Bundesanstalt für Geowissenschaften und Rohstoffe, Hannover
Federal Institute for Geosciences and Natural Resources

Environmental Geology

Handbook of Field Methods and Case Studies

This book is based on a joint research project partially funded by the Ministry of Education and Research (BMBF) of the Federal Republic of Germany through the Project Agency Forschungszentrum Karlsruhe GmbH, Water Technology and Waste Management Division (PTKA-WTE / FKZ 0261218).

Klaus Knödel
Gerhard Lange
Hans-Jürgen Voigt

Environmental Geology

Handbook of Field Methods and Case Studies

with contributions by
Thekla Abel, Sven Altfelder, Ulrich Beims, Manfred Birke, Norbert Blindow, Antje Bohn, Tilmann Bucher, Reiner Dohrmann, William E. Doll, Dieter Eisenburger, Hagen Hilse, Peter Herms, Bernhard Hörig, Bernhard Illich, Florian Jenn, Stephan Kaufhold, Claus Kohfahl, Franz König, Friedrich Kühn, Manja Liese, Harald Lindner, Reinhard Meyer, Klaus-Henrik Mittenzwey, Kai Müller, Mike Müller, Ranjeet Nagare, Michael Neuhaus, Claus Nitsche, Ricardo A. Olea, Hellfried Petzold, Michael Porzig, Jens Radschinski, Thomas Richter, Knut Seidel, Kathrin R. Schmidt, Dietmar Schmidt, Andreas Schuck, Anke Steinbach, Alejandra Tejedo, Andreas Tiehm, Markus Toloczyki, Peter Weidelt, Thomas Wonik, Ugur Yaramanci

with 501 Figures and 204 Tables

Springer

Editor:

BUNDESANSTALT FÜR GEOWISSENSCHAFTEN UND ROHSTOFFE

Stilleweg 2
30655 HANNOVER
GERMANY

ISBN 978-3-540-74669-0 **Springer Berlin Heidelberg New York**

Library of Congress Control Number: 2007937937

Springer is a part of Springer Science+Business Media
springeronline.com

Cover design: deblik, Berlin
Production: A. Oelschläger
Typesetting: Camera-ready by Claudia Wießner, BGR, Berlin Office

Printed on acid-free paper 30/2132/AO 543210

Table of Contents

1	**Introduction** *Klaus Knödel, Gerhard Lange & Hans-Jürgen Voigt*	**1**
2	**Preparatory Steps and Common Problems** *Klaus Knödel, Gerhard Lange & Hans-Jürgen Voigt*	**11**
2.1	**Placing of Orders and Order Handling**	**11**
2.2	**Collection and Use of Existing Data**	**12**
2.3	**Information Campaign and Permit Application**	**14**
2.4	**Mobilization and Demobilization**	**16**
2.5	**Land Surveying**	**17**
2.6	**Quality Assurance and Reporting**	**21**
3	**Remote Sensing** *Peter Herms, Bernhard Hörig, Friedrich Kühn, Dietmar Schmidt & Anke Steinbach,* with a contribution by *Tilmann Bucher*	**23**
3.1	**Aerial Photography** *Dietmar Schmidt & Friedrich Kühn*	**23**
3.1.1	Principle of the Methods	23
3.1.2	Applications	24
3.1.3	Fundamentals	25
3.1.4	Instruments and Film	27
3.1.5	Survey Practice	34
3.1.6	Interpretation of Aerial Photographs	36
3.1.7	Quality Assurance	43
3.1.8	Personnel, Equipment, Time Needed	46
3.1.9	Examples	47
3.2	**Photogrammetry** *Dietmar Schmidt, Peter Herms, Anke Steinbach & Friedrich Kühn*	**73**
3.2.1	Principle of the Methods	73
3.2.2	Applications	74
3.2.3	Fundamentals	74
3.2.4	Instruments	81
3.2.5	Survey Practice	82
3.2.6	Processing and Interpretation of Data	82
3.2.7	Quality Assurance	85
3.2.8	Personnel, Equipment, Time Needed	86
3.2.9	Examples	86

3.3 Nonphotographic Imaging from Aircraft and Space-borne Platforms **97**
Friedrich Kühn, Bernhard Hörig & Dietmar Schmidt, with a contribution by *Tilmann Bucher*
3.3.1 Principle of the Methods 97
3.3.2 Applications 99
3.3.3 Fundamentals 100
3.3.4 Instruments 115
3.3.5 Survey Practice 124
3.3.6 Processing and Interpretation of Data 125
3.3.7 Quality Assurance 134
3.3.8 Personnel, Equipment, Time Needed 136
3.3.9 Examples 137

4 Geophysics **161**
Norbert Blindow, Klaus Knödel, Franz König, Gerhard Lange, Harald Lindner, Reinie Meyer, Klaus-Henrik Mittenzwey, Andreas Schuck, Knut Seidel, Peter Weidelt, Thomas Wonik, Ugur Yaramanci, with contributions by *Dieter Eisenburger, Bernhard Illich, Ricardo A. Olea, Hellfried Petzold & Thomas Richter*
4.1 Magnetic Methods **161**
Klaus Knödel
4.1.1 Principle of the Methods 161
4.1.2 Applications 162
4.1.3 Fundamentals 162
4.1.4 Instruments 168
4.1.5 Survey Practice 170
4.1.6 Processing and Interpretation of the Measured Data 171
4.1.7 Quality Assurance 174
4.1.8 Personnel, Equipment, Time Needed 175
4.1.9 Examples 176
4.2 Gravity Methods **185**
Knut Seidel & Harald Lindner
4.2.1 Principle of the Methods 185
4.2.2 Applications 186
4.2.3 Fundamentals 186
4.2.4 Instruments 191
4.2.5 Survey Practice 192
4.2.6 Processing and Interpretation of the Measured Data 195
4.2.7 Quality Assurance 197
4.2.8 Personnel, Equipment, Time Needed 198
4.2.9 Examples 199

4.3	**Direct Current Resistivity Methods** *Knut Seidel & Gerhard Lange*	**205**
4.3.1	Principle of the Methods	205
4.3.2	Applications	207
4.3.3	Fundamentals	207
4.3.4	Instruments	215
4.3.5	Survey Practice	216
4.3.6	Processing and Interpretation of Measured Data	221
4.3.7	Quality Assurance	225
4.3.8	Personnel, Equipment, Time Needed	227
4.3.9	Examples	228
4.4	**Electromagnetic Methods** *Gerhard Lange & Knut Seidel*	**239**
4.4.1	Principle of the Methods	239
4.4.2	Applications	243
4.4.3	Fundamentals	243
4.4.4	Instruments	255
4.4.5	Survey Practice	260
4.4.6	Processing and Interpretation of Measured Data	263
4.4.7	Quality Assurance	266
4.4.8	Personnel, Equipment, Time Needed	269
4.4.9	Examples	270
4.5	**Ground Penetrating Radar** *Norbert Blindow,* with contributions by *Dieter Eisenburger, Bernhard Illich, Hellfried Petzold & Thomas Richter*	**283**
4.5.1	Principle of the Methods	283
4.5.2	Applications	285
4.5.3	Fundamentals	286
4.5.4	Instruments	297
4.5.5	Survey Practice	300
4.5.6	Processing, Presentation and Interpretation of the Measured Data	302
4.5.7	Quality Assurance	303
4.5.8	Personnel, Equipment, Time Needed	304
4.5.9	Examples	305
4.5.10	Special Applications and New Developments *Dieter Eisenburger*	316
4.6	**Seismic Methods** *Andreas Schuck & Gerhard Lange*	**337**
4.6.1	Principle of the Methods	337
4.6.2	Applications	340
4.6.3	Fundamentals	341
4.6.3.1	Propagation of Elastic Waves	341
4.6.3.2	Elastic Parameters and Seismic Velocities	343

4.6.3.3 Reflection, Transmission and Diffraction 347
4.6.3.4 Surface Waves 350
4.6.3.5 Seismic Resolution 352
4.6.4 Instruments 354
4.6.4.1 Seismic Sources 354
4.6.4.2 Seismic Sensors 359
4.6.4.3 Seismic Recording Instruments 362
4.6.5 Seismic Refraction Surveying 363
4.6.5.1 Principle of the Method 363
4.6.5.2 Survey Practice 364
4.6.5.3 Processing and Interpretation 365
4.6.5.4 Personnel, Equipment, Time Needed 369
4.6.6 Seismic Reflection Surveying 369
4.6.6.1 Principle of the Method 369
4.6.6.2 Survey Practice 373
4.6.6.3 Processing and Interpretation of Measured Data 376
4.6.6.4 Quality Assurance 382
4.6.6.5 Personnel, Equipment and Time Needed 383
4.6.7 Borehole Seismic Methods 384
4.6.7.1 Principle of the Methods 384
4.6.7.2 Applications 384
4.6.7.3 Fundamentals 385
4.6.7.4 Instruments 386
4.6.7.5 Survey Practice 386
4.6.7.6 Processing and Interpretation of Measured Data 387
4.6.7.7 Quality Assurance 387
4.6.7.8 Personnel, Equipment, Time Needed 388
4.6.8 Examples 388
4.7 Surface Nuclear Magnetic Resonance 403
Gerhard Lange, Ugur Yaramanci & Reinhard Meyer
4.7.1 Principle of the Method 403
4.7.2 Applications 404
4.7.3 Fundamentals 404
4.7.4 Instruments 408
4.7.5 Survey Practice 410
4.7.6 Processing and Interpretation of the Measured Data 413
4.7.7 Quality Assurance 415
4.7.8 Personnel, Equipment, Time Needed 416
4.7.9 Examples 416
4.8 Borehole Logging 431
Thomas Wonik with a contribution by *Ricardo A. Olea*
4.8.1 Principle of the Methods 431
4.8.2 Applications 433
4.8.3 Slimhole Logging Equipment and Logging Methods 434

4.8.3.1	Radioactivity Logging Methods	436
4.8.3.2	Electrical Methods	438
4.8.3.3	Electromagnetic Methods	440
4.8.3.4	Acoustic Methods	440
4.8.3.5	Optical Methods	441
4.8.3.6	Methods for Determining the Properties of Drilling Fluids (Fluid Logs)	441
4.8.3.7	Methods for Determining Borehole Properties	443
4.8.4	Survey Practice, Personnel, Equipment, Time Needed	443
4.8.5	Quality Assurance	447
4.8.6	Processing and Interpretation of the Logging Data and Examples	449
4.8.7	Expected Future Developments	470
4.9	**Geophysical In-situ Groundwater and Soil Monitoring** *Franz König, Klaus Knödel, Klaus-Henrik Mittenzwey & Peter Weidelt*	**475**
4.9.1	Principle of the Methods	475
4.9.2	Applications	477
4.9.3	Fundamentals	478
4.9.3.1	Environmental Parameters	478
4.9.3.2	Optical Spectroscopy	481
4.9.3.3	EM Monitoring	484
4.9.4	Instruments	488
4.9.5	Field work	492
4.9.6	Processing	493
4.9.7	Quality Assurance	494
4.9.8	Personnel, Equipment, Time Needed	494
4.9.9	Examples	495
5	**Geological, Hydrogeological, Geochemical and Microbiological Investigations** *Sven Altfelder, Ulrich Beims, Manfred Birke, Reiner Dohrmann, Hagen Hilse, Florian Jenn, Stephan Kaufhold, Klaus Knödel, Claus Kofahl, Manja Liese, Kai Müller, Mike Müller, Ranjeet Nagare, Michael Neuhaus, Claus Nitsche, Michael Porzig, Jens Radschinski, Katrin R. Schmidt, Andreas Thiem & Hans-Jürgen Voigt*	**507**
5.1	**Methods for Characterizing the Geological Setting** *Klaus Knödel, Kai Müller, Michael Neuhaus, Florian Jenn & Hans-Jürgen Voigt*	**507**
5.1.1	Geologic Field Observations *Florian Jenn, Klaus Knödel & Hans-Jürgen Voigt*	511
5.1.2	Trenching *Klaus Knödel & Hans-Jürgen Voigt*	519

5.1.3 Drilling 524
Florian Jenn & Hans-Jürgen Voigt
5.1.4 Direct Push Technology 540
Kai Müller, Michael Neuhaus & Hans-Jürgen Voigt
5.2 Methods for Characterizing the Hydrologic and Hydraulic Conditions 567
Ulrich Beims, Florian Jenn, Klaus Knödel, Manja Liese, Ranjeet Nagare, Claus Nitsche, Michael Porzig & Hans-Jürgen Voigt
5.2.1 Precipitation 569
Florian Jenn, Klaus Knödel, Manja Liese & Hans-Jürgen Voigt
5.2.2 Evaporation and Evapotranspiration 581
Florian Jenn, Klaus Knödel, Manja Liese & Hans-Jürgen Voigt
5.2.3 Runoff 590
Florian Jenn, Klaus Knödel, Manja Liese & Hans-Jürgen Voigt
5.2.4 Infiltration 603
Florian Jenn, Klaus Knödel, Manja Liese & Hans-Jürgen Voigt
5.2.5 Groundwater Recharge 612
Florian Jenn, Klaus Knödel, Manja Liese & Hans-Jürgen Voigt
5.2.6 Groundwater Monitoring 619
Florian Jenn, Claus Nitsche & Hans-Jürgen Voigt
5.2.7 Determination of Hydraulic Parameters 643
Ulrich Beims, Ranjeet Nagare, Claus Nitsche, Michael Porzig & Hans-Jürgen Voigt
5.2.7.1 Infiltrometer and Permeameter Tests 649
Florian Jenn, Ranjeet Nagare, Michael Porzig & Hans-Jürgen Voigt
5.2.7.2 Pumping Tests 681
Ulrich Beims
5.2.7.3 Laboratory Methods 711
Florian Jenn, Claus Nitsche & Hans-Jürgen Voigt

5.3 Methods for Characterizing the Geochemical and Microbiological Conditions 749
Sven Altfelder, Manfred Birke, Reiner Dohrmann, Hagen Hilse, Florian Jenn, Stephan Kaufhold, Klaus Knödel, Claus Nitsche, Kathrin R. Schmidt, Andreas Thiem & Hans-Jürgen Voigt

5.3.1 Sampling and Analysis of Groundwater and Surface Water 758
Claus Nitsche & Hans-Jürgen Voigt
5.3.1.1 Planning and Preparation of Work 759
5.3.1.2 Groundwater Sampling 768
5.3.1.3 Groundwater Analysis 778
Klaus Knödel
5.3.2 Sampling and Analysis of Soil, Rock, Stream and Lacustrine Sediments 785
Manfred Birke, Claus Nitsche & Hans-Jürgen Voigt
5.3.2.1 Planning and Preparation of Work 786
5.3.2.2 Sampling of Soil, Rock, Stream and Lacustrine Sediments 790
5.3.2.3 Analysis of Soil, Rock, Stream and Lacustrine Sediments 797
5.3.3 Sampling and Analysis of Soil Gas and Landfill Gas 807
Hagen Hilse & Hans-Jürgen Voigt
5.3.3.1 Planning and Preparation of Work 810
5.3.3.2 Sampling and Analysis of Soil Gas and Landfill Gas 811
5.3.4 Methods for Chemical Analysis used in Geochemical Investigations 816
Florian Jenn
5.3.5 Laboratory Methods for the Determination of Migration Parameters 823
Sven Altfelder with a contribution by *Claus Nitsche*
5.3.5.1 Basic Theory of Sorption 824
5.3.5.2 Basic Theory of Transport 830
5.3.5.3 Sampling and Preparation of Soil or Sediment for the Determination of Migration Parameter Values 831
5.3.5.4 Batch Tests 832
5.3.5.5 Column Experiments 838
5.3.5.6 Clays and Clay Minerals 849
Reiner Dohrmann
5.3.5.7 Cation Exchange Capacity 853
Reiner Dohrmann
5.3.5.8 Carbonates 862
Stephan Kaufhold
5.3.5.9 Iron and Manganese Oxides 867
Stephan Kaufhold
5.3.5.10 Organic Carbon 871
Klaus Knödel
5.3.6 Methods to Evaluate Biodegradation at Contaminated Sites 876
Andreas Thiem & Kathrin R. Schmidt
5.3.6.1 Microbial Processes in the Subsurface 877
5.3.6.2 Assessment Methods 883
5.3.6.3 Case Studies 892

5.4	**Interpretation of Geological, Hydrogeological, and Geochemical Results** *Florian Jenn, Claus Kofahl, Mike Müller, Jens Radschinski & Hans-Jürgen Voigt*	**941**
5.4.1	Statistical Methods *Florian Jenn & Hans-Jürgen Voigt*	941
5.4.1.1	Univariate Statistics	942
5.4.1.2	Multivariate Statistics	952
5.4.1.3	Time Series Analysis	958
5.4.1.4	Geostatistics and Interpolation of Spatial Data	959
5.4.1.5	Specific Tests of Hydrogeochemical Data	960
5.4.2	Conceptual Model *Hans-Jürgen Voigt & Jens Radschinski*	962
5.4.3	Groundwater Flow Modeling *Claus Kofahl*	1001
5.4.3.1	Fundamentals of Groundwater Flow Modeling	1002
5.4.3.2	Programs	1009
5.4.3.3	Guide for Construction and Use of a Groundwater Model	1014
5.4.4	Contaminant Transport Modeling *Mike Müller*	1020
5.4.4.1	Fundamentals of Transport Modeling	1021
5.4.4.2	Model Application	1041
6	**Integration of Investigation Results** *Thekla Abel, Manfred Birke, Antje Bohn, Klaus Knödel, Gerhard Lange, Alejandra Tejedo, Markus Toloczyki & Ugur Yaramanci*	**1053**
6.1	**Data Fusion** *Klaus Knödel, Marcus Toloczyki, Antje Bohn, Thekla Abel, Gerhard Lange & Alejandro Tejedo*	**1054**
6.1.1	Reprocessing and New Data Presentation *Klaus Knödel & Gerhard Lange*	1055
6.1.2	Geographic Information Systems *Marcus Toloczyki, Thekla Abel & Alejandra Tejedo*	1061
6.1.2.1	Fundamentals	1063
6.1.2.2	Hardware, Network, Software, and Manpower	1067
6.1.2.3	Data Acquisition and Analysis	1072
6.1.2.4	Examples	1075
6.1.3	Other Data Fusion Examples *Klaus Knödel & Antje Bohn*	1091
6.2	Joint Interpretation *Manfred Birke, Klaus Knödel, Gerhard Lange & Ugur Yaramanci*	1099

6.2.1 Qualitative and Semiquantitative Approach 1099
Klaus Knödel & Gerhard Lange
6.2.2 Quantitative Approach 1150
Manfred Birke, Klaus Knödel, Gerhard Lange & Ugur Yaramanci
6.2.2.1 Joint Quantitative Interpretation of Several Geophysical Measurements and Core Analysis Results 1050
Ugur Yaramanci & Gerhard Lange
6.2.2.2 Joint Inversion 1059
Klaus Knödel, Ugur Yaramanci & Gerhard Lange
6.2.2.3 Joint Interpretation Using Statistical Methods 1063
Manfred Birke

Glossary **1195**
Abbreviations **1319**
Units of Measure **1333**
SI Prefixes **1335**
None SI Units **1335**
Physical Constants **1336**
Mathematical Symbols and Constants **1337**
Subject Index **1339**

List of Authors

Dr. Thekla Abel
Bundesanstalt für Geowissenschaften und Rohstoffe
Stilleweg 2
D-30655 Hannover

Dr. Sven Altfelder
Bundesanstalt für Geowissenschaften und Rohstoffe
Stilleweg 2
D-30655 Hannover

Prof. Dr.-Ing. Ulrich Beims
GFI Grundwasserforschungsinstitut GmbH Dresden
Meraner Straße 10
D-01217 Dresden

Dr. Manfred Birke
Bundesanstalt für Geowissenschaften und Rohstoffe
Dienstbereich Berlin
Wilhelmstraße 25-30
D-13593 Berlin

Dr. Norbert Blindow
Westfälische Wilhelms-Universität
Institut für Geophysik
Corrensstraße 24
D-48149 Münster

Antje Bohn
Brandenburgische Technische Universität Cottbus
LS Umweltgeologie
Erich-Weinert-Straße 1
D-03406 Cottbus

Tilmann Bucher
Deutsches Zentrum für Luft- und Raumfahrt
Berlin-Adlershof
Einrichtung Optische Informationssysteme am Institut für Robotik und Mechatronik
Rutherfordstraße 2
D-12489 Berlin

Dr. Reiner Dohrmann
Bundesanstalt für Geowissenschaften und Rohstoffe
Stilleweg 2
D-30655 Hannover

Dr. William Doll
Research Leader
Batelle - Oak Ridge Operations
105 Mitchell Road, Ste 103
Oak Ridge, TN 37830

Dieter Eisenburger
Bundesanstalt für Geowissenschaften und Rohstoffe
Stilleweg 2
D-30655 Hannover

Dr.-Ing. Hagen Hilse
GICON – Großmann Ingenieur Consult GmbH
Tiergartenstraße 48
D-01219 Dresden

Peter Herms
Hansa Luftbild AG
Elbestraße 5
D-48145 Münster

Bernhard Hörig
Bundesanstalt für Geowissenschaften und Rohstoffe
Dienstbereich Berlin
Wilhelmstraße 25-30
D-13593 Berlin

Bernhard Illich
GGU Gesellschaft für
Geophysikalische Untersuchungen
Amalienstraße 4
D-76133 Karlsruhe

Florian Jenn
Brandenburgische Technische Universität Cottbus
LS Umweltgeologie
Erich-Weinert-Straße 1
D-03406 Cottbus

Dr. Stephan Kaufhold
Bundesanstalt für Geowissenschaften und Rohstoffe
Stilleweg 2
D-30655 Hannover

Dr. Klaus Knödel
Bundesanstalt für Geowissenschaften und Rohstoffe
Dienstbereich Berlin
Wilhelmstraße 25-30
D-13593 Berlin

Dr. Claus Kohfahl
FU Berlin
Institut für Geologische Wissenschaften
Fachrichtung Geochemie, Hydrologie, Mineralogie
Arbeitsbereich Hydrologie
Malteserstrasse 74-100, Haus B
D-12249 Berlin

Franz König
Bundesanstalt für Geowissenschaften und Rohstoffe
Dienstbereich Berlin
Wilhelmstraße 25-30
D-13593 Berlin

Dr. Friedrich Kühn
Bundesanstalt für Geowissenschaften und Rohstoffe
Stilleweg 2
D-30655 Hannover

Gerhard Lange
Bundesanstalt für Geowissenschaften und Rohstoffe
Dienstbereich Berlin
Wilhelmstraße 25-30
D-13593 Berlin

Manja Liese
Brandenburgische Technische Universität Cottbus
LS Umweltgeologie
Erich-Weinert-Straße 1
D-03406 Cottbus

Prof. Dr. Harald Lindner
TU Bergakademie Freiberg
Institut für Geophysik
Gustav-Zeuner-Straße 12
D-09599 Freiberg

Reinhard Meyer
Groundwater Sciences Research Group
Natural Resources and the Environment (NRE),
Council for Scientific and Industrial Research (CSIR)
PO Box 395
Pretoria, 0001

Dr. Klaus-Henrik Mittenzwey
OPTOSENS
Optische Spektroskopie und Sensortechnik GmbH
Rudower Chaussee 29 (IGZ)
D-12489 Berlin

Kai Müller
Boden- und Grundwasserlabor GmbH (BGD)
Tiergartenstrasse 48
D-01219 Dresden

Dr. Mike Müller
Ingenieurbüro für Grundwasser GmbH
Nonnenstrasse 9
D-04229 Leipzig

Ranjeet Nagare, M.Sc.
Brandenburgische Technische Universität Cottbus
LS Umweltgeologie
Erich-Weinert-Straße 1
D-03406 Cottbus

Dr. Michael Neuhaus
FUGRO Consult GmbH
Wolfener Straße 36 V
D-12681 Berlin

Dr. Claus Nitsche
Boden- und Grundwasserlabor GmbH (BGD)
Tiergartenstrasse 48
D-01219 Dresden

Ricardo A. Olea
Bundesanstalt für Geowissenschaften
und Rohstoffe, Dienstbereich Berlin
Wilhelmstraße 25-30
D-13593 Berlin

Now:
Eastern Energy Resorces
US Geological Survey
12201 Sunrise Valley Dr., MS956
Reston, VA20192

Dr. Hellfried Petzold
LAUBAG
Lausitzer Braunkohle AG
Knappenstraße 1
D-01968 Senftenberg

Michael Porzig
Brandenburgische Technische Universität Cottbus
LS Umweltgeologie
Erich-Weinert-Straße 1
D-03406 Cottbus

Jens Radschinski
Brandenburgische Technische Universität Cottbus
LS Umweltgeologie
Erich-Weinert-Straße 1
D-03406 Cottbus

Dr. Thomas Richter
Bo-Ra-Tec GmbH
Damaschkestr. 19a
D-99425 Weimar

Knut Seidel
Geophysik GGD
Gesellschaft für
Geowissenschaftliche Dienste
Ehrensteinstraße 33
D-04105 Leipzig

Now:
GGL Geophysik und
Geotechnik Leipzig GmbH
Bautzner Straße 67
D-04347 Leipzig

Kathrin R. Schmidt
DVGW - Technologiezentrum Wasser (TZW)
Karlsruher Straße 84
76139 Karlsruhe

Dietmar Schmidt
Hansa Luftbild AG
Elbestraße 5
D-48145 Münster

Dr. Andreas Schuck
Geophysik GGD
Gesellschaft für
Geowissenschaftliche Dienste
Ehrensteinstraße 33
D-04105 Leipzig

Now:
GGL Geophysik und
Geotechnik Leipzig GmbH
Bautzner Straße 67
D-04347 Leipzig

Anke Steinbach
Hansa Luftbild AG
Elbestraße 5
D-48145 Münster

Alejandra Tejedo
SEGEMAR Servicio Geológico Minero Argentino
Buenos Aires
Argentina

Dr. Andreas Tiehm
DVGW - Technologiezentrum Wasser (TZW)
Karlsruher Straße 84
D-76139 Karlsruhe

Dr. Markus Toloczyki
Bundesanstalt für Geowissenschaften und Rohstoffe
Stilleweg 2
D-30655 Hannover

Prof. Dr. Hans-Jürgen Voigt
Brandenburgische Technische Universität Cottbus
LS Umweltgeologie
Erich-Weinert-Straße 1
D-03406 Cottbus

Prof. Dr. Peter Weidelt
TU Braunschweig
Institut für Geophysik und Extraterrestrische Physik
Mendelsohnstraße 3
D-38106 Braunschweig

Dr. Thomas Wonik
GGA Institut für Geowissenschaftliche Gemeinschaftsaufgaben
Stilleweg 2
D-30655 Hannover

Prof. Dr. Ugur Yaramanci
TU Berlin, Fakultät 6
Fachbereich Bauingenieurwesen und Angewandte Geophysik
Ackerstraße 71-76
D-13355 Berlin

List of Reviewers

Dr. Ron D. Barker
Earth Sciences
The University of Birmingham
Edgbaston
Birmingham B15 2TT
United Kingdom

Jean Bernard
IRIS instruments
1 Avenue Buffon, BP 6007
45060 Orleans cedex 02
France

Dr. Manfred Birke
Federal Institute for Geosciences
and Natural Resources, DB Berlin
Wilhelmstraße 25-30
D-13593 Berlin

Prof. Dr. Hans Jürgen Burkhardt
Technical University Berlin
Dept. of Applied Geophysics
Ackerstrasse 71-76
D-13355 Berlin

Dr. Christoph Grissemann
Federal Institute for Geosciences
and Natural Resources, DB Berlin
Stilleweg 2
D-30655 Hannover

Dr. Martin Herold
ESA GOFC GOLD PROJECT OFFICE
Department of Geography
Friedrich Schiller University Jena
Loebdergraben 32
D-07743 Jena

Prof. Dr. Anatoly Legchenko
Institut de Recherche
pour la Developpement
LTHE, BP53
38041 Grenoble cedex 9
France

Prof. Dr. Ludwig Luckner
Dresdner Grundwasserforschungszentrum e.V.
Meraner Str. 10
D-01217 Dresden

Reinhard Meyer
Groundwater Sciences Research Group
Natural Resources and the Environment
CSIR
Pretoria, 0001
South Africa

Dr. Edgar Stettler
Council for Geoscience
Pretoria, 0001
South Africa

Now:
Thani Mineral Investment,
United Arab Emirates

Dr. Manfred Teschner
Federal Institute for Geosciences
and Natural Resources
Stilleweg 2
D-30655 Hannover

Prof. Dr. Peter Weidelt
Technical University Braunschweig
Institut of Geophysics und Meteorology
Mendelssohnstraße 3
D-38106 Braunschweig

Acknowledgements

This handbook is a product of the Thai-German Research Project WADIS (Waste Disposal: Recommendations for Site Investigations of Waste Disposal Sites and Contaminated Sites in Thailand, funded by the Ministry of Education and Research (BMBF) of the Federal Republic of Germany (grant no. 0261218) through its Project Management Agency Forschungszentrum Karlsruhe, Water Technology and Waste Management Division (PTKA-WTE). The editors and authors gratefully acknowledge the financial support by the German Federal Ministry of Education and Research (BMBF) and the scientific and administrative support by the Project Management Agency. Without the assistance of Prof. Dr. D. Mager, Federal Ministry of Economics and Technology, Mineral Resources and Geosciences Section, the project would not have been started and successfully carried out.

The WADIS project was approved by the Thai Cabinet on 30th May 2000 and the Project Agreement was signed in September 2000 in Bangkok between the Ministry of Industry and the German Federal Institute for Geosciences and Natural Resources (BGR). Strongly appreciated are, in particular, the support and efforts continually given to the research project by the management of the Department of Mineral Resources (DMR) and especially the director of the Environmental Geology and Geohazards Division, Khun Worawoot Tantiwanit, and his colleagues in Bangkok and in the Chiang Mai office.

An important aspect of the project work was the excellent team spirit among Thai and German partners. The joint Thai-German project group included the following institutions and companies:

DMR - Department of Mineral Resources, Bangkok	BGR - Federal Institute for Geosciences and Natural Resources, Hannover
CMU - Chiang Mai University, Chiang Mai	BTU - Brandenburg Technische Universität, Cottbus
ATOP Technology Co. Ltd., Bangkok	GGD - Geophysik GmbH, Leipzig
PCD - Pollution Control Department, Bangkok	BGD - Soil and Groundwater Laboratory GmbH, Dresden
Sky Eyes Co. Ltd., Bangkok	Hansa Luftbild GmbH, Münster
AZTEC Engineering Co. Ltd., Lampang	Gicon GmbH, Dresden
METRIX Associates Co. Ltd., Bangkok	Ingenieurbüro Sehlhoff GmbH, Vilsbiburg

The contributions and achievements of all partners in these institutions and companies are gratefully acknowledged.

This book is a joint venture of scientists in companies, universities, and institutions. Thanks are also due to the colleagues, companies and institutions outside the WADIS project group whose contributions and case studies enriched this handbook.

Dr. Christoph Grissemann of the Geophysics for Resource Management Section of the Federal Institute for Geosciences and Natural Resources conducted the final phase of the WADIS project. The editors would like to thank him for the excellent cooperation and beneficial discussions during the preparation of this handbook.

Particular thanks go to all reviewers (see List of Reviewers) for the thorough revision of manuscripts and numerous constructive comments for improvement.

Dr. R. Clark Newcomb has linguistically revised most of the contributions to this book. The manuscripts have benefited from his helpful comments and reviews. Henry Toms was an indispensable help in geoscientific terminology. The authors wish to thank both colleagues for their support.

Last but not least, the authors and editors are indebted to Claudia Wießner and Ingrid Boller for their careful preparation of figures and layout of the handbook. They contributed substantially to the design and quality of this book. Thanks also to Claudia Kirsch for her assistance in the handbook preparation.

Preface

As earth's population continues to grow and the detrimental aftereffects of industrialization and environmental negligence become more apparent, society has become more aware of, and concerned about, stewardship of the natural environment – water, soil, and air. Sustainable development has become more widely received and promoted in many parts of the world. The need is now critical for earth and environmental scientists and engineers to work together to implement technologies that can preserve our environment.

The Earth's population was 6.6 billion as of April 2007 according to the U.S. Census Bureau. This number is expected to rise to 9.4 billion by 2050. The population is increasing the demand for natural resources and energy, and increasing stress on the environment. Thus, protection of the environment and remediation of damage to the environment must be a priority. It is also important to develop procedures that will help to avert further damage to the environment and to recognize as early as possible the risks associated with changes in the environment.

Many methodologies and technologies have become more advanced in the past few decades, and new technologies and approaches have been developed, all to address the growing need for environmental assessment, monitoring, and remediation. As these technologies have grown, the need for interdisciplinary cooperation has also become more apparent. Specialists in remote sensing, geophysical methods, hydrogeology, geology, and geochemistry must maintain current awareness of developments within their sister disciplines in order to formulate effective overall approaches for environmental issues.

Too often, resolution of environmental problems is constrained by political or economic boundaries. In many parts of the world, standards for acceptable air, water, and soil quality must be compromised in order to meet more pressing human needs. It is therefore important that cost-effective interdisciplinary solutions be developed, in order to allow all nations and socio-economic groups to benefit from a clean, sustainable environment.

Economic development is ultimately limited by environmental quality. Water is needed for all aspects of life. All social and economic activities depend on having a reliable supply of high quality water. As populations grow and economic activity increases, many countries are rapidly confronted with the problem of water scarcity, which limits economic development. More than 1 billion people, mainly in the developing countries, lack an adequate supply of safe drinking water. By 2050, 25 percent of the people on Earth will live in countries in which water is permanently scarce. Contaminated drinking water is a major cause of disease and death in developing countries. An adequate water supply is a prerequisite for human existence, not only for drinking, but also for agriculture. Degradation of soils also poses a direct threat to food production in developing countries. The availability of arable land is decreasing. Population growth is complicating the situation. According to the

United Nations Environment Programme (UNEP), there are more refugees from a deteriorating environment than from war. Others warn that water quality will play a growing role in regional conflicts and wars. High quality water and uncontaminated soil must be given greater priority throughout the world.

This book, **Environmental Geology - Handbook of Field Methods and Case Studies**, is intended to enable progress toward these challenging goals. It provides a broad spectrum of investigatory methods in several disciplines to support cross-fertilization among experts in those disciplines and others. Methods that are treated in this book include remote sensing, geophysics, geology, hydrogeology, geochemistry, and microbiology. Most of the methods described in this handbook are available and used in developing countries. Information is provided about the principle of the method, possible applications, fundamentals, instruments, survey practice, processing and interpretation of the data, quality assurance, personnel, equipment, and time needed. Examples are given, as well as references and sources for further reading. Besides geoscientific methods, the procedures for stepwise site investigations are described, as well as common problems encountered in field operations.

This handbook is not intended to be used as a textbook, but instead as a reference, providing insights into the fundamentals, application and limits of methods. Interdisciplinary case studies from different parts of the world have been selected as examples for extrapolation to other geoenvironmental concerns. With this structure the handbook can be used as a practical guide for training students.

The descriptions of the methods and case studies also illustrate the advantages of interdisciplinary geoscientific site investigations to decision-makers who deal with environmental investigations. This applies to both remediation assessments and preventative measures. Thorough and knowledgeable application of such an approach will enhance its reliability, credibility, and value to future generations.

William E. Doll
Oak Ridge, 2007

1 Introduction

KLAUS KNÖDEL, GERHARD LANGE & HANS-JÜRGEN VOIGT

Increasing population density and industrialization are creating a high strain on the natural environment and resources of many countries. Therefore, precautionary measures to protect the environment and remedial action to repair the damages of the past have high priority. Resources to be protected are surface water and groundwater, soil and air. Hazards to these resources are landfills and industrial sites as well as mining facilities, including tailings, conditioning plants, and smelters, oil refineries, distribution facilities and pipelines, gas stations and other areas used by humans (e.g., military training sites).

Waste disposal, mining, and industrial sites are an absolutely necessary part of the infrastructure of an industrial society. Suitable new sites must be found for the disposal of waste and for mining and industrial facilities. It is often very difficult to obtain political approval and this is possible only if state-of-the-art methods are used to show that such sites have layers that can function as barrier, preventing entry of contaminants into the environment. Areas of both consolidated and unconsolidated rock can be suitable sites for landfills and industrial facilities.

Knowledge and experience with the disposal of waste and the operation of mines and industrial facilities in an ecologically nondetrimental way have been acquired only gradually during the past several decades. On the basis of this knowledge, numerous abandoned landfills, mining, and industrial sites must now be regarded as hazardous.

Impermeable layers at such sites are the most important barrier for impeding the spread of pollutants. It must be assumed that this geological barrier is always of importance, since the currently used techniques for preparing sites for landfills, mines and industrial plants will prevent the spread of pollutants for only a finite time.

A site investigation stepwise is usually carried out in at least two phases: (1) a orientating investigation and (2) a detailed investigation. A flow chart of the stepwise procedure for site investigations is shown in *Figure 1-1*. Some references to chapters of this book are given in the flow chart. Field and laboratory work as well as data processing, data presentation and interpretation of the data from individual methods are described in the chapters of Parts 3 to 5. Solutions to common problems are treated in Part 2. For more details of site investigation procedure see books with strategies and recommendations for site investigations (e.g., WILKEN & KNÖDEL, 1999, VOIGT et al., 2006).

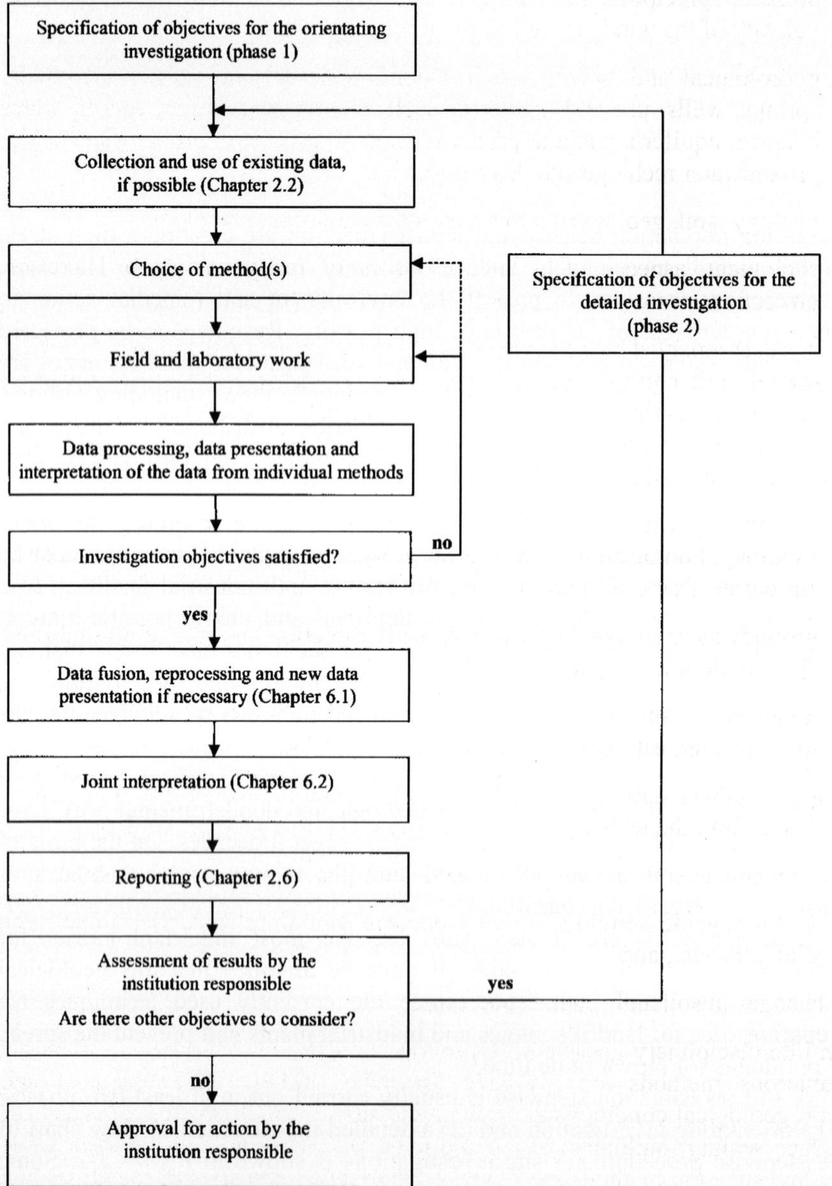

Fig. 1-1: Flow chart for site investigations

In the orientating investigation, the following information is obtained from maps and other archived data sources:

- topography, land use and vegetation, settlements, roads and railways,
- climate: precipitation, temperature, evapotranspiration, direction and velocity of the wind, as well as the frequency of strong winds,
- hydrological and hydrogeological conditions: streams, lakes and ponds, springs, wells, use and quality of surface and groundwater, runoff, water balance, aquifer/aquiclude properties and stratigraphy, groundwater table, groundwater recharge and discharge,
- geology: soil, geological structures, stratigraphy and lithology,
- ecological aspects: e.g., nature reserves, protected geotopes, water protection areas.

This is accompanied by a reconnaissance survey in the field and by a historical review of earlier use of the site (interviews of persons who lived or worked around the site during the time of mining or industrial operations).

The following aspects or parts of them must be taken into account for a detailed site investigation and assessment:

- geology: thickness and lateral extent of strata and geological units, lithology, homogeneity and heterogeneity, bedding conditions and tectonic structures, fractures, impact of weathering,
- groundwater: water table, water content, direction and rate of groundwater flow, hydraulic conductivity, value of aquifer,
- geochemical site characterization: chemical composition of soil, rocks and groundwater, estimation of contaminant retention,
- geotechnical stability: The geological barrier must be capable of adsorbing strain from the weight of a landfill, slag heap or industrial building.
- geogenic events: active faults, karst, earthquakes, subsidence, landslides,
- anthropogenic activities: mining damage, buildings, quarries, gravel pits, clay pits, etc., and
- changes in soil and groundwater quality.

An interdisciplinary geoscientific program is required for a site investigation. Numerous methods are available for such studies. The geological and hydrogeological conditions as well as the surface conditions (e.g., vegetation, surface sealing, buildings) at the site must be taken into consideration when the investigation methods are selected. There are some rules of thumb for site investigations: Start with the less expensive methods, expanding as necessary to more expensive methods for detailed investigations, i.e., remote sensing before geophysics, geology and hydrogeology before geochemistry and modeling. Mapping should be carried out before sounding and drilling, investigations of the area as a whole before point data. First, a representation of the data is made, it is then interpreted, and an assessment of the conditions at the site is made. Not only the geological structures immediately below a landfill, slag heap or industrial site has to be examined, but also the surrounding area. The investigation of the surrounding area must include that part of the geological barrier that is expected to be needed for contaminant retention and that part of the regional groundwater system that will possibly become contaminated. Each landfill, mining or industrial site, whether in operation or abandoned, is within a groundwater system of several tens of square kilometers. A general survey of this area has to be made. A detailed study must be carried out in an area of 0.1 - 1 km^2 around the site itself. The structures down to a depth of 50 m are relevant to a study of abandoned and planned waste disposal sites. It is often necessary to extend the investigations down to 150 m to obtain information on the groundwater system. In some cases, the geology and the groundwater system below groundwater protection areas also have to be investigated in order to assess their vulnerability to pollution.

The following methods and tools can be used to assess a geological barrier and the potential for the spread of contaminants: *Remote sensing methods* can provide geoscientific data for large areas in a relatively very short time. They are not limited by extremes in terrain or hazardous conditions that may be encountered during an on-site appraisal. In many cases aerial photographs and satellite images should be used to prepare a base map of the investigation area. Remote sensing methods can enable a preliminary assessment and site characterization of an area prior to the use of more costly and time consuming techniques, such as field mapping, geophysical surveys and drilling. The data obtained from satellite-based remote-sensing systems is best suited for regional studies as well as for detecting and monitoring large-scale environmental problems. However, for detailed site characterization, satellite data is sometimes of limited use due to relatively low spatial resolution. Mapping scales of 1 : 10 000 or larger are required for a detailed geoenvironmental assessment of landfills, mining, and industrial sites. High-resolution aerial photographs, airborne scanners, and some satellite-based remote-sensing systems provide data at the required spatial resolution (e.g., 70 cm and better). Aerial photographs made at different times can reveal the changes at sites suspected to be hazardous.

Geophysical methods are used to develop a model of the geology below the site, to locate fracture zones, to investigate the groundwater system, to detect and delineate abandoned landfills and contamination plumes, as well as to obtain information on the lithology and physical parameters of the ground. Necessary condition for a meaningful use of geophysical methods is the existence of contrasts in the physical parameter values (magnetization, susceptibility, density, electrical resistivity, seismic velocities, etc.) of soil and rock. The parameter values to be expected at the site must be considered before a geophysical survey is conducted. Geophysical methods supplement each other because they are sensitive for different physical parameters. Seismic methods are used to investigate structures and lithology. Electrical and electromagnetic methods are very sensitive to changes in electrolyte concentrations in the pore water. Ground-penetrating radar can be used in areas with dry, low-conductivity rocks. Both magnetic and electromagnetic mapping have proved useful for locating and delineating concealed landfills. Both methods are fast and easy to conduct, enabling large areas to be investigated in a short time. Seismic, dc-resistivity, electromagnetic and gravity methods are used to investigate groundwater systems on a regional scale. Geophysical surveys help find suitable locations for drilling groundwater observation wells and provide information between boreholes and for areas where it is impossible to drill. Well logging is absolutely necessary. Logging data are not only necessary for processing and interpretation of surface geophysical data but also as a bridge between geophysical surveys and hydrogeological modeling.

Geological and hydrogeological studies are used to investigate lithological structures, to determine the homogeneity of the rock, to locate fractures, to determine the permeability of the rock with respect to water, gases and various contaminants, to assess the mechanical stability of the ground, and to obtain data on the groundwater system. Flow and transport models must be developed to estimate groundwater recharge and the potential for groundwater contamination. The main tasks of geological and hydrogeological surveys are to gain information directly by examining outcrops, digging trenches and drilling boreholes, conducting hydraulic tests (e.g., pumping tests and tracer tests) in wells to determine hydraulic properties *in situ*. This work is augmented by geological mapping, examination of drill cores, construction and expansion of a network of groundwater observation wells. Rock, soil and groundwater samples are taken to determine physical, chemical, petrographic and mineralogical parameters. Special laboratory experiments can be carried out to estimate migration parameters and the texture of rock and soil samples. Data from cone penetration tests and other field and laboratory methods are used to assess the stability of the ground.

Geophysical and *geochemical methods* can be used to delineate and monitor operating and abandoned landfills as well as to determine the spread of contaminants. Geochemical methods are necessary to obtain information about the capacity of the rock to retain contaminants seeping from a landfill, as well as about the degradation of the hazardous substances.

The objectives of the site investigation must be well defined. Additionally important is for suitable methods to be chosen to accomplish the objectives. The sections “Possible Applications” and “Examples” in the chapters of Parts 3 to 5 of this book can aid the selection of suitable methods. Suitable methods are recommended in *Tables 1-1* to *1-3*, for example, for geological investigations, investigations of landfill waste bodies, and the evaluation of groundwater conditions. These tables should not be considered as a substitute for the advice of an experienced specialist.

This handbook is designed to provide geoscientific methods to investigate landfills and mining and industrial sites. It describes methods in the fields of remote sensing, geophysics, geology, hydrogeology, geochemistry, and microbiology. Most of the methods described in this volume are also available and used in developing countries. The descriptions provide information about the principle of the method, possible applications, fundamentals, instruments, survey practice, processing and interpretation of the data, quality assurance, personnel, equipment, time needed, examples, as well as references and sources for further reading. Most of the remote-sensing and geophysical methods are subdivided in this way (see Table of Contents). The geological, hydrogeological, and geochemical methods are subdivided in other ways, owing to the different way they are applied in site surveys.

The handbook is not intended to be used as a textbook, but, instead, to provide insights into the fundamentals, application and limits of methods, as well as case studies, selected as examples for extrapolation to other geoenvironmental concerns. This handbook cannot replace consultation with an experienced remote-sensing expert, geologist, geophysicist and/or chemist. Involvement of experts insures that the most up-to-date methods and techniques are applied. Prior starting a project, it is beneficial to first define the objectives and goals of the study so that the most suitable techniques and methods can be used.

The Federal Institute for Geosciences and Natural Resources (BGR), Germany, in cooperation with scientists from universities, research institutes, and industry, has carried out two projects entitled "Methods for the Investigation and Characterization of the Ground below Waste Disposal Sites" and “Recommendations for Site Investigations of Waste Disposal Sites and Contaminated Sites in Thailand” funded by the German Federal Ministry for Education and Research (BMBF). The primary objective of these studies was to increase the understanding of the characteristics of the ground below waste disposal sites using geoscientific methods. Eight volumes of the German Handbuch zur Erkundung des Untergrundes von Deponien und Altlasten (Handbook for Investigation of the Ground below Landfills and Sites Suspected to be Hazardous) (Springer-Verlag) were prepared in the first project. The handbook “Site Investigation Methods” presented here is a product of the second project. This book contains numerous examples from both of the above-mentioned projects.

Table 1-1: Geoscientific methods for site investigations: geology

1 - occasionally used for this purpose 2 - suitable method 3 - method is well suited to this application 4 - preferred method (can be more than one) for this application/objective	Remote Sensing					Geophysics								Geology and Hydrogeology							Mineralogy and Geochemistry				
METHODS	satellite imagery	aerial photography	hyperspectral scanner	radar methods	laser scanner	magnetics	gravity method	dc resistivy and electromagnetics	ground penetrating radar	refraction seismics	reflection seismics	SNMR	geophysical downhole logging	geological field work	hydrogeological field work	direct push technology	investigation in trenches	drilling and installation of wells	hydraulic and tracer tests	flow and transport modeling	lithological and mineralogical methods	chemical analysis	elution, batch, reactor and column test	soil gas analysis	geochemical modeling
APPLICATIONS/OBJECTIVES	suitability of the method for the respective application/objective																								
lateral delimitation of lithological units		1	1			1	1	4	2	3	2			3		2	2	2			1	2			
vertical delimitation of lithological units							1	4	2	3	3		3			1	2	3			1	2			
determination of lithological, petrophysical and hydrological properties of lithological units								3	1	2	2		4			1	2	3	4		3				
assessment of homogeneity of rock layers		1	1			2	1	4	1	2	1		3	1		1	2	4			3	2			
investigation of regional geological structures	2	2	2	1		1	2	3		4	4		1	2	2			2				2		2	
investigation of local geological structures	1	2	2		1	2	1	3	2	3	3		2	2	2	1	2	3				2		2	
determination of dip and strike of beds		3				1	1	2	1	2	3		3	1		1	2	3							
mapping of erosion channels	1	2	1		1	1	3	3	1	2	3			1	1	1	2	3							
determination of the depth to hard rock beneath unconsolidated cover rock, thickness of the weathering layer							2	3	1	4	3	3	3			1	2	4							
localization of faults, joints and unconsolidated zones in consolidated rock	1	3	1	1	2	2	2	3	1	3	4		2	2			2	2	1					3	
investigation of geological structures and areas of elevated permeability in unconsolidated rock	1	2	1		1		2	3	2	3	2	1	2			1	3	3			1			3	

Table 1-2: Geoscientific methods for site investigations: landfill waste body (both operating and abandoned)

	Remote Sensing					Geophysics								Geology and Hydrogeology							Mineralogy and Geochemistry				
1 - occasionally used for this purpose **2** - suitable method **3** - method is well suited to this application **4** - preferred method (can be more than one) for this application/objective **METHODS**	satellite imagery	aerial photography	hyperspectral scanner	radar methods	laser scanner	magnetics	gravity method	dc resistivy and electromagnetics	ground penetrating radar	refraction seismics	reflection seismics	SNMR	geophysical downhole logging	geological field work	hydrogeological field work	direct push technology	investigation in trenches	drilling and installation of wells	hydraulic and tracer tests	flow and transport modeling	lithological and mineralogical methods	chemical analysis	elution, batch, reactor and column test	soil gas analysis	geochemical modeling
APPLICATIONS/OBJECTIVES	suitability of the method for the respective application/objective																								
identification of concealed abandoned landfills		**3**	1			**4**	1	**3**	1	1			1	2	1	2	2	2				**3**			
investigation of anthropogenic structures below landfills		**4**				2	2			2	2							2							
determination of the thickness of the landfill, mapping the base of the landfill		**3**					1	**3**		**3**	**3**					1	1	2							
internal structure of the landfill		2	1			**3**	1	**3**	1	1						1	1	1				**3**			
investigation of temporal and spatial development of a landfill		**4**	1			1	1	1	1	1	1					1	1	1				2	2		
observation of destabilization processes		**3**																							
verification and delimiting contaminant emission		**3**						**4**							1		1	2				**4**		**3**	

Table 1-3: Geoscientific methods for site investigations: groundwater conditions

	Remote Sensing					Geophysics								Geology and Hydrogeology							Mineralogy and Geochemistry				
1 - occasionally used for this purpose 2 - suitable method 3 - method is well suited to this application 4 - preferred method (can be more than one) for this application/objective METHODS	satellite imagery	aerial photography	hyperspectral scanner	radar methods	laser scanner	magnetics	gravity method	dc resistivy and electromagnetics	ground penetrating radar	refraction seismics	reflection seismics	SNMR	geophysical downhole logging	geological field work	hydrogeological field work	direct push technology	investigation in trenches	drilling and installation of wells	hydraulic and tracer tests	flow and transport (inverse) modeling	lithological and mineralogical methods	chemical analysis	elution, batch, reactor and column test	soil gas analysis	geochemical modeling
APPLICATIONS/OBJECTIVES	suitability of the method for the respective application/objective																								
type of GW, water table depth, thickness of aquifers and aquitards (hydrostatigraphic units)								3	1	2	2	2	3	2			2	4	3		2	2			
permeability of unsaturated zone								1				3	3		1		1	2	4	2	3				
permeability and transmissivity of saturated zone								2	1			3	3		1	2		3	4	2	3				
effective porosity, degree of saturation, storage coefficient								1				3	3		1		1	2	4	2	3				
water table: unconfined/confined/artesian								2	1	3	1		3		3	3	1	3	2	2					
level and fluctuation of groundwater												2			4	1	1	3		2					
relation to surface water		2	1												3	1	1	2	2	2					
direction of flow, flow rates								2							1			2	3	2					
groundwater recharge, run-off, water balance															4			1		3		2			
hydrogeochemical conditions		2	2										2		3	3				2		4	3	3	3

The authors and editors of this volume believe that the method descriptions and case studies in this volume will illustrate the advantages of applying remote-sensing, geophysical, geological, hydrogeological, geochemical, and microbiological methods to decision-makers faced with their own environmental investigations. The consistent and knowledgeable application of methods will improve the timeliness, cost-effectiveness, and thoroughness of most environmental site assessments. Site investigation is active environmental protection. Money for effective site investigations is well invested. This is an investment in our future and that of our children.

References and further reading

VOIGT, H.-J., NÖLL, U., KNÖDEL, K., JENN, F., RADSCHINSKI, J., GRISSEMANN, C. & LANGE, G. (2006): Recommendations for Site Investigations of Waste Disposal Sites and Contaminated Sites in Thailand. Bangkok, Berlin, Hannover, Cottbus. (contact: u.noell@bgr.de)

WILKEN, H. & KNÖDEL, K. (1999): Handbuch zur Erkundung des Untergrundes von Deponien und Altlasten, Band7: Handlungsempfehlungen für die Erkundung der geologischen Barriere bei Deponien und Altlasten (Handbook for Investigation of the Ground below Landfills and Sites Suspected to be Hazardous, vol 7: Recommendations for the Investigation of Geological Barrier below Landfills and Sites Suspected to be Hazardous). Springer, Berlin.

2 Preparatory Steps and Common Problems

KLAUS KNÖDEL, GERHARD LANGE & HANS-JÜRGEN VOIGT

The procedures for investigating planned, operating, or abandoned landfills, mining or industrial sites and the available methods are introduced in Part 1. The geoscientific methods are described in more detail in Parts 3 to 5. Certain preparations are required for any site investigation using ground-based methods in order to ensure the efficient implementation of field operations.

2.1 Placing of Orders and Order Handling

The "rules" for commissioning site investigations differ from country to country, very often depending on national or international regulations (e.g., for World Bank or EU projects). The "rules" must be well formulated and known and accepted by both customer and the contractor. Invitations to tender must contain a specification of the investigation aims and detailed terms of references for the commissioned work. The bidding contractor has to prove that he is able to carry out the requested work. Before a bid is submitted the site should be inspected by all prospective contractors. Thus, the bidders can become acquainted with the local conditions and the customer can be questioned in detail about the proposed investigation. The situation at the site (e.g., noise sources, sealed ground surface) can hamper the application of some methods and method combinations. On the other hand, information about the terrain, accessibility and plant cover in the area to be surveyed are important for estimating the length of time needed of the field work and thus for cost calculations. Not only is information about the survey conditions important for conducting a geoscientific survey, but attention must be paid to operational safety. The opinion of an operational safety specialist may be needed to rule out any hazards for the field team. In the case of a military or abandoned ammunition site, a specialist for the removal of explosives has to be consulted. This will entail additional costs. It has to be kept in mind – that the lowest bid is not always the best one! The contract awarded to the best bidder should include a listing of the rights and duties of both customer and contractor, a description of the work to be carried out, quality assurance measures, time schedule for the work, including the delivery of the reports and data, as well as the terms of payment.

The quality of the work should be checked (e.g., well logging to check the placement of screens in groundwater observation wells) by the customer. If the customer is not able to do so, the geological survey and/or the environmental protection authorities will provide advice and can help with the quality assurance of such work. Quality assurance is discussed in Chapter 2.6 and the corresponding sections in the method descriptions. The same base map (see Chapter 2.2) and the same benchmarks at the site or near it have to be used for all work carried out at a site. It is very important that the local authorities and residents be informed about the field work (see Chapter 2.3). It is also important for the investigation results to be available and transparent to the local people. Details of the rights and duties of the customer and the contractor, as well as on operational safety, are not subject of this volume. The objective here is to sensitize authorities, customers and contractors to these problems.

2.2 Collection and Use of Existing Data

The most basic requirement for successful site investigations is the availability of maps and documents which facilitate a general overview of the investigation area and its surroundings and provide as much detailed information as possible.

Topographic maps

Topographic maps mainly display the morphology, the vegetation cover, the drainage system, and infrastructure. Scales of official maps vary between 1 : 100 000 and 1 : 10 000. A widely used scale is 1 : 50 000. Map content, however, can be rather outdated if map revisions are made only at long time intervals. While the 1 : 50 000 scale appears to be appropriate for an overview, site investigations usually require the most up-to-date maps at a considerably larger scale (1 : 10 000 to 1 : 2000). Such a large-scale map is needed as the base map for all field operations, as well as for the subsequent data documentation and interpretation. In many countries, large-scale topographic maps are not available. Therefore, base maps have to be prepared from aerial photographs or high resolution satellite images (e.g., IKONOS and QUICK BIRD). Cloud cover on the images should be less than 20 %. A ground check of the maps with a handheld GPS instrument can provide coordinates with a precision of about 10 m. This is sufficient for site investigations in most cases, however. If higher precision is necessary, maps constructed from large-scale aerial photographs (greater than 1 : 10 000) have to be geocoded and rectified by ground checks in the field (see Chapter 3.2).

Geoscientific maps

Geoscientific maps are a further necessary tool for site investigations. They generally depict a variety of information about geology, hydrogeology, mineral exploration, soil, natural hazards, and land use. Although the scale of the available geoscientific maps might not always be suitable to obtain the needed information in the necessary detail, the maps nevertheless place the investigation area in its regional context and permit the survey data to be interpreted with respect to its geological, hydrogeological and structural setting.

Aerial photographs and satellite images

Aerial photographs and satellite images are very useful tools for planning and implementing field operations. They provide the best overview and impression of the topography and infrastructure of the investigation area. In certain cases they allow a three-dimensional evaluation. They can provide information about lineaments, geomorphological structures, and flood-prone areas, as well as land use. Aerial photographs taken at different time are helpful for reconstructing the history of a site, often with immediate relevance to environment-related problems (see Part 3).

Point data compilation

An important preparatory step for a geoscientific site investigation and data interpretation is the collection of already existing information about the subsurface (desk studies). This task mainly comprises the compilation and assessment of available data from boreholes for groundwater or mineral exploration. These point data are a valuable source of information and are indispensable for the interpretation of geophysical results and calibration of hydrogeological models. Results of desk studies on the geology at the site should be verified in the field. Quarries, road cuts and other outcrops provide information about the stratigraphy, lithology and structures in the investigation area. The data should be easily accessible in an appropriately prepared and documented database (e.g., GIS).

Document review

For the investigation of abandoned landfills, industrial and mining sites, it is necessary to obtain the history of the site as completely as possible. For this purpose, a review of all documents on the site facilities, their construction, length of operations, as well as waste composition, quantity, and treatment methods. These documents will be normally available from the relevant authorities or the site owners. This information not only aids the interpretation of geophysical survey data and hydrogeological observations, but it is also essential for any assessment of the risk present at the site and for recommendations of remediation measures and further use of the site. As not all or even any such information may be available, it is highly useful to interview persons who lived or worked around the site during the time of landfill, mining or industrial operations.

2.3 Information Campaign and Permit Application

Investigation of an operating, abandoned and planned landfill, mining or industrial site requires the approval of the relevant authorities at the state, district and/or local levels and of the affected private property owners. The objectives of the investigations, the estimated duration, and the scope of the work, as well as the methods and their likely impacts, should be explained in detail to those affected before approval for the operations is requested. Recommendations and information from public and private sources should be integrated into the working concept.

As geophysical surveys and likely follow-up shallow drilling have to extend beyond the immediate investigation site, a large number of private and public landowners may be affected. Before field work is begun, permits must be applied for in order to ensure the work will be undisturbed during the site investigation. Before a permit is applied for, the customer should give the contractor a letter requesting support for the field work. The customer should also provide a site map showing the coordinates of survey lines and/or the boundary of the survey area. Depending on nature and extent of the investigation area, as well as the authorities and organizations concerned, e.g., municipal administrations, police stations and other relevant institutions must be informed on the impending field work (see check list in *Table 2-1*). If the investigation area includes nature, landscape, or water protection areas or other restricted areas, the governmental and/or private organizations involved must be provided information about possible impacts of the survey method(s) that will be used in order to obtain an exemption to the restrictions.

Table 2-1: Check list for the preparation of site investigations

customer	provides the contractor with a letter authorizing the implementation of the work
customer (if possible)	provides plans and maps
land surveying offices, land registry offices, map distributors	provides topographic and geological maps (also as digital maps), general and detailed maps, land register maps, and names of property owners, information about topographic benchmarks
district authorities	to be informed and, if necessary, authorization is requested
municipal authorities	to be informed and, if necessary, authorization is requested
local authorities	to be informed and, if necessary, authorization is requested
police stations if necessary	to be informed and, if necessary, authorization is requested
mining authorities if necessary	to be informed and, if necessary, authorization is requested
water management authorities if necessary	to be informed and, if necessary, authorization is requested
forestry offices if necessary	to be informed and, if necessary, authorization is requested
road authorities if necessary	to be informed and, if necessary, authorization is requested
railway operators if necessary	to be informed and, if necessary, authorization is requested
occupants/leaser and/or owners of the premises	provides name, address, designation of the premises involved, private roads that may be used, authorization to use the premises, description of any damages
utilities operators (long-distance and local operators may be different): – freshwater – waste water – electricity (including street lighting) – gas – telecommunications	provides information on the location of utility lines
specialists for the removal of unexploded ordnance (UXO), if necessary	would involve a subcontractor
contractor	secures office and lodgings in the field (address, telephone and fax numbers, number of rooms, price)
contractor	allocates human resources and equipment, hires local manpower if necessary
contractor	provides the field team with road maps showing the way to the investigation area and the meeting point

It is important to collect all available information on existing surface and subsurface installations (e.g., power lines, water mains, oil and gas pipelines, and sewer lines). If no official maps are available, local people generally know about these installations. As the locations of power lines and pipelines in an area often differ from the records provided by the operators, it is advisable to check the information on the maps with a cable/pipeline detector before starting the work. If public roads or railway tracks are crossed by survey cables it is necessary to discuss safety precautions with the road authorities or railway operators and procure written authorization. In some cases, the field investigations should be announced in local newspapers, stating time and area concerned.

When the ownership of the premises has been determined, the owners or occupants must be contacted. The persons affected by the planned survey are given leaflet containing information on the site investigation and a questionnaire to obtain information about the locations of drainage ditches and other drainage facilities, private roads, power and water supply lines etc. so that appropriate measures can be taken to avoid damage. In farming areas, the contractor should meet with property owners and/or leasers to discuss the avoidance of disruptions of the field work and damage to survey equipment by agricultural machines.

Immediately after termination of the survey, owners and occupants/leasers have to be contacted in order to assess any crop damage etc. Indemnity for crop damage depends on the season, on the anticipated harvest, and on the kind of damage to the land.

2.4 Mobilization and Demobilization

Field work generally starts with the mobilization of a survey party or a survey group and ends with its demobilization. This involves the following work and is part of the contract:

- Preparation of the survey equipment and transport to the survey site,
- transport of the working party to the survey area,
- establish a field office and preparation of the field work,
- disassembly and loading of the equipment at the end of the field work,
- transport of the working party back to the home base, and
- unloading and maintenance of the equipment.

The personnel and time required depends on the methods, the distance to the location, and the size of the investigation program.

2.5 Land Surveying

Additional geotechnical measures, for example, drilling and cone penetration tests, or supplementary geophysical investigations are often planned within the scope of site investigations. These are carried out on the basis of the results of a geophysical survey campaign. Thus, the exact determination of measuring points and profiles is very important in order to document consistently all investigations in a site map of an appropriate scale.

To properly interpret the geophysical data, it is often necessary to relate the surveyed profiles and measured points to the country's national topographic grid. The coordinates and elevation of benchmarks in or near the area to be surveyed have to be obtained from the land surveying office before the fixed points are surveyed at the investigation site. A map of the site and surroundings at a scale[1] appropriate for the aims of the survey should be made available by the customer. This will make it possible to compare the results of several contractors working in the same project area.

Position determination

The fixed points (benchmarks) near the investigation area must be inspected and their coordinates must be expressed in the units of the country's geodetic reference system (WGS 84, GRS 80, Universal Transverse Mercator Grid System - UTM, Gauss-Krueger system) so that the survey results can be related to the official maps.

Table 2-2: The required accuracy of the geophysical measuring points and boreholes as a function of the scale of the field and presentation maps

Scale of the field and presentation maps	Required accuracy
1 : 10 000	± 10 m
1 : 5 000	± 7.5 m
1 : 2 000	± 5 m
1 : 1 000	± 2 m
1 : 500	± 1 m

The required accuracy of the geophysical measuring points and boreholes, as shown in *Table 2-2*, depends on the working and presentation scale. The tolerable error in the relative positions measured within the survey area depends on the applied geophysical method. It must be smaller than ± 1 m. For quality assurance, the accuracy of the position and height measurements should be documented in the report.

[1] The scale should be chosen so that the density of measuring points in maps and profile plots is not less than three measuring points per centimeter.

If there are several benchmarks in the investigation area, the locations of the geophysical measuring points and profiles can be determined by "free positioning". The instrument (e.g., tachymeter) is positioned at a point where a line-of-sight view of the profiles to be measured geophysically and at least two, better three, fixed points is possible. The coordinates of this point can be determined from these measurements. The measurement to three or more benchmarks provides an indication of the precision of the coordinate determination for the standpoint. This method cannot be used if there is an insufficient number of fixed points. In this case, measurements must be made at several points along the profile and subsequently connected to the fixed points.

In some site investigations, a local base grid is used and permanent markers placed at three base points of the grid. The locations of these base points must be marked on the maps or sketch maps included in the technical and scientific reports. The locations of the measuring points and profiles are determined with respect to this base grid using measuring tapes, an optical square and ranging poles. If possible, the local grid can be connected to the country's grid.

Altimetry

The required precision for altimetry measurements is given in *Table 2-3* according to the purpose they are used for. If there are no geodetic points with an elevation value near the survey area, measurements to tie the local benchmarks with remote benchmarks must be carried out. The most exact method for this objective is geometric leveling, which, however, takes a considerable amount of time, especially in rugged terrain. Measurements must be made to at least two remote points as a control on each other. This can result in long topographic leveling paths.

Table 2-3: Required precision for altimetry measurements

Method/object	Precision
gravity methods	± 0.03 m, for special measurements ± 0.003 m
seismic methods	± 0.1 m
geoelectrics, geomagnetics	± 0.5 m
groundwater observation wells	± 0.01 m

The measurements can be made much more quickly if electronic tachymeters are used. The possibility of direct transfer of the data from the tachymeter to a computer also facilitates the data processing. The precision of the tachymeter measurements is less than that of geometric leveling, but sufficient for geophysical investigations. Because benchmarks often give only elevations or only coordinates, separate altimetry measurements are then necessary to tie the local benchmarks to the official grid.

Satellite supported positioning

The American Department of Defense completed the **Nav**igation **S**atellite **T**iming **a**nd **R**anging – **G**lobal **P**ositioning **S**ystem (NAVSTAR-GPS) in 1993. This system consists of 24 operating and 3 extra satellites, with at least 4 satellites being visible from any place on the Earth at any time. Each satellite transmits modulated navigation information on two carrier frequencies (L1, L2). This information includes the orbit parameters of the satellite (ephemerides), a refraction model, a clock parameter, as well as the orbital data (in simplified form) of all satellites in the space segment. Two positioning signals (C/A code[2], P code[3]) are modulated on frequency L1. With the P code modulated on L2 and the navigation information, the signal traveltime from the satellite to the receiver can be determined. The P code is encoded as a safeguard against falsification ("anti-spoofing", A-S) and is accessible only by authorized users, especially military. From the traveling time of the signals, a GPS receiver calculates its three-dimensional position (longitude, latitude and altitude). Longitude and latitude can be calculated from the data of at least three satellites and using the signals of a fourth satellite the altitude can also be determined.

"Selected Availability" (SA), which reduced the attainable precision by random signal falsification, was deactivated in May 2000. When that was done, measurements could be made by nonprivileged users with greater precision. Thus, using simple GPS receivers, a precision of ± 5 to 20 m can be reached for the latitude and longitude. The precision for altitude is generally lower by a factor of 1.5 to 2. Maximum precision, however, can only be attained with an optimum constellation of the satellites. How optimal the constellation is is indicated by the PDOP factors[4] of the GPS display. Multipath effects, generated at the walls of buildings, can lead to a decrease in precision.

Precision can be enhanced, for example, by using differential GPS (DGPS). The error in each satellite signal is calculated from the signals received by a GPS base station with coordinates known from a non-GPS source. The correction data is then transmitted to the portable DGPS station. The precision in the GPS measurements can be within a few centimeters when both the L1 and L2 carrier frequencies are used. There are three ways to transmit this information:

- Direct transmission is possible if the base station and the mobile station are only a few kilometers apart and there are no significant barriers between them. This is due to the limited range of the radio in the GPS instruments.

---

[2] C/A code stands for Clear/Acquisition, Clear/Access or Coarse/Access

[3] P code: Precise Code

[4] PDOP: Position Dilution of Precision: the smaller the value the better the results that may be expected.

- As the correction signal of a reference station, however, can be used for distances up to several hundred kilometers (depending on the required precision) powerful commercial transmitters are being increasingly used for transmission of the correction data.
- The third possibility is to send the correction signal via satellite. Some companies offer reference signals from geostationary satellites for a fee. Required is line-of-sight contact between the GPS instrument and the satellite. This kind of reference data transmission is of interest in non-European countries.

Use of satellite positioning for geoscientific purposes is a relatively new development. Besides the NAVSTAR-GPS and the Russian GLONASS systems, which can be used worldwide, the EU is installing the GALILEO system, for which more than 30 satellites will have been launched by 2008. Especially multifunctional "satellite positioning services" will increase the importance of satellite-supported positioning and navigation. These systems can be used without problems in open land, difficulties may occur in woodland or urban areas. A precision of about 1 cm to several millimeters is attainable by post-processing. The coordinates and altimetry data determined by GPS must be converted to the national or local frame of reference.

Personnel, equipment, time needed

How long it takes a surveying crew to execute a land survey depends on the precision required by the geoscientific methods that will be used. The survey progress is determined by topography, vegetation, visibility conditions, the distance to fixed points (benchmarks) and the survey method. High-performance GPS systems and electro-optical tachymeters are now available, allowing precise survey data with relatively few personnel. In favorable cases, e.g., dc resistivity soundings or gravity and magnetic surveys, a land survey can be carried out by the geophysical personnel together with the data acquisition. The personnel, equipment and time needed in comparison with geophysical measurements are given in *Table 2-4*.

Table 2-4: Personnel, equipment and time needed in comparison with geophysical measurements

	Survey personnel[5]	**Equipment needed**	**Time**[6] **required [in %]**
marking of points on the basis of the topography and map	0	measuring tape	5 - 10
marking of points on the basis of the topography and map and subsequent determination of position and elevation	1 - 2	position determination by tachymeter or GPS	20 - 50
preparation of map showing the positions and elevations of the profiles and points in the survey area	1 - 2	computer with appropriate software for maps	50 - 100

2.6 Quality Assurance and Reporting

Quality assurance is an important part of the field work, the data processing, the interpretation, and the reporting. It must be documented in field protocols as well as summarized in the reports. Quality assurance includes

- instrument checks before, at certain time intervals during, and after the measurements;
- careful maintenance and calibration of equipment;
- entry of all relevant occurrences and possible noise sources (e.g., passing cars during the geophysical measurements, heavy rain, storm, steel fences, power lines) in the field protocols;
- daily data check and plotting of results during the field campaign;
- repetition of a certain percentage of the measurements in order to estimate measurement errors;
- documentation of all steps and parameter values used in the data processing and interpretation.
- The report must contain all information needed to understand the report.
- The figures and maps must be clear, understandable, and convincing.

---

[5] Land survey personnel in addition to the personnel needed for the geophysical survey.

[6] Time needed for the land survey compared to the time needed for the geophysical survey. 100 % means that the same time is needed for the land survey as for the geophysical measurement.

- The conclusions of the report must answer the question whether and how far the tasks and targets were achieved, must list remaining problems, and must be understandable for non-experts.
- All measured data, relevant intermediate results, as well as figures and maps of the final report must be submitted to the customer in well documented form in both hardcopy and machine-readable form (e.g., CD or floppy disk) for later use as evidence in possible legal dispute and/or reinterpretation together with results of other investigations and, if necessary, with supplementary measurements.

For the specific aspects of quality assurance of methods, see the "Quality Assurance" sections in Parts 3 to 5.

3 Remote Sensing

PETER HERMS, BERNHARD HÖRIG, FRIEDRICH KÜHN, DIETMAR SCHMIDT & ANKE STEINBACH, with a contribution by TILMANN BUCHER

3.1 Aerial Photography

DIETMAR SCHMIDT & FRIEDRICH KÜHN

3.1.1 Principle of the Methods

Despite all technical progress in digital imaging, interpretations of standard aerial photographic images remain an important remote-sensing tool. Aerial photography can address a multitude of geoscientific questions and can be highly effective when used for logistics and planning (BÖKER & KÜHN, 1992). The cost of aerial photography is rather low. The data are informative, easy to manage, and the film does not require special image processing resources for analysis.

Cameras for aerial photography, also referred to as metric cameras and frame reconnaissance cameras, are analog to standard photographic cameras using lenses, shutters, and film. These cameras, however, are more expensive, and mechanically and electronically complicated since they need to adjust the orientation of the camera during data acquisition. Ideally, aerial photography equipment is coupled to a Global Positioning System (GPS) to ensure better precision and accuracy.

Photographic multispectral cameras have been widely used in planes and satellites in the 1970s and 1980s (COLWELL, 1983; KÜHN & OLEIKIEWITZ, 1983). Since then, the application of such cameras has decreased for several reasons: their complicated operation mode, spectral distortion depending on the angle of incidence on the camera lenses and filters, and the complicated and imprecise evaluation procedures. They have been replaced by multispectral digital imaging systems (see Chapter 3.3).

Although there are a variety of types of film available for aerial photography, three kinds are used in almost all situations. These commonly utilized films are panchromatic (black and white), color and color-infrared

(CIR) film. The latter one is able to serve most environmental and geological tasks.

Vertical aerial photographs provide a three-dimensional impression of an area. Interpretations of stereo-pair photographs require an overlap of 60 - 90 % between two images along the flight line. An overlap of approximately 35 % is needed between adjacent flight lines to allow for flight irregularities. As the distance between any two consecutive images is greater than the distance between the human eyes, the topography of the target area appears strongly enhanced and exaggerated when the images are viewed stereoscopically. This enhancement commonly provides detailed information for geological and environmental analysis. Aerial photographs can be used for, but are not limited, for detection of man-made features.

In addition, aerial photographs can often be analyzed to identify rock formations and types of soil, typical relief forms, distinctive vegetation types, drainage patterns and specific types of land use via changes in color or gray scale. In general, as the frequency of man-made features increases in an area, it becomes more difficult to extract geologic information from aerial photographs.

Even older aerial photographs provide proper spatial resolution and often document changes in an area of interest over a period of several decades. They allow the evaluation of geological and initial environmental situation. Archives of aerial photographs are maintained worldwide, e.g., in governmental survey offices, and can provide aerial photographs at low cost.

A thematic interpretation of a specific site using aerial photographs is usually carried out at scales between 1 : 2000 and 1 : 10 000. A topographic base map at the selected scale is necessary. The preparation of topographic maps is described in Chapter 3.2. Regional analyses (e.g., lineament analyses) are carried out at scales between 1 : 25 000 and 1 : 100 000.

3.1.2 Applications

- Chronological analysis of the development of land use, waste disposal, mining, and industrial sites and their surroundings,
- assessment of the previous geological and environmental situation at sites now covered by waste disposal, mining, and industrial facilities,
- search for seepage of water at the edges of landfills and mining waste heaps,
- locating springs and moisture anomalies,
- investigation of natural and artificial drainage systems,
- locating areas of high and low permeability to water,

- recognition and monitoring of areas of destabilization,
- investigation of vegetation vitality,
- assessment of surface water conditions,
- delineation of fractures and lineaments,
- inventory of sites suspected to be contaminated,
- mapping of land use patterns and biotops, and
- search for potential new waste disposal sites.

3.1.3 Fundamentals

Remote sensing utilizes electromagnetic (EM) radiation in the ultraviolet, visible, infrared, and microwave portions of the EM spectrum to obtain information about the Earth's surface with sensors carried by either an aircraft or satellite. Active and passive methods are used for remote sensing. Passive methods (see Chapters 3.1 - 3.3) use reflected solar radiation and radiation emitted from a surface. Active remote-sensing systems (Chapters 3.2 and 3.3) have their own source of radiation (e.g., radar or laser).

Interaction of electromagnetic radiation with the Earth's surface provides information about the reflecting or absorbing materials. Due to the small wavelengths of EM radiation used (nanometer to micrometer range) there is limited penetration of the targeted (optically thick) land surface objects. Given the proper atmospheric transmission, remotely sensed data represent the Earth surface conditions. Because EM radiation is the carrier of the remote-sensing information, it is important to define the different ranges of the EM spectrum. In the literature, it is difficult to find an unambiguous division of the electromagnetic spectrum, since it is of continuous nature. Divisions are commonly based on type of sensor and response of natural materials, etc. We have decided to use the divisions of the spectrum of ERB (1989). In addition to ERB's divisions, we will use a further subdivision of the **m**iddle **i**nfra**r**ed (MIR) into MIR-I and MIR-II (*Table 3.1-1*) for remote-sensing purposes.

The sun is the primary source of the Earth's incident EM radiation. The energy spectrum of the sun is almost identical to that of a black body at 5900 K. Solar radiation reflected from the Earth's surface is in the NUV to NIR-II (0.315 - 3.0 µm). However, there are gaps in this spectral range due to scattering, absorption, and reflection by gases, particles, and other atmospheric constituents. Major atmospheric absorption bands appear at 1.4 µm, 1.9 µm, and 2.5 - 3.0 µm resulting from atmospheric water vapor and carbon dioxide (KRONBERG, 1985; KUEHN et al., 2000). Little or no solar radiation reaches the Earth's surface at these wavelengths and they cannot be used for land remote-sensing purposes.

Table 3.1-1: Spectral range of electromagnetic radiation utilized by remote-sensing sensors; spectral divisions by ERB (1989)

Radiation	Abbreviation	Wavelength λ [in μm]
near ultraviolet	NUV	0.315 – 0.38
visible light	VIS	0.38 – 0.78
near infrared	NIR-I	0.78 – 1.4
	NIR-II	1.4 – 3.0
middle infrared	MIR	3.0 – 50.0
(thermal infrared)	MIR-I	3.0 – 5.5
	MIR-II	8.0 – 15.0
far infrared	FIR	50.0 – 1000
microwave (radar)	MW	1000 – 10^6

Solar radiation is either transmitted, absorbed or reflected at the surface. Reflections can be either direct or diffuse. The proportions of reflection, absorption, and penetration of the incident radiation at the Earth's surface depend on the physical, chemical, structural, and textural properties of the surface. Molecules and chemical bonds (which make up all materials, including soils, rocks, water, and plants) are characterized by a specific energy level. The absorption of the incoming radiation provides the energy required for the transition from one energy state to another. Remote-sensing systems record the reflected radiation that reaches the sensor after its interaction with the Earth's surface. Hence, the properties of the soil, rock, and other materials at the Earth's surface are indirectly recorded in the data or images.

The maximum energy of solar radiation is within the visible part of the spectrum (VIS) at 0.48 μm. The maximum radiation from the Earth is at about 9.7 μm. Thus, the best range for recording thermal radiation from the Earth's surface is between 8 and 12 μm.

Within this spectral range, slight temperature variations resulting from differences in the type of soil, soil moisture, or the presence of specific pollutants can be identified. Therefore, thermal scanners (Chapter 3.3) are designed to be most sensitive in the spectral range of 8 - 12 μm. To investigate hot surfaces (e.g., radiation from a blast furnace or the structure of a lava flow), a spectral range of 3 - 5 μm is more suitable.

Radiation in the visible and infrared portions of the spectrum is absorbed by certain constituents of the atmosphere. Major absorption bands lie between 5 and 8 μm and between 13 and 30 μm. These spectral regions are usually not used for land remote sensing. Applications of thermal remote sensing are described in Chapter 3.3.

Vertical aerial photographs are commonly produced at scales of 1 : 20 000 (sometimes 1 : 50 000) to 1 : 5000 for applications in environmental geology. Using a lens with a focal length of 150 mm, the flight altitude required to produce these scales is between 3000 m and 750 m above ground, respectively. Standard film size of 23 cm × 23 cm then will capture an area of 4600 m × 4600 m (at a scale 1 : 20 000) or 1150 m × 1150 m (scale 1 : 5000). The basic parameters of aerial photography, i.e., the scale (S) of the aerial photograph, the field of view (A) of the camera, and the ground resolution (R_g), can be calculated using the following equations (BORMANN, 1981a):

$$S = \frac{c_f}{h_g} = \frac{1}{S_n}, \tag{3.1.1}$$

$$A = \frac{h_g k}{c_f}, \tag{3.1.2}$$

$$R_g = \frac{h_g k}{R_L}, \tag{3.1.3}$$

where c_f is the focal length of the camera lens,
S_n scale constant,
h_g flight altitude above ground,
k film frame size (usually 23cm × 23 cm), and
R_L line resolution (lpi) of the film according to the manufacturer.

Equation 3.1.3 gives an estimate of the spatial resolution, R_g. To precisely calculate the resolution of a given "film–lens" system, a contrast modulation function has to be used which describes the relation between the degree of contrast in the terrain and in the image as a function of object size for a given spatial frequency domain (BORMANN, 1981b).

3.1.4 Instruments and Film

Aerial photographs are used for both thematic interpretation and photogrammetry. The equipment and quality standards are mainly the same. They differ only in the films used. The following equipment and materials are required for aerial photo surveys:

- Large format aerial cameras (*Table 3.1-2*) are available with different focal lengths (*Table 3.1-3*). Modern survey cameras (*Table 3.1-2*) are fully equipped for movement compensation (IMC – Image Motion Compensation). Blurring caused by the forward movement of the aircraft, can be compensated by FMC (Forward Motion Compensation). Fuzzy film images resulting from vibrations of the aircraft are prevented by using a gyroscope mount.
- The "Flight Management System" (corresponds to IGI's CCNS-4, T-FLITE of ZI/IMAGING or Ascot of LH-Systems), which was specially developed for survey flights on the basis of GPS, enables higher precision

in flight management and more accurate shutter release for the specified areas.

- DGPS (Differential Global Positioning System, see Chapter 2.5) receiver in the aircraft and on the ground, as well as GPS for terrestrial surveying.
- Suitable film (see below in this section).
- Equipment for the developing of black and white, color and/or CIR aerial film.

Table 3.1-2: Technical specifications for a large format aerial camera

	ZEISS RMK TOP	**LEICA RC 30**
manufacturer	ZI/Imaging	LH – Systems
format	23 cm × 23 cm	23 cm × 23 cm
film magazine capacity	152 m	152 m
exposure cycle time	1.5 s	1.5 s
weight	ca. 150 kg	ca. 150 kg

Table 3.1-3: Lens types for aerial cameras, ALBERTZ (2001)

Lens Type	**Focal length [cm]**	**Maximum angle**
narrow angle	61	33 gon (30 degrees)
normal angle	30	62 gon (56 degrees)
intermediate angle	21	83 gon (75 degrees)
wide angle	15	104 gon (94 degrees)
super wide angle	9	134 gon (122 degrees)

Depending on the task and the desired results, different kinds of film can be used for aerial photography:

Panchromatic black-and-white film

In many countries the most common film used for aerial photographs is black and white panchromatic film. This film is frequently applied for geodetic and cartographic purposes, but may also be used for specific thematic mapping tasks. The high spatial resolution allows for identification of minor fractures. Phenomena such as changes in rock and soil type, soil moisture, or secondary effects related to contamination can also be detected with this film. Stereoscopic techniques using pairs of aerial photographs allow an experienced interpreter to characterize objects at the surface of a site in detail. SCHNEIDER (1974), KRONBERG (1984) and CICIARELLI (1991) show a multitude of applications for panchromatic film. Updates of topographic maps are in many countries made with panchromatic black-and-white film (*Fig. 3.1-1*).

Fig. 3.1-1: Black-and-white aerial photograph taken by the Royal Thai Survey Department on November 11, 1995, showing the more than 35 m deep Nong Harn borrow pit near Chiang Mai in northern Thailand. The pit is being filled with waste from the west along a road to the bottom of the pit. The pit has steep slopes and was previously about 45 - 50 m deep. The western part of the pit floor is covered by more than 10 m of waste. The southern, western, and eastern slopes are covered by a plastic liner. The fine white lines on these slopes are ropes to hold down the strips of the liners. The northern slopes have the ropes too, but no plastic liner cover, as indicated by the brighter gray color and by the micro-topography of the uncovered slope.

Infrared film

Black-and-white film can be sensitized to extend beyond the visible range of light (VIS) into near-infrared wavelengths (NIR-I). This part of the near-infrared spectrum, also referred to as the photographic infrared, ranges from wavelengths of 0.78 µm to about 0.9 µm. EM radiation detected by near-infrared film is reflected solar radiation and should not be mistaken with thermal infrared radiation. The latter is the EM range between 8 and 15 µm and is used for thermal imaging. Due to the intense absorption of infrared radiation by water, near-infrared film is highly sensitive to changes in soil moisture content. Also, near-infrared film can be used to detect changes, extent, and state of vegetation cover. Healthy vegetation intensely reflects radiation in the same range. Color infrared film (CIR) has largely replaced black-and-white infrared film for investigations of damage to forests and other environmental conditions.

Color and CIR film

Color aerial film is used to depict a terrain in natural colors. Color film has three light-sensitive layers for the visible wavelengths of blue, green, and red. By substituting a near-infrared-sensitive layer for the blue-sensitive layer, a false-color photograph can be produced. This near-infrared-sensitive layer reacts to the intense reflection of NIR-I radiation by vegetation. As a result, if a reversal process is used for developing the film, the chlorophyll-rich vegetation will be represented by intensive red colors and low chlorophyll content by pale grayish red. Because there is less scattering of infrared radiation in the atmosphere than that in the visible range, CIR film (*Fig. 3.1-2*) yields sharper images with greater contrast from high altitudes than normal color film. The increase in contrast is enhanced by the use of a yellow filter to suppress the blue parts of light.

CIR film has proven indispensable for a broad range of thematic mapping objectives. Environmental mapping relies on CIR aerial photography because of its high sensitivity to changes in plant type and vitality. Methods for mapping the vitality of trees in the vicinity of a waste disposal site using CIR images are described in Section 3.1.6.

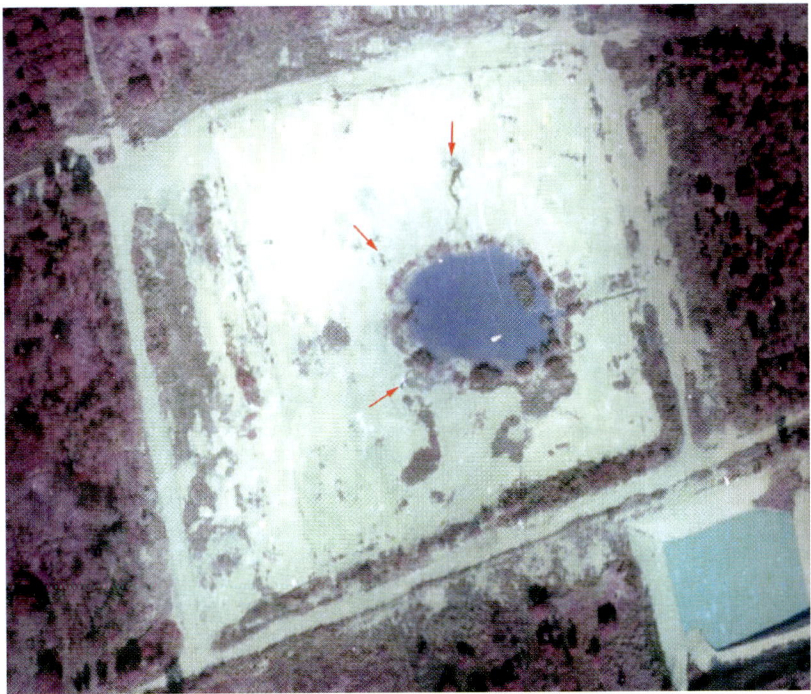

Fig. 3.1-2: CIR photograph, taken by Hansa Luftbild (German Air Survey) together with Sky Eyes (Thailand) on December 22, 2000, showing the refilled Nong Harn borrow pit near Chiang Mai in northern Thailand. The site is surrounded by a forest of several species of trees, recognizable by different colors and shapes of the crowns. Compaction of the 45 - 50 m thick waste in the pit creates a depression cone, visible by the blue round pond in the middle. Usually, water in CIR photographs appears dark or black. The blue color indicates relatively high turbidity. Chemical reactions within a landfill produce gas, which carries water with it as it ascends and escapes at the surface. The red arrows show the flow direction of surface water towards the pond at the surface.

Archival aerial photographs

Generally, any characterization of hazardous sites starts with an archival search for aerial photographs to reconstruct the historical development of the site and its surroundings. These photographs are commonly panchromatic black-and-white. Archives of aerial photographs are maintained by governmental survey offices, municipal administrative offices, commercial aerial photography firms, military institutions, private archives, and users of aerial photographs.

Fig. 3.1-3: World War II aerial photograph (March 24, 1945) of a destroyed industrial complex in the Dortmund area (Ruhr district, Germany). Damaged industrial plants, from which pollutants may have seeped into the soil, and bomb craters and pits, which may have served as dumps, are potential hazardous sites (courtesy of Luftbild-Datenbank, Ing.-Büro Dr. Carls, Estenfeld).

Most wartime aerial photographs were taken during World War II. More rarely, photographs from times before are available as well. Target Information Sheets exist for most wartime reconnaissance flights; these define the rationale for and the outcome of each bombing. In general, it is possible to use these sheets to deduce which potentially toxic substances seeped into the soil after bombing an ordnance plant or a chemical factory. These sites should be considered to be potentially hazardous. Although aerial photographs taken as part of a land survey usually cover large areas, military photographs are generally taken on linear traces with sudden direction changes (DECH et al., 1991). The archives of the American and British governments are additional sources of historical aerial photographs. There are also archives of aerial photographs which were taken during military actions by the American or British army since World War II. *Figure 3.1-3* shows an archival aerial photograph taken in 1945. In general, private consulting companies are specialized in archive searches for these wartime aerial photographs.

Oblique aerial photographs

Geo-environmental assessments may require oblique aerial photographs, when overviews of large areas are needed (*Fig. 3.1-4*). Oblique photographs may prove particularly valuable in the early phases of a project when the overall site is being characterized.

Fig. 3.1-4: Oblique aerial photograph taken on May 11, 1993, looking westward across the waste incineration facilities at Gallun, south of Berlin, Germany, (*foreground*) and the northern part of the Schöneiche landfill (*middle ground*) to the Schöneicher Plan landfill in the background. (Photo: F. BÖKER and F. KÜHN, BGR).

Interpretation instruments

The thematic evaluation and interpretation of aerial photographs can be carried out using a mirror stereoscope or a digital stereoplotter. Stereoscopes are available in various shapes and sizes.

3.1.5 Survey Practice

Important aspects in planning and carrying out an aerial photography survey are selection of the camera, objective and navigation system, scale, ground resolution and flight altitude, flight path and spacing of flight lines, selection of a suitable film, choice of season and time of day, weather conditions, and ground check and logistics. For scale, ground resolution and flight altitude see Section 3.1.3. State-of-the-art instruments and films are described in Section 3.1.4. Aerial photography surveys, which now are semi-automated with modern navigation and camera systems, require extensive sophisticated equipment and techniques. As already mentioned in Section 3.1.1, aerial photographs should be taken with 60 - 90 % overlap in the direction of flight for stereoscopic restitution to be carried out. An overlap of approximately 35 % is needed between adjacent flight lines to allow for flight irregularities (*Fig. 3.1-5*). Current aerial film has different spectral sensitivities and geometric resolution (see Section 3.1.4). The film must be transported from the flight area to the photo lab and developed strictly following manufacturer's instructions.

Ground control point coordinates for rectification of aerial photographs may be determined after a flight and the film has been developed in a separate GPS survey if no reference data set is at hand. For this task it is necessary to select and carefully mark control points in the aerial photographs as well as some additional points that can be used if the selected ones are inaccessible on the ground. During the ground survey it is helpful to carry out photographic documentation or sketches of the terrain at the control point to confirm the relationships between the images at later stages. Coordinates of ground control points must always be determined at ground level in order to avoid a shift between the on-site measurement and the control points on the aerial photograph (e.g., roofs and masts). Particularly suited for control points are geometric man-made features such as street intersection and trail crossings, junctions, bridges, and the corners of buildings and walls.

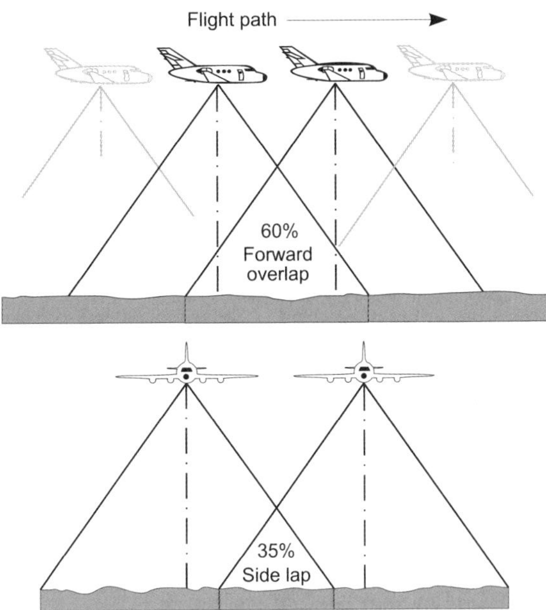

Fig. 3.1-5: A 60 % forward overlap and a 35 % flight-path overlap (sidelap) are necessary for effective stereoscopic evaluation (3-D) of photo pairs, after SCHNEIDER (1974)

The time of acquisition of the aerial photography is a critical factor in maximizing the value of aerial photographs for interpretation (BÖKER & KÜHN, 1992). Vegetation often covers geological features. Therefore, photographs for geologic interpretation should be taken during seasons with minimum vegetation. In temperate climate zones for example, aerial photographs taken in early spring after the snow has melted and the surface has dried up, and before the vegetation has fully emerged, are well suited for separating different kinds of soils and rocks and for detecting fractures and permeability anomalies. On the other hand, the existence of pollutants in specific areas may be recognized on the basis of certain vegetation patterns. In this case, and for mapping land use patterns and biotopes, it is better to use aerial photographs taken during the summer. Data about damages to vegetation caused by contamination should be acquired during the second half of the growing season in the late summer to early autumn. Taking photographs during this season avoids both the springtime vegetation burst, when small or moderate changes in vitality are hard to detect, and the rapid seasonal decay in late autumn.

In the climates of the subtropics and tropics, the alternation between the temperature and moisture conditions of rain and dry seasons determines the time of aerial photo acquisition. For investigations of barren soils, rocks, and geological structures the best time is at the end of the dry season, when the vegetation cover is at its minimum. If the vegetation itself is the object of evaluation, e.g., for land use or biotope mapping, forest inventory and vitality evaluation, the beginning of the dry season is the best time for the photographic survey. The vitality of the vegetation is commonly still high at this time and visibility from the air is good for taking aerial photographs.

Weather conditions are an important factor when acquiring images for geologic interpretation because soil moisture can greatly affect the quality of data interpretations. In some instances, it will be even useful to acquire data at a low altitude of the sun (10° to 30° above horizon), since low-relief forms appear enhanced by the more pronounced shadows on the ground. This increases the chance of detecting subtle geologic features, such as fractures. Consequently, the experience of the remote-sensing specialist is crucial in determining the appropriate strategy for acquisition and interpretation of aerial photographs.

The interpretation should be combined with at least two ground checks, one just before the survey – to establish a classification scheme, i.e., interpretation key – and one at the end of the survey.

3.1.6 Interpretation of Aerial Photographs

The most important image characteristics and landscape factors used in mapping vegetation, soil, rocks, and geologic structures from aerial photographs are:

- photographic gray scales,
- morphology,
- vegetation,
- drainage systems,
- geologic structures, and
- patterns related to land use.

Fig. 3.1-6 (next page): Generalized schematic of a landfill with a permeable base and without a drainage water collection system (top is a plan view and bottom is a cross section) showing structures, features, and properties that may be identified with remote-sensing techniques, depending on the site conditions (examples in Section 3.1.9)

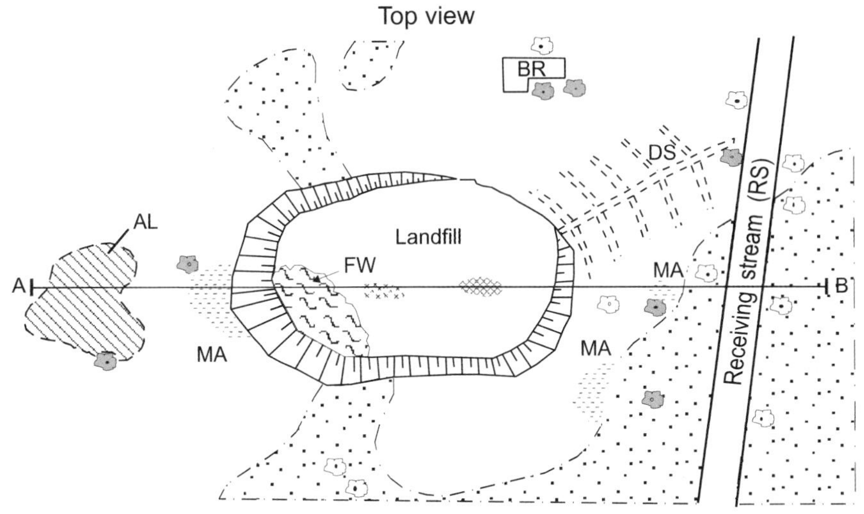

Cross - section A - B

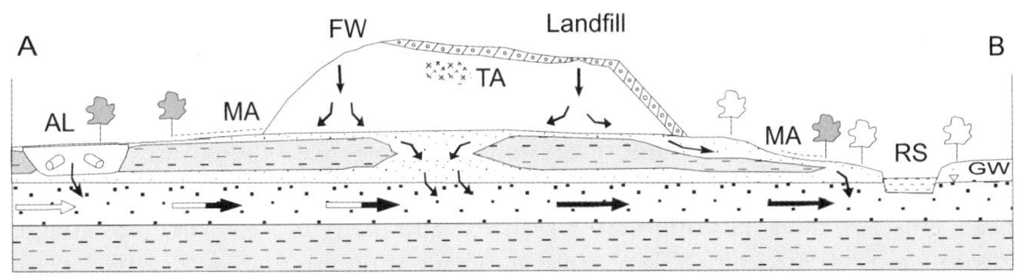

Legend

- Impermeable sediments
- Permeable sediments
- Gaps or weak points in the cover of the landfill
- Wet areas (MA), escape of leachate
- Receiving stream (RS)
- Leachate
- Drainage system (DS)
- Thermal anomaly (TA)
- Abandoned landfill (AL)
- Fresh waste (FW)
- Covering of landfill, e.g., loam stratum
- Building / settlement remains (BR)
- Tree (uncontaminated), no root contact with contaminated groundwater
- Tree (contaminated), root contact with contaminated groundwater

Groundwater:

- Clean groundwater
- Precontaminated groundwater
- Contaminated groundwater
- GW Groundwater surface of uppermost aquifer

These features are described, discussed and illustrated by MILLER & MILLER (1961), KRONBERG (1984) and SABINS (1996).

An evaluation of the environment using remote-sensing data requires consideration of further aspects (example in *Fig. 3.1-6*). Indicators of tectonic fractures, potential pathways for pollutant migration, and the water retention potential of soil layers, as well as anthropogenic features of a landscape, need to be considered. The most practical features for identifying anthropogenic components of a landscape in aerial photographs include the following (DODT et al., 1987):

- anomalous topography that may indicate waste disposal sites, mine spoil heaps, excavation work, etc.,
- anomalous gray-scale patterns that might indicate contaminated soils, filled-in depressions, spread of leachate from waste disposal sites, abandoned industrial sites, etc., and
- vegetation anomalies that might indicate contaminated soil and groundwater.

Conventional and digital processing methods are used for the interpretation of aerial photographs with digital processing having become more important in the last years.

Digital mapping and interpretation systems use the original aerial photographs, which are interpreted using a high quality stereoscope with a zoom lens, as well as computer hardware and software to calculate image coordinates and parallax for precise determination of distances and terrain elevations. Such systems allow mapping and GIS-related recording of terrain features.

Photogrammetric mapping systems use a pair of stereoscopic orthophotographs on the screen, where they can be viewed and interpreted through a special pair of glasses.

It has to be evaluated before starting a project, whether the objectives of the interpretation can be achieved with the available aerial photography. Generally, detrimental impacts on water, soil and vegetation can be better understood and evaluated when previous events or the initial conditions are known. Therefore, if human activity at the site occurred over a long period of time, historical photographs of the site and its surroundings should be incorporated as much as possible. Color or better CIR aerial photographs are especially suitable for such a site investigation.

Chronological development of waste disposal, industrial or mining sites

In order to assess the contamination potential of waste disposal, industrial or mining sites multi-temporal analyses of archival aerial photographs (*Figs. 3.1-7* and *3.1-8*) have been proven to be a useful tool. This method requires aerial photographs taken frequently over a relatively long period of time to investigate spatial and temporal changes in natural and man-made features. Historical photographs reveal the kind of activities at the site, where and how the waste disposal started, the type of waste (liquid, scrap, dust, etc.), and disposal of dangerous waste in barrels or as mud. These investigations are even more important when no documents or only fragmented records about the activities exist. Multi-temporal analyses of archival aerial photographs are also useful for an inventory of sites suspected to be hazardous. The results can be stored and updated in a geographical information system (GIS – see Part 6).

Evaluation of the former environmental and geological situation at the location of a current landfill

Many old waste deposits are in natural or excavated holes, such as sinkholes, caves with openings at the surface, abandoned quarries, and clay or gravel pits. Remote-sensing data predating a landfill can be used to estimate its present subsurface characteristics (*Figs. 3.1-9* and *3.1-10*). Under certain conditions, current remote-sensing data of the surroundings might be extrapolated to zones below the landfill, but archived aerial photographs are normally the best source of such information. Shallow geophysical methods can also be used to estimate the subsurface characteristics, but at greater expense and more limited capability to characterize the layers below the landfill. For the investigation of such sites the following aspects are relevant:

- surface characteristics prior to landfill operations,
- spatial distribution of impermeable and permeable soil and rock (to determine potential areas for spreading of leachate),
- formerly waterlogged areas now underlying present landfills,
- natural and/or artificial drainage systems that provide pathways from the waste disposal site into the surrounding area.

Springs, permeability and soil moisture anomalies, indications of water seepage, and natural and artificial drainage systems

Springs and moist areas associated with a lack of ground cover usually appear darker in black-and-white aerial photographs. In CIR images, areas of high moisture content can also be recognized by the associated vegetation shown as intense reddish colors. Signs of wetness on the surface may be associated with slight changes in topography or with the presence of impermeable, near-surface beds. Anomalous plant growth, discoloration of soil, or visibly wet areas at a landfill's edges indicate possibly contaminated seepage water and potential groundwater contamination (*Fig. 3.1-11*). Plant growth depends, among other factors, on the permeability of the soil and the underlying beds.

The lighter colored areas in *Figure 3.1-12* correlate with areas of poor plant growth due to the sandy soil and the permeable beds there. Precipitation and leachate from the neighboring landfill can migrate through this permeable material directly into the aquifer. If wet areas are detected near to a waste disposal site, as shown in *Figure 3.1-13*, chemical analysis of surface and groundwater samples is recommended. Natural surface and subsurface "channels", like frost cracks in glacial till and erosion trenches in boulder clay filled with sandy, permeable material can act as pathways for leachate from landfills, industrial and mining sites. An artificial drainage system and drainage ditches may have to be installed when the water flow is blocked by impermeable rocks and soils or when the water table is high (*Fig. 3.1-13*). During deposition of the waste, drainage ditches are often filled with permeable materials, such as excavated soil, building rubble, or waste. In contrast, irrigation ditches have to provide, for example, agricultural fields with additional water during dry seasons or when the water table is too deep for the plant roots to reach it. A dense network of irrigation ditches through cultivated land, in rice paddies for example, may later be covered by a landfill (*Fig. 3.1-14*). Waste disposal and other sites suspected to be hazardous have to be investigated for signs of artificial or natural drainage systems. The results of these investigations help to determine the pathways for the migration of contaminated leachate and identify potential hazards.

Recognition and monitoring of destabilization of the ground

Knowledge about the stability of the ground is necessary for risk assessments of waste disposal, industrial and mining sites. Recent and historical aerial photographs often contain indications for destabilization of the ground, e.g., subsidence caused by subrosion and mining, landslides or sinkholes. Terrain features indicating destabilization of the ground are fracturing, subsidence of the land surface, caving to the surface, slippage along bedding planes on slopes, breaks, as well as tensions and relaxation indications. These features can also be seen in the aerial photographs as changes in topography, soil moisture, roughness, water permeability and vegetation. If the slope angle of landfill or mining spoil heaps and/or other geotechnical parameters are not suitable, collapses of the deposited material can be expected. Such features can also be observed and monitored by aerial photography (*Fig. 3.1-15*).

Vitality of vegetation

Plants, in particular trees are very good indicators of the presence of hazardous substances in the air, soil, and groundwater if they are in contact with the contamination via roots or leaves. CIR film is a good tool for evaluating such damage because of its sensitivity to the near infrared (NIR-I) wavelengths. This part of the electromagnetic spectrum indicates the condition of vegetation that is not observable to the human eye. The appearance of a tree under environmental stress is different from that of a healthy tree. The red color of healthy vegetation in CIR aerial photographs results from the strong reflection of near-infrared radiation by the chlorophyll in leaves. In general, lower chlorophyll content, causes less reflection of NIR-I radiation. Thus, the color of unhealthy leaves and needles in the photographs will become less red in proportion to the degree of vegetation "greenness". Different tree species often show different color changes in CIR photographs in response to stress, e.g., damaged oaks and willows turn black, birches, alders, and willows turn brown, linden trees (*Tilia*) turn bluish-green, and silver willows become whiter. In addition, a decrease in tree vitality is also recognizable by the typical thinning of the crown, caused by the loss of leaves and small branches. In the photograph, a decrease in vitality turns a crown into a more open shape with visible branches. A vitality scale with 4 to 5 steps can be worked out for each species on the basis of changes in the color, interior structure and shape of the crown. These three criteria were developed during forest inventories using appropriate interpretation keys (MURTHA, 1972) and have also used in tropical forests (DE MILDE & SAYN-WITTGENSTEIN, 1973; TIWARI, 1975; MYERS, 1978). Such forest surveys in Thailand using aerial photographs have been described by TANHAN (1989). A color key for photo interpretation developed immediately after the aerial CIR photographs are taken can eliminate potential errors due to seasonal or weather-related changes in vitality. Areas with a shallow groundwater table, where the transport of the pollutants and contact with the roots is ensured for a long period of time and over a large area, are best suited for detection of contamination. In contrast, if a waste site is situated on top of a hill, where there is a deep groundwater table, the method might not work. Wind causing movement of the vegetation at the time the aerial photographs were taken reduces the applicability of the method. The following issues must be considered for an assessment of vitality:

- only trees of the same species and similar age are comparable and
- local and biological characteristics for differences in vitality (e.g., local dryness or moisture, parasites, diseases) should be excluded.

This is especially important in regions with a distinct separation of wet and dry seasons. Even during the dry season, changes in topography and associated groundwater levels, springs or different permeability of underlying deposits play an important role. Except for the evaluation of isolated trees, the analysis of a wood's edge may reveal trees which have been more intensively damaged by frequent exposure to toxic substances transported by wind than the trees deep inside the forest. Thus, by including information about prevailing wind direction, the toxic emissions may be traced back to their origin.

Suitable scale to evaluate trees in a CIR photograph is between 1 : 2000 and 1 : 10 000. Because shadows can disturb the interpretation, the sun should be high above the horizon during photo acquisition. *Figures 3.1-16* and *3.1-17* show the influence of contaminated water seeping from a landfill towards nearby trees in an area with a high groundwater level.

Assessment of surface water conditions

Surface water and groundwater from waste disposal, industrial, mining or agricultural sites may contain contaminants. Such contamination in receiving streams, ponds and lakes is indicated in CIR aerial photographs as cloudiness and changes in the color of surface water. Water bodies are normally black in CIR photographs since water absorbs most of the radiation. In this context, color photographs including shorter wavelengths of the visible blue and green are more suitable since they are less absorbed and penetrate deeper into the water surface. The *Fig 3.1-18* photograph indicates blue colors for water bodies emphasizing the stronger reflection in shorter wavelengths due to higher concentrations of insoluble contaminants and algae.

Lineaments

Fractures are often easily detected in aerial photographs (*Fig. 3.1-19*). They are recognizable as linear changes in topography, drainage patterns or patterns of vegetation, as well as by different colors of rocks and soils (MILLER & MILLER, 1961; KRONBERG, 1984 and SABINS, 1996). If morphologically expressed, fracture systems can be best observed in images obtained at low sun angles. Additionally, fractured rocks show signs of enhanced weathering and erosion. Based on information about fractures obtained from the evaluation of aerial photographs, further geophysical investigations can be assisted and better planned.

Biotope and land use mapping

Aerial photographs are a primary source for the mapping of biotopes and land use, because of their excellent spatial detail they provide. Because different kinds of plants can be better distinguished in CIR photographs than in normal color or even panchromatic aerial photographs, they are essential for this kind of problem. Such surveys are carried out worldwide for planning and regional natural resource management. *Figures 3.1-20* and *3.1-21* show an example of a poorly chosen location of an industrial plant and its disposal site immediately next to Esch-sur-Alzette in Luxembourg. The legend and key for interpretation depend on the legal requirements, standards for regional planning documents and the pattern of land use. The standard system for land use mapping in Thailand was designed by the Department of Land Development in 1982, based on black and white aerial photographs at a scale of 1:15 000 (WACHARAKITTI, 1982).

The mapping of biotopes differs from land use mapping. The botanical and forestry classification is more detailed. The quality of such an inventory depends on the expertise of the remote-sensing specialist to recognize all categories of the rather sophisticated classification system. Nevertheless, the interpretation results need to be confirmed by ground checks. The time required for interpretation depends on the complexity of the classification scheme. Classification systems for biotopes with hundreds of different types are common in Central Europe.

Search for new waste disposal sites

Remote-sensing data, particularly when combined with more traditional geologic and geophysical tools, can be used during the planning and development of new disposal sites. Information on the following features can be derived from aerial photographs:

a) impermeable rocks at the location of the planned waste disposal site,

b) existence of natural and artificial drainage systems,

c) substrate with a high permeability,

d) fractures,

e) shallow groundwater table at the site of interest, springs and marshy areas,

f) indications of landslides, and

g) previous human activities (e.g., former gravel pits, quarries, mining) which might have increased permeability of the underlying rocks.

3.1.7 Quality Assurance

For high quality aerial surveys in site investigation, the basic technical standards must be met (SCHWEBEL, 2001). In Germany, these standards are obligatory in accordance with the DIN standard (DIN 18740 - 1) and European standards at a similar level are in preparation (BALTASAVIAS & KAESER, 1999). The quality of site investigations by aerial photography depends on:

- the equipment and the technology used,
- the conditions when the aerial photographs are being taken, and
- the quality of the film processing and photo interpretation.

Equipment and techniques:

1. Large format aerial cameras must be recalibrated every two years in accordance with ISPRS guidelines (ISPRS = International Society of Photogrammetry and Remote Sensing).
2. The camera should be mounted on a gyroscope.
3. The focal length of the camera lens must be selected on the basis of the topography in the area of investigation. As a rule, the focal length should not exceed that of a wide angle lens.
4. Aerial photographs should be taken with full IMC (Image Motion Compensation) and FMC (Forward Motion Compensation).
5. High-precision DGPS must be used based on ground reference stations closer than 50 km from the project area.
6. High-resolution aerial film that yields at least 30 lp/mm (lp = visual line pair) must be used.
7. Storage and usage of film: Experience is needed if good results are to be obtained with film for aerial photography, especially CIR film. The film's age and how it is stored before and after exposure influence the quality of the photographs. The film should be stored according to the manufacturer's instructions (e.g., at -18 °C). A film stored at very low temperatures should be defrosted before use and should be kept cold during transport to the photo laboratory. The film must be developed following the manufacturer's instructions. For quality assurance, sensitive CIR film should be handled according to the VDI guidelines (VDI-GUIDELINE 3793, 1990).
8. Appropriate filters must be used to obtain images with optimum color differentiation. This means a blue filter is required for CIR film.
9. Photographic scales between 1 : 2000 and 1 : 10 000 are recommended.
10. Forward overlap of 60 % and a sidelap of 35 % are necessary for stereoscopic processing (*Fig. 3.1-5*).
11. It is important that the maximum deviation from the planned flight line does not exceed 100 - 150 m.
12. The preferred flight direction is north–south.

Conditions while taking aerial photographs:

1. A cloudless sky and a minimum visibility of about 10 km are required. There should not be any ground haze.

2. The timing of an aerial photo survey depends on the weather and seasonal development of the vegetation within the investigation area, thus, the objective of the survey determines the time of the survey.

Processing and interpretation:

1. For most thematic surveys, at least two groundchecks are necessary: one for preparing the interpretation key, the other one to confirm the results of the interpretation. The ground control points for the aerial photographs and in the terrain must be recorded with comprehensive documentation; the more detail the better.
2. For interpretation, aerial photographs have to be viewed stereoscopically. For most applications, misinterpretation can be expected if the photographs are viewed only two-dimensionally.
3. High-precision aerial photo scanners should be used.
4. A rectification program for elimination of geometric image shifts must be used.
5. If results from previous investigations or other methods exist, they should be incorporated into the interpretation. An interdisciplinary approach improves the recognition and understanding of details in aerial photographs.
6. Aerial photo interpretation is limited by the specifications of the film used in the aerial photo survey, by vegetation cover, and poor weather conditions.

3.1.8 Personnel, Equipment, Time Needed

	Personnel	Equipment	Time needed [d]
taking aerial photographs	2-3 technicians or specialists	special survey aircraft with flight management system, aerial metric camera, aerial film	1
film development	1 technician	1 roller transport processor	1
search for archived aerial photographs	1 archivist		3 - 14
scanning of the aerial photographs	1 technician	1 photogrammetric scanner	0.5
evaluation of development of a landfill, industrial or mining site	1 specialist	1 stereoscope 1 PC	2 - 5
evaluation of the available data about environmental conditions and geology of a landfill, industrial or mining site	1 specialist	1 stereoscope 1 PC	5 - 15
mapping of seepage water at the edges of landfills	1 specialist	1 stereoscope 1 PC	0.5
evaluation of springs and soil moisture anomalies	1 specialist	1 stereoscope 1 PC	0.5 - 5
investigation of natural and artificial drainage systems	1 specialist	1 stereoscope 1 PC	1 - 2
fracture analysis	1 specialist	1 stereoscope 1 PC	5 - 15
evaluation of destabilization of landfill or mining waste heap slopes	1 specialist	1 stereoscope 1 PC	1 - 5
mapping of vegetation vitality	1 specialist	1 stereoscope 1 PC	5 - 15
assessment of surface-water conditions	1 specialist	1 stereoscope 1 PC	1 - 2
mapping of land-use patterns and biotopes	1 specialist	1 stereoscope 1 PC	5 - 15
inventory of other suspected contaminated sites in the area around a landfill, industrial or mining site	1 specialist	1 stereoscope 1 PC	3 - 15
search for a new waste disposal site	1 specialist	1 stereoscope 1 PC	3 - 15
ground check	1 specialist	1 4WD vehicle	1 - 5
digitalization of one thematic map	1 technician	1 PC	0.5 - 5

3.1.9 Examples

Development of a waste disposal site

Two aerial photographs of the Schöneicher Plan waste disposal site near Berlin, Germany, are shown in *Figures 3.1-7* and *3.1-8a*. They were taken in 1992 and 1967, respectively. An interpretation of the 1967 photograph by KRENZ (1991) is given in *Figure 3.1-8b*. These aerial photographs allow a chronological investigation of the disposal site. The areas where waste was dumped and the technology used can be identified in good quality aerial photographs, and the risks of these waste deposits can be evaluated.

Fig. 3.1-7: Panchromatic aerial photograph (photomosaic) taken on May 15, 1992, showing the areal extent of the Schöneicher Plan waste disposal site (courtesy: Luftbildsammelstelle der Landesvermessung und Geobasisinformation Brandenburg)

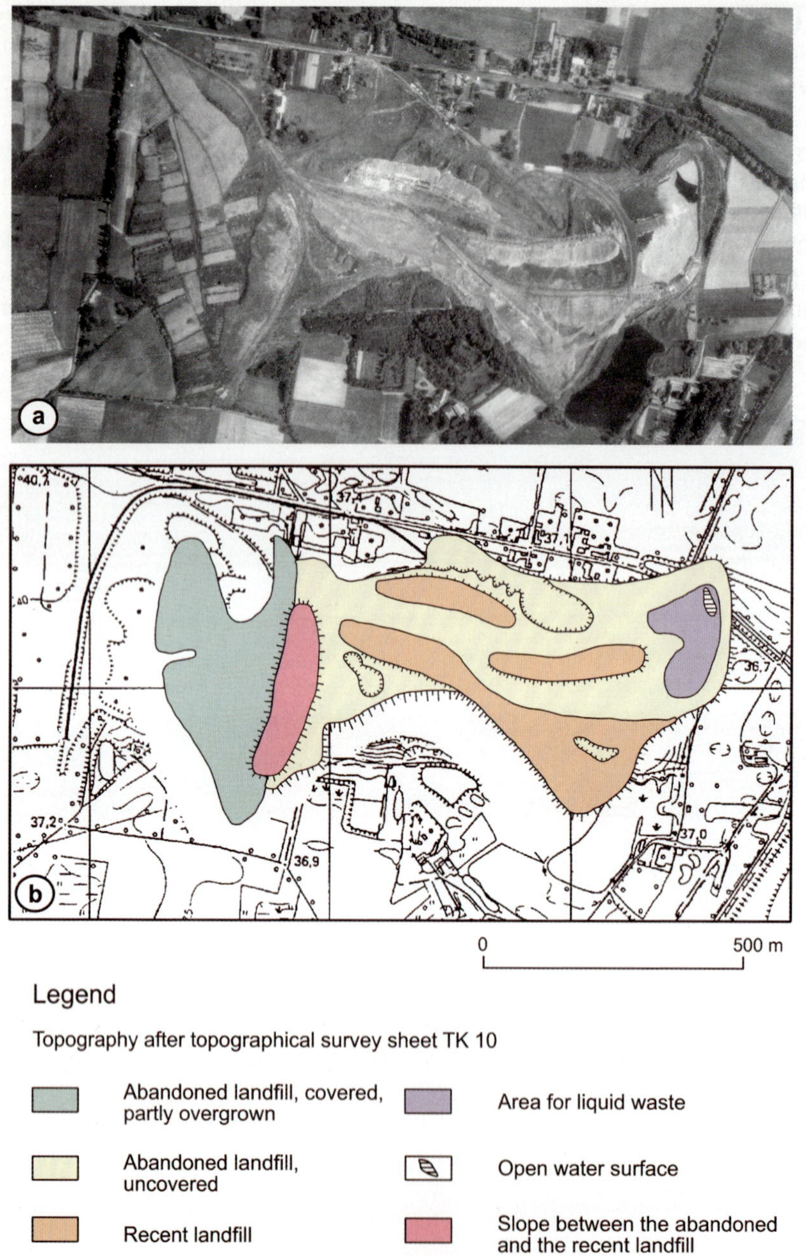

Fig. 3.1-8: (a) Aerial photograph taken on June 23, 1967; (b) Interpretation by KRENZ (1991) on a topographic base map from 1982 (source: Bundesarchiv (Federal Archives of Germany)

Evaluation of the previous environmental and geological situation at the location of an operating landfill

An aerial photograph from 1991 (*Fig. 3.1-9*) depicts an abandoned waste disposal site southeast of Ludwigslust in Mecklenburg-Vorpommern, Germany. The stereo-pair from 1953 (*Fig 3.1-10*) shows that the site was a gravel pit that has been filled with waste. Leachate from the waste site has direct access to the uppermost aquifer, which is exposed at the sides of the pit without the benefit of any natural filtering through soil. This would have occurred if the waste site had been placed on the natural surface instead of in a man-made depression.

The depth of the pit (and thickness of the waste), the areal extent of the old activities and the quantity of waste were determined by interpretation of the 1953 stereo-pair. If the archived photographs had not been available, the subsurface characteristics of the waste site could have been determined using shallow geophysical techniques, such as dc resistivity and shallow seismic methods, but at much greater expense and time.

Fig. 3.1-9: Aerial photograph (stereo-pair) taken on July 7, 1991, showing a disused waste disposal site (arrow) southeast of Ludwigslust in Mecklenburg-Vorpommern, Germany. The presently covered and partially overgrown waste site slightly rises above the surrounding terrain (printed with courtesy of the Amt für Geoinformationswesen der Bundeswehr in Euskirchen, Germany). Luftbild © AGeoBw, 1991 - Lizenz B - 7A004

Fig 3.1-10: Archive aerial photograph (stereo-pair) taken on May 28, 1953, showing a deep gravel pit (arrow) at the location of a now abandoned waste disposal site southeast of Ludwigslust (*Fig. 3.1-9*). Direct dumping of waste into the pit provides direct pathways for contaminated leachate into the uppermost aquifers (source: uve GmbH Berlin)

Seepage of water at the edge of a waste disposal site

Anomalous plant growth, discoloration of the soil, and visible wet areas at the edge of a landfill indicate seepage of water from the landfill and potential groundwater contamination at the site (*Fig. 3.1-11*). In general, water seeping from of a landfill is likely to be highly contaminated. Once leachate from a landfill is identified, further action is needed to characterize the leachate and control or restrict its movement.

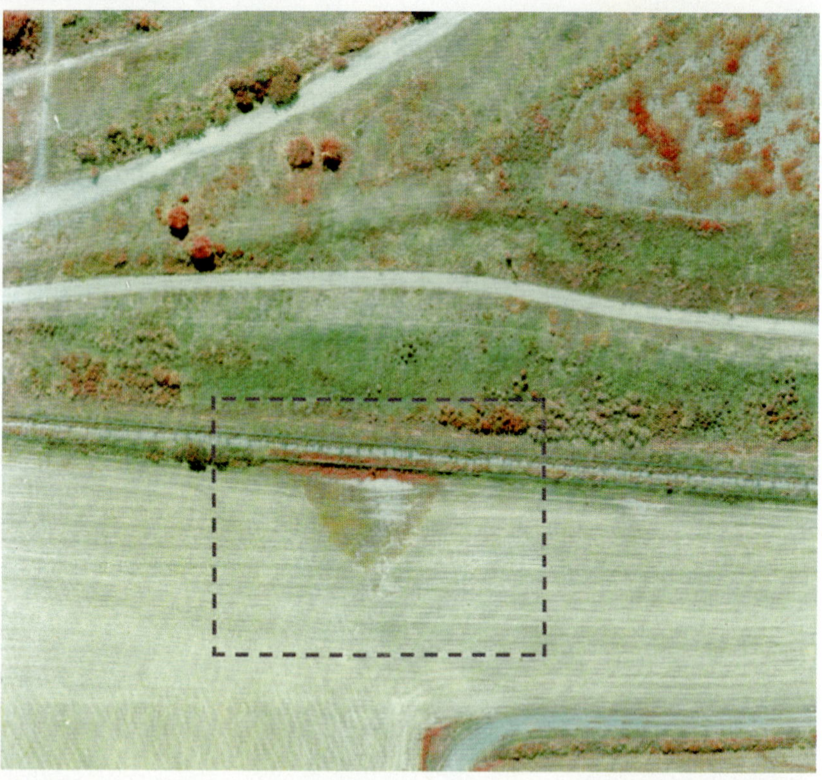

Fig. 3.1-11: CIR aerial photograph taken on August 2, 1990, showing signs of water leachate (within the dashed rectangle) near the edge of an abandoned disposal site which was transformed into a park in Berlin (aerial photograph: Eurosense GmbH, printed courtesy of the Senat Department for Urban Development, Berlin)

Areas of high and low permeability to water

Plant growth depends, among other factors, on the permeability of the soil and the underlying beds. The lighter colored areas in *Figure 3.1-12* correlate with areas of poor plant growth due to the sandy soil and the permeable beds there. Precipitation and leachate from the neighboring landfill can migrate through this permeable material directly into the aquifer.

Fig. 3.1-12: CIR aerial photograph taken on July 3, 1993, of an area west of the Schöneiche waste disposal site. Extensive sandy soil substrates responsible for impaired plant growth can be recognized by the lighter colors, (Aerial photograph: WIB GmbH for BGR).

Soil moisture anomalies, natural and artificial drainage systems

It is often a coincidence if archived historical aerial photographs can contribute to the clarification of site-specific problems. Seasonal conditions and the weather at the time the aerial photographs were taken determine their usability. Weather conditions during a reconnaissance flight in the Mittenwalde area south of Berlin on April 10, 1945 have been optimal (*Fig. 3.1-13*). New vegetation growth patterns delineated substrate and moisture conditions in the area at the time the data was acquired. Currently, the area of faint stripes recognizable in *Figure 3.1-13* is covered by the northern part of the Schöneiche landfill. The stripes may indicate long-term agricultural use (old field patterns) or the existence of abandoned drainage systems (see SCHNEIDER, 1974). Although no old drainage systems were documented in this area, it cannot be excluded that the faint linear features (*arrow*) indicate an old artificial drainage system. A ground check of such features is recommended if they are observed within the area of a landfill. An abandoned drainage system may still function and may transport waste leachate away from the landfill. In this particular case, damaged plants and slightly discolored water were observed in CIR aerial photographs of the areas where the feature joins the Gallun Canal (*Fig. 3.1-16*).

The area marked by the blue circle in *Figure 3.1-14* attracted attention during a ground check because of an abnormal phenotype of a large tree. At this time every other tree of the same species around the site had brown, ripe fruits with only small water content. But, the fruit on this tree were still green and juicy. The tree is at the top of the slope at the edge of the landfill. Closer inspections revealed water seeping from the slope even during the dry season. Interviews with the residents of a nearby village, together with the interpretation of the archival aerial photographs taken at the end of World War II, confirmed the existence of former irrigation ditches. Obviously the ditches are, at least to some extent, still functional beneath the covering waste.

Fig. 3.1-13: Wartime aerial photograph taken on April 10, 1945, showing cultivated land now covered by the northern part of the Schöneiche landfill (*within dashed line*). The high contrast reveals differences in moisture content and lithology. Faint stripes (*arrow*) suggest the presence of an old drainage system or long-term agricultural use (old field patterns). (Courtesy of Luftbilddatenbank, Ing.-Büro Dr. Carls, Estenfeld)

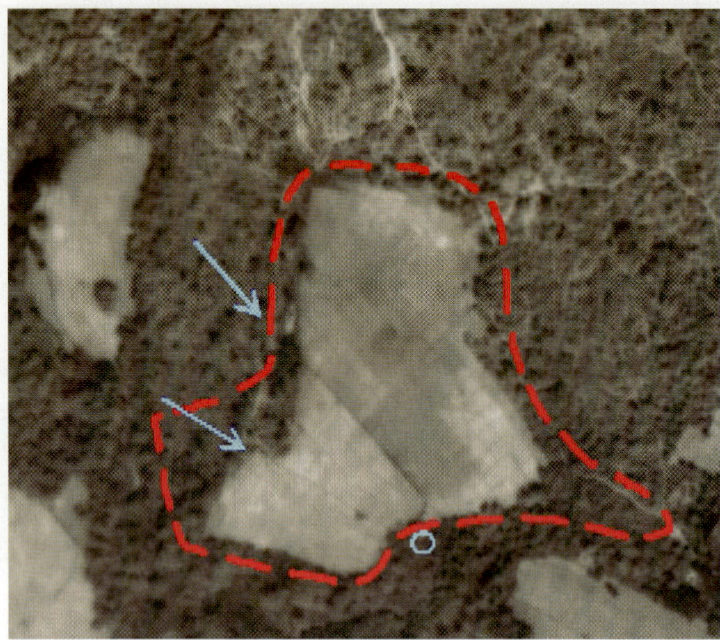

Fig. 3.1-14: Archive aerial photograph taken on February 1, 1945, showing a rice field near the city of Chiang Mai in northern Thailand. Irrigation ditches can be observed in the photograph within the rice field, which is now a waste disposal site. The ditches act as drainage system leading contaminants to the edge of the present-day landfill. *Dashed red line*: boundary of the waste disposal site. *Arrow*: direction of flow in the rice field's irrigation ditches. *Blue circle*: wet area at the edge of the waste site with assumed leachate. Archives research and acquisition by Luftbilddatenbank, Ing.-Büro Dr. Carls, Estenfeld, for BGR

Destabilization of mining spoil and tailings heap slope

Under certain geological conditions, the potential risk is not limited for soil and groundwater contamination. The stability of waste disposal sites and mining spoil and tailings heaps may be threatened as well. The stereo-pair of aerial photographs in *Figure 3.1-15* shows part of a tailings heap south of Magdeburg, Germany. This heap (the white area on the left) contains waste from potash production. A sinkhole caused by collapse in the potash mine appeared near the heap in 1975 (LÖFFLER, 1962 and BRÜCKNER et al., 1983). Stable conditions apparently prevail on the opposite side of the sinkhole from the heap (no concentric collapse fractures). However, the terrain surface is gradually collapsing along concentric fractures in the direction of the heap. Aerial photographs taken over a period of time can be interpreted to determine the stability of the area and predict additional collapse (KÜHN et al., 1997 and 1999).

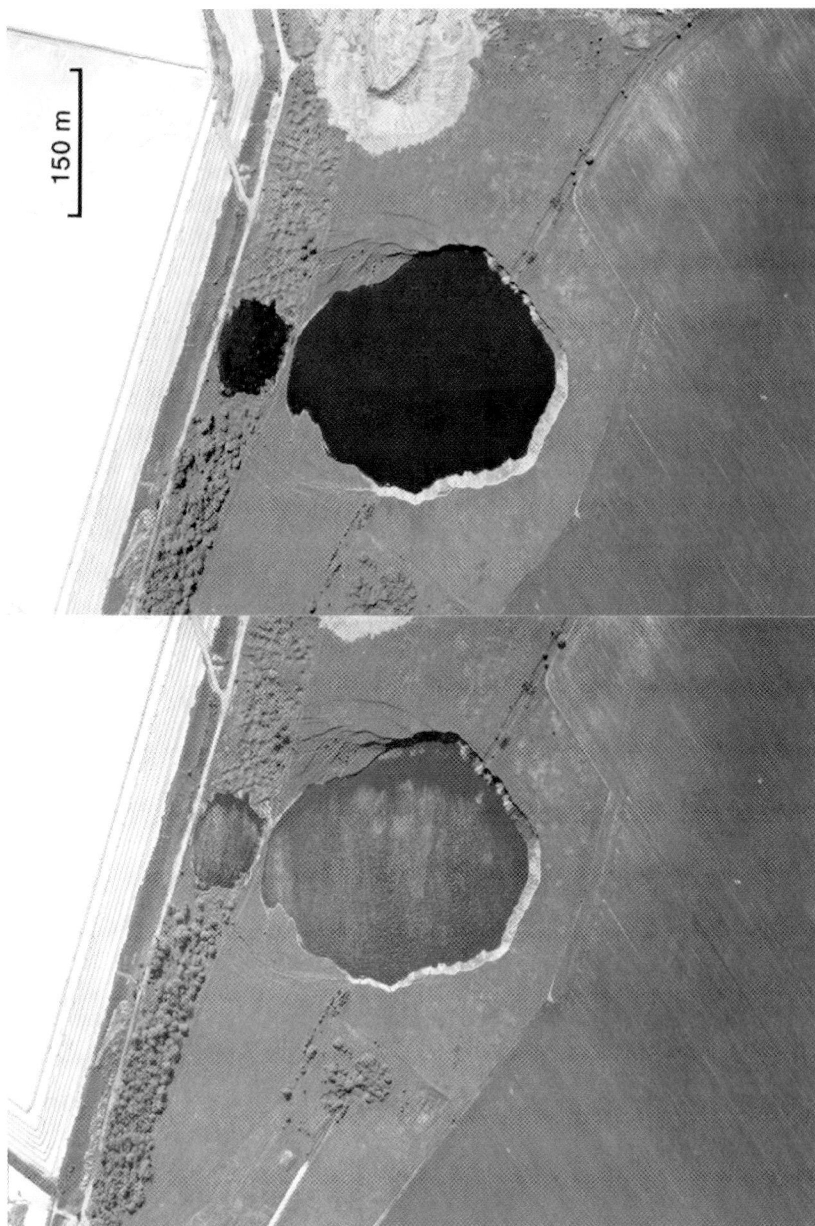

Fig. 3.1-15: Panchromatic aerial photographs (*stereo-pair*) can be interpreted to predict whether a waste heap will become unstable. Circular features at the edge of the collapse area (diameter about 270 m) indicate ongoing collapse in the direction of the heap (Photograph taken May 8, 1994, by Berliner Spezialflug, Luftbild GmbH for BGR).

Vitality of vegetation

A CIR photograph and map depicting the state of health of trees in the vicinity of the Schöneicher Plan and Schöneiche landfills are shown in *Figures 3.1-16* and *3.1-17*. The landfills are 0.5 m below and 1.5 m above the groundwater level, respectively. The uppermost, unconfined aquifer consists of sand and is in hydraulic contact with a number of streams. Runoff from both sites mainly passes through numerous ditches and finally flowing into the Notte Canal, to the north of the sites. The lack of an impermeable layer, e.g., clay, acting as a seal at the base of the landfills, allows the leachate to enter the aquifer and, thus, the surface drainage system. The CIR image shows the area north of the Schöneiche landfill with several rows of trees growing along both the open and silt-filled ditches. Most of the trees are willows, black poplars, and alders.

The map (*Fig. 3.1-17*) was compiled using only CIR aerial photographs from July 2, 1993, and ground check information. This map supports the earlier assumption that contamination spreads northward from the landfill. It can also be seen that the black poplars along the Gallun Canal northeast of the landfill are considerably stressed. This indicates that seepage water from the landfill infiltrates the ground via an old drainage system. However, an impact on the health of the black poplars by the neighboring waste incineration plant cannot be excluded.

Fig. 3.1-16: CIR aerial photograph taken on July 2, 1993, showing an area north of the Schöneiche landfill. The road along the northern edge of the landfill is just visible at the right bottom of the photo (photo taken by WIB GmbH Berlin for BGR).

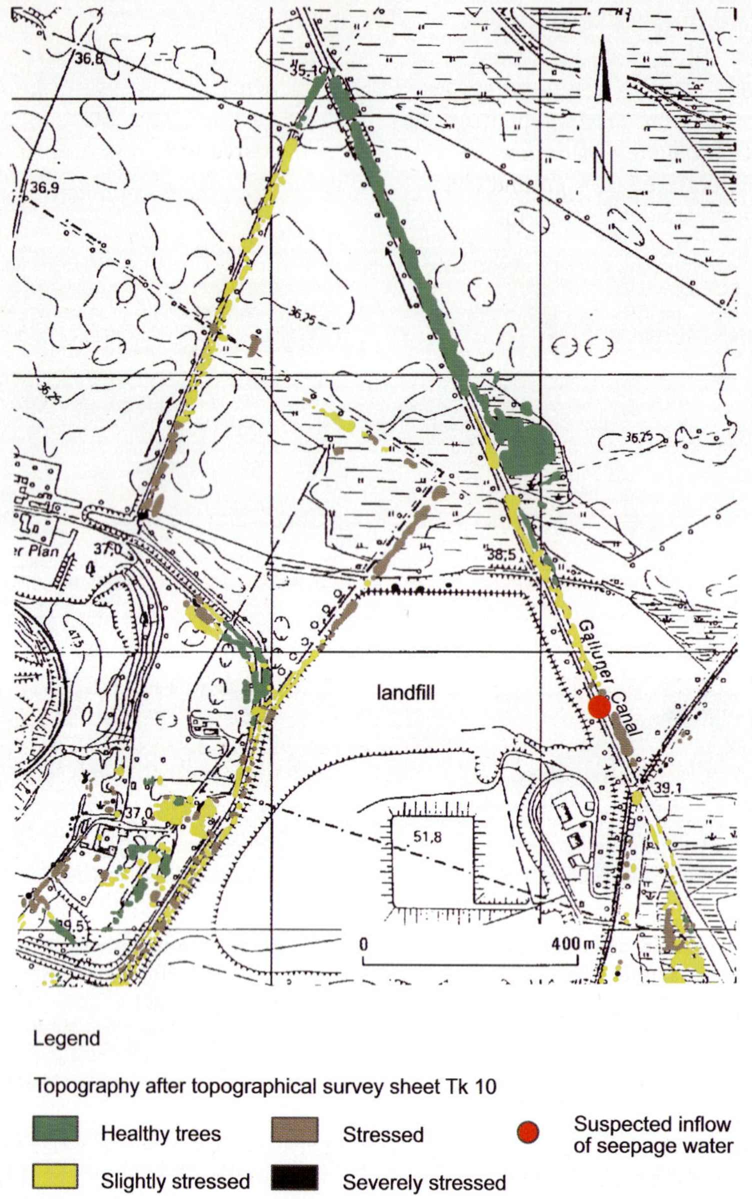

Fig. 3.1-17: Map of tree vitality near the Schöneicher Plan and Schöneiche landfills; mapped at a scale of 1 : 5000 (simplified)

Assessment of surface water conditions

In many cases, pollution of surface water is indicated by changes in the color of the water in the CIR photographs. The blue of the water body in the CIR aerial photograph of *Figure 3.1-18* indicates intensive turbidity. This turbidity was determined by a ground check to be caused by illegal discharge of liquid manure from a nearby farm.

Fig. 3.1-18: CIR aerial photograph of water-filled clay pits of a former brickyard north of Ketzin, Brandenburg, Germany. The blue color indicates the discharge of liquid manure (water bodies are normally dark or black in CIR photographs). Photograph taken July 28, 1990, by Berliner Spezialflug, Luftbild GmbH; printed with courtesy of FUGRO Consult GmbH, Berlin.

Fractures

Disturbed ground and fractures are often easy to detect in aerial photographs. The surface expression of such features is marked by changes in topography, drainage patterns, vegetation growth patterns, and rock and soil color differences.

A laser image (Chapter 3.3) and CIR aerial photograph of the same area are shown in *Figure 3.1-19*. The CIR aerial photograph was taken on April 21, 2000, and shows a field of young cereal *(Fig. 3.1-19B)*. Due to the spectral sensitivity of the film, the red color indicates healthy vegetation. In this photograph, the intensive red indicates advanced growth of the young plants. The vitality of the plants is due to the fertile soil, composed of thick loess, accumulated in open fractures. The red lines in the aerial photograph mark the individual fractures. Narrow linear depressions visible in the laser image *(Fig. 3.1-19A)* suggest that the ground is already subject to displacement. The two images in *Figure 3.1-19* corroborate the existence of a fracture zone, which was also verified in the field. Since this area is still used for farming, potential hazards to people, road traffic, and farm machinery have to be taken into consideration.

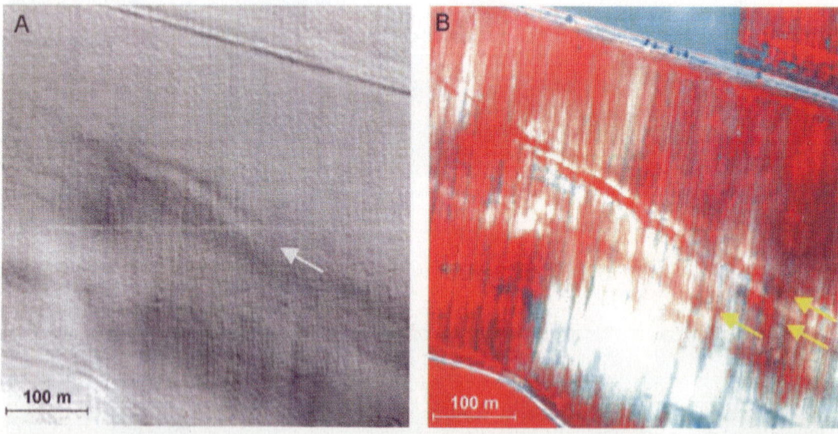

Fig. 3.1-19: "Laser Image" (A, shaded relief map – sun elevation: 45°, azimuth: 45°) and B Color-infrared (CIR) aerial photograph of the same part of a fracture zone north of the lake "Suesser See" near Eisleben, Germany. Faint linear features in the laser image and linear to curvilinear features in the CIR aerial photograph (yellow arrows) are surface expressions of fractures.

Mapping of land use pattern and biotopes

In many countries, land use maps are used to assist planning and decision-making at the regional scales. The CIR photograph in *Figure 3.1-20* depicts an industrial plant and its waste disposal site. The landfill was established for temporary storage of ferruginous dust and mud. Two mud ponds and a thin vegetation cover are present in the southern part of the landfill. No vegetation can develop where the dust was deposited. The land use interpretation in *Figure 3.1-21* documents the problematic location of the waste disposal site with respect to the nearby settlement. Residential buildings are less than 100 m away from this source of dust emission (red area). Dust transported by southerly and easterly winds may have negative affects on the health of the residents and the agriculture fields in the vicinity of the waste site. Proper information and planning prior to the establishment of a waste disposal site helps to avoid such problems.

Fig. 3.1-20: CIR aerial photograph of an industrial plant (lower right corner) and its waste disposal site (in the center) near the city of Esch-sur-Alzette in Luxembourg. The photo was taken on May 27, 1999, at a scale of 1 : 15 000 by Hansa Luftbild (German Air Survey), Münster

Fig. 3.1-21: Interpretation of aerial photographs with regard to land use as a basis for regional planning, Hansa Luftbild (German Air Survey), Münster

Land use mapping

Land use was mapped during an investigation of the abandoned Nong Harn landfill near Chiang Mai (northern Thailand) using the classification system of WACHARAKITTI (1982) adapted to the objectives of the site investigation and using 1 : 7500 CIR aerial photographs (*Fig. 3.1-22*). The classification system distinguishes six hierarchic levels of land categories. The first hierarchical level is labeled by letters, e.g., "a" for agricultural land, "u" for urban land, and "f" for forest land; the sublevels are labeled by numbers, separated by periods. If in one level an attribute cannot be assigned to an area, a blank is used instead of a number. The more attributes that can be recognized, the better the land use of the area can be specified.

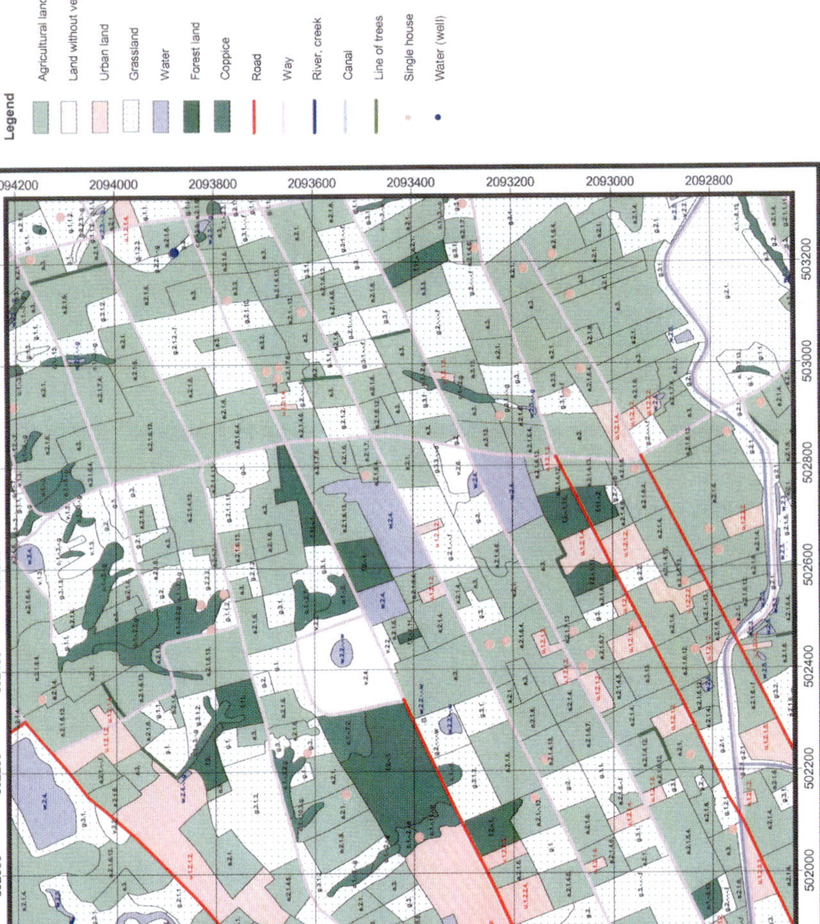

Fig. 3.1-22: Land use map of the Nong Harn area containing the codes of all objects recognized in the interpretation of aerial photographs

Inventory of sites suspected to be hazardous

An inventory of sites suspected to be hazardous in the surroundings of the abandoned Mae Hia landfill, Chiang Mai, Thailand was performed using historical aerial photography. The goal was to assess the different possible sources of contamination. Due to the poor records available, the interpretation scheme was limited to two classes. Sites and facilities with likely existence of hazardous substances have been assigned to class 1. Less dangerous sites were mapped as class 2. *Table 3.1-4* and *Figure 3.1-23* show the results of the inventory.

Table 3.1-4: Overview of the sites suspected to be hazardous in the Mae Hia investigation area

	Class 1	**Class 2**
Abandoned industrial sites		
motor vehicle garage	2	
gas/petrol station	2	
workshop in which wood is impregnated or is worked	1	
other		19
Waste deposits		
landfills	2	
small sites with buried and/or heaped waste		9
litter		89

Fig. 3.1-23 (next page): Sites in the Mae Hia area of northern Thailand suspected to be contaminated on the basis of interpretation of aerial photographs

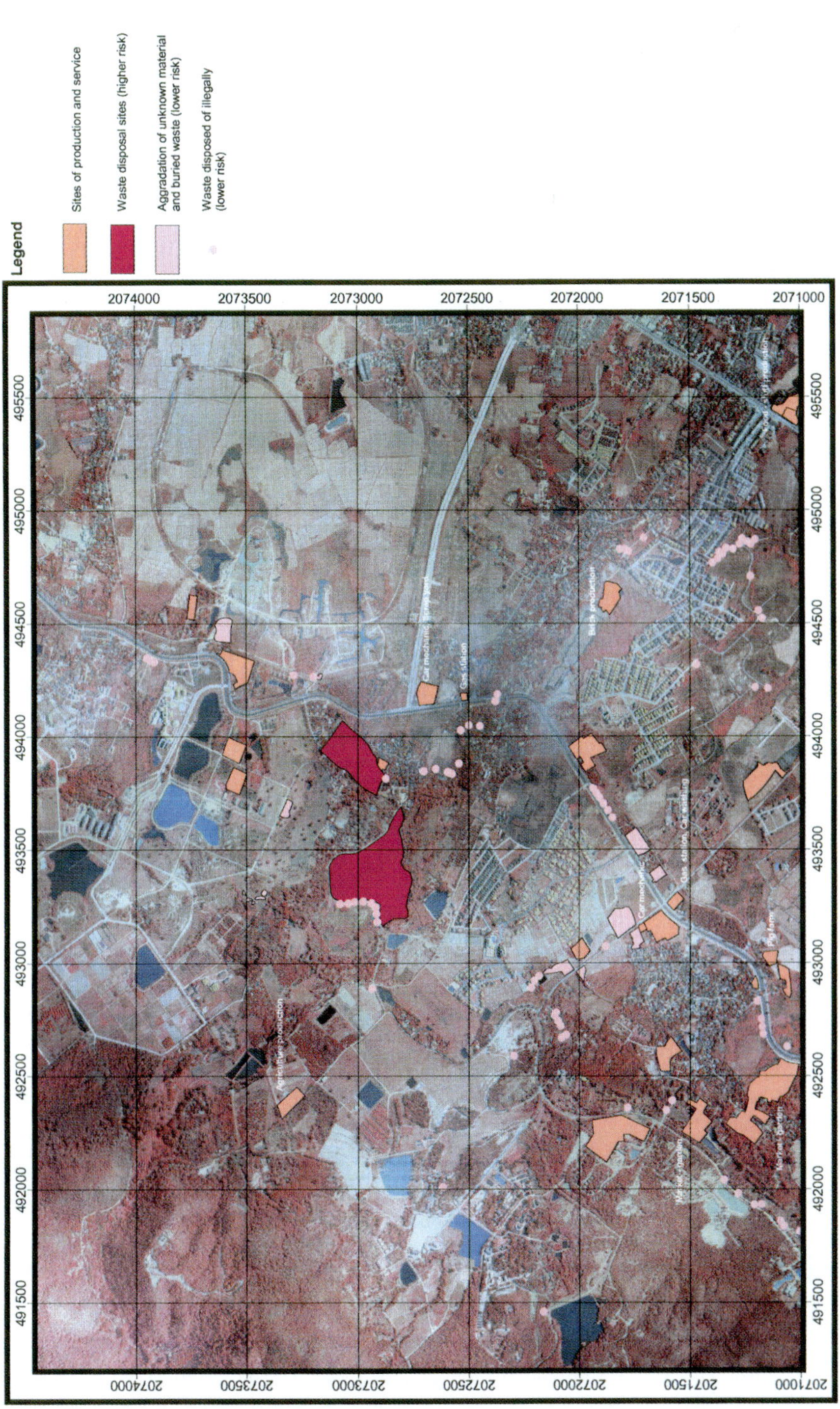

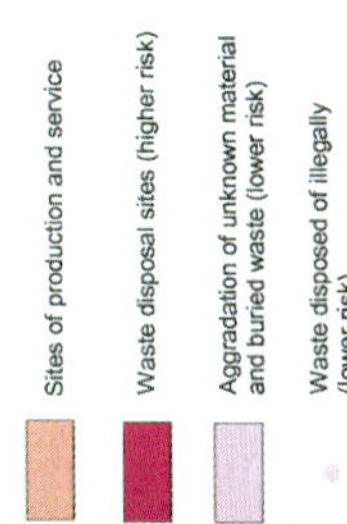

Legend
Sites of production and service
Waste disposal sites (higher risk)
Aggradation of unknown material and buried waste (lower risk)
Waste disposed of illegally (lower risk)
2074000
2073500
2073000
2072500
2072000
2071500
2071000
495500
495000
494500
494000
493500
493000
492500
492000
491500
Brick production
Gas station
Agriculture production
Pig farm
Market garden

References and further reading

ALBERTZ, J. (2001): Einführung in die Fernerkundung, Grundlagen der Interpretation von Luft- und Satellitenbildern. Wissenschaftliche Buchgesellschaft, Darmstadt.

AVERY, T.E., Berlin GL (1985): Interpretation of aerial photographs. Burgess Publ. Comp., Minneapolis.

BALTASAVIAS, P. & KAESER, C. (1999): Quality Evaluation of the DSW 200, DSW 300, SCAI and Ortho Vision Photogrammetric Scanners. Unpublished Report.

BARRETT, E.C. & CURTIS, L. F. (Eds.) (1974): Environment remote-sensing: applications and achievements. Edward Arnold Ltd., London.

BÖKER, F. & KÜHN, F. (1992): Zur Verwendung von Luftbildern bei der geologischen Kartierung. Zeitschrift für angewandte Geologie, **38**, 2, 80-85.

BORMANN, P. (1981a): Passive Fernerkundungssensoren im optischen Bereich des Spektrums. Vermessungstechnik, **29**, 2, 45-48.

BORMANN, P. (1981b): Was ist unter dem Auflösungsvermögen eines Fernerkundungssensors zu verstehen. Vermessungstechnik, **29**, 10, 331-335.

BORRIES, H.-W. (1992): Altlastenerfassung und -erstbewertung durch multitemporale Karten- und Luftbildauswertung. Vogel, Würzburg.

BRÜCKNER, G., KNITSCHKE, G., SPILKER, M., PELZEL, J., SCHWANDT, A. (1983): Probleme und Erfahrungen bei der Beherrschung von Karsterscheinungen in der Umgebung stillgelegter Bergwerke des Zechsteins in der DDR. Neue Bergbautechnik, **13**, 8, 417-422.

BRUNEAU, M. (1980): Mapping Landscape Dynamics in the Highlands and Lowlands of Northern Thailand. In: IVES, J. D., SABHASRI, S. & VORAURAI, P.: Conservation and Development in Northern Thailand, The United Nations University, Tokyo.

Center for Remote Sensing and Geographic Information Science (1999): Historical Aerial Photos and Maps for Documenting Changes over a Site. Michigan State University, www.crs.msu.edu.

CHUCHIP, K. (1997): Satellite Date Analysis and Surface Modeling for Land Use and Land Cover Classification in Thailand. Berliner Geographische Studien, **46**, Berlin.

CICIARELLI, J. A. (1991): A Practical Guide to Aerial Photography. Van Nostrand, Würzburg.

COLWELL, R. N. (ed.) (1983): Manual of Remote Sensing, **1**, Theory, Instruments and Techniques, **2**, Interpretation and Applications. American Society of Photogrammetry, Falls Church.

DECH, S. W., GLASER, R., KÜHN, F. & CARLS, H.-G. (1991): Ökologische Probleme durch Rüstungsaltlasten in der Colbitz-Letzlinger Heide. DLR-Nachrichten, **64.**

DE MILDE, R. & SAYN-WITTGENSTEIN, L. (1973): An Experiment in the Identification of Tropical Tree Species on Aerial Photographs. Proc. IUFRO Symp. S. 6.05, Freiburg 1973, 21-38.

DIN 18740-1 (1999): Qualitätsanforderungen an photogrammetrische Produkte, Teil 1: Bildflug und analoges Luftbild, Deutsches Institut für Normung e.V.

DODT, J., BORRIES, H. W., ECHTERHOFF-FRIEBE, M. & REIMERS, M. (1987): Zur Verwendung von Luftbildern und Karten bei der Ermittlung von Altlasten. Ruhr-Universität Bochum.

DOUGLAS, S. (1973): Terrain Analysis – A Guide to Site Selection Using Aerial Photographic Interpretation. Community Development Series, Dowden, Hutchinson & Ross Inc., Stroudsburg.

EDWARDS, G. J. (Ed.) (1988): Color Aerial Photography in the Plant Sciences and Related Fields, A Compendium 1967 - 1983, Falls Church.

EHRENBERG, M. (1991): Beprobungslose Altlastenerkundung. wlb. Wasser, Luft und Boden, **7-8**, 56-59.

ERB, W. (1989): Leitfaden der Spektroradiometrie. Springer, Berlin.

GLASER, R. & CARLS, H.-G. (1990): Kriegsluftbilder 1940 - 1945: Ein Hilfsmittel bei der Verdachtsflächenermittlung von Kriegsaltlasten und in der Umweltplanung. Laufener Seminarbeitr., **90.1**, 65-73, Laufen/Salzach.

GOETZ, A. F., BARRET, N. R. & ROWAN, L. C. (1983): Remote Sensing for Exploration: An Overview. Econom. Geol., **78/4**, 573-592.

GUPTA, R. P. (1991): Remote Sensing Geology. Springer, Berlin.

HAAS, R. (1992): Konzepte zur Untersuchung von Rüstungsaltlasten. Erich Schmidt, Berlin.

HEGG, K. M. (1966): A photo identification guide for the land and forest types of interior Alaska. U.S. For. Serv. North. For. Exp. Station, Research Paper NOR-**3,** 55.

HILDEBRANDT, G. (1996): Fernerkundung und Luftbildmessung: für Forstwirtschaft, Vegetationskartierung, und Landschaftsökologie. Herbert Wichmann, Heidelberg.

HOLZFÖRSTER, B. & TIEDEMANN, M. (1991): Die Bedeutung der Luftbildauswertung für die Erfassung von Rüstungsaltlasten am Beispiel Niedersachsen. In: THOMÉ-KOZMIENSKY (Ed.): Untersuchung von Rüstungsaltlasten. EF-Verlag für Energie und Umwelttechnik, Berlin.

HUBER, E. & VOLK, P. (1986): Deponie- und Altlastenerkundung mit Hilfe von Fernerkundungsmethoden. Wasser und Boden, **10**, 509-515.

KENNEWEG, H. (1994): Forest condition and forest damages – Contribution of remote sensing. Geo-Journal, **32**, 47-53.

KRENZ, O. (1991): Luftbildinterpretation der Deponiestandorte Schöneiche und Schöneicher Plan - Teilbericht zum BMFT-Projekt „Abfallwirtschaftliche Rekonstruktion von Altdeponien am Beispiel des Deponiestandortes Schöneiche“. Unpublished report.

KRONBERG, P. (1984): Photogeologie, eine Einführung in die Grundlagen und Methoden der geologischen Auswertung von Luftbildern. Enke, Stuttgart.

KRONBERG, P. (1985): Fernerkundung der Erde. Enke, Stuttgart.

KÜHN, F. & OLEIKIEWITZ, P. (1983): Die Nutzung der Multispektraltechnik zur Früherkennung von senkungs- und erdfallgefährdeten Gebieten. Z. angew. Geol., **29**, 6, 71-74.

KÜHN, F. & HÖRIG, B. (1995): Geofernerkundung, Handbuch zur Erkundung des Untergrundes von Deponien und Altlasten, **1,** Springer, Heidelberg.

KÜHN, F., HÖRIG, B. & BUDZIAK, D. (2004): Detecting unstable ground by multisensor remote sensing. Photogrammetrie – Fernerkundung – Geoinformation (PFG), **2**, 101-109.

KUEHN, F., KING, T., HOERIG, B. & PETERS, D. (2000): Remote Sensing for Site Characterization. Methods in Enviromental Geology, Springer, Berlin.

KÜHN, F., TREMBICH, G. & HÖRIG, B. (1997): Multisensor Remote Sensing to Evaluate Hazards Caused by Mining. In: Proceedings of the Twelfth International Conference on Applied Geologic Remote Sensing, 17-19 November, Denver, Colorado, ERIM, I, 425-432.

KÜHN, F., TREMBICH, G. & HÖRIG, B. (1999): Satellite and Airborne Remote Sensing to Detect Hazards Caused by Underground Mining. In: Proceedings of the Thirteenth International Conference on Applied Geologic Remote Sensing, 1-3 March, Vancouver, British Columbia, Canada, ERIM, II, 57-64.

LINTZ, J. & SIMONETT, D. S. (Eds.) (1976): Remote Sensing of Environment. Addison-Wesley, London.

LÖFFLER, J. (1962): Die Kali- und Steinsalzlagerstätten des Zechsteins in der Deutschen Demokratischen Republik. Freiberger Forschungsheft, Teil III, Sachsen-Anhalt, **C97**/III, Akademie Verlag, Berlin.

MILLER, V. & MILLER, C. F. (1961): Photogeology. McGraw-Hill, New York.

MURTHA, P. A. (1972): A Guide to Air Photo Interpretation of Forest Damage in Canada. Can. Forestry Service Publ., **1292**, 63.

MYERS, B. J. (1978): Separation of Eucalyptus species on the basis of crown colour on large scale colour aerial photographs. Austr. For. Res., **8**, 139-151.

ONGSOMWANG, S. (1994): Forest Inventory, Remote Sensing and GIS (Geographic Information System) for Forest Management in Thailand. Berliner Geographische Studien, Berlin.

PAINE, D. P. (1981): Aerial Photography and Image Interpretation for Resource Managment. Wiley & Sons, New York.

PETERS, D. C., HAUFF, P. L. & LIVO, K. E. (1995): Remote sensing for mine waste discrimination and characterization. In: CURRAN, P. J. & ROBERTSON, Y. C., comps., RSS 95: Remote Sensing in Action, Southampton, UK, Sept. 12 - 15, 1995, The Remote Sensing Society, 866-877.

SABINS, F. F. (1996): Remote sensing - Principles and interpretation. 3rd edition, Freeman, San Francisco.

SAYN-WITTGENSTEIN, L. (1960): Recognition of tree species on air photographs by crown characteristics. Com. Dept. For. Techn. Note 95.

SCHMIDT, D. (2000): Vitality of Trees. In KUEHN, F., KING, T., HOERIG, B. & PETERS, D. (Eds.): Remote Sensing for Site Characterization, Springer Verlag, Berlin, 91-93.

SCHNEIDER, S. (1974): Luftbild und Luftbildinterpretation. de Gruyter, Berlin.

SCHNEIDER, S. (1984): Angewandte Fernerkundung: Methoden und Beispiele. Vincentz, Hannover.

SCHWEBEL, R. (2001): Qualitätssicherung für Bildflug und analoges Luftbild durch neue DIN-Norm, Photogrammetrie, Fernerkundung, Geoinformation, **1**, 39-44.

SMITH, J. T. & ABRAHAM, A. (Eds.) (1968): Manual of Color Aerial Photography. American Society of Photogrammetry, Falls Church.

STRATHMANN, F.-W. (1993): Taschenbuch zur Fernerkundung. Wichmann, Karlsruhe.

TANHAN, S. (1989): Application of Aerial Photography for Land Right Cultivation. In: Remote Sensing and Mapping for Forest Management: Report of Project Course B.UNDP, FAO, Bangkok, 261-264.

TEEUW, R. M. (2007): Mapping hazardous Terrain using remote sensing. GSL Special Publications, London.

THORLEY, G. ed. (1975): Forest Lands: Inventory and Assessment. In: Manual of Remote Sensing 1975, **II**, American Society of Photogrammetry.

TIWARI, K. P. (1975): Tree species identification on large scale aerial photographs at new forest. Indian forest, **10**, 791-807.

VDI-Guideline 3793 (1990): On-site Determination of Vegetational Injuries Method of Aerial Photography with Color Infrared Film. Verein Deutscher Ingenieure, VDI-Handbuch Reinhaltung der Luft, **1**, Berlin.

WACHARAKITTI, S. (1982): Land Use Classification System or Land Use Design. Faculty of Forestry, Kasetsart University, Bangkok (Version Thai).

WEBER, H. (ed.) (1993): Altlasten: Erkennen, Bewerten, Sanieren. Springer, Berlin.

3.2 Photogrammetry

DIETMAR SCHMIDT, PETER HERMS, ANKE STEINBACH & FRIEDRICH KÜHN

3.2.1 Principle of the Methods

Photogrammetric techniques provide reliable measurements of geometric characteristics of Earth surface features from photographic images taken from remote sensing platforms. These remote measurements not only include the size, shape, and position of objects, but also their color or tone, texture, and spatial patterns and associations. Hence, such observations can be interpreted and analyzed with regard to their geoscientific information (Chapter 3.1) and with regard to their geometric structures (Chapter 3.2).

Photogrammetry techniques are used to prepare maps and digital elevation models (DEM) from remote-sensing data. Other tasks are the preparation of orthophotographs, photomosaics, terrain cross-sections, and vector data (e.g., the height of a terrain slope, length of a stream, volume of a waste heap or landfill). Most photogrammetric products derived from aerial photographs have been used for decades. These comprehensive experiences have developed to standards and rules assuring the quality and consistency of the mapping products (SCHWEBEL, 2001; STUTTARD & DOWMAN, 1998).

There are alternative approaches to traditional photogrammetry. For obtaining topography data airborne laser scanning has proven to be a successful method (Chapter 3.3). Optical-electronic cameras (Chapter 3.3) have also been developed for photogrammetric purposes. These sensor systems provide geometric resolution down to the centimeter range. Digital cameras that meet the requirements for photogrammetry processing are in an experimental stage but are expected to widely replace conventional aerial cameras in the near future (Chapter 3.3).

The major objective of photogrammetry is to relate the pixel coordinates measured by the sensor as exactly as possible to the geographic coordinates (longitude, latitude, height) of terrain points. This process involves the removal of distortions caused by the data acquisition system (aerial camera or scanner), perspective, and motion of the aircraft or the space-borne platform. This processing, called rectification of the image, requires the specifications of the camera and precise coordinates from ground control points (GCP). Two types of digital processing are used in photogrammetry:

- Geometric correction: This interpolation method uses northing and easting, (latitude and longitude respectively) of known terrain points and their corresponding image coordinates in the aerial photographs to obtain a "true-to-position" representation without consideration of location differences in the third dimension (topography).

- Orthorectification: A mathematical process in which transformation parameters, including the sensor specifications (interior orientation) and position during the acquisition of the photograph (exterior orientation), are used to obtain the geometric relationships in the aerial photographs and elevation data and thus remove all geometric distortions caused by the data acquisition process.

In both cases, ground checks are essential to determine the geometric relationship between point measurements in the photographs and the true location on the ground. Ground control points should be well defined in the photograph and easy to find and accessible on the ground. Points marked by reflective foil, fabric, plastic or paint, called ground targets, may be applied during the data acquisition to provide such control points if landscape characteristics (e.g., large uniform agricultural areas, steppe or desert) are not appropriate. The coordinates of the ground control points have to be surveyed on the ground if they cannot be accurately extracted from topographic maps. These field surveys are usually quickly and inexpensively accomplished using GPS technology (Chapter 2.5).

Orthophotographs, photomosaics, thematic maps, terrain cross-sections, and vector datasets derived from photogrammetry are widely used for site investigations in conjunction with other geological, hydrological, geophysical, and remote-sensing data. Topographic measurements acquired at different times can be used to observe changes of landfills, waste heaps and slopes suspected to be instable, as well as for planning and decision-making by governmental agencies and in the business world.

3.2.2 Applications

- Preparation of orthophotographs and photomosaics,
- development of topographic maps,
- derivation of digital elevation models (DEM),
- derivation of terrain cross-sections and vector data (e.g., determination of the height of a terrain slope, length of a stream, volume of a waste heap or landfill).

3.2.3 Fundamentals

The most common task for photogrammetry is the preparation of maps and comparable products, e.g., orthophotographs. While all points on a map are usually depicted at their true relative horizontal positions (parallel projection), an aerial photograph is distorted in a regular geometric manner. Because of these inherent geometric distortions aerial photographs cannot be directly used for mapping purposes. However, the geometric distortions can be removed or at least minimized with specific rectification procedures.

The geometric distortions result from the data acquisition process as the Earth surface is being viewed from a single point (central projection), and changes in terrain elevations, as well as, the instability of the aircraft causing dislocations of image pixel coordinates. The ladder, rather irregular distortions are minimized by compensation systems mounted on the sensor platform, e.g., Image Motion Compensation (IMC) and Forward Motion Compensation (FMC), and neglected in routine investigations.

Geometric distortions of vertical aerial photographs

Distortions resulting from a non-vertical position of the optical axis in vertical aerial photography (rarely greater than 5 gon (4.5°) and usually much smaller) is neglected in most cases (LILLESAND & KIEFER, 1994; ALBERTZ, 2001).

Because of the nature of central projection, any variation in terrain elevation will result in scale variations and displaced positions of ground objects in a vertical aerial photograph.

Corresponding to Equation (3.1.1) the scale of an aerial photograph is directly proportional to the focal length c_f of the camera lens and inversely proportional to the height h_g of the camera above ground. Therefore, aerial photographs taken over terrain of varying elevation will have a continuous range of scales associated with changes of h_g due to variations in terrain elevations.

The location displacement of a ground object in a vertical aerial photograph due to variations of terrain and/or object height (topography) and central projection is called relief displacement (SABINS, 1996; GUPTA, 2003). The amount of relief displacement d in an aerial photograph is directly proportional to the difference Δh in terrain elevation and/or the object height with respect to the reference plane (map plane), directly proportional to the radial distance r from the principal point, i.e. the optical center of the photograph, and inversely proportional to the height h_g of the camera above the terrain:

$$d = \frac{\Delta h\, r}{h_g}. \qquad (3.2.1)$$

The maximum displacement is at the corners of the photograph. *Figure 3.2–1* demonstrates how elevation differences are reflected in a horizontal displacement. Terrain depressions such as valleys and pits are shifted towards the principal point N of the image. In contrary, elevated areas such as hills, mountains, buildings, towers or walls are displaced away from the principal point.

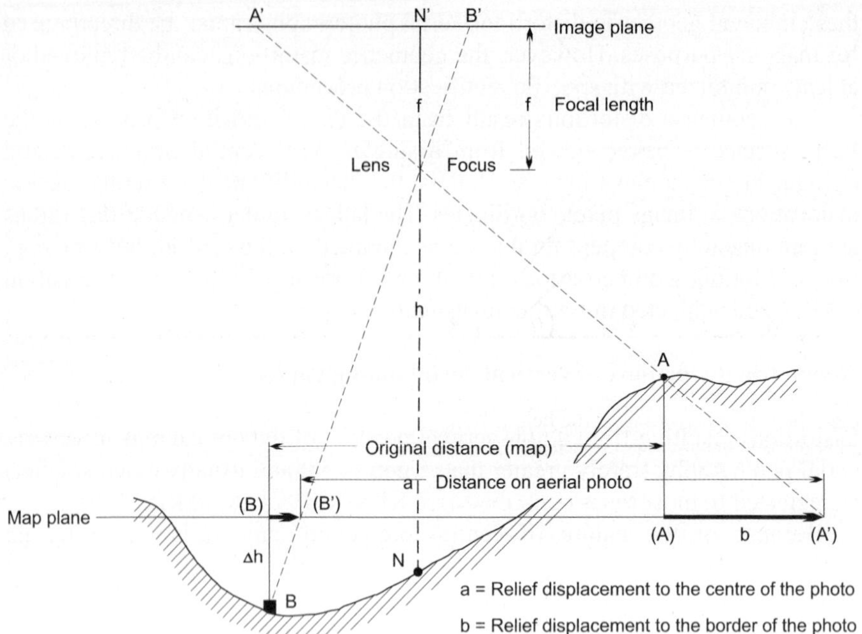

Fig. 3.2-1: Radial distortion caused by relief, SCHNEIDER (1974)

Determination of object heights from radial distortion or shadow length

Equation (3.2.1) can be used to determine the height of an object after solving for Δh. If the base and top of an object (building, tower, steeple, tree) are visible in the image, the object height Δh can be determined from the radial displacement $\Delta r'$ (*Fig. 3.2–2*) using Equation (3.2.2) (ALBERTZ, 2001).

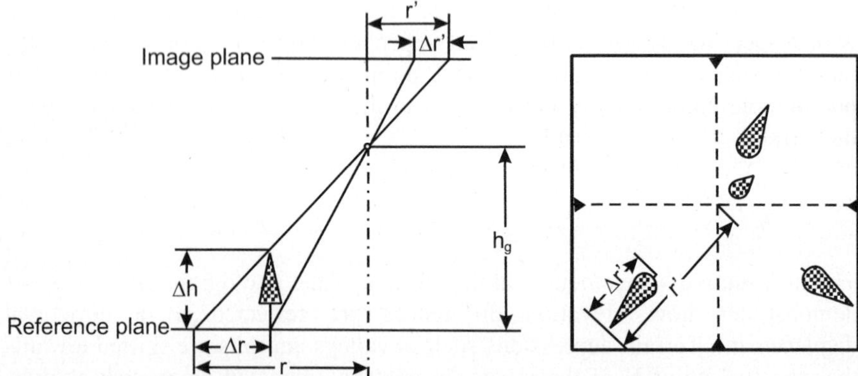

Fig. 3.2-2: Radial distortion due to differences in the height of the objects, ALBERTZ (2001)

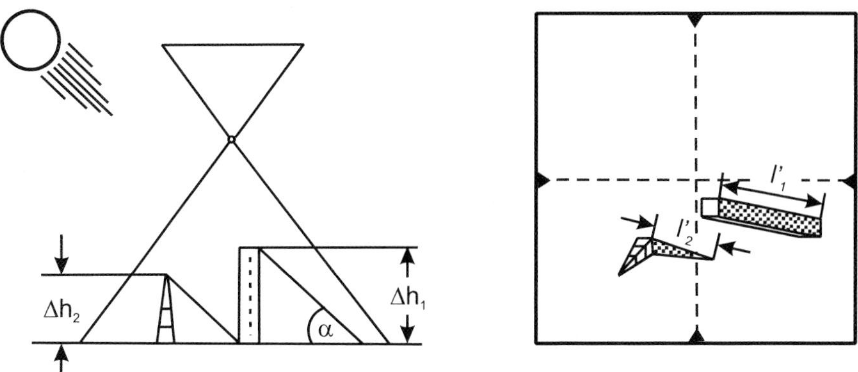

Fig. 3.2-3: Determination of the heights of objects from shadow length. Left: in the case of a known sun elevation angle α. Right: if the object height Δh_1 is known, ALBERTZ (2001)

$$\Delta h = \frac{\Delta r' h_g}{r'} \tag{3.2.2}$$

The height of the object can be estimated from the length l' of its shadow if the angle α with respect to the position of the sun (sun elevation angle) is known (*Fig. 3.2-3, left*) (SCHNEIDER, 1974; ALBERTZ, 2001). S_n is the scale constant (Equation 3.1.1).

$$\Delta h = l' S_n \tan \alpha \tag{3.2.3}$$

If the height Δh_1 of one object is known, the height Δh_2 of other objects (*Fig. 3.2-3, right*) can be determined using Equation (3.2.4) (ALBERTZ, 2001):

$$\Delta h_2 = \frac{l_2' \Delta h_1}{l_1'}. \tag{3.2.4}$$

Simple rectification of vertical aerial photographs

A simple rectification procedure can be applied in case of flat terrain. As a rule of thumb, a terrain can be regarded as flat if the elevation differences are less than the scale constant S_n (Equation 3.1.1) divided by 500. *Table 3.2-1* gives the elevation differences for which terrain can be regarded as being flat, and the resulting position errors as a function of map scale.

Table 3.2-1: Permissible elevation differences for which a terrain can be regarded being flat and resulting position errors as a function of map scale (LÜSCHER, 1944; SCHNEIDER, 1974)

Map scale	Permissible elevation difference in m	Resulting position error in m
1 : 1000	2	1
1 : 2000	4	2
1 : 5000	10	5
1 : 10 000	20	10
1 : 25 000	50	25
1 : 50 000	100	50

Graphical, optical-photographic, and digital image processing methods are available for rectification of aerial photographs in areas with flat terrain (ALBERTZ, 2001). The perspective relationship between the plane of the image and the ground surface can be unambiguously determined from four ground control points with known coordinates (e.g., from GPS measurements) that can be accurately located in the photograph. In digital image processing the relationship between the point locations on the image plane and on the ground surface can be expressed by two equations. The coefficients in the equations are determined from the ground control points and solved respectively.

Stereometric processing of vertical aerial photographs

Aerial photographs successively taken along a flight line usually overlap by approximately 60 %. The area of overlap allows for stereoscopic viewing, i.e., viewing the two images simultaneously using stereoscopic instruments. Viewed with the proper spacing of the images, the area is presented in three dimensional perspective. The photographs must have about the same scale, i.e. they must be taken from a similar sensor flight elevation. Satellites, e.g., IKONOS and QuickBird-2 have the capability to tilt the line-of-sight in either the along-track or the across-track direction to generate both along-track and across-track stereo pairs of photographs (GUPTA, 2003).

Stereoscopic viewing and measurement is based on the "principle of parallax", which is fundamental to photogrammetric processing. Parallax is the apparent displacement of the position of stationary objects caused by a shift in the position of the viewer, e.g., apparent location difference of a target in two consecutive, overlapping aerial photographs. In vertical aerial photographs, parallax displacements only occur parallel to the flight line. Distortions due to aircraft instabilities affect the compatibility of photographs for stereometric processing. *Figure 3.2-4* illustrates the concept of parallax and the geometric relations in stereoscopic viewing for overlapping vertical aerial photographs.

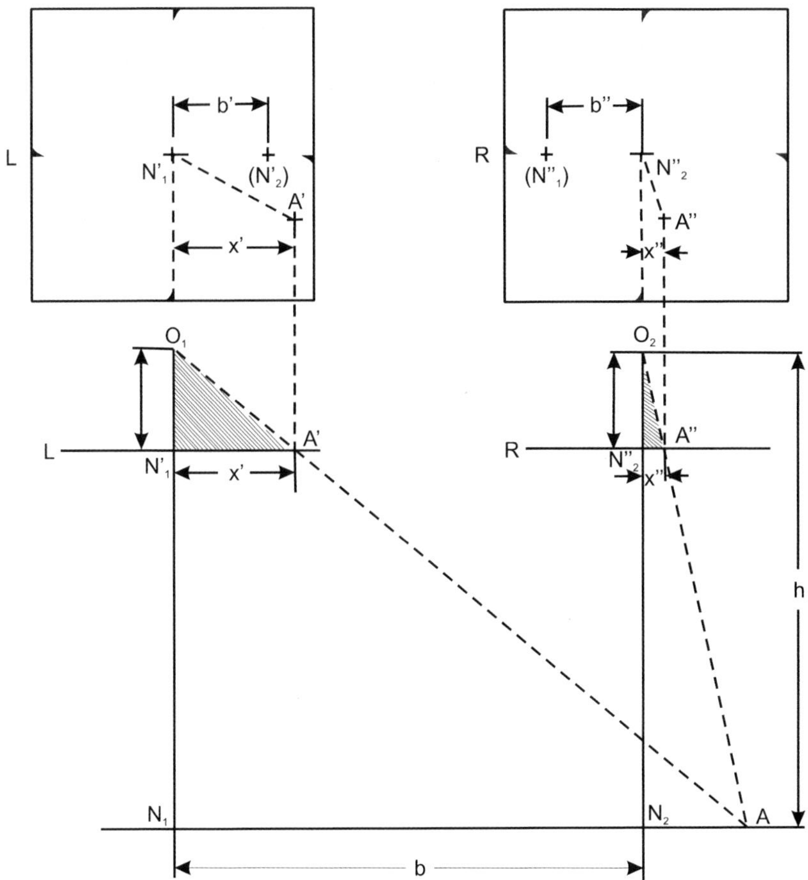

Fig. 3.2–4: Concept of Parallax: stereoscopic viewing of overlapping aerial photographs and geometric relationships, SCHNEIDER (1974)

For stereoscopic processing, the first photograph in the flight direction has to be aligned on the left side of the stereoscope and the overlapping photograph on the right side. The image centers (nadir points) N'_1 and N''_2 of the aerial photographs have to be determined and the nadir point in the left photograph is marked on the right one and vice versa (*Fig. 3.2-4*). The flight line axis can be found by connecting the nadir points N'_1 and N''_2.

Point A on the ground is represented in the stereo pair of photographs as A' and A". The distances from nadir points N'_1 and N''_2 to the image points A' and A" are x' and x''. The difference $p = x' - x''$ is called the parallax of point A. For the parallax p the Equation (3.2.5) can be established:

$$\frac{p}{f} = \frac{b}{h}. \tag{3.2.5}$$

The focal length of the camera lens is f, the height of the camera above a point on the ground is h. The horizontal distance b between the exposure points N_1 and N_2 is called the air base. Equation (3.2.5) can be solved with respect to h, as shown in Equation (3.2.6), and be used to estimate the elevation of the terrain:

$$h = \frac{f\,b}{p}. \tag{3.2.6}$$

The height Δh (in meters) of an object can be determined from measurements of the parallax difference Δp in the positions of the top and base of the object measured in hundredths of a millimeter:

$$\Delta h = \frac{h_0\,\Delta p}{b' - \Delta p}, \tag{3.2.7}$$

where b' is the horizontal distance between points N'_1 and (N'_2) in *Figure 3.2-4* measured in hundredths of millimeter range, and h_0 is the flight elevation above ground in meters.

Measurements of object height and elevation differences are made with stereoscopic instruments. Numerous instruments such as lens stereoscope, mirror stereoscope, and analytic data processing systems are available but differ in their technical capabilities. In stereometric processing the three-dimensional coordinates with respect to a reference plane can be calculated for any terrain point and/or object. These relative coordinates are transformed using ground control points into the (absolute) coordinates of the national or international topographic reference system[1]. Photogrammetric processing is predominantly carried out in a digital environment.

Automatic digital stereoscopic restitution is carried out in three steps:

1. Calculation of the spatial coordinates and parallax: An automatic recognition process for image correlation searches for and determines identical (corresponding) points in a pair of scanned stereoscopic images. Corresponding image points are identical terrain points in both images of a stereo pair. The spatial coordinates of these points and their parallax are calculated.
2. Introduction of ground control point coordinates and development of a true position model: The parallax values only reflect relative height differences. Thus, ground control point coordinates (from GPS measurements, topographic maps, etc.) are to be included in the dataset. They allow

[1] The most important reference systems for worldwide applications are the UTM-system (Universal Transverse Mercator system), the MGRS (UTMREF) for military operations and the UPS-system for polar regions.

determination of absolute heights and are used to derive a true position elevation model.

3. Interactive modifications: The elevation model is improved interactively by editing areas with no or incorrect data. Objects which are not required (e.g., houses, trees) are eliminated.

Further detail for photogrammetric processing is given, e.g., by HALLERT (1968); ALBERTZ & KREILING (1989); WOLF & WITT (2000); EGELS & KASSER (2001).

Digital elevation models (DEM) can also be prepared by digital stereoscopic restitution. *Figure 3.2-11* illustrates the results of the digital stereoscopic restitution of a landfill.

Differential rectification of vertical aerial photographs

If displacements caused by large elevation differences exceed a specified tolerance (see *Table 3.2-1*), differential rectification has to be applied to transform the data and correct for these distortions. This process eliminates displacement caused by topography resulting in an orthophotograph. Such rectification is more complex and more expensive than simple interpolation and requires detailed elevation data, typically an elevation model. Detailed elevation data are usually derived by digitizing contour lines from existing topographic maps. If such maps are not available or are insufficient, elevation information has to be determined by stereo photogrammetric restitution of the aerial photographs or by other methods such as airborne laser scanning. Ground control point coordinates and the camera specifications described in a calibration certificate (provided by the company that conducts the aerial photographic survey) are required in addition to elevation data.

3.2.4 Instruments

Common remote sensing instruments used in photogrammetry are described in Section 3.1.4. *Table 3.2-2* gives an overview of the technical specifications of photogrammetric scanners. Photogrammetric scanners are used for digitizing analog aerial photographs for digital image processing. In addition, airborne laser scanners are used as an alternative to traditional aerial photography for three-dimensional imaging of terrain.

Table 3.2-2: Types of photogrammetric scanners, modified after BALTSAVIAS & KAESER (1999)

Name	**LH Systems/DSW 500**	**ISM XL–10**	**ZI/Imaging PhotoScan 2001**
manufacturer	LH Systems, USA	Swissphoto Vermessung AG, Switzerland	ZI Imaging, Germany
scan pixel size [µm]	4 - 20	10 - 320	7 - 224
radiometric resolution [bit] (10 bit internal/ 8 bit output)	10/8 or 10	10/8	10/8
scanning format x/y [mm]	265/265	254/254	275/250
roll film width/length [mm/m]	35 - 241/152 manual, automatic	211 manual, automatic	245/150 manual, automatic

3.2.5 Survey Practice

Aerial photographic survey practice is described in Section 3.1.5 and survey practice of nonphotographic imaging systems is described in Section 3.3.5.

3.2.6 Processing and Interpretation of Data

Derivation of orthophotographs, photomosaics and topographic maps

Orthophotographs are determined from aerial photographs using a perspective projection where displacements from the tilt and topography have been removed. Once projected, orthophotographs have a consistent scale throughout the image and can be used as for mapping purposes. An assembly of several overlapping and/or adjacent aerial photographs, satellite or scanner images joined together to a continuous spatial representation is called a photomosaic.

The preparation of orthophotographs and photomosaics starts with the scanning of aerial photographs and ground checks of control points using GPS or other reference data. Aerial photographs with scales finer than or equal to 1 : 15 000 should be used. The resolution of photographs at coarser scales is usually too low. Optimal definition of the scanning resolution has to consider that at least four pixels are needed to clearly identify an object. As a rule of thumb, this means: scan resolution < (scale × desired resolution of an orthophotograph)/2, with the “scan resolution” (pixel size) in µm and the “desired resolution of an orthophotograph” in m. For example, if the orthophotograph is to resolve details of 0.5 m in size and the scale of the aerial photograph is 1 : 7500, then 1/7500 × 0.5 m = 66 µm. To adequately satisfying the rule of thumb, the pixel size for scanning should be 30 μm resulting in a square of four pixels that is 60 μm for a side.

In order to obtain sufficient accuracy, the aerial photograph should be rectified by simple rectification, stereometric processing or differential rectification (Section 3.2.3). The decision for which rectification method to be used depends on the elevation differences in the area of interest. The limits for simple rectification are given in *Table 3.2-1*. For areas with greater topographic differences, differential rectification with an elevation model is the appropriate method to prepare precise orthophotographs and thematic maps.

Rectification procedure

Digital image processing is commonly used in photogrammetry. In its basic form of understanding, an input image is processed by a transformation function into an output image. The processing can involve changes in the geometric relationships and the radiometric content of the image. Radiometric transformations change the gray values of the pixels, e.g., in order to improve the spectral quality or contrast of the image. The geometric projection of an image onto a horizontal reference plane is called rectification or restitution. The spatial relationship between image coordinates and reference plan locations (e.g., the Earth's surface) is determined by ground control points. These points are used to rectify the output image into a topographic reference system.

A difficult problem in the rectification is the determination of the transformation functions. Quadratic equations are suitable but require at least six control points to determine the equation coefficients. A location precision of less than one pixel can be obtained. Two approaches are commonly used to transform to original pixel values to the rectified image using these equations (ALBERTZ, 2001):

- Direct transformation: The location of each point in the input image is mapped to the output image and the value of the input pixel is assigned to the output pixel. Due to the geometric distortion of the input image this can lead to heterogeneous pixel occupancy in the output image. Interpolation to a regular pixel distribution is necessary after the transformation.

- Indirect transformation: For each pixel of the output image, an inverse transformation function is used to determine the location of the corresponding pixel in the input image to obtain the desired pixel data. The coordinates of data points in the input image will not map to pixel centers. This procedure is commonly used.

In this context, different rules have been introduced to obtain "suitable" estimations of gray values for the pixel matrix of the output image (GÖPFERT, 1991; MATHER, 1999; ALBERTZ, 2001):

- The nearest neighbor takes the pixel value of the input image closest to the calculated coordinate value.
- Bilinear interpolation: The relevant pixel value is calculated by linear interpolation from the values of the four closest pixels around the calculated coordinate value.
- Bi-cubic interpolation: interpolation using the 4×4 pixels neighborhood around the calculated coordinate value[2].

Before rectified aerial photographs can be merged to a photomosaic, the radiometric differences between the individual images in terms of brightness, contrast, and color are minimized by radiometric balancing. The available image processing software can be used for individual images or groups of images.

Digital elevation models derived from stereo processing of aerial photographs

A digital elevation model (DEM) can be produced by digital photogrammetric processing of overlapping aerial photographs or by other methods such as laser scanning or radar interferometry. The quality of the digital restitution of pairs of aerial photographs depends on the quality of the image correlation, as the most important part of stereoscopic restitution (FÖLLER & GERTLOFF, 1998). After scanning the aerial photograph and determination of the ground control points, the area of interest is divided into subareas with uniform terrain characteristics. The procedure for image correlation has to be selected for each subarea. Criteria for the selection of processing procedures are terrain texture, steepness of slopes, degree of urbanization, vegetation, terrain breaks, and raster width of the digitized image.

For the automatic processing of a pair of aerial photographs, a raster cell size of one meter has proven to be suitable. Smaller raster sizes do not increase accuracy, while larger grid cells can lead to a loss of both detail and small land features. An alternative consists of closely spaced elevation measurements along terrain breaks and a more widely spaced grid in areas

---

[2] More information for the gridding/interpolation algorithms is given in Part 6.

with only minor elevation changes. Before calculation of elevation contour lines, errors caused by disturbing objects such as trucks, single trees and hedges and forested areas should be eliminated interactively.

Digital elevation models derived from airborne laser scanning

Laser scanning (Chapter 3.3) can be used to prepare highly accurate digital elevation models. These models can also be combined with other products of photogrammetry, such as orthophotographs, photomosaics, topographic maps and DEM from stereometric processing.

The coordinates of the laser reflection points are calculated after data acquisition. These data are combined and compared with coordinate values from DGPS, the flight data, and the inertial navigation data from the aircraft. Disturbing objects, such as trees, buildings and cars, should be eliminated from the digital elevation model.

3.2.7 Quality Assurance

On the commercial market, aerial photogrammetry is influenced by pricing pressures with possible negative impacts on the quality. On the other hand, photogrammetric products are increasingly used by organizations which do not have photogrammetric know-how. Therefore, quality control standards are a necessity. A number of companies have developed quality management systems. Different aspects of quality assurance for aerial photograph surveys are described in Section 3.1.7.

3.2.8 Personnel, Equipment, Time Needed

	Personnel	Equipment	Time needed
aerial photography survey	2-3 technicians or specialists	special survey aircraft with a flight management system, aerial metric camera, aerial film	1 day
film development	1 technician	roller transport processor	1 day
scanning of two aerial photographs	1 technician	photogrammetric scanner	1 hour
terrestrial GPS/DGPS measurements (ca. 10 ground control points)	1 driver 1 technician or specialist	1 4WD vehicle, 1 GPS or DGPS	1 - 3 days
preparation of orthophotographs or aerial photomosaics	1 specialist	PC, plotter, software	1 day
preparation of digital elevation model using aerial photographs: for 10 km² for 100 km²	1 specialist	PC, plotter, software	 5 - 10 days 4 - 6 weeks
updating of existing map	1 specialist	PC, plotter, software	1 - 3 weeks
preparation of new topographic map	1 specialist	PC, plotter, software	4 - 8 weeks

3.2.9 Examples

Preparation of orthophotographic base maps from aerial photographs and IKONOS data

Up-to-date topographic maps (base maps) at a scale of 1 : 5000 or less are required for site investigations. Topographic maps at a scale of 1 : 50 000 published by the Royal Thai Survey Department are usually available for applications in Thailand and have proven to. However, these maps are insufficient as base maps for site investigations. Therefore, for the investigation of the area of the Dan Khun Thot landfill in the Nakhon Ratchasima province of Thailand, base maps had to be prepared from aerial photographs or high-resolution satellite images. IKONOS satellite images

(UTM projection, zone 47N, 1 m pixel size, datum WGS84, panchromatic) and aerial photographs at a scale of 1 : 15 000 from the Royal Thai Survey Department were used. Cloud coverage of the IKONOS images was 1 % for site 1 and about 20 % for other sites. A ground check of the prepared maps with a handheld GPS revealed deviations in the coordinates for site 1 between 3 and 13 m. This accuracy is sufficient for site investigations. The proposed waste disposal site 1 (*Fig. 3.2–5*, *red line*) is located approximately 17 km southwest of Dan Khun Thot with an area of about 135 000 m^2. The terrain is undulating with elevations between 235 and 265 m above sea level.

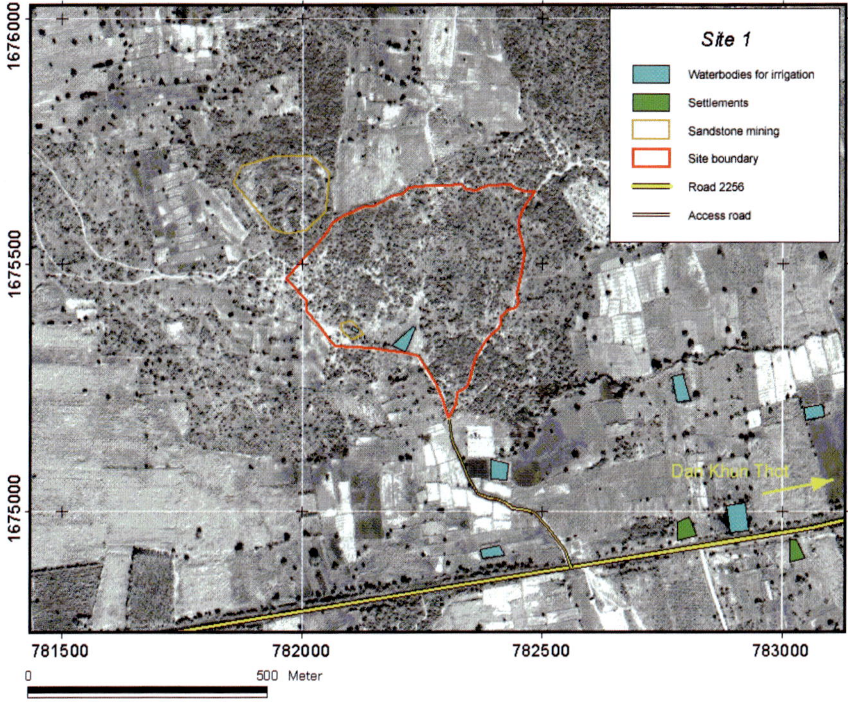

Fig. 3.2-5: Base map of the proposed Dan Khun Thot waste disposal site 1 (Thailand) prepared from IKONOS images from Nov. 27, 2002, coordinates: easting 0781750 to 0782750 and northing 1675000 to 1675750, Thai-Viet projection, UTM zone 47

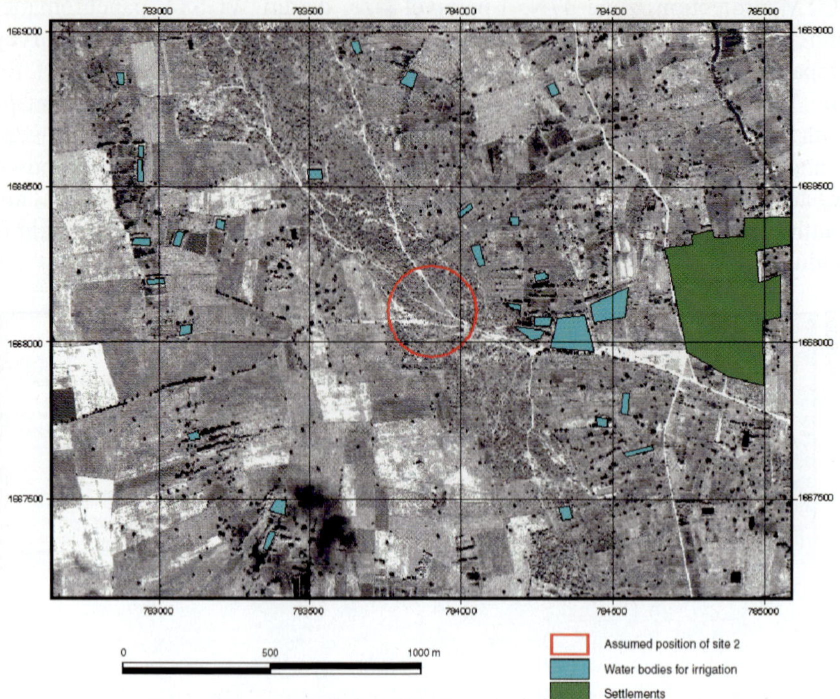

Fig. 3.2-6: Orthophotograph of the proposed Dan Khun Thot waste disposal site 2 (Thailand) prepared from aerial photographs, coordinates: easting 0782600 to 0785200 and northing 16672080 to 1669100, Thai-Viet projection, UTM zone 47

Figure 3.2-6 shows an example of an orthophotograph prepared from aerial photographs at a scale of 1 : 15 000.

Preparation of an aerial photomosaic and a topographic base map for site investigations

CIR aerial photographs at 1 : 7500 scale were acquired to provide a base map for the Mae Hia waste disposal site near Chiang Mai in northern Thailand and for remote-sensing site investigations. The aerial photomosaic (*Fig. 3.2-8*) is the result of rectifying and geo-referencing individual aerial photographs (*Fig. 3.2-7*), merging and joining them. First, the CIR images were scanned with a 28 µm pixel size. A differential rectification was carried out considering elevation differences in parts of the investigation area. Contour lines from the official 1 : 50 000 topographic map and DGPS control points were used for this purpose. Thailand's 1975 reference ellipsoid and the coordinate system of UTM zone 47 formed the geodetic reference system. The individual images and their joining were processed and merged as a mosaic.

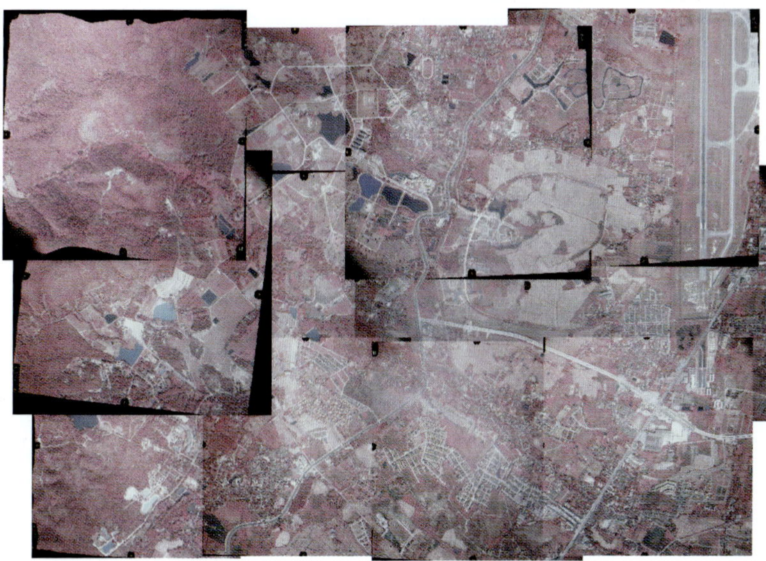

Fig 3.2-7: CIR-aerial photographs of the area around the Mae Hia waste disposal site in Chiang Mai, northern Thailand

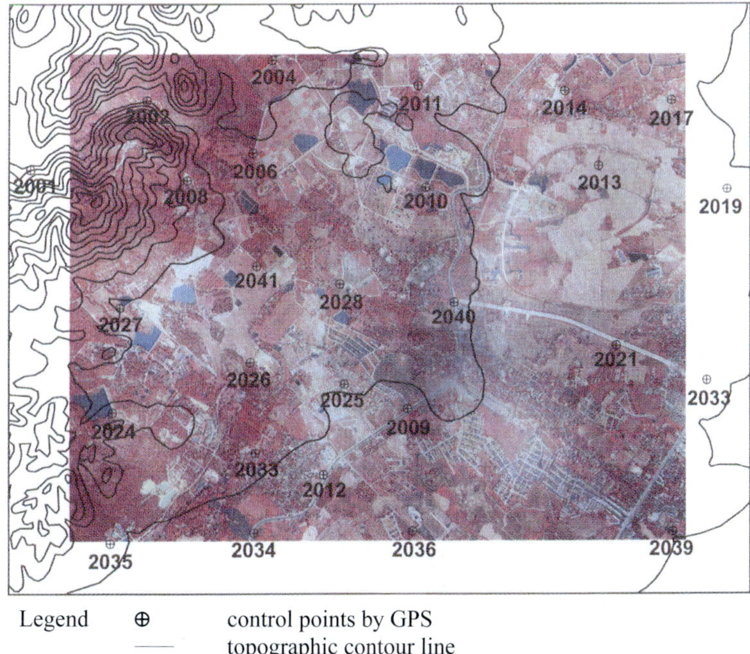

Legend ⊕ control points by GPS
—— topographic contour line

Fig 3.2-8: Rectified aerial photomosaic with contour lines and control points in the area of the Mae Hia waste disposal site

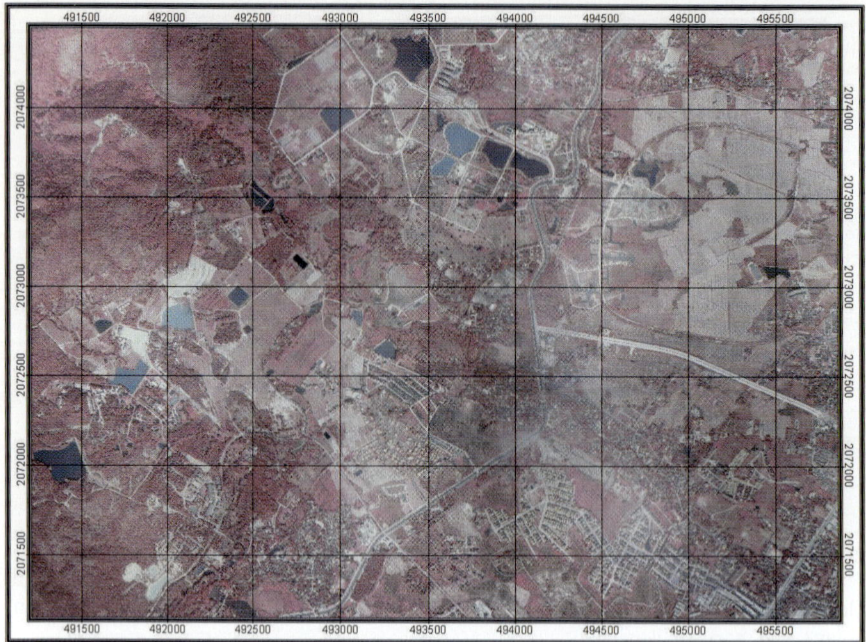

Fig 3.2-9: Aerial photographic map of the area around the Mae Hia waste disposal site

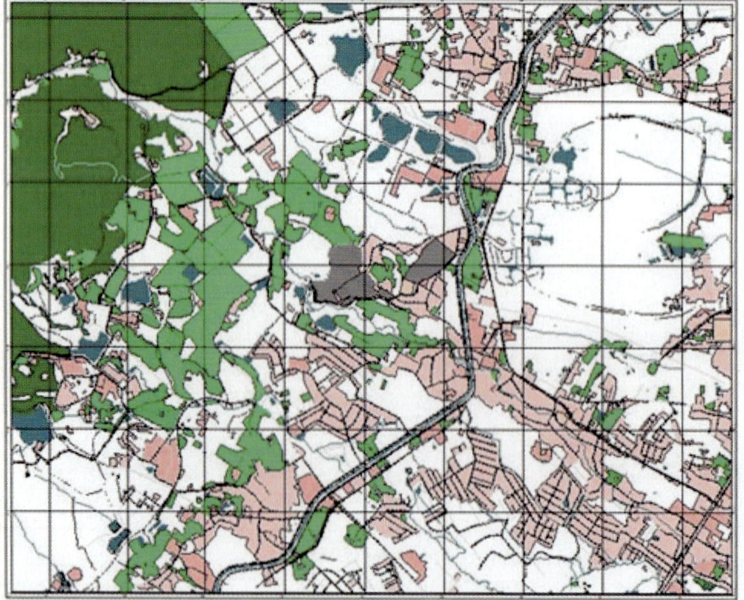

Fig 3.2-10: Topographic map of the area around the Mae Hia waste disposal site

The aerial photographic map of *Figure 3.2-9* was produced from the rectified and geo-referenced aerial photomosaic in *Figure 3.2-8*. On the basis of this aerial photographic map, a topographic base map (*Fig. 3.2-10*) at a scale of 1 : 5000 was prepared by on-screen digitizing and mapping using the GIS ArcView 3.2a interpretation software. This base map was used for the geological, geophysical, geochemical and remote-sensing site investigations of the Mae Hia waste disposal site. Only selected topographic elements, such as the perimeter of the disposal sites, streets, paths, rivers, canals, villages and towns, were plotted as a topographic base map on the maps with other geoscientific data.

Development and updating of a digital elevation model of a landfill

A digital stereo restitution of 1 : 5000 color aerial photographs was carried out to prepare a digital elevation model of a landfill. During digital stereo restitution, elevations were determined using a 15 m grid for most of study area and in more detail along terrain breaks (*Fig. 3.2-11*).

Aerial photographs were taken annually and used to update the elevation model of the landfill. This is done to document changes in the surface of the landfill, to estimate the increase in volume of waste, and observation of shrinkage (compaction). Monitoring the relief and changes in volume informs the operating agency about the condition of the landfill over a period of time and about how long the site can be used before the limits are reached, as well as, supports planning efforts for new necessary sites.

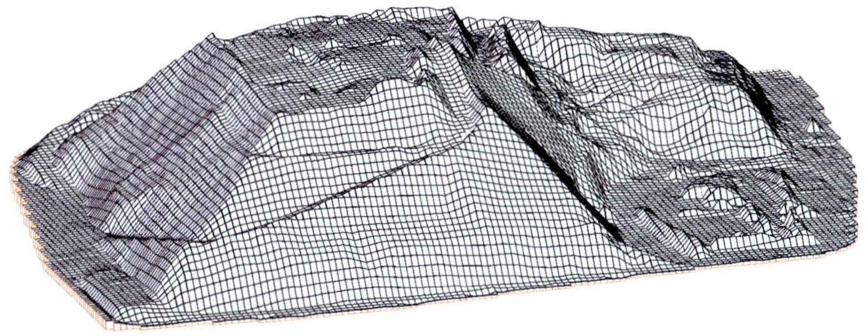

Fig 3.2-11: Perspective view of a digital elevation model of a municipal landfill in Germany, Hansa Luftbild (German Air Survey), Münster

Preparation of a regional digital elevation model from airborne laser scanning

Laser scanning observations can be used to monitor natural and manmade changes in relief. The area of investigation is located in a region of intense subrosion leading to caving to the surface and subsidence. Mining activities in the area have reinforced and intensified this process. A cone-shaped tailings and slag heap, rising about 150 m above the surrounding area, can be seen in the center of the derived digital elevation model (*Fig. 3.2-12*).

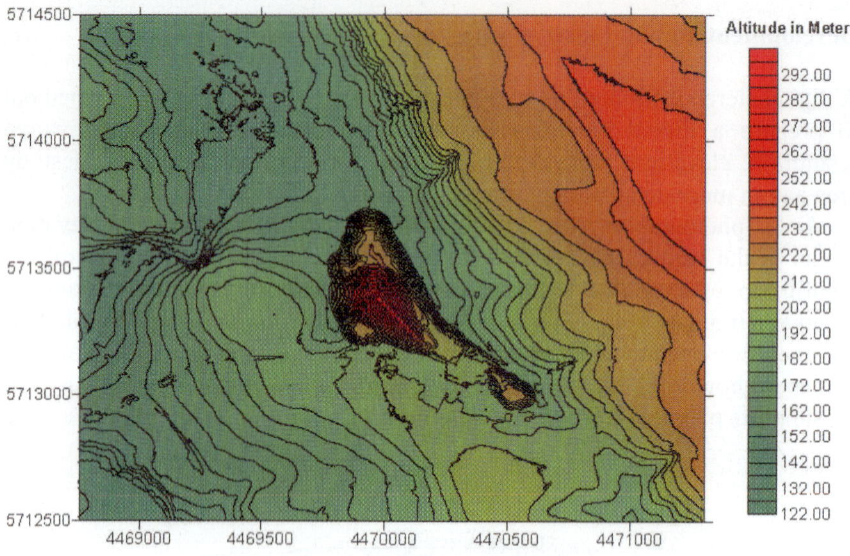

Fig 3.2-12: Regional digital terrain elevation contour model (system: Gauß-Krüger) of part of the Eisleben mining area in Germany from airborne laser scanning, BGR

References and further reading

ACKERMANN, F. (1995): Digitale Photogrammetrie – ein Paradigma-Sprung. Zeitschrift für Photogrammetrie und Fernerkundung, **63** (3), 106-115.

ACKERMANN, F. (1996): Verfahrenstechnik und Potential des Airborne Laser Scanning. Hansa Luftbild Symposium 1996: Airborne Laser Scanning – Ein neues Verfahren zur berührungslosen Erfassung der Erdoberfläche, Münster, 25.04.1996.

ALBERTZ, J. (2001): Einführung in die Fernerkundung, Grundlagen der Interpretation von Luft- und Satellitenbildern. Wissenschaftliche Buchgesellschaft, Darmstadt.

ALBERTZ, J. & KREILING, W. (1989): Photogrammetrisches Taschenbuch – Photogrammetric Guide; Wichmann, Karlsruhe.

BÄHR, H. P. & VÖGTLE, T. (1998): Digitale Bildverarbeitung. Wichmann, Heidelberg.

BALTASAVIAS, P. & KAESER, C. (1999): Quality Evaluation of the DSW 200, DSW 300, SCAI and Ortho Vision Photogrammetric Scanners. Unpublished Report.

BAUER , M. (1997): Vermessung und Ortung mit Satelliten. Wichmann, Karlsruhe.

BRIESE, C. & PFEIFER, N. (2001): Airborne Laser Scanning and Derivation of Digital Terrain Models. In: GRÜN & KAHMEN (Eds.): Optical 3-D Measurement Techniques, V, Wichmann, Karlsruhe, 80-87.

BURNSIDE, C. D. (1979): Mapping from Aerial Photographs. Granada, London.

CICIARELLI, J. A. (1991): A practical guide to aerial photography: with an introduction to surveying. Van Nostrand Reinhold, New York.

COLWELL, R. N. (1983): Manual of Remote Sensing. American Society for Photogrammetry, Falls Church (Virginia).

ECKER, R., KALLIANY, R. & OTEPKA, G. (1993): High quality rectification and image enhancement techniques for digital orthophoto production. Photogrammetric Week **93**, 142-155.

EGELS, Y. & KASSER, M. (2001): Digital Photogrammetry. Taylor & Francis.

FALKNER, E. (1995): Aerial Mapping: Methods and Applications. Lewis Publishers, Boca Raton.

FÖLLER, J. & GERTLOFF, K. H. (1998): Überwachung der Verformung von Deponieoberflächen mit Methoden der digitalen Photogrammetrie. Zeitschrift für Vermessungswesen, **9**, 287-294.

FRIESS, P. (1996): Systemkonfiguration und Anwendungsbereiche des Airborne Laser Terrain Mapping – System ALTM 1020, Hansa Luftbild Symposium 1996: Airborne Laser.

GIM International (2001): Product Survey on Airborne Laser Scanner - an Overview.

GÖPFERT, W. (1991): Raumbezogene Informationssysteme. Wichmann, Karlsruhe.

GUPTA, R. P. (2003): Remote Sensing Geology. Springer, Berlin.

HALLERT, B. (1968): Photogrammetry, Basic Principles and General Survey. Mc Graw Hill, New York.

HEIPKE, C. (1999): Digital Photogrammetric Workstations, GIM International, **13**, 1, 81.

HILDEBRANDT, G. (1996): Fernerkundung und Luftbildmessung: für Forstwirtschaft, Vegetationskartierung und Landschaftsökologie. Wichmann, Karlsruhe

HOFFMANN-WELLENHOF, B., LICHTENEGGER, H. & COLLINS, J. (2001): Global Positioning System. Theory and Practice, Springer, Wien.

HOSS, H. (1997): Einsatz des Laserscanner-Verfahrens beim Aufbau des Digitalen Geländehöhenmodelles in Baden-Würtemberg. Mitteilungen d. Deutschen Vereins f. Vermessungswesen, Heft 1.

KONECNY, G. (1997): From Photogrammetry to Spatial Information Systems. ITC Journal.

KONECNY, G. & LEHMANN, G. (1984): Photogrammetry. de Gruyter, Berlin.

KÖLBL, O. (1992): Popularisation of Photogrammetry. I. Arch. Ph. XXIX – **B2**, 601- 607.

KRAUS, K. (1993): Photogrammetry. Vol. 1: Fundamentals and Standard Processes, Dümmler, Bonn.

KRAUS, K. (1997): Photogrammetry. Vol. 2: Advanced Methods and Application. Dümmler, Bonn.

KRAUS, K. (2000): Photogrammetrie. Vol 3: Topographische Informationssysteme. Dümmler, Köln.

KUMM, W. (2000): GPS Global Positioning System. Delius, Klasing, Bielefeld.

LILLESAND, T. M. & KIEFER, R. W. (1994): Remote Sensing and Image Interpretation. 3rd edition, Wiley & Sons, New York.

LINDNER, W. (2003): Digital Photogrammetry – Theory and Applications. Springer, Berlin.

LÜSCHER, H. (1944): Kartieren nach Luftbildern. Berlin.

MATHER, P. M. (1999): Computer Processing of Remotely Sensed Images: An Introduction. Wiley & Sons, Chichester.

MIKHAIL, E. M., BETHEL, J. S. & MCGLONE, J. C. (2001): Raumbezogene Informationssysteme, Indroduction to Modern Photogrammetry. Wiley & Sons, Chichester.

REEVES, R. G. (1975): Manual of Remote Sensing. American Society for Photogrammetry, Falls Church (Virginia).

RITCHIE, W., WOOD, M., WRIGHT, R. & TAIT, D. (1988): Surveying and Mapping for Field Scientists. Longman.

SABINS, F. F. (1996): Remote Sensing: Principles and Interpretation. 3rd edition. Freeman and Co., New York.

SCHNEIDER, S. (1974): Luftbild und Luftbildinterpretation. de Gruyter, Berlin.

SCHWEBEL, R. (2001): Qualitätssicherung für Bildflug und analoges Luftbild durch neue DIN-Norm, Photogrammetrie, Fernerkundung. Geoinformation, **1**, 39-44.

STUTTARD, M. & DOWMAN, I. (1998): Guidelines for Quality Checking of Geometrically Corrected Remotely Sensed Imagery, Unpublished Report.

SCHWIDEFSKY, K. & ACKERMANN, F. (1976): Photogrammetrie – Grundlagen, Verfahren, Anwendungen. Teubner, Stuttgart.

SEEBER, G. (1989): Satellitengeodesie. de Gruyter, Berlin.

STEINER, T. D. & MERRIT, P. H. eds. (1998): Airborne Laser Advanced Technology, **338**, S P I E- International Society for Optical Engineering.

TEUNISSEN, P. J. G. & KLEUSBERG, A. (1998): GPS for Geodesie. Springer, Berlin.

VEGT, J. W. & HOFFMANN, A. (2001): Airborne Laser Scanning Reaches Maturity. Geoinformatics Sept. 2001, Emmelord, 32-39.

WOLF, P. R. & WITT, B. (2000): Elements of Photogrammetry. Mc Graw Hill, New York.

3.3 Nonphotographic Imaging from Aircraft and Space-borne Platforms

FRIEDRICH KÜHN, BERNHARD HÖRIG & DIETMAR SCHMIDT, with a contribution by TILMANN BUCHER

3.3.1 Principle of the Methods

Instead of using light-sensitive film, nonphotographic imaging systems detect the incoming electromagnetic (EM) radiation with semiconductor detectors or special antennas. While photography is limited to a spectral range from 0.3 to 0.9 µm, multispectral scanners (MSS) can operate in wavelength regions from 0.3 to approximately 14 µm (LILLESAND & KIEFER, 1994; GUPTA, 2003). This range includes radiation in the near ultraviolet, visible light, near, middle, and thermal infrared (*Table 3.1-1*). MSS can work in rather narrow spectral bands of a few nanometers. Thus, such remote sensing can focus on specific spectral features to identify detailed physical and chemical characteristics of the land surface objects of interest.

The basic operating characteristics and image geometry of multispectral and thermal scanners are similar. Both passive and active methods are used for nonphotographic imaging. Passive methods sense the natural electromagnetic radiation reflected or emitted from the Earth's surface. Active sensors contain their own source of EM radiation (e.g., laser or radar). They transmit a coherent signal and receive the reflected radiation to obtain geometric, geological, and environmental information from the Earth's surface.

The most important nonphotographic imaging systems for geoscientific site investigations are:

- opto-mechanical line scanners or whisk broom scanners,
- opto-electronic line scanners, using charge coupled devices (CCD) basic sensor unit and are also refered to as linear scanners or push broom scanners,
- digital cameras (CCD array),
- imaging spectrometers,
- microwave sensors, mainly radar-based systems, and
- laser scanners.

They are operated from aerial platforms and from space-borne platforms. During last decades, the technical development of nonphotographic imaging sensors has been and continues to be very rapid. The theoretical background and technical principles of nonphotographic remote-sensing systems are described in more detail in COLWELL (1983), KRONBERG (1985), GUPTA (1991 and 2003), ALBERTZ (2001), and SABINS (1996).

Technically, space-borne imaging systems operate similarly airborne systems. Differences arise from the longer distance between the sensor and the object on the Earth surface (affecting spatial resolution and observation noise), the methods of data transfer, the temporal resolution, and costs for operation and data. There are some limitations on the use of space-borne data. Because of the greater distance there is more atmospheric interference during image capture. This can result in more radiometric distortions in the image data, e.g., object boundaries can be blurred compared to aerial photos with similar spatial resolution (KERSTEN et al., 2000). For geologic applications, data acquisition is sometimes restricted to a particular season and the study area must be free of clouds. Thus, the chances of adequate data capture are somewhat reduced, because the Landsat Thematic Mapper, for example, covers the same site on the Earth only every 16 days, which may not coincide with optimum weather conditions. Bad weather results in a delay in the acquisition of the data or makes the use of archival material necessary.

Satellite images and aerial photographs can be interpreted using the similar analytical methods. Sophisticated processing and interpretation of nonphotographic data require special advanced software and appropriate skills of the human analysist. Satellite images differ from aerial photographs in that they normally have a lower spatial resolution with larger area coverage. The first Landsat sensors in the 1970s had pixel resolutions of about 80 by 80 m. The Landsat-TM sensors of the 1980s produced images with a spatial resolution of 30 by 30 m, and the panchromatic band of the Landsat-ETM+ (launched in 1999) even provides 15 by 15 m. Similar to aerial photographs, overlapping satellite images can be analyzed stereoscopically. Data from nonphotographic systems (*Landsat Thematic Mapper*, *SPOT*, *ERS-1*, *ERS-2*, *RADARSAT*, *IKONOS 2*, etc.) may be obtained in digital form on magnetic tape, optical disk, CD-ROM or as hard copy prepared according to standardized image processing.

Nonphotographic remote-sensing systems have the following basic advantages over conventional aerial photographic cameras:

- the ability to detect reflected or emitted radiation from objects on the Earth's surface over a wider wavelength range (near-infrared, thermal radiation, microwaves, e.g., radar),
- the ability to measure electromagnetic (EM) radiation in narrow bands simultaneously with the same optical system,
- higher radiometric and spectral resolution,
- the ability to directly record data in digitized form supporting more rapid data processing,
- electronically generated data are more amenable to calibration, and

- the ability to transmit data from the sensor to a ground-based receiving station providing immediate availability of the data.

Images from satellite-based remote-sensing systems are primarily used to obtain data for large areas. For a long time, the spatial resolution of a satellite-based remote-sensing system was usually insufficient to delineate and identify small objects on the Earth's surface. During the last several years, new satellite-based remote-sensing systems, e.g., IKONOS 2 and QUICKBIRD-3, have improved spatial resolution comparable to aerial photography. These high-resolution commercial satellite images can be used to generate or update topographical maps for a site of interest and for investigation of the region around the site being investigated. After rectification using a digital elevation model (DEM) and ground control points, IKONOS and QUICKBIRD data can be used to produce orthorectified images with a geometric accuracy of 1 - 5 m. However, airborne sensors with spatial resolutions of several centimeters are still more suitable for detailed investigations, e.g., the evaluation of the vitality of trees or the search for leachate water, springs, ground destabilization, thermal features. For multi-temporal analysis, aerial photographs are still indispensable (Chapter 3.1) due to the extensive historical archives. To obtain most appropriate information for a specific purpose, it is recommended to combine photographic and nonphotographic data and analyze them synergistically.

3.3.2 Applications

- Analysis of the development of waste disposal, mining, and industrial sites and their surroundings,
- preparation and/or updating of orthophotographs and topographic maps,
- preparation of digital elevation models,
- evaluation and assessment of environmental and geological conditions using methods similar to aerial photograph interpretations,
- assessment of previous geological and environmental conditions at a waste disposal, mining, and industrial site,
- search for seepage of water at the edges of landfills and heaps,
- localizing seepage, springs, and soil moisture anomalies,
- investigation of natural and artificial drainage systems,
- localizing high and low permeability areas for water,
- recognition and monitoring of destabilization processes (horizontal and vertical land movement),

- analysis of the vitality of vegetation,
- assessment of surface water conditions,
- delineation of fractures and lineaments,
- inventory of sites suspected to be contaminated,
- mapping of land use patterns and biotopes,
- search for new waste disposal sites,
- localizing heat sources inside landfills ("hot spots"),
- detection of gas emission and surveillance of gas emission ducts in landfills,
- assessment of mining disposals and associated weathering products in mining districts, and
- detection of hydrocarbons or materials containing hydrocarbons on the ground at oil fields, refineries, chemical factories, petrol depots, airports, coking plants, gas works, along or close to transport ways and near waste disposal sites.

3.3.3 Fundamentals

The term "image" is used as a general term for a picture representation of a scene or object recorded by a remote-sensing system. It is usually restricted to representations acquired by nonphotographic methods. Nonphotographic imaging systems consist of an optical component, a detector, a data recording system, and in space-borne applications a telemetric system for transmitting the data to the ground base stations. The optical component contains lenses and mirrors to collect the incoming radiation, and filters, prisms, and/or gratings to split the signal into different wavelengths. The detector includes devices transforming optical energy into electrical signals, such as photo-emissive detectors, photo-conductive detectors, and photodiodes. The frequently used charge-coupled devices (CCD) are photodiodes.

The efficiency of nonphotographic imaging systems and the possible applications depend on the following basic parameters and system-related conditions: The quality of images depends on the *aperture*, the opening in a remote-sensing system that admits electromagnetic radiation to the film or detector, and the *instantaneous field of view (IFOV)*, the solid angle in which the detector is sensitive to radiation. The IFOV is normally expressed as the cone angle β within which the incident energy is focused on the detector. The angle β is determined by the optical component and the size of detectors and is measured in radians. *Spatial or ground resolution* is directly related to the IFOV and describes the ground area sensed by one of the sensor/detector elements and determines smallest terrain features to be recognized in the resulting images. All energy (from the land surface) arriving at the detector within the IFOV contributes to the detector response at any instant. A larger IFOV results in more energy being caught by the detector. This improves the ability of sensor systems to discriminate gentle differences in radiation intensity (radiometric resolution). The smaller the IFOV gets, the smaller is the size of the minimum resolvable scene element and the finer the ground resolution, but with the less radiation arriving at the sensor. Thus, there is a trade-off between spatial and radiometric resolution.

A similar trade-off exists for width of spectral bands. In general, the *spectral resolution* is the ability of sensor systems to discriminate spectral differences and is determined by the number of spectral bands, their bandwidths, and the spectral region covered. The amount of radiation arriving at the sensor decreases for narrower spectral bands. Given a constant system noise, a lower amount of acquired radiation signal causes the *signal-to-noise ratio* to decrease and ultimately lowers abilities to analyze the spectral signal. The signal-to-noise ratio is an essential parameter for characterizing the capabilities of imaging spectrometers.

The *dwell time* is the time required for a detector IFOV to sweep across a ground resolution cell. The *angular field of view* is the angle subtended by lines from a remote-sensing system to the outer margins of the strip of terrain that is imaged by the system. The width of the strip of terrain that is imaged by a scanner system – called *ground swath* – is determined by the total *angular field-of-view (FOV)*.

Depending on the nonphotographic system, the images show *distortions*, i.e. changes in shape and position of objects with respect to their true shape and position on the Earth's surface. In general opto-electronic scanners have images with more simple geometric properties than opto-mechanical line scanners. Opto-mechanical line scanner images have a panoramic distortion. The constant angular velocity of the rotating scanner mirror results in a scale distortion perpendicular to the flight direction. The width of ground resolution cells (IFOV) increases with the lateral distance from the nadir to the edge of the swath. The image scale in the direction of flight is constant. Images taken by optical-electronic scanners have no panoramic distortion. In opto-mechanical line scanner images vertical features are displaced at right angles from the nadir line (one-dimensional relief displacement). Flight parameter distortions have to be corrected for both scanner types, particularly in the processing of images from airborne systems.

Opto-mechanical line scanners

Opto-mechanical line scanners sense the Earth's surface by scanning across a swath of land. This scanner system is often referred to as a "whisk broom" scanner or “across-track” scanner. Opto-mechanical line scanners have been frequently used in the last several decades because they can be operated multispectrally at a wide range of wavelengths from 0.3 to 14 µm. Opto-mechanical line scanners are used for both airborne applications, e.g., the Daedalus-AADS scanner (*Fig. 3.3-2*), and space-borne applications, e.g., MSS, TM, and ETM+ scanners on Landsat, MTI and ASTER (TIR part) satellites (Section 3.3.4). The principle of this type of scanner – consisting of opto-mechanical line scanning system, detector unit, and data recording system – is shown in *Figure 3.3-1*. It is similar for both aerial and space-borne applications. A moving mirror tilted at 45° is the basic element of an optical-mechanical line scanning system.

It scans the area by rotating or oscillating perpendicular to the flight direction. Due to the forward motion of the sensorcraft a two-dimensional data array (image) is obtained.

The size of the smallest discrete area indentifyable in an image (pixel) depends on the flight altitude and the IFOV. The spatial resolution of most scanners is between 1 and 2.5 mrad. A scanner with an IFOV of 1.5 mrad, equivalent to 0.086°, will scan with a pixel size of 1.5 m × 1.5 m at a flight altitude (h_g) of 1000 m. These figures are for the center of a scanned swath. As a rule of thumb, to be recognizable on a scanned image an object on the Earth's surface should have a minimum size about 2.8 times the pixel size (ALBERTZ, 2001). For the preceding example (pixel: 1.5 m × 1.5 m; h_g = 1000 m), an object would be at least 4.2 m across to be clearly identifiable. Airborne opto-mechanical line scanners are operated at flight altitudes between 300 and 12 000 m.

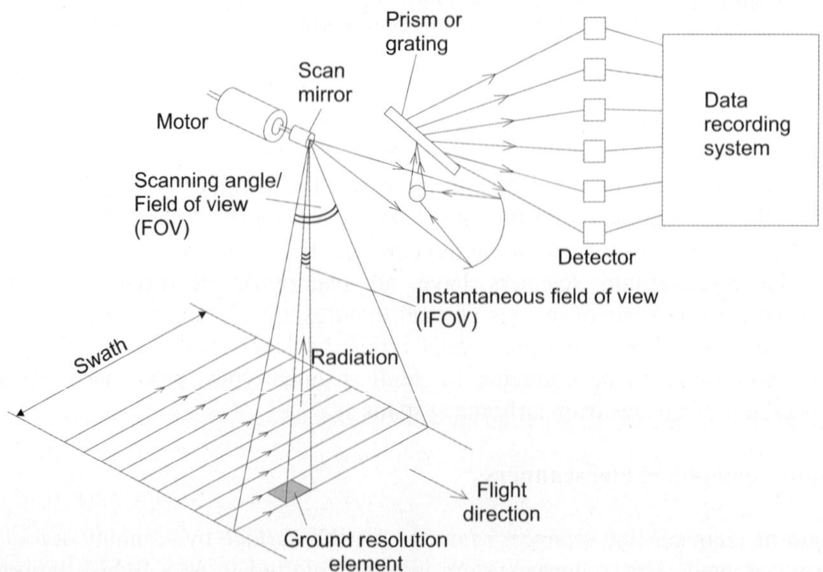

Fig. 3.3-1: Principle of opto-mechanical line scanners, modified after GUPTA (1991)

There often is a compromise between spatial resolution and swath width that has to be considered for mission planning. The margins of the scanned image usually show distortion of the pixels (panoramic distortion). They have to be geometrically corrected later. Because of inherent aircraft instability, geometric rectification of aircraft data is a major problem if the data needs to be georeferenced. Also, specific areas on the edges of the swath may be not acquired. With satellite-borne scanners, these distortions are usually irrelvant.

A scan line is the result of a single scan (moving mirror) across the width of the swath. The length of one scan line, i.e., the width of the swath, is defined by the angular field of view (FOV) of the scanner. This angle is usually between 90° and 120°. Due to the forward motion of the aircraft (or satellite), the terrain is scanned line by line resulting in a continuous image. Only if the synchronization is not properly adjusted overlaps or gaps will appear between the scan lines. The acquisition of high-quality images requires a perfect synchronization between the scan frequency of the mirror and the speed of the aircraft or satellite.

Fig. 3.3-2: Daedalus-AADS scanner image of the Münchehagen waste disposal site in Lower Saxony, Germany. The image was taken on July 6, 1989, by DLR Oberpfaffenhofen for BGR (channel 3 = red, 4 = green, and 5 = blue).

The reflected or emitted signal from the Earth surface reaches the scan mirror and passes through the optical system to the *detector unit*. The detectors transform the optical signals into electrical signals. The intensity of the electric signal largely depends on the intensity of the incoming radiation and, thus, on the spectral properties (absorption, reflection, or emission) of the ground surface. Multispectral scanners with hundreds of spectral channels have been developed and applied. The radiation is divided by a system of gratings and prisms into partial spectra and then passed to specialized detectors. Usually, it is necessary to use several different detectors because different parts of the electromagnetic spectrum between the visible part of spectrum and thermal radiation require different semiconductor crystals. It is common to employ Si detectors for the visible range and NIR-I and PbS detectors for the NIR-II part of the electromagnetic spectrum. InSb detectors for thermal radiation in the mid-infrared (MIR-II) range require cooling with liquid nitrogen (77 K). A temperature resolution on the order of 0.1 °C is attainable (LILLESAND & KIEFER, 1994). The electrical signals from the detectors, commonly measured as voltage, are stored digitally in the *data recording system*.

In addition to the relatively low ground resolution, the major disadvantage of optical-mechanical line scanners is their sensitivity to mechanical failure. Numerous moving parts must be adjusted very precisely and are subject to aging and wear. In addition, previously mentioned distortions related to movement of the aircraft and panoramic distortion at the edges of the scanned swath, require computer correction. Related processing and interpretation of the data requires access and capabilities for operating a digital image processing system.

To obtain optimum results, a geologic interpreter should be involved in the processing of the data so they are aware what processing has been done with the original data. Geographical and geological knowledge about the target study area and a well-focused, goal-orientated approach are prerequisites to a successful study. A more detailed explanation of the construction and performance of opto-mechanical scanners can be found in KRONBERG (1985), GUPTA (1991 and 2003), LILLESAND & KIEFER (1994), SABINS (1996) or ALBERTZ (2001).

Although being technologically outdated by opto-electronic scanners, the primary reason opto-mechanical line scanners are still in use is their ability to record thermal radiation. Furthermore, such sensors are quite effective including a larger number of spectral bands. In the course of technological development, however, it is assumed that future generations of opto-electronic scanners will be capable of routine data acquisition in the thermal wavelength ranges.

Information on the composition of surface materials and identification of temperature anomalies in and around a landfill can be obtained from *thermal scanner images*. The detected thermal properties of an object are determined both by the material of the object and by the character of the surrounding environment. The intensity of thermal radiation is dependent on both the surface temperature and the emissivity of the materials. The following physical and environmental parameters influence the thermal behavior of an object at the Earth's surface (after KRONBERG, 1985):

Physical parameters
- color
- composition
- surface properties
- density
- porosity or pore volume
- permeability
- water content

Environmental parameters
- topographical position in the field
- orientation of the surface with respect to the sun
- meteorological conditions
- microclimate
- humidity
- time of day, season
- type and extent of vegetation cover

Remotely sensed radiation temperatures have to be calibrated against temperatures measured at the ground surface. Thus, calculation of surface temperatures requires considerable field work; an effort only necessary for specific applications. Any given object in a thermal image can be identified as "warm" or "cold" relative to its surroundings (*Fig. 3.3-3*). Environmental influences can cause confusing overlap or even mask thermal radiation emitted from natural or artificial objects, creating false anomalies in the thermal images. The diurnal temperature cycles of two very different materials show that it is impossible to generalize about "warm" or "cold" characters of materials on the basis of their physical characteristics. According to GEBHARDT (1981), physical and environmental factors may act independently of each other or they may be synergetic or antagonistic to each other. In addition, the environmental factors are difficult or impossible to determine only on the basis of the images.

A landfill, mining or industrial site usually covers an area smaller than 1 km^2. Thus, highly sophisticated scanning systems designed for thermal mapping of larger areas may not be economically feasable in every case. Simple thermographic images often fit the requirements of site characterization best (*Fig. 3.3-3, Tables 3.3-2* and *3.3-3*).

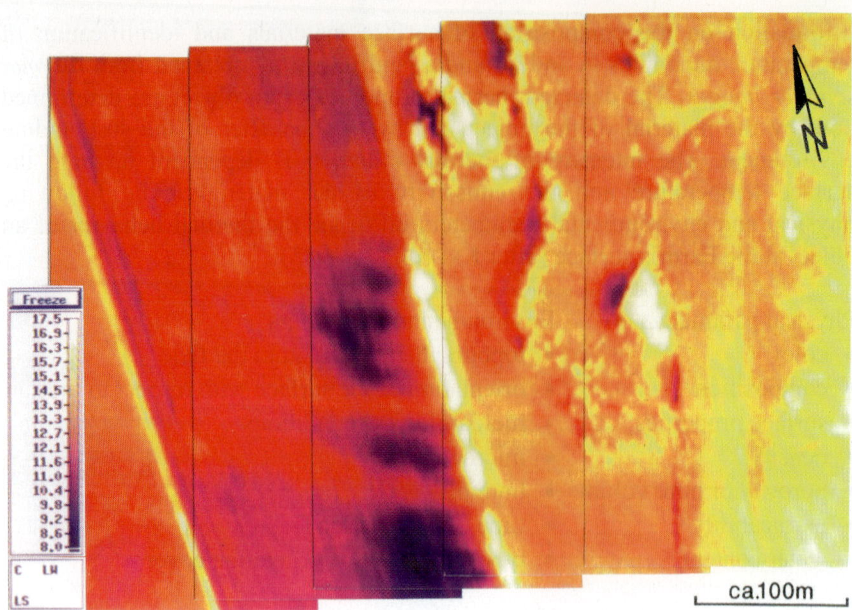

Fig. 3.3-3: Thermal image of the western rim of the Schöneicher Plan landfill and the adjacent area. The margin of the deposit (right) is warmer than the field to the left. Relief, groups of trees, and local winds generate exogenic thermal anomalies. Temperatures shown in °C (image taken on May 13, 1993, at 05:19 CET by F. BÖKER and F. KÜHN, BGR).

Optical-electronic scanners

Optical-electronic scanners are often referred to as CCD (Charge Coupled Device) linear array scanners, push-broom scanners, or along-track scanners. In principle, they are similar to photographic cameras. Instead of film, CCD arrays are installed in the focal plane. CCD arrays consist of a large number of detector elements, e.g., silicon-based photo detectors. Several thousand detectors may be placed on a single 1.5 cm^2 chip. Each individual spectral band, or channel, requires its own linear CCD array. This usually limits the number of spectral bands to be put on one sensor.

The CCD chips are mounted in the focal plane of an optical electronic scanner. The optical system focuses the incoming radiation on the surface of the light-sensitive chips. The radiation on each detector is integrated over a short time interval (dwell time) and transformed into electrical signals and recorded. The intensity of each signal depends on the spectral response of the related landscape feature. The detector chips are in a line perpendicular to the direction of flight. The swath is scanned due to the forward motion of the sensorcraft line by line in the manner a push-broom is used (*Fig. 3.3-4*). Opto-electronic scanners are operated from airborne platforms as well as from space.

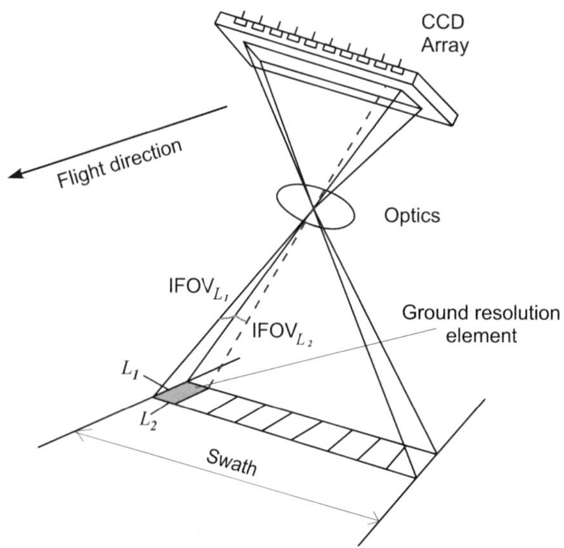

Fig. 3.3-4: Principle of an optical-electronic scanner (CCD linear array scanner, often called a push broom scanner), modified after GUPTA (1991)

The size of the ground resolution element of an opto-electronic scanner depends on the size of the detector cell as projected through the optics onto the ground surface. The ground resolution is, thus, dependent on the flight altitude h_g, the dwell time τ of each scan, and the velocity v of the aircraft. The ground resolution element is defined by a line L_1 parallel to the direction of flight (Equation 3.3.1) and L_2 perpendicular to this direction (Equation 3.3.2) as follows:

$$L_1 = v\tau \qquad (3.3.1)$$

$$L_2 = \frac{a\,h_g}{c_f}, \qquad (3.3.2)$$

where a is the distance between the centers of adjacent detector elements on the chip and c_f is the focal length of the optical system.

The major advantages of opto-electronic scanners compared to opto-mechanical scanners are their better reliability (due to the lack of moving parts), higher geometric precision along a scan line, and a longer durability. A longer dwell time enables a much stronger signal with better radiometric resolution and poorer spatial resolution. Disadvantage of this scanner type is the need to calibrate many more detectors. It is still difficult to produce chips with rows of detectors that are sensitive in the thermal infrared, are close enough together and can be individually cooled during flight. But, it is only a matter of time until the technology extends detection capability beyond the range of visible light (VIS) and parts of the near infrared (NIR-I and II) spectrum.

In multispectral opto-electronic scanners, the incoming radiation is splited into its spectral components by a grating, prism or color filters. It is possible to scan the surface over a selected wavelength interval by arranging several rows of detectors. The CASI scanner (described in the following section) is a multispectral recording system which exclusively uses an opto-electronic scanning system.

A more detailed explanation of the construction and performance of opto-electronic scanners can be found in GUPTA (1991 and 2003), LILLESAND & KIEFER (1994), SABINS (1996) or ALBERTZ (2001).

Digital cameras

The principle of remote-sensing systems known as digital cameras is quite similar to CCD linear array scanners (push-broom scanners) except that a two-dimensional array (matrix) is used instead of a linear array. At present array sizes of 4 k × 4 k to 9 k × 9 k are common. Digital cameras with 20 k × 20 k CCD matrix sensor chips will be available soon. It is expected that in the near future digital cameras will be used for most topographic mapping surveys. It is easier to develop digital elevation models from digital camera images than from aerial photographs. The spatial resolution of digital cameras depends on the number and size of the sensor elements on the chip, flight altitude, and optical sensor parameters (e.g., focal length and aperture). The spectral sensitivity of silicon sensor chips limits digital cameras to the spectral range 0.4 – 1.0 µm. Space-borne multispectral systems usually provide images in the four spectral bands: blue, green, red and near-IR (GUPTA, 2003).

The high resolution stereoscopic camera HRSC (*Table 3.3-12*) has the capabilities of both CCD linear array scanners and digital cameras. The HRSC contains nine CCD lines in the focal plane of the camera. Five lines are used to acquire panchromatic stereoscopic-images from different view angles and four lines are used to obtain multispectral data. Further examples of digital cameras are the Digital Modular Camera DMC (*Table 3.3-13*) and the ADS 40. They are suitable for photogrammetric surveys because their accuracy in the horizontal and vertical planes is of the order of centimeters, similar to a "metric" camera. Besides for airborne missions, CCD array technology is used on space platforms such as IKONOS, QuickBird, OrbView and ISI-EROS (GUPTA, 2003). An example of an IKONOS image is given in Section 3.2.9.

Imaging spectrometers

Airborne Imaging Spectrometers (AIS), also called hyperspectral scanners, operate on the principle of the push broom or whisk broom. The term hyperspectral is used to indicate the large number of contiguous spectral bands. Whereas opto-mechanical and opto-electronical scanners commonly used for multispectral sensing have 4 - 10 spectral channels with bandwidths of 100 - 200 nm, hyperspectral scanners record images of the Earth's surface in 50 - 300 narrow bands with bandwidths between 1 - 20 nm in the visible and near-infrared parts of the electromagnetic spectrum. Thus, a quasi-continuous spectral signature for each pixel or ground resolution cell can be obtained (GUPTA 1991 and 2003, SABINS 1996). Radiation-splitting optics, CCD linear, and two- and multi-dimensional detector arrays are used to measure the radiation intensity as described above.

Imaging spectrometry is based on reflectance spectroscopy – the study of the reflection, absorption and scattering of the solar radiation using laboratory and/or field spectrometers. The spectral features of minerals, rocks, vegetation types, manmade materials, water and dissolved organic compounds and environmental contaminants are recorded in digital spectral libraries. If these materials have unique and identifiable absorption features, they can be mapped by analysis of the imaging spectroscopy data. The wavelength region of interest is the portion of the electromagnetic spectrum in solar reflection band from 0.4 to 2.5 µm.

Airborne imaging spectrometers provide data with a ground resolution from 0.5 - 20 m. In general, pixels may represent a spectral mixed signal of various materials inside this elementary mapping area. This limits the ability to clearly identify specific materials on the basis of the pixel spectra, e.g., comparison with standard spectra from a reference library. The same is true for classification of the pixels to map materials from spectral data. However, there are a suite of image analysis techniques focused on sub-pixel spectral characterization of the material compositions, i.e. spectral unmixing or matched filter analysis.

In the 1990s, several German remote-sensing companies have explored the Canadian *Compact Airborne Spectrographic Imager CASI*. This sensor measures VIS to the NIR signals, i.e. 0.43 µm - 0.87 µm (*Fig. 3.3-5*). Theoretically, *CASI* contains 288 spectral channels, each with a bandwidth of 1.8 nm. Due to the large data volumes, not all of the 288 channels were used for data acquisition. The German Aerospace Center (DLR, Deutsches Zentrum für Luft- und Raumfahrt) has been developing and applying a digital airborne imaging spectrometer (*GER-DAIS-7915*) since 1995.

NASA's hyperspectral *AVIRIS* scanner (*Airborne Visible InfraRed Imaging Spectrometer*) has been used mainly by U.S. agencies and companies to investigate mining districts in North America and to evaluate the hazards of the mining waste to ground and surface water (PETERS & PHOEBE, 2000; KING et al., 2000).

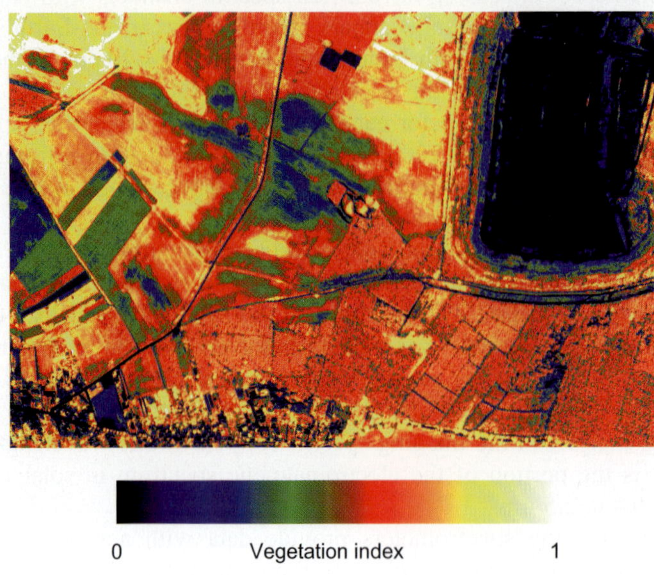

Fig. 3.3-5: *CASI* image taken on June 26, 1993, showing the SW corner of the Schöneiche landfill (*right*) and adjacent area. The image shows the vegetation index NDVI = (NIR - red) / (NIR + red) obtained using one recording channel each in the red and near infrared (NIR-I). The green/blue color of the zone in the center indicates poor growth on soil above permeable near-surface beds. (Images and image processing: WIB GmbH, Berlin, for BGR)

The Australian *HyMap*™ (**H**yperspectral **M**apping) system and its predecessor model *Probe-1* (*Table 3.3-16*) is known for its outstanding signal-to-noise-ratio of about 500:1, hence it provides clear and reliable spectral information. HÖRIG et al. (2001) and ELLIS et al. (2001) detected hydrocarbons and materials containing hydrocarbons on the ground using a *HyMap* (*Table 3.3-16*) or *Probe-1*. The *Ekwan* system (*Table 3.3-17*) is a new instrument with an even higher spatial and spectral resolution. Only the *Modular Optoelectronic Scanner MOS* of the German Aerospace Center (DLR) is operated on a satellite (ALBERTZ, 2001).

Imaging spectrometers are sophisticated and rather expensive instruments. They are usually produced as individual instruments or in small numbers by technical laboratories or specialized companies. Because of the rapid technological development of remote-sensing devices, the number of manufacturers and new instruments has increased considerably. A summary of current and planned imaging spectrometers is given in *Table 3.3-18*. The cost of acquisition and operation of modern and technically advanced imaging data remains high. If possible, cost and data should be shared among different users or agencies.

Microwave remote-sensing methods, mainly radar-based

Active and passive remote-sensing systems have been used for measuring microwave electromagnetic radiation in wavelengths of 1 mm to 1.0 m (*Table 3.1.1*).

Passive microwave sensor systems (radiometers) are commonly used to measure the intensity of natural radiation in the wavelength range 1 mm to 30 cm for mapping purposes (GUPTA, 2003). The signals recorded by passive microwave sensor systems represent surface temperature, emissivity, electrical and structural properties of the ground objects. Therefore, microwave radiometers can be used for geological and environmental investigations. However, the rather small amounts of natural spectral emissions in this spectral range only allow for coarse spatial resolution data to be acquired.

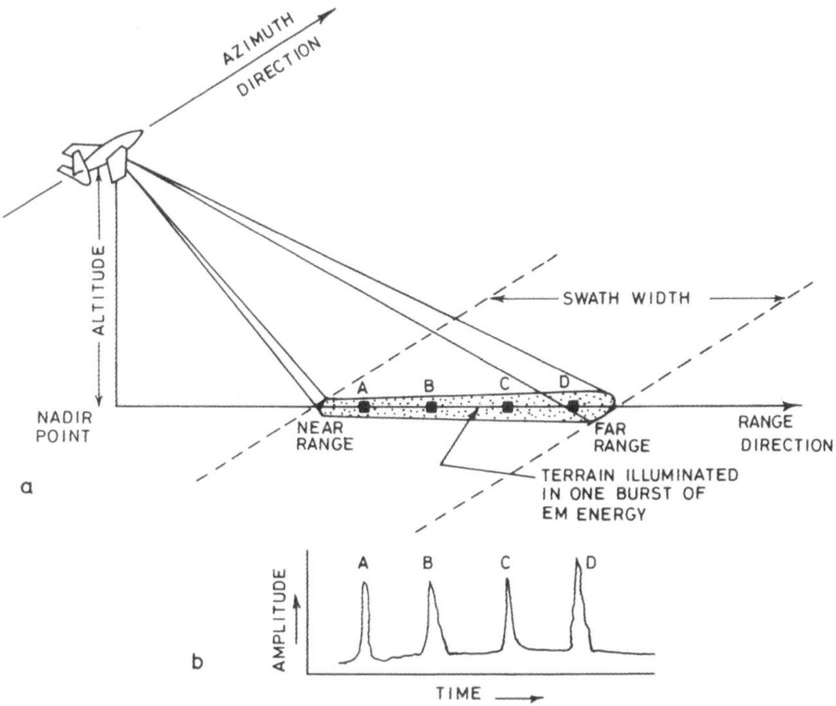

Fig. 3.3-6: Principle of side-looking airborne radar (SLAR), GUPTA (2003), (a) terrain illuminated with one transmitted radar impulse, (b) received back-scattered signal from the ground as an amplitude-time function

Most microwave remote sensing activities have been focused on active microwave systems particularly radar-based methods. *Radar* (acronym for Radio Detection and Ranging) illuminates an area with its self-produced coherent radiation in the cm-wavelength range. The long-wave electromagnetic radiation of radar penetrates clouds and is independent of natural radiation, e.g., can also be used at night. This all-weather and any-time capability is one of the distinct advantages of radar. Due to the oblique illumination angle of the transmitted radar signal to the Earth's surface (*Fig. 3.3-6*), radar images are especially useful for detecting slight changes in topography due to the radar shadowing effect. This property is useful for mapping topographically expressed faults. Radar measurements are also sensitive to differences in soil moisture content. Common disadvantages of radar data are the complexity of data acquisition, geometric distortions, usually lower spatial resolution than optical systems and challenges in data interpretation. *Figure 3.3-6* shows the principle of *side-looking airborne radar (SLAR)*.

In SLAR systems, a transmitter and receiver is mounted on an aircraft platform. A microwave pulse generated by the transmitter is obliquely send from the antenna to the Earth's surface illuminating narrow strips on the ground. The antenna is mounted on the side of the fuselage with a radiation direction perpendicular to the azimuth direction or flight direction and with a given look angle (*look angle* – angle between the line to the nadir from the antenna and the transmitted ray, *depression angle* – compliment of the look angle). The back-scattered signals (radar echoes) are acquired by the receiver antenna and converted to an amplitude-time function. Echoes from points on the ground closer to the antenna are recorded earlier as those further away. After the echo from the far range of the swath is received the next radar impulse is emitted. Strip by strip, the ground is scanned along the flight line, and a radar image is generated. In common radar systems the same antenna is alternately operated as transmitter and receiver antenna. The ground resolution of SLAR images is different in the range direction and the flight direction (azimuth). The range resolution is related to the duration of the transmitted radar pulse and to the look angle. Short pulses and a smaller look angle improve the range resolution. The resolution in the direction of flight (azimuth resolution) depends on the antenna characteristics and, therefore, on antenna length and on the wavelength of the transmitted radar wave. Shorter wavelengths and a longer antenna improve the resolution in the direction of flight. This resolution decreases from the near range to the far range. The possible antenna length is limited. With respect to azimuth resolution, radar systems are subdivided into *real-aperture radar (RAR)* and *synthetic-aperture radar (SAR)*. In RAR systems, the azimuth resolution is determined mainly by the physical length of the antenna. In SAR systems, the azimuth resolution is determined by processing Doppler shift data for multiple return pulses from the same object to create synthetically longer antenna. Consecutive antenna positions are treated as if they were elements of one large antenna. This results in a better azimuth resolution than obtained with RAR systems without physically changing the length of the antenna. Sophisticated algorithms are necessary for SAR data processing. Current aerial and space-borne systems use SAR technology. Besides ground resolution, radar wavelength, beam polarization, the most important parameters for SAR systems are look angle and swath width. *Table 3.3-1* gives the most frequently used radar bands. Microwaves with shorter wavelengths are more attenuated in the atmosphere. The penetration depth and the interaction of radar waves with materials on the Earth's surface are strongly influenced by the wavelength and polarization of the system. The backscatter signal depends on the dielectric properties (e.g., soil/plant moisture) and geometric characteristics such as roughness and shape of Earth surface types.

Table 3.3-1: Radar bands used in SLAR systems after GUPTA (2003); values in parentheses indicate the commonly used radar wavelengths

Radar band	Wavelength in cm	Frequency in GHz (10^9 cycles s^{-1})
Ka	0.8 - 1.1 (0.86)	40.0 - 26.5
Ks	1.1 - 1.7	26.5 - 18.0
Ku	1.7 - 2.4	18.0 - 12.5
X	2.4 - 3.8 (3.1)	12.5 - 8.0
C	3.8 - 7.5 (5.7)	8.0 - 4.0
S	7.5 - 15.0 (15)	4.0 - 2.0
L	15.0 - 30.0 (23.5)	2.0 - 1.0
P	30.0 - 100.0 (50)	1.0 - 0.3

The radar response signal is a complex number consisting of a real and an imaginary part. The amplitude and the phase of the signal can be calculated from its real and imaginary parts, respectively. Whereas conventional SAR imaging usually only use the amplitudes of the radar signals, *interferometric SAR (InSAR)* and polarimetric SAR (PolSAR) utilizes the phase measurements as well. The InSAR method does not use the phase information of single image for the interferograms but the phases of two images of the same ground scene recorded by SAR antennas at different locations and/or at different times. Detailed digital elevation models (DEM) can be derived from the phase differences of the interferograms. Differential interferometric SAR uses interferograms acquired over the same area at different times to detect displacements of the Earth's surface in the centimeter range due to earthquakes, volcanic events, landslides or land subsidence.

Additional information about the use of microwave remote sensing can be found in the literature (COLWELL, 1983; TREVETT, 1983; KRONBERG, 1985; WOODING, 1988; GUPTA, 1991 and 2003; SABINS, 1996).

Laser scanning

Airborne laser scanning is an active remote sensing system as well. A pulsed laser beam scans the Earth's surface in a strip across the flight line (*Fig. 3.3-7*). The pulse frequency and the speed of the aircraft determine the measured point density and the resulting mesh size of the data grid. An opto-mechanical scanner with rotating mirror guides the laser beam across the flight path. The path of the reflection points is a zigzag line due to the forward movement of the aircraft and the sideward oscillation of the scanner's mirror. Complicated data processing derives the coordinates (position and elevation) of each terrain point. Data collection is possible even during the night and under cloudy conditions. The current standard for image accuracy is about 15 cm horizontally and 5 - 50 cm vertically for a flying altitude of 1000 meters above ground during data acquisition (VEGT & HOFFMANN, 2001).

A single laser signal can be reflected before it reaches the Earth's surface, i.e. by vegetation such as grass, trees, and shrubs. Therefore, high-resolution surveys, e.g., for monitoring of land subsidence, should be carried out in spring before the vegetation begins to grow. Suitable ground control points should be selected on the basis of the survey objectives. The last step is a classification of the terrain characteristics (topography, vegetation, buildings, etc.).

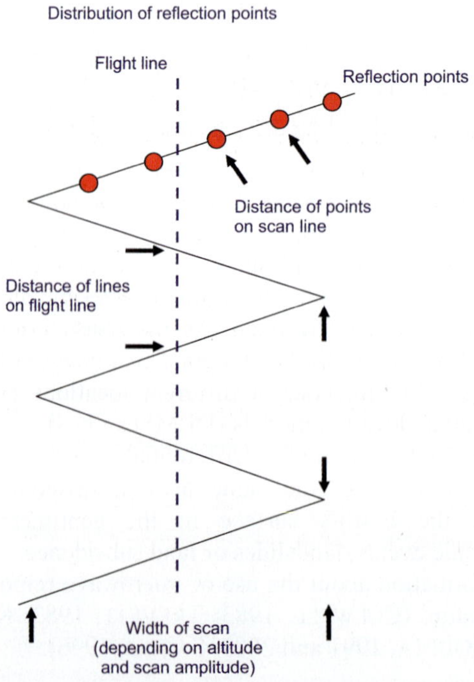

Fig. 3.3-7: Principle of data acquisition with an optical-mechanical laser scanner with a rotating mirror

Modern navigation technology, such as DGPS and INS (Inertial Navigation System), are used in order to estimate the precise aircraft position. Irregular movements of the aircraft (dipping, rolling, tilting) during the flight are recorded and stored in the INS unit. This information is used to calculate the coordinates of points in the image data along the flight path. The use of INS and DGPS guarantees a position and altitude precision in the order of 5 - 20 cm. An example of airborne laser scanning is given in *Fig. 3.2-11*.

3.3.4 Instruments

The specifications of selected airborne and space-borne nonphotographic remote-sensing systems are given in tabular form for the following categories:

- thermal scanners, *Tables 3.3-2, 3.3-3*
- satellite multispectral imaging systems, *Tables 3.3-4, 3.3-5, 3.3-6, 3.3-7, 3.3-8, 3.3-9*
- space-borne radar systems, *Tables 3.3-10, 3.3-11*
- airborne digital cameras, *Tables 3.3-12, 3.3-13*
- imaging spectrometers and hyperspectral scanners, *Tables 3.3-14, 3.3-15, 3.3-16, 3.3-17, 3.3-18* and
- airborne laser scanners, *Table 3.3-19*.

Field spectrometers operating in the spectral range 0.4 - 2.5 μm are used to provide basic information about the spectral properties of materials and to calibrate and validate remote sensing data. Airborne nonphotographic imaging systems include – besides the imaging sensor(s) – a navigation system and a gyroscopic platform to prevent tilting of the scanner. Usually, the navigation system is a differential global positioning system (DGPS) to determine the flight coordinates. An inertial navigation system (INS) is used to determine changes in aircraft attitude and to improve the position determination.

Table 3.3-2: Specifications of the thermography scanner *AGEMA 900* from AGEMA Infrared Systems, Sweden, (from documentation of the Engineering and Physical Sciences Research Council (EPSRC), Great Britain, and the University of Magdeburg)

spectral sensitivity	8 - 12 µm
filming speed	15 Hz
number of pixels	136 × 272
FOV	20° × 10° 10° × 5.0° 5.0° × 2.5° 2.5° × 1.25°
IFOV	1.28 mrad 0.64 mrad 0.32 mrad 0.16 mrad
temperature range	-30 °C to 1500 °C (2 000 °C with filter)
thermal sensitivity	0.08 °C at +30 °C

Table 3.3-3: Specifications of the thermography scanner *Thermo Tracer TH 1101* from NEC San-ei, Japan, (from documentation of the ebs GmbH, NEC San-ei distributor in Germany)

spectral sensitivity	8 - 13 µm
number of pixels	344 × 207
FOV	30° × 27°
IFOV	1.5 mrad
temperature range	-50 °C to 2000 °C
thermal sensitivity	0.1 °C

Table 3.3-4: Specification of the Landsat sensors, ALBERTZ (2001)

	Landsat 4 & 5 (1-3) Multispectral Scanning System (MSS)	**Landsat 4 & 5 Thematic Mapper (TM)**	**Landsat 7 Enhanced Thematic Mapper (ETM+)**
in operation	since 1972	since 1982	since 1999
altitude	705 km (915 km)	705 km	705 km
repeat cycle [d]	16 (18)	16	16
swath width	185 km	185 km	185 km
pixel size	79 × 79 m²	30 × 30 m²	30 × 30 m²
spectral channels	1 (4) 0.50 - 0.60 µm 2 (5) 0.60 - 0.70 µm 3 (6) 0.70 - 0.80 µm 4 (7) 0.80 - 1.10 µm	1 0.45 - 0.52 µm 2 0.52 - 0.60 µm 3 0.63 - 0.69 µm 4 0.76 - 0.90 µm 5 1.55 - 1.73 µm 7 2.08 - 2.35 µm	1 0.45 – 0.52 µm 2 0.52 - 0.60 µm 3 0.63 - 0.69 µm 4 0.76 - 0.90 µm 5 1.55 - 1.73 µm 7 2.08 - 2.35 µm
thermal channel		6 10.4 - 12.5 µm 120 × 120 m²	6 10.4 - 12.5 µm 60 × 60 m²
panchromatic channel			8 0.52 – 0.90 µm 15 × 15 m²

Table 3.3-5: Specification of the SPOT-Sensors, ALBERTZ (2001)

	SPOT HRV (XS-mode) multispectral	**SPOT HRV (P-mode) panchromatic**
in operation	since 1986	since 1986
altitude	832 km	832 km
repeat cycle [d]	26 *	26 *
pixel size	20 × 20 m²	10 ×10 m²
image size	60 × 60 km²	60 × 60 km²
spectral channels	1 0.50 - 0.59 µm 2 0.61 - 0.69 µm 3 0.79 - 0.89 µm	0.51 - 0.73 µm

SPOT 4 (launched 1998) is equipped with an additional spectral channel in short-wave infrared (1.58 – 1.75 µm) with a pixel size 20 × 20 m².
* Shorter repeat cycles are possible for certain areas by changing the look angle within +27° and -27°.

Table 3.3-6: Specifications of the Indian satellites IRS-IC and IRS-ID, ALBERTZ (2001)

	WiFS	**LISS-III**	**PAN**
launched	1996/1997	1996/1997	1996/1997
altitude	817 km	817 km	817 km
repeat cycle [d]	5	24	24 (5 days with a look angle of up to 26°)
resolution	188 m	23 m	5.8 m
swath width	810 km	142 km	70 km
spectral channels	0.62 - 0.68 μm 0.77 - 0.86 μm 1.55 - 1.75 μm	0.52 - 0.59 μm 0.62 - 0.68 μm 0.77 - 0.86 μm 1.55 - 1.70 μm resolution 70 m	0.50 - 0.75 μm

Table 3.3-7: Specifications of the OPS sensor system on Japan's Earth Resources Satellite-1 (JERS-1)

Spatial resolution	**Detector**	**Spectral range**	**Swath width**
18.3 × 24.2 m	CCD linear array	0.52 - 0.60 μm 0.63 - 0.69 μm 0.76 - 0.86 μm 0.76 - 0.86 μm * 1.60 - 1.71 μm 2.01 - 2.12 μm 2.13 - 2.25 μm 2.27 - 2.40 μm	75 km

* Forward viewing for stereo imaging.

Besides the optical sensor (OPS), an SAR system is operated on the same satellite.

Table 3.3-8: Specifications of the current high-resolution satellite sensors, ALBERTZ (2001) completed

	IKONOS 2	**Quick Bird 2**	**OrbView 3**
company	Space Imaging	Earth Watch Inc.	Orbital Image Corp.
launched	24.09.1999	18.10.2001	26.06.2003
altitude	680 km	600 km	470 km
orbit inclination	98°	66°	97°
repeat cycle [d]	14	20	16
minimal repeat cycle [d]	1 - 3	1.5 - 2.5	1 - 3
resolution (panchromatic)	0.82 m	0.82 m	1 m
band width	0.45 - 0.90 µm	0.45 - 0.90 µm	0.45 - 0.90 µm
resolution (multisp.)	4 m	3.28 m	4 m
swath width	11 km	22 km	8 km
spectral channels	0.45 - 0.52 µm 0.52 - 0.60 µm 0.63 - 0.69 µm 0.76 - 0.90 µm	0.45 - 0.52 µm 0.52 - 0.60 µm 0.63 - 0.69 µm 0.76 - 0.89 µm	0.45 - 0.52 µm 0.52 - 0.60 µm 0.62 - 0.70 µm 0.76 - 0.90 µm

Table 3.3-9: Specifications of the Advanced Spaceborne Thermal Emission and Reflection (ASTER) sensor on the Earth Observation Satellite (EOS-AM-1), GUPTA (2003)

	Spectral band	**Spectral range**	**Scanner type**	**Spatial resolution**	**Signal quantization**
VNIR					
green	1	0.52 - 0.60 µm	5 000 cells	15 m	8-bit
red	2	0.63 - 0.69 µm			
NIR	3	0.76 - 0.86 µm			
NIR	3B	0.76 - 0.86 µm			
SWIR	4	1.60 - 1.70 µm	2 048 cells	30 m	8-bit
	5	2.145 - 2.185 µm			
	6	2.225 - 2.245 µm			
	7	2.235 - 2.285 µm			
	8	2.295 - 2.365 µm			
	9	2.360 - 2.430 µm			
TIR	10	8.13 - 8.48 µm	OM	90 m	12-bit
	11	8.48 - 8.83 µm			
	12	8.90 - 9.25 µm			
	13	10.25 - 10.95 µm			
	14	10.95 - 11.65 µm			

Swath width: 60 km; band 3B is backward-looking for stereo viewing

Table 3.3-10: Satellite radar system specifications, ALBERTZ (2001)

	ERS	**RADARSAT-1**	**ENVISAT**
company	ESA (European Space Agency)	CSA/CCRS (Canadian Space Agency)	ESA (European Space Agency)
launched	1995 (ERS-2)	1995	2002
altitude	780 km	798 km	800 km
orbit inclination	98.5°	98.6°	98.5°
sensor system	AMI (Active Microwave Instrument)	SAR	ASAR (Advanced SAR)
mode	image mode	standard mode	image mode
resolution (panchromatic)	0.82 m	0.82 m	1 m
swath width	100 km	100 km	56 - 120 km
repeat cycle [d]	35	24	35
resolution	25 m	28 m	30 m
band	C (5.3 GHz/5.6 cm)	C (5.3 GHz/5.6 cm)	C (5.33 GHz/5.6 cm)
depression angle	~ 23°	20 - 49°	15 - 45°
polarization	vertical/vertical	horizontal/horizontal	vertical/vertical or horizontal/horizontal

Table 3.3-11: Specifications of the Shuttle Radar Topographic Mission (SRTM) dedicated to SAR interferometry for high-accuracy and high-resolution DEM, GUPTA (2003)

Shuttle		
launched	11 February 2000	
duration	11 days	
altitude	~ 275 km orbit	
SAR sensors		
	C-band	X-band
wavelength	5.6 cm	3.1 cm
frequency	5.3 GHz	9.6 GHz
look angle	30 - 60°	50 - 55°
swath width	225 km	50 km

Table 3.3-12: Specifications of the High-Resolution Stereo Camera *HRSC* of the DLR German Aerospace Center (from DLR documents)

	HRSC-A	HRSC-AX	HRSC-AXW
focal length	175 mm	151 mm	47 mm
FOV	38° × 12°	41° × 29°	30° × 79°
number of CCD lines	9 (4 colors)	9 (4 colors)	5 (2 colors)
pixel per line	5272	12 172	12 172
pixel size	7 μm	6.5 μm	6.5 μm
radiometric resolution	8 bit	12 bit	12 bit
max. scan frequency	450 lines s^{-1}	1640 lines s^{-1}	1640 lines s^{-1}
spectral resolution and viewing axis (model AX)		blue: 440 - 510 nm green: 520 - 590 nm red: 620 - 680 nm NIR: 780 - 850 nm PAN: 520 - 760 nm	-4.6° -2.3° 2.3° 4.6° 20.5°; -20.5°; 12°; -12°; 0°

Table 3.3-13: Specifications of the *Digital Modular Camera* of *ZI* Imaging, Germany (from documents of the manufacturer)

FOV cross × along	74° × 44°
panchromatic: resolution, lens system	13 500 × 8000 pixel, 4 × f = 120 mm
multispectral: channels, resolution, lens system	4 × RDG & NIR, 3000 × 2000 pixels, 4 × f = 25 mm
frame rate	2 s/image
radiometric resolution	12 bit

Table 3.3-14: Spectral recording channels and band widths of the *Daedalus AADS-1268* airborne scanner (after DLR documents)

Channel number	Spectral range	Channel range in μm
1	visible light	0.420 - 0.450
2		0.450 - 0.520
3		0.520 - 0.600
4		0.605 - 0.625
5		0.630 - 0.690
6		0.695 - 0.750
7	near infrared (I)	0.760 - 0.900
8		0.910 - 1.050
9	near infrared (II)	1.55 - 1.75
10		2.08 - 2.35
11	thermal infrared	8.5 - 13.0

Table 3.3-15: Selected specifications of the *CASI* scanner (from documentation of WIB GmbH, Berlin)

angular field of view (FOV)	35.4°
number of detectors per scan line	512
IFOV	1.21 mrad
spectral range	0.43 – 0.87 μm
total number of spectral recording channels available	288 with 1.8 nm bandwidth
maxium number of channels that can be simultaneously used	15

Table 3.3-16: Selected specifications of the *HyMap* hyperspectral scanner (from documentation of Integrated Spectronics, Australia)

Module	Spectral range	Bandwidth across module	Average spectral sampling interval	IFOV	Signal-to-noise ratio
VIS	0.45 - 0.89 μm	15 - 16 nm	15 nm	2.5 mrad along-track 2.0 mrad across-track	>500 : 1
NIR	0.89 - 1.35 μm	15 - 16 nm	15 nm		
SWIR1	1.40 - 1.80 μm	15 - 16 nm	13 nm		
SWIR2	1.95 - 2.48 μm	18 - 20 nm	17 nm		

Table 3.3-17: Selected specifications of the *Ekwan* hyperspectral imager (from documentation of Ekwan Technology Corporation)

IFOV	0.9 mrad
number of spectral channels	384
wavelength	370 to 2495 nm
signal to noise	> 400 : 1 (SWIR)
ground resolution	0.8 - 7 m
spectral bandwidth	5.25 nm to 6.25 nm
swath width	up to 2 km

Table 3.3-18: Examples of current and future imaging spectrometers, SLONECKER & WILLIAMS (2001)

Acronym	Name	Manufacturer	Bands	Spectral range [nm]
ASAS	Advanced Solid-State Array Spectrometer	NASA (Goddard)	62	400 - 1200
AIS-1	Airborne Imaging System 1	NASA (JPL)	128	800 - 1600 1200 - 2400
AISA	Airborne Imaging Spectrometer for Applications	SPECIM, Ltd.	286	400 - 1250
CIS	Chinese Imaging Spectrometer	Shanghai Institute of Technical Physics	91	-
Dais 7915	Digital Airborne Imaging Spectrometer	GER Corp	79	400 - 12000
IRIS	Infrared Imaging Spectroradiometer	ERIM	256	2000 - 15000
MIVIS	Multispectral Infrared and Visible Imaging Spectrometer	Daedalus	102	433 - 12700
VIMS-V	Visible Infrared Mapping Spectrometer	ASI	512	300 - 1050
AVIRIS	Advanced Visible and Infrared Imaging Spectrometer	NASA (JPL)	224	400 - 2500
HYDICE	Hyperspectral Digital Imagery Collection Experiment	NRL	210	400 - 2500
HYMAP	Airborne Hyperspectral Scanner	Integrated Spectronics	200	400 - 12000
MAS	MODIS AIRBORNE SIMULATOR	Daedalus	50	530 - 14500
MODIS	Moderate Resolution Infrared Spectrometer	NASA	36	400 - 14400
TRWIS III	TRW Imaging Spectrometer	TRW	384	300 - 2500
NEMO	Navy Earth Map	U.S. Navy	210	400 - 2500
Warfighter1	Observer Warfighter	U.S. Air Force	280	400 - 5000

Table 3.3-19: Airborne laser scanner specifications, GIM International (2001)

Name	ALTM 2033	Saab / TopEye	TopoSys® II	ALS 40
manufacturer	Optech Systems Corp.	TopEye, Saab	TopoSys, Saab	LH Systems
scanning principle	oscillating, Z-shaped, mirror	rotating, oscillating, Z-shaped, fibers mirror, optical	fibers	oscillating, mirror
data rate	max. 33 000 Hz	max. 7000 Hz	max. 83 000 Hz	max. 30 000 Hz
scan rate	max. 66 Hz	6.25 - 25 Hz	650 Hz	20 - 26 Hz
scan angle	max. ± 20°	max. ± 20°	max. ± 7°	max. ± 37.5°
max. height	6000 m	960 m	1600 m	6100 m
height accuracy	15 - 25 cm	< 10 cm	< 15 cm (depends on DGPS accuracy)	8 cm (depending on flight altitude)

3.3.5 Survey Practice

GEBHARDT (1981) discusses factors influencing and disturbing a *thermal survey*, e.g., wind, precipitation and clouds. Wind blurs thermal anomalies and can cause temperature effects that make the data unusable. Precipitation lowers thermal contrasts for a considerable period of time. Cloud cover causes atmospheric counter-radiation. This produces artifacts and reduces the contrasts between the anomalies on the ground. Vegetation prevents thermal evaluation of the soil and rocks and contributes its own thermal signature. The time after sunset is best suited for distinguishing thermal anomalies on the Earth's surface and buried objects (e.g., waste, "hot spots" inside tailing dams, gas emissions). Nighttime is preferred because the temperature is more constant than during the day ensuring more representative data and improved interpretations. Thermal contrast observed during the day is mostly due to different amount of exposure to solar radiation.

For investigation of geologic thermal phenomena, e.g., hot spots, thermal springs and gas emissions, temperatures slightly below zero are advantageous since thermal contrasts are enhanced. The time just after noon are most suitable for thermal localization of springs, seepage localities, drainage systems, irrigation ditches, etc. This is due to the high thermal inertia of water (*Fig. 3.3-8*), which causes these area to have a lower temperature than the area around it at this time. Higher evaporation and moisture-saturated air on top of a moist site can cause thermal minima as well.

The conditions necessary for a *digital camera survey flight* are similar to those for an aerial photographic survey flight (Chapter 3.1). The weather should be clear without clouds or haze. A sophisticated navigation system, e.g., DGPS, provides precise image-coordinates in the cm-range. This helps to avoid additional ground surveys to obtain the control point coordinates.

Imaging spectrometer measurements require high intensity solar radiation during the flight to allow sufficient signal for all channels of the instrument and thus improve the signal-to-noise ratio. Fog, mist, haze, smog, precipitation and all kind of clouds have to be avoided. Acquisitions near solar noon are the ideal time of day if there are no disturbing meteorological or man-made factors. Vegetation cover (season) should be avoided if the investigation targets are soils, rocks or their contamination. A calibration site and several additional sites should be carefully chosen before the flight. These are measured on the ground using a field spectrometer during the flight. The sites should be spectrally homogeneous (given the spatial resolution of the sensor) in order to obtain calibration spectra for the radiometric correction. The site has to represent a single possibly bright material and should be larger than one pixel of the remote-sensing image so calibration pixels contain a spectrum for only this material. The quality of such a standard field measurement determines the interpretation quality of the remote sensing data. For simplification, the calibration sites should be near the airport or on the premises of the client. Additional measurement of different objects within the area of interest further improves the data processing and ensures the reliability of the data interpretation.

Some manufacturers build *hyperspectral scanner* systems with spectral ranges from visible light to the middle thermal infrared. If the study involves a larger spectral range it has to be considered that thermal surveys have different requirements than a survey constrained to the 0.4 – 2.5 μm range.

For *high-resolution satellite sensors* (e.g., IKONOS, QuickBird), up-to-date images for selected areas can be acquired and obtained. Acceptable time slots and percentage of cloud cover must be given along with the data order. The time window should not be too short in order to increase the chances for good weather conditions when the satellite passes over the target area. The smaller the desired percentage of cloud cover given in the order, the higher the cost of the images. On the other hand, clouds and their shadows may obscure objects of interest.

3.3.6 Processing and Interpretation of Data

Sophisticated computers with an appropriate memory, powerful graphic cards, and large storage capabilities are an essential for the processing and interpretation of remote-sensing data. Another prerequisite is the availability of an effective software package for the following tasks:

- image rectification and restoration (preprocessing),
- image enhancement,
- image analysis,
- image classification,
- data merging, GIS integration and map preparation.

Examples of powerful and versatile remote-sensing software systems are:

- ENVI,
- ERDAS,
- Earth Resource Mapping / ER Mapper,
- PCI, and
- Tetracorder.

Exemplarily some features of the ENVI system are listed:

- Support of numerous data import and output formats,
- data preprocessing and radiometric calibration,
- image registration and orthorectification,
- interactive image enhancement,
- panchromatic, multi- and hyperspectral, and radar data analysis,
- spectral analysis and filtering,
- change analysis,
- mosaic and subset,
- supervised and unsupervised classification,
- integrated GIS features,
- annotation and map composition, and
- terrain analysis.

Preprocessing

Raw images from nonphotographic scanners contain systematic and nonsystematic geometric distortions. Sources of these distortions include variations in altitude, attitude, and velocity of the sensor platform, panoramic distortions, effects due to Earth's curvature and rotation, as well as relief displacement. In airborne applications, data from the DGPS and INS navigation system are used to eliminate distortions from variations in the altitude, attitude, and velocity of the aircraft. These correction procedures are not necessary for optical satellite data. Due to the small aperture and the high altitude of space-borne imaging systems the relief displacement can often be neglected. For geoscientific interpretation, the images from space-borne systems can be used directly as delivered by the provider.

Rectification and restoration of radar images requires sophisticated processing. A digital elevation model (DEM) is necessary for correction of topographic displacements. Preprocessing can be done by the provider of the satellite imagery. Satellite images are delivered and indexed in terms of path and row numbers for easy referencing and ordering. Web-based access allow the search in large satellite data archives. The images can be ordered in different qualities (e.g., standard and precision) as photographs (e.g., positive or negative image, false-color images) or as digital data.

Processing

Thermal scanner manufacturers provide specific and customized software systems for processing and interpretation of the data. Images from thermal scanners can be handeled with the same hardware and software used to record and store the data during the flight. This easy-to-use windows-based software permits the generation of most suitable thermal images using different processing techniques, such as enhancement of temperature patterns by stretching, plotting isotherms, determination of temperature along profiles and at point sites, and color coding of temperature intervals.

The processing of the data from *digital cameras* is described here using the example of the *HRSC* camera. The first processing step is always radiometric correction to minimize or at least reduce noise and to enhance the desired signal. The external orientation of the image lines is then computed using the post-processed data from a special flight management system that recorded the exact position of the aircraft every few seconds during the flight.

Using the same points derived from the different stereo channels and from overlapping image strips, the angular offset between the Inertial Navigation System (INS) and the camera axis can be calculated and the orientation of the adjacent image strips along the flight line can be corrected. Calculations of a digital elevation model can now be completed from the set of object points. The terrain model is needed to generate orthoimages correcting for topographic displacements. The orthoimages are then combined to orthoimage-mosaics. The final step is color processing to generate color composites (color, CIR) of the ortho images or ortho image mosaics. All data processing is fully automated.

The processing of data from *imaging spectrometers* always starts with a radiometric correction using laboratory standards. This step aims to calibrate the electric signal of the spectrometer with respect to the radiance of the incident radiation. When the sensor has been calibrated, an atmospheric and other radiometric correction are performed. An atmospheric correction is advised because of the sensor's sensitivity to changing atmospheric conditions, i.e. gases, water vapor, aerosols and other substances in the atmosphere. This correction is vital to suppress or eliminate this type of noise and to improve the signal-to-noise ratio. There are different models of atmospheric scattering and absorption available for this step. Further radiometric correction is necessary to match the signals measured by the imaging spectrometer to the spectrally acquired materials on the ground. After the radiometric correction, the absolute value of the radiance is converted into percentage reflectance of the incident radiation.

The next step is a geometric correction using the orientation data from INS and DGPS. The quality of the geometric correction can be improved by using the data from a second sensor with a higher ground resolution, like a digital camera.

Interpretation

Interpretations of nonphotographic imaging data and the use of the results for site investigations depend on the spectral range of the sensors used for imaging. The *solar reflection (SOR) region* from 0.3 to ~3 µm seems the best investigated part of the electromagnetic (EM) spectrum. Such data are easy to interpret in terms of directly observable objects and physical phenomena. All types of sensors used on aerial and space-borne platforms acquire data in this spectral range. But, the images from these systems differ in spatial and radiometric resolution. The use of multi-spectral data from the SOR region facilitates geological and environmental investigations and monitoring, such as topography and drainage patterns, identification of rock and soil types, moisture content, faults, and vegetation.

Given the thermal conditions of the Earth, EM radiation with wavelengths from 3 to 35 µm is emitted from the surface. The amount of energy an object on the ground radiates depends on its surface temperature and emissivity. Emissivity is a materials property. Previous investigations of waste disposal sites have shown that the thermal conditions depend on the time of day, time of year, and the weather prior to data collection; the usability and reproducibility of aerial photographs and thermal images are also related to these parameters. Qualifiers, such as warm or cold, can be used only together with a specific time of day because diurnal temperature variations differ for different materials (*Fig. 3.3-8*). Consequently, generalized statements regarding the thermal properties of materials in the Schöneiche landfill (e.g., EHRENBERG, 1991) may be misleading.

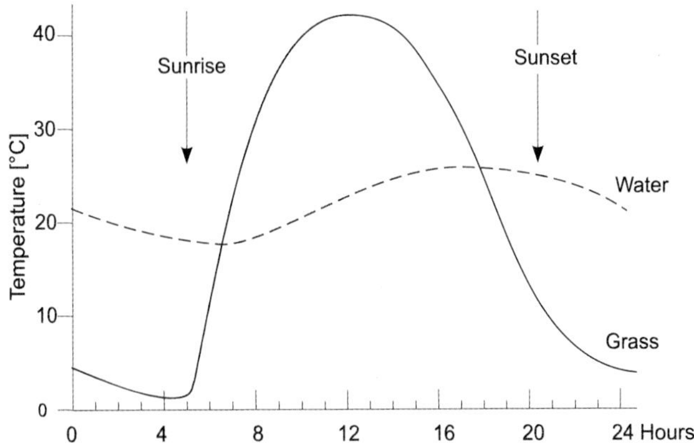

Fig. 3.3-8: Surface temperatures of water and grass during a diurnal cycle, LOWE (1969)

To maximize the usefulness of thermal remote-sensing data, surface temperatures of rocks, soils, vegetation, and waste should be determined using an array of ground sensors at the time of remote data take. Emissivities of samples may also be measured in the laboratory.

One objective of thermal imagery is the detection of heat sources within mine waste heaps ("hot spots"), for example produced by dumping of hot ashes and highly inflammable material (*Figs. 3.3-11, -12,* and *-13*). If the amount of smoldering material is large enough and the internal temperature is high enough, such phenomena can be detected even when covered up to several meters.

Fermentation of organic matter in a landfill produces heat that is transported to the surface by methane and other gases. Ducts are installed to remove these gases under controlled conditions (*Fig. 3.3-16*). Uncontrolled emissions are also possible and can be detected using thermal imagery.

Another application of thermography is the detection of buried waste. Because the heat capacity of waste is usually different than surrounding natural material, the waste is revealed by thermal observations (*Figs. 3.3-17* and *3.3-18*).

It is well known that the thermal inertia and heat capacity of water is different from those of soils and rocks. This information is used to detect natural and artificial features containing water. Such features can be springs and soil moisture seeps in areas with high groundwater levels or in areas with non-cohesive soil (*Figs. 3.3-19* and *3.3-20*), filled-in natural and artificial drainage systems (*Fig. 3.3-21*), seepage water, and fractures and faults in karst regions.

The combination of a DEM with thermal data can reveal indications of night time cold air fluxes and natural ventilation. A sufficient slope of the landfill and a sparse vegetation cover are required conditions for such processes to happen. Cold air flowing from the top of the landfill is indicated in night images by "cold" anomalies along the slope. Such cold air flows can possibly carry harmful substances that would then accumulate in depressions around the landfill.

Data from *digital cameras* can be interpreted independently or in combination with other remote-sensing data. Photogrammetric products (orthophotos, DEMs, orthophoto mosaics, topographic maps, etc., Chapter 3.2) or topographic base maps for field work and the presentation of results can be produced from digital camera images. The spectral channels of the camera permit a spectral composition similar to color and CIR photographs and the integration of the images with other data sets, e.g., with a DEM (*Fig. 3.3-24*). There are a range of applications like regional land-use mapping and mapping of biotopes to site investigations, for example, to evaluate the vitality of vegetation, assess surface water conditions, or search for seepage water, springs and moisture anomalies (Chapter 3.1). Yet, the spatial resolution of an analog aerial photograph, depending on the film granularity, is still better than the spatial resolution of most digital cameras.

Hyperspectral images particularly allow for identification of certain subpixel features. Most hyperspectral pixels have a mixed spectral signature, originating from different objects or materials. If this mixture can be detected in the pixel spectrum due to the presence of different spectral signatures, the spectral signatures can be assigned to different objects and their areal coverage (spectral unmixing).

In order to recognize and distinguish materials in the *imaging spectrometer data*, local measurements need to be made with a field spectrometer. This is necessary to determine whether the contaminated materials can be spectrally discriminated from the surrounding uncontaminated materials and thus mapped from the image data.

Minerals and mining waste exposed in open-pit mines, at mill sites, mine dumps, tailing and spoil heaps, etc. have to be identified and mapped to characterize large regional mining districts. LIVO (1994), JANSEN (1994) and PETERS & PHOEBE (2000) developed a method for this purpose using the AVIRIS system achieving a spatial resolution of 20 m in mining areas of North America. KING et al. (2000) discussed the impact of water and wind on mining districts. Sulfides constitute a significant part of the mineral paragenesis of such districts, as well as many other ore minerals and resulting weathering products. Pyrite and alunite generate sulfuric acid in contact with rain water seeping into the groundwater. Such acidic sites in mining districts can be detected by mapping the resulting mineral alunogen. Using specific software, the spectra of the image pixel are compared with standard spectra of the digital spectral libraries as provided by the USGS (U.S. Geological Survey). The spectral range for discriminating minerals is often focused on 2200 - 2300 nm. Visual examination and modern algorithms for spectral identification can be used by an experienced specialist. Such methods not only allow to identify pixels with the spectrum for a single material, but also pixels showing the spectra of two minerals, for example dolomite and calcite, kaolinite and hematite, jarosite and muscovite and other materials.

(a)

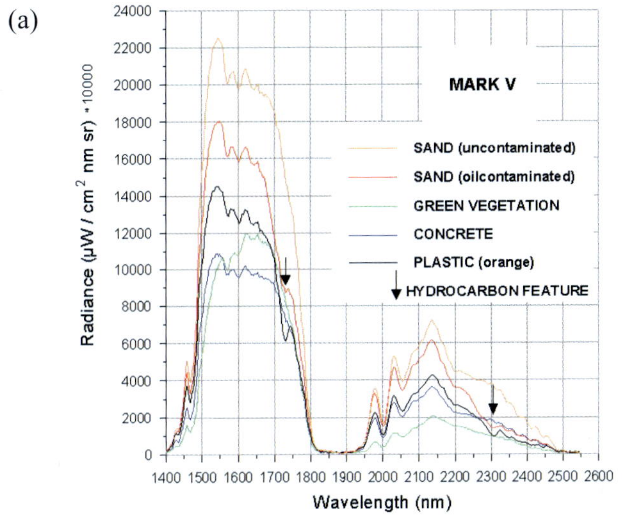

(b)

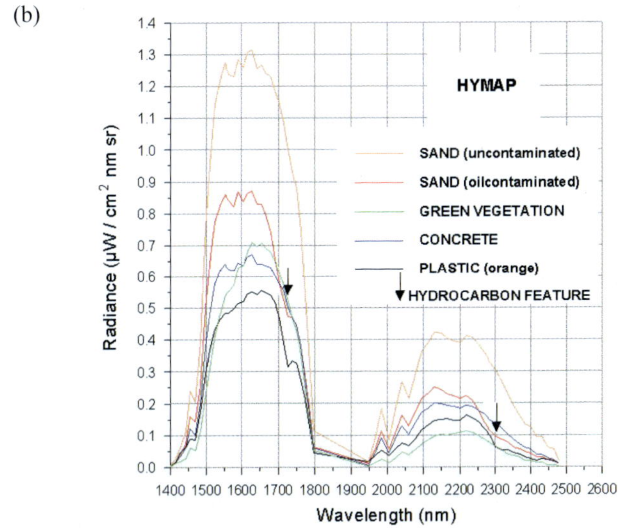

Fig. 3.3-9: Radiance spectra of MARK V laboratory spectrometer (a) and HyMap spectra (b) of reference areas on BGR premises

Tests with hyperspectral images in different mining areas with similar contaminated materials indicate that a simple transfer of the method is not possible. The different climatic, weathering, geomorpological and vegetation conditions in other regions and the mineral mixtures, man-made materials and vegetation cover typical for those regions point at the importance to collect spectroscopic data in the field and laboratory, hence regional spectral libraries as a basis for reliable calibration and interpretation.

The *HyMap* airborne imaging spectral system has been used to *detect hydrocarbons* (HÖRIG et al. 2001) (*Figs. 3.3.-9 and 3.3-22*). Hydrocarbons and any materials containing them are characterized by absorption maxima at wavelengths of 1730 and 2310 nm (CLOUTIS, 1989). The radiance spectra recorded by a *Mark V* laboratory spectrometer (*Fig. 3.3-9a*) had the same two absorption peaks as shown in the spectra of the airborne system recorded simultaneously (*Fig. 3.3-9b*). The 1730 nm feature is more pronounced. Better results for identifying the hydrocarbons were obtained using this absorption peak than with the 2310 nm peak. The interpretation uses at least three bands of the *HyMap* which cover the portion of the spectrum around these two peaks.

Fig. 3.3-10: Processed HyMap image showing materials containing hydrocarbon at the ground surface in Spandau, Berlin (BGR premises and oil-contaminated reference areas; A, artificial turf; T, race track; B, plastic roofs)

A test area in Spandau, Berlin is shown in *Figure 3.3-10*. Most of the surface materials in the images appear grey because spectral characteristics are similar in this part of the spectrum. Only the areas containing hydrocarbons appear colored due to their significant absorption features within the same narrow spectral range. Consequently, oil-contaminated soil reference areas (southern part of the picture, near BGR), plastic roofs of buildings (B), artificial turf (A), and athletics race tracks (T) appear pinkish. The intensity of the color of the area depends probably on the oil content.

ELLIS et al. (2001) used airborne hyperspectral imaging data from *Probe-1*, the preceding model of *HyMap* to detect several onshore oil seeps. Earlier mapped oil seeps were confirmed by these measurements and new sites were discovered. The "depth" of the absorption minimum in both the field and airborne spectra at 2310 nm correlates well with the amount of bitumen (or tar and asphalt).

The main objective of further investigation is the compilation of a spectral library of hydrocarbon maxima and minima, including spectra with different amounts of oil, tar, vegetation, soil and rock.

Using this approach, it is possible to map contaminations with oil, gasoline/petrol, kerosene, diesel, etc. even for large areas. Natural oil seeps, natural hydrocarbon-bearing sediments, e.g., oil shale or sandstone, and artificial plastic hydrocarbon materials, may be detected. No chemical laboratory method is able to provide such a spatial database as quickly, reliably and as inexpensively as this remote sensing method.

Radar images from various aerial and space-borne SAR sensors provide information about changes in the terrain slope on a decameter scale, of changes in surface roughness on a centimeter scale, and about electrical properties of the materials on or near the Earth's surface. Information about the soil moisture and mineral content of soil and rock can be derived from the dielectric properties. A large number of studies have shown that interferometric SAR (InSAR) provides very good data for detecting vertical ground displacement on a centimeter scale (*Fig. 3.3-25*). An overview of current InSAR applications for land subsidence detection is available at various websites, e.g., http://www.npagroup.com and http://www.atlsci.com.

3.3.7 Quality Assurance

To obtain high-quality results, an approriate remote-sensing system has to be selected given the objectives of the investigation. This decision depends on the spatial and radiometric resolution, the spectral range needed and the expected cost. This is the most important factor that determines the quality of the data and data products. In addition, the quality is determined by the following conditions:

- technological factors,
- conditions during data acquisition, and
- processing and interpretation.

Technological factors:

1. Calibration of the nonphotographic remote-sensing systems must be done before and during data acquisition following the instructions of the manufacturer.
2. INS (Inertial Navigation System) must be used for accurate determination of changes in aircraft attitude around all three axes and for correction the GPS data.
3. Data with full Image Motion Compensation (IMC) or Forward Motion Compensation (FMC) should be preferred if airborne systems are used.
4. High precision DGPS with a ground reference station closer than 50 km from the project area is strongly advised.
5. For laser scanning flights, DGPS and INS are essential in order to estimate the precise aircraft position at any point. A video camera should be used to record the path of the aircraft. A synchronization system is necessary to correlate the data with other data acquisition systems.

Conditions during data-sampling:

1. Weather:
   - Thermal scanner: no wind, clouds, mist, or precipitation, as well as complete clear conditions.
   - Digital camera: sunny, without clouds, mist or precipitation.
   - Imaging spectrometer: sunshine with no fog, mist, haze, smog, clouds, or precipitation, etc. to reduce the intensity of the sun's radiation.

2. Time of data acquisition:

    - Thermography: season with only little or no growing vegetation, during the night or in the early morning before sunrise following a sunny day.
    - Digital camera: depends on the investigation objectives, e.g., season with or without growing vegetation.
    - Imaging spectrometer: season with only little or no growing vegetation, the hours around high noon are preferable.

Processing and interpretation:

- Experience and references: The company or institution conducting the survey should provide references in carrying out remote-sensing surveys using state-of-the-art hardware and software.

- Integration of other data: Including other remote-sensing and non remote-sensing information increases the reliability and value of the results.

- Ground truth: Thermograpic and spectroscopic measurements are made on the ground at the time of the survey flight. If airborne digital camera data is acquired during the flight, additional ground truth data may be required for the interpretation.

3.4.8 Personnel, Equipment, Time Needed

	Personnel	Equipment	Time needed
data acquisition with thermography scanner	2 - 3 technicians or specialists	special survey aircraft with a flight management system and a thermography scanner	1 day
processing data from a thermography scanner	1 specialist	PC, plotter, software	1 - 3 days
interpretation of thermography data	1 specialist	PC, plotter, software	3 - 8 days
data acquisition with a digital camera	2 - 3 technicians or specialists	special survey aircraft with a flight management system and a digital camera system	1 day
processing data from a digital camera (HRSC)	1 specialist	several PCs, plotter, software	1 - 3 weeks
interpretation of data from a digital camera (HRSC)	1 specialist	PC, plotter, software	1 day to some weeks (depends on the task)
groundwork for imaging spectrometry	1 - 2 technicians or specialists	1 field spectrometer	1 - 3 days
data acquisition with an imaging spectrometer	2 - 3 technicians or /specialists	special survey aircraft with a flight management system, an imaging spectrometer and a digital camera	1 day
measurements at the calibration site for imaging spectrometry	1 - 2 technicians or specialist	1 field spectrometer	1 day
processing data of imaging spectrometer	1 specialist	PC, plotter, software	1 - 3 weeks
interpretation of imaging spectrometer data (mapping of hydrocarbons)	1 specialist	PC, plotter, software	1 - 3 weeks
data acquisition with an airborne laser scanner	2 - 3 technicians or specialists	special survey aircraft with a flight management system and laser scanning equipment	1 day
preparation of a digital elevation model by laser scanning (area 100 km^2)	1 specialist	PC, plotter, software	3 - 5 weeks

3.3.9 Examples

Waste disposal sites on a SPOT image

Fig. 3.3-11: Schöneiche and Schöneicher Plan waste disposal sites (*inside dashed lines*), south of Mittenwalde near Berlin, shown on a SPOT image taken on August 25, 1990 in panchromatic mode. Courtesy of FKP Ingenieurgesellschaft für Fernerkundung, Photogrammetrie, Kartographie und Vermessung mbH (Consulting Engineers in Remote Sensing, Photogrammetry, Cartography, and Surveying), Berlin

Preparation of a map of flood prone areas

Landsat 7 data from May 3, 2000, were used for mapping flood-prone areas in the Dan Khun Thot region of Thailand in order to classify preselected areas for a potential waste disposal site. Image quality was good with cloud coverage of about 20 - 25 %. The satellite remote sensing systems used scanners for 7 multi-spectral bands. Ground resolution was between 15 × 15 m and 60 × 60 m. The data were corrected for systematic and radiometric errors. Purpose of the processing was to provide the interpreter with enhanced imagery for mapping flooded areas. A combined approach of unsupervised and supervised classification was chosen to classify selected land cover types in the area. This classification divided the pixels into statistically defined classes for certain types of relevant land cover: standing water, swamps, elevated areas, plantations, soil salinization areas and settlements. A final map at a scale of 1 : 50 000 was derived (*Fig. 3.3-12*).

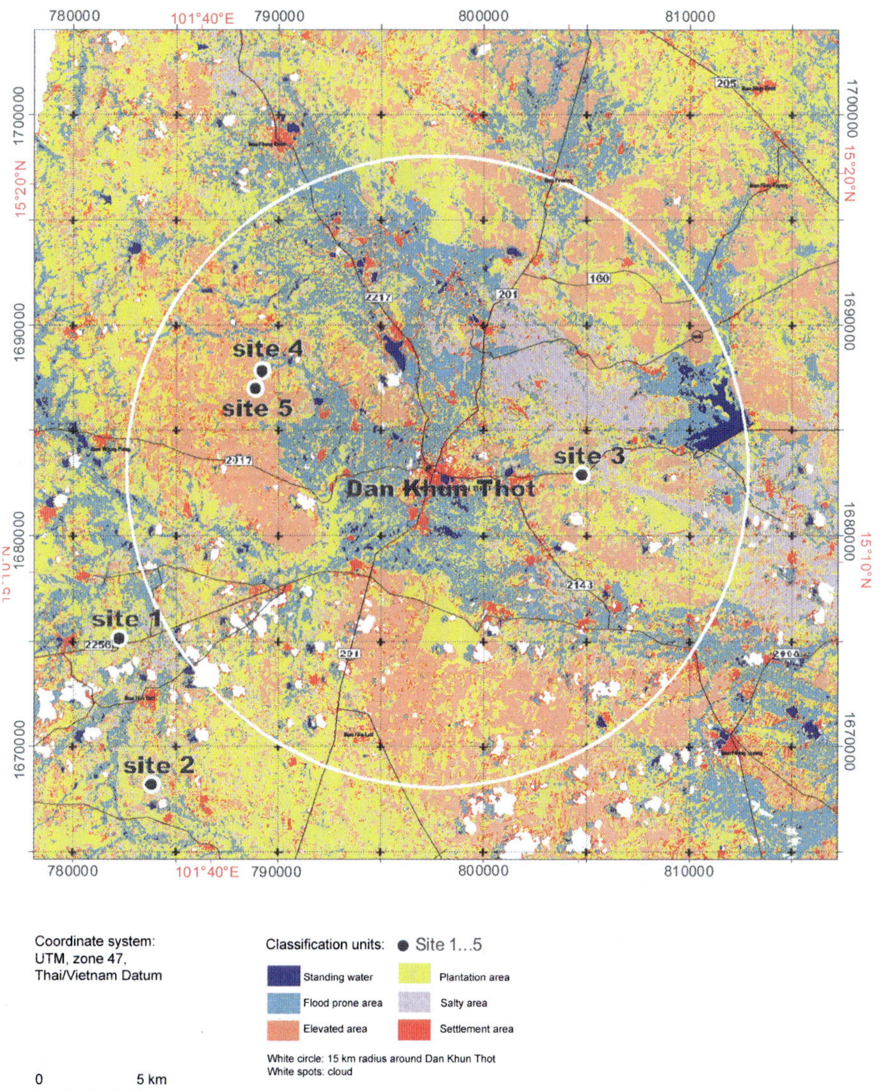

Fig. 3.3-12: Dan Khun Thot region of the Nakhon Ratchasima province of Thailand, map of flood-prone areas prepared using Landsat 7 data from May 3, 2000, showing the locations of proposed waste disposal sites

Localizing heat sources inside a mine spoil heap

Mining activities commonly result in changes to the landscape. Investigations are necessary when risks are identified that may threaten people, infrastructure and natural resources. The detailed black-and-white aerial photograph in *Figure 3.3-13* depicts parts of an abandoned open-pit mine in Thuringia, Germany. Mine dumps, mill tailings, tailing ponds, and abandoned mining equipment are visible. The mill tailings contain smoldering pyrite and carbon-rich schist. The smoldering is sustained by oxygen seeping through fissures and crevasses in the heap. This may lead to contamination of the groundwater if rainwater percolates through the heap (*Fig. 3.3-15*).

Fig. 3.3-13: Aerial photo taken April 24, 1995, showing a mining landscape in Thuringia, Germany. The dominant landscape structures include part of an abandoned open-pit mine (being filled with tailings), waste heaps, and an abandoned smelting plant (Source: Landesvermessungsamt Thüringen).

The thermal scanner image in *Figure 3.3-14*, taken at night, displays the area visible in the black-and-white aerial photograph of *Figure 3.3-13*. Areas of high temperature in the heap are indicated in the thermal image although the smoldering material is several meters deep inside the heap. The range of apparent surface temperatures (in °C) depicted in the image is indicated by the scale in the lower left corner. The yellow color denotes temperatures between 25 and 35 °C, the maximum. Monitoring of waste heaps with thermal systems provides information on the temporal and spatial extent of the oxidation zone. Rainwater percolating through smoldering pyrite and carbon-rich schist forms sulfuric acid, which will then contaminate the groundwater. These smoldering zones must be identified for effective remediation. The situation and the method are explained in *Figure 3.3-15*.

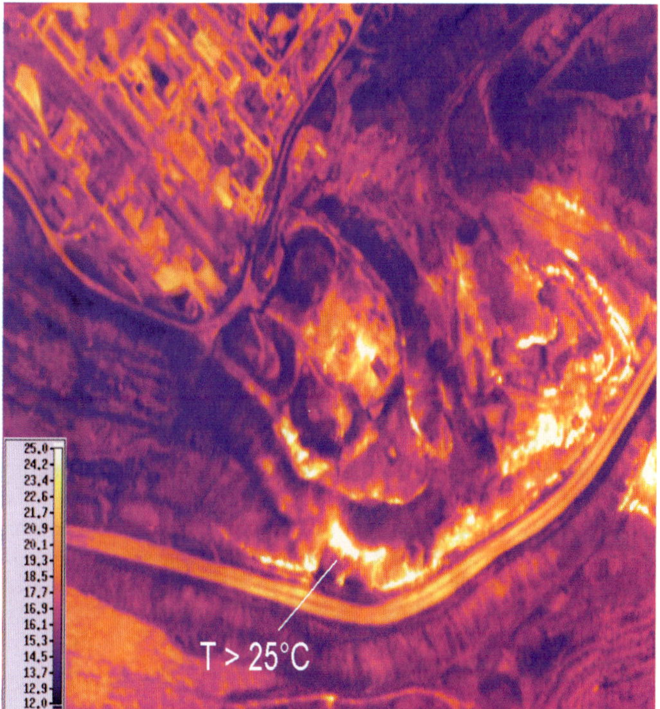

Fig. 3.3-14: Thermal image taken at 03:45 a.m. on May 23, 1994, of the abandoned mining site shown in *Fig. 3.3-13*. Temperature anomalies (bright) indicate areas of smoldering pyrite and carbon-rich schist measured through several meters of cover. These zones of buried smoldering material are marked by significant temperature anomalies at the ground surface. (Image: F. KÜHN, BGR).

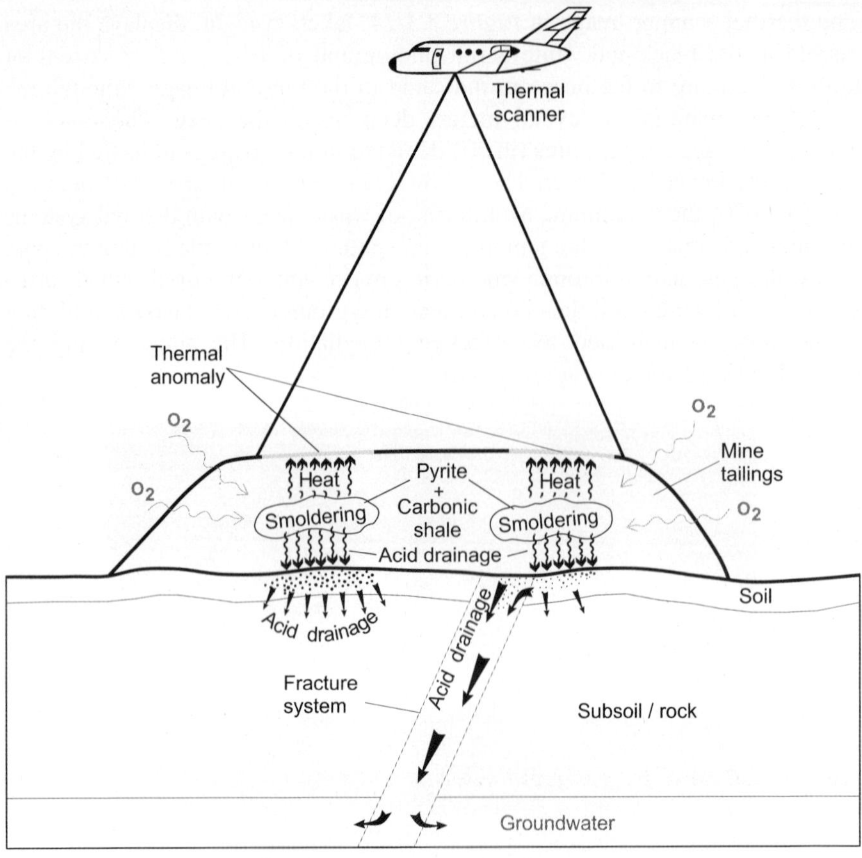

Fig. 3.3-15: Smoldering pyrite and carbon-rich schist generate heat within a mine waste heap. The smoldering is sustained by an influx of oxygen through fissures and crevasses in the heap. Sulfuric acid is produced, which may lead to groundwater contamination. The buried smoldering zones are easily detectable by thermal remote sensing (*Fig. 3.3-14*).

Gas emissions

Anaerobic conditions inside a landfill favor the production of methane and other gases by fermentation. In the northern part of the Schöneiche landfill (*Fig. 3.3-11*), a network of ducts is used to collect and disperse the gas. Where the gas ducts emerge, they are observable as "hot spots" in thermal images taken at night, when the landfill surface cools off rather rapidly. During the daytime, solar radiation heats the surface of the landfill and the ducts appear colder than the surrounding areas. These gas ducts are the only source of information about temperature anomalies and the decay processes within the landfill (*Fig. 3.3-16*).

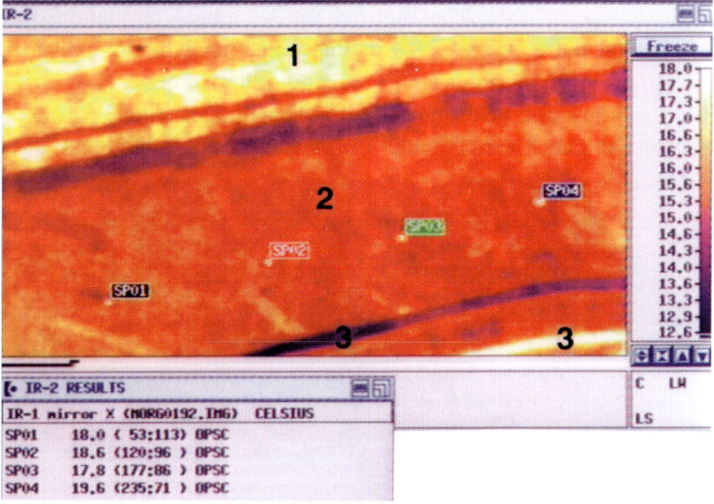

Fig. 3.3-16: A thermal image made at 5:15 CET on May 13, 1993, shows four emerging ducts (SP01 – 04) used for the dispersal of biogases. AGEMA 900 software for image processing and evaluation permits calculation of the "radiation temperature" at the gas ducts, which fluctuates between 17.8 and 19.6 °C. The temperature increase of 1.8 °C between SP03 and SP04 indicates a temperature increase within the landfill. (1) Slope, (2) covered waste, and (3) roads. (Image: F. BÖKER and F. KÜHN, BGR)

Abandoned landfills and areas of suspected contamination

The detection of soil and groundwater contamination raises questions concerning the temporal and spatial dispersal of contaminants. Because remediation can be expensive, and the polluter is legally required to cover the cost, the landfill operator will want to establish the actual sources of the contamination to avoid charges of clandestine waste disposal that often occur at large sites.

In 1992, the Working Group for Remediation and Management of Waste Disposal, Berlin, (Arbeitsgemeinschaft "Sanierungs- und Betriebskonzept/Abfallentsorgung Berlin") conducted a survey of abandoned waste disposal sites and other areas of suspected contamination near the Schöneiche and Schöneicher Plan landfills (*Fig. 3.3-11*). A number of current and abandoned waste disposal sites, as well as sewage and waste water irrigation systems, were observed and mapped.

The following example demonstrates how aerial photographs and thermal images are applied to detect and characterize an abandoned landfill. An elevated area, typical for landfills, can be identified between the Schöneiche and Schöneicher Plan sites in the center of the stereo-pair photographs in *Figure 3.3-17*. The area is near the southeast rim of a water-filled pit (black). It shows patterns typical of cropland. Additional information was provided by thermal data (*Fig. 3.3-18*): The thermal image reveal temperature anomaly patterns oriented more or less N-S in this area, most likely indicating near-surface changes of material (dark blue and red patches). NW–SE-oriented yellowish strips are due to the grain crop. The image helped define areas for taking samples to check the character of materials dumped in the area.

Fig. 3.3-17: CIR aerial photograph taken on July 3, 1993, (stereo-pair) showing an abandoned landfill (area in the center just below the body of water) between the Schöneiche and Schöneicher Plan landfills (Photo taken by WIB GmbH Berlin for BGR)

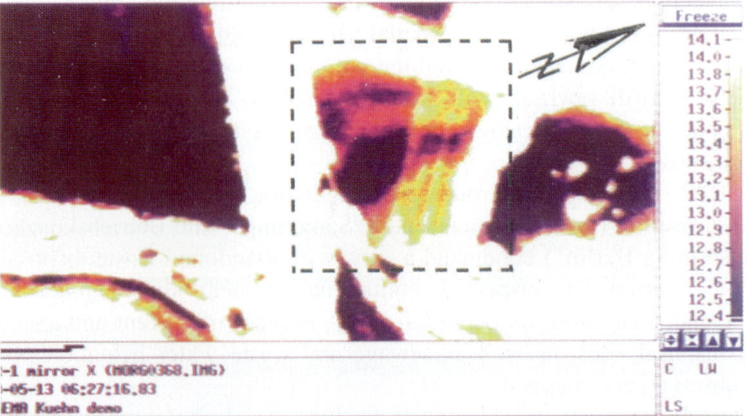

Fig. 3.3-18: Thermal image taken at 05:27 CET on May 13, 1993, showing the landfill (within dashed line) in *Fig. 3.3-17* and temperature anomalies resulting from a partial vegetation cover (yellow and orange stripes) and from soil (dark blue stripes trending roughly N–S). Image taken by F. BÖKER and F. KÜHN, BGR

Springs and soil moisture anomalies

As discussed previously, moist areas and springs are easily detected by remote-sensing techniques. Moist areas commonly develop at the surface due to the presence of perched groundwater or to a shallow water table. Geochemical sampling of moist areas may identify the sources of the moisture by their geochemical signatures.

The historical aerial photograph from 1974 in *Figure 3.3-19* shows an area about 100 m northwest of the Schöneiche landfill. A system of ditches can be seen that are partially filled with silt (D), indicating that this area must be permanent wetland.

This assumption is supported by thermal images of the site (*Fig. 3.3-20*). Signs of elevated water content can be seen in the thermal images. A "cold" (*dark blue*) zone indicates high moisture content in the topsoil. Electrical conductivities around 1200 μS cm^{-1}, which is about four times the normal value, were measured in groundwater samples from this zone. This suggests that mineralized groundwater contains leachate from the landfill about 200 meters north of the waste site.

Dark patches (P) in the aerial photograph (*Fig. 3.3-19*) also suggest surface inhomogeneity. A related ground check confirmed this feature to be an abandoned landfill. The extent to which the abandoned landfill contributes to the elevated groundwater mineralization is yet to be investigated.

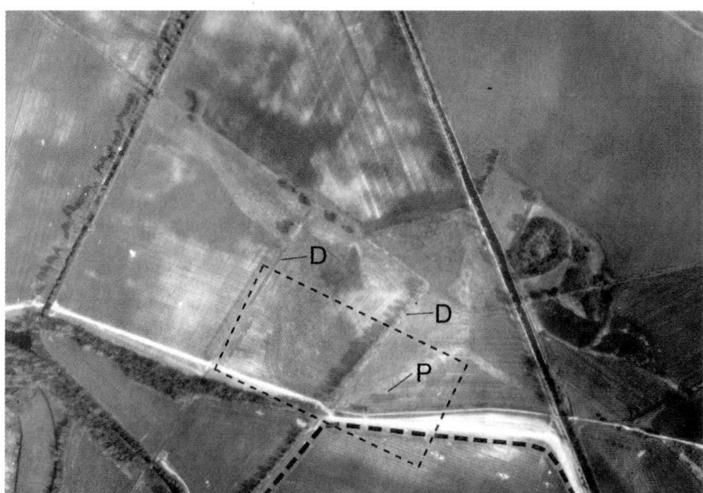

Fig. 3.3-19: Aerial photograph taken on April 20, 1974, of an area just north of the Schöneiche landfill. The dark gray areas indicate higher surface moisture content. The water has been drained via partially silted-up ditches (D). The area of *Fig. 3.3-20* is marked by the thin dashed line and the margin of the landfill is indicated by the thick dashed line. Relatively dark patches (P) indicate an abandoned landfill. Photo from Bundesarchiv (Federal Archives of Germany)

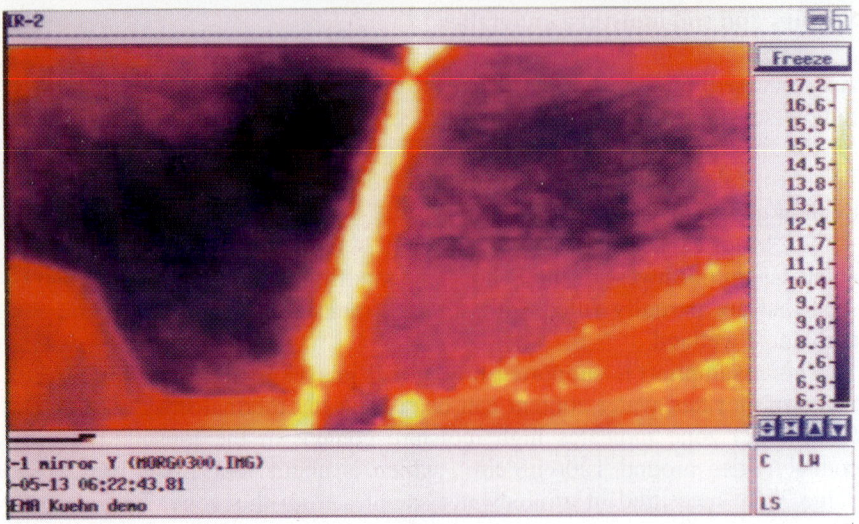

Fig. 3.3-20: Thermal image taken at 05:22 CET on May 13, 1993, of the area marked in *Fig. 3.3-19*. Dark areas correspond to low temperatures, indicating a higher soil moisture content (Image: F. BÖKER and F. KÜHN, BGR)

Natural and artificial drainage systems

A flat lowland adjacent to a waste disposal site is shown in a black-and-white aerial photograph in *Figure 3.3-21* (*top*). Geologically speaking, this area consists of humic sand, mud, and bog lime deposits. The aerial photograph was taken at the beginning of intensive plant growth and relatively high soil moisture content (mid-May). Weak features that may indicate an artificial drainage system (arrow) can be seen.

The thermal image (MIR-II) in *Figure 3.3-21* (*bottom*), taken under similar conditions, depicts the surface drainage more clearly because the drained ground has a higher temperature than the surrounding soil with a higher moisture content. In this case, the intensity of temperature anomalies are indicators of the efficiency of the drainage system. If there are signs of artificial or natural drainage or irrigation systems on or near a landfill, the site should be inspected to determine whether leachate from the landfill is drained by them.

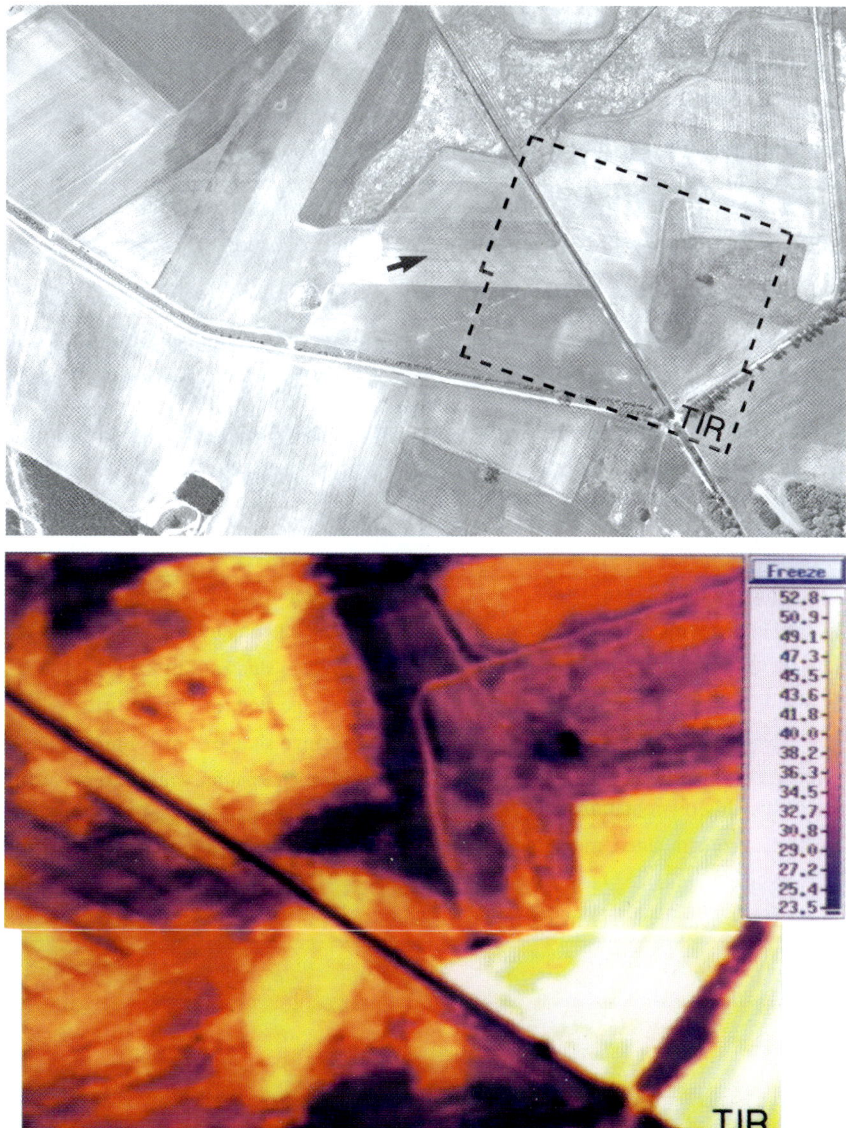

Fig. 3.3-21: Black-and-white aerial photograph taken May 15, 1992 (*top*), and a daytime thermal image, MIR-II (TIR), taken May 11, 1993 (*bottom*). The fishbone pattern of an artificial drainage system is weakly visible in the aerial photograph (arrow). This pattern is better emphasized in the thermal image. The soil is relatively dry and "warm" where the drainage system still functions, (aerial photo: Luftbildsammelstelle der Landesvermessung und Geobasisinformation Brandenburg; thermal image: F. BÖKER and F. KÜHN, BGR)

Identification and mapping of hydrocarbons and materials containing hydrocarbons

A case study performed to determine the possibility to identify and to map hydrocarbons and materials containing hydrocarbons. For this purpose reference areas were prepared containing different materials (*Table 3.3-20*). The ability to distinguish materials containing hydrocarbons from other materials is based on the absorption peaks of hydrocarbons that occur within a narrow range of the short-wave infrared (SWIR) spectrum (see Section 3.3.6).

Figure 3.3-22 shows the result of HyMap data analysis. In this part of the SWIR spectrum, non-hydrocarbons appear gray, e.g., uncontaminated sand and sewage sludge. Only the hydrocarbons appear colored, due to their significant absorption within this narrow portion of the spectrum. The reference areas containing hydrocarbons smaller than 4 m × 4 m are below the spatial resolution of the HyMap system and, thus, appear gray too. The black tarpaulin acts as a black body reflector.

Table 3.3-20: Reference areas on HyMap images shown in *Fig. 3.3-22*

1 - roofing felt (4 m × 4 m)	10 - sand, uncontaminated (4 m × 4 m)
2 - sand mixed with $CaCO_3$ (4 m × 4 m)	11 - sand, uncontaminated (6 m × 6 m)
3 - sand, slightly oil-contaminated[1)] (4 m× 4 m)	12 - sewage sludge (4 m × 4 m)
	13 - sewage sludge (4 m × 4 m)
4 - sand, highly oil-contaminated[2)] (4 m × 4 m)	14 - sewage sludge (4 m × 4 m)
5 - sand, highly oil-contaminated[2)] (2 m × 2 m)	16 - plastic tarpaulin, orange (6 m × 8 m)
6 - sand, highly oil-contaminated[2)] (1 m × 1 m)	17 - plastic tarpaulin, black (4 m × 8 m)
7 - sand, uncontaminated, moist (4 m × 4 m)	21 - pile of transparent plastic sheet
8 - sand, uncontaminated (1 m × 1 m)	
9 - sand, uncontaminated (2 m × 2 m)	

1) 25 mL oil per 1 kg sand
2) 100 mL oil per 1 kg sand

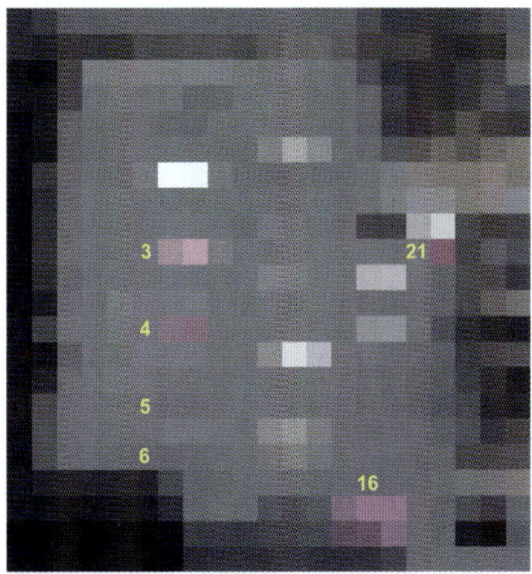

Fig. 3.3-22: Reference areas on processed and enlarged HyMap images (SWIR-1 bands 21/red, 26/green, 31/blue with linear stretching)

Identification and mapping of materials containing hydrocarbons by merging the data from two remote-sensing platforms (contribution by Tilmann Bucher)

Hyperspectral data (*HyMap*) can be processed together with data from a digital camera (*HRSC-A*) to merge high spatial and high spectral information. The HyMap system provides data for the location of hydrocarbons. The HRSC-A camera provides precise geometrically corrected images with a spatial resolution finer than 1 m (*Fig. 3.3-23*) and a DEM for 3-D visualization (*Fig. 3.3-24*).

Characteristic absorption features in the SWIR range can be used to map materials containing hydrocarbon (HÖRIG et al., 2001). This method was evaluated at the overburden heap of the Espenhain lignite mine located in the Central German Lignite Mining District in Eastern Germany. The HyMap and the HRSC-A data of this area have been recorded in August 1998, with a ground resolution of about 8 m for the HyMap data and 30 cm for the HRSC-A data (*Fig. 3.3-23*).

Figure 3.3-24 depicts the merge of the two data sets. A white triangle is clearly visible in the center of the lower picture, marking the hydrocarbon-bearing area. One side of the triangle is more than 300 m long. The DEM shows the dimensions of the dump in this mining district and the elongated crests; a deposition product of the swinging wings of the mining excavator. The crests are within the hydrocarbon-free and in the hydrocarbon-bearing area, thus, the contamination with hydrocarbons occurred after the deposition of the crests.

Fig. 3.3-3-23: Overlay of high-resolution HRSC-A data (spatial resolution 30 cm, center) with HyMap data (spatial resolution 8 m) from an area near Espenhain, Germany. Clouds (dark) cover parts of the area

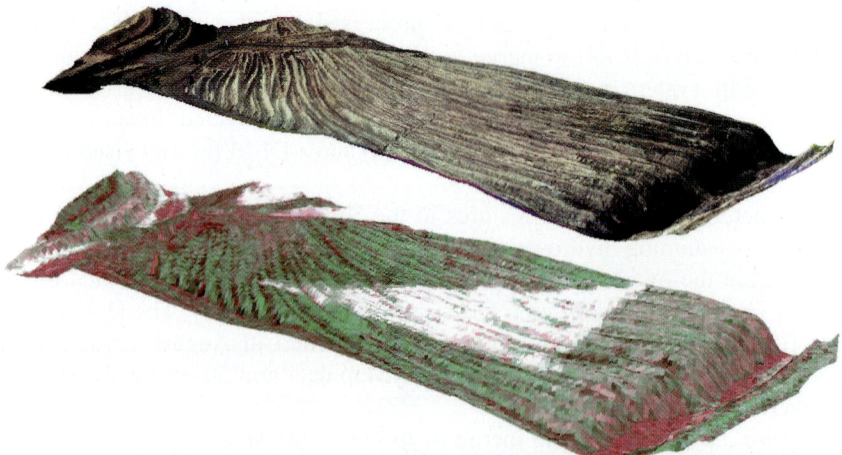

Fig. 3.3-24: An image produced for visualization by the merging of data from different sources. A high-resolution HRSC-A image (top) and a false color HyMap image (bottom) are draped over the HRSC-A DEM. The white areas (bottom) show strong indications of hydrocarbon absorption

InSAR-based land subsidence map for Bangkok, Thailand

Traditional leveling at benchmark stations are usually used to monitor land subsidence resulting from extraction of groundwater from the main aquifer in the Greater Bangkok area. Land subsidence maps derived from leveling show precise results for each benchmark station, but the areal distribution of subsidence is widely inaccurate. Furthermore, the leveling data were commonly collected over long periods of time depending on the size of the observation area and the number of benchmark stations. Thus, it is sometimes difficult to compare data from different stations with one another.

An InSAR-based land subsidence map showing the subsidence between February 20 and October 23, 1996, (left) is compared in *Figure 3.3-26* with the land subsidence map derived from conventional leveling in the Greater Bangkok area in 1995/1996. The InSAR-based land subsidence map is more accurate in terms of the two-dimensional representation of subsidence than the map made from benchmark leveling measurements. A combination of precise benchmark leveling and InSAR-based mapping provides the best results for the investigation of land subsidence. For details see KÜHN et al. (2004).

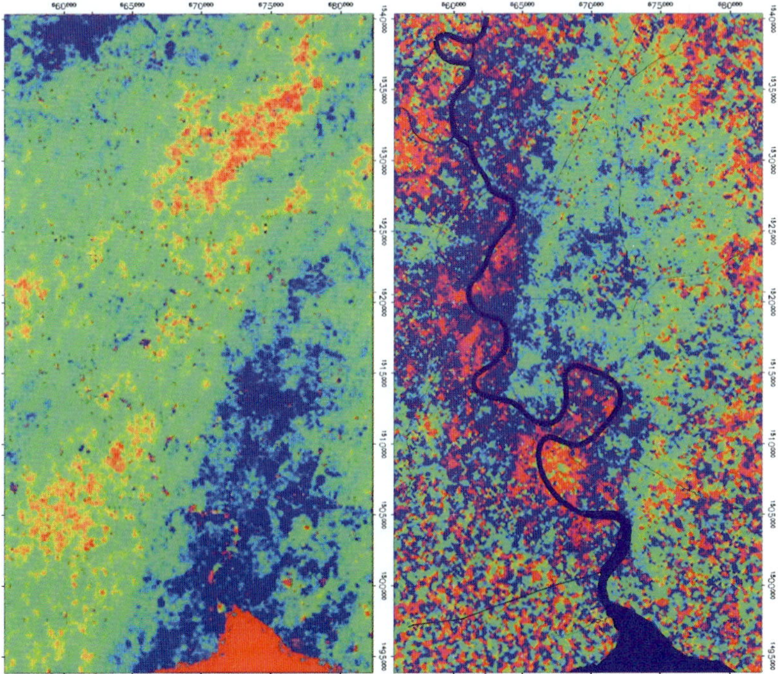

Fig. 3.3-25: (left) Interferogram calculated from ERS 1/2 data recorded on 21 and 22 December 1999 used for correction of the contribution of topography to the phase and (right) differential interferogram derived from ERS 1/2 data recorded on 20 February and 23 October 1996

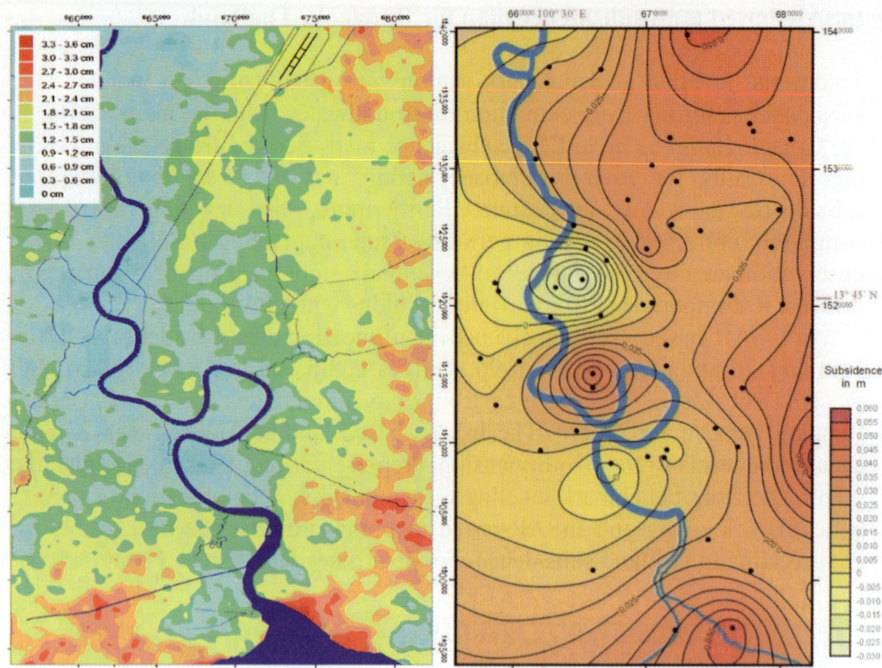

Fig. 3.3-26: InSAR-based land subsidence map showing subsidence between February 20 and October 23, 1996, (left) compared with the land subsidence map derived from conventional leveling in 1995/1996 by BONTEBAL (2001) (right)

References and further reading

ALBERTZ, J. (2001): Einführung in die Fernerkundung, Grundlagen der Interpretation von Luft- und Satellitenbildern. Wissenschaftliche Buchgesellschaft, Darmstadt.

ANGER, C. D., BABEY, S. K. & ADAMSON, R. A. (1990): A new approach to Imaging Spectroscopy. Proceedings of SPIE, **72**, 72-86.

AZCUE, J.M., MURDOCH, A., ROSA, F., HALL, G. E. M., JACKSON, T. A. & REYNOLDSON, T. (1995): Trace elements in water, sediments, porewater, and biota polluted by tailings from an abandoned gold mine in British Columbia, Canada. J. Geochem. Explor., **52**, 25-34.

BABEY, S. K. & ANGER, C. D. (1989): A Compact Airborne Spectrographic Imager (CASI). IGARSS Proceedings, **2**, 1028-1031.

BERGER, B. R. (1986): The geological attributes of Au-Ag-base metal epithermal deposits. In: ERICKSON, R., comp., Characteristics of mineral-deposit occurrences, USGS Open File Report, 82-795, 119-126.

BIANCI, R., CAVALLI, R. M., FIUMI, L., MARAINO, C. M. & PIGNATTI, S. (1997): Airborne remote sensing: Results of two years of imaging spectrometry for the study of environmental problems. In: Remote Sensing '96, Balkema, Rotterdam, 269-273.

BOLDT, C. M. K. & SCHEIBNER, B. J. (1987): Remote sensing of mine waste. U.S. Bureau of Mines Information Circular 9152.

BONTEBAL, M. (2001): Land Subsidence in Bangkok: An Overview of Changes in Land Subsidence over the last 25 Years. M. Sc. Thesis, Free University of Amsterdam, Netherlands.

BRITTON, B. (2002): Uses of Satellite Imagery for Planning or Site Selection, EOM, **11**, 2.

BROWN, R. L. (1991): Cripple Creek then and now. Sundance Publications, Ltd., Denver.

CHUCHIP, K. (1997): Satellite Date Analysis and Surface Modeling for Land Use and Land Cover Classification in Thailand. Berliner Geographische Studien, **46**, Berlin.

CLARK, R. N. (1999): Spectroscopy of Rocks and Minerals, and Principles of the Spectroscopy, In: RENCZ, A. N. (Ed.): Remote Sensing for the Earth Sciences, **1**, 3-58.

CLARK, R.N. (1999): Spectroscopy of Rocks and Minerals and Principles of Spectroscopy, In: A.N. RENCZ (Ed.). Manual of Remote Sensing, Chapter **1**, Wiley and Sons, New York, 3-58.

CLARK, R. N., GALLAGHER, A. J. & SWAYZE, G. A. (1990a): Material absorption band depth mapping of imaging spectrometer data using a complete band shape least-squares fit with library reference spectra. In: Proceedings of the Second Airborne Visible/Infrared Imaging Spectrometer (AVIRIS) Workshop, JPL Publication 90-54, 176-186.

CLARK, R. N., KING, T. V. V., AGER, C. & SWAYZE, G. A. (1995): Initial vegetation species and senescence/stress mapping in the San Luis Valley, Colorado using imaging spectrometer data. In: Summaries of the Fifth Annual JPL Airborne Earth Science Workshop, January 23.–26., R.O. Green, Ed., JPL Publication 95-1, 35-38.

CLARK, R. N., KING, T. V. V., KLEJWA, M., SWAYZE, G. & VERGO, N. (1990b): High Spectral Resolution Reflectance Spectroscopy of Minerals. J. Geophys. Res., **95**, 12653-12680.

CLARK, R. N., SWAYZE, G. A., KOCH, C., GALLAGHER, A. & AGER, C. (1992): Mapping Vegetation Types with the Multiple Spectral Feature Mapping Algorithm in both Emission and Absorption. In: Summaries of the Third Annual JPL Airborne Geosciences Workshop, **1**, AVIRIS Workshop, held in Pasadena, CA, on June 1.-5., 1992. JPL Publication 92-14, 60-62.

CLARK, R. N., SWAYZE, G. A., GALLAGHER, A., KING, T. V. V. & CALVIN, W. M. (1993): The U. S. Geological Survey, Digital Spectral Library, **1**, 0.2 to 3.0 µm, USGS, Open File Report 93-592, 1340, (Also published as a USGS Bulletin, 1300+pp, 1996).

CLARK, R. N., SWAYZE, G. A., LIVO, K. E., KOKALY, R. F., SUTLEY, S. J., DALTON, J. B., MCDOURAL, R. R., & GENT, C. A. (2003): Imaging spectroscopy: Earth and planetary remote sensing with the USGS Tetracorder and expert systems, J. Geophys. Res., **108** (E12), 5131, doi: 10.1029/2002JE001847, December, 2003. http://speclab.cr.usgs.gov/PAPERS/tetracorder

CLOUTIS, E. A. (1989): Spectral Reflectance Properties of Hydrocarbons: Remote Sensing Implications. Science, **245**, 165-168.

COLWELL, R. N. (Ed.) (1983): Manual of Remote Sensing, **1**, Theory, Instruments and Techniques, **2**, Interpretation and Applications. American Society of Photogrammetry, Falls Curch.

CONWAY, E. D. (1997): An Introduction to Satellite Image Interpretation. Johns Hopkins University Press.

COCKS, T., JENSSEN, R., STEWART, A., WILSON, I. & SHIELDS, T. (1998): The Hymap™ Airborne Hyperspectral Sensor: The System, Calibration and Performance. In: SCHAEPMAN, M. et al. Eds. 1st EARSeL Workshop on Imaging Spectroscopy, 37-42.

CRAWFORD, M. F. (1987): Preliminary Evaluation of Remote Sensing Data for Detection of Vegetation Stress Related to Hydrocarbon Microseepage: Mist Gas Field Oregon. Proc of 5th Thematic Conf on Remote Sensing for Exploration Geology, Environmental Research Institute of Michigan, **1**, 161-177.

CRAWFORD, G. A. (1995): Environmental improvements by the mining industry in the Sudbury Basin of Canada. J. Geochem. Explor., **52**, 267-284.

CURRAN, P. J. (1989): Remote sensing of foliar chemistry. Remote Sens. Environ., **30**, 271-278.

CURRAN, P. J., DUNGAN, J. L., MACLER, B.A. & PLUMMER, S. E. (1991): The effect of a red leaf pigment on the relationship between red edge and chlorphyll concentration. Remote Sens. Environ., **35**, 69-76.

CURRAN, P. J., DUNGAN, J. L., MACLER, B. A., PLUMMER, S. E. & PETERSON, D. L. (1992): Reflectance spectroscopy of fresh whole leaves for the estimation of chemical concentration. Remote Sens. Environ., **39**, 153-166.

CURRAN, P. J. & KUPIEC, J. A. (1995): Imaging Spectrometry: A New Tool for Ecology. In: DANSON, F. M. & PLUMMER, S. E. (Eds): Advances in environmental remote sensing. Wiley & Sons, Chichester, 71-88.

CURTISS, B. & USTIN, S. L. (1989): Parameters effecting reflectance of coniferous forests in the region of chlorophyll absorption. In: Proc of IGARSS '89, 12th Can Symp on Remote Sensing, Vancouver, BC.

DAVIS, A. D. & WEBB, C. J. (1995): Abandoned mines inventory and reclamation in the Black Hills of South Dakota. In: SCHEINER, B. J., CHATWIN, T. D., EL-SHALL, H., KAWATRA, S. K. & TORMA, A. E. (Eds): New remediation technology in the changing environmental arena. Littleton, Colorado. Society for Mining, Metallurgy, and Exploration, Inc., 27-33.

DE VOS, K. J., BLOWES, D. W., ROBERTSON, W. D. & GREENHOUSE, J. P. (1995): Delineation and evaluation of a plume of tailings derived water, Copper Cliff, Ontario. Mining and the Environment, Proceedings, Sudbury, 673-682.

DIAS, N. W. (2002): Applying Red-Edge & WaterAbsorption Geometry Analyses to Estimated Hardwood Forest Structure and Biomass by Using Hyperspectral Data, EOM, **11**, 2.

DOUGLAS, W. J. (1995): Environmental GIS – Applications to industrial facilities. Lewis Publishers, Boca Raton, FL.

EHRENBERG, M. (1991): Beprobungslose Altlastenerkundung. wlb Wasser, Luft und Boden, **7-8**, 56-59.

Ekwan Technology Corporation: www.ekwantech.com

ELLIS, J. M., DAVIS, H. H. & ZAMUDIO, J. A. (2001): Exploring for onshore oil seeps with hyperspectral imaging. Oil and Gas Journal, **99**, 10, 49-58.

ERB, W. & Endres, L. (1989): Leitfaden der Spektroradiometrie. Springer, Berlin.

FRICKER, P., SANDAU, R. & WALKER, A. S. (1999): Digital photogrammetric cameras: possibilities and problems. In: FRITSCH, D. & SPILLER, R. (Eds.): Photogrammetric Week '99, Wichmann Verlag, Heidelberg, 71-82.

GAFFEY, S. J., MCFADDEN, L. A., NASH, D. & PIETERS, C. M. (1993): Ultraviolet, visible, and nearinfrared reflectance spectroscopy: Laboratory spectra of geologic materials. In: PIETERS, C. M. & ENGLERT, P. A. J. (Eds.): Remote Chemical Analysis: Elemental and Mineralogical Composition, Cambridge University Press, 43-77.

GAMBA, P. and B. HOUSHMAND (2002): Joint analysis of SAR, LIDAR and aerial imagery for simultaneous extraction of land cover, DTM and 3D shape of buildings, International Journal of Remote Sensing, **23**, 20, 4439-4450.

GATES, D. M., KEEGAN, H. J., SCHLETER, J. D. & WEIDNER, V. R. (1965): Spectral properties of plants. Applied Optics, **4/1**, 11-20.

GEBHARDT, A. (1981): Thermografie, Anwendungen bei der geophysikalischen Naherkundung. Freiberger Forschungshefte, C367, Dt. Verlag f. Grundstoffindustrie, Leipzig.

GIM International (2001): Product Survey on Airborne Laser Scanner - an Overview.

GLAESER, C. (1989): Beiträge zur Anwendung der Multispektraltechnik für die Lösung geowissenschaftlicher Aufgaben. Dissertation B, Martin-Luther-Universität, Halle-Wittenberg, Textband.

GOODCHILD, M. F., STEYAERT, L. T., PARKS, B. O., JOHNSTON, C., MAIDMENT, D., CRANE, M. & GLENDINNING, S. (Eds.) (1996): GIS and environmental modeling. Progress and research issues: GIS World Books, Fort Collins, Colorado.

GRAHAM, D. F., ST-ARNAND, E. L. & RENCZ, A. N. (1994): Canada Geologic Survey Monitors Mine Tailings, Disposal sites with Landsat. EOS Magazine, 38-41.

GUNN, J. M. (ed.) (1995): Restoration and Recovery of an Industrial Region. Springer, New York.

GUPTA, R. P. (1991): Remote Sensing Geology. Springer, Berlin.

GUPTA, R. P. (2003): Remote Sensing Geology. Springer, Berlin.

HAUFF, PH. L. (1993): Spectral Reflectance Properties of Oil and Hydrocarbon-Bearing Rocks and Sediments. Application Note, **1**, 1993, Spectral International Inc., Lafayette, CO.

HEIER, H. (1999): Application and markets for digital airborne cameras. In: FRITSCH, D. & SPILLER, R. (Eds.): Photogrammetric Week '99, Wichmann Verlag, Heidelberg, 43-49.

HÖRIG, B., KÜHN, F., OSCHÜTZ, F. & LEHMANN F. (2001): HyMap hyperspectral remote sensing to dedect hydrocarbons. Int. J. Remote Sensing, **22**, 8, 1413-1422.

HORNSBY, J. K., BRUCE, B. & MACKENZIE-GREIVE, G. (1989): Monitoring Vegetation Regrowth on Placer Mine Tailings, Bonanza Creek, Yukon Territory. In: Proc of Int Symp on Remote Sensing, 2518-2521.

HUNT, G. R. & SALISBURY, J. W. (1976): Visible and nearinfrared spectra of minerals and rocks: XII. Sedimentary Rocks. Mod. Geol., **5**, 211-217.

HUNT, G. R. (1977): Spectral signature of particulate minerals in the visible and near infrared. Geophysics, **42**, 501-513.

IRVINE, J. M., STAHL, G., ODENWELLER, J., SMYRE, J. L., EVERS, T. K., DALE, H. & KING, A. L. (2000): Thermal Remote Sensing to Detect Buried Waste Material (Oak Ridge, U.S.A.). In: KUEHN, F., KING, T., HOERIG, B. & PETERS, D. (Eds.): Remote Sensing for Site Characterization, Springer Verlag, Berlin, 96-105.

JANSEN, W. T. (1994): The Mapping of Mineral Distributions Using Remotely Sensed Hyperspectral Images and Standard Spectral Libraries. Int. Symposium on Remote Sensing and GIS for Site Characterizations – Applications and Standards, ASTM, San Francisco, Jan. 27.-28. 1994.

JOHNSON, A. I., PETTERSSON, C. B. & FULTON, J. L. (Eds.) (1992): Geographic information systems (GIS) and mapping – Practices and standards: ASTM, Publication STP, **1126**, 346.

KERSTEN, T., BALTSAVIAS, E., SCHWARZ, M. & LEISS, I. (2000): IKONOS-2 CARTERRA™ GEO - erste geometrische Genauigkeitsuntersuchungen in der Schweiz mit hochaufgelösten Satellitendaten. VPKOL 8, 2000.

KING, D. J. (1993): Digital frame cameras: the next generation of low cost remote sensing sensors. In: Proc of ASPRS Biennial Workshop on Photography and Videography in the Plant Sciences, Logan, Utah.

KING, T. V. V. & RIDLEY, W. I. (1987): Relation of the spectroscopic reflectance of olivine to mineral chemistry and some remote sensing implications. J. Geophys. Res., **92**, 11457-11469.

KING, T. V. V. & CLARK, R. N. (1989): Spectral Characteristics of Chlorites and Mg-Serpentine Using High-Resolution Reflectance Spectroscopy. J. Geophys. Res., **94**, 13997-14008.

KING, T. V. V. (Ed.) (1995): Environmental Considerations of Active and Abandoned Mine Lands: Lessons From Summitville, Colorado. USGS Bulletin 2220. 38.

KING, T. V. V., CLARK, R. N. & GREGG, A. S. (2000): Applications of Imaging Spectroscopy Data: A Case Study at Summitville, Colorado. In: KUEHN, F., KING, T., HOERIG, B. & PETERS, D. (Eds.): Remote Sensing for Site Characterization, Springer, Berlin, 164-185.

KOKALY, R. F., CLARK, R. N. & LIVO, K. E. (1987): Mapping the biology and mineralogy of Yellowstone National Park using imaging spectroscopy. In: Summaries of 7th Annual JPL Airborne Earth Science Workshop, R.O. Green, ED., JPL 97-21, **1**.

KOSCHMANN, A. H. (1949): Structural control of the gold deposits of the Cripple Creek district, Teller County, Colorado. USGS Bulletin 955-B, 60.

KRONBERG, P. (1985): Fernerkundung der Erde. Enke, Stuttgart.

KRUSE, F. A., HAUFF, P. L., DIETZ, J., BROCK, J. C. & HAMPTON, L. R. (1989): Characterization and mapping of mine waste at Leadville, Colorado using imaging spectroscopy. Boulder, Center for the Study of Earth from Space, U of Colorado, EPA Contract No. 68-01-7251, **2**.

KÜHN, F. & HÖRIG, B. (1995): Environmental Remote Sensing for Military Excercise Places. - Remote Sensing and GIS for Site Characterizations: Applications and Standards, ASTM STP 1279, American Society for Testing and Materials, 5-16.

KÜHN, F. & HÖRIG, B. (1995): Handbuch zur Erkundung des Untergrundes von Deponien und Altlasten, **1**, Geofernerkundung. Springer, Berlin.

KÜHN, F. & OLEIKIEWITZ, P. (1983): Die Nutzung der Multispektraltechnik zur Früherkennung von senkungs- und erdfallgefährdeten Gebieten. Z. angew. Geol. **29**, 6, 71-74.

KÜHN, F., TREMBICH, G. & HÖRIG, B. (1997): Multisensor Remote Sensing to Evaluate Hazards Caused by Mining. In: Proceedings of the Twelfth International Conference on Applied Geologic Remote Sensing, 17.-19. November, Denver, Colorado, ERIM, **I**, 425-432.

KÜHN, F., TREMBICH, G. & HÖRIG, B. (1999): Satellite and Airborne Remote Sensing to Detect Hazards Caused by Underground Mining. In: Proceedings of the Thirteenth International Conference on Applied Geologic Remote Sensing, 1.-3. March, Vancouver, British Columbia, Canada, ERIM, **II**, 57-64.

KUEHN, F., KING, T., HOERIG, B. & PETERS, D. (2000): Remote Sensing for Site Characterization. Methods in Enviromental Geology. Springer, Berlin.

KÜHN, F., MARGANE, A., TATONG, T. & WEVER, T. (2004): InSAR – Based Land Subsidence Map for Bangkok, Thailand. Z. Angew. Geol. 1/2004, 74 -81.

LILLESAND, T. M. & KIEFER, R. W. (1994): Remote Sensing and Image Interpretation. 3rd edition, Wiley & Sons, New York.

LINTZ, J. & SIMONETT, D. S. (Eds.) (1976): Remote Sensing of Environment. Addison-Wesley, London.

LIVO, K. E. (1994): Use of remote sensing to characterize hydrothermal alteration of the Cripple Creek area, Colorado. Colorado School of Mines, M. Sc. Thesis T-4613 (unpubl.), Golden.

LOWE, D. S. (1969): Optical sensors. In: Principles and applications to earth resources surveys. CNS and Univ. of Michigan, Paris, 73-136.

LYON, J. G. & MCCARTHY, J. (Eds.) (1995): Wetland and environmental applications of GIS. Lewis Publishers, Boca Raton, Florida.

LYON, R. J. P. (1994): Weathering and desert varnish in arid terrains, In: Proceedings of the First International Airborne Remote Sensing Conference and Exhibition, Strasbourg, France, Sept. 11.-15., 1994. Ann Arbor, Environmental Research Institute of Michigan, **1**, 257-268.

MCCARTHY, F., CHENG, P. & TOUTIN, T. (2000): Case Study of Using IKONOS Imagery in Small Municipalities. EOM **11**.

MCGREGOR, R. G., BLOWES, D. W. & ROBERTSON, W. D. (1995): The Application of chemical Extractions to Sulfide Tailings at the Copper cliff Tailings Area, Sudbury, Ontario. Mining and the Environment, Sudbury 95, Proceedings, Sudbury, 1133-1142.

MCCANN, M. E. & KING, A. (1995): Case Study of Advanced Technologies for Hydrological Site Characterization. Mining and the Environment, Sudbury 95 Proceedings, Sudbury, 667-671.

MEER, F. D. van der & JONG, S. M. de (Eds.) (2001): Imaging spectrometry: basic principles and prospective applications / Kluwer Academic.

MEINEL, G., REDER, J. & NEUBERT, M. (2001): IKONOS- Satellitenbilddaten und ihre Klassifikation – ein erster Erfahrungsbericht. Kartographische Nachrichten, **1**, 2000, 40-46.

MILLER, V. & MILLER, C. F. (1961): Photogeology. Mc Graw-Hill Book Company Inc. New York, 1961.

MILLER, J. R., HARE, E. W., HOLLINGER, A. B. & STURGEON, D. R. (1987): Imaging spectroscopy as a tool for botanical mapping. In: Vane, G. (Ed): Imaging Spectroscopy II, International Society for Optical Engineering, Bellingham, WA, 108-113.

MILTON, N. M., AGER, C. M., EISWERTH, B. A. & POWER, M. S. (1989): Arsenic and selenium-induced changes in spectral reflectance and morphology of soybean plants. Remote Sens. Environ., **30**, 263-269.

MUNTS, S. R., HAUFF, P. L., SEELOS, A. & MCDONALD, B. (1993): Reflectance spectroscopy of selected base-metal bearing tailings with implications for remote sensing. In: Proc of 9th Thematic Conf on Geol Remote Sensing, Pasadena, CA, Feb. 8.–11., 1993: Ann Arbor, Environmental Research Institute of Michigan, 567-578.

MUSSOKOWSKI, R. (1983): A Technique for Mapping Environmental Change - using Digital Landsat Data.- COSPAR, Advances in Space Research, **8**, 103-107.

MURRAY, I. & WILLIAMS, P. C. (1987): Chemical principles of near-infrared. In: WILLIAMS, P. & NORRIES, K. (Eds): Near Infrared Technology in the Agricultural and Food Industries, American Association of Cereal Chemists, St. Paul, MN, 17-37.

ONGSOMWANG, S. (1994): Forest Inventory, Remote Sensing and GIS (Geographic Information System) for Forest Management in Thailand. Berliner Geographische Studien, Berlin.

PARKINSON, C. L. & MCGUIRE, A. (Eds.) (1997): Using Color-Coded Satellite Images To Examine The Global Enviroment. University Science Books.

PETERS, D. C. & PHOEBE, L. H. (2000): Multispectral Remote Sensing to Characterize Mine Waste (Cripple Creek and Goldfield, U.S.A). In: KUEHN, F., KING, T., HOERIG, B. & PETERS, D. (Eds.): Remote Sensing for Site Characterization, Springer, Berlin, 113-164.

PISCHEL, P. & NEUKUM, G. (1999): The High Resolution Stereo Camera HRSC-A – Digital 3-D Image Acquisition; Photogrammetric Processing and Data Evaluation. In: Proceedings, Joint Workshop "Sensors and Mapping from Space 1999", Institut für Photogrammetrie und Ingenieurvermessung, Universität Hannover, **18**, 1999.

RICHARDS, I. A. (2005): Remote Sensing Digital Image Analysis. 4th ed., Springer, Berlin.

ROBERTS, D. A., ADAMS, J. B. & SMITH, M. O. (1993): Discriminating Green Vegetation, Non-Photosynthetic Vegetation and Soils in AVIRIS Data, Remote Sensing of Environment, **44** (2/3), 255-270.

ROBERTS, D. A. & HEROLD, M. (2004): Imaging spectrometry of urban materials. In: KING, P., RAMSEY, M.S. and G. SWAYZE, (Eds.): Infrared Spectroscopy in Geochemistry, Exploration and Remote Sensing, Mineral Association of Canada, Short Course Series Volume 33, London, Ontario, 155-181. URL: http://www.ncgia.ucsb.edu/ncrst/research/pavementhealth/urban/imaging_spectr ometry_of_urban_materials.pdf

ROCK, B. A. (1988): Comparison of in-situ and airborne spectral measurements of the blue shift associated with forest decline. Remote Sensing Environ, **24**, 109-127.

ROWAN, J. S., BARNES, S. J. A., HETHERINGTON, S. L., LAMBERS, B. & PARSONS, F. (1995): Geomorphology and pollution – the environmental impacts of lead mining, Leadhills, Scotland. J. Geochem. Explor., **52**, 57-65.

SABINS, F. F. (1996): Remote sensing - Principles and interpretation. 3rd. edition, Freeman, San Francisco.

SALOMONS, W. (1995): Environmental impact of metals derived from mining activities. Processes, predictions, prevention. J. Geochem. Explor., **52**, 5-23.

SENGUPTA, M. (1993): Environmental impacts of mining. Lewis, Boca Raton, Florida.

SINGHROY, V. H. (1992): Spectral Characterization of the Boreal Forest Species Associated with Geochemical Anomalies. International Geological Congress, Abstracts, Kyoto, Japan, **3**, 980.

SINGHROY, V. H. (1995): Spectral characterization of vegetation at mine tailings. Mining and the Environment, Sudbury 95 Proceedings, Sudbury, 193-200.

SINGHROY, V. H. & KRUSE, F. (1991): Detection of metal stress in the Boreal Forest species using the 670 μm chlorophyll band. ERIM, 8th Thematic Conf on Geologic Remote Sensing, 361-372.

SLONECKER, E. T. & WILLIAMS, D. J. (2001): Imaging spectroscopy for detecting fugitive environmental contaminants. Enviromental Protection Agency Report.

SMITH, R. B. (2001): Indroduction to Hyperspectral Imaging, MicroImages Inc.

STEWART, K. C. & SEVERSON, R. C. (Eds.) (1994): Guidebook on the geology, history, and surface-water contamination and remediation in the area from Denver to Idaho Springs, Colorado. USGS Circular 1097, 55.

STRATHMANN, F.-W. (1993): Taschenbuch zur Fernerkundung. Wichmann, Karlsruhe.

STRUHSACKER, D. W. (1995): The importance of waste characterization in effective environmental planning, project design and reclamation. In: SCHEINER, B. J., CHATWIN, T. D., EL-SHALL, H., KAWATRA, S. K. & TORMA, A. E. (Eds): New remediation technology in the changing environmental arena: Littleton, Colorado, Society for Mining, Metallurgy, and Exploration, Inc., 19-25.

TANDY, B. C. & AMOS, E. (1985): Airborne thermal infrared linescan in geology. Proc of Intern Symp on Remote Sensing of Environment, 4th Thematic Conf "Remote Sensing for Exploration Geology", San Francisco, California, 1. - 4. April.

TANG, L., DOERSTEL, C., JACOBSON, K., HEIPKE, C. & HINZ, A. (2000): Geometric accuracy potential of the Digital Modular Camera, IAPRS, XXXIII, Amsterdam.

THEILEN-WILLIGE, R. (1993): Umweltbeobachtung durch Fernerkundung. Enke, Stuttgart.

THOMPSON, T. B., TRIPPEL, A. D. & DWELLEY, P. C. (1985): Mineralized veins and breccias of the Cripple Creek District, Colorado. Econom Geol, **80**, 1669-1688.

TREVETT, J. W. (1983): Imaging radar for resource surveys. Chapman & Hall, London

TOUTIN, T. & CHENG, P. (2000): Demystification of IKONOS, EOM **9**, 7.

U.S. Forest Service (1993): Acid drainage from mines on the national forests: a management challenge. Program Aid 1505, 12.

USTIN, S. L., MARTENS, S. N., CURTISS, B. & VANDERBILT, V. C. (1994): Use of high spectral resolution sensors to detect air pollution injury in conifer forests. In: FENSTERNMAKER, L. A. (Ed.): Remote sensing applications for acid deposition. EPA publication CR81400201, 72-85.

VANE, G., GREEN, R., CHRIEN, T., UNMARK, H., HANSEN, E. & PORTER, W. (1993): The airborne visible/infrared imaging spectrometer (AVIRIS). Remote Sensing Environ., **44**, 127-143.

VANGRONSVELD, J., STERCKX, J., VAN ASSCHE, F. & CLIJSTERS, H. (1995): Rehabilitation studies on an old non-ferrous waste dumping ground. Effects of revegetation and metal immobilization by beringite. J. Geochem. Explor., **52**, 221-229.

VARSHNEY, P. & ARORA, M. (2004): Advanced Image Processing Techniques for Remotely Sensed Hyperspectral Data. Springer, Berlin.

VEGT, J. W. & HOFFMANN, A. (2001): Airborne Laser Scanning Reaches Maturity. Geoinformatics, Sept. 2001, Emmelord, 32-39.

WOODING, M. G. (1988): Imaging radar applications in Europe, illustrated experimental results (1978-1987). ESA TM-01, Noordwijk.

ZILIOLI, E., GOMARASCA, M. A., & TOMASONI, R. (1992): Application of terrestrial thermography to the detection of waste-disposal sites. Remote Sensing Environ, **40**, 2, 153-160.

4 Geophysics

NORBERT BLINDOW, KLAUS KNÖDEL, FRANZ KÖNIG, GERHARD LANGE, HARALD LINDNER, REINIE MEYER, KLAUS-HENRIK MITTENZWEY, ANDREAS SCHUCK, KNUT SEIDEL, PETER WEIDELT, THOMAS WONIK, UGUR YARAMANCI, with contributions by DIETER EISENBURGER, BERNHARD ILLICH, RICARDO A. OLEA, HELLFRIED PETZOLD & THOMAS RICHTER

4.1 Magnetic Methods

KLAUS KNÖDEL

4.1.1 Principle of the Methods

Everywhere on the Earth there is a natural magnetic field which moves a horizontally free-moving magnetic needle (magnetic compass) to magnetic north. The magnetic field is a vector field, i.e., it is described by its magnitude and direction. The magnetic field consists of three parts: the main field, a fluctuating field, and a local anomaly field.

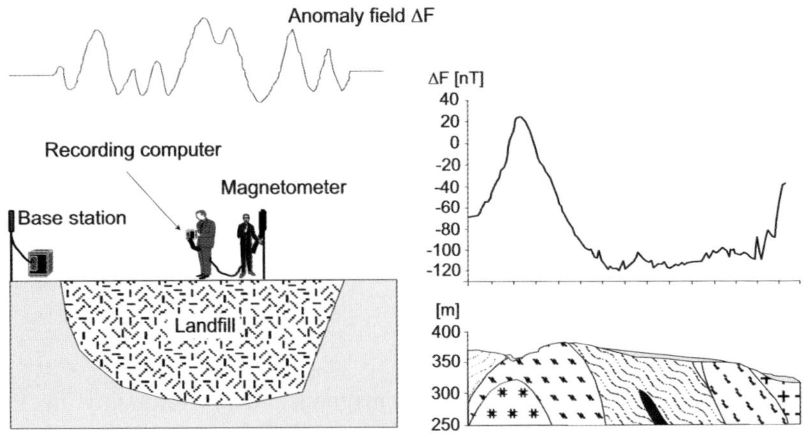

Fig. 4.1-1: Typical applications of the magnetic method, left: search for concealed waste disposal sites containing magnetic materials, right: investigation of the geological structures

The main field, whose origin is within the Earth, varies very slowly with time (years to decades) and is superimposed by a rapidly varying (in fractions of seconds to days) field component, whose origin is outside the Earth (external field, time-varying field). In addition to these geomagnetic field components, there is an almost constant local anomaly field ΔF, which results from the magnetization of material in the upper crust. Not only do geological bodies, consisting of, for example, basalt, metamorphic rocks, and some ore deposits, cause local magnetic anomalies, but also metal objects at legal and illegal waste disposal sites (*Fig. 4.1-1*).

The values of the magnetic flux density are 25 000 to 60 000 nT for the main field, 0.1 to several 100 nT for the fluctuating field, and up to several 1000 nT for the local anomaly field.

The local anomaly field is estimated from magnetic data and conclusions are made about the sources. To obtain the local anomaly field, the main field and the time-varying field must be eliminated from the measured total field. The local anomaly field is represented on an isoline map, (pseudo 3-D plot) or along profiles. The sources of the magnetic anomalies are interpreted from these maps and, in some cases, from model calculations for two and/or three-dimensional models.

4.1.2 Applications

- Search for concealed waste disposal sites containing magnetic materials,
- localization of concentrations of barrels suspected of containing hazardous waste in a landfill,
- determination of the parts of a landfill with a high proportion of building rubble,
- search for unexploded ordnance (UXO),
- investigation of the geology below planned landfills and industrial or mining sites in areas of igneous and metamorphic rocks, and
- delineation of faults, particularly in areas of igneous and metamorphic rocks.

4.1.3 Fundamentals

The International Association for Geomagnetism and Aeronomy in 1973 resolved that geomagnetic field measurements shall be reported in tesla, i.e., the SI units (International System of units) of magnetic flux density $\boldsymbol{B}$. One tesla (T) is equal to 10^9 nanotesla (nT). Although all magnetometers are now calibrated in units of magnetic flux density, the term "magnetic field" has been retained for the measured quantity. For this reason, the strength of the total magnetic field, the time-varying field, or the local anomaly field is always understood as magnetic flux density. In air, the relationship between magnetic flux density $\boldsymbol{B}$ and magnetic field strength $\boldsymbol{H}$ is approximated as follows: $\boldsymbol{B} \approx \mu_0 \boldsymbol{H}$. The most widely used magnetometers in applied geomagnetics measure the intensity F of the total field $\boldsymbol{F}$. ΔF is the intensity of the local anomaly field, where $\Delta F = F - F_0$. ΔF is a good approximation of the projection of $\boldsymbol{\Delta F}$ on the main field $\boldsymbol{F_0}$.

The local magnetic anomaly field is due to local and regional differences in the magnetization of rocks (*Table 4.1-1*) and other magnetic materials. A distinction is made between induced (i) and remanent (r) magnetization. These add vectorially: $\boldsymbol{M} = \boldsymbol{M}_\mathbf{i} + \boldsymbol{M}_\mathbf{r}$.

The induced magnetization $\boldsymbol{M}_\mathbf{i}$ is proportional to the magnetic field strength $\boldsymbol{H}$:

$$\boldsymbol{M}_\mathbf{i} = \kappa \boldsymbol{H} = \kappa \boldsymbol{B} / \mu_0 \tag{4.1.1}$$

$$\boldsymbol{B} = \mu_0 \mu \boldsymbol{H} = \mu_0 (1 + \kappa) \boldsymbol{H} = \mu_0 (\boldsymbol{H} + \boldsymbol{M}_\mathbf{i}) \tag{4.1.2}$$

In magnetically anisotropic material μ and κ are tensors. The magnetic susceptibility κ and the relative magnetic permeability μ are properties of magnetic materials (*Fig. 4.1-2*). The susceptibility is sometimes required in cgs units for model calculation programs: $\kappa_{\text{cgs}} \equiv (1/4\pi)\kappa_{\text{SI}}$.

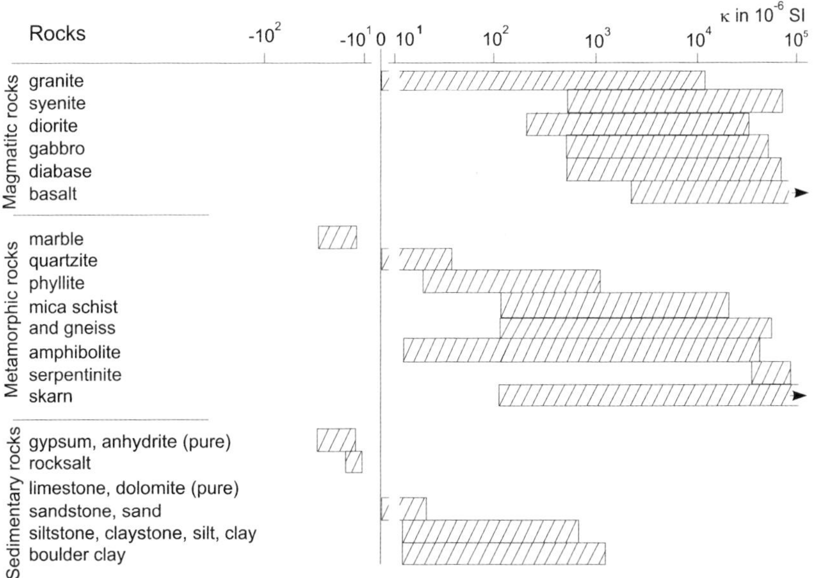

Fig. 4.1-2: Ranges of values for the magnetic susceptibility κ of different rocks, compiled by SCHÖN (1983)

According to BREINER (1973), the magnetic susceptibility of most iron and steel objects is between 10 and 130 (SI). Stainless steel is practically non-magnetic. The magnetic susceptibility of corroded scrap metal is 40 000 - 750 000 10^{-6} (SI).

Table 4.1-1: Magnetization (in A m-1) of the most important rock types, Bosum (1981)

	M_i	mean M_i	M_r	mean M_r
Ultrabasic igneous rocks				
pyroxenite	0.01 - 0.45	0.1 - 0.2		
serpentinite	0.5 - 10	0.75	0.2 - 0.8	0.6
Basic igneous rocks				
basalt	0.01 - 100	0.02 - 1.5	0.01 - 200	0.05 - 3
diabase	0 - 3.5	0.04 - 0.9	0 - 4	0.05 - 0.9
gabbro	0.05 - 3		0.2 - 30	
Intermediate and acidic igneous rocks				
diorite	0 - 1.5		0 - 0.003	
porphyry	0 - 0.3			
granite	0 - 2.2	0.03 - 0.5	0 - 9	
Metamorphic rocks				
amphibolite	0 - 1.3		0 - 1.6	
gneiss	0 - 1.8	0.01 - 0.75	0 - 2	0.01 - 1.4
Sedimentary rocks	0 - 0.4		0 - 0.4	
Magnetic ores				
magnetite	20 - 1000	20	30 - 1000	
pyrrhotite	0.3 - 50			2
hematite	0 - 0.9		0.1 - 0.25	
chromite	0.01 - 2		0.02 - 30	

Conventional methods yield an apparent susceptibility κ of the source body as a whole, which is related to the susceptibility κ_M of the magnetic substances in that body as follows: $\kappa = (p\ \kappa_M) / (1 + N\ \kappa_M)$, where p is the proportion of magnetized grains in the rock and N is a demagnetization factor. N is dependent on the shape of the grains: It equals 1/3 for homogeneous, magnetically isotropic spheres.

The magnetization of a rock or material that remains in the absence of an external field is called remanent magnetization, $\boldsymbol{M}_r$, which depends on the content of ferrimagnetic matter and on the formation conditions.

A sphere with homogeneous magnetization is the simplest model of a magnetic geological or artificial body. The magnetic field (i.e., magnetic flux density) of a homogeneously magnetized sphere is equal to the field of a dipole at the center of the sphere when the distance r between source and point of measurement is larger than the radius R and the magnetic moment $m = (4\pi / 3)\, R^3\, \boldsymbol{M} = V\boldsymbol{M}$, where V is the volume of the sphere. From this, it follows that for all spheres for which $R < r$ and which have the same $R^3 M$ value and the same center, the same magnetic anomaly is observed (equivalence principle of potential theory).

If (a) the center of a sphere with a radius R lies at a depth d ($d > R$), (b) I is the inclination[1] of the magnetic field at this site, (c) α is the angle between magnetic north and the x-axis, and (c) the origin ($x = y = z = 0$) of the geophysical coordinate system lies directly above the center of the sphere, then the anomaly of the total field δF can be estimated as follows (LINDNER & SCHEIBE, 1978):

$$\delta F(x) = \underbrace{\left\{\frac{\frac{4}{3}\pi R^3 \kappa F_0}{4\pi}\right\}}_{factor} \frac{3d^2 \sin^2 I + 3x^2 \cos^2 I \cos^2 \alpha - 3xd \sin^2 I \cos \alpha - x^2 - d^2}{(x^2 + d^2)^{5/2}} \qquad (4.1.3)$$

Taking the magnetic anomaly at the Earth's surface above the center of the sphere at $x = 0$ into consideration, it can be seen that the magnetic field directly above the source body decreases with the third power of the depth d. The local anomaly field depends not only on the size and magnetization of the source body but also to a large degree on its depth. Small near-surface source bodies in landfills cause a strongly varying local anomaly field. These local variations smooth rapidly with increasing height h above the surface (1 to 4 m).

For further formulas to calculating δF anomalies of model bodies with a simple geometry, see TELFORD, GELDART & SHERIFF (1990). *Figure 4.1-3* shows the typical appearance of the anomalies of the total field intensity for inclinations $I = 15°$ and $30°$ (e.g., South East Asia) and $I = 67.5°$ (e.g., Central Europe). These anomalies, which are due to induced magnetization, are superimposed by the anomaly field due to the remanent magnetization. The measured value ΔF is the sum of these two local anomalies.

The anomalies of the total magnetic field can be calculated using Equation (4.1.3.) To estimate the maximum value expected for the anomaly, set $x = 0$ and $I = 90°$ in Equation (4.1.3.) This yields

[1] The inclination is the angle between the vector of the main magnetic field and the horizontal.

$$\delta F_{\max} = \frac{\frac{3}{4}\pi R^3 2\kappa F_0}{4\pi d^3}. \tag{4.1.4}$$

Introducing the parameter "effective susceptibility" κ_{eff}, we can define the magnetic moment m_c as follows:

$$m_c = \frac{3}{4}\pi \kappa_{\text{eff}} R^3 F_0 . \tag{4.1.5}$$

For a spherical body at a depth *d*, the maximum value of the anomaly of the total field intensity is

$$\delta F_{\max} = \frac{m_c}{4\pi d^3}. \tag{4.1.6}$$

Mean values for the normal magnetic field in the Berlin area are $F_0 = 49\,000$ nT and $I = 67.5°$. Values of $\kappa = 100$ (SI) can be assumed for the magnetic susceptibility of iron and steel and 6600 - 8100 kg m^{-3} (mean 7800 kg m^3) for their density. *Figure 4.1-4* shows a nomogram for Equation (4.1.6). This nomogram shows that at the Earth's surface the local anomaly field of a model body containing 100 kg iron at a depth of 5 m has an maximum value of 50 nT.

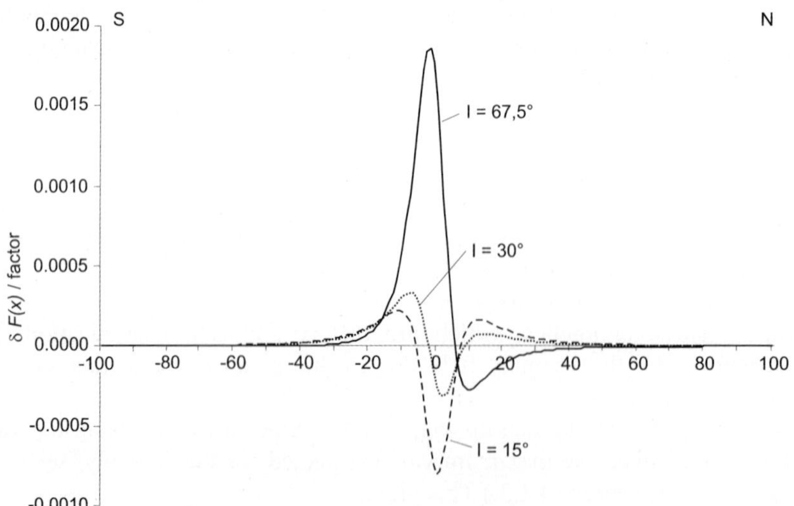

Fig. 4.1-3: Anomaly (induced part) of the total field intensity for a sphere with a radius *R* and a homogeneous magnetization, at $x = 0$, at a depth $d = 10$ m, for inclination $I = 15°$, 30°, 67.5° and $\alpha = 0$

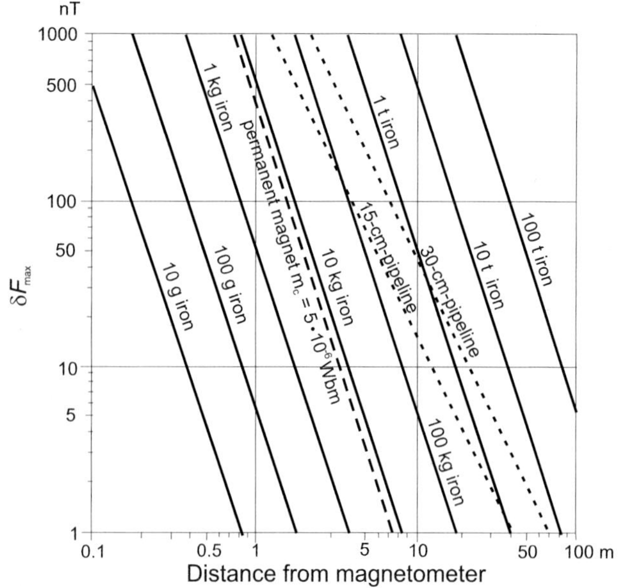

Fig. 4.1-4: Nomogram for estimating the maximum anomalies δF_{max} from iron objects with $m_c = 5 \cdot 4\pi \cdot 10^{-8}$ Wb m per kilogram after BREINER (1973); taken from MILITZER et al. (1986). Note that δF_{max} can change by up to a factor 5, depending on the hardness of the iron, the inclination, and the direction of the remanent magnetization vector

Table 4.1-2 can be used to estimate the magnetic anomaly of barrels before the measurements are carried out.

Table 4.1-2: Size and mass of steel containers for waste disposal and the maximum magnetic fields δF_{max} according to Equation (4.1.6) at a depth *d*. The magnetic moment $m_c = 5 \cdot 4\pi \cdot 10^{-8}$ Wb m per kilogram

		Local anomaly field δF_{max} in nT		
		d = 1 m	*d* = 3 m	*d* = 5 m
capacity outside diameter length sheet metal thickness mass (empty)	30 - 220 L 290 - 610 mm 440 - 950 mm 0.5 - 1.2 mm 2.6 - 22.6 kg	130 - 1130	5 - 42	1 - 9

4.1.4 Instruments

Geomagnetic fields are measured with fluxgate magnetometers, proton-precession magnetometers, and optically pumped magnetometers. Fluxgate magnetometers measure the components of the magnetic field (e.g., vertical intensity) and its gradients. Proton-precession magnetometers and optically pumped magnetometers measure total field intensity.

Fluxgate magnetometers utilize the nonlinearity of the magnetization curve $\boldsymbol{B}$ as a function of $\boldsymbol{H}$ of the high permeability magnetic material of the coil core. The core is magnetized to saturation by an ac field in order to measure the component of the magnetic field that is parallel to the core axis. Continuous measurement is possible. At a walking pace, 1 - 8 readings can be taken per profile meter. Thus, the high density of measurement points necessary for environmental studies can be obtained within an acceptable time. Fluxgate magnetometers measure 1 or 3 components of the geomagnetic field or they measure the vertical gradient of the vertical component. State-of-the-art fluxgate magnetometers have the following specifications:

sensitivity	0.1 nT
precision	±1 nT
dynamic range	±2000 or 20 000 nT
response time	20 ms
temperature range	-10 to +40 °C
temperature drift	0.1 nT/°C.

The most widely used magnetometers in applied magnetics are *proton-precession magnetometers*. They measure the precession frequency of protons in the geomagnetic field after a polarization field is switched off. The precession frequency is proportional to the field strength of the total magnetic field. The *Overhauser* effect (transfer of the spin moment of electrons to protons by electromagnetic waves with a frequency selected for a specific chemical compound) is used to overcome the disadvantage of discontinuous readings of the proton-precession magnetometers.

State-of-the-art proton-precession magnetometers have the following specifications:

sensitivity	0.1 nT
precision	±1 nT
dynamic range	18 000 to 110 000 nT
gradient tolerance	6000 nT/m
response time	0.5 to 3 s
temperature range	-40 to +55 °C.

Almost all proton-precession magnetometers have gradient probes for measuring the vertical gradient of the total field. The distance Δr between the sensors for the gradient measurement is 0.5 - 2 m. Δr must be less than 1/5 - 1/10 the distance r of the gradient probes from the source body (BREINER, 1973). This condition is certainly fulfilled when geological structures are being explored, but not always for investigations of landfills. Not all gradient probes measure the gradients simultaneously by both sensors. The measurement of the total fields by the two sensors 1 - 2 s apart can lead to incorrect gradient measurements in areas where the magnetic field changes rapidly or if there is movement of the probes, for instance, by the wind. Gradiometers have the following advantages: higher sensitivity with respect to near-surface targets and elimination of the fluctuating part of the geomagnetic field and regional trends.

Proton-precession magnetometers with a response time of 0.5 s make essentially continuous measurements possible if the magnetometer is carried at walking speed along a profile. Proton-precession magnetometers are also available as base stations for recording variation of the total field. The time between two readings can be chosen between 5 s and 60 min.

Optically pumped magnetometers (also called alkali vapor magnetometers or optical absorption magnetometers) are based on the Zeeman effect, the splitting of the spectral lines in the Earth's magnetic field. The distance between the split spectral lines is proportional to the magnetic field strength. The optical pumping technique is used instead of spectrometric methods. Rubidium, caesium or potassium vapor is used in optical absorption cells. Therefore, such magnetometers are referred to as rubidium or caesium, or potassium vapor magnetometers. The sensitivity of this type of magnetometer is about two orders of magnitude higher than that of proton-precession magnetometers. This sensitivity is needed only for special investigations. The more sensitive optically pumped magnetometers, with which measurements can be made more rapidly, are more expensive than proton-precession magnetometers.

All modern magnetometers give direct readings of the magnetic field values in nanoteslas. A total of 10 000 - 100 000 readings can be stored in a datalogger for subsequent downloading to a computer. Nowadays magnetometers integrated with differential GPS (Global Positioning Systems) are available, which makes it unnecessary to use a regular survey grid (TAMIR L KLAFF, 1999). Particularly for unexploded ordnance (UXO) investigations, the combined use of magnetic and electromagnetic sensors integrated with differential GPS can be useful (BARROW et al., 1996).

4.1.5 Survey Practice

The objectives must be known in detail for optimum planning and execution of a magnetic survey. Also of importance is knowledge of the local geology and the situation in the field (e.g., accessibility, benchmarks, sources of noise). The parameters and equipment for the survey are chosen on the basis of this information:

1. Planning of the survey grid:
   a) location and size of the study area (possibly on the basis of test measurements),
   b) line spacing,
   c) spacing of measurement points,
   d) orientation of the profiles,
   e) consideration of utility lines (water, gas, electricity, telephone),
   f) choice of an undisturbed base point and if necessary further reference and control points,
   g) minimum number of points for replicate measurements,
   h) sequence of measurements.
2. measured parameter (e.g., total field intensity and/or vertical gradient of the total field intensity, vertical gradient of the vertical component),
3. precision,
4. distance of the sensors from the Earth's surface,
5. use of a base station,
6. type of magnetometer.

The survey area should be large enough that the non-anomalous area surrounding the target area is also included. For landfills, the area surveyed must be larger than the planned landfill.

In the normal case, magnetic measurements are carried out along profiles. The profiles should be laid out perpendicular to the strike of the geological structures. If the strike is known, a line spacing of 2 to 5 times the distance between measurement points is often used. Grids with same spacing for profiles and measurement points lead to the most reliable results. If the magnetic field fluctuates irregularly and the line spacing is too large, the anomalies will indicate an erroneous strike direction. An example of this is given by HAHN et al. (1985) (pg. 126), who recommends that the spacing of profiles and measurement points be about one-third of the expected depth of the target(s). For investigations of landfills, the grid spacing should be 5 - 10 m. The spacing can be reduced for more detailed measurement of source bodies of interest and for exact determination of the boundaries of concealed landfills. The search for concealed barrels normally requires a grid spacing of less than 1 m. In such cases, quasi-continuous measurements with two readings per second or per meter are useful.

Measurement of the total field anomaly ΔF is often impossible near electric railways. However, simultaneous sensor readings for gradient measurements still produce useful results in this case. Iron water pipes disturb gradient measurements more strongly than ΔF measurements (LINDNER et al., 1984). Therefore, particularly in areas in which there are magnetic disturbances, it is recommended that the total field and gradient be measured simultaneously.

Since the local magnetic anomaly field decreases approximately with the third power of the distance of the probe from the target, the influence of small near-surface bodies decreases quickly with increasing height of the sensor above the Earth's surface. This effect is used to eliminate the anomalies resulting from these near-surface bodies and to smooth the measured local anomaly fields. Depending on the objective of the survey, the sensors are placed 0.3 - 4 m above the Earth's surface. To avoid interference fields ≥ 1 nT, a distance of at least 30 m should be kept from motor vehicles and 150 m from the steel towers of power lines (HAHN et al., 1985). See Section 4.1.9 for examples of parameter values for various survey objectives.

4.1.6 Processing and Interpretation of the Measured Data

The gradient values do not require any further processing after measurement (i.e., main field and variations are already eliminated). But the following corrections of the total field must be carried out to determine the local anomaly field ΔF:

Corrections for diurnal variations: Variations of the geomagnetic field are best measured at a base station within the survey area which automatically records the total field, for example, every 10 s. These values can be used to eliminate the time-varying field from all measurements. The time-varying field can also be largely eliminated by frequent control measurements (tie-line method).

Normal field and elevation corrections: The normal field can be described as a function of geographical longitude and latitude, altitude above sea level and time (e.g., January 1994 = Epoch 1994.1). Tables of the normal field or computer programs using the International Geomagnetic Reference Field (IGRF) are employed to calculate the normal field. Owing to irregular secular variations of the main field, the IGRF is updated every five years. Because the survey area for a site investigation is normally small, the differences in measurement elevation are small and the measurements are made over a short period of time, one normal-field value F_0 can be used for the entire survey area. A single base station in a neighboring undisturbed area is also sufficient.

Terrain correction is not necessary for the objectives discussed here. In special cases, a *drift correction* is necessary for fluxgate magnetometer measurements.

For qualitative investigations, such as the search for and delineation of abandoned industrial and landfill sites, a correction for diurnal variation or for normal field and elevation is not often done, because anomalies caused by abandoned landfills also are clearly visible in the total field.

The anomalies of the total field and/or of the vertical gradients are represented on a contour map and/or a set of profiles. Color or grey shading increase the readability of contour maps. Three-dimensional plots are normally suitable only for a general impression of the results. The delineation of underground structures is too inaccurate with pseudo 3-D maps. Trend removal, filtering, field continuation, second-derivative analysis, Fourier analysis, pole reduction, and reduction to the magnetic equator can be helpful for the interpretation of the data, making the underground structures more clearly visible on the anomaly maps. For details, see TELFORD et al. (1990).

Information about the subsurface can often be derived directly from isoline maps and profiles by qualitative interpretation by a geophysicist. Some guidelines to qualitative interpretation of magnetic profiles and maps are listed in *Table 4.1-3*.

The geometry, depth, magnetic susceptibility (and remanent magnetization) of the magnetic material (e.g., geological structures, wrecked cars or barrels, UXOs) can be estimated by quantitative interpretation. Magnetic effects of simple bodies are depicted, for example by TELFORD et al. (1990) and REYNOLDS (1997). These depictions are useful in order to get an understanding of possible causes of magnetic anomalies.

As rule of thumb, the half-width of the anomaly corresponds approximately to 1 to 3 times the depth of the center of the model body. The depth d of a model body can also be estimated from measurements of the total field anomaly ΔF and the vertical gradient dF/dz (BREINER, 1973). The following relation is valid at the maximum of the anomaly: $d = (-n\Delta F) / (dF/dz)$, where $n = 3$ for dipole sources, $n = 2$ for a horizontal cylinder, and $n = 1$ for a narrow vertical dike. For this equation, only an assumption about the type of model body is necessary and not about the magnetic susceptibility. Other simple methods for depth estimation of magnetically effective structures in the subsurface as Peters‘ Half-Slope method (PETERS, 1949) and Parasnis‘ method (PARASNIS, 1986) are described in REYNOLDS (1997).

Forward modeling and inversion programs for 2, 2.5, and 2.75-D magnetic and/or gravity models are current practice. Three-dimensional structures can be modeled in special cases. Model calculations can also take into account the remanent magnetization. Inversion of measured data cannot unambiguously yield the correct model, owing to the principle of equivalence. The number of possible models must be reduced on the basis of geological knowledge and/or by comparison with the results of other methods. In any case, it is essential that the field data used for modeling is adequate with respect to data density and quality.

Table 4.1-3: Guidelines to qualitative interpretation of magnetic profiles and maps, modified after REYNOLDS (1997). © John Wiley and Sons Ltd. Reproduced with permission.

Taken into consideration	Observation	Indication for
segments of a profile and areas of maps	magnetically quiet field	near surface rocks with a low magnetic susceptibility κ
	magnetically noisy field	near surface rocks with a moderate to high magnetic susceptibility κ
anomaly	Wavelength	short => near-surface feature
		long => deep-seated feature
	positive or negative amplitude	intensity of magnetization
profiles and maps	anomaly structure* and shape	dip and dip direction
		induced magnetization if the anomaly shows a minimum to the north and a maximum to the south in the northern hemisphere and vice versa in the southern hemisphere; if this is not the case, it implies significant remanent magnetization present
profiles and maps	magnetic gradient	possible contrast in $\kappa$and/or magnetization direction
maps	linearity in anomaly	possible strike of magnetic feature
	dislocation of contours	lateral offset by fault
	broadening of contour interval	downthrow of magnetic rocks

*i.e., positive peak only, negative peak only or doublet positive and negative peaks

Recent developments in interpretation of magnetic data are target characterization algorithms designed for determination of an optimum source location of geologic structures as well as location and apparent size of buried ferrometallic objects, such as drums, pipes and UXOs (REID et al., 1990; YAGHOOBIAN et al., 1993; REYNOLDS, 1997). The most common technique is the Euler deconvolution. This method is both a geometry finder and a depth estimator. Euler's homogeneity equation relates the magnetic field and its gradient components to the location of magnetic sources with the degree of homogeneity expressed as a structural index. The structural index is a measure of the rate of change with the distance of the field from the source and is directly related to the source dimension. The advantages of the Euler deconvolution are that it can be applied to large gridded data sets, no particular geological model is necessary, and Euler's equation is insensitive to magnetic inclination, declination and remanence. Geologic constraints are imposed by choice of the structural index. In addition to the Euler deconvolution, a 3-D analytic signal is used for magnetic interpretation (ROEST et al., 1992; REYNOLDS, 1997). The analytic signal can be used to estimate magnetic contrast and approximate depth, while the Euler deconvolution can give a more detailed structural interpretation at depth. The absolute value of the analytic signal is defined as the square root of the squared sum of the two horizontal and the vertical derivatives of the total magnetic field. Maxima of the analytic signal indicate locations of magnetic contrasts. The depth of a magnetic source can be estimated from the shape of the analytic signal.

The objective of a magnetic survey for investigating a site must be a map which summarizes the results for geophysicists as well as for nonspecialists in a clear and understandable form.

4.1.7 Quality Assurance

The general aspects of quality assurance are given under this term in the glossary. The following measures are important for quality assurance in a magnetic survey:

- Check that the survey personnel are not carrying any magnetic material and that the sensor is not dirty.
- Check direction dependence of the readings ("heading effects"), i.e., with the sensor always held by the technician on the same side of his/her body, making readings in all four cardinal directions.
- Check for measurement reproducibility and noise: Equipment check, e.g., for low battery voltage, loose sensor cables, low sensor fluid level, exceedance of gradient tolerance, and sources of noise (from electric railways, power lines, etc.).
- Record observations in the area, e. g., sources of disturbance (steel towers for power lines and iron fences).

- Plot the readings at the end of a field day so that errors and disturbances can be recognized and, if necessary, measurements can be repeated.
- Take replicate measurements: Approximately 2 - 5 % of the readings should be repeated.
- Recognized errors must be removed by repeating the measurements. Obscuring errors by "data processing" is inadmissible.

4.1.8 Personnel, Equipment, Time Needed

	Personnel	Equipment	Time needed
mobilisation and demobilisation	depends on the distance to the survey area		
topographic survey	s. Chapter 2.5		
measurements	1 geophysical technician (+1 assistant)	1 4WD vehicle, 1 magnetometer or gradiometer, (1 base station), 1 PC	a = 1 - 5 m L = 800 - 2500 N/d a = 10 m L = 600 - 900 N/d quasi continuous measurement of the gradient with 1 - 8 readings/s L = up to 15 000 N/d
data processing, interpretation, reporting	1 geophysicist (+ 1 assistant)	1 PC with plotter, printer, and software	1 - 2 days are necessary for each day in the field

a = grid spacing, N = number of readings, d = 10-hour work day, L = N/d

4.1.9 Examples

Search for and delineation of an concealed landfill

A pit about 8 m deep, dug to obtain material for a freeway, was filled with waste until 1979. The landfill was then covered with a layer of soil 1 m thick. Today, the landfill is not visible in the field. The abandoned landfill is in a ca. 50 m thick layer of Triassic sandstone. The magnetic method was used to determine the exact location of the landfill. Survey parameters:

size of the study area	400 × 400 m
line spacing	5 m
spacing of the measurement points	5 m
number of measurement points	6581
duration of the survey	9 days
personnel	1 geophysical technician
physical parameters	total field intensity and vertical gradient
distance of the sensors from the ground surface	2.0 and 1.3 m
equipment	G 856-AX
base station magnetometer	G 856 (sampling rate 10 s).

The concealed landfill stands out clearly in the anomalies of both total field intensity ΔF (*Fig. 4.1-5*) and vertical gradient (*Fig. 4.1-6*). The delineation of the landfill area is clearest in the isoanomaly plot of the vertical gradient (*Fig 4.1-6*). Calculations based on the ΔF values yielded depths of less than 5 m for the magnetic materials and susceptibility values indicative of iron. There is a high-voltage power line along the 400E grid line with a steel tower at 400E, 520N. The vertical gradients are more strongly disturbed by the power line than the ΔF isoanomalies. The anomalies in the eastern part of the survey area are caused by building rubble containing iron.

Figure 5.4-25 shows the modeling of the magnetic anomaly related to the Nong Harn landfill near Chiang Mai, Thailand.

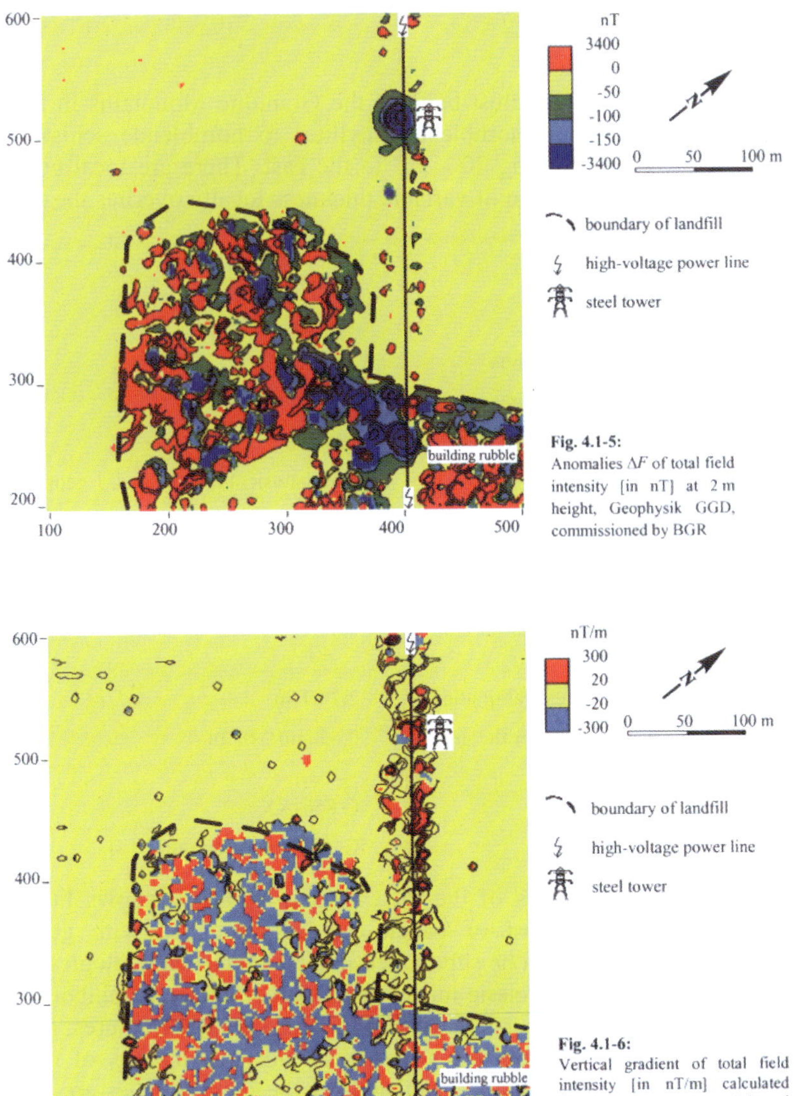

Fig. 4.1-5: Anomalies ΔF of total field intensity [in nT] at 2 m height, Geophysik GGD, commissioned by BGR

Fig. 4.1-6: Vertical gradient of total field intensity [in nT/m] calculated from measurements at 2.0 and 1.3 m height, Geophysik GGD, commissioned by BGR

Lithology and tectonic structure below a landfill in an area of igneous and metamorphic rocks

The site is in the "outer schist belt" of the Granulite Mountains in Saxony, Germany. The phyllites, hornblende phyllites to hornblende schists strike approximately N - S dipping 20 - 50° to the east. There are local bodies of carbonate rocks. Loess loam of variable thickness locally overlies these rocks. Survey parameters:

size of the study area	1400 m × 2000 m
line spacing	10 m
spacing of measurement points	10 m
number of measurement points	26 854
duration of the survey	33 days
personnel	2 geophysical technicians, 1 geophysicist
number of measurements at repetition points and check points	447
physical parameters	total field intensity and vertical gradient
standard error in the field ΔF at 2.05 m above ground level	2.3 nT
standard error for the vertical gradient	2.2 nT/m
distance of the sensors from the ground surface	0.65 and 2.05 m
equipment	G 856-AX
base station magnetometer	G 856 (sampling rate 10 s)

The map of the anomalies of the total field *(Fig. 4.1-7)* shows linear and concentric structures, most of which can be correlated with geological sources. These sources can be subdivided into (i) concealed geological bodies (probably consisting of gneissic mica schists, serpentinites, amphibolites, and phyllites) that have no clear relationship to near-surface structures, (ii) near-surface rocks and magnetic intrusive rocks, and (iii) outcrops of magnetic material. The parameter values calculated for the model bodies are given in *Fig. 4.1-8*. Furthermore, expensive studies of the geological barrier can be targeted more precisely, reducing the cost of such investigations.

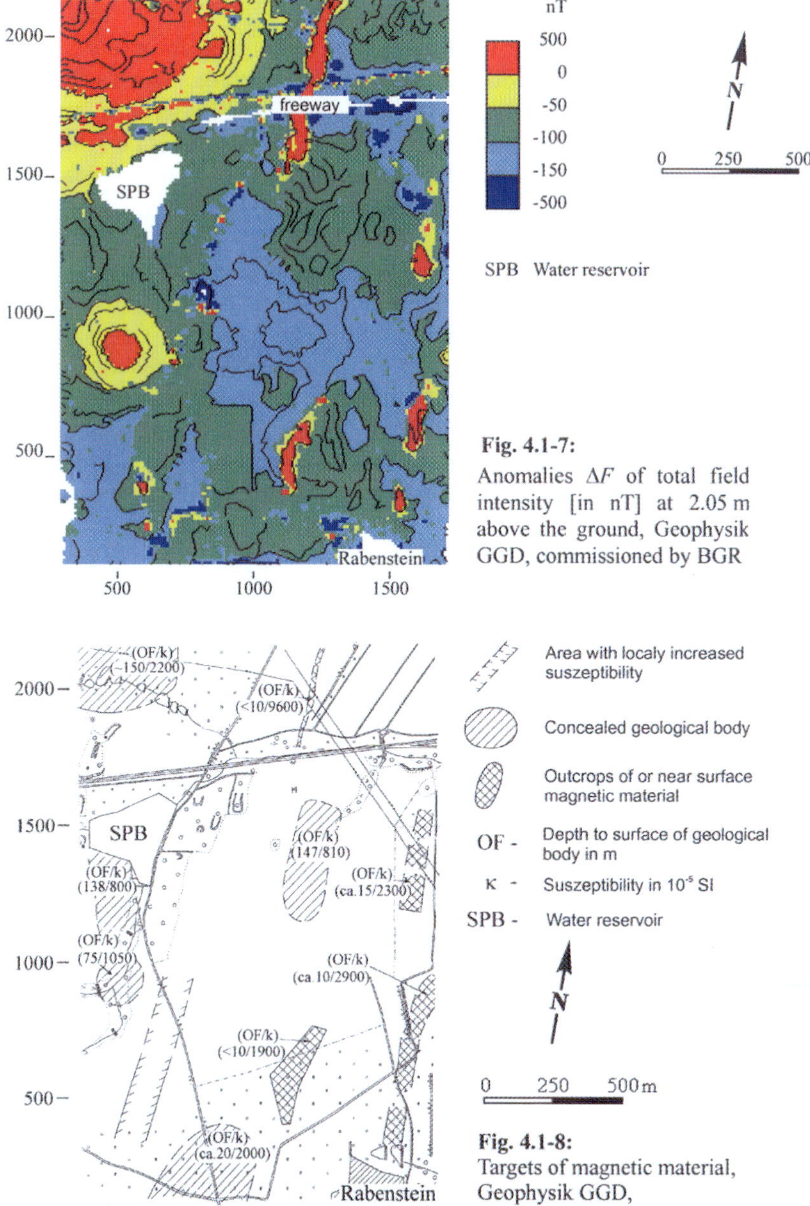

Fig. 4.1-7:
Anomalies ΔF of total field intensity [in nT] at 2.05 m above the ground, Geophysik GGD, commissioned by BGR

Fig. 4.1-8:
Targets of magnetic material, Geophysik GGD, commissioned by BGR

Search for unexploded ordnance (UXO)

Magnetic and electromagnetic techniques have been used for many years to detect unexploded ordnance concealed in the subsurface. The use of both methods for the same investigation site leads to more reliable and less ambiguous results. In the past "mag and flag" methods, i.e., magnetic measurements without data collection and indications flagged for excavation, were commonly used. Nowadays, surveys with data collection, sophisticated processing and interpretation are state-of-the-art.

On the UXO test field of the Berlin police department, different types of UXO and other targets, such as steel girders, a steel ball, and metal containers were buried. Measurements with a magnetic gradiometer *(Fig. 4.1-9)* and with a transient electromagnetic probe *(Fig. 4.1-10)* reveal the concealed targets. In addition to the location and depth, the weight of the targets was estimated. In this example, depth estimations on the basis of magnetic anomalies are more reliable than those from EM 61 data. Due to remanent magnetization, the estimated mass is often too small. Survey parameters:

size of the study area	25 m × 50 m	25 m × 50 m
line spacing	0.50 m	0.50 m
spacing of measurement points	0.25 m	0.195 m
number of measurement points	10 250	13 100
duration of the survey	1 day	1 day
personnel	1 technician, 1 geophysicist	1 technician, 1 geophysicist
parameters	vertical gradient of vertical component of the magnetic field	electromagnetic response
equipment	FM 36	EM 61

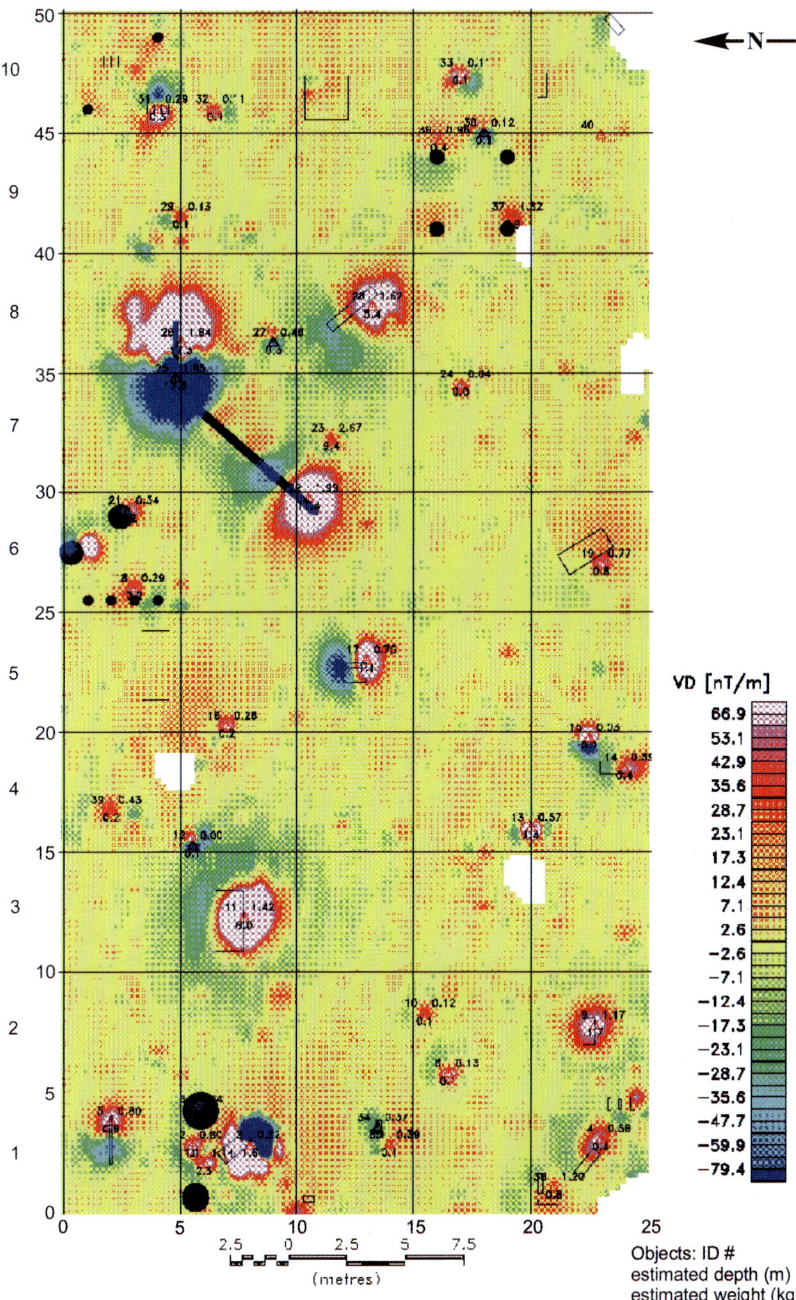

Fig. 4.1-9: UXO test field of the Berlin police department, vertical gradient of the vertical component of the magnetic field [in nT/m], including interpretation, courtesy of Büro für Geophysik Lorenz, Berlin by permission

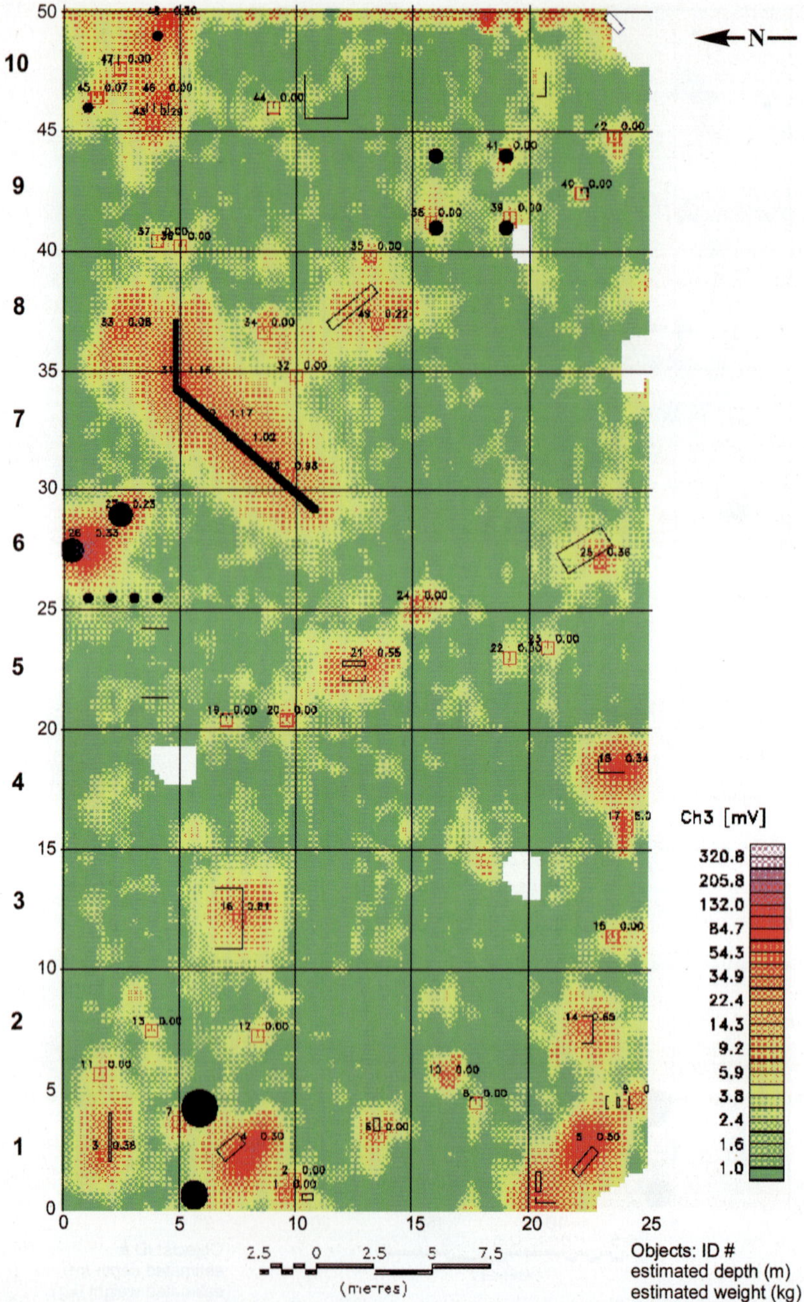

Fig. 4.1-10: UXO test field of the Berlin police department, electromagnetic response of EM 61 [in mV], including interpretation, courtesy of Büro für Geophysik Lorenz, Berlin by permission

Parameters and units

	Symbol	Units
magnetic field strength	$\boldsymbol{H}$	A m^{-1}
magnetic flux density or magnetic induction	$\boldsymbol{B}$	tesla (T) = V s m^{-2} 1 nT = 10^{-9} T 1 T = 1 Wb m^{-2}
magnetic permeability in a vacuum	μ_0	$\mu_0 = 4\pi\ 10^{-7} \cdot$ V s $A^{-1}m^{-1}$
relative magnetic permeability	μ	dimensionless
magnetic susceptibility	κ	dimensionless
total magnetic field	$\boldsymbol{F}$	nT
main field (also called normal field, reference field)	$\boldsymbol{F_0}$	nT
local anomaly field	$\boldsymbol{\Delta F}$	nT
inclination	I	degrees
magnetization	$\boldsymbol{M}$	A m^{-1}
magnetic dipole moment	$\boldsymbol{m}$	A m^2
magnetic moment after Coulomb	m_c	Wb m

References and further reading

BARROW, B., KHADR, N., DIMARCO, R. & NELSON, H. H (1996): The Combined Use of Magnetic and Eletromagnetic Sensors for Detection and Characterization of UXO. SAGEEP 1996, 469-477.

BOSUM, W. (1981): Anlage und Interpretation aeromagnetischer Vermessungen im Rahmen der Erzprospektion. Geol. Jb., **E 20**, 3-63.

BREINER, S. (1973): Applications manual for portable magnetometers. GeoMetrics, Sunnyvale, CA.

HAHN, A., PETERSEN, N. & SOFFEL, H. (1985): Geomagnetik, In: BENDER, F. (Ed.): Angewandte Geowissenschaften, **II**: Methoden der Angewandten Geophysik und mathematische Verfahren in den Geowissenschaften, Enke, Stuttgart, 57-155.

LINDNER, H. & SCHEIBE, R. (1978): Die Berechnung von δg- und δT-Anomalien für regelmäßige homogene Störkörper. Gerlands Beitr. Geophys., **87,** 29-45.

LINDNER, H., MAURITSCH, H., MILITZER, H., RÖSLER, R., SCHEIBE, R., SEIBERL, W., WALACH, G. & WEBER, F. (1984), In: MILITZER, H. & WEBER, F. (Eds.): Angewandte Geophysik, **1**: Gravimetrie und Magnetik, Springer, Vienna, and Akademie-Verlag Berlin.

MILITZER, H. & SCHEIBE, R. (1981): Grundlagen der angewandten Geomagnetik. Freiberger Forschungsheft C 352, VEB Dt. Verlag für Grundstoffindustrie, Leipzig.

MILITZER, H., SCHÖN, J. & STÖTZNER, U. (1986): Angewandte Geophysik im Ingenieur- und Bergbau, 2. Aufl., Enke, Stuttgart.

PARASNIS, D. S. (1986): Principles of Applied Geophysics, 4th ed., Chapman & Hall, London.

PETERS, L. J. (1949): The direct approach to magnetic interpretation and its practical application. Geophysics, **14**, 290-320.

REID, A. B., ALLSOP, J. M., GRANSER, H., MILLETT, A. J. & SOMERTON, I. W. (1990): Magnetic interpretation in three dimensions using Euler deconvolution. Geophysics, **55**, 80-91.

REYNOLDS, J. M. (1997): An Introduction to Applied and Environmental Geophysics. John Wiley & Sons, Ltd., Chichester.

ROEST, W. R., VERHOEF, J. & PILKINGTON, M. (1992): Magnetic interpretation using the 3-D analytic signal. Geophysics, **57**, 116-125.

SCHÖN, J. (1983): Petrophysik, Enke, Stuttgart.

TAMIR L KLAFF, P. G. (1999): The Application of an Integrated Magnetometer/Global Positioning System for an Unexploded Ordnance Investigation. SAGEEP, 793-801.

TELFORD, W. M., GELDART, L. P. & SHERIFF, R. E. (1990): Applied Geophysics, 2nd ed., Cambridge University Press, Cambridge, 62-135.

YAGHOOBIAN, A., BOUSTEAD, G. A. & DOBUSH, T. M. (1993): Object Delineation Using Euler's Homogeneity Equation, Location and Depth Determination of Buried Ferro-Metallic Bodies. SAGEEP, 613-632.

4.2 Gravity Methods

KNUT SEIDEL & HARALD LINDNER

4.2.1 Principle of the Methods

Gravity is defined as the force of mutual attraction between two bodies, which is a function of their masses and the distance between them, and is described by Newton's law of universal gravitation. An effect of gravity is observed when the fruit from a tree falls to the ground. The gravity field at each location on Earth consists of a global field which is superimposed by a local anomaly field. In a gravity survey, measurements are made of the local gravity field differences due to density variations in the subsurface. The effects of small-scale masses are very small compared with the effects of the global part of the Earth's gravity field (often on the order of 1 part in 10^6 to 10^7).

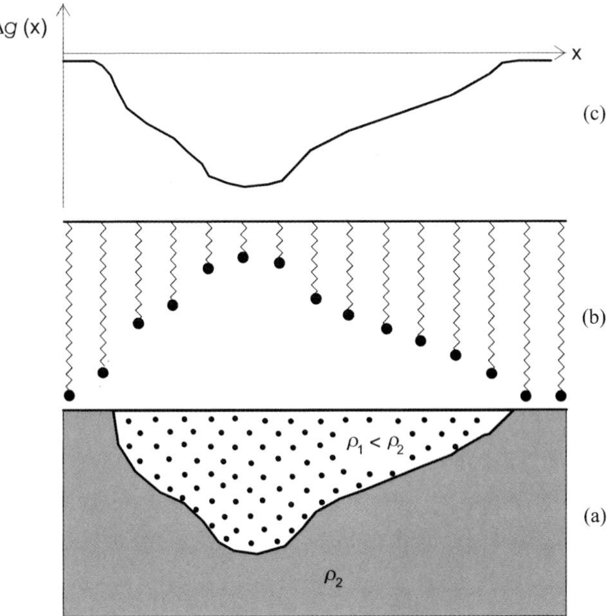

Fig. 4.2-1 a-c: Principle of a gravity measurement: (a) the model shows a geological structure with a density ρ_1 embedded in material with a higher density ρ_2, (b) the spring with a small mass at the end of it changes length with changes in the gravity field, (c) the measurement results are plotted to document the gravity anomaly $\Delta g(x)$

Highly sensitive gravimeters are necessary for measuring such variations in gravitational attraction accurately. Special data processing and interpretation techniques (Section 4.2.6) are used to interpret the shape and amplitude of the anomalies in terms of subsurface geological or anthropogenic structures. Gravity measurements can be performed on land, at sea, and in the air. For environmental problems, land measurements are generally made. A necessary condition for the application of this method is the existence of density contrasts. The schematic in *Fig. 4.2-1* shows a structure with a density ρ_1 embedded in material with a higher density ρ_2. This could be a channel in a boulder clay layer filled with sand or gravel or a pit filled with waste. Because of the negative density contrast ($\rho_1 - \rho_2 < 0$), the resulting gravity anomaly Δg is negative too. Considering a gravimeter as basically a mass on a spring, the amplitude of the gravity anomaly is a function of the expansion or contraction of the spring, the geological situation in *Fig. 4.2-1a* will cause the length of the spring to decrease above the anomalous structure.

4.2.2 Applications

- Structural investigation of landfill, industrial or mining sites and their surroundings,
- obtaining structure and thickness of unconsolidated sediments,
- lithological subdivision of the subsurface, particularly in areas of unconsolidated rocks,
- detection of lithological and structural changes as well as fractures in consolidated rock,
- detection of cavities,
- locating concealed waste dumps,
- estimation of the thickness and/or mean density of waste deposits,
- detection of density inhomogeneities inside a waste dump (a rare application), and
- gravity data are also used to provide constraints in the interpretation of seismic data.

4.2.3 Fundamentals

Each point *P* on, above and below the Earth's surface is affected by the gravitational (or gravity) field *g*(*P*). This is a natural potential field like, for example, the magnetic field. The gravity field is measured in units of acceleration [m s^{-2}] [1]. This field is mainly caused by the attraction between masses of the Earth and an arbitrary mass at point P with the coordinates x, y, z. The gravitational force F between two point masses m and m' with the distance r between them is:

$$F(P) = \mathrm{G}\frac{m m'}{r^2} \qquad (4.2.1)$$

where the gravitational constant G = 6.672 · 10^{-11} m^3 kg^{-1} s^{-2}. The effect of the gravitational field g of m on m' is derived from Equation (4.2.1):

$$g_{\mathrm{E}}(P) = \mathrm{G}\frac{m}{r^2}. \qquad (4.2.2)$$

This effect is called gravitational acceleration. The mass m of a body is given by the product of its density ρ and volume V. In addition to the gravitational acceleration of the Earth $g_{\mathrm{E}}(P)$, point P is affected by centrifugal acceleration $g_{\mathrm{C}}(P)$ and by the gravitational attractions of mainly the moon and the sun $g_{\mathrm{T}}(P)$, the variations of which are called tides. These tides vary with respect to place and time. The centrifugal acceleration $g_{\mathrm{C}}(P)$ is due to the rotation of the Earth and depends on the latitude φ of the point P. The gravitational field at point P is:

$$g(P) = g_{\mathrm{E}}(P) + g_{\mathrm{C}}(P) + g_{\mathrm{T}}(P). \qquad (4.2.3)$$

This formula describes the global gravity field. The calculation of the individual components of the global field is described in Section 4.2.6. The anomalous gravity field Δg caused by density inhomogeneities of geological or anthropogenic structures is superimposed on the global field. Equation (4.2.2) can be used to calculate Δg if the mass m is replaced by the anomalous mass $\Delta m = \Delta\rho\, V$, where $\Delta\rho$ is the density contrast between the inhomogeneity and the surrounding material, V is the volume of the inhomogeneity. Consequently, the gravity value $g(P)$ at point P is calculated from the theoretical gravity field value γ_0, the tidal effect and the anomalous field:

$$g(P) = \gamma_0(P) + g_{\mathrm{T}}(P) + \Delta g(P). \qquad (4.2.4)$$

For geophysical surveys, only gravity anomalies $\Delta g(P)$ related to density inhomogeneities in the subsurface are of interest.

---

1 In the SI system, the unit of gravity acceleration is 1 µm s^{-2} (= 10^{-6} m s^{-2}), called a gravity unit (g.u.), but in practice the older unit 1 mGal (= 1 milliGal = 10^3 µGal = 10 µm s^{-2} = 10^{-5} m s^{-2}) is often used. Thus 0.1 mGal = 1 g.u.

Petrophysical basis

Gravity measurements are only useful if the density of the target is significantly different from the density of the surrounding material. Therefore, it is important to have an idea of the typical density values of the materials in the area to be investigated. The density values of selected materials, including typical waste, are given in *Table 4.2-1.*

Table 4.2-1: Typical densities of selected sedimentary, metamorphic and igneous rocks, and waste, ∅ = mean value

		Density in g cm^{-3} (10^3 kg m^{-3})																		
	∅	**1.0**	**1.1**	**1.2**	**1.3**	**1.4**	**1.5**	**1.6**	**1.7**	**1.8**	**1.9**	**2.0**	**2.1**	**2.2**	**2.3**	**2.4**	**2.5**	**2.6**	**2.7**	**2.8**
Sediments																				
fine sand, dry	1.5				■	■	■	■	■											
fine sand, wet	1.8								■	■	■	■								
medium sand	1.7						■	■	■	■	■	■								
gravel, dry	1.7						■	■	■	■	■	■								
gravel, wet	2.0						■	■	■	■	■	■	■	■	■					
silt	1.7			■	■	■	■	■	■	■										
clay, dry	1.7				■	■	■	■	■	■										
clay, wet	1.9								■	■	■	■	■	■						
limestone	2.55										■	■	■	■	■	■	■	■	■	■
sandstone	2.35							■	■	■	■	■	■	■	■	■	■	■	■	
rock salt	2.2												■	■	■	■	■	■		
Metamorphic and igneous rocks																				
granite	2.65																■	■	■	
granodiorite	2.7																	■	■	
basalt	3.1																			⇒
gneiss	2.7																	■	■	■
Waste																				
organic	1.25			■	■															
high ash content	1.3				■															
domestic, mixed	1.5					■	■	■												
high rubble content	1.8									■										

The density of waste material is almost always less than that of the surrounding rock, especially consolidated rock. Thus, there is a negative gravity anomaly, sometimes called a gravity minimum or gravity low, above a landfill. If no density data are available in the investigation area, a few cone penetration tests (in unconsolidated material) or geophysical borehole logs (e.g., gamma-gamma logs, sometimes called density logs) can help to obtain basic data.

Gravity anomalies of simple structures

Calculation of the gravity anomalies of simple structures is useful in different phases of a gravity survey. In the project planning phase this can help to decide whether gravity data would be useful and which anomalies are to be expected. Based on this knowledge, the parameters and equipment for the survey can be selected.

In the interpretation phase, the measured gravity anomalies are first interpreted qualitatively, but the main aim of the interpretation is to obtain information about the size and depth of the structures of interest. To obtain such a quantitative interpretation, it is necessary to calculate the theoretical gravity anomaly that would be caused by a geological structure and to compare the result with the measured data. Geological structures usually have a complicated geometry and their effect on the gravity field cannot be sufficiently described by Equation (4.2.2). Therefore, Equation (4.2.2) is replaced by the integral expression (4.2.5).

For all points $P(x, y, z)$ on the Earth's surface ($z = 0$) the gravity anomaly caused by an anomalous body with a density difference $\Delta\rho$ to the host material is:

$$\Delta g(x, y, 0) = \mathrm{G}\,\Delta\rho \int_V \frac{z'}{r^3} dx'dy'dz' \qquad (4.2.5)$$

where $r^2 = (x - x')^2 + (y - y')^2 + z'^2$; x, y, z are the coordinates of the point of measurement and x', y', z' are the source points of the body. The gravity anomalies of several structures (forward modeling) can be calculated by using different integration paths in Equation (4.2.5) corresponding to the boundaries of the source body. In general, the modeling can be done for 2-D or 3-D structures:

Two-dimensional structures:	The length L of the source body (y'-direction) is quasi-infinite, i.e., much larger than its width W in the x' and z'-direction ($L \geq 4W$), permitting integration only across its cross-section F in the x, z-plane.
Three-dimensional structures:	The source body is finite in the x', y' and z'-directions; integration is over the volume V of the body as shown in Equation (4.2.5).

Simple examples of both types of structures are a horizontal cylinder and a sphere. These shapes can easily be used for a rough calculation of the gravity anomalies of similar geological structures. Assuming a horizontal cylinder and a sphere with a radius R and the center of the structure at a depth d ($d > R$) on the z-axis and the density contrast $\Delta\rho$ with respect to the host material, then the gravity anomalies of these bodies at the Earth's surface $z = 0$ can be calculated as follows (LINDNER & SCHEIBE, 1978):

$$\Delta g(x)_{\text{cylinder}} = 2\pi R^2 \mathrm{G}\,\Delta\rho \frac{d}{x^2 + d^2} \tag{4.2.6}$$

and

$$\Delta g(x)_{\text{sphere}} = \frac{4\pi}{3} R^3 \mathrm{G}\,\Delta\rho \frac{d}{(x^2 + d^2)^{3/2}}. \tag{4.2.7}$$

According to (4.2.6) and (4.2.7) the maximum or minimum of the gravity anomaly is at $x = 0$, above the center of a homogeneous body.

Figure 4.2-2 shows the anomalies of horizontal cylinders of same size at different depths. As can be seen in *Fig. 4.2-2* the anomaly amplitude decreases with increasing depth. On the other hand, the width of the anomaly increases with increasing depth.

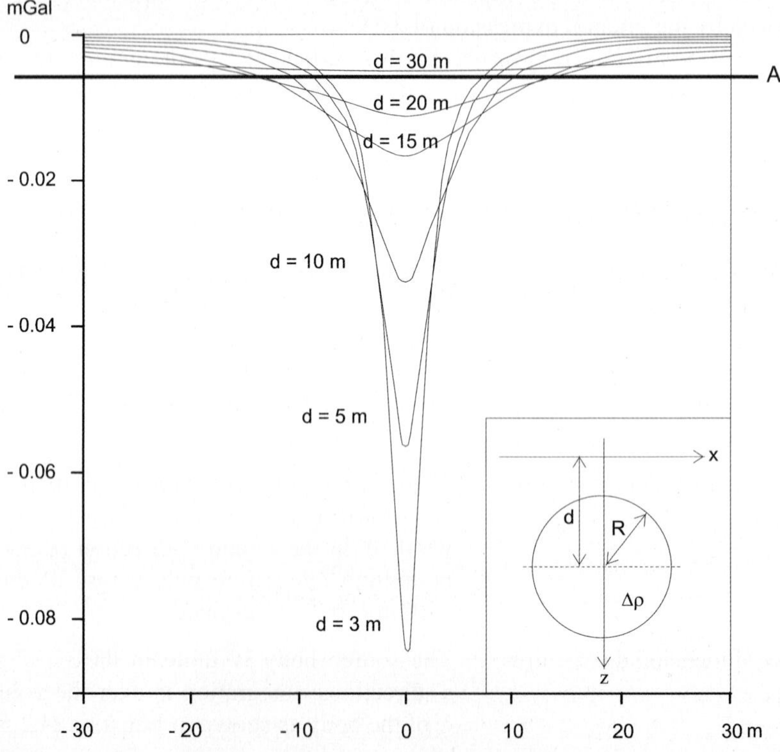

Fig. 4.2-2: Gravity anomaly of a horizontal cylinder calculated for different depths d, density contrast between the cylinder and its surroundings $\Delta\rho = -1.0 \times 10^3$ kg m^{-3}, $R = 2$ m, center of the cylinder at $x = 0$ m, A – detection limit of the gravity measurements as defined by the precision

A collection of formulas for other simple models is given by TELFORD et al. (1990), and LINDNER & SCHEIBE (1978). Such simple models can be used only for a rough interpretation of complicated geological structures. For more than a rough interpretation, software has to be used for two-dimensional and/or three-dimensional modeling (GÖTZE & LAHMEYER, 1988; BOULANGER & CHOTEAU, 2001). The latter is rather expensive and requires considerable experience in gravity modeling (Section 4.2.6).

Ambiguity of the results

The gravity field is a potential field. Therefore, it is not possible to obtain unequivocal information about depth, extent and/or density from gravity measurements alone (owing to the principle of equivalence). This ambiguity can be reduced or even be removed by inclusion of the following additional information when the gravity data is interpreted:

- all available geological information about the survey area,
- reliable density data of the materials expected,
- inclusion of the results of other geophysical methods (e.g., seismic and geoelectric methods).

Thus, the most useful results of gravity methods are obtained within the scope of an integrated geophysical survey (SILVA et al., 2002).

4.2.4 Instruments

The gravity field is measured by a gravimeter. Gravimeters used in a gravity survey can only measure the difference in gravity between two observation points, not the absolute values. Therefore, relative gravity values are measured in a survey, i.e., the measured data at single stations is related to a reference station with a known gravity value. The results of the local survey are linked to a larger gravity network in this way. If there is no reference station available in the investigation area, one station can be defined as reference station with an arbitrarily chosen value. The absolute values are generally not important for environmental surveys.

All instruments used in field geophysics employ some type of static spring-mass system, i.e., the gravitational force on a mass in the gravimeter is balanced by a spring, and the change in length of the spring is measured. Most gravimeters are of the astatic type to increase the precision. In this type of gravimeter, the amount necessary to balance the mass to zero is determined. There are two main spring systems: The first is based on a concept developed by S. P. Worden in 1947 and uses quartz springs (e.g., the Sodin and Sharpe gravimeters); the second system, developed by LaCoste & Romberg, uses metal springs. Most instruments house their systems in a vacuum and have thermostats to control temperature drift, keeping the drift to a minimum and increasing the precision (CHAPIN et al., 1999; CARBONE & RYMER, 1999).

Gravimeters with manual adjustment of the length of the spring and automated gravimeters with an electronic feedback system are available. The instruments with manual compensation require experienced operators and do not allow digital data storage. But, these instruments are less expensive than automated gravimeters, which use electrostatic systems to adjust the length of the spring. The measured gravity values, time, instrument tilt, drift, etc. are stored in RAM and can be transferred to a PC.

State-of-the-art automated gravimeters (e.g., CG-5, Scintrex Ltd.; Graviton EG, LaCoste & Romberg) have the following specifications:

sensitivity	0.001 to 0.005 mGal
precision in the field	± 0.005 to 0.01 mGal
dynamic range without resetting	8000 mGal
measuring time per station	2 - 3 minutes (in areas with low seismic noise)
temperature range	-40° to + 45° or 0 to 55°C

Instruments with manual adjustment (e.g., Sodin 410, W. Sodin Ltd.; LCR G or D from LaCoste & Romberg) have a precision of 0.005 to 0.010 mGal. This is sufficient for most survey purposes, but for the detection of cavities, instruments with a higher precision should preferably be used.

4.2.5 Survey Practice

Planning of a survey

Planning of a survey is an important step in exploration because the selection of appropriate survey parameters strongly influences the quality of the results. The following survey parameters are to be determined:

1. *Size of the survey area*: The survey area should be larger than the target (e.g., a covered dump site) in order to record the entire gravity anomaly, which is larger than the target itself, and to allow the definition of the background, or regional field.

2. *Selection of profile and station spacing*: If possible, the measurements should be done at almost equidistant grid points. If the measurements can only be made along profiles, the distance between the profiles should not be more than 2 - 4times the distance between stations along a profile. The station spacing depends on the information needed and on the depth and

size of the expected structure. For the detection of cavities or local density inhomogeneities the expected anomaly should be covered at least by five stations. The magnitude and areal extent of the expected anomalies can be calculated using the simple models mentioned in Section 4.2.3 or by sophisticated model calculations.

3. *Selection of a location for a reference station*: Because the gravity value depends on the elevation, the gravity value and station elevation must be measured for each field station. At least one station must serve as reference for both measurements. This station should be in a seismically quiet area and it should not be possible for its value to change during the survey. In large survey areas, it is useful to establish several reference stations.

Determination of coordinates and elevation

Determination of the coordinates and elevation of a gravity station is an indispensible part of a gravity survey. The values for all stations are to be related to the reference station. The measurements can be done by leveling, by electro-optical tachymetry, or by GPS (Global Positioning System). The advantage of the latter two is that the coordinates of the station can be determined simultaneously with the elevation.

In the case of levelling, the position of the stations has to be determined using a tape measure. For small areas and small station spacings (≤ 25 m), a regular grid can be advantageous. For larger station spacings, the locations of the stations are drawn on a good quality topographic map and the coordinates are determined by digitization. The selection of the appropriate and most efficient method depends on the size and the topographic conditions of the survey area. In any case, the elevation of a station should be determined with a precision better than $\pm$ 0.05 m (in order to obtain a precision of the gravity value better than 0.01 mGal). See also Chapter 2.5.

Gravity survey

All gravity measurements at the field stations are related to a single reference station or to one of the reference stations of the base net (if established). The tie to the reference station(s) can be obtained by either

single-loop measurement: all field stations are measured once, or

double-loop measurement: all field stations are measured twice.

The double-loop method requires almost double the effort of the single-loop method but it is useful if very high precision is desired (e.g., when underground cavities are being searched for).

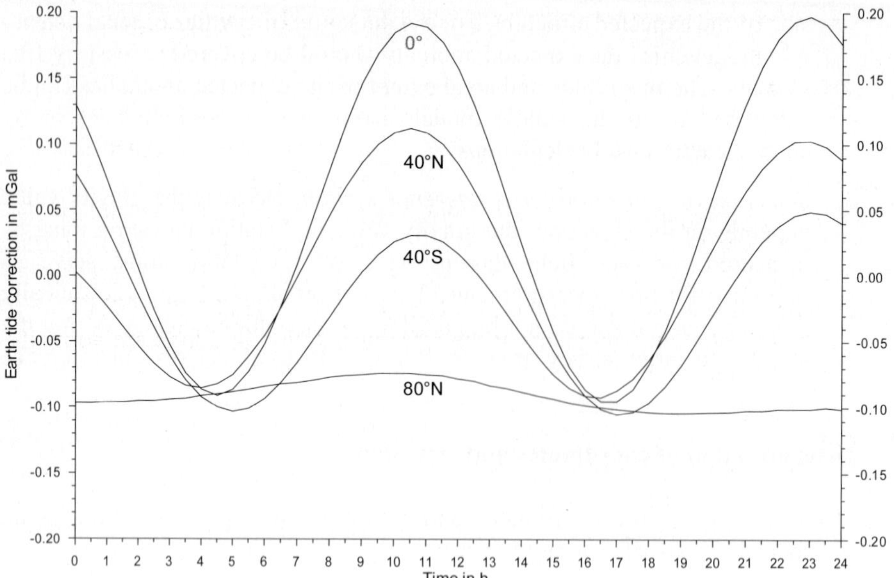

Fig. 4.2-3: Earth tide correction for different latitudes along 15° E on 4th October 2002; the tidal curve is superimposed by the drift curve of the instrument (the daily drift of a good quality instrument should be less than 0.02 mGal)

The measurement at the reference station has to be done at least at the beginning and at the end of each loop. As the measured value varies with time due to the Earth tide (*Fig. 4.2-3*) and the drift of the instrument, both of these parameters have to be determined. Normally, only a theoretical value is calculated for the Earth tide correction. Instrument drift must be determind by repeated measurement at the base station within a certain time interval. This interval depends on the quality and specifications of the gravimeter and can be between 30 and 240 minutes. If the Earth tide correction cannot be calculated, a tie-back to a base station on a hourly basis can be done to record and correct the superimposed influence of the Earth tide and the instrument drift. For quality control, some stations (5 - 10 %) of previous loops should be measured repeatedly in each loop.

4.2.6 Processing and Interpretation of the Measured Data

Corrections

The gravity value is influenced by the elevation and the coordinates of the gravity station, the time of the measurement, and the surrounding morphology. Therefore, it is necessary to carry out several corrections. The aim of the corrections is to make the measurement at a single station comparable to the results at the other stations and to remove from the gravity value all known influences that are not due to the investigated structure (LEFEHR, 1991).

The repeated measurements at the reference station in a loop are normally different from the first measurement in the loop. This is due to the effect of tides and instrument drift. These effects have to be eliminated using the internal software of the gravimeter or PC software. When the measurement at the reference station is repeated at least every 45 - 60 minutes, the differences from the first measurement are due to instrument drift and tidal changes; only when the measurements are repeated can the correction of both effects be carried out in one step. The result of this data processing is the drift and tidal corrected gravity value Δg.

The resulting value due to unknown subsurface structures is called the *Bouguer* anomaly $\Delta g_0''$ and it is calculated as follows:

$$\Delta g_0'' = \Delta g - \gamma_0 + \delta g_{FA} - \delta g_{Boug} + \delta g_{Top} \qquad (4.2.8)$$

The main corrections are described in *Table 4.2-2*. The interpretation is based on the Bouguer anomaly calculated using Equation (4.2.8). The data can be shown along profiles as a measured curve or plotted as a contour map. In most cases, the data have to be interpolated on a regular grid before plotting. A colored contour map improves the readability of the gravity map.

Table 4.2-2: The main gravity corrections

Symbol	Name	Description	
γ_0	normal gravity (theoretical gravity field)	$= \gamma_e (1 + \beta \sin^2 \varphi - \beta_1 \sin^2 2\varphi)$ in mGal where (GRS67) $\gamma_e = 978031.8$ mGal $\beta = 0.005\,3024$ $\beta_1 = 0.000\,0059$	 (GRS80)[2] $\gamma_e = 978032.7$ mGal $\beta = 0.005\,3024$ $\beta_1 = 0.000\,0058$
δg_{FA}	free air correction	$= 0.3086\, h$	in mGal
δg_{Boug}	Bouguer correction	$2\pi\, G = 0.04192\, \rho\, h$	in mGal
δg_{Top}	topographic (terrain) correction	As the Bouguer correction assumes a flat, horizontal area around the station, in hilly or mountainous areas it is necessary to correct for the effect of the masses in the surroundings that are not at the same elevation as the station. The topographic relief in the immediate vicinity of the station or survey area may require special surveying. This correction is calculated using special software. A template or a zone chart can also be used.	

h - station elevation in m
ρ - density in g cm^{-3} or in 10^3 kg m^{-3}
φ - latitude of the gravity station
$G = 6.672 \times 10^{-11}$ m^3 kg^{-1} s^{-2} - gravitational constant

Interpretation

Separation of the regional trend from the residual gravity anomaly: A Bouguer anomaly contour map contains all the gravity effects of both deep sources (the regional part of the gravity field) and shallow sources (the local or residual part). For a targeted interpretation, it is useful to separate the regional and the local parts of the field, in particular to make the anomalies caused by shallow sources more easily recognized. In general, the regional trend (field) is determined by various methods (e.g., graphical and smoothing techniques, gridding methods, filtering, HINZE (1988); TELFORD et al. (1990), or by modeling known structures). The regional field is subjectively choosen and it represents the gravity field that would have been measured if the gravity expression of the anomalous structure being investigated was not present. This regional field is subtracted from the Bouguer gravity and the residual field comprises mainly the local anomalies. The quality of the field separation

[2] The Geodetic Reference System 1980 (GRS80) was adopted by the International Association of Geodesy (IAG) during the General Assembly 1979 as reference system for size, shape, and gravity field of the Earth for geodetic, geophysical, astronomical and hydrographic applications. It replaced the Geodetic Reference System 1967 (GRS67). For principal parameters of GRS80 see
http://dgfi2.dgfi.badw-muenchen.de/geodis/REFS/grs80.html.

strongly influences the interpretation of the data, especially if the residual field is used as reference field for modeling the geological structures. Maps depicting regional or residual fields are both used for qualitative interpretation (PAWLOWSKI, 1995). Local field maps are important for the identification of tectonic structures in a survey area.

Modeling: The main aim of interpretation is to provide quantitative information about the geology below the surface, i.e., depth, extent and petrophysical properties of structures (ABDERRAHMAN et al., 2001). Gravity modeling is necessary to obtain such information. Depending on the geology, 2-D, 2.5-D, 2.75-D or 3-D modeling is carried out.

2-D modeling is carried out along a profile if the structure (perpendicular to the profile direction) is much larger than it is wide. Each such structure along the profile is approximated by polygonal cross-sections. The calculated field is compared with the measured field and the model is modified until the calculated data sufficiently fits the measured data. 2.5-D or 2.75-D modeling will increase the accuracy of the modeling if the length of the structure is not quasi-infinite, i.e., less than four times the width.

The most accurate modeling of the subsurface geology is done with 3-D modeling. Programs for inverting surface gravity data to derive a 3-D distribution of density contrast are also available. Geological units are constructed from triangular or rectangular elements and thus, very complex structures can be included in the model (GÖTZE & LAHMEYER, 1988; LI & OLDENBURG, 1998). The disadvantage of this type of modeling is that it requires special software, time-consuming model preparation and therefore, is often too expensive for small projects. Independent on the type of modeling, all available information (drill cores, borehole logs, density data, geological knowledge) has to be included in the modeling process to produce the most reliable model of the subsurface (BOULANGER & CHOTEAU, 2001; BARBOSA et al., 2002).

4.2.7 Quality Assurance

The general aspects of quality assurance are given in the glossary. The following measures are important for quality assurance in a gravity survey:

- Coordinates and elevation of the stations have to be determined with the required precision (Section 4.2.5). The errors in coordinates and elevation of observation stations have to be documented.
- Documentation of the coordinates and elevations of all gravity reference stations and benchmarks. Reference stations should be permanently marked in the field.

- The gravimeters should be calibrated at least once a year at base stations with at least 80 mGal difference. The drift of the gravimeters and function tests should be documented.
- To determine the error of the gravity measurements, repeated measurements should be made in different loops at 5 - 10 % of the gravity stations.
- The quality of the data should be checked during the field work, permitting errors to be detected so that measurements can be repeated at these stations, if necessary.
- The corrections are to be documented. In addition, information about datum planes and the density values used for reductions is necessary.
- Uncertainties in the interpretation, in particular in the modeling, should be given.

4.2.8 Personnel, Equipment, Time Needed

	Personnel	**Equipment**	**Time needed**
mobilization and demobilization	depends on the distance to the survey area		
measurements: position, elevation	1 surveyor 1 assistant	1 instrument for surveying , 1 4WD vehicle	a = 5 - 10 m L = 200 N/d
			a = 200 m L = 30 N/d
gravity survey	1 operator (1 assistant)	1 gravimeter, 1 4WD vehicle 1 PC, (e-mail for data transfer)	a = 5 - 10 m L = 80 - 120 N/d
			a = 200 m L = 30 N/d
data processing interpretation report	1 geophysicist (1 technical assistant)	1 PC with plotter, printer and software; (e-mail for data transfer)	3 - 4 days for each day in the field

a = station spacing, N = number of stations, d = 10-hour working day, L = N/d

4.2.9 Examples

Investigation of a concealed waste dump in an abandoned quarry

This example describes the investigation of a concealed waste dump in a former sandstone quarry in southern Germany. The objective was to determine the location and size of the landfill.

Survey parameters:

size of the investigation area	12 000 m^2
number of stations	353
repeated stations	10 %
station spacing	5 m, at the margin 10 m
duration of the survey	6 days
personnel	1 geophysical technician, 1 assistant
measured physical parameter	Δg
equipment	Sodin 410,
corrections and reductions	drift correction
	free air correction
	normal gravity (relatively calculated)
	Bouguer reduction using a density of
	$2.0 \cdot 10^3$ kg m^{-3}

The Bouguer anomaly map of the area under investigation is shown in *Fig. 4.2-4a.* The gravity low in the southeastern part of the area is assumed to coincide with the location of the former quarry, now filled with waste and covered with soil. The edge of the landfill can be identified clearly, both laterally and with depth. The gravity high east of the gravity low is caused be sandstone. To the north and west the thickness of the weathered material covering the sandstone increases. An estimate of the depth of the waste is given in *Fig. 4.2-4b* and *c*. For modeling, the residual field is calculated by subtracting an almost horizontal linear trend from the Bouguer anomalies. Depending on the density contrast assumed, the maximum depth of the former quarry is between 9 m and 20 m. As the density difference and the depth of the waste are not known, this ambiguity inherent in gravity data can only be removed by further investigation, e.g., by boreholes in the area of the gravity anomaly or by additional geophysical investigations such as seismics or geoelectrics. Other examples of gravity modeling of waste pits are given in *Fig. 5.4-24.*

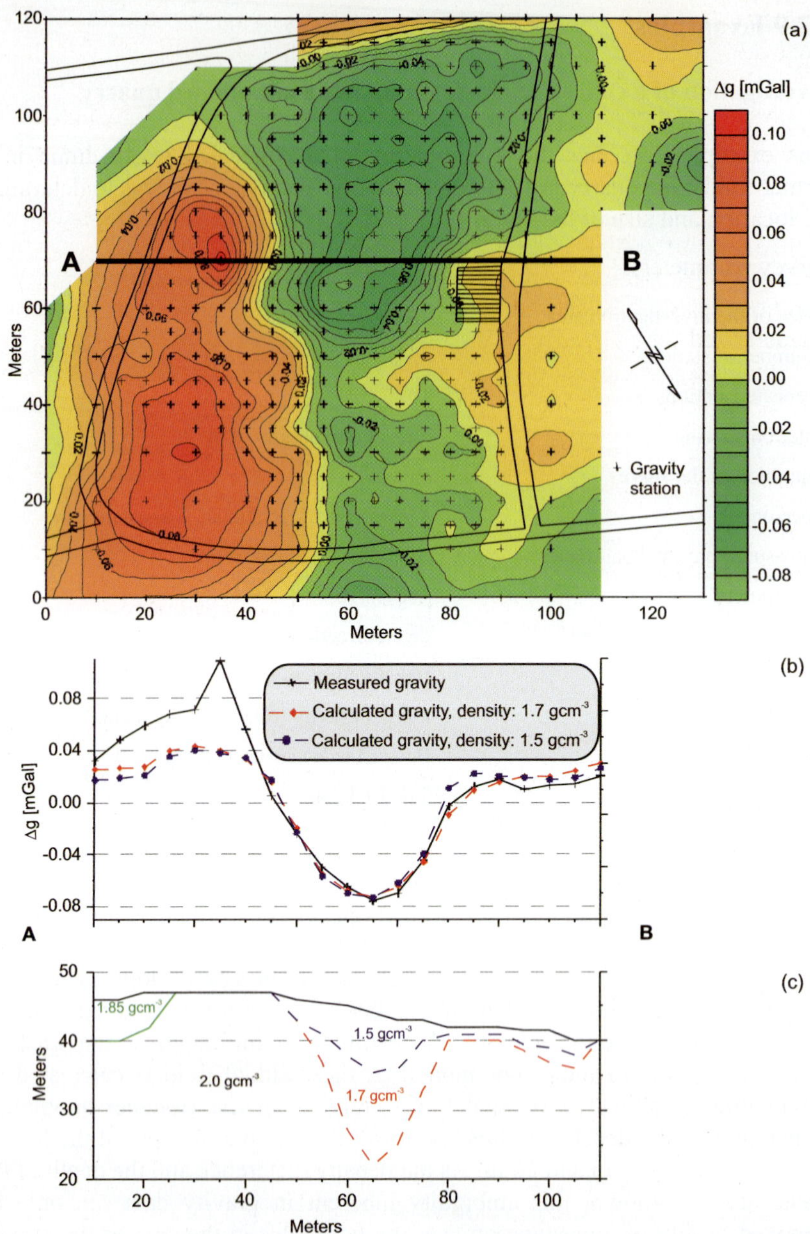

Fig. 4.2-4: Gravity survey of a concealed waste dump in a former quarry, (a) Bouguer anomaly map, (b) comparison of the measured and calculated gravity anomalies of the profile A - B for the two models shown in (c)

Investigation of structures in the area around an open pit mine in Sarawak, Malaysia

A short time after the re-opening of an open-pit mine directly south of the town of Bau in Sarawak, there was some subsidence and sinkholes appeared. In order to establish a correlation between the sinkholes, subsidence, and geological structures such as faults and other types of contacts, staff of the Geological Survey of Malaysia, Sarawak (GSMS), assisted by German consultants, carried out a gravity survey of the entire municipal area of Bau. Additionally some seismic lines were surveyed. The structure in this area is characterized by nearly vertical contacts between limestone and unconsolidated materials (e.g., mudstone, tailings). The limestone is karstified and it was thought that collapsing solution cavities were responsible for the sinkholes.

Survey parameters:

size of the investigation area	≈ 1.5 km^2
number of stations	658
repeated stations	46
station spacing	40 - 50 m
duration of the survey	13 days
personnel	1 geophysical technician, 1 assistant
measured physical parameter	Δg
equipment	LCR Model G
corrections	drift correction
	free air correction
	normal gravity (formula: GRS67)
	Bouguer correction using a density of
	$2.0 \cdot 10^3$ kg m^{-3}
precision of corrected data	0.018 mGal

The main results of the survey are shown in *Fig. 4.2-5*. This residual map was prepared by applying a high-pass wavelength filter with a cutoff wavelength of 1 km to the Bouguer anomaly data. The gravity highs (yellow to red) coincide with areas where the overburden above the limestone has a very small thickness. Outside the areas of these gravity highs, the soil is generally thicker. Several local gravity lows (e.g., in the NW, N and SE part of the survey area) coincide with areas in which shale has been mapped. In these gravity-low areas the shale can be expected to be very thick.

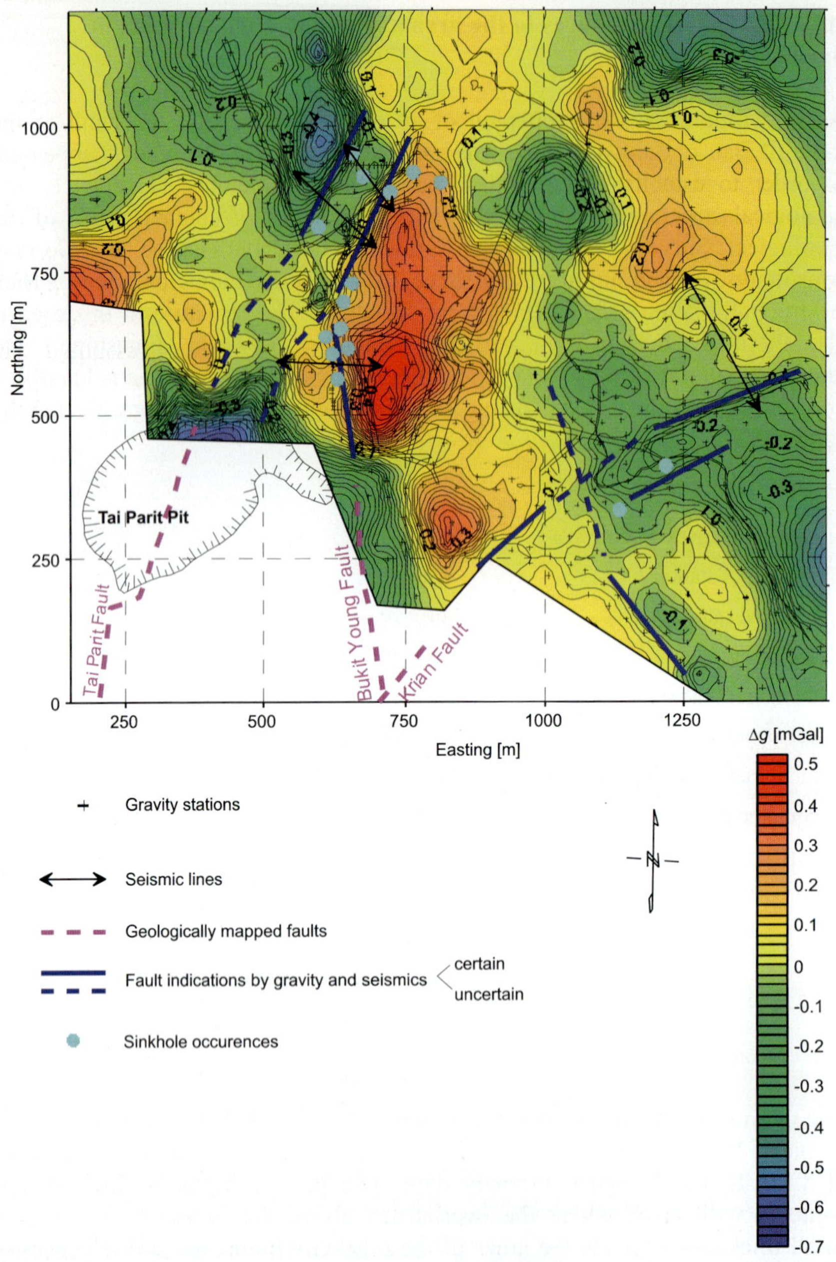

Fig. 4.2-5: Gravity survey of structures in the area around an open-pit mine in Sarawak, Malaysia, contour map of the residual gravity field obtained using a high-pass wavelength filter with a cutoff wavelength of 1 km

Besides the information about variation in overburden thickness, the gravity data showed the locations of several faults (some of them as extensions of known faults) and provided a detailed outline of the structures in the survey area. These structures were later confirmed by a seismic survey. The main geological structure, starting from the northern edge of the pit, trends NNE-SSW. Most of the larger sinkholes in the town are along this structure, in particular to the crossing of two faults (Tai Parit fault and Bukit Young fault). This zone west of the strongest gravity high coincides with a zone where the depth to the limestone abruptly increases (high gradients in the gravity anomaly) and whose steep edges are in a direct contact with unconsolidated material. On the basis of the geophysical results, it is now assumed that groundwater moves along the faults and in the karst system. At the contact of the limestone with the unconsolidated material, the latter is removed by water producing cavities, which finally collapse.

Another example of a structure map based on residual gravity is given in *Fig. 5.4-21.*

Parameters and units

	Symbol	Units
gravitational acceleration	g	µm s^{-2}, mGal, gravity units (g.u.)
density	ρ	kg m^{-3} (g cm^{-3}) 1 g $cm^{-3} \equiv 10^3$ kg m^{-3}
gravitational constant	G	6.672×10^{-11} m^3 $kg^{-1}s^{-2}$
mass	m	kg

References and further reading

ABDELRAHMAN, E-S., M., EL-ARABY, T. M. EL-ARABY, H. M. & ABO-EZZ, E. R. (2001): A new method for shape and depth determinations from gravity data. Geophysics, **66**, 1774-1780.

ARAFIN, S. (2004): Relative Bouguer anomaly. The Leading Edge, 850-851.

BARBOSA, V. C. F., SILVA, J. B. C. & MEDEIROS, W. E. (2002): Practical applications of uniqueness theorems in gravimetry: Part II – pragmatic incorporation of concrete geologic information. Geophysics, **67**, 795-809.

BOULANGER, O. & CHOTEAU, M. (2001): Constraints in 3D gravity inversion. Geophys. Prosp., **49**, 265-280.

CARBONE, D. & RYMER, H. (1999): Calibration shifts in a LaCoste-and-Romberg gravimeter: comparison with a Scintrex CG-3M. Geophys. Prosp., **47**, 73-83.

CHAPIN, D. A., CRAWFORD, M. F. & BAUMEISTER, M. (1999): A side-by-side test of four land gravity meters. Geophysics, **64**, 765-775.

GIBSON, R. I. & MILLEGAN, P. S. (1998): Geologic applications of gravity and magnetics: Case histories. SEG, Tulsa.

GÖTZE, H.-J. & LAHMEYER, B. (1988): Application of three-dimensional interactive modelling in gravity and magnetics. Geophysics, **53**, 1096-1108.

HINZE, W. J. (1988): Gravity and magnetic methods applied to engineering and environmental problems. Proceedings of the "Symposium on the Application to Engineering and Environmental Problems", March 28 - 31, Golden, Colorado, 1-107.

HINZE, W. J. (2003): Bouguer reduction density, why 2.67?. Geophysics, **68**, 1559-1560.

LEFEHR, T. R. (1991): Standardization in gravity reduction. Geophysics, **56**, 1170-1178.

LI, Y. & OLDENBURG, D. W. (1998): 3-D inversion of gravity data. Geophysics, **63**, 109-119.

LI, X. (2001): Vertical resolution: Gravity versus vertical gravity gradient. The Leading Edge, 901-904.

LITINSKY, V. A. (1989): Concept of effective density: key to gravity depth determinations from sedimentary basins. Geophysics, **54**, 1474-1482.

LINDNER, H. & SCHEIBE, R. (1978): Die Berechnung von δg- und δT- Anomalien für regelmäßige homogene Störkörper. Gerlands Beitr. Geophys., Leipzig, **87**, 29-45.

MILITZER, H. & WEBER, F. (1984): Angewandte Geophysik. **1**. Springer, Wien.

PARASNIS, D. S. (1997): Principles of Applied Geophysics. Chapman & Hall, London.

PAWLOWSKI, R. S. (1995): Preferential continuation for potential-field anomaly enhancement. Geophysics, **60**, 390-398.

SHARMA, B. & GELDART, L. P. (1968): Analysis of gravity anomalies using Fourier transforms. Geophys. Prosp., **16**, 77-93.

SILVA, J. B. C., MEDEIROS, W. E. & BARBOSA, V. C. F. (2002): Practical applications of uniqueness theorems in gravimetry: Part I - constructing sound interpretation methods. Geophysics, **67**, 788-794.

Simon Fraser University, Vancouver (2003): Earth-sciences courses. www.sfu.ca/earth-sciences/courses/new207.

TALWANI, M. & EWING, M. (1960): Rapid computation of gravitational attraction of three-dimensional bodies of arbitrary shape. Geophysics, **25**, 203-225.

TELFORD, W. M., GELDART, L. P. & SHERIFF, R. E. (1990): Applied geophysics (2nd edn.). Cambridge University Press, Cambridge, New York, Port Chester, Melbourne, Sydney.

4.3 Direct Current Resistivity Methods

KNUT SEIDEL & GERHARD LANGE

4.3.1 Principle of the Methods

Direct current (dc) resistivity methods use artificial sources of current to produce an electrical potential field in the ground. In almost all resistivity methods, a current is introduced into the ground through point electrodes (C_1, C_2) and the potential field is measured using two other electrodes (the potential electrodes P_1 and P_2), as shown in *Fig. 4.3-1*. The source current can be direct current or low-frequency (0.1 - 30 Hz) alternating current. The aim of generating and measuring the electrical potential field is to determine the spatial resistivity distribution (or its reciprocal - conductivity) in the ground. As the potential between P_1 and P_2, the current introduced through C_1 and C_2, and the electrode configuration are known, the resistivity of the ground can be determined; this is referred to as the "apparent resistivity".

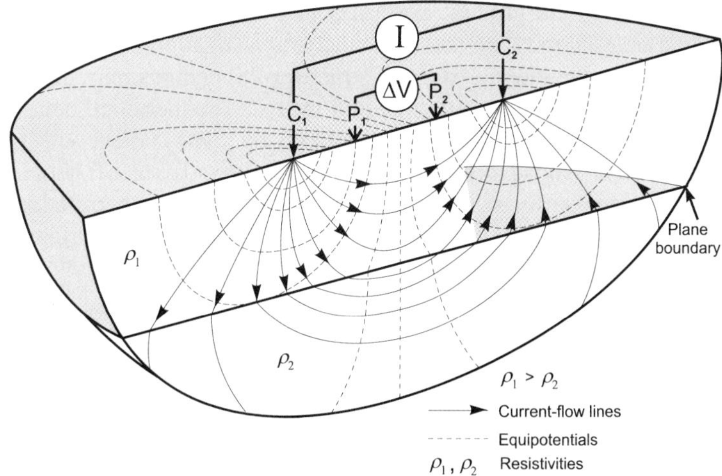

Fig. 4.3-1: Principle of resistivity measurement with a four-electrodes array

Resistivity measurements may be made at the Earth's surface, between boreholes or between a single borehole and the surface. With special cables, measurements can be made underwater in lakes, rivers and coastal areas. The following basic modes of operation can be used:

- profiling (mapping),
- vertical electrical sounding (VES),
- combined sounding and profiling (two-dimensional resistivity imaging),
- three-dimensional resistivity survey (3-D resistivity imaging), and
- electrical resistivity tomography (ERT).

Profiling methods use fixed electrode spacings to detect lateral resistivity changes along a profile down to a more or less constant investigation depth, which is governed by the electrode spacing. The results are normally interpreted qualitatively. Contour maps or profile plots of the measured apparent resistivities allow delineation of lateral boundaries of geogenic structures and anthropogenic features (e.g., waste dumps and contamination plumes), as well as hydrogeological conditions.

The main aim of sounding methods is to determine the vertical distribution of the resistivity in the ground. Several soundings at regular spacings along profiles and/or randomly in the area under investigation will also provide information about the lateral extent of structures. Soundings may be made for investigation depths up to several hundred meters. The measured data may be interpreted both qualitatively and quantitatively. The latter will provide resistivity models whose layer boundaries are boundaries of geoelectrical layers but not necessarily of lithological layers. For a better correlation to the geology, geoelectrical data should be correlated with borehole logs or the results of other geophysical methods, such as reflection and refraction seismic sections.

Sounding and profiling can be combined in a single process (2-D resistivity imaging) to investigate complicated geological structures with strong lateral resistivity changes. This combination provides detailed information both laterally and vertically along the profile and is the most frequently applied technique in environmental studies. 2-D inversion yields a two-dimensional distribution of resistivities in the ground.

Three-dimensional (3-D) resistivity surveys and ERT measurements provide information about complex structures. In current practice they do not play an important role in geophysical site investigations, as they are still very time consuming and expensive. Some convincing examples in small survey areas show that new and effective 3-D techniques are being developed, including data acquisition and interpretation (Dahlin et al., 2002).

4.3.2 Applications

- Investigation of lithological underground structures,
- estimation of depth, thickness and properties of aquifers and aquicludes,
- determination of the thickness of the weathered zone covering unweathered rock,
- detection of fractures and faults in crystalline rock,
- mapping of preferential pathways of groundwater flow,
- localization and delineation of the horizontal extent of dumped materials,
- estimation of depth and thickness of landfills,
- detection of inhomogeneities within a waste dump,
- mapping contamination plumes,
- monitoring of temporal changes in subsurface electrical properties,
- detection of underground cavities,
- classification of cohesive and non-cohesive material in dikes, levees, and dams.

4.3.3 Fundamentals

Physical basics

A point electrode introducing an electrical current I will generate a potential V_{r} at a distance r from the source. If both source and measuring points are at the surface of a homogeneous half-space with resistivity ρ, this potential is given by:

$$V_{\mathrm{r}} = \frac{\rho I}{2\pi r} \tag{4.3.1}$$

In the case of a four-electrode array (*Fig. 4.3-1*) consisting of two current electrodes (C_1, C_2) that introduce a current $\pm I$, the potential difference ΔV between the potential electrodes P_1 and P_2 can be calculated as follows:

$$\Delta V = \rho I \left[\frac{1}{2\pi} \left(\frac{1}{r_1} - \frac{1}{r_2} - \frac{1}{r_3} + \frac{1}{r_4} \right) \right], \tag{4.3.2}$$

where $r_1 = C_1P_1$, $r_2 = C_1P_2$, $r_3 = C_2P_1$, and $r_4 = C_2P_2$.

Replacing the factor in square brackets by $1/K$, we obtain the resistivity of the homogeneous half-space as follows:

$$\rho = K\frac{\Delta V}{I}. \tag{4.3.3}$$

The parameter K, the configuration factor or geometric factor, can be easily calculated for all practical configurations. *Table 4.3-1* gives K parameters for arrays most commonly used in field surveys. For inhomogeneous conditions, Equation (4.3.3) gives the resistivity of an equivalent homogeneous half-space. For this value the term apparent resistivity ρ_a is introduced, which is normally assigned to the center of the electrode array. In a given four-electrode array, the current and potential electrodes are interchangeable. This concept (the principle of reciprocity) can be used in multi-electrode systems to improve the signal-to-noise ratio.

There are many different types of electrode arrays (also called spreads or configurations) although in practice only a few of them are used. The most common are compiled in *Table 4.3-1*. Each array has its advantages and disadvantages with regard to depth of investigation, resolution of horizontal and vertical structures, sensitivity to lateral changes in resistivity (lateral effects) and inhomogeneity, to depth of targets, to dip and topography, etc. Depending on the array characteristics and the objective, the appropriate array has to be selected for each survey. Field logistical qualities are also essential for a survey. A comprehensive assessment of electrode arrays is given by WARD (1990).

The type of electrode array selected for a resistivity survey plays a role with regard to resolution and depth of investigation. For resistivity soundings (VES), the vertical resolution refers to how thin a layer can be detected in a layer sequence. In recent years 2-D resistivity measurements with a multielectrode system have become increasingly important, especially for environmental applications (BARKER, 1981; DAHLIN, 1996). The vertical and horizontal resolution of dipping structures and lateral resistivity distribution is of interest in the case of a 2-D survey.

At the same time as the new measuring techniques, 2-D inversion schemes have been developed using the increasing capacity of personal computers (DEY & MORRISON, 1979a; BARKER, 1992). A sensitivity matrix is an important part of the inversion algorithms. For a given electrode array, this matrix is used to evaluate the contribution of the spatial elements of the subsurface model to the measured apparent resistivity. Sensitivity matrices are very helpful for the understanding of measured data and, especially for the planning a survey, as they allow conclusions to be drawn about the resolution capabilities and investigation depths of different electrode arrays.

For the case of a homogeneous half-space, some examples of sensitivity matrices for commonly applied electrode arrays are given in *Fig. 4.3-2*. Each of these examples shows that there are areas of positive as well as of negative sensitivity. In an area of negative sensitivity, there will be a decrease in the measured apparent resistivity when the structure in that area has a higher resistivity than the surrounding material. Negative sensitivities always occur in the space between a current and a potential electrode, whereas the sensitivity is always positive between two current electrodes and between two potential electrodes (*Fig. 4.3-2*). The spatial sensitivity generally decreases with increasing distance from the electrodes (FRIEDEL, 2000).

The plots in the left side of *Fig. 4.3-2* show sensitivities for small electrode spacings, whereas those on the right side show sensitivities for a larger electrode spacing. The different arrays can be assessed on the basis of these sensitivity plots. It can be seen, for example, in plots (*c*) and (*d*) that the sharp, almost vertical boundary between positive and negative sensitivities is the reason why pole-dipole arrays have a very good lateral resolution. The fairly high resolution of dipole-dipole measurements in the case of small targets results from a structured sensitivity distribution (*Fig. 4.3-2e* and *f*). But it can also be seen that inhomogeneity near the electrodes strongly influences the measured apparent resistivity. Experience shows that the results of Schlumberger arrays are mainly determined by the ground conditions below the potential electrodes, as *Fig. 4.3-2g* and *h* confirm. Finally, the fairly smooth sensitivity distributions shown in plots (*i*) and (*j*) explain the comparatively low lateral resolution of a Wenner array. But this array is advantageous if the investigation area is subject to electromagnetic noise.

Two examples of dipole-dipole-measurements with electrodes in boreholes are given in *Figs. 4.3-2k* and *l*. *Figure 4.3-2m* shows the sensitivity of a Wenner array in the x-y-plane at the surface. It can be seen that differences in the rock to the side of the profile will also influence the measured data. This is very important to know if misinterpretation is to be avoided in areas with a complicated 3-D geology.

In general, the geology of interest has a 3-D resistivity distribution. Normally, this would require a 3-D resistivity survey. But this is, as already mentioned, very time consuming and expensive at present. In many practical cases the problem can be reduced to a 2-D or even a 1-D case, but it must always be kept in mind that the measurements can be influenced by objects outside the arrays (SPITZER, 1995).

In the 2-D case it is assumed that the resistivity of the ground varies only in the vertical and one horizontal direction. There is no resistivity variation in the second horizontal direction (strike direction). As a consequence, survey profiles should run perpendicular to the strike of such structures.

Table 4.3-1: Some electrode arrays for dc resistivity measurements

Electrode array	Electrode configuration	Configuration factor
Wenner Wenner α Lee	C_1 P_1 O P_2 C_2; a a a	$K = 2\pi a$
Schlumberger	C_1 P_1 P_2 C_2; n·a a n·a	$K = \pi\ n(n+1)\,a$ $n > 3$
dipole-dipole axial dipole Wenner β	C_1 C_2 P_1 P_2; a n·a a	$K = \pi\ n(n+1)(n+2)\,a$
pole-dipole half Schlumberger Hummel	C_1 P_1 P_2 $C_2\infty$; n·a a	$K = 2\pi n\ (n+1)\,a$ $n > 3$
pole-pole	$C_2\infty$ C_1 P_1 $P_2\infty$	$K = 2\pi a$
gradient	C_1 P_1 P_2 C_2; a; y; x; $\overline{C_1C_2}/3$ $\overline{C_1C_2}/3$ $\overline{C_1C_2}/3$; Δ Δ; Δ > 10 · a	$K = 2\pi \left[\frac{1-X}{(Y^2+(1-X)^2)^{\frac{3}{2}}} + \frac{1+X}{(Y^2+(1+X)^2)^{\frac{3}{2}}} \right]^{-1}$ $(X = \frac{x}{\Delta};\ Y = \frac{y}{\Delta})$
surface-borehole dipole-dipole pole-dipole pole-pole	$C_2\infty$ C_2 C_1 Brg1; $P_2\infty$ P_2 P_1 Brg2	
in boreholes dipole-dipole pole-dipole pole-pole	$C_2\infty$ C_2 C_1 Brg; P_1 P_2 $P_2\infty$	

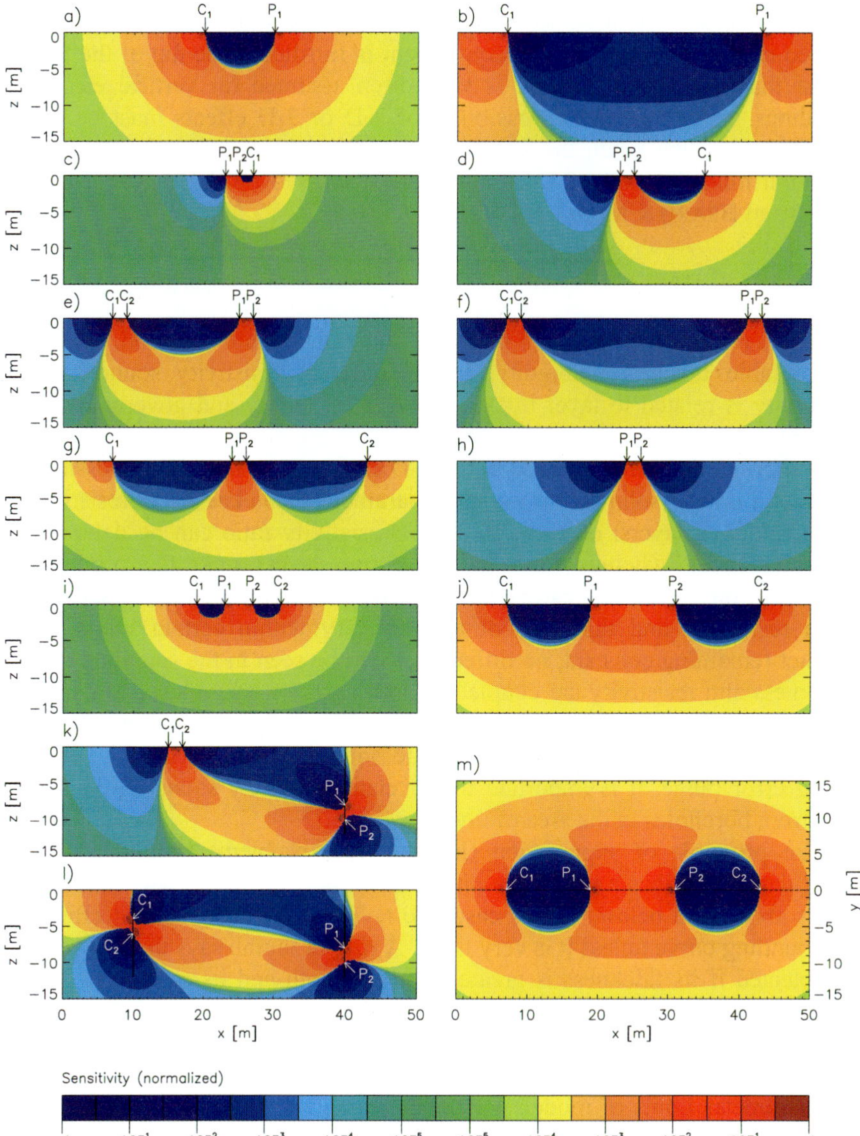

Fig. 4.3-2: Sensitivity of several electrode arrays, examples for the 2-D-sensitivity distribution in a homogeneous half-space: (*a,b*) pole-pole, (*c,d*) pole-dipole, (*e,f*) dipole-dipole, (*g,h*) Schlumberger array, (*i,j*) Wenner array, (*k,l*) dipole-dipole with electrodes in boreholes, (*m*) Wenner array but x-y-plane of the sensitivity. All plots are normalized with respect to the maximum of each matrix. (courtesy of S. FRIEDEL, University of Leipzig, for details on computation, see FRIEDEL, 2000).

In the 1-D case, it is assumed that the ground corresponds to a horizontally layered model. In practice, this requirement is fulfilled if the dip of the layers is less than 10°. The 1-D case is very often assumed for vertical electrical soundings. In any case, the possibility of 2-D or 3-D effects must be taken intro consideration when the sounding survey is planned. The limits of 1-D inversion modelling of a 2-D situation are discussed by SCHULZ & TEZKAN (1988) and BEARD & MORGAN (1991).

Ambiguity of results

The dc resistivity method, being a potential field method, exhibits considerable inherent ambiguity. In the case of resistivity sounding, the ambiguity is related to layer thickness, layer resistivity and governed by the principles of equivalence and suppression.

The principle of equivalence in the 1-D case means that it is impossible to arrive at a unique solution for the layer parameters (thickness and resistivity). For conductive layers, only the thickness/resistivity ratio can be determined (S-equivalence), whereas for highly resistive layers only the product of thickness times resistivity can be determined (T-equivalence). This means that when resistivity and thickness of a layer vary within certain limits, their product remaining constant, no differences can be seen in the sounding curve. Thickness and resistivity are coupled in both cases of equivalence and cannot independently determined. Because this principle results in different equivalent layer models which all fit the sounding curve within a selected fitting error range, it is important to correlate resistivity sounding data with data from boreholes in the area of investigation.

The principle of suppression (hidden layer problem) applies when an intercalated layer has a resistivity intermediate between the resistivities of the layers above and below. In this case, the layer has an insignificant effect on the sounding curve unless it is very thick. A rule of thumb says that a layer can be detected if its thickness is greater than its depth and its resistivity differs from the cover layer. By modeling of sounding data, a series of thin layers will mostly be represented by only one layer and a mean resistivity.

There are still problems with the determination of the confidence intervals and parameter limits of models for the 2-D and 3-D cases. In 2-D inversion, ambiguity is influenced by several factors: the structure of the grid used to approximate the geological structures, limited data density, and errors in the data. Not to be neglected is the sensitivity distribution of the electrode configuration used (SPITZER & KÜMPEL, 1997; FRIEDEL, 2000; *Fig. 4.3-2*). 2-D equivalence of smooth and sharp boundary inversion in the investigation of simple structural models (e.g., vertical fault, graben, horst) using Wenner array is discussed by OLAYINKA & YARAMANCI (2002). Inversion using sharp boundaries (block inversion) indicates that - for good quality data with sufficiently high density - the range of 2-D equivalence is relatively narrow.

For equivalent solutions the misfit between the observed data and the calculated data is very small in these cases. Models that are incorrect can be readily identified on the basis of very large data misfits.

Petrophysical basics

Resistivity methods are useful only if the resistivity of the target differs significantly from the resistivity of the host material. For the successful planning of any geoelectrical survey and for the interpretation of the data, it is very important to know the resistivities of the materials in the study area. Resistivities of some selected materials, including typical domestic and industrial waste, are listed in *Table 4.3-2.*

Table 4.3-2: Resistivities for geological and waste materials

Material	**Resistivity (in Ωm)**	
	minimum	**maximum**
gravel	50 (water saturated)	$>10^4$ (dry)
sand	50 (water saturated)	$>10^4$ (dry)
silt	20	50
loam	30	100
clay (wet)	5	30
clay (dry)		>1000
peat, humus, sludge	15	25
sandstone	<50 (wet, jointed)	$>10^5$ (compact)
limestone	100 (wet, jointed)	$>10^5$ (compact)
schist	50 (wet, jointed)	$>10^5$ (compact)
igneous and metamorphic rock	<100 (weathered, wet)	$>10^6$ (compact)
rock salt	30 (wet)	$>10^6$ (compact)
domestic and industrial waste	<1	>1000 (plastic)
natural water	10	300
sea water (35‰ NaCl)	0.25	
saline water (brine)	<0.15	

Pore fluids considerably reduce the resistivity of porous sediments (*Table 4.3-2*). The resistivity of a rock that is saturated with highly mineralized water can be significantly lower than given in *Table 4.3-2.* Sandy sediments containing highly mineralized pore water can have the same resistivity as clay. In such cases it is difficult to distinguish between sand and clay layers on the basis of resistivity alone. Data from a combination of methods (e.g., induced polarization, seismics, ground penetrating radar) should be used to identify subsurface lithology and structures (see next section).

Induced polarization as a complementary method

In some environmental investigations and research (e.g., VANHALA et al., 1992; WELLER et al., 1999), it has been found that induced polarization (IP) surveys can provide additional or even more detailed information about a landfill and the lithology of the host material than a resistivity survey. Originally developed to explore of minerals, the IP method based on the ability of metallic minerals in a rock with a very low conductivity to take an electrical charge and on differences in ion concentrations in the pore water or at the pore surfaces (SUMNER, 1976). After an excitation current pulse (time-domain method) or a sinusoidal current (frequency-domain method) has been introduced in the ground, the measured voltage will not instantly return to zero, but shows a slow material-dependent decay.

In the time-domain method, the chargeability is given by the integrated amplitude of the decay curve (in mV sV^{-1}) normalized with respect to the excitation voltage after the excitation current is switched off. Chargeability is a useful parameter for distinguishing between kinds of materials. Material in a landfill normally has a higher chargeability but lower resistivity than the surrounding rocks. Galvanic sludge has a high chargeability. For this reason IP measurements can be used to detect galvanic sludge in landfills. In addition, clay, which is the most important material for geological barriers, shows a higher chargeability than sandy material.

In the frequency-domain, subsurface resistivity is measured as a function of the excitation current frequency. Induced polarization causes the apparent resistivity measured at high frequencies to be smaller than the apparent resistivity measured at low frequencies. If the apparent resistivity ρ_a is measured at two frequencies $\omega_1 < \omega_2$, the induced polarization is expressed by the frequency effect (FE):

$$FE = \frac{|\rho(\omega_1)| - |\rho(\omega_2)|}{|\rho(\omega_2)|} . \tag{4.3.4}$$

In the case of spectral induced polarization (SIP), the resistivity of the ground is measured at 15 - 20 frequencies from 0.1 Hz to 1000 Hz. The resulting amplitude and phase spectra are interpreted with regard to the electrical properties of the ground. Research (e.g., WELLER & BÖRNER, 1996) have shown that SIP measurements may be used to determine petrophysical parameter values, such as porosity or coefficient of permeability.

For IP measurements the same four-electrode arrays as for dc resistivity surveys apply. In every case, the IP method determines the apparent resistivity as well as the chargeability (or frequency effect, FE) of the subsurface material. For IP surveys, more sophisticated equipment is needed than for resistivity surveys. Multi-channel instruments have proved useful for logistical and economic reasons. To obtain high quality data it is favorable, if not essential, to use non-polarizing electrodes to measure the voltage. IP measurements take at least twice the amount of time needed for dc resistivity measurements.

Although there are many possible applications of IP methods, they are not standard in a site investigation because of their high survey cost, the sophisticated equipment required and, last but not least, because of the lack of knowledge regarding the application of such methods to solve environmental problems. Therefore, this method is only briefly mentioned here. For further reading the reader it referred to SUMNER (1976), TELFORD et al. (1990) and recent research publications.

4.3.4 Instruments

The basic equipment for dc resistivity measurements consists of a transmitter, receiver, the power supply, electrodes, and cables. Transmitter, receiver and power supply may be integrated in a single unit or may be separated. In practice, different types of current are used:

- direct current (dc),
- low-frequency alternating current (ac < 30 Hz), as well as
- pulsed square wave dc with changing polarity (on$^+$ - off - on$^-$).

Most modern instruments use pulsed square wave dc or low-frequency sinusoidal ac to avoid electrode polarization. Transmitters produce either a constant voltage or a constant input current. In any case, it is advantageous if the transmitted current and voltage are measured and the data stored together with the resistivity data.

Transmitter power ranges from 10 W to more than 3 kW. The output voltage can range from < 50 to 2000 V and the output current from 1 mA to 10 A. The current is automatically stabilized and the cycle timing (on$^+$ - off - on$^-$ ...) is also automatically controlled. Depending on the output power to be generated, the appropriate power supply has to be chosen (battery pack, car battery or generator).

Receivers should have an input resistance of >10 MΩ and equipped with notch filters for 50 Hz and/or 60 Hz power-line rejection. The precision of the voltage readings of modern instruments is about 1 μV and that of the measured resistivities is about ±1 % within the operating temperature range from -20°C to 70°C. Most instruments offer signal enhancement by stacking and filtering as well as automatic self-potential tracking and suppression. Some receivers are able to also measure chargeability (Section *4.3.3*).

It is advantegeous if the equipment can be used for mapping/profiling and soundings, as well as 2-D and 3-D measurements. An additional switching unit is necessary for 2-D and 3-D measurements to activate the electrodes individually. For some instruments a laptop is required for data acquisition, display and storage. The data acquisition software must be able to take into consideration the electrode configuration, electrode spacing and other survey parameters.

Stainless steel electrodes seem to be best choice for the galvanic coupling of the electrodes with the ground, as they are durable and have fairly low self-potentials, when placed in the ground. To obtain a sufficiently low contact resistance, the electrodes should be about 1 cm in diameter and at least 0.5 m long. For vertical electrical sounding with very large electrode spacing, a bundle of several metal rods can be used as current electrode in order to improve the galvanic coupling. In very dry ground the contact can be improved by pouring water around the electrodes.

The type of cable used depends on the kind of measurement. Plastic- or rubber-insulated single-core cable with a cross-section of about 1.5 mm^2 is sufficient for profiling and sounding. Multi-core cables are used for 2-D/3-D resistivity measurements. Two types of operation are common. The first one uses "passive" electrodes connected to a switching box via a cable containing 20 or more cores. The other type of operation use "active" electrodes: A cable containing only a few cores – which are used to transmit current and communications, as well as measure voltage – is connected to an addressable box on each electrode. The addressable box is used to switch the electrode between active and passive modes. In practice, the first type is easier to handle, but the cable becomes very heavy for large basic electrode spacings (> 5 m). Crosstalk between cores can influence data acquisition in low-resistivity environments, and then it is better to use cables with separately shielded cores. The advantage of the second type is that the crosstalk is reduced because the cores all have a larger cross-section. Moreover, it is easier to increase the number of electrodes and to use basic electrode spacings of 10 m and more. For 3-D resistivity surveys multichannel instruments are essential to reduce data acquisition time to an acceptable level.

4.3.5 Survey Practice

Planning a survey

Before a survey is started, it is necessary to select the optimal kind of measurement (mapping/profiling, sounding or 2-D imaging). Such a decision is influenced by geology and topography in the study area, by the size of the area, and last but not least by economic considerations. An early discussion between the geophysicist and customer about the different geophysical, technical and economic parameters is recommended. For data processing and interpretation it is necessary to collect all available information about drilling results, the local geology and hydrogeology (particularly depth to the water table and groundwater flow direction) and potential sources of cultural noise (metal pipes, power lines, industrial objects). It is always very helpful if resistivity values for the materials of the geological structure to be investigated and information about possible contaminants in the soil and groundwater are available. In many cases, forward modeling of the expected situation can help to determine the most suitable method.

It is essential to select the most appropriate electrode array for the problem at hand. Each array has advantages and disadvantages, which should be weighted against each other. Resolution and depth of investigation of different arrays have been investigated by several researchers (e.g., ROY & APPARAO, 1971; EDWARDS, 1977; BARKER, 1989; WARD, 1990). They all show that both the Schlumberger array and the Wenner array have very good vertical resolution, but due to their symmetric electrode geometry, a lower lateral resolution than asymmetric configurations, e.g., pole-dipole or dipole-dipole. Wenner arrays provide the best results under noisy conditions. Experience, confirmed by sensitivity matrix analysis (*Fig. 4.3-2*), suggests that the dipole-dipole array has the best resolution with regard to the detection of single objects (e.g., cavities, sand and clay lenses).

The investigation depth of commonly used arrays is, as a rule of thumb, in the range of $L/6$ to $L/4$, where L is the spacing between the two outer active electrodes. The dipole-dipole array, which offers the best depth of investigation (as long as the dipoles are small compared to the distance between them), has the disadvantage of a comparatively low signal-to-noise ratio (WARD, 1990). Therefore, to obtain good quality data, the electrodes must be adequately coupled to the ground and the equipment must have a high sensitivity.

Further important points in the planning of a survey are size and location of the survey area. The survey area should be somewhat larger than the target (e.g., a concealed landfill) in order to properly determine its boundaries. The spacing of profiles and stations depends on the lateral resolution required to evaluate a certain geological situation - the more detailed the information needed, the smaller the interval between measurement stations. However, it must be remembered that the resolution decreases with increasing depth.

Influence of noise and topography

Resistivity measurements in populated areas can be disturbed by various noise sources, such as grounded metal fences, underground metal pipes and cables, power lines, electric corrosion protection for pipelines or leakage currents generated by industrial facilities, streetcars and trains. Especially underground metal pipes can produce anomalies which influence the interpretation quite seriously (VICKERY & HOBBS, 2002). If it is unavoidable to measure in such an area, it is better for the profiles to be perpendicular to the pipes than parallel to them. Another type of noise of natural origin, for example, induction effects of magnetic storms or atmospheric electrical discharges (sferics), can also significantly influence the measurements (HOOGERVORST, 1975).

In many cases, the results of resistivity measurements may be considerably affected by near-surface conditions. It is important to take resistivity changes in the soil cover into account if it is necessary to compare a series of measurements (e.g., soundings) carried out at different times and meteorological situations. Variations in resistivity in the uppermost layer (caused by differences in near-surface moisture content as a result of alternating dry and wet periods, soil temperature, plant water uptake, etc.) can have a significant impact on the sounding curve and the results of the profiling.

Another factor which adversely affects the results is topography (FOX et al., 1980). In a high-relief area, apparent resistivities cannot be calculated using the configuration factors given in *Table 4.3-1* since these *K* values are for a horizontal ground surface. The terrain relief may have a strong influence on shaping the equipotential surfaces and, therefore, the *K*-factor has to be modified to take the terrain geometry into account. This is an important step in the interpretation of 2-D or 3-D resistivity measurements, which requires the topography along the profile to be surveyed by leveling. The results of leveling have to be taken into account as a first step in the 2-D modeling process.

Mapping and profiling

Mapping/profiling methods provide information about the lateral resistivity distribution within a certain depth range. For environmental problems they are mainly used to delineate the boundaries of concealed waste dumps or to locate contamination plumes. In the survey, an electrode array whose parameters are kept fixed is moved along profiles (profiling) or, if possible, on a regular grid (mapping). Thus, the investigation depth varies only slightly with changes in the subsurface resistivity distribution. Because of their operational advantages, Wenner, Schlumberger, and dipole-dipole arrays are the most commonly used (WARD, 1990; MILSOM, 1996). The geometric parameters of an array have to be adjusted to obtain the investigation depth necessary to achieve the project goal. These parameter values can be optimized using the results of several vertical electrical soundings carried out at well-chosen points in the survey area. The survey grid and the profile lines have to be established before starting the survey. One disadvantage of a mapping and profiling survey is the need for manpower – at least 3 to 4 people are necessary. Another problem is the long cables that have to be moved in the field to achieve large investigation depths. For this reason, electromagnetic methods have largely replaced resistivity profiling and mapping in recent years. These methods provide comparable results in many cases requiring less time and less personnel, i.e., they are cost-saving (see Chapter 4.4). An alternative to profiling is 2-D resistivity imaging. Although the costs is often almost the same as for conventional profiling, this method provides much more detailed information along the profile in both the lateral and vertical directions (BARKER, 1992; DAHLIN, 1996).

Vertical electrical sounding

Vertical electrical soundings (VES) has been the most important tools in resistivity surveys for a long time, and for deep investigation depths (>100 m) they still are, although transient electromagnetic soundings (TEM) are being increasingly used for investigation of these depth ranges (see Chapter 4.4). VES is also used for near-surface environmental problems, but in this field 2-D resistivity measurements are becoming increasingly important. Soundings should preferably be used to investigate horizontally layered ground. They provide information about layer thicknesses and resistivity. The apparent resistivities allow conclusions to be drawn with regard to lithological parameters and the composition of the pore fluid (e.g., mineralization of groundwater). If the resistivity data can be correlated with drilling results or other a-priori information, the accuracy of the parameter values derived from 1-D inversion can be better than 5 %.

Soundings should be carried out on an almost regular grid or along profiles. The result of one sounding on a profile can be used as input for the 1-D inversion of the next sounding on that profile. A geoelectrical cross-section can be produced from the results of the soundings.

Because of its high vertical resolution, the Schlumberger array is preferably used for sounding. To obtain information about resistivity as a function of depth, the current electrode (C_1, C_2) spacing[1] L is increased stepwise (geometric sounding). It has proven to be useful to increase the current electrode spacing in logarithmic steps (optimum: 6 - 8 steps per decade), so that the points appear roughly equally spaced when plotted on log-log graph paper. The following sequence has proved useful:

$L/2$ = 1, 1.3, 1.8, 2.4, 3.2, 4.2, 5.6, 7.5, 10, 13, 18, ... 100, 130, 180, ...1000 m.

The potential electrode (inner electrodes) spacing l is increased (as long as the ratio $l/L < 1/3$) only if the measured voltage becomes too small and, therefore, the signal-to-noise ratio becomes too low. In this case, some of the measurements must be repeated (overlap readings), starting with the two current electrode spacings before the potential became too low. Overlapping segments of the sounding curve are obtained, which have to be shifted so that

[1] For the electrode spacing L often the symbol AB is used.

in the overlapping range they are superimposed. This procedure is necessary because increasing the potential electrode spacing often causes the resistivity to change, due to inhomogeneity in the ground near to the potential electrodes or causes the coupling with the ground to change. Data quality should be checked already in the field by plotting the sounding curves on log-log graph paper. This way, the operator is able to react to data outliers, for example, resulting from insufficient galvanic coupling of the electrodes with the ground or geometric errors.

The sounding sites (coordinates of the array centre, direction of the electrode spread) have to be documented in the field notebook. If there is rough terrain, the elevation of each sounding position has to be determined by leveling.

2-D resistivity surveys

In the geoelectrical investigation of environmental problems, 2-D resistivity surveys (2-D imaging) have played an increasingly important role in the last few years. The advantages of 2-D measurements are their high vertical and lateral resolution along the profile, comparatively low cost due to computer-driven data acquisition, which means only a small field crew is needed (one operator and, for basic electrode spacings ≥ 5 m, one assistant). Owing to the performance capabilities of the available instruments, 2-D surveys are, in general, limited to investigation depths down to about 100 m. 2-D measurements are often the most suitable geophysical method for solving environmental problems, which are frequently related to 2-D or 3-D resistivity features.

Many different instruments have been developed for 2-D measurements, usually multi-electrode systems, sometimes multi-channel ones. All systems use multi-core cables (see Section 4.3.4). The procedure for measurements with a Wenner electrode array is shown in *Fig. 4.3-3*.

A multi-core cable with equidistant “takeouts” for connecting the electrodes to the cores is placed along the profile. The basic electrode spacing may be either the distance between the takeouts or a shorter distance. In any case, the electrodes should be positioned with equal spacings along the profile. For high quality data, it is important to ensure good galvanic coupling of the electrodes with the ground. Data acquisition is begun after the multi-core cables are connected to the switching box and the electrode contacts have been checked. The measurements are controlled by the microprocessor-driven resistivity meter or by a computer (BARKER, 1981).

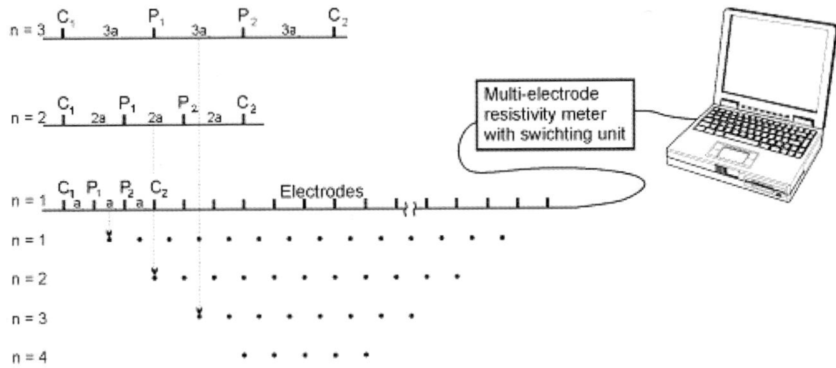

Fig. 4.3-3: Setup for a 2-D resistivity measurement with a Wenner electrode array

Several depth levels may be measured by increasing the electrode spacing stepwise as shown in *Fig. 4.3-3.* The apparent resistivities are plotted as a function of location along the profile and electrode separation. This 2-D plot is called a "pseudosection". For the Wenner electrode array, the location on the profile is given by the center of the array and the depth ("pseudodepth") is given by the spacing of the current electrodes. The "pseudodepth" corresponds to the depth of investigation ($\approx a/2$).

Normally, two multi-core cables are used for each measurement. If the profile is longer than the total length of these two cables, the first cable is placed at the end of the second one and a new measurement is made. The survey can be carried out more efficiently if a third cable is available, as this could be laid out ahead while the measurement using cables 1 and 2 is being made.

To starting and end points and the points where the profile changes direction define the profile. To avoid geometrical errors, these bends should not exceed 15 degrees. In rough terrain the elevation of several prominent points along the profile should be determined.

4.3.6 Processing and Interpretation of the Measured Data

Mapping / profiling

The mapping and profiling results are presented as colored contour maps of apparent resistivity or as profile plots, respectively. Profiling results should be shown as maps only if the distance between the profiles is not larger than five times the station interval along the profile, especially where resistivities vary considerably between the profiles. If there is rugged terrain in the survey area, a topographical correction is necessary before the final maps are made (FOX et al., 1980). The mapping data are interpreted qualitatively. If some resistivity values in the area are known, the lithology in the survey area can be derived, lateral changes can be delineated (e.g., boundaries of concealed landfills), and information about contamination plumes obtained.

Vertical electrical sounding

The results of a resistivity sounding survey can be interpreted both qualitatively and quantitatively. Vertical electrical soundings can provide information about the areas between boreholes. If soundings are more or less regularly distributed in the survey area, apparent resistivity maps can be made for different electrode spacings C_1C_2. This type of map is similar to that obtained in a mapping survey, although the station spacings are commonly larger. In many cases they provide an idea about the resistivity distribution in the area. If soundings are carried out along profiles and the distance between soundings is not too large, the results can be displayed as contoured pseudosections of apparent resistivities. "Pseudodepths" correspond to approximately half the spacing of the current electrodes and, therefore, are preferably presented on a logarithmic scale. This gives a rough visual impression of the geological structures below the profile. The preferred quantitative interpretation of resistivity soundings requires information about the local geology and experience using the method, including knowledge of its limitations. Several computer programs are available for the inversion, allowing data processing, such as editing of data sets and sounding array parameters or shifting of overlapping segments to produce continuous sounding curves (see Section 4.3.5). Modern computer programs allow modelling of the individual segments in a Schlumberger sounding. The interpretation of a sounding curve to derive a layered resistivity model which fits the measured data well is carried out in three steps (ZOHDY, 1989; SANDBERG, 1993).

In the first step, a reasonable starting model has to be created. Experienced geophysicists can derive the number of layers and an initial estimate of the resistivities and thicknesses of the layers directly from the sounding curve. The influence of layer resistivity and thickness on the curve shape is checked by doing both forward and inverse calculations. Important for the interpretation is that the program allows masking of noisy data points and incorporate of a-priori information about the local geology from well logs and other geophysical methods.

The second step involves automatic inversion to confirm and refine the starting model. If model parameter values are known (layer thickness and/or resistivity), they can be held constant. If some parameters do not vary during the inversion, the remaining parameters are more reliably estimated. A least-squares fit of the model curve to the edited data is then made until a selected fitting error is reached. Due to the equivalence basic (see above), several models may fit the sounding curve within the fitting error. Therefore, an inversion program should be able to determine the range of equivalence for a certain error threshold. Equivalence analysis makes it possible to generate a set of equivalent layered models that are also valid solutions of the inverse problem. They fit the data nearly as well as the best-fit model, but deviate from this within a defined error range. With this knowledge it is then possible in a third step to obtain a model which is geologically more plausible than the best-fit model, but which is still acceptable within the error limits. An example is given in *Fig. 4.3-4*.

A by-product of the equivalence analysis is a „parameter resolution matrix". This triangular matrix provides an indication as to how well the layer parameters (thickness and resistivity) are resolved. Owing to equivalence, the two parameters cannot be resolved independently for all layers (Section 4.3.3). The resolution matrix indicates when this may be expected (INTERPEX Ltd., RESIX-IP user´s manual, 1993).

The best way to obtain a realistic interpretation of the soundings is to correlate soundings with lithological borehole logs or well logging data. As a rule, the interpretation is an iterative process because results of the individual soundings always have to be adjusted to the results of neighbouring soundings. The final result of a quantitative interpretation can be presented as a geological model (cross-sections) derived from the sounding data. Contour maps of the depth to layer boundaries or of the thickness of different layers can be constructed from the results of several soundings regularly distributed in the survey area. It has to be mentioned that such layer boundaries do not necessarily correspond to lithological boundaries - they represent significant changes in the electrical properties. This can also occur within a lithological layer (e.g., a change in the mineralization of the groundwater in an aquifer).

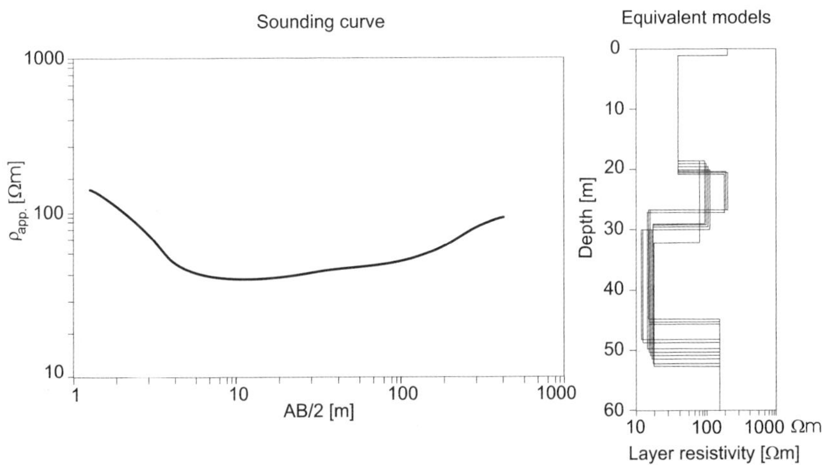

Fig. 4.3-4: Graphical representation of equivalent models. Each of these models fits the measured data equally well.

2-D resistivity surveys

For many environmental problems it is preferable to carry out 2-D resistivity measurements instead of simple soundings and/or profiling. The aim of a geoelectrical survey at the Earth surface is to provide detailed information about the lateral and vertical resistivity distribution in the ground. As the principles applied are similar to those of computer tomography in medicine, such geoelectrical investigations are also called impedance tomography or 2-D resistivity imaging.

The first – qualitative - result of a 2-D resistivity survey is a pseudosection along the profile (EDWARDS, 1977). Pseudosections display the apparent resistivity as a function of location and electrode spacing, which is indirectly, related to the depth of investigation of the array (ROY & APPARAO, 1971; BARKER, 1989; SHERIFF, 1991). *Figure 4.3-3* shows the principle behind the generation of a Wenner pseudosection.

Pseudosections provide an initial picture of the subsurface geology. A 2-D inversion of the measured data is necessary for the final interpretation. This process transforms the apparent resistivities and "pseudodepths" into a 2-D model. In recent years, several computer programs have been developed to carry out such inversions (SHIMA, 1990; BARKER, 1992; LOKE & BARKER, 1995 and 1996). Depending on the algorithm used, the result is a smoothed layer model or a block model showing sharp layer boundaries, comparable to the result of 1-D inversion.

The model used for inversion consists of a number of rectangular cells (blocks). The size of these cells is determined either automatically as a function of the electrode spacing, or manually by the user. In general, the size of the cells increases with increasing depth and towards the beginning and the end of a profile, due to decreasing spatial sensitivities in these areas (see *Fig. 4.3-2*).

All inversion programs require a reasonable starting model (initial estimate). This model is either determined by the program itself or user defined (BARKER, 1989; ZOHDY, 1989). The final model is calculated in an iteration process which includes step-by-step forward modeling and inversion. The program compares the forward modeling results with the measured data, the quality of the fit is determined and the parameter values for the next iteration step are specified. This process is continued until the maximum number of iterations or a selected deviation (error) is reached. The modeling program must allow interactive modification of parameters. For rough terrain, the program should be able to include topographic correction of measured data. Recent programs allow the user to place the electrodes in non-standard positions (e.g., non-regular spacing, electrodes in boreholes or underwater). As in the 1-D inversion of vertical soundings it is advantageous if a-priori information can be included.

In spite of the various capabilities of different programs, the final assessment of inversion results has to be done by the user, i.e., the geophysicist. Due to the limited number of values and precision of the data, all results are ambiguous (Section 4.3.3). This ambiguity can only be reduced by including all available information from boreholes and other geophysical surveys. As the inversion will not normally provide sharp layer boundaries but resistivity gradients (smooth inversion), data for a vertical electrical sounding in areas with almost horizontal layers can be extracted from the 2-D data set and inverted as 1-D vertical soundings. Generally, if gradual structural transitions are indicated by gradual transitions in the resistivity cross-section, a smooth 2-D inversion should be carried out; if the structures are known to have sharp boundaries, then block inversion should be employed.

The final result of a 2-D resistivity survey is a cross-section of the calculated rock resistivities along the profile line. This cross-section should include the structural interpretation of the resistivity data (e.g., the groundwater table or the boundaries of a waste dump). If the spacing between profiles is not too large, horizontal resistivity sections for different depths can also be derived from the data.

4.3.7 Quality Assurance

The following measures are important for quality assurance in a dc resistivity survey. During the survey the following aspects should be checked in the field:

- coupling of the electrodes to the ground (contact resistance),
- drifting self-potentials,
- the parameters (voltage, current) employed for the measurement,
- stability and precision of the measurements (e.g., by repeated readings),
- leakage currents from large power consumers, e.g., industrial plants or electric railways (industrial noise),
- influence of electromagnetic coupling if long cables are used,
- influence to topography, and
- the influence of meteorological factors.

The field notebook documentation should include:

- the instrument employed and the parameters used for the measurements (array type and geometry),
- a topographic map (including major topographic features, such as roads, buildings, rivers and lakes) showing the location of profiles and measurement stations; this description shall enable the reestablishment of the survey grid,
- information about the location of buried objects (cables, metal pipes, etc.),
- rapid changes in the relief, topographic survey results,
- information about visible sources of anomalies (e.g., iron fences, waste dumps, wet and dry areas, changes in the composition of the ground surface), and
- weather conditions.

Attention must be paid to the following aspects of data processing and interpretation:

- The raw and processed data must be delivered to the client in digital form. It must also be archived so that it is available for later reprocessing.
- The client must be provided with a description of the processing parameters and software (e.g., for data correction, data filtering and/or smoothing).
- A technical report should be submitted as part of the final report.
- The results have to be presented in a form that can be understood by the client. The presentation must also provide an explanation of uncertainties in the interpretation and recognized problems in the modeling, particularly from a geophysicist's point of view.

4.3.8 Personnel, Equipment, Time needed

	Personnel	Equipment	Time needed
mobilization and demobilization	depending on the distance to the survey area		
		1 4WD vehicle or van	
mapping: Schlumberger $P_1P_2 / 2 < 3$ m $P_1P_2 / 2 > 3$ m Wenner $C_1C_2 > 15$ m	 1 operator, 2 assistants; 1 operator, 3 assistants; 1 operator, 3 assistants	 resistivity meter, cables, electrodes	 $C_1C_2 / 2 < 30$ m: 500-800 *N*/d $C_1C_2 / 2 > 30$ m: 300-500 *N*/d 400-700 *N*/d
soundings (VES): Schlumberger $C_1C_2 < 130$ m $C_1C_2 \geq 130$ m	 1 operator, 1-2 assistants; 1 operator, 2-3 assistants	resistivity meter with appropriate power supply (high capacity for deep VES), cable reel, electrodes, (optional walkie-talkie)	 12 - 15 VES 5 - 12 VES
2-D resistivity survey: (imaging, tomography) basic electrode spacing: $a < 5$ m $a \geq 5$ m	 1 operator 1 operator 1 assistant	resistivity meter with switching unit, multi-core cables, passive or active electrodes, 1 laptop, (optional: e-mail for data transfer)	about 1500-1800 datum points per day $a = 2$ m: 300 - 500 m/d $a = 5$ m: 700 - 900 m/d (16 depth levels for both)
data processing, interpretation, reporting	1 geophysicist (1 technical assistant)	1 PC with plotter, printer and software; (optional: e-mail for data transfer)	2 - 3 days for each day in the field

a = basic electrode spacing, N = number of stations, d = 10-hour working day, C_1C_2 current electrode spacing, P_1P_2 potential electrode spacing

4.3.9 Examples

Investigation of a concealed landfill in a former sand pit

This example describes the investigation of a covered waste dump in a former sand pit in Germany. The objective was to determine the location and size of the dump site, as well as the distribution of cohesive and non-cohesive soil. A combination of resistivity sounding and mapping was used. Survey parameters:

- Size of the survey area (part of it shown in *Fig. 4.3-5*) 300 m × 500 m
- mapping (Schlumberger array)
  - spacing of current electrodes C_1C_2 20 m
  - spacing of potential electrodes P_1P_2 2 m
  - spacing of profiles 20 m
  - station spacing 10 m
  - total number of stations 815
- soundings (Schlumberger array)
  - max. spacing of current electrodes 360 m
  - number of soundings 5
- duration of the survey 10 days
- personnel 1 technician, 2 assistants
- measured physical parameter apparent resistivity.

The mapping results are shown as a contour map of apparent resistivity ρ_a in the upper part of *Fig. 4.3-5*. The results clearly mark the location of the concealed landfill, as the resistivities of the waste materials are much lower than those of sediments in the surroundings. Although the resistivity map suggests that a contamination plume might be moving to the southeast away from the dump, the resistivity soundings show that cohesive material (boulder clay) is responsible for these low resistivities, while the higher values represent sand and gravel. Because the distance between resistivity soundings is relatively small, a geoelectrical cross-section could be produced; this cross-section shows the landfill is 2 to 4 m deep, mostly covered by non-waste material (*Fig. 4.3-5)*. Sounding 2 and 13 show a sandy aquifer overlain by a layer of cohesive material, which is in turn overlain by a cover layer of sand. The upper sandy layer increases in thickness northwards along the profile. It was advantageous to correlate the sounding results with borehole data in the study area, thus improving the reliability of the interpretation.

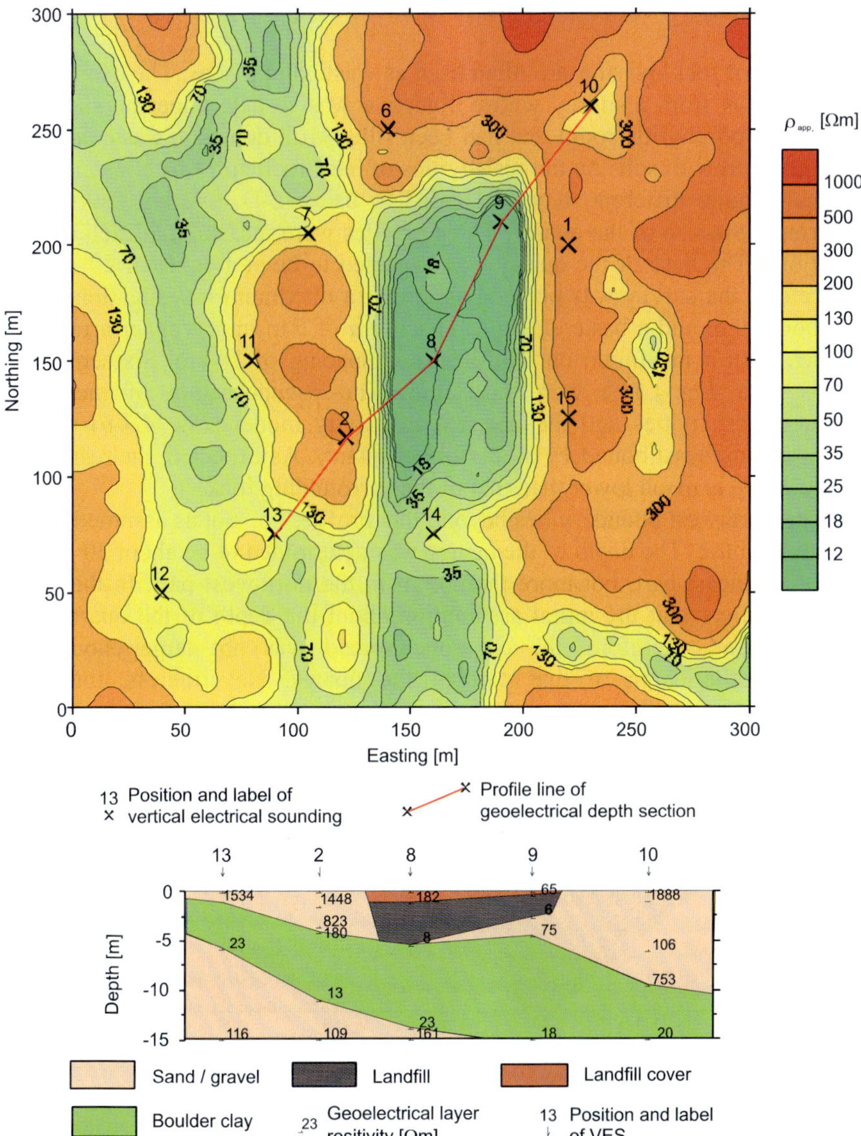

Fig. 4.3-5: dc resistivity survey of a concealed landfill in a former sand pit. *Top*: mapping with Schlumberger array, ρ_a with an electrode spacing $C_1C_2/2 = 10$ m. *Bottom*: geoelectrical cross-section derived from Schlumberger resistivity soundings

Investigation of a concealed landfill in a former quarry

Former hard rock quarries are often used as waste dump sites. The example in *Fig. 4.3-6* is from a survey whose objective was to determine the size of a quarry (in porphyritic rock) that had been filled with domestic waste and was now covered with soil. As the resistivities of hard rock and waste material are normally quite different, it was decided to use a 2-D resistivity survey to determine the size of the quarry. Three profiles were measured at each of five locations. The base of the quarry was expected to be at a depth of 20 - 30 m. Therefore, the survey was planned assuming a maximum investigation depth of about 40 m and a basic electrode spacing a of 5 m in a Wenner array was chosen with a maximum current electrode spacing of 240 m. Sixteen depth levels (see *Fig. 4.3-3*) were measured. The pseudosection of measured apparent resistivities and the inverted resistivity model (*Fig. 4.3-6*) clearly show the southeast boundary of the former quarry. As expected, the resistivity of the waste is much lower than that of the surrounding rock.

The northwest boundary lies beyond this profile, i.e., not as assumed in the planning phase. The depth of the quarry was determined to be about 20 - 25 m in the southeast part, but more than 35 m in the northwest part. In the latter case it is clear that the initial assumption about the depth of the quarry was incorrect and, therefore, the electrode spread was not large enough to determine the depth of that part of the quarry. Nevertheless, this 2-D resistivity survey has proved that it can be very useful for solving such problems.

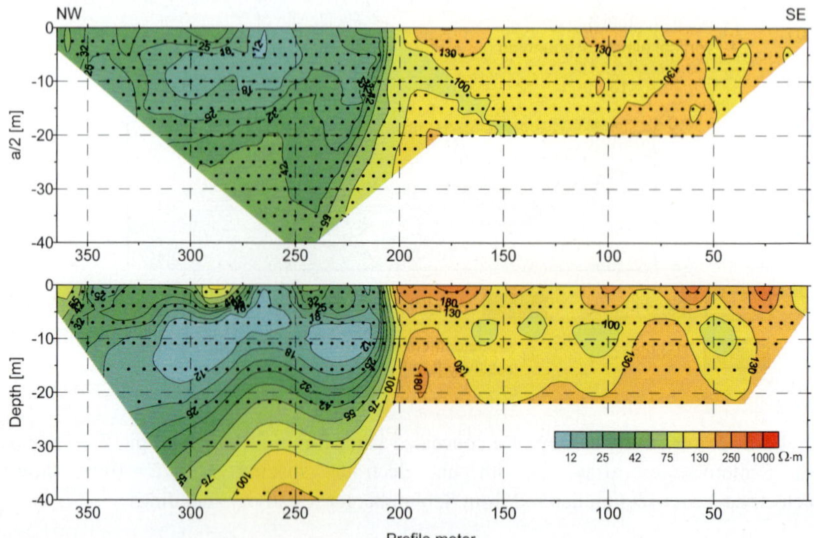

Fig. 4.3-6: dc resitivity survey of a concealed landfill in a former quarry. *Top*: pseudosection of apparent resitivity. *Bottom*: calculated resistivity cross-section.

Investigation of the geological structures in an Alpine valley

A geophysical survey was carried out to determine the geological structures in an Alpine valley in order to assess the environmental hazard emanating from an abandoned, derelict chemical plant in the Austrian Alps. As the study area is in a valley, information about the depth to the bedrock and classification of the overlying unconsolidated material was of interest. Reflection seismics and geoelectrical surveys were conducted for this project. The survey results were correlated with data obtained from several boreholes drilled as part of the project.

To reach an investigation depth of about 100 m, resistivity soundings using Schlumberger array with a maximum electrode spacing C_1C_2 of 840 m were carried out, mostly along three profiles. Before the survey, it was assumed that the bedrock would have a significantly higher resistivity than the overlying unconsolidated sediments, but sounding results did not confirm this. The data from the first borehole showed that the phyllite in the top part of the bedrock was strongly weathered to a low resistivity material. Consequently, it was clear that the boundary between the unconsolidated sediments (mainly sand and gravel) and the underlying bedrock is marked by gradually decreasing resistivities. *Figure 4.3-7* shows the measured sounding curves along one of the profiles, together with the cross-section derived from them. The low resistivity of the bedrock can be clearly recognized in the sounding curves. The resistivity model shown in the lower part of *Fig. 4.3-7* was constructed after 1-D inversion of the sounding data. The bedrock surface dips from the margin of the valley in the west towards the valley center at the east end of the profile. This result is confirmed by the seismic data. The geoelectrical survey provides additional information about the aquifer (sand and gravel) above the bedrock and the layer of more cohesive material (fine sand, silt) at shallow depth. This layer could prevent entry of contaminants into the groundwater. Although data along the individual profiles correlate well, the large distances between profiles, resulting from the urban nature of the survey area, precluded construction of accurate maps of bedrock topography. Despite these limitations, the geophysical results provide a valuable contribution to an understanding of the geological structures in the survey area.

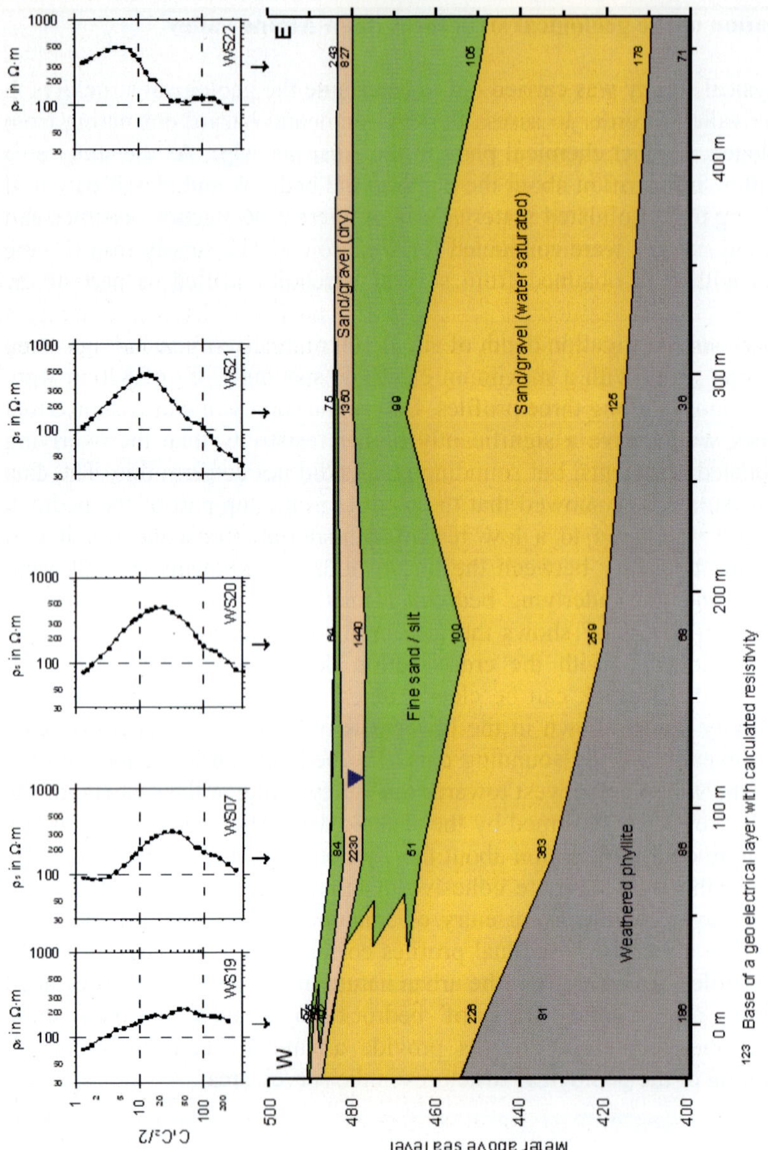

Fig. 4.3-7: Investigation of geological structures by dc resistivity sounding (VES). *Top:* sounding curves (Schlumberger array). *Bottom:* resistivity model as derived from VES with lithological interpretation

Investigation of the site of an old military fortification

Tunnels which were part of a 19th century military fortification (of which nothing remains above ground) were discovered during site studies for redevelopment in a German city. Part of the tunnels were accessible and a considerable amount of explosive material was found, although the fortification was thought to have been completely destroyed at the end of World War II. These explosives were extremely hazardous not only for the development of this area but also could easily contaminate the groundwater. Because the exact location of the old fortification was not known, a geophysical survey was conducted to locate the tunnels which had not been completely destroyed. The tunnels extended radially from the main building. They were 1.3 - 2.2 m high and 0.7 - 1.5 m wide and 5 - 30 m long. They were covered by ground approximately 3 m thick.

Several methods (magnetics, gravity measurements, ground penetrating radar (GPR) and 2-D resistivity imaging) were tested before the survey. As it provided the most convincing results, the resistivity method was chosen for the survey. For economic reasons, the 2-D measurements were carried out with the Wenner electrode array.

For the 2-D resistivity survey using a Wenner array, the following parameters were used for the 150 m × 150 m area:

basic electrode spacing *a*	1 m
number of levels	12
profile spacing	5 m
number of profiles	40
total profile length	5280 m
survey duration	18 days
personnel	1 geophysical technician, 1 assistant
measured parameter	apparent resistivity

The upper part of *Fig. 4.3-8* displays calculated resistivities from the inversion results of all profiles for a depth of about 4.3 m. The higher resistivities (red) mark the location of the remaining tunnels. The geoelectrical results were confirmed by excavation and showed that the initial estimates of tunnel locations were inaccurate. The survey, therefore, provided a means of correcting the preliminary map of tunnel locations.

The lower part of *Fig. 4.3-8* shows the vertical section of the 2-D resistivity profile C-D over a known tunnel in order to look for indications of tunnels extending perpendicular to it. The inversion reveals several high resistivity zones at a depth of ≥ 2.5 m. These high resistivity zones are assumed to be tunnels.

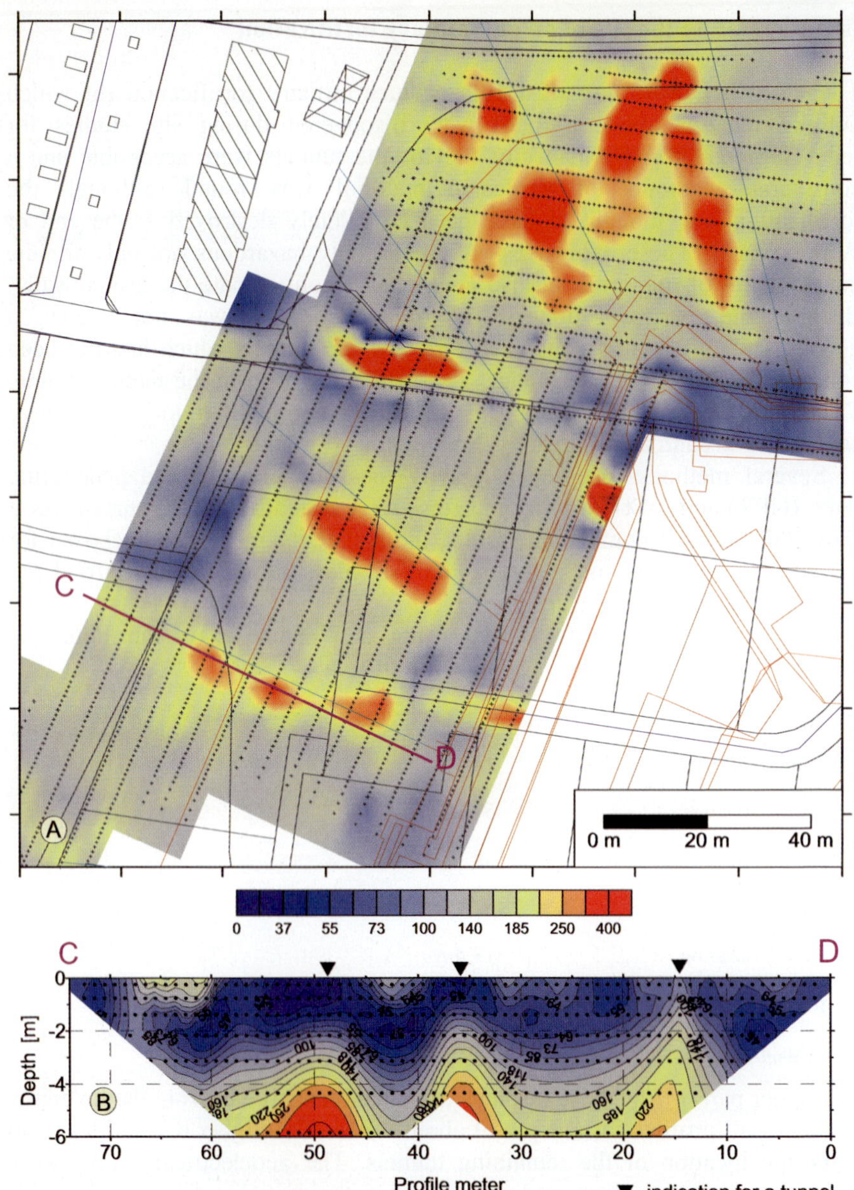

Fig. 4.3-8: Investigation of the site of an old military fortification by a 2-D resistivity survey. *Top*: map of calculated resistivities for a depth of about 4.3 m. *Bottom*: calculated vertical resistivity section for a profile C-D marked in the map

Further examples of dc resistivity surveys are given in *Figures 5.4-22*, *5.4-23*, *6.2-7*, *6.2-16*, *6.2-28*, *6.2-29* and *6.2-37a*.

Parameters and units

	Symbol	Units
electrical potential	U	V
electrical current	I	A
resistivity	ρ	Ωm
apparent resistivity	ρ_a	Ωm
conductivity	σ	mS m^{-1}
apparent conductivity	σ_a	σ_a [mS m^{-1}] = 1000/ρ_a [Ωm]

References and further readings

BARKER, R. D. (1981): The offset system of electrical resistivity sounding and its use with a multicore cable. Geophys. Prosp., **29**, 128-143.

BARKER, R. D. (1989): Depth of investigation of collinear symmetrical four-electrode arrays. Geophysics, **54**, 1031-1037.

BARKER, R. D. (1992): A simple algorithm for electrical imaging of the subsurface. First Break, **10**, 2, 52-62.

BEARD, L.P. & MORGAN, F.D. (1991): Assessment of 2D-resistivity structures using 1D-inversion. Geophysics, **56**, 874-883.

DAHLIN, T. (1996): 2D-resistivity surveying for environmental and engineering applications. First Break, **14**, 275-283.

DAHLIN, T., BERNITONE, C. & LOKE, M. H. (2002): A 3-D resistivity investigation of a contaminated site at Lernacken, Sweden. Geophysicis, **67**, 1692-1700.

DEY, A. & MORRISON, H. F. (1979a): Resistivity modelling for arbitrarily shaped two-dimensional structures. Geophys. Prosp., **27**, 106-136.

DEY, A. & MORRISON, H. F. (1979b): Resistivity modelling for arbitrarily shaped three-dimensional structures. Geophysics, **44**, 753-780.

EDWARDS, L. S. (1977): A modified pseudosection for resistivity and induced polarisation. Geophysics, **42**, 1020-1036.

FOX, R. C., HOHMANN, G. W., KILLPACK, T. J. & RIJO, L. (1980): Topographic effects in resistivity and induced-polarization surveys. Geophysics, **45**, 75-93.

FRIEDEL, S. (2000): Über die Abbildungseigenschaften der geoelektrischen Impedanztomographie unter Berücksichtigung von endlicher Anzahl und endlicher Genauigkeit der Meßdaten. Dissertation, Universität Leipzig. Berichte aus der Wissenschaft. Shaker, Aachen.

HABBERJAM, G. M. (1979): Apparent resistivity observations and the use of square array techniques. Geoexploration Monographs, Series 1, **9**. Bornträger, Berlin.

HOOGERVORST, G. H. T. C. (1975): Fundamental noise affecting signal-to-noise ratio of resistivity surveys. Geophys. Prosp., **23**, 380-390.

INTERPEX Ltd. (1993): RESIX-IP user´s manual. Golden Colorado.

KAROUS, M. & PERNU, T. K. (1985): Combined sounding-profiling resistivity measurements with three-electrode arrays. Geophys. Prosp., **33**, 447-459.

KOEFOED, O. (1979): Geosounding principles **1**: resistivity sounding measurements. Elsevier, Amsterdam.

LILE, O. B., BACKE, K. R., ELVEBAKK, H. & BUAN, J. E. (1994): Resistivity measurements on the sea bottom to map fracture zones in the bedrock underneath sediments. Geophys. Prosp., **42**, 813-824.

LOKE, M. H. & BARKER, R. D. (1995): Least-squares deconvolution of apparent resistivity pseudosections. Geophysics, **60**, 1682-1690.

LOKE, M. H. & BARKER, R. D. (1996): Rapid least-squares inversion of apparent resistivity pseudosections using a quasi-Newton method. Geophys. Prosp., **44**, 131-152.

MARIN, L. E., STEINICH, B., JAGLOWSKI, D. & BARCELONA, M. J. (1998): Hydrogeologic site characterization using azimuthal resistivity surveys. JEEG, **3**, 179-184.

MILSOM, J. (1996): Field Geophysics. Geological Society of London handbook. Open University Press and Halsted Press. Wiley and Sons, Chichester.

MORRIS, M., RONNING, J. S. & LILE, O. B. (1997): Detecting lateral resistivity inhomogeneities with the Schlumberger array. Geophys. Prosp., **45**, 435-448.

NOWROOZI, A. A., HORROCKS, S. B. & HENDERSON, P. (1999): Saltwater intrusion into the freshwater aquifer in the eastern shore of Virginia: a reconnaissance electrical resistivity survey. Journal of Applied Geophysics, **42**, 1-22.

OGILVY, R., MELDRUM, P., CHAMBERS, J. & WILLIAMS, G. (2002): The Use of 3D Electrical Resistivity Tomography to characterise Waste Leachate Distribution within a closed Landfill, Thriplow, UK. Journal of Environmental and Engineering Geophysics, **7**, 11-18

OLAYINKA, A. I. & YARAMANCI, U. (2000): Assessment of the reliability of 2D-inversion of apparent resistivity data. Geophys. Prosp., **48**, 293-316.

OLAYINKA, A. I. & YARAMANCI, U. (2002): Smooth and sharp-boundary inversion of two-dimensional pseudosection data in presence of a decrease in resistivity with depth. Europ. J. Environm. Engin. Geophys., **7**, 139-165.

OLDENBURG, D. W. & LI, J. (1999): Estimating depth of investigation in dc resistivity and IP surveys. Geophysics, **64**, 403-416.

RAMIREZ, A., DAILY, W., BINLEY, A., LABRECQUE, D. & ROELANT, D. (1996): Detection of leaks in underground storage tanks using electrical resistance methods. JEEG, **1**, 189-203.

ROY, A. & APPARAO, A. (1971): Depth of investigation in direct current methods. Geophysics, **36**, 943-959.

SANDBERG, S. K. (1993): Examples of resolution improvement in geoelectrical soundings applied to groundwater investigations. Geophys. Prosp., **41**, 207-227.

SCHULZ, R. & TEZKAN, B. (1988): Interpretation of resistivity measurements over 2D-structures. Geophys. Prosp., **36**, 962-975.

SHARMA, S. P. & KAIKKONEN, P. (1999): Appraisal of equivalence and suppression problems in 1D EM and DC measurements using global optimization and joint inversion. Geophys. Prosp., **47**, 219-249.

SHERIFF, R.E. (1991): Encyclopedic Dictionary of Exploration Geophysics, 3rd edn. SEG, Tulsa, OK.

SHIMA, H. (1990): Two-dimensional automatic resistivity inversion technique using alpha - centers. Geophysics, **55**, 682-694.

SPITZER, K. (1995): A 3D-finite difference algorithm for DC resistivity modelling using conjugate gradient methods. Geophys. J. Int., **123**, 903-914.

SPITZER, K. & KÜMPEL, H.-J. (1997): 3D FD resistivity modeling and sensitivity analyses applied to a highly resistive phonolitic body. Geophysical Prospecting, **45**, 963-982.

SPITZER, K. (1998): The three-dimensional DC sensitivity for surface and subsurface sources. Geophys. J. Int., **134**, 736-746

SUMNER, J.S. (1976): Principles of Induced Polarization for Geophysical Exploration. Elsevier, Amsterdam.

TELFORD, W. M., GELDART, L. P. & SHERIFF, R. E. (1990): Applied geophysics, 2nd edn., Cambridge University Press, Cambridge.

VANHALA, H., SOININEN, H., KUKKONEN, I. (1992): Detecting organic chemical contaminants by spectral-induced polarization method in glacial till environment. Geophysics, **57**, 1014-1017.

VICKERY, A. C. & HOBBS, B. A. (2002): The effect of subsurface pipes on apparent-resistivity measurements. Geophys. Prosp., **50**, 1-13.

WARD, S. H. (1990): Resistivity and induced polarization methods. In: WARD, S. H. (Ed.): Geotechnical and environmental geophysics, **I**: Review and Tutorial. Society of Exploration Geophysicists, Tulsa, Oklahoma, 147-189.

WATSON, K. A. & BARKER, R. D. (1999): Differentiating anisotropy and lateral effects using azimuthal resistivity offset Wenner soundings. Geophysics, **64**, 739-745.

WELLER, A. & BÖRNER, F.D. (1996): Measurements of spectral induced polarization for environmental purposes. Environmental Geology, **27**, 329-334.

WELLER, A., FRANGOS, W., SEICHTER, M. (1999): Three-dimensional inversion of induced polarization data from simulated waste. Journal of Applied Geophysics, **41**, 31-47.

ZOHDY, A. A. R. (1989): A new method for the automatic interpretation of Schlumberger and Wenner sounding curves. Geophysics, **54**, 245-253.

4.4 Electromagnetic Methods

GERHARD LANGE & KNUT SEIDEL

4.4.1 Principle of the Methods

Electromagnetic inductive methods provide an excellent means to obtain information about electrical ground conductivities. They can be classified as natural field methods and controlled source methods. The well-known natural field method magnetotellurics, used since the 1950s employs fluctuations of the Earth's magnetic field ranging from 10^{-5} seconds to several hours to study the distribution of the conductivities with depth (CAGNIARD, 1953). The electromagnetic exploration methods described in this Chapter are based on the use of electromagnetic fields generated by controlled sources.

Easy to operate, active methods are preferably used in environmental geophysics. Dual coil systems (with a transmitter and a receiver coil) have been commercially available since the 1970s. Used in controlled-source methods, they are important tools for investigating shallow ground structures. They can be subdivided into two basic types:

- frequency-domain electromagnetic methods (FEM) and
- time-domain or transient electromagnetic methods (TEM).

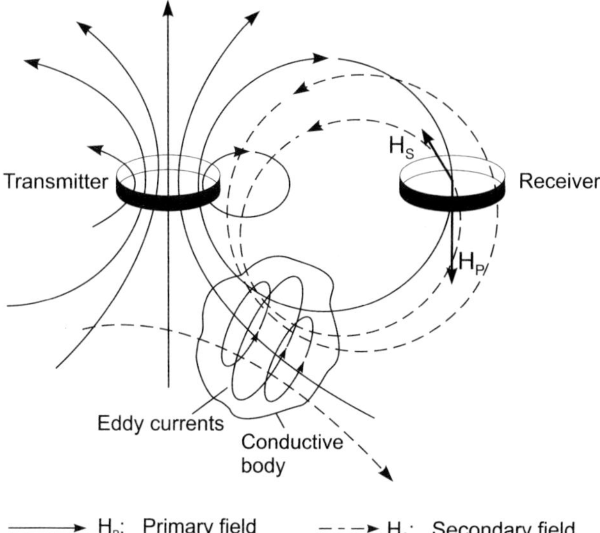

Fig. 4.4-1: Principle of electromagnetic induction methods

The principle of the most commonly used FEM systems is shown in *Fig. 4.4-1.* A transmitter coil, continuously energized with a sinusoidal audio-frequency current, forms a magnetic dipole. Its primary magnetic field induces very weak eddy currents in the conductive ground. These eddy currents in turn generate a secondary magnetic field that is of the same frequency but with a different phase and of lower amplitude than the primary field. The primary and secondary magnetic fields are superimposed on each other and the resultant field is detected by the receiving coil. To obtain the ground conductivity information, the weak secondary field has to be separated from the primary field.

The secondary magnetic field strength depends on the electrical conductivity of the ground, on the transmitter-receiver coil figuration, and on the operating frequency. Although these relationships are expressed by a complicated function, a linear relationship between ground conductivity and the secondary magnetic field can be obtained by choosing an appropriate coil spacing and frequency. Operation using a coil spacing and frequency pair that yields such a linear relationship is called operation at "low induction numbers (LIN)". LIN ground conductivity meters allow the conductivity to be determined directly (McNeill, 1980).

Time-domain electromagnetic (TEM) soundings are being increasingly used in environmental and groundwater surveys (Hoekstra & Blohm, 1990). Investigation depths of commercially available equipment range from about 5 m to 3000 m. Like FEM techniques, TEM methods use a dipole magnetic source field. The waveform of this primary field is not continuous as in FEM methods, but a modified square wave (*Fig. 4.4-2*). By abruptly turning off a constant current in the ungrounded transmitter loop, a transient electromagnetic field is produced. According to Faraday's law, this rapid change induces eddy currents in nearly horizontal circles in the ground below the loop, which in turn create a secondary magnetic field. With increasing time after transmitter current switch-off, the induced fields have penetrated deeper into the Earth by "diffusion".

The decay of the secondary magnetic field caused by the eddy currents is recorded during the transmitter´s off-time, i.e., in absence of the primary field (*Fig. 4.4-2c*). The decay rate as a function of the ground conductivity and the time, after the transmitter current has been switched off, determines the investigation depth. As the current density maximum migrates to greater depths (*Fig. 4.4-6*), the influence of conductive near-surface structures on the signal response decreases. This behavior is important as near-surface inhomogeneities are often the main cause of poor data quality in other electrical sounding methods.

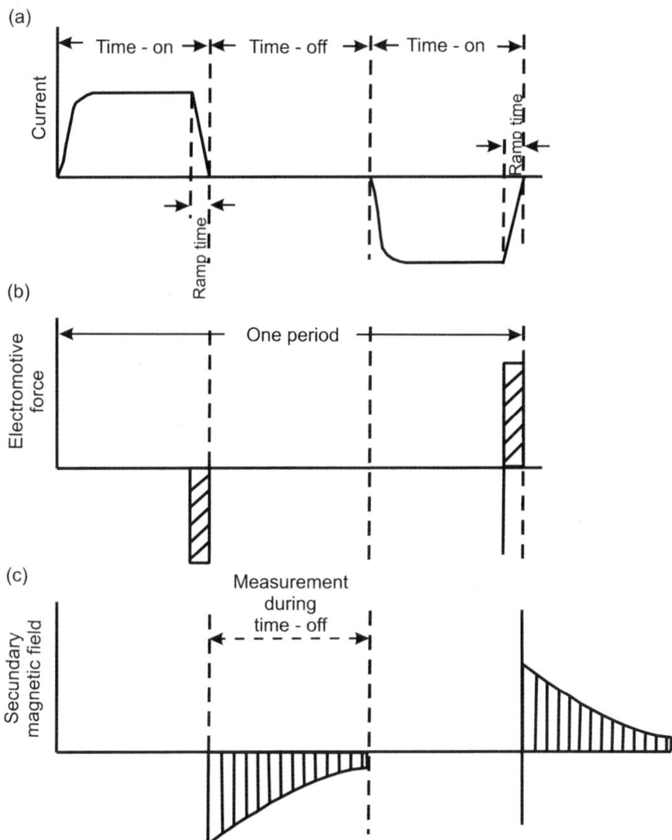

Fig. 4.4-2: Typical waveforms occurring during a TEM measurement: (a) a current in transmitter loop, (b) induced electromotive force caused by current, (c) secondary magnetic field caused by eddy currents, HOEKSTRA & BLOHM (1990)

FEM and TEM methods are available for land-based and airborne surveys and for borehole logging (DYCK, 1991). For small-scale environmental investigations surface measurements are usually chosen. Due to inductive coupling, these methods are preferably used in areas with high-resistivity cover layers (e.g., arid areas, areas of nonreinforced concrete or asphalt, exposed bed rock, or frozen ground), where direct-current resistivity methods fail due to poor galvanic coupling. Because of the induction principle, EM methods are best suited for detecting highly conductive targets in a low-conductivity host environment (e.g., clay layers embedded in sandy material and buried metal objects).

There are three kinds of measurements: mapping, profiling, and sounding. Profiling provides information about the variation in lateral conductivity within a certain depth range. This depth range is directly proportional to the transmitter-receiver coil separation and is inversely proportional to the operating frequency and ground conductivity. FEM methods are mainly used for profiling as they are not as time consuming as TEM methods. The interpretation of profiling data delivers mainly qualitative results. A very fast, high-sensitivity TEM mapping technique has been developed to detect both ferrous and nonferrous metallic targets, such as unexploded ordnance (UXO) (see example in Section 4.1.9). Measurements of the secondary magnetic field decay in different time gates allow the characterization of detected metallic objects (McNeill & Bosnar, 1996; Snyder et al., 1999).

FEM soundings are done by changing the frequency and/or the coil separation (parametric/geometric sounding). TEM techniques generally provide better sounding data than FEM techniques. Their depth of investigation is directly proportional to the shut-off time, to the recording time interval, and to the loop size (Spies, 1989). To obtain TEM data for both near-surface and deep depth ranges, different loop sizes have to be used. A quantitative interpretation is obtained from the sounding data by modeling and inversion. Until now 1-D inversion is the most commonly applied interpretation.

The very low frequency (VLF), the VLF-resistivity (VLF-R), and the radiomagnetotellurics (RMT) methods are passive methods which employ the electromagnetic fields of remote radio transmitters. Distributed around the world, VLF transmitters are primarily used for military communications; RMT utilizes the frequencies of normal broadcast stations. In contrast to these methods, which utilize electromagnetic waves from stationary transmitters, the equipment for controlled source audiomagnetotellurics (CSAMT) includes a portable transmitter.

The transmitter frequencies are between 10 to 30 kHz for VLF/VLF-R and up to 2.4 MHz for RMT. Because of the usually very large distances to the source, the transmitted electromagnetic waves can be considered as plane waves. When these waves strike the ground, they are partly reflected, another part enters the ground, inducing eddy currents there. To obtain information about the conductivity of the ground from VLF-measurements, selected magnetic components of the coupled EM field and, in case of VLF-R, electric field components are measured. At least two stations transmitting within a narrow frequency range should be used for a survey. A disadvantage of VLF methods is that the transmitter stations are not always available. The main use of the VLF/VLF-R methods is for mapping with only a qualitative interpretation of the data. RMT, when used as a multifrequency method, provides sounding capabilities and, due to the higher frequency range, RMT data allow modeling of 1-D, 2-D and simple 3-D near-surface situations. Because the equipment for VLF/VLF-R and RMT methods is light-weight and easy to operate, these methods can also be used to locate conductive geological and man-made structures in low conductivity host material and to delineate their extent and strike (e.g., mapping of fractures in consolidated rock, boundaries of landfills and leachate plumes) (MC NEILL & LABSON, 1991; TURBERG et al., 1994; TURBERG & BARKER, 1996).

EM surveys are less time consuming and require less manpower than dc resistivity surveys. As the transmitter-receiver array necessary to investigate a certain depth range is much smaller than the equivalent dc array, EM methods provide a better lateral resolution and a higher survey efficiency. The disadvantage of EM methods is their lower vertical resolution and higher sensitivity to electromagnetic noise compared to dc resistivity.

4.4.2 Applications

Typical applications for EM methods in near-surface geophysics are:

- delineation of concealed or abandoned landfills and sometimes determination of landfill thickness,
- mapping of pollution plumes in groundwater (if conductive),
- mapping and monitoring of saltwater intrusion,
- geological mapping (soil/rock types, lithology, faults, fracture zones),
- groundwater surveys,
- detection of clayfilled karst cavities,
- detection of metallic objects (both magnetic and nonmagnetic), and
- UXO-detection.

4.4.3 Fundamentals

Electrical conductivity in rocks

The propagation of electromagnetic fields within the Earth is determined by the bulk conductivity, dielectric permittivity, and magnetic permeability of the medium. Very high-frequency methods, such as ground penetrating radar (Chapter 4.5), are strongly affected by magnetic and dielectric properties. For low-frequency electromagnetic fields, as considered here, conduction currents are important. There are three types of conduction.

Electron conduction

In rock materials containing free electrons (e.g., metals, sulfides), the current is based on electron conduction. As most rock-forming minerals (silicates, oxides) are insulators, this type of conduction can be neglected for this discussion.

Electrolytic conductivity

The conductivity of porous rocks, especially unconsolidated sediments, strongly increases if they are fluid-saturated. In this case, the electrical current arises from the flow of ions in the formation fluid through the effective pore space. Isolated fluid-filled pores have only a negligible impact on the bulk conductivity, but only a thin continuous layer of fluid at the grain surfaces can increase the conductivity of an unsaturated medium by 4 - 5 orders of magnitude.

The conductivity of fully or partially saturated sediments is predominantly determined by their porosity and the conductivity of the pore fluid. An empirical equation commonly used to quantify the conductivity of clean sandy sediments (no shale) is defined as Archie´s law (1942):

$$\sigma_0 = \frac{\Phi^m}{a}\sigma_w S^n \qquad (4.4.1a)$$

where

σ_0 electrical conductivity of the sediment (in mS m^{-1}, µS cm^{-1}),
σ_w electrical conductivity of the pore fluid (in mS m^{-1}, µS cm^{-1}),
ϕ effective porosity (in %)
S fraction of pores containing water,
n „saturation exponent“ (often $n \approx 2$),
m cementation exponent (for most sedimentary rocks $1.3 < m < 2.5$), and
a empirical parameter, depending on the type of sediment $0.5 < a < 1$.

For fully fluid-saturated material is $S = 1$. In partially saturated material, the insulating influence of air in the pore space decreases the conductivity.

Interface conductivity

A third type of electrical conductivity results from excess ions in a diffuse double layer (Helmholtz double layer) at the boundary between solid and pore fluid. To take this interface conductivity into account, Archie´s Equation (4.4.1a) is modified by introducing an interface conductivity term σ_{q0}:

$$\sigma_0 = \frac{\Phi^m}{a}\sigma_w S^n + \sigma_{q0} \,. \qquad (4.4.1b)$$

The interface conductivity (excess conductivity) is a frequency dependent component of the measured conductivity and has a real and an imaginary part. Its real part is directly proportional to the porosity Φ, and its imaginary part is directly proportional to the inner surface area of the sediment. This allows the hydraulic properties to be determined from conductivity measurements carried out at different frequencies (see Chapter 4.3).

The geoelectrical ground parameters are the same for both electromagnetic and dc resistivity surveys. In dc resistivity surveys, the unit of ohm × meter (Ω m) is used, while in EM surveys the unit is siemens per meter (S m^{-1}), for practical reasons millisiemens per meter (mS m^{-1}). Variations in resistivity and conductivity may result from changes in porosity, degree of saturation with water, salinity, and clay content. Resistivity values of various materials are given in *Table 4.3-2* (Chapter 4.3). To convert the resistivity values ρ in the table into conductivity σ, note that $\sigma = 1 / \rho$, i.e., a conductivity of $\sigma = 1$ mS m^{-1} corresponds to a resistivity of $\rho = 1000\ \Omega$ m.

Frequency-domain electromagnetic

Small two-coil systems, so called loop-loop systems, are used for land-based FEM surveys. For transmitter-receiver separations greater than five loop diameters, the loop can be treated as a magnetic dipole. Vertical magnetic dipoles result when coils are horizontal and horizontal magnetic dipoles result when the coils are vertical (*Fig. 4.4-5*). Eddy currents are generated by induction in the ground, which give rise to a secondary magnetic field superimposed on the primary field. The resulting total field has the same frequency as the primary field but differs in intensity, direction and phase. The total field is described by a complex-valued function consisting of both real and imaginary parts. The real part, which has the same phase as the primary field, is called in-phase. The imaginary part, which shows a 90-degree phase difference from the transmitted signal, is called out-of-phase, outphase, or quadrature. To obtain information about the conductivity, the magnetic part of the resulting electromagnetic (EM) field detected by the receiver coil is recorded.

Calculations of EM fields for inhomogeneous ground conditions are rather complicated. To explain the induction and measuring process, a simplified representation of an EM system is given in *Fig. 4.4-3*. A detailed explanation of interactions in the system consisting of transmitter loop (S_1), receiver loop (S_3) and conductive ground (S_2) can be found in several papers (WEST & MACNAE, 1991; TELFORD et al.,1990).

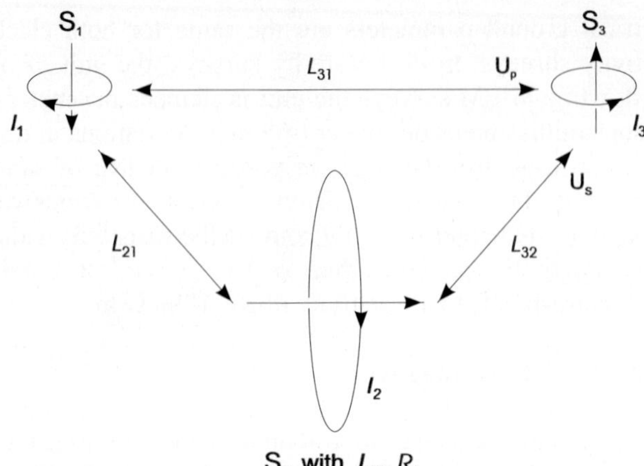

Fig. 4.4-3: Electromagnetic mutual coupling of conductive ground (S_2) to transmitter loop (S_1) and receiver loop (S_3), simplified representation of complicated interactions, L_{ij} are the mutual inductances, WEST & MCNAE (1991)

Important for the understanding of EM effects is the response function $f(Q)$. This complex-valued function describes the electromagnetic response from conductive ground structures to an electromagnetic excitation field:

$$f(Q) = \frac{Q^2 + \mathrm{i}Q}{1 + Q^2}. \qquad (4.4.2)$$

The argument of $f(Q)$, called the response parameter Q or induction number, depends on the operating frequency, conductivity of the ground and the target, and on the size of the target and the geometry of the transmitter and receiver loops:

$$Q = \mu\sigma\omega l^2$$

where

μ	magnetic permeability,
σ	conductivity,
$\omega = 2\pi f$	angular frequency (s^{-1}),
f	operating frequency, and
l	parameter for the geometry of the target and/or loop-system.

The response parameters for simple models are given in *Table 4.4-1*. If one of the variables changes, then one or more of the others must be changed to keep Q constant.

Tab. 4.4-1: Response parameters for simple models

Model	**Response parameter Q**	
homogeneous half-space	$\mu\sigma\omega r^2$	r – coil separation
infinite horizontal plate	$\mu\sigma\omega dh$	d – plate thickness h – height of loop above plate
semi-infinite vertical plate	$\mu\sigma\omega dr$	d – plate thickness r – coil separation
disc	$\mu\sigma\omega da$	a – disc radius d – disc thickness
sphere	$\mu\sigma\omega a^2$	a – sphere radius

In the plate and disc models, the product of conductivity σ and thickness d is called conductance.

Although higher frequencies result in a stronger response for a given ground situation and system geometry, increasing the frequency will lead to a reduction in penetration depth. Hence, when multifrequency instruments are used, it is important to select a frequency (or frequencies) that will give both an optimal response and desired penetration depth.

The behavior of the complex-valued response function $f(Q)$ for a homogeneous ground is shown in *Fig. 4.4-4*. For values of $Q < 1$ and $Q > 1$ the imaginary part (quadrature) and the in-phase part prevail, respectively. When $Q << 1$, the contribution of the real part to the measured signal is negligible and the information about the ground conductivity is mainly in the quadrature component. This is important for understanding the principle of the low-induction-number (LIN) instruments.

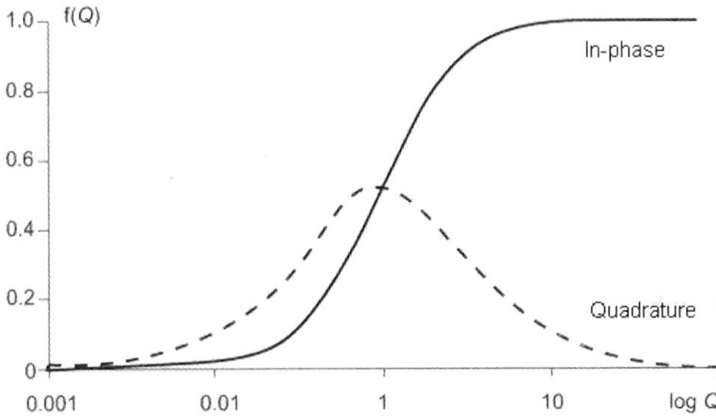

Fig. 4.4-4: Complex response function of a loop-loop system to conductive ground, GREINWALD (1985)

Skin depth

For every survey situation it is important to have an idea of the depth to which the transmitted electromagnetic signals penetrate. Induced eddy currents consum energy and attenuate the fluctuating field strength, thus reducing penetration. The depth at which the amplitude of a plane wave has been attenuated to 37 % (1/e) of its original value is called the skin depth δ, also called effective depth (SHERIFF, 1991). For a homogeneous ground the skin depth δ is given by the following numerical equation:

$$\delta[\mathrm{m}] = \sqrt{\frac{2}{\sigma\mu\omega}} = 503\sqrt{\frac{1}{\sigma\left[\mathrm{Sm}^{-1}\right]\cdot f[\mathrm{Hz}]}} = 503\sqrt{\frac{\rho[\Omega\mathrm{m}]}{f[\mathrm{Hz}]}} \tag{4.4.3}$$

However, this relationship is only valid for plane waves from large distances to the source (far-field conditions), as employed by the magnetotelluric, VLF, and RMT methods. For the small two-loop FEM systems, which are operated with short loop separations, the penetration depth is less than that derived from Equation (4.4.3) due to the difference in the wave-front geometries. But, there is no general rule to determine the depth of investigation of two-loop FEM systems. Generally, the penetration depth increases with increasing loop separation and decreasing frequency. Although REYNOLDS (1997) states that “a realistic estimate of the depth, to which a conductor would give rise to a detectable EM anomaly is $\approx \delta/5$”, the use of general rules for depth estimation is not recommended.

Loop configuration

The loop configuration strongly influences the depth of penetration, resolution and sensitivity of EM systems. The most common source-receiver configurations used in FEM methods are shown in *Fig. 4.4-5.*

In practice, mainly horizontal coplanar (HLEM, Slingram) and vertical coplanar loop configurations (VLEM) are used. Only a few instruments allow measurements with configurations c) and d) (e.g., APEX MaxMin). Of prime importance is the depth of penetration and the resolution of an FEM system used (VERMA & SHARMA, 1995). HLEM systems have the greatest penetration depth range, whereas VLEM systems have the highest sensitivity to near-surface layers. This means the strongest contribution to the measured signal is from these layers (*Fig. 4.4-9*).

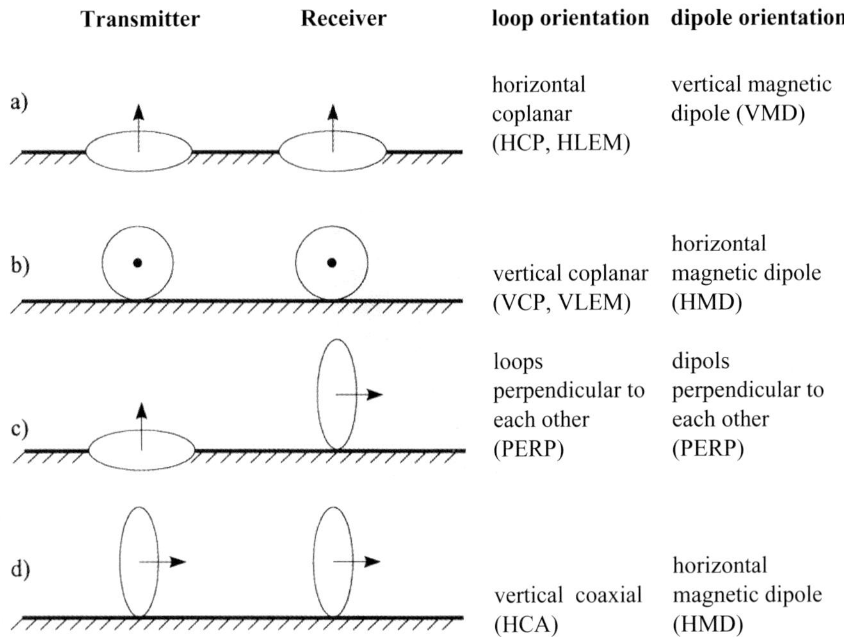

Fig. 4.4-5: Common loop configurations of small loop dipole-dipole systems, after SPIES & FRISCHKNECHT (1991)

Electromagnetic conductivity measurements

Determination of conductivity (or its reciprocal, resistivity) from electromagnetic measurements is much more complicated than from dc resistivity data (DAS, 1995). Therefore, the measured values of out-of-phase and in-phase components are frequently presented on profiles or maps for a qualitative interpretation of the locations of conductivity anomalies.

During the early 1980s the company GEONICS developed and constructed instruments which allowed the measurement of conductivity directly. The principle of these "ground conductivity meters" (GCM) is the low induction number (LIN). As shown in *Fig. 4.4-4*, the quadrature component dominates the signal response when $Q << 1$; moreover, it is directly proportional to the ground conductivity. Q can be expressed in terms of skin depth δ as follows:

$$Q = r / \delta$$

Thus, it can be shown that the LIN condition is fulfilled if the coil separation r is very small compared to the skin depth δ (MCNEILL, 1980).

Let $H_{z,p}$ be the vertical component of the primary magnetic field in free space and H_z the vertical component of the measured field over a homogeneous conductive half-space. Then for a vertical magnetic dipole source (VMD) the mutual coupling ratio $H_z/H_{z,p}$ is given by (SPIES & FRISCHKNECHT, 1991):

$$\frac{H_z}{H_{z,p}} = \frac{2}{(\gamma r)^2}\left[\left(9 + 9\mathrm{i}\gamma r - 4\gamma^2 r^2 - \mathrm{i}\gamma^3 r^3\right)e^{-\mathrm{i}\gamma r} - 9\right] \tag{4.4.4}$$

For a horizontal dipole (vertical coplanar loops, HMD), the tangential component H_Φ has to be taken into consideration, resulting in the following mutual coupling ratio:

$$\frac{H_\Phi}{H_{\Phi,p}} = \frac{2}{(\gamma r)^2}\left[3 + \gamma^2 r^2 - \left(3 + 3\mathrm{i}\gamma r - \gamma^2 r^2\right)e^{-\mathrm{i}\gamma r}\right] \tag{4.4.5}$$

In terms of the induction number Q and skin depth δ, the term γr is

$$\gamma r = \sqrt{2\mathrm{i}}\frac{r}{\delta} = \sqrt{2\mathrm{i}}Q.$$

When $Q \ll 1$ and, correspondingly, $|\gamma r| \ll 1$, both mutual coupling ratios reduce to

$$\frac{H_z}{H_{z,p}} \cong \frac{H_\Phi}{H_{\Phi,p}} \cong \frac{\mathrm{i}\mu_0\omega\sigma r^2}{4}. \tag{4.4.6}$$

On the basis of Equation (4.4.6), the apparent conductivity σ for both HLEM as well as VLEM measurements can be directly derived from the quadrature component:

$$\sigma = \frac{4}{\mu_0\omega r^2}\left(\frac{H}{H_p}\right)_{\text{quadrature component}}. \tag{4.4.7}$$

Some instruments (e.g., EM31, CM-031; EM38, CM-138 and EM34-3) utilize this concept by using fixed combinations of frequency and coil separation and a proper calibration of the instrument. But, this kind of conductivity determination fails when metallic objects are present (high induction number due to the extremely high conductivity of metals) or if there are very strong lateral conductivity changes.

Time-domain electromagnetics

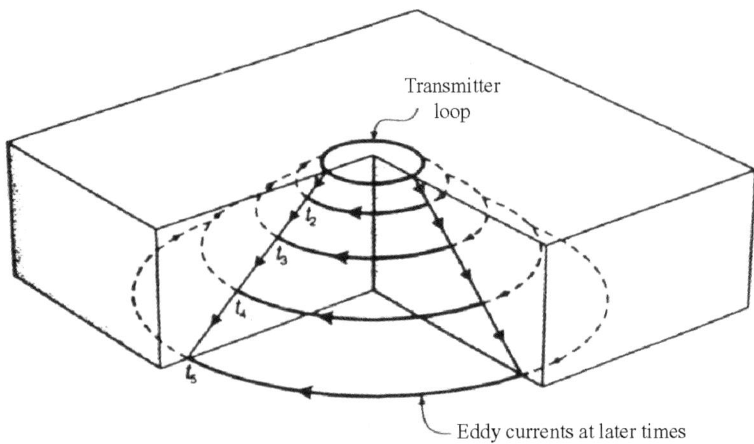

Fig. 4.4-6: Propagation of circular eddy current systems in a homogeneous half-space at successive time intervals, REYNOLDS (1997). © John Wiley and Sons Ltd. Reproduced with permission.

TEM surveys are conducted with ungrounded square or circular loops lying on the ground. The transmitter loop is energized with short pulses of direct current, which is turned off as rapidly as technically possible after a specified "time-on" period (*Fig. 4.4-2*). The time it takes for the current to return to zero after being turned off mainly depends on the loop size (PELLERIN et al., 1994). For loops with side lengths ≤ 100 m this length of time ("ramp time") is less than 5 µs. When the transmitter is turned off, the primary field rapidly decays, eddy currents are generated in conductive ground material below the loop concentrically around the vertical axis of the horizontal loop. Similar to the movement of "smoke rings" they "diffuse" downwards, expanding outward with time and dissipate due to ohmic losses. The decaying currents induce new electromagnetic fields, which cause new currents at even greater depth. This "smoke ring-like diffusion" is illustrated in *Fig. 4.4-6*. The "diffusion" depth is the time-domain equivalent of the skin depth in FEM methods. The depth of investigation of the TEM method is about half of the diffusion depth. The velocity v_m of the downward movement of the induced currents is given by (NABHIGHIAN & MACNAE, 1991):

$$v_m = \frac{2}{\sqrt{\pi \mu \sigma t}} \quad [\mathrm{m\ s^{-1}}] \qquad (4.4.8)$$

where

σ conductivity [S m^{-1}]
t propagation time [s]
μ magnetic permeability [Vs A^{-1} m^{-1}].

The diffusion depth z at time t can be calculated as follows:

$$z = \sqrt{2t/(\mu\sigma)} \quad [\mathrm{m}]. \tag{4.4.9}$$

Unlike FEM methods, the signal decay is recorded when the primary field is not present. Because there is no crosstalk between the primary field and the secondary field, TEM measurements can use very high transmitter powers, thus allowing areas with high overburden conductivities to be surveyed, where penetration depth of FEM measurements is limited by the skin depth and/or current "channeling". The very weak signal response, measured in nV m^{-2}, requires multiple excitation and stacking of several hundreds to thousands of responses to improve the signal-to-noise ratio in order to obtain good quality data. A typical TEM response, which is a function of conductivity and time, can be divided into early, intermediate, and late stage. The response curve is measured as receiver output voltage at a succession of up to thirty narrow time gates starting from a few microseconds and progressively increasing to tens or hundreds of milliseconds after the transmitter current turn-off. This is comparable to the measurement of the IP transient in time-domain IP systems (see Chapter 4.3).

TEM survey configurations

Differences in the sensitivity of different loop geometries and loop sizes are one of the most useful aspects of TEM methods. Although there is a wide variety of different loop configurations, as shown in *Fig. 4.4-7,* the most common survey configuration consists of a multi-turn horizontal receiver coil located in the center of a single-turn squared transmitter loop (an in-loop configuration). The depth of investigation is proportional to the magnetic moment $\boldsymbol{m}$ of the transmitter loop, which is the product of current intensity and loop area. Because of limited transmitter current intensities, an increase in the depth of investigation can only be obtained by increasing the loop area. Side lengths as small as 5 to 10 m are appropriate for shallow depths. Small targets can be more exactly located with an in-loop configuration, while large targets are often better delineated with fixed transmitter-loop, moving receiver-loop layouts. To take full advantage of the differing sensitivities of loop configurations, the TEM equipment used for high-resolution surveys should allow rapid and easy changes of loop geometry (MAULDIN-MAYERLE et al., 1998).

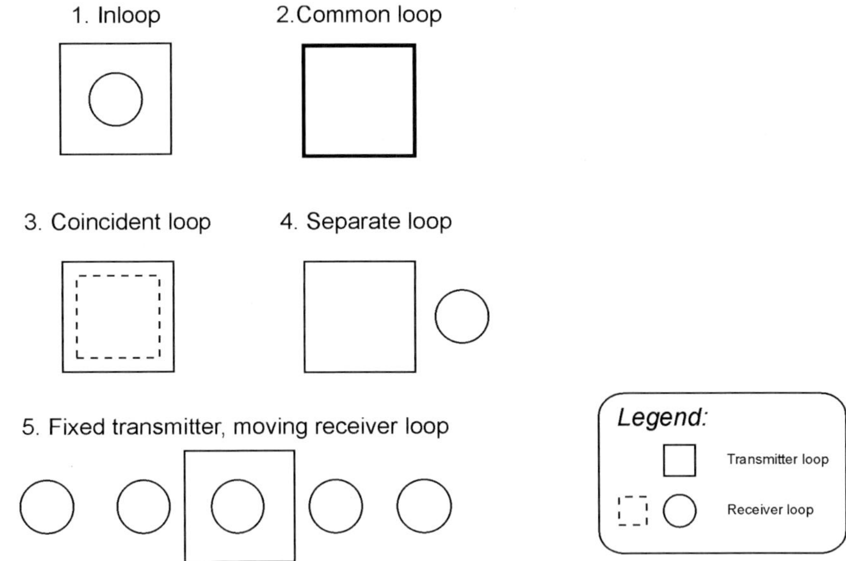

Fig. 4.4-7: Common layout configurations of transmitter and receiver loops for TEM surveys

Very low frequency (VLF, VLF-R) and radiomagnetotelluric (RMT) method

Electromagnetic fields used in VLF, VLF-R and RMT are transmitted by remote broadcasts. They consist of coupled electrical and magnetic fields, oscillating perpendicular to the direction of propagation. McNeill & Labson (1991) show that almost all magnetic VLF field anomalies are caused by the accompanying subsurface horizontal electrical field. This electrical field creates "current channeling" into conductive targets embedded in a more resistive host. The secondary magnetic fields coupled with these currents generate anomalous magnetic field components, which are detected with a VLF-receiver measuring the vertical magnetic component. As VLF is not a controlled source method, there is no reference signal like in the above-described loop-loop methods. Readings of a single field component can be strongly influenced by the source field and they are, therefore, meaningless without calibration. At least one further component must be selected as a reference to compare amplitudes and phases. The most suitable VLF reference is a horizontal magnetic field component, as this approximates the source signal the best (Milsom, 1996; Berktold, 2005). The measured vertical magnetic component is normalized with respect to the maximum value of the horizontal magnetic field. To avoid the influence of strong temporal variations in the source field, which mostly occur during sunrise and sunset, VLF surveys should not be carried out at such times. The depth of penetration is estimated to be about two-thirds of the skin depth, corresponding to equation (4.4.3). Depending on the frequency used, the depth of penetration is limited to the top 10 m for resistivities ≈10 Ωm, but reaches more than 100 m for resistivities >>1000 Ωm (bedrock, dry sands). In areas where a thick conductive overburden exists, as a result of chemical weathering in, for example, tropical regions, leads to a reduction in depth penetration.

The maximum VLF response is recorded when the strike of the target (e.g., a vertical conductive fracture) is same as the direction of the transmitter, decreasing rapidly with increasing deviation from this direction. The curves for the in-phase and quadrature phases of the vertical magnetic component measured on a profile perpendicular to steeply dipping sheet-like conductors cross above the conductive target (*Fig. 4.4-8*). Conductors with a strike perpendicular to the transmitter direction are not well coupled and could be overlooked. Therefore, the directions of profile lines and directions to the transmitter stations should always be depicted in VLF interpretation maps.

In the VLF resistivity method (VLF-R), one horizontal electrical field component and the horizontal magnetic field component orthogonal to this are measured. In analogy to magnetotellurics, an apparent ground resistivity can be derived from these two values. A parameter which frequently shows more detailed structural features than the magnetic field values is the phase difference between the horizontal electric and magnetic field components.

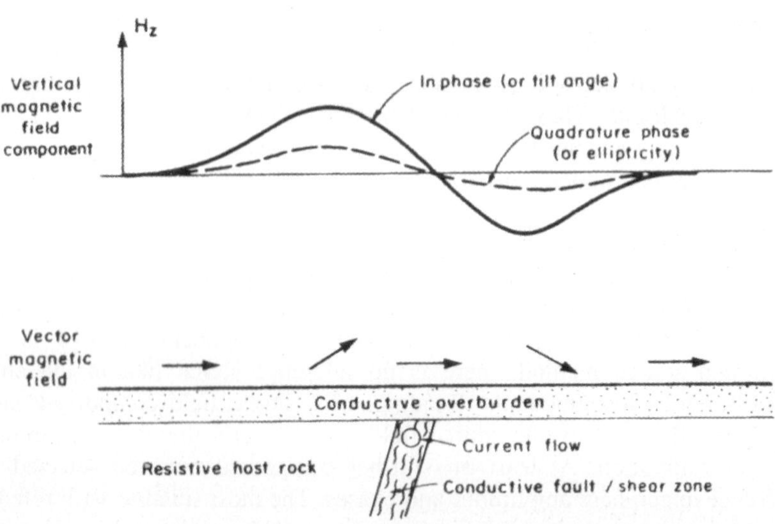

Fig. 4.4-8: Vertical magnetic field component over a steeply dipping conductor, MCNEILL (1990)

Therefore, this parameter should be used in any more detailed interpretation. Both apparent resistivity and phase are automatically calculated by the VLF-R instrument and stored on an internal data storage medium. Above conductivity anomalies whose strike differs from that of the transmitter-receiver direction, the electrical field in the strike direction and the magnetic field component perpendicular to that should be measured. An advantage of VLF-R technique is that terrain relief has little influence on the measurements.

Radiomagnetotellurics (RMT) covers a wider range of frequencies than the VLF-R method. The use of frequencies up to 240 kHz makes soundings possible. This is a considerable advantage over VLF/VLF-R, which can be used only for mapping or profiling. Large areas can be surveyed by RMT mapping and/or sounding within a short time. RMT data reflect lateral and vertical conductivity structures with high resolution. 1-D and 2-D data inversion methods are available (TEZKAN et al., 2000).

VLF and RMT measurements can typically be done by one person. As they can be carried out quickly, they can be used to assess the suitability of an area under investigation before planning a more expensive conventional EM survey. A comprehensive description of VLF and VLF-R is given by MCNEILL & LABSON (1991). For further information about RMT refer to the publications of TURBERG et al. (1994) and TURBERG & BARKER (1996).

4.4.4 Instruments

Electromagnetic methods offer the broadest variety of instrumental systems used in geophysics and different instruments are used in North America, Europe, Asia, and Australia. EM instruments can be divided as follows:

- FEM: frequency-domain methods, single and multifrequency instruments, LIN instruments,
- TEM: time-domain EM, VETEM (very early time domain EM),
- VLF/VLF-R, RMT: very low frequency/very low frequency resistivity and radiomagnetotellurics, and
- CSAMT: controlled source audiomagnetotellurics.

In environmental investigations there is a need for instruments that are easy to operate and allow shallow depth ranges to be rapidly investigated with a high efficiency. FEM and TEM equipment fulfill these requirements. CSAMT is increasingly being used for hydrogeological investigations. Although VLF and RMT are very effective methods with low manpower requirements, they are of minor importance at present. The main reason for not using VLF more frequently is the unreliable availability of VLF stations.

Frequency-domain instruments

All FEM instruments used for environmental exploration employ transmitter and receiver loops with diameters of less than 1 m. Transmitter and receiver are normally connected by a shielded reference cable to carry the phase-reference signal for the primary field compensation. Some instruments allow voice communication via this cable. The Swedish name "Slingram" is sometimes used for the horizontal-loop method. In some instruments for near-surface investigations, the two coils are mounted at a fixed separation in one unit. The measurements are made at one or at several frequencies. The internal data storage medium must be capable of storing at least one day's measurements.

Instruments allowing variable coil separations are employed for exploration of structures deeper than 10 m. Ground conductivity meters (GCM) operate with a fixed combination of coil separation and frequency. Other instruments allow a more or less free combination of these parameters and, therefore, can be more easily adapted to the survey objectives Most instruments are designed to be used with at least two loop configurations: horizontal coplanar or vertical coplanar (*Figs. 4.4-5* and *4.4-9*). The configuration selected mainly depends on the depth of investigation required to solve the problem at hand.

A large number of EM instruments are available, thus any list of instruments will surely be incomplete. However, the most widely used instruments used for environmental investigations are given in *Table 4.4-2*.

Table 4.4-2: Specifications and applications of selected FEM instruments

Name and specifications	**Application and special features**
Geonics EM 38 and EM 31 **Geofyzika CM-138 and CM-031** fixed frequency f and intercoil spacing r; coil figuration horizontal/vertical coplanar EM 38(CM-138): f=14.6 kHz, r = 1 m EM 31(CM-031): f= 9.8 kHz, r = 3.66 m measured parameters: conductivity (mS m^{-1}) and in-phase response (ppm)	Mapping of shallow depth (0 - 1.5 m or 0.5 - 6.0 m); due to the fixed transmitter-receiver separation, there are no errors due to inexact coil separation measurement; depth penetration and sensitivity can be easily changed by changing coil orientation; one-person operation.
Geonics EM 34-3 3 frequency f_i and coil spacing r_i pairs; coil figuration horizontal/vertical coplanar; f_1 = 6.4 kHz; r_1 = 10 m f_2 = 1.6 kHz; r_2 = 20 m f_3 = 0.4 kHz; r_3 = 10 m measured parameter: apparent conductivity	Mapping and limited sounding capabilities for investigation depths of about 6 - 60 m; measurements require two persons.
Apex MAXMIN up to 10 frequencies (110 - 56 320 Hz) and coil separations from 5 m to 400 m can be combined; four coil configurations are possible, standard is horizontal coplanar; measured parameters: in-phase/out-of-phase component and apparent conductivity	Mapping, sounding and imaging with exploration depths of about 5 - 400 m; greatest variability of all FEM instruments; frequency as well as geometric soundings are possible; measurements require two persons.
Geophex GEM 300 allows up to 16 user-defined frequencies between 330 Hz and 20 kHz to be measured simultaneously. Fixed coil spacing r = 1.3 m; measured parameters: apparent conductivity (in mS m^{-1}) and in-phase response (in ppm)	Mapping of near-surface depths. By using multiple frequencies the user can select the best results for interpretation; one-person operation.
Geometrics STRATAGEM EH4 combination of magnetotellurics (MT) and CSAMT. Frequency range 0.1 Hz to 92 kHz; portable dual-loop vertical transmitter antenna; orthogonal electric and magnetic fields are recorded to provide tensor impedance for investigating apparent resistivity/conductivity structures.	Sounding and high-resolution 2-D imaging with investigation depths of 5 to 500 m; deeper structures are imaged using natural MT-signals.

Time domain instruments

In TEM instruments, the transmitter and receiver can be installed in one unit or in separate units. If separated, the components must be synchronized via reference cable. The TEM source consists of a rapid-turn-off, high-current transmitter with a loop on the ground as antenna. The current is cycled on and off in pulses. The transient signal decay is recorded during the current off-time in multiple time windows (gates). The number of gates varies from 20 up to 30. Traditionally, TEM has been used for the exploration of several hundred meters deep ground and, therefore, time intervals monitored on the decay curve typically vary from 50 μs to hundreds of microseconds. For near-surface investigations, time windows vary from microseconds to some milliseconds.

The ramp time (time the current takes to return to zero after being turned off) depends on the loop size, number of turns, and strength of the current. Ramp times for the deep penetration depths required for ore exploration range between 20 - 120 μs. As mainly shallow depths have to be surveyed in environmental investigations, the ramp time must be very short. Recent improvements in electronics allow transmitter ramp times of < 1 μs in conjunction with small loops (< 10 m × 10 m) and very high sampling rates, increasing the amount of information for modeling and interpretation. Instruments with loop diameters of about 1 m have been developed for special applications such as UXO detection. Instruments mounted on a cart provide multiple measurements with high spatial resolution. Because eddy currents induced in metallic objects show longer decay times than those in the host rocks, metallic targets appear above the background (McNeill & Bosnar, 1996; Snyder et al., 1999).

One of the latest instruments designed to investigate conductivity and dielectric properties of the very shallow depth range is the USGS VETEM (Very Early Time domain EM). This instrument operates in the frequency range between commercial time-domain systems and ground penetrating radar (GPR) to fill the gap between GPR and the upper limit of traditional TEM depth range (Pellerin et al., 1994; Smith et al., 2000).

Table 4.4-3: Specifications and possible applications of selected TEM instruments

Name and specifications	Application and special features
PROTEM /TEM47 Separate transmitter and receiver units with synchronization via reference cable; for shallow exploration 5 m × 5 m single- or 8-turn transmitter loop; shut-off times 0.5 µs and up to 30 time gates allow highest resolution at shallow depths.	Shallow ground characterization, locating conductive contaminant plumes, saline groundwater intrusion or lithological and structural profiling; exploration depths up to 150 m.
Zonge NanoTEM / ZEROTEM Very fast turn-off TEM system with separate transmitter / (multichannel) receiver. Transmitter-loop sides 5, 10, ... 100 m; all three magnetic field components (Hx, Hy, Hz) can be detected using a three-axis receiver loop (side length 1 m); shut-off time 1.5 µs allows investigation depths of less than 2 m.	The system offers both deep sounding metal detection as well as resistivity sounding capabilities; waste site, groundwater characterization and lithological and structural features; UXO and metallic objects detection.
EM 61 / EM 61-HH / EM 63 TEM metal detectors to locate both ferrous and nonferrous objects; target characterization and discrimination (material, size, shape) based on the signal decay rate in multiple time gates; fixed transmitter and receiver loop mounting; GPS compatible for real-time positioning.	Underground metal tank and UXO detection; single 200 L (55 gal.) metal drums at depths > 3 m are detected; relatively insensitive to nearby noise sources (e.g., powerlines, fences, buildings); depth to target estimated from the width of the response; one-person operation.
VETEM (prototype) Square transmitter and receiver loops with 0.76 m side length; transmitter shut-off time <10 ns; time window 10 to 1000 ns after shut-off; rapid stacking for signal/noise improvement; high density imaging of the 1 - 10 m depth range, also in conductive terrain.	Investigation of landfills by high resolution imaging of conductive and dielectric targets, as well as in conductive host material; multidimensional VETEM data processing software with graphical interfaces.

VLF/VLF-R instruments

By using induction coils VLF/VLF-R instruments measure the vertical magnetic component and one or two (mutually perpendicular) horizontal magnetic components of the electromagnetic field emanating from distant radiostations or portable VLF transmitters. Additionally, most types are able to record one or two components of the induced electric field E_x and/or E_y (parallel and perpendicular to the profile direction, respectively) by means of the voltage between two electrodes in the ground. Signals from up to three transmitting stations are recorded simultaneously. State-of-the-art instruments display the measured data in numerical or graphic mode on LCDs and store data on data storage media. The signal distortion of field components shown on the display of the instrument in the field provides the first indication of conductive structures (e.g., water-bearing fracture zones). The most commonly used instruments are the GEONICS EM 16/EM 16R, ABEM WADI and SCINTREX ENVI. Lists of VLF transmitter stations around the world are given in BERKTOLD (2005) and in the International Frequency List of ABEM (http://*www.abem/se/vlf/vlf-freq.pdf*).

4.4.5 Survey Practice

The first step in an electromagnetic survey is to select an appropriate technology to accomplish the survey objectives. The selection of the method depends mainly on following parameters:

- required information (e.g., lateral or vertical conductivity distribution, fracture detection),
- depth of investigation,
- size of the area under investigation,
- geological conditions, and
- terrain conditions and logistics in the survey area.

Mapping

For environmental and engineering applications, mostly shallow depths need to be surveyed. The lateral distribution of the ground conductivity can be best investigated with a mapping or profiling survey. Due to their fast and easy operation, FEM ground conductivity meters operating in the low induction number range are preferably used to investigate depths down to about 30 m. Mapping is carried out along profiles with a constant station spacing depending on the required lateral resolution. If detection of fractures in solid rock is required, the station spacing should not exceed one half of the intercoil spacing. The distance between profile lines should be approximately the same as the coil spacing. The profile spacing can be increased for economic reasons, but it should not exceed five times the station spacing to allow reliable correlation of anomaly patterns on neighboring profiles. In general, profile lines should be perpendicular to the strike of geological structures. If the geological conditions are not known, it could be useful to carry out the survey along profiles in two perpendicular directions.

Although the depth of exploration depends on the coil separation and the frequency used (SPIES, 1989), it is also important to select an appropriate coil configuration. Either horizontal coplanar (HCP) or vertical coplanar (VCP) coil configurations can be selected with most instruments. Depending on the coil configuration the current strength in the ground varies with depth, indicating that different configurations have different sensitivities. The functions that express these different sensitivities quantify the contribution of conducting material at different depths to the measured signal. The sensitivities of horizontal coplanar and vertical coplanar coil systems are shown in *Fig. 4.4-9*.

In the case of vertical coils, near-surface material will contribute most of the measured value. For horizontal coils, the maximum contribution originates from material at depths of about 0.4 times the coil separation. Therefore, horizontal coils (HCP) should be used for the mapping of relatively great depths. For depth ranges greater than 30 m, instruments allowing coil separations of 40 m and more have to be used. In this case, the frequencies must be carefully selected to fulfill the low-induction-number criterion.

Mapping is also successfully applied for the detection of metal objects (UXO detection) and in locating cables or metal pipes to about 5 m depth. Besides FEM instruments, specialized TEM instruments, such as the EM61(63) product series and the NanoTEM have proved to be suitable. The measurements are carried out along a regular grid with small spacing (< 1 m) and allow locating both ferrous and nonferrous objects together with target characterization and discrimination.

For mapping of fractures and faults, the VLF and RMT methods yield good results and are an alternative to the active loop-loop methods. The advantage of the VLF methods is a very rapid survey while a disadvantage is its dependence on appropriate transmitter stations.

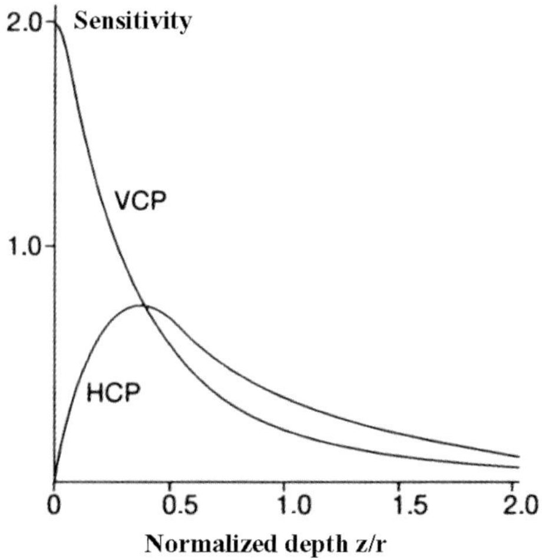

Fig. 4.4-9: Sensitivity of horizontal coplanar (HCP) and vertical coplanar coil systems (VCP); the depth z is normalized relative to the coil separation r, McNEILL (1980)

Transmitters with sufficient transmitter power must be present in an appropriate direction from the structures to be surveyed. While the western hemisphere is covered by many stations, there are only a few stations in the eastern hemisphere. Sometimes, transmitters can be off-air and the survey has to be interrupted or repeated. That could restrict and delay or even prevent conducting a survey.

Soundings

Sounding methods are used to investigate the vertical distribution of ground conductivity. Electromagnetic soundings offer advantages especially in arid regions, as they do not need galvanic ground contact. Since the EM configuration required for a certain depth of investigation is much smaller than that of the corresponding resistivity array, EM survey results have a higher lateral resolution. Experience has shown that the disadvantage of EM soundings is their lower vertical resolution for shallow depths than that obtained with resistivity soundings. Different penetration depths in FEM soundings are achieved by either changing the frequency (parametric sounding) or the coil separation (geometric sounding) or a combination of both. For sufficient resolution, at least four frequencies must be measured per frequency decade (SPIES & FRISCHKNECHT, 1991). There are only a few instruments suitable for such soundings, e.g., APEX MaxMin-10 (with ten frequencies from 110 Hz up to 56 360 Hz) and the CSAMT instrument STRATAGEM. Depending on the ground conductivity, the maximum depth of penetration can be assumed to be in the range of 0.5 to 1.5times the coil separation. A good quality sounding requires the coil separation to be kept as accurately as possible. The sounding result is assigned to the midpoint of the transmitter and receiver coil configuration.

Topography can have considerable influence on the data. In contrast to the VCP mode, the HCP mode, particularly the in-phase component, is very sensitive to the relief. If the slope of the ground between transmitter and receiver position exceeds 10 degrees, it has to be measured (e.g., using an inclinometer). An appropriate program must be used to correct for the terrain effects. When the HCP mode is used for profiling and sounding-profiling, a decision has to made in rough terrain whether the coils are held horizontal or parallel to the slope. The former is easier to operate, but requires appropriate corrections.

TEM soundings with small loop sizes (≤ 20 m) are being increasingly used for shallow investigation depths. TEM soundings with larger loops are the best choice for the depth ranges below 100 m. *Table 4.4-4*, which shows the depth penetration in dependence on the ground resistivity, provides a guideline for planning TEM soundings.

Table 4.4-4: Maximum depth penetration of a TEM sounding, GEONICS PROTEM

Square-loop side length [m]	**Resistivity of the top layer [Ωm]**		
	25	100	400
10	50 m	70 m	95 m
40	90 m	120 m	160 m
80	120 m	160 m	220 m

In addition to selected loop size, appropriate time gates for the recording of the decay curve have to be chosen, as the depth penetration is also a function of time. In practice, square loops are laid out as these are easier to handle than circular loops. The loop corners have to be positioned precisely.

Due to the high sensitivity of EM methods to metal objects and to electromagnetic disturbances, it is important to know as much as possible about sources of EM noise in and around the survey area (e.g., metallic fences, cables, pipelines, railway tracks, transmitter stations). If practicable, the profile should be perpendicular to elongated noise sources, e.g., cables and power lines.

4.4.6 Processing and Interpretation of Measured Data

Mapping

Normally, FEM instruments which directly provide conductivity values are employed for mapping. The data is plotted as profile plots or as contour plots. All plots should be accompanied with sufficient information for the client to be able to locate anomalies and understand the results. The plots are the basis for a qualitative interpretation of the lateral conductivity distribution (e.g., location and delineation of abandoned landfills, detection of contamination plumes, changes in lithology or water content). When vertical or inclined fractures are encoutered, the conductivities calculated from quadrature data on the assumption of a homogeneous or horizontally layered ground may be erroneous. For this reason, instruments providing in-phase and quadrature data are indispensable when fractures are present. Profile plots or maps of these components show distinct anomalies indicating the locations of faults and fracture zones. The typical shape of the profile or contour plot when a fault is crossed is shown in *Fig. 4.4-10*. This behavior can be used for a quantitative estimation of position, depth and dip of vertical or near-vertical fractures.

The shape of such an anomaly will be deformed if the profile is not perpendicular to the conductor and the station spacing is $\leq 0.5\ r$ (where r is the coil separation). Inversion software or type curves derived from modeling are used to interpret the data (KETOLA & PURANEN, 1967). If topography affects the data, corrections have to be made before interpretation.

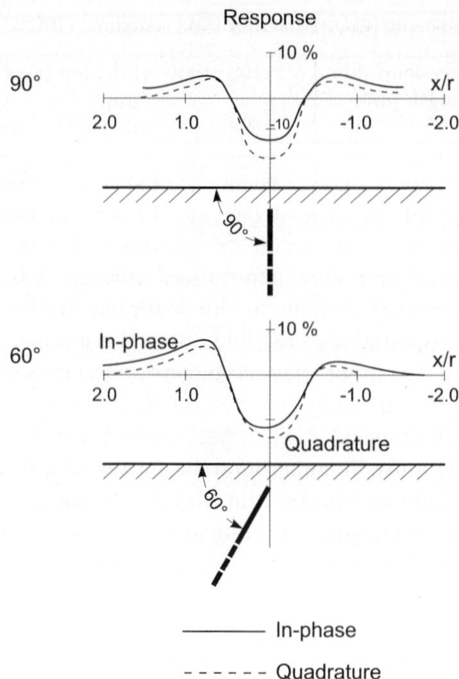

Fig. 4.4-10: The influence of the dip of a thin conductive plate on the in-phase and quadrature components, VOGELSANG (1991). The dip causes the curves to be asymmetric.

In the case of VLF surveys, corrections are necessary due to varying transmitter signal strength as well as terrain effects (VALLÉE et al., 1992; EBERLE, 1981). The interpretation can be supported by data filtering. The Fraser or Karous-Hjelt (averaging) filter is used to shift the maximum of a thin dipping sheet anomaly to above the thin sheet (cross-over point). When this is done, fault zones are easier to recognize (FRASER, 1969; KAROUS & HJELT, 1983; GUERIN et al., 1994).

TEM metal detectors, e.g., the EM61(63) line, NanoTEM and VETEM, can be used to locate underground storage tanks, buried drums, pipelines or for UXO detection and produce data that can be plotted as contour maps. The depth to the target and the type of target can be estimated with the software used with these instruments.

Soundings

Electromagnetic soundings provide information about the vertical conductivity distribution. As in dc-soundings, the data is presented in the form of a layered ground model. For FEM and TEM soundings, only 1-D inversion software is commercially available. The electromagnetic modeling and analysis program "EMMA" can be used to learn 1-D modeling of electrical and EM data. It is available online at no charge (AUKEN, et al., 2002). In recent years, numerous investigations have been conducted to improve inversion algorithms and measurement techniques to enable 2-D and 3-D EM surveys, but up to now, the application of such software has been limited to research projects (OLDENBURG, 2001).

As FEM soundings can be carried out by varying the frequency and/or the coil spacing, it is recommended to select inversion software that can handle both types of soundings. The result of 1-D inversion is a horizontal layer model of layer thickness and resistivity/conductivity. The 1-D inversion of FEM soundings is mostly limited to 3 or 4 layers. The results of sounding-profiling or of soundings along a profile are presented as geoelectrical cross-sections or as contour maps of layer boundaries. It is also possible to produce horizontal sections of the calculated resistivity for a given depth. It should however be noted that geoelectrical layers are not necessarily identical to lithological layers and that boreholes for calibration are desirable. Before the inversion is begun, it is necessary to check the quality of the data. It is especially important to recognize noise and 2-D/3-D effects (i.e., on the shape and amplitude of the response function). Abrupt lateral changes in geology at the sounding location strongly affect the inversion results. Moreover, erroneous interpretation can result from placement of the coils at incorrect separation distances. The in-phase component, which is strongly affected by such separation errors, must be zero for short coil separations at low frequencies (e.g., when using the MaxMin instrument with <100 m separation at 110 Hz). Separation errors are identified by fluctuating in-phase values. Similar effects appear along slopes even if the straight line distance between the coils is correct. Such terrain effects have to be recognized already in the field. Incorrect coil separation should be corrected for by using appropriate software before plotting and interpretation.

Two TEM soundings with different loop sizes and time windows (early time and late time curves) have to be carried out if information about both shallow and deeper depth ranges is required. In many cases, 1-D inversion would be sufficient, as the lateral resolution of TEM soundings is very high; in any case better than for dc resistivity soundings, due to the comparably small layout. The data processing steps depend on the instruments used. In general, the following steps are necessary for 1D-inversion:

- Transient voltages are normally measured within given time gates. If necessary, these voltages are normalized with respect to the given transmitter current, the area of the receiving loop and the gain applied to different channels.
- Replicate data are checked for errors and averaged. Different time intervals are combined into a single data set.

- Decay voltages are plotted versus decay time to check the data quality. The interpretation must be made with reservation if ground chargeability (IP-effects) or 3-D structures influence the data.
- The following equation (NABIGHIAN & MACNAE, 1991) provides an initial estimate of the apparent resistivity ρ_a:

$$\rho_a = \left\{ \frac{\pi (\mu_0 / t)^5 r^8}{400} \left| \frac{I}{U} \right|^2 \right\}^{1/3} \tag{4.4.10}$$

where

I - current [A],
μ_0 - permeability of a vacuum,
r - loop radius [m] (loop area has to be converted into equivalent circular area),
t - time [s], and
U - measured voltage [V].

If soundings are made along profiles, this approximation can be used to produce pseudosections of apparent resistivity.

- The final step is the 1-D inversion, which yields a horizontal layered model with layer thicknesses and resistivities/conductivities. The software for 1-D inversions normally allows interactive and/or automatic fitting of model curves to the measured data.

The presentation of TEM results is similar to that for FEM data. Soundings along profiles are presented as geoelectrical cross-sections, including topography, or as contour maps of the interpreted layer boundaries. It is also possible to produce horizontal slices of the calculated resistivity for selected depths.

4.4.7 Quality Assurance

The principles of quality assurance are given in the Chapter 2.6 and in the Glossary. In particular, EM surveys should avoid

- metal fences, cars, buildings containing reinforced concrete,
- areas where an accumulation of metal at the surface is present, and,
- power lines, telephone lines, metal pipes in or above the ground and transmitters for radio, cell phones or radar.

as these could produce noise that will impact negatively on the data. Potential noise sources that cannot be avoided should be documented.

Important practical aspects for FEM surveys are:

- Avoid low battery voltage, poor cable connections, incorrect calibration (e.g., zeroing of the instrument).
- Extreme care should be taken that the coils are placed at the correct distance from each other (one of the main sources for error!).
- Check that the coils are exactly oriented (use a level).
- Check data visually during acquisition (fluctuating data may due to noise).
- The distance between the coil and the ground must not be changed during measurements in HCP (HLEM) mode. If done for whatever reason, this has to be considered in the interpretation.
- In rugged terrain (slope >10°), determine the slope for topographic corrections.
- Take into account that lateral inhomogeneities in geology will influence sounding results.
- Carry out at least one daily check of data quality so that erroneous measurements can be repeated.

Important for TEM surveys:

- Avoid low battery voltage, damaged cable insulation of the loops, poor cable connections.
- Care should be taken that the transmitter loop size is correct.
- Improve the signal/noise ratio by repeating measurements and averaging the data.
- When metal detection instruments are used, metallic objects should not be carried near the coil system.
- The best quality control during a survey is to make each sounding twice.

Important for VLF-surveys:

- If possible, do not measure just a single profile but several parallel profiles.
- Select a radio station of sufficient signal strength in a direction as near as possible perpendicular to the strike of the target.
- If the strike of the target is unknown, use two transmitters in different directions.
- Information about topography and potential noise sources must be recorded.

- Avoid profiles parallel to elongated conductors (power lines, cables, etc.).

The survey documentation should include

- the instruments and the measuring parameters,
- a topographic map showing roads, buildings, water bodies, the position of profiles and stations,
- information about the location of buried, disturbing obstacles (cables, metal pipes etc.),
- abrupt changes in the topography, results of the leveling survey,
- information about obvious features in the landscape (e.g., wet and dry areas, changes in surface materials, illegal waste deposits), and
- meteorological conditions.

4.4.8 Personnel, Equipment, Time Needed

	Personnel	Equipment	Time needed
mobilization and demobilization	depends on the distance to the survey area		
		1 4WD vehicle	
mapping/profiling: fixed coil systems (EM38, EM31, EM61, GEM300)	1 operator/ geophysicist (1assistant)	EM instrument, cables, data logger, 1 PC	a = 0.2 - 2 m: pd = 1 - 3 m L = 1000 - 5000 N/d
variable coil systems (e.g., EM34, MaxMin)	1 operator 1-2 assistants		a = 5 - 20 m: pd = 10 - 50 m L = 200 - 1000 N/d
VLF/VLF-R	1 operator	VLF instrument, electrodes for VLF-R, 1 PC	depends on the terrain conditions and number of frequencies L = 200 - 300 N/d
sounding: FEM	1 operator / geophysicist 1-2 assistants	FEM instrument (transmitter, receiver) cables, data logger, 1 PC	20 - 100 soundings/d, depending on the terrain conditions, and the distance between soundings
TEM	1 operator 1-2 assistants	TEM instrument (transmitter, receiver) loops, cables, data logger, 1 PC, geodetic equipment for loop layout	5 - 40 soundings/d, depending on transmitter loop size and loop configuration, terrain conditions, and distance between soundings
CSAMT	1 operator 1-2 assistants	CSAMT instrument	30 - 60 soundings/d, depending on frequency and depth range
data processing, interpretation, reporting	1 geophysicist (+ 1 assistant)	1 PC with plotter, printer, and software	1 - 2 days are necessary for each day in the field

a = station spacing, pd = profile distance, N = number of stations, d = 10-hour-day, L = N/d

4.4.9 Examples

Site investigation for a planned landfill in Germany

For the site investigation of a planned landfill near Chemnitz, Germany, an electromagnetic mapping survey was carried out to determine the tectonic structures at the site, and the lithology and thickness of the unconsolidated sediments. Survey parameters:

Size of the investigation area	1700 m × 2000 m
measuring mode	VCP (VLEM)
coil separation	20 m
line spacing	30 m
station spacing	10 m
number of stations	8456
duration of the survey	20 days (10 h/d) incl. surveying
personnel	1 geophysical technician (receiver), 1 assistant (transmitter)
physical parameter	apparent conductivity
equipment	GEONICS 34-3/XL + Polycorder.

The survey parameters were chosen on the basis of test measurements. The depth of investigation of the selected configuration was between 12-15 m. VCP mode was chosen instead of HCP mode owing to its easier operation in rough terrain and because no terrain correction is needed. The survey results are shown in *Fig. 4.4-11* (*Top*) as a contour map of apparent conductivity. The lateral conductivity distribution correlates with sediment thickness and a clear structural picture of the area could be gained.

A dc resistivity mapping survey (Schlumberger array, fixed electrode spacing of AB/2 = 50 m) was carried out together with the EM survey for comparison. The plot for the selected profile *Fig. 4.4-11* (*Bottom*) shows essentially almost the same anomalies as those in the EM data. But because the current flow is parallel to the lamination of the schist in the EM measurements and perpendicular in the dc resistivity measurements, the mean apparent resistivity determined by the EM method is here about 40 % lower than obtained with the dc method. The principal advantage of the EM survey was its considerably higher productivity with less personnel than needed for the resistivity method.

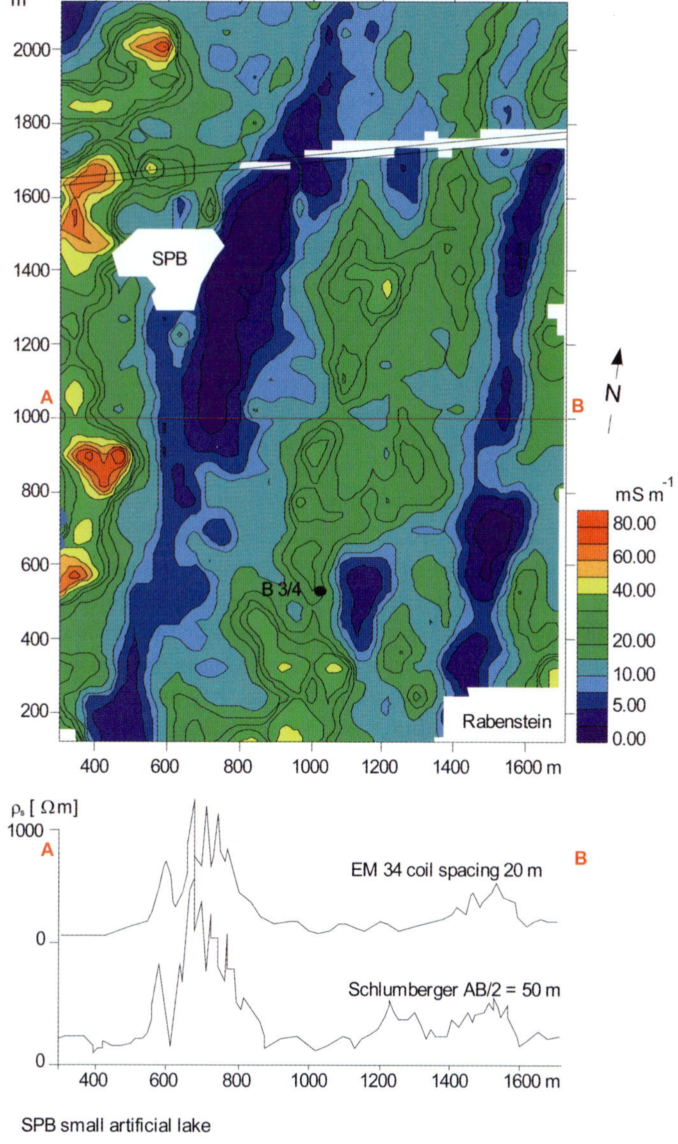

Fig. 4.4-11: Results of conductivity mapping survey using the EM34-3 instrument. *Top*: contour map of apparent conductivity. *Bottom*: comparison of apparent resistivities from EM 34-3 measurements with dc resistivity mapping using a Schlumberger array for the profile A - B shown on the map. Geophysik GGD commissioned by BGR

Exploration of an abandoned landfill

An electromagnetic survey was carried out at the Eulenberg test site near Arnstadt, Germany, using the multifrequency instrument APEX MaxMin. The EM method was used to determine whether it would provide results comparable with those of the usually used dc resistivity method. Survey parameters:

Size of the investigation area	1000 m × 1000 m
measuring mode	HCP
coil separation	40 m
line spacing	40 m
station spacing	10 m
number of stations	1531 (N-S), 1592 (E-W)
number of frequencies	10
measuring time	4 minutes per station
personnel	1 operator, 2 assistants
equipment	APEX MaxMin I-10
measured parameters	inphase, outphase, apparent conductivity.

Because data was recorded for all ten available frequencies, there was a large amount of data to be processed. The apparent conductivity calculated for 7040 Hz by the MaxMin computer was selected for the contour map in *Fig. 4.4-12*. This map shows essentially the same results as the dc resistivity survey in this area. In the southern part of the survey area, the Middle Triassic limestone, covered by a thin overburden, is characterized by low conductivities. The conductivity increases towards the north, where there is weathered Upper Triassic claystone. Two conspicuous linear anomalies in the northeastern part of the investigation area are caused by underground cables. The landfill itself is clearly distinguished from the surrounding area by a very high apparent conductivity.

In the profile plot of *Fig. 4.4-13*, the in-phase and quadrature components of five selected frequencies are shown along an E-W profile ($y = 450$ m in the local grid coordinates). The landfill boundary is indicated by typical EM anomalies at about profile meter 450, where it can be seen that at especially the highest frequency ($f = 28\,160$ Hz) the in-phase component decreases rapidly, and the quadrature component increases. The local anomaly at profile meter 110 is caused by an underground cable.

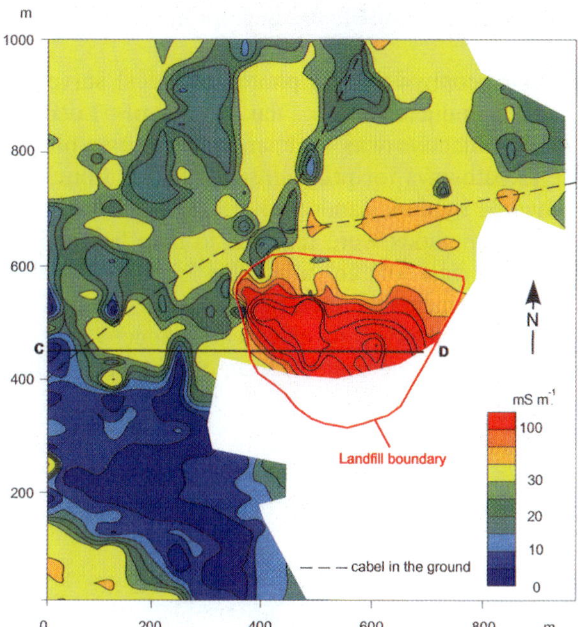

Fig. 4.4-12: Apparent conductivity map. APEX MaxMin instrument, frequency 7040 Hz, coil separation r = 40 m, Geophysik GGD, commissioned by BGR

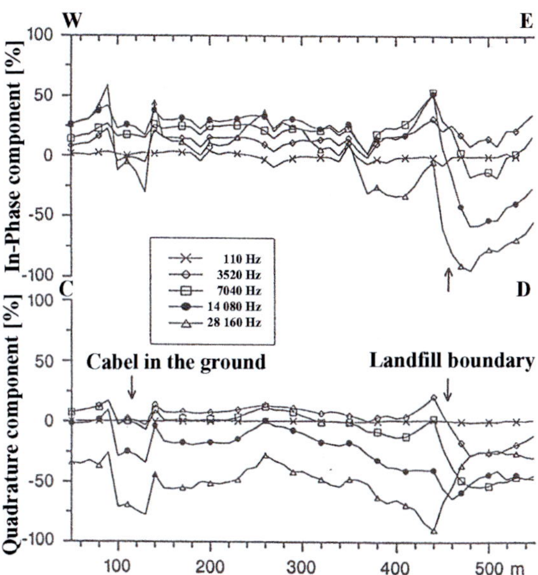

Fig. 4.4-13: Profile plot of in-phase and quadrature component for five selected frequencies, Geophysik GGD, commissioned by BGR

Investigation of a planned landfill site in Namibia

A combined ground geophysical and photo-geological survey of a potential landfill site was conducted near the town of Luederitz, Namibia (Section 6.2.1). The objective was to determine fractures present in bedrock that might serve as pathways for pollutants originating from the waste. Since the bedrock is covered by dune sands and weathering debris in some parts of the area, geophysical methods were required to map fractures hidden by the cover sediments. As part of an integrated ground geophysical survey electromagnetic measurements were carried out using an APEX MaxMin I-10 multifrequency system. Changes in electrical conductivity as derived from these measurements were interpreted as an indication of the presence of fractures. Survey parameters:

Size of the investigation area	450 m × 700 m
measuring mode	HCP
coil separation	40 m
line spacing	20 m
station spacing	10 m
number of stations	1250
frequencies	3520, 7040 and 14 080 Hz
duration of the survey	10 days (10 h/d) incl. positioning (no GPS)
personnel	1 operator, 2 assistants
equipment	APEX MaxMin I-10
measured parameters	inphase, outphase, apparent conductivity

In order to ensure a high degree of spatial resolution, and to be able to focus the EM survey on identifying shallow conductive features underlying the sand, survey parameters were determined beforehand by a series of tests. Three out-of-phase and apparent conductivity maps were interpreted under the assumption that due to weathering, increased conductivity would reflect fractured bedrock. Apparent conductivities calculated by the MaxMin computer are derived from the out-of-phase values basing on the low induction number principle (Section 4.4.3). Of the three frequencies used, data at 7040 Hz yielded the best results (*Fig. 4.4-14*).

Linear trends can be clearly seen in the data. Most of the detected fractures occur in the western part of the survey area, whereas the eastern part has almost none and is characterized by uniformly low out-of-phase values. This conclusion is in agreement with the results of the interpretation of the aerial photographs (Section 6.2.1). The long narrow trends on the out-of-phase and the apparent conductivity maps are interpreted as fracture zones.

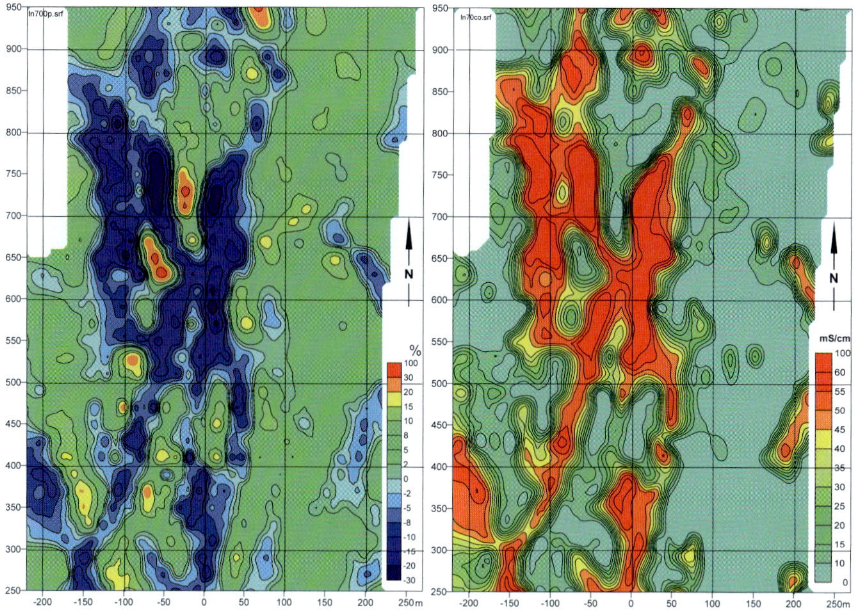

Fig. 4.4-14: Planned landfill site at Luederitz, Namibia, results of the APEX MaxMin multifrequency electromagnetic measurements. *Left*: Out-of-phase values for the 7040 Hz frequency. *Right*: Apparent conductivity map as derived from the out-of-phase values left

TEM soundings for hydrogeological exploration

To assess the hydrogeology in the area around a former salt mine, a TEM sounding survey was conducted to determine the location of the saltwater and freshwater boundary. The TEM method was selected as the depth of penetration should be at least 200 m and strong lateral changes in material were expected. Therefore, a method with a high lateral resolution was required. As the survey area conditions did not guarantee a successful TEM survey at first, a test profile was measured. Survey parameters:

length of the profile	8 000 m
loop size	200 m × 200 m
loop configuration	coincident loops
station spacing	200 - 600 m
number of stations	20
duration of the survey	4 days (10 h/d) plus 1 day for testing
personnel	1 geophysicist/operator, 2 assistants
equipment	SIROTEM II and III
physical parameter	transient voltage (decay curve).

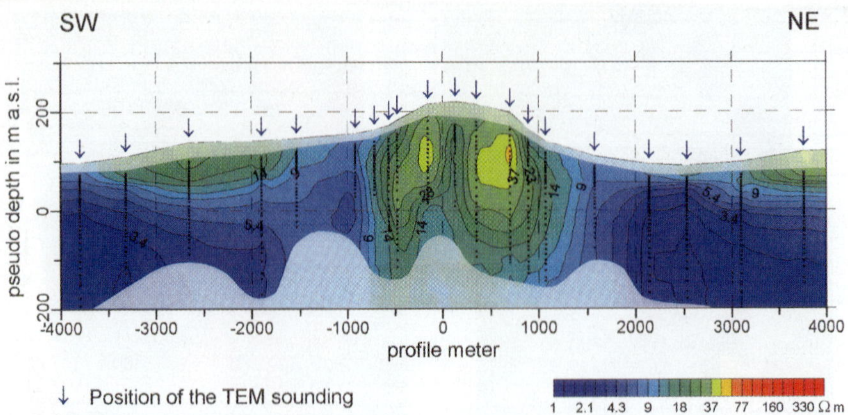

Fig. 4.4-15: Pseudosection of apparent resistivity derived from TEM soundings, courtesy of GSF GmbH

The test profile results are presented in *Fig 4.4-15* as a pseudosection of apparent resistivity. The grey hatching at the top and the bottom of the section denote the depth ranges not covered by the soundings. Differences in the exploration depths result mainly from differences in the quality of the data from the individual soundings. The apparent resistivity was calculated using Equation (4.4.10) and the pseudodepth is based on Equation (4.4.9), taking about the half of the "diffusion depth" as the optimum depth of exploration.

The pseudosection reflects the topography and dots mark the pseudodepths determined for each sounding. The center of the profile is higher than at the ends, pushed up by the ascending salt of the diapir. All strata seen outcropping at the flanks of the diapir in the field show a steep dip.

The extremely low resistivities ($< 6\ \Omega$m) clearly show where mineralized groundwater can be expected. Contrary to expectations, the saltwater reaches almost to the surface south and north of the hill. The depths of the relatively high resistivities on the south side of the diapir correlate quite well with location of the saltwater/freshwater interface found in boreholes. The same situation can be probably assumed for the north side. Considering the steep vertical resistivity gradients, the TEM soundings provided good lateral resolution. 1-D inversion was carried out for the soundings, which improved the depth data. However, in this case, 2-D inversions would have been better, but 2-D inversions programs were not commercially available that time.

Parameters and units

	Symbol	Units
coil separation (distance between FEM receiver and transmitter)	r	m
magnetic field intensity	$\boldsymbol{H}$	A m^{-1}
magnetic moment	$\boldsymbol{m}$	A m^2
frequency	f	Hz, s^{-1}
angular frequency	$\omega = 2\pi f$	Rps, s^{-1}
magnetic permeability of free space	μ_0	$4\pi \cdot 10^{-7}$ Vs $A^{-1}m^{-1}$
magnetic permeability	μ	Vs $A^{-1}m^{-1}$
relative magnetic permeability	$\mu_r = \mu/\mu_0$	dimensionless
conductivity (apparent conductivity)	σ (σ_a)	Sm^{-1} (1 mS m^{-1} ≡ 1000 Ωm)
resistivity (apparent resistivity)	ρ (ρ_a)	Ωm
response parameter (or induction number)	$Q = \mu\sigma\omega l^2$	dimensionless
current	I	A
time	t	s
voltage	U	V

References and further reading

ARCHIE, G. E. (1942): The electrical resistivity log as an aid in determining some reservoir characteristics. Transactions Americ. Inst. Mineral. Met., **146**, 54-62.

AUKEN, E., NEBEL, L., SORENSEN, K., BREINER, M., PELLERIN, L. & CHRISTENSEN, N. B. (2002): EMMA-a geophysical training and education tool for electromagnetic modeling and analysis. JEEG, **7**, 57-68. Software free available from www.hgg.au.dk.

BARTEL, L. C., CRESS, D. H. & STOLARCZYK, L. G. (1997): Evaluation of the electromagnetic gradiometer concept for detection of underground structures - theory and application. JEEG**, 2**, 127-136.

BERKTOLD, A. (2005): VLF, VLF-R und Radiomagnetiotellurik. In: KNÖDEL, K., KRUMMEL, H. & LANGE, G. (Eds.) (2005): Handbuch zur Erkundung des Untergrundes von Deponien und Altlasten. Band 3 Geophysik, 2nd edn. Springer, Berlin.

BUSELLI, G. & CAMERON, M. (1996): Robust statistical methods for reducing sferics noise contaminating transient electromagnetic measurements. Geophysics, **61**, 1633-648.

CAGNIARD; L. (1953): Basic theory of the magneto-telluric method of geophysical prospecting. Geophysics, **18**, 605- 635.

CHRISTIANSEN; A. V. & CHRISTENSEN, N. B. (2003): A quantitative appraisal of airborne and ground-based transient electromagnetic (TEM) measurements in Denmark. Geophysics, **68**, 523-534.

DAS, U. C. (1995): Apparent resistivity curves in controlled source electromagnetic sounding directly reflecting true resistivities in a layered earth. Geophysics, **60**, 53-60.

DAS, K., BECKER; A. & LEE, K. H. (2002): Experimental validation of the wavefield transform of electromagnetic fields. Geophys. Prosp., **50**, 441-451.

DYCK, A. V. (1991): Drill-hole electromagnetic methods. In: NABIGHIAN, M. N. (Ed.): Electromagnetic methods in applied geophysics – applications part B. Society of Exploration Geophysicists, Tulsa, 881-930.

EBERLE, D. (1981): A method of reducing terrain relief effects from VLF-EM data. Geoexploration, **19**, 103-114.

FRASER D. C. (1969): Contouring of VLF-EM data. Geophysics, **34**, 958- 967.

FRISCHKNECHT, F. C., LABSON, V. F., SPIES, B. R. ANDERSON, W. L. (1991): Profiling methods using small sources. In: NABIGHIAN, M. N. (Ed.): Electromagnetic methods in applied geophysics, **2**, part A. Society of Exploration Geophysicists, Tulsa, 105-270.

GEONICS Limited (2001): www.geonics.com/tdem.html#TEM47.

GOLDMAN; M., TABAROVSKY, L. & RABINOVICH, M. (1994): On the influence of 3-D structures in the interpretation of transient electromagnetic sounding data. Geophysics, **59**, 889-901.

GREINWALD, S. (1985): Wechselstromverfahren. In: BENDER, F. (Ed.): Angewandte Geowissenschaften, **II**: Methoden der Angewandten Geophysik und mathematische Verfahren in den Geowissenschaften. Enke, Stuttgart, 352-387.

GUERIN, R., TABBAGH, A. & ANDRIEUX, P. (1994): Field and/or resistivity mapping in MT-VLF and implications for data processing. Geophysics, **59**, 1695-1712.

HOEKSTRA, P. & BLOHM, M. W. (1990): Case histories of time-domain electromagnetic soundings in environmental geophysics. In: WARD, S. H. (Ed.): Geotechnical and environmental geophysics, **II.** Society of Exploration Geophysicists, Tulsa, 1-15.

KAROUS, M. & HJELT, S. E. (1983): Linear filtering of VLF dip angle measurements. Geophys. Prosp., **31**, 782-794.

KAUFMAN; A. A. (1989): A paradox in geoelectromagnetism and its resolution, demonstrating the equivalence of frequency and transient domain methods. Geoexploration, **25**, 287-317.

KAUFMAN, A. A. & HOEKSTRA, P. (2001): Electromagnetic soundings. Methods in Geochemistry and Geophysics, **34**. Elsevier, Amsterdam.

KETOLA, M. & PURANEN, M. (1967): Type curves for the interpretation of slingram (Horizontal Loop) anomalies over tabular bodies. Report of Investigations, No. 1. Geological Survey of Finland, Otaniemi.

MAULDIN-MAYERLE, CARSON, N.. M. & ZONGE; K. L. (1998): Environmental application of high resolution TEM methods. Proceedings of the 4th Meeting of the Environmental and Engineering Geophysical Society. Barcelona, September 14-17, 829-832.

MCNEILL, J. D. (1980): Electromagnetic terrain conductivity measurements at low induction numbers. GEONICS Technical Note TN6.

MCNEILL, J. D. (1988): Advances in electromagnetic methods for groundwater studies. In: Proceedings of the Symposium on the Application of Geophysics to Engineering and Environmental Problems (SAGEEP). March 28-31, 1988, Golden, Colorado, 251-348.

MCNEILL, J. D. (1990): Use of electromagnetic methods for groundwater studies. In.: WARD, S. H. (Ed.): Geotechnical and environmental geophysics, **I**: Review and tutorial. Society of Exploration Geophysicists, Tulsa, 191-218.

MCNEILL, J. D. (1991): Advances in electromagnetic methods for groundwater studies. Geoexploration, **27**, 65-80.

MCNEILL, J. D. & LABSON, V. F. (1991): Geological mapping using VLF radio fields. In: NABIGHIAN, M. N. (Ed.): Electromagnetic methods in applied geophysics, **2** (applications), part B, investigations in Geophysics, Society of Exploration Geophysicists, Tulsa, Oklahoma, 521-640.

MCNEILL, J. D. & BOSNAR, M. (1996): Application of time domain electromagnetic techniques to UXO detection. Paper presented at the Williamsburg UXO Conference, March 1996.

MEJU, A. M. (1998): A simple method of transient electromagnetic data analysis. Geophysics, **63**, 405-410.

MILSOM, J. (1996): Field geophysics. Open University Press, Milton Keynes and Halsted Press, Wiley, New-York.

NABIGHIAN, M. N. & MACNAE, J. C. (1991): Time-domain electromagnetic prospecting methods - In: NABIGHIAN, M. N. (Ed.): Electromagnetic methods in applied geophysics, **2**, part A. Society of Exploration Geophysicists, Tulsa, 427-520.

OLDENBURG, D. W. (2001): University of British Columbia - geophysical inversion facility inversion and modelling of applied geophysical electromagnetic data (IMAGE). Consortium: http//www.geop.ubc.ca/gif/research/image/index.html

PALACKY, G. J. (1991): Application of the multifrequency horizontal-loop EM-method in overburden investigations. Geophys. Prosp., **39**, 1061-1082.

PATRA, H. P. & MALLICK, K. (1980): Geosounding principles, **2**. Time-varying geoelectric soundings. Elsevier, Amsterdam.

PELLERIN; L., LABSON, V., PFEIFER, C. (1994): VETEM a very early time electromagnetic system. Proceedings of the symposium on the application of geophysics to engineering and environmental problems, SAGEEP, Boston, March 1994, 795-802.

REYNOLDS, J. .M. (1997): An introduction to applied and environmental Geophysics. John Wiley & Sons Ltd., Chichester.

SINHA, A. K. (1990): Interpretation of ground VLF-EM data in terms of vertical conductor models. Geoexploration, **26**, 213-231.

SMITH, D. VON G., WRIGHT, D. L., ABRAHAM, J. D. (2000): Advances in very early time electromagnetic (VETEM) system data analysis and image processing. Proceedings of the Symposium on the Application of Geophysics to Engineering and Environmental Problems, Arlington, February 20-24, 2000, 469-475.

SNYDER, D. D., MACINNES, S., URQUHART, S. & ZONGE, K. L. (1999): Possibilities for UXO classification using characteristic modes of the broad-band electromagnetic induction response. Paper presented at "A new technology conference on the science and technology of unexploded ordnance (UXO) removal and site remediation". Outrigger Wailea Resort, Maui, Hawaii, November 8-11, 1999.

SHERIFF, R. E. (1991): Encyclopedic dictionary of exploration geophysics. 3rd edn. Society of Exploration Geophysicists, Tulsa.

SPIES, B. R. & EGGERS, D. W. (1986): The use and misuse of apparent resistivity in electromagnetic methods. Geophysics, **51**, 1462-1471.

SPIES, B. R. (1989): Depth of investigation in electromagnetic sounding methods. Geophysics, **54**, 872-888.

SPIES, B. R. & FRISCHKNECHT, F. C. (1991): Electromagnetic Sounding. In: NABIGHIAN, M. N. (Ed.): Electromagnetic methods in applied geophysics, **2**, part A. Society of Exploration Geophysicists, Tulsa, 285-425.

TABBAGH, A., BENDERITTER, Y., ANDRIEUX, P., DECRIAUD, J. P. & GUERIN, R. (1991): VLF resistivity mapping and verticalisation of the electric field. Geophys. Prosp., **39**, 1083-1097.

TELFORD, W. M., GELDART, L. P. & SHERIFF, R. E. (1990): Applied geophysics, 2nd edn., Cambridge University Press, Cambridge.

TEZKAN, B., HÖRDT, A. & GOBASHY, M. (2000): Two dimensional inversion of radiomagnetotelluric data: selected case histories for waste site exploration. Journal of Applied Geophysics, **44**, 237-256.

TURBERG, P., MÜLLER, I. & FLURY; B. (1994): Hydrogeological investigation of porous environments by radio magnetotelluric resistivity (RMT-R 12-240 kHz). J. Appl. Geophysics, **31**, 133-143.

TURBERG, P. & BARKER, R. (1996): Joint application of radiomagnetotelluric and electrical imaging surveys in complex subsurface environments. First Break, **14**, 105-112.

VALLÉE, M. A., CHOUTEAU, M. & PALACKY, G. J. (1992): Effect of temporal and spatial variations of the primary signal on VLF total-field surveys. Geophysics, **57**, 97-105.

VERMA, S. K. & SHARMA, S. P. (1995): Focused resolution of thin conducting layers by various dipole EM systems. Geophysics, **60**, 381-389.

VOGELSANG, D. (1991): Elektromagnetische Erkundung grundwasserführender Strukturen. Geol. Jb., **E 48**, Hannover, 283-308.

VOZOFF, K. (1988): The magnetotelluric method. In NABIGHIAN, M. N. (Ed.): Electromagnetic methods in applied geophysics, **2**, application, part B. Society of Exploration Geophysicists, Tulsa, 641-711.

WON, I. J., KEISWETTER, D. A., FIELDS, R. A. & SUTTON, L. (1996): GEM-2: A new multifrequency electromagnetic sensor. JEEG, **2**, 129-137.

WITTEN, A., WON, I. J. & NORTON, S. (1997): Imaging underground structures using broadband electromagnetic induction. JEEG, **2**, 105-114.

WARD, S. H. & HOHMANN, G. W. (1987): Electromagnetic theory for geophysical applications. In: NABIGHIAN, M. N. (Ed.): Electromagnetic methods in applied geophysics - theory, **1**. Society of Exploration Geophysicists, Tulsa, 131-311.

WEST, G. F. & MACNAE, J. C. (1991): Physics of the electromagnetic induction exploration method. In: NABIGHIAN, M. N. (Ed.): Electromagnetic methods in applied geophysics, **2**, Part A. Society of Exploration Geophysicists, Tulsa, 5-45.

WRIGHT, D. L., SMITH, D. V. & ABRAHAM, J. D. (2000): A VETEM survey of a former munitions foundry site at the Denver Federal Center. Proceedings of the symposium on the application of geophysics to engineering and environmental problems, SAGEEP, Arlington, February 20-24, 2000, 459-468.

YANG, C.-H., TONG, L.-T. & HUANG, C.-F. (1999): Combined application of dc and TEM to sea-water intrusion mapping. Geophysics, **64**, 417-425.

ZACHER, G., TEZKAN, B., MÜLLER, I., NEUBAUER, F. M. & HÖRDT, A. (1996): Radiomagnetotellurics: a powerful tool for waste site exploration. European Journal of Environmental and Engineering Geophysics, **1**, 139-159.

4.5 Ground Penetrating Radar

NORBERT BLINDOW with contributions by DIETER EISENBURGER, BERNHARD ILLICH, HELLFRIED PETZOLD & THOMAS RICHTER[1]

4.5.1 Principle of the Methods

Ground penetrating radar[2] (GPR) is an electromagnetic pulse reflection method based on physical principles similar to those of reflection seismics. It is a geophysical technique for shallow investigations with high resolution which has undergone a rapid development during the last two decades (cf. e.g. GPR Conference Proceedings 1994 to 2006). There are several synonyms and acronyms for this method like EMR (electromagnetic reflection), SIR (subsurface interface radar), georadar, subsurface penetrating radar and soil radar. GPR has been used since the 1960s with the term radio echo sounding (RES) for ice thickness measurements on polar ice sheets. The method was first applied by STERN (1929, 1930) in Austria to estimate the thickness of a glacier. GPR has been increasingly accepted for geological, engineering, environmental, and archaeological investigations since the 1980s.

In its simple time domain form, electromagnetic pulses are transmitted into the ground. A part of this energy is reflected or scattered at layer boundaries or buried objects. The direct and reflected amplitudes of the electric field strength E are recorded as a function of traveltime.

Reflections and diffractions of electromagnetic waves occur at boundaries between rock strata and objects that have differences in their electrical properties. Electric permittivity ε and electric conductivity σ are the petrophysical parameters which determine the reflectivity of layer boundaries and the penetration depth. Because the magnetic permeability μ is approximately equal to μ_0 ($= 4\pi \cdot 10^{-7}$ V s A^{-1} m^{-1}) for most rocks except ferromagnetics, only the value of μ_0 has to be considered in calculations.

---

1 BERNHARD ILLICH: Localization of objects, Delineation of a mineralized fault in consolidated rocks, Investigations of concrete constructions, Examination of masonry structures, and Investigation of residual foundations; HELLFRIED PETZOLD: Investigation of a domestic waste site in a refilled open-pit mine; THOMAS RICHTER: Radar tomography to assess the ground below buildings; DIETER EISENBURGER: Special Applications and New Developments.

2 Radar is an acronym for Radio Detection And Ranging, used since the mid-1930s in aerial and naval navigation.

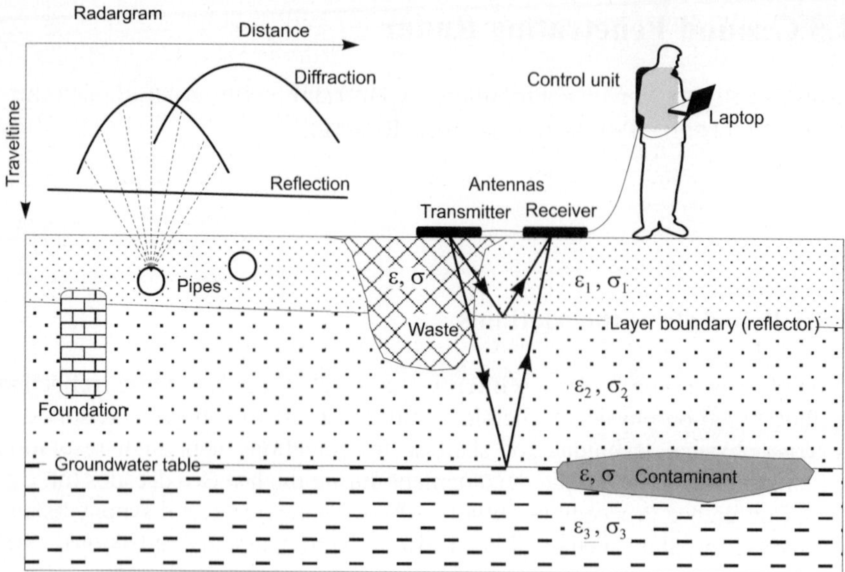

Fig. 4.5-1: Principle of the method

Broadband dipole antennas are normally used for transmission and reception of the signals. Frequencies between 10 and 1000 MHz are used for geological and engineering investigations. For materials testing, frequencies higher than 1000 MHz are also useful. Placing the antennas flat on the ground provides the best ground coupling and alters the antenna characteristic significantly compared to an antenna in air (focusing effect). A high pulse rate enables measurements to be made quasi-continuous by pulling the antennas along a profile. Using an antenna-configuration like in *Fig. 4.5-1*, several kilometers can be surveyed per day. If the underground conditions are favorable, the advantages of the method are that it is non-invasive with high horizontal and vertical resolution providing profiling results in realtime in the form of radargrams on a monitor or plotter. In many cases, a preliminary interpretation is possible in the field.

The center frequency f_m (spectral maximum of the usually broad pulse spectrum) is chosen on the basis of the task at hand and the properties of the materials being investigated. Unconsolidated rocks and soils have an average permittivity $\varepsilon_r = 9$, which means that frequencies of 10 - 1000 MHz correspond to wavelengths λ of 10 - 0.1 m. High frequencies, i.e., short wavelengths, provide a higher resolution. On the other hand, due to higher absorption and scattering, signals with higher frequencies have less penetration depth than that with lower frequencies. GPR is particularly suitable for materials with higher resistivities, such as dry sand with low clay content or consolidated rocks. In these cases GPR is the geophysical method with the highest resolution (achieving the centimeter range) for subsurface imaging.

Ground penetrating radar can be used from the Earth's surface (Section 4.5.9), in single boreholes, between boreholes, from drifts in mines, from an aircraft or helicopter (Sections 4.5.9 and 4.5.10), and from space (see remote sensing techniques, Chapter 3.3). As in reflection seismics (Chapter 4.6), correlation and frequency domain variations (the chirp and stepped frequency methods, see Section 4.5.10) are used besides the time domain (pulsed) method.

4.5.2 Applications

- Determination of location (including depth), orientation, size, and shape of underground metal and plastic pipelines, cables, and other buried manmade objects (e.g., barrels and foundations),
- detection of cavities within rock masses,
- investigation of karst structures,
- subsidence investigations,
- investigation of sediment and soil structures, distinguishing between homogeneous and inhomogeneous areas,
- lake and riverbed sediment mapping,
- geological investigations of glacial deposits,
- location of faults, joints, and fissures in consolidated rock,
- location of clay lenses, ice wedges, small peat deposits, etc.,
- determination of the depth to water table in gravel, sand, and sandstone, mapping of the aquifer base (in the case of water with low conductivity),
- determination of the condition of landfill lining systems,
- determination of rock structure in salt mines being investigated for use as a waste repository,
- detection and monitoring of contamination plumes,
- estimation of soil moisture content,
- permafrost investigations,
- glaciological investigations (e.g., ice thickness mapping, determination of internal glacier structures),
- identification of buried landmines and unexploded ordnance (UXO),
- road pavement analysis,
- testing integrity and moisture content of building materials,

- testing of concrete and checking the location of reinforcement bars ("rebars") in it, and
- localization of concealed structures and objects before and/or between archaeological excavations.

The method fails if the top layers contain good-conducting material (e.g., moist clay or silt, saline water, ferrous slag).

4.5.3 Fundamentals

Wave propagation, velocity, and absorption

The propagation of electromagnetic waves in rock is similar to the propagation of seismic waves. There are several basic differences, however, that have consequences for the practical application in the field and for the processing of the data. Velocity and absorption of electromagnetic waves are highly dependent on frequency. More than for seismic waves, this frequency dependence (dispersion) causes changes in the pulse shape during propagation and reflection and diffraction at boundary planes.

Important and simple solutions of Maxwell's equations are harmonic plane waves, e.g. a transverse electric field $E(t,z)$ propagating in z-direction (v. HIPPEL, 1954):

$$\boldsymbol{E}(t,z) = E_0\, e^{(\mathrm{i}\omega t - \gamma z)} \qquad [\mathrm{V\,m^{-1}}] \tag{4.5.1}$$

with the angular frequency $\omega = 2\pi f$ $[\mathrm{s^{-1}}]$ and
the propagation constant $\gamma = \alpha + \mathrm{i}\,\beta$ $[\mathrm{m^{-1}}]$. (4.5.2)

The magnetic field component is strongly coupled with the electric field. It is perpendicular to the electric field vector and – like the electric field component – oscillates perpendicular to the direction of propagation.

The following parameters are needed to describe the propagation of radar waves:

- permittivity $\varepsilon^* = \varepsilon' - \mathrm{i}\varepsilon''$;
- relative permittivity $\varepsilon_r^* = \varepsilon^*/\varepsilon_0$,
  where $\varepsilon_0 = 8.8544 \cdot 10^{-12}\ \mathrm{A\,s\,V^{-1}\,m^{-1}}$;
- permeability $\mu^* = \mu' - \mathrm{i}\mu''$,
  $\mu^* \approx \mu_0 = 4\pi \cdot 10^{-7}\ \mathrm{VsA^{-1}m^{-1}}$ for most rocks;
- and the loss angle δ, defined by the loss tangent

$$\tan\delta = \frac{\varepsilon''}{\varepsilon'} = \frac{\sigma}{\omega\varepsilon'} = \frac{1}{\omega\varepsilon'\rho} . \qquad (4.5.3)$$

Thus, $\tan\delta$ is directly proportional to conductivity σ and inversely proportional to the frequency f. The loss tangent is the ratio of conduction to displacement currents. Velocity and absorption of electromagnetic waves are nearly independent of frequency when $\tan\delta < 0.5$. When $\tan\delta > 2$, there is considerable dispersion and the energy is propagated mainly by diffusion. In Equation (4.5.3) the conductivity σ is generally frequency dependent and consists of a dc component and an ac component due to losses of the displacement currents. The dc conductivity can be estimated from geoelectrical sounding data (Chapter 4.3). The attenuation coefficient α, which is the real part of the propagation constant γ in Equation (4.5.2), and the phase constant β, which is the imaginary part, are calculated as follows:

$$\alpha = \frac{\omega}{c_0}\sqrt{\frac{\varepsilon'_r}{2}\left(\sqrt{1+\tan^2\delta}-1\right)} \quad [\mathrm{m}^{-1}] \qquad (4.5.4)$$

and

$$\beta = \frac{\omega}{c_0}\sqrt{\frac{\varepsilon'_r}{2}\left(\sqrt{1+\tan^2\delta}+1\right)} \quad [\mathrm{m}^{-1}] \qquad (4.5.5)$$

with the speed of light in vacuum $c_0 = 2.998 \cdot 10^8$ m s^{-1} = 0.2998 m ns^{-1}. The absorption coefficient $\alpha' = 8.686\ \alpha$ [dB m^{-1}] is usually used instead of the attenuation coefficient[3] α.

The propagation or phase velocity v of the radar waves is determined from the spacing of planes with the same phase: $v = \omega/\beta = \lambda f$ [m ns^{-1}] with the wavelength $\lambda = 2\pi/\beta$ [m]. The velocity v in a low loss material, i.e., $\tan\delta \ll 1$, is in good approximation given by

$$v \approx \frac{c_0}{\sqrt{\varepsilon'_r}} .$$

The electromagnetic properties of a dielectric can also be described by the complex characteristic impedance Z^* (the ratio of electric to magnetic field strength)

---

[3] The unit bel (symbol B) as a logarithmic measure of ratios of power levels is used mostly in telecommunication, electronics and acoustics. Commonly used is the decibel (dB) equal to 0.1 B. The decibel value is given by $10 \log_{10}(P_1/P_2)$, where P_1 and P_2 are the power levels (energy, power density, etc.). If the field parameters (voltage, current intensity, etc.) are used instead of power levels, the decibel value is given by $20 \log_{10}(P_1/P_2)$. The decibel is a dimensionless unit like percent. The attenuation coefficient is often given in neper (1 NP = 8.686 dB). The phase constant is in rad/m (1 rad ≈ 57.3°).

$$Z^* = Z' + \mathrm{i}Z'' = \sqrt{\frac{\mu^*}{\varepsilon^*}} \tag{4.5.6}$$

where μ^* is the complex magnetic permeability and
ε^* is the complex permittivity.

The penetration depth can be estimated if the absorption coefficient α' is known. The depth of discontinuities in the rock can be calculated from traveltimes of the signals and the propagation velocity v.

High frequency properties of rocks and liquids

Characteristic values of the relative permittivity ε'_r (real part), the conductivity σ, the propagation velocity v, and the absorption coefficient α' are given in *Table 4.5-1* for various media. The values for sediments and consolidated rocks are average values for a large number of samples. The propagation velocity v, which is important for interpreting depth, is governed to a large degree by the water content of the rock, owing to the high value of the real part of the permittivity ε'_r of water ($\varepsilon'_r = 80$). Moreover, the ionic content of the water influences the conductivity and thus the absorption and penetration depth in moist or water-saturated rock. At frequencies greater than 100 MHz, absorption heavily increases due to Debye relaxation of the water molecules (a property which is utilized in the GHz range in microwave ovens). Owing to dispersion, these parameter values are a function of frequency.

Table 4.5-1: Relative permittivity ε'_r, electric conductivity σ, velocity v, and absorption coefficient α' of several materials at 100 MHz (modified after DAVIS & ANNAN, 1989). The values for v are typical for possible ranges of ε'_r. The values for ice are for 60 MHz (JOHARI & CHARETTE, 1975). The values for oil are from BEBLO (1982).

Material	ε'_r [dimensionless]	σ [mS m^{-1}]	v [m ns^{-1}]	α' [dB m^{-1}]
air	1	0	0.2998	0
distilled water	80	0.01	0.033	0.002
fresh water	80	0.5	0.033	0.1
sea water	80	000	0.01	1000
dry sand	3 - 5	0.01	0.15	0.01
water-saturated sand	20 - 30	0.1 - 1	0.06	0.03 - 0.3
silt	5 - 30	1 - 100	0.07	1 - 100
clay	5 - 40	2 - 1000	0.06	1 - 300
limestone	4 - 8	0.5 - 2	0.12	0.4 - 1
shale	5 - 15	1 - 100	0.09	1 - 100
granite	6	0.01 - 1	0.12	0.01 - 1
dry salt	≈ 6	0.001 - 0.1	0.125	0.01 - 1
ice	3.18	0.01	0.168	0.02
oil, asphalt	2 - 3	0.01	0.19	0.01

Reflection, transmission and diffraction during wave propagation

At the boundary between two media (1) and (2) with different electrical properties, an arriving electromagnetic wave is both reflected and refracted. Since in the far-field[4] of the transmitting dipole these waves appear as plane waves, the spatial relations can be determined using the equations for refraction as in seismics (Section 4.6.3.4). Amplitudes are calculated using Equation (4.5.1). The coefficient of reflection r (reflected amplitude of an incident wave) is given by

$$r = \frac{Z_2 \cos\phi - Z_1 \cos\psi}{Z_2 \cos\phi + Z_1 \cos\psi}, \tag{4.5.7}$$

and the refracted (transmitted) part t by

$$t = \frac{2 Z_2 \cos\phi}{Z_2 \cos\phi + Z_1 \cos\psi},$$

where ϕ is the angle of incidence, ψ is the angle of refraction, v_1 and v_2 are the wave velocities in the two media, and Z_1 and Z_2 are the electrical wave impedances. When the incidence is perpendicular to the boundary plane (i.e., $\psi = \phi = 90°$), Equation (4.5.7) reduces to

$$r = \frac{Z_2 - Z_1}{Z_2 + Z_1},$$

which under low loss conditions ($\tan\delta \ll 1$ and $\mu_i^* \approx \mu_0$) can be expressed as:

$$r \cong \frac{\sqrt{\varepsilon'_{r1}} - \sqrt{\varepsilon'_{r2}}}{\sqrt{\varepsilon'_{r1}} + \sqrt{\varepsilon'_{r2}}} \cong \frac{v_2 - v_1}{v_2 + v_1}.$$

Because the impedances Z_i are complex values (Equation (4.5.6)), r and t are also complex, even for incident waves perpendicular to the boundary plane. Reflection and transmission of electromagnetic waves at the boundary between two strata with different electrical properties (i.e., there is a change in $\tan\delta$) always involves deformation of the wavelet[5] (*Fig. 4.5-2*). This is a significant difference from reflection seismics.

---

[4] The distance from the transmitting antenna to where the transition from the near-field to the far-field occurs depends on the wavelength of the electromagnetic signal, the antenna parameters, and the electromagnetic properties of the ground. This distance is greater than one wavelength (OLHOEFT, 2004). The immediate vicinity of the transmitting antenna is called "near-field". In the far-field, spherical waves can be treated to a good approximation in the same way as plane waves. Some radar systems remove the near-field effects by filtering. However, the near-field response can be used to detect incipient desiccation cracks in clay, to locate land mines, to study soil compaction, etc. (OLHOEFT, 2004).

[5] A wavelet is a pulse that consists of only a few oscillation cycles.

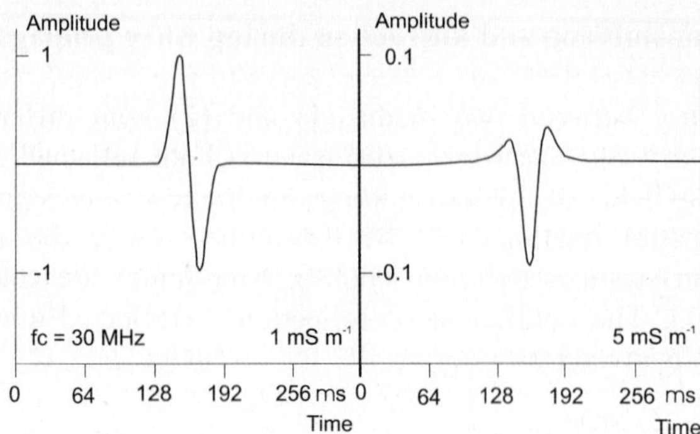

Fig. 4.5-2: Wavelet before (*left*) and after (*right*) perpendicular reflection at a boundary plane at which the conductivity changes from 1 to 5 mS m^{-1}

When incident to boundaries, electromagnetic waves behave in a complicated geology (e.g., thin layers, lamellae, gradient zones) similarily to seismic waves (Section 4.6.3). Multiple reflections at the ground surface are not significant for GPR, because there is usually considerable absorption by the soil and rock and only up to about 10 % of the transmitted energy is reflected by the ground surface.

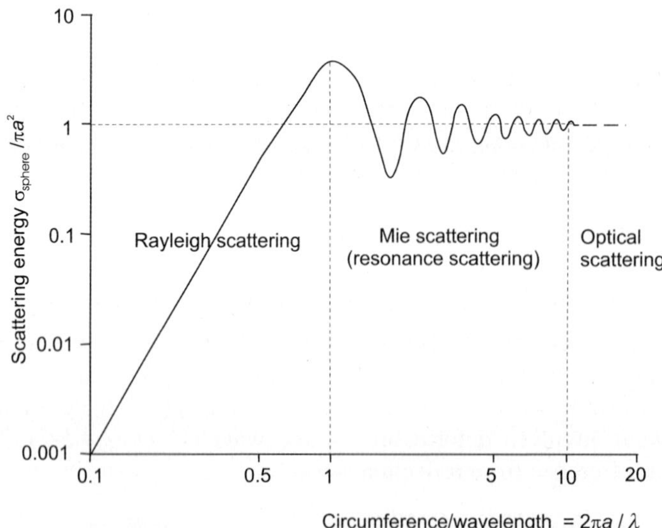

Fig. 4.5-3: Scattering from a conductive sphere with a radius *a*, SKOLNIK (1970)

As in seismics, diffraction occurs at discontinuities of reflectors (interrupts, faults with a shift) and objects whose dimensions are small compared to the wavelength. As can be seen in *Fig. 4.5-3*, the amplitude of the waves diffracted from a spherical body increases in the Rayleigh zone until the wavelength reaches approximately the same size as the diffractor. To reduce the disturbing influence of inhomogeneities in the rock (geological noise, also called clutter), a low operating frequency should be selected such that the wavelength is considerably larger than the size of the inhomogeneities. A compromise has to be made between this and the achievable resolution.

In order to localize diffractors with an irregular shape or diffractions being much longer than they are wide (the 2-D case, e.g., pipelines or cables) the dipole axis of the antennas (polarization of the electric field) should be orientated parallel to the target[6]. If the orientation of the target is not known, measurements must be made along orthogonal profiles.

Horizontal dipole on the boundary plane between two half-spaces

In most cases, horizontal electrical dipole antennas lying flat on the ground are used for transmitting and receiving the high-frequency GPR pulses. The transmission pattern of a dipole lying at the air/ground boundary plane is considerably different from that of a dipole in a quasi-infinite space. The characteristic of a horizontal dipole under far-field conditions is shown in *Fig. 4.5-4* for two directions (perpendicular and parallel to the plane of incidence). It can be seen that due to the strong contrast of electrical properties effective coupling is achieved by simply laying the dipole antenna flat on the ground.

The antenna pattern $t_A(\phi)$ of a short dipole (Hertz dipole) with a polarization direction perpendicular to the plane of incidence can be approximately described in the far-field using equations for geometrical optics (ANNAN et al., 1975). The following equation can be used for the lower half space:

$$t_A(\phi) = \frac{2\cos\phi}{\cos\phi + \sqrt{\frac{1}{\varepsilon_r^*} - \sin^2\phi}} \qquad (4.5.8)$$

[6] Polarization means that the field vector points in a particular direction. The most common commercial GPR systems use linearly polarized antennas. Linear polarized means that the electric fields of the transmitter and receiver antennas are oriented parallel to each other, parallel to the ground surface, and moved perpendicular to the electric field direction.

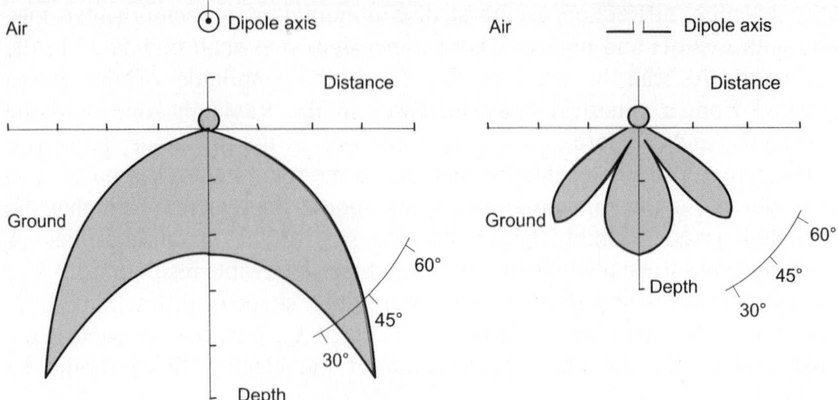

Fig. 4.5-4: The antenna patterns of a horizontal dipole lying on the air/ground interface oriented perpendicular (*left*) and parallel (*right*) to the plane of incidence, TSANG et al., 1973

The function $t_A(\phi)$ is complex when transmission angles ϕ are larger than the critical angle ϕ_c, i.e., $\phi > \phi_c = \arcsin(1/\varepsilon_r^*)$, even for real values of the permittivity. This means that there is a phase change of the transmitted and received wavelet for these transmission angles.

If the transmitting and receiving antennas are identical, the square of the single antenna pattern $T_A(\phi) = t_A^2(\phi)$ gives the amplitude and phase patterns of the system as a whole.

Wave paths, traveltimes, and amplitudes

The propagation of radar waves can be described by an idealized representation of rays as in optics and seismics. A simple two-layer model for the GPR method requires four wave paths and traveltime curves. The GPR ray path scheme is shown in *Fig. 4.5-5*.

Two direct waves with different phase velocities and amplitudes travel along the ground surface: the air wave and the ground wave (BAÑOS, 1966; CLOUGH, 1976). Since the air wave travels with the largest possible velocity for electromagnetic waves – the velocity of light in a vacuum – it can be used to determine time zero (like the time break in seismics). The velocity in the uppermost stratum is determined from the ground wave. Changes in the direct waves indicate changes in the uppermost stratum (e.g., moisture content, type of rock).

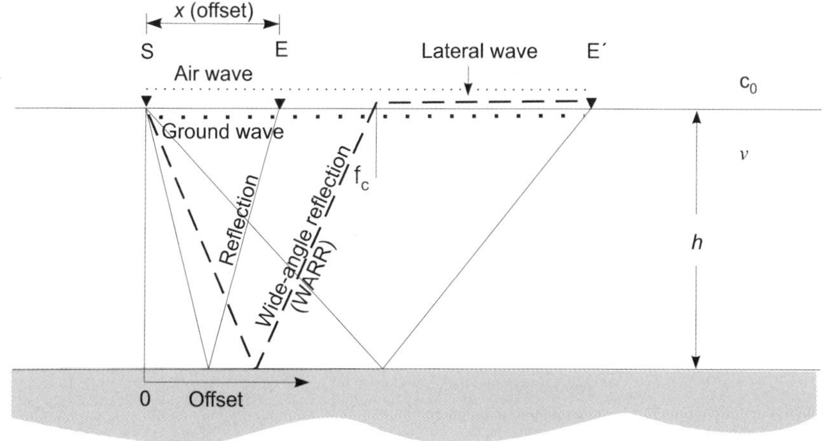

Fig. 4.5-5: GPR ray paths – principle sketch

The traveltimes of the air wave t_a and the ground wave t_g are given by

$$t_a = \frac{x}{c_0} \qquad \text{and} \qquad t_g = \frac{x}{v_g},$$

where x is the distance from transmitter to receiver dipole [m],
v_a is the velocity of the air wave ($\approx c_0 \approx 0.3$ m ns^{-1}), and
v_g is the velocity of the ground wave [m ns^{-1}].

Like in reflection seismics, the traveltime t_r of reflected waves is given by the hyperbola:

$$t_r = \frac{1}{v}\sqrt{x^2 + 4h^2}\,. \tag{4.5.9}$$

Because v is always less than c_0, a lateral wave is generated at the critical angle $\phi_c = \arcsin(v/c_0)$, analog to the head wave in refraction seismic. This wave propagates in air parallel to the ground surface. The critical angle ϕ_c is related to the critical distance x_c, which is given by

$$x_c = \frac{2hv}{\sqrt{c_0^2 - v^2}}\,.$$

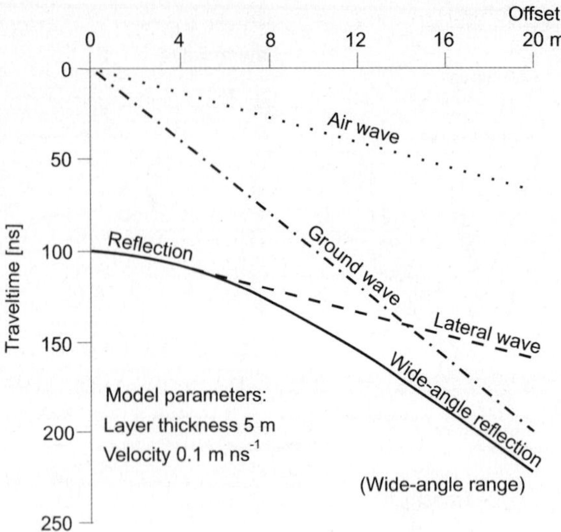

Fig. 4.5-6: Traveltime diagram for the horizontal two-layer case

The traveltime of the lateral wave is given by

$$t_1 = \frac{x}{c_0} + 2h\sqrt{\frac{1}{v^2} - \frac{1}{c_0^2}} \,.$$

Refracted waves are seldom observed, because in most cases the velocity decreases with depth. A traveltime diagram of the different wave types is shown in *Fig. 4.5-6*.

The field strength of the direct wave E_a (tangential component parallel to the dipole) decreases in the far-field with the square of the distance from the source. The field strength of the ground wave E_g is diminished in addition by absorption and scattering.

$$E_a \approx \frac{1}{x^2} \qquad \text{and} \qquad E_g \approx \frac{1}{x^2}\exp(-\alpha x).$$

The field strength of the lateral wave (analog to seismics) is given by (BREKOVSKIKH, 1980)

$$E_1 \approx \frac{1}{\sqrt{x}\left(\sqrt{x - x_c}\right)^3} \qquad \text{for} \qquad x > x_c,$$

and that of the reflected wave in the far-field by

$$E_r \approx \exp(-\alpha s) s^{-1} r(\phi) T_A(\phi) = AGrT, \tag{4.5.10}$$

where s is the path length ($s = \sqrt{x^2 + 4h^2}$),
ϕ is the angle of incidence (ϕ = arctan ($x/2\ h$)),
A is the absorption und scatter (frequency dependent),
α is the attenuation coefficient [m^{-1}],
G is the geometry, spherical divergence,
r is the reflection coefficient for plane waves derived from the amplitudes, and
$T_A(\phi)$ is the antenna pattern.

The factors of proportionality A, G, and T are derived from the field strengths in the near-field of the transmitter dipole and the modification of the antenna pattern by coupling and shielding.

In the case of diffractors, the following equation (e.g., JANSCHEK et al., 1985) is a suitable alternative to Equation (4.5.10) in the zero-offset case ($x = 0$) for waves diffracted back towards the source:

$$P_r = P_t q_s \frac{G^2 \lambda^2}{(4\pi)^3 h^4} \exp(-4\alpha h), \tag{4.5.11}$$

where P_t is the transmission power [W],
P_r is the power of the field at the receiver [W]
q_s is the scatter cross-section (area of the first Fresnel zone) [m^2],
G is the antenna gain (relative to that of a spherical dipole),
λ is the wavelength of the center frequency [m], and
h is the distance between the antenna and the diffractor [m].

Vertical and horizontal resolution

Resolution is a measure of the ability to distinguish between signals from closely spaced targets. In ground penetrating radar the resolution depends on the center frequency (or wavelength, which is proportional to the pulse period) and the bandwidth as well as on the polarization of the electromagnetic wave, the contrast of electrical parameters (mainly conductivity and relative permittivity), and the geometry of the target (size, shape, and orientation). Important are also the coupling to the ground, the radiation patterns of the antennas (especially the diameter of the first Fresnel zone[7]), and the noise conditions in the field. As a rule of thumb, the vertical resolution is theoretically one-quarter of the wavelength $\lambda = v f^{-1}$, where v is the velocity of the electromagnetic wave in the medium (see *Table 4.5-1*) and f is the center frequency.

[7] See Section 4.6.3.5 Equation (4.6.8) and *Fig. 4.6-7*.

Example: The wavelength λ is equal to 1.2 m for a limestone with $v = 0.12$ m ns^{-1} and a 100 MHz antenna ($1/f = 10$ ns); the resolution will be on the order of 0.3 m. For very good conditions, the resolution can be one-tenth of the wavelength and for unfavorable conditions one-third of the wavelength. Because the velocity is lower in wet rock than in dry rock (e.g., sand, see *Table 4.5-1*), the resolution will be better in wet materials. Due to the above-described effects in real Earth materials, the actual resolution will be in the order of one wavelength. The horizontal resolution is thoroughly discussed in Section 4.6.3.5 for the case of reflection seismics. It has been shown that objects can be separated laterally if their distance is larger than the diameter d_F of the first Fresnel zone. For example, a GPR system operating at $f_m = 200$ MHz in an environment with $v = 0.1$ m ns^{-1} has a wavelength of $\lambda = 0.5$ m. The horizontal resolution at a depth $h = 4$ m will be $d_F = (2\,\lambda\,h)^{1/2} = 2$ m, which is much poorer than the vertical resolution of better than 0.5 m.

Estimation of penetration depth and amplitude

To avoid dispersion, the operating frequency should be chosen so that $\tan\delta < 0.5$ (see Equation 4.5.3). Solving Equation (4.5.3) for frequency, we obtain

$$f_m \geq \frac{36\,000}{\rho\varepsilon_r'},$$

where f_m is the center frequency [MHz],
ρ is the electrical (dc) resistivity [Ωm], and
ε_r' is the real part of the relative permittivity (see *Table 4.5-1*).

If no other information is available, the value of the absorption coefficient α' can be taken from *Table 4.5-1* or, using Equation (4.5.3), estimated from the resistivity ρ provided by dc resistivity measurements or taken from *Table 4.3-2* (Chapter 4.3). The values for the parameters ω and ε_r' are inserted in Equation (4.5.4), with 0.5 for $\tan\delta$. Equation (4.5.4) then reduces to

$$\alpha' \approx \frac{1640}{\rho\sqrt{\varepsilon_r'}}, \tag{4.5.12}$$

where α' is in dB/m.

For successful application of the method, wave absorption along the way from the ground surface to the reflector and back to the surface should not exceed 40 - 60 dB. Experience has shown that otherwise spherical divergence, reflection and dispersion loss, geological and technical noise will take up the remaining dynamic range. For example: a center frequency > 40 MHz is selected for ground with a resistivity of 90 Ωm and $\varepsilon'_r = 10$. A value of 5.8 dB/m is obtained for the absorption coefficient using Equation (4.5.12). Thus, the maximum depth of penetration is about 5 m ($h_{max} \approx 60$ dB/$2\alpha'$). A more exact value can be obtained using Equations (4.5.10) and (4.5.11) if the system parameters (bandwidth, sensitivity, output of the pulse generator, antenna efficiency, effective area of the antennas, spectral distribution of the noise, and the high-frequency properties of the ground) are known.

4.5.4 Instruments

GPR systems consist of a pulse generator, an antenna for transmission of high-frequency electromagnetic waves, a second antenna to receive the direct and reflected impulses or a switch for switching between transmission and reception if only one antenna[8] is to be used, and a receiver which converts the received signals to be recorded and displayed. These components are designed and arrayed differently by different producers of radar equipment, but the functionality is generally the same.

Antennas

Broadband antennas are needed to transmit and receive short electromagnetic impulses. Conventional broadband systems with directional antenna gain, as used for radio and television reception in the VHF and UHF ranges (e.g., logarithmic-periodic antennas), are unsuitable for single pulses. A broad bandwidth is generally achieved by damping electrical dipoles.

Several antenna designs have proved to be useful, e.g., linear dipoles with hyperbolic resistive loading (see WU & KING, 1965 for theoretical treatment) and combinations of butterfly and loop dipoles with empirically determined resistivity damping (*Fig. 4.5-7*). Antennas with loading according to WU & KING have a signal loss of 20 dB relative to an undamped dipole. The radiated wavelet is approximated by the time-derivative of the excitation function. The frequency spectrum of a damped loop dipole is narrower, the waveform is similar to a Ricker wavelet (Section 4.6.3.3), and the amplitude loss per antenna pair is about 10 dB.

---

[8] The use of a single antenna for both transmission and reception is called monostatic radar; the use of two antennas is called bistatic radar, with the results assigned to the midpoint between the two antennas. Most GPR investigations can be carried out with monostatic or bistatic systems. For some applications, e.g., wide-angle reflection and refraction (WARR) a bistatic system is necessary.

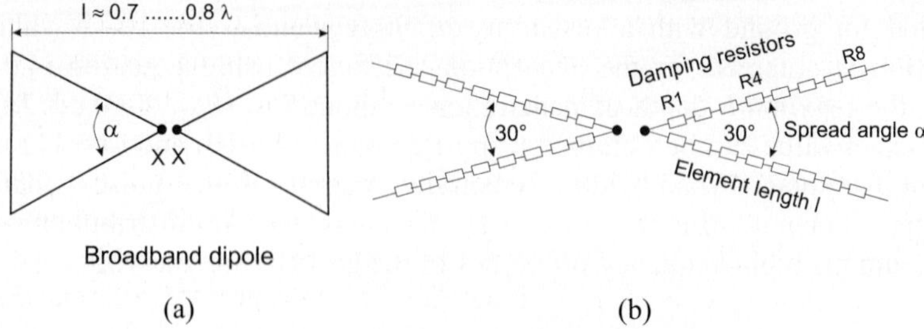

Fig. 4.5-7: Schematic of examples of broadband antennas: (a) butterfly dipole, ROTHAMMEL (1991); (b) V-dipole with resistive loading, e.g., BLINDOW (1986)

When radar data is interpreted it is necessary to keep in mind that the source pulse is longer than one wavelength and may have a complex waveform. The downgoing wavelet is modified by ground coupling and attenuation effects[9]. Therefore, the reflection is also complex. It consists of more than one wavelet.

Typical antenna patterns are shown in *Fig. 4.5-4* for different dipole orientations. When measurements are made within buildings, under power lines and trees, etc., "air reflections" are obtained from reflectors and diffractors in the half-space above the antennas. To suppress these disturbances, absorber and/or metal screens can be placed above the antennas, especially at center frequencies above about 100 MHz (i.e., when compact dipoles are used). When these are used, however, the antenna patterns are changed. The antennas are usually completely enclosed and the manufacturer usually gives only the central frequency (which is usually for use in the air and thus too high for georadar) and a note that a screening effect will take place. Borehole antennas often are due to the limited space simple or damped dipoles. To obtain a directional effect, the dipole is exentrically embedded in a dielectric material; another possibility is to use a loaded loop antenna.

Pulse generators

Pulse generators for producing short, high-energy impulses for use in the field are often build up as cascade generator with transistor switches using the avalanche effect[10]. In principle, a series of parallel capacitors are charged and then discharged to the antenna in a cascade by the transistors switches (PFEIFFER, 1976). Pulses with nanosecond rise times and amplitudes of up to 2 kV can be produced with frequencies of more than 100 kHz. The typical pulse length of a transmitted electromagnetic wave is < 20 ns, depending on

[9] Attenuation is the process of transformation and loss of energy of the propagating electromagnetic wave. The energy loss through geometric effects in the wave propagation, such as scattering, geometric spreading, multipathing, etc., is called apparent attenuation.

[10] Avalanche effect is a process in solids analog to gaseous discharge where few injected electrons release many further electrons like an avalanche.

the antenna frequency and type. It is important that the transmitted signal is generated and repeated with a high accuracy. The timing accuracy within radar systems used for geophysics is usually ±1 ns. The repetition rate is typically 50 000 times per second.

For measurements with small offsets it is important that no further signal is produced by the pulse generator between pulses. Such signals would appear as arrival times with either constant or variable traveltimes. Recently developed MOS switches and other high-voltage devices now provide alternatives for pulse generation.

Receiver systems

The time window for georadar measurements ranges from several 10 ns (for travel paths of several meters) to 50 μs (e.g., for travel paths of 4 km in ice). Depending on the frequency used, the measurements are made with sampling intervals of down to less than nanoseconds. Either analog or digital systems with a high dynamic range can be used. For reasons of weight and power consumption, mobile systems are usually designed to take only one sample for each transmitted pulse as the antennas are drawn along the profile (i.e., sequential measurements). The principle is discussed, for example, by PFEIFFER (1976). This is not a disadvantage for continuous measurements, since achievable pulse rates are very high and the pulse generators normally have a long lifetime.

The sequentially received high-frequency signals are converted to audio waves so that they can be digitized and recorded. The achievable dynamic range of good sampling systems is 80 - 90 dB. The sensitivity (without stacking) under ideal conditions depends on the thermal noise in the input and the noise factor of the receiver. The effective thermal noise potential U_{eff} (also called Nyquist noise) is given by the following equation:

$$U_{eff} = \sqrt{4kTR\Delta f}$$

where k is the Boltzmann constant ($1.3805 \cdot 10^{-23}$ A s K^{-1}),
T is the absolute temperature [K],
R is the antenna impedance (electrical resistivity) [Ω], and
Δf is the band width [Hz].

Example: For $R = 50\ \Omega$, $T = 300$ K, and $\Delta f = 400$ MHz, the thermal noise potential $U_{eff} = 36\ \mu V$. Thus, the resolution of the digitizing system should be about 10 μV. The maximum range is then about 0.65 V for a 16 bit system, which is sufficient in most cases.

To avoid oscillation of the system as a whole, the antennas must be matched to the input of the receiver, and propagation of high-frequency electromagnetic waves in metal cables must be prevented, especially between the transmission and receiver antennas. The GPR systems of some manufacturers use fiber optic cables for this reason. High-quality digital storage oscilloscopes (DSOs) and other fast A/D converters make it possible to replace the manufacturers "black box" analog sampling systems with systems with known specifications. Because these digital systems require only one "shot" per time trace, it would be possible to use high output pulse generators with a lower repetition rate. For surveys of glaciers, there are a number of prototype digital systems available (e.g., JONES et al., 1989; WRIGHT et al., 1990). Since most systems have only one to four channels, multiple coverage measurements like those in reflection seismics are done sequentially. Normally, georadar measurements are carried out with single coverage at contant antenna offset.

Performance factor

The system dynamics is given by some manufacturers as the ratio of the peak voltage of the pulse generator and the minimum recordable voltage of the receiver in dB (performance factor PF). Performance factors of 130 to 140 dB are typical for existing systems. Because antenna efficiency, ground coupling, and spectral content of the broadband transmitter pulse are not specified, the PF is of little use in practice. It could not in any case be used to calculate the depth of penetration, etc. More important for good results are the dynamic range of the receiver, adaptation of the pulse generator and antennas to the ground material, clean excitation pulses, and avoidance of high-frequency oscillations in the cables.

4.5.5 Survey Practice

There are several procedures used for GPR surveys: radar reflection profiling, wide-angle reflection and refraction (WARR), common-midpoint soundings (CMP), and radar tomography. See Section 4.5.10 for special applications. For radar reflection profiling one or two antennas are moved over the ground while simultaneously measuring the traveltimes of the reflected radar pulses, as is done in seismic reflection profiling (see Chapter 4.6). This method is the most frequently used for GPR surveys. For WARR soundings, the transmitter is kept at a fixed location and the receiver is moved away from it. For CMP soundings, both the transmitter and receiver are moved simultaneously away from a fixed midpoint. This makes it possible to determine the velocity-depth function. WARR/CMP soundings can be carried out only with bistatic antenna systems. For radar tomography (trans-illumination) applications, transmitter and receiver antennas on opposite sides of the volume to be investigated (e.g., between boreholes or for the non-destructive testing of walls and pillars) are moved as shown in *Fig. 4.5-13*.

As for all geophysical methods, the objective of the field work must be clearly defined for the planning and execution. Important for an assessment of success of the method are the depth to the groundwater table, the kind of cover sediments (good-conducting, cohesive material or low conductivity, noncohesive material), and the location of buildings, buried and above-ground cables and other buried objects that are not investigation targets. Visible objects should be marked on the base map, as well as distances and dimensions. When ground penetrating radar measurements are interpreted, the influence of temperature, precipitation and chemicals present must be taken into account. Therefore, it is necessary to note the weather conditions and any observations of pollution during the measurements.

The suitability of an area or target object for georadar can be checked with a test measurement or by geoelectric sounding, from which the absorption can be estimated from the electrical resistivity of the ground. Unsuitable are areas with moist clay and those paved with slag.

To calibrate the GPR data, test measurements should be made along a profile with known underground conditions that are typical of the area. The target object (buried object, reflection horizon) should be recognizable in the test data. The operating frequency (and thus the resolution), the distance between the transmission and receiver antennas (constant offset for simple coverage measurements), and the distance between profiles and measurement points must be selected on the basis of the objectives. The useable dynamic range is negatively influenced by a small offset; if the offset is too large for investigating shallow depths, direct and lateral signals will overlie the reflections. As a rule, the offset should be about a fourth to a fifth of the expected depth to the reflector or diffractor. The spacing of the measurements should be a compromise between the necessary resolution and the fastest possible speed of measurement. For measurements of an entire area, the guidelines for reflection seismic should be used (see Chapter 4.6).

All GPR systems provide possibilities for filtering the data during acquisition in order to sharpen the signal waveform. Both high-pass and low-pass filters are used for this purpose. A rule of thumb given by REYNOLDS (1997) says that the filter settings should kept as broadband as possible so that potentially valuable signals are not removed during the acquisition phase. Digital systems have gain-setting options and a stacking function to optimize data quality.

4.5.6 Processing, Presentation and Interpretation of the Measured Data

Filtering is the first step in post-acquisition data processing. For many applications, this is sufficient to prepare the data for presentation and interpretation. Data-processing packages are available for a more thorough analysis. Most of the steps for a seismic survey can be directly applied to a georadar survey. Normally, however, multiple coverage measurements are seldom made. Hence, CMP stacking, one of the most important processing steps in seismics, is not usually done. There are also differences in the application of deconvolution. The complex reflection and transmission coefficients and dispersion effects during propagation cause wavelet deformation. Because the wavelets are not minimum-phase, but have a "mixed-phase" character, the deconvolution method used in seismics cannot be applied directly. A data adaptive deconvolution is necessary in GPR.

Radargrams often show a complicated picture, due to a large number of diffraction hyperbolas, which require considerable experience to interpret. Usually the location of the diffraction center is indicated by the apex of the hyperbola (e.g., the location of a buried pipeline or cable, *Fig. 4.5-8*). If there is little absorption, the migration programs used for reflection seismic can be used to obtain a good mapping of diffractors with considerable improvement in resolution.

If there are reflection horizons in the survey area, the velocity-vertical traveltime function can be obtained from CMP or well logging measurements, analogously to reflection seismic. The analysis of diffraction hyperbolas is suitable only for a rough estimate of velocity, since location and shape of the diffractor are often not sufficiently known. An exception is diffraction hyperbolas from profiles perpendicular to buried pipelines or cables. These velocities can be used to calculate depth from traveltimes.

Due to the high data density from profile measurements, the data are often displayed compressed in grey-scale or raster plots instead as individual data points. In the presentation of GPR data it is necessary to give the traveltime and a distance scale, as well as the time-zero point, the antenna offset, the velocity-depth function, and/or a depth scale derived from it. The identification of the horizons and objects should be well founded. To avoid erroneous interpretation by the client, system noise (e.g., cable waves, which appear in the radargram as parallel stripes at constant time intervals) should be labeled.

There are two types of areal surveys:

- Use of profile spacing larger than half the wavelength of the central frequency for reconnaissance surveys: Such a survey is suitable for following subhorizontal horizons or to map areas with varying cover layer properties. The data is processed like for single profiles and presented on a map, e.g., horizon depth relative to MSL or special features in the cover layer.

- 3-D measurements (without spatial aliasing[11]): The amount of data, the necessary processing steps, and the possibilities for representation are comparable to 3-D seismics. Special programs are available for processing the data.

In some cases it is difficult to distinguish between significant reflections and multiple events, extraneous reverberations, off-section ghosts, etc. In such cases the interpretation of georadar data can be checked by modeling. Appropriate modeling software has been developed for the 2-D and 3-D cases (CAI & MCMECHAN, 1995; GOODMAN, 1994). The structural modeling in these programs is done by ray tracing according to optical principles. The dynamic aspects of wave propagation (energy conditions and signal form) influenced by absorption, divergence, reflection, and transmission are usually taken into consideration in these programs. Calculations using the FDTD (finite-difference time-domain) method have become more powerful during the last decade (e.g., RADZEVICIUS et al., 2003; LAMPE et al., 2003).

4.5.7 Quality Assurance

General guidelines for quality assurance are given in Chapter 2.6. For georadar, the following points should be observed:

- Diffractors and reflectors (trees, powerlines, roofs, etc.) above the antennas must be documented.
- Recognizable changes in the cover layer and indications (e.g., manhole covers and hydrants) of buried pipelines or cables should be documented. Cable detectors should be used.
- Survey parameter values appropriate to the survey objectives (operating frequency, antenna offset, location of the profiles, and orientation of the antennas relative to the profiles, especially when pipelines or cables are the target) should be selected.
- All survey parameter values and equipment settings should be recorded in the field and deviations from normal documented. Profile position and parameter values not recorded in file headers must be noted during the survey in files for this purpose.

---

[11] Spatial aliasing occurs when the data density is too small and can lead to overlook of local anomalies.

- If the measurements are affected by disturbances in the functioning of the apparatus, or erroneous or nonoptimal parameter values are obtained, the profile is to be resurveyed before continuing the survey.
- The kind of calculations used to estimate velocities for the depth calculations are to be noted: CMP (comparable to WARR = wide-angle reflection and refraction), diffraction hyperbolas, well logging, measurement of the permittivity of samples, calibration with respect to known reflectors or objects (e.g., from exposures), or estimation from experience.
- The data is to be delivered to the client in well documented form as legal evidence and for possible reinterpretation together with the results of other methods.

4.5.8 Personnel, Equipment, Time Needed

	Personnel	Equipment	Time needed
mobilization and demobilization	depends on the distance to the survey area		
topographic survey	see Chapter 2.5		
field work	1 geophysicist (or technician) 1 assistant	1 4WD vehicle or station wagon, 1 georadar system, various types of antennas (in pairs) and accessory equipment (tape measure, survey wheel, or other instrument for measuring distance, etc.)	for $a = 0.02$ - 1 m, 200 - 20 000 m d^{-1}, depending on the conditions in the field, the objectives, and the type of antennas
data processing, interpretation, and report preparation	1 geophysicist (1 or 2 assistants)	PC or workstation with > 256 MB RAM and > 40 GB hard disk; radar or seismic software, printer (plotter)	1 - 2 days are necessary for each day in the field

a is the distance between measurement points, d = 10-h workday

4.5.9 Examples

Localization of objects

Georadar can be used to localize both metallic and nonmetallic pipelines, cables, and other objects. Possibilities for application exist also in the field of archeology and in non-destructive testing. The size of the target is normally smaller than the wavelength of the radar wave, so that characteristic diffraction curves are obtained. The shape of the diffraction curves from pipelines and cables depends on the direction of polarization of the propagating wavefield and the properties of material it passes through. Due to changes in propagation velocity, backfill, e.g., in cable and pipeline trenches, creates additional pull-ups or push-downs[12], which can indicate the presence of the object. The resolution of closely spaced pipes is not very high. Cohesive or inhomogeneous ground negatively affects the determination of the location of an object. The antennas must be located near (about 10 cm) or at the ground surface. The interpretation of a radargram is based mainly on the regocnition of diffractions. 3-D measurements have proved useful for complicated cases, e.g., where linear targets cross each other. An example is shown in *Fig. 4.5-8*, in which the locations of the pipelines can be clearly seen, marked by arrows. The section A - A' is running perpendicular to pipe L, which is indicated by a diffraction hyperbola. Section B - B' extends along pipe L, indicated by reflections. The kind of pipe (metallic, plastics) can be identified by analyzing the intensity of reflections and from previous knowledge. A second example is given in *Fig. 4.5-9*, which shows buried vault structures below a church.

[12] Pull-ups or push-downs are sudden rises or depressions in a reflection event. They are observed when there is a large difference between the velocities inside a buried object and the surrounding material (or the air/soil boundary in the case of a cavity) and may not be interpreted as faults.

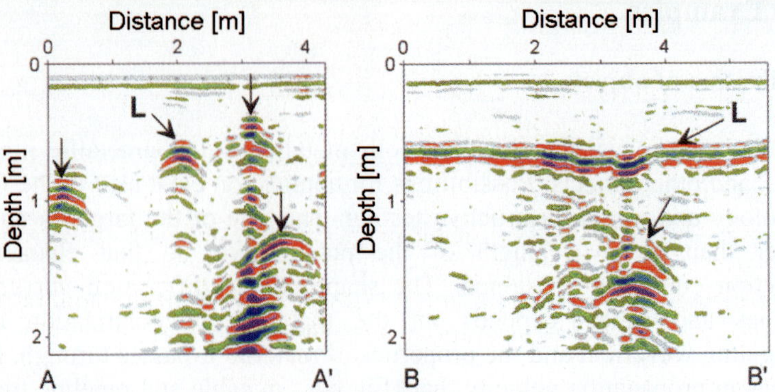

Fig. 4.5-8: Radargrams showing the location of pipes, *left*: section A - A', *right*: section B - B' perpendicular to A – A', GGU, Karlsruhe, Germany

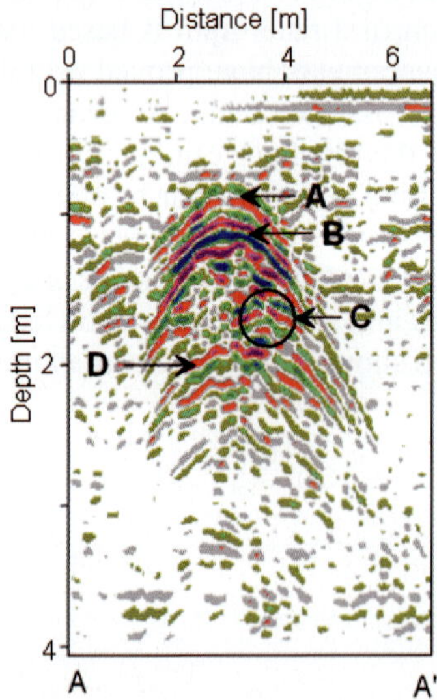

Fig. 4.5-9: Radargrams showing the location of buried objects, A: top edge of a vault, B: bottom edge of the masonry, C: object in the cavity, D: bottom of the cavity, GGU, Karlsruhe, Germany

Quaternary geology and hydrogeology

The location of the groundwater table can often be clearly observed in radargrams from coarse sands, gravel and porous limestone layers, caused by the large difference in impedance between unsaturated and saturated rock. Reflections are observed at layer boundaries when both layers have distinctly different water contents. The depth of penetration in saturated rock depends strongly on the conductivity of the water – from a few decimeters to several tens of meters. In the example in *Fig. 4.5-10*, the base of the aquifer at 28 m depth, mapped in parts of the profile, was confirmed by drilling.

Survey setup:

system	sampling system, opto-electronic transmission and 2 kV pulser (developed by the University of Münster),
antennas	dipoles with resistive loading, approx. 50 MHz,
sampling interval	1 ns, 20fold stacking,
measuring speed	3 km h^{-1}, wheel triggering, 5 traces m^{-1},
personnel	1 operator, 1 assistant,
processing	frequency filtering, time-dependent amplitude control.

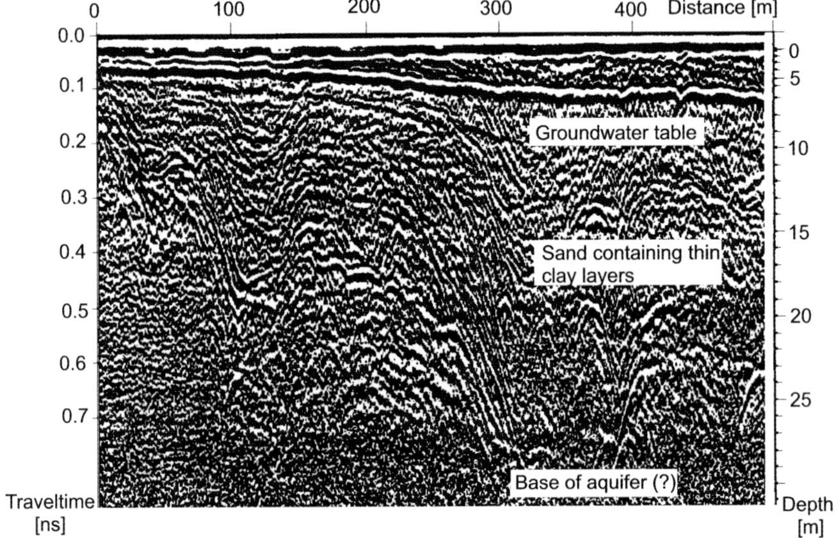

Fig. 4.5-10: Radargram showing the groundwater table, layer boundaries, and the base of the aquifer in glacial sediments in the Lüneburg Heath, Institut für Geophysik, Westfälische Wilhelms-Universität Münster (1992)

Delineation of a mineralized fault in consolidated rocks

GPR has a large penetration depth in consolidated, unweathered, dry rocks, like dolomite due to high electrical resistivity. In contrast, mineralized, clay-filled, and water-bearing faults have electrical properties (resistivity and permittivity) that are different from those of the surrounding rocks. Such faults are good reflectors for the radar waves. *Fig. 4.5-11* shows a radargram recorded during the investigation of a fault containing a lead and zincblende mineralization in a dolomite mine. The depths were calculated from the traveltimes of the reflected signals and the velocity of electromagnetic waves in the dolomite. The velocity was determined in two ways: from a CMP sounding and from diffraction hyperbolae.

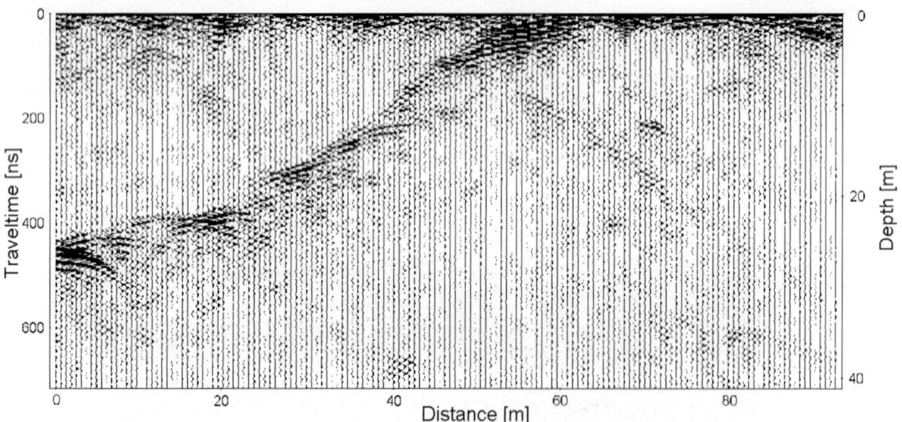

Fig. 4.5-11: Radargram recorded in an investigation of a mineralized fault in a dolomite, frequency 100 MHz, GGU, Karlsruhe, Germany

Investigation of a domestic waste site in a refilled open-pit mine

Georadar measurements were performed along with geoelectric resistivity measurements in the area of exhausted open-pit lignite mines. An example of these measurements on the surface of a refilled pit is shown in *Fig. 4.5-12*. When pumping to keep the open pit dry was switched off, the groundwater table rose in the stripped overburden filling the pit. Beginning at the edge of the pit at the left, the reflection from the water table can be clearly seen. Between 160 and 215 m from the edge of the pit, the location of a covered domestic waste landfill in a hole in the refilled overburden can be recognized in both the radargram and the resistivity curve. The base of the waste – dumped without a base seal – is below the water table, making it possible for pollutants to enter the groundwater and eventually surface water bodies. The measurements were made in continuous mode from a 4WD vehicle.

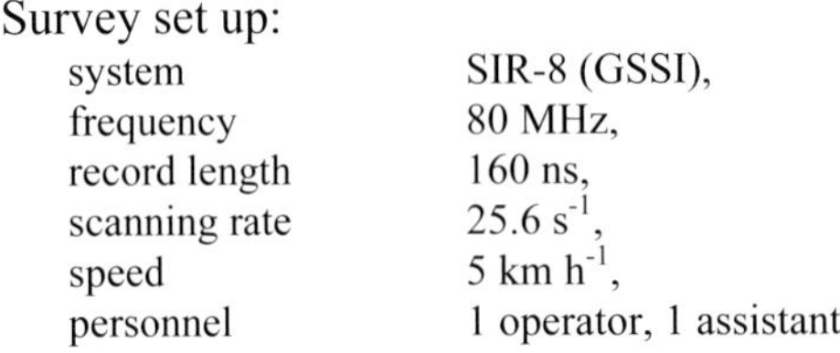

Survey set up:

system	SIR-8 (GSSI),
frequency	80 MHz,
record length	160 ns,
scanning rate	25.6 s^{-1},
speed	5 km h^{-1},
personnel	1 operator, 1 assistant.

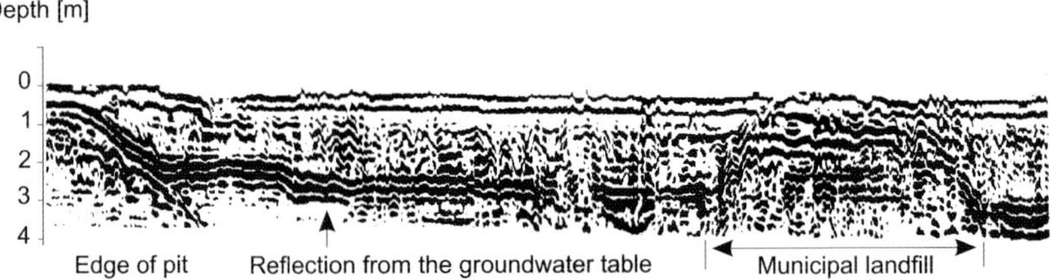

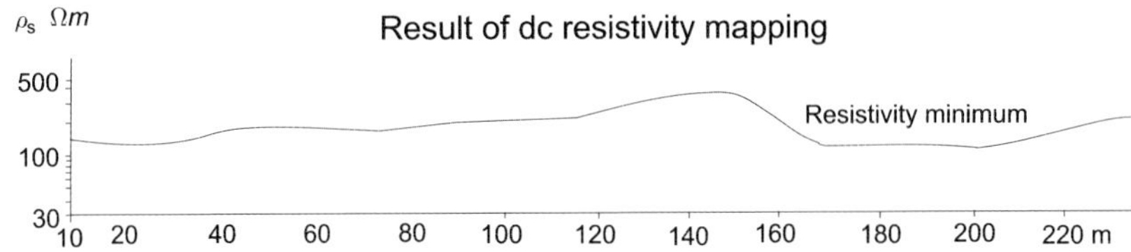

Fig. 4.5-12: Domestic waste landfill at the edge of an exhausted lignite open-pit mine, Lausitzer Braunkohlen AG, Arbeitsgruppe Geophysik

Radar tomography to assess the ground below buildings

Radar velocity tomography in boreholes can be used to assess foundations, even below existing buildings. In gypsum and other evaporates, solution processes (subrosion) change the properties and structure of the ground. The objective of the site survey was to investigate the rock structures and properties below an old church. *Fig. 4.5-13* shows the measurement scheme used in radar velocity tomography. Transmitter and receiver are placed in separate boreholes and the traveltime of the radar wave is determined for each raypath. The velocity distribution (*Fig. 4.5-14, left*) can be calculated from the traveltime and the known distance using a tomographic inversion algorithm. The geological cross-section (*Fig. 4.5-14, right*) was derived from the velocity distribution of the radar waves. The cross-section contains information about the structure and the rocks in the ground below the church.

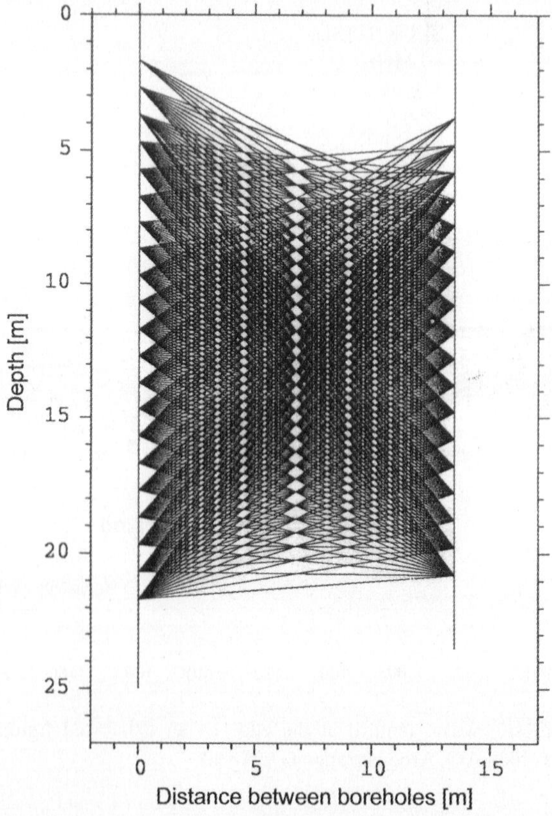

Fig. 4.5-13: Radar velocity tomography, ray pattern for several transmitter and receiver positions between two boreholes, HALLEUX & RICHTER (1994)

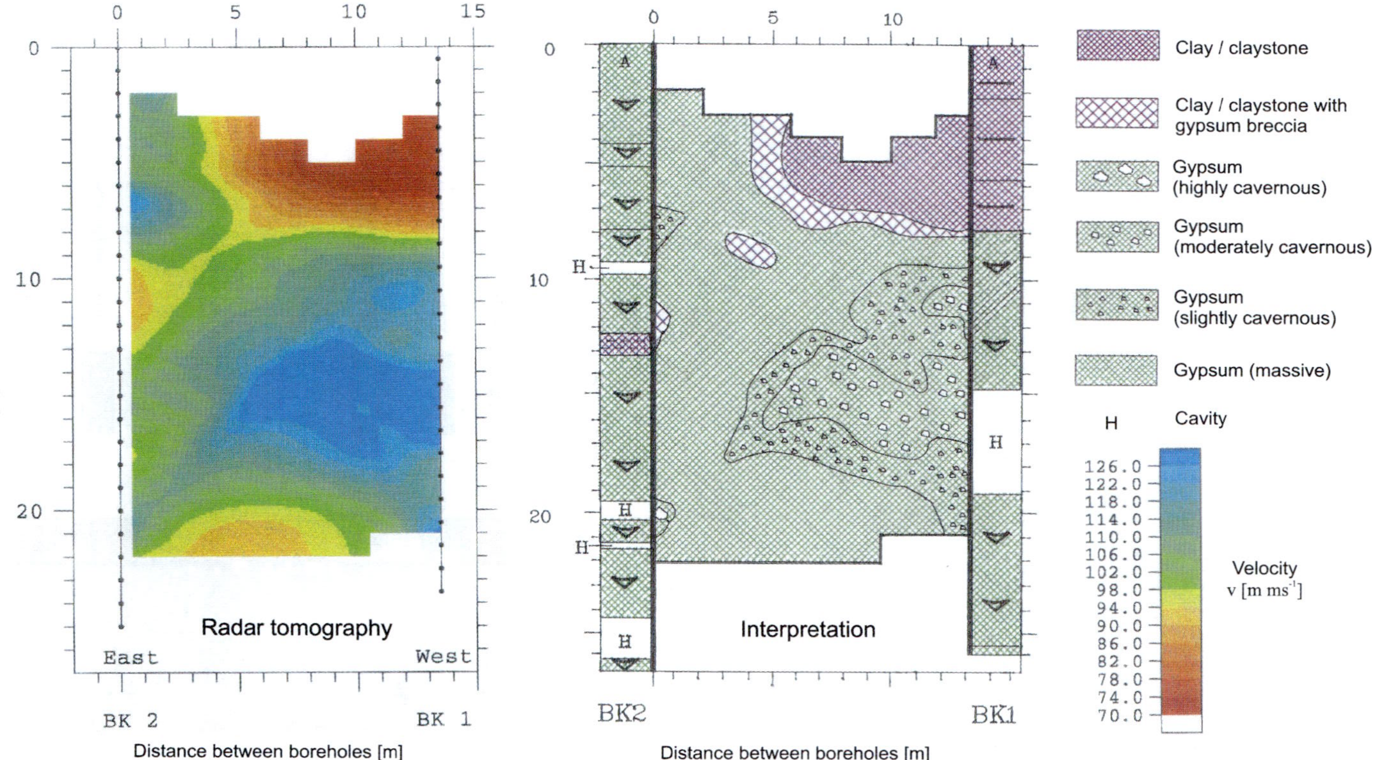

Fig. 4.5-14: Radar velocity tomography, *left*: velocity of the radar waves in the ground between two boreholes, BK1 and BK2, determined using the measurement scheme shown in *Fig. 4.5-13*. *Right*: geological cross-section derived from the radar data, HALLEUX & RICHTER (1994)

Investigations of concrete constructions

Ground penetrating radar is an excellent method for quality control of concrete constructions (*Fig. 4.5-15*). Defects in concrete usually cannot be concealed by rebars if the profiles cross the rebars (steel rods).

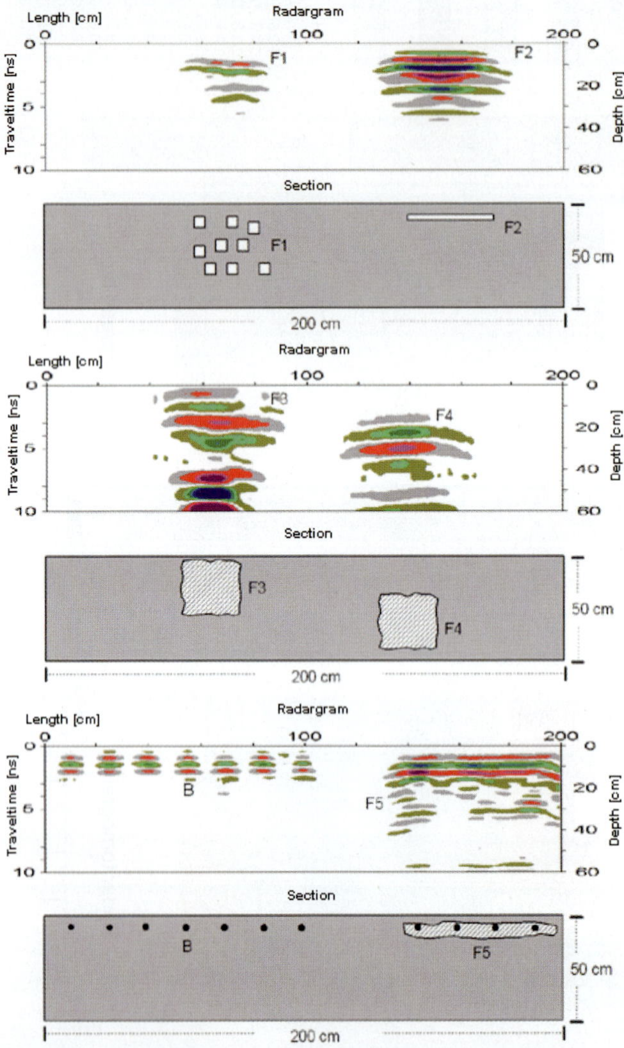

Fig. 4.5-15: Radargram recorded for quality control of concrete constructions, cube-shaped voids (F1) and an air gap (F2), gravel nests at various depths (F3 and F4), uncorroded reinforcement bars (rebars) (B) and corroded rebars in a gravel nest (F5) in a concrete plate, GGU, Karlsruhe, Germany

Examination of masonry structures

Retaining walls are often used to stabilize embankments. But moving water or other processes (e.g., burrowing animals) change the structure and destabilize the construction. *Fig. 4.5-16* shows an example of a GPR investigation of a retaining wall. Several parallel, vertical profiles were measured on the wall. Due to the wetness and salt content of the masonry, a 200-MHz frequency antenna was used. The thickness of the wall, structures in the wall, as well as cavities were determined by the measurements.

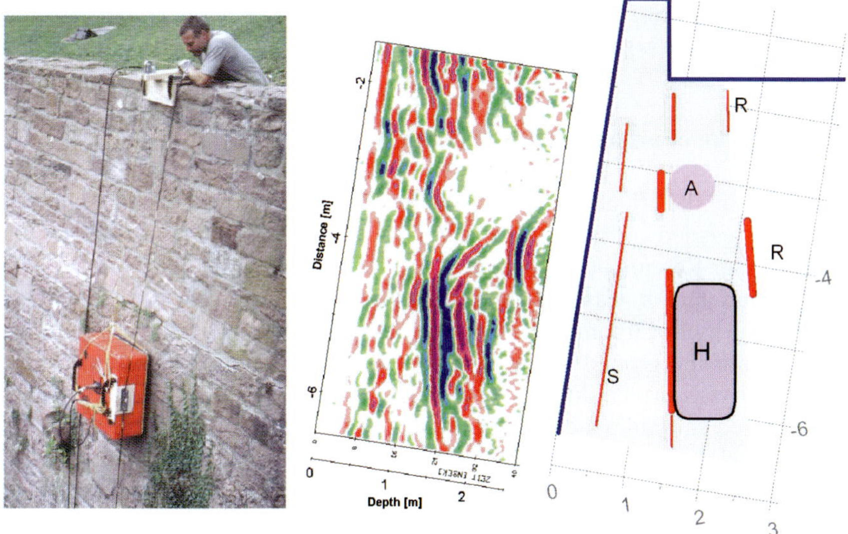

Fig. 4.5-16: Investigation of a retaining wall, *left*: photo from the measurements, *middle*: radargram of a vertical profile, *right*: section showing the result of the interpretation of the GPR data calculated using $v = 0.13$ m ns^{-1}; H cavity, A inhomogeneity, R interior side of the wall, S internal boundary within the wall, GGU, Karlsruhe, Germany

Investigation of residual foundations

Residual foundations impedes the reuse of sites. GPR measurements on a grid of profiles or several parallel profiles with small line spacing are useful to reveal construction obstacles in the ground. For an areal representation, the strength of radar signal amplitudes is estimated on all profiles from the radargrams for a given traveltime. From this, a time slice representation (*Fig. 4.5-17*) is plotted. The time slice can be assigned to an approximate depth. Dark shading indicates high signal amplitudes from residual foundations.

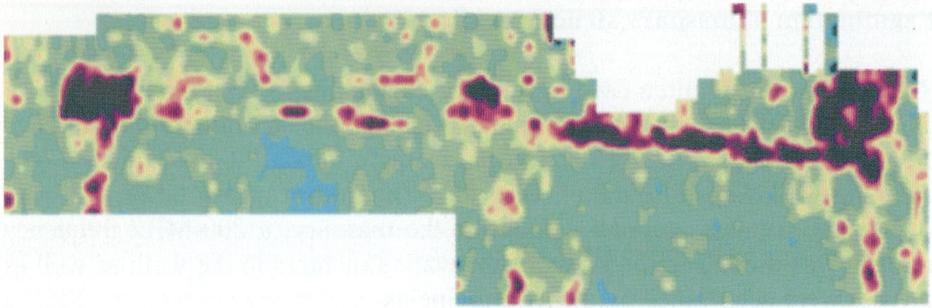

Fig. 4.5-17: Areal representation (time slice, plan view) of radar signals backscattering for a depth of about 1 m below ground surface, GGU, Karlsruhe, Germany

Identification of organic contaminants

To determine the effectiveness of GPR for identifying organic contaminants, a test was carried out in 1991 by the Canadian Waterloo Center for Groundwater Research (BREWSTER et al., 1992; BREWSTER & ANNAN, 1994). They found that it was possible to observe the movement of a plume of chlorinated hydrocarbons in a sandy aquifer. Perchloroethylene (770 L) was injected into the ground in the center of a 9 m × 9 m area surrounded by a sheet steel wall driven down to the clay layer at a depth of 3.3 m to prevent uncontrolled entry into the surrounding groundwater.

A Pulse-Ekko-IV system (200 MHz) was used before and 16 and 917 hours after injection with a 1-m profile spacing and 5 cm spacing between measurement points along a profile. Radargrams along a profile through the center of the test area are shown in *Fig. 4.5-18*. In the radargram before injection (top), the top of the clay layer appears as a high-energy reflection at about 110 ns. Weaker reflections within the aquifer are caused by laminations in the sand. The inclined reflections are diffractions from the steel walls. After 16 hours (middle), reflections at about 1 and 2 m depth can be recognized. These correlate with elevated PCE concentration. The plume front is already differentiated at this depth but has not spread much laterally. After 917 hours (bottom), strong reflections in the center above the clay layer can be seen, showing that the front has reached the clay layer and is spreading laterally.

In another experiment (GREENHOUSE et al., 1993), the "Borden Experiment", the contaminant front was indicated by "bright spots" (local, elevated amplitudes caused by abrupt changes in reflection properties) in the radargrams. More studies need to be made to determine the practical value of such measurements. A possible application would be the monitoring of sanitation projects.

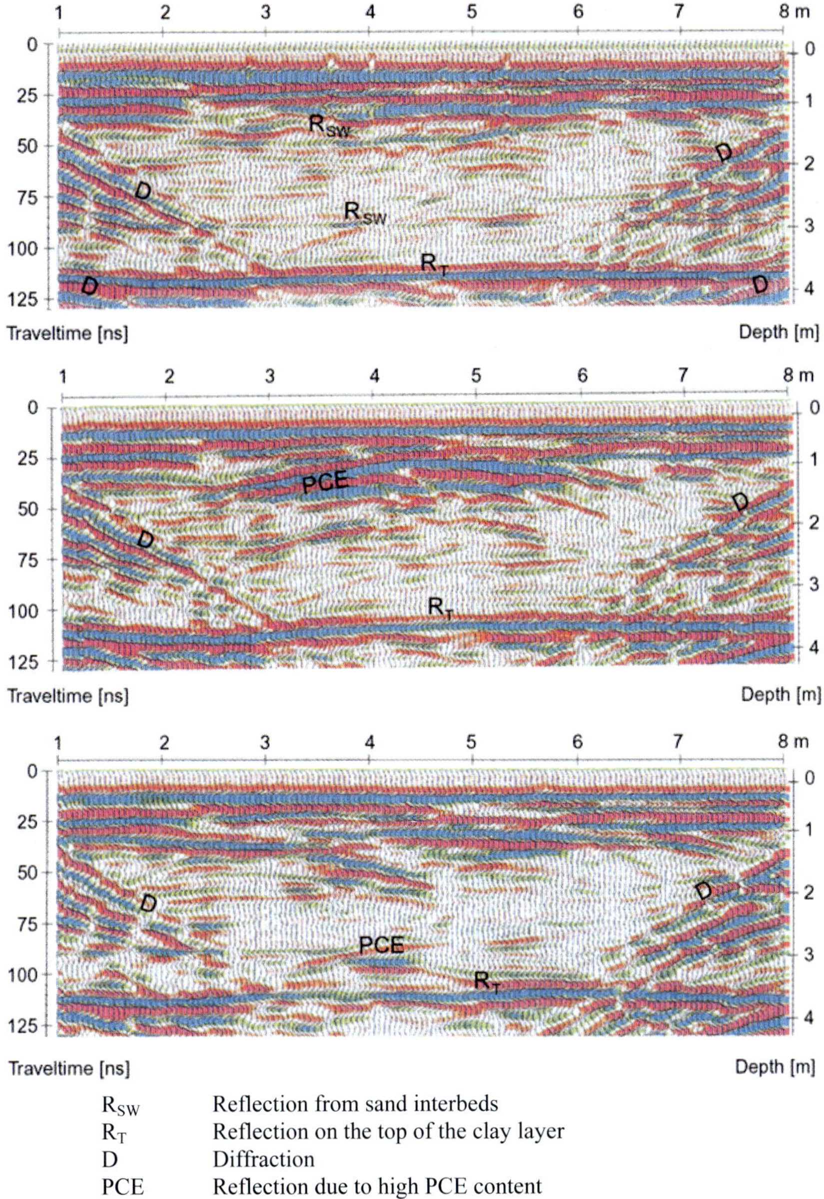

R_{SW}	Reflection from sand interbeds
R_T	Reflection on the top of the clay layer
D	Diffraction
PCE	Reflection due to high PCE content

Fig. 4.5-18: Radargram from the dense nonaqueous-phase liquids (DNAPL) experiment of Brewster et al. (1992), top: before the experiment; middle: 16 hours after the beginning of the percolation of perchloroethylene (PCE) into the ground; bottom: 38 days later, modified after Brewster et al. (1992)

4.5.10 Special Applications and New Developments

DIETER EISENBURGER

Radar surveys in underground workings

There is a need to deposit dangerous materials safely in underground repositories. Because technical barriers only provide a limited degree of security, it is important to find natural geological barriers which are effective and safe in the long term. GPR surveys can be conducted to investigate suitable host rocks. The use of electromagnetic waves for surveying directly in underground workings goes back over 30 years. GPR is a non-destructive method and logistically simpler and more economical to carry out than other exploration methods used for the geological investigation of repositories. Underground GPR surveying methods provide spatial information on the location of structures and heterogeneities by using directional receiving antennas. Such GPR surveys are carried out along profiles in drifts or boreholes. The difficulty with GPR reflection surveys is the extraction of spatial information from a linear antennas array moved along a straight profile.

If it is assumed that a reflection surface is in an optically favorable position with respect to the survey profile, then it is possible to calculate the distance to a reflection point (Pn) on the reflecting surface for each survey point if the traveltime and velocity are known (*Fig. 4.5-19*). The exact position within the reflection plane[13] and the angle λ is determined by migration[14] along the profile. However, the angle and distance are not sufficient to determine the precise position of a point in space because a second angle α is required. This is determined from measurements taken along a cross-section perpendicular to the drift axis or by using direction-receiving antennas. For the migration process the wave front method is used because it is able to keep the radial angle α from the survey in this process. This is a robust method which has the advantage of still being applicable even if the distance between survey points does not fulfill the Nyquist condition[15]. If the precise position of the associated reflection points on a reflection surface has been determined, it is possible to construct an element surface for each of these points. If the

[13] The reflection plane is defined by the transmitter/receiver position and the reflection point.

[14] Migration is a process by which events on a radargram are mapped to their true spatial position. It requires a knowledge of the velocity distribution along the raypath.

[15] The Nyquist condition is fulfilled when there are more than two samples per cycle (Nyquist rate) for the highest frequency of the spatial waveform.

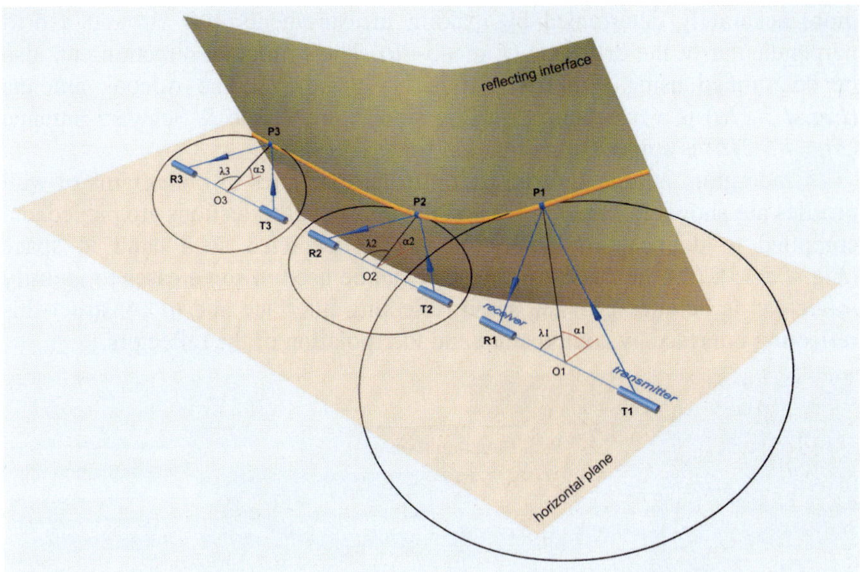

Fig. 4.5-19: Determination of the location of a reflection point: principle

element surfaces of all points are joined, a band in space is generated, visualizing the spatial position of the reflecting surface.

GPR surveys in drifts are carried out, if possible, on four profiles: on the floor of the drift, on the two walls and on the roof (*Fig. 4.5-20*).

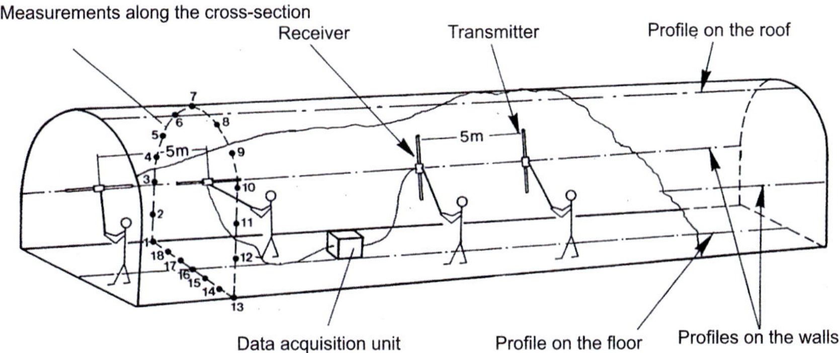

Fig. 4.5-20: Configuration of GPR measurements in a drift

It is recommended that this survey is carried out immediately after cutting the drift to prevent the influence of subsequent installations such as cables, pipes, equipment, etc., influencing the survey. The focusing effect of the denser medium (in this case, the surrounding rock) means that reflections from the direction of the profile (floor, wall, roof) are preferentially received. The azimuthal direction from which the individual reflections are received are more accurately determined by making measurements 360° around a drift perpendicular to the drift axis (*Fig. 4.5-20*). The azimuthal direction can also be determined using a direction finder (*Fig. 4.5-21*). The Adcook antenna (*Fig. 4.5-21a*) is a minimum direction finder; the Rohde & Schwarz antenna (*Fig. 4.5-21b*) is a maximum direction finder.

A radargram is prepared for each profile (examples of radargrams of wall profiles are shown in *Fig. 4.5-22*) and the individual reflections are picked and compiled to delineate reflectors, which are depicted as a band in space (*Fig. 4.5-23*). Geological considerations can be used in some cases to identify reflectors as belonging to the same reflecting interface and to construct this reflecting interface by combination and interpolation of the reflectors.

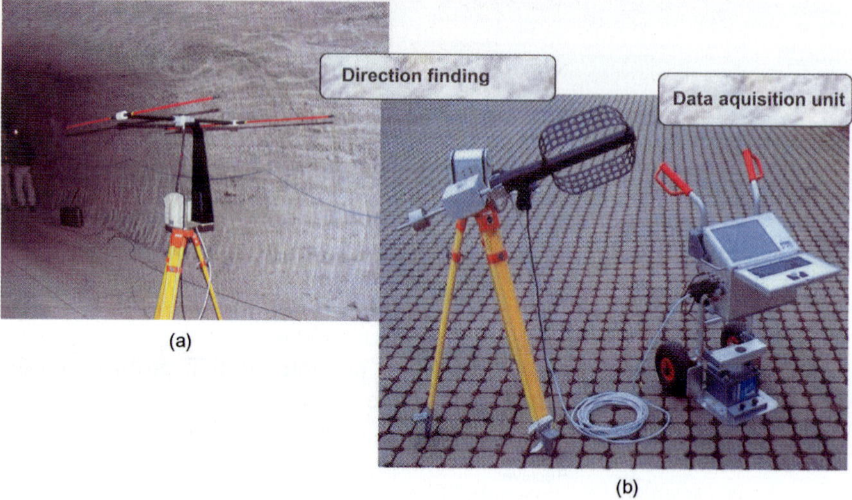

Fig. 4.5-21: GPR surveying equipment with direction finding equipment

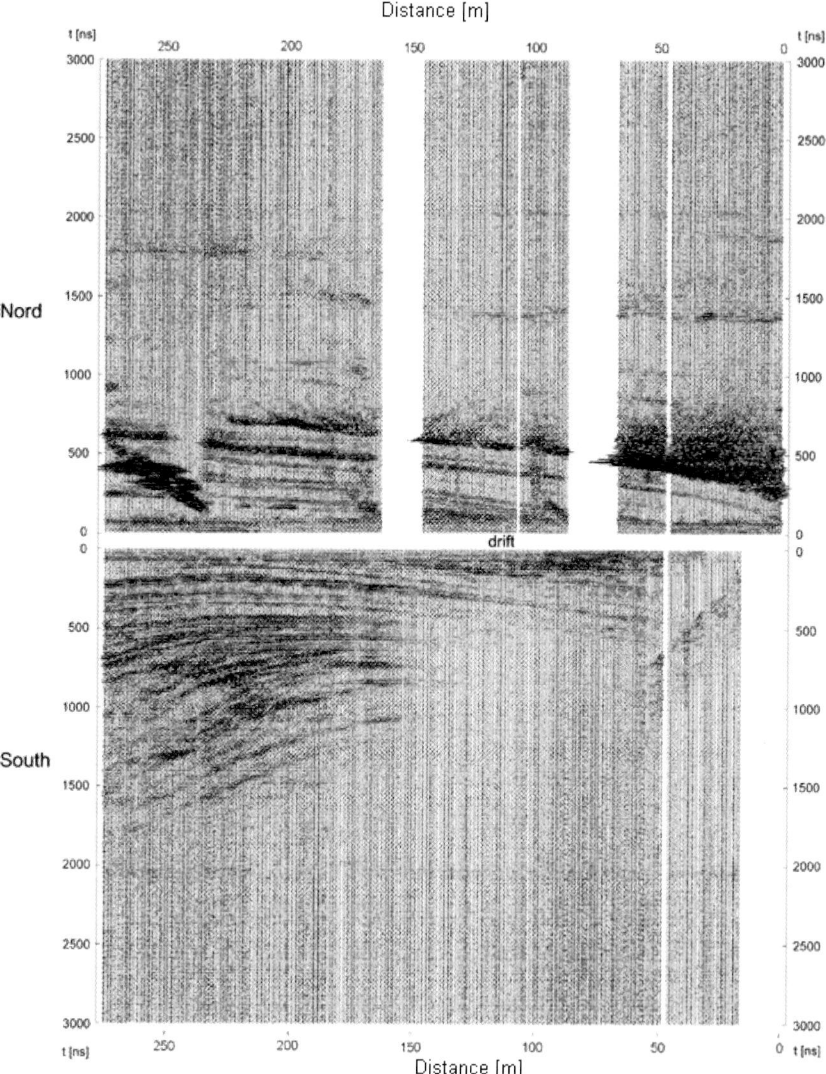

Fig. 4.5-22: Radargrams from the north and south wall profiles along a directional drift

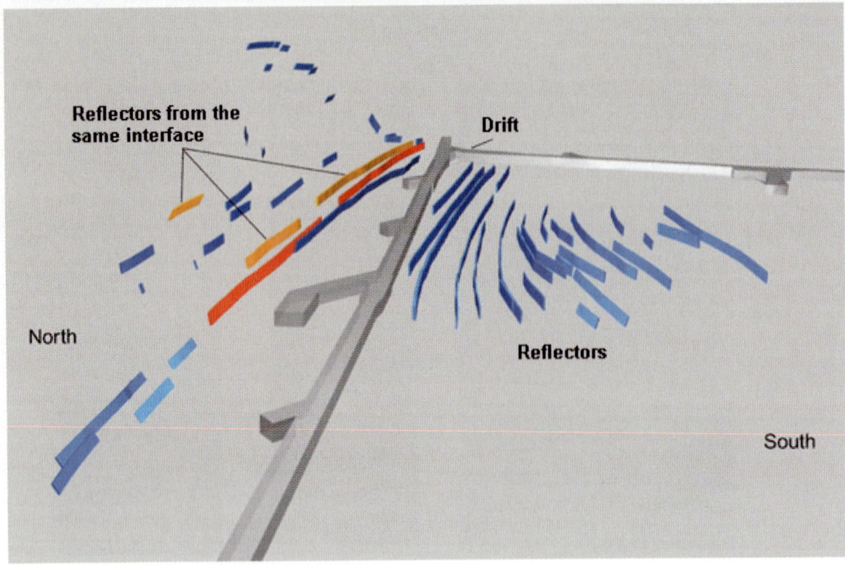

Fig. 4.5-23: Spatial evaluation of the reflectors

GPR single borehole logging

GPR borehole logs are carried out in boreholes drilled underground as well as those drilled from the surface. The GPR borehole logging system described here is based on the pulsed radar method described above. It consists of a transmitter with a dipole as the transmission antenna and a receiver with a direction-sensitive receiving antenna, plus the necessary digitizing and recording unit (EISENBURGER et al., 1993). The receiving antenna consists of two orthogonal loop antennas which can be connected to form a dipole. The pulse-generating transmission electronics is built into the transmission antenna. The use of direction-sensitive receiving antennas allows to determine the spatial position of reflection interfaces from a single borehole. Logging is carried out with the tool held at discrete points. The signals received by the two loop antennas, as well as by the dipole formed by them are recorded and used to determine the direction from which a reflection is received using Equation (4.5.13). The dipole signal is used to resolve the ambiguity of the angle derived from the two loop antennas. The angle determined remains stable for the whole time a reflection signal is received (*Fig. 4.5-25*) and it is a gage for good quality reflections.

$$\tan\alpha_{1,2} = \frac{1}{2\Sigma x_i y_i}\left[-\left(\Sigma x_i^2 - \Sigma y_i^2\right) \pm \sqrt{\left(\Sigma x_i^2 - \Sigma y_i^2\right)^2 + 4\left(\Sigma x_i y_i\right)^2}\right] \quad (4.5.13)$$

Several amplitudes are taken from a wavelet recorded with loop 1 (x_i) and loop 2 (y_i). In addition to the calculated angle (α) from the loop signals, the precise position of the tool needs to be known. For this reason, the measurement also includes the depth in the borehole and the distortion of the tool around the roll angle (β). The reference position for the roll angle in vertical boreholes is magnetic north, and in the case of horizontal boreholes, the vertical-up position. The absolute direction from which a reflection is received can be calculated from the angle α using (4.5.13) and the roll angle β (*Fig. 4.5-24*). The method described above can be used to determine the location of the reflectors from the traveltime and the calculated receiving direction. *Figures 4.5-26* and *4.5-27* show an example of a radar log using a directional antenna in a vertical borehole.

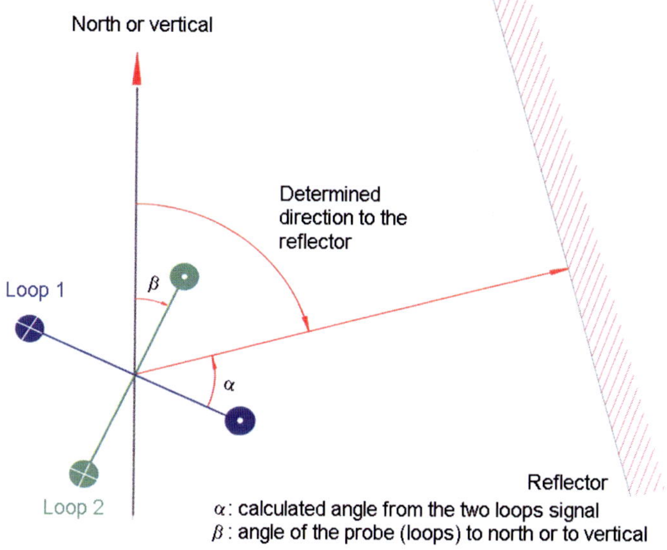

Fig. 4.5-24: Principle of determination the azimuthal direction of reflections received in boreholes

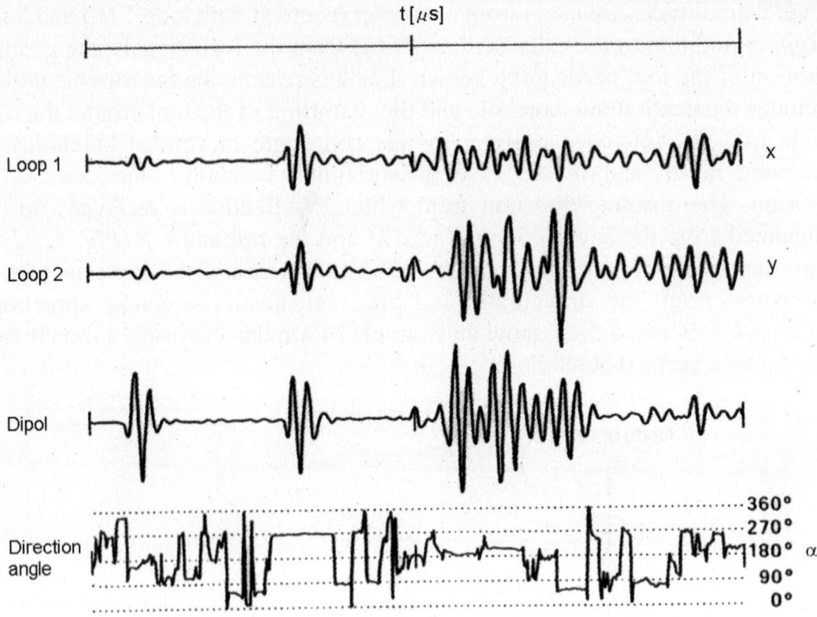

Fig. 4.5-25: Example of the determination of the azimuthal direction from which a reflection is received

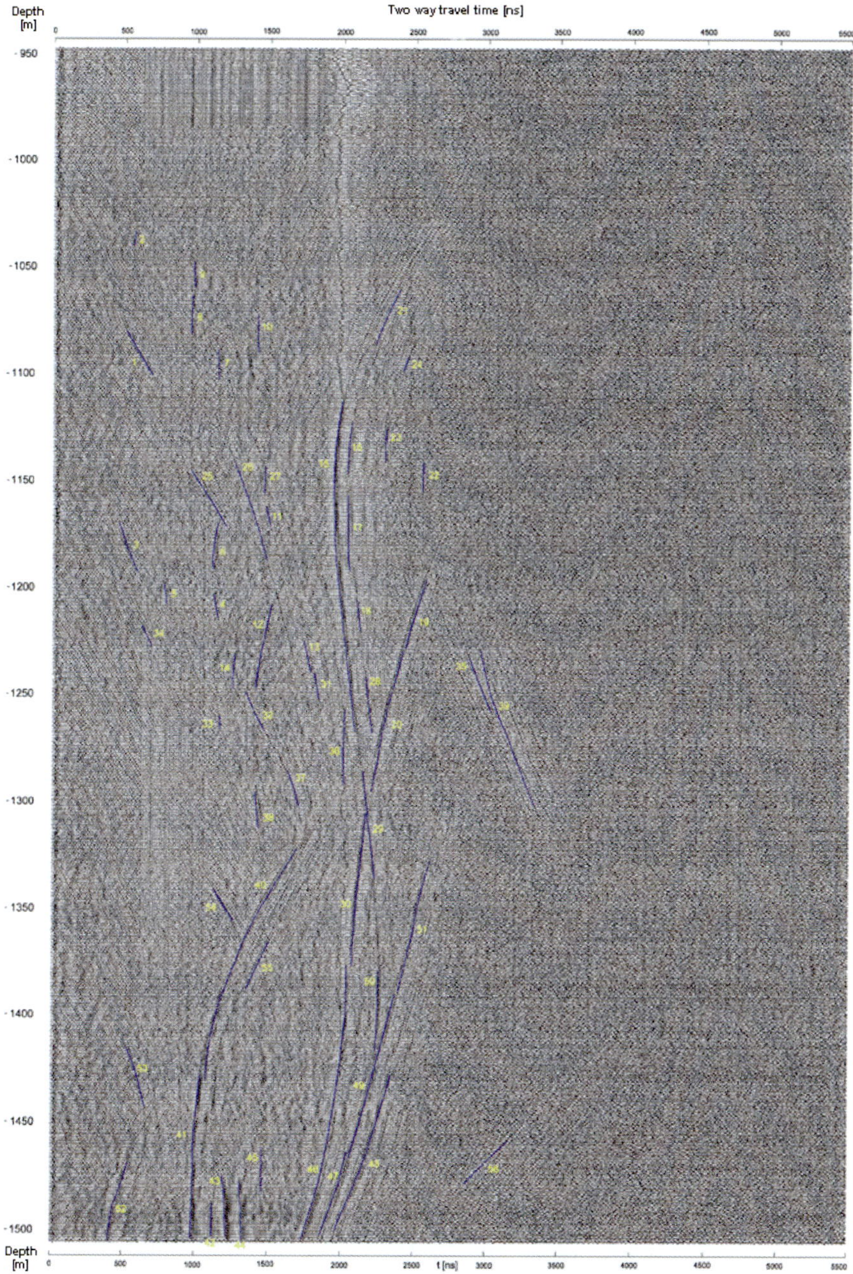

Fig. 4.5-26: Radargram of a vertical borehole log with the marked reflectors

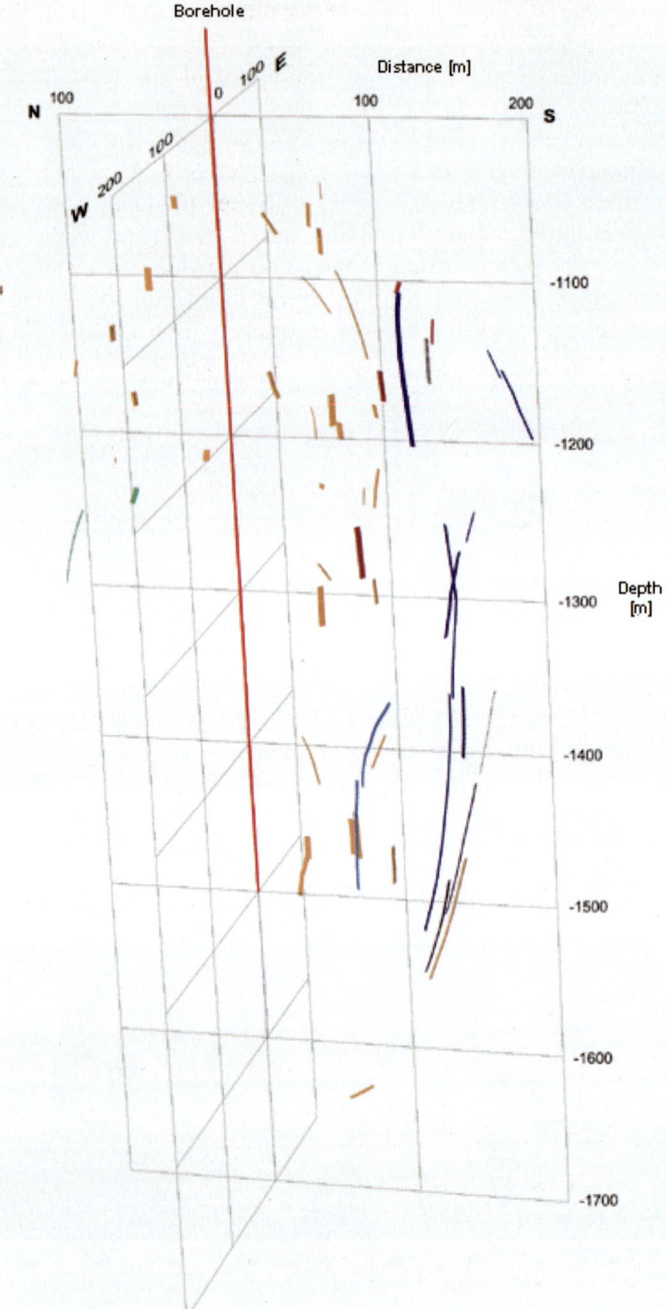

Fig. 4.5-27: Perspective diagram showing reflectors determined using a direction-sensitive GPR borehole log

The reflectors determined in this way represent reflection horizons which usually correspond to geological boundaries. This method allows to obtain spatial information on the geological structure of the rocks surrounding a single borehole. This information is used to create a three-dimensional geological model and for the much more precise and better characterization of geological barriers than possible with other methods. All of this information increases the safety and efficiency of the planning of underground repositories. In already existing repositories, e.g., caverns for the storage of hydrocarbons, this borehole logging method can help to acquire additional information on the rock around the caverns – information which is a very valuable supplement on the often inadequately investigated geology, and which, therefore, makes a contribution to improve the economic efficiency and safer operation of a storage cavern.

Helicopter-borne surveys with stepped-frequency radar (SFR)

In the past, helicopter-borne GPR systems have been restricted to the use of pulse-radar technology to determine the thickness of polar glaciers. Because of its limited resolution, this method is only of limited suitability for the geological investigation of shallow structures. Stepped-frequency radar (SFR) is a better alternative radar method for shallow geological investigations.

Figure 4.5-28 shows how classic pulse-radar works. Transmission pulses (TX) propagate from the transmission antenna and are received by the receiving antenna after a time delay corresponding to the distance to the target.

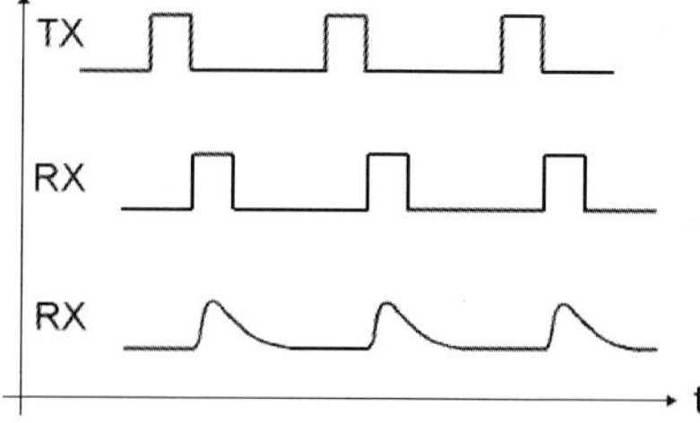

Fig. 4.5-28: Scheme of pulse radar, showing the transmission pulse (TX) and the receiving pulse (RX)

In *Fig. 4.5-28*, the upper received signal (RX) represents an idealized situation. The received signal shown in the lower part of the diagram shows the typically distorted measured pulse. The vertical resolution depends on the transmission pulse width. Therefore, a system with improved resolution requires a reduced pulse width. Because the transmitted energy corresponds to the "pulse area" (i.e., amplitude of the pulse integrated over the pulse width), a reduction of pulse width requires an enlargement of transmitted peak power in order to maintain the pulse energy decisive for the penetration depth. Limits to the achievable vertical resolution arise from technical limits to transmission strength, regulatory limits, or the technical inability to reduce the pulse width.

In contrast to classical pulse radar systems, SFR systems operate with amplitude-continuous radar signals (IIZUKA et al., 1984). The signal bandwidth being required for the desired radar resolution is generated sequentially instead of providing the instantaneous complete spectrum of the pulse radar case. *Figure 4.5-29* illustrates the basic SFR modulation scheme. The radar transmitter sequentially provides signals stepping through the desired frequency range. Depending on the application this could be done, for instance, in linear steps as depicted in the figure. In the receiver section, both phase and amplitude measurement of the echo signal is performed. The results are fed into a data processing, the well-known IFFT (Inverse Fast Fourier Transform), for example. The output is a pulse that can be compared to that of a pulse radar. It contains the range information of the target, while its pulse width, the radar range resolution, is related to the bandwidth of the applied radar spectrum.

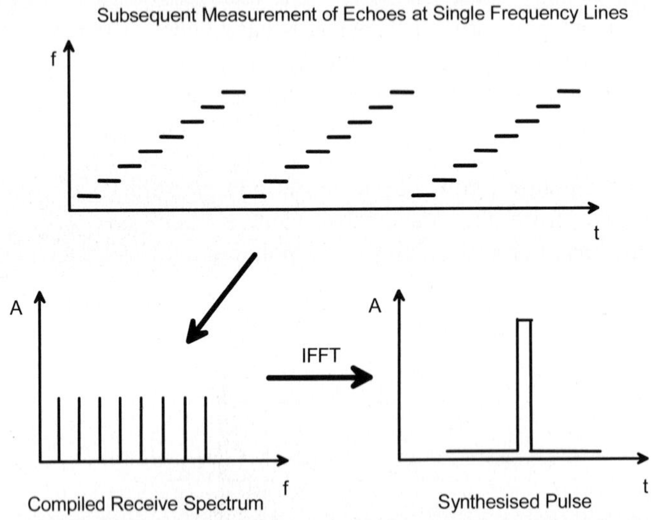

Fig. 4.5-29: Scheme of stepped-frequency radar (SFR), calculation of a synthetic pulse from several received echo lines

The advantages of the SFR can briefly be summarized as follows: low instantaneous bandwidth, high sensitivity, high penetration depth, low sensitivity to radio frequency - interferences, low power consumption, high resolution with respect to measurement frequency, low output data rate (saves memory, allows for high dynamic range AD-converters), reduced wideband antenna problems. Most of the advantages are directly related to the continuous wave operation and the low instantaneous bandwidth of the SFR. The second key factor resulting from the sequential operation principle, is the unique possibility for powerful instrument calibration in both, the frequency as well as in the time domain. The influence of the overall calibration is finally reflected by the achieved radar range resolution performance that is close to theory.

This radar method produces better resolution and penetration depths by controlling the range of the frequencies used (*Fig. 4.5-30*). And because the stepped-frequency technique only requires very low power of the transmitter, it is suitable for surveys where the potential interference with other equipment is to be avoided. A helicopter-borne SFR system (*Fig. 4.5-31*) makes it possible to quickly investigate large areas and can also be used to fly over dangerous or poorly accessible ground.

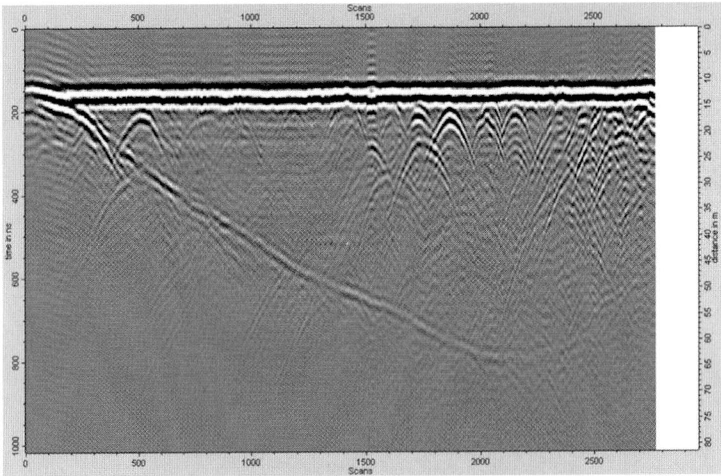

Fig. 4.5-30: Example of a stepped-frequency radar line

Fig. 4.5-31: SFR helicopter system in operation

Physical parameters and units

Parameters	Symbols	Units
electric field strength	$\boldsymbol{E}$	$V\ m^{-1}$
power	P_t, P_r	W, $m^2\ kg\ s^{-3}$
frequency	f	Hz
angular frequency	$\omega = 2\pi f$	s^{-1}, rad s^{-1}
induction constant of a vacuum	μ_0	$4\pi 10^{-7}\ V\ s\ A^{-1}\ m^{-1}$
permittivity of a vacuum	ε_0	$8.854\ 10^{-12}\ A\ s\ V^{-1}\ m^{-1}$
velocity of light in a vacuum	c_0	$2.998\ 10^8\ m\ s^{-1} = 0.2998\ m\ ns^{-1}$
Boltzmann constant	k	$1.3805 \cdot 10^{-23} A\ s\ K^{-1}$
phase velocity	v	$m\ s^{-1}$, $m\ ns^{-1}$
permittivity	$\varepsilon^* = \varepsilon' - i\varepsilon''$	$A\ s\ V^{-1}\ m^{-1}$
relative permittivity	$\varepsilon_r = \varepsilon / \varepsilon_0$	dimensionless
characteristic impedance (electrical impedance)	Z^*	Ω
propagation constant	$\gamma = \alpha + i\beta$	m^{-1}
absorption coefficient	α'	dB m^{-1}
electrical resistivity	ρ	Ωm
electrical conductivity	σ	$S\ m^{-1}$, $mS\ m^{-1}$
tangent of the loss angle δ	$\tan\delta = \frac{\varepsilon''}{\varepsilon'} = \frac{\sigma}{\omega\varepsilon'}$	dimensionless

References and further reading

AHRENS, T. J. (Ed.) (1995): A handbook of physical constants. Am. Geophys. Union, Washington, 3 vols.

ANNAN, A. P., WALLER, W. M., STRANGWAY, D. W., ROSSITER, J. R., REDMAN, J. D. & WATTS, R. D. (1975): The electromagnetic response of a low loss, 2-layer, dielectric earth for horizontal electric dipole excitation. Geophysics, **40**, 285-298.

BALANIS, C. A. (1996): Antenna theory: analysis and design. Wiley.

BAÑOS, A. (1966): Dipole radiation in the presence of a conducting halfspace. Pergamon Press, New York.

BEBLO, M. (1982): Elektrische Eigenschaften. In: ANGENHEISTER, G. (Ed.): LANDOLT - BÖRNSTEIN: Zahlenwerte und Funktionen aus Naturwissenschaften und Technik. Neue Serie **V**, **1b**, 254-261, Springer, Berlin.

BLEIL, U. & PETERSEN, N. (1982): Magnetic properties: in Landolt-Bornstein numerical data and functional relationships in science and technology: group V, geophysics and space research, **1b**, physical properties of rocks. ANGENHEISTER, G. (Ed.), Springer, Berlin, 308-432.

BLINDOW, N. (1986): Bestimmung der Mächtigkeit und des inneren Aufbaus von Schelfeis und temperierten Gletschern mit einem hochauflösenden elektromagnetischen Reflexionsverfahren. Dissertation, Institut für Geophysik, Westfälische Wilhelms-Universität Münster.

BLINDOW, N., ERGENZINGER, P., PAHLS, H., SCHOLZ, H. & THYSSEN, F. (1987): Continuous profiling of subsurface structures and groundwater surface by EMR methods in Southern Egypt. Berliner Geowiss. Abh. (A) **75.2**, 575-627.

BOHIDAR, R. N. & HERMANCE, J. F. (2002): The GPR refraction method. Geophysics, **67**, 1474-1485.

BREKOVSKIKH, L. M. (1980): Waves in layered media, 2nd edn. Academic Press, New York.

BREWSTER, M. L., ANNAN A. P. & REDMAN, J. D. (1992): GPR Monitoring of DNAPL migration in a sandy aquifer. In: Fourth international conference on ground penetrating radar. Geological Survey of Finland, Special Paper **16**, 185-190.

BREWSTER, M. L. & ANNAN, A. P. (1994): Ground-penetrating radar monitoring of a controlled DNAPL release: 200 MHz radar. Geophysics, **59**, 1211-1221.

BREWSTER, M. L., ANNAN, A. P., GREENHOUSE, J. P., KUEPER, B. H., OLHOEFT, G. R., REDMAN, J. D. & SANDER, K. A. (1995): Observed migration of a controlled DNAPL release by geophysical methods: Ground Water. **33**, 977-987.

BRISTOW, C .S. & JOL, H. M. (2003): Ground penetrating radar in sediments. Geological Society Publication 211, London.

CAI, J. & MCMECHAN, G. A. (1995): Ray-based synthesis of bistatic ground-penetrating radar profiles. Geophysics, **60**, 87-96.

CARCIONE, J. M. & SERIANI, G. (2000): An electromagnetic modelling tool for the detection of hydrocarbons in the subsoil. Geophys. Prosp., **48**, 231-256.

CARMICHAEL, R. S. (Ed.) (1982): Handbook of physical properties of rocks. CRC Press, Boca Raton, 3 vols.

CLOUGH, J. W. (1976): Electromagnetic lateral waves observed by earth sounding radars. Geophysics, **41**, 1126-1128.

CONYERS, L. B. & Goodman, D. (1997): Ground-penetrating radar: an introduction for archaeologists. Altimira.

DANIELS, D. J., GUNTON, D. J. & SCOTT, H. F. (1988): Introduction to subsurface radar. IEE Proceedings F, **135** (F,4), 278-320.

DANIELS, J. J. (1989): Fundamentals of ground penetrating radar. Proceedings of the Symposium on the Application of Geophysics to Engineering and Environmental Problems, SAGEEP 89, Golden, Colorado, 62-142.

DANIELS, J. J. (1996): Surface Penetrating Radar. The Institution of Electrical Engineers, London.

DANIELS, J. J. (2004): Ground Penetrating Radar – 2nd edn. The Institution of Electrical Engineers, London.

DANIELS, J. J. (1989): Fundamentals of ground penetrating radar. Proceedings of the Symposium on the Application of Geophysics to Engineering an Environmental Problems, SAGEEP 89, Golden, Colorado, 62-142.

DAVIS, J. L. & ANNAN, A. P. (1989): Ground penetrating radar for high-resolution mapping of soil and rock stratigraphy. Geophys. Prosp., **37**, 531-551.

DOOLITTLE, J. A. (1982): Characterizing soil map units with the ground-penetrating radar. Soil Surv. Horizons, **23**, 4, 3-10.

DOOLITTLE, J. A. (1983): Investigating Histosols with the ground-penetrating radar. Soil Surv. Horizons, **24**, 3, 23-28.

DOUGLAS, D. G., BURNS, A. A., RINO, CH. L. & MARESCA, J. W. (1992): A study to determine the feasibility of using a ground-penetrating radar for more effective remediation of subsurface contamination. Risk Reduction Engineering Laboratory Office of Research and Development U. S. Environmental Protection Agency, Cincinnati, Ohio, Report EPA/600/R-92/089.

EISENBURGER, D., SENDER, F. & THIERBACH, R. (1993): Borehole Radar- An Efficient Geophysical Tool to Aid in the Planing of Salt caverns and Mines. Seventh Symposium on Salt, **I**, 279-284, Elsevier, Amsterdam.

FORKMANN, B. & PETZOLD, H. (1989): Prinzip und Anwendung des Gesteinsradars zur Erkundung des Nahbereichs. Freiberger Forschungshefte, **C 432**, Dt. Verlag für Grundstoffindustrie, Leipzig.

FORKMANN, B. & PETZOLD, H. (1991): Gesteinsradar - Prinzip und Anwendungsmöglichkeiten. Z. angew. Geol., **37,** 25-30.

GEVANTMAN, L. H. (Ed.) (1981): Physical properties data for rock salt. National Bureau of Standards Monograph 167.

GOODMAN, D. (1994): Ground-penetrating radar simulation in engineering and archaeology. Geophysics, **59**, 224-232.

GPR 1994, Proceedings of the Fifth International Conference on Ground Penetrating Radar; 12-16 June, 1994, Kitchener, Ontario Canada.

GPR 1996, Proceedings of the Sixth International Conference on Ground Penetrating Radar; September 30- October 3, 1996, Tohoku Japan.

GPR 1998, Proceedings of the Seventh International Conference on Ground Penetrating Radar; 27-30 May, 1998, Lawrence, Kansas USA.

GPR 2000, Proceedings of the Eighth International Conference on Ground Penetrating Radar; 23-26 May, 2000, Gold Coast, Australia.

GPR 2002, Proceedings of the Ninth International Conference on Ground Penetrating Radar; April 29- May 2, 2002, Santa Barbara, California, USA.

GPR 2004, Proceedings of the Tenth International Conference on Ground Penetrating Radar; 21 – 24 June, 2004, Delft, The Netherlands.

GPR 2006, Proceedings of the Eleventh International Conference on Ground Penetrating Radar; June 19-22, 2006, Columbus, Ohio, USA.

GREENHOUSE, J., BREWSTER, M., SCHNEIDER, G., REDMANN, D., ANNAN, P., OLHOEFT, G., LUCIUS, J., SANDER, K. & MAZELLA, A. (1993): Geophysics and solvents: the Borden experiment. The Leading Edge, 261-267.

HALLEUX, L., FELLER, P., MONJOIE, A. & PISSART, R. (1992): Ground penetrating and borehole radar surveys in the Borth salt mine (FRG). In: Fourth International Conference on Ground Penetrating Radar. Geological Survey of Finland, Special Paper **16**, 317-321.

HALLEUX, L. & RICHTER, T. (1994): Radar tomography for shallow engineering geological investigations. Poster presentation on the Fifth International Conference on Ground Penetrating Radar; June 12–16, 1994, Kitchener, Ontario, Canada.

HANSEN, V. W. (1989): Numerical solution of antennas in layered media. Wiley.

V. HIPPEL, A. R. (1954): Dielectrics and waves. M. I. T. Press, Cambridge, Massachusetts.

HUGGENBERGER, P., MEIER, E. & PUGIN, A. (1994): Ground-probing radar as a tool for heterogeneity estimation in gravel deposits: Advances in data processing and facies analysis. Applied Geophysics, **31**, 171-184.

IIZUKA, K., FREUNDORFER, A. P., WU, K. H., MORI, H., OGURA, H. & NGUYEN, V.-K. (1984): Step-frequency radar. J. Appl. Phys., **56**, 9, 2572-2583.

JANSCHEK, H., MAURITSCH, H., RÖSLER, R. & STEINHAUSER, P. (1985): Hochfrequenzmethoden. In: MILITZER, H. & WEBER, F. (Eds.): Angewandte Geophysik, **2**, Geoelektrik-Geothermik-Radiometrie-Aerogeophysik. Springer Wien, Akademie Berlin, 151-173.

JOHARI, G. P. & CHARETTE, P. A. (1975): The permittivity and attenuation in polycrystalline and single-crystal ice Ih at 35 and 60 MHz. J. Glac., **14**, 293-303.

JONES, F. H. M., NAROD, B. B. & CLARKE, G. K. C. (1989): Design and operation of a portable impulse radar. J. Glac., **35**, 143-147.

KING, R. W. P. (1980): Antennas in matter: fundamentals, theory, and applications: M.I.T. Press, Cambridge, Massachusetts.

KRAUS, J. D. & Marhefka, R. J. (1988): Antennas, 2nd edn.: McGraw-Hill, New York.

KRAUS, J. D. (1991): Electromagnetics, 4th edn.: McGraw-Hill, New York.

LAMBOT, S. & GORITTI, A. (Eds.) (2007): Special Issue on Ground Penetrating Radar. Near Surface Geophysics, **5**, 1, 5-82.

LAMPE, B., HOLLIGER, K. & GREEN, A. G. (2003): A finite difference time-domain simulation tool for ground-penetrating radar antennas. Geophysics, **68**, 971-987.

LEPAROUX, D., GIBERT, D. & COTE, P. (2001): Adaptation of prestack migration to multi-offset ground penetrating radar (GPR) data. Geophys. Prosp., **49**, 374-386.

LUCIUS, J. E., OLHOEFT, G. R., HILL, P. L. & DUKE, S. K. (1992): Properties and hazards of 108 selected substances - 1992 edn: U.S. Geological Survey Open-File Report 92-527.

MCMECHAN, G. A., LOUCKS, R. G., MESCHER, P. & ZENG, X. (2002): Characterization of a coalesced, collapsed paleocave reservoir analog using GPR and well-core data. Geophysics, **67**, 1148-1158.

MEHLHORN, H., RICHTER, TH. & BAND, S. (1988): Beitrag zu theoretischen Aspekten der Anwendung des Radarverfahrens zur Ortung von punkt- oder flächenartigen Zielen. Neue Bergbautechnik, **18**, 176-178.

MEYERS, R. A., SMITH, D. G., JOL, H. M. & PETERSON, C. D. (1996): Evidence for eight great earthquake-subsidence events detected with ground-penetrating radar. Willapa Barrier, Washington. Geology, **24**, 99-102.

MILITZER, H. & WEBER, F. (1987): Angewandte Geophysik, **3**, Seismik. Springer, Berlin, Akademie Berlin.

MILLER, E. K. (Ed.) (1986): Time-domain measurements in electromagnetics. Van Nostrand, New York.

MOREY, R. M. (1998): Ground penetrating radar for evaluating subsurface conditions for transportation facilities. NAS/NRC/TRB NCHRP Synthesis Report 255.

MUNDRY, E. (1991): Numerische Modelluntersuchungen zur Reflexion hochfrequenter elektromagnetischer Wellen im Salzgestein. Geol. Jb., E **48**, Hannover, 259-282.

NOON, D. A., LONGSTAFF, D. & YELF, R. J. (1994): Advances in the development of step frequency ground penetrating radar. Proceedings of the Fifth International Conference on Ground Penetrating Radar. Kitchener, Ontario, June 12-16, 117-131.

OLHOEFT, G. R. (1988): Interpretation of hole-to-hole radar measurements. In: Proceedings of the Third Technical Symposium on Tunnel Detection, January 12-15, 1988, Golden, CO, 616-629.

OLHOEFT, G. R. (1992): Geophysical detection of hydrocarbon and organic chemical contamination. In: BELL, R. S., (Ed.): Proceedings on Application of Geophysics to Engineering, and Environmental Problems, Oakbrook, IL: Society of Engineering and Mining Exploration Geophysics, Golden, CO, 587-595.

OLHOEFT, G. R. & CAPRON, D. E. (1994): Petrophysical causes of electromagnetic dispersion. Proceedings of the Fifth International Conference on Ground Penetrating Radar. Kitchener, Ontario, June 12-16, 145-152.

OLHOEFT, G. R., POWERS, M. H. & CAPRON, D. E. (1994): Buried object detection with ground penetrating radar. In: Proc. of Unexploded Ordnance (UXO) detection and range remediation conference, Golden, CO, May 17-19, 1994, 207-233.

OLHOEFT, G. R. (1998): Electrical, magnetic and geometric properties that determine ground penetrating radar performance. In: Proc. of GPR'98, 7th Int'l. Conf. On Ground Penetrating Radar, May 27-30, 1998, The Univ. of Kansas, Lawrence, KS, USA, 177-182.

OLHOEFT, G. R. (2000): Maximizing the information return from ground penetrating radar. J. Applied Geophys., **43**, 175-187.

OLHOEFT, G. R. (2004): www.g-p-r.com. Webpage on Ground Penetrating Radar with a GPR Tutorial, a Bibliography and links to GPR Manufactures.

PFEIFFER, W. (1976): Impulstechnik. Hanser, München.

PIPAN, M., FORTE, E., GUANGYOU, F. & FINETTI, I. (2003): High resolution imaging and joint characterization in limestone. Near Surface Geophysics, **1**, 39-55.

RADEZEVICIUS S. J., CHEN C. C., PETERS L. & DANIELS J. J. (2003): Near-field dipole radiation dynamics through FDTD modelling. Journal of Applied Geophysics, **52**, 75-91.

REYNOLDS, J. M. (1997): An Introduction to Applied and Environmental Geophysics. John Wiley & Sons Ltd., Chichester.

ROTH, F., VAN GENDEREN, P. & VERHAEGEN, M. (2004): Radar Scattering Models for the Identification of Buried Low-Metal Content Landmines. Proceedings of the Tenth International Conference on Ground Penetrating Radar; 21–24 June, 2004, Delft, The Netherlands.

ROTHAMMEL, K. (1991): Antennenbuch. 10th edn., Franck-Kosmos, Stuttgart.

SKOLNIK, M. I. (1970): Introduction to radar systems. McGraw-Hill, New York.

SLOB, E. & YAROVOY, A. (Eds.) (2006): Special Issue on Ground Penetrating Radar. Near Surface Geophysics, **4**, 1, 5-75.

STERN, W., (1929): Versuch einer elektrodynamischen Dickenmessung von Gletschereis. Gerl. Beitr. zur Geophysik, **23**, 292-333.

STERN, W., (1930): Über Grundlagen, Methodik und bisherige Ergebnisse elektrodynamischer Dickenmessung von Gletschereis. Z. Gletscherkunde, **15**, 24-42.

SZERBIAK, R. B., MCMECHAN, G. A., CORBEANU, R., FORSTER, C. & SNELGROVE, S. H. (2001): 3-D characterization of a clastic reservoir analog: From 3-D GPR data to a 3-D fluid permeability model. Geophysics, **66**, 1026-1037.

TAYLOR, B. N. (1995): Guide for the use of the international system of units (SI). NIST Spec. Publ. 811, 1995 ed., USGPO, Washington, DC. (http://www.nist.gov).

THIERBACH, R. (1974): Electromagnetic reflections in salt deposits. J. Geophys., **40**, 633-637.

THYSSEN, F. (1985): Erkundung oberflächennaher Strukturen und Eigenschaften mit dem elektromagnetischen Reflexionsverfahren. In: HEITFELD, K.-H. (Ed.): Ingenieurgeologische Probleme im Grenzbereich zwischen Locker- und Festgesteinen. Springer, Berlin, 597-609.

TRONICKE, J., DIETRICH, P., WAHLIG, U. & APPEL, E. (2002): Integrating surface georadar and crosshole radar tomography: A validation experiment in braided stream deposits. Geophysics, **78**, 1516-1523.

TURNER, G. & SIGGINS, A. F. (1994): Constant *Q*-attenuation of subsurface radar pulses. Geophysics, **59**, 1192-1200.

TSANG, T., KONG, J. A. & SIMMONS, G. (1973): Interference patterns of a horizontal electric dipole over layered dielectric media. J. Geophys. Res., **78**, 3287-3300.

TILLARD, S. (1994): Radar experiments in isotropic and anisotropic geological formations (granite and schists). Geophys. Prosp., **42,** 615-636.

VAN OVERMEEREN, R. A. (1994): Georadar for hydrogeology. First Break, **12**, 401-408.

WRIGHT, D. L., HODGE, S. M., BRADLEY, J. A., GROVER, T. P. & JACOBEL, R. W. (1990): A digital low-frequency, surface-profiling ice-radar system. J. Glac., **36**, 1112-1121.

WU, T. T. & KING, R. W. P. (1965): The cylindrical antenna with nonreflecting resistive loading, IEEE Trans. Antennas and Propagation AP-13, 369-373.

WYATT, D. E. & TEMPLES, T. J. (1996): Ground-penetrating radar detection of small-scale channels, joints and faults in the unconsolitated sediments of the atlantic coastal plain. Environmental Geology, **27**, 219-225.

ZENG, X. & MCMECHAN, G. A. (1997): GPR characterization of buried tanks and pipes. Geophysics, **62**, 797-806.

ZENG, X., MCMECHAN, G. A. & XU, T. (2000): Synthesis of amplitude-versus-offset variations in ground-penetrating radar data. Geophysics, **65**, 113–125.

4.6 Seismic Methods

ANDREAS SCHUCK & GERHARD LANGE

4.6.1 Principle of the Methods

The basic principle of all seismic methods is the controlled generation of elastic waves by a seismic source in order to obtain an image of the subsurface. Seismic waves are pulses of strain energy that propagate in solids and fluids. Seismic energy sources, whether at the Earth's surface or in shallow boreholes, produce wave types known as:

- body waves, where the energy transport is in all directions, and
- surface waves, where the energy travels along or near to the surface.

Two main criteria distinguish these wave types from each other - the propagation zones and the direction of ground movement relative to the propagation direction.
Of prime interest in shallow seismics are the two types of body waves:

- P-waves (primary, longitudinal or compressional waves) with particle motion parallel to the direction of propagation, and
- S-waves (secondary, shear or transverse waves) with particle motion perpendicular to the direction of propagation – when particle motion is in the vertical plane they are referred to as SV-waves, and SH-waves when the particle motion is in the horizontal plane.

Surface waves, often considered to be a source of noise, contain valuable information about material properties of the shallow ground. Their use is on the increase in engineering studies.

The velocity of seismic waves is the most fundamental parameter in seismic methods. It depends on the elastic properties as well as bulk densities of the media and varies with mineral content, lithology, porosity, pore fluid saturation and degree of compaction. P-waves have principally a higher velocity than S-waves. S-waves cannot propagate in fluids because fluids do not support shear stress.

During their propagation within the subsurface seismic waves are reflected, refracted or diffracted when elastic contrasts occur at boundaries between layers and rock masses of different rock properties (seismic velocities and/or bulk densities) or at man-made obstacles. The recording of seismic waves returning from the subsurface to the surface allows drawing conclusions on structures and lithological composition of the subsurface. By measuring the traveltimes of seismic waves and determining their material specific velocities, a geological model of the subsurface can be constructed.

In a seismic survey elastic waves are generated by different energy sources (e.g., explosives, vibrator, weight-drop, sledgehammer). The seismic response is simultaneously recorded by a number of receivers (geophones, seismometers). These are positioned along straight profile lines (2-D seismics) or over an area in 3-D seismic surveying and connected to a seismograph. The signals of individual geophones or groups of geophones are recorded by a seismograph, processed and displayed in seismic sections to image the subsurface structure.

The classical seismic methods are refraction seismics (Section 4.6.5) and reflection seismics (Section 4.6.6). Applications of seismic methods on inshore waters or in boreholes are also standard practice (*Fig. 4.6-1*).

In refraction seismics "head waves" (or Mintrop waves) are used, which arise at the interface between two layers when refraction at the critical angle occurs. The critical angle is that angle of ray incidence for which the refracted ray moves along the contact surface (*Fig. 4.6-12*). This requires an increase in velocity with depth. If a layer of lower seismic velocity underlies a layer of higher velocity (velocity reversal), no refraction at the critical angle occurs and the layer cannot be detected (hidden layer). Such conditions may cause ambiguity in the interpretation. In the refraction analysis the traveltimes of first arrivals on recorded seismograms are commonly used. The first arrivals have to be identified, picked and presented in traveltime curves (plots of arrival times vs. source-receiver distance). Main aim of refraction surveys is to determine the depths to layers, the refractor topography (dip of layers) and layer velocities. To extract this information from traveltime curves, some methods have been developed (e.g., HAGEDOORN, 1959; PALMER, 1981; VAN OVERMEEREN, 2001) and user-friendly software packages, running on PCs are available to assist in the data interpretation.

Reflection seismics is the most important method to prospect for oil and gas at greater depths. During the 1980s the importance of the seismic reflection technique in shallow investigations, typically less than 50 m, has increased (e.g., DAVIES et al., 1992; HILL, 1992b; KING, 1992). It must however, be pointed out that the application of reflection seismics to shallow investigations, is not merely a case of transferring the seismic technology used so successfully in prospecting for hydrocarbons. The unprocessed field records are mostly characterized by poor signal-to-noise ratios that result in some special problems for shallow seismic surveys. Because of the short traveltimes, source generated noise, surface waves, first arrivals of direct waves and refracted waves dominate and genuine reflections are masked (*Fig.4.6-15*).

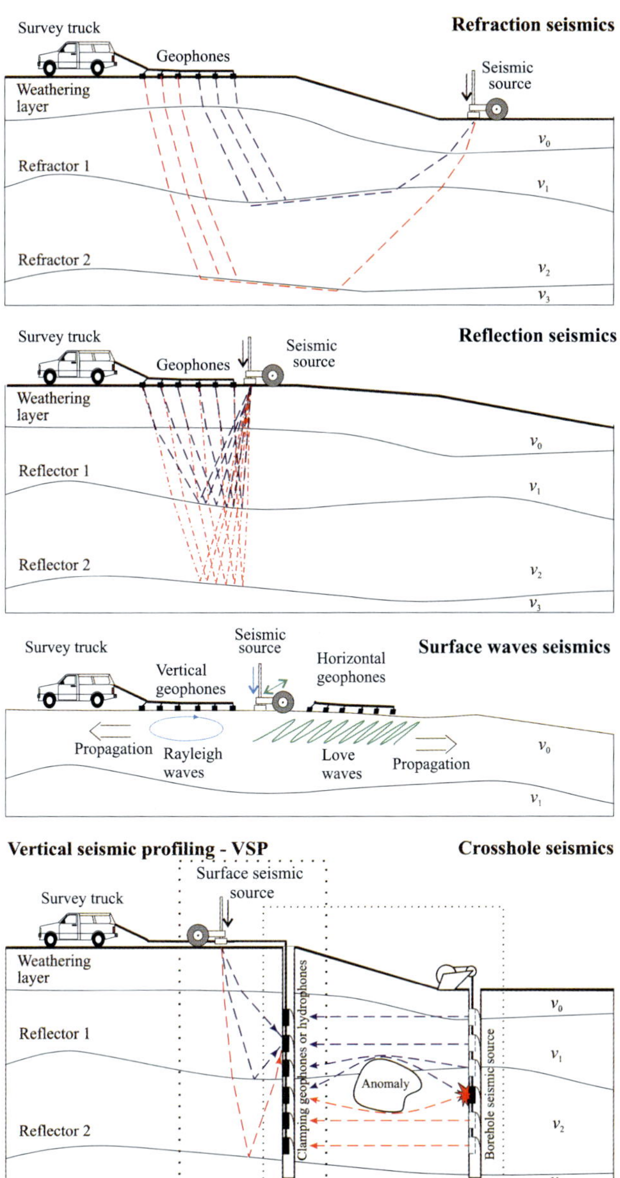

Fig. 4.6-1: Principles of seismic methods. Dashed lines show raypaths of seismic waves from the source to receivers; v_0, v_1, v_2, v_3 are seismic velocities in the respective layers. The returning signal is recorded by geophones at distances much greater (refraction seismics), at distances less than the investigation depth (reflection seismics) and at distances mostly equal to the investigation depth (seismic surface waves). In VSP and crosshole seismics the geophones (or hydrophones) are located in boreholes and the source is positioned at the surface (VSP) or in boreholes (crosshole seismics).

Thus, both acquisition and standard processing as performed in conventional seismic reflection methods cannot be transferred by simply using a "scaling down" approach. Some possible sources of misinterpretation resulting from the complex wavefields appearing in shallow reflection seismic records are described by STEEPLES & MILLER (1998).

The surface wave method is based on analyzing the variation of seismic velocities with frequency (dispersion). Dispersion of surface waves occurs when near surface velocity layering is present (Section 4.6.3.4). Since the late 1990s the method has been successfully used in many applications, as for example underground cavity detection and the delineation of abandoned waste sites (e.g., PARK et al., 1999; HAEGEMAN & VAN IMPE, 1999; LEPAROUX et al., 2000; SHTIVELMAN, 2002).

4.6.2 Applications

The main applications with respect to waste disposal and contaminated sites, civil engineering and foundation studies, and in the search for groundwater can be summarized as follows:

- investigation of regional and local geological structures,
- detection of faults, joints and other zones of weakness,
- identification of shallow stratigraphic sequences,
- determination of depth to bedrock and relief of bedrock surface,
- investigation of layer boundaries, in particular with respect to groundwater-bearing formations,
- distribution and thickness of weathering layer and of erosion channels,
- delineation of areas of different lithology and facies,
- mapping the topography of the base of disposal sites,
- investigations below sealed surfaces,
- investigation of ground deformations caused by mining and subrosion,
- determination of elastic parameters (Poisson's ratio, shear modulus, elastic bulk modulus, Lamé parameter, etc. – see *Table 4.6-1*), especially rock competence for geotechnical application,
- localization of manmade subsurface structures (e.g., buildings, tanks, foundations), and
- localization of cavities and voids.

4.6.3 Fundamentals

4.6.3.1 Propogation of Elastic Waves

During their propagation within the ground elastic waves disturb the medium causing particle motion as well as volume and shape changes. A mathematical expression of particle displacement ψ and wave velocity v as functions of space (x, y, z) and time t is the wave equation. The wave equation is based on the theory of elastic continua which relates deformation (strain) and stress. Equation (4.6.1) describes the wave propagation in a homogeneous isotropic elastic medium, where ψ can be a vector or a scalar, e.g., a component u of the particle displacement:

$$\nabla^2\psi = \frac{\partial^2\psi}{\partial x^2} + \frac{\partial^2\psi}{\partial y^2} + \frac{\partial^2\psi}{\partial z^2} = \left(\frac{1}{v^2}\right)\frac{\partial^2\psi}{\partial t^2}. \tag{4.6.1}$$

The simplest solution of the wave Equation (4.6.1) is that of a planar wave with harmonic ground motion

$$\psi(t) = A\cos(\phi + \omega t) \tag{4.6.2}$$

with A being the amplitude, ω the angular frequency ($2\pi f$) and ϕ the phase of the wave.

Adding an additional phase term kz delivers a generalized form of Equation (4.6.2) in the z direction of wave propagation (1-dimensional case):

$$\psi(t) = A\cos(\phi + \omega t + kz), \tag{4.6.3}$$

where k is the wavenumber ($k = 2\pi/\lambda$) with λ being the wavelength, and z the distance along the wave propagation. However, planar harmonic waves are idealized waveforms which can be used as approximation only in a very large distance to the source. Solutions of the wave equation for several types of wavefront geometries (cylindrical waves, spherical waves, etc.) can be found, for example, in TELFORD et al. (1990).

For very small deformations of a homogeneous isotropic elastic medium strain is directly proportional to stress (Hooke's law) - an assumption which is valid for wave propagation in many geological materials. The relations between stress (force per unit area) and strain (change of volume and shape) are described by the general elasticity tensor, which contains up to 21 independent elastic constants for anisotropic media. These elastic constants are expressed in terms of each other, P- and S-wave velocities (v_p and v_s), and bulk density ρ (*Table 4.6.1* and SHERIFF, 1991). The simplest assumption is seismic wave propagation in an isotropic medium. In this case only two elastic constants, noted as Lamé's constant λ (lambda constant) and μ (shear modulus), exist and the solution of the wave equation yields two different phase velocities $v_p = \sqrt{\lambda + 2\mu/\rho}$ and $v_s = \sqrt{\mu/\rho}$. Both phase velocities are independent of the direction of wave propagation and identical with the group velocity[1].

The wave traveling in an isotropic medium at the higher velocity v_p (P-wave) is polarized parallel to the direction of wave propagation. Waves with lower velocity v_s (S-wave) can be split into a wave vertically polarized (SV-wave) and a wave horizontally polarized (SH-wave) perpendicular to the direction of propagation. P-waves cause changes of shape and volume, whereas S-waves only cause changes to the shape of the medium through which they are transmitted (*Fig. 4.6-2*).

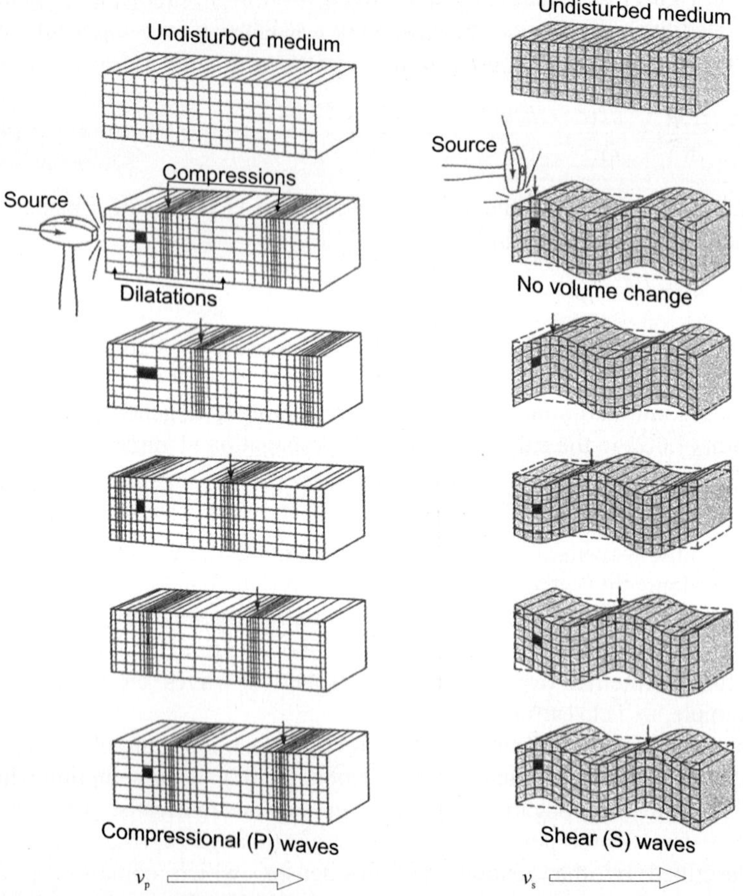

Fig. 4.6-2: Particle motion of a compressional wave (*left*) and a shear wave (*right*), when passing through an elastic material

[1] Phase velocity is the velocity of any phase (peak or trough) of an event correlated on a seismic record. Group velocity is the velocity with which the seismic energy in a wavetrain travels. Because of dispersion the phase velocity may differ from the group velocity.

In an inhomogeneous medium both wave types are coupled. Seismic energy, traveling partly as a P-wave and partly as an S-wave, is converted from one to the other when reflection or refraction occurs at oblique incidence on a boundary in the elastic properties (seismic boundary). Since mode conversion is small for low incident angles, converted waves become more prominent for large source-receiver distances (offsets). In the most frequent case of an anisotropic medium (e.g., fine layered ground), the velocities of P-, SV- and SH-waves are different and depend on the direction of wave propagation (HELBIG, 1994). Phase and group velocity are not identical and the P- and S-waves are polarized orthogonal to each other, but the polarization, generally, does not coincide with the direction of wave propagation.

For spherical waves radiating from a small source (point source) at or near to the Earth's surface, wavefronts propagate outward and away from their point of origin (*Fig. 4.6-8*). Their seismic energy is distributed over a sphere with an area $a = 4\,\pi\,r^2$ (r being the source distance). This results in a geometrical wave energy decay inversely proportional to the square of the distance ($1/r^2$). The geometrical decay is independent of the wave frequency and called spherical or geometric divergence.

A decrease in amplitude during wave propagation is also caused by energy loss due to the non-elastic behavior of real media. This phenomenon, known as absorption or intrinsic attenuation, is caused by viscous energy dissipation, e.g., by movement of fluids in the pore space of sediments. Absorption is strongly dependent on wave frequency. The amplitudes of waves with higher frequencies are attenuated more than those of waves with lower frequencies. Therefore, for low frequencies and short source distances, spherical divergence influences the wave amplitude more than absorption. However, with increasing distance and higher frequencies absorption dominates.

In a first order approximation energy loss with increasing distance due to absorption can be regarded as being exponentially. If the medium is characterized by an absorption coefficient (or attenuation factor) $\alpha = \alpha\,(f)$, the relation between the source energy E_0 and the actual energy E at a distance x from the source is $E = E_0\,e^{-\alpha x}$. Hence, the decrease in amplitude A or energy E is described in a logarithmic scale and expressed in decibel:
decrease = $10 \log (E_0/E) = 20 \log (A_0/A)$ [dB].

4.6.3.2 Elastic Parameters and Seismic Velocities

Elastic parameters

For small deformations of an elastic medium the relation between stress and deformation is described by Hooke's law (Section 4.6.3.1). Depending on the deformation (compressional or shear) the proportionality between stress and strain in the linear range is specified by elastic moduli (*Fig. 4.6-3*,

Table 4.6-1). In addition, a relation exists between elastic moduli and the velocity of seismic waves. Elastic (dynamic) parameters and velocities as derived from non-destructive testing by seismic measurements are used to describe soil mechanical properties in ground engineering, for example for the construction of large buildings. Their determination is standardized and described in recommendations of the American Society for Testing Materials (ASTM), the British Standards Institution and the German Society for Non-Destructive Testing (DGZfP).

Table 4.6-1: Interrelations between elastic constants, P- and S-wave velocities, and bulk density ρ for isotropic media

		λ, μ	k, μ	E, σ	v_p^2, v_s^2
Lamé's constant	λ	λ	$k - \frac{2}{3}\mu$	$\frac{E\sigma}{(1+\sigma)(1-2\sigma)}$	$\rho(v_p^2 - 2v_s^2)$
shear modulus	μ	μ	μ	$\frac{E}{2(1+\sigma)}$	ρv_s^2
bulk modulus	k	$\lambda + \frac{2}{3}\mu$	k	$\frac{E}{3(1-2\sigma)}$	$\rho\left(v_p^2 - \frac{4}{3}v_s^2\right)$
Young's modulus	E	$\mu\frac{3\lambda + 2\mu}{\lambda + \mu}$	$\frac{9k\mu}{3k + \mu}$	E	$\rho V_p^2 \frac{3v_p^2 - 4v_s^2}{v_p^2 - v_s^2}$
Poisson's ratio	σ	$\frac{\lambda}{2(\lambda + \mu)}$	$\frac{3k - 2\mu}{6k + 2\mu}$	σ	$\frac{\frac{v_p^2}{v_s^2} - 2}{2\frac{v_p^2}{v_s^2} - 2}$
P-wave velocity	v_p^2	$\frac{\lambda + 2\mu}{\rho}$	$\frac{k + \frac{4}{3}\mu}{\rho}$	$\frac{E(1-\sigma)}{\rho(1+\sigma)(1-2\sigma)}$	v_p^2
S-wave velocity	v_s^2	$\frac{\mu}{\rho}$	$\frac{\mu}{\rho}$	$\frac{E}{2\rho(1+\sigma)}$	v_s^2

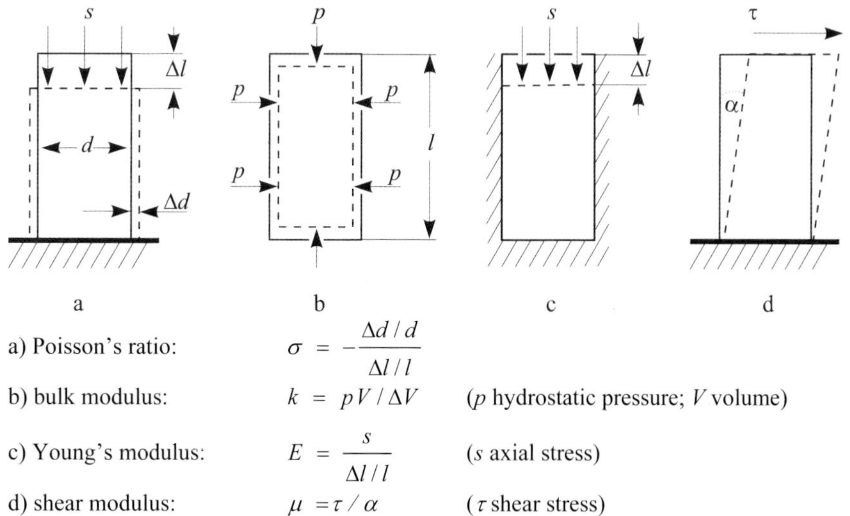

Fig. 4.6-3: Illustration of elastic deformation of a cylindrical isotropic sample under different stress conditions and mechanical constraints, BERCKHEMER (1990)

Seismic velocities

The most important relationship for the joint interpretation of P- and S-wave sections is the velocity-ratio v_p/v_s. As v_p/v_s is directly proportional to the traveltime-ratio t_s/t_p, it can be directly derived from traveltime differences in P- and S-wave sections. For a Poisson's ratio $\sigma = 0.25$ the velocity-ratio $v_p/v_s = \sqrt{3}$. This often holds for consolidated sediments and solid rocks. For shallow unconsolidated and slightly consolidated sediments, the v_p/v_s-ratio varies between 2 and 12. The simultaneous observation of both P- and S-waves allows drawing conclusions on lateral changes in the lithology, and on the type and degree of fluid saturation (BACHRACH et al., 1998; MAVKO et al., 1998). Whereas P-waves provide information on lithology and fluid content, S-waves contain information about lithology.

Table 4.6-2 gives a compilation of velocities and bulk densities of unconsolidated sediments, rocks, and fluids.

Table 4.6-2: Seismic velocities and bulk densities of unconsolidated sediments, selected rocks, and fluids, FERTIG (2005)

Material	**Velocity** v_p [m s^{-1}]	**Velocity** v_s [m s^{-1}]	**Bulk density** ρ [g cm^{-3}]
clay	500 - 2800	110 - 1500	1.25 - 2.32
-sandy clay	2000 - 2750		
-loamy clay	500 - 1900	440 - 1080	0.76 - 1.57
sand	100 - 2000	100 - 500	1.80 - 2.05
-dry	100 - 600		2.33 - 2.80
-wet	200 - 2000		1.50 - 2.00
-saturated	1300 - 1800		1.80 - 2.05
-coarse	1835		2.03
-fine grained	1740		1.98
gravel	180 - 1250		1.95 - 2.20
-wet	750 - 1250		1.95 - 2.20
weathering layer	100 - 500		1.20 - 1.80
marly limestone	3200 - 3800		2.65 - 2.73
marlstone	1300 - 4500	715 - 2250	
siltstone	1900 - 2000		2.35 - 2.73
sandstone	800 - 4500	320 - 2700	2.30 - 2.55
-loose	1500 - 2500	575 - 1101	1.80 - 2.40
-compacted	1800 - 4300	672 - 1023	2.22 - 2.69
-siliceous	2200 - 2400		
quartzite	5800		2.64
schistose quartzite	5500		2.65
lime	3000 - 6000		
-chalky	3560		
-fine grained	4680		
-crystalline	5500		2.67
marble	5100		2.66 - 2.70
chalk	1800 - 3500		2.00 - 2.57
limestone	2000 - 6250	1800 - 3800	1.75 - 2.88
dolomite	2000 - 6250	2900 - 3740	1.75 - 2.88
gypsum	1500 - 4600	750 - 2760	2.31 - 2.33
anhydrite	4500 - 6500	750 - 3600	2.15 - 2.44
halite	4500 - 6500	2250 - 3300	2.15 - 2.44
bituminous coal	1600 - 1900	800 - 1140	1.25 - 1.84
brown coal	500 - 1800		1.20 - 1.50
air (depending on temperature)	310 - 360		$1.29 \cdot 10^{-3}$
petroleum	1035 - 1370		0.92 - 1.07
water	1430 - 1590		0.98 - 1.01
saltwater (seawater)	1400 - 1600		1.01

4.6.3.3 Reflection, Transmission and Diffraction

To describe the seismic wave propagation it is important to know the behavior of seismic waves at an interface. When an advancing planar wavefront[2] arrives at a boundary between two layers of different elastic properties, every point at this boundary acts as a source of a secondary spherical wave (Huygens' principle). These secondary waves overlap and new wavefronts defined by the envelope to all the secondary waves are formed - a reflected wavefront in the upper halfspace and a transmitted (refracted) wavefront in the lower halfspace. For oblique incidence converted waves arise, traveling partly as P-waves and partly as S-waves.

The ratio between reflected and transmitted energy and, hence, the ratio of the wave amplitudes, depends on the elastic properties of both layers, and is expressed as their acoustic impedance. The acoustic impedance of a layer is defined by its seismic velocity multiplied by the bulk density. Based on the contrast in acoustic impedances at a seismic boundary, reflection and transmission coefficients can be calculated. These values define that part of energy which is reflected or transmitted. A reflection coefficient of value "1" (theoretically) means that all incident seismic energy is reflected. To calculate the reflection and transmission coefficients at an interface it is necessary to match boundary conditions for the displacement vector $\boldsymbol{u}$ (x, t) and the stress tensor.

A simple example to illustrate the calculation of reflection and transmission coefficients is the case when a planar wave incidents normal to an interface between two rigidly connected solid media (*Fig. 4.6-4*). Keeping the boundary conditions for displacement and stress one can state: The reflection coefficient R (reflectivity) for displacement amplitudes u_0 of the incident wave and u_r of the reflected wave is:

$$\frac{u_r}{u_0} = R = \frac{\rho_2 v_2 - \rho_1 v_1}{\rho_2 v_2 + \rho_1 v_1}. \tag{4.6.4}$$

The transmission coefficient T for the displacement amplitude u_T of the transmitted wave is:

$$\frac{u_T}{u_0} = T = \frac{2\rho_1 v_1}{\rho_1 v_1 + \rho_2 v_2} = 1 - R. \tag{4.6.5}$$

---

[2] A wavefront is that surface over which the phase of a wave is the same. At large distances from the source a wavefront can be considered as planar. As it is simpler to consider wave propagation in terms of rays, analog to the ray-geometric optics, in the following rays and raypaths are considered. A ray is the normal to the wavefront.

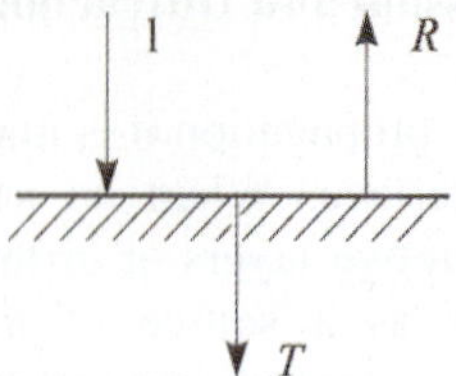

Fig. 4.6-4: Reflection and transmission coefficients R and T for a planar interface between two half-spaces with the acoustic impedances $I_1 = \rho_1 v_1$ and $I_2 = \rho_2 v_2$, respectively. The incident wave I (here shown as a ray normal to the wavefront) has the normalized amplitude of value 1.

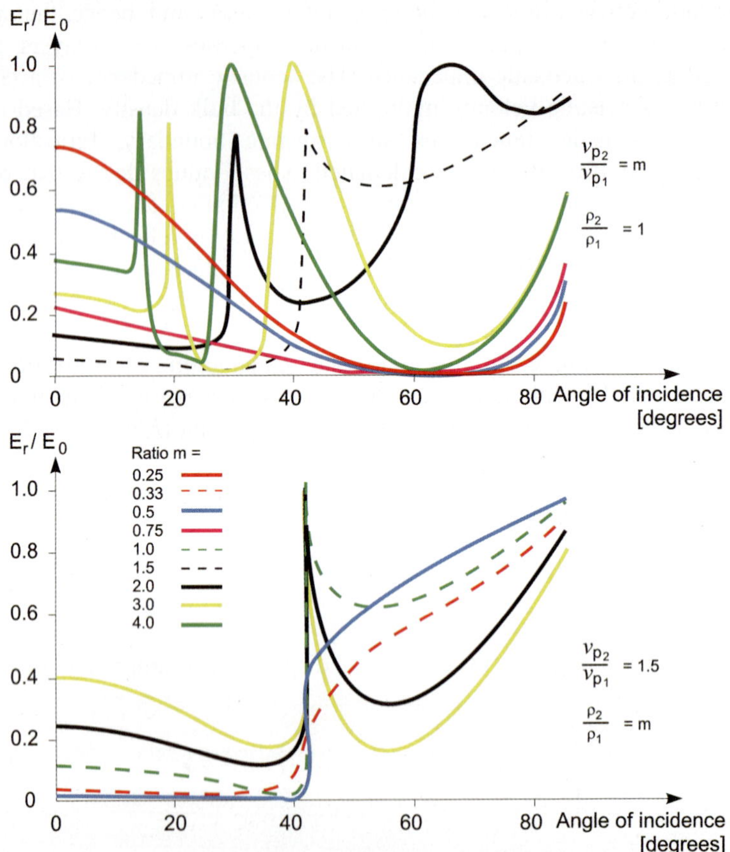

Fig. 4.6-5: Ratio of reflected energy to incident energy E_r/E_0 as a function of the incident angle of a P-wave raypath impinging on the planar interface between medium 1 and 2. In the upper picture the ratio of bulk densities ρ_1/ρ_2 is kept constant and the velocity ratio varies. In the lower picture the velocity ratio remains constant and the density ratio varies. In both cases Poisson's ratio $\sigma = 0.25$.

However, this only holds for normal incidence of a ray. In the general case reflection and transmission coefficients vary with the angle of incidence. Angle-dependent reflection and transmission coefficients for elastic plane waves at a non-slip horizontal boundary are given by the Zoeppritz equations (AKI & RICHARDS, 1980; SHUEY, 1985; www.crewes.org). In *Fig. 4.6-5* energy ratios are presented which are proportional to the square of displacement amplitudes. Due to the general boundary conditions for the incidence of a P-wave or S-wave, both reflected and transmitted P- and S-waves are generated - so called converted waves. They allow, for example, obtaining S-wave information in cases where S-waves are primarily not generated by the seismic source used (see radiation pattern in *Fig. 4.6-8*).

The angle of the reflected or transmitted wave at oblique incidence can be easily calculated using Snell's law:

$$p = \frac{\sin(i_\alpha)}{v_\alpha} = \text{const.} \tag{4.6.6}$$

with p being the ray parameter (slowness), i_α the angle of the incident, reflected or transmitted ray and v_α the (P- or S-wave) velocity of the medium in which the wave is propagating. Snell's law is the fundamental principle for all ray-geometry observations and describes the reflection and refraction angles between two half-spaces with different velocities. Due to the velocity contrast at the interface the transmitted wave is refracted and changes its direction of propagation. Note that the reflection and refraction angles are independent from bulk density.

For example, a P-wave propagating in medium 1 with the velocity v_{p1} incidents on a planar interface separating medium 1 and 2 under the angle i_{p1}. Due to the above described boundary conditions, four waves arise: one reflected P-wave, one reflected SV-wave, one transmitted P-wave and one transmitted SV-wave (*Fig. 4.6-6*).

The reflected waves propagate with the velocities v_{p1} and v_{s1} of medium 1. The refracted waves propagate with the velocities v_{p2} and v_{s2} of medium 2. Using Snell's law (4.6.6) the angles of reflected and refracted rays can be calculated:

$$\frac{\sin(i_{p1})}{v_{p1}} = \frac{\sin(i_{p2})}{v_{p2}} = \frac{\sin(i_{s1})}{v_{s1}} = \frac{\sin(i_{s2})}{v_{s2}} = p = \text{const}. \tag{4.6.7}$$

The premises for reflection and transmission of waves are valid only for planar and adequately extensive interfaces. They are not applicable to the area around faults and in case of truncated reflectors, isolated objects or layers wedging out. If the extension of such structures is equal or less than the wavelength, the energy will neither be reflected nor refracted, but diffracted. Diffractions are explained by Huygens' principle. When the incident wavefront strikes a point (diffractor), this point is considered to be the source of a secondary spherical wave.

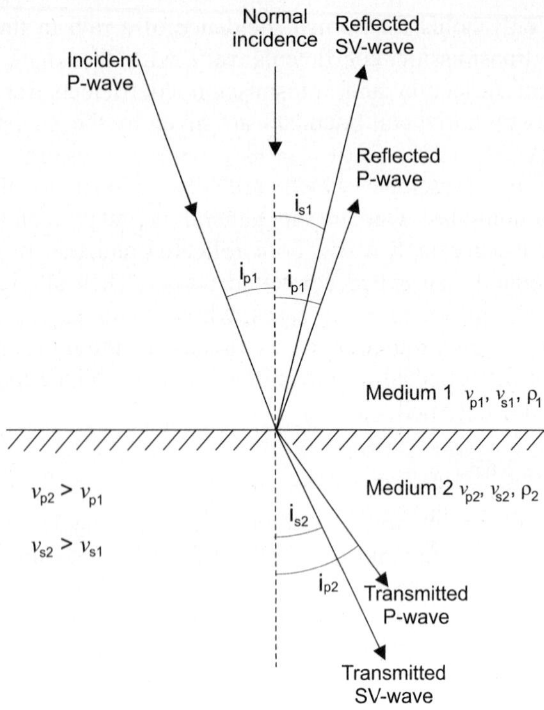

Fig. 4.6-6: Reflection, refraction and mode conversion of a P-wave incident at a plane interface

Diffractions play an important role in the mapping of reflectors consisting of several elements of limited extent (fault and fracture zones, horst and graben structures, etc.) or of single objects like cavities, boulders, and man made subsurface objects. Identifying diffracted waves on seismic records may be difficult, since the reflected, transmitted, and diffracted waves interfere with each other.

4.6.3.4 Surface Waves

In a homogeneous isotropic full-space only body waves exist, and are transmitted as compressive waves with velocity v_p and as shear waves with velocity v_s. If the medium is bounded by the Earth's surface (stress free surface) different types of surface waves (for example Rayleigh, Love, Lamb waves) are also present. Surface waves (boundary waves) are usually characterized by low velocity, low frequency, and large amplitude. They propagate directly from source to receiver, are confined to the surface (interface), and depending on the wavelength their amplitude decreases with depth. At receiver stations close to the source location they tend to obscure the events of reflected waves, because of their often very large amplitudes. In conventional seismics they are merely regarded as coherent noise ("ground roll"). For the investigation of near-surface layers, however, they are an important source of information (e.g., HERING et al., 1995; SOCCO & STROBBIA, 2004).

Rayleigh waves appear along a free surface and result from interfering P- and SV-waves. Particle motion in a homogeneous medium is retrograde[3] elliptical in the vertical plane and the wavefront is approximately cylindrical. In contrast to spherical body waves (Section 4.6.3.1) their wave amplitudes decay with $1/\sqrt{r}$, r being the distance from the source. Within a homogeneous half-space Rayleigh waves are not dispersive. Their velocity v_R is slightly less than the velocity of an S-wave ($v_R \approx 0.92\ v_s$ for Poisson's ratio $\sigma = 0.25$ – see *Table 4.6-1*). However, when a thin layer with lower velocity overlays a homogeneous half-space (e.g., the weathering layer), Rayleigh waves become dispersive. In this case the propagation velocity depends on the frequency or wavelength. Waves with longer wavelengths penetrate deeper into the ground. Waves with shorter wavelengths are sensitive to near-surface material. The dispersive behavior allows determining the thickness and S-wave velocity of near-surface layers. XIA et al. (2004) developed methods for display and inversion of Rayleigh waves. The multichannel analysis of surface waves (MASW) includes acquisition, extraction, and inversion of high frequency Rayleigh waves. By inversion of dispersion curves (phase velocities as a function of source-receiver distance), taking into account P-wave velocity, thickness and bulk density, S-wave velocities of near-surface layers can be calculated (XIA et al., 1999). Practical experiences, however, show that severe inhomogeneities in the near-surface layer have an adverse effect on the success of the method.

Love-waves (also Q-waves) occur when a free surface and a deeper interface are present and the shear body wave velocity in the near-surface layer is lower than that in the underlying layer. As a surface seismic channel wave their particle motion is parallel to the Earth's surface and perpendicular to the propagation direction (as for SH-waves). The velocity of Love waves is intermediate between that of the S-wave velocities of the adjacent layers. They always show dispersive behavior, but usually travel faster than Rayleigh waves. The dispersion of Love-waves can be used to determine the thickness of near-surface layers. Because Love-waves are not generated by pure P-wave sources, they can only be used for subsurface investigation when an SH-wave source is applied (*Figs. 4.6-1* and *4.6-8*).

[3] The particle movement is opposite (counterclockwise) to the propagation direction.

4.6.3.5 Seismic Resolution

In a seismic context, resolution can be defined as the ability to differentiate between two geological features that are close to each other. A distinction must be made between vertical and horizontal (or lateral) resolution, but they are not independent of each other - improving one automatically improves the other.

Resolution is determined by the wavelength λ of the seismic signal and the target depth z. Short wavelengths (higher frequencies) provide better resolution than longer ones. The wavelength is directly proportional to the wave velocity v in the corresponding material and inversely proportional to the (dominant) frequency f of the seismic signal: $\lambda = v / f$.

Vertical resolution implies how thin a layer can be to allow the separate detection of both its upper and lower boundary. Due to the relatively short near-vertical raypaths reflection seismics offers highest vertical (and horizontal) resolution of seismic methods. The vertical resolution is within the range of $\lambda/4$ and $\lambda/8$ (WIDESS, 1973). Up to this value two reflections can be separated and the true layer thickness can be obtained. When the thickness d of a layer is $<\lambda/8$ it can be detected because of constructive interference (superposition) of the top and bottom reflections, but its true thickness cannot be determined.

Horizontal resolution is the ability to separate single objects on a reflector and is determined using the first Fresnel zone. As the wave cone spreads from the source point it becomes increasingly delayed from the original signal. The first Fresnel zone is that part of the wave cone where most of the signal power is accumulated (*Fig. 4.6-7*). As a rule, reflections from two neighboring objects at a plane reflector on a 2-D seismic section can only be observed individually if their distance is larger than the diameter of the first Fresnel zone. The Fresnel zone diameter d_F is estimated as:

$$d_F = 2 r_F \approx 2 \sqrt{\frac{1}{2} z \lambda} \, . \qquad (4.6.8)$$

The Fresnel zone diameter d_F is usually much larger than the trace spacing on a seismic section. An important question that requires attention during planning and interpretation of CMP sections[4] (see Sections 4.6.6.1 and 4.6.6.3) is: How independent is the signal of one local feature from another feature at the same reflector? At least two samples (points) per apparent (horizontal) wavelength must be obtained in order to recognize single features.

---

[4] CMP (common-midpoint) is in multichannel seismic data acquisition the point on the surface halfway between source and receiver. The CMP method exploits the redundancy of multiple reflector coverage to enhance the quality of reflections by stacking multiple traces corresponding to the same CMP. The display of CMP stacked seismic traces (channels) along a profile is a CMP section. The trace spacing in stacked sections is half of the receiver spacing at the surface (Section 4.6.6.2).

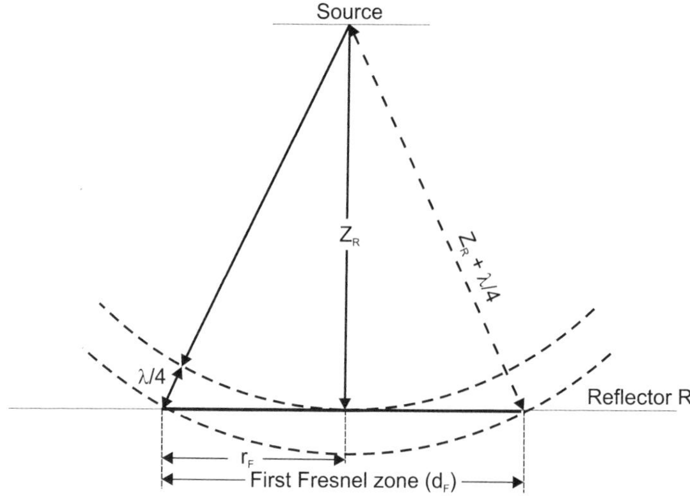

Fig. 4.6-7: Diagram of the first Fresnel zone

This is illustrated by a shallow reflection seismic survey carried out to determine structural and lithological details at depth z = 20 m. The layer velocity v at that depth is 1600 ms^{-1} and the dominant frequency of the reflection event is f = 80 Hz (resp. $\lambda = v/f$ = 20 m). Inserting these parameters in Equation (4.6.8) yields a first Fresnel zone diameter of about 28 m. Using a geophone spacing Δx = 5 m results a CMP-distance of 2.5 m. This means that 12 traces are required to provide visual independence between two structural/lithological features on the target horizon. On the traces in between, detailed information is obscured in the reflection events. For higher velocities and lower dominant frequencies the first Fresnel zone is broader.

Seismic migration (Section 4.6.6.3) allows the improvement of the lateral resolution by minimizing the size of the first Fresnel zones on successive reflectors. Using 2-D migration the first Fresnel zone is reduced in the profile direction only. Theoretically, the first Fresnel zone reduces to a circle of diameter $\lambda/4$ by applying 3-D migration (LINDSEY, 1989).

For a good vertical (temporal) resolution an impulse signal of short duration, corresponding to a broad bandwidth of frequencies, is required. This must already be taken into account when selecting a suitable seismic source (Section 4.6.4.1). Usually, signals generated by an impulsive seismic source are characterized by an amplitude maximum at the beginning of the signal. Such signals are called minimum-phase[5]. From the interpreters view,

[5] Minimum-phase, zero-phase, mixed-phase: characterization of the shape of basic seismic signals (wavelets). In minimum-phase signals the energy is concentrated in the front end of the pulse, whereas non-causal zero-phase wavelets are symmetrical about zero time. To improve the (temporal) resolution, non-zero-phase signals are converted to zero-phase signals.

however, a zero-phase signal is ideal because it provides less distortion and sharper definition of subsurface details. To achieve the best temporal resolution, minimum-phase signals are converted to zero-phase signals during seismic data processing. Deconvolution as a step in seismic processing also increases temporal resolution (Section 4.6.6.3).

4.6.4 Instruments

Seismic data acquisition systems basically consist of three main components:

- seismic source,
- seismic receivers (sensors, geophones), and
- a recording unit (seismograph).

The refracted or reflected seismic signals are received by geophones that are connected to a multicore seismic cable with contact positions (referred to as "take-outs") located at increasing distances from the source. The seismic signals received at each geophone are then transmitted to the recording unit by means of the multicore cable. These cables often consist of segments that can be connected to each other and are then moved along the seimic profile. The objective of a seismic survey (depth to target, resolution and detail required, site logistics, etc.) will dictate decisions regarding issues such as distance between geophones, shotpoint pattern, and the length of the profile. These have to be decided during the planning stage and before embarking on a seismic survey (STEEPLES & MILLER, 1988; STONE, 1994).

4.6.4.1 Seismic Sources

A seismic source is a device to generate seismic wave energy. For it to be used for shallow seismic surveys it has to comply with the following basic requirements:

- generate sufficient energy in order to record a measurable signal,
- produce signals with a frequency bandwith high enough to resolve individual layers,
- emit consistent waveforms, not corrupted by acoustic noise,
- the source signature must be consistent, and
- easy to operate with no major license restrictions.

There are two types of seismic sources:

- impulsive sources, stressing the surrounding media by a sudden release of energy (sledgehammer, weight drop, rifle sources, surface projectile impacts, in-hole shotguns), and
- vibrator swept sources, radiating a controlled low energy wavetrain generated by sinusoidal vibration of continuously varying frequency.

Seismic sources are placed in boreholes or at the Earth's surface. Explosives are mostly used in boreholes and generate a spherical radiated pattern of P-waves. Noninvasive seismic sources act as a vertical or horizontal force on the Earth's surface. This results in a typical radiation pattern of the seismic energy for each wave type (*Fig 4.6-8*). Every seismic source at the Earth's surface simultaneously generates P- and SV-waves, but only horizontal acting forces generate SH-waves.

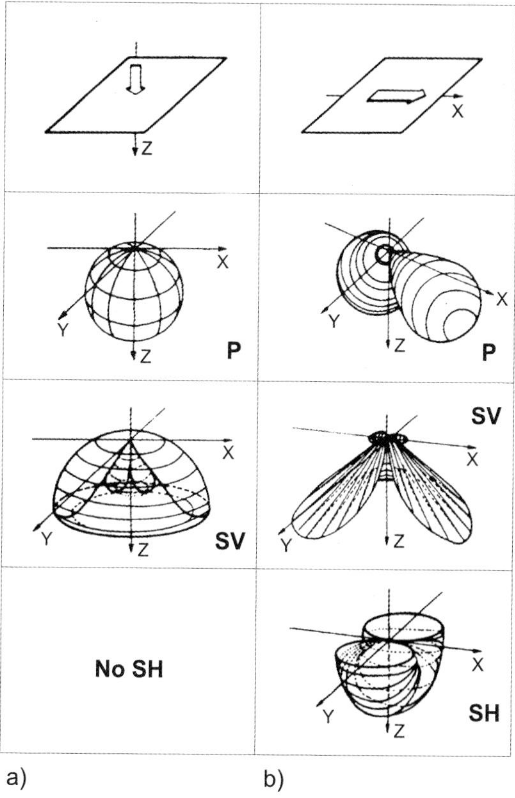

Fig. 4.6-8: Radiation pattern of a vertical and horizontal point force acting on a half-space (Earth's surface). The Poisson's ratio is approximately 0.25, KÄHLER & MEIßNER (1983).

The technical specifications of different seismic sources vary, including energy levels generated and the frequency characteristics of the impulses (MCCANN et al., 1985; KNAPP & STEEPLES, 1986; LINDSEY, 1991; DOLL et al., 1998). The investigation depth, subsurface conditions and desired resolution amongst others, should all be considered when selecting a suitable seismic source for a particular survey application (VAN DER VEEN et al., 2000). Factors which control the seismic penetration depth are the amount of energy generated by the source, its coupling to the ground and the geologic conditions, in particular energy absorption in different rock and soil types, thickness of the weathering layer, and depth to the water table. The resolution of subsurface structures depends on the frequency spectrum of the source. The broader the frequency range and the higher the upper frequency, the better is the resolution power. The effective penetration depth, however, depends on the individual ground conditions being present. In every survey situation there is a trade-off between source energy generation and spectral properties. The frequency content of the source signal is dependent on primary energy and ground coupling. In general, coupling to soft material results in low frequencies and large amplitudes whereas coupling to hard material produces high frequencies but small amplitudes. Therefore, changes of source coupling conditions have to be avoided during surveys. When using impulsive sources the following rule of thumb generally applies: with increasing source strength the center frequency in the frequency spectrum decreases resulting in a poorer vertical resolution.

Explosive sources

Explosives are normally detonated in shallow boreholes (referred to as shot holes) to improve the coupling to the ground and to minimize surface damage. By varying the charge mass or detonation velocity, the exploration depth as well as frequency characteristics of the impulse are adjustable. Explosive sources in shot holes generate almost purely P-waves. However, the storage and transportation of explosives is hazardous and special permission is normally required for the use of explosives in seismic surveys. Gas emissions are generally present at waste dumps, and often at other sites suspected to be hazardous. Therefore, non-explosive impulsive seismic sources are preferred for seismic surveys at such sites.

Non-explosive impulsive sources

For shallow exploration depths down to some tens of meters a sledgehammer (3 or 6 kg) struck against a steel or aluminium baseplate is a simple and very cost-effective impulsive seismic source. Its signal contains a broad frequency spectrum allowing a high subsurface resolution. Because of the low energy generated by hammer impacts, several recordings (normally 4 - 8) have to be stacked at each source point (vertical stacking), in order to improve the signal-to-noise (S/N) ratio. For this reason the scale of a seismic reflection survey using a sledgehammer source is limited by the manpower being available (KEISWETTER & STEEPLES, 1995).

Weight drops are commonly used non-explosive seismic sources. A mass (typically weighing 10, 50 or 250 kg) can be lifted to various heights (1 - 3 m) by a hydraulic system. It is then released and dropped onto a steel baseplate. The moment of impact is precisely determined by a sensor on the baseplate which triggers the recording of the seismic signal. Even more energy is generated if an accelerated weight drop (e.g., EWG III, based on sling shot technology) is used. As is the case with the sledgehammer method, the recordings of several drops at every source point are stacked. Because a repeatable signature is obtained from a weight drop, the S/N ratio is significantly enhanced by vertical stacking. Another advantage is the limited surface damage caused during the survey. Generating up to about 9500 Nm of energy, accelerated weight drops are suitable sources for investigations as deep as 1500 m (HERBST et al., 1998; DOLL et al., 1998). Airguns and sparkers are another form of non-explosive impulsive sources used for surveying on water and in boreholes. Airguns discharge compressed air into the water, while sparkers utilize an electrical discharge as energy source.

Vibrator sources

Vibrators are rapid and convenient low energy sources which emit controlled sinusoidal wavetrains of continuously varying frequency lasting a few seconds (sweeps). For shallow investigations specially designed minivibrators fixed to a pick-up or a trailer are normally used (GHOSE et al., 1998). As they generate low-amplitude signals, they can be applied in densely populated areas where other sources are not applicable because of possible structural damage they may cause. Energy generated by mini-vibrators can penetrate the subsurface down to a few hundred meters. Their common frequency range of 10 - 550 Hz allows a high vertical (temporal) resolution. Portable mini-vibrators weighing 60 - 70 kg and producing sweeps of 20 - 1500 Hz are restricted to investigation depths of less than 100 m. Using vibrator sources a processing step is necessary that transfers the long recorded signal ("vibrogram") into a compressed signal - the "correlogram". A correlogram looks similar to a seismic record that would be obtained with an impulsive source. To produce a correlogram, the recorded signal is cross-correlated with the radiated sweep as reference signal. The correlation process corresponds to zero-phase filtering (*Fig. 4.6-9*) and leads to an enhanced temporal (vertical) resolution in the resulting correlogram (Section 4.6.3.5, BUTTKUS, 2000 and REYNOLDS, 1997).

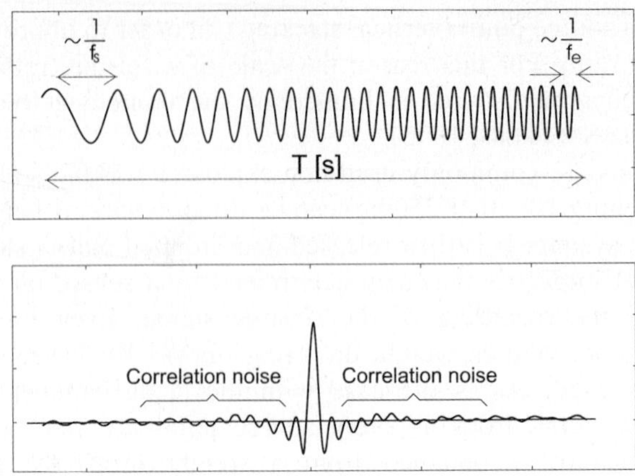

Fig. 4.6-9: *Top*: Sweep with its typical parameters: sweep length T (e.g., 6 s), start frequency f_s (20 Hz), end frequency f_e (300 Hz); *Bottom*: sweep after auto-correlation. The side lobes of the maximum peak can be minimized by tapering the sweep signal.

All non-explosive sources described above are acting with a vertical force on the Earth's surface. According to *Fig. 4.6-8* this generates not only P-waves, but also SV-waves with a different radiation pattern, and surface waves. P-wave energy is radiated preferably vertically into the ground, whereas SV-wave energy is directed mainly at an angle of about 45 degrees. In order to record P-waves, the survey layout has to be designed in such a way that reflections from steep angles are recorded. This necessitates the use of short offsets (*Fig. 4.6-1*). Serious problems may arise using short offsets due to the generation of high-energy surface waves. P-wave signals can be obscured by surface waves arriving simultaneously at the receivers.

S-wave sources

S-waves can be polarized in different ways – where the particle motion is horizontal (SH-wave) and where particle motion is vertical (SV-wave). In principle, every P-wave source can be applied for the generation of SV-waves, but difficulties arise from the simultaneous generation of P-waves obscuring the SV-wave signal. In site investigation SH-waves are preferred as they do not convert to P- or SV-waves during reflection or transmission (*Fig. 4.6-6*). For interpretation purposes it is important to note that SH-waves are easier to identify in the seismogram than SV-waves. SH-waves can be generated by horizontal strikes with a sledgehammer against a vertically buried board/metal baseplate, against the wall of a trial hole or using a pendulum or rotating hammer. The Shover method (EDELMANN, 1981) allows SH-wave generation by vibration orthogonal to the polarization using a second (vertical) vibrator with opposite polarity in parallel.

The use of SH-wave sources offers an elegant possibility to suppress noise. For this purpose the wave generation is alternately carried out in opposite directions and the corresponding records are stacked with reversed sign. This simple procedure essentially improves the signal-to-noise ratio.

Due to their lower velocities S-waves allow a higher vertical resolution than P-waves. However, S-wave signals generated with a certain source are mostly of lower frequency than P-waves. Compared to P-waves the vertical resolution could be further increased when S-wave signals of the same frequency content are used at the site under investigation. But, using this fact during a full scale survey to determine lithological information (Section 4.6.3.2), only vibrator sources offer a realistic chance of success.

4.6.4.2 Seismic Sensors

The sensors used in land seismic surveys to convert the ground movement caused by seismic waves into a recordable electrical signal (voltage), are called geophones (receivers, seismometers). The most frequently used type is the moving-coil geophone, which contains a coil that is suspended by springs inside the magnetic field of a permanent magnet (*Fig. 4.6-10*). Contact between the ground and the geophone is achieved by inserting the spike attached to the base of the geophone vertically into the ground. When a seismic wave causes a relative movement between the magnetic field of the permanent magnet and the coil, a voltage is generated in the coil by induction. As the coil can only move in one direction, usually vertical, the geophone only senses the component of ground motion along the coil axis.

The voltage induced is proportional to the velocity of the coil movement. Typical magnitudes of the vertical coil movement with respect to the magnet are of the order of ± 0.1 mm. The springs and mass of the coil form an oscillating system which is characterized by a natural frequency. This system is electronically damped by a shunt resistor and usually the frequency range above the resonant frequency is used to record the signal resulting from true ground motion caused by the wave passage. The natural frequency of the mechanical system and the damping factor h determine the frequency characteristic of a geophone (*Fig. 4.6-11*). If the damping is very small (underdamped), a system will oscillate at the natural frequency. At critical damping ($h = 1$) the coil will very quickly return to rest. Most geophones are slightly underdamped, often having optimum damping with $h = 0.60 - 0.66$ (see *Fig. 4.6-11* and SHERIFF, 1991).

In seismic reflection surveys a large number of geophones are required. For hydrocarbon exploration surveys geophones with 8, 10, or 14 Hz natural frequency are commonly used, while high resolution investigations require 40 Hz, or 60 Hz geophones. For very shallow-target studies, 100 Hz geophones are usually selected (HILL, 1992a).

The signal from each geophone or geophone group registered in one channel of a seismograph is displayed as one trace on the seismogram. The distance between geophones is also referred to as trace spacing. When using short geophone spacings during shallow seismic surveys, a single geophone per channel is usually used. Sometimes geophones are connected in specific patterns in order to provide a summed output to reduce the influence of poorly coupled detectors and to reduce the coherent noise generated by the seismic source, traffic, powerlines, trains or radio transmitters. To ensure sufficient energy transfer to the geophones proper ground coupling is essential. Specific attention therefore has to be paid to the careful installation in a nearly upright position of every single geophone (KROHN, 1984; VAN DER VEEN & GREEN, 1998). Lower sensitivity, higher distortion and the generation of spurious resonances are some of the undesired results when the geophones are tilted at high angles.

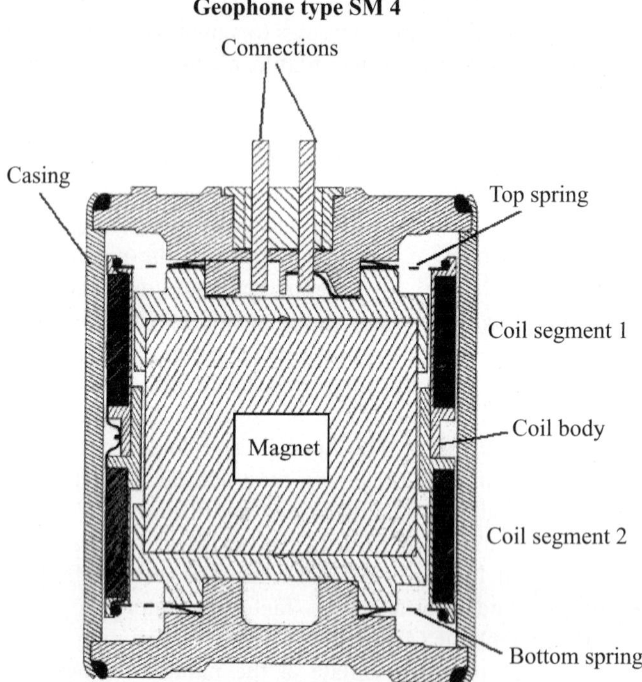

Fig. 4.6-10: Principle of a geophone (see also SHERIFF, 1991)

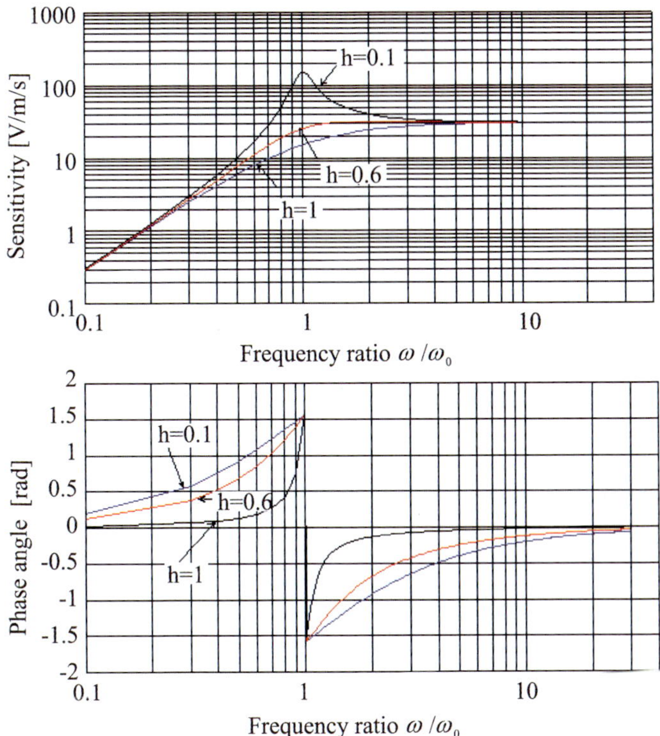

Fig. 4.6-11: Amplitude and phase characteristics of a geophone for different damping factors *h*, as function of the ratio frequency ω to natural frequency ω_0

As described in Section 4.6.3.1 seismic waves are composed of three-dimensional ground movements. Usually, only vertical and/or horizontal components of the ground movement are recorded. For the measurement of horizontal components, especially for SH-waves, geophones with a coil system turned by 90 degrees are used. A complete three-dimensional recording can be realized by three-component geophones which are an assembly of three orthogonal geophones - one vertical and two horizontal.

Geophones often receive rough treatment and are, therefore, constructed as shock-resistant. They must operate under a wide variety of pressure, temperature and moisture conditions and still provide reliable results. The rough treatment geophones are often subjected to, can damage the sensitive mechanical systems and thereby affecting their output characteristics. Geophones should therefore be calibrated at regular time intervals or before used in high resolution surveys. During calibration the geophone parameters as specified by the manufacturer shall be adhered to.

For seismic surveys carried out on lakes and rivers or in swamps and marshlands, hydrophones[6] are used. Their electrical output is generated by the piezoelectric effect produced by water pressure variations and is directly proportional to the acceleration of the ground motion.

4.6.4.3 Seismic Recording Instruments

Seismic recording systems consist of an analog amplifier, an analog-digital converter (ADC) and a storage unit. Before digitizing the geophone output (referred to as sampling a seismic trace) the signal has to be amplified without distortion. In high-resolution seismic reflection surveying multi-channel recording systems (up to 100 channels and more) are commonly used. Using a multiplexer, groups of 6, 12 or 24 channels are often connected to the same amplifier, thus securing uniform amplification of the seismic signals. After amplification the analog-digital converter transforms the continuous analog signal into a series of discrete numbers with a constant sample interval Δt. The sampling interval is a critical parameter for the digitization of seismic data because it determines the highest resolvable frequency of the digital signal. This frequency, known as Nyquist frequency f_N, is computed as $f_N = (2\,\Delta t)^{-1}$. For example, if a seismic signal is sampled at every 1 millisecond it preserves all frequencies up to 500 Hz. To avoid aliasing[7] during AD conversion, the analog data must be passed through an anti-aliasing filter prior digitization to suppress all frequencies higher than the Nyquist frequency. The dynamic range of the digitization depends on the word length measured in bits of the ADC. By over-sampling the highest recorded frequency, the distance between two samples becomes very small. In this case it is sufficient to record only the sign of the amplitude change (Delta-Sigma-technique). Hence, not every analog value has to be measured. The signal can be reconstructed by a series of amplitude changes. Usually, a 24 bit-length word is used, thus preserving a dynamic range of approximately 108 dB for a 125 Hz signal over-sampled at a rate of 32 kHz. A sufficiently instantaneous dynamic range of the equipment is required to allow shallow reflections to be recorded in the presence of high-amplitude (ground roll) noise (STEEPLES & MILLER, 1998). Finally, the digitized signal is stored on the hard disk of the seismic recorder in a standardized word length of 32 bit. For further processing the data are transferred to magnetic tape or optical disk in a standard format recommended

---

[6] Seismic receiver which is sensitive to changes in pressure, in contrast to a geophone which is sensitive to ground motion.

[7] Aliasing occurs when the analog seismic trace is not sampled often enough and produces unwanted low frequencies.

by the Society of Exploration Geophysicists Technical Standards Committee (e.g., SEG 2, SEG D or SEG Y).

4.6.5 Seismic Refraction Surveying

4.6.5.1 Principle of the Method

The seismic refraction method uses the process of critical refraction (see Section 4.6.1) to determine the depth and dip of layer boundaries as well as the velocity of layers. According to Snell's law (4.6.6) an incident P-wave is split in both refracted and reflected P- and S-waves at each layer boundary (*Fig. 4.6-6*). When the energy is transmitted to a layer with higher velocity the refraction angle is greater than the incidence angle. A critically refracted wave arises when the transmitted wave emerges at the critical angle i_c. This wave propagates along the layer boundary in the lower layer with its higher velocity (*Fig. 4.6-12*). According to Huygens' principle the interaction of the critical refracted wave with the layer interface produces secondary waves that are transmitted as wavefronts back to the surface. These are known as head waves. The head waves can be registered with geophones at the Earth's surface.

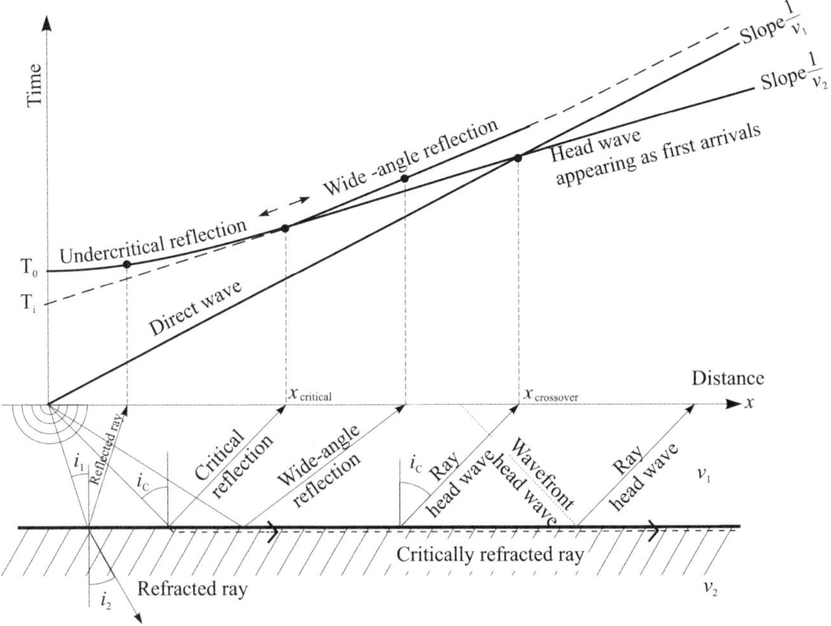

Fig. 4.6-12: Refraction seismics: raypaths and associated traveltime curves for a horizontal two-layer model

In refraction seismogram sections the arrivals of head waves appear beyond a certain (critical) distance $x_{critical}$ only, but are overlain by the events of subcritical and overcritical reflections at the layer boundary. The arrivals of refracted waves are, therefore, only clearly recognizable as first breaks in the seismogram at geophone offsets greater than the crossover distance $x_{crossover}$ (*Fig. 4.6-12*). At the crossover distance the direct wave and the refracted wave arrive at the same time. $x_{crossover}$ will always be greater than twice the refractor depth. The time difference between the time break (time of the wave generation at the source point) and the arrival of the refracted wave, presented in a time-distance graph (traveltime curve), is a linear function of the offset as well as velocity and dip of the layer boundary. The intercept time T_i is the interpolated intersecting point of this linear function to an offset equal to zero and directly related to the depth of the refractor below the shotpoint. The velocity in the (lower) refracting layer is derived from the reciprocal slope of the traveltime curve (*Fig. 4.6-12*).

But, this velocity is not necessarily a true velocity. In the case of dipping layers the velocity value derived from the traveltime curve is an apparent velocity. For shots updip[8] the apparent velocity is higher and for shots downdip the apparent velocity is less than the true velocity. To determine true velocities, depth, and dip of layers, at least two overlapping traveltime curves derived from forward and reverse direction shots (shot updip and shot downdip) are necessary. In the modeling procedure Equations (4.6.9 - 4.6.11) are used.

Because a critical angle only exists if the underlying layer has a higher velocity than the overlying layer, the application of the seismic refraction method is restricted. A layer of lower velocity below a layer with higher velocity cannot be detected (hidden-layer, shadow zone).

4.6.5.2 Survey Practice

Typical problems that can be solved by this method are the determination of depth to bedrock or water table, thickness of a geologic barrier or waste deposits, as well as thickness and velocity of the weathering layer (to determine static corrections for seismic reflection surveys). Theoretically, the method is not restricted with respect to the number of layers to be investigated. In practice, however, not more than 3 - 5 layers can normally be resolved in a multilayer Earth.

As shown in *Fig. 4.6-12* seismic refraction surveying is carried out along straight profile lines. A seismic source is normally located at one end of the profile and the arrivals are detected by geophones regularly spaced along the

[8] Updip: the refractor dips from the shotpoint towards the geophones.

profile. The geophone locations are routinely surveyed relative to the source point to precisions of ±15 cm in height and ±10 cm in the horizontal position.

When dipping layers occur every shot must be reversed. Because refracted arrivals are only recorded as first breaks in a seismogram beyond the crossover distance, a survey line should be 5 to 10 times as long as the investigation depth. However, the profile length necessary is dependent on the velocity variation with depth. For small-scale refraction surveys on waste-disposal or construction sites, spread lengths < 100 m are normally sufficient. A weight drop energy source normally provides sufficient energy to achieve the set of objectives, and two or three operators can easily handle all the necessary equipment.

In addition to the first arrivals used in traditional refraction surveys the CMP-refraction method uses also amplitude and frequency characteristics (GEBRANDE, 1986). A roll-along shooting pattern is started at the beginning of the geophone spread progressively moving forward to the next recording station. The first arrivals of refracted waves obtained are assigned to common midpoints (CMP) located half way between the individual source and detector points. In this way a CMP-traveltime curve is obtained with the slope of every individual branch being the reciprocal of the apparent refractor velocity (*Fig. 4.6-12*). It is important that the spread is designed in such a way that for every CMP a well-sampled traveltime curve can be constructed (at least 3 sample points for each branch).

An areal coverage by the seismic refraction method is possible in cooperation with 3-D shallow seismic reflection surveys (BÜKER et al., 1998a). To compute static corrections (refraction statics) the method utilizes the first arrivals of reflection seismograms (see *Fig. 4.6-14 right*). This allows the determination of a near-surface velocity field and to derive a 3-D depth model for the refracting boundaries.

4.6.5.3 Processing and Interpretation

There are various methods available for the processing and interpretation of refraction survey data (e.g., HALES, 1958; HAGEDOORN, 1959; PALMER, 1986). While the standard kinematic methods are restricted to the interpretation of refracted arrivals in small-scale refraction surveys (usually utilizing first breaks only), the refraction convolution section (RCS) uses amplitudes and first-arrival traveltimes in a full-trace processing to generate time sections similar to that of reflection seismics (PALMER, 2001a,b,c)

The intercept-time method is a quick and efficient tool, however, limited with respect to more complex subsurface conditions. It allows an exact determination of the seismic velocities and layer thicknesses under following assumptions:

- the subsurface is composed of a series of layers separated by planar interfaces,
- the dip of the layer interfaces is moderate (≤ 10 degrees),
- the seismic velocities within each layer are uniform and increase with depth, and
- no cross-dip occurs along the profile line so that all raypaths are restricted to the vertical plane beneath the profile.

If the velocities within the individual layers are not constant or if the top of a layer undulates, only a rough estimation of layer velocity and layer thickness is possible. Therefore, the intercept-time method is most suitable in the case of nearly horizontal layering and as a reconnaissance survey method for the planning of further fieldwork. The traveltimes of the first arrivals are plotted against the distance of the receivers (geophones) from the source. For the direct wave and every velocity layer being present in the investigated depth interval, a different slope is obtained from the resulting traveltime curve. In the case of horizontal layering the velocity of each individual layer is determined by the reciprocal of the relevant slope (*Fig. 4.6-12*). For dipping layers or lateral velocity changes the layer velocity is calculated from the apparent velocities determined by reverse shooting at each end of the geophone spread. The depths of the individual refractors can be calculated from the intercept time derived from each slope.

$$t_{\mathrm{d}} = \frac{\sin(i_{\mathrm{c}} + \alpha)}{v_1} x + \frac{2 z_{\mathrm{d}} \cos(i_{\mathrm{c}})}{v_1} \tag{4.6.9}$$

$$t_{\mathrm{u}} = \frac{\sin(i_{\mathrm{c}} - \alpha)}{v_1} x + \frac{2 z_{\mathrm{u}} \cos(i_{\mathrm{c}})}{v_1} \tag{4.6.10}$$

These two formulae represent the traveltime curves of the critical refracted wave for a 2-layer-case with dipping planar interfaces where t_{d} is the traveltime of the shot "downdip", z_{d} the depth below this shotpoint perpendicular to the refracting interface, t_{u} is the traveltime of the shot "updip", z_{u} the depth below this shotpoint perpendicular to the refracting interface, i_{c} the critical angle, α the layer dip and v_1 the velocity of the direct wave in the first layer. As these formulas are equations of a straight line

with slope $m_{\mathrm{d,u}} = \sin(i_{\mathrm{c}} \pm \alpha)/v_1$ and

intercept time $T_{\mathrm{id,u}} = 2z_{\mathrm{d,u}} \cos(i_{\mathrm{c}})/v_1$,

it is easy to determine the values of velocity v_2, depth z and dip α by "reading" the traveltime curves. The dip α is given by

$$\alpha = \frac{1}{2}\left(\arcsin(v_1 m_{\mathrm{d}}) + \arcsin(v_1 m_{\mathrm{u}})\right). \tag{4.6.11}$$

With i_c the critical angle as

$$i_c = \frac{1}{2}\left(\arcsin\left(v_1 m_d\right) - \arcsin\left(v_1 m_u\right)\right),$$

using Snell's law and the velocity v_2 of the second layer

$$v_2 = \frac{v_1}{\sin\left(i_c\right)},$$

the depth below the shotpoints $z_{u,d}$ results as

$$z_{u,d} = \frac{v_1 T_{iu,d}}{2\cos\left(i_c\right)}.$$

The Generalized Reciprocal Method (GRM) is a generalization of the intercept-time method for planar refractors (PALMER, 1981). Besides the depth below the shotpoints, GRM calculates the depth at every geophone station (*Fig. 4.6-13*). This allows the consideration of changes in depth and velocity along the refractor with high resolution (SJÖGREN, 2000). Other direct inversion methods such as the plus-minus method (HAGEDOORN, 1959), Hales' method (HALES, 1958) and the wavefront method (THORNBURGH, 1930; ROCKWELL, 1967) are applicable to curved refractors.

Because velocity and depth are closely connected in the traditional refraction seismic interpretation, their independent determination from refraction data alone is restricted. Calculated models are always ambiguously over a relatively broad range of velocity-depth pairs. To constrain refraction models, the refraction convolution section (RCS) uses the information content of amplitudes. Any lateral changes in refractor wavespeed should have an associated amplitude expression (PALMER, 2001b).

In addition to these direct inversion methods seismic iterative forward modeling or refraction tomography can also be applied. In general, these methods require a starting model for subsequent iterative calculations. The new model resulting from these calculations is compared with the observed data in order to modify the model parameters. Based on these modifications the iterative modeling process is repeated until a sufficient similarity of the calculated model to the observed data is reached.

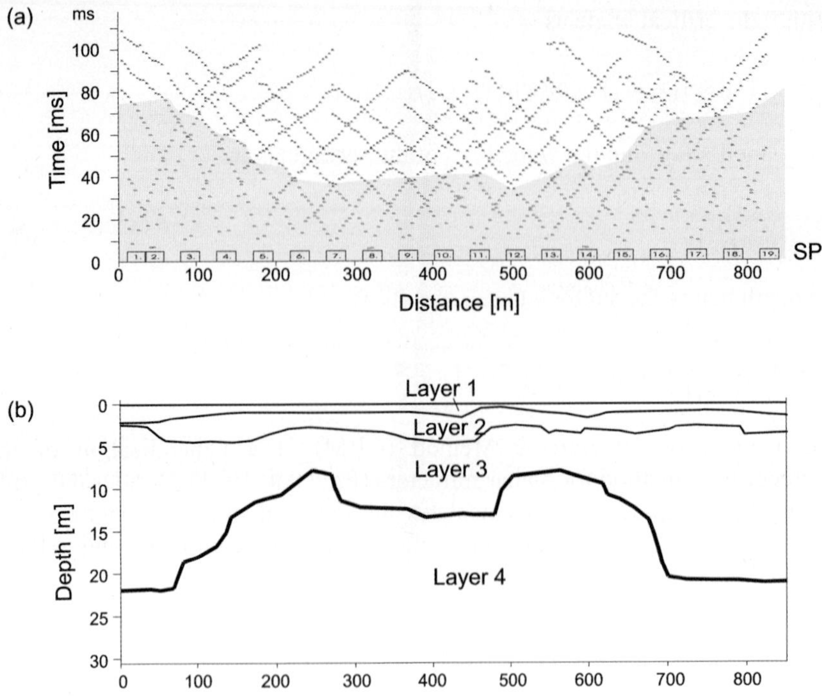

Fig. 4.6-13: Refraction seismics - example of using GRM, (a) Traveltime curves as derived from "picked" first arrivals - forward and reverse shots at source points 1 - 19: dots in traveltime curves represent geophone positions, (b) Layered subsurface model as derived from GRM analysis

Refraction tomography (NOLET, 1987; STEWART, 1991; LANZ et al., 1998) allows the localization of lateral velocity anomalies in the subsurface (caused by, for example, manmade alterations or due to fault and fracture zones). Compared to standard inversion techniques a significantly higher resolution of these anomalies is accomplished. In refraction tomography the traveltimes of diving waves are used. Diving waves arise due to continuous refraction in high velocity-gradient zones with continuously increasing velocity. The diving wave rays are bent in a way that they return to the Earth's surface without the presence of a distinctive refractor. To determine the velocity field from the observed data, the plane of investigation is subdivided (gridded) into rectangular cells, which are characterized by the model parameter velocity. A prerequisite for the correct inversion is that each cell of the model is crossed by several seismic raypaths. Forward modeling is performed with a ray-tracing algorithm and the model modifications are calculated by an algebraic reconstruction technique (e.g., SIRT, Simultaneous Iterative Reconstruction Technique).

4.6.5.4 Personal, Equipment, Time Needed

Service	Personal	Equipment	Time needed
refraction seismic survey	1 operator (= party chief) for recording 1 shooter for weight drop, sledgehammer or explosives 1 temporary worker for geophone layout	1 4WD-vehicle for transport of material and crew 1 seismograph (24 – 96 channels recommended) 1 seismic source (weight drop or sledgehammer and baseplate or explosives) geophones and cables for at least 3 spreads	depending on terrain conditions 4 - 6 spreads (40 – 80 shots) per day (1 day = 10 working hours)
processing, interpretation, reporting	1 geophysicist temporarily 1 drawing assistant	1 PC with processing software	2 - 3 days per field day

4.6.6 Seismic Reflection Surveying

4.6.6.1 Principle of the Method

The seismic reflection method observes elastic waves reflected from interfaces where contrasts in acoustic impedance occur (reflectors). Because the value of the reflection coefficient (4.6.3.3) is much smaller than "1" for most sedimentary layers, a certain part of the energy of the incident wave is transmitted to deeper layers. Although the energy of seismic signals is very weak and interfered with noise, the penetration depth of reflection seismics can be as much as a few kilometers, and having a higher resolution compared to other geophysical methods. The principle of reflection seismics is illustrated in *Fig. 4.6-14* for a layered Earth model with horizontal and dipping interfaces, showing raypaths and the resulting seismogram.

For normal incidence onto a reflector a ray is reflected into itself. The arrival time T_0 in the apex of the hyperbola of a certain reflection which would be observed for a vertical raypath (vertical time, zero-time) is of special significance for the interpretation of CMP reflection seismic data. Based on vertical times T_{0i} the reflector topography in the subsurface model can be imaged in time domain.

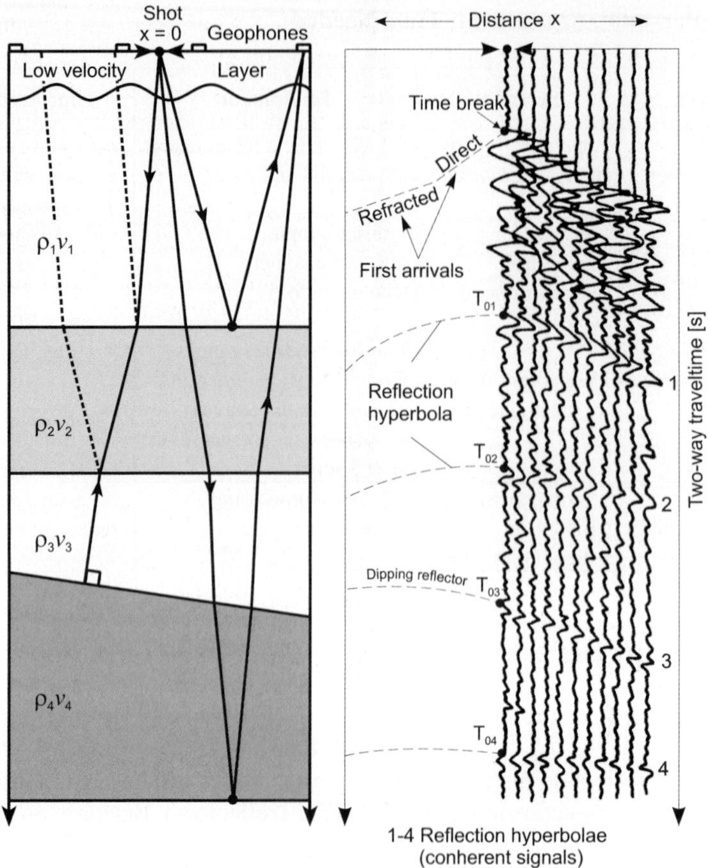

Fig. 4.6-14: *left*: Layered model showing a single source point (shot) and selected seismic raypaths, *right*: Seismogram of a shot gather belonging to the model on the left. Reflections can be identified as coherent signals. ρ_i /v_i are densities/velocities of the respective layers, T_{01}…T_{04} are the zero-offset times (zero-times) of the reflections, Beck (1981)

As in refraction seismics, geophone sets positioned along a line (2-D) or over an area (3-D), source point and receiver points are mostly not coincident. The increasing distance to the source (offset) results in different reflection arrivals at the geophones. Time shifts of the reflection arrivals at different receiver positions to the zero-offset time T_0 for horizontal layering are called normal moveout "NMO" (*Fig.4.6-18*). Time shifts caused by dipping reflectors are called dip moveout (DMO). For a horizontal plane reflector at depth z, the traveltime curve $T(x)$ of the reflected wave is a hyperbola with the parameters offset x, layer velocity v and zero-time $T_0 = 2\,z/v$

$$T^2(x) = T_0^2 + \frac{x^2}{v^2}. \tag{4.6.12}$$

Figure 4.6-15 Top shows a real central shot seismogram in conventional reflection seismics for oil and gas prospecting. Several reflections as hyperbolic coherent events are clearly visible. In the central part of the seismogram events with low velocities, high amplitudes and low frequencies mask the reflections and complicate the exact determination of their zero-times T_0. These high-energy waves are surface-waves (Section 4.6.3.4), also called ground roll, and because of its tendency to mask reflection events they are considered as a major source of noise in shallow reflection seismics.

This problem is further illustrated by a seismogram of a central shot in a near-surface reflection seismic survey with up to 600 ms two-way traveltime (*Fig. 4.6.15 Bottom*). The edged area I shows the time-/depth-range typical for the majority of shallow seismic surveys. Seismic noise, such as surface waves and air waves, dominate and reflections cannot be recognized here. A clear separation between reflections and noise occurs only in area II which is described as the "optimum window". The "optimum window" is used in a special shallow reflection seismic technology (PULLAN et al., 1991; WHITELEY et al., 1998).

In a layered subsurface the rays between source and geophones are refracted and reflected several times. Therefore, the path of a ray reflected at a certain layer is complicated to trace (*Fig. 4.6-14*). In this case the traveltimes can be approximated as a hyperbola only for small offsets in relation to the reflector depth. The velocity v in Equation (4.6.12) has to be replaced by the rms (root-mean-square) velocity v_{rms}

$$v_{rms} = \sqrt{\frac{\sum_{i=1}^{n} v_i^2 \cdot \Delta t_i}{t_n}}. \qquad (4.6.13)$$

The rms-velocity is determined from the inverse slope of squared time-distance curves at $x = 0$ using

$$v_{rms}^2 = \left(\frac{\Delta t^2(x^2)}{\Delta x^2}\right)^{-1}$$

based on a horizontally layered model described by the interval velocities v_i and thicknesses of the layers, expressed in zero-traveltimes Δt_i perpendicular to these horizontal layers (with $t_n = \sum_{i=1}^{n} \Delta t_i$). On the other hand, interval velocities can be iteratively derived from these rms velocities by the formula of DIX (1955):

$$v_{int}^2 = \left(v_{rms,n}^2 \sum_{i=1}^{n} t_i - v_{rms,n-1}^2 \sum_{i=1}^{n-1} t_i\right) / \left(t_n - t_{n-1}\right)$$

$$= \left(v_{rms,n}^2 t_n - v_{rms,n-1}^2 t_{n-1}\right) / \left(t_n - t_{n-1}\right) \qquad (4.6.14)$$

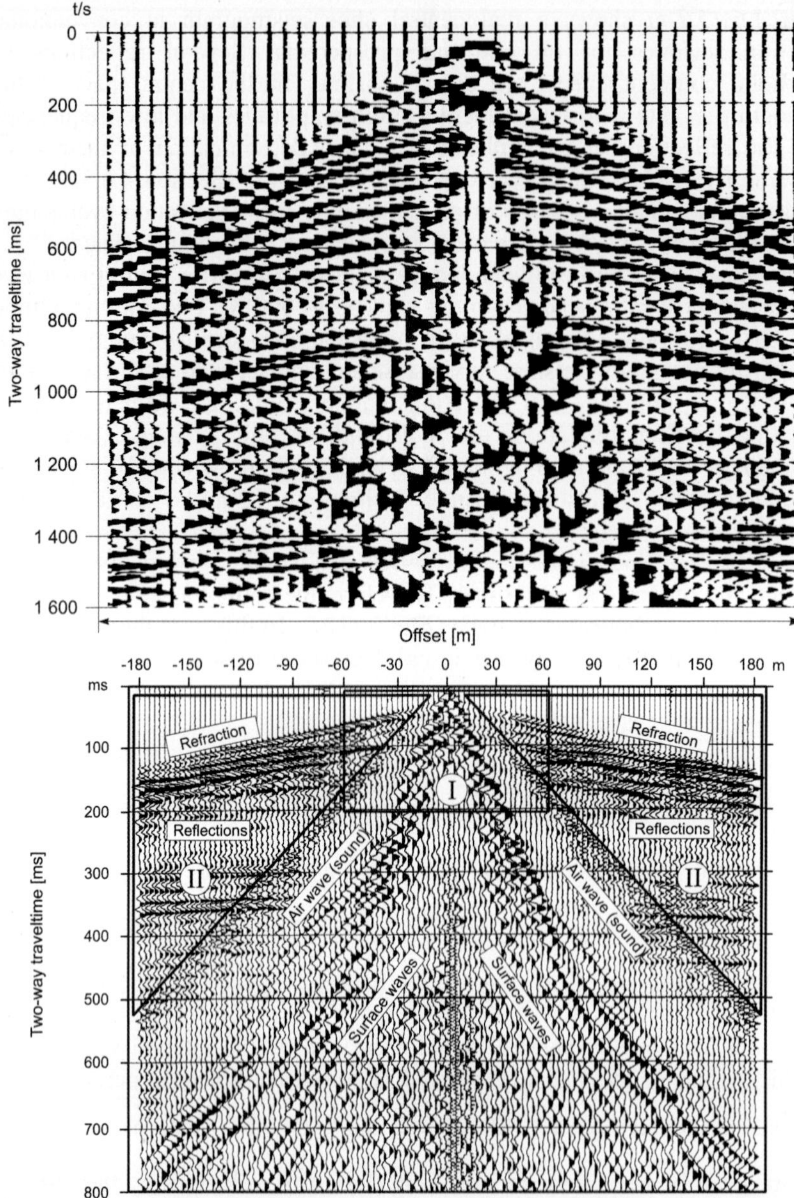

Fig. 4.6-15: Comparison of seismic reflection data for a deep hydrocarbon prospecting survey and a shallow survey. *Top*: Conventional seismogram recorded for a two-way reflection traveltime of up to 1.6 sec showing distinct reflection hyperbolas and the noise cone in the center, KEARY & BROOKS (1991), *Bottom*: Seismogram of a central shot showing a typical wavefield in shallow seismic surveys. Area II represents the "optimum window" where reflection events are clearly visible.

4.6.6.2 Survey Practice

In shallow reflection seismic surveys the investigation depths range from approximately 5 meters to a few hundred meters, and seldom up to one kilometer. P-waves are most commonly used, but in special investigations S-waves are also used. Collecting shallow seismic data requires the use of high frequencies, and therefore, appropriate seismic sources and high-frequency geophones are needed (KNAPP & STEEPLES, 1986; HILL, 1992a).

Problems arise from significant changes of the seismic velocity within the shallow depth interval. In addition, the seismic records can be affected by strong noise such as ground roll, air-coupled waves, direct waves, refractions and diving waves, diffracted waves and multiples[9]. Moreover, seismic measurements are disturbed by ground motion due to wind, rain (JEFFERSON et al., 1998), human activities (industry, traffic), and electromagnetic noise from powerlines and electric cables.

In reflection profiling (two-dimensional surveying) all source points and receivers are aligned along preferably straight lines. Only when profiles run perpendicular to the prominent geologic strike, the reflected raypaths can be assumed to travel in the profile plane. During the survey the source point and its associated geophone spread are progressively moved forward along the profile line to achieve lateral (multi-) coverage of the underlying geological section. *Figure 4.6-16* shows various source-geophone arrangements:

a) split spread (split dip): the geophone array is arranged symmetrically on both sides of the source (typical seismic records see *Fig. 4.6-15*)

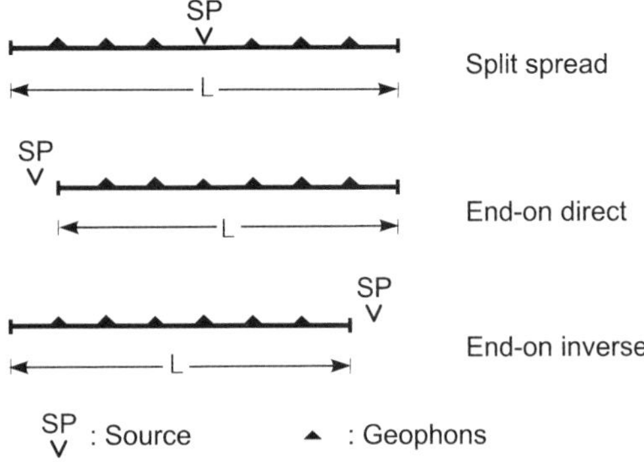

Fig. 4.6-16: Commonly used source-geophone arrangements (spreads) for reflection seismic data acquisition

---

[9] In comparison to primary reflections, which are reflected only at one reflector before received at the surface, multiples (or secondary reflections) are reflected more than once at several reflectors. They can be suppressed by CMP stacking and filtering. In conventional seismics multiples are a noteworthy source of misinterpretation.

b) end-on direct: the geophones are arranged in-line at one side of the source and in the direction of the source movement, and

c) end-on inverse: the geophones are arranged in-line at one side of the source but in the opposite direction of the source movement.

When the spread is stepwise moved by an amount less than half the total spread length, it is possible to record reflections due to each source at multiple geophones and, reversed using each geophone to record from multiple shotpoints. This means that most of the reflecting points in the ground are multifold covered by different traces. Nowadays, the application of the multiple coverage principle (common-midpoint method, CMP) is standard practice in reflection seismic surveying. The stacking of traces corresponding to the common midpoint (CMP stack or horizontal stack) permits effective noise reduction to enhance the signal-to-noise ratio (including multiples suppression) and provides a seismic section showing traces that would have been recorded by a coincident source and receiver point (vertical time section). Such a CMP section represents a geologically interpretable image of the subsurface (*Figs. 4.6-22* and *4.6-23*). To achieve this all reflections from source-geophone configurations with the same CMP are combined before stacking. The set of traces that have a common midpoint is referred to as a common midpoint gather (CMP gather). In the case of horizontal layering all seismic traces of a distinct CMP gather originate from the common reflecting point. When the reflectors dip at a slight angle the different traces of a CMP gather originate from different, but closely spaced, reflection points within a small area around the midpoint depth location. In this case the CMP concept holds, especially considering that some corrections are applied during the data processing (Section 4.6.6.3). The fold C (the number of traces that have been added together) on a 2-D CMP profile is calculated using the formula:

$$C_{2D} = \frac{n}{2} \cdot \frac{\Delta x}{\Delta SP} \tag{4.6.15}$$

with n being the number of geophones per spread, Δx the geophone spacing and ΔSP the distance between the source points. Thus, with a 96-channel spread, a geophone spacing of 1 m and a source point distance of 2 m the resulting coverage is 24-fold (see Section 4.6.8 for examples).

An important aspect of survey practice is the wavefield sampling in time and space. To avoid spatial and temporal aliasing as a source of ambiguity in the signal frequency content and, sometimes giving a false picture of ground motion, the data have to be sampled at sufficiently dense intervals (STEEPLES & MILLER, 1998). The temporal sample interval depends on the highest frequency present in the seismic signal (Section 4.6.4.3). At the same time it should be kept in mind that the absorption of waves with higher frequencies is stronger than the absorption of waves with lower frequencies (Section 4.6.3.1). In a shallow seismic survey temporal sample intervals chosen as 1 ms or 0.5 ms and shorter are typical, thus preserving frequencies in the digitized signal of 500 Hz or 1 kHz, respectively.

To ensure sufficient spatial sampling of the wave field the geophone spacing has to be choosen in consideration of the surface relief, the dip of the target horizons, and the size of the first Fresnel zone (Section 4.6.3.5). Experience shows that, using Equation (4.6.8), the optimum geophone spacing ranges between 1/6 to 1/12 of the first Fresnel zone diameter of the shallowest horizon to be investigated. To avoid aliasing in wavenumber filtering ($k = \omega/v$, k being the wavenumber or spatial frequency), the spatial sample interval Δx should be equal to the space converted temporal sample interval Δt using $\Delta t = \Delta x/v$.

As a rule of thumb: the length of an individual geophone spread should be equal to the maximum investigation depth. The deeper the target zone, the more channels are required. A seismic recording system for CMP multichannel reflection profiling should have at least 48 channels for an adequate subsurface coverage, but seismographs with 96 or more channels are frequently employed.

Using a set of several closely-spaced parallel or intersecting survey lines a three-dimensional image of the subsurface can be approximated. The line spacing depends on the maximum investigation depth and the specific problem to be solved. In structurally complex survey areas, however, a much higher resolution of the subsurface geology can be achieved by three-dimensional seismic reflection surveying. A 3-D seismic survey provides a "true image" of the subsurface in three dimensions thus avoiding artifacts of 2-D seismic surveying if, for example, reflections could originate outside of the vertical plane.

In the case of 3-D seismic surveys the field data are collected in such a way that a ground volume is sampled and the recording is not restricted to raypaths in the vertical plane below a single profile line as in the 2-D case. For data acquisition on land a cross array with source points and receivers being distributed over an orthogonal grid of in-lines and cross-lines is often employed. The survey area is subdivided into several strips, each strip consisting of some geophone lines. The source points are regularly distributed on lines running perpendicular to these strips. *Figure 4.6-17* illustrates the principle of 3-D seismic surveying. Detailed descriptions how to design a 3-D seismic survey are given by STONE (1994), BÜKER et al. (1998a), and YILMAZ (2001). Generally, 3-D surveying requires more equipment, a larger survey crew and more sophisticated processing techniques.

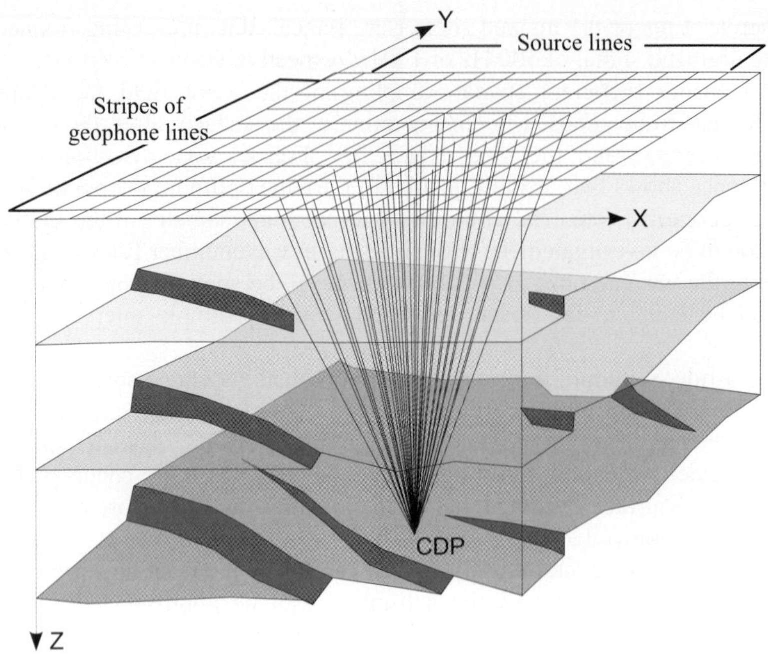

Fig. 4.6-17: Principle of a 3-D seismic reflection survey: *x* is the direction of the geophone lines, *y* is the direction of the source point lines, CDP is a common depth point covered by several seismic raypaths between selected geophone stations and source points

4.6.6.3 Processing and Interpretation of Measured Data

A detailed discussion of seismic processing techniques is beyond the scope of this book. The fundamentals of seismic processing are presented in great detail in DAVIES & KING (1992) and YILMAZ (2001). The main problem in shallow seismic processing is to enhance reflections by suppressing unwanted energy. STEEPLES & MILLER (1998) summarize special problems of shallow seismic processing and interpretation as follows: "Processing and interpreting near-surface reflection data correctly often requires more than a simple scaling-down of the methods used in oil and gas exploration. For example, even under favorable conditions, separating shallow reflections from shallow refractions during processing may prove difficult, if not impossible. Artifacts emanating from inadequate velocity analysis and inaccurate static corrections during processing are at least as troublesome when they emerge on shallow reflection sections as they are on sections of petroleum exploration. Consequently, when using shallow seismic reflection, an interpreter must be exceptionally careful not to misinterpret as reflections those many coherent waves that may appear to be reflections or not. Evaluating the validity of a processed, shallow seismic reflection section therefore requires that the interpreter has access to at least one field record and, ideally to copies of one or more of the intermediate processing steps to corroborate the interpretation and to monitor for artifacts introduced by digital processing".

In the following a brief overview about the different steps in a typical seismic data processing sequence (standard processing flow) is given:

- **Generation of the survey geometry**

  Numbering of shot and geophone positions, consideration of shotpoint and geophone spacing in absolute or relative coordinates; assignment of geophone point and shotpoint geometry, CMP-binning (sorting data into data sets for discrete areas according to midpoint locations), etc.

- **Compiling database and trace headers**

  Location, date, seismic data format, sampling interval, trace length and other information are written to a database and in the trace headers.

- **Editing of records and traces**

  The quality of the recorded raw seismic data often varies. Typical factors affecting quality are: no signals, very high noise level, interference with mono-frequency signals, spiky signals, wrong polarity. Traces which are affected by these factors can be filtered, their polarity can be reversed or the appropriate traces can be deleted (trace removal) in order to ensure a homogeneous dataset with high quality.

  Shallow reflections are often masked by direct waves, refractions, air waves or ground roll. If these waves cannot be suppressed or minimized by CMP stacking, frequency or *f-k* filtering[10] and spectral balancing[11], their amplitudes can be muted (BAKER et al., 1998). A prominent feature of reflections used to distinguish between reflections from layers near the surface and low frequency noise, is that they should have a frequency content close to that of the direct wave and early refractions.

- **Static corrections (statics)**

  Statics are corrections applied to field data to compensate for the influence of variations in shot and geophone point elevation as well as of low-velocity layers near to the surface. Data are corrected to a selected datum plane. The static corrections can be calculated using the results of refraction seismics or up-hole measurements. If information on the low-velocity layer is lacking, a velocity is chosen from experience to calculate the statics.

---

[10] Filter using frequency *f* and wavenumber *k* to attenuate unwanted events with certain apparent velocity or dip (also called velocity filter, fan filter, pie-slice filter).

[11] Seismic data sets are spectral balanced if all records have approximately the same frequency spectrum. Spectral balancing of input records is done in different ways: deconvolution, band-pass filtering, spectral averaging, etc.

Deriving effective statics is more difficult for shallow seismic surveys than for conventional seismics as the frequencies are normally higher and near-surface reflections tend to be less continuous. FREI (1995) introduced a procedure for field statics across rugged terrain where the target depth is of the same order as relief variations over the spread length.

- **True amplitude recovery and gain recovery**

The decrease in amplitude of seismic waves caused by spherical divergence and absorption (Section 4.6.3.2) has to be equalized. Time variable gain or automatic gain control (AGC) are typical operations to set the data to a distinct (relative) amplitude level.

- **Velocity analysis and dynamic corrections**

To determine the dynamic corrections (normal-moveout correction, NMO), reflection hyperbolas have to be corrected for the time delay of an offset trace using Equations (4.6.12) and (4.6.13). The objective is to determine stacking velocities (rms-velocities) which will result in the highest signal-to-noise ratio of the final stacked seismic section used for the geological interpretation. The stacking velocity depends on traveltime, position along the profile and the dip of layers. As the velocity often changes drastically within very short distances in shallow seismic surveys, it is recommended to generate a series of constant-velocity CMP stacks with a broad range of velocities for densely spaced CMPs (velocity scan). To derive the best NMO velocity from the constant-velocity gathers (velocity panels), a cross-correlation technique has been developed (TANER & KOEHLER, 1969). Using these velocities primary reflection events on offset traces are moved to the traveltime T_0 in the apex of the reflection hyperbola and, as they are now in-phase, they can be stacked properly to enhance the signal (*Fig. 4.6-18*).

- **Common-midpoint stacking**

Stacking is adding or averaging seismic signals. Common-midpoint stacking (CMP stacking) is used to improve the signal-to-noise ratio and to attenuate multiples. The NMO corrected traces from different records, due to rays reflected at positions near the common depth point, are summed up (stacked) and displayed as one trace at normal incidence (zero offset). The result is a composite record in time domain which reflects the structural image underneath the survey line (*Fig. 4.6-18*). Strictly speaking, the NMO-procedure is only applicable under near horizontally layered conditions. When dipping reflectors are stacked the reflection points are smeared. In these cases a special dip moveout processing routine (DMO) can be applied to correct for the smearing effects (see below).

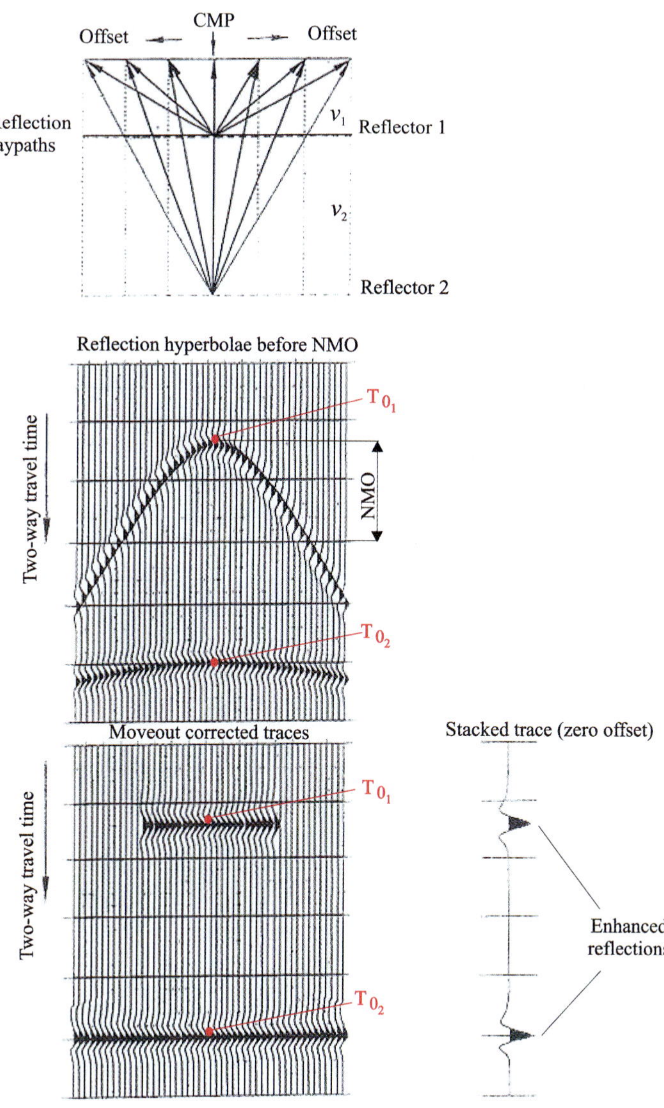

Fig. 4.6-18: Principle of normal-moveout (NMO) correction and CMP stacking

- **Deconvolution**

When traveling through the subsurface, seismic signals undergo a filtering process which changes their waveshape. Dissipative properties of the ground and intrinsic attenuation act as lowpass filters that decrease the high frequencies of the source signal. This reduces the resolution power. The main goal of deconvolution (inverse filtering) is to enhance the resolution of reflected events. This is achieved by recovering high signal frequencies, equalizing amplitudes within a certain frequency band, by producing zero-phase wavelets or by minimizing variations between seismic records using the knowledge of the source waveform (signature deconvolution). Another important goal, especially in conventional seismics, is the attenuation of multiples by predictive deconconvolution.

- **Migration**

Seismic migration was originally developed to move reflection events from dipping interfaces to their "true subsurface reflection points", to focus diffraction effects and to correct amplitude distortions caused by curved reflectors (GRAY et al., 2001; FAGIN, 1998). The moving of reflection events from their recording point position to their reflection point position by seismic migration is demonstrated by *Fig. 4.6-19*. There has to be distinguished between different migration algorithms: integral and finite-difference solutions or frequency-wavenumber implementations. To carry out the migration processing appropriate parameters (input data, migration velocities) have to be provided and depending on the complexity of the geological structure under investigation an adequate migration strategy has to be choosen: post-stack or pre-stack migration, time or depth migration. For shallow seismic investigations usually 2-D post-stack migration is used. According to some authors migration may produce in most cases only minor changes to very shallow reflection data and may even be unnecessary for low velocities (<1000 m s^{-1}) and gently dipping reflectors. However, the sharpening of horizontal resolution by migration may have advantages even in surveys of areas without complex structures (DAVIES & KING, 1992). Nevertheless, with the powerful computers available today, migration should be carried out to test whether or not significant displacement of seismic energy is visible. To migrate seismic reflection data properly, high-quality reflections are always required (BÜKER, 1998).

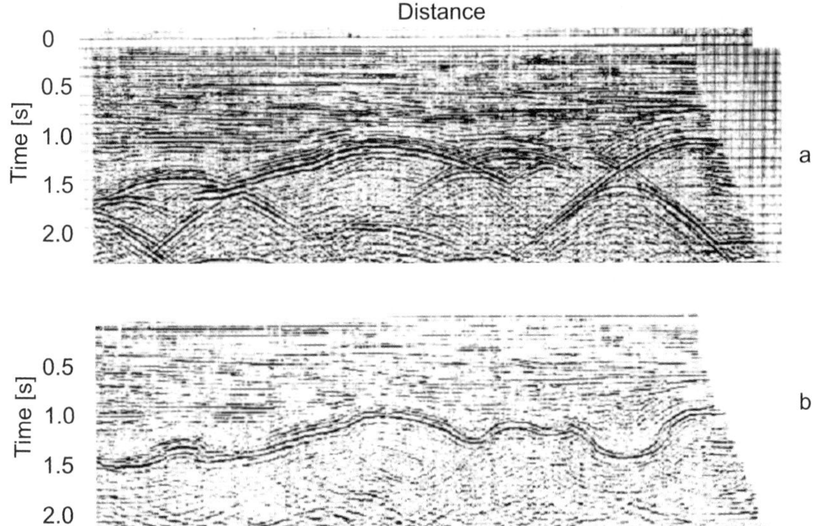

Fig. 4.6-19: Effect of migration in seismic time sections, (a) the stacked unmigrated time section shows bow tie reflections which result from narrow syncline structures, (b) time migration process moves the bow ties to their correct subsurface position (example modified from SFU-website)

Besides these basic processing steps listed above several processing operations can be applied pre-stack or post-stack to enhance the signal-to-noise (S/N) ratio and to improve the seismic resolution:

- frequency filtering (high-pass, low-cut, band-pass or notch filter) to suppress coherent (non-reflection) noise,
- two-dimensional frequency-wavenumber domain filtering (*f-k* filtering, velocity filtering) to suppress certain types of waves (e.g., ground roll) according to their apparent velocities,
- spectral whitening to increase the high frequency content of seismic signals which should result in a higher resolution,
- residual statics to enhance the lateral coherency of reflections, and
- dip moveout processing (DMO) applied to NMO corrected pre-stack data to compensate for the effects of (steeply) dipping reflectors. The DMO procedure harmonizes the stacking velocity for events corresponding to the same rms velocity regardless of reflector dips. DMO corrects for the reflection-point smear that results when dipping reflections are stacked (BRADFORD et al., 1998). Including a full range of dips leads to an enhanced stacking quality and to an improved seismic image.

After processing and structural interpretation of the stacked and/or migrated seismic sections in the time domain, the data are transformed into the depth domain using seismic velocity information in the form of a "velocity field". To design this "velocity field" data from several sources of information should be included. Refraction seismic profiles provide velocity information about the shallow layers approximately down to the base of the weathering layer or to the groundwater table. Stacking velocities computed during the velocity analysis are useful to show the tendency of the velocity to increase or to decrease with depth. By optimizing the velocity analysis in order to obtain a low noise final stacked section with coherent reflections, the use of only the corresponding stacking velocities for the time-depth conversion may result in a poor depth model (AL CHALABI, 1994; ROBEIN, 2003). The most reliable velocity information is derived from checkshots or vertical seismic profiles in boreholes (Section 4.6.7). Both measurements provide traveltime curves which allow calculating exact interval velocities. A very important element for the estimation of plausible interval or average velocities is to correlate certain reflections to geologic horizons known from boreholes.

The interpretation of seismic data in geological terms is the objective and leads to the final product of each seismic survey. Boreholes located in or near to the survey area allow the calibration of seismic sections to the geology. Reflections have to be linked to geologic horizons and traced along the profiles or mapped over an area in order to identify anticlines, synclines, faults, pinch-outs, unconformities, domes, bedrock surface, channels or other structures (BÜKER et al., 1998; SHTIVELMAN et al., 1998; WIEDERHOLD et al., 1998). The interpreter has to evaluate the wavefield character along a seismic profile, pay attention to changes in the reflection pattern across faults, do the phase correlation of reflections at intersecting profiles and check the closing of profile loops (ties). The checking of geological models for reliability regarding layer thickness, throw or dip requires the services of an experienced geophysicist or geologist.

4.6.6.4 Quality Assurance

Before commencing with a seismic survey, the seismic lines have to be fixed in the field by means of a land survey and documented on a map of suitable scale.

As a first step, all survey equipment should be regularly checked for proper performance prior to starting the field campaign. Sampling interval, registration length, recording filters and pre-amplification are critical parameters that have to be chosen carefully by the operator-in-charge. To optimize all these parameters, a field operator with extensive experience is necessary.

In order to test the parameters of the seismic source (if possible, testing different types of sources), the filters and pre-amplifiers applied and geophone arrays, a full day of on-site noise analyses is normally performed before the actual surveying campaign starts. For this purpose energy is generated at a fixed source point and the geophone spread with very short spacing is moved to progressively larger offsets from the source point; or the source points are moved while the geophones are kept fixed (walkaway testing).

To acquire seismic data with a high signal-to-noise ratio the field operator has to control the quality of the recordings continuously during the survey. In particular, sufficient vertical stacking rates and proper coupling of the geophones to the ground have to be ensured. All survey steps should be documented in a survey protocol that allows the assignment of each record to the corresponding source location if required at a later stage. To control the quality of the records and to prepare the pre-processing (in particular by the calculation of statics), field processing and on-site interpretation are further required. For the transfer of data to the seismic processing system, accepted digital data formats such as the SEG standards SEG 2, SEG A, SEG D or SEG Y are used (SHERIFF, 1991). Usually, a supervisor is in control during a larger field survey. Permanent training of the staff ensures that only state-of-the-art techniques are employed.

4.6.6.5 Personnel, Equipment, Time Needed

Survey	Personnel	Equipment	Time needed
2-D reflection seismic survey	1 operator (= party chief) for seismic recording 1 shooter for weight-drop or sledgehammer 1 line checker 1 - 4 temporary workers for geophone layout	1 recording van, 1 or 2 4WD-vehicles for equipment and crew transport 1 seismograph (48 - 144 channels recommended) 1 seismic source (weight drop or sledgehammer and baseplate) geophones and cables for at least 3 spreads roll-along switch	depending on terrain conditions 80 - 150 shot points per day (1 day = 10 working hours)
processing, interpretation, reporting	1 geophysicist for processing, 1 geo-physicist or geologist for interpretation), 1 drafting assistant (at times)	1 PC or workstation with processing and interpretation software	2 - 4 days per field day

4.6.7 Borehole Seismic Methods

4.6.7.1 Principle of the Methods

Borehole seismic methods use elastic waves generated and recorded at various depths in a borehole using sources or detectors clamped to the borehole wall. They are carried out both solely for investigating the subsurface in close proximity of a borehole and as a complementary method to seismic reflection surveys. To determine layer velocities, the traveltimes of P- and/or S-waves between two points in conjunction with measurements of bulk densities are recorded. This enables, for example, the determination of elastic moduli (Section 4.6.3.2). The installation of shotpoints and detectors close to the geologic units of interest allow the investigation of complicated structures with significantly enhanced resolution and can provide greater structural/lithological detail. Furthermore, the observation of elastic waves at fixed depths in a borehole allows the determination of seismic velocities (GALPERIN, 1985; HUNTER et al., 1998; WONG, 2000).

4.6.7.2 Applications

Depending on the survey configuration, the following borehole seismic methods are used (*Fig. 4.6-20*):

A: both seismic sources (SP) and detectors (DP) are positioned in the same borehole (= single well imaging),

B1: seismic sources and detectors in different boreholes (= borehole-to-borehole method, crosshole imaging),

B2: recording the response of detectors at various depths in a borehole to shots inside the borehole, at the wellhead or at an offset point (= tomography),

C: vertical seismic profiling (VSP), offset VSP, walkaway VSP (moving source VSP), check shot (well shooting), and reversed VSP.

The positions of source and detectors can be interchanged (reversed VSP). By means of check shots and VSP information about seismic velocities of the various geologic units is obtained, so that surface seismic sections can be calibrated geologically. Also, by determining the reflection properties of particular horizons identified in the borehole section, the detection and discrimination of multiple reflections is possible. Borehole seismic methods are also used as stand-alone surveys where the investigation of geologic structures in the vicinity of a borehole is of interest. Using single well imaging, complicated or steep dipping geologic structures close to the borehole can be investigated, whereas crosshole imaging provides information

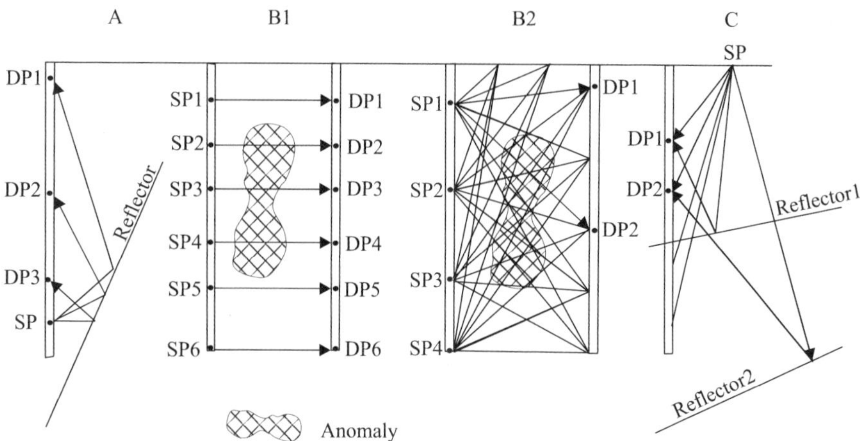

Fig. 4.6-20: Principles of borehole seismic methods: A) single well imaging; B1) cross well imaging; B2) cross well seismic tomography; C) vertical seismic profiling

about the velocity field in the space between two boreholes (GOUDSWAARD et al., 1998; WONG, 2000).

4.6.7.3 Fundamentals

Compared to (surface) reflection seismic surveying borehole measurements have the advantage that the raypath is usually shorter and to a far lesser extent influenced by unconsolidated near-surface sediments. Because the receivers are located in the borehole, noise caused by wind, rain or human activities is significantly reduced. For these reasons higher frequencies can be used and seismic resolution is increased. In case of steeply dipping boundaries less effort to obtain a suitable survey configuration is also required.

However, due to the physical conditions in a borehole environment, signals may be obscured by waves propagating through the drilling fluid or along the borehole wall (fluid waves, tube waves, Stoneley waves) and in cased holes along the casing pipe itself. While tube waves and fluid waves can be suppressed by proper detector coupling to the borehole wall, the latter are best avoided when conducting the measurements in an open hole. In cased holes the voids between pipe and host rock should be filled with cement or clay pellets in order to ensure better coupling of the casing pipe to the borehole wall.

4.6.7.4 Instruments

Sources

To generate the seismic energy at the surface in a VSP survey, the same seismic sources as described in Section 4.6.4.1 can be applied. Because of the high signal-to-noise ratio suitable sources are sledgehammers for shallow depths up to 100 m or weight drops (e.g., EWG III) for exploration depths down to 1000 m. The fold of energy stacking is in the range from 4 to 8. To investigate depths greater than 1000 m, a dynamite or vibrator source is required. For wave generation in boreholes either impulsive downhole sources such as (piezoelectric) sparkers, mechanical hammers and explosives (blasting caps) or vibrators can be used (GIBSON & PENG, 1994; WONG, 2000; DALEY & COX, 2001). The source type determines the range of dominant frequencies.

Detectors

When recording seismic signals in a borehole, special detectors are necessary. To record only the direct wave (e.g., for a check shot survey) in a less than 500 m deep water filled borehole, cables consisting of 12 - 24 hydrophone detectors with 1 - 4 m spacing can be used (MILLIGAN et al, 1997; WONG, 2000). Measurements in dry holes or wells deeper than 500 m require seismometers properly coupled to the borehole wall by means of mechanical, electro-mechanic, or hydraulic systems. Three-component systems usually serve as receivers, but in the case of shallow depths, one-component accelerometers also yield satisfactory results. As in surface seismic reflection surveys the data is recorded by a digital recording system.

4.6.7.5 Survey Practice

Borehole seismic measurements should be carried out starting from the lowest recording level working upward in order to secure accurate recording depths. When reaching the deepest recording level, the recording tool has to be pressed against the borehole wall and the cable be released to avoid energy transmission along the cable (cablebreaks). Only when the wire-line is at rest, the generation of seismic waves should be started. In order to suppress tube and pipe waves, the distance of a surface seismic source from the wellhead should be at least 4 m for shallow recording depths and 15 m for depths greater 100 m.

4.6.7.6 Processing and Interpretation of Measured Data

The processing sequence comprises:

- pre-processing, depthwise ordering (arranging) of the records,
- orientation of horizontal components, suppressing of disturbing waves and noise,
- deconvolution (signal shaping), and
- separation and visualization of downgoing and upcoming events. To clearly recognize reflections in the vertical seismic profile they are aligned horizontally by trace shifting.

The interpretation consists of:

- calculation of time-depth relationships derived from traveltimes of the direct wave,
- calculation of interval and average velocities for check shot and VSP surveys,
- tomographic analysis to create tomograms of crosshole first arrival traveltimes (traveltime tomography), and
- discrimination of primary waves from multiple reflections and correlation to stratigraphic units.

As a result of this processing and interpretation sequence a model describing the seismic velocities and the structural conditions around the borehole is obtained. Conclusions about the lithological characteristics in the vicinity of the borehole can be derived in this way, while the accuracy of the time-depth conversion of surface seismic reflection data can be improved.

4.6.7.7 Quality Assurance

Borehole seismic surveys can be affected by the following shortcomings:

- time-lag between the moment of the seismic pulse generation (zero time) and the start of recording,
- erroneous depths because of tool creep due to the weight of the tool, and
- noise due to cable waves, fluid waves or tube waves.

In order to detect and eliminate recording time-lags, a control geophone is installed in the borehole at a constant distance below the source. Wrong recording levels become obvious when the measurement is repeated or by overlapping detector arrays. Cable waves and tube waves can largely be prevented from being recorded by proper coupling of the receiver to the borehole wall and if the wireline is released during recording. Tube waves can in general, not be suppressed, but because of their high propagation velocities they arrive prior to the direct wave. To avoid misinterpretations they have to be identified.

4.6.7.8 Personnel, Equipment, Time Needed

The staff required to conduct borehole seismic surveys usually consists of a survey engineer and one or two assistants operating the source and the receiver tool. One to two vehicles are needed for the transport of the equipment, while the recording instruments, cable drums and winches are usually installed on a special recording truck. In addition, a seismic source capable of providing sufficient energy is needed to generate reflections from the required investigation depths. A pulley fixed to either a derrick or tripod is required to lower the downhole equipment into the borehole. Depending on the desired recording depth, on average 10 - 25 minutes are needed for each recording level or each section covered by a hydrophone array. For comparison purposes: to conduct a borehole seismic survey at 10 m depth intervals to a depth of 300 m using a hydrophone array, will take about 3 hours, while 6 hours are required to cover the same depth interval using three-component receivers.

4.6.8 Examples

Refraction tomography at a waste disposal site

The first example is from an old limestone quarry near Arnstadt in Germany, now filled with industrial and domestic waste (see also Section 6.2.1). The waste body, showing rugged surface topography, conceales the original quarry boundaries and geological structures. The geophysical survey at this location consisted of different geophysical methods including a combined seismic reflection and refraction survey. The objective of this survey was to delineate the outline as well as the floor contours of the quarry and to locate an underground building constructed for the production of secret weapons during the World War II.

A suitable method to describe smallscale lateral and vertical inhomogeneities in the subsurface is the refraction (transmission) tomography technique. Here the data set of a reflection seismic profile measured across the waste body was used. This data satisfies the principal conditions for transmission tomography:

1. The subsurface is crossed by an adequately dense set of raypaths, and
2. first arrivals must be of a high quality to resolve small ground features.

Survey parameter values

seismograph	Bison DIPF 9048	source	EWG III, acceler. weight drop
channels	48	vertical stacks	1 - 3
sampling rate	0.5 ms	shotpoint distance	4 m
record length	250 ms	shotpoint at channel	24 (central shot)
high-pass filter	60 - 500 Hz	geophone type	L40 A-1 Mark Products
notchfilter	none	geophone offset	-92... + ...96 m
pre-amplification	0 - 60 dB	geophone interval	4 m
		geophones/channel	1

In transmission tomography the subsurface is divided (gridded) into cells (here 2 m × 2 m). Each cell is assigned a specific velocity value (starting model). Traveltimes are calculated by tracing rays through the model. These traveltimes are compared with the observed values and the resulting model is iteratively modified to minimize residuals. The model calculation was carried out with the Simultaneous Iterative Reconstruction Technique (SIRT). The starting model for this calculation was derived from the results of conventional refraction seismics, interpreted with GRM (*Fig. 4.6-21*, *Bottom*).

To fulfill the high precision criteria for tomography, a manual "picking" of first arrivals was necessary. In this way a precision of 1 ms on average was achieved. The difference of modeled and observed arrivals after 10 iterations of the curved-ray-modeling was 2 ms on average, thus within the criteria set for picking precision. *Fig. 4.6-21* (*Top*) shows the resulting tomogram. The former quarry is recognizable between stations 65 - 85 as a trough structure with steep flanks. The maximum thickness of the deposited waste material, characterized by velocities of 300 - 500 m s^{-1} is of the order of 20 m.

This example illustrates that complicated structures can be mapped in detail with refraction tomography. The result is in good agreement with other sophisticated methods such as prestack Kirchhoff depth migration (PASASA et al., 1998). The conventional GRM modeling poorly resolved smallscale structures because of the inhomogeneity in lateral stratification. Large velocity gradients cannot be resolved correctly by GRM. The refraction tomography however, represents the "real" conditions with high detail, depending, of course, on inherent spatial resolution (cf. Section 4.6.3.5)

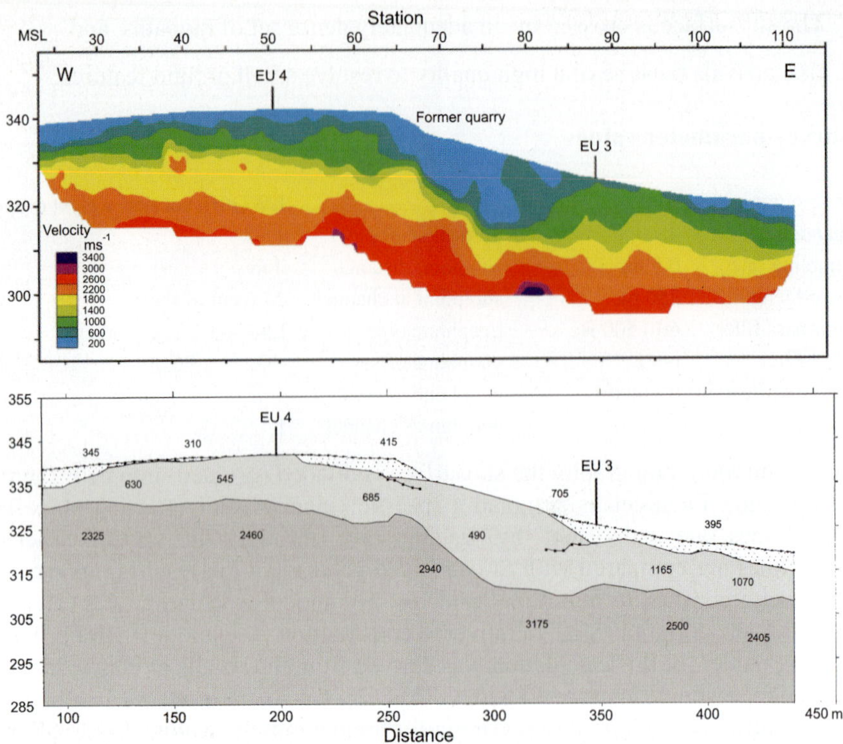

Fig. 4.6-21: Refraction tomography at a waste disposal site, *Top*: tomogram showing subsurface layering and contours of the former quarry, *Bottom*: refraction seismic interpretation using GRM (starting model), EU 3 and EU 4 are boreholes

Mapping of faults by shallow reflection seismics

A high resolution reflection seismic survey was done on the area around a former landfill to investigate the local and regional groundwater system. The landfill at Arnstadt/Germany (see example refraction tomography and Section 6.2.1) is located within a dominant regional fault system, accompanied by numerous fractures and joints. Karst phenomenons occur in the limestone of the Muschelkalk formation, covered by clayey sediments of Keuper. It was suspected, that contaminants spread out from the landfill along tectonic faults and thereby causing groundwater pollution. To perform a risk assessment, additional information on the fault pattern and the hydraulic conductivity of faults are essential. The reflection seismic method is an accepted way of mapping faults and for describing fault architecture (SHTIVELMAN et al., 1998). 2-D reflection seismic measurements in the investigation area were carried out along 17 lines totaling 12.5 km. The topography of the survey area is rugged in places. Data acquisition with the conventional CMP-technique was carried out.

Survey parameter values

seismograph	BISON 9048	source	EWG III, sledgehammer
channels	48	vertical stacks	1 - 8
sampling rate	1 ms	shotpoint interval	6 m / 4 m
record length	500 ms	shotpoint at channel	24 (central shot)
high-pass filter	60 – 250 Hz	geophone interval	6 m / 4 m
notchfilter	out	geophone type	L40 A-1 Mark products / 60Hz
pre-amplification	0 - 60 dB	geophones/channel	1
CMP-fold	24	geophone offset	-138…144 m / -92…+96 m

Processing was carried out using Landmark`s ProMAX[12] with AGC (automatic gain control), building geometry files, static corrections (statics) to an inclined reference datum above profile line, CMP sorting, bandpass filtering (25/35 - 120/150 Hz), muting, determining NMO velocities by velocity scan, NMO-correction, CMP stacking and post-stack bandpass filtering, coherence filtering to emphasize reflection events and to suppress noise, static corrections to a horizontal reference datum, time-to-depth conversion using velocities from database or refraction seismics, and plotting of stacked sections. *Figure 4.6-22* presents two interpreted time sections. On both sections distinct reflection patterns appear. Reflected events can be observed between two-way traveltimes (TWT) less than 25 ms and 250 ms.

Clear indications of the presence of tectonic faults in the above seismic sections are: changes in the dip of layers, phase splitting, jumps (reversals) in the phase correlation and short diffractions. By careful interpretation of all seismic sections a detailed fault and block pattern was derived and jointly interpreted with the results of geological mapping, other geophysical methods, and aerial photograph interpretation.

---

[12] Landmark`s ProMAX® is a frequently used system for seismic data processing. It includes a complete suite of applications for 2-D, 3-D and VSP.

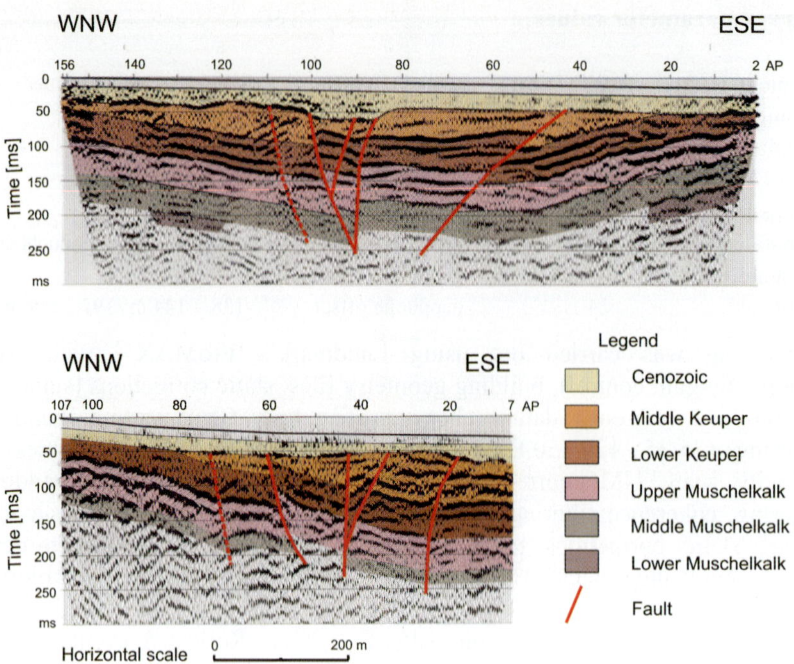

Fig. 4.6-22: Mapping faults with reflection seismics - interpreted seismic sections

Shallow reflection seismic survey to investigate the subsoil of a potential waste disposal site with changes in seismic facies

An integrated geophysical survey was conducted about 50 km north of Nakhon Ratchasima (Khorat) in NE Thailand to study the geological barrier of a potential waste disposal site. This area, located on the Khorat plateau, is characterized by increased land salinization. Saline soil is the origin of various environmental problems, such as shortage of freshwater and soil degradation. The geological causes for salinization are still unclear.

Along four lines in the study area a combined reflection/refraction seismic survey was conducted. The top of the evaporites (horizon A) appears as a marker horizon on all seismic sections at a depth range from 75 - 100 m. Different reflection configurations in the overburden indicate different seismic facies units within the clastic rocks. While the reflections in the eastern part can probably be attributed to intercalating basin sediments, only a few distinct reflections can be detected in the western part. Discontinuous events, low amplitudes and low frequencies indicate a relatively monotonous sedimentary succession in this area. West and east of borehole DKT 3-1 paleochannels (PCS), formed as a result of fluvial erosion, are present. The transition zone between the areas of different sedimentation, where flexures can be observed, is a possible zone of weakness for ascending saline water (*Fig. 4.6-23*).

Survey parameter values of the combined reflection/refraction seismic

seismograph	BISON 24096	source	sledgehammer
channels	96	vertical stacks	2 - 8
sampling rate	0.5 ms	shotpoint interval	3 m
record length	500 ms	shotpoint at channel	48 (central shot)
geophone type	SM4 (20 Hz)	geophone offset	-141…144 m
geophones/channel	1	CMP-coverage	48
geophone spacing	3 m		

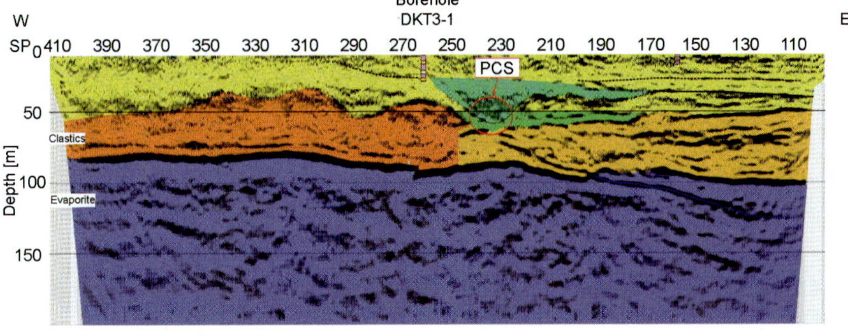

Fig. 4.6.23: Interpreted seismic depth section of line DKT-03

Processing using Landmark`s ProMAX® was carried out with input of field data and conversion into 32-bit format, geometry assignment, CMP-binning, database generation, quality control (QC) of geometry and seismic data, notch filtering of 50 Hz and harmonics, interactive trace editing, calculation and application of statics based on first arrivals (refraction statics), trace muting, true amplitude recovery, spectral whitening, interactive velocity analysis, residual statics, NMO correction and CMP stacking, finite difference (FD) time and depth migration, final band pass filtering and coherence filtering, and plotting of final time and depth sections.

Vertical seismic profiling (VSP) for the determination of elastic moduli

To determine Young´s modulus E (stretch modulus) and the shear modulus μ (see *Table 4.6-1*) of the shallow ground surrounding a borehole, a vertical seismic profile was measured to obtain P- and S-wave data (*Fig. 4.6-24*) The aim of these measurements was to assess the foundation state for the construction of a new building. VSP surveys provide in situ parameter values which complement geotechnical laboratory experiments on soil and drill core samples.

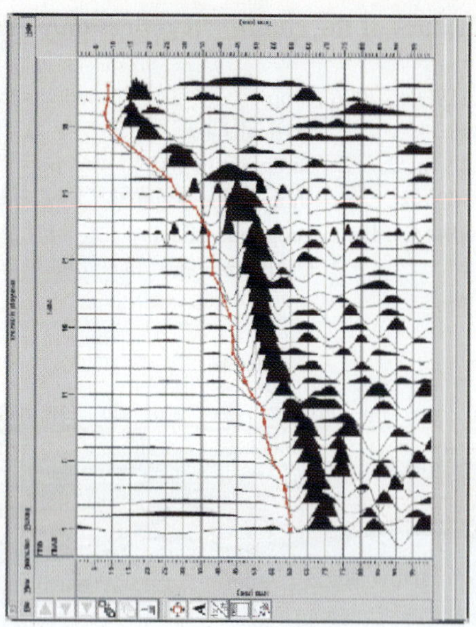

Fig. 4.6-24: Seismogram recorded in the borehole using a chain of 24 horizontal geophones - first breaks of SH-waves (red)

At a distance of 4 m from the wellhead on the Earths' surface, P-waves as well as vertically polarized shear waves (SV) were generated by vertical hammer blows on a metal base plate. Horizontally polarized shear waves (SH-waves) were excited by hammer blows onto the side of a special shear wave rake. Both the wavetrains of SH- and P-waves are recorded by a chain of 24 horizontal geophones placed in a 17 m deep borehole with a spacing of 0.5 m. The advantage of using a geophone chain is the very fast data acquisition because of simultaneous wave recording at 24 depth levels. Alternatively to the geophone chain a 3-component borehole seismic tool can be used *(Fig. 4.6-25)*. This tool is coupled against the casing by clamps. The borehole is cased with a 75 mm diameter plastic pipe. The casing is coupled to the borehole wall by "clay pellets" and filled with freshwater. An advantage of using a 3-component tool is that P- and S-waves can be distinguished not only by their traveltime but also by their polarization.

Fig. 4.6-25: VSP equipment, left: tripod with cable and registration car, right: geophone chain (black cable), 3-component borehole tool (metallic rod) and seismograph

Survey parameter values

seismograph	ABEM Mark VI	shotpoint offset	4 m
channels	48	vertical stacks	5 - 20
sampling rate	0.1 ms	geophone type	SG 10 horizontal
record length	410 ms	number of geophones	24
source type	sledgehammer	geophone interval	0.5 m
source direction	vertical and horizontal (subtraction of SH^+ and SH^--waves)		
	depth interval	0.5 - 16.5 m	

Specific elastic moduli are calculated by determining the velocity of direct P- and SH-waves (cf. Section 4.6.3.2) according to the following procedure:

- Pick the first breaks of P- and SH-waves (*Fig. 4.6-24*).
- Calculate vertical traveltimes.
- Display traveltime curves vs. depth.
- Compute P- and SH-wave velocities by regression analysis.
- Calculate Young's modulus E and shear modulus μ.

To determine the dynamic moduli, information about bulk density as a function of depth is required. These values can be determined by well logging (e.g., density log and neutron-neutron-log for water content) or by laboratory tests on rock samples. The following table shows the final result. The depth ranges for calculating the elastic modulus are selected according to the geological model as derived from drilling results.

Depth range [m]	v_p [m s^{-1}]	v_s [m s^{-1}]	ρ [t m^{-3}]	μ [MPa]	E [MPa]
0 - 1.5	(566)	(400)	2.10	(336)	(336)
2.0 - 4.5	274	157	2.00	49	124
5.0 - 15.5	2009	348	2.25	272	809
16.0 - 16.5			2.40	291	863

Values in brackets are probably erroneous because it was not possible to separate P- and SH-waves at the shallow depth ranges. At the depth range from 5.0 to 15.5 m the P-wave velocity is quite high due to the high bulk density of water saturated gravel, whereas the SH-wave velocity is not affected. That results in a rather high velocity ratio ($v_p/v_s > 5$). This value is, however, not unusual for unconsolidated sediments. At a depth range below 16 m no velocities can be determined. Hard rock was encountered by drilling at this depth and it has to be assumed that the v_p/v_s ratio would decrease to a value of ~ 2.

References and further reading

AKI, K.I. & RICHARDS, P.G. (1980): Quantitative seismology: Theory and methods, vol. **1**. Freeman, New York.

AL-CHALABI, M. (1994): Seismic velocities – a critique. First Break, **12**, 589-596.

AVSETH, P., MUKERIJ, T. & MAVKO, G. (2005): Quantitative seismic interpretation: Applying rock physics tools to reduce interpretation risk. Cambridge University Press.

BACHRACH, R., DVORKIN, J. & NUR, A. (1998): High resolution shallow-seismic experiments in sand. Part I and II. Geophysics, **63**, 1225-1240.

BAKER, G. S., STEEPLES D. W. & DRAKE, M. (1998): Muting the noise cone in near-surface reflection data: An example from southeastern Kansas. Geophysics, **63**, 1332-1338.

BAKER, G. S., STEEPLES D. W. & SCHMEISSNER, C. (2001): The effect of seasonal soil-moisture conditions on near-surface seismic reflection data quality. First Break, **20**, 1, 35-41.

BECK, A. E. (1981): Physical principles of exploration methods. Mac Millan, London.

BERCKHEMER, H. (1990): Grundlagen der Geophysik. Wiss. Buchgesellschaft, Darmstadt.

BERRYMAN, J. G. (1999): Tutorial - Origin of Gassmann´s equations. Geophysics, **64**, 1627-1629.

BRADFORD, J. H., SAWYER, D. S., ZELT, C. A. & OLDOW, J. S. (1998): Imaging a shallow aquifer in temperate glacial sediments using seismic reflection profiling with DMO processing. Geophysics, **63**, 1248-1256.

BROUWER, J. & HELBIG, K. (1998): Shallow high-resolution reflection seismics: Vol. **19** in: HELBIG, K. & TREITEL, S. (Eds.): Handbook of Geophysical Exploration, Seismic Exploration. Elsevier, Amsterdam.

BÜKER, F. (1998): Mapping the shallow subsurface using 2-D and 3-D reflection seismic techniques. Dissertation ETH Zürich, Switzerland.

BÜKER, F., GREEN, A. & HORSTMEYER, H. (1998a): Shallow 3-D seismic reflection surveying: Data acquisition and preliminary processing strategies. Geophysics, **63**, 1434-1450.

BÜKER, F., GREEN, A. & HORSTMEYER, H. (1998b): Shallow seismic reflection study of a glaciated valley. Geophysics, **63**, 1395-1407.

BUTTKUS, B. (2000): Spectral analysis and filter theory in applied geophysics. Springer, Berlin.

Consortium for Research in Elastic Wave Exploration Seismology: www.crewes.org

CORSMIT, J., VERSTEEG, W.H., BROUWER, J. H. & HELBIG, K. (1987): High resolution 3D reflection seismic on a tidal flat: Acquisition, processing and interpretation. First Break **6**, 9-23.

DALEY, T. & COX, D. (2001): Orbital vibrator seismic source for simultaneous P- and S-wave crosswell acquisition. Geophysics, **66**, 1471-1480.

DAVIES, K. J., BARKER, R. D. & KING, R. F. (1992): Application of shallow reflection technique in hydrogeology. The Quarterly Journal of Engineering Geology, **25**, 207-216.

DAVIES, K. J. & KING, R. F. (1992): The essentials of shallow reflection data processing. The Quarterly Journal of Engineering Geology, **25**, 191-206.

DIEBOLD, J. B. & STOFFA, P. L. (1981): The traveltime equation, tau-P-mapping and inversion of common midpoint data. Geophysics, **46**, 238-245.

DIX, C. H. (1955): Seismic velocities from surface measurements. Geophysics, **20**, 68-86.

DOLL, W. E., MILLER, R. D. & XIA, J. (1998): A non-invasive shallow seismic source comparison on the Oak Ridge Reservation, Tennesse. Geophysics, **63**, 1318-1331.

EDELMANN, H. A. K. (1981): Shover shear-wave generation by vibration orthogonal to the polarization. Geophysical Prospecting, **29**, 541-549.

EEGS (2005): Special Issue Seismic Surface Waves. Journal of Environmental & Engineering Geophysics, **10**, 67-234.

ETH Zuerich-website: www.aug.geophys.etzh.ch/teach.htm

FAGIN, S. (1998): Model-based depth imaging. Course note series, no. **10**, SEG Tulsa.

FERTIG, J. (2005): Geschwindigkeits- und Dichtewerte in Sedimenten. In: KNÖDEL, K. KRUMMEL, H. & LANGE, G.: Handbuch zur Erkundung des Untergrundes von Deponien und Altlasten, vol. **3**, Geophysik. Springer, Berlin.

FREI, W. (1995): Refined field static corrections in near-surface reflection profiling across rugged terrain. The Leading Edge, 259-262.

GALPERIN, E. I. (1985): Vertical seismic profiling and its exploration potential. D. Reidel, Dordrecht.

GEBRANDE, H. (1986): CMP-Refraktionsseismik. In: DRESEN, L., RÜTER, H. & BUDACH, W. (Eds.): Seismik auf neuen Wegen, 6th Mintrop-Seminar, Unikontakt, Ruhr-Universität Bochum, 191-206.

GHOSE, R., NIJHOF, V., BROUWER, J., MATSUBARA, Y., KAIDA, Y. & TAKAHASHI, T. (1998): Shallow to very shallow, high-resolution reflection seismic using portable vibrator system. Geophysics, **63**, 1295-1309.

GIBSON, R. L. JR. & PENG, C. (1994): Low- and high-frequency radiation from seismic sources in cased boreholes. Geophysics, **59**, 1780-1785.

GRAY, S. H., ETGEN, J., DELLINGER, J. & WHITMORE, D. (2001): Seismic migration problems and solutions. Geophysics, **66**, 1622-1640.

GREGORY, A. R. (1976): Fluid saturation effects on dynamic elastic properties of sedimentary rocks. Geophysics, **41**, 895-921.

GOUDSWAARD, J. C. M., TEN KROODE, A. P. E., SNIEDER, R. K. & VERDEL, A. R. (1998): Detection of lateral velocity contrasts by cross-well traveltime tomography. Geophysics, **63**, 523-533.

HAGEDOORN, I. G. (1959): The plus-minus method of interpreting seismic refraction sections. Geophysical Prospecting, **7**, 158-182.

HAEGEMAN, W. & VAN IMPE, W. F. (1999): Characterization of disposal sites from surface wave measurements. JEEG, **4/1**, 27-33.

HALES, F. W. (1958): An accurate graphical method for interpreting seismic refraction lines. Geophysical Prospecting, **6**, 285-294.

HELBIG, K. (1994): Foundations of Anisotropy for Exploration Seismics. In: Handbook of Geophysical Exploration. Seismic Exploration, vol. **22**.

HERING A., MISIEK R., GYULAI A., ORMOS T., DOBROKA, M. & DRESEN L. (1995): A joint inversion algorithm to process geoelectric and surface wave data. Geophysical Prospecting, **43**, 135-156.

HERBST, R., KAPP, I., KRUMMEL, H. & LÜCK, E. (1998): Seismic sources for shallow investigations: A field comparison from Northern Germany. Journal of Applied Geophysics, **38**, 301-317.

HILL, I. A. (1992a): Field techniques and instrumentation in shallow seismic reflection. The Quarterly Journal of Engineering Geology, **25**, 183-190.

HILL, I. A. (1992b): Better than drilling? Some shallow seismic reflection case histories. The Quarterly Journal of Engineering Geology, **25**, 239-248.

HUNTER, J. A., PULLAN, S. E., BURNS, R. A., GOOD, R. L., HARRIS, J. B. PUGIN, A. SKVORTSOV, A. & GORIAINOV (1998): Downhole seismic logging for high-resolution reflection surveying in unconsolidated overburden. Geophysics, **63**, 1371-1384.

JEFFERSON, R. D., STEEPLES, D. W., BLACK, R. A. & CARR, T. (1998): Effects of soil-moisture content on shallow seismic data. Geophysics, **63**, 1357-1362.

KÄHLER, S. & MEIßNER, R. (1983): Radiation and receiver pattern of shear and compressional waves as function of Poisson's ratio. Geophysical Prospecting, **31**, 421-435.

KEAREY, P. & BROOKS, M. (1991): An Introduction to Geophysical Exploration. Blackwell, Oxford.

KEISWETTER, D. A. & STEEPLES, D. (1995): A field investigation of source parameters for the sledgehammer. Geophysics, **60**, 1051-1057.

KING, R. F. (1992): High-resolution shallow seismology: history, principles and problems. The Quarterly Journal of Engineering Geology, **25**, 177-182.

KNAPP, R. W. & STEEPLES, D. W. (1986): High resolution common depth point seismic reflection profiling: Instrumentation / Field acquisition parameter design. Geophysics, **51**, 276-294.

KROHN, CH. E. (1984): Geophone ground coupling. Geophysics, **49**, 722-731.

LANZ, E., MAURER, H. & GREEN, A. (1998): Refraction tomography over a buried waste disposal site. Geophysics, **63**, 1414-1433.

LEHMANN, B. (2007): Seismic traveltime tomography for engineering and exploration application. EAGE Publication BV, Houten.

LAVERGNE, M. (1989): Seismic methods. Editions Technip, Paris.

LEPAROUX, D., BITRI, A. & GRANDJEAN, G. (2000): Underground cavity detection: A new method based on seismic Rayleigh waves. Europ. J. Environm. Engin. Geophys., **5**, 33-53.

LINDSEY, J. P. (1989): The Fresnel zone and it´s interpretive significance. The Leading Edge, **8**, 33-39.

LINDSEY, J. P. (1991): Seismic sources I have known. The Leading Edge, **10**, 47-48.

MCCANN, D. M., ANDREW, E. M. & MCCANN, C. (1985): Seismic sources for shallow reflection surveying. Geophysical Prospecting, **33**, 943-955.

MAVKO, G., MUKERJI, T. & DVORKIN, J. (1998): The rock physics handbook. Cambridge University Press.

MILLER, K. C., HARDER, S. H., ADAMS, D. C. & O´DONNEL, T. JR. (1998): Integrating high-resolution refraction data into near-surface seismic reflection data processing and interpretation. Geophysics, **63**, 1339-1347.

MILLIGAN, P. A., RECTOR III, J. A. & BAINER, R. W. (1997): Hydrophone VSP imaging at a shallow site. Geophysics, **62**, 842-852.

NOLET, G. (Ed.) (1987): Seismic tomography: with applications in global seismology and exploration geophysics. D. Reidel, Dordrecht.

PALMER, D. (1981): An introduction to the generalized reciprocal method of seismic refraction interpretation. Geophysics, **46**, 1508-1518.

PALMER, D. (1986): Refraction seismics. In: HELBIG & TREITEL (Eds.): Handbook of geophysical exploration. Seismic exploration, **13**, Geophysical Press, London-Amsterdam.

PALMER, D. (2001a): Imaging refractors with the convolution section. Geophysics, **66**, 1582-1589.

PALMER, D. (2001b): Resolving refractor ambiguities with amplitudes. Geophysics, **66**, 1590-1593.

PALMER, D. (2001c): A new direction for shallow refraction seismology: integrating amplitudes and traveltimes with the refraction convolution section. Geophysical Prospecting, **49** (Special Issue in Memory of J.G. HAGEDOORN), 657-673.

PARK, C. B., MILLER, R. D. & XIA, J. (1999): Multichannel analysis of surface waves. Geophysics, **64**, 800-806.

PASASA, L., WENZEL, F. & ZHAO, P. (1998): Prestack Kirchoff depth migration of shallow seismic data. Geophysics, **64**, 1241-1247.

PULLAN, S. E., MILLER, R. D., HUNTER, J. A. & STEEPLES, D. W. (1991): Shallow seismic reflection survey – CDP or "optimum offset". 61st Ann. Internat. Meeting., Soc. Expl. Geophys., expanded abstracts, 576-579.

REYNOLDS, J. M. (1997): An Introduction to Applied and Environmental Geophysics. John Wiley & Sons Ltd., Chichester.

ROBEIN, E. (2003): Velocities, time-imaging and depth-imaging in reflection seismics. Principles and methods. EAGE Publications BV, Houten.

ROCKWELL, D. W. (1967): A general wavefront method. In: MUSGRAVE A. W. (Ed.): Seismic refraction profiling. Soc. Explor. Geophys., Tulsa, 363-415.

SCHLUMBERGER Oilfield Glossary: www.glossary.oilfield.slb.com

SEG technical standards: www.seg.org/publications/tech-stand/

SHERIFF, R. E. (1980): Nomogram for Fresnel-zone calculation – short note. Geophysics, **45**, 968-972.

SHERIFF, R. E. & GELDART, L. P. (1982): Exploration seismology, **1**, history, theory and data acquisition. Cambridge University Press, Cambridge.

SHERIFF, R. E. (1991): Encyclopedic dictionary of exploration geophysics. Geophysical reference series, **1**. Society of Exploration Geophysicists, Tulsa.

SHTIVELMAN, V., FRIESLANDER, U., ZILBERMAN, E. & AMIT, R. (1998): Mapping shallow faults at the Evrona playa site using high-resolution reflection method. Geophysics, **63**, 1257-1264.

SHTIVELMAN, V. (2002): Surface wave sections as a tool for imaging subsurface inhomogeneities. Europ. J. Environm. Engin. Geophys., **7**, 121-138.

SHUEY, R.T. (1985): A simplification of the Zoeppritz equations. Geophysics, **50**, 609-614.

SIMON FRACER UNIVERSITY (SFU) website: www.sfu.ca/earth-sciences/courses

SJÖGREN, B. (2000): A brief study of applications of the generalized reciprocal method and of some limitations of the method. Geophysical Prospecting, **48**, 815-834.

SOCCO, L. V. & STROBBIA, C. (2004): Surface-wave method for near-surface characterization: a tutorial. Near Surface Geophysics, **2**, 165-186.

STEEPLES, D. W. & MILLER, R. D. (1988): Seismic reflection methods applied to engineering, environmental, and ground-water problems. Proceedings of the Symposium on the Application of Geophysics to Engineering and Environmental Problems (SAGEEP). March 28 – 31, Golden, Co., 409-461.

STEEPLES, D. W., GREEN, A. G., MCEVILLY, T. V., MILLER, R. D., DOLL, W. E. & RECTOR, J. W. (1997): A workshop examination of shallow seismic reflection surveying. The Leading Edge, 1641-1647.

STEEPLES, D. W. & MILLER, R. D. (1998): Avoiding pitfalls in shallow seismic reflection surveys. Geophysics, **63**, 1213-1224.

STEEPLES, D. W. (2005): Shallow seismic methods. In: RUBIN, Y. & HUBBARD, S. S. (Eds.): Hydrogeophysics. In: Water Science and Technology Library, vol. **50**, 215-251. Springer, Dordrecht.

STEWART, R. R. (1991): Exploration seismic tomography: fundamentals, course notes series, **3**. S.N. DOMENICO (Series ed.), Society of Exploration Geophysics, Tulsa.

STONE, D. G. (1994): Designing seismic surveys in two and three dimensions. Geophysical Reference Series **5**. SEG, Tulsa.

TANER, M. T. & KOEHLER, F. (1969): Velocity spectra – digital computer derivation and applications of velocity functions. Geophysics, **34**, 859-881.

TELFORD, W. M., GELDART, L. P., SHERIFF, R. E. & KEYS, D. A. (1990): Applied Geophysics. Cambridge University Press.

THORNBURGH, H. R. (1930): Wavefront diagram in seismic interpretation. Bull. Am. Ass. Petr. Geol., **14**, 185-200.

VAN DER VEEN, M. & GREEN, A. G. (1998): Land streamer for shallow seismic data acquisition: Evaluation of gimbal-mounted geophones. Geophysics, **63**, 1408-1413.

VAN DER VEEN, M., BUNESS, H. A., BÜKER, F. & GREEN, A. (2000): Field comparison of high-frequency seismic sources for imaging shallow (10 - 250 m) structures. JEEG, **5/2**, 39-56.

VAN OVERMEEREN, R. A. (2001): Hagedoorn´s plus-minus method: The beauty of simplicity. Geophysical Prospecting, **49**, 687-696.

WIDESS, M. B. (1973): How thin is a thin bed? Geophysics, **38**, 1176-1180.

WHITELEY, R. J., HUNTER, J. A., PULLAN, S. E. & NUTALAYA, P. (1998): "Optimum offset" seismic reflection mapping of shallow aquifers near Bangkok. Geophysics, **63**, 1385-1394.

WIEDERHOLD, H., BUNESS, H. A. & BRAM, K. (1998): Glacial structures in northern Germany revealed by a high-resolution reflection seismic survey. Geophysics, **64**, 1265-1272.

WONG, J. (2000): Crosshole seismic imaging for sulphide orebody delineation near Sudbury, Ontario, Canada. Geophysics, **65**, 1900-1907.

XIA, J., MILLER, R. D. & PARK, C. B. (1999): Estimation of near-surface shear-wave velocity by inversion of Rayleigh waves. Geophysics, **64**, 691-700.

XIA, J., MILLER, R. D., PARK, C. B., IVANOV, J., TIAN, G. & CHEN, C. (2004): Utilization of high-frequency Rayleigh waves in near-surface geophysics. The Leading Edge, 753-759.

YILMAZ, Ö. (2001): Seismic data analysis. In: COOPER, M. R. and DOHERTY, S. M. (Eds.): Seismic data analysis, 2 Volumes, SEG, Tulsa.

YOUNG, R. (2005): A lab manual of seismic reflection processing. EAGE publications.

4.7 Surface Nuclear Magnetic Resonance

GERHARD LANGE, UGUR YARAMANCI & REINHARD MEYER

4.7.1 Principle of the Method

Surface nuclear magnetic resonance[1] (SNMR) is a novel non-invasive geophysical tool with which the water content of an aquifer and other geohydrological parameters of an aquifer can be determined directly. In contrast to SNMR, other geophysical techniques, e.g., dc resistivity soundings (VES), electromagnetics, seismics, and ground penetrating radar only provide information on the lithological character of aquifers and aquicludes indirectly.

The SNMR method utilizes the nuclear properties of water. Because hydrogen nuclei possess a magnetic moment (spin), they are aligned with the Earth's magnetic field. It is possible to excite them with an external magnetic field and to measure the signal response resulting from precession of the protons after the external magnetic field is switched off. Since the amplitude of the SNMR signal is related to the number of excited hydrogen nuclei (protons), the technique can be used to estimate the water content of soil and rock. Information about other hydraulic properties, such as pore size and hydraulic conductivity can also be derived from this signal.

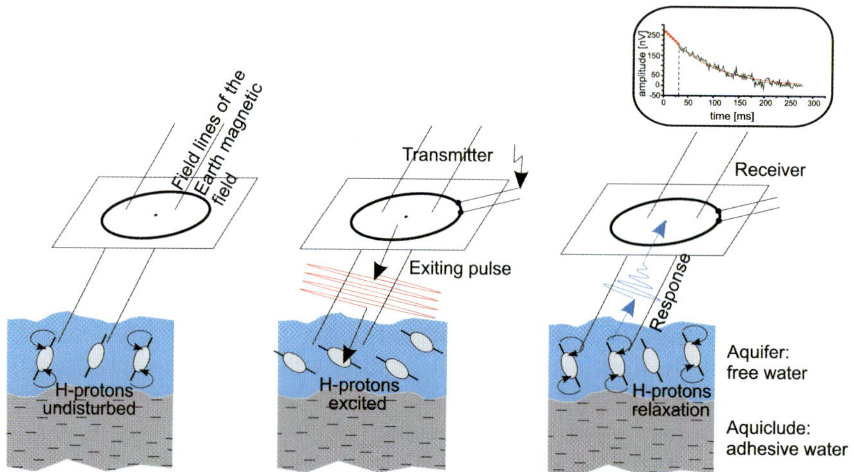

Fig. 4.7-1: Principle of the surface nuclear magnetic resonance method

[1] Some authors use the terms magnetic resonance sounding (MRS) or proton magnetic resonance (PMR) for this surface geophysical method.

The first high-precision studies of nuclear magnetic resonance signals from hydrogen nuclei were made by BLOCH and PURCELL in 1946. For this scientific achievement both scientists received the Physics Nobel Prize in 1952. Since then, this technique has found widespread application in chemistry, physics, medicine and borehole geophysics.

In contrast to the NMR methods applied in the laboratory, which use two strong orthogonal magnetic fields generated by the NMR instrument, SNMR uses the Earth's magnetic field and an external magnetic field generated in a large loop on the Earth's surface. In the 1960s, VARIAN and BARRINGER (1968) had the initial idea of SNMR, namely to use the principle of the proton-precession magnetometer (see Section 4.1.4) to develop a tool for prospecting for water. But not until the 1980s were effective instruments developed. They were initially used by Russian scientists (SEMENOV et al., 1988) to search for primary aquifers. Extensive testing of SNMR worldwide under different geological conditions began in the early 1990s.

4.7.2 Applications

- Direct detection of groundwater,
- distinguish between hydrogeological units (aquifers and aquicludes),
- determination of the water content of primary aquifers,
- estimation of hydraulic conductivity,
- determination of the amount of water in the vadose zone,
- estimation of water content and hydraulic properties of karstic and fractured aquifers (research activities).

4.7.3 Fundamentals

Hydrogen nuclei of water molecules have a magnetic moment μ. They can be described in terms of a spinning charged particle. Generally, μ is aligned parallel to the local magnetic field B_0 of the Earth. When another magnetic field – the excitation field – is applied, the axis of the proton is deflected, owing to the torque applied to the spinning protons of the nuclei (*Fig. 4.7-1*). When the excitation field is removed, the protons generate a magnetic field as they become realigned along B_0 (relaxation) while precessing around B_0 with the angular frequency

$$\omega_L = \gamma B_0 = 2\pi f_L \tag{4.7.1}$$

where γ = 0.267 518 Hz nT^{-1} is the gyromagnetic ratio and f_L is the Larmor frequency for hydrogen protons.

Field measurements are made using a large circular or square loop on the surface of the Earth. To excite the protons in the ground, an alternating current

$$i(t) = i_0 \cos(\omega_L t) \tag{4.7.2}$$

with the Larmor frequency f_L (between 800 and 3000 Hz) is passed through this loop for a limited time τ, attaining a pulse moment (excitation intensity) $q = i_0\ \tau$. To satisfy the resonance condition, the frequency of the excitation pulse has to be very close to the local Larmor frequency.

When the current in the transmitter loop is switched off, a voltage $e(t)$ is induced in the loop, now acting as receiver loop, by the relaxation of the protons (SEMENOV et al., 1982; SCHIROV et al., 1991):

$$e(t) = E_0 \exp\left(-t/T_2^*\right) \cos\left(\omega_L t + \varphi_0\right). \tag{4.7.3}$$

The excitation pulse shown on the left side of the graph in *Fig. 4.7-2* has a carrier frequency equal to f_L. The relaxation signal is recorded after the excitation pulse is switched off and a dead time of 30 ms has lapsed.

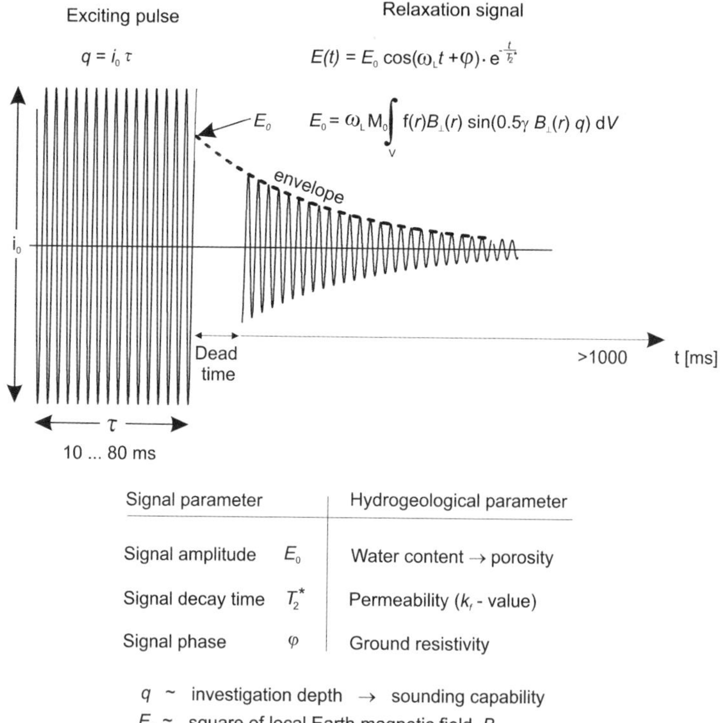

Fig. 4.7-2: Record of the transmission of the excitation pulse and the relaxation signal in an SNMR measurement and the hydrogeologic information provided by the signal parameters

The exponential decay of the SNMR signal reflects the relaxation of the hydrogen protons in the groundwater. The relationship between the signal parameters and hydrogeological parameters is given in the lower part of the figure.

The initial amplitude E_0 of the signal is related to the water content and can range between a few nanovolts to millivolts:

$$E_0 = \omega_L M_0 \int_V f(r) B_\perp(r) \sin\left(0.5\gamma B_\perp(r) q\right) dV , \tag{4.7.4}$$

where $M_0 = 3.29 \cdot 10^{-3} B_0$ [J T^{-1} m^{-3})] is the nuclear magnetization (magnetic moment of the unit volume dV of water), $f(r)$ is the volume fraction of water in a unit volume dV at the source point r. $B_\perp(r)$ is the component of the excitation field perpendicular to the static magnetic field B_0. In a conductive medium, $B_\perp(r)$ consists of the primary field of the loop and the induced field. Note that the argument of the sine function, $0.5\gamma B_\perp(r) q = \Theta$, in Equation (4.7.4) is the angle of deflection of the magnetic moment of the protons from B_0.

The decay time T_2^* – the relaxation time – can be of the order of a few milliseconds for water saturated fine-grained material up to > 1000 ms for karst. Since relaxation takes place more rapidly at grain surfaces, the relaxation time is related to the mean pore size and, therefore, grain size of the material. Empirical relationships show that clayey material, including sandy clay, usually has decay times of less than 30 ms, whereas that of sand is of the order of 60 - 300 ms, gravel 300 - 600 ms and pure water about 1000 ms (*Fig. 4.7-3*). Several authors (SCHIROV et al., 1991; LIEBLICH et al., 1994; YARAMANCI et al., 1999) have shown that SNMR measurements can detect only free water in the pores. Adhesive water on the pore surfaces has very short relaxation times and cannot be detected due to the inevitable dead time (*Fig. 4.7-2*).

The phase φ_0 of the relaxation signal is related to the phase of the excitation signal in Equation (4.7.3). If the conductivity of the ground is negligible, $B_\perp(r)$ is in phase with the excitation current ($\varphi_0 = 0$). If the ground is highly conductive, a secondary magnetic field, superimposed on the primary field is induced. This modifies the amplitude and phase of the field $B_\perp(r)$. Therefore, φ_0 is an indicator of the conductivity of the ground. The influence of conductivity on the signal amplitude reduces the investigation depth, as is the case for all electromagnetic methods. A highly conductive layer above the aquifer decreases the amplitude of the SNMR signal, the same layer below the aquifer increases it.

The resolution and precision of the method depend on the $B_\perp(r)$ gradient and, therefore, decrease with depth. A strong current i_o and/or long τ is needed to excite the protons at greater depths (as long as $\tau << T_2$*). Increasing q increases the depth of investigation. Thus, the method has sounding capabilities similar to those of the electromagnetic and dc resistivity methods.

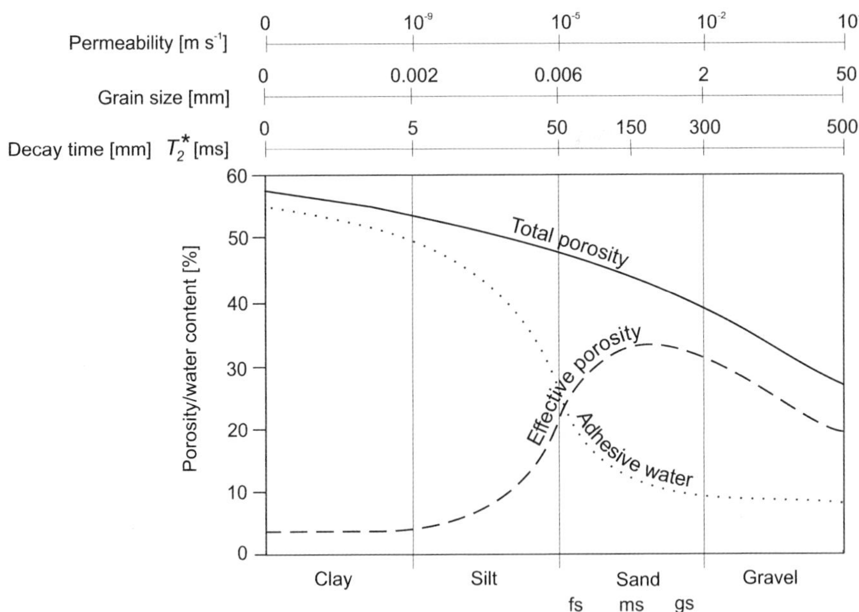

Fig. 4.7-3: Top: Relationship of grain size to SNMR decay time, after SCHIROV et al. (1991) and permeability, after HÖLTING (1992); bottom: Relationship between porosity and adhesive water content of unconsolidated sediments, after DAVIS & DE WIEST (1966)

Important for the use of SNMR in all parts of the world is the fact that the signal amplitude E_0 is approximately proportional to the square of the total intensity of the Earth's magnetic field B_0. The total intensity ranges from > 60 000 nT in the polar regions to < 25 000 nT in South America (*Fig. 4.7-4*). Thus, the precession signal amplitudes for the same hydrogeological conditions in different regions can differ considerably, as shown in *Fig. 4.7-4* by a comparison of model sounding curves for Germany and Namibia assuming the same aquifer and the same loop configuration.

The influence of loop configuration and loop area on the signal amplitude in SNMR soundings at a test location is shown in *Table 4.7-1.*

Table 4.7-1: Maximum signal amplitudes for SNMR soundings with different loop configurations, measured at a test location in Germany. Parameters: total magnetic intensity: 48 750 nT; field inclination: 62°, depth of primary aquifer: 20 - 80 m; effective porosity: 30 %.

Loop configuration	**Loop area**	**Max. signal amplitude E_0**
circular loop, d = 150 m, 1 turn	17 670 m^2	2890 nV
circular loop, d = 100 m, 1 turn	7855 m^2	1320 nV
circular loop, d = 75 m, 2 turns	8835 m^2	1310 nV
circular loop, d = 50 m, 2 turns	3927 m^2	540 nV
figure-8, loop, d = 50 m, 1 turn	3927 m^2	320 nV

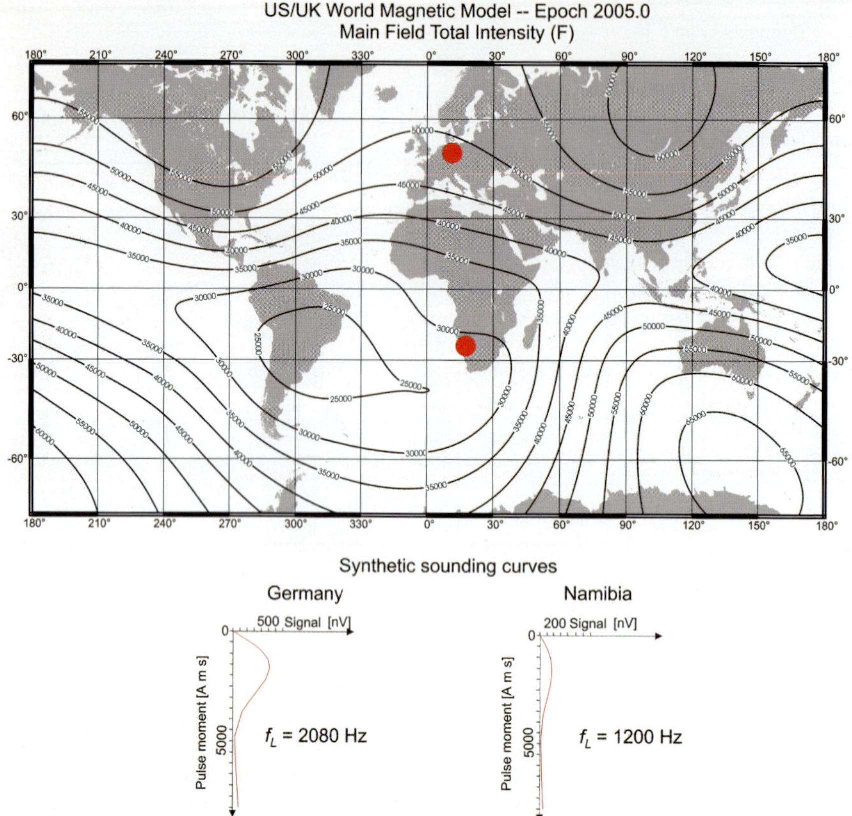

Fig. 4.7-4: The total intensity of the Earth's magnetic field and consequences for SNMR soundings at different geographical locations. Aquifer: depth 15 - 30 m, effective porosity 10 %. Loop geometry: circular, 150 m in diameter. (Magnetic data from http://www.ngdc.noaa.gov/seg/WMM/image.shtml)

4.7.4 Instruments

Worldwide there are currently only three manufacturers of single- and two manufacturers of multichannel SNMR measuring systems: The single-channel systems are known as "AQUATOM" and "HYDROSCOPE" (Russian), and "NUMIS" (French), while the multichannel systems are referred to as NMR MIDI (German) and VistaClara MRS (American). In principle, all these instruments consist of a transmitter and a receiver (Tx/Rx), a dc/dc converter, and transmitting/receiving loops. All systems are computer controlled. Car batteries provide the energy to generate high-energy pulses of up to 450 A/4000 V.

The most widely used commercially available system is the NUMIS^{Plus} made by IRIS Instruments which has the following main features:

- modular design, thus portable, allowing measurements in rough terrain (e.g., desert dunes and mountains),
- two dc/dc-converter units to energize the transmitting loop: one unit for depths down to 100 m, two units to extend the depth range down to 150 m,
- a Tx/Rx unit, to generate the excitation pulses and record the resulting signal,
- a computer (laptop), to receive the raw data, and process, and display the data for quality control, and store the values for inversion and interpretation, and
- special tuning units for measurements at low magnetic field strength (e.g., in southern Africa and South America).

Table 4.7-2: NUMIS^{Plus}: technical specifications

Units	Specifications
dc/dc–converter	
power supply	12 V, 60 Ah (2/4 car batteries)
working capacity	6 - 8 h on average
output	± 380 V, 0.5 A (1 dc/dc converter) ± 430 V, 1 A (2 dc/dc converters)
transmitter (Tx)	
frequency range	800 - 3000 Hz
maximum output	4000 V, 450 A
pulse duration	20 - 80 ms
pulse moment	100 - 30 000 A ms
receiver and tuning units (Rx)	
A/D –converter	14 bit
sampling frequency	4times the local Larmor frequency f_L
calibration	with a phase reference
band-pass filter width	100 Hz
recording interval	10 - 1000 ms
digital stacking capabilities	noise and signal stacking
transmitter/receiver loop	
4 to 6 segments on cable reels	75 - 150 m length each, 6 - 24 mm² cross-section

4.7.5 Survey Practice

Modeling before surveying

Before starting an SNMR survey in an unknown area, model sounding curves should be calculated in order to assess the feasibility and usefulness of the measurements. For this purpose, total intensity and inclination of the Earth's magnetic field in the survey area, as well as hydrogeological information (expected depth of the water table, thickness and porosities of the aquifers and aquicludes) are required. If the location is characterized by highly conductive rocks, information about electrical conductivity is important to know. Forward modeling with these data provides the magnitudes of the signal response that may be expected. This is important for the interpretation of the measured SNMR signals (*Fig. 4.7-5*).

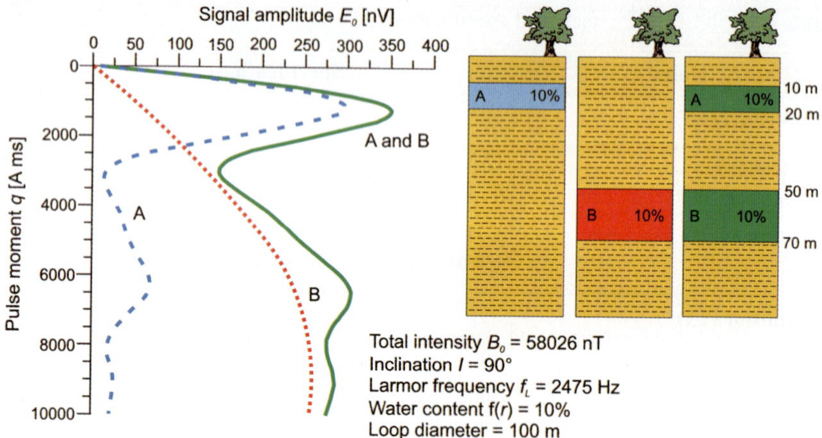

Fig. 4.7-5: SNMR sounding curves for simple hydrogeological situations. Rule of thumb: the broader the maximum, the thicker the aquifer and/or its depth

Loop configuration

The survey is started with the installation of the transmitter/receiver loop. A loop consists of 4 to 6 cable sections, each 75 -150 m in length. Loop shape and dimension affect the depth of investigation and the complexity of the loop configuration. The maximum area for a given length of cable is obtained by a circle, but if the site is covered by trees or bushes it may be easier to lay out a square loop.

In order to overcome the problem of man-made electromagnetic noise from powerlines or other sources, a figure-8 loop can be used as a noise-attenuating antenna (TRUSHKIN et al., 1994). In this case, two circular or square loops are connected in series, but with opposite polarity. (i) The axis of the figure-8 has to be aligned parallel to a linear noise source, or (ii) the normal to the axis of the figure-8 to a point source must pass through the center of the figure-8. The noise in each of the loops induces equal voltages of opposite polarity which cancel each other while the signal induced in each loop will add to the other. These two effects will increase the signal-to-noise ratio by a factor of as much as 5 to 20. The disadvantage of a figure-8 layout is that the loop area is smaller, compared with a full circle or square. As investigation depth is directly proportional to the loop diameter, a reduced depth of investigation is thus attained. A rule of thumb for estimating the SNMR investigation depth is: Maximum investigation depth is < loop diameter (see *Fig. 4.7-9*).

Table 4.7-3: SNMR loop configurations and properties

Configuration	Signal	S-N-ratio	Investigation depth	Resolution
circular loop	1	3	1	1
square loop	2	3	2	2
figure-8	3	1	3	3

range: 1 – best; 2 – medium; 3 – poor

Data acquisition

Data acquisition is computer controlled. The selection of parameters in the set-up is decisive for data quality. To satisfy the resonance condition it is necessary to transmit an excitation pulse whose frequency is equal to the Larmor frequency f_L, corresponding to the local Earth's magnetic field (*Fig. 4.7-2*). The excitation pulse frequency should not deviate more than ± 1 Hz from the calculated f_L. A larger difference may cause distortion of the decay curve, which in turn causes erroneous E_o and T_2* values. To obtain preliminary information about ω_L, the magnetic field B_0 is measured with a proton magnetometer. The frequency f_L [Hz] is determined from the magnetic field B_0 [nT] as follows: $f_L = 0.042\,58\ B_0$. Because of inhomogeneities in the local magnetic field this value often needs to be adjusted to an effective frequency, which is displayed on the computer screen during acquisition. If there is a deviation of more than 1 Hz, especially for the peaks in the sounding curve, the sounding should be repeated with a new frequency value. To avoid a deviation of more than 1 Hz, a test sounding with a small number of pulse moments (5 - 6) should be carried out before the sounding is started, to determine the correct frequency.

The investigation depth of a sounding is increased by increasing the pulse moment q. The pulse moment q is the product of current intensity i_o and pulse duration τ, which can be between 10 - 80 ms (*Fig. 4.7-2*). A pulse moment of 40 ms has proven to be useful for normal surveys. Between 4 and 40 q-values can normally be selected, but 16 is an optimum number. Values of q between

30 A ms and 30 000 A ms are automatically selected by the equipment software and can be modified, added or removed interactively.

SNMR is very sensitive to external electromagnetic noise. Because the signals are generally weak - the signal amplitude is frequently several orders of magnitude lower than the ambient noise - all measuring systems have stacking capabilities to increase the signal-to-noise ratio. This is very important for soundings near industrial locations, but also in rural environments, where motors, generators and powerlines cause noise problems. Besides technical sources of noise, natural sources may also cause noise corruption, for example, distant thunderstorms, lightning, or current channeling in the ground (LEGCHENKO et al., 1997).

The number of stacks for the selected pulse moment depends on the signal-to-noise ratio. Because the stacking number fundamentally influences the duration of a sounding, it has to be selected as a compromise between quality and duration. There is no absolute rule, but *Table 4.7-4* gives recommendations for this parameter value.

SNMR systems allow the signal decay to be recorded up to about 1000 ms. This corresponds to decay times for water in, for example lakes and rivers (100 % water saturation). Since both increasing of the recording interval and the stacking number influence the acquisition time, the recording interval should be in the range of maximum decay times typical for aquifer rock at the sounding location. A guide level can be taken from *Figure 4.7-3*. As a rule of thumb, use about 250 ms for primary aquifers and >500 ms for karst aquifers. When doing soundings at a new site, start with a relatively long time interval and adjust as the survey proceeds.

A geoelectrical sounding should be carried out to obtain information about the subsurface structure and conductivity in order to improve the SNMR inversion. Experience has shown that when a dc vertical electrical sounding (VES) is conducted simultaneously with an SNMR sounding, the two methods do not influence each other. Therefore, no additional time is needed and the electrical sounding can be set up and carried out while the SNMR sounding runs automatically.

Table 4.7-4: Recommended number of stacks (IRIS Instruments)

Ambient noise	Signal amplitude [nV]		
[nV]	30	100	300
200	64	32	16
500	128	64	32
1000	256	128	64

4.7.6 Processing and Interpretation of the Measured Data

The aim of the inversion is to transform a sounding curve – a data set of signal amplitude (E_0) versus pulse moment (q) – into a hydrogeological section showing water content as a function of depth for a horizontally layered model (*Figs. 4.7-5* and *4.7-6*).

The inversion algorithms currently available for routine application are one-dimensional (for horizontally layered primary aquifers). In these algorithms, the subsurface is modeled as a half-space subdivided into thin layers, all with the same thickness. The signal resulting from a given pulse intensity is calculated as the sum of the signals from all of the layers in the half-space. This is done for several pulse moments. The contribution of each layer to the signal for each pulse moment is entered in a matrix ("contribution matrix") for the given loop configuration, taking into account the total intensity and inclination of the Earth's magnetic field, and the electrical conductivity of the ground. The water content of each layer is adjusted to obtain the least-squares best fit of the model to the observed data. As the contribution matrix can be calculated in advance, the inversion is not very time consuming and, therefore, suitable for on site interpretation of the data.

Other inversion algorithms use a random search scheme (Monte-Carlo type) and allow smooth or block inversion. Incorporation of a-priori information (e.g., layer boundaries obtained by other methods) and SNMR decay times is also possible (MOHNKE & YARAMANCI, 2002). Aquifer characterization can be improved by joint inversion of SNMR and VES data. Because these two methods are based on different physical principles, the information they provide supplements the information from the other, making the structural and hydrogeological interpretation more reliable (HERTRICH & YARAMANCI, 2002). Several processing steps have to be taken before the inversion is started:

- If the asymptotic part of the decay curve is too noisy, it will be necessary to use only a portion of it instead of the entire recorded interval. The recorded signal length should be close to the maximum T_2*.
- Depending on the noise level, the raw data must be filtered (moving average) to smooth the relaxation curves using a constant time window between 5 to 40 ms.
- Matrix regularization: the matrix inversion of noisy data can be unstable and yield a non-smooth result. To avoid this, a matrix regularization can be carried out that minimizes the difference in water content between adjacent layers.

These processing steps can significantly affect the inversion results (YARAMANCI et al., 1998). For a detailed interpretation, the electrical conductivity must be taken into account. This parameter can have a major influence on the SNMR signal. Experience and forward modeling show that a "normal" electrical conductivity has practically no influence. But a layer with a resistivity ρ_s <100 Ωm can attenuate the signal, causing errors in the depth estimate, and water content values for aquifers below this layer will be too low.

The results of the inversion are plotted showing the water content as a function of depth; the decay times are shown using color or grey scale (*Fig. 4.7-6*). Owing to the decreasing sensitivity to the water content with depth, there is a better resolution for shallow layers than for deeper ones (*Fig. 4.7-9*). This is taken into account by increasing the thickness of the layers with depth on semi-logarithmic scale. When this is done, there is no significant difference between the resolution of the SNMR and geoelectric methods. In the case of SNMR, the product of water content and layer thickness is then constant. This means that the total quantity of water is determined very well, which is the most important information needed for determining whether an area is of sufficient hydrogeological interest for test-drilling.

From several field tests under different geological conditions it is known that the decay time T_2* not only depends on the pore geometry, but also on the magnetic properties of the rocks. Rocks with high magnetic susceptibilities may cause inhomogeneities in the local magnetic field. In such cases the above-discussed empirical relationship between pore size and relaxation time cannot be used. For a given geological situation, the SNMR results should be compared with information from a borehole (LEGCHENKO & BEAUCE, 1999).

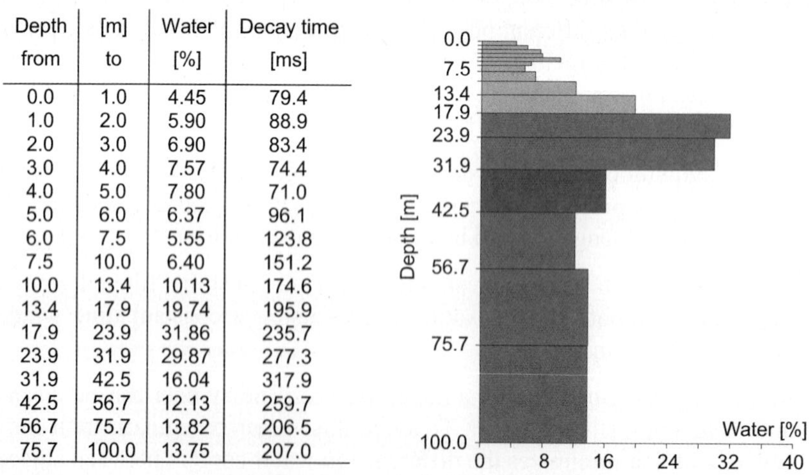

Depth from	[m] to	Water [%]	Decay time [ms]
0.0	1.0	4.45	79.4
1.0	2.0	5.90	88.9
2.0	3.0	6.90	83.4
3.0	4.0	7.57	74.4
4.0	5.0	7.80	71.0
5.0	6.0	6.37	96.1
6.0	7.5	5.55	123.8
7.5	10.0	6.40	151.2
10.0	13.4	10.13	174.6
13.4	17.9	19.74	195.9
17.9	23.9	31.86	235.7
23.9	31.9	29.87	277.3
31.9	42.5	16.04	317.9
42.5	56.7	12.13	259.7
56.7	75.7	13.82	206.5
75.7	100.0	13.75	207.0

Fig. 4.7-6: Example of a smooth matrix inversion using the NUMIS software

4.7.7 Quality Assurance

- Before SNMR soundings are started in an unknown area, the total intensity of the Earth's magnetic field should be determined at several points or profiles in the survey area. This provides information about inhomogeneities in the local magnetic field.
- Soundings in the vicinity of powerlines, cables, and magnetic objects, e.g., drilling rigs and boreholes with metallic casing, should be avoided.
- The loop must be laid out accurately.
- If a figure-8 loop is used, its axis has to be aligned parallel to any linear source of noise.
- Noise sources inside the transport vehicle must be taken into consideration. Air conditioning systems and even the external power supply of the computer may generate considerable electromagnetic noise.
- The data acquisition should be monitored on the screen. If there is too much noise, the stacking number has to be increased.
- If the selected pulse frequency deviates more than ± 1 Hz from the "effective frequency" or if there are outliers then the pulses within a given range should be repeated.
- Combined application with other geophysical methods is recommended to improve the SNMR inversion and interpretation. In any case, a geoelectrical sounding (VES) should be carried out to determine the electrical conductivity of the ground.

4.7.8 Personnel, Equipment, Time Needed

	Personnel	Equipment	Time needed
mobilization and demobilization	depends on the distance to the survey area		
topographic survey	s. Chapter 2.5		
measurements	1 geophysical technician (1 – 3 assistants)	1 SNMR system, 4 - 6 cable reels (each 75 - 150 m), 2 - 4 car batteries (12 V, > 48 Ah), 1 laptop, 1 proton magnetometer, measuring tapes, surveying equipment 1 or 2 4WD-vehicles with sufficient space for the equipment (approx. 1.5 m × 2 m), geoelectrical sounding equipment	Depends on the noise conditions, which determines the stacking number n and recording time, as well as the accessibility of the location: n: 16 - 64, 4 - 5 soundings/d*; n: 64 - 128, 3 soundings/d; n: 200 - 400, 1 - 2 soundings/d. A dc vertical electrical sounding (VES) can be carried out during the computer-controlled acquisition of the SNMR data.
processing, interpretation, reporting	1 geophysicist	1 PC with plotter, printer, and software	1 - 2 days are necessary for each day in the field

d* 10-hour workday

4.7.9 Examples

Haldensleben/Germany test site

The first SNMR measurements in Germany were conducted in 1997 at the Haldensleben site to check the suitability of the method in populated areas (YARAMANCI et al., 1999). The results from vertical electrical sounding (VES), well logging and core analysis were available to determine the geoelectrical layering, porosity, hydraulic conductivity, and free and adhesive water content. The lithological log of the reference borehole shows a sandy sequence down to a depth of 46 m (vadose zone and upper aquifer), underlain by glacial till, and a second aquifer beginning at a depth of 60 m. Average porosity of the sandy material of the upper aquifer as derived from the impulse neutron gamma log (ING.W) is about 35 %. The hydraulic conductivity values between 10^{-4} and 10^{-3} m s^{-1} obtained from borehole data are similar to those estimated from the SNMR decay times. In the VES results, the vadose zone appears as an extremely high resistivity layer, although the total water content is around 15 % (*Fig. 4.7-7 bottom, left*).

Comparison of SNMR data with borehole data provides clear evidence that the water content determined by SNMR represents the free (or mobile) water content. Taking into account the 5 - 10 % adhesive water derived from the grain-size distribution, the SNMR result is in good agreement with the ING.W log, which gives the total water content (in the case of saturation this corresponds to total porosity – see *Fig. 4.7-3*).

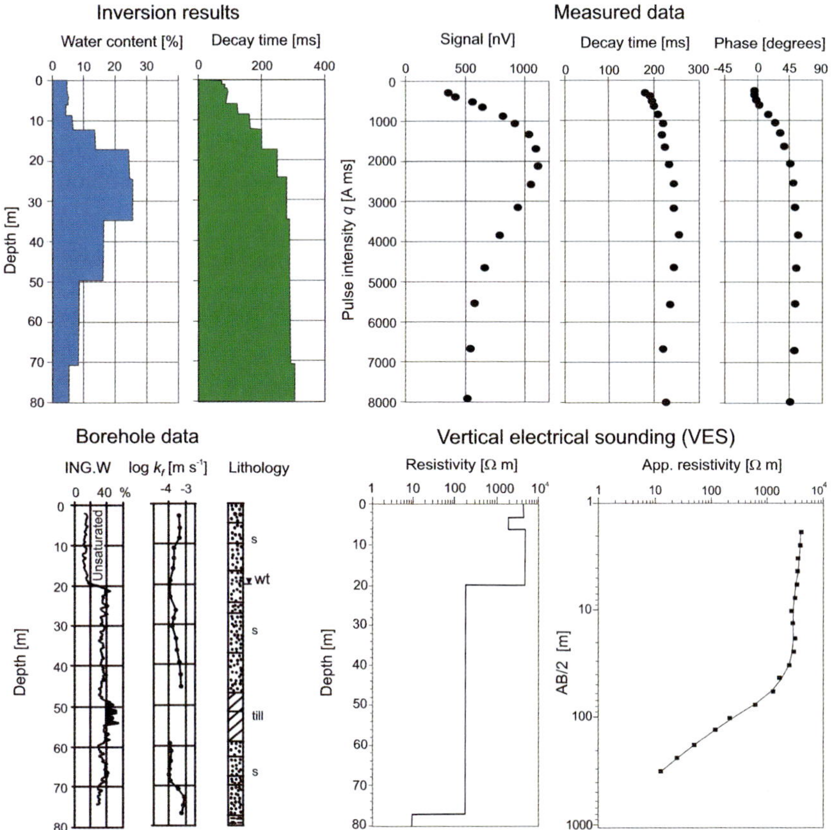

Fig. 4.7-7: Comparison of SNMR results with borehole and VES data from the Haldensleben test site

Survey parameter values:

total intensity	48 757 nT; magnetic field inclination: 67°
loop-parameters	circular, 1 turn, diam. = 100 m, cable section: 24 mm^2
pulse frequency	2076 Hz; pulse length: 40 ms
pulse number	16; stacking rate: 24 (low noise)
acquisition time	2 h.

Haldensleben – deep aquifer below a thick vadose zone

The results of the first soundings at the Haldensleben site already showed the potential of SNMR to detect free water in the unsaturated (vadose) zone. However, there were still doubts whether the determined water contents were artefacts of the measuring and/or the inversion process.

To detect water in the vadose zone excluding artefacts caused by "smearing" effects due to measuring and inversion, seven soundings with different loop-types (*Fig. 4.7-9*) were measured at a borehole drilled into an unconfined aquifer overlain by a 48 m thick vadose zone. The objectives of these soundings were:

- to determine whether water in the vadose zone could be detected, and
- to investigate the depth of investigation and resolution problem of SNMR.

Of the seven soundings, all of which were of high quality, one was selected for comparative studies. In *Fig. 4.7-8* the interpretation based on the inversion of SNMR data is compared to the lithological borehole description and well logging data (gamma-ray, short- and long-normal resistivity logs). Within the vadose zone, at the depth range of approx. 10 - 25 m, the presence of thin layers with a water content of about 3 %, were proven with all soundings. These water accumulations could be explained as perched water held up by a low permeability layer preventing water to percolate downward into the underlying unsaturated zone. The presence of two low permeability or cofining layers consisting of fine-grained (silty, marly) material in the sandy vadose zone is indicated by the increased values in the gamma-log and lower values in the resistivity-logs at corresponding depths. Between the depth of 25 m and the water level at 48.6 m the SNMR result shows no clear indication of the presence of water. A more detailed forward modeling approach, however, confirmed the presence of on average a 2 % water content in the upper 25 m. This low water content is reflected in the ascending branch (low pulse moments) of all seven sounding curves as a deviation from the "waterless" curve.

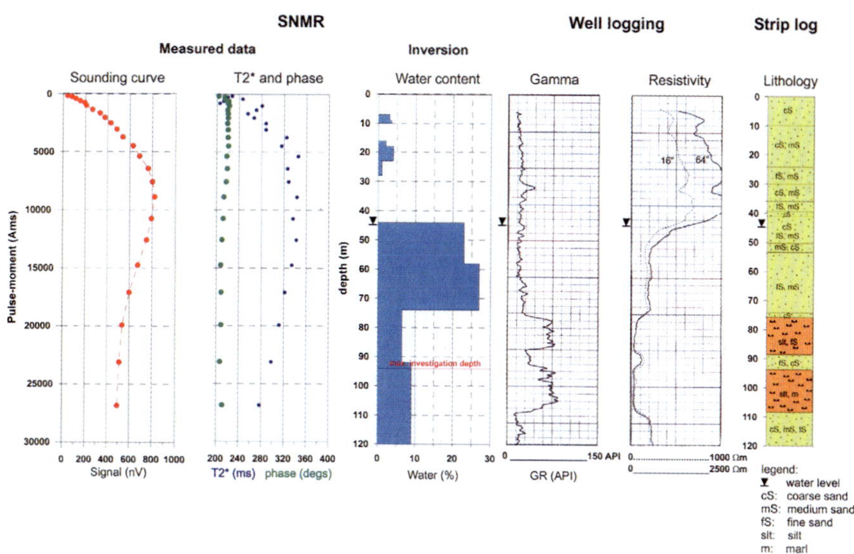

Fig. 4.7-8: Measured data and inversion result of the SNMR sounding compared to strip log lithology and well logs in the nearby borehole

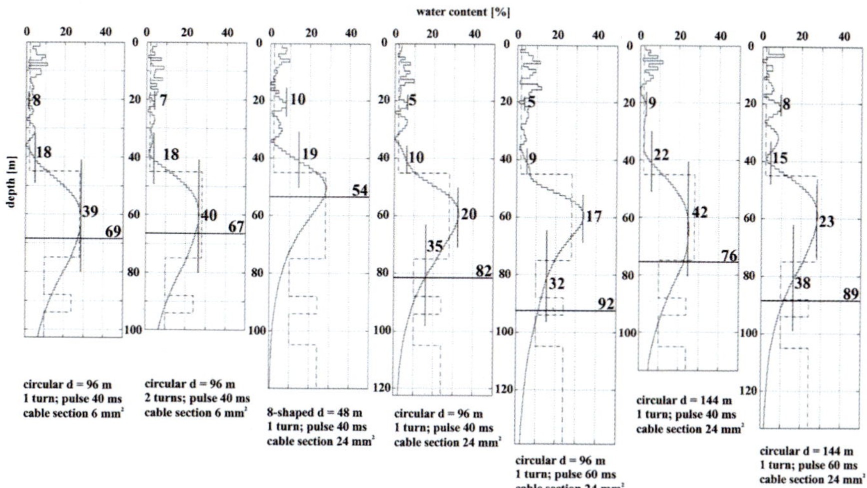

69 investigation depth [m]; | 39 vertical resolution [m]

-------- water content as derived from borehole data

Fig. 4.7-9: Investigation depth and vertical resolution as derived from loop-configuration tests at Haldensleben/deep aquifer. The values are determined by singular value decomposition (SVD) of the data resolution matrix of the measured data. The resistivity model is derived from well logging data in *Fig. 4.7-8,* MÜLLER-PETKE et al. (2006)

Although the lithologic profile (*Fig. 4.7-8, right*) does not show any significant changes within the sandy aquifer material, there is a transition zone below the water table showing a stepwise change in the SNMR water content. Between the depths of 45 and 58 m the interpreted water content is 23 % and below that it is 27 %. Normally, such differences in adjacent layers may result from the smooth inversion in the context of the equivalence principle. However, the increase to a 27 % water content corresponds to a gradual decrease in the short- and long-normal resistivity values. To understand this behavior more detailed investigations are required.

The transition at 74 m depth to the aquiclude consisting of silt and marl, can be recognized as a drastic reduction in the SNMR-water content from 27 % to 7 %. Because of the reduced resolution power with depth (*Fig. 4.7-9*), the thin coarse sand interlayer (cS) at 88 to 94 m within the aquiclude cannot be resolved. Detailed modeling with singular value decomposition (SVD) of the data resolution matrix confirmed the maximum depth of investigation for the particular loop configuration used (circular, $d = 96$ m, max. pulse moment 26 850 A ms) to be 92 m and a layer resolution of $\approx$ 35 m at that depth interval (MÜLLER-PETKE et al., 2006).

Survey parameter values:

total intensity	49 033 nT; magnetic field inclination: 67°
loop-parameters	circular; 1 turn, diam. 96 m; cable section 24 mm^2
pulse frequency	2087.8 Hz; pulse length: 60 ms
pulse number	24; stacking rate: 36
acquisition time	2.5 h.

Omdel/Namibia - groundwater below saline soil

The VES data indicates a very low resistivity layer down to about 27 m at the Omdel site which could be interpreted as representing an aquifer (*Fig. 4.7-10*). However, this seems to be in contradiction to the SNMR results, which show free water only below this depth. The low resistivity at shallow depth is caused by saline soil. High soil salinity is due to salts dissolved from the sediment matrix by percolating water and absorbed on the grain surfaces of sandy material. This absorbed water is not visible in the SNMR signal. A higher layer resistivity below 27 m depth indicates the presence of freshwater. This is unambiguously recognized by SNMR, even though the water content is unusually low. The SNMR result is verified by information from a nearby well, where the water table was measured as 29.2 m depth below surface.

The measured decay time-depth profile at this site shows some unexpected features. Despite the absence of free water, high decay times are measured at shallow depth. These are believed to be an indication of artifacts and needs more research in order to understand the cause. Decay times of around 80 ms for the aquifer, corresponding to fine-grained sand, indicate a relatively high proportion of adsorbed water and a low proportion of free water. This explains the low water content derived from the signal amplitudes of the sounding curve.

At this location the groundwater of the Omdel aquifer is artificially recharged. This example shows that using geoelectrical data (VES) only, an incorrect assessment of the aquifer distribution is made. By incorporating SMNR soundings the regional extent of the aquifer can be determined and a reliable estimate of the water content of the aquifer can be made.

Survey parameter values:

total intensity	29 222 nT; magnetic field inclination: -63°
loop-parameters	figure-8, 1 turn, diam. 75 m; cable section 24 mm^2
pulse frequency	1244.3 Hz; pulse length: 60 ms
pulse number	16; stacking rate: 300 (high noise)
acquisition time	9 h.

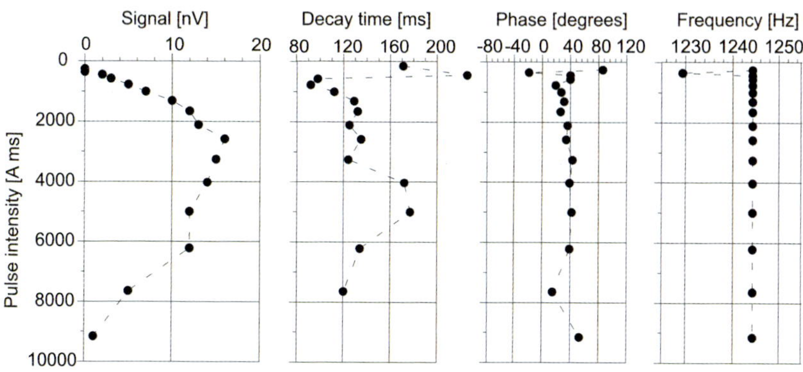

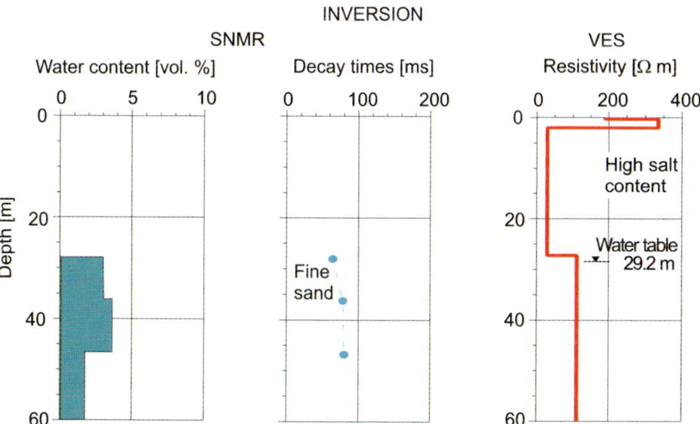

Fig. 4.7-10: Comparison of SNMR sounding data and VES data from the Omdel site in Namibia.

South African geohydrology and challenges to the SNMR technique

South African hydrogeology is dominated by fractured rock aquifers which are present over approximately 80 % of the surface area of the country. Important primary aquifers are found along the coastal areas while others are associated with ephemeral river systems. Because of the dominance of fractured rock aquifers, the main objective of the South African SNMR experimental programme is to test, evaluate and further develop the method under different fractured rock aquifer conditions.

Three challenges confront SNMR experiments in Southern Africa:

- A low Earth's Magnetic field intensity. SNMR measurements in regions of low Earth's magnetic field intensity are a special challenge. Because of the square proportionality of the SNMR signal to the geomagnetic field strength (approximately 26 000 nT to 28 000 nT), the signal response in southern Africa is only approximately 1/3 of that under European conditions for the same hydrogeological situation (see *Fig. 4.7-4*).
- Low water content of the fractured rock aquifers and therefore low signal strength. Typical fractured basement rock aquifers have, at best, a water content of only a few percent, and therefore the signal strength, which is directly proportional to the water content, is significantly lower than those typically observed in unconsolidated sand aquifers where a water content of 20 - 30 % is not uncommon.
- High natural and man-made electromagnetic noise levels. The initial experimental measurements on low water content basement fractured rock aquifer conditions were rather disappointing due to a very low S/N ratio (in places lower than 1) and special measures are required to improve the S/N. High noise levels are currently mitigated by using double turn transmitter/receiver loops with a noise suppressing reference loop system; but nevertheless large stacking numbers are required to obtain smooth sounding curves. This can result in recording times of several hours and sometimes more than a day per sounding.

To confirm the viability of the technique under low Earth magnetic field strengths, a follow-up SNMR experimental programme included not only low water content fractured basement rock aquifers, but also soundings on alluvial high water content unconsolidated primary aquifers. Two examples are presented below: a coastal aquifer system consisting of two unconsolidated to semi-consolidated sand aquifers separated by a clay/limestone layer along the Cape West Coast north of Cape Town (*Fig. 4.7-11*), and an intensely fractured quartzite of Silurian age associated with the Cape Fold Belt present along the South African south coast (*Figs. 4.7-12 a,b*).

High water content sand aquifer, Langebaan aquifer, Cape West Coast, South Africa

Figure 4.7-11 illustrates the SNMR effect of an upper approximately 20 m thick unconfined sand aquifer, separated from a lower confined aquifer, comprising of sands and gravel by an ~15 m thick aquiclude (calcrete and clay layer). The lower aquifer is underlain by the Darling granite pluton at a depth of 77 m.

Two inflection points on the sounding curve indicate the presence of two aquifers. The large pulse moment (~2800 A ms) at which the maximum signal amplitude is reached is an indication of increased depth and thickness of the lower aquifer. Average porosity of about 20% over the depth range of 40 m to 70 m is in agreement with borehole and other hydrogeological data. The confining layer between the upper and lower aquifer is clearly identified by the inversion. This example illustrates that the SNMR-technique can be used to map the largely unknown lateral extent of this layer and could significantly reduce costly drilling. The regional extent of this confining layer is of great importance in the formulation of the geohydrological model of the area.

Survey parameter values:

total intensity	26 140 nT; magnetic field inclination: -65°
loop-parameters	circular; 2 turns; diam. 96 m; cable section: 6 mm^2
pulse frequency	1113.6 Hz; pulse length: 40 ms
pulse number	24; stacking rate: 72
acquisition time	4.5 hours.

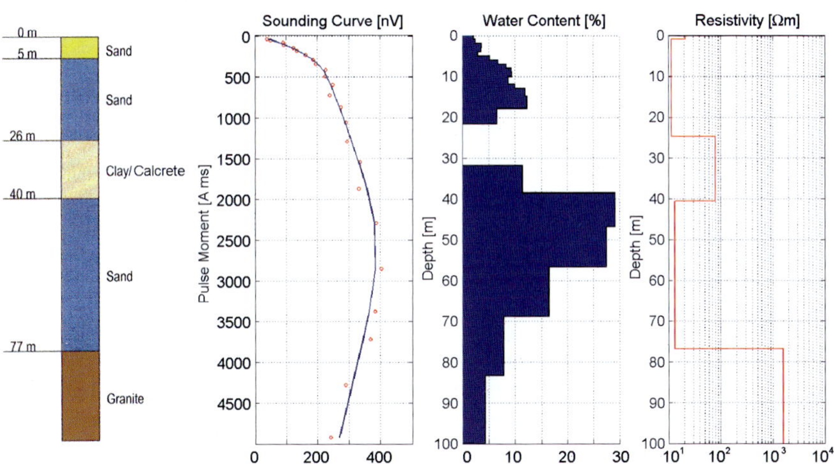

Fig. 4.7-11: SNMR sounding Muishondfontein, Langebaan Road aquifer, Cape West Coast, South Africa

Steenbras dam site, Western Cape Province, South Africa

Figure 4.7-12 (*a*) is a photo taken at the sounding site showing the intense fracturing in the investigation area. The site is located in a steeply dipping north-easterly striking synclinal structure (just south of the dam) which coincides with a major and extensive fracture zone extending for approximately 30 km to the NE. The main fracture and fault directions visible on an aerial photograph are SE-NW and SSE-NNW together with numerous other fracture directions of lower intensity. The sounding position was selected to be at the intersection of some of these major fractures. Unfortunately no borehole information is yet available, but the site has been earmarked as a potential drilling target as part of a major water supply project for the City of Cape Town. *Figure 4.7-12* (*b*) illustrates the expected nature of the fracturing at depth.

The expected hydrogeological profile for the site is however, essentially a two layer case: The upper few meters consist of clean quartzitic sand being the weathering product of the quartzite followed by a highly fractured quartzite extending to a depth in excess of 100 m.

(a) (b)

Fig. 4.7-12: (a) Photograph of the Steenbras survey area showing the intensely faulted and fractured nature of the rock formations, (b) Close up view illustrating the degree of fracturing in the quartzite

The SNMR sounding curve and the 1-D inversion interpretation is shown in *Fig. 4.7-13*. Due to high noise levels and instrument malfunction, data recording had to be extended over two days. The end result was a high quality sounding curve which raises the question of quality data versus production. As the evaluation of the applicability of the SNMR technique in fractured rocks is part of the research project, the emphasis was on quality data rather than production. The presence of the two inflection points shown by the sounding curve are reflected in the interpretation as representing a double aquifer condition where the upper aquifer is separated from the lower by a thin, apparently semi-confined layer. The interpretation predicts an up to 3 % porosity for the deeper aquifer between depths of 25 m and about 70 m. The presence of the presumably semi-confined layer at around 15 m is unexpected and will be investigated further as it has important implications in terms of the environmental constraints governing the abstraction of groundwater in this ecological highly sensitive area.

The two South African examples illustrate that the SNMR technique, despite the low Earth's magnetic field intensity and high natural noise levels, can be used effectively to investigate low water content fractured aquifers in Southern Africa. In most instances however, it does require the use of noise suppression techniques and longer recording periods to produce high quality sounding curves. As more experience is gained and improved techniques for noise suppression are developed, a compromise between data quality and interpretation resolution will be established that should eventually result in shorter data recording times.

Survey parameter values:

total intensity	25 994 nT; magnetic field inclination: -65°
loop-parameters	Figure-8, 2 turns, diam. = 96 m, cable section: 6 mm^2
pulse frequency	1103.3 Hz; pulse length: 40 ms
pulse number	20; stacking rate: 300 (using noise threshold)
acquisition time	12 h.

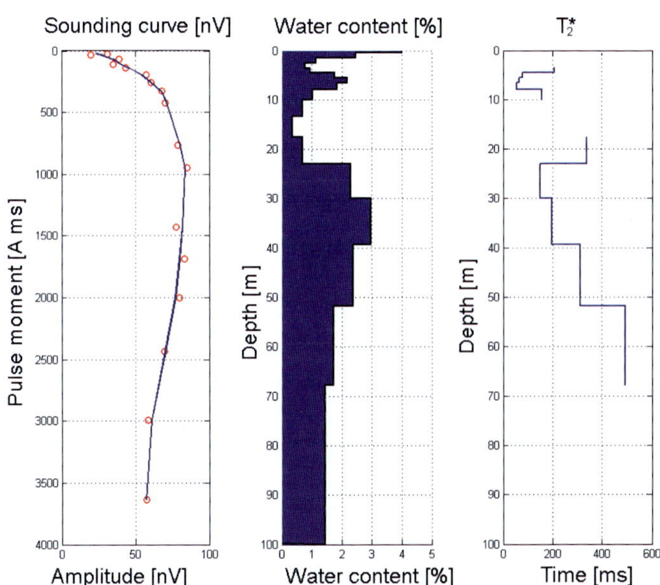

Fig. 4.7-13: SNMR sounding recorded at the Steenbras site and showing the interpreted water content and the T_2^* time distribution with depth based on the 1-D inversion

Parameters and units

Parameters	Symbols	Units
Larmor frequency	f_L	Hz
gyromagnetic ratio	γ	0.267 518 Hz nT^{-1}
total magnetic field of the Earth	B_0	nT (1 nT = 10^{-9} T)
transmitted magnetic field perpendicular to the static field	$B_\perp(r)$	nT
time	t	ms
alternating current in the loop	$i(t)$	A
duration of alternating current	τ	ms
amplitude of current	i_0	A
excitation intensity - pulse moment	q	A ms
voltage in the loop	$e(t)$	V
initial amplitude	E_0	nV (1 nV = 10^{-9} V)
relaxation time	T_2*	ms
phase	φ_0	degrees
nuclear magnetization	M_0	J T^{-1} m^{-3}
water content	$f(r)$	vol %
hydraulic conductivity	k_f	m s^{-1}

References and further reading

BALTASSAT, J.-M., LEGCHENKO, A., AMBROISE, B., MATHIEU F., LACHASSAGNE, P. & WYNS, R. (2005): Magnetic resonance sounding and resistivity characterisation of a mountain hard rock aquifer: the Ringelbach Catchment, Vosges Massif, France. Near Surface Geophysics, **3**, 267-274.

BARRINGER, A. R. & WHITE, J. F. (1968): Groundwater survey method and apparatus. US Patent 3, 398, 355. 3.

BEAUCE, A., BERNARD, J., LEGCHENKO, A. & VALLA, P. (1996): Une nouvelle méthode géophysique pour les études hydrogéologiques: l'application de la résonance magnétique nucléaire. Hydrogéologie, **1**, 71-77.

BRAUN, M., HERTRICH, M. & YARAMANCI, U. (2005): Complex inversion of MRS data. Near Surface Geophysics, **3**, 155-164.

DAVIS, S. N. & DE WIEST, R. J. M. (1966): Hydrogeology. Wiley & Sons, NewYork.

DUNN, K.-J., BERGMAN, D. J. & LATORRACA, G. A. (2002): Nuclear Magnetic Resonance, Petrophysical and logging applications. In: Seismic Exploration, **32**, Pergamon Press, Amsterdam, London.

GOLDMAN, M., RABINOVICH, B., RABINOVICH, M., GILAD, D., GEV, I. & SCHIROV, M. (1994): Application of the integrated NMR-TDEM method in groundwater exploration in Israel. Journal of Applied Geophysics, **31**, 27-52.

GUILLEN, A. & LEGCHENKO, A. (2002): Inversion of surface nuclear magnetic resonance data by an adapted Monte Carlo method applied to water resource characterization. Journal of Applied Geophysics, **50**, 193-205.

HERTRICH, M. (2005): Magnetic Resonance Sounding with separated transmitter and receiver loops for the investigation of 2D water content distributions. PhD thesis. TU Berlin.

HERTRICH, M., BRAUN, M. & YARAMANCI, U. (2005): Magnetic Resonance Soundings with separated transmitter and receiver loops. Near Surface Geophysics, **3**, 141-154.

HERTRICH, M., BRAUN, M. & YARAMANCI, U. (2006): High resolution 2D inversion of separated loop Magnetic Resonance Sounding (MRS) surveys. Proceedings of the Symposium on the Application of Geophysics to Engineering and Environmental Problems (SAGEEP), Seattle/USA.

HERTRICH, M. & YARAMANCI, U. (2002): Joint inversion of Surface Nuclear Magnetic Resonance and Vertical Electrical Sounding. Journal of Applied Geophysics, **50**, 179-191.

HÖLTING, B. (1992): Hydrogeologie. Enke, Stuttgart.

HUNTER, D. & KEPIC, A. (2005): SNMR signal contribution in conductive terrains. Exploration Geophysics, **36**, 73-77

IRIS Instruments: NUMIS Proton Magnetic Resonance System, Operation Manual. Orleans, France.

KIRSCH, R., YARAMANCI, U., & HÖRDT, A. (2006): Geophysical characterisation of aquifers. In: KIRSCH, R. (Ed.): Groundwater Geophysics – a tool for hydrogeology. Springer, Berlin, 439-457.

KENYON, W. E. (1992): Nuclear magnetic resonance as a petrophysical measurement. Int. Journal of Radiat. Appl. Instrum., Part E, Nuclear Geophysics, **6**, 153-171.

LACHASSAGNE, P., BALTASSAT, J., LEGCHENKO, A. & DE GRAMOT, H. M. (2005): The links between MRS parameters and the hydrogeological parameters. Near Surface Geophysics, **3**, 259-266.

LANGE, G., HERTRICH, M., KNÖDEL, K. & YARAMANCI, U. (2000): Surface-NMR in an area with low geomagnetic field and low water content – a case history from Namibia. Proceedings of the 6th Meeting of Environmental and Engineering Geophysics, European Section. September 3-7, 2000, Bochum.

LANGE, G., MOHNKE, O. & GRISSEMAN, C. (2005): SNMR measurements in Thailand to investigate low porosity aquifers. Near Surface Geophysics, **3**, 197-204.

LANGE, G., MEYER, R., DIBATTISTA, M., & MÜLLER-PETKE, M. (2006): Loop configuration experiments on multiple layered primary aquifers in Germany and South Africa. MRS2006 Proceedings, 69-72, Madrid-Tres Cantos, Oct. 25-27. www.igme.es/MRS2006/tech_program.html

LEGCHENKO, A., BEAUCE, A., GUILLEN, A., VALLA, P. & BERNARD, J. (1997): Natural variations in the magnetic resonance signal used in PMR groundwater prospecting from the surface. European Journal of Environmental and Engineering geophysics, **2**, 173-190.

LEGCHENKO, A. V. & BEAUCE, A. (1999): Surface Proton Magnetic Resonance method: What do users get? Proceedings of 5th Meeting of the Environmental and Engineering Geophysics, European Section. September 6 – 9, 1999, Budapest.

LEGCHENKO, A. V., BALTASSAT, J. M., BEAUCE, A. & BERNARD, J. (2002): Nuclear magnetic resonance as a geophysical tool for hydrogeologists. Journal of Applied Geophysics, **50**, 21-46.

LEGCHENKO, A. V., BALTASSAT, J.-M., BOBACHEV, A., MARTIN, C., ROBAIN, H. & VOUILLAMOZ, J.-M. (2004): Magnetic resonance soundings applied to the characterization of aquifers. Groundwater, **42**, 436-473.

LEGCHENKO, A. & SHUSHAKOV, O. A. (1998): Inversion of surface NMR data. Geophysics, **63**, 75–84.

LEGCHENKO, A. & VALLA, P. (2002): A review of the basic principles for proton magnetic resonance sounding measurements. Journal of Applied Geophysics, **50**, 3-19.

LIEBLICH, D. A., LEGCHENKO, A., HAENI, F. P. & PORTSELAN, A. A. (1994): Surface nuclear magnetic resonance experiments to detect subsurface water at Haddam Meadows, Connecticut. SAGEEP, 717-736.

LUBCZYNSKI, M. & ROY, J. (2004): Magnetic resonance sounding (MRS): New method for groundwater assessment. Groundwater, **42**, 291-303.

MEYER, R., LANGE, G., DIBATTISTA, M. & SOLTAU, L. (2006): Fractured rock aquifer MRS in high noise, low water content and low frequency environments, South Africa. MRS2006 Proceedings, 85-88, Madrid-Tres Cantos, Oct. 25-27. www.igme.es/MRS2006/tech_program.html

MOHNKE, O. & YARAMANCI, U. (2002): Smooth and block inversion of surface NMR amplitudes and decay times using simulated annealing. Journal of Applied Geophysics, **50**, 163-177.

MOHNKE, O. & YARAMANCI, U. (2005): Interpretation of relaxation signals using multi-exponential inversion. Near Surface Geophysics, **3**, 165-186.

MÜLLER-PETKE, M., HERTRICH, M. & YARAMANCI, U. (2006): Analysis of magnetic resonance sounding kernels concerning large scale applications using singular value decomposition (SVD). In: Proceedings of SAGEEP, Seattle/USA.

PACKARD, M. & VARIAN, R. (1954): Free Nuclear Induction in the Earth's Magnetic Field. Physical Review, **93**, 941.

PLATA, J. & RUBIO, F. (2002): MRS experiments in a noisy area of a detrital aquifer in the south of Spain. Journal of Applied Geophysics, **50**, 83-94.

PURCELL, E. M., TORREY, H. C. & POUND, R. V. (1946): Resonance absorption by nuclear magnetic moment in a solid. Physical Review, **69**, 37-38.

ROY, J. & LUBCZYNSKI, M. W. (2003): The magnetic resonance sounding technique and its use for groundwater investigations. In: Hydrogeology Journal: IAH, **11**, 455-465.

SEMENOV, A. G., PUSEP, A.Y. & SCHIROV, M. (1982): Hydroscope - an installation for prospecting without drilling. USSR Academy of Sciences, Novosibirsk (in Russian).

SEMENOV, A. G., BURSHTEIN, A. I., PUSEP, A.Y. & SCHIROV, M. (1988): A device for measurement of underground mineral parameters (in Russian). USSR Patent 1079063.

SCHIROV, M., LEGCHENKO, A. & CREER, G. (1991): A new direct non-invasive groundwater detection technology for Australia. Exploration Geophysics, **22**, 333-338.

SHUSHAKOV, O. A. (1996): Groundwater NMR in conductive water. Geophysics, **61**, 998-1006.

SUPPER, R., JOCHUM, B., HÜBL, G., RÖMER, A. & ARNDT, R. (2002): SNMR test measurements in Austria. Journal of Applied Geophysics, **50**, 113-121.

TRUSHKIN, D. V., SHUSHAKOV, O. A. & LEGCHENKO, A. V. (1994): The potential of a noise-reducing antenna for surface NMR groundwater surveys in the Earth's magnetic field. Geophysical Prospecting, **42**, 855-862.

TRUSHKIN, D. V., SHUSHAKOV, O. A. & LEGCHENKO, A. V. (1995): Surface NMR applied to an electroconductive medium. Geophysical Prospecting, **43**, 623-633.

VALLA, P. & LEGCHENKO, A. (2002): One-dimensional modelling for proton magnetic resonance sounding measurements over an electrically conductive medium. Journal of Applied Geophysics, **50** (1-2), 217-229.

VALLA, P. & YARAMANCI, U., (Eds.) (2002): Special Issue on Surface Nuclear Magnetic Resonance. Journal of Applied Geophysics, **50**, 229.

VARIAN, R. H. (1962): Ground liquid prospecting method and apparatus. US Patent 3019383.

VOUILLAMOZ, J.-M., DESCLOITRES, M., TOE, G. & LEGCHENKO, A. (2005): Characterization of crystalline basement aquifers with MRS: Comparison with boreholes and pumping tests data in Burkina Faso. Near Surface Geophysics, **3**, 204-214.

VOUILLAMOZ, J.-M., DESCLOITRES, M., BERNARD, J., FOURCASSIER, P. & ROMAGNY, L. (2002): Application of integrated magnetic resonance sounding and resistivity methods for borehole implementation - A case study in Cambodia. Journal of Applied Geophysics, **50**, 67-81.

WARSA, W., MOHNKE, O. & YARAMANCI, U. (2002): 3-D modeling of Surface-NMR amplitudes and decay times. In: Water Resources and Environmental Research, ICWRER, 209-212.

WEICHMAN, P. B., LAVELY, E. M. & RITZWOLLER, M. H. (2000): Theory of surface nuclear magnetic resonance with applications to geophysical imaging problems. Physical Review E **62** (1, Part B), 1290-1312.

WEICHMAN, P. B., LUN, D. R., RITZWOLLER, M. H. & LAVELY, E. M. (2002): Study of surface nuclear magnetic resonance inverse problems. Journal of Applied Geophysics, **50**, 129-147.

YARAMANCI, U. & HERTRICH, M. (2006): Magnetic resonance sounding. In: KIRSCH, R. (Ed.): Groundwater Geophysics – a tool for hydrogeology. Springer, Berlin, 253-273.

YARAMANCI, U., KEMNA, A., VEREECKEN, H. (2005): Emerging Technologies in Hydrogeophysics. In: RUBIN Y. & HUBBARD S. (Eds.): Hydrogeophysics. Springer, Dordrecht, 467-486.

YARAMANCI, U., LANGE, G. & HERTRICH, M. (2002): Aquifer characterisation using Surface NMR jointly with other geophysical techniques at the Nauen/Berlin test site. Journal of Applied Geophysics, **50**, 47-65.

YARAMANCI, U., LANGE, G. & KNÖDEL, K. (1998): Effects of regularisation on Surface NMR measurements. Proceedings of the 60th conference of EAGE, 10-18.

YARAMANCI, U., LANGE, G. & KNÖDEL, K. (1999): Surface NMR within a geophysical study of an aquifer at Haldensleben (Germany). Geophysical Prospecting, **47**, 923-943.

YARAMANCI, U. & LEGCHENKO, A. V. (Eds.) (2005): Special Issue Magnetic Resonance Sounding - aquifer detection and characterization. Near Surface Geophysics, **3**, 119-222.

4.8 Borehole Logging

THOMAS WONIK with a contribution by RICARDO A. OLEA

4.8.1 Principle of the Methods

Boreholes yield direct information about the structure of the ground below the surface. Drill cores show the lithology at the drill site, and physical parameter values and petrographic data can be obtained from them in the laboratory. For technical and economic reasons, however, drill cores are not taken from every borehole. Borehole logging (also called well logging or downhole logging) provides in-situ information in addition to the direct geological information. Geophysical borehole logging is the process of measuring physical, chemical, and structural properties of penetrated geological formations using tools that are lowered into a borehole on a wireline cable. The measurements are electrically transmitted via cable to the surface where they are recorded in digital format as a function of depth. The physical properties of the rock strata and a lithological classification can be derived from the data. Measurements on the cores themselves are also sometimes made. Borehole logging has certain advantages with respect to core measurements. For example, in a hole where there is limited core recovery, the depth of the incomplete cores can be uncertain; logs provide a continuous depth record of formation properties. The in-situ nature of the downhole measurements is an important factor because measurements on recovered cores are affected by the fact that the material is no longer under the pressure and temperature conditions that exist at depth (HARVEY & LOVELL, 1998).

Since the first geophysical borehole logs were made more than seventy years ago, a number of probes have been developed to measure in a borehole nearly every possible physical parameter. The probes used in hydrocarbons exploration have diameters of 10 - 15 cm, weigh several hundred kilograms, and are stable under the temperature and pressure conditions at depths of several kilometers. Sophisticated methods have been developed to interpret the logging data, mostly for assessing a prospect for hydrocarbons and characterizing the reservoir sediments (SERRA, 1984 & 1986; WESTERN ATLAS INTERNATIONAL, 1985; SCHLUMBERGER EDUCATIONAL SERVICES, 1987 & 1989a & b; LUTHI, 2001). An example of the application of these methods to crystalline rock is the Continental Deep Drilling Program (KTB) (e.g., BRAM & DRAXLER, 1995). The successful application of borehole logging to marine geology within the scope of the Ocean Drilling Projects (ODP) has been summarized by GOLDBERG (1997).

Light, easily moved equipment (slimhole equipment) is used for boreholes in near-surface sediments. Many exploratory boreholes are completed as groundwater observation wells.

Borehole geophysical logging methods are used to determine

- the physical parameters density, electrical resistivity, velocity of elastic waves, dielectric constant,
- the geometric parameters caliper, dip, azimuth, orientation of joints, and
- the condition of the well: casing, cementation, clay barriers, screens.

Figure 4.8-1 shows the main components of slimhole logging equipment. The downhole logs are rapidly collected, are continuous with depth, and provide in-situ property values with a high sampling resolution. They can be used for many purposes. The most important aspects are that geophysical well logs can be interpreted in terms of the formation's stratigraphy, lithology, and mineralogy. They provide information on the geology and the groundwater system. Logs also provide the major link between borehole and surface geophysical measurements. In order to obtain the maximum benefit from geophysical logging methods it is necessary to understand the basic principles of the many tools commonly used and the most effective log analysis techniques.

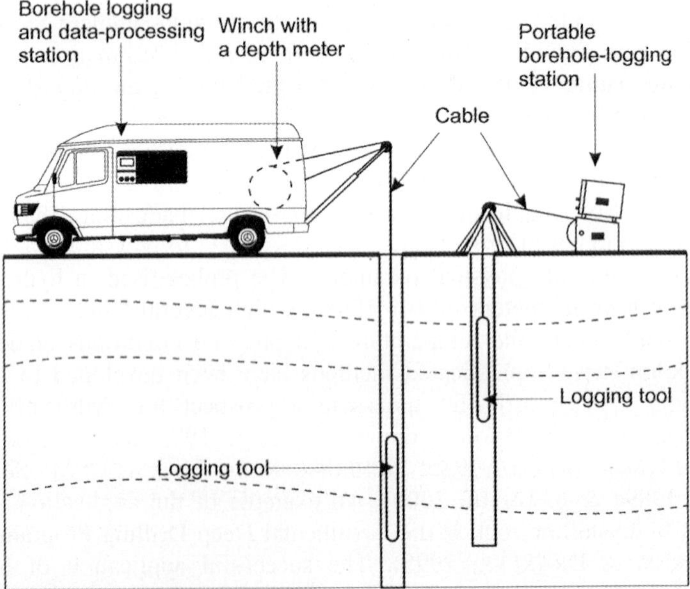

Fig. 4.8-1: Main components of a slimhole logging equipment: surface unit, winch and cable, and logging tools

Although the reasons for using well logging methods in deep boreholes are different from those in shallow boreholes, some of the methods developed for deep boreholes can be used. Measurements in shallow boreholes, normally less than 100 m deep, are often made in unconsolidated sediments. Problems are encountered in the interpretation of the data that cannot be solved with the methods developed for hydrocarbons exploration. The shallow boreholes – with diameters of 3.5 - 6.0 cm (slimholes) – require different probes. There are fewer problems with seals and temperature, but the smaller size makes it necessary to redesign the electronic components. The special requirements of shallow boreholes has led to the development of a special field called "slimhole" well logging. The equipment, field application, and interpretation methods have been described, for example, by HALLENBURG (1987 & 1992), KEYS (1990), YEARSLEY & CROWDER (1990), SAMWORTH (1992). The following applications of geophysical well logging methods in shallow boreholes have been dealt with by a large number of publications over the last several years: geological investigations (DOVETON & PRENSKY, 1992), determination of physical properties (HEARST et al., 2000), hydrogeological investigations (REPSOLD, 1989; MARES et al., 1994; JORGENSEN & PETRICOLA, 1995), and environmental investigations (TAYLOR et al., 1990; KEYS, 1997; KRAMMER, 1997). This chapter focuses on the use of geophysical well logging methods for geoscientific investigations in shallow boreholes and wells (slimhole logging).

4.8.2 Applications

- Lithological classification and documentation of the strata penetrated by a borehole on the basis of geophysical well logs,
- correlation of the strata penetrated by a borehole with neighboring boreholes,
- determination of the dip and strike of strata, detection of small-scale disturbances, e.g., faults, joints, and karst phenomena,
- determination of parameter values (e.g., density and electrical conductivity) needed for the evaluation of geophysical survey data obtained at the ground surface,
- determination of hydrogeological and geotechnical parameter values from geophysical parameters (e.g., density, clay content, porosity, degree of saturation with water),
- determination of water influx and efflux in wells,
- inspection of the condition of the well at the time of well completion and later times, including hydraulic efficiency, and

- in-situ determination of physico-chemical parameter values of groundwater.

See also *Table 4.8-2*.

4.8.3 Slimhole Logging Equipment and Logging Methods

Logging equipment consists mainly of three components: a surface unit, winch and cable, and logging tools (*Fig. 4.8-1*). The surface unit controls the logging, including the movement of the probe in the borehole; it provides the energy supply to the probe and records, displays, and stores the data. The energy is supplied through the cable, which also conveys the electrical signals between the probe and the surface unit. The depth at which a measurement is made is given by a gauge on the winch.

Two types of logging equipment are used for shallow boreholes, differing mainly with respect to the type of winch. Portable equipment is needed for rough terrain, with a winch with about 300 m of single-core cable. For deeper holes, it is best to use a truck or van in which the surface unit and a winch with about 1000 m of cable are installed. Three kinds of logging tools are used: (a) the probe is lowered free in the hole, (b) the probe is centered in the hole by centralizer, and (c) the sensors in the tool are pressed against the borehole wall. The tools and the principles behind them are discussed in the following sections. The different tools are not named according to any particular system: some are named on the basis of the parameter measured, others according to the principle by which the measurement is made, and still others on the basis of the geometry of the probe or the trade name. The most common methods are given in *Table 4.8-1*. Further information about the individual methods is given, for example, by RIDER (1996), KEYS (1997), ZSCHERPE & STEINBRECHER (1997) and FRICKE & SCHÖN (1999). PRENSKY (2002) describes recent developments in logging technology.

Table 4.8-1: Overview of the most common well logging methods (after FRICKE & SCHÖN, 1999)

Method	**Principle**	**Log**	**Abbreviation**	**Measured parameter [derived parameters]**
radioactivity logging methods				
a) "passive" gamma measurement	measurement of the natural gamma radiation - sum - spectral	gamma-ray log, spectral gamma-ray log	GR GRS	counting rate (cps, API-units) [Th, U, K concentrations (%, ppm)]
b) "active" gamma-gamma measurement	measurement of the gamma-radiation using the Compton effect	gamma-gamma log, density log	GG	counting rate (cps) [density (g cm^{-3})]
neutron measurement	measurement of the neutron radiation on collision with an atom; measurement of the radiation resulting from neutron capture	neutron-neutron log neutron-gamma spectrometry (a log for each element determined)	NN, INN EBS	[neutron porosity (%, WU)] [element concentration (ppm, %)]
electrical methods				
self-potential	measurement of the self-potential	self-potential log	SP	potential (mV)
resistivity	measurement of the electrical resistivity	conventional resistivity measurement	EL, ML	electrical resistivity (Ωm)
		focused electric log	FEL, DLL, MLL	electrical resistivity (Ωm)
		dipmeter	DIP	electrical resistivity (Ωm)
electromagnetic methods				
inductive measurement	measurement of the electrical conductivity and magnetic susceptibility	induction log, susceptibility log	IL SUSZ	electrical conductivity (mS m^{-1}) magnetic susceptibility (dimensionless)

Table continued next page.

Method	Principle	Log	Abbreviation	Measured parameter [derived parameters]
acoustic methods				
a) transmission refraction	measurement of the ultrasound waves refracted from or guided waves along the borehole wall	acoustic log	AL	traveltime (µs m^{-1})
	measurement of the attenuation of the ultrasound waves in the cement annulus fill	cement bond log	CBL	attenuation (dB m^{-1})
b) reflection	measurement of the ultrasound waves reflected from the borehole wall or casing	borehole televiewer	BHTV	entire wavelet
optical methods				
	video	borehole TV	OPT	
methods for determining the properties of drilling fluids (fluid logs)				
	measurement of temperature, electrical conductivity and hydrochemical parameters	temperature log, salinity log, drilling fluid log (environment log)	TEMP SAL MIL	temperature (°C) electrical conductivity (mS m^{-1}) pH, redox potential (mV), O_2 concentration (%)
	measurement of vertical flow	impeller flowmeter	FLOW	revolutions per time unit [flow velocity]
	measurement of the transparency of the water	photometric turbidity measurement	PMT	[groundwater flow velocity]
methods for determining borehole properties				
	measurement of the diameter and deviation from the vertical	caliber log, borehole deviation log	CAL DEV	caliber (mm) azimuth, dip (°)

WU: water units, calibration against water as opposed to the API method

4.8.3.1 Radioactivity Logging Methods

Radioactivity logging methods measure (total and/or spectral measurement) the natural gamma-radiation (gamma-ray log) or the secondary gamma or neutron radiation produced by a primary radiation source (gamma-gamma log, neutron-neutron log, neutron-gamma log) (IAEA, 1999). A review of radioactivity logging methods is given by MEYERS (1992).

The natural gamma-radiation measured with a *gamma-ray logging tool* (GR) is from the natural ^{40}K in the ground and the isotopes of the uranium and thorium decay series. These isotopes occur naturally in clay, making it possible to distinguish between sand and clay layers and to estimate the clay content. Uranium ores are characterized by sharp peaks and high counting rates in the GR log. A *spectral gamma-ray probe* (GRS) permits a distinction between different minerals containing radio-isotopes by measuring the proportion of the gamma-radiation from the ^{40}K, ^{238}U and ^{232}Th decay series. This is done with a scintillation detector (e.g., bismuth-germanium oxide). After calibration with a standard, the concentrations of K, U, and Th can be given in percent or ppm. In order to compare measurements made with differently designed probes, the calibration is done in API (American Petroleum Institute) units according to international convention. An API unit is 1/200 of the difference in measured activity of a layer with low radioactivity and a layer with a higher radioactivity in the reference well on the campus of the University of Houston.

Gamma-ray logs can be made in dry and cased boreholes. The attenuating influence of the casing can be corrected for. The vertical resolution of a gamma log depends on the length of the scintillation crystal, normally about 20 cm long. The lateral distance of detection, normally 15 - 20 cm, is influenced by the density of the rock. Ninety percent of the measured γ-radiation is from this interval.

At the bottom of a *density probe* (γ-γ log, GG) there is a gamma-ray source (usually Cesium-137); at the top of the probe there is a γ detector (e.g., a sodium iodide scintillation counter), shielded from direct radiation from the source below by a lead cylinder. The gamma-radiation emitted by the source is scattered by the atoms of the surrounding rock partially adsorbed, depending on the density of the rock (Compton effect). Some of the scattered radiation is deflected back to the detector and recorded. The porosity of the rock can be derived from the measured density of the rock if the densities of the rock matrix and the pore fluid are known. Density and porosity are important parameters for assigning a lithology to the strata penetrated by the borehole. The radioactive source and the detector are 15 - 50 cm apart in the probe. Depending on the distance between the source and the detector, the investigation depth is 5 - 10 cm. The vertical resolution (20 - 60 cm) also depends on the source-detector spacing.

A *neutron-neutron tool* (NN or INN) contains a neutron source and one or more neutron detectors. The neutron source is either a neutron-emitting radioactive isotope, e.g., californium-252, (NN) or a neutron generator (INN). In an INN tool, an accelerated deuterium beam is directed at a tritium target to produce neutrons with an energy of about 14 MeV. There are between 10 and 20 hits per second. These fast neutrons loose energy when they collide with the nuclei of the atoms of the surrounding rock and are registered by the detectors as thermal and/or epithermal neutrons. Because the energy transfer is the most effective when the neutrons collide with hydrogen nuclei, which have the same mass, the counting rate is inversely proportional to the water content and porosity of the rock. The neutron source and the detectors are separated by a neutron barrier (lead shield) to suppress direct radiation. The counting rate is calibrated with a material with a known porosity and expressed as neutron porosity. Water content and the lithology of the rock can be derived from the neutron porosity values. The influence of the borehole itself is corrected for by simultaneously measuring with two detectors at different distances from the source. The influence of the borehole on the measurement can be checked by calibrating the tool using a model constructed according to API specifications with the measurement expressed in neutron porosity (RIDER, 1996) or using a model with a known hydrogen content with the measurement expressed in water units (WU). The depth of investigation depends mainly on the hydrogen content of the rock, i.e., on the porosity and water content. It varies from about 10 cm in water-saturated, porous rock to as much as 100 cm in dry rock. The high hydrogen contents in clay and coal cause values like those of porous rock. The detector spacing is larger than 30 cm. The vertical resolution of an NN or INN tool is about 50 cm.

A *neutron-gamma tool* (EBS, neutron-gamma spectrometry) also emits fast neutrons from a neutron generator into the surrounding rock. Gamma-rays are emitted as a result of inelastic collision of the neutrons with atomic nuclei, as well as neutron capture, which are then detected by a scintillation counter and recorded in a spectral log. The gamma energy spectrum is also recorded just before and after each neutron pulse. The energy spectra are characteristic of the chemical elements emitting the gamma-rays. The concentrations of Si, Ca, C, and O can be determined from the spectrum resulting from inelastic collisions of the neutrons. The elements H, Fe, Si, and Ca can be recognized in the neutron-capture spectrum, permitting relative concentrations in the rock to be estimated. Complicated calibration and interpretation procedures are necessary to derive the mineral composition of the rock (e.g., FANG et al., 1996; JACOBSEN & WYATT, 1996; LOFTS et al., 1995).

4.8.3.2 Electrical Methods

Electrical well logs can be made only within water or drilling fluid in a well. They are used in open holes to determine the electrical resistivity of the rock, which together with other physical parameters can be used to derive a lithological log for the borehole (MAUTE, 1992; SPIES, 1996). Some electrical well logging tools measure the self-potential; others measure the resistivity using one of several electrode configurations.

A *self-potential tool* (SP) measures the natural electrical potential between an electrode at the ground surface and an electrode in a drilling-fluid-filled borehole. This natural potential is caused by electrochemical processes occurring between different fluids (in this case the drilling fluid and the groundwater). The measurement is often disturbed by electrical noise. Prerequisite for an interpretable SP log is a distinct difference between the resistivities of the drilling fluid and the formation pore water together with an alternating sand/clay sequence with a distinct difference between the potentials of the sand and clay layers. The relative change in potential with depth is plotted.

In *conventional resistivity logging* (electric log, EL, and microlog, ML), the resistivity of the rock is measured using a four-electrode array, analog to dc resistivity surveys at the ground surface (Chapter 4.3). A constant current is introduced into the rock between two current electrodes in the logging tool. The potential measured between two other electrodes (potential electrodes) is proportional to the electrical resistivity of the rock. The tools used for conventional resistivity logging have the following lengths (current electrode spacings): small normal (L = 10 - 50 cm), large normal (L = 50 - 200 cm), 16 inch normal (L = 40 cm), and 64 inch normal (L = 160 cm). The measured value is called the "apparent resistivity" and is dependent on the size of the borehole, the adjacent rock and the overlying and underlying rock. The "true" resistivities of the rock can be derived from the apparent resistivities using master curves. The measured resistivity logs are symmetric. The curve deviates strongly from the true value when the layer thickness h is less than 5 L and a reverse curve when $h / L < 1$. The depth of investigation and vertical resolution are determined by the length L of the tool. The depth of investigation and vertical resolution are inversely proportional. The ratio of the formation resistivity R_t to the resistivity of the drilling fluid R_m also influences the results. Useable results are obtained when $R_t / R_m > 1$. Micrologs have a small depth of investigation, due to the small electrode spacing, and are used to investigate the rock immediately next to the borehole. The depth of investigation is 3 - 5 cm and the vertical resolution is 2 - 5 cm.

A *laterolog* (focused electrolog, FEL) is made with a tool that uses additional electrodes, called guard or bucking electrodes, to focus the current to nearly right angles to the logging tool, considerably increasing the lateral investigation depth and vertical resolution over those of the conventional logging method. Thin layers (down to 20 cm) can be detected with this tool. In a dual laterolog tool (DLL), two focusing systems are used with different investigation depths. The system with the smaller penetration depth mainly measures the rock next to the borehole that has been disturbed by the drilling and has been penetrated by the drilling fluid (invaded zone). The system with the greater penetration depth is influenced mainly by the undisturbed rock further from the borehole. In a microlaterologging tool (MLL), the electrodes have a small spacing and are pressed against the borehole wall during the measurement, so that there is no drilling fluid between the tool and the rock. The vertical resolution is about 5 cm and the investigation depth is as much as 10 cm. Laterologging tools can be used to detect damage to plastic casing in a well.

A *dipmeter tool* (DIP) is used to determine the dip and dip direction of the layers penetrated by a borehole. The tool records at least three rock resistivity values measured using an electrode array pressed against the borehole wall, similar to a MLL measurement, while the deflection of the tool arms is measured for caliber determination (Section 4.8.3.7). The orientation of the tool is determined from the data from three orthogonal magnetic field sensors. Layers with thicknesses of 1 - 1.5 cm are detected.

4.8.3.3 Electromagnetic Methods

Electromagnetic well logging methods can be used in both dry wells and those containing water or drilling fluid. In contrast to electrical methods, these methods can be used in boreholes with plastic casing. The parameters electrical conductivity and susceptibility can be determined using an induction tool or a susceptibility tool, respectively. Both parameters can be used for lithological classification of the rock sequence.

Induction tools (IL) are used to determine electrical conductivity/resistivity of the rock surrounding a borehole. A transmitter coil generates an alternating magnetic field around the borehole, which in turn induces electrical eddy currents that are proportional to the conductivity of the rock. An induction tool usually contains two coil systems with different coil spacings and thus different investigation depths. The investigation depth depends on the conductivity of the rock and is 60 - 100 cm for a dual induction log. The vertical resolution is about 150 cm for low conductivities and about 50 cm for higher conductivities. If the differences in conductivity are large, the resistivity curve recorded by focused systems overshoots the proper value when the logging tool passes the layer interface. In highly conductive rock, the signal is attenuated, due to the skin effect.

The same principles are used by a *susceptibility tool* (SUSZ). Besides magnetic markers in a borehole, metal parts that have been lost in a borehole can also be found.

4.8.3.4 Acoustic Methods

An *acoustic tool* or sonic tool (AL) is used to determine the velocity of sound in the rock surrounding the borehole. Acoustic well logging tools contain one or two ultrasound generators and several receivers in a linear array. The value for the ultrasound traveltime is an average for the distance between the transmitter and the receiver(s). The acoustic signal is recorded as a wavelet from which the traveltimes of the compression and shear waves are calculated. Acoustic well logs can be made only within water or drilling fluid in a well. They are used in open holes for lithological classification of the rock sequence, as well as for detecting joint and fracture zones. If the velocity in the rock matrix is known, the porosity of the rock can be calculated. The depth of investigation is 1 - 2 cm, the vertical resolution 60 cm. By increasing the distance between the ultrasound generator and receiver (e.g., up to 200 cm), the investigation depth is increased and the vertical resolution decreased. There are problems with the method in shallow boreholes in unconsolidated sediment if the traveltime in the drilling fluid is shorter than in the rock.

Acoustic tools are used to inspect the annulus cementation in wells (*cement bond log*, CBL). The signal amplitude is a measure of the quality of the cementation: high attenuation indicates a good cement bond. This means the wave energy is conducted very well via contact of the casing with the cement and of the cement with the rock. If the contact of the cement with the casing and/or rock is poor, the wave energy is conducted mainly along the casing from the transmitter to the receiver, i.e., there is little attenuation and the signal amplitude is high. The main aspects of the acoustic method are discussed by PAILLET et al. (1992).

An acoustic image of the borehole wall can be produced with *borehole televiewer* (BHTV). A rotating sonic generator transmits 250 ultrasound pulses per rotation with 3 - 6 rotations per second (corresponds to a resolution of 1.5°). Traveltime and amplitude of the received signal are recorded and displayed in false color to represent the borehole walls relative to north. The image has a very high resolution, permitting fractures, joints, and fracture zones to be recognized together with their spatial orientation.

4.8.3.5 Optical Methods

An image of the borehole wall can be obtained directly using a *video camera* (OPT), permitting a qualitative assessment of the borehole wall or casing. The depths of screens in the casing can be checked and problems with the casing can be determined. Together with direction and distance measuring systems, the size of penetrated cavities can be determined (GOCHIOCO et al., 2002). A video camera can be used only in dry boreholes or in clear water; visibility also depends on the lens and available light.

4.8.3.6 Methods for Determining the Properties of Drilling Fluids (Fluid Logs)

The *temperature* (TEMP) and *electrical resistivity* (SAL, for salinity) of the drilling fluid are usually measured together with a single logging tool. A thermister is used to register the temperature. The electrodes of the SAL probe are in a mini-four-point configuration with relatively small electrode spacings of about 50 mm. To avoid eddies in the drilling fluid, this is usually the first well log to be carried out in a borehole or well, and in contrast to normal logging, the log is recorded while the tool is lowered in the well. Electrical resistivity cannot be measured above the groundwater table, and therefore it is easy to determine when the probe enters the water or drilling fluid. The salinity of the water can be determined from the SAL readings. The tool is calibrated in the laboratory with a standard salt solution. The data from an SAL probe are often used to correct the values obtained with the electrical logging tool (Section 4.8.3.2). The combination of the TEMP and SAL logs provides an indication of vertical movement of the water in the well. The logs often show irregularities – anomalies – at depths at which inflow or efflux of water occurs in borehole with no casing, as well as where leaks in the casing of a well occur. If the temperature of the surrounding rock undisturbed by the drilling process is needed, it is necessary to wait until the temperature has returned to the natural state. A rule of thumb is to wait as long as time taken by the drilling. Calculations for the equilibration time have been published by BULLARD (1947).

A *multi-parameter probe* (MIL) measures pH, oxygen concentration, and redox potential of the drilling fluid. It can be used to monitor water quality (e.g., contamination). See also Chapter 4.9.

The vertical flow velocity of water in a borehole is measured with a *flowmeter* (FLOW). The water flowing through the meter turns a propeller, whose revolutions are recorded. Movement of the water caused by movement of the tool is corrected for during processing of the data. The measurement is strongly influenced by the diameter of the borehole and for this reason is normally done in well with casing. The tool can be used only in clear water, because the high viscosity and suspended matter in drilling mud prevents proper functioning. The pumping rate or artesian flow rate must be adequately high to determine the level(s) at which there is inflow of groundwater and its contribution to the total amount of water pumped. A pump must be installed in the well before the FLOW measurements are made. An electromagnetic flowmeter is being increasingly used (MOLZ & YOUNG, 1993).

The turbidity of the water is measured with a *photometric probe* (PMT). With this tool, the amount of fine-grained material washed into the well, even if it cannot be seen with the naked eye, can be determined. By adding a dye tracer, the groundwater flow rate can also be determined.

A *sample collecting tool* (SAMP) is used to take groundwater samples from a given depth, keeping it at the original pressure. Samples of about 1 L are collected in a hermetically sealed cylinder. The sampling tool is lowered to the specified depth, a valve is opened, allowing water to flow into the cylinder, and the valve is closed again so that when the tool is raised again the sample remains at the pressure at the sampling depth.

4.8.3.7 Methods for Determining Borehole Properties

Caliper logging tools (CAL) have one to four arms with springs to hold the ends of the arms against the borehole walls. Caliper data are used for installation of casing and cementation. They are also used for the evaluation of the data from nearly all other logging tools (as a correction parameter).

The deviation of a borehole from the vertical and the direction (DEV) is determined using a *deviation tool*. This is necessary information for proper casing installation or filling of the annulus. A compass is used to determine direction of deviation, a plumb line to determine the angle of deviation. The tool can be used only in a borehole in which the casing is plastic or there is none at all, i.e., in "open" holes.

4.8.4 Survey Practice, Personnel, Equipment, Time Needed

The utility of geophysical well logs strongly depends in the well construction, casing, drilling fluid, borehole diameter, and whether the borehole is drilled into consolidated or unconsolidated rocks.

The objectives of the different well logging methods are given in *Table 4.8-2*. This table can be used to select the methods that can be applied for a given problem and well conditions. It can be seen in *Table 4.8-3* that most of the methods cannot be used in wells with PVC casing. For each borehole/well the methods must be selected that will best fulfill the objectives. A borehole can seldom be logged over its entire length. For example, some methods can be used only in fluid-filled parts of a well or where there is no casing.

The cost of well logging is determined primarily by how long it takes: Not only the actual logging time, but also the idle time (time the logging crew has to wait for the borehole to be ready for logging, and the time the drilling crew has to wait for the logging to be completed) costs money. One method of saving time is to use multi-parameter tools, with which several parameters can be measured at the same time. For this purpose, several tools are lowered in "strings". It must be taken into consideration, however, that not all of the parameters can be measured down to the bottom of the borehole. Another cost factor is the rate at which the measurements can be made. The logging engineer will choose a logging speed to minimize acquisition time while maintaining an acceptable resolution and signal-to-noise ratio. The best data is obtained at the slowest speed. Consequently, it is recommended to run the tool as slowly as possible. Typical logging speeds are between 1 and 10 m min^{-1} (*Table 4.8-3*).

Tab. 4.8-2: The most important logging methods and their purposes

<table>
<tr><th rowspan="2">Logging method</th><th colspan="3">Objectives</th></tr>
<tr><th>Borehole in unconsolidated rock</th><th>Borehole in consolidated rock</th><th>Wells</th></tr>
<tr><td rowspan="2">GR, GRS</td><td colspan="3">lithological classification of the rock sequence, determination of the clay content, determination of radioactivity anomalies</td></tr>
<tr><td colspan="2">recognition of aquicludes</td><td>recognition of annular clay seals</td></tr>
<tr><td rowspan="3">GG</td><td colspan="3">lithological classification of the rock sequence</td></tr>
<tr><td colspan="2">determination of rock density and porosity</td><td rowspan="2">recognition of leaks in the clay seals and density distribution in the annulus, determination of casing thickness</td></tr>
<tr><td></td><td>identification of disturbed zones</td></tr>
<tr><td rowspan="2">NN, INN</td><td colspan="3">lithological classification of the rock sequence, determination of water content and porosity of the rock</td></tr>
<tr><td>determination of fine-grained fraction</td><td></td><td>determination of water in the annulus</td></tr>
<tr><td>EBS</td><td colspan="3">determination of the concentration of some chemical elements</td></tr>
<tr><td>SP</td><td colspan="2">lithological classification of the rock sequence, determination of formation-water resistivity, estimation of clay content</td><td></td></tr>
<tr><td rowspan="2">EL, ML, FEL, DLL, MLL</td><td colspan="2">lithological classification of the rock sequence, depth of invasion, calculation of true layer resistivities</td><td>identification of well screens, testing of casing seals</td></tr>
<tr><td></td><td>identification of joint and fracture zones, and weathering zones</td><td></td></tr>
<tr><td>DIP</td><td colspan="2">determination of the dip and strike of penetrated strata</td><td></td></tr>
<tr><td rowspan="2">IL</td><td colspan="3">lithological classification of the rock sequence, determination of the degree of water saturation of the rock</td></tr>
<tr><td colspan="2"></td><td>identification of electrical anomalies in the annulus</td></tr>
<tr><td rowspan="2">SUSZ</td><td colspan="3">lithological classification of the rock sequence, detection of metals</td></tr>
<tr><td colspan="2">recognition of magnetic minerals</td><td>recognition of annulus fill materials</td></tr>
</table>

Logging method	Objectives		
	Borehole in unconsolidated rock	**Borehole in consolidated rock**	**Wells**
AL, CBL	lithological classification of the rock sequence, determination of the elastic properties, and porosity of the rock		cementation check
		identification of joint and fracture zones	
BHTV		identification of fractures and joints	
OPT		identification of fractures and joints	casing check
TEMP	determination of the depth to the water table, localization of water influx and outflow, determination of the temperature distribution		
			recognition of leakage at casing joints, of circulation in the annulus fill, hydraulic effectiveness of the clay seals in the annulus
SAL	determination of the depth to the water table, localization of water influx and outflow, determination of the mineralization of the water		
	correction factor for determination of the resistivity of the rock		
MIL	localization of water influx and outflow, monitoring water quality		
FLOW	localization of water influx and outflow, identification of changes in the hydraulic conditions		
FMT			turbidity of water, determination of water flow velocity in tracer tests
SAMP	collection of water samples under the original pressure		
CAL	determination of the diameter and volume of the borehole, identification of downhole obstructions, correction factors for other physical parameters		identification of obstructions, casing joints and defects, locations of screens
		indirect detection of joints	
DEV	deviation of the borehole from the vertical and its direction		

Table 4.8-3: The most important logging methods and their possible applications depending on drilling mud and casing used: + measuring possible without limitations; * measuring possible with limitations; - measuring either not possible or impracticable.

Logging tool symbol	Parameter	Logging speed (m min^{-1})	Open hole, with drilling fluid	Steel casing, with drilling fluid	PVC casing, with drilling fluid	no casing, dry
GR	natural gamma-radiation	5	+	+	+	+
GRS	spectral natural gamma-radiation	1 - 3	+	+	+	+
GG	density	3 - 5	+	-	*	-
NN, INN	neutron porosity	3 - 5	+	-	*	-
EBS	chemical elements	1 - 2	+	-	-	-
SP	self potential	10	*	-	-	-
EL, ML	electrical resistivity (multiple point methods)	10	+	-	-	-
FEL, DLL, MLL	electrical resistivity (focused methods)	10	+	-	*	-
DIP	strike and dip	3 - 5	+	-	-	-
IL	electrical conductivity	10	+	-	+	+
SUSZ	magnetic susceptibility	5 - 10	+	-	+	+
AL	acoustic velocity (open hole)	3 - 5	+	-	-	-
CBL (cement-bond log)	acoustic velocity (cased hole)	3 - 5	-	+	-	-
BHTV (borehole televiewer)	acoustic velocity	1 - 3	+	-	-	-
OPT	visual image of the borehole wall	2 - 4	*	*	*	+
TEMP	temperature of the drilling fluid	10	+	+	+	-
SAL	electrical resistivity of the drilling fluid	10	+	+	+	-
MIL	pH, oxygen concentration, redox potential of the drilling fluid	3 - 5	+	+	+	-
FLOW	vertical flow rate of the fluid in the borehole	3 - 5	+	*	*	-
PMT	turbidity	1 - 3	*	*	*	-
SAMP	sample collection		+	+	+	-
CAL	borehole diameter	10	+	-	-	+
DEV	borehole geometry	10	+	-	+	+

The time needed for the actual logging can thus be calculated for a particular borehole and problem. A rule of thumb says that the total time needed is the actual logging time times 2. This factor takes into consideration the time needed to lower the tool to the bottom of the borehole before the logging is started, the time necessary to wait for temperature equilibrium of the electronic equipment (or crystals), and the time needed for setting up and dismantling of the equipment (tripod, diverter pulleys, laying out of cables, etc.). There are additional costs for mobilization and demobilization and for the surveying to determine the coordinates and elevation of the borehole. The well logging can be carried out by one geophysical engineer and one assistant. One geophysicist and possibly one assistant are needed for the processing, interpretation, and reporting. The processing will require a PC with the appropriate software and a plotter. The cost of equipment and services tends to be much lower than for petroleum logging. The cost of logging is typically a small percentage of the cost of drilling a borehole.

It is important to select the proper sequence in which the measurements are made. The following criteria should be taken into consideration:

- If temperature is to be measured, it should be measured first so that the drilling fluid is not first mixed by the lowering and raising of the other logging tools. Also, a temperature logging tool is relatively inexpensive, and its loss in the case of difficult conditions in the borehole would not be as serious as would be the case if one of the more "sophisticated", expensive tools were lost.
- If there is the danger of caving of the borehole walls during logging, the type of tool (free-hanging or wall contact) and the cost of the tool should be taken into consideration when the sequence of measurements is selected.
- Because it can be used for many purposes in both open and cased holes, a gamma-ray log is particularly important and is often one of the first measurements.

The safety measures required when radioactive sources are used must also be taken into consideration. Most of the standard density devices use 0.2 - 0.4 GBq (50 - 100 mCi) sources, which must be used in a limited access area and by an operator licensed for work with radioactive materials.

4.8.5 Quality Assurance

Repeated sections of a log allow to prove that there is no drift or artifact in the measurements. Repeatability, however, is not sufficient to demonstrate the quality of a well log. A number of factors influence the quality of logging data (THEYS, 1991). A major factor is the quality of the equipment – the detectors and logging tools. For example, high quality sensors for deviation tools have a precision of ± 0.2°. This means an important decision about the quality of the logging data is made when the logging tool is purchased. The quality of the data is also affected by how carefully the sensors are calibrated. A distinction is made between the initial calibration (recording of a calibration curve) and control calibrations. An example of the latter: The gamma-ray source in a density tool is usually cesium-137, which has a half-life of 30 years. The "aging" of the source must be taken into account by control measurements in test pits or in cylinders of a material with a known density.

The measured values are influenced not only by the rock, but also by the geometry of the borehole, the drilling fluid, the well completion, and the position of the tool in the hole (centering). The correction terms for these factors are called environmental corrections (caliper corrections, stand-off corrections, and corrections for mud properties). The complexity of the environmental factors that affect log sensors has led to the development of numerous empirical curves for determining their influence on the measured geophysical parameters (e.g. SCHLUMBERGER EDUCATIONAL SERVICES, 1989b; WESTERN-ATLAS INTERNATIONAL, 1985). Only when the environmental corrections have been applied, are data obtained that really represent the properties of the rock. The quality of these corrections has a direct influence on the quality of the data on the rock properties.

If the measurements are made in several passes through the borehole ("runs"), the "zero depth" must be the same for each run. To take possible deviations in the starting depth from run to run into consideration, as well as any stretching of the cable, a gamma-ray detector is included in each tool so that it is possible to correlate depths of the different logs. This processing step also takes differences in logging speed into consideration, for example caused by the tool becoming stuck in the hole (yoyo effect).

The logging speed depends on the logging method and has a large influence on the quality of the data. Typical logging speeds are given in *Table 4.8-3*. The sampling interval also influences data quality. It is typically 10 cm, with a vertical resolution of about 50 cm. Some tools have a higher sampling rate and resolution; e.g., the borehole televiewer can acoustically image sub-cm-scale features.

In total, the quality of the data and information content of geophysical well logs are dependent on the measurement conditions, the investigation interval, and the depth resolution. The measurement conditions are determined primarily by the drilling method, the borehole geometry (diameter, borehole wall, caving), the content of the borehole (water, air, drilling fluid) and its physical properties (density, electrical resistivity, pH, properties that affect the slowing and capture of neutrons), the properties and size of the infiltrated zone (open borehole), and the type and properties of the well completion (fully cased, screens, annulus filling, gravel, clay, cement).

The well logging methods – which are based on different physical principles – together with a sensor configuration adapted to the measurement conditions, can be used to obtain data from a limited, irregular rock volume.

The vertical and radial extent of this volume is influenced by the following factors:

- the borehole diameter and the physical properties of the content of the borehole (borehole fluid and well completion materials) and surrounding rock,
- the ratio of the borehole diameter to the diameter of the tool and the position of the tool in the borehole,
- the tool design (detector size, electrode spacing, transmitter-receiver spacing, radioactive source-detector spacing).

Thus, each tool has a characteristic depth resolution and an average radial depth of investigation under the given conditions, which are discussed in Section 4.8.3.

4.8.6 Processing and Interpretation of the Logging Data and Examples

Normally, one value for every ten centimeters is stored for each logging method. Field plots are prepared for the standard methods for immediate delivery to the client. The header of the log contains the identifying and descriptive data for the borehole and the logging method. Color can be used to distinguish between plots for different methods in a single graph. Color or hatching can be used to emphasize anomalies. The proper scale for each curve has to be included.

The data are interpreted in terms of the material composition and structure of the rock penetrated by the borehole: the lithology of the profile and, for example, the clay content from the gamma log or the porosity from the gamma-gamma log. Crossplots are used to obtain lithological details, such as porosity and degree of water saturation. Methods are available to estimate target parameters from one or more measured parameters on the basis of deterministic or probabilistic relationships. Pattern recognition methods are used for qualitative or semi-quantitative interpretation, especially for the identification of fractures and joints in the dip log and acoustic televiewer data.

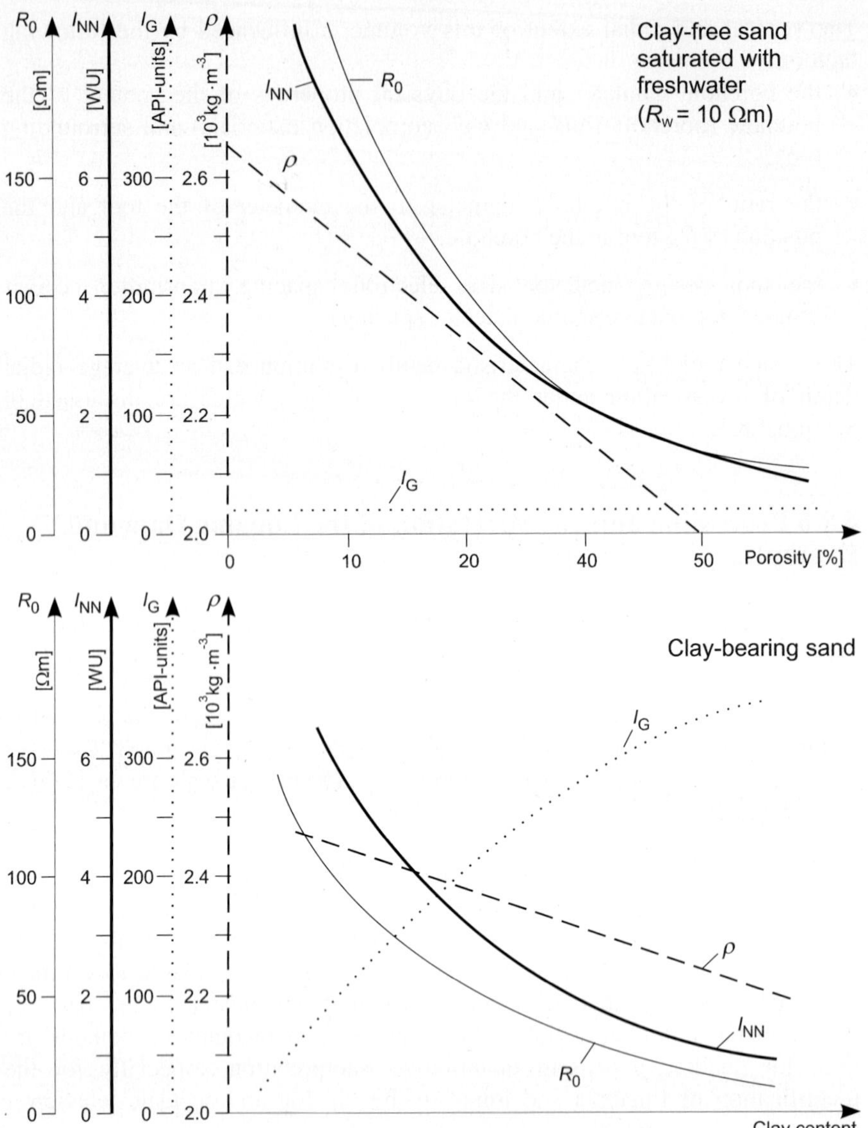

Fig. 4.8-2: Natural gamma radiation IG, neutron-neutron flux density INN, electrical resistivity R_0, and density ρ as functions of porosity and clay content for a clay-free, freshwater-saturated sand and a clay-bearing sand, respectively; R_w is the electrical resistivity of the groundwater.

The penetrated rock contains pores and solid material. The pores can be saturated with water or partially saturated. The solid material normally consists of several different minerals. The geophysical parameters are very sensitive to the physical properties and proportions of the rock components, the pore structure, and the pore filling. *Figure 4.8-2* shows natural gamma radiation I_G, neutron-neutron flux density I_{NN}, electrical resistivity R_0, and density ρ (gamma-gamma log) as functions of porosity and clay content for a clay-free, freshwater-saturated sand and a clay-bearing sand, respectively. Joint interpretation of several well logging parameters is necessary for a reliable assignment of the lithology of the penetrated strata and for reliable estimates of physical and other parameter values.

Lithological classification and documentation of the rock penetrated by a borehole

The primary objective of geophysical well logging is to obtain geological information from the data. Characteristic physical rock parameters, especially several in combination, can be used to determine or confirm the lithology of the rocks penetrated by the borehole. The most useful parameters for this are: gamma-radiation, density, porosity, electrical resistivity, magnetic susceptibility, acoustic velocity, and caliper (Sections 4.8.3.1 - 4.8.3.4 and 4.8.3.7). Increasing the number of parameters measured either increases the accuracy of the interpretation due to redundancy or can be used to determine further constituents of the rock. A plot of the typical responses of a set of hypothetical geophysical logs to different rock types is useful for deriving lithological profiles. This is shown for sedimentary rocks in *Fig. 4.8-3* and for igneous and metamorphic rocks in *Fig. 4.8-4*.

The idealized logs are based on experience with different kinds of rocks, as shown for example in *Fig. 4.8-4*: Porosity tends to be very low in igneous and metamorphic rocks, where groundwater flows only along fissures in the rock. Granite tends to be more radioactive than granodiorite and gabbro, but when weathered or altered, the gamma-ray signal may decrease. Pegmatites may be highly radioactive due to the inclusion of rare-earth minerals. Large fractures may be indicated by any of the logs. The caliper log may be misleading in steeply dipping fractures, as each of the arms makes a separate excursion into the fracture at different depths. A BHTV log is essential to correctly interpret the response of other logs to fractures.

An example of a lithological log derived from well logs made in unconsolidated rock at the Schöneiche test site near Berlin is shown in *Fig. 4.8-5*. The borehole was drilled with a nominal diameter of 180 mm using drilling mud. The resistivity and neutron-neutron logs indicate thick layers of sand at 32.8 - 61.2 m and 80.2 - 98.7 m in the borehole. The gamma-ray, resistivity, and density logs indicate layers of silt at 11.2 - 23.7 m and 61.2 - 80.2 m. The susceptibility log and microlaterolog have a vertical resolution in the centimeter range.

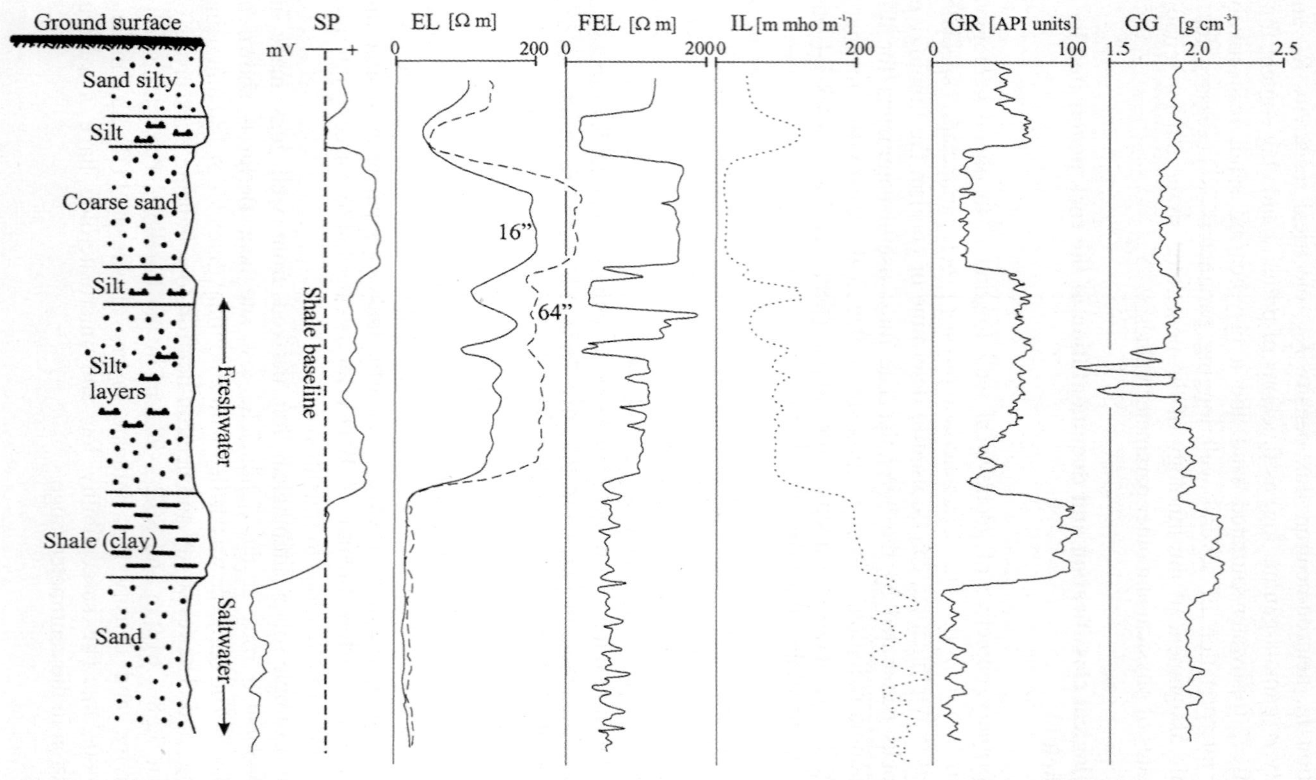

Fig. 4.8-3: Typical geophysical downhole logs for a sequence of sedimentary rocks, REPSOLD (1989). SP: self-potential log, EL: electric log 16- and 64-inch normal, FEL: focused electric log, IL: induction log, GR: gamma-ray log, GG.D: gamma-gamma-density log

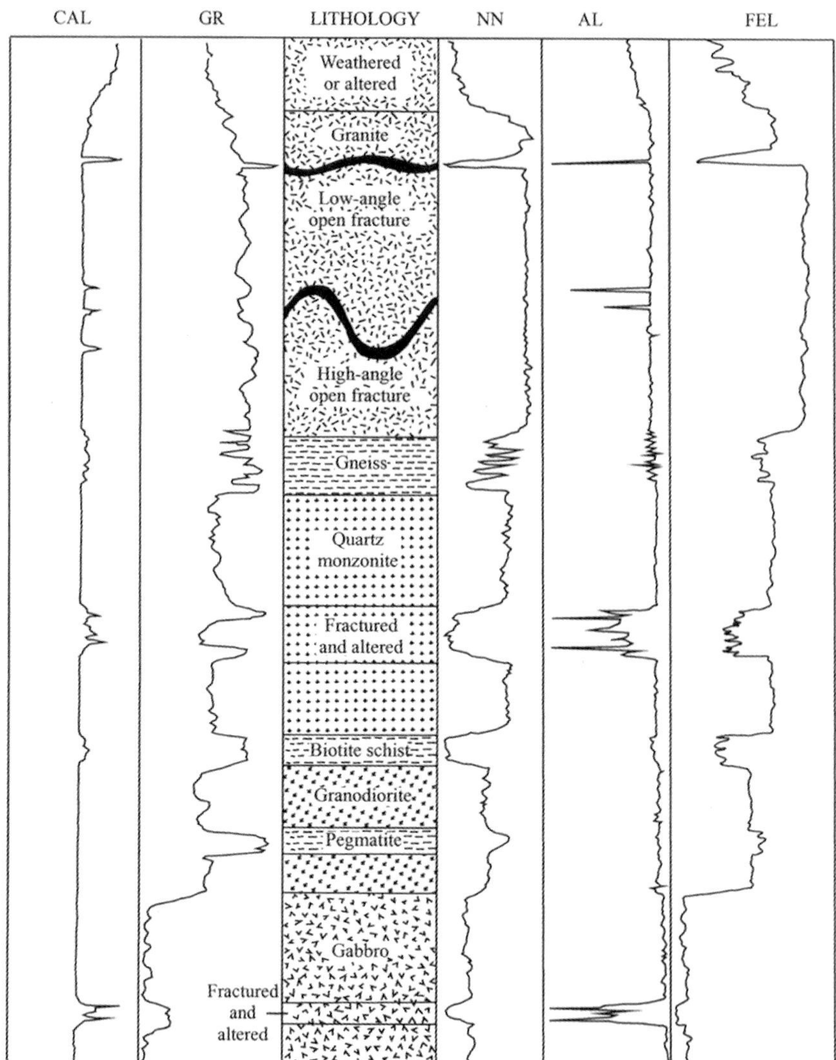

Fig. 4.8-4: Typical geophysical downhole logs for various crystalline rocks, KEYS (1997). All measurement units increase to the right. CAL: caliper log, GR: gamma-ray log, NN: neutron-neutron log, AL: acoustic log, FEL: focused electric log

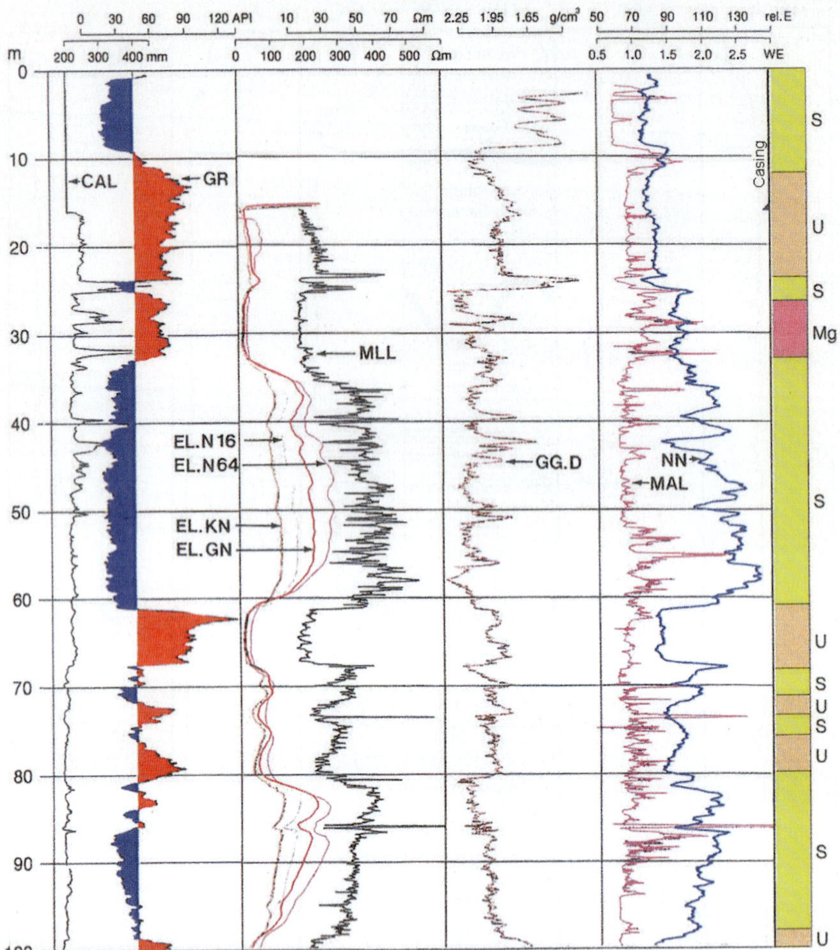

Fig. 4.8-5: Simplified lithology derived from borehole logs in unconsolidated sediments at Schöneiche in Brandenburg, Germany, ZSCHERPE & STEINBRECHER (1997). The following logs were made in the open borehole: caliper (CAL) in mm; gamma-ray (GR) in API units; electrical resistivity 16" normal (EL.N 16), 64" normal (EL.N 64), 25 cm normal (EL.KN, short normal), 100 cm normal (EL.GN, long normal); microlaterolog (MLL) in Ωm; gamma-gamma density (GG.D) in g cm^{-3} = 10^3 kg m^{-3}; neutron-neutron (NN) (calibrated with respect to water (WE=WU=water units)); and magnetic susceptibility (MAL) in relative units; S: sand; U: silt, Mg: marl

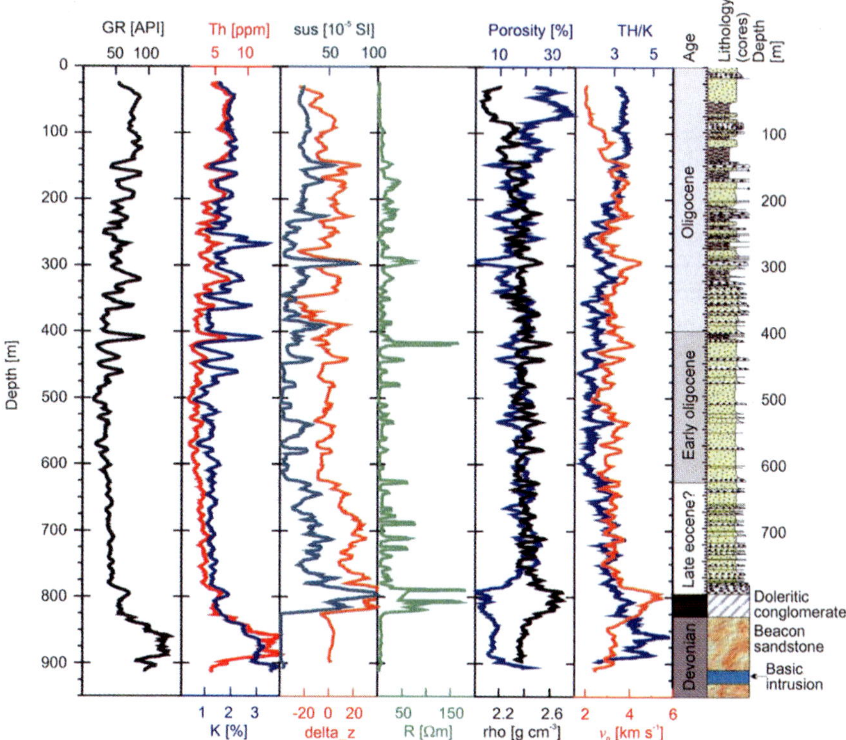

Fig. 4.8-6: Core-derived lithology and downhole logs for the Cape Roberts CRP-3 borehole in Antarctica, BÜCKER et al. (2001). GR: gamma-ray; Th: thorium; K: potassium; sus: susceptibility; delta_z: anomaly of vertical component of magnetic field; R: resistivity; porosity; rho: density; Th/K: thorium/potassium ratio; v_p: P-wave velocity

A further example of the interpretation of well logs is shown in *Fig. 4.8-6*. The borehole (CRP-3 at 77.007°S; 163.719°E) was drilled in the northern McMurdo Sound near Cape Roberts (Ross Sea, Antarctica) to a depth of about 900 m below the seafloor (BÜCKER et al., 2001). Glacially influenced marine sediments (sand, silt, and gravel) of early Oligocene and earliest Eocene age were encountered down to 790 m below surface. Below that depth, a coarse doleritic conglomerate and Devonian Beacon Sandstone were drilled. Lithology, grain size, and diagenesis can be identified on the basis of the geophysical signatures in the borehole logs. The physical properties of the Tertiary and Devonian sediments differ considerably, especially in the susceptibility, porosity, and gamma-ray logs. The Tertiary sediments are not as homogeneous as might be assumed from the visual core description. On the basis of the borehole logs, they can be subdivided into three units, indicating three source areas. Log Unit I, from 0 - 144 m, is characterized by generally very homogeneous log responses, particularly for the gamma-ray, susceptibility, and K/Th logs. The resistivity, density, and velocity logs show a downward increase, whereas porosity generally decreases. Log Unit II, from 144 to 630 m below surface, is characterized by bimodal responses for all logs, with one mode similar to Log Unit I. Alternations on a 10 - 30 m scale represent sedimentary changes. A basal conglomerate with high resistivity and density, low porosity, and high susceptibility marks most sequence boundaries. Some fining-upward sequences from clean sand to muddy sand and silt are visible in the log responses. Log Unit III, from 630 to 790 m, is lithologically relatively homogeneous, with low gamma-ray values, but elevated Th/K values and high susceptibilities. Porosity, density, and velocity are heterogeneous, with variations similar to those in Log Unit II: low porosities in the conglomerates and well-cemented sandstones, with high-porosity sandstones also present. Except for the conglomerates, grain-size changes are not as evident in the gamma-ray and susceptibility logs as in Log Unit II.

Interpretation of the well logs provides, for example, the thicknesses of homogenous layers for hydrogeological modeling. In addition to the identification of aquifers and aquitards, the lithological logs often aid decisions about the completion of the well.

Manual correlation of the strata penetrated by a borehole with neighboring boreholes

An example of the correlation of the geophysical well logs of neighboring boreholes is shown in *Fig. 4.8-7*. The borehole logs reflect the lithological units in the boreholes and can be correlated.

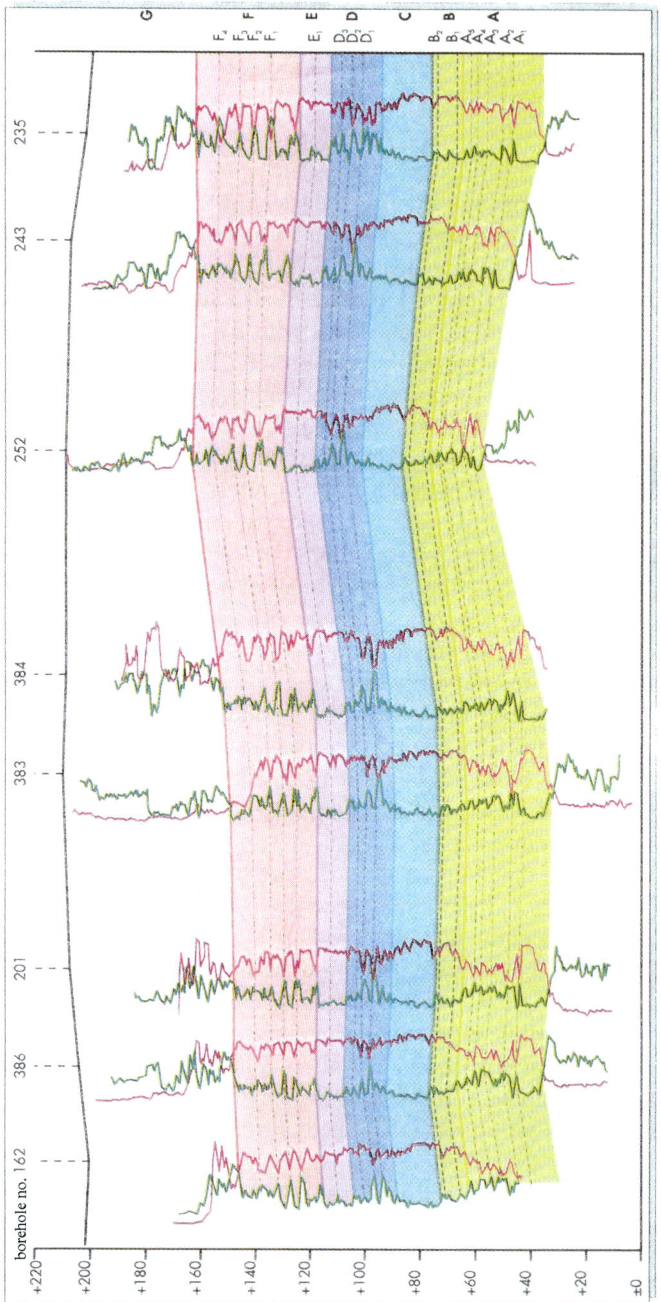

Fig. 4.8-7: Correlation of geophysical well logs made in the Berzdorf lignite district, Germany, ZSCHERPE & STEINBRECHER, (1997), green: natural gamma-activity (GR); red: gamma-gamma density (GGD)

Geologic modeling by computer well-log correlation

Contribution by RICARDO A. OLEA

Although manual correlation of logs by visual inspection continues to be the prevailing method for correlating logs, computer correlation offers several advantages. Computer correlation is consistent, in the sense that exactly the same criteria are used to correlate all the intervals, and repetition of a correlation using the same parameters and data will always yield the same results. Further advantages over manual correlation are the high resolution of the results and availability of cross-sections in electronic form. In this example the computer program CORRELATOR was used (OLEA & SAMPSON, 2002; OLEA, 2003a). The program allows the geometry of the subsurface layers to be examined, along with some of the petrophysical attributes. The method used by the program is a mathematical systematization of the skills gained by experienced geologists. As in manual correlations, the program requires two logs per well. In one of these logs, the measurement of a parameter is shown that directly relates with the amount of shale, e.g., natural gamma-ray radiation. The second log (called correlation log) can be of any petrophysical property with high vertical resolution, e.g., resistivity or neutron log. The program correlates intervals, not individual values and selects an optimum length for such intervals. CORRELATOR uses the shale log to discriminate intervals of clean and shaly horizons to reduce the number of intervals in a well that may correlated with any given interval in another well. The program uses the correlation log to compare the pattern of variation in the logs. The underlying presumption is that such variations are the result of changes through time in the depositional environment. Two intervals are considered lithostratigraphically equivalent when the product of their correlation and shale coefficients is at a maximum. This product is called the weighted correlation coefficient. As in manual correlation, hiatuses, erosion, and faulting are events that cannot be eliminated or ignored. Hence, sometimes an interval in a well does not have an equivalent interval in another well. A clear correlation will have a weighted correlation coefficient close to 1.0 and two totally unworkable intervals will have a weighted correlation coefficient close to zero.

For the two wells at Nong Harn, Chiang Mai, Thailand (see also Section 5.4.2), the gamma-ray, resistivity, neutron, and density logs are available. The density logs were discarded because of the high noise level. The logs were digitized during logging at 1 cm intervals. Due to the resolution of the tools the digitization interval was changed to 10 cm by ignoring 9 out of 10 successive readings. The resolution of the resistivity logs is significantly lower than that of the neutron and gamma-ray logs, a condition that in this case is an advantage in Nong Harn. The considerable heterogeneity of the sediments and the absence of any markers preclude clear correlation of the gamma ray and neutron logs. The resistivity logs, however, provide better clues about lithostratigraphic equivalences because they have fewer features. The disadvantage of low-resolution logs is that they do not allow many correlations. Consequently, the work in this study was organized in two stages. First, the gamma-ray and resistivity pair of logs was used to define the general geological architecture between wells; and then the results were used for the correlation of the gamma-ray logs with the neutron logs.

Figure 4.8-8 (a) displays all tie lines between the wells that have a reliability of 0.4 or higher, reliability that is provided by the weighted correlation coefficient in a set uses the resistivity log as correlation log. Color is used to denote the quality of the correlation. In *Fig. 4.8-8 (b)* the correlation log is the neutron log. The results from *Fig. 4.8-8 (a)* were used to start the interpretation of *Fig. 4.8-8 (b)*, whose preparation would have been more uncertain without this assistance. The color of the lines has been used to represent the amount of clay calculated from the gamma-ray log using a clay line of 350 units and a clean line at the minimum reading. *Figure 4.8-8 (c)* shows the lateral and vertical variations in lithology, which in this case is reduced to clay as one class and all coarser fractions as the other. The two wells at Nong Harn suggest a high clay content in the uppermost 30 meters, underlain by cleaner and tilting sediments. The high frequency in the gamma-ray and neutron logs is due to highly heterogeneous sediments rather than noise. Particularly for the shallow sediments, the geometry of the layering is subparallel to the land surface, which suggests that deposition was governed by a topography that did not change much during the time that it took to accumulate the top 30 meters. On the basis of two seismic sections measured in the vicinity, geophysicists suggest that the bottom portion of the colluvial sediments may be faulted. With just two wells, it is difficult to confirm or discard the existence of a fault between the two wells 30 m below the surface. Note also that the graphs have a 5fold vertical exaggeration, so the dip is not as pronounced in the subsurface as it looks in the figures.

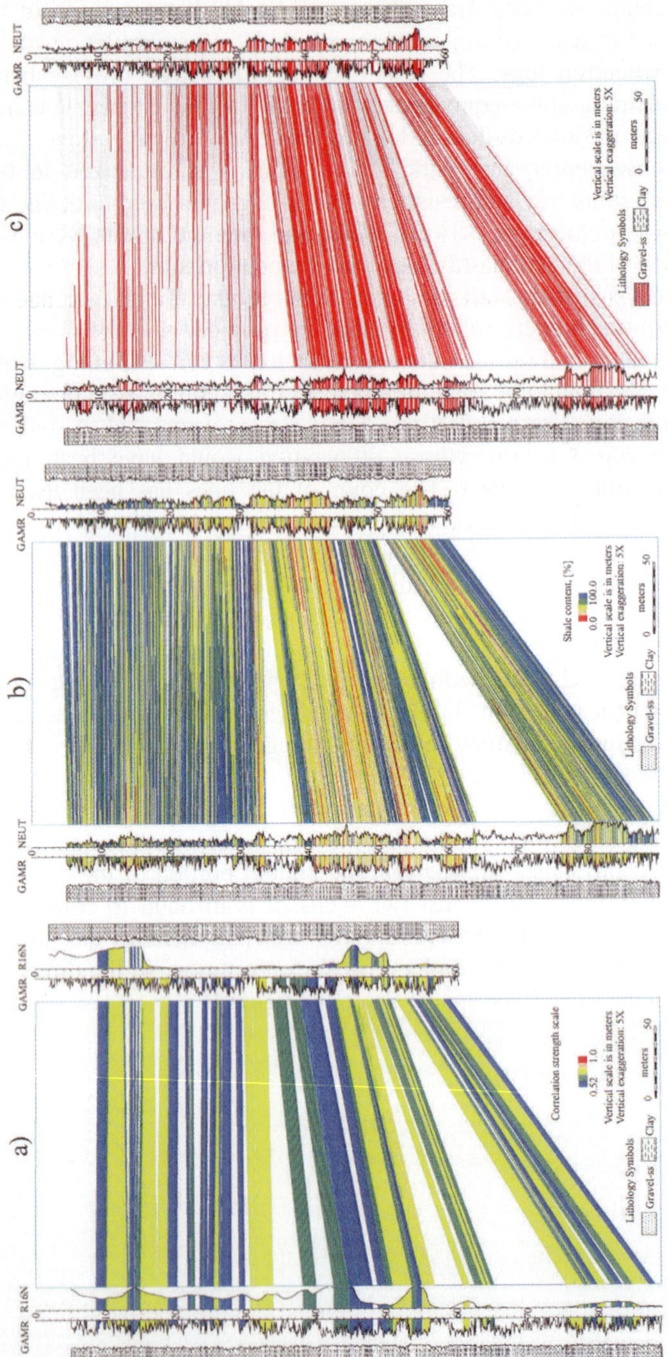

Fig. 4.8-8: Nong Harn, Chiang Mai, Thailand, (a) cross-section through wells 1a and 2a, lithostratigraphic correlations with a weighted correlation coefficient > 0.4 derived from the gamma-ray and resistivity logs, (b) shale percentage derived from the gamma ray and neutron logs, (c) lithology derived from the gamma ray and neutron logs

Estimation of petrophysical parameter values

Many geoscientific problems, particularly hydrogeological ones, can be solved by deriving petrophysical properties from well logs. A distinction is made here between direct and indirect well logging methods. In direct well logging methods, the desired parameter values are obtained directly from the well logs. For example, density values are taken directly from the GG log (Section 4.8.3.1). In indirect well logging methods, one or more parameters are logged and plotted. The desired rock parameter is then calculated from these logs. Analysis of geophysical well logs requires a knowledge of petrophysics (e.g., CARMICHEL, 1989; SCHÖN, 1996). Because rocks normally consist of a number of different minerals, their physical properties are determined by the proportions and properties of these minerals, their distribution in the rock and how they are bound. If the rock contains fluids (either liquids or gases), the properties of the rock are also influenced by the properties and distribution of the fluids within the rock.

The possibilities for determining the different petrophysical parameters from geophysical well logs are shown in *Table 4.8-4*. The number of geological models can be considerably reduced by quantitative data for parameters such as clay content, density, porosity, permeability, and storativity.

Table 4.8-4: Possibilities for determining petrophysical parameter values from geophysical well logs (after BLM, 1996)

Estimated parameters	Well logging method									
	SP	**FEL**	**IL**	**SUSZ**	**AL**	**GR**	**GRS**	**GG**	**NN**	**EBS**
density								D		
electrical resistivity		D	D							
dielectric constant			D							
magnetic susceptibility				D						
compression wave velocity					D					
shear wave velocity					D					
damping constant					D					
modulus of elasticity					B			B		
porosity		I	I					I,B	D,B	
water content									D	
degree of water saturation		B	B						D	
clay content	I					I	I			
clay minerals							I			I
ionic content of the pore water	I,B	I,B	I,B							
K, U, and Th content							D			
Al, Si, C, O, Ca, Fe content							I			D

D direct measurement of the parameter,
I indirect determination of the parameter values via petrophysical models,
B determination of the parameter values from the results of several logging methods

The often empirical methods for deriving petrophysical parameter values from well logs can be illustrated here with the estimation of shale volume from natural-gamma logs (Section 4.8.3.1). Although the gamma ray value for shale or clay varies enormously, in any area or borehole, the values for pure shale tend to be constant. The following equations provide the simplest way to calculate the volume of shale or clay V_{sh} from the radioactivity index *GRI*.

V_{sh} [%] = *GRI* [%],

where

GRI [%] = $(GR - GR_{min})/(GR_{max} - GR_{min})$ and

GR_{max} = 100 % shale, GR_{min} = 0 % shale, i.e., a clean formation.

This method tends to overestimate shale content in the intermediate ranges. Thus, modifications of this linear relationship have been proposed on the basis of empirical correlations by LARIONOV (1969) for pre-Tertiary (consolidated) rocks and Tertiary (unconsolidated) rocks. Other equations are available for other rocks, e.g., FRICKE & SCHÖN (1999) have published a statistical relationship between gamma-ray values and clay content for Rotliegend sandstone in northern Germany. *Figure 4.8-9* shows the relationship between radioactivity index, *GRI*, and shale or clay volume, V_{sh}.

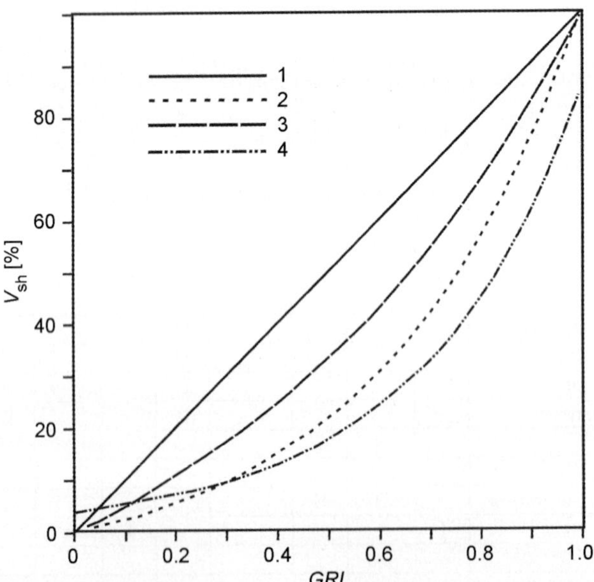

Fig. 4.8-9: The radioactivity index, *GRI*, as a function of shale or clay volume, V_{sh}, FRICKE & SCHÖN, (1999), 1: linear relationship 2: Tertiary (unconsolidated) rocks (LARIONOV, 1969) 3: pre-Tertiary (consolidated) rocks (LARIONOV, 1969) 4: Rotliegend sandstone in northern Germany (FRICKE & SCHÖN, 1999)

The shale or clay content V_{sh} (GR) in *Fig. 4.8-10* was calculated from the GR log in the Nieder Ochtenhausen borehole in northern Germany using the formula of LARIONOV (1969) for Tertiary sediments (WONIK, 2001). The clay content above a depth of 101 m increases from about 3 % in the Quaternary cover to 20 % in the Miocene sediments. Below 101 m the clay content increases rapidly to a maximum of 70 %, reducing again to about 30 % at 125 m. Grain-size analyses of the < 2 µm fraction taken from core samples are shown for comparison (HINDEL et al., 2001). They also determined the mineral composition by XRD (x-ray diffraction) analysis. The proportions of the clay minerals muscovite/illite, kaolinite, smectite and chlorite were added and are shown as V_{sh} (MKSC). The curves for the three methods show considerable agreement between 35 and 115 m. Below 115 m, the clay contents determined by grain-size analysis and XRD are similar (50 - 70 %) in contrast to the V_{sh} (GR) values of about 30 %. The geochemical and mineralogical composition of the Langenfeldian and Reinbekian stages differs considerably from that of the overlying Gramian stage (KUSTER, 2001). The factor that influences the natural gamma radiation the most, and thus the calculated potassium, thorium, and uranium concentrations, is the humus content. A high TOC content was determined in the Reinbekian stage (HINDEL et al., 2001), a very fine-grained material with a high clay mineral content, but in which K, Th and U have been replaced by nonradioactive elements. The gamma-ray log is a simple and good tool for estimating clay content, but the section below 115 m is an example that shows that automatic calculation from the GR log can lead to the wrong conclusions.

The gamma-gamma log (Section 4.8.3.1) measures the bulk density of rocks σ_b. Rock densities range from 1200 kg m^{-3} (= 1.2 g cm^{-3}) for soils and lignite to 3300 kg m^{-3} (= 3.3 g cm^{-3}) for peridotite (SCHÖN, 1996). The density of rocks strongly depends on their mineral composition and on the proportion of cavities. The total mass M of a volume V is given by

$$M = V\sigma_m\left(1 - \Phi\right) + V\sigma_f\Phi,$$

where Φ (porosity) is the percentage of the total volume V that is cavities, σ_m is the density of the solid components of the rock, and σ_f is the density of the fluid cavity fill.

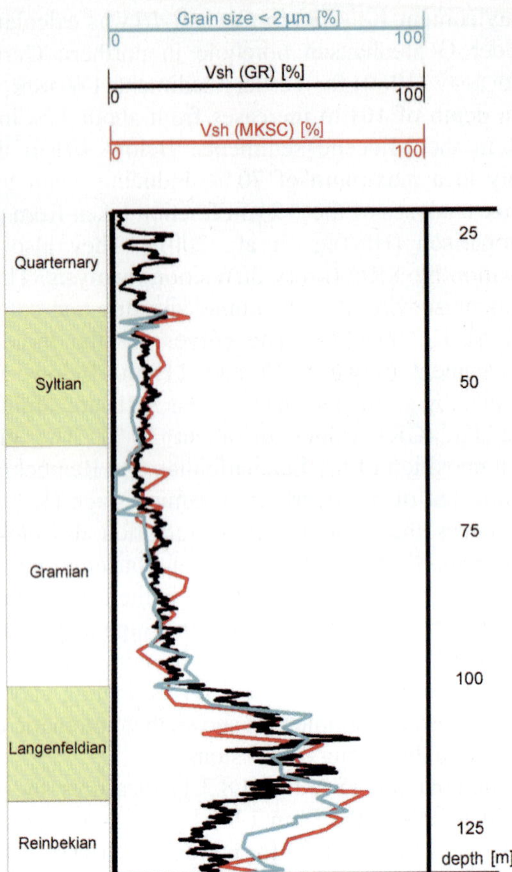

Fig. 4.8-10: Comparison of the clay content [%] in the Upper and Middle Miocene in the Nieder Ochtenhausen borehole in NW Germany derived from the results of three methods, WONIK (2001): grain-size analysis by sieving to determine the < 2 µm fraction (HINDEL et al., 2001); calculation of the V_{sh} from the GR log; and calculation from XRD data (V_{sh} (MKSC), see text) (HINDEL et al., 2001)

The bulk density σ_b is given by

$$\sigma_b = \sigma_m - (\sigma_m - \sigma_f)\Phi .$$

Solving this equation for porosity, we obtain

$$\Phi = (\sigma_m - \sigma_b)/(\sigma_m - \sigma_f) .$$

The influence of cavities on the density of igneous and most metamorphic rocks can usually be neglected ($\sigma_b \rightarrow \sigma_m$). In this case the variation of density indicates variation of the mineral composition of these rock types. Cavities and their fill strongly influence the bulk density of unconsolidated rock and porous rock. Combining the results of the gamma-gamma (σ_b) and the neutron-neutron (Φ) tools and making assumptions about the fluid density, σ_f, it is possible to calculate the matrix density, σ_m, which provides a strong indication of the lithology.

The three logging tools density, neutron-neutron (Section 4.8.3.1), and acoustic (Section 4.8.3.4) can be used to determine porosity. If only one of the tools is used, assumptions are made about the lithology and fluid properties required to estimate porosity. These assumptions must be made on the basis of local knowledge or other measurements. If more than one logging tool is used, the combined data is normally sufficient to determine porosity quite precisely.

The first attempts to derive permeability values from well logs were made in the 1950s using porosity and resistivity data (Sections 4.8.3.1 and 4.8.3.2). A qualitative permeability index can be indirectly calculated from gamma-ray, conductivity, and acoustic logs (Sections 4.8.3.1, 4.8.3.3, and 4.8.3.4). Permeability can be determined using the nuclear magnetic resonance method (NMR) by making certain assumptions (see Chapter 4.7). There is as yet no permeability logging tool for the slimhole range. It is expected that such a tool will be developed soon (Section 4.8.7).

Spectral gamma-ray logs (Section 4.8.3.1) can be used to distinguish between different clayey sediments on the basis of their U, Th, and K content, providing indirect information about their contaminant retention capacity, i.e., how well the groundwater is protected from contamination. Acoustic logs (Section 4.8.3.4) via the compression modulus provide indirect information about the storage coefficient of the rock.

Determination of water influx and efflux in wells

Horizons from which there is inflow into or outflow from the well are determined by measuring vertical flow in the well with a flowmeter (Section 4.8.3.6). These measurements also permit determination of the proportion of the total yield contributed by each horizon. Self-potential, temperature, and drilling fluid resistivity measurements (Sections 4.8.3.2 and 4.8.3.6) can also be used for this purpose. Other methods (e.g., PMT, Section 4.8.3.6) using tracers allow determination of groundwater flow rates.

Derivation of information about the rock structure

Information about rock textures can be obtained by well logging, especially in consolidated rocks. Caliper logs (Section 4.8.3.7) indicate the locations of caving of the borehole wall, thus indirectly indicating hydraulic flow paths. Borehole deviation logs (Section 4.8.3.7) provide information about changes in the direction of the borehole, and thus about the present stress field. Indirect information about fractures can be obtained by determining Th anomalies from spectral gamma-ray logs (Section 4.8.3.1). Dip and strike of rock strata can be estimated from dipmeter logs (Section 4.8.3.2). Dipmeter logs are also useful for detecting fractures (i.e., hydraulic pathways). Fractures and the dip of strata can be determined with excellent accuracy using a borehole televiewer (Section 4.8.3.4) in the ultrasound range. Programs are available for calculating dip angles from the data from the last two methods named. An example is shown in *Fig. 4.8-11* from the Fürstenwald borehole near Kassel, Germany. The borehole penetrated fine-grained sandstone with claystone and siltstone intercalations at a depth of 204 - 207 m (Lower Triassic). In this section the dip is 25° - 30° to the north and northwest, increasing to about 45° at 204.5 m. This data permits the drill core to be oriented.

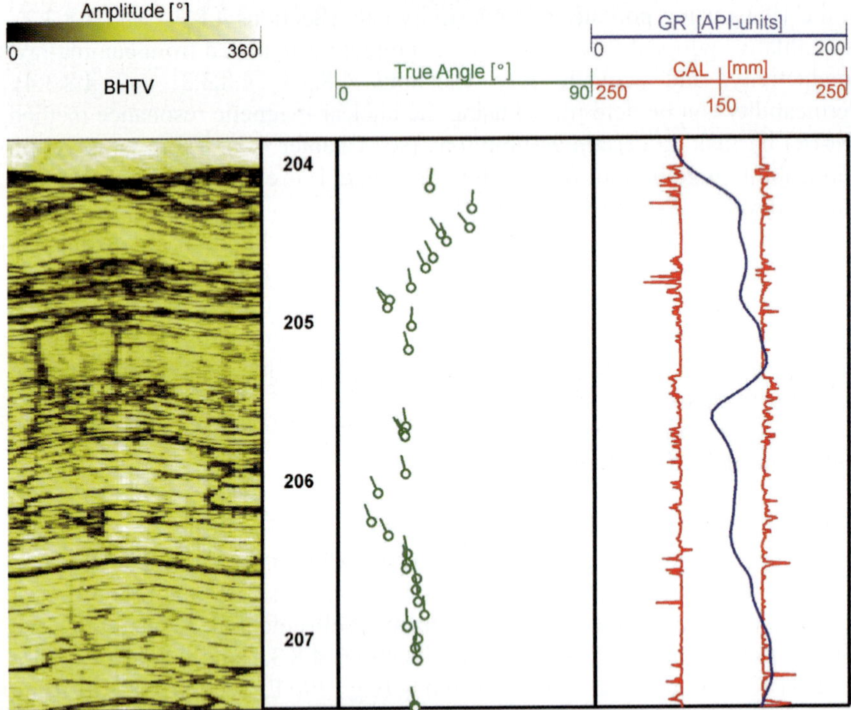

Fig. 4.8-11: Acoustic image (BHTV) of the depth interval 203.8 - 207.5 m in the Fürstenwald borehole near Kassel (central Germany). The lithology consists of fine-grained sandstone with claystone and siltstone intercalations of Lower Triassic age. The deviation from the vertical in this interval is 1° WNW. GR: gamma-ray log, CAL: caliper log, True Angle: direction and dip angle of fractures and other structures in the borehole wall

Inspection of well completion and of technical conditions of a well

Well logs can be used to check the condition of a well. Damage to PVC casing, for example, can be easily determined via focused electrologs (Section 4.8.3.2). Caliper logs (Section 4.8.3.7) and videocameras (Section 4.8.3.5) can also be used to check the condition of a well. Several parameters are logged to check the annulus fill, e.g., the depth and condition of the clay barriers and the cementing. For this purpose, natural gamma-ray logs, magnetic susceptibility, density, and neutron-neutron logs (Sections 4.8.3.1 and 4.8.3.3) provide the best information. Acoustic logs are used to check cementing (Section 4.8.3.4). Electromagnetic methods (Section 4.8.3.3) have proved to be useful for identifying metal scrap forced into the rock during drilling. An example of well logs for checking the condition of a groundwater observation well is shown in *Fig. 4.8-12*. The well was drilled with a diameter of 150 mm and reamed out to 250 mm. Well completion was carried out using several different materials to fill the annulus for testing and calibration purposes. The well has a 5" diameter PVC casing with screens at depths of 13 - 20 m and 24 - 37 m. The following materials were used for the annulus fill: Dämmer-Suspension™, Quellon™, gravel, clay spheres, drill cuttings, Witterschlicker Clay™, Compactonit™, and a cement-bentonite-monazite-sand mixture.

The well logs provide an unambiguous determination of the annulus fill. The Quellon swelling clay can be clearly identified in the gamma-ray and susceptibility logs. The Compactonit swelling clay (69.0 - 75.5 m) can be identified in the neutron-neutron, gamma-gamma, and susceptibility logs. The strata boundary of the Witterschlicker Clay can be identified in the gamma-gamma and neutron-neutron logs. The gravel can be recognized in the neutron-neutron log together with gamma-gamma log.

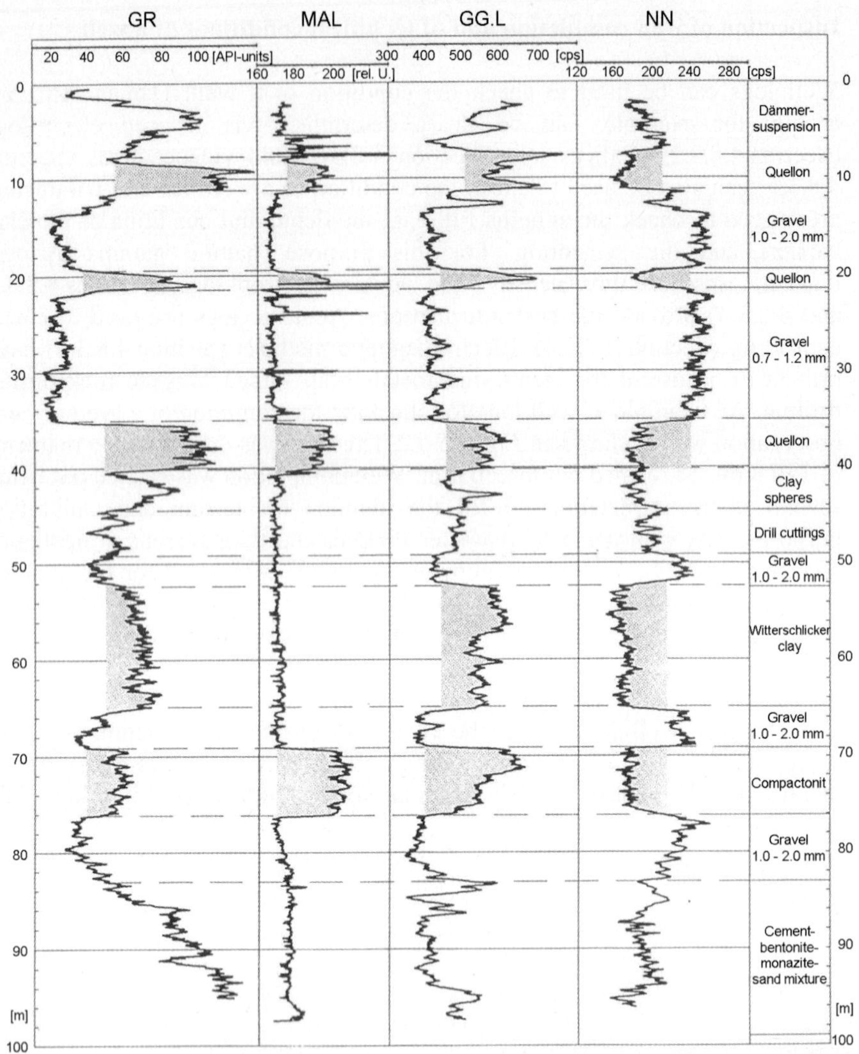

Fig. 4.8-12: Borehole logs testing the well completion of a groundwater observation well, ZSCHERPE & STEINBRECHER (1997): The different materials used for filling the annulus are indicated by the following logs: GR (gamma log in API units), MAL (magnetic susceptibility in relative units), GG.L (gamma-gamma log in cps), NN (neutron-neutron log in cps)

In-situ determination of physico-chemical parameter values of groundwater

The physical and chemical properties of groundwater can be determined by taking samples at specific depths or by measuring temperature, electrical resistivity, pH, redox potential, and degree of oxygen saturation (Section 4.8.3.6). Determination of the freshwater/saltwater boundary is an important application of drilling fluid resistivity logging. This cannot be done, however, immediately after the drilling of the well or if a saline drilling fluid was used. In such cases, it may be possible to determine the location of the freshwater/saltwater boundary from the electrical resistivity/conductivity of the rock (Sections 4.8.3.2 and 4.8.3.3). The logging tools used for this purpose have a relatively large lateral depth of penetration of about 100 cm and the data thus also reflect the electrical properties of the pore water in the sediment. A DLL log (Section 4.8.3.2) made in a borehole about 5 km from the coast near Cuxhaven is shown in *Fig. 4.8-13*. The electrical conductivity of the drilling mud was 0.33 S m^{-1}. A transition zone separating freshwater and saltwater is recognizable between 50 and 60 m in the DLL logs.

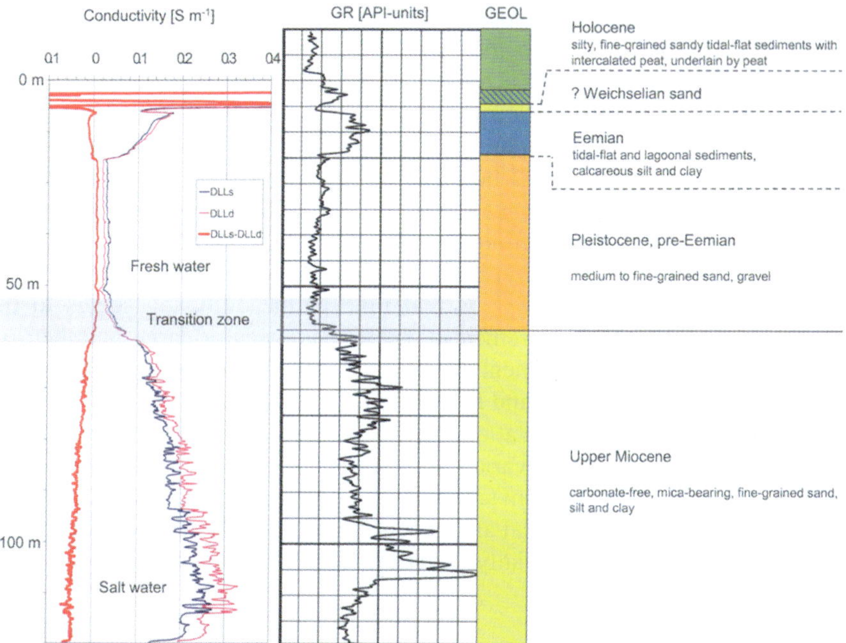

Fig. 4.8-13: Freshwater/saltwater interface in the Cuxhaven borehole in northern Germany in conductivity logs DLL; the electrical conductivity of the drilling fluid was 0.33 S m^{-1}; DLLs: shallow dual laterolog; DLLd: deep dual laterolog; GR: gamma-ray log

Contamination of the groundwater can be easily detected if the electrical resistivity is changed by the contaminant. This is the case for brine and wastes from the metal processing industry. Organic substances do not ionize sufficiently to change the resistivity of the water, especially at low concentrations. Chemical and optoelectro sensors can be used to determine concentrations of dissolved substances (e.g., nitrate). Such sensors are becoming increasingly important for identifying and delimiting contamination plumes originating from old landfills and other contaminated sites. Well logs can be used for monitoring water quality in a well or well field. See also Chapter 4.9.

Further information on the processing and interpretation of the logging data is given, for example, by ALBERTY (1992), BASSIOUNI (1994) and DOVETON (1994).

4.8.7 Expected Future Developments

Many of the well logging methods developed for oil exploration have also been applied in shallow wells with a delay of 5 - 10 years. Improvements can be expected in imaging and scanning methods (LOVELL et al., 1999), but also new developments that permit the logging of parameters that have not been possible to log up to now. The new methods will make it possible to better answer numerous geoscientific questions. It can also be expected that new methods will open new questions. Two methods that have not been possible to use in the slimhole range of borehole diameters can be mentioned here: (i) formation microscanner: Detailed, high-resolution spatial data providing structural and textural information could be obtained from the measurement of electrical resistivity with a large number of electrodes (LOVELL et al., 1999). (ii) nuclear magnetic resonance: A real time permeability log seems to be possible from NMR measurements. There has been a large number of publications on the development of such a method, for example, KENYON (1997), COATES et al. (1999) and DUNN et al. (2002).

The application of interpretation methods for hydrocarbons exploration and basement rocks (e.g., multivariate statistics) to shallow boreholes opens further possibilities for problem solving. Combination of borehole geophysical methods with other geophysical and geochemical methods may be expected to be increasingly applied, especially to hydrogeological problems.

References and further reading

ALBERTY, M. W. (Ed.) (1992): Wireline Methods. In: MORTON-THOMPSON, D. & WOODS, A. M. (chief Eds.): Development Geology Reference Manual. AAPG Methods in Exploration Series, 10, 142-194, The American Association of Petroleum Geologists, Tulsa, USA.

BASSIOUNI, Z. (1994): Theory, measurement and interpretation of well logs. Society of Petroleum Engineers Textbook Ser. 4. Richardson, Texas.

BLM (1996): Petrophysik für die Analyse geophysikalischer Bohrlochmessungen. Messen im Untergrund, **4**, 1-11, Gommern.

BRAM, K. & DRAXLER, J. K. (1995): Basic Research and Borehole Geophysics (Final Report). Borehole Logging in the KTB-Oberpfalz HB Interval 6013.5 - 9101.0 m. KTB Report **94-1**, 453.

BÜCKER, C., JARRARD, R. D., NIESSEN, F. & WONIK, T. (2001): Statistical Analysis of Wireline Logging Data of the CRP-3 Drillhole (Victoria Land Basin, Antarctica). Terra Antartica, **8**, 4, 491-505.

BULLARD, E. C. (1947): The Time Necessary for a Borehole to Attain Temperature Equilibrium. Mon. Notes R. Astron. Soc, **5**, 127-130.

CARMICHEL, R. S. (1989): Physical properties of rocks and minerals. CRC Press, Boca Raton.

COATES, G. R., XIAO, L. & PRAMMER, M. G. (1999): NMR Logging: Principles and Applications. Haliburton Energy Services, Houston.

DOVETON, J. H. (1994): Geologic Log Analysis Using Computer Methods. AAPG Computer Applications in Geology, **2**.

DOVETON, J. H. & PRENSKY, S. E. (1992): Geological Applications of Wireline Logs: A Synopsis of Developments and Trends. The Log Analyst, **33**, 3, 286-303.

DUNN, K. J., BERGMAN, D. J. & LATORRACA, G. A. (2002): Nuclear Magnetic Resonance – Petrophysical and Logging Applications. Pergamon, Amsterdam.

FANG, J. H., KARR, C. L. & STANLEY, D. A. (1996): Transformation of Geochemical Log Data to Mineralogy Using Genetic Algorithms. The Log Analyst, **37**, 2, 26-31.

FRICKE, S. & SCHÖN, J. (1999): Praktische Bohrlochgeophysik. Enke, Stuttgart.

GOCHIOCO, L. M., MAGILL, C. & MARKS, F. (2002): The Borehole Camera: An Investigative Geophysical Tool Applied to Engineering, Environmental, and Mining Challenges. The Leading Edge, **21**, 5, 474-477.

GOLDBERG, D. (1997): The Role of Downhole Measurements in Marine Geology and Geophysics. Review of Geophysics, **35**, 3, 315-342.

HALLENBURG, J. K. (1987): Geophysical Logging for Mineral and Engineering Applications. Penn Well Books, Tulsa, Oklahoma.

HALLENBURG, J. K. (1992): Nonhydrocarbon Logging. The Log Analyst, **33**, 3, 259-269.

HARVEY, P. K. & LOVELL, M. A. (Eds.) (1998): Core-log Integration. Geol. Soc. Spec. Publ., **136**, London.

HEARST, J. R., NELSON, P. H. & PAILLET, F. L. (2000): Well Logging for Physical Properties. 2nd edn., Wiley & Sons, Chicester.

HINDEL, R., KOCH, J., RÖSCH, H., WEHNER, H. & ZIMMERLE, W. (2001): Geochemische und petrographische Untersuchungen an Sedimenten aus der Forschungsbohrung Nieder Ochtenhausen. In: MEYER, K.-J. (ed.): Forschungsbohrung Nieder Ochtenhausen – Ein Beitrag zur Miozän-Stratigraphie in NW-Deutschland. Geol. Jahrbuch, **A 152**, 71-100.

IAEA (International Atomic Energy Agency) (1999): Nuclear Geophysics and its Applications. International Atomic Energy Agency, Technical Reports Series, **393**, Vienna.

JACOBSEN, L. A. & WYATT, D. F. (1996): Elemental Yields and Complex Lithology Analysis From the Pulsed Spectral Gamma Log. The Log Analyst, **37**, 1, 50-64.

JORGENSEN, D. G. & PETRICOLA, M. (1995): Research Borehole-Geophysical Logging in Determining Geohydrologic Properties. Ground Water, **33**, 4, 589-596.

KENYON, W. E. (1997): Petrophysical Principles of Applications of NMR Logging. The Log Analyst, **38**, 2, 21-43.

KEYS, W. S. (1990): Borehole Geophysics Applied to Ground-water Investigations. Techniques of Water-resources Investigations of the United States Geological Survey, Book 2, Chapter E2, Washington.

KEYS, W. S. (1997): A Practical Guide to Borehole Geophysics in Environmental Investigations. Lewis Publishers, Boca Raton, Florida.

KRAMMER, K. (1997): Geophysikalische Bohrlochmessungen bei der Erkundung von Deponiestandorten. In: BEBLO, M. (Ed.): Umweltgeophysik. Ernst, Berlin.

KUSTER, H. (2001): Forschungsbohrung Nieder Ochtenhausen: Bemerkungen zum Ansatzpunkt, zur Lithologie und Paläogeographie. In: MEYER, K.-J. (Ed.): Forschungsbohrung Nieder Ochtenhausen – Ein Beitrag zur Miozän-Stratigraphie in NW-Deutschland. Geol. Jahrbuch, **A 152**, 15-38.

LARIONOV, W. W. (1969): Radiometrija skwaschin. Nedra Verlag, Moscow.

LOFTS, J. C., HARVEY, P. K. & LOVELL, M. A. (1995): The Characterization of Reservoir Rocks Using Nuclear Logging Tools: Evaluation of Mineral Transform Techniques in the Laboratory and Log Environments. The Log Analyst, **36**, 2, 16-28.

LOVELL, M. A., WILLIAMSON, G. & HARVEY, P. K. (Eds.) (1999): Borehole Imaging: Applications and Case Histories. Geol. Soc. Spec. Publ., **159**, London.

LUTHI, S. M. (2001): Geological Well Logs – Their Use in Reservoir Modeling. Springer, Berlin.

MARES, S., ZBORIL, A. & KELLY, W. E. (1994): Logging for the Determination of Aquifer Hydraulic Properties. The Log Analyst, **35**, 6, 28-36.

MAUTE, R. E. (1992): Electrical logging: State-of-the-art. The Log Analyst, **33**, 3, 206-227.

MEYERS, G. D. (1992): A Review of Nuclear Logging. The Log Analyst, **33**, 3, 228-238.

MOLZ, F. J. & YOUNG, S. C. (1993): Development and Application of Borehole Flowmeters for Environment Assessment. The Log Analyst, **34**, 1, 13-23.

OLEA, R. A. (1999): Geostatistics for Engineers and Earth Scientists. Kluwer Academic Publishers

OLEA, R. A. & SAMPSON, R. J. (2002): CORRELATOR 5.2 – computer program and user manual: Open–File Report No. 2002-51, Kansas Geological Survey, Lawrence, Kansas.

OLEA, R. A. (2003a): High-resolution characterization of subsurface geology by computer assisted correlation of wireline logs: Proceeding of the 12th Annual Meetings of the Gesellschaft für Geowissenschaften held in Husum, 74-77.

OLEA, R. A. (2003b): Geologic modeling of colluvial sediments in Thailand by computer well-log correlation. Unpublished report Bundesanstalt für Geowissenschaften und Rohstoffe, Berlin.

PAILLET, F. L., CHENG, C. H. & PENNINGTON, W. D. (1992): Acoustic-waveform Logging-advances in Theory and Application. The Log Analyst, **33**, 3, 239-258.

PRENSKY, S. (2002): Recent Developments in Logging Technology. Petrophysics, **43**, 3, 197-216.

REPSOLD, H. (1989): Well Logging in Groundwater Development. Internat. Contributions to Hydrogeology, **9**; Verlag Heinz Heise, Hannover.

RIDER, M. (1996): The Geological Interpretation of Well Logs. 2nd edn., Whittles Publishing, Caithness, Scotland.

SAMWORTH, J. R. (1992): Quantitative Open-hole Logging With Very Small Diameter Wireline Tools. SPWLA 33rd Annual Logging Symposium, Paper NN.

SCHLUMBERGER EDUCATIONAL SERVICES (1987): Schlumberger Log Interpretation Principles/Applications, Houston TX.

SCHLUMBERGER EDUCATIONAL SERVICES (1989a): Schlumberger Log Interpretation Principles II, Houston TX.

SCHLUMBERGER EDUCATIONAL SERVICES (1989b): Schlumberger Log Interpretation Charts, Houston TX.

SCHÖN, J. (1996): Physical Properties of Rocks: Fundamentals and Principles of Petrophysics. Elsevier, Amsterdam.

SERRA, O. (1984): Fundamentals of Well-log Interpretation. Part 1: The Acquisition of Logging Data. Developments in Petroleum Science, **15A**, Elsevier, Amsterdam.

SERRA, O. (1986): Fundamentals of Well-log Interpretation. Part 2: The Interpretation of Logging Data. Developments in Petroleum Science, **15B**, Elsevier, Amsterdam.

SPIES, B. R. (1996): Electrical and Electromagnetic Borehole Measurements: A Review. Surveys in Geophysics, **17**, 517-556.

TAYLOR, K., HESS, J. & WHEATCRAFT, S. (1990): Evaluation of Selected Borehole Geophysical Methods for Hazardous Waste Site Investigations and Monitoring. Final Report, United States Environmental Protection Agency, EPA/600/4-90/029, Las Vegas, Nevada.

THEYS, P. P. (1991): Log Data Acquisition and Quality Control. Éditions Technip, Paris.

WESTERN ATLAS INTERNATIONAL (1985): Atlas Log Interpretation Charts, Houston TX.

WONIK, T. (2001): Ergebnisse bohrlochgeophysikalischer Messungen in der Forschungsbohrung Nieder Ochtenhausen. In: MEYER, K.-J. (Ed.): Forschungsbohrung Nieder Ochtenhausen – Ein Beitrag zur Miozän-Stratigraphie in NW-Deutschland. Geol. Jahrbuch, **A 152**, 1-512.

YEARSLEY, L. E. & CROWDER, R. E. (1990): State-of-the-art Borehole Geophysics Applied to Hydrology. Canadian/American Conf. on Hydrology, Sept. 18-20, 1990.

ZSCHERPE, G. & STEINBRECHER, D. (1997): Bohrlochgeophysik. In: KNÖDEL, K., KRUMMEL, H. & LANGE, G. (Eds.): Handbuch zur Erkundung des Untergrundes von Deponien und Altlasten, **3**, Geophysik, Springer, Berlin, 789-896.

4.9 Geophysical In-situ Groundwater and Soil Monitoring

FRANZ KÖNIG, KLAUS KNÖDEL, KLAUS-HENRIK MITTENZWEY & PETER WEIDELT

4.9.1 Principle of the Methods

The area around both vulnerable sites (e.g., waterworks) and hazardous sites (e.g., landfills and industrial plants) have to be monitored for contaminants. The integrity of contaminant barriers and the success of rehabilitation of contaminated sites must also be monitored. The spread of contaminants occurs via three paths: water, soil, and air. Of these, groundwater and surface water are the most significant.

Current practice is to take water samples from observation wells and soil and gas samples at vulnerable and hazardous sites at regular intervals for chemical analysis on site and/or in the laboratory. It is also current practice to use a multi-parameter probe monitoring several environmental parameters in wells, such as temperature, electrical conductivity, pH, redox potential and oxygen concentration in the water. The repeated visits to the observation wells are personnel intensive and hence expensive. Further disadvantages are the uncertainties deriving from the sample collection, transport, and the chemical analysis, in addition to the masses of data produced. The advantages are that numerous substances can be analyzed and that the chemical analyses can be used in court. All of these procedures yield point data. To adequately monitor a site, numerous sampling points are necessary.

A new technological development permits continual in-situ monitoring over long periods of time. The concept is based on a combination of measurements made with multi-parameter probes and an optical sensor system to obtain point data together with an electromagnetic (EM) system to obtain spatial data.

Multi-parameter probe is an instrument for measuring several environmental parameters at the same time, such as electrical conductivity, temperature, pH, redox potential, and oxygen concentration. The information about the quality of groundwater and surface water that can be derived from environmental parameter values is discussed in Section 4.9.3.1.

Electromagnetic monitoring system: Electrical conductivity is a sensitive parameter for monitoring contaminants from landfills or a freshwater/saltwater interface. The real and imaginary components of the electrical conductivity are functions of the frequency of the transmitted electromagnetic waves. The water and electrolyte content can be derived from the conductivity function.

This function is determined for selected frequencies from the measured signal attenuation and phase shift between the transmitter and receiver antennas.

Optosensor system: This sensor system permits simultaneous measurement of transmission/absorption, scatter, fluorescence, and refraction. (a) Transmission: Molecules absorb radiation energy and convert it to another form (e.g., heat or radiation at another frequency). The amount of absorbed energy depends on the concentration of the substance. The concentration of substances such as phenols, BTEX, dyes, explosives, and nitrate can be determined in this way. Humic acids also absorb light and can serve as an indicator of a pulse of seepage water from a landfill. (b) Radiation can also be scattered by molecules and particles. The extent of scatter depends on concentration and particle size and provides a measure of turbidity and macromolecule content. (c) Fluorescence is the emission of radiation after excitation of a molecule by absorption of radiation energy. The intensity of fluorescence is proportional to the concentration of the fluorescing substances. Because few substances fluoresce, e.g., PAH, this parameter is selective for certain groups of substances. (d) When radiation enters a liquid, it changes its velocity and direction. This is called refraction. The index of refraction depends on the liquid and the substances in it. Refraction measurements can be used to determine substances that do not absorb in the UV/VIS/NIR spectral range, e.g., salts and fatty acids.

The cost and uncertainties of monitoring can be reduced by using a permanently installed in-situ system. Such a system reduces the frequency of on-site visits to take samples and the subsequent chemical analyses. The disadvantages of the customary system – uncertainties deriving from the sample collection, transport, and the chemical analysis, in addition to the masses of data produced – are avoided or reduced. In addition, it is possible to react more quickly to the data obtained online. The reliability of the data is increased by the combination of point and spatial data.

A schematic of such an on-site monitoring system is shown in *Fig. 4.9.1.* To avoid the influence of the system components (probes, optosensors, and EM system) on each other, the three components are activated separately. The multi-parameter probes and the optosensors can be installed together in a 3" borehole. Transmitters and receivers for the EM system are placed in other boreholes to avoid disturbances from the cables of the probes and optosensors.

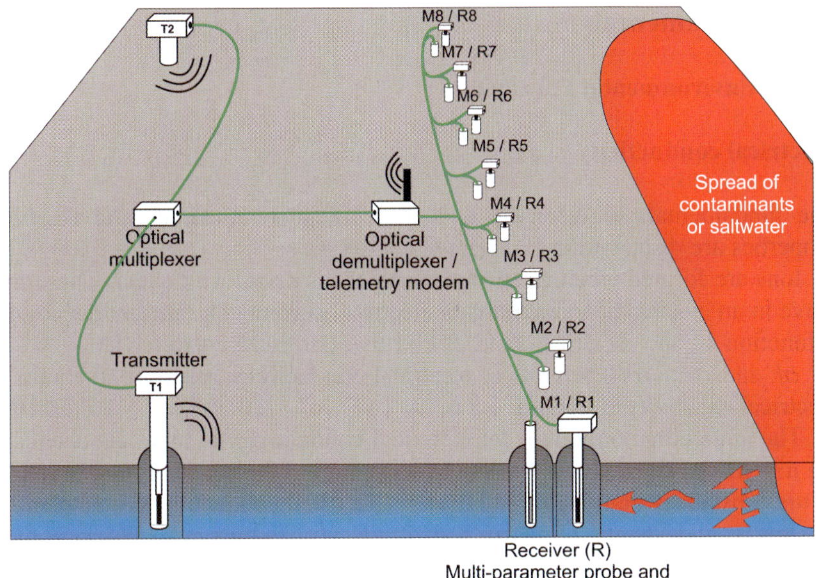

Fig. 4.9.1: Schematic of a permanently installed on-site monitoring system

4.9.2 Applications

- Monitoring of vulnerable sites (e.g., waterworks and water protection areas),
- monitoring of hazardous sites (e.g., landfills, industrial areas, and mines),
- monitoring of freshwater/saltwater interfaces,
- monitoring of well water,
- monitoring of waste water,
- monitoring of the rising groundwater after lignite mining is terminated, and
- monitoring of the integrity of contaminant barriers and the success of rehabilitation of contaminated sites.

4.9.3 Fundamentals

4.9.3.1 Environmental Parameters

Electrical conductivity

The fundamentals of electrical and electromagnetic methods and electrical properties are discussed in Chapters 4.3 and 4.4.

Ions are formed when electrolytic substances dissolve in water. These ions move in an electric field, creating an electrical current. The current density I is a function of the electrical field E and the electrical conductivity σ, where $I = \sigma E$ and $\sigma = 1/\rho$. The units of electrical conductivity are S m^{-1}, the units of electrical resistivity ρ are Ωm. 1 S m^{-1} = 1 Ω^{-1} m^{-1} = 10^3 mS m^{-1} = 10^4 µS cm^{-1}.

The units commonly used for electrical conductivity in hydrogeochemistry are µS cm^{-1}, because the values then correspond to the concentration in mg L^{-1} of the substances dissolved in the water. Electrical conductivity is temperature dependent and always normalized to 25 °C. Tables are available to convert values measured at another temperature to 25 °C (HÖLTING, 1989). This is normally done automatically by the software delivered with environmental parameter monitoring systems.

Pure water has an electrical conductivity of $5.483 \cdot 10^{-2}$ µS cm^{-1}, owing to its auto-ionization. The following conductivities can be expected for water (HÖLTING, 1989; JORDAN & WEDER, 1995; MATTHESS, 1990):

Source of water	**Conductivity** [µS cm^{-1}]
distilled or deionized water	1 - 10
rainwater	5 - 50
freshwater (drinking water)	30 - 2000
groundwater	30 - 5000
stream or lake water	100 - 5000
mineral water, medicinal water	10 000 - 20 000
sea water	45 000 - 55 000
oilfield water	> 100 000

The limit for drinking water in the German drinking water regulations of 1990 is 2000 µS cm^{-1}; the limit in the EU guidelines of 1980 is 400 µS cm^{-1}.

The electrical conductivity of an aqueous solution is a function of the following parameters (for details, see BERKTOLD, 1982 & VOIGT, 1990):

- the concentration of the electrolyte and its degree of ionization,
- interaction between the ions (with increasing concentration, this can lead to a maximum conductivity, followed by decreasing conductivity),
- ionic charge,

- mobility of the ions in solution,
- density, viscosity, dielectric constant, pressure, and temperature.

Temperature

Groundwater, as well as lake and river water, has an annual temperature "wave" down to a specific depth. The depth at which the annual temperature becomes nearly constant (fluctuation less than 0.1 °C) is called the damping depth. It is at about 15 m for groundwater in the temperate zone. The amplitude of fluctuation depends on the vegetation cover: less below a forest than below grassland. Below the damping depth, the temperature increases (20 - 40 m/°C) with increasing depth, depending on the geology. The temperature in the damping depth corresponds to the mean annual temperature of the locality. In addition, groundwater temperature is influenced by groundwater recharge, groundwater flow, dissolution, and anthropogenic factors (e.g., exothermic reaction in a landfill).

pH

Pure water undergoes self-ionization to form H^+ (actually H_3O^+) and OH^- ions in equilibrium with each other. At 22 °C these ions have a concentration of 10^{-7} M (M = moles per liter). If an acid or base is added to the water, the concentration of hydrogen and hydroxide ions changes. When acids add hydrogen ions, the concentration of hydroxide is decreased; when bases add hydroxide ions, the hydrogen ion concentration is decreased. The equilibrium constant for this is given by the following formula:

$$K_{\mathrm{w}} = c\left(\mathrm{H}^+\right) f\left(\mathrm{H}^+\right) c\left(\mathrm{OH}^-\right) f\left(\mathrm{OH}^-\right), \tag{4.9.1}$$

where K_w is the temperature-dependent equilibrium constant for the self-ionization of water, $c\,(H^+)$ and $c\,(OH^-)$ are the concentrations of the hydrogen and hydroxide ions in mol L^{-1}, and $f(H^+)$ and $f(OH^-)$ are the activity coefficients. The product of $c\,(H^+)$ and $f(H^+)$ is called the hydrogen ion activity or apparent hydrogen ion concentration $a\,(H^+)$. The pH is defined as the negative logarithm of the hydrogen ion activity:

$$\mathrm{pH} = -\log c\left(\mathrm{H}^+\right) f\left(\mathrm{H}^+\right). \tag{4.9.2}$$

pH values lie between 0 and 14, i.e., apparent hydrogen ion concentrations between 10^0 and 10^{-14} M. Neutral water, i.e., when H^+ and OH^- have equal concentrations, has a pH = 7. Acidic solutions have a pH < 7, basic solutions have a pH > 7. The pH value is less temperature-dependent than the electrical conductivity of water. The solubility of many substances is a function of pH. The significance of pH in hydrogeochemistry is discussed, for example, in

APPELO & POSTMA (1996), DREVER (1988), HÖLTING (1989), JORDAN & WEDER (1995), MATTHESS (1990) and VOIGT (1990).

Groundwater with no anthropogenic contaminants has a pH of 5.5 to 8 (in exceptional cases as high as 11.6, see HÖLTING, 1989). River and lake water has a similar range. Rainwater has a pH between 3.0 and 9.8, varying greatly with time and place (CARROLL, 1962). Sea water has a pH of 8.

Redox potential

The redox potential E_h is a measure of the activity of the oxidizing and reducing substances dissolved in the water. The redox potential is measured between an inert electrode (usually platinum) and a standard hydrogen electrode. The redox potential of a redox reaction is given by the Nernst equation:

$$E_h = E_0 + \frac{R\,T}{n\,F} \ln\left(\frac{a_{ox}}{a_{red}}\right), \tag{4.9.3}$$

where E_0 is the standard potential [V] under standard conditions 25 °C, 10^5 Pa, $a_{ox} = a_{red} = 1$,
a_{ox}, a_{red} are the activities of the oxidizing and reducing substances,
n is the number of electrons transferred in the reaction,
T is the temperature [K],
R is the gas constant (= 8.3143 J K^{-1} mol^{-1}), and
F is the Faraday constant (= 9.64867 A s mol^{-1}).

Standard potentials for reactions of interest in hydrogeochemistry are given by, for example, APPELO & POSTMA (1996), HÖLTING (1989), MATTHESS (1990), SCHÜRING et al. (2000), and VOIGT (1990). The Nernst equation is expanded for pH-dependent reactions (MATTHESS, 1990). The significance of pH in hydrogeochemistry is discussed, for example, in HÖLTING (1989), MATTHESS (1990), SCHÜRING et al., (2000) and VOIGT (1990).

Oxygen content

The concentration of oxygen (in mg L^{-1}) in both groundwater and lake and river water is one of the most important factors affecting the quality of the water for living organisms and the chemical reactions that occur in the water. Oxygen is physically dissolved in water, rather than chemically (like CO_2 or HCl, for example). The source of the dissolved oxygen is the atmosphere, where it has a mean concentration of 20.95 vol%. Its concentration in soil air is normally less than that in the atmosphere. The equilibrium concentration between the gas phase and the aqueous phase is temperature dependent and dependent from the partial pressure of the oxygen in the gas phase, and the type and concentration of the other dissolved substances. Oxygen not only enters groundwater and surface water directly from the atmosphere or soil air, but also via precipitation.

Chemical reactions in the unsaturated zone of the soil, in surface water and groundwater consume oxygen (MATTHESS, 1990). The extent to which this occurs is reflected by the oxygen saturation index: the ratio (in percent) of the oxygen concentration to the theoretical saturation concentration at the ambient pressure and temperature. Good quality drinking water has an oxygen concentration near the saturation concentration. The redox potential is a function of oxygen concentration.

4.9.3.2 Optical Spectroscopy

Transmission/absorption

When light is passed through a fluid, its intensity is decreased. The ratio of the intensity I after passing through the fluid to the intensity I_0 of the entering light is termed "transmission" T:

$$T = I/I_0 . \qquad (4.9.4)$$

The decrease in intensity is caused by substances in the fluid that absorb or scatter the light. Substances that absorb light are dissolved in the fluid, e.g., aromatic and aliphatic hydrocarbons. Substances that scatter light are suspended and colloidal particles. The transmission is then given by the following equation:

$$T = e^{-(\alpha+\beta)d} \qquad (4.9.5)$$

where α is the absorption coefficient, β is the scattering coefficient, and d is the distance traveled by the light in the fluid.

This equation is valid for ideal conditions. Under natural conditions, e.g., in groundwater, microbes and minerals form a film on the lens and mirrors that are in contact with the water during the measurement. To correct for this, the equation must be extended by an extinction term E_B:

$$T = e^{-(\alpha+\beta)d-E_B} \qquad (4.9.6)$$

To eliminate the extinction term, the transmission is measured through two paths of different lengths, where the transmission for the longer path T_L is given by

$$T_l = e^{-(\alpha+\beta)d_l-E_B} , \qquad (4.9.7)$$

where d_l is the longer distance (e.g., 0.1 m), and the transmission of the shorter path T_s is given by

$$T_s = e^{-(\alpha+\beta)d_s - E_B} \tag{4.9.8}$$

where d_s is the shorter distance (e.g., 0.01 m).

It is assumed that the film affects the lens and mirrors uniformly (fouling). The corrected transmission T_c is then obtained by dividing Equation (4.9.7) by Equation (4.9.8):

$$T_c = T_1 / T_s = e^{-(\alpha+\beta)(d_1 - d_s)} \tag{4.9.9}$$

To determine the concentration of absorbing substances, Equation (4.9.9) is solved for the absorption coefficient α:

$$\alpha = \frac{1}{(d_1 - d_s)} \ln(1/T_c) - \beta . \tag{4.9.10}$$

For a single absorbing substance, α is directly proportional to its concentration:

$$\alpha = \tilde{\alpha}\, C_A \tag{4.9.11}$$

where $\tilde{\alpha}$ is a constant which depends on the wavelength and the absorbing substance.

The scattering coefficient β is less substance specific than α. As a rough approximation, it can be assumed to be linear with respect to wavelength with a slightly negative slope. There are two possibilities for elimination of the influence of scattering:

- measurement at an absorption maximum and at an absorption minimum, followed by dividing the one by the other;
- measurement of the intensity of the scattered light, calculation of β, and calculation of α using Equation (4.9.10).

Scattering

The intensity of the scattered light I_s is given by the following equation:

$$I_s = I_0 H \frac{\beta}{\alpha+\beta}\left(1 - e^{-(\alpha+\beta)d}\right) \tag{4.9.12}$$

where I_0 is the intensity of the light as it enters the fluid, H is a constant (for the instrument used for the measurement), and d is the path length. It can be seen that the absorption coefficient α is needed to calculate the intensity of scattering. Moreover, the scattering is also influenced by the film fouling the lens and mirrors. Inserting Equation (4.9.7) into Equation (4.9.12), the following equation is obtained:

$$I_s = I_0 H \beta f(T_1) \tag{4.9.13}$$

where

$$f(T_1) = \frac{(1 - T_1)d}{\ln(1/T_1)}.$$

Equation (4.9.13) is then solved for β:

$$\beta = \frac{I_s}{I_0 H f(T_1)} \quad (4.9.14)$$

β is directly proportional to the concentration of the matter responsible for the scattering:

$$\beta = \tilde{\beta} C_s \quad (4.9.15)$$

$\tilde{\beta}$ is a substance-specific constant. The intensity of scattering is dependent on the concentration of the particles, their diameter and their shape. It is often used as a measure of turbidity.

Fluorescence

The intensity of fluorescence is given by an equation similar to that for scattering:

$$I_F = I_0 H \frac{Q_F \alpha_F}{\alpha + \beta}\left(1 - e^{-(\alpha+\beta)d}\right) \quad (4.9.16)$$

where Q_F is the fluorescence quantum yield; α_F is the absorption coefficient of the fluorescing substance, which is a component of the total absorption coefficient α; β is the scattering coefficient; d is the path length; and H is a constant for the apparatus. Like for scattering, insertion of Equation (4.9.7) into Equation (4.9.16) yields the following equation for the fluorescence intensity:

$$I_F = I_0 H Q_F \alpha_F f(T_1) \quad (4.9.17)$$

where

$$f(T_1) = \frac{(1 - T_1)d}{\ln(1/T_1)}.$$

Equation (4.9.17) is then solved for α_F:

$$\alpha_F = \frac{I_F}{I_0 H Q_F f(T_1)} \quad (4.9.18)$$

α_F is directly proportional to the concentration of the fluorescing substance:

$$\alpha_F = \tilde{\alpha}_F C_F \tag{4.9.19}$$

where $\tilde{\alpha}_F$ is a substance-specific constant which depends on the wavelength.

Refraction

There are many substances that cannot be detected by absorption, fluorescence, or scattering in the UV-VIS spectral range, e.g., ethanol, salt, sugar, and fatty acids. Refraction is a possibility for detecting these substances. The index of refraction n_F of a liquid is determined from the intensity of light I_{SPEC} reflected from a glass/liquid interface as follows:

$$I_{SPEC} = \left(\frac{n_G - n_F}{n_G + n_F} \right)^2 \tag{4.9.20}$$

where n_G is the index of refraction of the glass in contact with the liquid. I_{SPEC} is a normalized parameter with values between 0 and 1. The following values illustrate the range that can be expected:

glass/water:	0.00360 = 0.360 %
glass/ethanol:	0.00205 = 0.205 %
glass/air:	0.04000 = 4.000 %
aluminum/air:	0.85000 = 85.00 %.

Solving Equation (4.9.20) for n_F, we obtain

$$n_F = \frac{n_G \left(1 - I_{SPEC}^{1/2}\right)}{1 + I_{SPEC}^{1/2}} . \tag{4.9.21}$$

The parameter n_F is directly proportional to the index of refraction n_N of a substance dissolved in the liquid and to its concentration C_N:

$$n_F = f\left(C_N n_N\right). \tag{4.9.22}$$

Because the index of refraction of water at a given temperature is known, any deviation of n_F from this value is due substances dissolved in the water.

4.9.3.3 EM Monitoring

The field of the borehole antenna used in the electromagnetic monitoring system can be approximated by the field of an electrical dipole with an infinitesimally short length l in a homogenous space with a conductivity σ and a permittivity ε. This approximation does not take into consideration (a) that the dipole has a finite length with an undetermined current distribution,

(b) that the dipole is electrically insolated and only inductively coupled to its surroundings, or (c) the geometry of the antenna and the borehole.

The following equations are used for the case of harmonic excitation of the antenna with a frequency f and a time factor $e^{i\omega t}$, where $\omega = 2\pi f$:

$$\nabla \times \boldsymbol{E} = -i\omega\mu_0 \boldsymbol{H}, \tag{4.9.23}$$

$$\nabla \times \boldsymbol{H} = (\sigma + i\omega\varepsilon)\boldsymbol{E} + \boldsymbol{I} l \delta^3(\boldsymbol{r} - \boldsymbol{r}_0)\hat{z}, \tag{4.9.24}$$

where

$\boldsymbol{E}$	electric field vector,
$\boldsymbol{H}$	magnetic field vector,
$\nabla \times$	rotation (curl) operator (nabla × operator),
i	imaginary unit,
μ_0	magnetic permeability in a vacuum $\mu_0 = 4\pi \times 10^{-7}$ Vs A^{-1} m^{-1},
ω	angular frequency,
$\boldsymbol{I}$	current in the dipole,
l	length of the electric dipole,
σ	electrical conductivity,
$\varepsilon = \varepsilon_0 \varepsilon_r$	permittivity,
ε_0	permittivity of a vacuum = 8.854×10^{-12} As V^{-1} m^{-1},
ε_r	relative permittivity,
$\hat{z}$	vertical unit vector,
$\boldsymbol{r}_0$	position vector of the dipole,
$\boldsymbol{r}$	observation point, and
$\delta^3(\cdot)$	three-dimensional Dirac Delta function.

By elimination of $\boldsymbol{H}$, we obtain

$$\nabla \times \nabla \times \boldsymbol{E} = -k^2 \boldsymbol{E} - i\omega\mu_0 I l \delta^3(\boldsymbol{r} - \boldsymbol{r}_0)\hat{z}, \tag{4.9.25}$$

where

$$k^2 = i\omega\mu_0(\sigma + i\omega\varepsilon). \tag{4.9.26}$$

Equation (4.9.25) is solved using the Hertz vector Π, which has only one component in the z-direction,

$$\Pi = \Pi \hat{z} \tag{4.9.27}$$

to obtain

$$\boldsymbol{E} = -k^2 \Pi + \nabla\nabla \cdot \Pi \tag{4.9.28}$$

and

$$\boldsymbol{H} = (\sigma + i\omega\varepsilon)\nabla \times \Pi. \tag{4.9.29}$$

Π satisfies the following equation:

$$\nabla^2 \Pi = k^2 \Pi - \frac{Il}{\sigma + i\omega\varepsilon} \delta^3 (\boldsymbol{r} - \boldsymbol{r}_0). \tag{4.9.30}$$

Solving this equation, we obtain

$$\Pi = \frac{Il e^{(-kR)}}{4\pi(\sigma + i\omega\varepsilon) R}, \quad R = |\boldsymbol{r} - \boldsymbol{r}_0|. \tag{4.9.31}$$

Using cylindrical coordinates and placing the dipole at the origin of the coordinate system, we obtain the following field components from Equations (4.9.27), (4.9.28), (4.9.29), and (4.9.31):

$$E_r = \frac{Il(3 + 3u + u^2) r z e^{-u}}{4\pi(\sigma + i\omega\varepsilon) R^5}, \tag{4.9.32}$$

$$E_z = \frac{Il\{(3 + 3u + u^2) z^2 - (1 + u + u^2) R^2\} e^{-u}}{4\pi(\sigma + i\omega\varepsilon) R^5} \tag{4.9.33}$$

$$H_\phi = \frac{Il(1 + u) r e^{-u}}{4\pi R^3} \tag{4.9.34}$$

where

$$R = \sqrt{r^2 + z^2} \text{ and } u = kR. \tag{4.9.35}$$

The field can be subdivided on the basis of the value of the dimensionless parameter u:

For the near-field, $|u| \ll 1$,

$$E_r = \frac{3Ilrz}{4\pi\sigma R^5}, \tag{4.9.36}$$

$$E_z = \frac{Il(3z^2 - R^2)}{4\pi\sigma R^5}, \tag{4.9.37}$$

$$H_\phi = \frac{Ilr}{4\pi R^3}; \tag{4.9.38}$$

and for the far-field, $|u| \gg 1$,

$$E_r = +\frac{Ilk^2 r z e^{-u}}{4\pi(\sigma + i\omega\varepsilon) R^3}, \tag{4.9.39}$$

$$E_z = -\frac{I l k^2 r^2 e^{-u}}{4\pi(\sigma + i\omega\varepsilon)R^3}, \tag{4.9.40}$$

$$H_\phi = \frac{I l k r e^{-u}}{4\pi R^2}. \tag{4.9.41}$$

For the present application, the near-field equations are of minor importance, because they are for the dc field of a grounded dipole antenna, and the borehole antenna is an insolated antenna without any galvanic contact to its surroundings.

The following table gives typical values for $k = \alpha + i\beta$ [m^{-1}] for selected frequencies f, conductivities σ and relative permittivities ε_r:

(a) for $\varepsilon_r = 1$

f [kHz]	0.003 S m^{-1}	0.01 S m^{-1}	0.03 S m^{-1}	0.1 S m^{-1}
10	0.0112 + i 0.0112	0.0199 + i 0.0199	0.0353 + i 0.0353	0.0628 + i 0.0628
32	0.0199 + i 0.0199	0.0353 + i 0.0353	0.0628 + i 0.0628	0.1117 + i 0.1117
100	0.0353 + i 0.0354	0.0628 + i 0.0628	0.1117 + i 0.1117	0.1987 + i 0.1987
316	0.0627 + i 0.0630	0.1116 + i 0.1118	0.1986 + i 0.1987	0.3533 + i 0.3534
1000	0.1108 + i 0.1127	0.1981 + i 0.1992	0.3530 + i 0.3536	0.6281 + i 0.6285

(b) for $\varepsilon_r = 10$:

f [kHz]	0.003 S m^{-1}	0.01 S m^{-1}	0.03 S m^{-1}	0.1 S m^{-1}
10	0.0112 + i 0.0112	0.0199 + i 0.0199	0.0353 + i 0.0353	0.0628 + i 0.0628
32	0.0198 + i 0.0199	0.0353 + i 0.0354	0.0628 + i 0.0628	0.1117 + i 0.1117
100	0.0350 + i 0.0356	0.0627 + i 0.0630	0.1116 + i 0.1118	0.1986 + i 0.1987
316	0.0611 + i 0.0646	0.1108 + i 0.1127	0.1981 + i 0.1992	0.3530 + i 0.3536
1000	0.1024 + i 0.1220	0.1932 + i 0.2043	0.3502 + i 0.3565	0.6266 + i 0.6301

For the relatively low frequencies that are used, a quasi-stationary approximation can be used in which displacement current is neglected. The real and imaginary parts of k are approximately equal in this approximation. The increase in permittivity from 1 to 10 in the above tables shows how little permittivity affects the value of k. The tables also show that in the range of interest for f, σ, and ε_r, the parameter $u = kR$, where R = 10 - 100 m, ranges from $|u| \ll 1$ to $|u| \gg 1$. Thus, there is no simple asymptotic equation like Equations (4.9.36) to (4.9.41) that can be used to calculate the conductivity σ.

Assuming far-field conditions for the field strengths E_{z1} and E_{z2} measured in two boreholes at points (r_1, z_1) and (r_2, z_2) such that $|u_1| \gg 1$ and $|u_2| \gg 1$, we obtain the following quasi-stationary approximation from Equation (4.9.40):

$$\sigma = \frac{2}{\omega\mu_0}\left\{\frac{1}{R_2 - R_1}\ln\frac{|E_{z2}|R_2^3/r_2^2}{|E_{z1}|R_1^3/r_1^2}\right\}^2 \tag{4.9.42}$$

When $z_1 = z_2 = 0$, Equation (4.9.42) can be simplified as follows:

$$\sigma = \frac{2}{\omega\mu_0}\left\{\frac{1}{r_2 - r_1}\ln\frac{r_2|E_{z2}|}{r_1|E_{z1}|}\right\}^2 . \tag{4.9.43}$$

It is very difficult to calculate the electrical field strength at the point of transmission. The reason for this is the quasi-simultaneous overlap of near- and far-field components. Because the near-field is reactive out-of-phase, its energy is idle in a standing wave on the antenna. It is better to use the dipole moment $I\,l$ of the transmitter, which is proportional to the field strength of the receiver.

If the transmitter is operated in resonance mode, the electrical conductivity of the soil can be calculated from the dipole moment of the transmitter and the field strength E_z of the receiver. In this case, the reactive power of the transmitter has no influence on the field propagating away from the antenna in the conducting surroundings. Hence, the conductivity of the soil can be calculated using Equation (4.9.40) for the far-field. It follows for the absolute values $|E_z|$ and $|I|$ for $z = 0$:

$$\sigma = \frac{2}{\omega\mu_0}\left\{\frac{1}{r^2}\ln\left(\frac{4\pi r|E_z/I|}{\omega\mu_0 l}\right)\right\}^2 . \tag{4.9.44}$$

In order to use this equation, the antenna current must be measured.

4.9.4 Instruments

The monitoring system described here consists of multi-parameter probes, optosensors, and an EM system. These three components can be used for many applications in different combinations. All of the system components can be installed in boreholes with a diameter of 5 cm or more and at water depths of up to 100 m (10 bar). The probes are corrosion resistant and the cables are resistant to chemical and biological attack. To maintain operation for long periods of time, all system components are designed to use very little power and are equipped with a sleep-mode, in which even less current is needed. Data acquisition is with a data logger (12 analog channels, modulation range can be either unipolar or bipolar, 12 bit, ≥ 512 kByte memory, RS-232 interface converter, realtime clock). Operating temperature range is 0 - 60 °C. The sensors must function in both directions and may not be damaged by aggressive substances.

The specifications for the *multi-parameter probe* are given in *Table 4.9.1*. The probe contains sensors for temperature, electrical conductivity, pH, redox potential, as well as a piezoresistor pressure sensor for measuring the depth of the water. The oxygen sensor is optional.

Table 4.9.1: Specifications of the multi-parameter probe

Sensor	Range	Resolution	Lifetime
hydraulic pressure (compensated for atmospheric pressure)	0 - 100 m water column	2.5 cm	≥ 5 years
temperature	0 - 60 °C	0.03 °C	≥ 5 years
conductivity	0 - 1 mS cm^{-1} 0 - 100 mS cm^{-1} 0.5 - 500 mS cm^{-1}	0.001 mS cm^{-1} 0.05 mS cm^{-1} 0.25 mS cm^{-1}	≥ 5 years
pH	1 - 12	0.02 pH	≤ 2 years*
redox potential	± 2000 mV	1 mV	≤ 5 years*
dissolved oxygen	0.01 - 20 mg L^{-1}	5 µg L^{-1}	≤ 5 years*

* with maintenance at specific intervals

Electrical conductivity is measured using four special carbon electrodes that adsorb very little water. pH and redox potential are measured using a common reference electrode. Conductivity and pH are referred to the standard temperature of 25 °C using appropriate software.

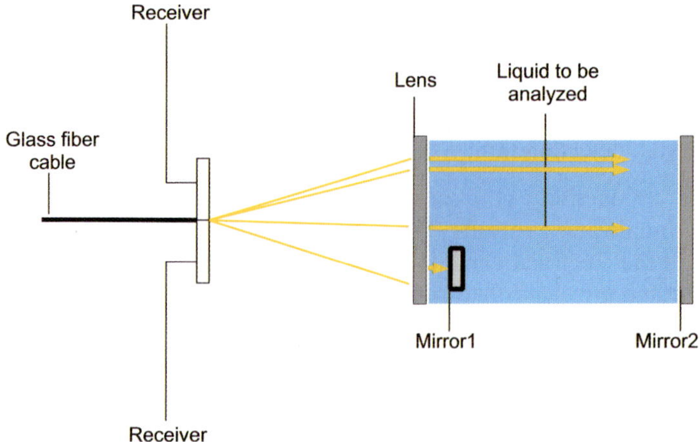

Fig. 4.9.2: The principle of the optosensor monitoring system

The principle of the optosensor monitoring system is illustrated in *Fig. 4.9.2*. A pulse of light (210, 370, 880 and 950 nm) from a xenon lamp is passed through a glass fiber cable to a lens and into the liquid to be analyzed. The light is collimated in the liquid where it is reflected from two mirrors back to the lens. The lens directs the reflected light to a different optoelectronic sensor for each mirror. The intensity of the reflected light is registered by the sensor. Because the path lengths are different, the intensities are different, from which a transmission value is calculated (corrected for the fouling film on the lens and mirrors) as described in Section 4.9.3. Two other sensors register the scattered and fluorescent light. The sensor for fluorescence has an edge filter to suppress scattered light. Because the intensity of the scattered light is generally much higher than fluorescent light, a filter is not needed for the sensor for scattered light. Optionally, the light refracted at the lens/liquid boundary can also be registered.

The sensors can indicate that a contaminant is present, but they cannot replace analysis in the laboratory. Just as important as the detection of contamination is detection of a change in concentrations. The following substances can be detected although it cannot be distinguished which one is present:

- total PAH: naphthaline, acenaphthene, fluorene, phenanthrene, anthracene, chrysene, benzo(k)fluoranthene, benzo(a)pyrene, benzo(g,h,i)perylene; the detection limits depend on the solvent and are lower than 5 μg L^{-1},
- total BTEX: benzene, toluene, ethyl benzene, xylene; the detection limits depend on the solvent and are lower than 25 μg L^{-1},
- total aliphatic hydrocarbons: cis- and trans-dichloroethene, trichloroethene, tetrachloroethene; the detection limits depend on the solvent and are lower than 50 μg L^{-1},
- nitrate: 0.01 - 2.0 μg L^{-1},
- suspended matter: 0.05 - 10 μg L^{-1},
- total salts: detection limit > 0.1 %.

Electrical conductivity is a sensitive parameter for monitoring contaminant load in groundwater and soils. The *electromagnetic monitoring system* detects changes in the electrical conductivity – and thus changes in the water content of the soil and the electrolyte concentration in the soil water or groundwater – by measuring signal attenuation and optionally phase shifts between the transmitter and receiver antennas (which are in different boreholes). Good-conducting soils are a problem owing to the strong signal attenuation in them, limiting the area that can be monitored. In addition, external (both natural and man-made) electromagnetic disturbances (mainly from the atmosphere) are present nearly everywhere, also limiting the size of the area that can be monitored.

The following requirements are placed on an EM monitoring system:

- Only one transmitter antenna system should be used for the frequency range from 10 kHz to 10 MHz.
- The transmitter antenna must radiate an optimum power into the soil.
- The vertical radiation pattern should show no zeros.
- The system has to operate in the far-field.
- Vertically polarized waves must be used.

- The antenna efficiency must be optimal.
- Disturbances in the transmission band must be a minimum.
- The receiver must have a high sensitivity and large dynamic range.
- The receiver must have wide-band capability.
- Secondary radiation by the receiver antenna must be low.

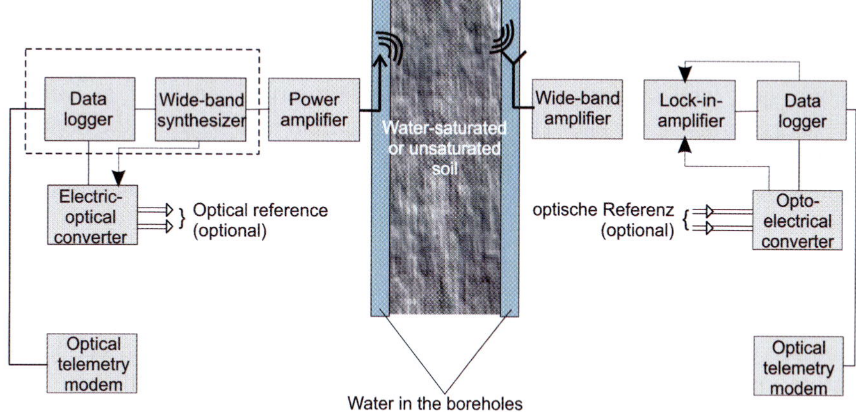

Fig. 4.9.3: Circuit diagram of the electromagnetic monitoring system

As shown in *Fig. 4.9.3*, the main components of the EM transmitter are two wide-band synthesizers, a data logger, a power amplifier, an optical computer interface, and an electrical/optical converter for the reference signals of the synthesizers. Quartz oscillators have been used before to obtain high stability of the discrete frequencies that are used. The wide-band synthesizer permits any frequency between 10 kHz and 10 MHz to be used. Frequency stability with very little phase jitter is attained by the use of such a module.

The main components of the receiver are a wide-band amplifier for the antenna and a lock-in amplifier suitable for installation in a borehole. The lock-in module can automatically measure amplitude and phase in the frequency range from 10 kHz to 10 MHz. This means that manual adjustment of the maximum reading is not necessary, as is the case with the normal lock-in voltmeter. The software of the data logger adjusts the amplification of the lock-in amplifier so that the sensitivity of the amplifier is a maximum.

The sensitivity of the amplifier is controlled by the system software to optimize the depth of penetration and reliability of the EM monitoring system. The software adjusts the antenna parameters according to the electrical conductivity of the soil. The transmission of the signal via glass fiber cables avoids disturbance by electromagnetic noise. This design of the instrument makes it possible to determine conductivity changes across the entire frequency range from 10 kHz to 10 MHz with a single transmitting antenna.

The influence of the water in the borehole is minimized by the design of the transmitter and receiver antennas. The system can be used in water-saturated soil as well as in the unsaturated zone. The data is transmitted by telemetry to the office that is to analyze the data.

4.9.5 Field work

Prerequisite for the field work is the knowledge of the coordinates of the observation well, its type of completion, the geological sequence, and any geophysical well logging and pumping tests that have been done, as well as the accessibility of the site. The first step in the field is to measure the depth to the water table. Depth profiles are then made of temperature, electrical conductivity and pH in each of the wells in order to determine the optimum depth to install the probes and antennas, taking into consideration the well parameters mentioned above. The spacing of the transmitter and receiver antennas is determined mainly by the electrical conductivity of the soil and groundwater. It is desirable to take groundwater samples for analysis in the laboratory. The multi-parameter probes and optosensors must be calibrated before installation; the EM system must be optimized after installation. The multi-parameter probes are calibrated using commercially available standard solutions.

The calibration of the DOC monitor OS100-1S (DOC – dissolved organic carbon) is described here as an example of the calibration of the optosensors. The HA-tech standard (a solution of humic acids), for example, can be used to calibrate the sensor for humic substances and mineral oil. A series of dilutions of the HA-tech reference with precisely known concentrations is prepared. A concentration range of 0 - 5 mg L^{-1} is recommended for water with low DOC values. The DOC concentration in each of these diluted solutions is determined chemically. The intensity of the reflected light is measured for each solution with the OS100-1S monitor. Two equations are derived from the resulting plots of the chemically determined DOC concentration versus the opto intensity and DOC concentration versus the percent dilution of the standard: The first function can be used to determine DOC directly from the sensor signal. The second function can be used to check the calibration curve without having to be determined DOC again in the laboratory. This second function is for the standard used to calibrate the instrument. If a standard other than HA-tech is used, the calibration curve will probably be different. For monitoring, however, the standard used is not important. Any change during the monitoring due to contamination with mineral oil or a decrease in mineral oil concentration will be registered. If there is any change in the opto-signal for DOC (corrected for fouling), a water sample must be taken and analyzed in the laboratory.

For the EM system, the transmitter antenna and up to three receiver antennas are installed in boreholes 15 to 100 m apart, depending on the electrical conductivity of the ground. The receiver antennas are placed deeper than the transmitter antenna to take optimal advantage of the radiation pattern. Multi-parameter probes and/or optosensors can be installed in neighboring boreholes. Each probe is connected to a data logger for data acquisition. Each data logger is connected to a interface multiplexer at the mouth of the borehole so that each of the probes/sensors can be individually selected. The multiplexer has two optical buses for communication with a transmitter or receiver, two RS485 buses for a multi-parameter probe and/or optosensor, and an RS232 bus for sending commands to the probes/sensors and for transferring data to a PC or a telemetry modem.

The resonance frequency of the EM transmitter is determined after it is installed in the borehole. At the resonance frequency, the transmitter draws a maximum current, i.e., the power level is a maximum. The power level of the transmitter is then set to zero, after which the zero potential of the receiver is determined. Next the transfer function for the frequency range from 10 kHz to 10 MHz is derived in order to be certain that the resonance frequency was correctly established.

The transmitter and receiver are then connected by glass fiber cable to provide a reference signal. The times that the multiplexer, the transmitter, and receiver are activated are set in the real-time clock system. At all other times, the system is in sleep mode.

At selected times, the resonance frequency is transmitted and the electric field parallel to both the transmitter and receiver boreholes is measured. Changes in the attenuation of the signal are registered and the electrical conductivity of the ground is calculated from the electrical field components.

4.9.6 Processing

The data stored by the data loggers are transferred to a PC on site or transmitted by telemetry to the office where they are to be evaluated. They are first stored in a format that is compatible with the data processing and graphics programs that will be used. The fundamentals of the evaluation procedures are discussed in Section 4.9.3. The values measured by the multi-parameter probes are converted to temperature, electrical conductivity, pH, redox potential, oxygen concentration, and water level. Optical parameter values are calculated from the measured values for absorption, fluorescence, scattering, and refraction for the different wavelengths and distances to the mirrors. Concentrations are then calculated using special algorithms and calibration curves. It is important to correct for films on the lenses and mirrors. The electrical conductivities between the transmitters and receivers are calculated using Equations (4.9.43) and (4.9.44).

The plots of the measured parameter values provide information about the quality of the groundwater and surface water as a function of time. Joint interpretation of the parameter values as described in Section 4.9.3 provided information about the complex hydrogeochemical processes occurring in the water and about any spreading of contaminants.

4.9.7 Quality Assurance

Aspects of quality assurance that are independent of the procedure or method used are dealt with in Chapter 2.6. The following aspects are of special importance for a permanently installed on-site monitoring system:

- calibration of the probes before installation and at certain intervals thereafter,
- safeguarding the installation from damage in the field,
- field records (recording of special occurrences),
- cleaning of the multi-parameter probes, optosensors, and cables after each downhole logging,
- monitoring of the battery voltage in the probes.

4.9.8 Personnel, Equipment, Time Needed

	Personnel	**Equipment**	**Time needed**
mobilization and demobilization		1 van or off-road vehicle	depends on the distance to the survey area
installation of the monitoring systems	1 geophysicist and 1 technician	probe(s) (transmitter, receiver, multi-parameter probe, optosensor), cable, multiplexer, telemetry modules, tripod, laptop computer	differs from case to case
measurements and processing	1 geophysicist and 1 technical assistant	1 PC, printer, plotter, software	differs from case to case
interpretation and preparation of the report	1 geophysicist and 1 technical assistant		differs from case to case

4.9.9 Examples

Monitoring of a contaminated site at Stassfurt, Germany, using multi-parameter probes and EM monitoring system

The center of the industries that processed the potash and other salts from the mines in the Stassfurt region was south of the Bode River in the Leopoldshall part of the city (GUT, 2001). The "Aktiengesellschaft Chemische Fabrik Concordia zu Leopoldshall" was founded here in 1872 between the present-day streets Bernburg and Hohenerxleben. At first, gravel was quarried; the eventual depth of the pits is assumed to have been 5 m. During this time, a potassium chloride plant was built to produce fertilizer. The waste products were dumped in the exhausted gravel pits. This was the start of the present Concordia landfill. Other plants were built for processing the mined material for other products and their waste products were also dumped in the gravel pits. Production decreased after World War II and some of the plants were closed down. By 1970, little production remained and regular dumping ceased. The waste heap was later recontoured, partially covered with soil and rubble, and the drainage ditch between the two sides was filled. Over the years, the following materials were dumped in the pits: salts (sodium chloride and some kieserite and magnesium chloride), siliceous sediments, clay, ash, soil and rubble (as cover material). It can be assumed that faulty/defective batches and contaminated rubble were also dumped here. Particularly the soluble salts in the waste can be leached into the groundwater. Subsidence, especially in the middle and western parts, indicates that leaching is occurring. Electrical conductivities in groundwater observation wells north and south of the landfill as high as 60 000 $\mu S\ cm^{-1}$ have been measured.

The study area is above the NW–SE-striking Stassfurt anticline. The gypsum cap rock of this halokinetic structure is overlain by Quaternary sediments (sand, gravelly sand, silt and loess) of considerably varying thickness. This is in turn overlain by waste (mostly residues from the chemical production and rubble) up to 5 m thick. The groundwater table in the Quaternary sediments is at a depth of 5 - 8 m. Groundwater flow is to the NE in the direction of the Bode River.

Environmental parameter values measured by a multi-parameter probe in groundwater observation well AH2 at a depth of 16 m are plotted in *Fig. 4.9.5* together with changes in electrical conductivity determined by the EM monitoring system with the transmitter in borehole AH7 at a depth of 12 m and the receiver in AH5 at a depth of 16 m. The locations of these boreholes are shown in *Fig. 4.9.4*. Several measurements were made each day from 27 August to 25 December 2001 (117 days); the mean was calculated for each day and transmitted by telemetry at regular intervals to the BGR branch office in Berlin.

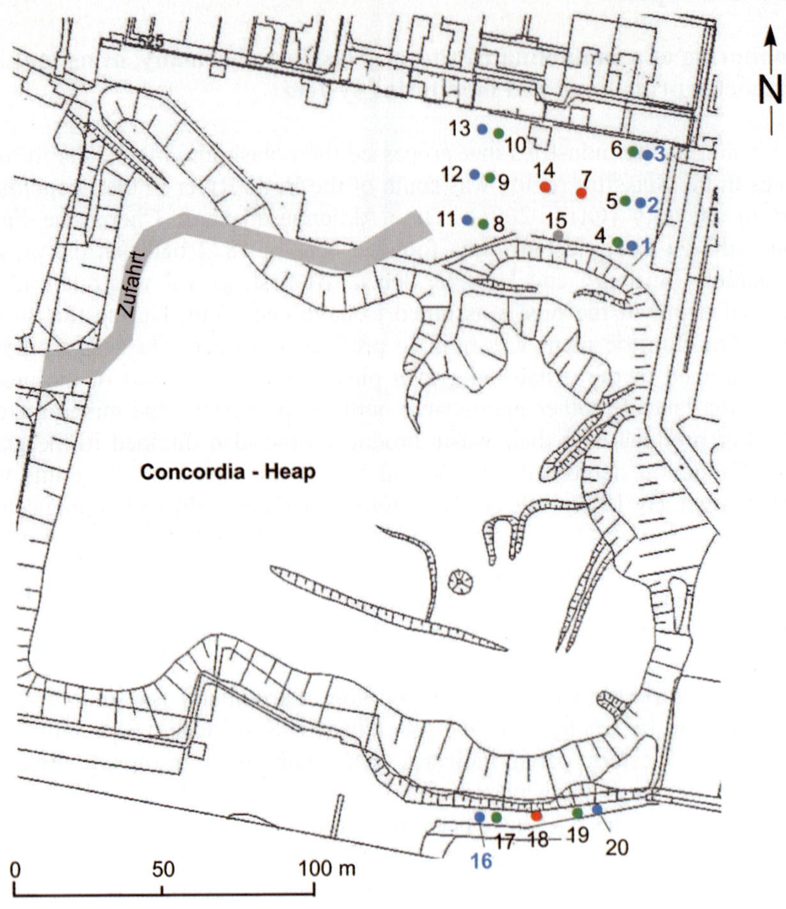

Fig. 4.9.4: Locations of the monitoring system boreholes at the Concordia landfill in Stassfurt, Germany

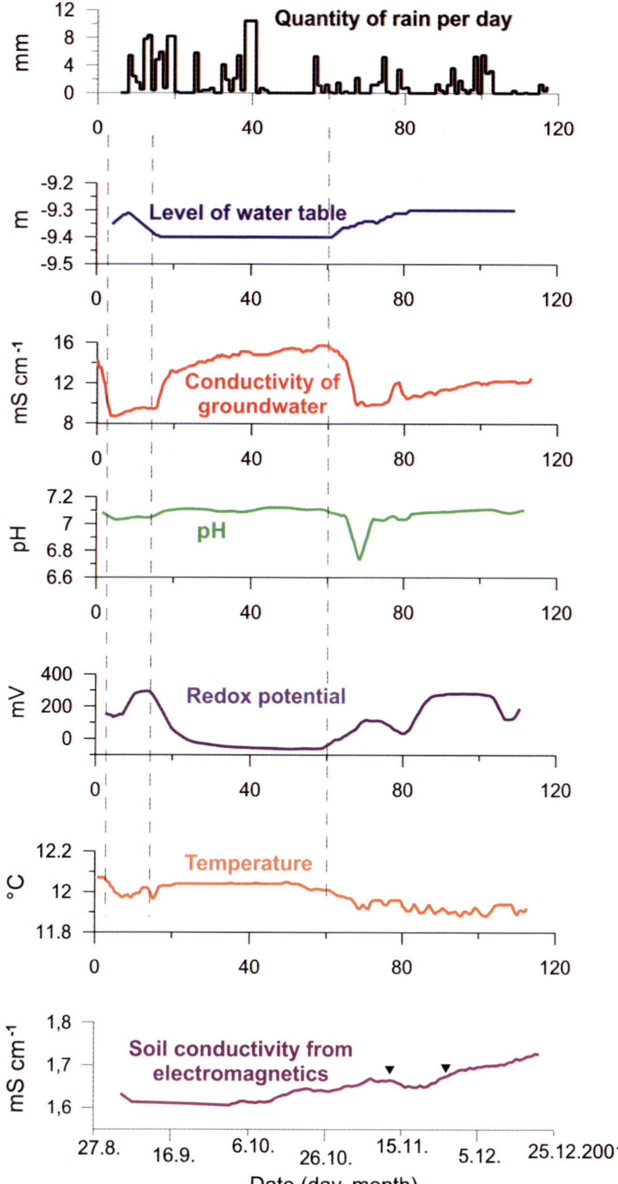

Fig. 4.9.5: Environmental parameter values as a function of time at the Concordia landfill in Stassfurt, Germany

Two events can be recognized in the curves for the environmental parameters that correlate with the changes in the depth to the water table in borehole AH2: A drop in conductivity, increasing redox potential, and falling temperature indicate inflow of colder water with a higher oxygen concentration and lower conductivity. This could be due to groundwater recharge due to heavy rains. With the present time series the origin of the water cannot be determined with certainty. The continual increase in electrical conductivity of the soil and groundwater seen in the EM data during the same period of time was interrupted by a minimum with a time lag relative to the drop in conductivity of the groundwater. The size of this change is within the expected range. Data acquired over a longer period of time – at least one hydrological year – will allow a more extensive interpretation.

Measurements with different transmitter/receiver configurations on the north and south sides of the landfill show that conductivities are higher on the south side (side of groundwater inflow, the groundwater flows northeast).

Relatively large fluctuations in the electrical conductivity of the soil with respect to both time and location were observed on the north side of the landfill. The EM data are in good agreement with the geoelectric measurements (NIEDERLEITHINGER, 2001).

An example of depth profiles of data from the multi-parameter probes is shown in *Fig. 4.9.6.* The borehole AH11 is on the north side of the landfill. The conductivity profile shows three groundwater layers. The low conductivity in the upper layer can be due to groundwater recharge. The values in the middle are can be due to efflux from the landfill. The high conductivities at the bottom of the profile indicates ascending saline water.

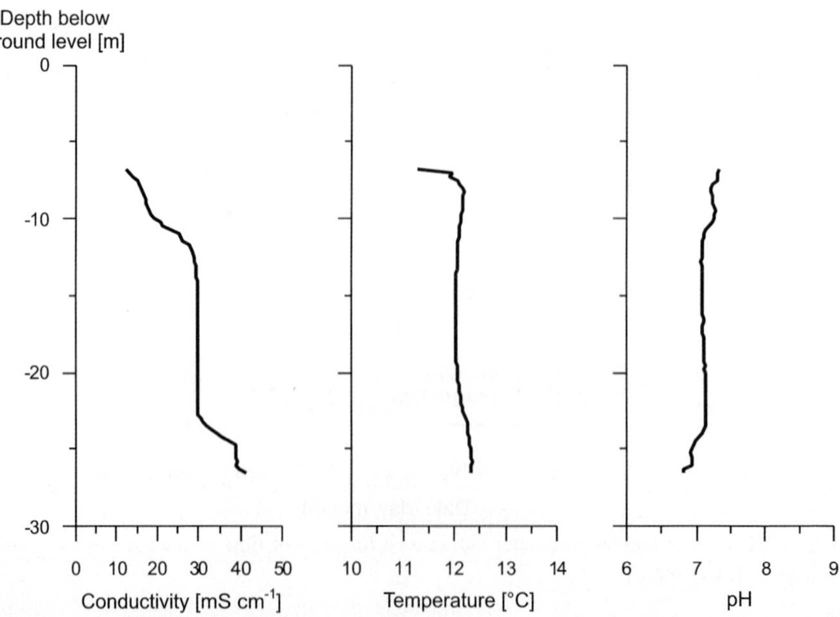

Fig. 4.9.6: Depth profiles for conductivity, temperature and pH in groundwater observation well AH11 at the Concordia landfill in Stassfurt, Germany, in January 2002

The layering can also be recognized in the pH and temperature profile. The data provide a basis for targeted sampling and an understanding of the groundwater dynamics and the kinetics of the chemical reactions.

Monitoring DOC in well water using optosensors

It is often necessary to monitor well water for contamination with oil. Oil and oil products are complex mixtures, mainly hydrocarbons. In the soil, some hydrocarbons are decomposed by microbial activity, others are metabolized to other substances, and still others remain unchanged. Entry of soluble compounds into the groundwater increases the DOC value (DOC – dissolved organic carbon). Humic matter is also formed from organic matter in the soil by microbial activity. Humic matter is a complex mixture of complex organic compounds. The low-molecular-weight, water-soluble humic compounds also increase the DOC value of the groundwater. The DOC value is a suitable parameter for monitoring contamination of groundwater with oil, permitting a waterworks to take appropriate measures at an early stage.

The optical DOC monitoring system OS 100-1S is particularly suitable for detecting very low DOC values (0 - 2 ppm). An OS 100-1S was installed in a well for the drinking water supply for a year at a depth of 30 m as well as in a well in an aquifer downstream from a landfill.

The OS 100-1S measures the absorption at a wavelength of 254 nm (at which aromatic hydrocarbons have an absorption maximum) and at a wavelength at which aromatic hydrocarbons do not absorb (for a turbidity correction).

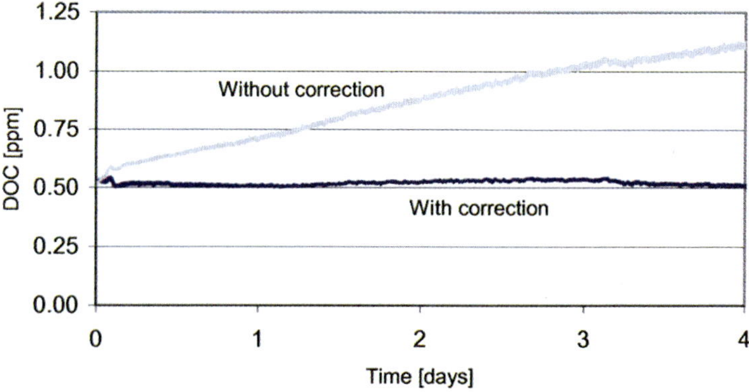

Fig. 4.9.7: DOC in drinking water well as a function of time with and without correction for the film on the lens and mirrors

Measurements are made in the groundwater during each pulse of the xenon lamp. Between flashes, the “dark” signal is measured. At the same time, the intensity of the reference (primary) signal is measured. Each value of the groundwater measurement is corrected for the "dark" signal, fouling, and turbidity and normalized with respect to the reference signal. The need to correct for fouling is illustrated in *Fig. 4.9.7* and discussed in Section 4.9.3. It can be seen that without the correction for fouling the values will drift.

Monitoring chemical oxygen demand and turbidity in waste water using optosensors

One way to assess waste water is to measure the content of organic matter and turbidity. The parameter chemical oxygen demand (COD) is a proxy for content of organic matter. The COD and turbidity values determine the treatment process and may not exceed certain legally defined limits. Optical sensors are used for on-line monitoring of a sewage line and of the waste water from a textile plant. For waste water, the absorption and scattering measurements are made at four wavelengths in the UV/VIS/NIR spectrum. The sensor is equipped with a device for pneumatically removing the film that forms on the lens and mirrors.

The chemical oxygen demand of the homogenized sample and the filtered sample as well as the turbidity are calculated from the measured data. The algorithm can be used for both domestic sewage and waste water from a textile plant. The large fluctuation in COD values for untreated sewage is shown in *Fig. 4.9.8* as a function of time. These values were confirmed by analysis in the laboratory.

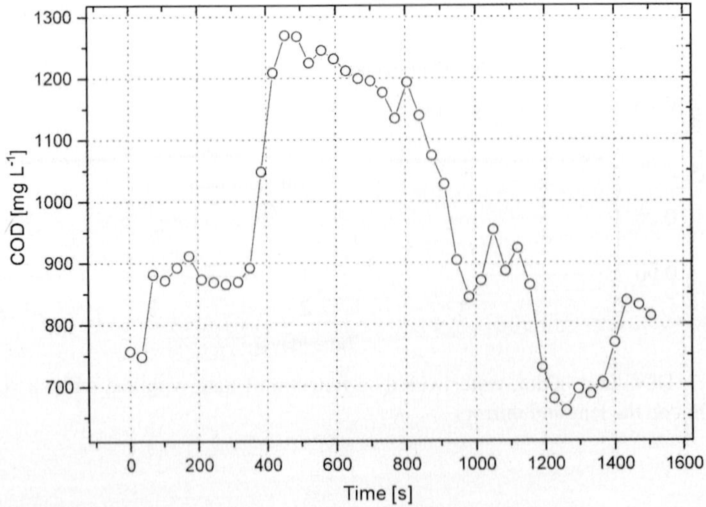

Fig. 4.9.8: COD values for untreated sewage as a function of time

Monitoring oil concentration in industrial waste water using optosensors

Environmental protection laws prescribe the maximum allowed concentration of oil in waste water when it enters the environment. Optical sensors used for on-line monitoring for compliance measure absorption and scattering at four wavelengths. The sensor is equipped with a device for removing the film that forms on the lens and mirrors, e.g., pneumatically. The algorithm used to interpret the measured data has to be sensitive to changes in oil concentration and insensitive to changes in total organic carbon (TOC) and turbidity.

Absorption in the UV spectrum of industrial waste water during a typical day is shown in *Fig. 4.9.9*. The large changes within a short time are mainly caused by changes in TOC and turbidity. The oil concentration varies little during a day (*Fig. 4.9.10*).

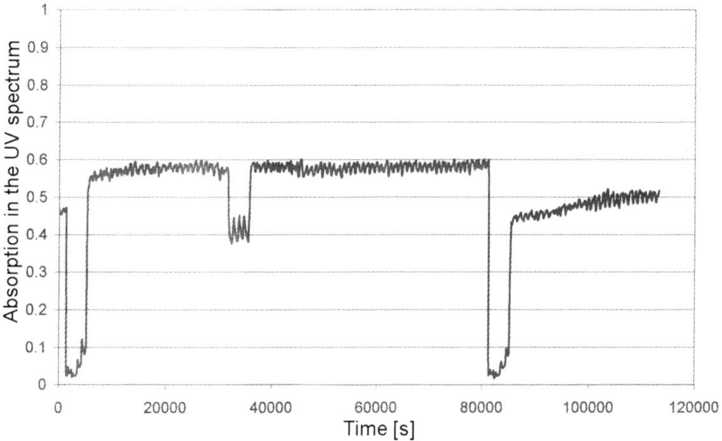

Fig. 4.9.9: Absorption in the UV spectrum by industrial waste water as a function of time

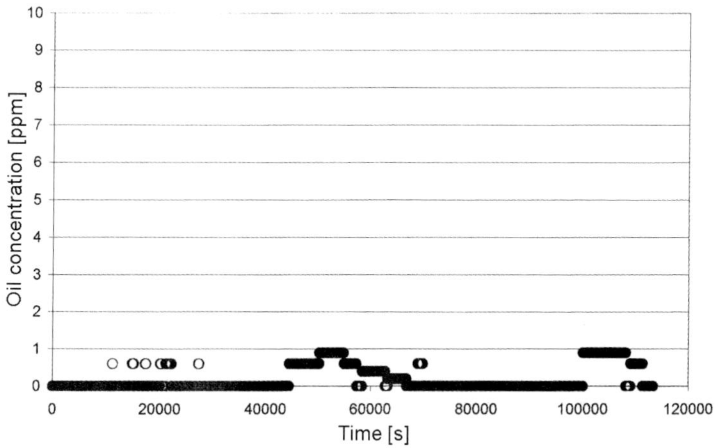

Fig. 4.9.10: Oil concentration in industrial waste water as a function of time

Monitoring groundwater quality below a gas coking plant using optosensors

To determine a depth profile of water quality on the grounds of an old gas coking plant in Berlin, optosensors were used in groundwater observation wells to measure absorption and fluorescence at several UV and VIS wavelengths. Absorption was measured over two path lengths so that both very low concentrations as well as high concentrations could be measured. Water samples were taken and analyzed in the laboratory. Comparison of the two sets of data showed that the optical measurements correlate with concentrations of alkylphenols, cyanides, PAH, BTEX, and the sum of PAH, naphthaline, alkylphenols, BTEX, and alkyl cyanides.

Absorption at 205 nm is shown in *Fig. 4.9.11* to be a linear function of the total concentration of PAH, naphthaline, alkylphenols, BTEX, and cyanides. All of these substances absorb at 205 nm. It can be seen that this wavelength is suitable for monitoring the total concentration of these substances. A depth profile is shown in *Fig. 4.9.12* for the total concentration of these substances in three wells as calculated from the linear function in *Fig. 4.9.11*. The three profiles differ considerably.

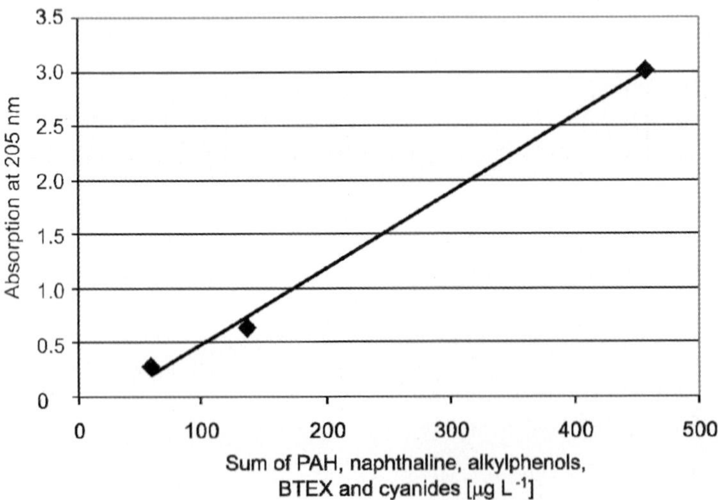

Fig. 4.9.11: Absorption at 205 nm as a function of the total concentration of PAH, naphthaline, alkylphenols, BTEX, and alkylcyanides in a groundwater observation well on the grounds of a gas coking plant in Berlin

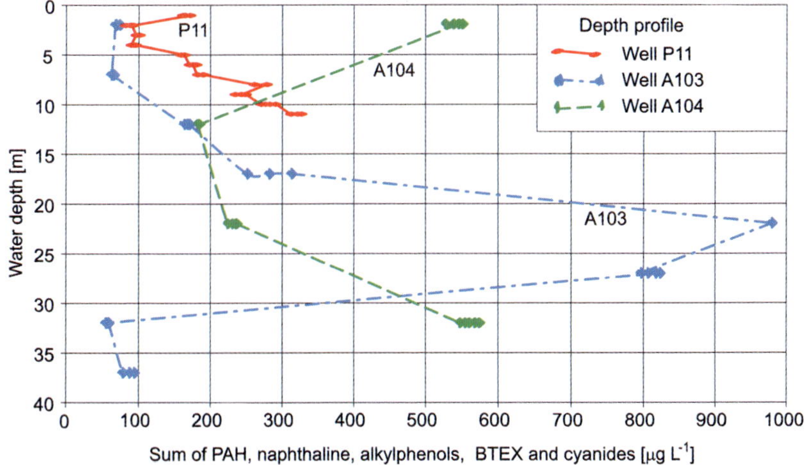

Fig. 4.9.12: Depth profiles of the total concentration of PAH, naphthaline, alkylphenols, BTEX, and cyanides in several groundwater observation wells on the ground of a gas coking plant in Berlin

Parameters and units

	Symbol	Units
electrical conductivity	σ	$\mu S\ cm^{-1}$
temperature	T	°C
pH	pH	–
redox potential	E_h	mV
oxygen concentration	O_2	$mg\ L^{-1}$

References and further reading

APPELO, C. A. J. & POSTMA, D. (1996): Geochemistry, Groundwater and Pollution. Balkema, Rotterdam/Brookfield.

BALDINI, F. & FALAI, A. (2000): Characterizations of an Optical Fibre pH Sensor with Methyl red as Optical Indicator. Optical Sensors and Mikrosystems, Edited bei Marte Llucci et al., Kluwer Academic/Plenum Publishers, New York 2000.

BERKTOLD, A. (1982): Elektrische Leitfähigkeit von Lösungen reiner Salze und von natürlichen Wässern. In: LANDOLT - BÖRNSTEIN: Zahlenwerte und Funktionen aus Naturwissenschaft und Technik, Neue Serie, Gruppe V: Geophysik und Weltraumforschung, Bd. **1**, Physikalische Eigenschaften der Gesteine, Teilband b. Springer, Berlin, 262-275.

BÜHLER, H. & INGOLD, W. (1982): Redoxmessung, Grundlagen und Probleme. Firmenschrift, Wissenschaftlich Technische Werkstätten GmbH, Weilheim.

CARROLL, D. (1962): Rainwater as a chemical agent of geologic processes - a review. US Geol. Surv. Water Supply Pap., **1535-G**, Washington, D.C.

DEGENER, R. & HONOLD, F. (1994): Kalibrierverfahren für die pH-Messung. Abwassertechnik, H. **3**, Bauverlag Wiesbaden.

DREVER, J. I. (1988): The Geochemistry of Natrural Waters. Prentice Hall, Englewood Cliffs.

FLACHENECKER, G. (1970): Optimierung von transistorischen Empfangsantennen. Habilitationsschrift, Technische Universität München.

GABILLARD, R. DEGAUQUE & WAIT, J. (1971): Subsurface Electromagnetic Telecommunication – A Review. IEEE Transactions on Communication Technology, Vol. COM. **19**, No. 6.

GUT Gesellschaft für Umweltsanierungstechnologien mbH (2001): Historisch-genetische Recherche Deponie 30 "Concordia-Halde" und deren Umfeld in Staßfurt. Gutachten im Auftrag der BTU Cottbus.

HÖLTING, B. (1989): Hydrogeologie, Einführung in die Allgemeine und Angewandte Hydrogeologie. Enke, Stuttgart.

IIZUKA, KEIGO (1962): An Experimental Study of the Insulated Dipole Antenna Immersed in a Conducting Medium. IEEE Transactions on Antennas and Propagation, 518-532.

JORDAN, H. & WEDER, H.-J. (1995): Hydrogeologie, Grundlagen und Methoden, Regionale Hydrogeologie: Mecklenburg-Vorpommern, Brandenburg und Berlin, Sachsen-Anhalt, Sachsen, Thüringen. Enke, Stuttgart.

KING, R. W. P., SMITH, G. S., OWENS, M. & TAI TSUN WU (1981): Antennas in Matter. Fundamentals, Theory and Applications. Copyright by the Massachusetts Institute of Technology.

KRAUS, S. (1996): Xerogelstrukturen als Sensormatrices – Eignungsuntersuchungen durch Immobilisierung pH-sensitiver Indikatoren. Forschungszentrum Karlsruhe, Technik und Umwelt, Wissenschaftliche Berichte FZKA 5707.

LINDENMEIER, H. (1974): Die transistorisierte Empfangsantenne mit kapazitiv hochohmigem Verstärker als optimale Lösung für den Empfang niedriger Frequenzen. ntz, **27**, 11.

LINDEMEIER, H. (1977): Kleinsignaleigenschaften und Empfindlichkeit einer aktiven Breitbandempfangsantenne mit großem Aussteuerungsbereich. ntz, **30**, 1.

MAIER, O. (2001): Analogsignale digital gesteuert. Elektronik, **1**.

MATTHESS, G. (1990): Die Beschaffenheit des Grundwassers. Borntraeger, Berlin.

MEINKE, H. (1969): Aktive Empfangsantennen. Internationale Elektronische Rundschau, **23**, 6, 141-144.

NIEDERLEITHINGER, E. (2001): Bericht Geophysikalische Messungen auf der Halde Concordia, Staßfurt. Büro für Geophysik Lorenz, Berlin, im Auftrag der BGR.

OLAF, S. (1989): Elektromagnetische Feldausbreitung in leitfähigen Medien. Diss. TU Clausthal.

RAUSCHER, C. (2000): Grundlagen der Spektrumanalyse. Rohde & Schwarz, 1. Auflage.

SCHÜRING, J., SCHULZ, H. D., FISCHER, W. R., BÖTTCHER, J. & DUIJNISVELD, W. H. M. (Eds.) (2000): Redox Fundamentals, Processes and Applications, Springer, Berlin.

SEBA Hydrometrie GmbH (1997): Firmenschrift über Drucksensoren. Kaufbeuren.

TIETZE, U. & SCHENK, Ch. (1996): Halbleiterschaltungstechnik. 10. Aufl., Springer, Berlin.

VOIGT, H.-J. (1990): Hydrogeochemie, Eine Einführung in die Beschaffenheitsentwicklung des Grundwassers. Springer, Berlin.

WAIT, J. & CAMBELL, L. L. (1953): The Fields of an Electric Dipole in a Semi-Infinite Conducting Medium. Journal of Geophysical Research, **58**, 1.

WAIT, J. R. & FULLER, J. A. (1973): Transmission Line Theory for an Insulated Linear Antenna in a Fluid- or Air-Filled Borehole. Appl. Physics, **1**, 311-316.

WOLFBEIS, S. (1991): Fiber optic chemical sensors and biosensors. Vol. II, CRC Press.

5 Geological, Hydrogeological, Geochemical and Microbiological Investigations

SVEN ALTFELDER, ULRICH BEIMS, MANFRED BIRKE, REINER DOHRMANN, HAGEN HILSE, FLORIAN JENN, STEPHAN KAUFHOLD, KLAUS KNÖDEL, CLAUS KOHFAHL, MANJA LIESE, KAI MUELLER, MIKE MÜLLER, RANJEET NAGARE, MICHAEL NEUHAUS, CLAUS NITSCHE, MICHAEL PORZIG, JENS RADSCHINSKI, KATRIN R. SCHMIDT, ANDREAS THIEM & HANS-JÜRGEN VOIGT

5.1 Methods for Characterizing the Geological Setting

KLAUS KNÖDEL, KAI MÜLLER, MICHAEL NEUHAUS, FLORIAN JENN & HANS-JÜRGEN VOIGT

Site investigations and assessment require a thorough understanding of the geology of the site. The following aspects or parts thereof must be taken into account: stratigraphic sequence, thickness and lateral extent of strata and other geological units, lithology, homogeneity and heterogeneity, bedding conditions, tectonic structures, fractures, and impact of weathering. Information about landforms (geomorphology), earthquake risk, activity of faults, land sliding, subsidence and caving to the surface as a result of mining and karst must also be collected. Planning of an investigation must take into account accessibility, whether the surficial rock is unconsolidated or consolidated and the investigation methods must be appropriate for the geological/hydrogeological conditions.

The investigation must focus not only on the immediate site (e.g., the actual landfill area), but also on the surrounding area. The geological surroundings are described as that area around a landfill or site suspected to be hazardous that can be assumed to be affected by possible spreading of pollution with a degree of probability greater than zero and whose contaminant retention capacity will be exploited. This includes that part of the regional groundwater system affected by possible contamination or in which the contamination has been reduced either by dilution and other processes to a level of trace or background values. The lateral and vertical extent of this area is specified depending upon the site conditions by way of plausibility considerations and experience. As a rule of thumb: Each landfill or site suspected to be hazardous is situated in a regional groundwater system that covers an area of several tens of km^2 around the immediate site. The regional groundwater system must be included in the site assessment and therefore, information about this system must be collected. The local groundwater system to be investigated in detail usually covers an area of 0.1 to 1 km^2. Relevant to site investigations is a depth range from the surface to 50 m, which may often be extended to about 150 m in order to better understand the regional structures (stratigraphy and tectonics) as well as the regional groundwater system.

The extent and thickness of the lithological units are determined using geological methods (e.g., geological mapping, excavations, drilling, direct push sounding, see Sections 5.1.1 to 5.1.4) in combination with geophysical and remote sensing methods (Parts 3 and 4). The lithological units are identified on the basis of their mineralogical composition, color, grain size, texture, and other physical properties.

Stratigraphy is the branch of geology that deals with the original sequence of deposition, age relationships, mineralogical composition, fossil content, and distribution of strata. Lithostratigraphy deals with the change in rock type both vertically and laterally, reflecting changing deposition environments, known as facies change. Chemostratigraphy is based on changes in the relative proportions of trace elements and isotopes within and between lithologic units. Biostratigraphy is based on the fossil content of the strata. Strata at locations containing the same fossil fauna and flora are correlatable in time. The stratigraphic information can be presented in a stratigraphic column.

The first step in a geological site investigation is to interpret the situation depicted on the geological map, if available. A geological and geophysical field check is made and the interpretation of the geology is revised and/or supplemented.

The *lithological, petrophysical and hydraulic properties* of the relevant lithological units must be established (Sections 5.2.7 to 5.2.7.3). These properties can be determined from well cores and other samples, laboratory analyses, geophysical borehole logging (Chapter 4.8), hydraulic well tests, and surface geophysical measurements. The *contaminant retention capacity* of the ground must also be estimated (Section 5.3.5.7). Jointing and fault zones which may be hydraulically effective and may provide pathways for water and material transport must be given special attention.

The *homogeneity/heterogeneity* of the hydrogeologically relevant layers must be determined to obtain information about the spatial variability of the hydraulic properties. The homogeneity of the subsurface can be investigated using geophysical methods, including borehole logging, by taking suitable samples from outcrops, trenches, and boreholes, as well as direct push soundings.

The number and spatial distribution of the boreholes, sampling and geophysical sounding points depend on many things. If little initial information is available, a regular rectangular, square or triangular grid with a node spacing of 25 m is appropriate, 15 m for relatively small areas and up to 100 m for very large areas. The grid can be reduced to a spacing of 5 m in the second investigation phase. A triangular grid pattern requires fewer points to cover an area than a square grid pattern does. Random grids are best suited where regular structures can be assumed in the underground.
The following properties are important for an assessment of the homogeneity/heterogeneity of the ground on a small scale (1 - 100 cm) with respect to

- texture, grain size, mineral content, density, water content, possibly contamination,
- strength (compressibility, shear strength),
- porosity and degree of saturation with water, and
- permeability.

These properties can be determined among other methods directly by analysis of samples in the laboratory. Soil samples become increasingly difficult to obtain with increasing depth. Soil mechanical tests in the laboratory are expensive and time consuming. For this reason, soundings are made to gain additional information. The sounding results cannot be unequivocally interpreted without calibration on the basis of samples and laboratory tests. Sampling and sounding methods are, therefore, complementary.

Sample collection influences the mechanical and hydraulic properties of the rock. Even the most careful sampling techniques do not totally avoid such interference. To a certain extent it is, however, possible to recreate original stress states on laboratory specimens. In the case of all laboratory tests, it should be borne in mind that the sample size is small to very small compared with the rock mass being assessed and the (often decisive) inhomogeneities which may be present. Mixed samples, consisting of material taken from several grid points, are often used for grain-size analysis, determination of water content, etc. The heterogeneity/homogeneity of hard rock is often determined on the basis of drill cores.

Other aspects that need to be taken into consideration for an assessment of the subsurface are *bedding* and *tectonic structures*. The tectonic structures are important for interpreting the formation and movement of the Earth's crust. Sedimentary rocks are deposited under the influence of water, wind and gravity, generally as even, parallel layers. The bedding created during rock formation can be changed, for example, by extension or compression. In the case of plastic deformation, these result in folds, flexures and salt domes. In the case of deformation with fracturing, the result is displacement, with the formation of faults. The subsurface can have faults formed at different times and by different mechanisms, and having different strike directions and dip angles and extending to different depths, not necessarily to the surface.

In order to assess the effectiveness of the ground as a geological barrier below a landfill or other site suspected to be hazardous it is necessary to investigate the bedding and tectonic structures, and to understand how they were formed. For an assessment of underground pathways for water, for example, it is crucial to know whether a fault zone was created by compression or extension. A broad spectrum of geoscientific methods are employed for this.

The investigation of the distribution, thickness and homogeneity/heterogeneity of the geological units as described above is inseparable from the investigation of the tectonic structures. Structures, such as dipping layers, faults, shear zones, jointing and cavities, are often visible in drill cores. Their spatial position (attitude) can often be determined locally with the support of geophysical borehole logging. In the case of steeply inclined layers, the borehole(s) should be inclined in order to avoid false interpretation. Seismic and other geophysical measurements help delineate subsurface structures. The results of the geophysical measurements, trenches, direct push soundings and boreholes are used to develop a three-dimensional model of the site.

Geomorphology is the study of the Earth's surface features (landforms), the processes that form them, and the geological structures beneath. Geomorphologists try to understand why landscapes look the way they do, landform dynamics, and to predict changes through a combination of field observation, physical experiments, and numerical modeling. Landforms evolve in response to a combination of natural and anthropogenic processes. The landscape can be changed through tectonic processes and volcanism. Erosion produces sediment that is transported and deposited elsewhere within the landscape or off the coast. Landscapes are also changed by subsidence and uplift, either due to tectonics or physical changes in underlying sedimentary deposits.

Recommendations with regard to the suitability of methods for certain applications/objectives in a detailed geological site investigation are given in *Table 1-1*. This table should not be viewed as a substitute for the advice of an experienced specialist. The methods should be selected so that optimal investigation results are obtained in the most economical way. In many countries there are already standards and regulations for different kinds of geological investigations, their quality requests and their documentation. An overview of the common drilling methods, their applicability criteria, usual diameter of the borehole, sample quality criteria, and advantages of the methods, as well as typical sources of errors is given in *Tables 5.1-2* and *5.1-3*. A summary of the most important geophysical logging methods, their purposes and measured parameters is given in *Tables 4.8-2* and *4.8-3* as well as their possible applications depending on drilling mud and casing used.

5.1.1 Geological Field Observations

FLORIAN JENN, KLAUS KNÖDEL & HANS-JÜRGEN VOIGT

Compilation of the available geological information (Chapter 2.2), geological field observations (e.g., geological mapping), and the evaluation of the data are steps in geological modeling of a site or an area under investigation. The geological model forms the basis for a conceptual hydrogeological model.

The first step in a site investigation should be to describe the study area relative to a *major structural feature*, if there is one. *Figure 5.1-1* shows the location of the investigation area of the abandoned Eulenberg landfill near Arnstadt, Thuringia, Germany, relative to the regional tectonic structure of Thuringia. The main strike direction is NW-SE. The site is in the area of the regional Gotha-Arnstadt-Saalfeld fault zone. This fault zone dominates the geological structures in the investigation area and is hydrogeologically important.

The next step is to visit the *outcrops* and other rock exposures in the study area. Outcrops are most important sources of information for geological mapping. *Figure 5.1-3* shows an exposure in a road cut used in geological mapping of the Eulenberg area near Arnstadt (*Fig. 5.1-2*). The most obvious feature of an outcrop of sedimentary rock is the strata. Although the rocks may be tilted or folded, the sediments are almost always originally laid down in horizontal layers which extended continuously in all directions. The outcrop is also used to determine the strike and dip directions as well as the properties of the joints.

Geological mapping provides information on the geological setting on the basis of field investigations supplemented by laboratory and office work. A geological map is a two-dimensional graphical depiction of the rock units and structures in an area using lines, symbols, patterns, and colors. The composition of the rocks and the age of the units is commonly given. A geological map is prepared by combination of field observations of location and lithology. The lithology is determined on the basis of hand specimens, outcrops and, if possible, boreholes. A geological cross-section (*Figs. 5.1-5* and *5.1-8*) is sometimes useful to understand the continuity of rock units between outcrops and at depth. What is shown on the map depends on the objectives of the site investigation. A bedrock map (British term: solid map), for instance, shows the rock structure beneath the surficial deposits, such as alluvium. A normal geological map (British term: drift map), on the other hand, shows the cover rock as well as any bedrock where it crops out. The map scale determines the level of detail on the map: The larger the scale the more information that can be shown. Geological maps for a site investigation are normally made at a scale of 1 : 5000 to 1 : 10 000. An example of a detailed geological map prepared in an investigation of an abandoned landfill site is shown in *Figure 5.1-2*. The scale is 1 : 5000.

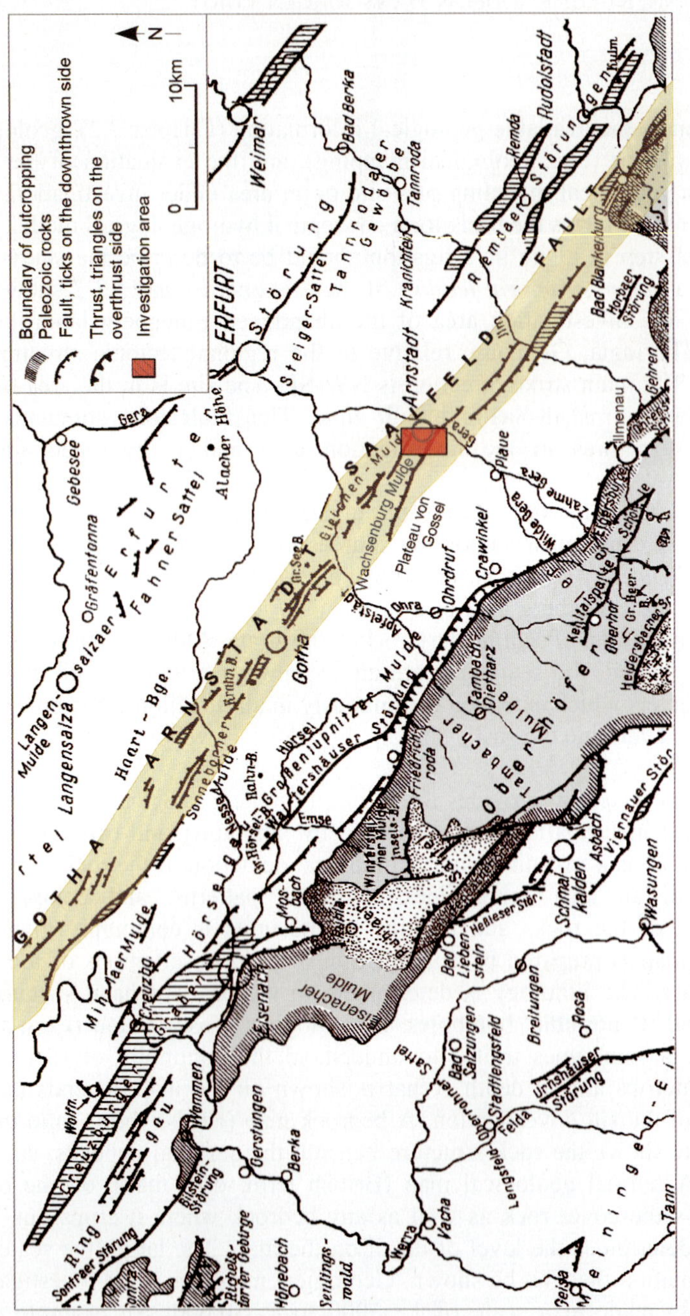

Fig. 5.1-1: Simplified tectonic map of Thuringia, Germany

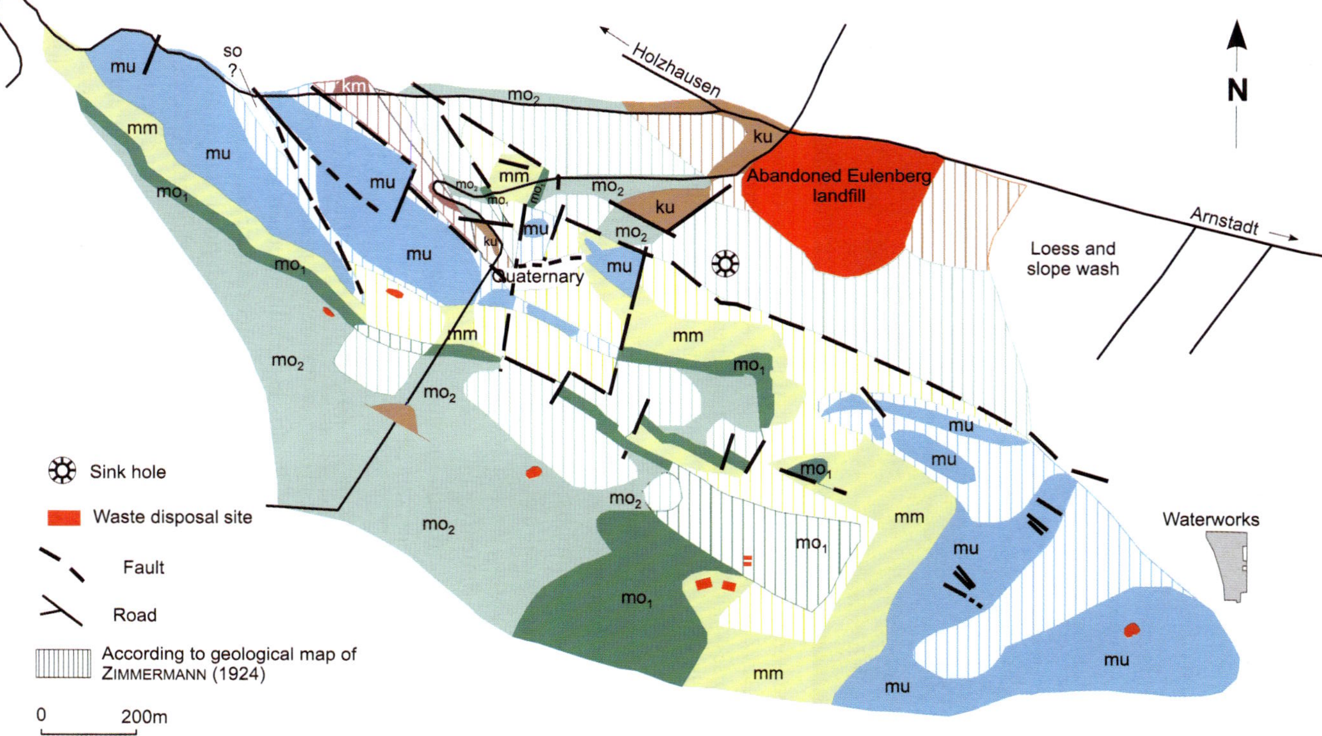

Fig. 5.1-2: Abandoned Eulenberg landfill near Arnstadt, Thuringia, Germany: detailed geological map, geological formation symbols are given in *Table 5.1-1*, SCHMIDT (1993)

Fig. 5.1-3: Abandoned Eulenberg landfill near Arnstadt, Thuringia, Germany: Exposure of the Terebratulid Bed Zone (mu) next to the road from Arnstadt to Bittstädt. In some places the material is coarse oolitic limestone and to a large percentage limonitic. NW-SE-trending joints contain calcite up to 0.5 cm thick. Photograph from SCHMIDT (1993).

The following supplies and equipment are needed for geological mapping: a large-scale topographic base map of the area under investigation, aerial photographs if possible, clipboard or map board, compass, hammer, and field notebook. Digital geological map databases and geological mapping software has been introduced during the last several years and are frequently used. The production of three-dimensional computerized representations of the subsurface is a relatively recent method in geology. For this a geological formation is generally represented using a three-dimensional array of relatively small subdivisions, or cells. Each cell in the model is assigned a rock type. The locations of the major formation boundaries, faults, folds, and unconformities are incorporated. The preparation of digital geologic models is computationally effortful. The method is especially used for the representation and analysis of oil and gas fields and groundwater resources.

A sequence of sedimentary rocks may be subdivided into a number of lithostratigraphic units of various sizes. A lithostratigraphic unit is defined as a body of rock that has the same lithological characteristics and stratigraphic position (position in the rock sequence). The smallest lithostratigraphic rock unit is a bed or layer. A formation is a set of similar layers, and is the fundamental unit of stratigraphy. Formations are identified by lithological characteristics and stratigraphic position and are mappable at the Earth's surface or traceable in the subsurface. It is not necessarily a geolocical time unit. Formations have formal names, generally derived from the stratigraphic position, geographic locality where the unit was first recognized and described and the dominant rock type (e.g., Leine Rock Salt in the Upper Permian). A set of similar or related formations may be combined to a group or subdivided into members. Boundaries of lithostratigraphic units are placed where the lithology (or rock type) changes.

When geologists study a sequence of sedimentary rocks they measure the thickness of each bed, record the physical, mineralogical, and paleontological characteristics of the rock, and note the nature of the interfaces between beds. These data are used to produce a *stratigraphic column*. Standard lithologic symbols are used to denote the rock types: e.g., *Figs. 5.1-4* to *5.1-7*. The lithology of each layer is denoted with lithologic symbols. Formation name, often abbreviated, and other information about the rocks are also noted (*Fig. 5.1-4*).

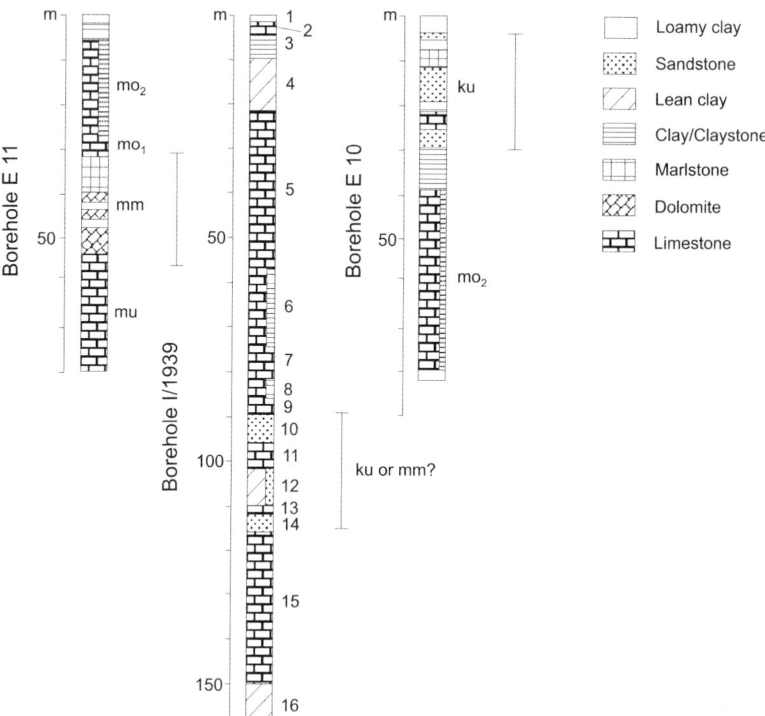

Fig. 5.1-4: Abandoned Eulenberg landfill near Arnstadt, Thuringia, Germany: drilling results, SCHMIDT et al. (2003)

Table 5.1-1: Standard stratigraphic column for the area of Arnstadt, Thuringia, Germany, SCHMIDT et al. (2003)

Stratigraphic unit (German)	Stratigraphic unit (English)	Symbol	Thickness [m]
Mittlerer Keuper	Middle Keuper	km	> 27
Unterer Keuper	Lower Keuper	ku	40-45
Oberer Muschelkalk	Upper Muschelkalk	mo	55-65
Ceratitenschichten	Ceratite Beds	moC	50-60
Trochitenkalk	Trochite Limestone	moT	4-7
Mittlerer Muschelkalk	Middle Muschelkalk	mm	25-55 in parts only 10
Unterer Muschelkalk	Lower Muschelkalk	mu	100-105
Schaumkalkzone	Aphrite Zone	muS	6-7
Oberer Wellenkalk	Upper Wellenkalk	muWO	18-22
Terebratulazone	Terebratulid Zone	muT	4-5
Unterer/Mittlerer Wellenkalk	Lower/Middle Wellenkalk	muWU/muWM	60
Oberer Buntsandstein	Upper Bunter	so	130
Myophorienfolge	Myophoria Formation	soM	17-24
Röt	Roethian		110
(davon Steinsalz)	(Roethian rock salt beds)	(soSANA)	(10)
Mittlerer Buntsandstein	Middle Bunter	sm	190
Unterer Buntsandstein	Lower Bunter	su	310
Zechstein	Zechstein	z	
Aller-Folge	Aller Formation	z4- z8	12
Leine-Folge	Leine Formation	z3	25
(davon Steinsalz)	(Leine rock salt beds)	(z3Na)	(<2)
Staßfurt-Folge	Stassfurt Formation	z2	42
(davon Steinsalz)	(Stassfurt rock salt beds)	(z2Na)	(<5)
Werra-Folge	Werra Formation	z1	160?
(davon Steinsalz)	(Werra rock salt beds)	(z1Na)	(<150)

After geologists have studied the *lithology* in a particular area and grouped the rocks into formations, the next step is to work out the age relationships: the *stratigraphy*. The sequence and timing of events are now extremely important. It is necessary to add faults and folds, dikes, unconformities, age relationships, and other complexities to the lithological representation. These tend to further complicate the picture. The stratigraphic information can be presented in several ways. One of the most useful is the stratigraphic column: a diagram or list of a chronological sequence of rock units arranged so that the oldest occurs at the bottom. For regional studies, geologists study the stratigraphy of as many separate areas as they can, prepare a stratigraphic column for each, and combine them in an attempt to understand the regional geological history of the area. A stratigraphic column allows other geologists who have never seen the local rocks to understand the geology of that area. There are several key pieces of information in a stratigraphic column. First, it shows the major rock units. It also shows the types of rocks these units are composed of in two ways. First, by the name of the unit itself, and also by the symbol used (dots, for example, tell a geologist that the rock type is a sand or sandstone). Lastly, it shows the relative thickness of the individual rock units (*Figs. 5.1-4* and *5.1-7*). For areas with large lateral variation in layer thickness it is useful to prepare a standard stratigraphic column with thickness ranges as shown in *Table 5.1-1*.

When stratigraphic columns have been drawn for all outcrops and/or boreholes in the area (*Fig. 5.1-4*), the rock units must be correlated or traced between outcrops and boreholes. This is done on the basis of hand specimens, drill cores, geophysical well logs, and other geophysical field measurements. This lithologic correlation is carried out under the hypothesis that units in two widely separated sequences are equivalent. Examples are given in *Figs. 5.1-5* and *5.1-8*. Marker beds, distinctive sequences of beds, bed thicknesses, and unconformities can all be used for correlation between sections. Marker beds have some unusual, distinguishing features which allows them to be readily identified. Beds may change laterally in thickness or lithology as a result of differences in the sedimentation rate or depositional environment (lateral facies changes). Unconformities are layer interfaces which represent a gap in the geological record because of either erosion or nondeposition.

A geological cross-section is a diagram showing the formations and structures in a vertical plane as if the ground were cut open and viewed from the side. Geological cross-sections are interpretative, since the underground relationships can generally not be observed directly. Only in areas of deep canyons or high mountains can natural cross-sections be observed. In other cases, drill holes or geophysical surveys provide data from which cross-sections can be constructed. The visualization of stratigraphic and structural relationships is often best accomplished in a geological cross-section. Other graphical renditions are fence diagrams, and block diagrams. Special software is used to draw fence diagrams and block diagrams. Geological cross-sections, fence diagrams, and block diagrams are often constructed with vertical exaggeration. Examples of geological cross-sections through areas with consolidated and unconsolidated rocks are shown in *Figs. 5.1-5* and *5.1-8*.

The abandoned Eulenberg landfill site (see also Section 6.2.1) consists mostly of consolidated rock; relevant for the investigation of the Schöneiche-Mittenwalde area (see also Section 6.2.1), however, is the unconsolidated rock. The strategy is different for the investigation of the two areas. Outcrops of bedrock are rare in Schöneiche-Mittenwalde area making it possible to use direct push technology, which is considerably less expensive than drilling in that area. In both cases geophysics can help to reveal the structures in the subsurface (examples in Section 6.2.1). Examples of schematic standard stratigraphic columns and geological cross-sections in an area of unconsolidated rock (Schöneiche-Mittenwalde) are shown in *Figs. 6.2-24* and *6.2-25*.

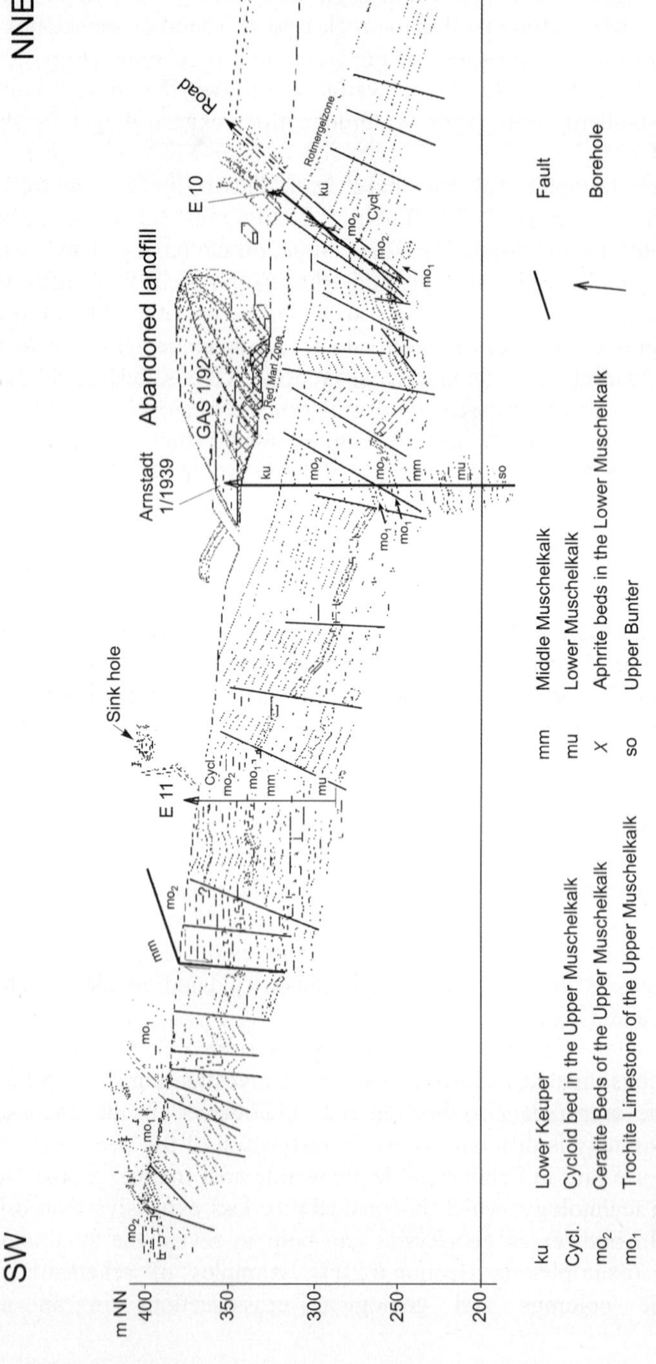

Fig. 5.1-5: Abandoned Eulenberg landfill near Arnstadt, Thuringia, Germany: geological cross-section, SCHMIDT et al. (2003)

5.1.2 Trenching

KLAUS KNÖDEL & HANS-JÜRGEN VOIGT

An exposure that can be used to investigate the top several meters of soil and rock can be made by digging a trench. This is usually done with an excavator, seldom by hand. Trenches offer a cost-effective means of obtaining an exact record of the lithological succession and geological structure in unconsolidated and consolidated rocks above the groundwater table down to a depth of several meters. Trenches allow the taking of samples to determine mechanical and hydraulic properties of rocks, e.g., compressibility, shear strength, and hydraulic conductivity.

The trenches can be dug at relatively arbitrary locations in areas of unconsolidated rock in order, for example, to investigate the depth and the extent of the weathering of boulder clay or to describe the bedding conditions in an area of sand cover. It is easy to construct such trenches and to install shoring in them.

In hard rock areas the walls of the trench are generally stable without the need for shoring. Problems in such areas are associated with excavation of the trench depending on the strength of the rock. Trenches in consolidated rock can also be useful not only for investigating the zone of weathering and zones of fracturing (e.g., faults), but also for mapping bedding planes and thinning out of beds as well as for investigating thin interbeds. Attention should also be paid to the presence of water-logged zones and, impervious and semipervious beds.

In many countries, attention must be paid to excavation standards, such as OSHA's[1] excavation standards. Due to these standards all excavations or trenches four feet (1.22 m) or greater in depth must be appropriately benched, shored, or sloped. The measures depend on the soil types found in the excavation.

The competent person in charge of the excavation shall determine by visual test and/or manual tests (e.g., thumb penetration test, dry strength test, plasticity or wet thread test) what type the soil is.

Spoil must be placed no closer than 2 feet (0.61 m) from the surface edge of the excavation. Everyone on an excavation site must wear a hard hat and not allowed to work under loads being lifted or moved by heavy equipment used for digging or lifting. Trenches four feet (1.22 m) or more in depth shall be provided with a ladder. Trenches left open overnight must be fenced and barricaded. Equipment for controlling standing water and water accumulation

[1] Occupational Safety & Health Administration (OSHA), U.S. Department of Labor. OSHA's mission is to assure the safety and health of America's workers by setting and enforcing standards. www.osha.gov

must be provided. Special measures must be taken if hazardous and/or toxic material is expected. For prevention of caving, a 10-foot-deep trench (ca. 3 m) would have to be sloped depending on the soil types. The excavated soil is mostly used to backfill trenches. Topsoil deposited separately can be used for recultivation. Settlement of backfilled trenches must be expected and be taken into account.

An example of a trenching survey at the abandoned Eulenberg landfill near Arnstadt, Germany, is shown in *Figs. 5.1-6* and *5.1-7. Figure 5.1-8* shows another example of a trenching survey and analysis at the same abandoned landfill in order to investigate whether a fault beneath the landfill detected by reflection seismics can be a pathway for leachate. The most prominent feature in seismic section EU4 (*Fig. 5.1-8 top*) is a three-phase reflection from the surface of the Muschelkalk (Middle Triassic). This reflection is offset beneath the northern part of the landfill. This is an indication of a fault. The result of the trenching survey is shown in the middle of *Fig. 5.1-8*. The tectonic model shown at the bottom of *Fig. 5.1-8* depicts an extension phase in the Mesozoic and a compression phase during the Mesozoic-Cenozoic. It was concluded that the fault is not "open" and is probably not a pathway for contaminant spreading.

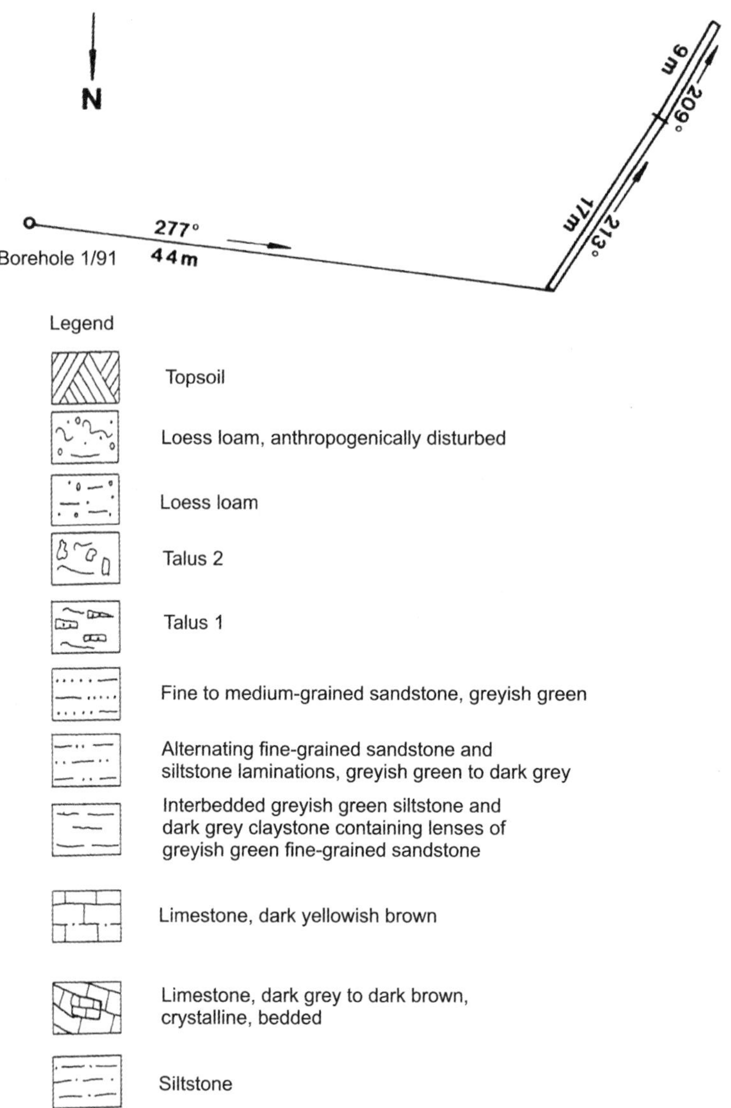

Fig. 5.1-6: Site sketch and legend of the trenching shown in *Fig. 5.1-7*, GEOS Ingenieurbüro GmbH, Jena, commissioned by BGR

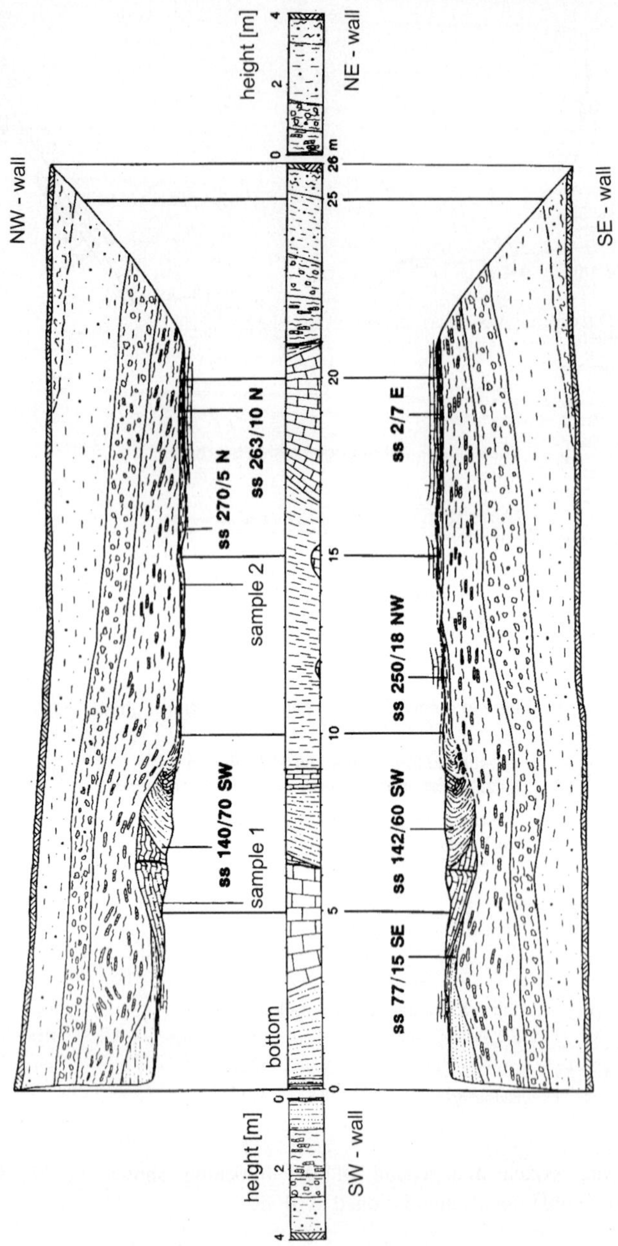

Fig. 5.1-7: Results of trenching at the abandoned Eulenberg landfill near Arnstadt, Thuringia, Germany, GEOS Ingenieurbüro GmbH, Jena commissioned by BGR

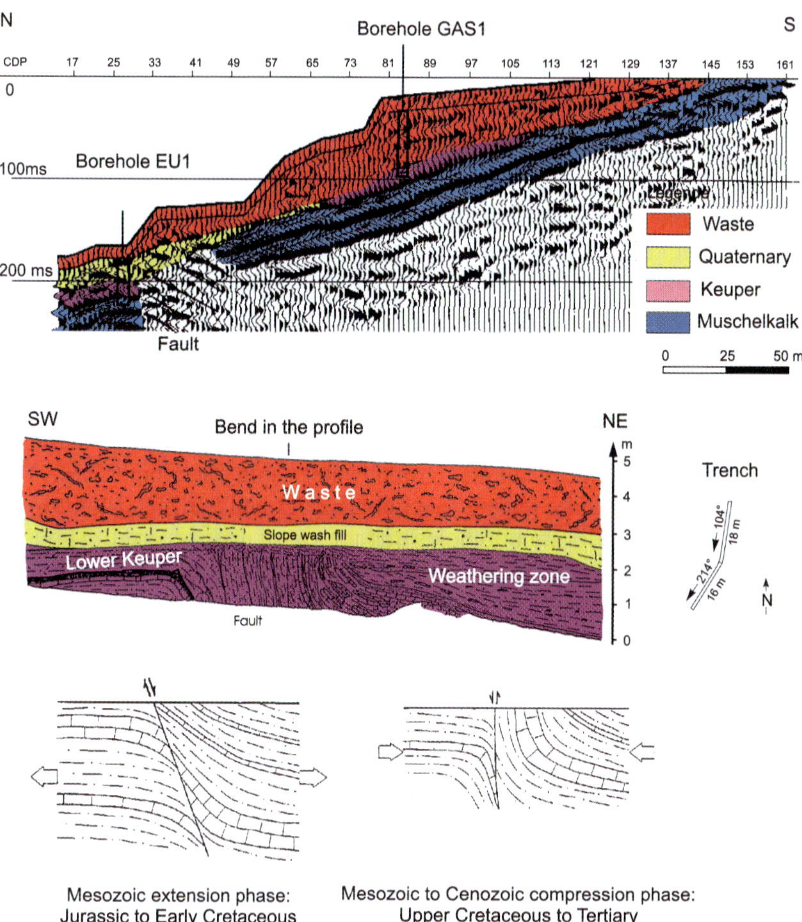

Fig. 5.1-8: Reflection seismic profile EU4, results of trenching and tectonic interpretation of trenching results at the abandoned Eulenberg landfill near Arnstadt, Thuringia, Germany, Geophysik GGD, Leipzig and GEOS Ingenieurbüro GmbH, Jena commissioned by BGR

5.1.3 Drilling

FLORIAN JENN & HANS-JÜRGEN VOIGT

Drilling is the process of making a circular hole with a drill or other cutting tool. Samples can be obtained from the drill cuttings or by coring during the drilling. Boreholes are used to obtain detailed information about rock types, mineral content, rock fabric and the relationships between rock layers at selected locations. Boreholes can also be used as monitoring wells, test wells and production wells in hydrogeological investigations (DÜRING, 1983). In some cases boreholes are plugged back to the surface after core sampling or logging, but in most cases they are used as monitoring wells. Monitoring wells are drilled to different depths of the aquifer to obtain information about the spatial distribution of contaminants and changes over time. They are also used for long-term measurements of the groundwater table (Section 5.2.6). Test wells are used for hydraulic testing (Sections 5.2.7.1 and 5.2.7.2). Monitoring, test, and production wells are boreholes that are cased and screened. Although production wells are normally used for water supply, at contaminated sites they can be used for precautionary (e.g., to prevent contaminant spreading) and/or remediation measures (pump and treat systems). The locations for drilling are selected using information obtained by geological, geophysical and/or geochemical methods. Three principal drilling methods are widely used for shallow-depth boreholes, depending on the type of information required and/or the rock types being drilled:

- cable tool method,
- auger drilling, and
- rotary drilling.

Either manual or engine-driven methods can be used for drilling. There are several variations of each principal method. Each of these methods has advantages and disadvantages, depending on the objectives of the investigation and the geological conditions. It is obvious that no drilling method is universal for all geological conditions and financial budgets. The main criteria for the selection of the most suitable drilling method are a reliable investigation of the subsurface down to the required depth, collection of good quality samples, and cost efficiency. Several considerations must be taken into account:

- What type of rock (unconsolidated and/or consolidated formations) is to be drilled through.
- In the case of hard rock, drilling tools will need cooling and lubrication.

- Rock cuttings and debris must be removed.
- Unconsolidated rock will require support to prevent the hole from collapsing.

It is also advised to take into consideration the possible subsequent installation of groundwater observation wells. The chosen method should not have any long-term influence on the characteristics of the groundwater. Before drilling is started, the place, objective and depth of each borehole must be stated in the plans for the field investigation.

The applicability of the different drilling methods is given in *Table 5.1-2*. An overview of the frequently used drilling methods, the usual diameter of the borehole, sample quality criteria, advantages, and typical sources of error is given in *Table 5.1-3*.

Table 5.1-2: Applicability criteria for the different drilling methods, LAPHAM et al. (1995)

Applicability criteria	**Drilling methods**									
	hollow-stem augering	solid-stem augering	bucket augering	hand augering	direct rotary with mud	direct rotary with air	reverse circulation rotary	cable tool (percussion)	jet wash and jet percussion	driven wells (percussion)
inexpensive	x	x		x					x	x
fast	x	x		x	x	x			x	x
excellent mobility	x	x		x				x	x	x
steel casing required								x		x
drilling fluid required					x	x	x		x	
limited to shallow depths	x	x	x	x					x	x
good formation samples	x		x	x			x	x		
poor formation samples		x			x	x			x	
no formation samples										x
gravel packing or grouting difficult to impossible		x		x					x	x
permits multiple-well completion in single hole	x		x		x	x	x	x		
can drill in most types of formations					x	x	x	x		
possibility of cross contamination	x	x			x	x	x		x	
limited casing diameter can be used	x	x		x					x	x

Table 5.1-3: Overview of frequently used drilling methods, ITVA (1996). Sample quality categories: 1 - very good, 2 - good, 3 - mediocre, 4 - unsatisfactory, 5 - unusable

Method	Diameter [mm]	Sample quality category	Advantages	Limitations	Sources of error
manual drilling	15-80	5 or worse	very fast, low cost	not always representative, max. drilling depth 2 m	contamination of borehole walls, inaccurate geological description, loss of volatile compounds
small-scale percussion drilling	35-80	at best 2 in cohesive soil, up to 3 in noncohesive soil	low cost, fast, can also be used in basements, houses, etc.	core loss due to deformation, caving, degassing	inaccuracies in geological description, borehole wall contamination, loss of volatile compounds
percussion core drilling	80-300	2 in consolidated sediment, 2-3 in unconsolidated sediment	yields good cores, applicable even below groundwater	significant heating occurs in hard soils, mixing of material when loose sediment is removed	loss of volatile compounds, displacement of contaminants
rotary core drilling	65-200	at best 4 above groundwater, 5 below groundwater	low cost, large sample volumes	retrieved samples are disturbed, significant heating occurs in hard soil	mixing of sample material, loss of fine material below groundwater surface
grab drilling	400-2500	3 above groundwater, 5-4 below groundwater	large borehole diameters and large sample volumes, even in very coarse material	inaccurate lithological profiles, disturbed samples, disposal	mixing of sample material, loss of fine material below groundwater surface
liner coring	80-200	1–2	exact profiles, no entry of air or water into the samples, protection from degassing	elaborate method, easily affected by obstructions	interactions with liner material

Detailed information about the various drilling techniques is given in the literature or in national standards, for example DRISCOLL (1986), DÜRING (1983), SHUTER & TEASDALE (1989), SCHREINER & KREYSING (1998), DIN 4020 and 4021, and EPA (1990).

Cable tool method

Cable percussion (cable tool) drilling is the oldest, simplest, most reliable, and economical technology available for drilling water wells (*Fig. 5.1-9*). It can be used to drill any material – from soft sands and clays to hard rock like granite. It requires no mud, mud pits, auxiliary pumps or chemicals. The drilling depth is only limited by the length of the wire cable. The equipment consists of a tripod with pulley, strong rope or cable, heavy drill stem with a drill bit or bailer, and a drum or pitman at the other end of the rope. By using the drum to alternately pull the rope taut and let it go slack, the drill bit is raised and allowed to fall (a pitman can also be used for this purpose). The resulting cuttings are periodically brought to the surface (normally after each drilled meter) with the bailer or with a suction pump (*Fig. 5.1-10*). Undisturbed material is not obtained, but it is easy to determine the lithology from the mixed sample from each meter. In unconsolidated rock the sample can be used for grain-size analysis and for laboratory migration tests. Another kind of percussion technique with core sampling is shown in *Fig. 5.1-11*.

The cable tool method is used in the vadose zone and in the groundwater zone. The major disadvantage of cable tool drilling is its slow penetration rate. Its slowness limits the use of cable tool drilling to boreholes of less than 250 m depth. Other limitations of cable tool rigs are that they can only drill vertical wells, not angled or horizontal ones, and that long casing strings may be difficult to retrieve in some formations without special equipment.

A cable tool rig can be operated by one person. A helper is required for safety reasons and to handle the tools. A geologist is also employed at most drilling sites to log samples.

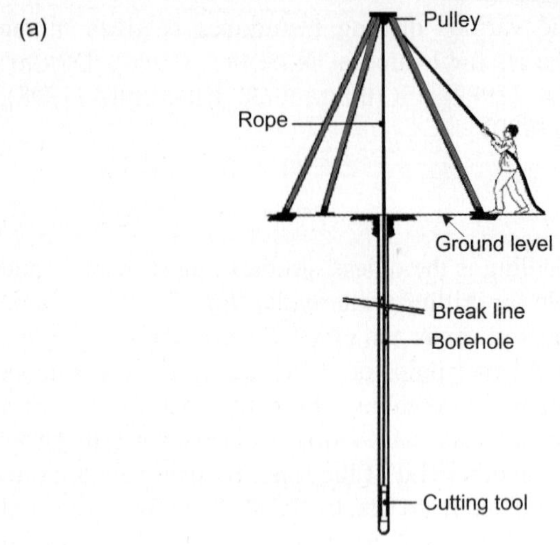

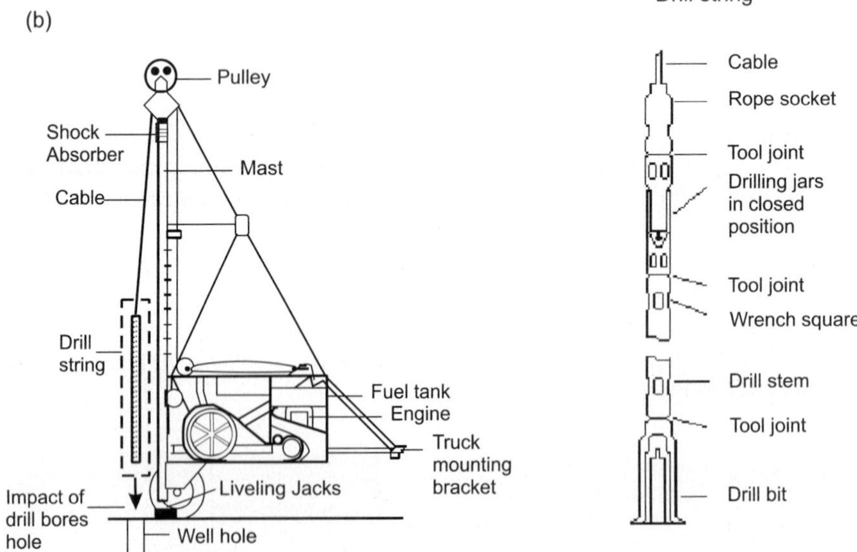

Fig. 5.1-9: Principle of cable tool drilling (a) manual version, ELSON & SHAW (2005), (b) engine-driven version, http://w3.pnl.gov:2080/WEBTECH/voc/cabtool.html

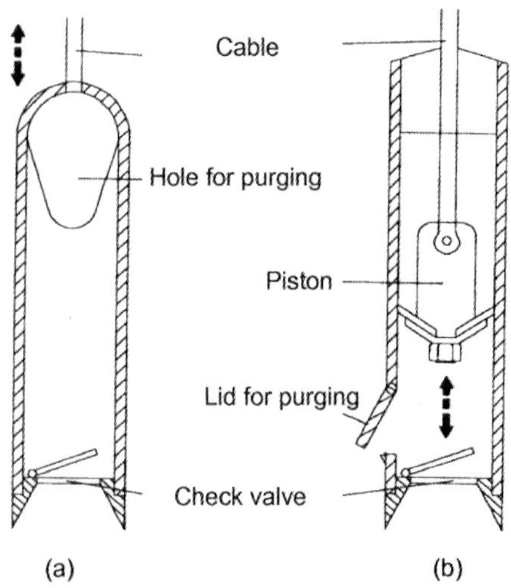

Fig. 5.1-10: (a) Cable tool bailer and (b) suction pump, Altlastenhandbuch des Landes Niedersachsen (1997)

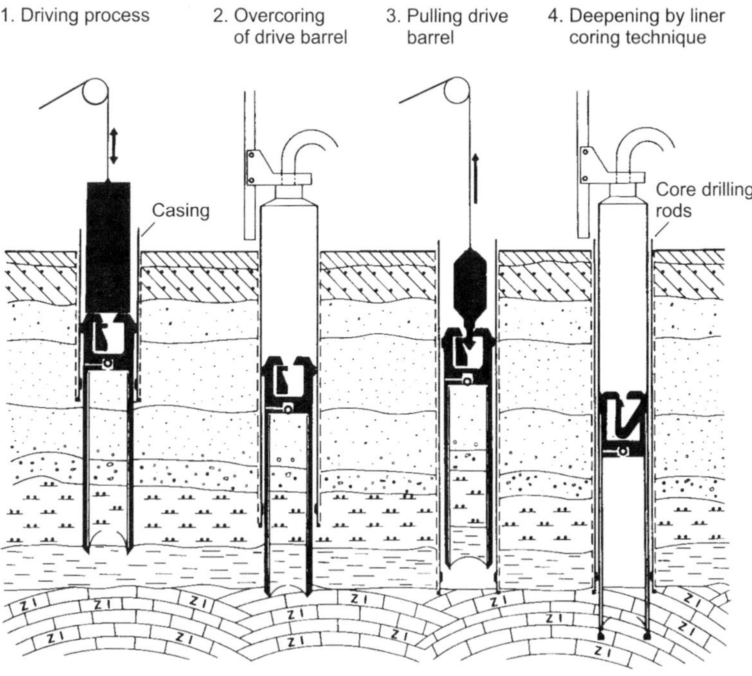

Fig. 5.1-11: Percussion drilling with core sampling, System Prakla-Seismos/Celler Brunnenbau, SCHREINER & KREYSING (1998)

Percussion drilling is a so-called “dry” drilling method. But that does not mean no water is used. All “dry” drilling methods need water in the borehole to protect against base failure when the bailer or the core sampler is pulled out. Experience shows that an average of 400 liters of additional water is used per drilled meter when using cable tool or hollow-stem auger techniques. This large amount of water use can significantly change the groundwater quality. Therefore, after installation a monitoring well must be carefully developed (Section 5.2.6).

Auger drilling

An auger is a drilling tool designed so that the cuttings are continuously carried by helical grooves on the rotating drill pipe to the top of the hole during drilling. Hand-auger drilling (*Fig. 5.1-12*) is suitable only for unconsolidated deposits and drilling to a shallow depth. It is inexpensive but slow compared to other methods. Problems can occur with unstable rock formations. Water is needed for dry holes. There are two types of augers: solid-stem augers and hollow-stem augers, see *Figure 5.1-13*.

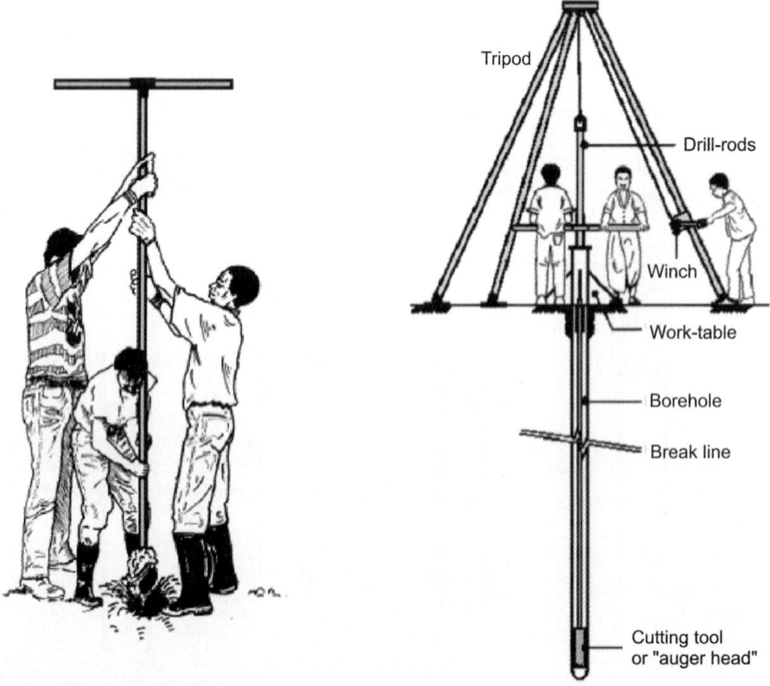

Fig. 5.1-12: Hand-auger drilling, ELSON & SHAW (2005)

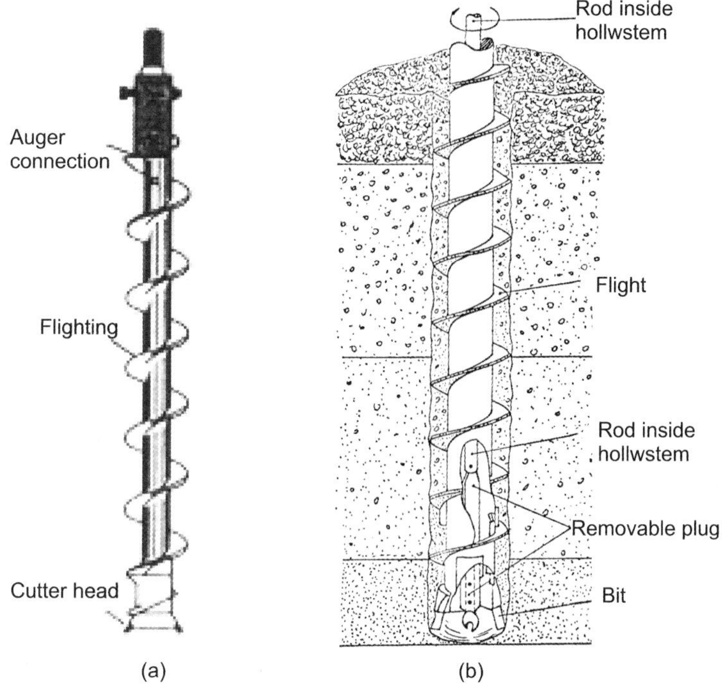

Fig. 5.1-13: (a) Solid-stem auger and (b) hollow-stem auger drilling tools, after SARA (1993); SCALF et al. (1981)

There are different types of solid-stem augers (bucket, spiral and others) which convey the rock material to the surface. This method allows drilling through unconsolidated or semiconsolidated materials such as sands, gravel, silt and clay. Normally, this method can be used to drill boreholes down to a depth of 40 m and up to 600 mm in diameter. On the basis of the cuttings transported to the surface, it is possible to determine the geological profile to within 50 cm resolution. The cuttings can be used for grain-size analysis and for geochemical laboratory analyses (batch tests). The hollow-stem auger technology allows undisturbed sediment samples to be taken. Several different sampling tools are used with hollow-stem augers. An example is shown in *Fig. 5.1-14*. The advantages of hollow-stem auger drilling are

- caving of the borehole walls is prevented,
- high drilling rate,
- allows well installation (up to 2 inches),
- easy sampling and documentation of sediments.

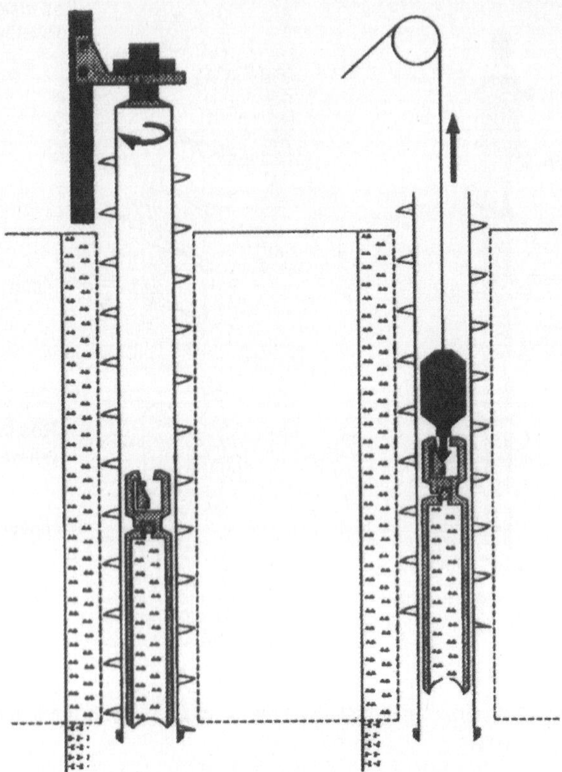

Fig. 5.1-14: Core sampling using a hollow-stem auger, HOMRIGHAUSEN (1993)

The auger method is the preferable method for drilling on waste or contaminated sites, but it is not recommended for water wells.

Rotary drilling

Jetting (*Fig. 5.1-15*) is a simple drilling method for unconsolidated rock. Water is pumped down the center of the drill stem, emerging from the borehole together with the cuttings. The cutting and washing out of the rock is facilitated by the up-and-down motion and rotation of the drill string. Either a foot-powered treadle pump or a small motor pump can be used to force the water into the borehole. This method can be used above and below the water table. Boulders can prevent further drilling.

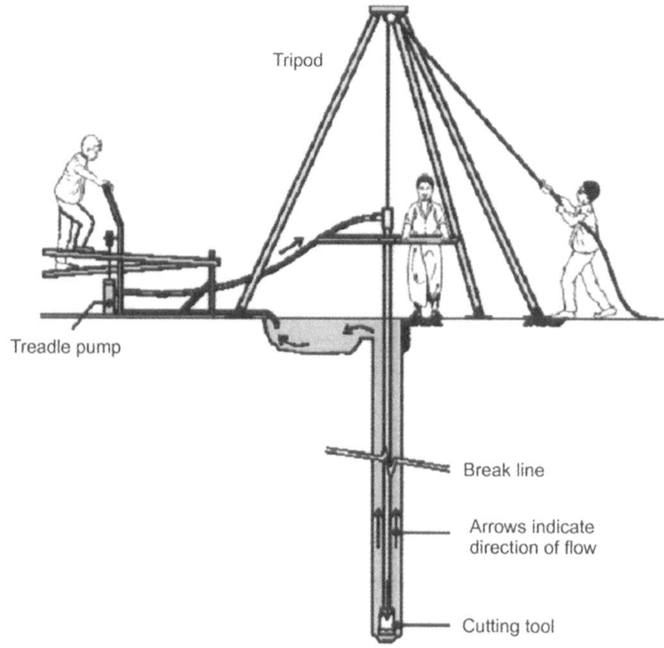

Fig.: 5.1-15: Jetting, ELSON & SHAW (2005)

Rotary drilling is one of the most common drilling methods. In this method, the entire drill string is rotated at the surface to turn the drill bit, and cuttings are removed from the hole by a circulating fluid. Water, mud or air is used as drilling fluid to

- cool and lubricate the drill bit,
- flush the drill cuttings up and out of the borehole,
- support and stabilize the borehole wall to prevent caving in, and
- to seal the borehole wall to reduce fluid loss.

Rotary-percussion drilling (*Fig. 5.1-16a*) is a fast method for drilling very hard rock, such as granite, and to penetrate gravel beds above and below the water table to a depth of about 300 m. A rapid action pneumatic hammer bit is forced through the rock by compressed air. As in air-rotary drilling, the rock cuttings are brought to the surface by the returning air. Rotation of 10 - 30 rotations per minute ensures that the borehole is straight and circular in cross-section. The method requires experience to operate and maintain the equipment.

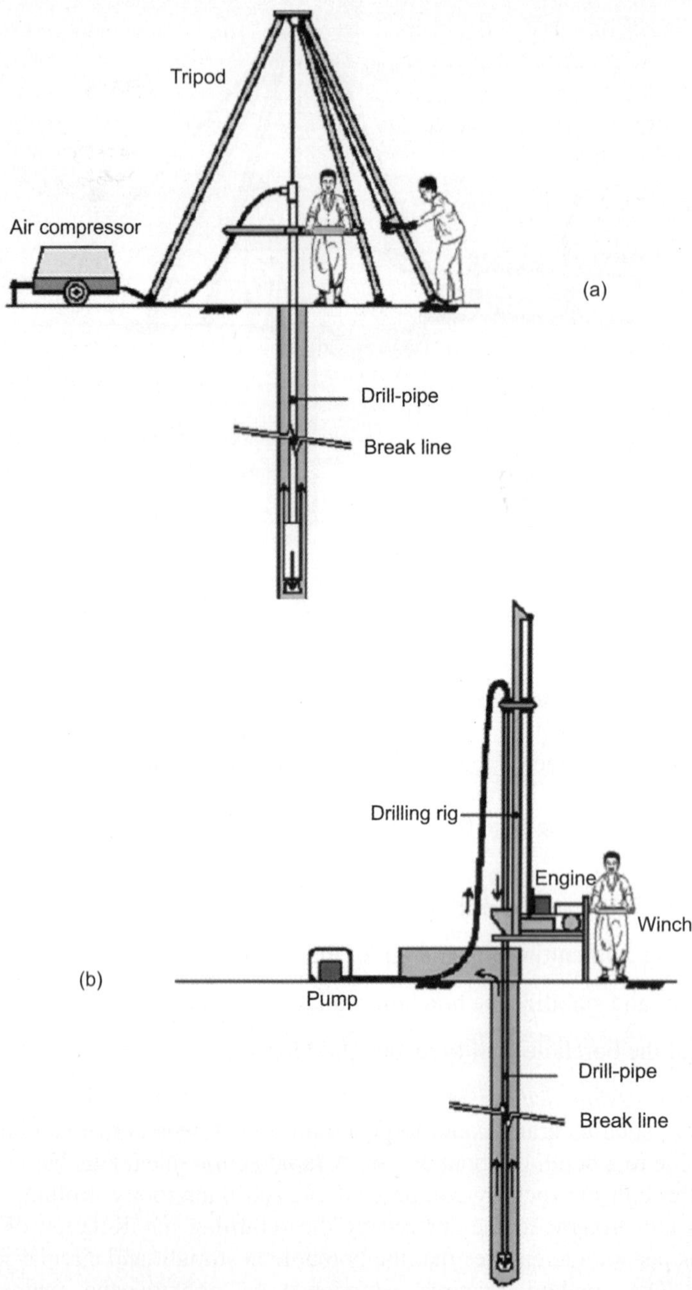

Fig.: 5.1-16: (a) Principle of rotary-percussion drilling, (b) principle of rotary drilling with flush, ELSON & SHAW (2005)

Air-rotary drilling is a fast and economical method. Compressed air is forced down the drill pipe and, as it returns to the surface, it carries with it the rock cuttings made by the rotating bit. This method is used in soft rock materials to a depth of about 25 m.

Most rock formations can be rapidly drilled by *rotary drilling with flush* (*Fig. 5.1-16b*). The drill stem and bit are rotated to cut the rock. Air, water, or drilling mud is pumped down the drill stem to flush out the cuttings and debris. The velocity of the flush in the borehole annulus must be sufficient to lift the cuttings to the surface. Water and mud support unstable borehole walls. Drilling is possible above and below the water table and the method can be used to depths of several hundred meters.

In *diamond core drilling*, a pipe encrusted with industrial diamonds at one end is used to drill through the rock, with a core of rock remaining in the center of the pipe. This core is recovered and provides detailed information about the rock.

There are two basic types of *mud or water rotary drilling*: the direct and reverse methods (*Fig. 5.1-17*). In the direct rotary method, the drilling fluid is pumped down through the drill pipe and out of the bit. The fluid then flows upwards in the annular space between the hole and the drill stem, carrying the cuttings in suspension to the surface into the mud pit. In reverse circulation rotary drilling, the fluid and its load of cuttings flow upwards inside the drill pipe and are discharged by a suction or vacuum pump into the mud pit.

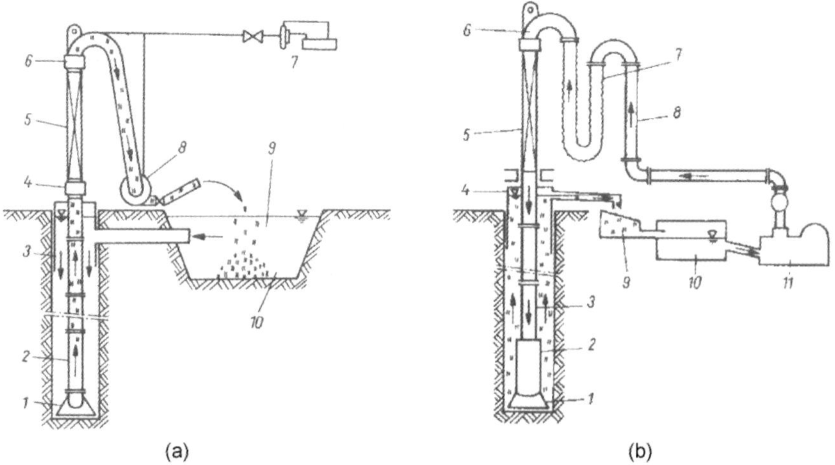

Fig.: 5.1-17: (a) Principle of reverse mud rotary drilling and (b) principle of direct mud rotary drilling, DÜRING (1983). (a): 1 - drilling bit, 2 - drill pipe, 3 - protection casing, 4 - drive table, 5 - kelly, 6 - swivel, 7 - vacuum pump, 8 - suction pump, 9 - mud or water, 10 - mud pond, (b): 1 - bit, 2 - heavy collar, 3 - drill pipe, 4 - protection casing, 5 - kelly, 6 - swivel, 7, 8 - mud pipe, 9 - sieve, 10 - mud pit, 11 - mud pump

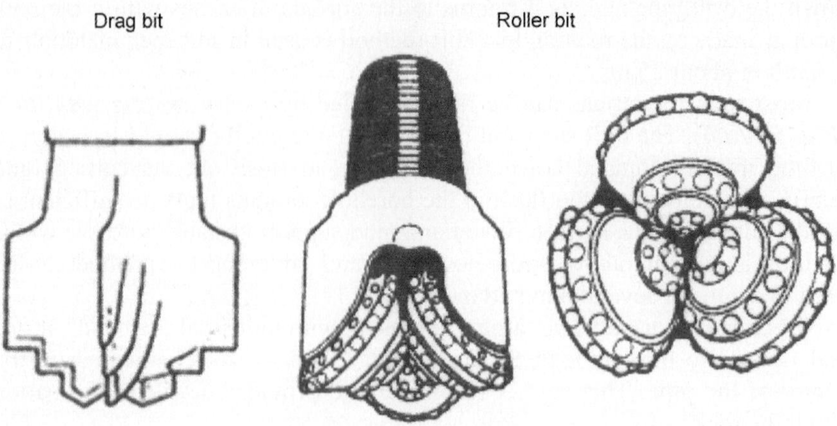

Fig.: 5.1-18: Two types of drill bits, LIFEWATER CANADA (2004)

Figure 5.1-17 shows the most common reverse rotary drilling method. A more specialized reverse technique is, for example, the counterflush method, which is used for small-diameter (2") boreholes in unconsolidated sediments. The down-hole-casing advance method allows drilling and casing of the borehole to be done at the same time. It is appropriate for drilling in semi-consolidated and hard rocks, but not in loose sediments. This method and air rotary drilling have the least impact on the quality of the surrounding groundwater.

Three types of bits are used for rotary drilling of water wells: a drag (blade or fishtail) bit, a roller (tricone) bit (*Fig. 5.1-18*) and a reamer bit. Reamer bits are used to widen a borehole for well installation.

The most common drilling mud is a suspension of bentonite (sodium montmorillonite clay) in water, but it can be advantageous under certain circumstances (depending on availability, geological conditions, etc.) to use other artificial or natural mixtures, such as Tixoton (calcium montmorillonite), Revert (a synthetic polymer), or mixtures of local clay. After the fluid is mixed, sufficient time must be allowed to elapse to insure complete hydration of the clay prior to its being circulated into the borehole (DRISCOLL, 1986). It is very important to have a strictly controlled fluid scheme during the entire drilling. The weight, viscosity, mud losses, and pumping rate must be monitored continuously and conditions adjusted accordingly. On the one hand, the drilling mud must be thick enough to bring up the cuttings, but on the other hand, if it is too thick, it will be difficult to pump and the cuttings will not settle out in the mud pit. And if mud is too thin, excessive migration of the mud into aquiferous layers can occur, changing the geochemical conditions. *Figure 5.1-19* shows the relationship between the loss of drilling mud and mud weight for different mud mixtures.

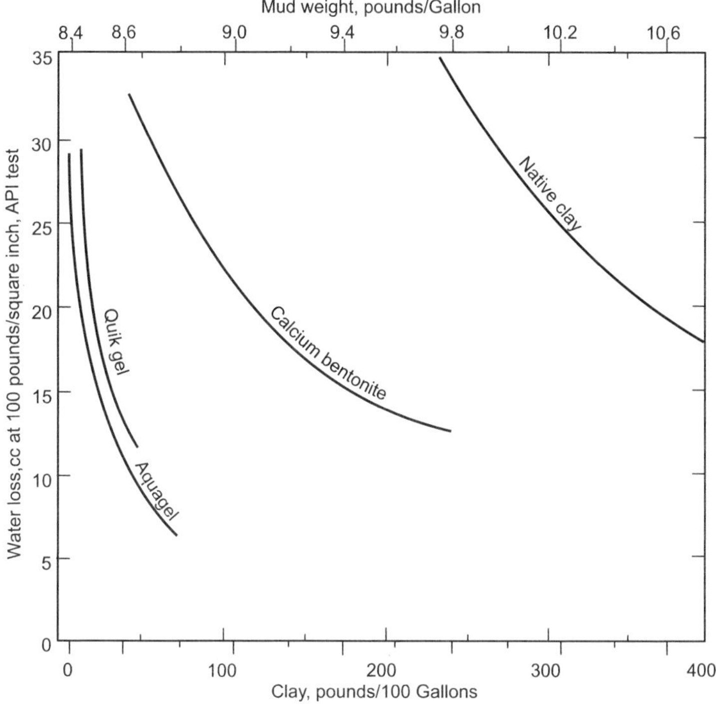

Fig.: 5.1-19: Mud weight versus drilling mud loss, SHUTER & TEASDALE (1989)
Conversion factors: 100 pounds/square inch 7.03×10^4 kg m^{-2}, pound/gallon (US) 0.1198×103 kg m^{-3}, cc = cm^3

Cuttings brought to the surface by rotary drilling cannot be used for lithological interpretation of the borehole. They can only provide an indication of changing geological units. Hence, rotary drilling operations must be always coupled with open-hole geophysical logging. A *standard program of geophysical logging* should include the following logging methods:

- caliper log,
- natural gamma ray log,
- gamma-gamma log (density log),
- neutron-neutron log (water content, porosity),
- resistivity log (16-inch normal, 64-inch normal),
- focused electrical log, and
- salinity-temperature log.

For detailed information about geophysical borehole logging, see Chapter 4.8.

At least one cored borehole is recommended at new landfill sites for each hectare of landfill area. This recommendation, however, does not take into consideration the difference between unconsolidated and consolidated rock, nor the very different geological and hydrogeological conditions at different locations and also cannot be applied to abandoned landfill sites. A larger number of cored boreholes would only be deemed necessary at locations with unconsolidated rock (in general, sites with Quaternary and Tertiary rocks). The boreholes must be backfilled prior to the construction of a new landfill; in the case of observation wells this would involve expensive over-drilling since the gravel fill would also have to be removed. Drilling boreholes in hard rock is more effortful than in unconsolidated rock and hence also more expensive. This is the reason why the number of boreholes is often kept to a minimum. From a geological point of view this is justified since it is valid to assume that the rocks will – as a consequence of their geological genesis – not have any sizable lateral inhomogeneities within the area of a landfill that are likely to remain undetected. A further factor which greatly influences cost is the drilling depth. Boreholes drilled into unconsolidated rock should generally have depths of 10 - 15 m, with individual boreholes drilled to greater depths depending upon the geological conditions. Boreholes in hard rock should normally be drilled down to 20 - 30 m, in individual cases also to greater depths, for example when deeper aquifers are to be investigated. In the case of unconsolidated rock, direct push testing can also be a suitable method for obtaining information from the subsurface. Geophysical methods can provide information between the boreholes in order to reduce costs.
Routine macroscopic processing of drill cores consists of the following:

- petrographic core description,
- photographic documentation,
- representation of the preliminary geological profile,
- description of the geological structures visible in the core,
- input of the data into a database, and
- sample archiving.

Secondary drill core processing includes – depending upon the scope of the site investigation – a selection of other sedimentological, petrographic, petrophysical, mineralogical and geochemical laboratory evaluations in which information can be gathered to focus on various aspects, including

- content and distribution of mineral and rock components,
- presence and content of clay minerals,
- total carbonate as well as the individual carbonate phases,
- differences in mineral content in different parts of the core,

- organogenic components,
- formation fluids,
- gas and oil traces, contamination,
- texture, bedding and deformation,
- grain-size distribution,
- particle sizes in the micron and submicron range,
- sediment matrix and cement,
- porosity and permeability,
- pore-size distribution,
- fissures, joints, slickensides,
- rock and grain density, and
- stress effects.

The core evaluation should be based both upon well proven standard, as well as on nondestructive core testing methods, such as radiometric density determination, transmission computer tomography, electrical resistance tomography, acoustic tomography, magnetic susceptibility measurement.

After the logging and before the installation of a well, the drilling mud must be replaced by clean water and the mud-cake on the borehole wall must be reamed out. When drilling has been completed at a contaminated site, the drilling equipment must be decontaminated and the contaminated materials (soil, mud, water, etc.) must be disposed of properly. A six-step procedure for the decontamination of well installation equipment and materials after ALLER et al. (1989) may be used:

1. Select a location for the decontamination procedures:
   - Avoid spilling decontamination fluids at the drilling site.
   - Prepare a clean area for cleaned equipment.
2. Select equipment requiring decontamination.
3. Determine the frequency of decontamination of the equipment.
4. Select the cleaning technique and type of cleaning solutions to be used for decontamination. The following procedure is standard for cleaning well-installation equipment and materials:
   - Wash outside (and inside where applicable – for example, well casing and screen) of equipment and materials used during well installation using a low-sudsing, nonphosphate detergent.

- Complete decontamination procedure with high-pressure steam-cleaning using potable tap water.

5. Contain residual contaminants and cleaning solutions, if necessary, and dispose of according to regulations.
6. Plan to collect some quality-control samples to evaluate the effectiveness of decontamination procedure (for example, a sample of the rinse water that was used to steam-clean or remove all residues and additional samples of rinse water taken from the equipment after it has been decontaminated).

5.1.4 Direct Push Technology

KAI MÜLLER, MICHAEL NEUHAUS & HANS-JÜERGEN VOIGT

Direct push methods are used for investigation of soils and sediments by driving, pushing, and/or vibrating small-diameter hollow steel rods into the ground. Their use is limited to unconsolidated sediments and semi-consolidated sediments and is generally not possible in consolidated rocks.

An enormous variety of equipment is available and – depending on the survey objectives and selected methods – ranges from 15 kg manual hammers to hydraulic pressure systems mounted on a heavy-duty truck. Examples for direct push equipment are shown in *Figs. 5.1-20* and *5.1-21.* See also percussion drilling in Section 5.1.3. For cone penetration testing (CPT), the mass of the truck (10 to 30 t) is used as a counterweight to the hydraulic pressure unit used to achieve depths of > 20 m. In contrast to the CPT method, a percussion hammer is used for depths of < 20 m by hammering rods into the ground or by vibration under pressure. In this case the mass of the truck can be considerably smaller than for CPT (usually less than 4 t). Because the trucks are smaller, the percussion hammer method can be used even within buildings, as well as in difficult terrain. Moreover, such tools can be rammed into the ground at an angle of up to 37.5° from the vertical. The rods used for the CPT method are 1 m long. They have male threads on one end and female threads on the other. The rods used for the percussion method are 1.5 - 3.5 m long.

There are two types of rod systems: single rod and cased. The most common type of rod systems is the single rod. The diameter of the rods is typically 1 inch (2.54 cm), but can range from 0.5 to 2.125 inches (1.27 to 5.40 cm). Cased systems, also called dual-tube systems, have an outer tube, or casing, and a separate inner sampling rod. The casing can be advanced simultaneously with the inner rod or immediately afterwards. Samples can be collected without removing the entire string of rods from the ground. The outer tube diameter is typically 2.4 inches (6.10 cm), but can range between

1.25 and 4.2 inches (3.18 to 10.67 cm). Single-rod systems are easier to use (EPA SUPERFUND, 2005).

If a soft layer is underlain by a hard one, it is possible that the rods will bend. For this reason, deviation from the vertical is automatically registered in the CPT method by an inclinometer in the tool (cone with sensors at the lower end of the first rod). The rods and tools must be decontaminated when they are removed from the ground. In the CPT method this is done with hot water or steam automatically while the rods are withdrawn.

For the percussion method, the U. S. EPA considers 180 m to be possible within an 8-hour working day. For the CPT method, 80 - 150 m daily is considered reasonable (HEITMANN, 2000).

CPT (sometimes called Dutch CPT) has been continuously developed and used since the 1930s to determine soil mechanical parameter values for engineering geology purposes. In the USA, direct push technology has been used by EPA since the end of the 1970s to investigate contamination within the scope of the Superfund Program and has become standard procedure (ASTM D5778-95, D6001-96, D6187-97, D6282-98). Guidelines for the use of these methods for investigating and monitoring waste sites are being prepared within the scope of the COST 629 program of the European Union.

Direct push testing is described here as an example of cone penetration testing. CPT can provide a highly detailed, 3-dimensional picture of the subsurface in less time than needed by traditional methods (e.g., boreholes). Various tools are used to identify lithology, stratigraphy, depth to the groundwater table and the capillary fringe, as well as to determine geotechnical parameter values and detect the presence of contaminants. There are also tools for collecting soil, water, and soil gas samples.

Logging methods (e.g., gamma, neutron-neutron) for small diameter holes (slim holes) provide very detailed information about the strata.

Special tools are available for detecting different types of contaminants, e.g., petroleum hydrocarbons, polycyclic aromatic hydrocarbons (PAH), phenols, volatile halogenated hydrocarbons (VHC), and benzene, toluene, ethylbenzene, xylene (BTEX). Individual compounds cannot be identified and the results are only semiquantitative. X-ray fluorescence (XRF) and laser-induced breakdown spectroscopy (LIBS) are used in direct push testing to identify inorganic substances (e.g., heavy metals).

In addition to the above-described tools, video tools are used to recognize very thin soil and contamination layers. The direct push technology can also be used for small-diameter (i.e., less than 2 inches) temporary or permanent monitoring wells, install remediation equipment, for soil vapor extraction wells and air sparging injection points (EPA SUPERFUND, 2005).

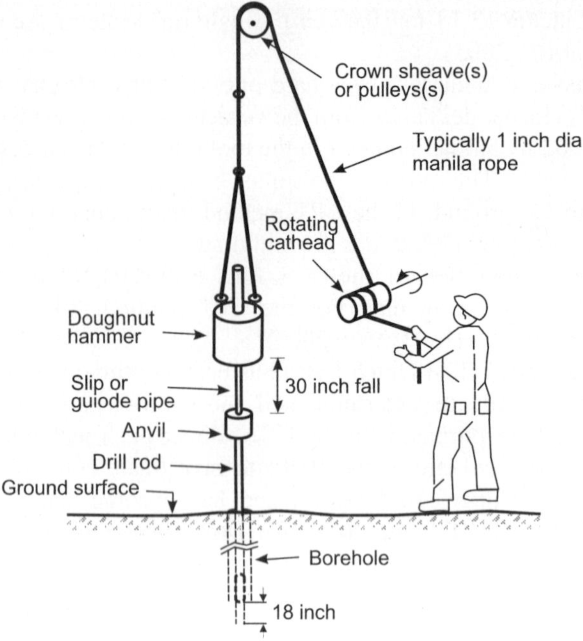

Fig. 5.1-20: Equipment for a standard penetration test

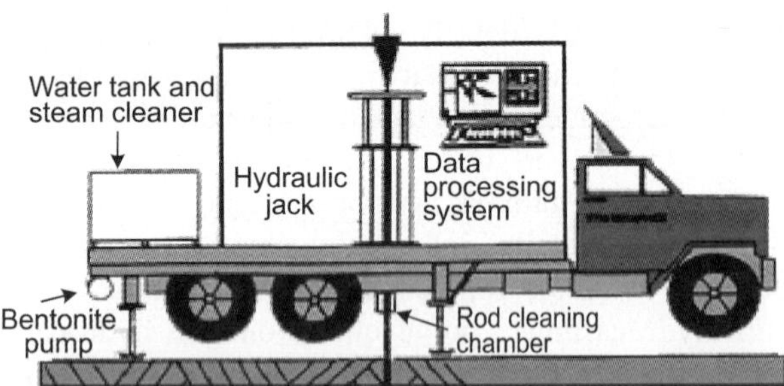

Fig. 5.1-21: Equipment for cone penetration testing mounted on a heavy-duty truck

Owing to the small diameter of the tools, the effect on the soil ecosystem is little. To avoid migration of any contaminants, the resulting hole is filled with bentonite when the testing is completed. Because there is no drilling, there are no drill cuttings to be disposed of, which would be expensive in the case of contaminants.

The following kinds of tools are discussed in this section:

- geological/geotechnical tools,
- tools for detecting types of contaminants, and
- sampling tools.

Most of the tools are connected to the lower end of a rod or rod string. The tool is pushed into the ground at a constant rate and measurements are made continuously. In some cases the CPT is carried out in two steps: In the first step, a geological/geotechnical tool is pushed to the final depth and the corresponding parameter values are measured. In the second step, various geophysical tools are lowered within the hollow rods and measurements are made as in geophysical borehole logging (*Fig. 5.1-22*).

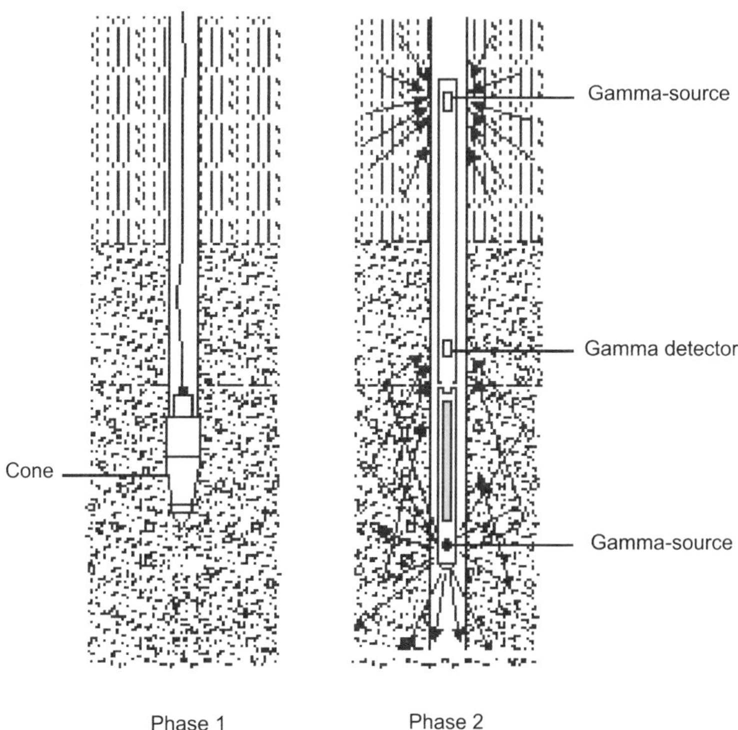

Fig. 5.1-22: Cone penetration testing with measurement of cone resistance and sleeve friction in the first phase and measurement of natural gamma activity and gamma-gamma density in the rod string in the second phase, FEJES et al. (1997)

Geological/geotechnical and geophysical tools

CPT systems consist of a thrust machine capable of 20 to 200 kN thrusting force and a counterweight system (the rig and truck), a cone penetrometer (tool) as well as recording equipment. The principle of a cone penetrometer is shown in *Fig. 5.1-23*. The currently used type of cone with a friction sleeve is shown in *Fig. 5.1-24*. The accepted reference is a cone penetrometer that has a cone with a 10 cm^2 base area and an apex angle of 60 degrees and is specified as the standard in the International Reference Test Procedure (IRTP, 2001). The following parameters are measured with geological/geotechnical tools (cone penetrometers): cone (tip) resistance, sleeve friction, inclination, depth, and in some cases pore water pressure. The most commonly measured parameters are either cone resistance and sleeve friction, or cone resistance and total force on the penetrometer in the CPT method. Cone resistance, sleeve friction and pore water pressure are the commonly measured parameters in the CPTU method. Inclination and depth are measured with both methods. Additional measurement of the pore water pressure at one or more locations on the penetrometer surface gives a more reliable determination of stratification, soil type and mechanical soil properties than standard CPT.

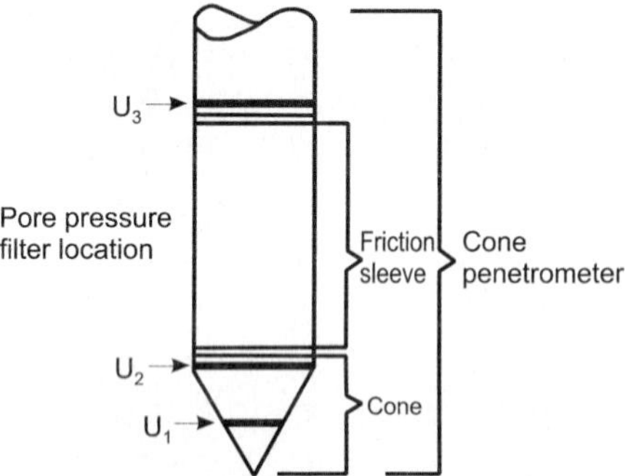

Fig. 5.1-23: Principle of cone penetrometer, U_1, U_2 and U_3 are filter locations for pore water pressure measurements, BROUWER (2002)

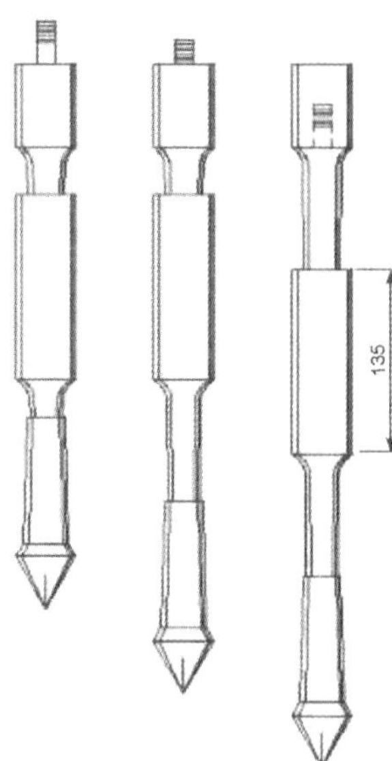

Fig. 5.1-24: Begemann-type cones with friction sleeve, SANGLERAT (1972)

There are two types of cone penetrometers: subtraction cones and compression cones.

- Subtraction cones measure the total force on the penetrometer (sleeve + tip) and the cone resistance. The sleeve friction is calculated by subtracting the cone resistance from the total force. There is no upper limit for the sleeve friction of the subtraction cone. The only limit is on the total force on the penetrometer.

- Compression cone measures the cone resistance and the sleeve friction separately. This results in a lower maximum value (1 MPa) for the sleeve friction. Specifications for cone penetrometers with 10 and 15 ton capacity are given in *Tables 5.1-4*, *5.1-5* an *5.1-6* as an example.

Table 5.1-4: CPT parameter values of subtraction cones, BROUWER (2002)

Parameter	10 tons	15 tons
apex angle of cone	60°	60°
diameter	35.7 mm	43.8 mm
projected area of cone	1000 mm^2	1500 mm^2
length of friction sleeve	134 mm	164 mm
area of friction sleeve	15 000 mm^2	22 500 mm^2
max. force on penetrometer	100 kN	150 kN
max. cone resistance q_c (if $f_s = 0$)	100 MPa (100 kN)	100 MPa (150 kN)
max. sleeve friction f_s (if $q_c = 0$)	6.6 MPa (100 kN)	6.6 MPa (150 kN)
diameter of push rods	36 mm	36 mm

Table 5.1-5: CPT parameter values of compression cones, BROUWER (2002)

Parameter	**10 tons**	**15 tons**
apex angle of cone	60°	60°
diameter	35.7 mm	43.8 mm
projected area of cone	1000 mm^2	1500 mm^2
length of friction sleeve	134 mm	164 mm
area of friction sleeve	15 000 mm^2	22 500 mm^2
max. force on penetrometer	100 kN	150 kN
max. cone resistance q_c (if $f_s = 0$)	100 MPa (100 kN)	100 MPa (150 kN)
max. sleeve friction f_s	1 MPa (15 kN)	1 MPa (22.5 kN)
diameter of push rods	36 mm	36 mm

Table 5.1-6: General parameter values of CPT systems, modified after BROUWER (2002)

rate of penetration	20 ± 5 mm s^{-1}
max. inclination of the rod string	15°
sensors	cone resistance sleeve friction inclination depth
	pore water pressure (only for CPTU) with a max. of 1 MPa, 2 MPa or 5 MPa
position of thrust machine	at least 1 m from previous CPT, or at least 20 times the diameter of a previous borehole
verticality of the thrust machine	± 2°
distance between measurements	max. 20 mm
calibration	every 3 months or 3000 m of soundings
field checks	see *Table 5.1-7*
max. wear	angle of cone: 60 ± 5°, sleeve: 35.7 ± 0.4 mm

Recommendations for maintenance routines to assure the quality of investigation results are given in *Table 5.1-7*.

The friction ratio R_f [%] can be used to identify the soil type. Graphs for soil classification using cone resistance and local sleeve friction or cone resistance and friction ratio, respectively, are given in *Figs. 5.1-25* and *5.1-26*.

$$R_f = (f_s / q_c)\ 100, \tag{5.1.1}$$

where f_s = unit sleeve friction [MPa] and q_c = cone resistance [MPa].

Table 5.1-7: Recommended maintenance routines (IRTP, 2001, Table A1.1; taken from BROUWER, 2002)

Checking Routine	Start of project	Start of test	End of test	Every 3rd month
verticality of thrust machine		x		
penetration rate		x		
safety functions	x			x
push rods	x	x		
wear	x	x	x	
gaps and seals	x	x	x	
filter	x	x	x	
zero drift		x	x	
calibration	x			x
function control	x			x

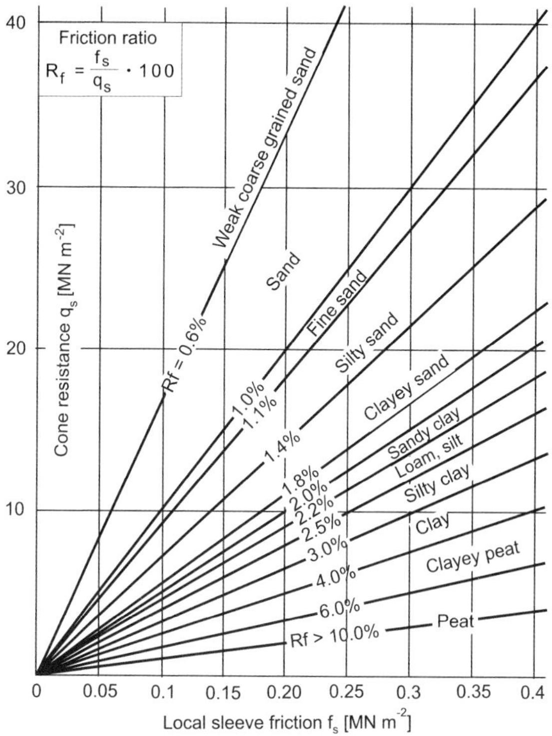

Fig. 5.1-25: Soil classification using cone resistance and local sleeve friction from CPT measurements, based on FUGRO documents after, WEISS (1990)

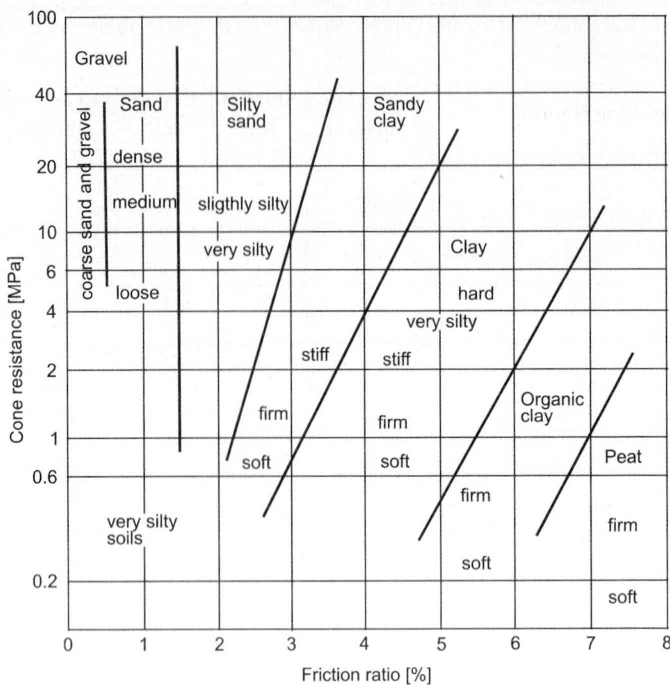

Fig. 5.1-26: Soil classification using cone resistance and friction ratio from CPT measurements, BROUWER (2002)

A lithological classification scheme for sediments based on cone resistance, natural gamma activity, and gamma-gamma density data is shown in *Fig. 5.1-27*. Examples of CPT results are given in *Figs. 5.1-28* and *5.1-29*. Besides rock type (lithology), other geotechnical parameter values, such as soil density, in-situ stress conditions and shear strength parameters, can be determined by CPT and CPTU. Most CPT equipment can also be used with other sensors, for example to measure electrical conductivity or the traveltime and amplitude of acoustic waves generated by a seismic source (e.g., a hammer striking a steel plate on the ground) to provide supplementary information.

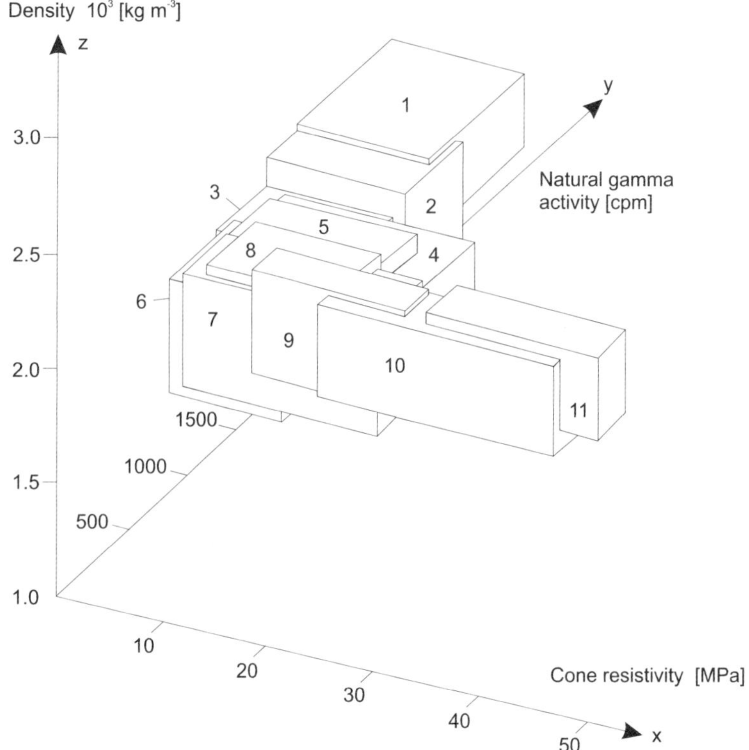

Fig. 5.1-27: Lithological classification of sediments on the basis of cone resistance, natural gamma activity, and gamma-gamma density data, FEJES et al. (1997): 1 - clay; 2 - lean clay to silty clay; 3 - clayey silt to silt; 4 - fine-sandy silt; 5 - silty fine sand; 6 - fine sand; 7 - silty sand; 8 - sand; 9 - coarse sand; 10 - gravelly sand to gravelly coarse sand; 11 - sandy gravel to gravel

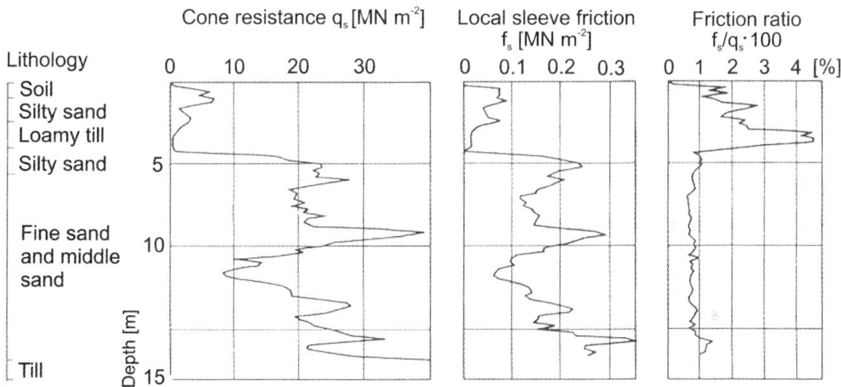

Fig. 5.1-28: Lithological profile based on the interpretation of cone resistance and local sleeve friction from CPT measurements, WEISS (1990)

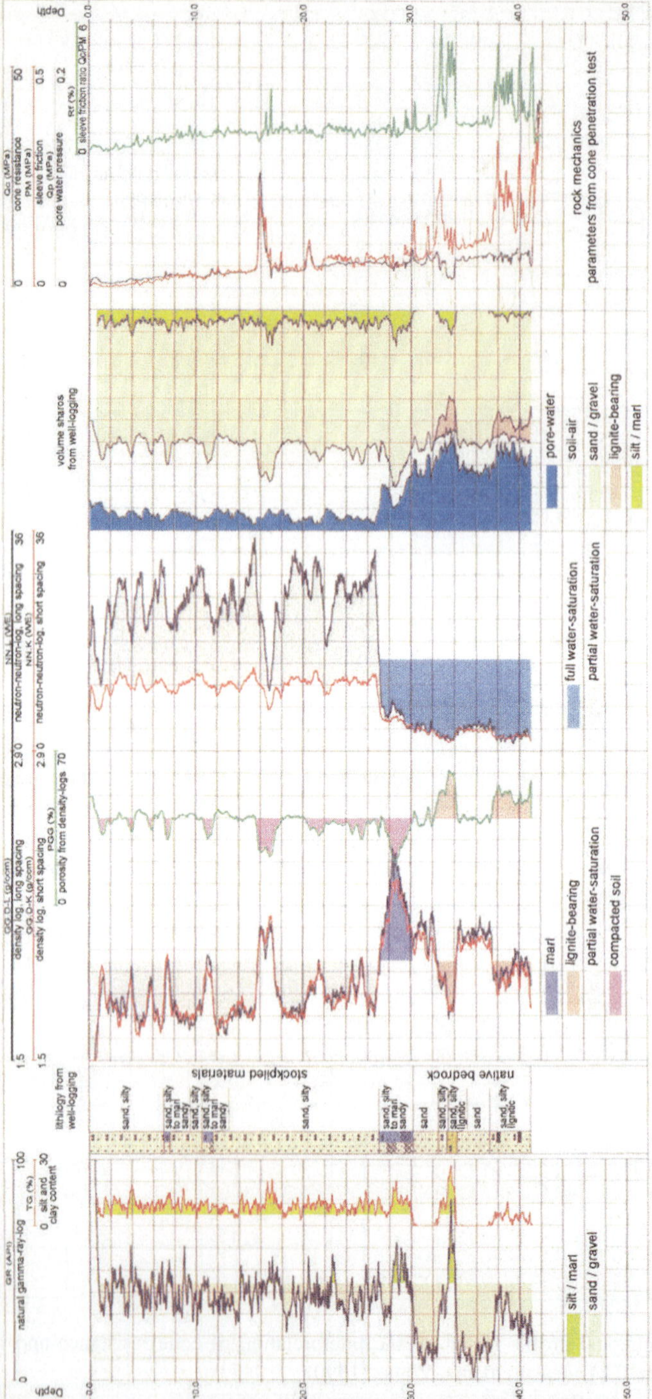

Fig. 5.1-29: Determination of geological parameter values in a lignite mine waste heap: cone resistance, sleeve friction and pore water pressure from CPT, and natural gamma activity, gamma-gamma density and neutron-neutron measurements from geophysical logging within CPT rods, courtesy of BLM Bohrlochmessung Storkow, Brandenburg, Germany

Other examples of cone penetration tests with geophysical logging are given in *Figs. 6.2-29* and *6.2-33* in Part 6.

Tools for detecting types of contaminants

Special tools are available for detecting different types of contaminants, e.g., petroleum hydrocarbons, volatile halogenated hydrocarbons (VOC) and BTEX (benzene, toluene, ethylbenzene, xylene). Individual compounds cannot be identified and the results are only semiquantitative.

Membrane interface probes (MIP) are used to detect the presence of volatile substances in nearly real time. A field laboratory is required in the CPT vehicle or in a separate vehicle if the percussion drilling method is used. Flameionization detectors and photoionization detectors are needed in the laboratory.

As the tool is pressed into the ground, a heating element in the tool heats the surrounding ground to about 130 °C (BRACKE, 2002), volatilizing substances with a low boiling point. The gases are drawn into the tool across a heated hydrophobic membrane made, for example, of teflon and metal, in a stainless steel housing and transported at a rate of about 40 mL min^{-1} through the drilling rods to the flame-ionization and photo-ionization detectors by a carrier gas (nitrogen or helium). A schematic diagram of a membrane interface probe is shown in *Fig. 5.1-30*.

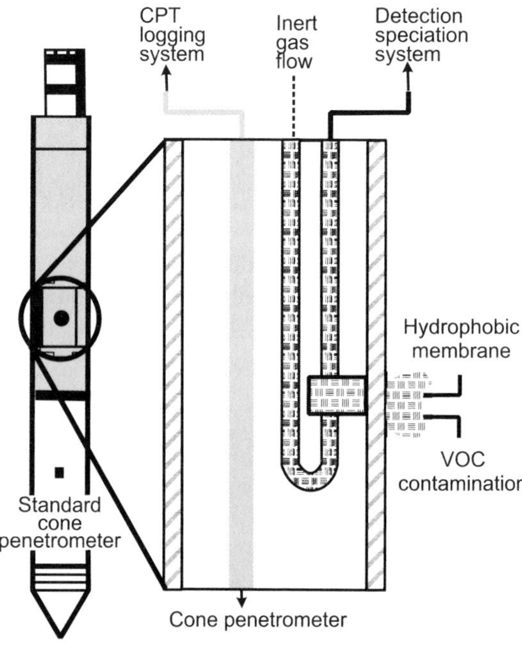

Fig. 5.1-30: Sample collection with a MIP tool, courtesy of FUGRO Consult

Individual compounds can be identified by also passing the collected gas through a gas chromatograph. A halogen-sensitive detector (XSD) for direct-push technology is developed for identification of different volatile organic halogen compounds (BRACKE, 2002).

The Rapid Optical Screening Tool (ROST®) and the FMG tool (field tool for detecting fluorescent substances) use lasers to induce fluorescence of certain compounds, e.g., aromatic hydrocarbons, PAHs, phenols, and BTEX. Fluorescence is the emission of radiation after excitation of a molecule by absorption of radiation energy (see also 4.9.3.2). The intensity of fluorescence is proportional to the concentration of the fluorescing substance. A UV laser beam or other light source is passed through a fiber glass cable to the tool, where it passes through a sapphire window into the soil surrounding the tool. The substances are determined from the frequencies of the UV fluorescence light (KRAM et al., 2001). The UV light from a mercury lamp, for example, causes aromatic hydrocarbons to fluoresce. A second fiber glass cable carries the emitted fluorescence to the sensors in the CPT vehicle, where the intensity of the light is measured, displayed on an oscilloscope, and recorded.

The ROST® tool can be operated in one of two modes: (i) The intensity of the fluorescence light is plotted as a function of depth, holding the wavelengths of the laser beam and of the recorded fluorescence constant. This is done to determine the location of the contamination (*Fig. 5.1-31*). (ii) The intensity spectra of the fluorescence are plotted 3-dimensionally as a function of depth and wavelength while the wavelength of the laser remains constant. The wavelengths of the fluorescence vary depending on the detected substances. The type of contaminant can be identified from the spectra if the characteristic 3-D plot („fingerprint") is known for that type of substance (*Fig. 5.1-32*).

Estimation the proportions of dissolved hydrocarbons, hydrocarbons in the oil phase and the residual phase are facilitated by spectral measurements (*Fig. 5.1-33*).

The detection limits of the ROST® tool are several mg contaminant per kg soil and several tens of $\mu g\ L^{-1}$ for groundwater, depending on the type of soil and the type of contaminant (HEITMANN, 2000). For example, the detection limit for petroleum hydrocarbons in sandy soil is about 100 ppm. Because clay soils have a large surface area, part of the induced fluorescence is reabsorbed before reaching the tool, reducing the intensity of the measured fluorescence. PAH compounds and additives (e.g., colored substances) in motor fuel and oil cause detection limits to be higher because they dampen the intensity of the fluorescence.

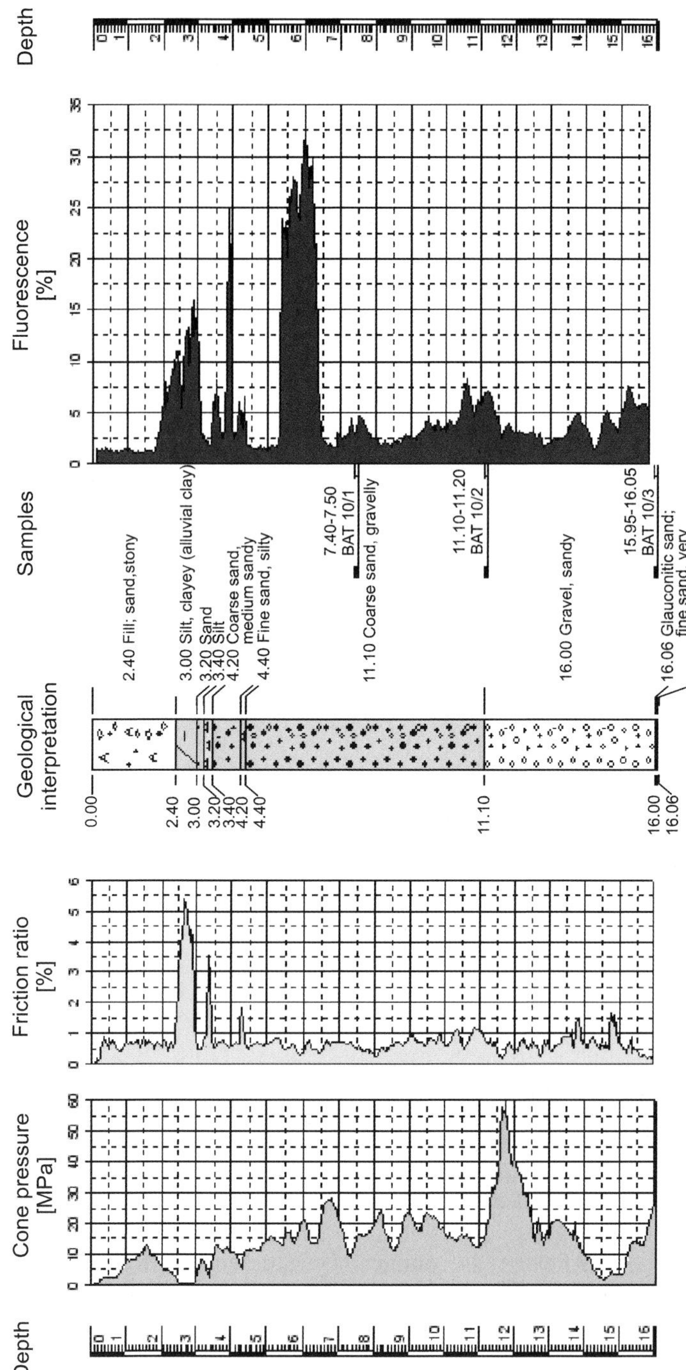

Fig. 5.1-31: Logs from ROST® and geotechnical tools, courtesy of FUGRO Consult

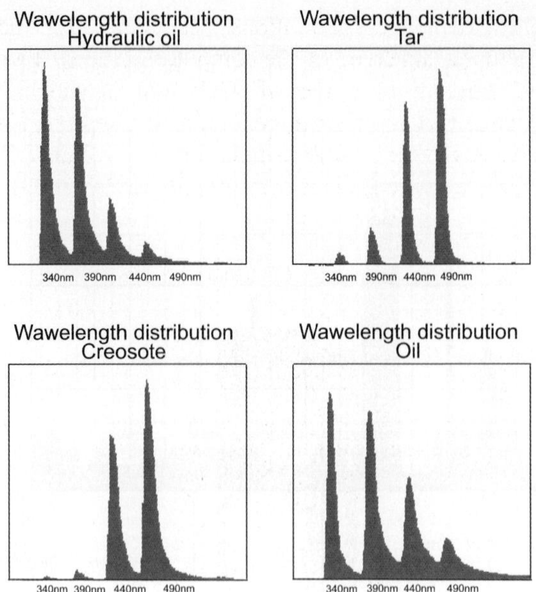

Fig. 5.1-32: Spectra of hydraulic oil, tar, creosote, and waste oil, courtesy of FUGRO Consult

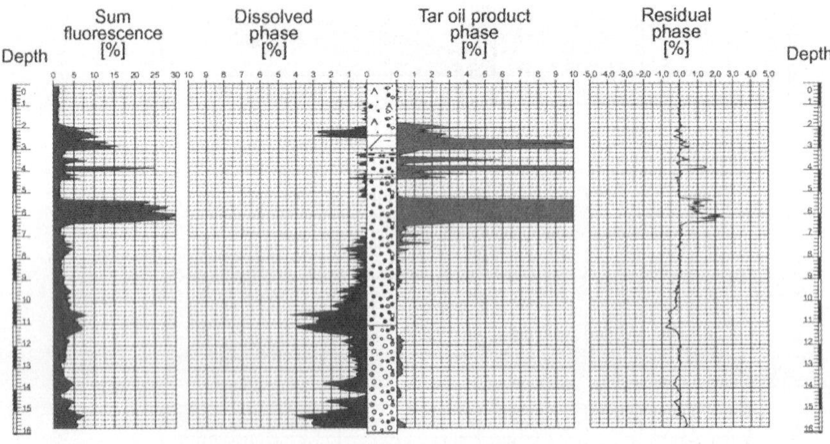

Fig. 5.1-33: Estimation of sum fluorescence, dissolved phase, product phase and residual phase by ROST measurements, courtesy of FUGRO Consult

X-ray fluorescence (XRF) and laser-induced breakdown spectroscopy (LIBS) are used in direct push testing to identify inorganic substances (e.g., heavy metals). In XRF, X-rays cause atoms to fluoresce at a typical wavelength characteristic of the fluorescing element. Thus, measurement of the wavelength and intensity of the fluorescence provides information about the elements present and their concentrations. The X-ray fluorescence spectrum registered by the tool is analyzed by the computer in the CPT vehicle.

A schematic diagram of a typical XRF tool is shown in *Fig. 5.1-34*. Because the distance that X-rays can penetrate is only 0.1 - 1 mm, there must be direct contact between the soil particle and the window in the tool. The detection limit for heavy metals is about 100 ppm (EPA, 2003).

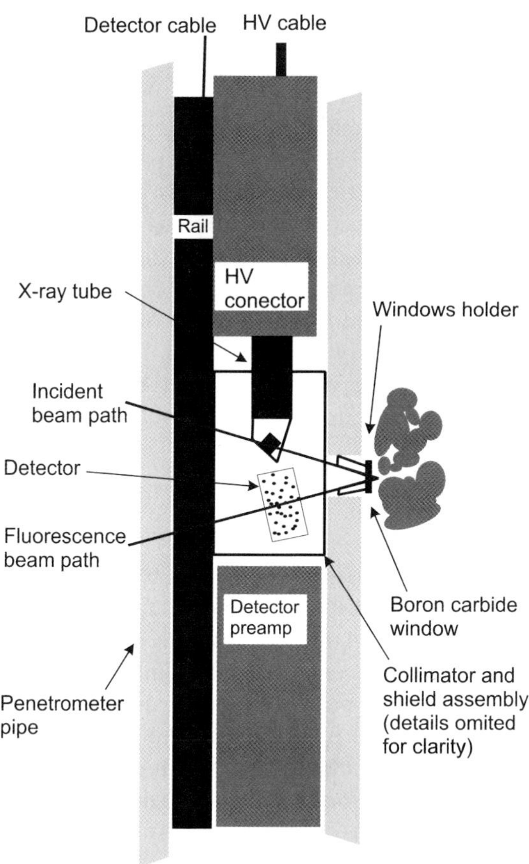

Fig. 5.1-34: XRF tool, modified after EPA (2003)

In the LIBS (Laser Induced Breakdown Spectroscopy) method for detecting heavy metals, a high-energy pulsing laser is used to produce a high-temperature plasma in the soil. This plasma emits pulses of light with wavelengths characteristic of the elements present. The wavelengths and their intensities registered by the tool are converted to an electrical signal, which is analyzed by the computer in the CPT vehicle. In general, LIBS has lower detection limits than XRF, for example 10 - 40 ppm for lead. Values obtained for chromium between 30 and 1200 ppm, for example, are in very good agreement with the values determined in the laboratory (EPA, 2003). The LIBS method, however, cannot be used below the groundwater table. In contrast, the XRF method can be used in both the saturated and unsaturated zones. XRF and LIBS are normally used together with geotechnical tools. The surface texture and inhomogeneity of the soil particles and the type of soil influence both methods. A particular problem is a heterogeneous distribution of heavy metals in the soil, because in contrast to laboratory analyses, the heavy metals cannot be separated from the soil particles.

In addition to the above-described tools, video tools are used to recognize very thin soil and contamination layers. Video tools contain miniature cameras with objectives that can magnify by up to factor 100 so that particles as small as 20 μm in diameter can be observed. Light emitting diodes provide the necessary light. Sometimes laser-induced fluorescence is used to make contaminated particles visible.

Sampling tools

Sample collection is also possible with direct-push methods. When the lithology and the type and distribution of contaminants have been determined using geotechnical and contaminant-detecting tools, soil, soilgas, and groundwater samples can be taken at selected depths. Only the basic aspects of sample collection with direct-push technology are discussed in this section with respect to the objectives of the investigation. There is a multitude of instruments available on the market, which will not be gone into here.

There are two types of *soil samplers*: sealed and unsealed samplers. Unsealed samplers are driven open into the ground to the desired sample depth. Sealed samplers are driven closed into the ground, opened at the selected depth, and then driven further, filling the liner (*Fig. 5.1-35*). The liner is made of brass, stainless steel, acrylic plastic, teflon, or PVC. The tool is then withdrawn to the surface, the liner containing the sample is removed and sealed. Sealed samplers help prevent spreading of contamination. Such samples can be directly used later for laboratory tests (e.g., percolation tests, migration tests, but also for determination of hydraulic conductivity) and chemical analysis.

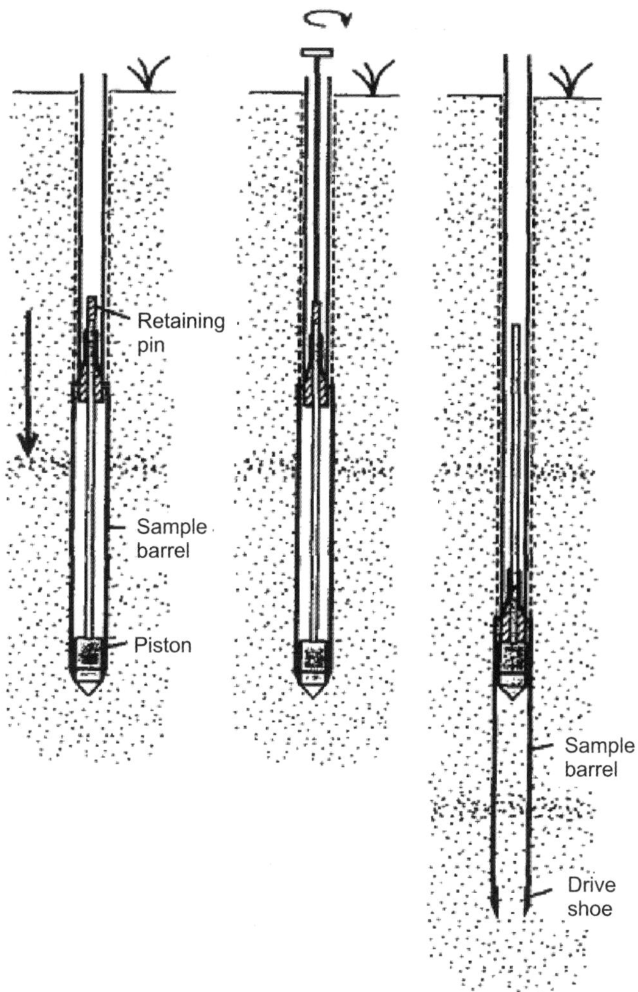

Fig. 5.1-35: Sample collection with a sealed sampler, modified after EPA (1997)

The MOSTAP soil sampler for cone penetration testing and the Geoprobe soil sampler for percussion drilling are examples of sealed soil samplers used worldwide. The thermal-desorption sampler (TDS), designed for detecting volatile organic compounds in soil down to depths of 30 m, cannot be used together with geotechnical tools.

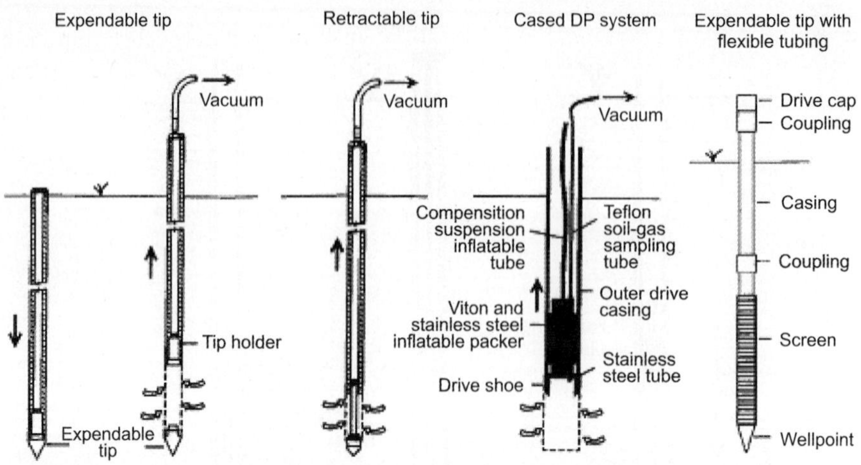

Fig. 5.1-36: Tools for sampling soil air, modified after EPA (1997)

There are four methods for *collecting soil gas samples* with direct-push technology (*Fig. 5.1-36*). A tool with a disposable tip is driven to the selected depth and then withdrawn several centimeters (leaving the tip behind). The valve of an evacuated headspace vial is opened to collect a gas sample, or a sample is drawn through tubing in the drilling rod to a container at the surface. In the first case, the tool is brought back to the surface and the gas sample removed. The tool is decontaminated and driven with a new tip to a greater depth in the same hole (if it is stable). The previously used tip remaining in the hole is pushed aside in the process, which can be a problem in clay.

If a disposable tip cannot be used owing to the presence of clay, a tool with a retractable tip is a possible alternative. The tool is driven to the desired depth, the tip extended, the sample collected, and the tip retracted again. The main problem with this procedure is that if the tip is not completely retracted into the tool, it may prevent withdrawal of the tool to the surface.

Another type of tool has screens (e.g., of stainless steel or PVC) through which soil gas samples can be continuously drawn as the tool is driven into the ground. The disadvantage of this tool is that the screen can drag contaminated material to greater depths or become clogged with fine material, affecting sample collection at those depths (EPA, 1997). Gas that remains in the tubing can also contaminate the next sample.

The danger of dragging contamination to different depths is reduced if casing is driven into the ground together with the cone penetration tool. The soil-air sample can then be collected using teflon tubing to the surface or a headspace vial can be used. Samples can be collected at different depths without removing the tool. Sample collection takes longer with this method than with the above-described tools.

Soil-air sampling tools are available that can be installed in direct-push holes for continuous monitoring (BRACKE, 2002). These have either a screen of stainless steel or HDPE (0.5 m length, 1 inch in diameter, 0.145 mm hole size). Filter packs of quartz sand sealed with bentonite and a tubing system permit the collection of samples at several levels.

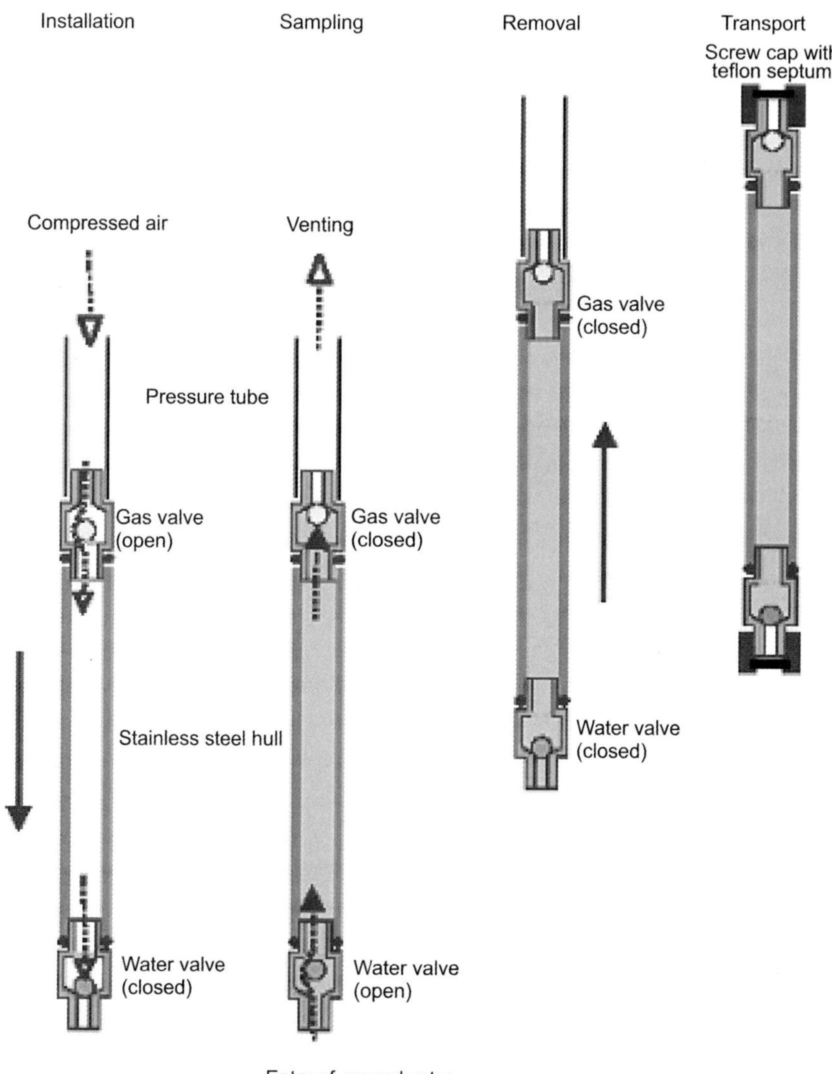

Fig. 5.1-37: Pneumatic HYDROP® tool of ECOS Co. (BRACKE, 2002)

Two methods are available for *collecting groundwater samples*. Which one is used depends on the kind of contamination. If the groundwater is contaminated with substances with elevated boiling points, it can be raised to the ground surface with a membrane pump (up to 9 m) or with a hydrolift pump. Groundwater contaminated with volatile substances should be collected without contact with the atmosphere to avoid escape of the volatile substance. In this case, the tool contains a evacuated container. A valve is opened when the tool is in the groundwater, the valve is closed and the tool is brought back to the surface.

One tool used in Germany (BRACKE, 2002) consists of a tube with a ball valve at either end (*Fig. 5.1-37*). The bottom valve is kept closed by elevated air pressure while the tool is lowered in the hole. At the level the sample is to be taken, the air pressure is lowered, allowing water to rise in the tube, closing the ball valve at the top. The tube is then raised to the surface to remove the sample.

In another tool, called BAT® Enviroprobe, a sleeve covering the screen is retracted when the tool is at the selected depth, allowing water into a chamber closed at the top with a septum. A glass vial under vacuum and closed with a septum is then lowered so that a double-ended hypodermic needle breaks through both septums, allowing water to enter the glass vial (*Fig. 5.1-38*). Relatively small samples are collected with both sampler types (500 mL maximum).

When groundwater samples are collected in sealed containers, the in-situ pressure in the aquifer is retained (ASTM D 6282). That volatile substances are retained when groundwater samples are collected using direct push technology is a considerable advantage over the collection of samples by pumping. It is practical to collect groundwater samples using direct push technology only down to a hydraulic conductivity of 10^{-4} cm s^{-1} (typical of silt). At lower conductivities, it is better to install a permanent monitoring tool.

Groundwater sampling tools designed for direct push technology can be easily and inexpensively used for groundwater monitoring. If the soil material is cohesive enough, the cone penetration tool and rod can be replaced with tubing (PVC, for example) with screened intervals and a slightly smaller diameter than the push tool. If protective tubing is used around the push rod, filter and seal packs can be installed at the screened intervals. This improves the hydraulic conductivity and keeps fine material away from the screen (DRISCOLL, 1986). The relatively large diameter of the protective tubing, however, limits the depth to which the push tool can be driven. Because observation wells prepared with direct push technology are smaller than conventional ones, the sample volumes are smaller and longer times are needed to collect the samples. Comparison of results from the two kinds of installations show very good agreement. The lack of filter packs results, however, in a higher turbidity of the sample and a lower hydraulic conductivity. Volatile organic substances, on the other hand, can hardly be detected in conventional observation wells. It has been shown that there is very little loss of VOC in samples from direct-push observation wells.

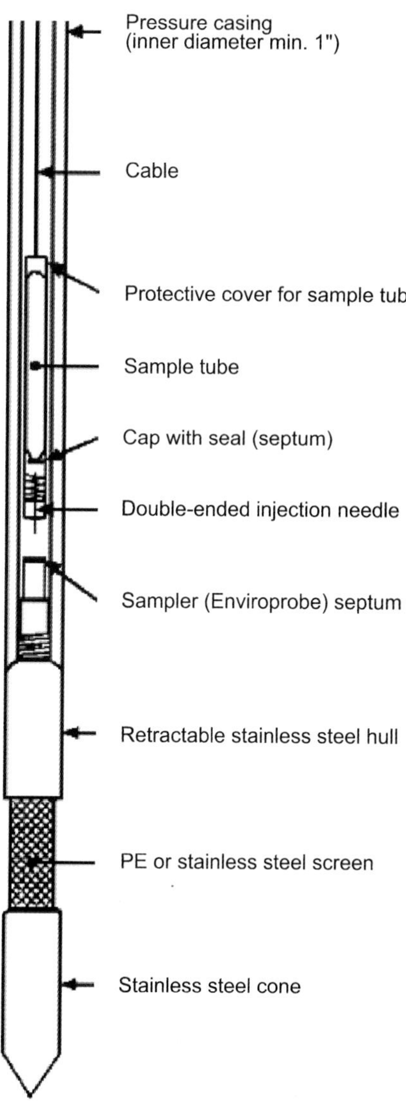

Fig. 5.1-38: BAT® Enviroprobe, courtesy of FUGRO Consult

Grouting of direct push holes

Open direct-push (DP) holes provide a path for vertical migration of contaminants. Several methods have been developed to avoid this (*Fig. 5.1-39*). The holes are usually filled with bentonite, which is very fine-grained clay and therefore, has a very low hydraulic conductivity.

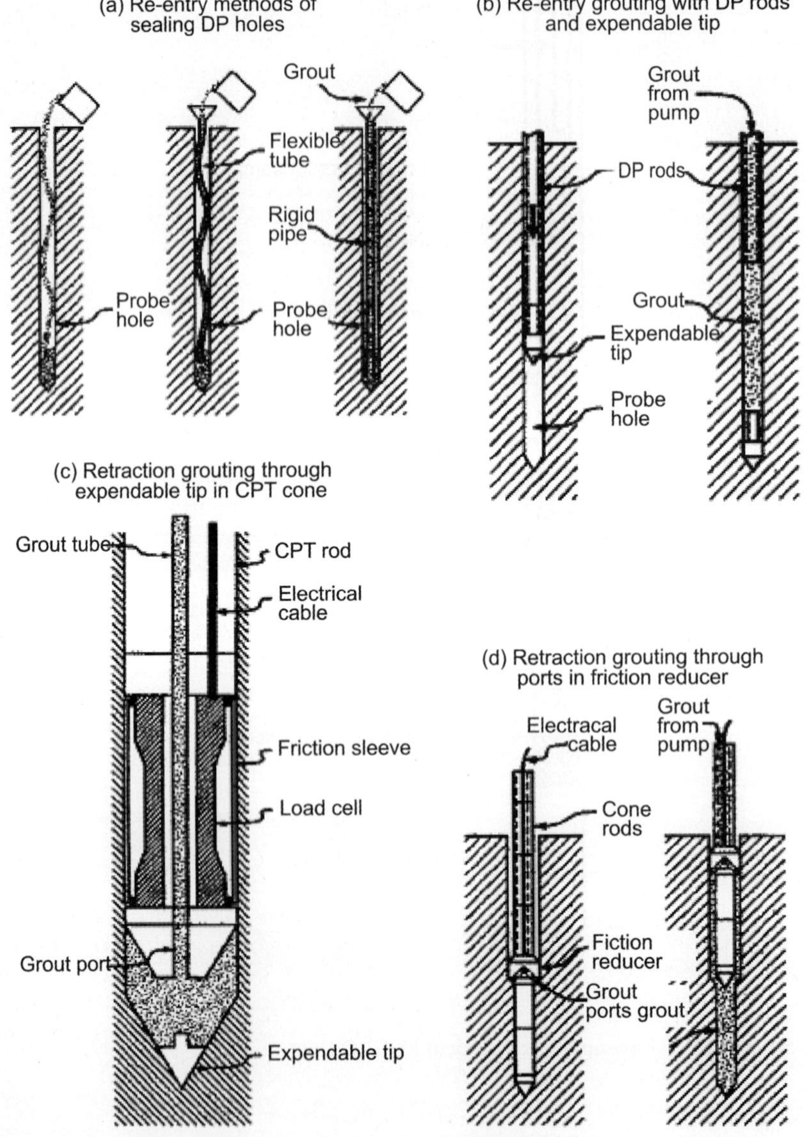

Fig. 5.1-39: Methods for direct-grouting push holes, EPA (1997)

References and further reading

ALLER, L., BENNETT, T. W., HACKETT, G., PETTY, R. J., LEHR, J. H., SEDORIS, H., NIELSON, D. M. & DENNE, J. E. (1989): Handbook of suggested practices for the design and installation of ground-water monitoring wells. National Water Well Association, Dublin, Ohio.

ALTLASTENHANDBUCH DES LANDES NIEDERSACHSEN (1997): Niedersächsisches Landesamt für Ökologie und Niedersächsisches Landesamt für Bodenforschung (Ed.). Springer, Berlin.

ASTM D 5778 (1995): Standard Test Method for Performing Electronic Friction Cone and Piezocone Penetration Testing of Soils. Reapproved 2000. American Society for Testing and Materials, West Conshohocken, Pennsylvania, USA.

ASTM D 6001 (1996): Standard Guide for Direct-Push Water Sampling for Geoenvironmental Investigations. Reapproved 2002. Superseded by: ASTM D 6001 (2005). Standard Guide for DirectPush Water Sampling for Geoenvironmental Investigations.

ASTM D 6187 (1997): Standard Practice for Cone Penetrometer Technology Characterization of Petroleum Contaminated Sites with Nitrogen Laser-Induced Fluorescence. Reapproved 2003.

ASTM D 6282 (1998): Standard Guide for Direct Push Soil Sampling for Environmental Site Characterizations. Reapproved 2005.

BRACKE, R. (2002): Direct-Push Technologien – Der Einsatz von DP-Verfahren für Geomonitorings und baugrundtechnische Untersuchungen. Ecos, Aachen.

BROUWER, J. J. M. (2002): Guide to Cone Penetration Testing – On shore and near shore. www.conepenetration.com/cpt3.html#test

DIN 4020 (2003): Geotechnische Untersuchungen für bautechnische Zwecke. Geotechnical investigations for civil engineering purposes. Beuth, Berlin.

DIN 4021 (1990): Aufschluss durch Schürfe und Bohrungen sowie Entnahme von Proben. Soil; exploration by excavation and borings; sampling. Note: To be replaced by DIN EN ISO 22475-1 (2004-08). Beuth, Berlin.

DIN 4022: Baugrund und Grundwasser; Benennen und Beschreiben von Boden und Fels. Subsoil and groundwater; classification and description of soil and rock. Teil 1 (1987): Schichtenverzeichnis für Bohrungen ohne durchgehende Gewinnung von gekernten Proben im Boden und im Fels. Part 1 (1987). Borehole logging of soil and rock not involving continuous core sample recovery. Note: To be replaced by DIN EN ISO 224751 (2004-08), DIN ISO 14689 (1997-01, t). Teil 3 (1982). Schichtenverzeichnis für Bohrungen mit durchgehender Gewinnung von gekernten Proben im Boden. Part 3 (1982). Borehole log for boring in soil (loose rock) by continuous extraction of cores. Note: To be replaced by DIN EN ISO 22475-1 (2004-08). Beuth, Berlin.

DIN EN ISO 22475-1, (Draft standard), Publication date: 200408: Geotechnical investigation and testing – Sampling by drilling and excavation methods and groundwater measurements – Part 1: Technical principles for execution (ISO/DIS 224751:2004); German version prEN ISO 22475-1:2004. Beuth, Berlin.

DIN EN ISO 22475-1, (Draft standard), Publication date: 200408: Geotechnical investigation and testing – Sampling by drilling and excavation methods and groundwater measurements – Part 1: Technical principles for execution (ISO/DIS 224751:2004); German version prEN ISO 22475-1:2004. Beuth, Berlin.

Driscoll, F. G. (1986): Groundwater and Wells, 2nd edn., St. Paul, Minnesota: Johnson Filtration Systems Inc.

Düring, P.-H. (1983): Geologische Bohrungen, Bd. 1 Bohrtechnik und Technologie. Deutscher Verlag für Grundstoffindustrie, Leipzig.

Elson, B. & Shaw, R. (2005): Simple drilling methods. WEDC Loughborough University Leicestershire LE11 3TU UK.
www.lboro.ac.uk/well/resources/technical-briefs/43-simpledrilling-methods.pdf

EPA (1990): Handbook of Suggested Practices for the Design and Installation of Ground-Water Monitoring Wells.

EPA (1997): Expedited Site Assessment Tools For Underground Storage Tank Sites: A Guide for Regulators. Office of Underground Storage Tanks (OUST). Also available online: www.epa.gov/OUST/pubs/sam.htm

EPA (2003): Field Analytic Technologies Encyclopedia. Office of Superfund Remediation and Technology Innovation (OSRTI). URL: http://fate.clu-in.org
Last Updated: January 2003.

EPA Superfund (2005): Direct Push Technologies.
www.epa.gov/superfund/programs/dfa/dirtech.htm

Fejes, I., Szabdvary, L. & Verö, L. (1997): Geophysikalische Penetrationssondierungen. In: Knödel, K., Krummel, H. & Lange, G.: Handbuch zur Erkundung des Untergrundes von Deponien und Altlasten, Band **3**: Geophysik. Springer, Berlin.

Gore, P. J. W. (2003): http://gpc.edu/~pgore/geology/historical_lab/stratigraphy.php

Heitmann, R. (2000): Anwendung von elektrischen Drucksonden mit laserinduzierter Fluoreszenzmessung (MIF) und Membrane Interface-Probe (MIP) zur Untersuchung kontaminierter Standorte. FUGRO Consult GmbH, Markkleeberg.

Homrighausen, R. (1993): Bohrungen für die Erkundungen von Altlasten, Industriestandorten und Deponien. bbr, **44**, 10, 2-7.

IRTP (2001): International Reference Test Procedure (IRTP) for the Cone Penetration Test (CPT) and the Cone Penetration Test with pore pressure (CPTU). Report of the International Society for Soil Mechanics and Geotechnical Engineering (ISSMGE).

ITVA (1996): Aufschlussverfahren zur Probengewinnung für die Untersuchung von Verdachtsflächen und Altlasten. Ingenieurtechnischer Verband Altlasten, Fachausschuss F2 „Probennahme“ Arbeitshilfe.

KRAM, M. L., KELLER, A. A., ROSSABI, J. & EVERETT, L. G. (2001): DNAPL Characterization Methods and Approaches, Part 1: Performance Comparisons. Ground Water Monitoring and Remediation. **21,** 4, 109-123.

LAPHAM, W. W., WILDE, F. D. & KOTERBA, M. T. (1995): Groundwater data-collection protocols and procedures for the National Water-Quality Assessment Program: Selection, installation, and documentation of wells, and collection of related data. USGS Open-File Report 95-398.

LIFEWATER CANADA (2004): Drilling and Well Construction Manual. Lifewater Canada: http://www.lifewater.ca/ndexdril.htm

LUCKNER, L. & SCHESTAKOW, W. M. (1991): Migration processes in the soil and groundwater zone. Lewis, Boca Raton, Florida.

LUNNE, T., ROBERTSON, P. K. & POWELL, J. J. M. (1997): Cone Penetration Testing in Geotechnical Practice. Blackie Academic & Professional. London. ISBN 0751 40393 8.

MEIGH, A. C. (1989): Cone Penetration Testing, Methods and Interpretation. Butterworths, London.

OSU Trenching/Shoring Manual (2005): www.pp.okstate.edu/ehs/manuals/Trench.htm#soil

SANGLERAT, G. (1972): The Penetrometer and Soil Exploration. Elsevier, Amsterdam.

SARA, M. N. (1993): Standard Handbook for Solid and Hazardous Waste Facility Assessments. Lewis, Boca Raton, Florida.

SARA, M. N. (2003): Site Assessment and Remediation Handbook. 2nd ed., Lewis, Boca Raton, Florida.

SCALF, M. R., McNabb, J. F., Dunlap, W. F., Cosby, R. L. & Fryberger, J. (1981): Manual of Groundwater Sampling Procedures. National Water Well Association, Worthington, Ohio.

SCHMIDT, J. (1993): Der Untergrund der Altlast „Eulenberg“ bei Arnstadt – ein tektonisches Modell und Beiträge zur Umweltgeologie. Bergakademie Freiberg, Diplomarbeit.

SCHMIDT, J., WUNDERLICH, J. & BÖHME, O. (2003): Fallbeispiel Eulenberg/Arnstadt – Geologie. In: LANGE, G. & KNÖDEL, K.: Handbuch zur Erkundung des Untergrundes von Deponien und Altlasten, Bd. **8**, Erkundungspraxis. Springer, Berlin, 547-595.

SCHREINER, M. & KREYSING, K. (Ed.) (1998): Handbuch zur Erkundung des Untergrundes von Deponien und Altlasten, Bd. 4: Geotechnik, Hydrogeologie (Manual for investigation of the subsoil of waste sites and hazardous sites, vol. 4: Geotechnics, Hydrogeology). Springer, Berlin.

SHUTER, E. & TEASDALE, W. E. (1989): Application of drilling, coring, and sampling techniques to test holes and wells. In: U.S. Geological Survey Techniques of Water-Resources Investigations, book **2**, chap. F1.

WEISS, K. (1990): Baugrunduntersuchung im Feld. In: SMOLTCZYK, H. (Ed.): Grundbautaschenbuch, **1**, Ernst, Berlin.

5.2 Methods for Characterizing the Hydrologic and Hydraulic Conditions

ULRICH BEIMS, FLORIAN JENN, KLAUS KNÖDEL, MANJA LIESE, RANJEET NAGARE, CLAUS NITSCHE, MICHAEL PORZIG & HANS-JÜRGEN VOIGT

Hydrology is science of the behavior of water in the atmosphere, on the Earth's surface, and in the subsurface, while the hydrogeology studies the groundwater and its relationship to the geologic environment. Both points of view are important in a site investigation and assessment. Basic in hydrology is the hydrologic cycle, also known as the water cycle, i.e. the continuous circulation of water between the atmosphere, land, surface water, groundwater, and plants, through condensation, precipitation, runoff, infiltration, evaporation, transpiration, groundwater flow, storage and seepage. *Figure 5.2-1* shows the elements of the hydrologic cycle. The hydrologic cycle supplies terrestrial organisms with freshwater and is therefore a source of life. The system powered by solar radiation determines essentially the climate.

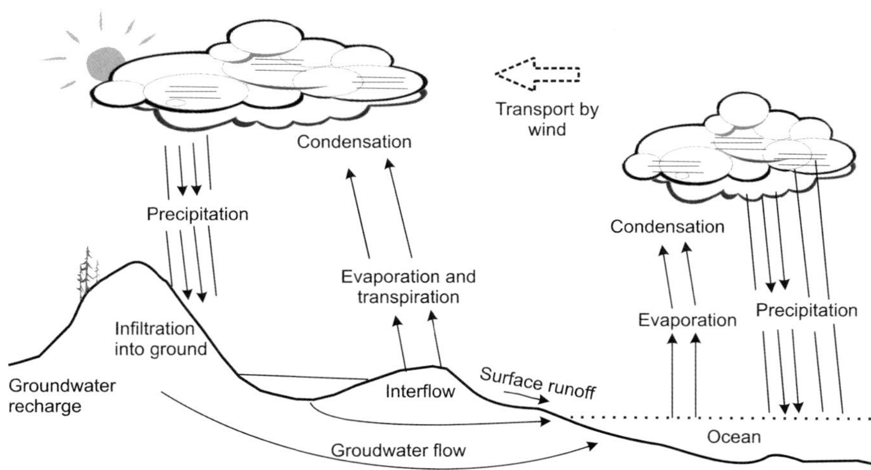

Fig. 5.2-1: Elements of hydrologic cycle, modified after WEIGHT & SONDEREGGER (2001)

Climate is the general pattern of weather conditions for a region over a long time period (at least 30 years), taking into account temperature, precipitation, humidity, wind, and other phenomena. The main influence governing the climate of a region is its latitude. The influence of latitude on climate is modified by one or more secondary influences including position relative to land and water masses, altitude, topography, prevailing winds, ocean currents, and prevalence of cyclonic storms. A broad latitudinal division of the Earth's surface into climatic zones based on global winds includes the equatorial zone, characterized by high temperatures with small seasonal and diurnal change and heavy rainfall; the subtropical, including the trade-wind belts and the horse latitudes, a dry region with uniformly mild temperatures and little wind; the intermediate, the region of prevailing westerly winds that because of several secondary influences, displays wide temperature ranges and marked changeability of weather; and the polar, a region of short summers and long winters, where the ground is generally perpetually frozen (permafrost). The transitional climate between those of the subtropical and intermediate zones, known as the Mediterranean type, is found in areas bordering the Mediterranean Sea and on the west coasts of continents. It is characterized by mild temperatures with moderate winter rainfall under the influence of the moisture-laden prevailing westerly winds and dry summers under the influence of the horse latitudes or the trade winds. In addition to hydrological and hydrogeological information climatic data must be collected for site investigation. Precipitation not only recharges the groundwater, but also forms leachate in landfills and mobilizes soluble substances in contaminated sites. Air pressure influences emissions of soil gas. For example, low air pressure promotes the venting of landfill gas. Direction and intensity of wind determines dust and gas emissions from landfills, industrial and mining sites into the surrounding areas.

Most important for characterizing the hydrologic conditions of an area is the water balance equation:

$$P = ET + R_d + R_b \pm \Delta S, \tag{5.2.1}$$

where P is precipitation, ET is evapotranspiration, R_d is direct runoff (overland flow and interflow), R_b is baseflow, and ΔS is change in groundwater storage due to lateral in/outflow.

5.2.1 Precipitation

FLORIAN JENN, KLAUS KNÖDEL, MANJA LIESE & HANS-JÜRGEN VOIGT

Precipitation is any form of water falling to Earth as part of the weather and is thus a major element of the hydrologic cycle. Of special interest to the hydrologist are the form, quantity (depth), spatial distribution and temporal variability of precipitation in the area under investigation. Depending on temperature, precipitation reaches the Earth surface as rain, snow, hail, sleet, mist, or dew. Rain is in non-polar regions the most common form of precipitation. Snow is the most common form of solid precipitation, occurring at air temperatures between - 40 and + 5 °C, most frequently around 0 °C (HÖLTING, 1992). From the hydrogeological point of view the slowly melting snow is important for groundwater recharge. The quantity of other precipitation forms is often wrongly neglected in the water balance. The mean yield of dew in Central Europe is in the order of 25 - 30 mm a^{-1}, while in the tropics (e.g., on the Loango coast of West Africa) 3 mm of dew can form in one night (MATTHES & UBELL, 1983). In the low mountain range of Central Europe the water deposition from mist droplets on trees and bushes can reach three times the quantity of precipitation forms measured in rain gauges (WECHMANN, 1964).

A basic understanding of *factors affecting the formation of precipitation* (*Table 5.2-1*) is needed for the interpretation of precipitation data. Precipitation occurs if air masses with a given water vapor content undergo changes in pressure and temperature leading to the saturation point (dew point) being exceeded. At the saturation point the maximum concentration of water vapor the air can hold at a given temperature and pressure is reached. When the temperature is lower than the dew point, the excess water condenses from the gaseous (vapor) to the liquid state. This process is called condensation and is the reverse of vaporization (evaporation), i.e. the change from liquid to gas. The condensation of water vapor into clouds and precipitation as the temperature drops or the atmospheric pressure rises is part of the hydrologic cycle. Condensation requires not only saturation, but also the presence of solid condensation nuclei (e.g., particulate products of combustion, volcanic ash, salt particles) around which water droplets or ice/snow crystals can form. Air pollution facilitates condensation.

Some of the precipitation evaporates as it falls to the Earth's surface. In extreme cases all of the precipitation has evaporated before the Earth's surface is reached.

Cooling by lifting of air masses is the most important process in the formation of precipitation. There are three main types of precipitation that result from different lifting mechanisms: cyclone precipitation, orographic precipitation and convective precipitation.

Cyclone precipitation is associated with a cyclone. Most cyclone precipitation can be classed as frontal precipitation. Frontal rainfall occurs at both warm fronts and cold fronts. At a warm front the lighter warm air rides up over stationary cold air. As it rises it also cools owing to the drop in pressure (adiabatic process). Moisture in the air condenses to form clouds, to eventually fall to the Earth as precipitation. At a cold front warm air is forced up by moving cold air and the same process occurs.

Table 5.2-1: Factors affecting precipitation, PONCE (2005)

Factor \ Effect on	Moisture availability	Condensation	Coalescence
1. latitude[A]	yes	-	-
2. global and mesoscale ocean currents[B]	yes	yes	-
3. atmospheric currents[C]	yes	yes	-
4. proximity to moisture source[D]	yes	-	yes
5. relative continental position[E]	yes	yes	-
6. season[F]	-	yes	-
7. presence of orographic barriers[G]	-	yes	-
8. land surface condition (texture, color, moisture content)[H]	yes	yes	-
9. natural/anthropogenic atmospheric particulates[I]	-	-	yes

[A] The climate is tropical, temperate, or polar; a function of the Hadley and Ferrell cells.
[B] For example, the ENSO (El Niño Southern Oscillation).
[C] For example, monsoon-related precipitation, such as that of the Bay of Bengal, India.
[D] Ocean or large inland lake (coastal precipitation); presence of salt particles (aerosols) from the ocean.
[E] Continental location with respect to one or more moisture sources (for example, in Arizona, convergence of Pacific and Gulf moisture); lifting through horizontal convergence; frontal lifting.
[F] Summer, fall, winter, spring; determines radiation balance; seasonally driven atmospheric cooling.
[G] Mountain ranges, which act as barriers to the movement of continental air masses, with upwind effects (orographic lifting).
[H] Determined by land-surface albedo, which conditions the near surface-radiation balance and makes possible thermal lifting.
[I] Through volcano eruptions or fires, which increase the load atmospheric particulates, favoring the downwind formation of precipitation.

Orographic precipitation is rain or snow that is generated when a moisture-bearing air mass is forced to rise over hills or mountain ranges, causing them to cool and precipitate their moisture. It is especially evident on the windward side of mountains. Often, the leeward side is drier (rain shadow).

Convectional rainfall occurs when the air is heated, usually by the land below it and consequently rises, cooling as it does so, causing the water vapor in it to condense. This kind of precipitation is most commonly found in tropical areas.

The precipitation intensity is given in terms of volume per unit of time. The volume is usually given as the depth to which the precipitation would accumulate if it fell on a horizontal flat surface, e.g., in inches or mm. Rainfall of 1 mm corresponds to 1 L of water per m^2 or 1000 m^3 per km^2. Mean annual precipitation at the Earth's surface is 973 mm a^{-1} (BAUMGARTNER & REICHEL, 1975). The highest annual precipitation is observed in tropical mountains, where more than 10 000 mm a^{-1} has been measured, while desert areas receive less than 100 mm a^{-1} and sometimes less than 20 mm a^{-1}.

Rainfall intensity is described as weak (< 2.5 mm h^{-1}), moderate (2.6 - 7.5 mm h^{-1}) or heavy (> 7.5 mm h^{-1}). Precipitation intensity is of particular significance from the hydrological/hydrogeological point of view. Surface runoff is greater for short heavy rains while groundwater recharge is greater in the case of moderate continuous rain (MATTHES & UBELL, 1983; HÖLTING, 1992). The highest amounts of precipitation usually occur within short rainfall events. The highest daily precipitation observed in the Mediterranean area is 400 - 700 mm d^{-1} and the highest observed in tropical areas and monsoon and typhoon areas is > 1200 mm d^{-1} (MATTHES & UBELL, 1983).

For the hydrological characterization of an area, the mean annual precipitation is an important parameter. *Figures 5.2-2* and *5.2-3* show as examples the distribution of mean annual precipitation in Germany and Thailand for a period of 30 years. Of special interest is the spatial variability of the mean annual precipitation related to geographic units.

The amount of precipitation depends especially on the latitude and distance from the moisture source. For instance a monsoon coming from the sea brings more precipitation than a monsoon from inland. Moreover, precipitation varies in a region over the year. *Figure 5.2-4* shows the annual distribution of monthly mean precipitation in Germany (Cottbus, State of Brandenburg) and Thailand (Chiang Mai Station).

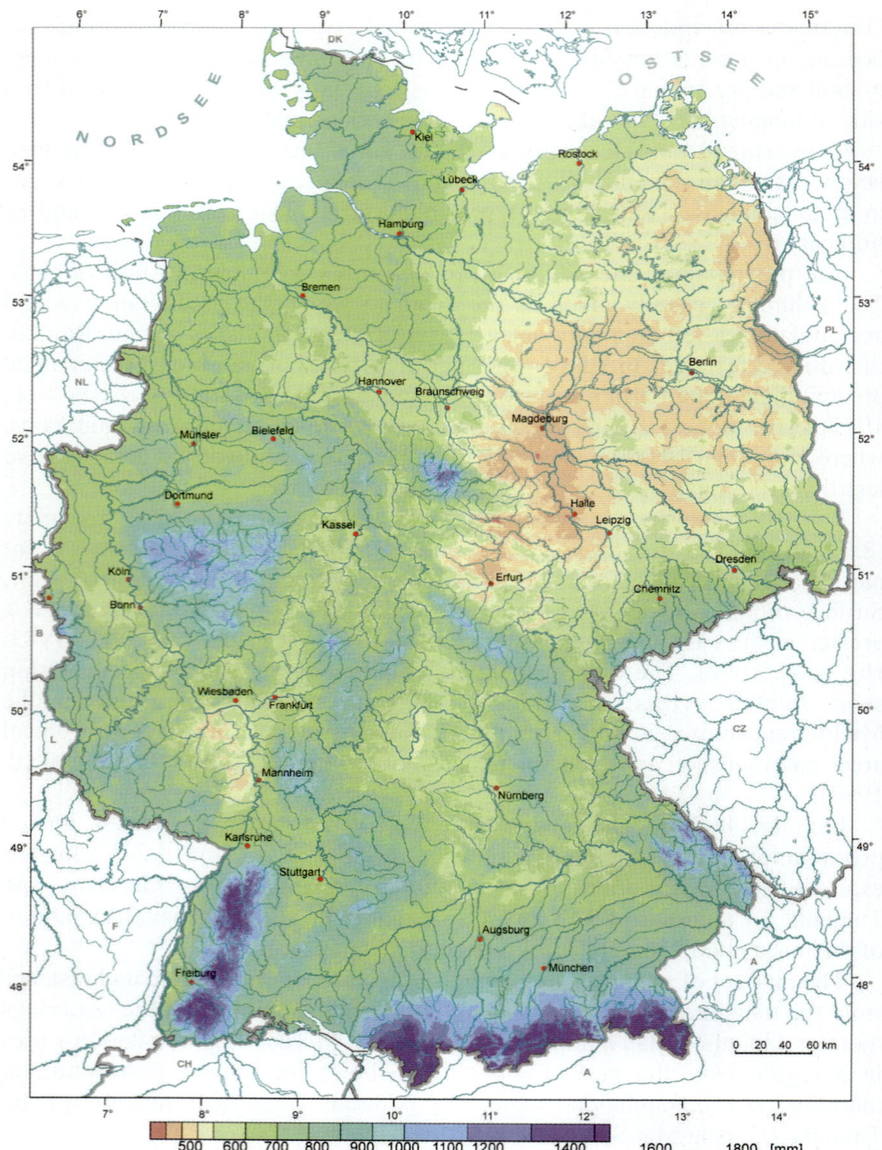

Fig. 5.2-2: Mean annual precipitation in Germany for the period 1961 - 1990, www.klimadiagramme.de

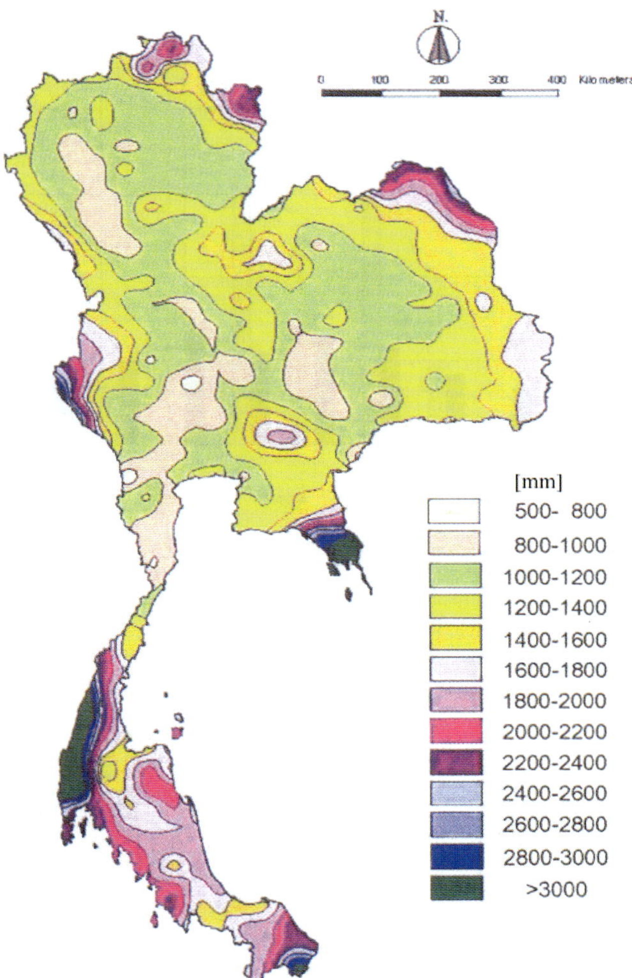

Fig. 5.2-3: Mean annual precipitation in Thailand for the period 1969 - 1998, Department of Meteorology, Thailand (1999)

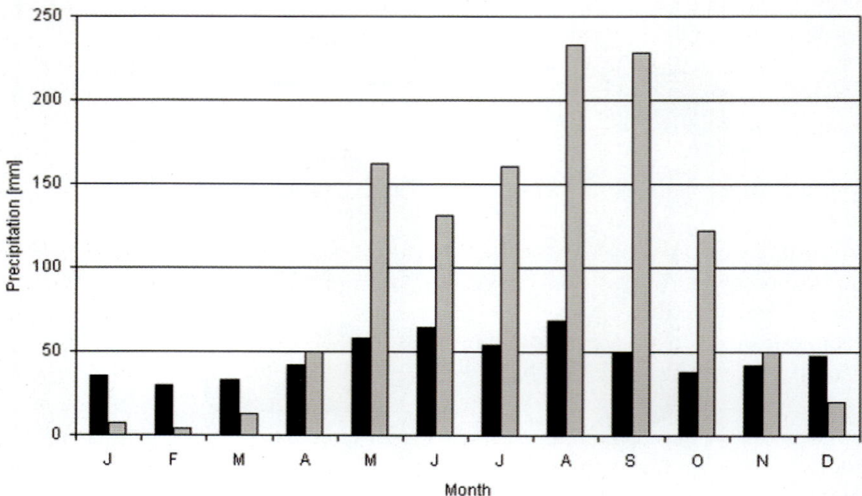

Fig. 5.2-4: Monthly average precipitation of the climatological normal period 1961 - 1990 for Cottbus, Germany (black) and Chiang Mai, Thailand (gray), www.klimadiagramme.de/Deutschland/cottbus.html and www.klimadiagramme.de/Asien/chiangmai.html

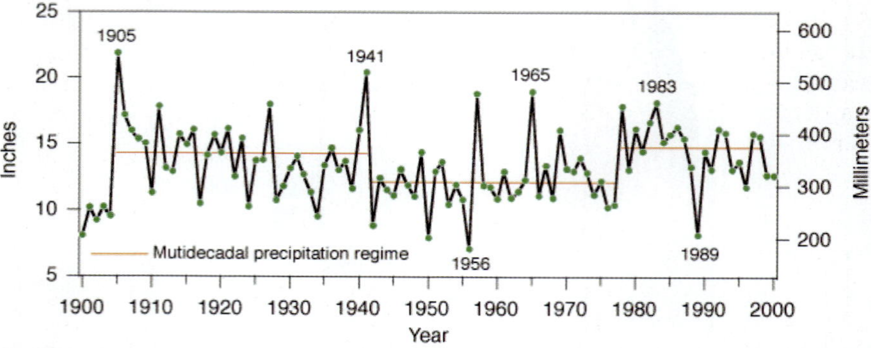

Fig. 5.2-5: Average annual precipitation recorded at 97 weather stations on the Colorado Plateau from 1900 to 2000. Horizontal lines show the average precipitation for three multidecadal precipitation regimes, HEREFORD et al. (2005).

The long-term variability of precipitation must also be taken into account for the hydrogeological investigation and assessment of an area. *Figure 5.2-5* shows the long-term variability of annual precipitation on the Colorado Plateau as an example (HEREFORD et al., 2005). The Colorado Plateau is an arid to semiarid region with mean annual precipitation of 136 to 668 mm a^{-1} and a median precipitation value of 300 mm a^{-1}. Mean annual precipitation, based on analysis of daily records from 97 long-term weather stations, varied considerably during the 20th century (*Fig. 5.2-5*). Three multi-decadal wet and dry regimes are recognizable in the precipitation time series: 1905-1941, 1942-1977, and 1978-1998. HEREFORD et al., 2005 show that the precipitation variability on the Colorado Plateau is linked spatially and temporally with events in the Pacific Ocean.

Measurement of precipitation

The amount of precipitation can be measured at selected points by recording and non-recording rain gauges (also spelled rain gages), as well as within an area by radar measurements. A rain gauge is an instrument that collects precipitation and measures the accumulated volume during a certain period of time. The precipitation depth is equal to the accumulated volume divided by the collection area of the gauge. So that measurements within an observation network can be compared, standardized gauges must be used and set up in a standardized way (same height above ground, etc.). The values obtained are point values. Several examples of rain gauges are shown in *Fig. 5.2-6*.

Rain gauges are open containers that catch and accumulate the total precipitation. The U.S. Weather Bureau (now: National Weather Service) *nonrecording standard rain gauge*, for instance, is a cylinder, 8 inches (20.32 cm) in diameter, corresponding to a collecting area of 324.3 cm^2. Rain gauges after Hellmann (used e.g., in Germany, *Fig. 5.2-6a*) have a collecting area of 200 cm^2. Rain gauges should be mounted 1 m above the ground at a site uninfluenced by obstacles (e.g., trees, buildings). Solid and liquid precipitation is collected with the same gauge. The gauges must be read daily at a defined time (e.g., in Germany at 7 a.m., or in the United Kingdom at 9 a.m.).

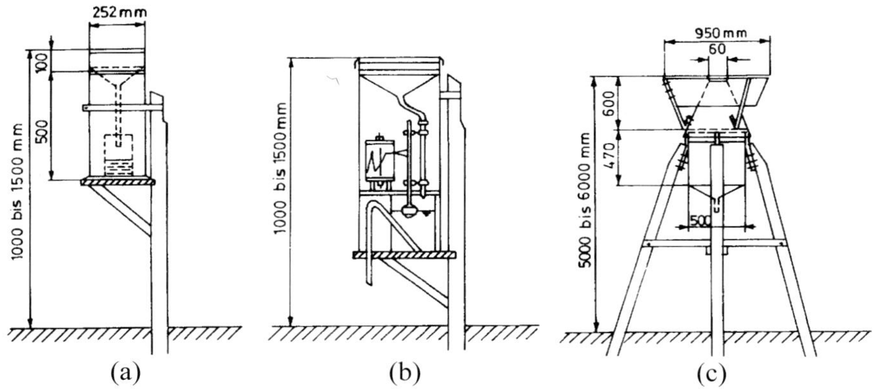

Fig. 5.2-6: Rain gauges: (a) Hellmann sampler, (b) recording rainfall gauge, (c) totalizator rain gauge for measurements over a long period of time, DRACOS (1980) & WECHMANN (1964)

Larger gauges (collecting area 500 cm^2) can be installed in mountainous areas above 1000 m MSL. Snow is melted using calcium chloride ($CaCl_2$) and the snow water is measured.

Recording gauges do not need to be visited to record the amount of precipitation collected. They can be equipped with a telemetry unit and are useful at remote, rarely visited or poorly accessible sites. An example is the *tipping bucket rain gauge*. The rain enters the gauge and is directed into a small bucket that holds a defined amount (e.g., 2 mL) of water. When the bucket is filled, it tips over bringing the second bucket under the funnel. Each tip generates an electric impulse, which is recorded and is a measure of rain intensity in increments of 0.1.

Float gauges and *weighing gauges* measure accumulated precipitation by continuously recording, the change in the height of the float or the accumulated weight of water. *Totalizators* collect precipitation for a long period of time (monthly, half-yearly or yearly).

Precipitation measurements are simple in principle. Nevertheless, difficulties occur in the determination of the amount. Errors are mainly due to effects of wind, ignoring trace amounts, evaporation from the gauge, water adhering to the sides of the gauge, and poor gauge exposure. Wind/turbulence error is usually around 5 %, but can be as large as 40 % in high winds (WILSON & BRANDES, 1979). The precision is 10 to 15 % for rain and up to 40 % for snow (BRUCE & CLARK, 1966). Measurements must be made for at least one year, including both dry and wet seasons.

The minimum density of one gauge per 25 km^2 is recommended, considering that some heavy rain events (e.g., thunderstorms) may cover only an area of 20 km^2 (HISCOCK, 2005). Higher rain gauge densities may be necessary in mountainous areas, where orographic effects cause rainfall variations over short distances. The quality of precipitation data increases with an increase in rain gauge density.

Weather radar can detect any type of hydrometeors[1] (e.g., rain, snow, clouds, dew, and frost) in atmosphere, therefore it can be used to determine the amount of precipitation in a specific area. An electromagnetic radar pulse (Chapters 3.3 and 4.5) is used to determine the reflection (echo) of the hydrometeors. In general, the intensity of the echo is a measure of the precipitation intensity. The distance from the radar site to the precipitation area can be measured by the traveltime between transmission of the radar pulse and the arrival of the echo. The spatial coordinates of the precipitation can be determined from the direction and distance measurements. These measurements can be disturbed by interference from trees and buildings (ground clutter), wind drift of particles, drop size distributions, and type of storm. Precipitation measurements by radar can cover an area instead of single point measurements by gauges on the ground. The radar estimates of the

[1] Any form of liquid or solid water in the atmosphere.

intensity of precipitation must be calibrated by measurements with traditional rain gauges. Therefore, the joint use of radar and gauges leads to more accurate estimates of precipitation (WANIELISTA et al., 1997; LEGATES, 2004).

The U. S. National Weather Service operates numerous WSR-88D (formerly NEXRAD) weather radar systems in the United States (HUNTER, 1996; HARDEGREE et al., 2003; LEGATES, 2004). These radar systems use the reflectivity of S-band microwaves (10 cm wavelength) to provide 1-hourly gridded precipitation estimates with high spatial resolution (4 km × 4 km). HARDEGREE et al. (2003) compared 150 rain gauges in six watersheds with radar precipitation estimates using three years of data. In 5 of 6 cases, the radar underestimated ground precipitation and in all cases, the gauge network detected precipitation for a much longer time than the radar. Similar tests were carried out, for example, in Thailand (CHOKNGAMWONG et al., 2005) with comparable results. HUNTER (1996), GERMANN & JOSS (2001) and LEGATES (2004) describe possible causes of errors in radar precipitation estimates and possibilities to improve them.

Hydrologic requirement for precipitation data

For hydrologic purposes, single point precipitation measurements as well as areal estimates of precipitation are used. For site characterization, single point precipitation data may be used if they are measured within a reasonably close distance (≤ 15 km) from the site under investigation (SARA, 1993) and there are no local factors affecting precipitation (e.g., orographic barriers) in the area.

In numerous cases (e.g., in hydrological modeling) estimation of areal precipitation (e.g., for a catchment area) is necessary. A number of techniques are available to convert single-point precipitation measurements to areal estimates. An example for a catchment area is given in *Figs.5.2-7* to *5.2-10* and *Table 5.2-2*.

The simplest method is to calculate the *arithmetic mean* of the amounts of precipitation recorded for all the rain gauges in the area. The arithmetic mean is acceptable in relatively flat areas where rainfall does not vary drastically with locality (THOMPSON, 1999).

Another method is to use *Thiessen polygons* (*Fig. 5.2-8*) to define the zone of influence of each rain gauge in the area. The polygons are constructed by first drawing lines between adjacent gauges. Next, the perpendicular bisectors of these connecting lines are drawn. The perpendicular bisectors are then joined to form the polygon boundaries. The areas of the polygons are determined with a planimeter or by digitizing the map and calculating the polygon areas. While this method assigns weights based on the location of the station in the basin and the relative density of the station network, it cannot account for the effects of elevation on precipitation.

In *isohyetal analysis* (*Fig. 5.2-9*), lines of equal precipitation (isohyets) are drawn based on point measurements with rain gauges. Then the precipitation in each grid area is estimated. The values of all the grid areas are added and the sum is divided by the number of grid areas in the investigation area to estimate the mean precipitation. It is assumed that the effects of elevation on precipitation are taken more into account by isohyetal analysis than in other methods of estimation of areal precipitation, because the isohyets follow considerably the topography (e.g., orographic barriers).

The *MAPX WSR 88-D estimate* (*Fig. 5.2-10*) uses the 1-hourly gridded precipitation weather radar estimates on a 4 km × 4 km grid. The mean areal precipitation for a catchment area is calculated by averaging the grid point estimates. *Table 5.2-2* gives a comparison of results of several mean areal precipitation estimates for an example catchment area.

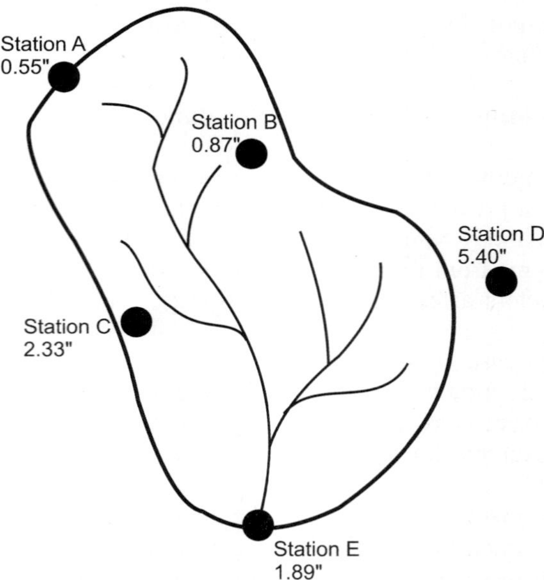

Fig. 5.2-7: Calculation of the mean areal precipitation for an example catchment area (Station A: 0.55" = 14 mm, Station B: 0.87" = 22 mm, Station C: 2.33" = 59 mm, Station D: 5.40" = 133 mm, Station E: 1.89" = 48 mm), after National Weather Service, Arkansas-Red Basin River Forecast Center (2003)

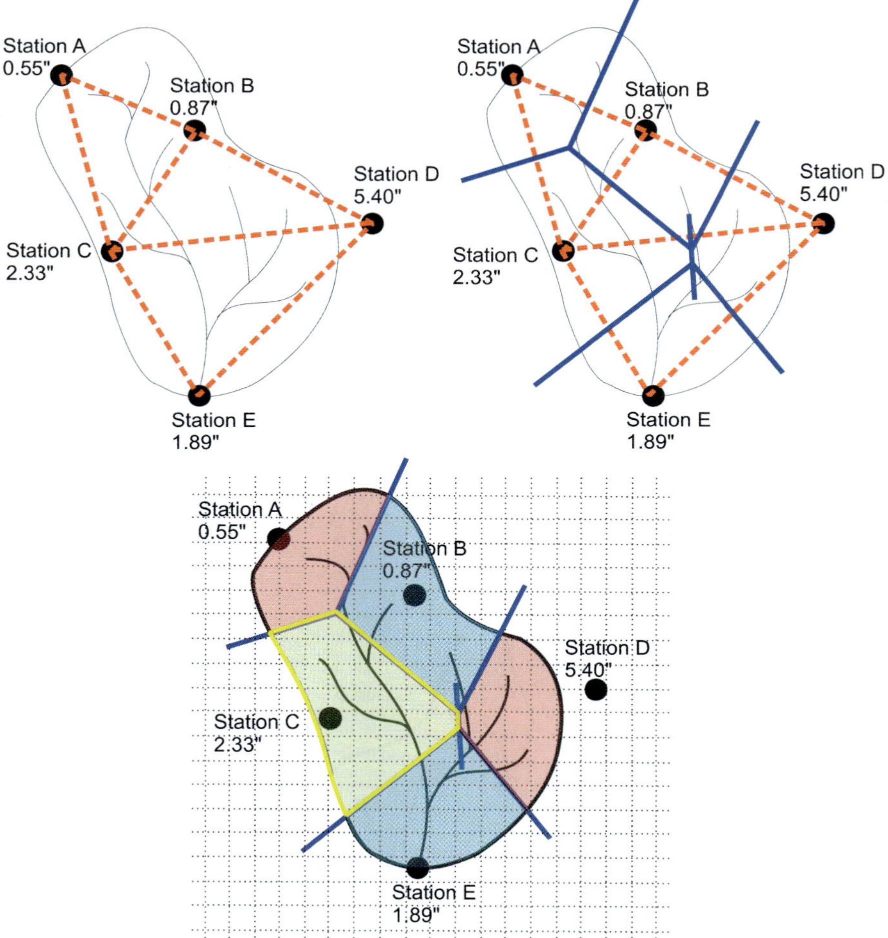

Fig. 5.2-8: Calculation of mean areal precipitation for an example catchment area, Thiessen polygons, after National Weather Service, Arkansas-Red Basin River Forecast Center (2003)

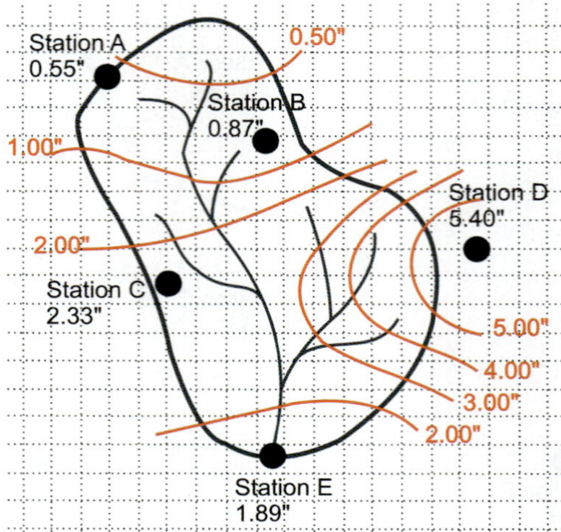

Fig. 5.2-9: Calculation of mean areal precipitation for an example catchment area, isohyetal analysis, after National Weather Service, Arkansas-Red Basin River Forecast Center (2003)

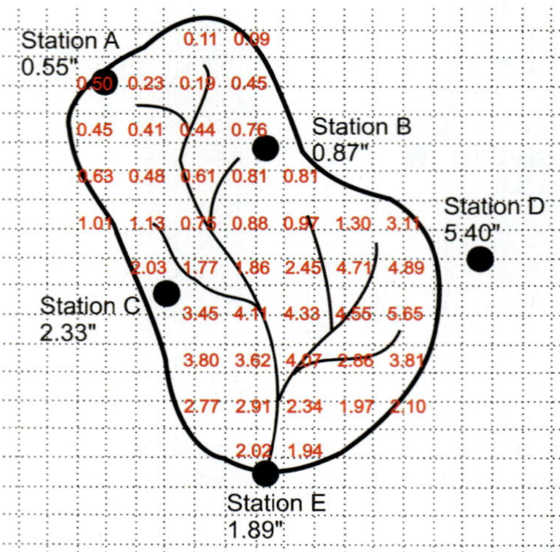

Fig. 5.2-10: Calculation of mean areal precipitation for an example catchment area, MAPX WSR 88-D estimate, after National Weather Service, Arkansas-Red Basin River Forecast Center (2003)

Table 5.2-2: Results of several mean areal precipitation estimates for an example catchment area, after National Weather Service, Arkansas-Red Basin River Forecast Center (2003)

Calculation method	Result [inches]	Result [mm]	Figure
arithmetic mean	2.21	56	no figure
Thiessen polygons	2.03	52	5.2-8
isohyetal analysis	1.90	48	5.2-9
MAXP WSR 88-D estimate	2.00	51	5.2-10

5.2.2 Evaporation and Evapotranspiration

FLORIAN JENN, KLAUS KNÖDEL, MANJA LIESE & HANS-JÜRGEN VOIGT

Evaporation is the conversion of a liquid to a vapor at a temperature below its boiling point. It is the opposite of condensation. The term is also used for the quantity of water that is evaporated. Water evaporates from a variety of surfaces, such as lakes, streams, soils and wet vegetation. Transpiration is the process by which water in plants, absorbed through the roots from the soil, is transferred into the atmosphere from the plant surface, principally from the leaves. Almost all water taken up by the plants is lost by transpiration and only a tiny fraction (about 1 %) is used within the plant in the growth process. The phenomena evaporation and transpiration are often combined to evapotranspiration (ET). Evapotranspiration is the water discharged to the atmosphere by evaporation from the soil and surface-water bodies and by plant transpiration. The term is also used for the volume of water lost through evapotranspiration.

The evapotranspiration rate is normally measured in millimeters per time unit. The time unit can be a day, month or year. The amount of water lost by evapotranspiration is expressed in units of water depth. Therefore, it can be converted to volume per unit area (e.g., 1 mm d^{-1} is equivalent to 10 m^3 ha^{-1} d^{-1}).

The rate of evapotranspiration from the Earth's surface is governed by the following factors:

- Energy availability: Energy is provided first of all by solar radiation and to a lesser extent by the ambient temperature. The more energy that is available, the greater the rate of evapotranspiration.
- Vapor pressure gradient (humidity gradient), i.e. the difference between the water vapor pressure at the surface at which evaporation or transpiration is occurring and that of the surrounding atmosphere. The rate at which water

vapor enters the atmosphere increases when the vapor pressure of the atmosphere decreases (the air becomes drier).

- Wind speed at the Earth's surface: If the wind across the evaporation or transpiration surface replaces the air containing the vapor from evaporation or transpiration with drier air, the rate of evapotranspiration is increased.
- Water availability: Evapotranspiration cannot occur if water is not available.
- Types and condition of vegetation and land use: When there is little vegetation, water is predominately lost by evaporation from the soil, but once the vegetation is well developed and completely covers the soil, transpiration becomes the main process. When a crop is planted, nearly 100 % of ET comes from evaporation; when the ground is fully covered, more than 90 % of ET comes from transpiration (ALLEN et al., 1998). Plants with deep reaching roots can more constantly transpire water. Woody plants keep their structure over long winters while herbaceous plants largely die back during the winter in seasonal climates, and will contribute almost nothing to evapotranspiration in the spring as they regenerate themselves. Conifer forests tend to have much higher rates of evapotranspiration than deciduous forests. This is because their needles have a much larger surface area.
- The soil water content, the ability of the soil to conduct water to the roots, waterlogging and soil water salinity are other factors that affect the evapotranspiration.

Hydrologists distinguish between actual evapotranspiration and potential evapotranspiration. The term "actual evapotranspiration" is used for evapotranspiration that actually occurs, while potential evapotranspiration is the evapotranspiration that would occur if a sufficient water source is available. Potential evapotranspiration is called "reference crop evapotranspiration" or "reference evapotranspiration" (ET_o) by agricultural experts. The concept of reference evapotranspiration was introduced to study the evaporation caused by climatological effects independently of crop type, crop development, management practices, and soil factors if water is abundantly available. The reference surface is a hypothetical grass reference crop with specific characteristics (ALLEN et al., 1998). Examples of actual evapotranspiration and potential evapotranspiration are given in the *Figs. 5.2-11* to *5.2-13*.

Actual evapotranspiration is usually less than precipitation due to percolation or runoff loss. Exceptions are areas with a high groundwater table, where capillary rise can move water to the ground surface and lakes whose tributary/tributaries from areas with elevated precipitation (e.g., mountains). Due to the ability of soil to hold water and a short supply of energy there is normally a time shift between precipitation and evapotranspiration.

Potential evapotranspiration is usually calculated from climate data. It also depends on the surface type (e.g., ocean, lake, bare soil, crop), soil type, and type of vegetation. Often a value for the potential evapotranspiration is calculated for a climate station on a reference surface, conventionally short grass, and is converted to other areas by multiplying by a surface coefficient. If the potential evapotranspiration is greater than the precipitation, then the soil will dry out unless it is irrigated. On a global scale, most of evapotranspiration of water at the Earth's surface occurs in the subtropical oceans.

The mean evapotranspiration value for Germany is 532 mm a^{-1}. Depending on soil and land use, actual evapotranspiration varies from less than 350 mm a^{-1} in urban areas and the higher parts of the Alps, the Erzgebirge mountains and other upland regions, to values above 700 mm a^{-1}, primarily in the Upper Rhine Plain (Oberrheinebene) and areas with a groundwater table close to land surface (BMU, 2003; WINNEGGE & MAURER, 2002).

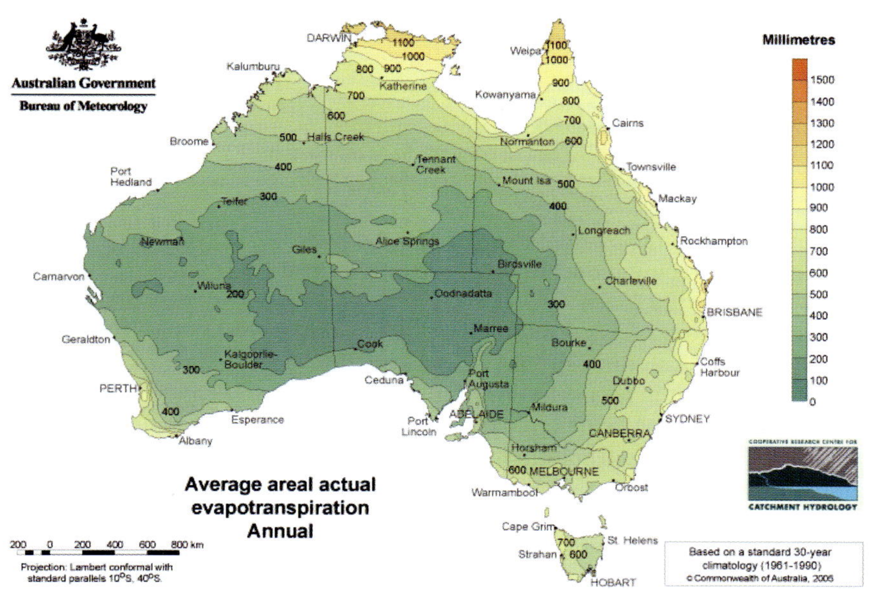

Fig. 5.2-11: Annual areal actual evapotranspiration in Australia based on a 30 year climatology (1961 to 1990), Bureau of Meteorology of Australia (2001), www.bom.gov.au/cgi-bin/climate/cgi_bin_scripts/evapotrans/et_map_script.cgi

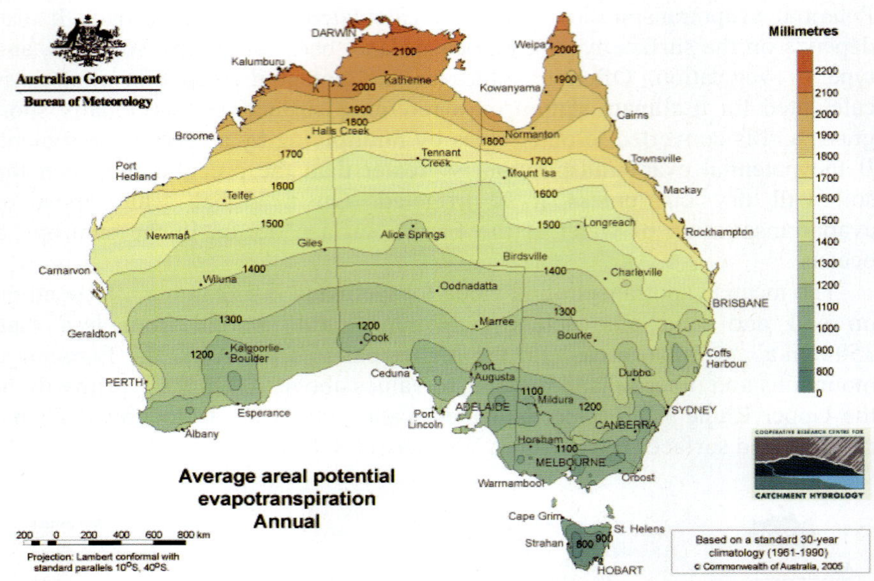

Fig. 5.2-12: Annual areal potential evapotranspiration in Australia based on 30 a year climatology (1961 to 1990), Bureau of Meteorology of Australia (2001), www.bom.gov.au/cgi-bin/climate/cgi_bin_scripts/evapotrans/et_map_script.cgi

Evapotranspiration is an important part of the hydrologic cycle (*Fig. 5.2-1*). A large part of precipitation, for example in Germany 62 % (HÖLTING, 1992) and in Australia 90 % (WANG et al., 2001), is returned to the atmosphere by evapotranspiration. This water does not contribute to the groundwater recharge and is a significant loss of water in the water budget of a catchment area. Therefore, it is necessary to estimate the value of ET as well as possible. It is difficult to exactly determine the ET value due to the complexity of the factors influencing evaporation and transpiration. Consequently numerous approximation methods have been developed.

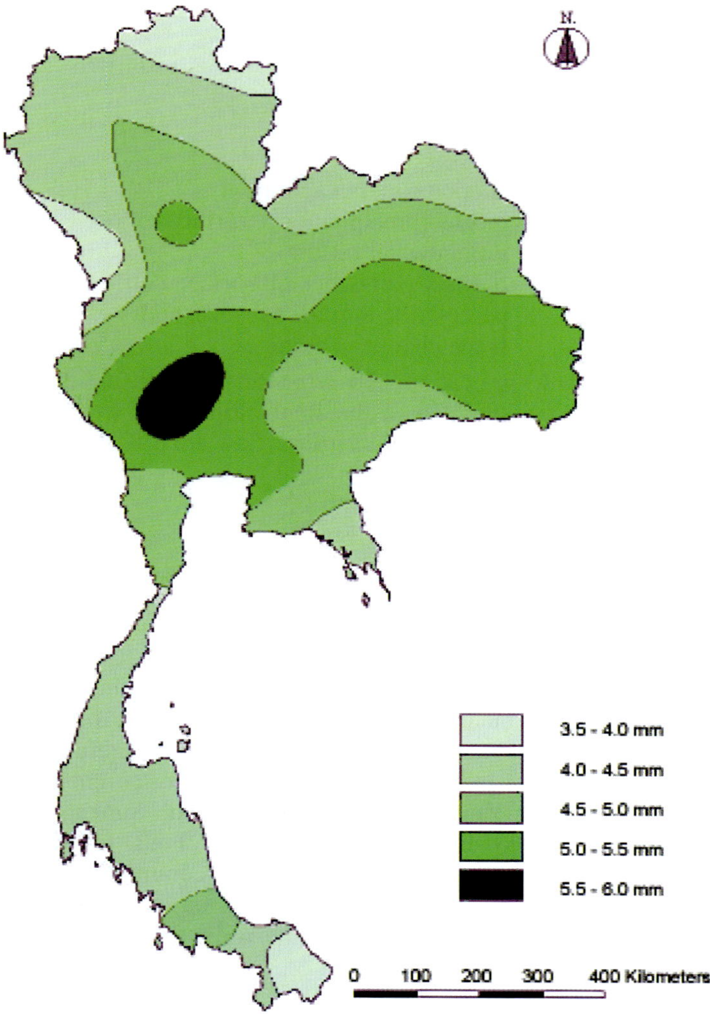

Fig. 5.2-13: Mean annual evapotranspiration in Thailand in last 30 years in mm d^{-1}, the Meteorological Department of Thailand (1999), www.ldd.go.th/FAO/z_th/thmp131.htm

Water budget equation

The open water evaporation E_o from lakes can be estimated from the following mass-balance equation:

$$E_o = I - O \pm S, \tag{5.2.2}$$

where E_o is open water evaporation,
$I = P + I_S + I_G$ inflow (precipitation + surface inflow + groundwater inflow),
$O = O_S + O_G + D_P$ outflow (surface + subsurface outflow + deep percolation into the subsoil), and
S is the change in storage.

This method has limited application. All errors in the measurement of the individual components are summed in the final estimate of evaporation. In addition, some of the components like subsurface flow are difficult to measure (WATSON & BURNETT, 1993).

Evaporation tank or pan

Evaporation can be measured with several different devices. But, the amount of evaporation measured with different devices is usually not the same. The U.S. National Weather Service evaporation pan called a Class A pan is accepted as an international standard. A Class A evaporation pan is cylindrical with a diameter of 47.5 inches (1206.5 mm) and a depth of 10 inches (254 mm) made of stainless steel. The pan is placed on a carefully leveled, wooden base on the ground in a grassy location and is often enclosed by a chain link fence to prevent animals drinking from it. The Sunken Colorado pan (square pan) and the British standard tank, both of which are 1.83 m × 1.83 m and a water depth 0.61 m, are also commonly used. Tank evaporimeters tend to give more accurate measurements of true evaporation, because tanks are less affected by heating of the walls than pans (HISCOCK, 2005).

Pan evaporation is a measurement that combines the effects of temperature, humidity, solar radiation, and wind. Evaporation is measured daily as the depth of water decreases in the pan. The evaporation rate can be measured by manual readings or with an evaporation gauge with analog output. Since the amount of evaporation is a function of temperature, humidity, wind, and other conditions, the maximum and minimum temperature of the water and the amount of air passage are normally recorded along with the evaporation. Water level sensors, water temperature sensors, weather sensors, and data-logging capabilities can be part of an evaporation monitoring system. The measurement day begins each morning at 7 o'clock with the pan filled to exactly two inches (50.8 mm) from the top of the pan. After 24 hours the remaining quantity of water is measured and the amount of pan evaporation per time unit E_{pan} (mm d^{-1}) is obtained from the difference between the two measured water depths. If precipitation occurs in the 24-hour period, it is measured simultaneously and taken into account in calculating the evaporation. Sometimes precipitation is greater than evaporation, and measured amounts of water must be removed from the pan. The potential evapotranspiration ET_o is calculated by multiplying E_{pan} by the pan coefficient K_{pan}:

$$ET_o = K_{pan} \times E_{pan} \tag{5.2.3}$$

For the Class A evaporation pan, K_{pan} varies depending on season and location from 0.35 to 0.85. The average K_{pan} is 0.70.

Determination of actual evapotranspiration with lysimeter

Lysimeters are tanks filled with either disturbed or undisturbed soil buried with tops flush with the ground level. The soil in them should be planted with same type of vegetation as the surrounding area. Two types of lysimeters are available, weighing (*Fig. 5.1-14*) and nonweighing. In precision weighing lysimeters, the water loss is directly measured by the change of mass. Evapotranspiration can be obtained with an accuracy of a few hundredths of a millimeter, and small time periods such as an hour can be considered (ALLEN et al., 1998). In nonweighing lysimeters the ET for a given time period is determined from the difference between the total water input and the drainage water, collected at the bottom of the lysimeters. Nonweighing lysimeters should be used to determine the actual evapotranspiration only if the change in soil moisture is negligible.

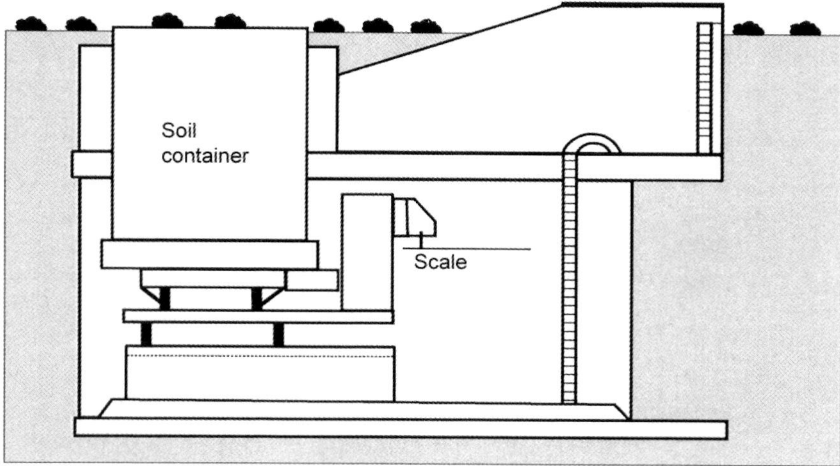

Fig. 5.1-14: Diagram of a weighing lysimeter, DYCK & PESCHKE (1989)

Precipitation must be measured simultaneously. The evapotranspiration can be estimated with the following water-balance Equation:

$$ET = P + IR - DP \pm \Delta S, \tag{5.2.4}$$

where ET is the evapotranspiration per time unit,
P precipitation,
IR irrigation water added to the lysimeter,
DP deep percolation (excess water drained from the bottom of the lysimeter), and
ΔS soil moisture storage.

As lysimeters are difficult and expensive to construct, operate and maintain, their use is limited to research purposes (ALLEN et al., 1998).

A nonweighing lysimeter in combination with a neutron probe can be used to determine evapotranspiration. A nonweighing lysimeter is used to determine deep percolation and the neutron probe measures the soil moisture. When all other parameters (rainfall, irrigation, surface and subsurface inflow and outflow) are measured or reduced to zero, the *ET* can be determined, using the water-balance Equation (5.2.4).

Thornthwaite equation

In the Thornthwaite equation the mean monthly temperature is used to estimate potential evapotranspiration.

$$ET = 1.6\left(\frac{10t}{TE}\right)^{a} \tag{5.2.5}$$

$$a = 0.49239 + 0.01792\ TE \tag{5.2.6}$$

$$TE = \sum_{i=1}^{12}\left[\left(t_i/5\right)^{1.514}\right], \tag{5.2.7}$$

where ET is evapotranspiration (cm per month),
t mean monthly temperature (°C), and
TE Thornwaite´s temperature efficiency index.

This equation usually has to be adjusted for the month and latitude.

Blaney-Criddle equation

This method for calculating potential evapotranspiration is based on the monthly mean temperatures and the monthly mean percentage of daylight hours per day. It can be used if no locally measured data on pan evaporation are available. The Blaney-Criddle equation (BROUWER & HEIBLOEM, 1986) is

$$ET = p\,(0.46\,T_{mean} + 8), \qquad (5.2.8)$$

where ET evapotranspiration (mm d^{-1}) as the average for a period of one month,
T_{mean} mean daily temperature (°C), and
p mean daily percentage of annual daytime hours.

The mean annual percentage of daylight hours for the latitudes 0° to 60° N and S and the months of the year can be taken from Table 12 of BROUWER & HEIBLOEM (1986).

The method only provides a rough, order-of-magnitude estimate. Especially in windy, dry, sunny areas, ET is underestimated by up to about 60 %, while in calm, humid, clouded areas, the ET is overestimated by up to about 40 % (BROUWER & HEIBLOEM, 1986).

Turc equation

The Turc equation give approximate values of actual evapotranspiration averaged over a year based on mean annual precipitation and mean annual air temperature:

$$ET = \frac{P}{\left(0.9 + \left(P/L\right)^2\right)^{1/2}}, \qquad (5.2.9)$$

where ET evapotranspiration in mm per year,
P mean annual precipitation in mm,
L $= 300 + 25\,T + 0.05\,T^3$, and
T mean annual air temperature in °C

This equation can be used for almost all climates.

FAO Penman-Monteith method

The Penman-Monteith method is recommended by the Food and Agriculture Organization of the United Nations (FAO) as the sole standard method with the reason that the method is able to predict the reference crop evapotranspiration ET_o correctly for a wide range of locations and climates. The reference crop is defined "as a hypothetical crop with an assumed height of 0.12 m having a surface resistance[2] of 70 s m^{-1} and an albedo of 0.23, closely resembling the evaporation of an extension surface of green grass of uniform height, actively growing and adequately watered..." (ALLEN et al., 1998). For the calculation of ET_o using Equation (5.2.9) after ALLEN et al., (1998), standard climatological records of solar radiation (sunshine), air temperature, humidity and wind speed are required:

[2] aerodynamic resistance

$$ET_o = \frac{0.408\Delta(R_n - G) + \gamma\frac{900}{T + 273}u_2(e_s - e_a)}{\Delta + \gamma(1 + 0.34u_2)}, \qquad (5.2.10)$$

where ET_o reference evapotranspiration [mm d^{-1}],
R_n net radiation at the crop surface [MJ m^{-2} d^{-1}],
G soil heat flux density [MJ m^{-2} d^{-1}],
T mean daily air temperature 2 m above the ground [°C],
u_2 wind speed 2 m above the ground [m s^{-1}],
e_s saturation vapor pressure [kPa],
e_a actual vapor pressure [kPa],
$e_s - e_a$ saturation vapor pressure deficit [kPa],
Δ slope of the vapor pressure curve [kPa °C^{-1}],
γ psychrometric constant [kPa °C^{-1}].

For more details, examples, parameter tables, and how this method can be applied when there is little data, see ALLEN et al., (1998).

Other methods

The use of Landsat images define crop distribution and thermal NOAA-AVHRR images to determine spatial variation in temperature has been proposed by HURTADO et al. (1997) for estimating actual evapotranspiration. Actual daily evapotranspiration can be estimated with this with an error of no more than 0.9 mm d^{-1}.

5.2.3 Runoff

FLORIAN JENN, KLAUS KNÖDEL, MANJA LIESE & HANS-JÜRGEN VOIGT

After precipitation and evapotranspiration the third component of the water balance of a catchment area is runoff. Runoff is that part of precipitation that flows toward streams on the surface of the ground or within the ground. Runoff may be classified according to how quick it appears after rainfall or snow melts (quick flow or baseflow) or according to location (surface runoff, interflow, or groundwater runoff). Total runoff (R) is composed of overland flow (R_{Su}), interflow (R_I), and groundwater runoff baseflow (R_B):

$$R = R_{Su} + R_I + R_B \qquad (5.2.11)$$

Overland flow is that portion of precipitation, snow melt, or irrigation water in excess of what can infiltrate the soil surface or be stored in small surface depressions. It is a major transporter of non-point source pollutants in rivers, streams, and lakes. Rain that falls on the stream, usually less than 1 % of runoff, is often included in this parameter. This channel precipitation is only significant where there is a large area of wetlands, lakes or during floods, but it always occurs. That portion of rainwater or snowmelt which cannot infiltrate the soil moves as overland flow over the land surface toward stream channels. This overland flow is expressed as discharge per unit width of slope. If the intensity of the rain is less than the infiltration rate, all precipitation water will infiltrate and no overland flow occurs. Three types of overland flow are distinguished:

- Hortonian overland flow,
- saturation overland flow, and
- overland flow due to surface crusting.

Hortonian overland flow occurs if the rain intensity exceeds the infiltration rate and the surplus water becomes runoff. Initially, depth and velocity are very small and sheet flow prevails. When the velocity increases, rill flow (flow in small rivulets) occurs. When velocity and depth of overland flow increase, soil erosion results and contaminant transport increases. It is more frequent in arid areas, where the vegetation is sparse and the infiltration rate is low.

Saturation overland flow occurs if the soil is water-saturated. This may happen after a long precipitation event or if the groundwater table is shallow, for example adjacent to a stream. This is more often the case in humid areas.

Overland flow due to surface crusting occurs when the surface becomes dry and crusted. This is mostly the case in arid areas, where the soil dries out (among other things because there is no vegetation cover protecting it) during the dry season and becomes encrusted.

Interflow is that portion of rainfall or snowmelt that infiltrates into the soil and moves laterally through the unsaturated zone during or immediately after a precipitation event and discharges into a stream or other body of water. This usually takes longer to reach a stream than surface runoff. Rapid interflow occurs as macropore flow, for example, through root canals and tunnels made by animals. Overland flow dominates in arid areas with poor vegetation and relatively thin soil layer. In humid areas with vegetation and a thick permeable soil layer, interflow predominates. Overland flow and interflow represent the surface runoff or quick flow or direct runoff of a stream catchment area.

Groundwater runoff (baseflow) is that part of runoff which has passed into the ground, has become groundwater, and has been discharged into a stream as spring or seepage water. Baseflow is an important source of streamflow between rainstorms. It is estimated by employing a hydrograph separation method in which long-term streamflow records are reconstructed so that most of the stormwater flow is subtracted from those records to theoretically arrive at the groundwater contribution to the stream. This is also referred to as dry-weather flow. In catchment areas with permeable sediments, e.g., sand, limestone, or sandstone, the baseflow component is a large fraction of the total runoff. The opposite is true in catchment areas with clay-rich sediments or outcropping hard rock.

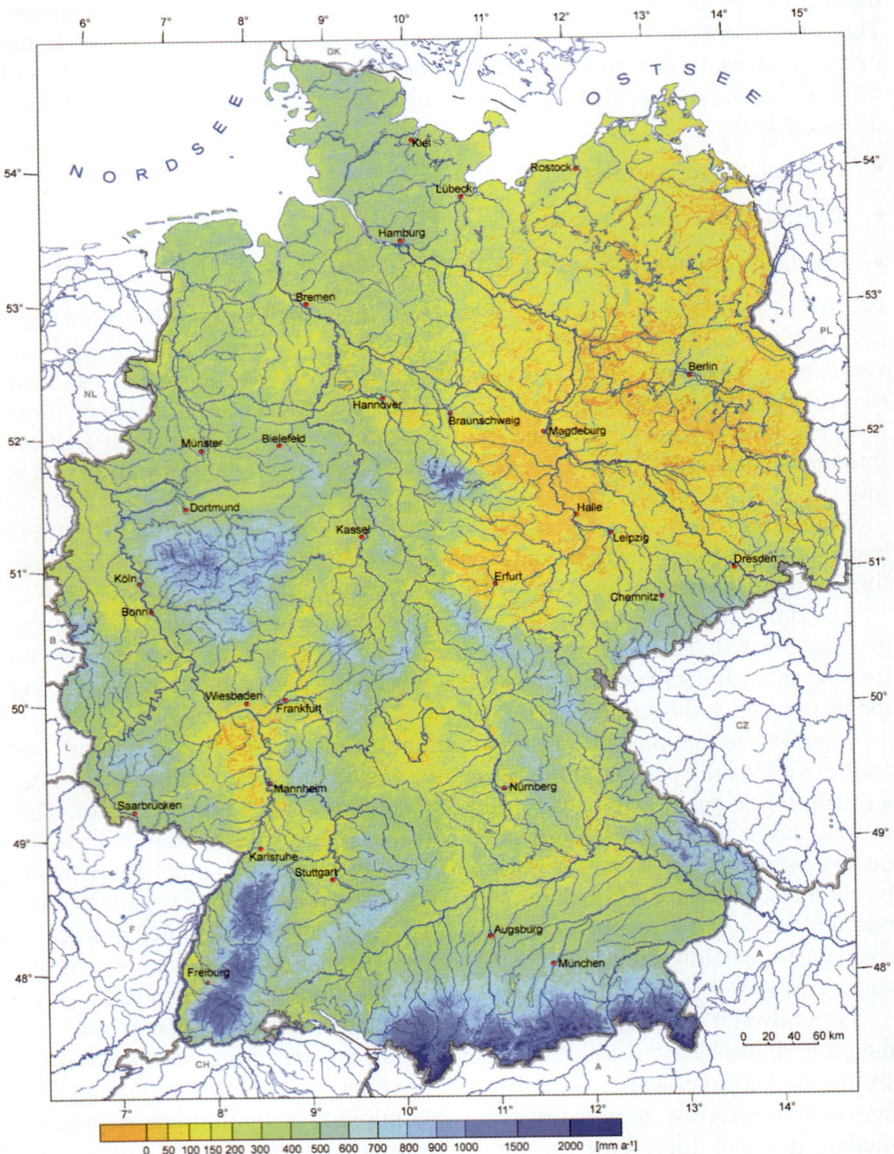

Fig. 5.2-15: Mean annual runoff depth for the period 1961 - 1990 in Germany with grid cells representing 1 km^2 each, BMU (2003)

Mean annual runoff in Germany is about 38 % of the mean annual precipitation (HÖLTING, 1992). The mean annual runoff in Germany for the period 1961 - 1990 is shown in *Fig. 5.2-15* with grid cells representing 1 km^2 each. The values are less than 100 mm a^{-1} in northeastern Germany and more than 2000 mm a^{-1} in the higher regions of the Alps (WINNEGGE & MAURER, 2002).

Table 5.2-3 gives typical values for continental runoff. Surface runoff sends 7 % of the land based precipitation back to the ocean (PIDWIRNY, 2005).

Table 5.2-3: Continental runoff values, LVOVITCH (1972), quoted from PIDWIRNY (2005)

Continent	**Runoff per year [mm a^{-1}]**
Europe	300
Asia	286
Africa	139
North and Central America	265
South America	445
Australia, New Zealand and New Guinea	218
Antarctica and Greenland	164

The following factors affect runoff:

- type of precipitation (rain, snow, sleet, etc.),
- intensity, amount, and duration of a rainfall event,
- distribution of rainfall within the watershed,
- direction of storm movement,
- antecedent precipitation and resulting soil moisture,
- other climatic conditions that affect evapotranspiration, such as temperature, humidity, solar radiation, and wind,
- land use,
- vegetation,
- soil type,
- soil sealing,
- drainage area,
- shape and orientation of the catchment area,
- topography, elevation and slope,
- drainage network patterns,
- ponds, lakes, reservoirs, sinks, etc. in the basin, which prevent or alter runoff from continuing downstream.

Determination of streamflow

Streamflow (Q) is the amount of water leaving the catchment area, i.e. surface runoff, interflow and baseflow. But these three kinds of flow are difficult to separate. The flow is now analyzed as a function of time, not the flow path (see Section "Runoff separation by hydrograph analysis"). Therefore, streamflow is determined as the volume of water passing a given point per time unit in a stream channel, usually expressed in cubic meters per second ($m^3\ s^{-1}$), smaller flows in liters per second ($L\ s^{-1}$). Streamflow is a function of depth, width, and velocity of the water in a channel. The streamflow is not constant. During a precipitation event in temperate areas it increases and reaches a maximum after the precipitation event (*Fig. 5.2-20*). Then it devreases until only baseflow occurs. In arid regions no runoff exists during the season of no precipitation. Ephemeral streams are typical in these regions because there is no baseflow. The variability of runoff during the year in temperate regions is low in comparison to arid regions. Moreover, runoff varies from year to year. During a dry period the soil dries out, and soil moisture decreases. The soil moisture is replenished by the precipitation event that follows. In spite of high precipitation, runoff can be low (because runoff depends on soil moisture, infiltration capacity and precipitation intensity).

Several methods can be used to determine streamflow. Some frequently used methods are described here. The simplest method for estimating small flows is by direct measurement of the time to fill a container of known volume. If the flow of water can be diverted into a pipe so that it is discharged under pressure and the pipe is discharges vertically upwards, the height to which the jet of water rises above the end of the pipe can be measured and the rate of flow can be calculated using the appropriate formula (HUDSON, 1993).

In another method Q is calculated from the average flow velocity (v) and the cross-section area (A) of the channel:

$$Q = A\,v. \qquad (5.2.12)$$

A simple possibility for estimating the velocity is to measure the time taken for a floating object or a strongly colored dye to travel a measured distance downstream. The velocity is not the same at all places in the stream. It is slower at the sides and bottom, and faster on the surface. Taking 0.8 of the surface velocity as measured by the float gives an approximate value for the average velocity (HUDSON, 1993). In turbulent streams a tracer can be added to the stream at a constant measured rate and samples are taken at points downstream. The dilution is a function of the rate of flow which can be calculated. This method is particularly useful when the cross-section of the stream is difficult to determine (ice-covered streams, sand-channel streams or boulder-filled mountain streams).

More accurate determination of velocity can be obtained with a current meter. A current meter is a device in hydrology for measuring the speed and direction or speed alone of flowing water. Current meters can be mechanical, electric, electromagnetic, acoustic, or combination thereof. Commonly used current meters are vertical-axis Price or Woltmann current meters (*Fig. 5.2-16*), and the Pygmy meter for shallow channels.

For a selected location in a stream, velocity is measured with a current meter at 0.2 and 0.8 of the depth *D* in several vertical section of the stream (at 0.6 in shallow streams). The streamflow is calculated from the mean velocity of each section and the area of that section using Equation (5.2.12) as shown in *Fig. 5.2-17*.

Streamflow can be monitored by continuously measuring the flow rate with a current meter and the height of the water surface (called stage) above an arbitrarily established datum plane with a water-level recorder. The results are shown on a stage-discharge curve, which is a graph of flow volume versus depth.

The relationship between stage and flow rate is affected by the irregularities and changes in the channel resulting from erosion and deposition. For this reason gauging the flow in a natural stream can never be precise. More reliable estimates can be obtained when the flow is passed through a standard structure, such as a weir or flume, constructed for this purpose. Master discharge tables or graphs can be be used to directly obtain the flow in a standard weir or flume. Nonstandard structures must be calibrated after installation.

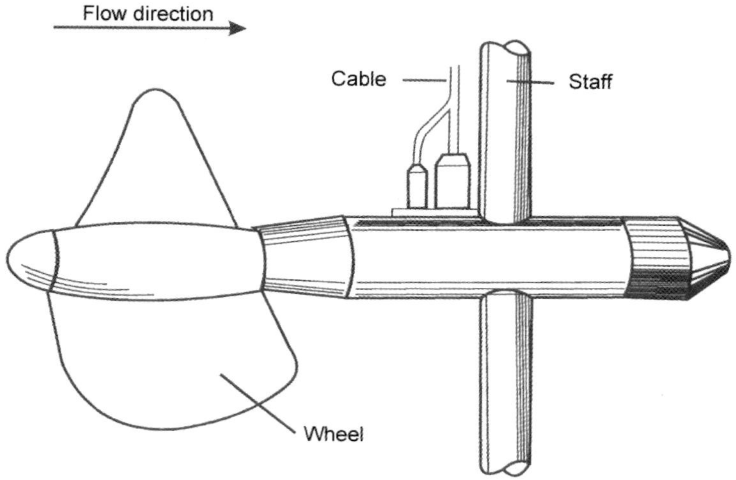

Fig. 5.2-16: Woltmann current meter, LAWA (1991)

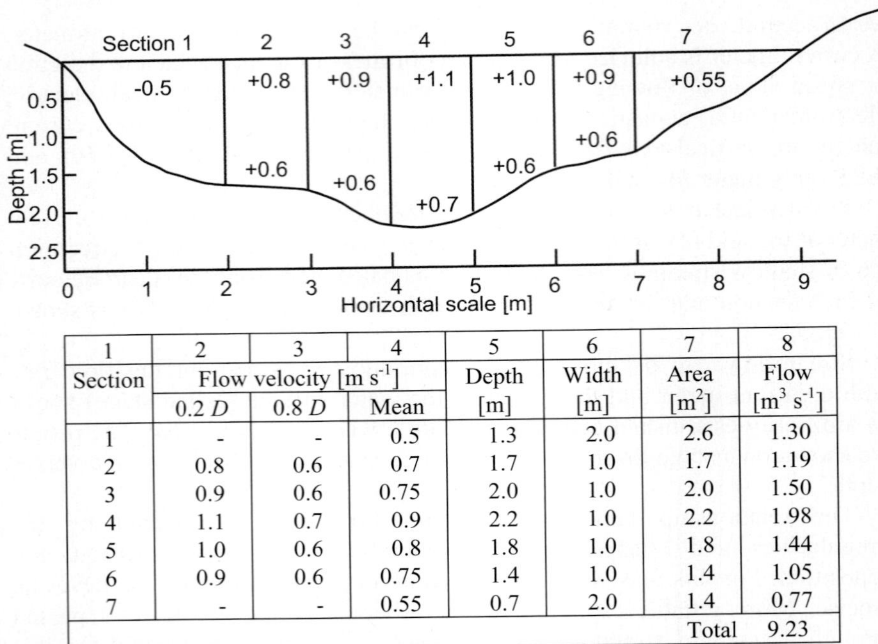

1	2	3	4	5	6	7	8
Section	Flow velocity [m s^{-1}]			Depth	Width	Area	Flow
	0.2 *D*	0.8 *D*	Mean	[m]	[m]	[m^2]	[m^3 s^{-1}]
1	-	-	0.5	1.3	2.0	2.6	1.30
2	0.8	0.6	0.7	1.7	1.0	1.7	1.19
3	0.9	0.6	0.75	2.0	1.0	2.0	1.50
4	1.1	0.7	0.9	2.2	1.0	2.2	1.98
5	1.0	0.6	0.8	1.8	1.0	1.8	1.44
6	0.9	0.6	0.75	1.4	1.0	1.4	1.05
7	-	-	0.55	0.7	2.0	1.4	0.77
						Total	9.23

D is the depth of the stream at the mid-point of each section.

Fig. 5.2-17: Estimating streamflow from current meter measurements, HUDSON (1993)

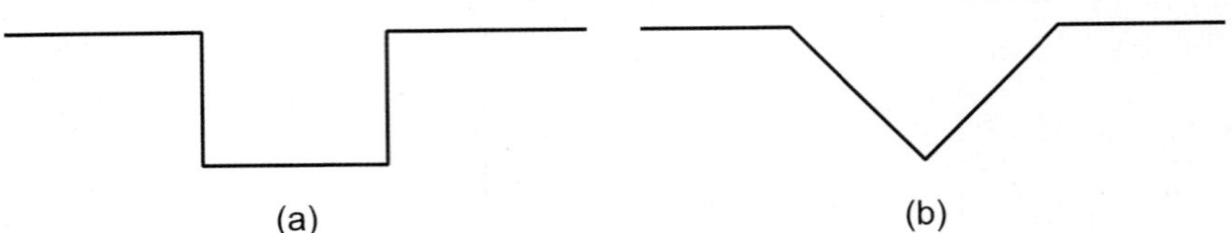

Fig. 5.2-18: Examples for weirs, (a) rectangular-notch weir, (b) 90° V-notch weir

A weir is a low overflow type dam built across a stream to raise its level or to direct or regulate flow. It can be used to measure the discharge by having the water flow over a specifically designed spillway (FETTER, 2001). Flumes are channel-type structures.

The V-notch and the rectangular-notch weirs (*Fig. 5.2-18*) are the two most common types of weirs. They are portable and simple to install in either temporary or permanent positions. The rectangular notch is more suitable for large flows because the width can be easily adapted to the expected flow. There must be an approach channel on the upstream side to remove any turbulence and ensure that the water approaches the notch smoothly. For accurate measurements the width of the approach channel must be 8times the width of the notch, and it must extend upstream for 15times the depth of flow over the notch and the notch must have a sharp edge on the upstream side. To determine the depth of flow through the notch, a staff gauge is placed in the approach channel well back from the notch so that it is not affected by the drawdown curve as the water approaches the notch. The zero of the scale on the staff is set level with the lowest point of the notch (HUDSON, 1993).

The flow rate over a weir can be estimated using the following empirical equations (FETTER, 2001) or from the *Table 5.2-4*:

rectangular-notch weir: $Q = 1.84\,(\mathrm{L} - 0.2\,H)\,H^{2/3}$, (5.2.13)

90° V-notch weir: $Q = 1.379\,H^{5/2}$, (5.2.14)

where Q is flow [$\mathrm{m}^3\ \mathrm{s}^{-1}$],
L length of weir crest [m], and
H height of the backwater above weir crest [m].

The equations are empirical and not subject to dimensional analysis.

Table 5.2-4: Flow rates over a rectangular-notch weir with end contractions and over a 90° V-notch weir, USDI (1975) quoted after HUDSON (1993)

Rectangular-notch weir		90° V-notch weir	
head [mm]	**flow [L s^{-1} per meter of crest length]**	**head [mm]**	**flow [L s^{-1}]**
30	9.5	40	0.441
40	14.6	50	0.731
50	20.4	60	1.21
60	26.7	70	1.79
70	33.6	80	2.49
80	40.9	90	3.34
90	48.9	100	4.36
100	57.0	110	5.54
110	65.6	120	6.91
120	74.7	130	8.41
130	84.0	140	10.2
140	93.7	150	12.0
150	103.8	160	14.1
160	114.0	170	16.4
170	124.5	180	18.9
180	136.0	190	21.7
190	146.0	200	24.7
200	158.5	210	27.9
210	169.5	220	31.3
220	181.5	230	35.1
230	193.5	240	38.9
240	205.5	250	43.1
250	218.5	260	47.6
260	231.0	270	52.3
270	244.0	280	57.3
280	257.5	290	62.5
290	271.0	300	68.0
300	284.0	350	100.0
310	298.0		
320	311.5		
330	326.0		
340	340.0		
350	354.0		
360	368.5		
370	383.5		
380	398.0		

Open channel flow

The semi-empirical Manning equation, first presented in 1889 by the Irish engineer Robert Manning, is the most commonly used equation to analyze water flow in channels and culverts where the water is open to the atmosphere, i.e. not flowing under pressure. The flow Q can be determined from the average velocity v of the water flowing in the channel and the cross-section area of the channel A according to Equation (5.2.12). The velocity of water flowing in an open channel is affected by the following factors:

- the channel gradient,
- roughness of the channel, causing frictional resistance to the moving water in the stream bed, and
- shape of the channel, which influences the frictional resistance.

The Manning equation takes into account these parameters to determine the velocity of water flowing in an open channel:

$$v = \frac{1}{n} M^{2/3} S^{1/2} \qquad (5.2.15)$$

The parameter M takes the shape of the channel into account and is called the hydraulic radius of the channel. It is defined as the cross-sectional area divided by the wetted perimeter, which is the width of the bed and sides of the channel which are in contact with the water. Channels with the same cross-section area can have different values for the hydraulic radius M (*Fig. 5.2-19*). The hydraulic radius thus has units of length.

$$M = \frac{A}{P}, \qquad (5.2.16)$$

where Q is flow [$m^3\ s^{-1}$],
v average velocity [$m\ s^{-1}$],
A cross-section area of the channel [m^2],
M hydraulic radius [m],
S slope of the channel [$m\ m^{-1}$],
P wetted perimeter, which is often approximated by the top width of the channel or by top width of the channel + 2 times the depth of the channel [m], and
n Manning's coefficient of roughness.

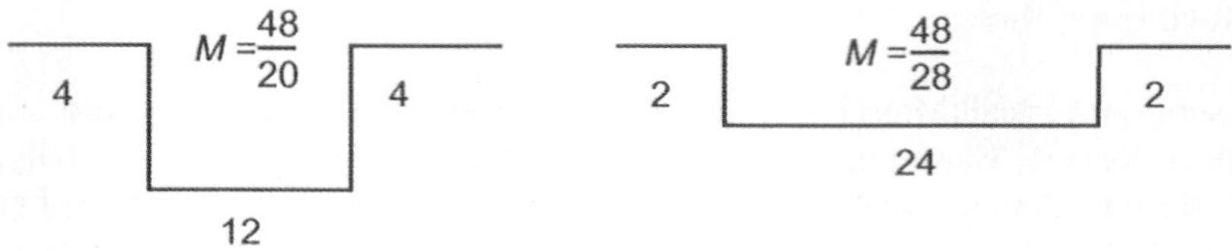

Fig. 5.2-19: Examples of channels with the same cross-sectional area that have different values for the hydraulic radius *M*

Table 5.2-5 gives examples of Manning's roughness coefficients n.

Table 5.2-5: Examples of Manning's roughness coefficients *n* for streams/channels, FETTER (2001)

mountain streams with rocky beds	0.04 - 0.05
winding natural streams with weeds	0.035
natural streams with little vegetation	0.025
straight, unlined Earth canals	0.020
smoothed concrete	0.012

Runoff separation by hydrograph analysis

The runoff in a catchment area consists of direct runoff and baseflow as shown in *Fig. 5.2-20a*. A hydrograph from a rainfall event (shown schematically in *Fig. 5.2-20b*) can be used to estimate the amounts of direct runoff and baseflow. A hydrograph is graphical representation showing for a given point on a stream the discharge, stage (depth), velocity, or other property of water as a function of time. Infiltration and percolation during and after a rainstorm recharges the groundwater, increasing the rate of baseflow, as shown in *Figs. 5.2-20b* and *5.2-21*. For a catchment area of 100 km^2, maximum baseflow is reached 3 to 4 days after a precipitation event (RICHTER & LILLICH, 1975). After infiltration, the baseflow declines, as shown in the baseflow depletion curve in *Fig. 5.2-20b*. Baseflow R_B can be estimated from hydrographs as shown in *Figs. 5.2-20b* and *5.2-21*. R_B can be calculated as a mean value Q_{mean} from single measurements Q_i made during dry periods and by integration of the area below the dry-weather flow line. The area above the dry-weather flow line is assigned to direct runoff. Other hydrograph separation methods are available (RICHTER & LILLICH, 1975; HÖLTING, 1992; HISCOCK, 2005).

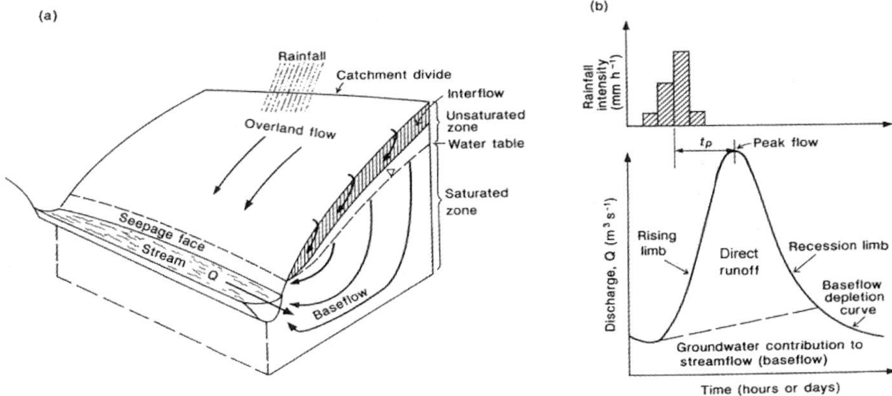

Fig. 5.2-20: Runoff in a catchment area: (a) runoff components, (b) schematic hydrograph from a rainfall event and its interpretation as described in the text (HISCOCK, 2005)

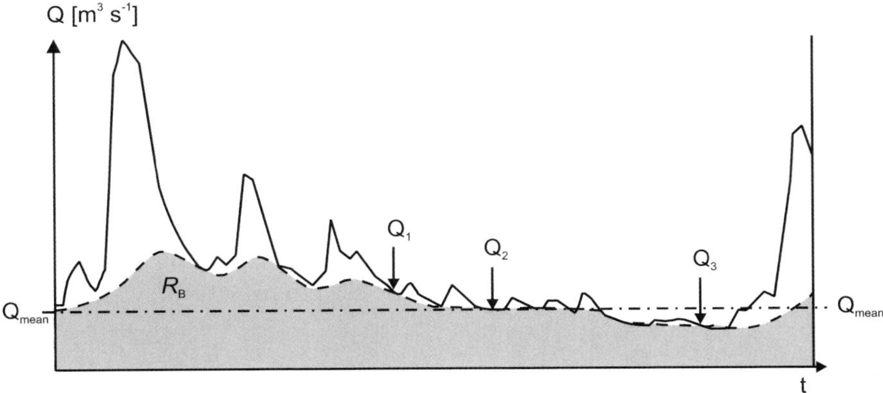

Fig. 5.2-21: Hydrograph of discharge Q as a function of time t of a stream and estimation of baseflow R_B as a mean value Q_{mean} from single measurements Q_i made during dry periods and by integration of the area below the dry-weather flow line (gray area), RICHTER & LILLICH (1975)

Maximum rate of runoff

The probable maximum rate of runoff in a catchment area can be estimated using the following equation (HUDSON, 1993):

$$Q = \frac{c}{360} I\,A\,, \tag{5.2.17}$$

where Q is rate of runoff [$m^3\ s^{-1}$],
I intensity of rain [$mm\ h^{-1}$],
A catchment area in hectars, and
c runoff coefficient [dimensionless].

The rain intensity I can be taken from local rainfall records. It is also necessary to take into account the longest time taken by surface runoff to reach the outlet from any point in the catchment area. The coefficient c is a measure of the proportion of the rain which becomes runoff. Values for different surface soils are given in *Table 5.2-6*.

Table 5.2-6: Values of the runoff coefficient c, SCHWAB et al., (1981), quoted after HUDSON (1993)

Topography and vegetation	Soil texture		
	open sandy loam	**clay and silt loam**	**tight clay**
woodland			
flat 0 - 5 % slope	0.10	0.30	0.40
rolling 5 - 10 % slope	0.25	0.35	0.50
hilly 10 - 30 % slope	0.30	0.50	0.60
pasture			
flat	0.10	0.30	0.40
rolling	0.16	0.36	0.55
hilly	0.22	0.42	0.60
cultivated			
flat	0.30	0.50	0.60
rolling	0.40	0.60	0.70
hilly	0.52	0.72	0.82
urban areas	30 % *	50 % *	70 % *
flat	0.40	0.55	0.65
rolling	0.50	0.65	0.80

*of area impervious

Probable runoff quantity of a catchment area

The probable volume of runoff or yield of a catchment area can be calculated for an individual storm using the formula of the U. S. Soil Conservation Service (HUDSON, 1993):

$$Q = \frac{(P - 0.2S)^2}{P + 0.8S}, \qquad (5.2.18)$$

where Q is runoff [mm],
P storm rainfall [mm], and
S amount of rainfall which can soak into soil during the storm [mm].

Table 5.2-7 gives values of S [mm] for the water yield Equation (5.2.18). Intermediate values may be used.

Table 5.2-7: Values of S [mm] for the water yield Equation (5.2.18), USDA-SCS (1964), quoted after HUDSON (1993)

Soil type	**Number of days since last storm which caused runoff**		
	more than 5	2 to 5	less than 2
good permeability, for example, deep sands	150	75	50
medium permeability, for example, sandy clay loams and clay loams	100	50	25
low permeability, for example, clays	50	25	25

5.2.4 Infiltration

FLORIAN JENN, KLAUS KNÖDEL, MANJA LIESE & HANS-JÜRGEN VOIGT

Infiltration is the process by which water at the ground surface passes into the unsaturated, or vadose zone, as well as the volume of that water. More than two thirds of the world's precipitation on land infiltrates into the ground (SERRANO, 1997). Infiltration provides the water for plant growth and almost all water in the groundwater reservoirs. Infiltration is also important for the spreading of contaminants. Therefore, an understanding of infiltration and water movement through the unsaturated zone is very important in hydrology and hydrogeology. Infiltration is governed by two forces: gravity and capillary action. Temporal and spatial factors greatly affect how water moves through the subsurface. Infiltration depends on the characteristics of the ground surface, such as slope, vegetation cover, land use, soil properties such as soil texture, porosity, soil water content, soil chemistry, as well as meteorological factors, such as antecedent rain events, precipitation intensity, depth of ponding, etc. Secondary pores (e.g., shrinkage cracks, root courses, earthworm courses) are also of a high importance.

During a rain storm event, the precipitation divides into infiltration, depression storage (ponding), and overland flow (Section 5.2.3) as shown in *Fig. 5.2-22*. Infiltration begins when the precipitation reaches the ground. Overland flow does not begin until the available depression storage is exhausted. Overland flow can continue past the termination of precipitation. Infiltration will continue as long as there is any water in the depression storage, usually past the period of overland flow.

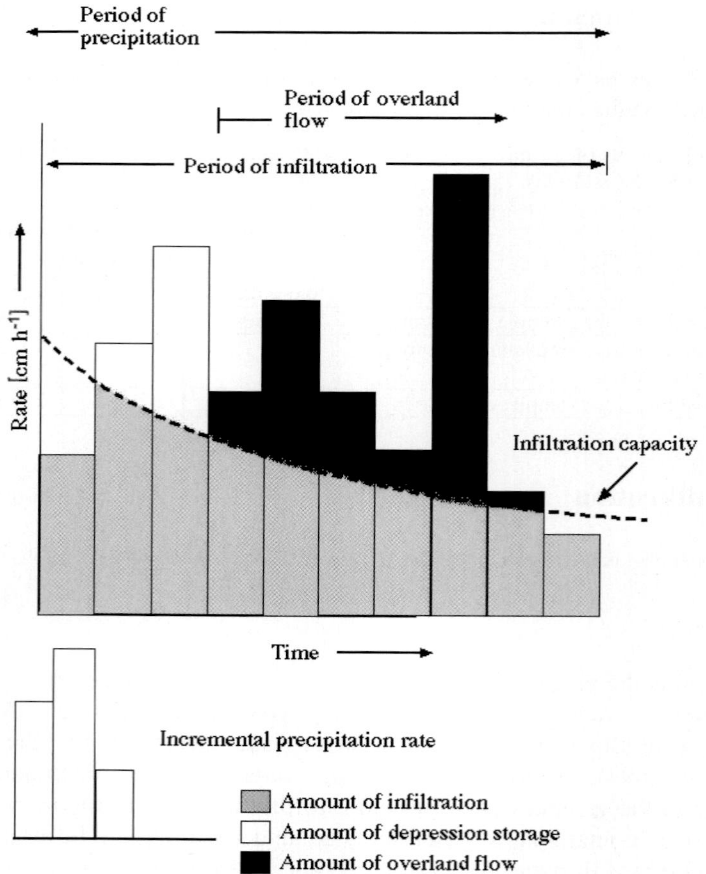

Fig. 5.2-22: Incremental precipitation rate and its dissociation into amounts of infiltration, depression storage (ponding), and overland flow, http://jan.ucc.nau.edu/~doetqp-p/courses/env302/lec4/LEC4.html

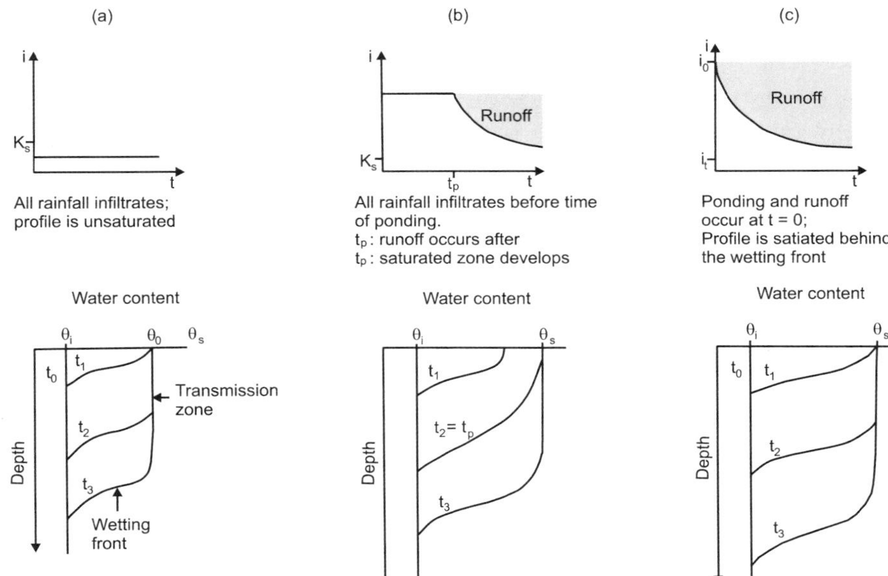

Fig. 5.2-23: Schematic of infiltration rate dependency on rain intensity i and profiles of water content for (a) low, (b) moderate, and (c) intense precipitation, K_s is the saturated hydraulic conductivity, R recharge, θ_i initial water content, θ_0 water content behind the wetting front, θ_s saturated water content, STEPHENS (1996)

Four infiltration scenarios are distinguished:

- Low precipitation: The rainfall intensity is less than the magnitude of saturated hydraulic conductivity[3] of the soil. All rainfall infiltrates, no ponding and no overland flow occur (*Fig. 5.2-23a*). The infiltration rate is equal to the water input rate and is less than the infiltration capacity.
- Moderate precipitation: Rain falls on dry soil with a constant rate. The rainfall rate exceeds the magnitude of saturated hydraulic conductivity of the soil. The soil initially takes up the rainwater as rapidly it falls. All rainwater infiltrates until saturation of the uppermost soil layer is reached. Then ponding and overland flow occur and the infiltration rate decreases to the value of infiltration capacity (*Fig. 5.2-23b*).
- Intense precipitation (saturation from above): When rainfall intensity is high relative to the saturated hydraulic conductivity of the soil, ponding and runoff occur instantly and the infiltration declines exponentially and asymptotically to a near constant value corresponding to the magnitude of saturated hydraulic conductivity of the soil (*Fig. 5.2-23c*). The infiltration rate is equal to the infiltration capacity.

---

[3] Saturated hydraulic conductivity is a quantitative measure of a saturated soil's ability to transmit water when subjected to a hydraulic gradient.

- Saturation from below: The water table has risen to or above the surface, ponding, no infiltration occurs.

Figure 5.2.23 also schematically shows the water content profiles and the downward propagation of the wetting front of infiltrating water during a rainfall event. The wetting front is diffuse in clay (due to high capillarity) and sharp in sand and, in addition, lateral inhomogeneities of soil physical properties cause an uneven front.

The actual infiltration rate is difficult to measure because of the variety of factors it depends on. Analog to the approach in modeling evapotranspiration, in which the two concepts actual "evapotranspiration" and "potential evapotranspiration" are used, the models of actual infiltration rate and infiltration capacity are used. The actual infiltration rate is the quantity of water on a soil surface that enters the soil in a specified time interval. The unit is depth of water per unit time, for example mm h^{-1} or cm s^{-1}. Infiltration capacity is the maximum rate at which infiltration can occur under given conditions. It is a function of soil moisture content. The unit is depth of water per unit time, for example mm h^{-1} or cm s^{-1}. The actual infiltration rate is less than the infiltration capacity if no ponding/flooding occurs and is the minimum between infiltration capacity and precipitation rate.

The infiltration rate is usually measured for ponded water. The infiltration rate depends on the ponding depth, primarily at early time of the precipitation event. It is adequate to use the term "ponded infiltration rate". The ponded infiltration rate curve is also affected by the antecedent water content. The dryer the soil, the higher is the initial infiltration rate.

Various researchers have modeled the infiltration curve as shown in *Fig. 5.23c*. One model is the Horton equation for infiltration capacity (HORTON, 1940):

$$f_p = f_c + (f_0 - f_c)e^{-kt}, \qquad (5.2.19)$$

where f_p infiltration capacity [m s^{-1} or mm h^{-1}],
t time [s or min],
f_0 initial (i.e. maximum) infiltration capacity [m s^{-1} or mm h^{-1}],
f_c equilibrium infiltration capacity (asymptotic infiltration capacity) [m s^{-1} or mm h^{-1}],
k a constant, representing the rate of decrease of infiltration capacity [s^{-1} or min^{-1}].

The parameters f_0, f_c, and k can be determined from infiltration measurements. For rough estimates of infiltration capacity the parameters from *Table 5.2-8* can be used.

Table 5.2-8: Parameter estimates for Horton's infiltration model, ASCE (1996)

Soil and cover	f_0 **[mm h^{-1}]**	f_c **[mm h^{-1}]**	k **[min^{-1}]**
standard agriculture (bare)	280	6 - 220	1.6
standard agriculture (turfed)	900	20 - 290	0.8
peat	325	2 - 29	1.8
fine sandy clay (bare)	210	2 - 25	2.0
fine sandy clay (turfed)	670	10 - 30	1.4

The most widely referenced model is Philip's transient infiltration equation for very shallow ponded water (PHILIP, 1957; STEPHENS, 1996):

$$i = \frac{1}{2} s t^{-1/2} + K_o , \qquad (5.2.20)$$

where i infiltration rate [m s^{-1}],
t time since infiltration began [s],
s sorptivity of the soil [cm s$^{-1/2}$],
K_o hydraulic conductivity of the uppermost soil layer [m s^{-1}],

Sorptivity is a soil hydraulic parameter that includes several factors and represents the capacity of soil to take up infiltrating water early during a precipitation event, such as the initial water content of the soil. Philip's infiltration equation can be used to predict the transient infiltration rate.

Soil can be classified by its texture (see Section 5.2.7.3). The texture is determined by the proportions of clay, silt and sand after particles larger than sand have been removed. Clay is defined as soil particles with a diameter less than 0.002 mm. Silt is soil particles between 0.002 and 0.05 mm in diameter and sand is particles between 0.05 and 2 mm in diameter. Particles greater than 2 mm in diameter are classified as gravel. If a significant portion of the soil is greater than 2 mm, the term "gravelly" or "stony" is added to the soil description.

The grain-size distribution must be determined for the classification of a soil. Several soil hydraulic properties depend on soil texture. The size of the pores through which water can flow is approximately equal to the grain size. Therefore, the pore-size distribution can be derived from the grain-size distribution. Typical curves of accumulated infiltration for different soil textures and land use are shown in *Fig. 5.2-24*.

Water flows in both the unsaturated, or vadose zone and in the saturated zone from areas of high potential energy to areas with low potential energy. The total soil-water potential consists of two parts: soil-water potential and elevation potential. For the soil-water potential, capillary and adsorptive forces are the most important. When the soil-water potential is expressed in energy units per unit weight of soil, it is called pressure head ψ (STEPHENS, 1996). It is also called suction or tension head and is measured with a tensiometer as water column.

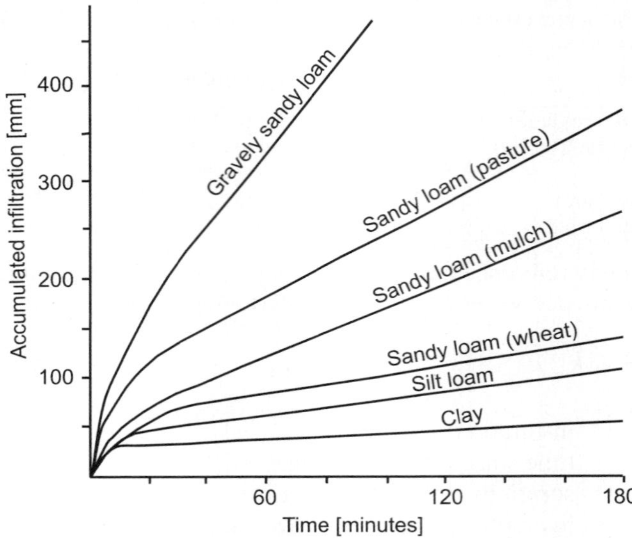

Fig. 5.2-24: Typical infiltration curves for different soil textures and land use, McLAREN & CAMERON (1990) quoted after MAF (2002)

Pressure head and hydraulic conductivity govern the rate of unsaturated flow in a porous media (soil, sediment, bedrock) according to Darcy's law. Both pressure head and hydraulic conductivity are related to the water content. The relationship between pressure head (soil water potential) and water content is shown in *Fig. 5.2-25* for material with different textures. The water content of soil at a given pressure head depends on its wetting history. This effect is called hysteresis and is depicted in *Fig. 5.2-26*. When the soil wets from an initially dry condition, the soil behaves according to the main wetting curve. The behavior of drying from complete saturation is depicted as main drying curve. The intermediate paths on the soil-water characteristic curve, bounded by the main wetting curve and main drying curve, are called scanning curves. The factors contributing to hysteresis are capillary effects and the effect of entrapped air in the soil pores.

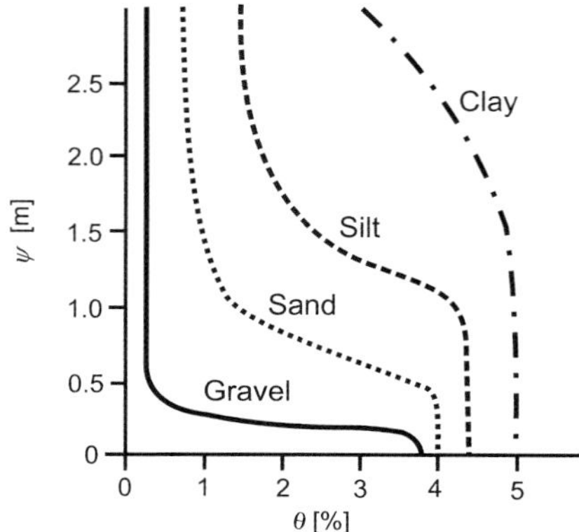

Fig. 5.2-25: Effect of texture on the soil-water characteristic curve; ψ is the pressure head, θ is water content (STEPHENS, 1996)

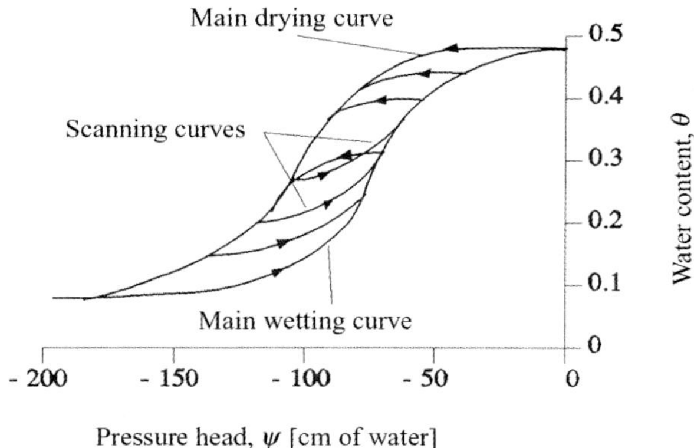

Fig. 5.2-26: Water content versus pressure head (SERRANO, 1997)

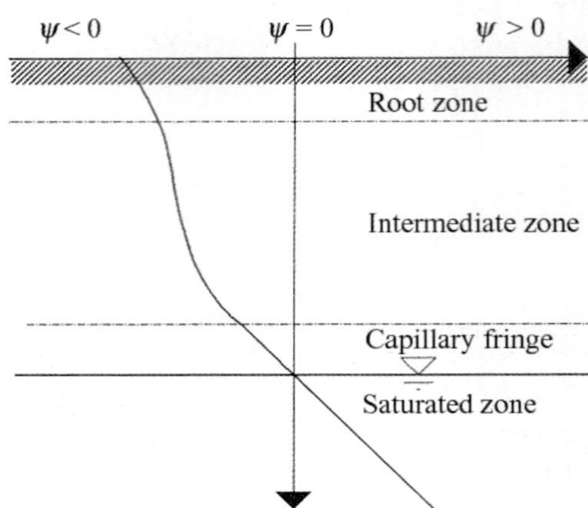

Fig. 5.2-27: Pressure head as a function of depth for the unsaturated and saturated zones, SERRANO (1997)

An idealized representation of the pressure head as a function of depth is shown for a typical soil profile in *Fig. 5.2-27*. The pressure head at an unconfined water table is zero, and in the zone of saturation the pressure head is positive. In the unsaturated zone the pressure head is negative. The effect of layering on water-content and pressure-head profiles under hydrostatic equilibrium is shown in *Fig. 5.2-28*.

Infiltration ceases when no water from precipitation, lakes, ponds, and streams is available to move through the soil surface into the soil. But movement of water within the soil does not stop when infiltration ceases. After a period of infiltration, water moves in the vadose zone as drainage and redistribution. Drainage in the uppermost vadose zone occurs when the water content decreases over time (*Fig. 5.2-29a*). Water behind the wetting front can move upward in response of evapotranspiration or it can move downward due to gravity and other potential gradients resulting from capillary, imbibitional and adsorptive forces (*Fig. 5.2-29b*). This process is called redistribution. Water percolating downward through the vadose zone can reach the water table and contribute to groundwater recharge.

According to isotope measurements in various regions, the mean infiltration velocity for Northern Germany is about 1 m a^{-1} (HÖLTING, 1992).

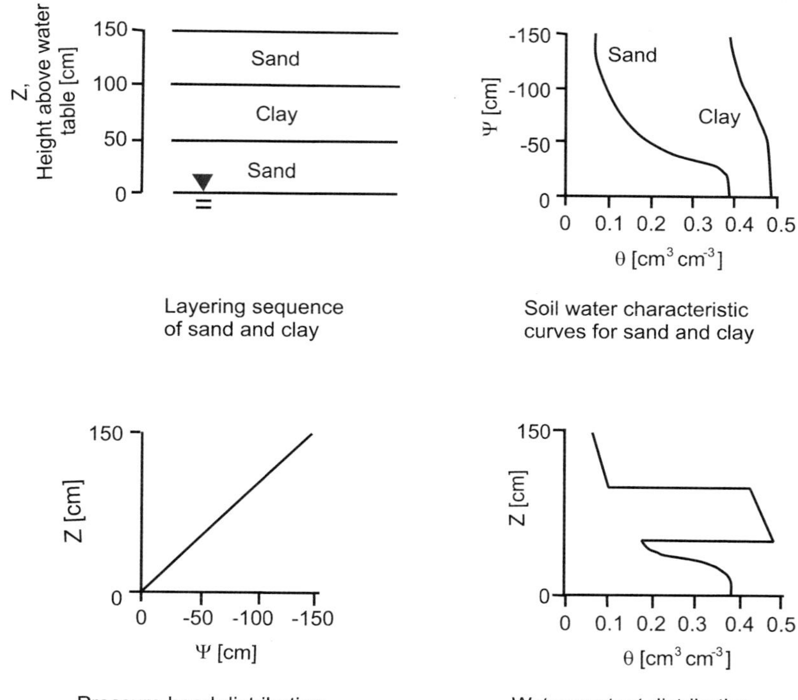

Fig. 5.2-28: Effect of layering on water-content and pressure head profiles under hydrostatic equilibrium, ψ pressure head, θ water content, STEPHENS (1996)

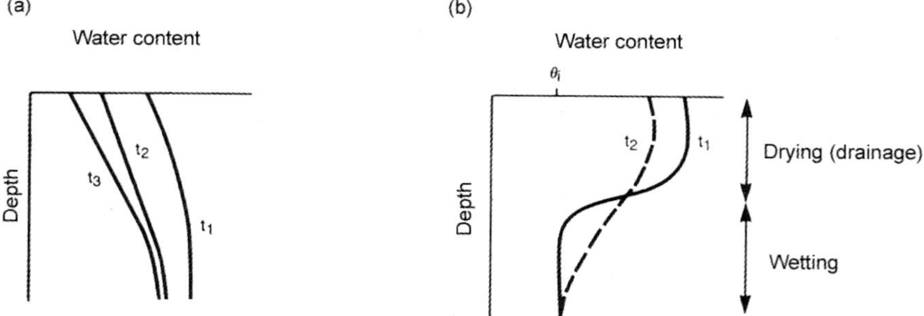

Fig. 5.2-29: Drainage (a) and redistribution (b) following a period of infiltration, t_1, t_2 and t_3 time, where θ_i is the initial water content, STEPHENS (1996)

Infiltration is usually quantified by water budget analysis, laboratory and field measurements, and/or calculation based on soil hydraulic properties. Various analytical models are used in soil science to estimate the annual infiltration rate based on the water balance equation. The infiltration rate is calculated from the difference between the annual precipitation and the sum of direct runoff and evapotranspiration. Examples of such models in Germany are ABIMO (GLUGLA & FÜRTIG, 1997), GROWA (WENDLAND et al., 2001), and the RENGER-WESSOLEK (RENGER & WESSOLEK, 1996). Lysimeters can also be used to study infiltration. For their operation and limitations, see Section 5.2.2 and BREDENKAMP et al. (1995). Other methods are described in Section 5.2.7.

5.2.5 Groundwater Recharge

FLORIAN JENN, KLAUS KNÖDEL, MANJA LIESE & HANS-JÜRGEN VOIGT

Groundwater recharge is the process of water infiltrating through the vadose zone and reaching the groundwater surface, thus adding water to the aquifer. The amount of this water is also called "groundwater recharge". Groundwater recharge occurs naturally as the net gain from precipitation or runoff. Injection of water into an aquifer through wells is one form of artificial recharge. Another method is to direct runoff into a collecting pond with a permeable bottom; this is often done in areas with hardpan. The groundwater recharge rate is reported in $mm\ a^{-1}$ or $L\ s^{-1}\ km^{-2}$ and the volume of groundwater recharge is calculated from the recharge rate multiplied by the land area under consideration and is given in $m^3\ s^{-1}$ or $L\ s^{-1}$.

The land area where recharge occurs is called the recharge area or recharge zone. Areas of one to more than hundreds of square kilometers can be involved. A recharge area is particularly vulnerable to any pollutants that may be in the water. Surface sealing, e.g., asphalt parking areas, and buildings, reduces the amount of water that can enter the aquifer from the surface.

Groundwater recharge is classified on the basis of the source of water and by the pathway the water takes to enter the saturated zone. Two principal types of recharge are recognized:

- *Direct recharge* is defined as "water added to the groundwater reservoir in excess of soil moisture deficits and evapotranspiration, by direct vertical percolation of precipitation through the unsaturated zone" (SIMMERS, 1997). Infiltration and soil water seepage are the main processes of groundwater recharge. As described in Section 5.2.4, how much water infiltrates depends, among other factors, on climatological aspects, vegetation cover, slope of the Earth's surface, soil composition, depth to the water table, the presence or absence of confining beds. Groundwater recharge depends on vegetation cover and is promoted by flat topography, permeable soil, a deep water table and the absence of confining beds. For groundwater modeling, it is necessary to recognize the variability of recharge within the area under consideration (example in *Fig. 6.2-21*). Direct recharge is also referred to as diffuse recharge. Diffuse recharge normally occurs over large areas, such as a watershed. Changes in land use, for example deforestation, have an impact on groundwater recharge.

- *Indirect recharge* can occur, for example, as localized recharge from water in natural depressions, ponds, channels and losing streams (influent streams), playas, arroyos and wadis or by interflow, vertical leakage across aquitards and underflow from adjacent aquifers. Losing streams, arroyos, wadis, and ponds play an important role for indirect recharge in arid and semiarid areas. Mountain front recharge takes place as unsaturated and saturated flow in fractured rocks and as infiltration along channels flowing across alluvial fans (STEPHENS, 1996). Recharge from water mains, septic tanks, sewers and drainage ditches in urban areas must also be taken into account for indirect recharge. Local recharge occurs when water infiltrates from perennial and/or ephemeral channels, lakes, etc. and can be modeled as a two- or three-dimensional process. The magnitude of infiltration depends on channel geometry, wetted perimeter, depth and duration of water flow, water temperature, clogging layers on the channel floor, and the hydraulic properties of the unsaturated zone. On the basis of streamflow gauge measurements in seven major channels in the Tucson Basin in southeastern Arizona, BURKHAM (1970) found that about 70 % of the runoff into the ephemeral channels was infiltrated. The channel infiltration along a particular reach can be highly variable in time and space. Irrigation can also contribute to groundwater recharge.

In humid climates, recharge mostly occurs by diffuse infiltration and percolation of precipitation and can amount to as much as 20 to more than 50 % of the mean annual precipitation (STEPHENS, 1996). Infiltration and percolation of precipitation does not occur continually over the hydrologic year. It depends on the distribution and intensity of precipitation events, land use, vegetation periods, evapotranspiration, antecedent saturation of soil, etc. In Central Europe groundwater recharge starts between October and December and lasts until March to April. From April to October, the main vegetation period, precipitation contributes little to the recharge (MATTHESS & UBELL, 1983). Groundwater discharge takes place as baseflow to streams (Section 5.2.3).

The often repeated statement that there is no direct recharge by rainfall through the unsaturated zone in arid regions is based on a fallacy. In arid and semi-arid climates precipitation occurs as intense summer thunderstorms and frontal showers in the winter and spring. In a summary of recharge studies in arid and semi-arid areas, STEPHENS (1996) makes the following statement, together with values for such areas in tabular form: "Regardless of the precipitation and evapotranspiration rates, there is the potential for recharge where the infiltration or redistributed water penetrates below the plant root zone. Winter precipitation, especially in a series of storms, is most likely to contribute to deep soil-water movement and recharge. During the winter storm periods when evapotranspiration is at a minimum, a soil-water budget over the wet period is likely to indicate that infiltrated water would penetrate below the root zone and potentially could become recharge. A few studies illustrate that annual mean precipitation (*P*) and potential evapotranspiration or gross lake evaporation (*E*) are not adequate indicators of whether recharge occurs in areas of low annual precipitation."

The volume of groundwater recharge as well as its spatial and time distribution must be known to assess, for example, the capacity of an aquifer for municipal water supply, the feasibility of dewatering a mine pit, the risk of groundwater contamination, the effectiveness of proposed remedial actions, as well as the risk of land subsidence and/or saline water intrusion as undesirable consequences of groundwater withdrawal.

Groundwater recharge cannot be determined by direct measurement. It is estimated or determined using indirect methods, either in local (point) investigations or in areal investigations for typical soil areas with different land use (hydrotops, local areas within a catchment area with uniform hydrological conditions) and/or catchment area studies. Common methods are

- measurements with soil lysimeters,
- water balance estimation,
- evaluation of groundwater hydrographs,
- stream gauging,
- soil-waterflow modeling,
- estimations using groundwaterflow models, and
- chemical methods.

Measurements with soil lysimeters

Lysimeters (Section 5.2.2) can be used to estimate the part of precipitation available for groundwater recharge. One problem of this method of recharge estimation is the transfer of point lysimeter data to an area. Other problems of lysimeter measurements are discussed in Section 5.2.2.

Water balance estimation

For hydrotop or catchment areas groundwater recharge GWR can be estimated using the groundwater budget Equation (5.2.1):

$$GWR = P - ET - R_d - \Delta S, \tag{5.2.22}$$

where P is precipitation, R_d is the surface runoff and the groundwater discharged from the area, ET is evaporation and transpiration loss, and ΔS is the change in groundwater storage. The annual groundwater recharge of flat areas without substantial surface runoff can be estimated by a simplified equation:

$$GWR = P - ET. \tag{5.2.23}$$

But it must kept in mind that R_d, and ΔS are neglected by this equation.

For different hydrotops the following specified estimation methods are used in Germany:

- BAGAROV & GLUGLA (GLUGLA & FÜRTIG 1997)
- LIEBSCHER & KELLER (1979)
- DÖRHÖFER & JOSOPAIT (1980)
- RENGER & WESSOLEK (1996).

Evaluation of groundwater hydrographs

Changes in groundwater level (Δh) can be monitored using groundwater hydrographs. These changes can be correlated with the precipitation in the same time period. If recharge is to be estimated from groundwater hydrographs, a long time period must be evaluated and the "period of reserve" and the "period of consumption" have to be considered separately. Most important is the qualitative and quantitative correlation between precipitation and the rise of water table in the period of reserve (MATTHESS & UBELL, 1983). The period of reserve is defined as the time when precipitation exceeds evaporation. This is the case, for example, during rainy season in Southeast Asia or in the winter in Europe. In this period a linear correlation between the rise of the water table and the precipitation is observed and can be plotted with the precipitation values on the x-axis and the rise in groundwater level on the y-axis as shown in *Fig. 5.2-30.*

In addition to the amount of precipitation, the rise in groundwater level depends on the amount of evaporation, depth to the groundwater table, specific yield, soil texture, surface runoff, and the change in soil moisture. These factors influence the slope of the best-fit straight line and the axis intercept (P_1). This line characterizes average conditions. The scatter in the values is due to variations in precipitation.

The precipitation can be divided into two parts: P_1 and P_2. P_1 corresponds to that part of the precipitation during which no rise in groundwater is observable, i.e., the water is evaporated, is stored in the upper layer as soil water storage (ΔSW), or becomes direct runoff (R_d).

$$P_1 = ET + R_d + \Delta SW. \tag{5.2.24}$$

The other part of rainfall ($P_2 = P - P_1$) is the effective precipitation for groundwater recharge ($P_2 = P_{eff} = GWR$), causing the rise in groundwater table. The steeper the line in the correlation plot of *Fig. 5.2-30*, the more precipitation is necessary to cause the same rise of groundwater table. The slope α is determined by the specific yield (S_y).

$$\tan\alpha = \frac{P_2}{\Delta h} = S_y. \tag{5.2.25}$$

The correlation line is given by the following equation:

$$P = S_y\,\Delta h + P_1. \tag{5.2.26}$$

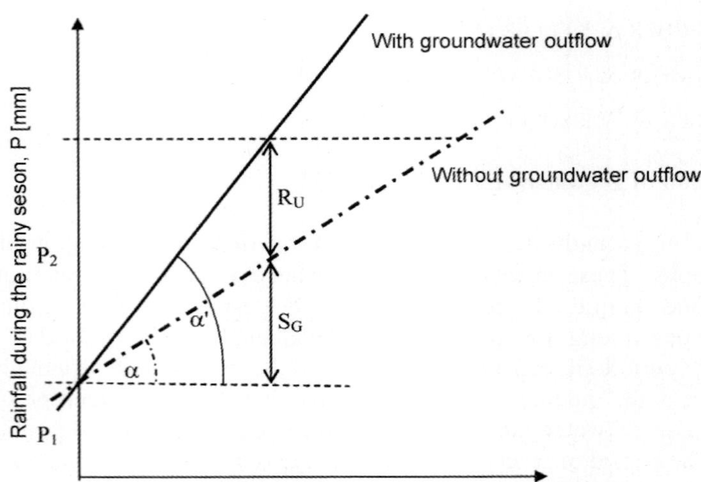

Fig. 5.2-30: Correlation between precipitation and rise of groundwater level with and without groundwater outflow, MATHESS & UBELL (1983)

If there is no groundwater outflow, the specific yield of the recharged aquifer can be estimated as described above from the slope of the line diagram. If there is outflow, however, groundwater recharge GWR by precipitation (i.e., P_2) can be further divided into groundwater storage S_G and groundwater outflow R_G (see *Fig. 5.2-30*, solid line):

$$GWR = P_2 = S_y \, \Delta h = S_G + R_G \qquad (5.2.27)$$

It is not easy to determine specific yield. If the specific yield is known, S_G and R_G can be calculated. Guide values for different materials are given in *Table 5.2-9*. The specific yield may vary due to inhomogeneities in the ground. A representative value must be found.

Table 5.2-9: Specific yield for various materials, BELL (1998)

Material	Specific yield [%]
gravel	15 - 30
sand	10 - 30
dune sand	25 - 35
sand and gravel	15 - 25
loess	15 - 20
silt	5 - 10
clay	1 - 5
till (silty)	4 - 7
till (sandy)	12 - 18
sandstone	5 - 25
limestone	0.5 - 10
shale	0.5 - 5

Stream gauging

Groundwater discharge is estimated by baseflow separation and by stream hydrograph analysis as described in Section 5.2.3.

Soil-waterflow modeling

The groundwater recharge rate can be estimated by soil-water flow modeling using measurements of water content and water tension as a function of depth and time. This provides detailed information about the percolation in unsaturated zone and groundwater recharge. But the measurements and calculations are difficult.

Estimations using groundwater flow models

Groundwaterflow modeling (see Section 5.4.3) starts with a conceptual model (preparation of the model of the aquifer-aquitard system, specification of boundary conditions and hydrological stresses), followed by the design of the numerical model, preparation of model inputs (initial conditions, boundary conditions, hydrogeological parameters, and hydrological stresses), model calibration and validation, and the scenario simulation. The calibration of the groundwaterflow model can be used to estimate the groundwater-recharge rate compatible with the measured groundwater data (e.g., level of the groundwater table and hydraulic gradient in the model area). This value of the groundwater-recharge rate can be compared with estimations by other methods.

Chemical methods

The chloride-profile method or chloride-budget method is frequently used in arid and semiarid areas to determine the water balance in the unsaturated zone (BREDENKAMP et al., 1995; EDMUNDS et al., 1990). But it can also be useful under humid climatic conditions for a general (provisional) groundwater recharge estimate (VOIGT, 1990). It must be kept in mind that the chloride concentration typically decreases with increasing distance inland from the sea.

The method is based on the assumption that chloride in the infiltration water originates only from precipitation and is subsequently concentrated by evapotranspiration on the way from the Earth's surface to the groundwater table. Chloride behaves as a conservative tracer during soil-water transport. The annual groundwater recharge rate is estimated from

$$R = \frac{(P - R_d)C_{input}}{C_{gw}} = \frac{(P - R_d)}{K_v}, \tag{5.2.28}$$

where R is the groundwater recharge rate [mm a^{-1}],
P precipitation [mm a^{-1}],
R_d direct runoff [mm a^{-1}],
C_{input} chloride concentration in rainwater [mg L^{-1}],
C_{gw} chloride concentration in the groundwater [mg L^{-1}], and
K_v evapotranspiration coefficient or infiltration coefficient.

The quantity $P - R_d$ is called the effective annual precipitation. For humid conditions in northeastern Germany, SCHLINKER (1969) found the infiltration coefficients for various soil substrates given in *Table 5.2-10.*

Table 5.2-10: Infiltration coefficients K_v for various soil types, SCHLINKER (1969)

Soil types	K_v
sand with low content of humus	4
sand with high humus content	5
sand, loamy	6.7
moraine (till), sandy	10
moraine (till), loamy	20

The problem in this method is the determination of chloride input (concentration in rainfall, irrigation water and the amount deposited by dry deposition). This requires either a sampling program, which is expensive, or indirect estimation of the chloride concentration of rainfall, because the chloride concentration in precipitation is different in different weather conditions (VOIGT, 1990).

In-situ sampling of isotopes, for example, tritium, chlorine-36, oxygen-18 and deuterium, is also used to estimate groundwater recharge (STEPHENS, 1996; HÖLTING, 1992). The sampling must be carried out below the root zone. It is an advantage that these methods facilitate the integration of hydrologic events over long periods (decades to thousands of years). Sometimes it is possible to investigate the infiltration history on the basis of chemical and isotopic depth profiles (HISCOCK, 2005).

5.2.6 Groundwater Monitoring

FLORIAN JENN, CLAUS NITSCHE & HANS-JÜRGEN VOIGT

Groundwater monitoring includes the periodic or continuous measurement of groundwater level and/or of groundwater quality data at selected points. This is done for site characterization, assessment of the situation at the site and prediction of future changes. Groundwater monitoring requires the existence and/or installation of high quality groundwater monitoring wells at suitable locations.

For the planning of groundwater monitoring, six questions must be answered (LAUTERBACH & VOIGT, 1993):

Why? Objectives of the monitoring must be defined.

Who? Who is to conduct the monitoring?

What? What parameter values must be measured?

Where? Where are the monitoring wells to be located?

When? Is the monitoring to be carried out at specified time intervals or are measurements to be made continuously?

How? What monitoring technique/equipment is to be used?

Monitoring strategies

LAUTERBACH & LUCKNER (1999) distinguish between *information-oriented monitoring* and decision-oriented monitoring. The task of information-oriented monitoring is to evaluate changes in quantity and quality of regional groundwater resources. Several parameters are measured to monitor groundwater quality. The parameters depend on the location of wells in the groundwater catchment area. Most wells are monitored for common groundwater constituents, such as calcium, sodium, and iron. Other wells are monitored primarily for nitrate, chloride, organic solvents and/or gasoline additives. The groundwater level and groundwater quality data can be used for groundwater modeling and predicting groundwater supply. Governmental agencies and/or water supply utilities are responsible for this type of monitoring.

Monitoring of groundwater quantity has been conducted for a long time in Europe. The oldest groundwater monitoring network has been in operation since 1845 in France. Most of the networks for groundwater quantity monitoring were installed at the beginning of the 20th century. The average length of records is between 30 and 50 years. Monitoring of groundwater quality has been carried out in most European countries since the 1970s and 1980s. A first overview of "Groundwater Monitoring in Europe" was published by the European Environmental Agency (EEA) in their Topic Report No. 14 in 1996 (KOREIMANN et al., 1996).

Examples of the objectives of *decision-oriented monitoring* are:

- detection of the spreading of contamination from different emission sources, e.g., landfills, wastewater plants, industrial sites, mining heaps, agricultural activities, and effectiveness check of remediation measures;
- determination of the influence of various technical measures (e.g., pump and treat) on the groundwater table (NILLERT & THIERBACH, 1999; PRZYBILSKI & RECHENBERG, 1999);
- scientific studies of processes at work in the unsaturated or saturated zone (AARGAARD et al., 2004);
- studies of the distribution of dissolved compounds or colloids in groundwater by tracer experiments (ENGLERT et al., 2000; HERFORT et al., 1998; LE BLANC et al., 1991).

A flow chart of the planning of decision-oriented monitoring, e.g., for landfills or sites suspected to be hazardous, is given in *Fig. 5.2-31*.

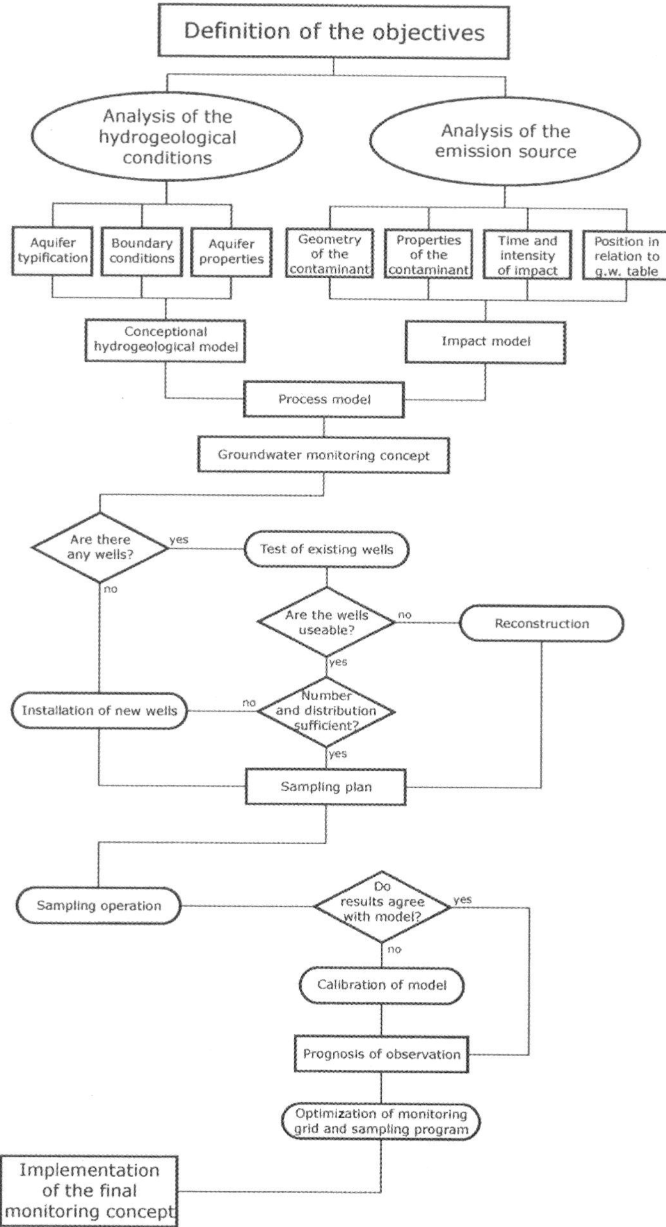

Fig. 5.2-31: Flow chart of decision-oriented monitoring

In contrast to information-oriented monitoring systems, decision-oriented monitoring systems must always be based on an analytical or numerical model of the hydrogeological processes (process model) at the site consisting of a conceptual model of the hydrogeological conditions and a model of the impact, for example, of the source of contamination (see also Section 5.4.4). A scheme for evaluation of the contamination source(s) has been developed (VOIGT, 1990), as well as one for monitoring hydrogeological conditions (VOIGT et al., 2000). The model of the processes must show why monitoring is to be carried out and what has to be investigated. It must also provide an answer to the most important question for decision-oriented monitoring: "Where do the monitoring wells have to be installed?" But, before new wells are drilled, the locations of existing wells must be determined. Recommendations for testing the suitability of existing wells are described later in this section.

Next, the "when" and "how" the groundwater samples are to be taken must be determined. State-of-the-art sampling methods and equipment, the determination of monitoring parameters (indicator substances), and sampling intervals (with respect to both time and depth) are given in Section 5.3.1. The first sampling and analysis operation must answer the question "Do the results agree with the process model?" If the results do not match, the model must be recalibrated and if necessary the grid of monitoring wells must be optimized. A prediction of future changes can be given based on the calibrated process model.

Current practice of groundwater quality monitoring is to take water samples from observation wells, as well as soil and gas samples at vulnerable and hazardous sites at regular intervals for chemical analysis on site and/or in the laboratory. For sampling, it is necessary to drill and install groundwater observation wells at suitable locations and of sufficient diameter to recover sufficiently large water samples. It is also current practice to use multi-parameter probes in wells to monitor several environmental parameters, such as temperature, electrical conductivity, pH, redox potential and oxygen content of the water.

In order to investigate the geogenic and anthropogenic background of the groundwater it is necessary to install one or two wells in the inflow area of the landfill or site suspected to be hazardous (reference wells, so-called A-wells, *Figs. 5.2-32* and *5.2-33*). The location should be selected such that no substances from the landfill or site suspected to be hazardous can migrate to these wells.

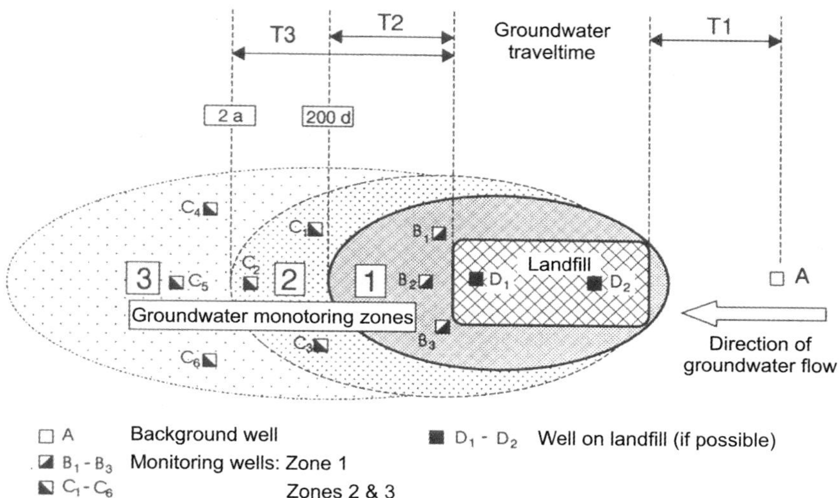

Fig. 5.2-32: Groundwater monitoring at landfills and sites suspected to be hazardous, principle schematic for arranging groundwater observation wells in the monitoring zone (view from above), for further notes see text, DÖRHÖFER et al., (1991)

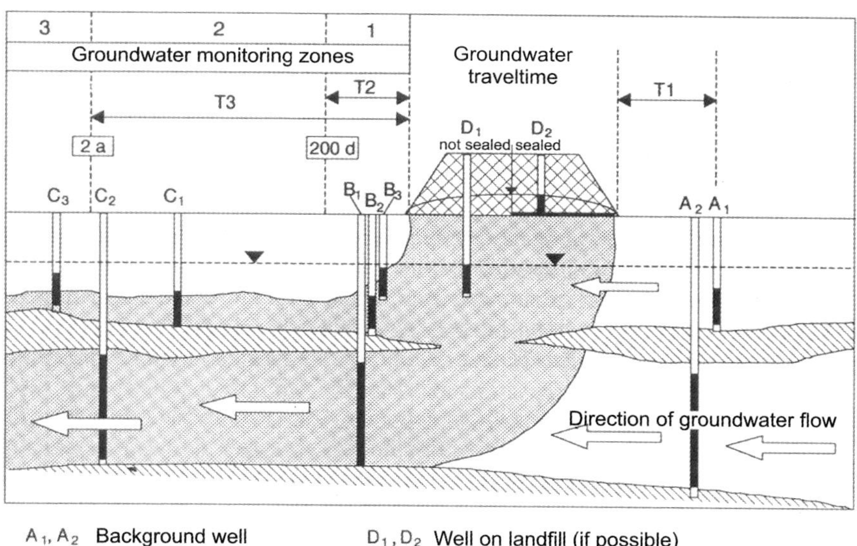

Fig. 5.2-33: Groundwater monitoring at landfills and sites suspected to be hazardous, principle schematic for arranging groundwater observation wells in the monitoring zones (hydrostratigraphic section), for further notes see text, DÖRHÖFER et al. (1991)

The outflow of the area to be monitored, as depicted in *Fig. 5.2-32*, is divided into three zones. The size of zone 1 (inner monitoring zone) should be such that the time the groundwater will flow to the margin of this zone from the landfill or site suspected to be hazardous is 200 days. The period of 200 days was selected so that the normal semiannual groundwater sampling provides early detection of changes. To determine the 200-day line, the displacement velocity is calculated and then the flow time.

Depending upon the size of the landfill or site suspected to be hazardous, several wells are installed within zone 1. These wells (depicted in *Figs. 5.2-32* and *5.2-33*) are designated as B-wells (interior monitoring zone) and are crucial in early detection of contamination. They should, therefore, be located as close as possible to the source of contamination (e.g., landfill).

Zone 2 (exterior monitoring zone) extends from the 200-day line out to the 2-year line. Several wells may be installed in this zone also (designated C-wells in *Figs. 5.2-32* and *5.2-33*) to monitor the spreading of substances over a broad area. Additional monitoring wells in a third zone (additional monitoring zone, zone 3) are required only if contaminants can be expected in the groundwater outside of zone 2.

The success of the monitoring will ultimately depend on the position and length of the screen installed in the well. If the object under observation is located above a thick aquifer, it is necessary to determine the vertical distribution of contaminants prior to installation of the screens. If, for example, the aquifer is fully penetrated by the observation well and the screen covers the entire thickness of the aquifer, the water sample is a mixed sample from the entire depth range. However, since substances will be entering the monitoring well from only one part of the aquifer, it is possible that the concentration of the substance in the sample will be so diluted that it no longer provides a basis for an assessment of the substance load migrating from the landfill or site suspected to be hazardous. It is difficult to decide on the depths of the screens in the case of landfills with a base seal system, since in this case the substances may enter the aquifer as a dot-like source that cannot be easily localized. The screens should be installed where the substances are expected. In thin aquifers the screens should be installed in segments where the highest permeabilities are present. In thick aquifers the wells should be installed as multiple groundwater observation wells (well clusters with screens at different depths or multi-level wells) to obtain data about the vertical distribution of the contaminant plume. The hydrogeological model, which should already exist for the location, is of major importance when the positions of the well screens are decided.

The selection of parameters depends on the composition of the material in the landfill or site suspected to be hazardous and the solubility of the substances. Other factors that should be taken into account are the geogenic and anthropogenic background load in the groundwater, which can be determined from samples taken from the A-well(s) in the inflow area. During the orientation phase it is necessary to carry out a comprehensive analysis of the chemical content of the groundwater. Later the extent of the analyses can be reduced. It may be possible to analyze only those substances or key parameters that are particularly obvious in the seepage water from the source or which have particular toxicological relevance. For the following types of waste should be analyzed for the given key parameters:

- waste from human settlements: dissolved organic carbon (DOC), chemical oxygen demand (COD), ammonia, potassium, iron, manganese, boron, adsorbed organic halogen (AOX),
- building waste: electrical conductivity, ammonia, calcium, chloride, sulfate, AOX.

The scope of the parameters as well as the frequency of investigation should be adapted to the determined load. The greatest effort will be associated with contaminated sites without seal systems. In the case of landfills installed according to current engineering practice, the investigations can be restricted to key parameters and longer time intervals.

Repeated visits to the observation wells are personnel intensive and hence expensive. Further disadvantages are the uncertainties deriving from the sample collection, transport, and the chemical analysis, in addition to the masses of data produced. The advantages are that numerous substances can be analyzed and that the results of chemical analyses can be used in court. All of these procedures yield point data in space and time. To adequately monitor a site, numerous sampling points in space and time are necessary.

A new technological development permits continual in-situ monitoring over long periods of time. The concept is based on a combination of measurements made with multi-parameter probes and optical sensor systems to obtain point data together with an electromagnetic (EM) system to obtain spatial data (Chapter 4.9). To avoid the influence of the system components (probes, optosensors, and EM system) on each other, the three components are activated separately. The multi-parameter probes and optosensors can be installed together in a 3-inch borehole. Transmitters and receivers for the EM system are placed in other boreholes to avoid disturbances from the cables of the probes and optosensors.

The cost and uncertainties of monitoring can be reduced by using a permanently installed in-situ system. Such a system reduces the frequency of on-site visits to take samples and the subsequent chemical analysis. The disadvantages of the customary system – uncertainties deriving from the sample collection, transport, and the chemical analysis, in addition to the masses of data produced – are avoided or reduced. In addition, it is possible to react more quickly to the data obtained online. The reliability of the data is increased by the combination of point and spatial data.

Installation of groundwater monitoring wells

Groundwater monitoring wells provide direct access to the groundwater system. The monitoring system should preserve the in-situ conditions as well as possible. Therefore, minimum standards are necessary in each of the following categories:

- drilling methods,
- materials,
- installation, and
- well development.

First, the drilling method must be selected. The decision is based on the requirements on the monitoring well, the hydrogeological conditions, the quantity of water required, the depth and diameter needed, and the budget of the project. The drilling methods are described together with their advantages and disadvantages in Section 5.1.3. The two major drilling procedures used in monitoring well installation are hollow stem augers and rotary drills. The most important decision is whether drilling fluid will be used and if so, what kind. When feasible, drilling procedures that do not introduce water or other liquids into the borehole should be used. If the introduction of drilling fluids cannot be avoided, safety precautions must be used to avoid contamination of the zone to be monitored.

Components of a monitoring well are the casing, well screen, centralizers, annular sealant, and gravel pack. During storage, assembly, and installation, all materials used in the construction of a monitoring well must be free of contaminants that could affect the monitoring of the groundwater system. PVC is standard for pipes and well screens. All PVC pipes should be threaded to avoid the volatile organic compounds contained in PVC glues, which could contaminate the system. If non-aqueous-phase liquids (NAPLs) or contaminants of high concentration which react with PVC are to be monitored, HDPE or stainless steel should be used instead of PVC.

To insure proper installation of the monitoring well, the following procedure is recommended: The screen interval is selected after the borehole has been drilled and logged. The well screen and the well casing are lowered directly within the hollow stem during a hollow stem auger installation. During a rotary-drilled well installation, potable water must be employed to dilute the drilling mud to ensure ease of installation. After the screen has been positioned and centered in the borehole, the gravel pack is put in place from the base of the well to 20 - 50 cm above the top of the well screen. The gravel pack should consist of material with a grain size on the order of the grain size of the surrounding aquifer. If the filter gravel is very coarse, a counterfilter of finer-grained gravel should complete the filter gravel on either end. Bentonite is then placed on top of the filter/counterfilter 2 - 5 m thick to prevent entry of water from the surface or overlying layers into the filter and the aquifer. The remaining borehole is then either completely grouted, or filled with sand, sealing off aquicludes with bentonite (i.e. filling material corresponds to the aquifer-aquiclude structure). On top of the borehole, another bentonite or grout seal (and optionally a concrete slab on the ground) protects against intrusion of rain or surface water. Finally, protective casing and locking well caps must be installed to insure that the monitoring well is protected from vandalism and accidental damage. Well development removes the fine-grained material from both the gravel pack and aquifer material adjacent to the borehole mostly by pumping to improve the hydraulic efficiency of the well. The installation and well development is called borehole completion.

Depending on the type of screen installation, monitoring wells are divided into

- fully penetrating wells,
- multi-screen wells,
- "nested" groundwater wells,
- cluster or group of monitoring wells, and
- special wells with equipment for sampling at preselected depths.

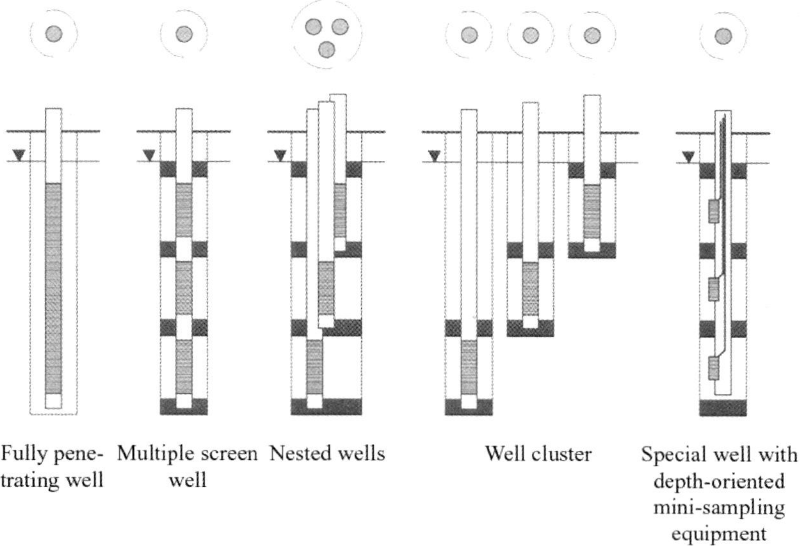

Fig. 5.2-34: Types of monitoring wells, after DVWK-Merkblatt 245 (1997)

The different types of monitoring wells are shown in *Fig. 5.2-34*. Until about 1995, fully penetrating or multiple-screen wells were constructed for decision-oriented monitoring in most of West Germany and also in most European countries. *Fully penetrating wells* have screens installed over the entire thickness of the aquifer. This is done to obtain a hydraulically weighted average water sample of the whole aquifer or, using double (straddle) packers, to obtain a sample from a specified depth. Studies conducted by KALERIS (1992) and BARCZEWSKI et al. (1993) show that it is practically impossible to hydraulically separate the different screened zones or the packer intervals (*Fig. 5.2-37*).

The principal differences in construction and the resulting hydraulic conditions in the different types of wells and their downstream areas are illustrated in *Fig. 5.2-35*. Figure *5.2-36* shows the hydraulic conditions in and around a fully penetrating groundwater well. In this example, the pump is placed just at the upper edge of the screen. The structure of the aquifer system is identified on the basis of the hydraulic conductivities k_1 to k_4. The flow resulting from pumping for sampling is denoted by an upward arrow. The figure also shows the velocity distribution. The black arrows indicate layers with high groundwater inflow. Vertical mixing takes place in the annular space of the gravel pack, thus depth-oriented sampling is not possible.

Moreover, in the intervals between two groundwater sampling events vertical exchange may take place within the gravel pack, the screen, and the well casing. This shows that fully penetrating wells lead to mixing of groundwater from vertically different conditions in the aquifers and particularly affects the downstream area. The same effect is found in wells with leaky casings.

It is possible to carry out sampling at specified depths in aquifers using inline packers and/or multi-packers (BARCZEWSKI et al., 1993; LERNER & TEUTSCH, 1995), however, it does not prevent the cross contamination in the area around the monitoring well (mainly downstream). *Figure 5.2-37* shows, for example, the influence of different packer systems on the sampling results in a fully penetrating well. In the opinion of the authors, these systems should be used only for indicating the qualitative layering of groundwater and for installation purposes and to carry out vertical groundwater profiling.

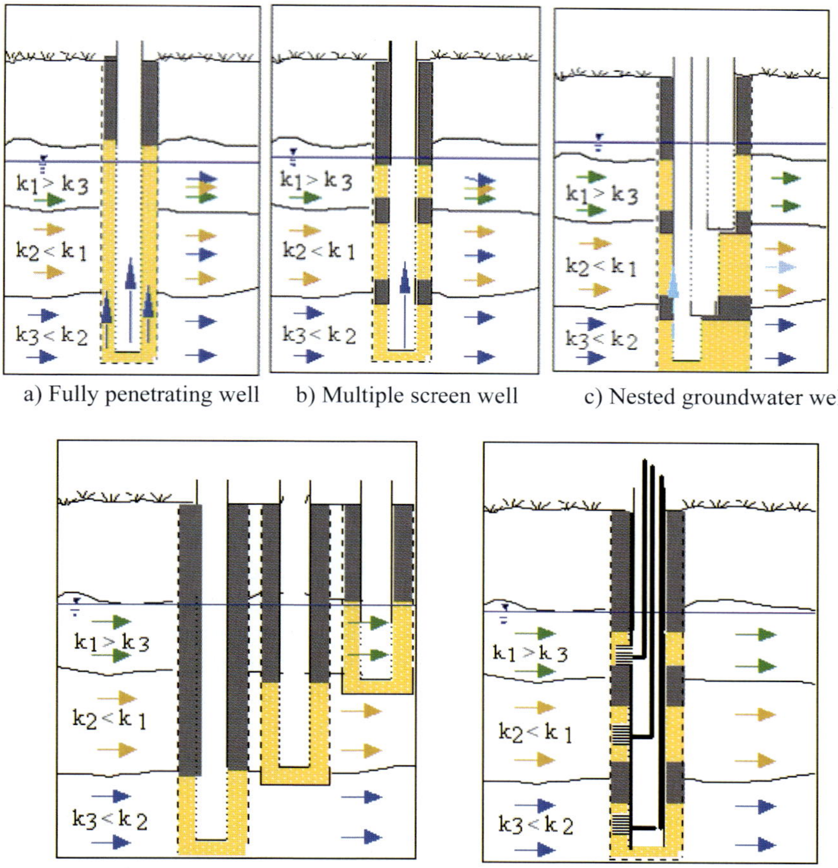

a) Fully penetrating well b) Multiple screen well c) Nested groundwater well

d) Group of monitoring wells e) Special well with depth oriented sampling equipment

Fig. 5.2-35: Hydraulic conditions for different well types

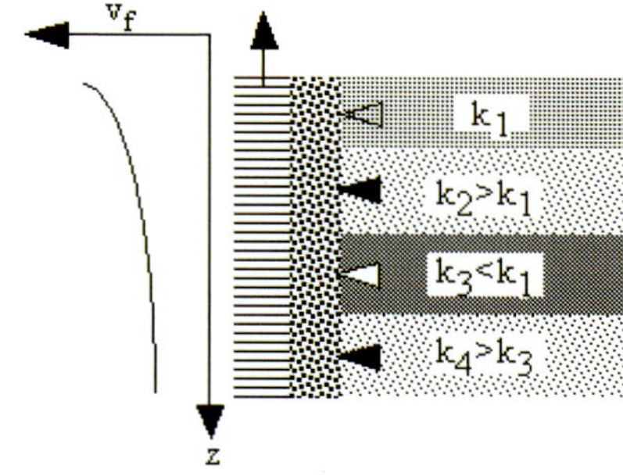

Fig. 5.2-36: Hydraulic conditions in and around a fully penetrating monitoring well during sampling, v_f: velocity distribution with depth z

Multiple screen wells are constructed with a single casing and screens at different depths depending on the hydrogeological profile. The different screen sections are separated by grouting seals so that there is no hydraulic connection between the different aquifer layers through the annular space between the casing and the sediments. But inside of the casing there can still be mixing of water from different layers (for example mixing of contaminated water by vertical convection). Accordingly, the use of multiple screens monitoring wells requires inline packer and/or multi-packers to be installed for the entire period of monitoring to prevent hydraulic connection between the individual aquifer layers, avoiding vertical mixing of groundwater of different quality from these aquifers. Studies of the effectiveness of different types of wells (e.g., DEHNERT et al., 2001) also describe the risk of mixing of contaminated water downstream of fully and multiple screen wells. So these types of monitoring wells are not recommended for sampling of groundwater at specific depths.

In *nested wells*, several individual well casings are installed in a borehole and the individual screens (usually 1 to 2 m long) are isolated from one another (HUDAK, 1998). As compared with multiple screen monitoring wells, properly constructed nested monitoring wells have the advantage that the different aquifer layers are not hydraulically connected, therefore, the mixing of water of different quality is avoided. Despite this advantage, the large borehole diameter required for this kind of construction and the disturbances in underground that it may cause during drilling as well as the changed groundwater conditions in the screened depth are the key issues to be considered. These factors frequently result into longer purging times, which can be considerably reduced by installing stationary packer pump systems. A further practical problem in construction of the nested monitoring wells is that of sealing the space between the individual groundwater screens. Since very small annular spaces cannot be always effectively sealed, it demands a lot of effort on the part of the drilling company to meet the specified requirements.

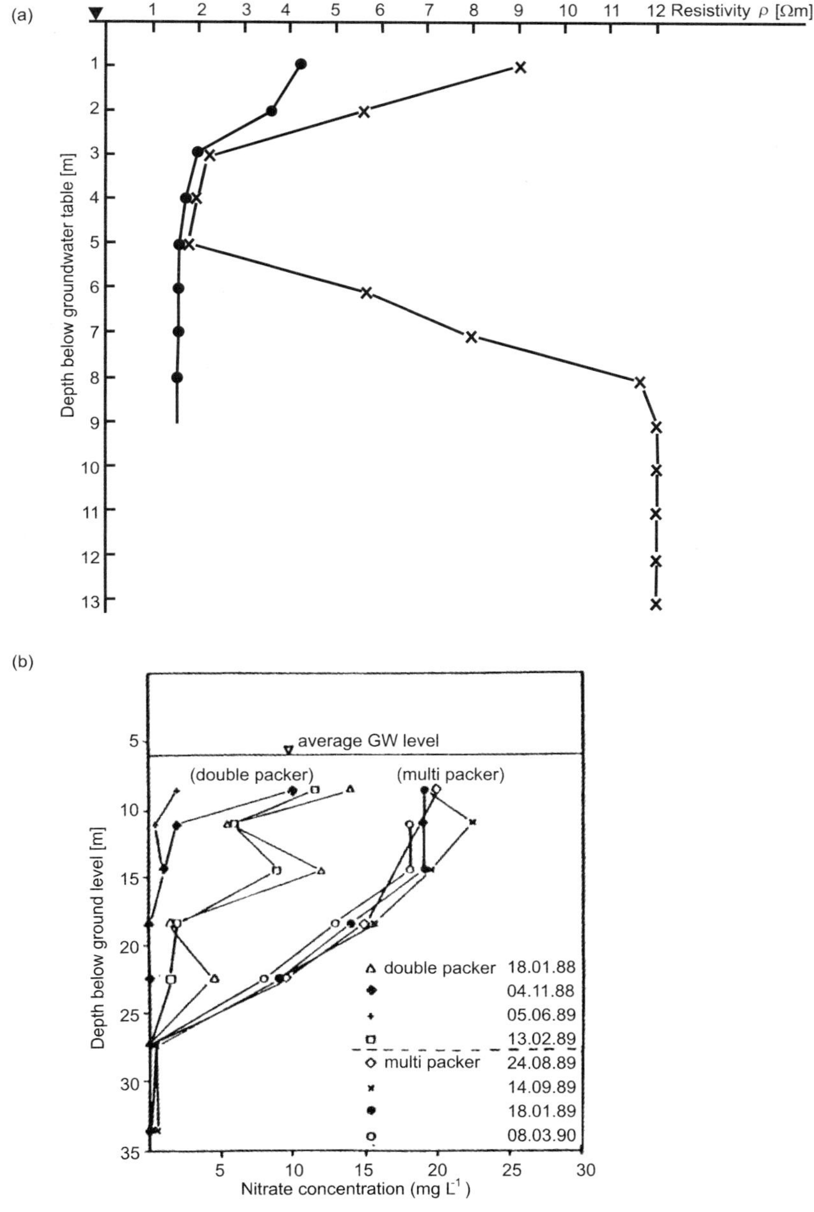

Fig.: 5.2-37: Influence of packer types on sampling results in a fully penetrating well, (a) Influence of a fully penetrating well on the electrical conductivity (resistivity) distribution during sampling, × resistivity from CPT sampling, • resistivity from sampling with double packers. (b) Influence of packer types on sampling results in a fully penetrating well, BARCZEWSKI et al. (1993)

Groups/clusters of monitoring wells are built with a single casing in each borehole and the main aim is to achieve no hydraulic connection between individual aquifer layers. This means that the groundwater tables of every single aquifer layer is independent of the other aquifers. However, due consideration must always be given to disturbance in the underground that may be caused during drilling closely spaced boreholes. The search for improvement in establishing groups of measuring points led to development of a special type of monitoring wells as explained below.

In *special wells with equipment for sampling at preselected depths*, mini-gauging screens are arranged at different depths and are hydraulically separated by impervious seals. In comparison with the well types already discussed, a minimization of changes in original conditions in the screened depth range is obtained by their direct arrangement in the transient area between the gravel pack and the original subsurface material.

The periodic exposure of the uppermost screen to the air is a common problem in monitoring wells. It is a function of the dimensions of the screen and the position of the groundwater table. Such exposure may lead to changes in the condition of groundwater in the exposed screened depth due to long-term biochemical reactions (e.g., clogging with iron precipitate). In special wells with equipment for sampling at specified depths, this effect is counterbalanced owing to the small dimensions of the screen. Screen and sample systems in special wells have non-return valves that additionally help prevent a change in original conditions in the sampling area. This is not the case in the monitoring wells already mentioned. Typical examples of special wells are

- multi-level screen (e.g., from the company Pumpen-Boese),
- SGM percolation-water and groundwater measurement systems (LUCKNER et al., 1992),
- BAT system (TORSTENSSON, 1984), and
- multi-screen monitoring wells and measuring point groups having stationary packer pump systems.

Suitability tests of existing groundwater monitoring wells

A preliminary survey to determine whether there are any monitoring wells that may be suitably integrated in the planned monitoring network is always the first step during planning a monitoring program. Any such existing well must satisfy the following minimum requirements in order to be considered for integration in the monitoring network:

- must be supported by documentation (e.g., information about coordinates, documentation of the installation design, lithological profile, hydrogeological monitoring data),

- must not be damaged or filled, and
- must have been in use during the entire last 10 years.

To determine whether an existing well can be used as a groundwater monitoring well for quality observation, it is necessary to examine the well on the basis of the criteria given in *Table 5.2-11* and check that it satisfies all the criteria.

Before selecting any existing groundwater monitoring well, it is of fundamental importance to check whether the screened section of this well is properly placed with respect to the area of observation. The best option is to construct a geohydraulic model coupled with a particle tracking model in order to understand the relationship between the sampled depth in the well and the source of contamination. As a first approximation, the groundwater equipotential lines (isohypses) constructed for the aquifer should be used. This approach only takes advective transport into consideration, i.e. hydrodynamic dispersion and the interaction between the contaminants and aquifer material are neglected.

To ensure the proper functioning of a monitoring well (or to ensure that the depth of sampling in the well represents the actual conditions in the aquifer), the well casing must be checked for leakage over the entire length. Borehole television or geophysical borehole methods (Chapter 4.8) can be used to check for leakage of the casing, at the joints of the pipes or in the screen. When that has been done, pumping or infiltration tests can be conducted. *Figure 5.2-38* shows the results of standard borehole logging to determine the location and quality of the grouting in a monitoring well. This includes the following logs:

- caliper,
- natural gamma,
- neutron log,
- density log,
- single point resistivity.

Table 5.2-11: Checklist for determining the suitability of an existing monitoring well

		Yes	No
I	**Preselection by evaluation of the available inventory documents**	**Yes**	**No**
1.	Are the well and the screen(s) placed at a depth to be monitored, and is the monitoring well type suited for the investigation objective?	⇓	⇒
2.	Is the casing diameter larger than 2.5 inches and if the screen is at a depth > 50 m, is the diameter larger than 4 inches?	⇓	⇒
3.	Does the location of the clay seal guarantee a good representative groundwater sample without influence from other aquifers penetrated by the well?	⇓	⇒
4.	Do the property rights allow safe access to the well; or, alternatively, is it possible to establish such conditions with reasonable effort?	⇓	⇒
II	**On-site visual evaluation**		
5.	Is the groundwater monitoring well easily accessible, or can present obstacles easily be removed?	⇓	⇒
6.	Does the casing material consist of stainless steel, HDPE or hard PVC, or another comparable plastic?	⇓	⇒
7.	Is the groundwater monitoring well externally intact and does the visual appearance match the inventory data? (Close fit of protective casing to the ground, subsidence, cracks, gaps, etc., deformation of casing, deviation of casing from the perpendicular, leaky or missing cover cap, waste deposition in the vicinity of the well.)	⇓	⇒
8.	Does the sounded depth match the stated depth within 0.5 m and is it possible to insert and remove a dummy sampler (e.g., a dummy pump) without problems?	⇓	⇒
III	**On-site technical evaluation**		
9.	Is the screen, after pumping it clean, sufficiently hydraulically connected to the aquifer? (necessary for pump or infiltration testing)	⇓	⇒
10.	Can the tightness of the casing be confirmed by performing a water pressure test in the casing after installing a packer above the screen?	⇓	⇒
11.	Can the position and effectiveness of the clay seals be confirmed (e.g., by geophysical GR, GG, or NN logging), and do they correspond to the stated well construction (see item 3 above)? (If necessary, inspection with a camera to check condition and location of screen section.)	⇓	⇒
12.	Was the initial sampling successful? (Checking of the analysis results for representativity of the monitoring well, deviations require geochemical/biological explanation.)	⇓	⇒
	The monitoring well is **suitable** for groundwater quality monitoring; provide well ID sheet with basic data and inspection record.		
	The monitoring well is **not suitable** for groundwater quality monitoring.		

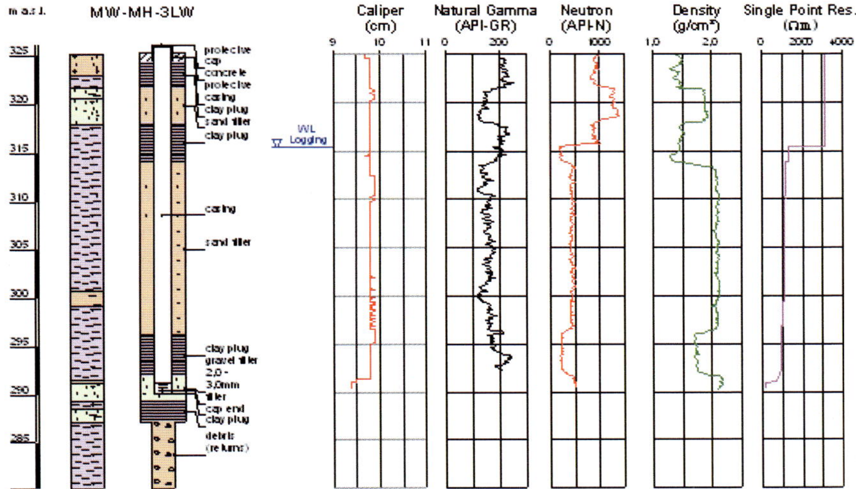

Fig. 5.2-38: Standard borehole logging to determine the condition of a monitoring well

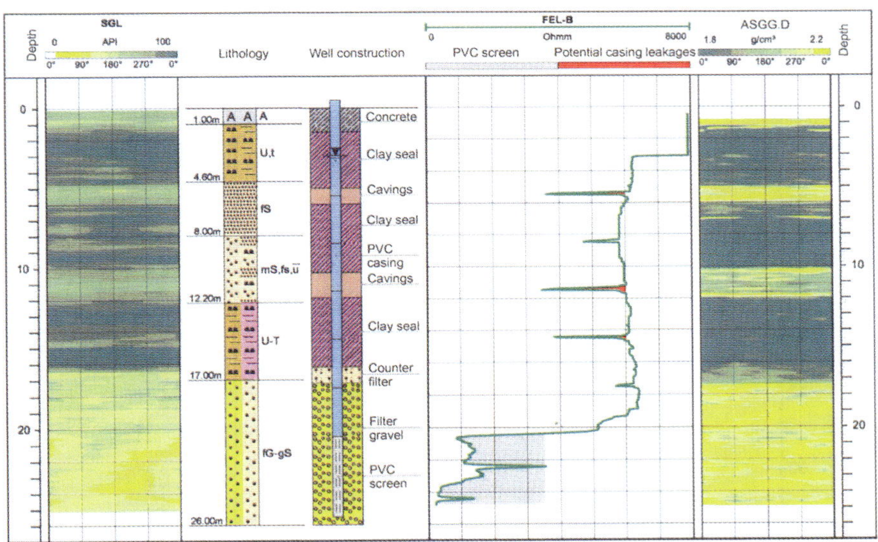

A	made ground
U, t	clayey silt
fS	fine sand
mS, fs, ū	medium sand, fine-sandy, very silty
U-T	silt - clay
fG-gS	fine gravel - coarse sand

Fig. 5.2-39: Results of FEL, SGL and ASGG.D logs in a monitoring well, for further information see text, BAUMANN & PETZOLD (2005). Logs by Bohrlochmessung - Storkow GmbH.

Single point resistivity in combination with a temperature log is extensively used to detect entry of water from unscreened aquifers into the well. These influences can be also detected by focused electric log (FEL) as shown in *Fig. 5.2.-39*. The combination of FEL log with segmented gamma-ray log (SGL) and annular space gamma-gamma density scanner (ASGG.D) give a better overview about the state of the well. These scanning logs – unlike the standard methods – help in indicating the quality of grouting seal in the whole annular space around the casing.

In addition to the above-mentioned methods, flowmeter measurements can be used to evaluate the hydraulic connection of the well screen to the aquifer and the extent of clogging. *Figure 5.2-40* shows an example where only a small part of the screen is connected with the aquifer, because most of it is clogged.

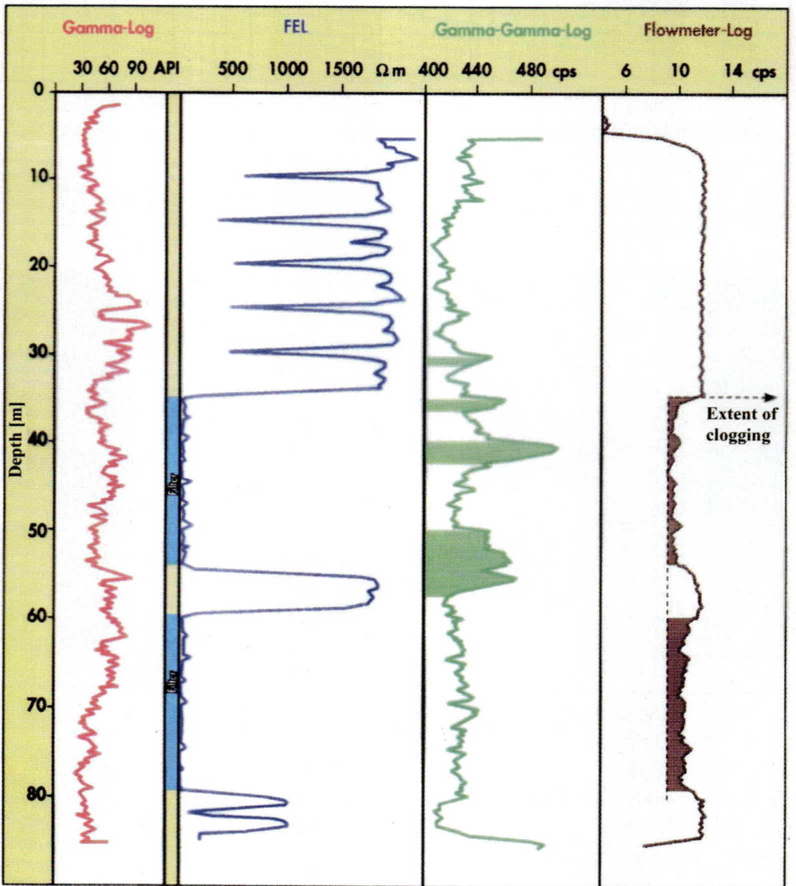

Fig. 5.2-40: Check of the screen function by borehole geophysical methods, BAUMANN & PETZOLD, (2005). Logs by Bohrlochmessung - Storkow GmbH.

The results of borehole geophysical methods are an important part of the characterization of the suitability of an existing well for monitoring. The multiple screen well, shown in *Fig. 5.2-41* was characterized as not suitable because the grouting seals between the screened sections failed to prevent a hydraulic connection between the screens.

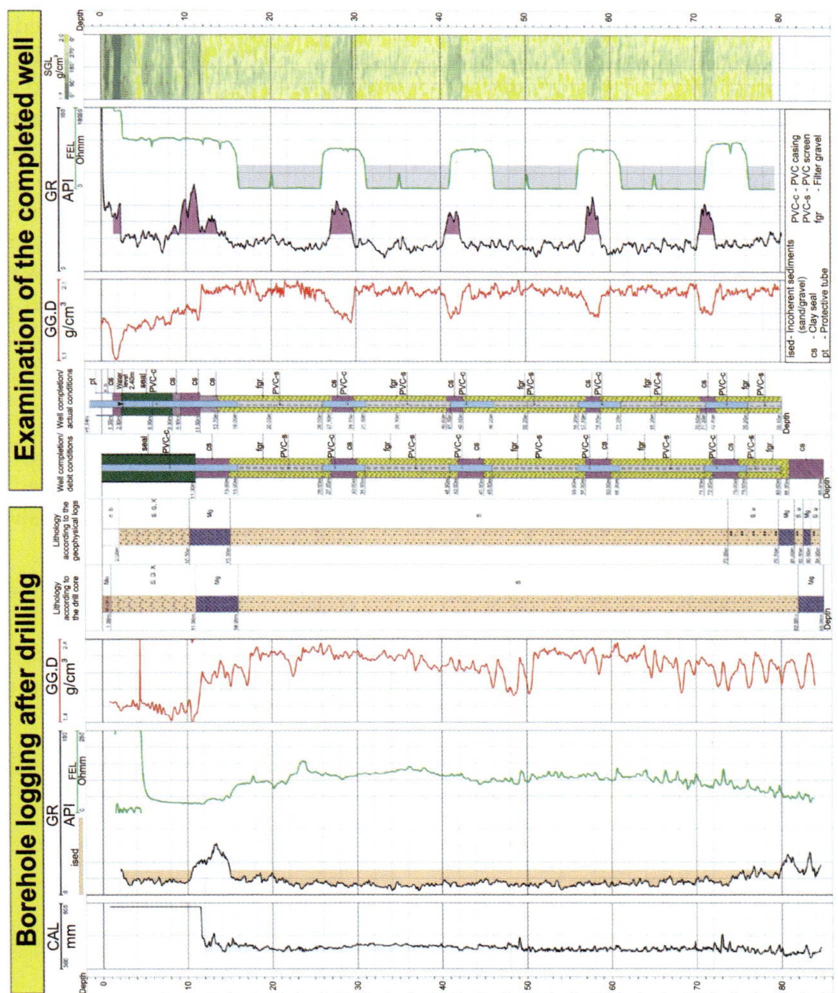

Fig. 5.2-41: Example of checking a multiple-screen monitoring well by geophysical logging, BAUMANN & PETZOLD (2005). Logs by Bohrlochmessung - Storkow GmbH.

Simplified pumping or infiltration tests are used to determine the connection between the screen and the surrounding aquifer in order to establish the effectiveness of the monitoring well. The water used in these tests should be taken from the aquifer being monitored. The fundamentals of such types of infiltration tests and pumping tests are described in Section 5.2.7.2.

These tests are not a replacement for more complete geohydraulic pumping tests, but only a rough method for quick estimate of parameter values based on simplified equations for idealized flow conditions. They yield an estimate of permeability (k) at a single point, since only a very limited area around the borehole is hydraulically stressed. These methods work particularly satisfactorily in quick investigations for an initial estimate of parameter values of small areas.

A first sampling operation can take place if all tests of the inspection scheme (points 1 to 11 in *Table 5.2-11*) are positively evaluated and, therefore, representative groundwater samples can be taken from the groundwater monitoring well. If the monitoring well cannot be included in the concerned monitoring net, then it is suggested to try to reconstruct it.

Identification sheet of monitoring wells

If the well is determined to be suitable for the monitoring grid, then an identification sheet (ID sheet) for the monitoring well should be prepared in cooperation with the governmental agency responsible for such. The sheet must contain the following information:

- lithological profile of the borehole,
- diagram of the well construction,
- results of the well tests, e.g., depth of the pump, pumping duration, pumping rate, drawdown,
- operating characteristics, e.g., pH, electrical conductivity, redox potential, oxygen content and temperature,
- stamp and signature of the assigned engineer's office and/or laboratory.

An example of such an ID sheet is given in *Figs. 5.2-42* and *5.2-43*.

Monitoring Well ID Sheet

MW-MH-2A

(Monitoring Well Mae Hia 2A)

General Information

Easting	493751
Northing	2073322
Elevation Welltop	329.49 m above MSL
Drilling Company	Aztec Engineering Co., Ltd.
Completed on	26.01.2002
Drilling Method	Direct Rotary
Borehole Diameter	279 mm

Initial Sampling

Depth of Pump	28.0 m
Duration	75 min
Flow Rate	5.8 L min^{-1}
Pumped Volume	440 L
Drawdown	3.77 m

Characteristic Parameters

Temperature	29.3 °C
El. Conductivity	548 μS cm^{-1}
pH	7.28
Eh	-111 mV
Oxygen Content	0.1 mg L^{-1}

Signature of Client

Maintenance Operations			
Date			
Regeneration Method			
K According to Well Test (in m d^{-1})			
Repairs			
Executing Company			
Remarks			
Signature of Client			

Fig. 5.2-42: Example of monitoring well ID sheet: front page

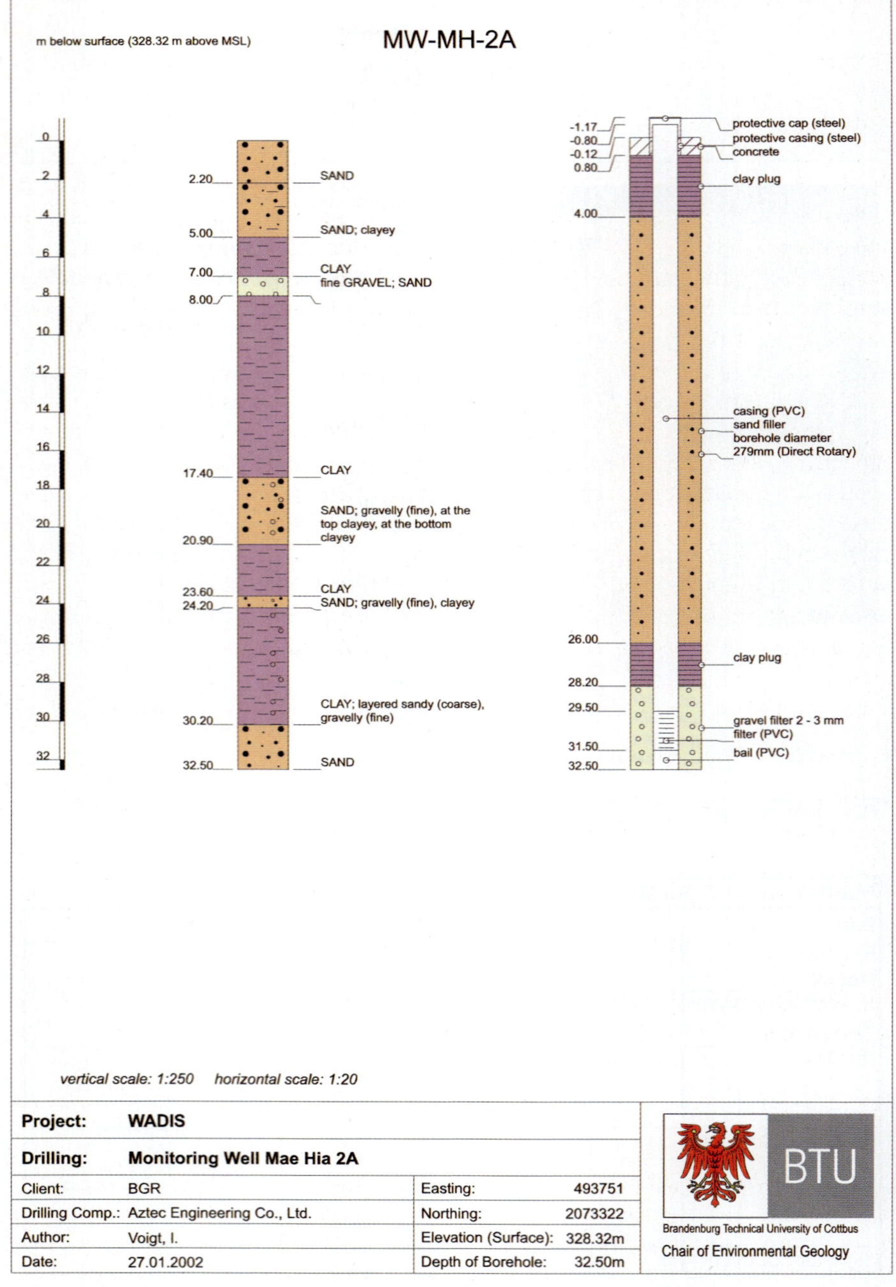

Fig. 5.2-43: Example of monitoring well ID sheet: back page

Equipment for groundwater monitoring

Groundwater monitoring instruments are designed to obtain manual or automatic water-level measurements, water samples and recordings of various water parameter values.

Water-level measurements are the principal source of information about the hydrologic stresses acting on an aquifer as well as about groundwater recharge and discharge. Water-level meters are used for manual measurements and water-level monitors are used for automatic continuous measurements. Water-level meters are devices that measure the depth to the water in a well or borehole. It is frequently made of a non-stretch measuring tape with a probe on the end of the tape, which completes an electric circuit when it contacts water. Tapes are marked in mm or 0.01 ft. This kind of water level-meter is also called a sounding light.

Hydrostatic water-depth probes, submersible pressure transducers and capacitive water depth probes are used for long-term measurements of water depths in standard ranges of 1, 2, 5, 10 and 20 meters. The resolution of water-level measurements depends on the measurement range. The water-depth probes are sometimes combined with temperature probes for the 0 to 60 °C range. Ultrasonic sensors can be used for water-level determinations where instruments cannot be submerged in the liquid, e.g., frozen water or corrosive liquids. The probes can be operated in combination with data loggers and telemetry.

Two types of water-level monitors are used:

- absolute water-level monitors, and
- differential water-level monitors with automatic barometric pressure compensation (also called vented water-level loggers).

Storms and elevation changes cause changes in barometric pressure. Since 1 hPa or mbar equals 1 cm of water, considerable changes in water level can be expected due to changes in barometric pressure[4].

Absolute water-level monitors measure changes in both water level and barometric pressure. To correctly interpret the data from this type of monitor, an external barometric pressure monitor is required. The true water-level reading is calculated from water-level monitor and external barometric-pressure monitor data. The overall accuracy of measurements made using an

[4] The mean value of the barometric pressure at sea level is 1013.25 hPa. The barometric pressure decreases rapidly with altitude according to an exponential function. The decrease near the Earth's surface is about 1 hPa per 8 m. A diurnal and annual variation as well as a variation due to storm events also occurs. Global extreme values of barometric pressure at sea level are 869.9 hPa and 1085.7 hPa. The largest drop (98 hPa) of barometric pressure within 24 hours was observed within the hurricane Wilma in October 2005 (http://de.wikipedia.org/wiki/Luftdruck).

absolute water-level device is lower than those made using a differential water-level monitor, because the error of both the water-level monitor and barometric pressure sensor must be considered.

Differential water-level monitors have vented cable to automatically compensate for barometric pressure changes. Therefore, direct readings of water-level changes are available in the field and no post-processing is necessary.

Interfacemeters measure the depth and thickness of floating or sinking hydrocarbon products (both LNAPLs and DNAPLs) in groundwater with an accuracy of 1 mm or 0.01 ft. A conductivity sensor is used to sense the water-level, an optical infrared refraction sensor is used to detect the hydrocarbon layer.

Submersible multi-parameter water-quality sensors are equipped with sensors for elec. conductivity, pH, redox potential, temperature and dissolved oxygen (Section 4.9). *Turbidity probes* are also used for water quality monitoring. The described monitoring well instruments normally fit inside 2-inch wells. They can also be used in streams, springs, ponds and lakes.

Meteorological measurements (e.g., precipitation depth, temperature, air pressure) are necessary for the evaluation and assessment of groundwater monitoring data. These measurements can be also carried out automatically and if necessary telemetrically.

Periodic *groundwater sampling and analysis* is conducted to monitor changes in the concentrations of chemical substances in groundwater. Different sampling intervals are used: Some wells are sampled quarterly and others are sampled annually, depending on the substances that are being monitored. Samples must be representative of in-situ conditions of the aquifer or surface water being monitored. Groundwater sampling and analysis are described in detail in Sections 5.3.1 and 5.3.4.

Determination of the direction of groundwater flow

The groundwater surface (piezometric surface) and the direction of groundwater flow can be determined from the mapping of water-level elevations in wells drilled into the specific aquifer. At least three groundwater observation wells (called a hydrogeological triangle) are required for this purpose, as shown in *Fig. 5.2-44*. The measured groundwater levels related to a common datum (mean sea level or a local datum) are plotted with their well position on a diagram made to scale. Then, lines of equal water level are drawn at selected levels between the three or more observation wells. If measurements in more than three groundwater observation wells are available, contour maps can be calculated as described in Section 6.1.1. Due to the anisotropy of aquifers, the contour lines are normally not straight lines as shown in *Fig. 5.2-44*. But in any case the groundwater flow direction is perpendicular to the contour lines.

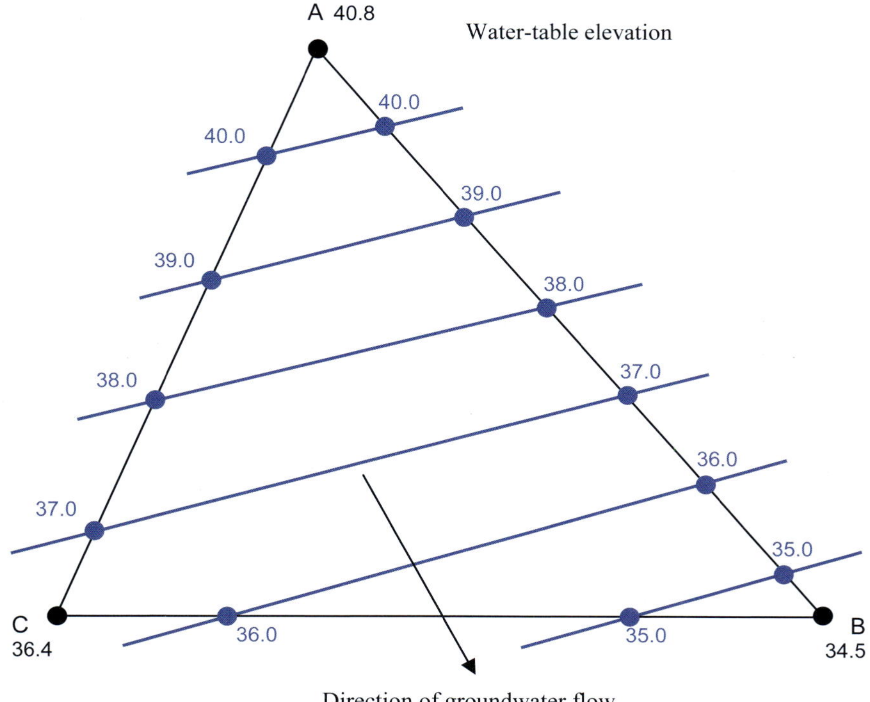

Fig. 5.2-44: Determination of groundwater flow direction using a hydrogeological triangle

5.2.7 Determination of Hydraulic Parameters

ULRICH BEIMS, RANJEET NAGARE, CLAUS NITSCHE, MICHAEL PORZIG & HANS-JÜRGEN VOIGT

In terms of hydrogeology, the subsurface consists of

- aquifers,
- aquitards, and
- aquicludes.

Aquifers are permeable geological strata that can both store and transmit water in significant quantities. *Aquitards* are geological units that have a low hydraulic conductivity relative to the units above and below it. *Aquicludes* are geological units that are incapable of transmitting significant quantities of groundwater under ordinary hydraulic gradients; such geologic units have a hydraulic conductivity < 10^{-9} m s^{-1}.

The following zones in an aquifer are distinguished on the basis of the degree of saturation with water:

- unsaturated or vadose zone,
- capillary fringe, and
- saturated zone.

The *unsaturated zone* or vadose zone is the zone between the land surface and the top of the saturated zone, i.e. the water table. In this zone some of the pore spaces are filled with air and some are filled with water. The *capillary fringe* is that part of the unsaturated zone immediately above the water table which contains water due to capillary rise. It is sometimes used synonymously with capillary space or zone of capillarity. How high the water will rise above the water table as a result of capillarity depends mainly on the size of the interstices (which is determined by the grain size) of the soil or rock. The *saturated zone* is the zone below the water table where every available space (pores and fractures) is filled with water and the fluid pressure is greater than atmospheric pressure.

Aquifers are classified as confined or unconfined. A *confined aquifer* is bounded above and below by an aquiclude or aquitard (confining bed) and whose piezometric surface is above the base of the confining layer. The water in a well tapping a confined aquifer will rise above the base of the confining layer. *Unconfined aquifers* (phreatic aquifers) are aquifers in which the surface of the water is at atmospheric pressure. The water in a well tapping an unconfined aquifer will rise only to the level of the groundwater table.

The type of the hydrogeological unit affects what hydraulic parameters the occurrence, the amount, and flow of water control. The following parameters are preferred to characterize aquifers, aquitards, and aquicludes:

Hydraulic conductivity k_f (sometimes denoted as K) is defined as the volume of groundwater that will move within a unit of time under a unit hydraulic gradient through a unit cross-sectional area that is perpendicular to the direction of flow. Hydraulic conductivity is preferred to the term "permeability". Hydraulic conductivity is the permeability to water only, whereas (intrinsic) permeability is a general property of the matrix only, applicable to different fluids. For a particular soil or rock layer, the hydraulic conductivity may not be the same in the horizontal direction as in the vertical direction. Hydraulic conductivity is defined by Darcy's law.

The hydraulic conductivity depends on the intrinsic permeability of the material and on the degree of water saturation. *Saturated hydraulic conductivity*, k_f, describes water movement through a saturated medium. The values of saturated hydraulic conductivity in soil and rock vary within several orders of magnitude, depending on the material (*Table 5.2-12*).

Table 5.2-12: Saturated hydraulic conductivity (k_f) for typical fresh groundwater conditions, modified from BEAR (1972)

<table>
<tr><td>k_f (cm s^{-1})</td><td>10^{2}</td><td>10^{1}</td><td>$10^{0}=1$</td><td>10^{-1}</td><td>10^{-2}</td><td>10^{-3}</td><td>10^{-4}</td><td>10^{-5}</td><td>10^{-6}</td><td>10^{-7}</td><td>10^{-8}</td><td>10^{-9}</td><td>10^{-10}</td></tr>
<tr><td>k_f (ft d^{-1})</td><td>10^{5}</td><td>10,000</td><td>1,000</td><td>100</td><td>10</td><td>1</td><td>0.1</td><td>0.01</td><td>0.001</td><td>0.0001</td><td>10^{-5}</td><td>10^{-6}</td><td>10^{-7}</td></tr>
<tr><td>relative permeability</td><td colspan="4">pervious</td><td colspan="4">semi-pervious</td><td colspan="5">impervious</td></tr>
<tr><td>aquifer</td><td colspan="5">good</td><td colspan="4">poor</td><td colspan="4">none</td></tr>
<tr><td>unconsolidated sand & gravel</td><td colspan="2">well sorted gravel</td><td colspan="3">well sorted sand or sand & gravel</td><td colspan="4">very fine sand, silt, loess, loam</td><td colspan="4"></td></tr>
<tr><td>unconsolidated clay & organic</td><td colspan="4"></td><td colspan="2">peat</td><td colspan="3">layered clay</td><td colspan="4">fat / unweathered clay</td></tr>
<tr><td>consolidated rocks</td><td colspan="4">highly fractured rocks</td><td colspan="3">oil reservoir rocks</td><td colspan="2">fresh sandstone</td><td colspan="2">fresh limestone, dolomite</td><td colspan="2">fresh granite</td></tr>
</table>

The *unsaturated hydraulic conductivity* is a function of the water content of the material. As a material dries out, the connected wet pathways through the media become smaller, the hydraulic conductivity decreasing with lower water content in a very non-linear way. The unsaturated hydraulic conductivity may be a fraction of saturated hydraulic conductivity. This is one of the main complications which arises in studying the vadose zone.

Darcy's law states that the discharge of groundwater Q through a porous medium is proportional to the product of saturated hydraulic conductivity k_f, the cross-sectional area of flow A and the hydraulic gradient J normal to that area:

$$Q = k_f A J \tag{5.2.29}$$

It is assumed that the flow is laminar and steady-state. The hydraulic conductivity k_f depends on the internal structure of the medium the water flows through and the viscosity of the water. If the density or viscosity are differing from normal groundwater conditions (e.g. by high temperatures, high salt concentrations, or if considering a different fluid, like oil) it may be necessary to adapt the k_f value, based on intrinsic permeability of the matrix and the fluid properties.

Porosity is the percentage of open pore space in soil, rock or other materials. It is the ratio of volume of open pores and fractures to the total volume of the material. The porosity is determined from cores or from borehole logs.

Effective porosity is the portion of pore space in a saturated permeable material where movement of water takes place. It is almost eqal to total porosity, and greater than drainable porosity/specific yield (SARA, 1993).

Transmissivity, is a measure of the capability of the entire thickness of an aquifer to transmit water. Transmissivity T is related to the average hydraulic conductivity k_f of the saturated part of aquifer by the relationship: $T = k_f\, b$, where b is the thickness of the saturated part of aquifer. Transmissivity is expressed in units of $m^2\ s^{-1}$. When the saturated thickness of an unconfined aquifer changes, the transmissivity also changes.

Storativity S is the volume of water that an aquifer releases from or takes into storage per unit surface area of the aquifer per unit change in head. It is also called storage coefficient and is equal to the product of specific storage and aquifer thickness. Storativity is vastly different in unconfined and confined aquifers. In confined aquifers, the change in volume is caused only by elastic compression/expansion of water and rock structure. The corresponding storativity values are on the order of 10^{-6} to 10^{-4}. In the unconfined case, the filling/draining of pore volume (i.e. specific yield) dominates the value of storativity, which is then on the order of 10^{-2} to 10^{-1}. In an unconfined aquifer, the storativity therefore is equal to the specific yield.

Specific storage is the volume of water released from storage per unit volume of a saturated porous medium per unit decrease in the hydraulic head. It has units of m^{-1}.

Specific yield is defined as the ratio of the volume of water that will drain from a porous medium (drainable porosity) by gravity to the volume of the porous medium (see also effective porosity).

Field capacity is the amount of water soil can hold against the force of gravity after saturation and runoff (usually 1 to 3 days after rainfall). This water is used by plants during the growing season. In soil science, the term field capacity is used, whereas in hydrogeology the same property is called specific retention (compare specific yield).

Infiltration rate is the rate that water enters a soil surface within a unit time interval. The unit is depth of water per unit time, for example mm h^{-1} or cm s^{-1}.

Volumetric water content is the volume of water per volume of soil (i.e. solid material and pores).

Figure 5.2-45 gives an overview of hydraulic test methods used in site characterization. Besides these field methods for the determination of hydraulic parameters, laboratory methods can also be used (Section 5.2.7.3).

Table 5.2-13 shows which methods can be used for the determination of different hydraulic parameters.

Figure 5.2-46 gives an overview of the characteristics of hydraulic test methods and the conditions under which they can be used.

Table 5.2-13: Methods used for the determination of hydraulic parameters

Parameter	Method						
	double-ring infiltrometer test	Guelph permeameter test	borehole permeameter test	slug/bail test	pulse test	pumping test	laboratory tests
hydraulic conductivity	X	X	X	X	X		X
transmissivity					X	X	
storativity				X		X	

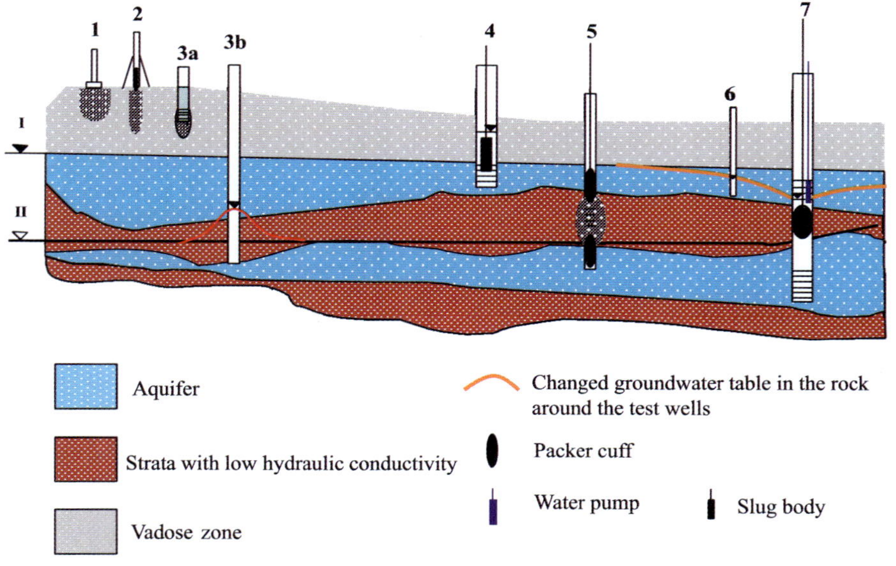

I Groundwater table in the upper unconfined aquifer

II Potentiometric surface of confined or partially confined aquifer

1 Double-ring infiltrometer

2 Guelph permeameter

3a Screened borehole permeameter (vadose zone)

3b Open bottom-end borehole permeameter (saturated zone)

4 Slug/bail test

5 Pulse test with double packer in open borehole

6 Piezometer (for pumping test)

7 Pumping test in upper screen of the pumped well, lower screened section is sealed by a single packer

Fig. 5.2-45: Overview of hydraulic test methods used for site characterization

Test method		Double ring infiltrometer	Guelph permeameter	Borehole permeameter test	Slug/bail test	Pulse test	Pumping test
test duration[1)]	with packer				1 - 2 h	1 min	10 h[3)]
	without packer	1 - 5 h	0.5 - 2 h	> 8 h	5 - 10 h		100 h[3)]
lateral extension[1)]	lateral confined				10 m	0.3 m	30 m
	lateral unconfined[2)]				0.5 - 2 m		1 - 5 m
diameter of test borehole			2 to 3 inches	in tests without packers the diameters of depthsounder and pressure transducer are the limiting factors			
depth of application		down to 1 m	down to 3.15 m	no restrictions by test method, but by geological and financial conditions as well as drilling method			
method applicable in - drilling casement - open borehole - tempoary casing		 No No No	 No Yes No	 Yes Yes Yes	 Yes Yes Yes	 Yes Yes No	 No Yes Yes

1) for $k_f = 10^{-7}$ m s^{-1}, rock mass of 5 m, test well diameter of 125 mm, compressibility of 2 x 10^{-8} m^2 N^{-1}

2) for unconsolidated rock the lower value and for tight formations the higher value should be assumed

3) time includes recovery phase

Fig. 5.2-46: Characteristics of hydraulic test methods and their applicability, modified from GARTUNG & NEFF (1999)

5.2.7.1 Infiltrometer and Permeameter Tests

FLORIAN JENN, RANJEET NAGARE, MICHAEL PORZIG & HANS-JUERGEN VOIGT

All hydraulic tests work on the principle of reaction to a change in pressure head. Infiltrometers and permeameters are instruments for determining hydraulic parameters of soil, sediment and rock by measuring the discharge through the material when a known hydraulic head is applied. Infiltrometer and permeameter tests can be subdivided into three groups on the basis of the method of causing a change in the head (*Table 5.2-14*).

Table 5.2-14: Classification of hydraulic tests

Method for initial changing the head in the well	Test methods
addition of water at the ground surface	• double ring infiltrometer (DRI)
gradual continual addition or removal of water in or from borehole	• Guelph permeameter (GP) • constant head borehole permeameter test (BP test) • variable head borehole permeameter test (BP test) • cone penetration technology (CPT)[1)]
sudden addition or removal of water in or from borehole	• slug/bail test (SB test) • pulse test

[1)] for details see Section 5.1.4

Test method selection

The flow chart in *Fig. 5.2-47* can be used to select the appropriate hydraulic test methods on the basis of the hydrogeology of the site. The objectives, the subsurface conditions, and the monetary budget govern the selection of the test method used at a site to be investigated. It is important for the person in charge to know the exact reason for the investigations and whether the tests being used to determine the hydraulic conductivity are suitable. Field tests can be used in combination with laboratory methods.

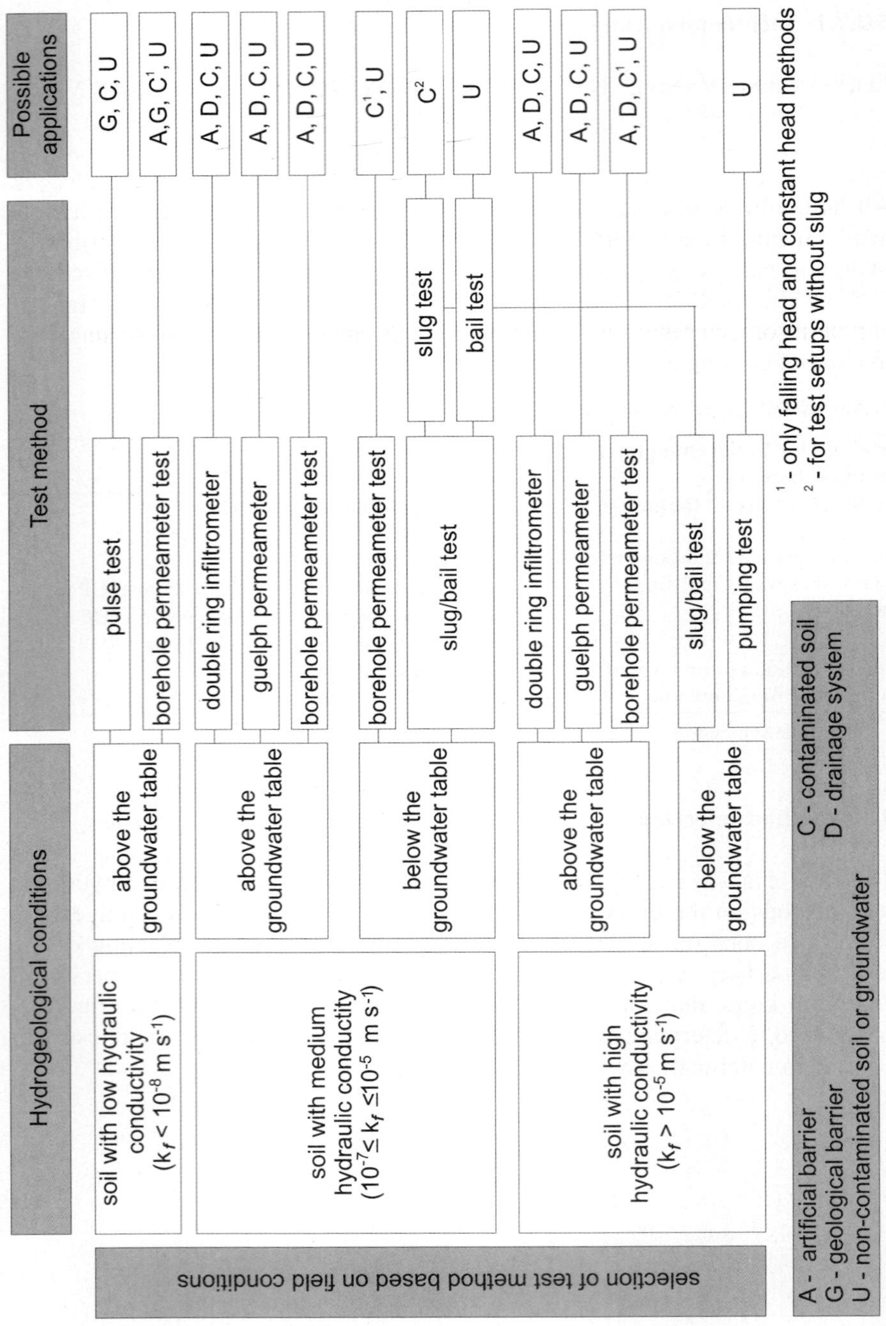

Fig. 5.2-47: Flow chart to aid selection of a hydraulic test method on the basis of the hydrogeology of the site

Test preparation

Drilling a borehole is the first step for all of the listed test methods, except for the double ring infiltrometer test. The borehole tests can be conducted either in cased or uncased boreholes. The boreholes have to be close to cylindrical shape in order to obtain reliable results. Hence, it is necessary to select a suitable drilling method (Section 5.1.3). It should be noted that only dry drilling methods may be used. Common problems that arise are

- deviation of the borehole from cylindrical shape,
- nonuniform borehole walls (as a result of caving) in the case of uncased boreholes,
- alteration of the borehole wall due to the construction and operation of the well, which may result in decreased hydraulic conductivity (skin effect). Hydraulic tests conducted in a borehole with skin effect will not reflect the actual hydraulic parameters of the surrounding material (YANG & GATES, 1997).

The stability of the borehole wall in uncased boreholes has to be ensured. A temporary casing is highly recommended in unconsolidated rock. In the case of consolidated rock it is recommended that uncased boreholes are used for the hydraulic tests.

Test setup and execution

Each hydraulic test method is suitable only for certain geological and/or hydrogeological conditions. If the geological and/or hydrogeological conditions are complicated, a combination of test methods may be necessary. It may happen that a test method is found unsuitable during execution. In such a case, it can be necessary to modify or substitute the selected test method by a more appropriate one. The coordinates of each test point must be determined and recorded when the test is set up.

Most of the standard test setups include pressure transducers. These instruments can be replaced by a transparent tube with a scale and sealed at the top. The head can be increased to a certain level and the decline directly observed. This kind of setup is particularly suitable for tests at shallow depths (2 - 5 m). In order to compare the results with other tests, groundwater temperature and barometric pressure should be recorded. Modern pressure transducers are able to monitor these parameters.

Water added in tests must be from the same groundwater at the location of the well, if possible, in order to avoid geochemical reactions in the aquifer. It is important to repeat the test at least three times in the same depth interval. This ensures a high accuracy of the final results. If the borehole was not developed properly, then the repetitions themselves help to develop the borehole and the results improve with each repetition. Moreover, if the discrepancy in the results during the repetitions is too large, then the test must be terminated to find out the reasons for the discrepancies. In addition, two or more different initial heads should be used during testing each well. A series of tests in which the initial head varies by a factor of at least two should be conducted. However, the first and last tests should have the same initial head in order to detect a possible skin effect.

General Assumptions

If not described otherwise, it should be generally understood that the tests and their evaluation methods are based on the following assumptions:

- Darcy's law is applicable,
- flow is assumed to be laminar, and
- the aquifer is not affected by external energy (e.g., man-made pressure changes).

Further assumptions, especially in case of borehole tests, are that

- the aquifer is homogeneous with isotropic properties and is quasi infinite in areal extent,
- the aquifer has a uniform thickness,
- the aquifer is horizontal, and
- the well fully penetrates the aquifer and is open through its entire extent.

Possible discrepancies in the final results could be attributed to these assumptions not being fulfilled, since it is seldom possible to assume that all of the above conditions are met in the field during actual testing.

Double-ring infiltrometer (DRI)

Infiltrometers are simple, officially approved, inexpensive instruments for determining the infiltration rate and hydraulic conductivity of soil or surface sediment layer. Single-ring and double-ring infiltrometers and disc permeameters are commonly used for this purpose. Single- and double-ring infiltrometers only measure flow under ponded (saturated) conditions. Double-ring infiltrometer measurements are particularly suitable for relatively uniform fine-grained soils of low plasticity and moderate to low resistance to ring penetration. The measuring range is $k_f = 10^{-8}$ to $5 \cdot 10^{-5}$ m s^{-1} (GARTUNG & NEFF, 1999). This test method may be carried out at the ground surface on bare soil or with vegetation in place or in a pit with a specified depth depending on the investigation objectives but not below the groundwater table or even below a perched water table.

Single-ring infiltrometer measurements are carried out by driving a single ring into the soil, leaving the top of the ring about ten centimeters above the ground. Water is poured into the ring so that either the amount of water within the ring is always held constant with a Mariotte bottle (*see Fig. 5.2-49*) (constant head) or the water is allowed to drop with time (falling head). The operator records how much water infiltrates into the soil within a given time period. The rate at which water enters the soil is proportional to the hydraulic conductivity of the soil or sediment layer.

A *double-ring infiltrometer* kit consists of two concentric rings with a lid or other cover, a wood block used as an anvil, one or two Mariotte bottles, sledge hammer, shovel and spade, stopwatch, carpenter's level, steel measuring tape, and a means to determine the location of the test point. The outer ring is, for example, 60 cm in diameter; the inner ring 30 cm. Each Mariotte bottle is calibrated and is graduated from 0 - 3000 mL in 50 mL subdivisions. The volume of water needed to maintain a specified level within specified time intervals is recorded together with the time.

When a suitable site has been selected, the coordinates of the site have been determined, and the soil surface prepared, trial pits should be dug in the test area in order to determine the soil type. The outer ring is driven into the soil using a wood block to absorb the blows from the sledgehammer. When the outer ring is in place, the inner ring can be centered inside and driven into the ground the same way. The rings should be inserted into the ground deep enough to prevent the water in the ring from flowing laterally at the surface. A depth of about 5 - 10 cm is usually adequate. Sometimes the rings are set into a bentonite slurry in order to reduce the potential for preferential flow along the ring walls (STEPHENS, 1996). The tops of the outer and the inner rings should be at the same level all the way around. The soil surrounding the rings should be free of significant disturbances. If significant disturbances are observed, the rings should be removed, relocated and reset using a technique that will avoid disturbance of the soil. The purpose of using two rings is to create one-dimensional vertical flow of water from the inner ring, as this simplifies the analysis of data. The outer ring helps force the vertical flow of water from the inner ring into the ground. If water flows in one dimension at a steady rate, the infiltration rate is approximately equal to the saturated hydraulic conductivity. Presaturation of the test site can be helpful to minimize the disturbance of the soil surrounding the rings, to reduce the time required for the test, and make it easier to drive the rings to equal depths. Both rings are then filled with water. The rings should be covered appropriately to prevent evaporation. Mariotte bottles or water-filled columns are used to maintain a constant pool of water in the rings. The outer ring can also be supplied with water manually. It is important to keep the water level equal in both rings. The Mariotte bottle (*Fig. 5.2-49*) may be replaced by a plastic column with a scale (*Fig. 5.2-48a*). *Figure 5.2-50* shows a computer-controlled test setup.

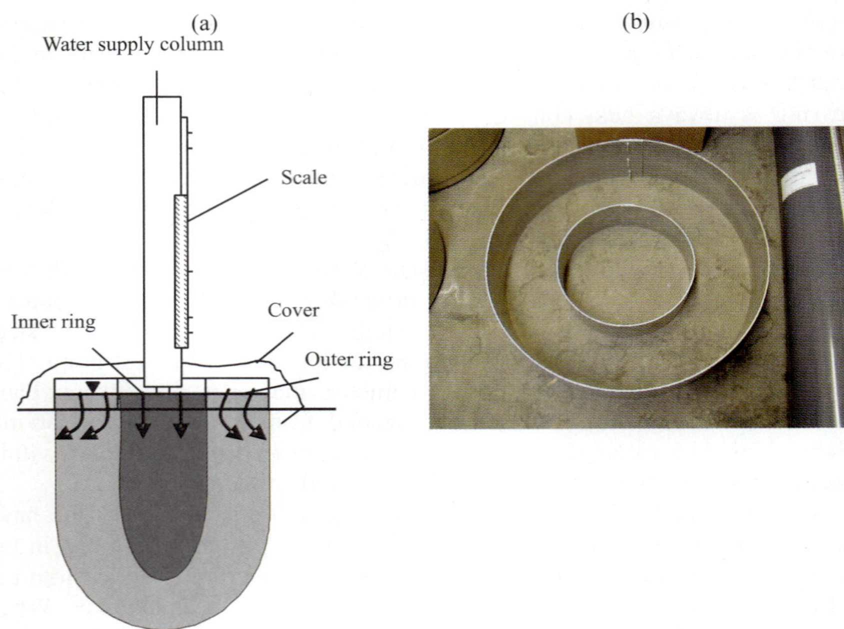

Fig. 5.2-48: Double ring infiltrometer: (a) principle of the method, (b) rings (photograph courtesy of the Technical University of Dresden)

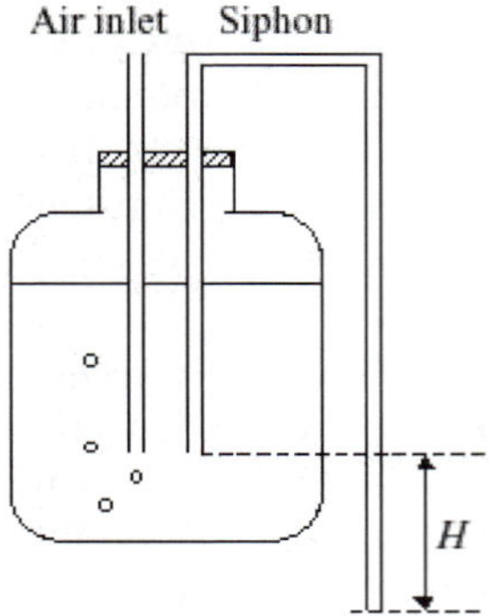

Fig. 5.2-49: Mariotte bottle supplying water at constant head, e.g., for infiltrometer. If the entrance to the siphon is at the same depth as the bottom of the air inlet, then it will always supply the water at atmospheric pressure and will deliver a flow under constant head H, regardless of the changing height of water within the reservoir.

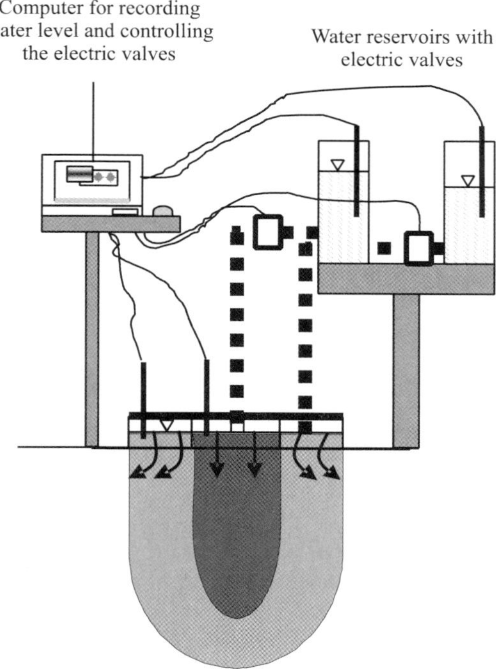

Fig. 5.2-50: Computer controlled DRI test setup

The water level in the column or the Mariotte bottle of the inner ring is recorded at regular intervals during the test. These measurements are continued until steady-state conditions are reached, i.e., when the water level decrease in the Mariotte bottle or column is constant and the infiltration rate slowly converges to an estimated well-determined linear function. The appropriate reading frequency has to be determined from experience and may be more frequent for soil or sediment with a higher hydraulic conductivity.

The steady-state flow rate through the inner ring is used to determine the saturated hydraulic conductivity of the material being tested using Darcy's law (5.2.29), which can be reformulated for vertical, saturated hydraulic conductivity k_f [m s^{-1}] as follows:

$$k_f = Q_\mathrm{i} / \pi\, r_\mathrm{i}^2, \qquad (5.2.30)$$

where Q_i [m^3 s^{-1}] is the steady-state flow rate through the inner ring, assuming that the flow within this ring is essentially one-dimensional downward at a unit hydraulic gradient, and r_i [m] is the radius of the inner ring. Sorptivity of the soil/sediment can be determined from data from the unsteady phase at the beginning of the test. Double-ring infiltrometer tests take several minutes to hours depending on soil/sediment type.

It should, however, be noted that 100 % saturation is never reached in these tests due to the various processes that occur in the unsaturated zone, different soil types, and residual soil air. Experience shows that the measured k_f value is 50 - 75 % less than the actual k_f value (REYNOLDS & ELRICK, 1986).

Some disadvantages associated with DRI method are that the rings are very bulky and heavy to move, and that the water levels in both rings have to be adjusted to be equal only by trial and error, which can be time consuming. Tests may take long time due to the fact that the soil mass enclosed within the rings has to be saturated first before the steady rate is attained. It also requires a flat, undisturbed surface, which sometimes is not available. During the experiment it is sometimes necessary to refill the Mariotte bottles. To do this, the tap has to be turned off and this disrupts the experiment.

Guelph permeameter (GP)

The Guelph permeameter is a widely used constant-head borehole tool for quick and accurate determination of in-situ hydraulic conductivity, soil sorptivity, and matrix flux potential of all types of soil (REYNOLDS & ELRICK, 1985). It is used for soil/sediment in the hydraulic conductivity range of 10^{-2} to 10^{-6} cm s^{-1}. GP measurements are carried out in an uncased 2-inch cylindrical borehole in the unsaturated soil zone from a normal operating depth of 15 - 75 cm to a maximum depth of 315 cm. A Guelph permeameter can also be used to investigate the changing water transmission characteristics of the soil with increasing depth in the same borehole. GP measurements are made in a shallow borehole. The borehole is successively deepened and GP measurements are carried out at the desired depths. *Figure 5.2-51* shows the principle of the method.

A water reservoir, essentially an in-hole Marriotte bottle constructed of concentric transparent plastic tubes, produces a steady flow which creates a saturated volume in the moist, but unsaturated soil surrounding the borehole. The standard equipment includes a tripod, permeameter, well auger, well development and cleaning tools, collapsible water container, hand vacuum pump, and accessories for extending the depth at which the permeameter can be used, ring attachments that allow ring infiltrometer measurements using 10 - 20 cm diameter rings, and a tension adapter attachment which allows a tension infiltrometer extension for measurements under tensional and very low head conditions.

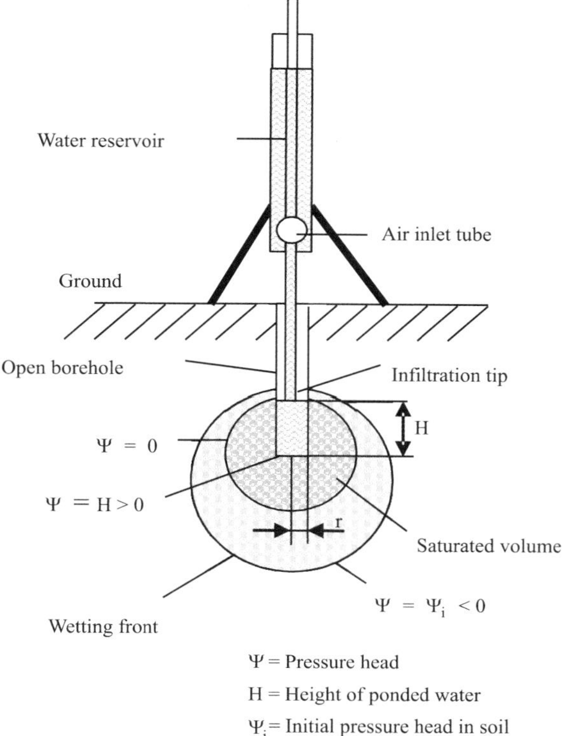

Fig. 5.2-51: Guelph permeameter, principle of the method, modified from GIAKOUMAKIS & TSAKIRIS (1999)

The Guelph permeameter provides point measurements in a heterogeneous three-dimensional soil environment. Therefore, it is the task of the people in charge to decide on the number of repetitions needed to obtain a representative estimate of the average hydraulic conductivity. This can be judged on the basis of the size of the area, the soil type, degree of heterogeneity, and the objectives of investigation. Soil profile descriptions and soil survey reports greatly enhance the understanding of the data obtained with a GP.

The field procedure can be summarized as follows:

1. Drill a borehole of required depth. Make sure that the borehole has been developed properly in order to avoid any discrepancies in the results due to the skin effect.
2. Insert the permeameter into the borehole with the water reservoir fully filled.
3. Start the measurement by raising the air inlet tube out of the outlet.

4. Set the head H to the desired value by adjusting the height of the air inlet tube.
5. Adjust the rate of decrease of the reservoir level until a steady rate r is attained.
6. The steady-state flow rate Q can be calculated by multiplying the steady-state rate r by the cross-sectional area of the reservoir A.
7. The saturated hydraulic conductivity k_f, sorptivity S_o, and soil matrix flux potential ϕ_m can then be estimated as described below.

It is important to support the weight of the water reservoir in order to avoid sinking of the permeameter tip. An accidental sinking of the outlet tip results in underestimation of the hydraulic conductivity due to lowering of the head and smaller effective infiltration area (BAGARELLO, 1997). This can be prevented by using a tripod to stabilize the GP. Furthermore, it is recommended to install a temporary borehole casing during measurements in unconsolidated materials.

The test data can be analyzed using REYNOLDS' solution of the RICHARDS equation (REYNOLDS et al., 1985) for steady-state flow Q out of a well:

$$2\pi H^2 k_f + C\pi r^2 k_f + 2\pi H\phi_m = CQ \qquad (5.2.31)$$

Rearranging this equation for k_f we obtain

$$k_f = \left(\frac{CQ - 2\pi H\phi_m}{2\pi H^2 + C\pi r^2}\right), \qquad (5.2.32)$$

where H is the steady-state head of the water in the borehole,
r well radius,
ϕ_m matrix flux potential,
C a dimensionless proportionality constant dependent on H/r; *see Fig. 5.2-52* for values.

Equation (5.2.31) is solved for k_f and ϕ_m using simultaneous equations or a least-squares approach.

The matrix flux potential ϕ_m can be calculated using an empirical approach for the hydraulic conductivity-pressure relationship using the formula (ELRICK et al., 1990):

$$\phi_m = \frac{k_f}{\alpha}, \qquad (5.2.33)$$

where k_f saturated hydraulic conductivity,
α empirical parameter; see *Table 5.2-15* for values.

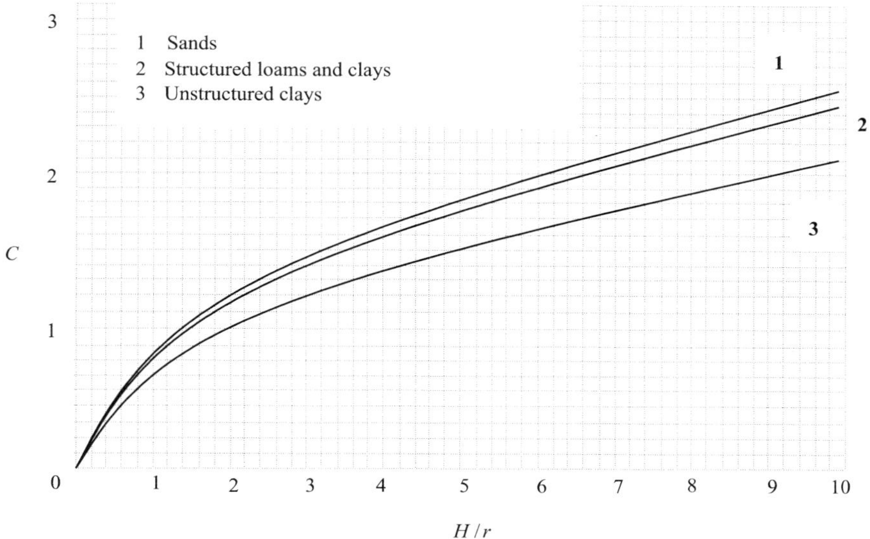

Fig. 5.2-52: Graph for determining the correction factor *C* for different soils, DURNER (2002)

Table 5.2-15: Values for the parameter α depending on soil type, ELRICK & REYNOLDS (1992)

Soil type	α [m^{-1}]
coarse sands, highly structured soils	36
most structured soils and medium sands	12
unstructured fine-textured soils	4
compacted clays	1

Advantages of the method are that the equipment can be transported, assembled and operated easily by one person. Measurements can be carried out in 0.5 to 2 hours depending on the soil type and requires only about 2.5 L of water. Disadvantages are the low maximum depth of application (315 cm). Furthermore, the initial cost and the maintenance costs could be decisive for the decision of using a GP for investigations. Though not confirmed, it is reported that increasing diameters of the test borehole generate higher k_f values. According to REYNOLDS & ELRICK (1985), this can be reflected as an auger or sample size effect (BAGARELLO, 1997).

Borehole permeameter (BP)

Borehole permeameter tests can be used to determine hydraulic conductivity k_f in all types of soil/sediment. This is the oldest and the most widely used method worldwide. BP test methods include a wide range of test designs

which differ to some extent with respect to theory, procedure, apparatus and methods of data analysis. A BP test can be conducted only at the bottom of a borehole. Hence, it is necessary either to interrupt the drilling at the required depth and conduct the BP test, or to drill a set of temporary test boreholes with different depths. The latter is costly and time-consuming, but allows the correct distribution of k_f values to be obtained over a greater lateral extent. The method is recommended for formations with k_f values between 10^{-5} and 10^{-9} m s^{-1} (GARTUNG & NEFF, 1999). *Figure 5.2-53* gives an overview of different BP test procedures.
Two basic techniques are used in borehole permeameter tests:

- In the *variable head test* the rate of rise or fall of head relative to the initial groundwater table is determined as a measure of hydraulic conductivity, and
- in the *constant head* test the rate required to maintain a constant head in the borehole is used to determine hydraulic conductivity.

The BP test can be carried out under unsaturated conditions with a constant head or with a falling head, as well as under saturated conditions with a constant head, falling head, or rising head. The constant head BP test is the most common method. Variable head BP tests are normally used in soil/sediment with a relatively low hydraulic conductivity. Tests with constant head and falling head are suitable for both saturated and unsaturated conditions. The rising head test can be conducted only under saturated conditions, i.e. below the groundwater table. This test is supposed to minimize the impact of skin effects.

A further variant of this method involves injecting water into a packered borehole test section at a predetermined externally applied additional pressure. The injection rate is adjusted to maintain a predetermined pressure head. The steady-state rate required to maintain the pressure head is then used to calculate the hydraulic conductivity. This test is conducted while maintaining a constant head; therefore, it can be classified as a constant head BP test (HEITFELD & HEITFELD, 1992; TRÖGER & POIER, 2000). Accurate results can be obtained even using low pressure heads.

In combination with percussion core drilling, BP tests can be conducted down to depths of around 15 m (SCHREINER & KREYSING, 1998). When rotary or circulation drilling methods are used, this depth is limited only by equipment and financial aspects.

Figure 5.2-54 shows the typical test setup for a BP test with and without packers. Packers are especially recommended for consolidated rocks or to separate different sections within a multi-screen well.

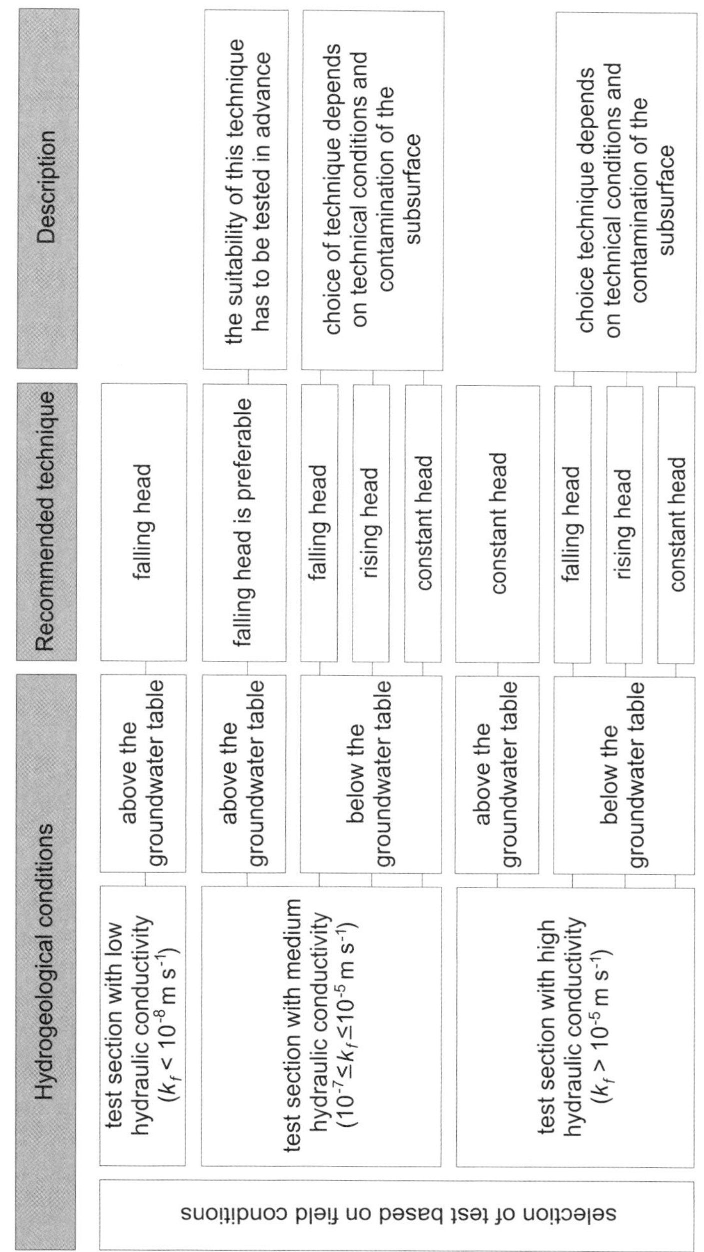

Fig. 5.2-53: Overview of suitable BP test procedures classified according to site conditions and area of application

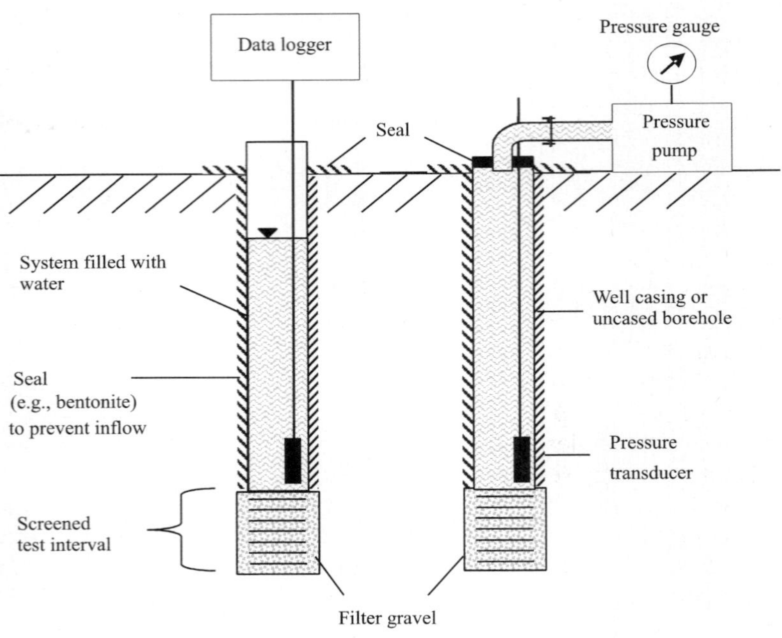

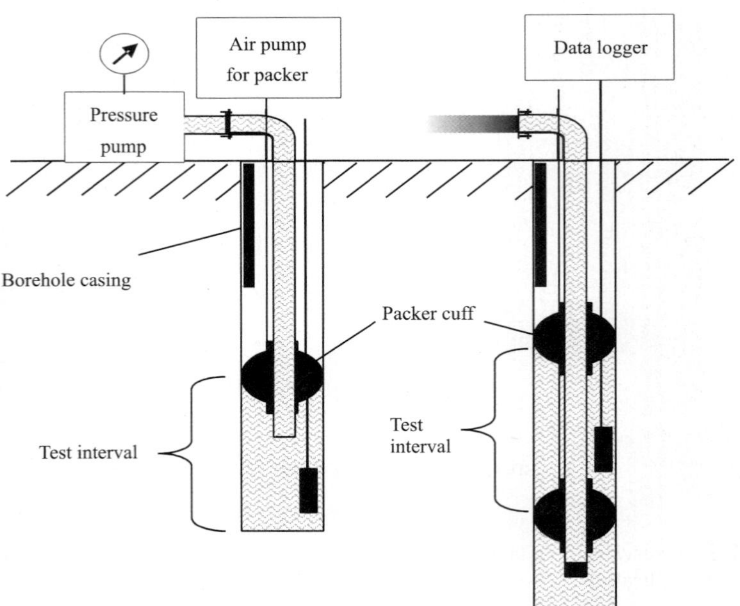

Fig. 5.2-54: Borehole permeameter tests, schematic diagram of the method

Simple equipment, such as a temporary casing, a water-level meter and a stopwatch are the basic equipment for BP tests. In addition, an electric pump may be required to extract water (for a rise test). To automate data collection, it is recommended to use a pressure transducer with data logger. A winch for lowering and lifting sensors, packers, etc. is desirable for BP tests. Water consumption in these tests depends on the properties of the test sections. Enough water must be supplied during the test to ensure proper results. The temporary casing used in a BP test is usually a plastic tube, which can be reused. The bottom of the well consists of a gravel filter with grain sizes between 2 and 4 mm.

In practice, the test setups suggested by the Earth Manual (USBR, 1974) and HEITFELD et al. (1979) are the most commonly used. *Figure 5.2-55* shows possible test configurations. Setups (*a*), (*b*), and (*c*) in *Fig. 5.2-55* are open bottom-end setups for interrupted drilling. Drilling can be continued after the test is carried out. Setup (*a*) is suitable for tests in normal borehole casings, percussion core boreholes, or hollow augers. Setups (*b*) and (*c*) involve pulling up the borehole casing several centimeters (depending on the thickness of the strata) to create a cavity for conducting the test (*Fig. 5.2-55c*). In unconsolidated rock, the cavity is filled with gravel to stabilize the borehole (*Fig. 5.2-55b*). With these three test setups, it is possible to establish an alternating drilling/testing program. In permanent wells, the test setups shown in *Fig. 5.2-55d* and *e* are the most common. Usually the screen section is enclosed in a gravel pack. Standardized equipment and the same test setup should be used for each test within a test series to obtain comparable results for k_f values. For accurate measurements under both saturated and unsaturated conditions, it is recommended to use the test setup with an open bottom-end screened section (*Figs. 5.2-55b*, *c*, and *d*). The test setup shown in *Fig. 5.2-55b* can be used in the unsaturated zone, but the cost and the loss of gravel should be considered. It is recommended to seal the space between the borehole walls and the casing above the screen section, preferably by grouting.

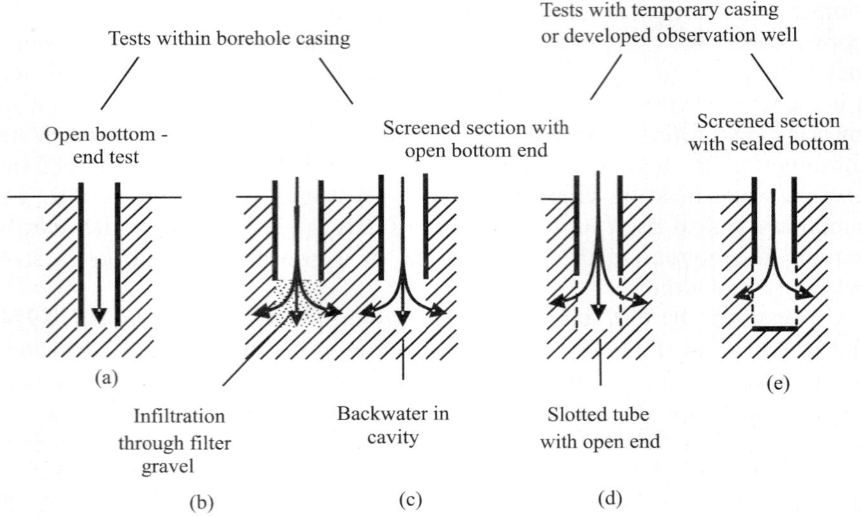

Fig. 5.2-55: Possible BP test configurations

Field procedure for a BP test:

1. Select proper locations for test points on the basis of data as derived from geophysical and borehole investigations.
2. Drill boreholes at the selected locations to the desired depths using the most suitable drilling method.
3. Install a well casing or temporary casing and seal the space between the casing and the borehole walls and at the ground surface to prevent inflow of surface water (*Fig. 5.2-54*).
4. To ensure proper results, the length of the screen should be no more than 20% of the thickness of the tested stratum (*Figs. 5.2-60* and *5.2-61*). In open bottom-end tests, the distance from the bottom of the casing to both the top and bottom of the test stratum should be more than ten times the casing radius (*Fig. 5.2-62*).
5. An existing developed observation well can also be used for a BP test.
6. Measure the static water level in the borehole/well. This should be done only after the oscillations of the water table caused by the installation of the test equipment have completely stopped.
7. There are two procedures for the variable head BP test: falling head and rising head. Water is added (falling head) or extracted (rising head) from the borehole to create a selected head (usually 2 - 5 m). The recovery of the head to the initial level of the water table is then observed by measuring water levels at specified time intervals. The level can then be plotted

against time. The hydraulic conductivity can be calculated from the resulting curve. This procedure is repeated several times to obtain a reliable average of hydraulic conductivity. The constant head BP test involves creating a head (usually 3 - 5 m) and then adding water to the well at a rate required to keep this head constant, recording the amount added and the time. The rate required to maintain a constant head in the well is a measure of hydraulic conductivity of the test section.

8. The data obtained in the test is analyzed by one of the methods of analysis explained below (*Figs. 5.2-56* to *5.2-62*).

Field procedure using packers:

1. Drill a borehole of the required specifications.
2. Install the test setup as shown in *Fig. 5.2-54b*. Keep the pressure valve open during installation. Close the valve and fill the test pipe with water until a predetermined pressure is reached. The space above the packer should be continuously checked for leakage. One good way to prevent leakage is to use two packers, above and below the test section.
3. Open the valve and start logging. In constant head tests, vary the rate at which water is pumped into the packered section through the pipe in order to maintain the predetermined pressure.
4. Use the steady-state rate or the change in head to calculate the hydraulic conductivity k_f of the stratum.
5. From this pressure-quantitative measurement, the k_f value can be obtained.

Due to the broad range of possible interpretation methods, one single evaluation procedure recommendation cannot be given. *Figures 5.2-56* and *5.2-59* give different equations for calculating k_f for different field conditions. The evaluation methods are divided into two parts according to the reference used: HEITFELD (1998) and Earth Manual (USBR, 1974). The following parameters are used in *Figs. 5.2-56* to *5.2-62* and in the equations:

Q steady-state flow rate of water added,
r_i inner radius of casing,
r_o radius of borehole, or outer radius of casing, as appropriate,
Tu difference between groundwater table and head in casing at time t,
H constant head in casing,
L length of screen section,
h_1 head in casing at time t_1,
h_2 head in casing at time t_2,
t_1 initial time after adding of water,
t_2 time of reaching equilibrium.

All parameters should be measured in the field.

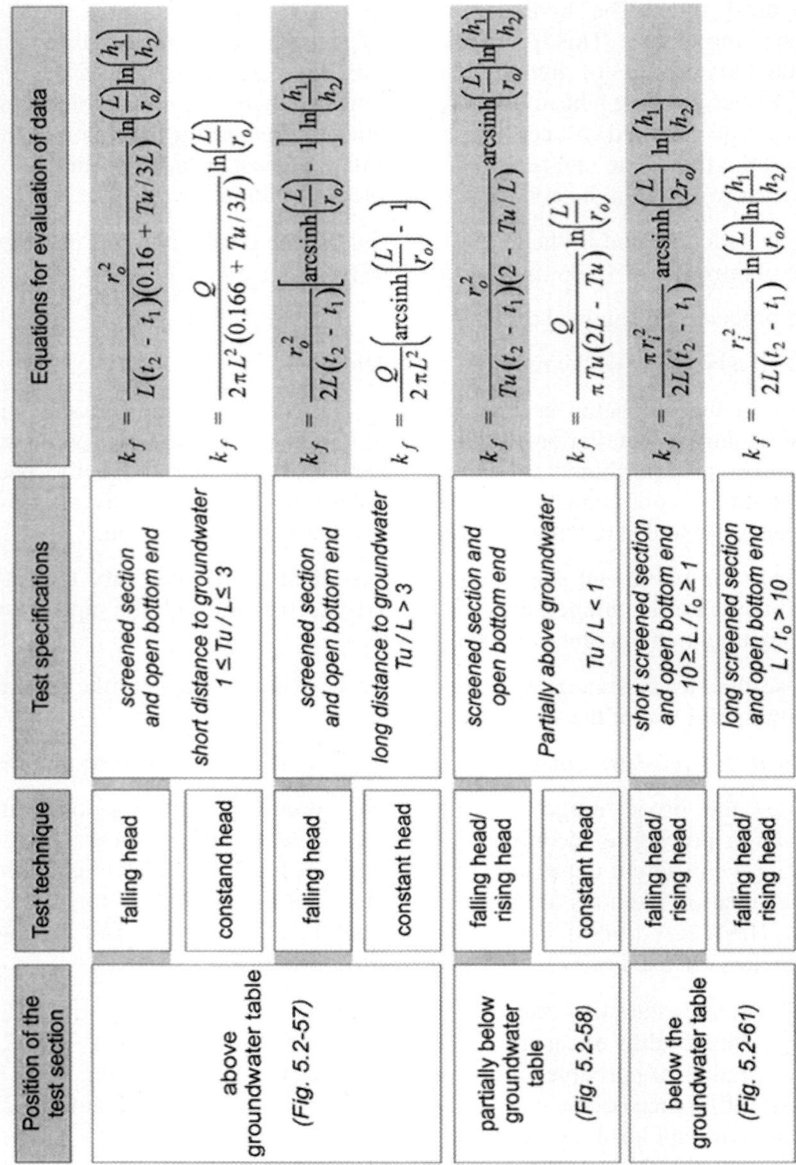

Position of the test section	Test technique	Test specifications	Equations for evaluation of data
above groundwater table	falling head	*screened section and open bottom end*	$k_f = \frac{r_o^2}{L(t_2 - t_1)(0.16 + Tu/3L)} \ln\left(\frac{L}{r_o}\right) \ln\left(\frac{h_1}{h_2}\right)$
	constand head	*short distance to groundwater* $1 \leq Tu/L \leq 3$	$k_f = \frac{Q}{2\pi L^2 (0.166 + Tu/3L)} \ln\left(\frac{L}{r_o}\right)$
(Fig. 5.2-57)	falling head	*screened section and open bottom end*	$k_f = \frac{r_o^2}{2L(t_2 - t_1)} \left[\operatorname{arcsinh}\left(\frac{L}{r_o}\right) - 1\right] \ln\left(\frac{h_1}{h_2}\right)$
	constant head	*long distance to groundwater* $Tu/L > 3$	$k_f = \frac{Q}{2\pi L^2} \left(\operatorname{arcsinh}\left(\frac{L}{r_o}\right) - 1\right)$
partially below groundwater table	falling head/ rising head	*screened section and open bottom end*	$k_f = \frac{r_o^2}{Tu(t_2 - t_1)(2 - Tu/L)} \operatorname{arcsinh}\left(\frac{L}{r_o}\right) \ln\left(\frac{h_1}{h_2}\right)$
(Fig. 5.2-58)	constant head	*Partially above groundwater* $Tu/L < 1$	$k_f = \frac{Q}{\pi Tu(2L - Tu)} \ln\left(\frac{L}{r_o}\right)$
below the groundwater table	falling head/ rising head	*short screened section and open bottom end* $10 \geq L/r_o \geq 1$	$k_f = \frac{\pi r_i^2}{2L(t_2 - t_1)} \operatorname{arcsinh}\left(\frac{L}{2r_o}\right) \ln\left(\frac{h_1}{h_2}\right)$
(Fig. 5.2-61)	falling head/ rising head	*long screened section and open bottom end* $L/r_o > 10$	$k_f = \frac{r_i^2}{2L(t_2 - t_1)} \ln\left(\frac{L}{r_o}\right) \ln\left(\frac{h_1}{h_2}\right)$

Fig. 5.2-56: Recommended evaluation methods according to HEITFELD (1998)

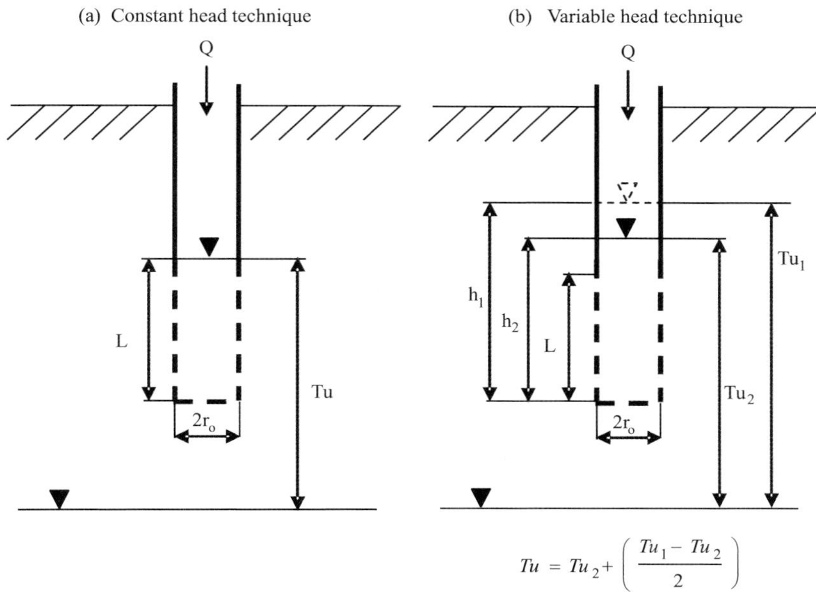

Fig. 5.2-57: BP test with test section above the groundwater table according to HEITFELD (1998)

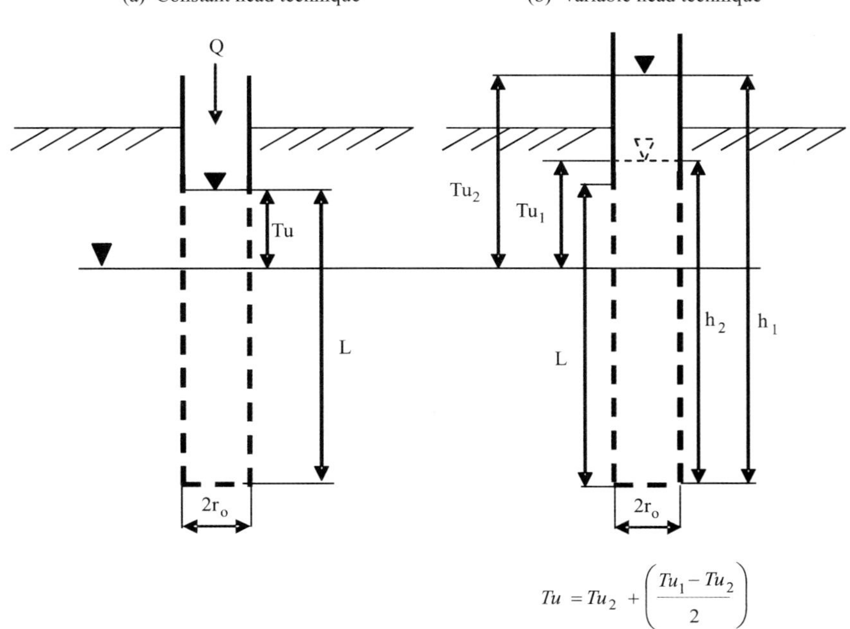

Fig. 5.2-58: BP test with test section partially within the groundwater zone according to HEITFELD (1998)

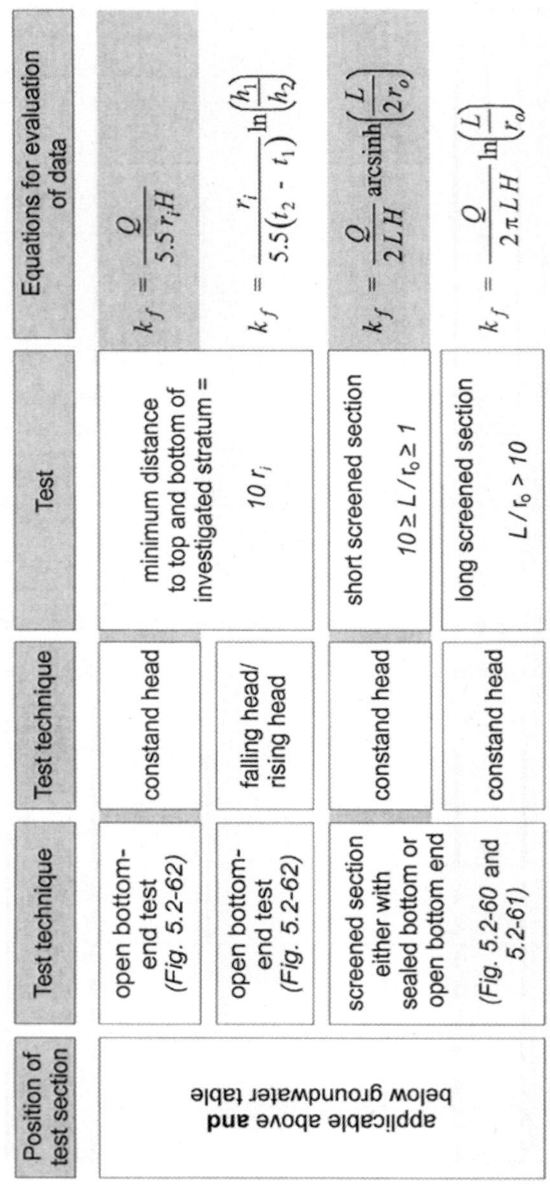

Position of test section	Test technique	Test technique	Test	Equations for evaluation of data
applicable above and below groundwater table	open bottom-end test *(Fig. 5.2-62)*	constand head	minimum distance to top and bottom of investigated stratum =	$k_f = \frac{Q}{5.5\, r_i H}$
	open bottom-end test *(Fig. 5.2-62)*	falling head/ rising head	$10\, r_i$	$k_f = \frac{r_i}{5.5(t_2 - t_1)} \ln\left(\frac{h_1}{h_2}\right)$
	screened section either with sealed bottom or open bottom end *(Fig. 5.2-60 and 5.2-61)*	constand head	short screened section $10 \geq L / r_o \geq 1$	$k_f = \frac{Q}{2LH} \operatorname{arcsinh}\left(\frac{L}{2r_o}\right)$
		constand head	long screened section $L / r_o > 10$	$k_f = \frac{Q}{2\pi LH} \ln\left(\frac{L}{r_o}\right)$

Fig. 5.2-59: Recommended evaluation methods according to Earth Manual (USBR, 1974), Method E-18

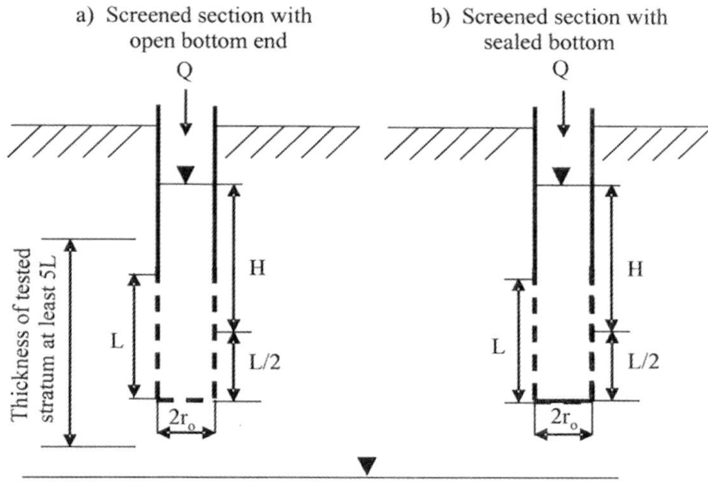

Fig. 5.2-60: BP test with test section above the groundwater table according to Earth Manual (USBR, 1974)

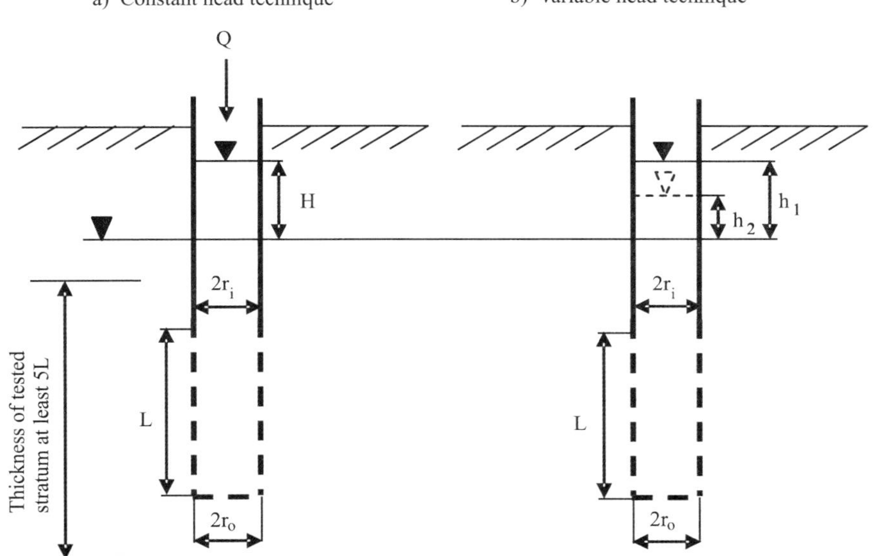

Fig. 5.2-61: BP test with test section below the groundwater table according to Earth Manual (USBR, 1974)

(a) Above groundwater table

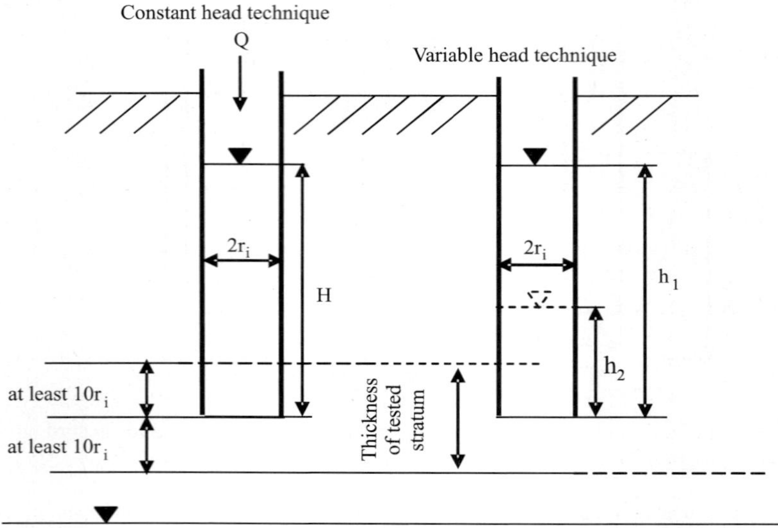

(b) Below groundwater table

Constant head technique

Q

Variable head technique

H

2r$_i$

2r$_i$

h$_1$

h$_2$

at least 10r$_i$

at least 10r$_i$

Thickness of tested stratum

Fig. 5.2-62: Open bottom-end BP test according to Earth Manual (USBR, 1974)

BP tests provide highly accurate results. The technical effort and the cost of carrying out the test are relatively low. BP tests are one of the most established methods for determining the hydraulic conductivity of an aquifer. There are numerous methods available for analysis of BP test data. Except for the rising head method, BP tests are a natural choice for investigating contaminated sites. The considerable time needed for testing formations with a low hydraulic conductivity is the only factor which may prompt a change to other methods, for example pulse tests (see section below). It is important to know the thickness of the strata being tested in advance. This helps to decide the penetration depth of the borehole. A problem may arise when the method proposed in the Earth Manual (USBR, 1974) is used and is mainly associated with the installation of the test section.

Slug/bail test (SB test)

The SB test is a variable-head single-borehole aquifer test method to determine aquifer parameters such as hydraulic conductivity and storativity. A piece of metal, called a "slug", with a defined volume is swiftly submerged in a well and the recovery of the water table to its original level is plotted as a function of time (called "slug insertion phase"). After recovery, the slug is quickly removed and the recovery of the water table is again plotted as a function of time.

The same effects can also be achieved by adding or bailing a defined volume of water to or from a well. There are several other methods to achieve the effect of sudden addition or removal of specific volume of water to or from a well during a slug/bail test. One recent method uses positive or negative gas pressure (e.g., air or nitrogen) to change the water table level in a well. See FETTER (2001) and BUTLER (1997) for detailed descriptions.

The SB test is applicable only under saturated conditions and is recommended for formations with k_f values of 10^{-5} to 10^{-9} m s^{-1} (GARTUNG & NEFF, 1999). The water table in cased wells should be above the screened section. In uncased boreholes, flow into the borehole above the water table should be prevented by suitable measures such as packers (consolidated rocks) or a temporary casing.

Figure 5.2-63 shows two possible test setups. The packer method (*Fig. 5.2-63b*) is preferred for cased boreholes with relatively small diameters and several screened sections, but can also be used in uncased boreholes in consolidated rocks. In this kind of setup the test intervals should be separated using single or double packers. A pump is necessary during the extraction phase of the method shown in *Fig. 5.2-63b*. The generator for supplying electricity and the winch for lowering and lifting tools into and from the well are not depicted in *Fig. 5.2-63*, but are indispensable.

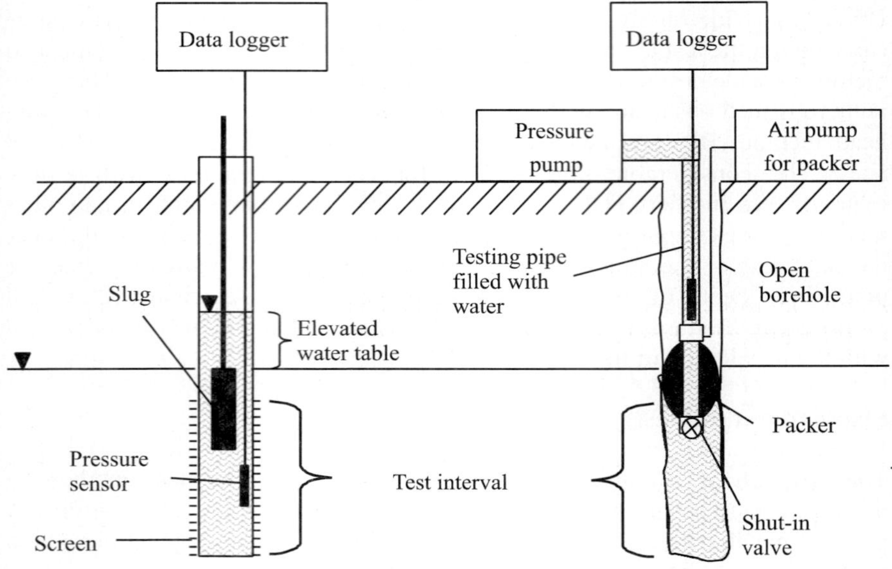

Fig. 5.2-63: SB test, setups for the two possible techniques of the method

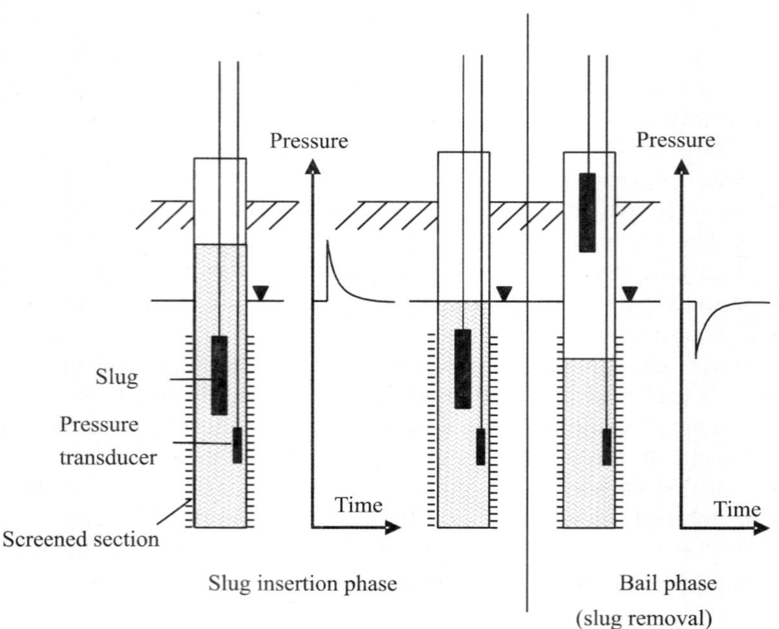

Fig. 5.2-64: Schematic diagram of a SB test and the resulting pressure as a function of time

The procedure for slug and bail tests may differ with respect to the method of changing the water level, but they are still the same in principle (*Fig. 5.2-64*).

1. Drill and develop a borehole to the required depth.
2. Install the test equipment to be used for the selected method.
3. Measure the static water level, i.e., the initial water level.
4. In the *first method* a metal slug is lowered into the well below the groundwater table. This is termed the "slug insertion phase". The addition of a known volume of water to the well and the lowering of a slug below the water table level are equivalent.
5. Measure the decrease in water level as a function of time until it once again reaches the initial water level.
6. Similar results can be obtained by removing the slug or withdrawing a known volume of water from the well and measuring the recovery of the water table over time to its initial level. This is termed the "bail phase".
7. The data are analyzed to determine the desired parameter values.

In the *second method*, a single or double packer is set at the selected depth and the shut-in valve is closed. The test pipe is then filled with water and the valve is opened. The decrease in the water table is measured as a function of time and recorded by a data logger. The slug-withdrawal phase starts by setting the packer at a desired depth. Once the packer is in place, the shut-in valve is closed and water in the pipe is pumped out completely. The valve is opened and the recovery of the water table to its initial level is observed using a pressure transducer.

The Bouwer-Rice method (BOUWER & RICE, 1976) is the most widely used method for slug/bail test data analysis. It is simple and yields highly accurate, reproducible results. It is suitable for both fully and partially penetrating wells. Although it was originally developed for unconfined aquifers, it can also be used in confined aquifers if the top of the well screen is in some distance below the bottom of the upper confining layer (FETTER, 2001).

Figure 5.2-65 shows the geometry of the borehole for the Bouwer-Rice method, where

r_c radius of well casing,
r_w radius of borehole,
L length of well screen,
E distance from the initial water table to the bottom of the test section,
D distance from the initial water table to the bottom of the tested aquifer,
$H(t)$ difference from initial head at time t.

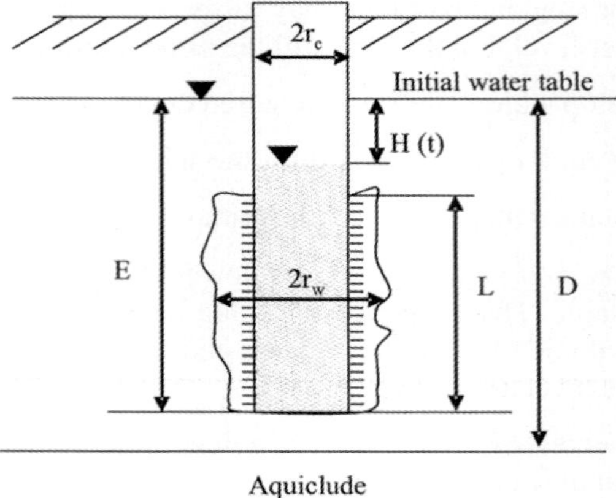

Fig. 5.2-65: Geometry of the borehole for Bouwer-Rice analysis method, BOUWER & RICE (1976)

The Bouwer-Rice equation to calculate the field-saturated hydraulic conductivity is

$$k_f = \frac{r_c^2 \ln\left(\frac{R_e}{r_w}\right)}{2L} \frac{1}{t} \ln\left(\frac{H_o}{H_t}\right), \tag{5.2.34}$$

where H_o drawdown at time 0,
H_t drawdown at time t,
R_e radial distance from the well over which the head is dissipated.

It is not possible to state exactly the value of R_e which applies to a particular well. BOUWER & RICE (1976) have also presented a method to estimate the dimensionless ratio R_e/r_w:
when $E < D$,

$$\ln\left(\frac{R_e}{r_w}\right) = \left[\frac{1.1}{\ln\left(\frac{E}{r_w}\right)} + \frac{(A+B)\ln\left(\frac{D-E}{r_w}\right)}{\frac{L}{r_w}}\right]^{-1}, \tag{5.2.35}$$

when $E = D$,

$$\ln\left(\frac{R_e}{r_w}\right) = \left[\frac{1.1}{\ln\left(\frac{E}{r_w}\right)} + \frac{C\,r_w}{L}\right]^{-1} . \qquad (5.2.36)$$

A, *B*, and *C* are empirical dimensionless parameters which can be obtained from the curve shown in *Fig. 5.2-66*.

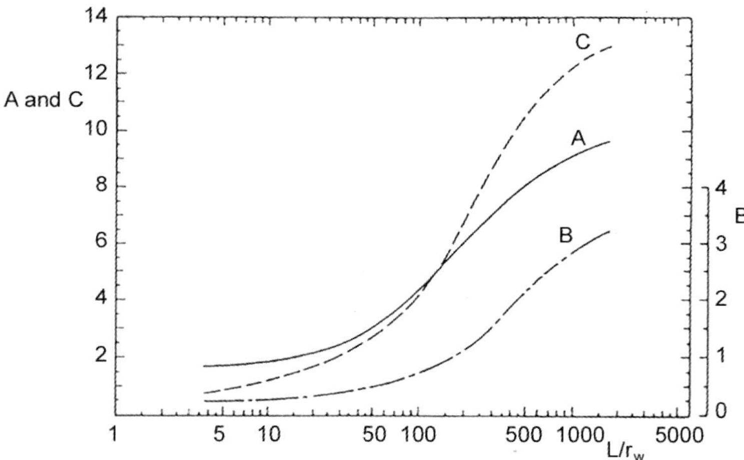

Fig. 5.2-66: Curves for determining the constants *A*, *B*, and *C*, BOUWER & RICE (1976)

The change in water level in the borehole is continuously recorded. This data is plotted semi-logarithmically versus time (*Fig. 5.2-67*). The values of H_0 and H_t can be determined from this curve. Although the terms H_0 and H_t are used in the equation, they merely represent drawdown values at two times. The k_f value can also be determined by substituting drawdown values at any other corresponding points in time during the recovery period.

When it is necessary to obtain very accurate results, the pattern of recovery has to be considered in order to select a suitable method for analysis. The water level may recover to the initial level in a smooth manner. This response is called overdamped (*Fig. 5.2-68*). An overdamped response is observed mainly in aquifers that contain fine-grained to sandy material (SARA, 2003).

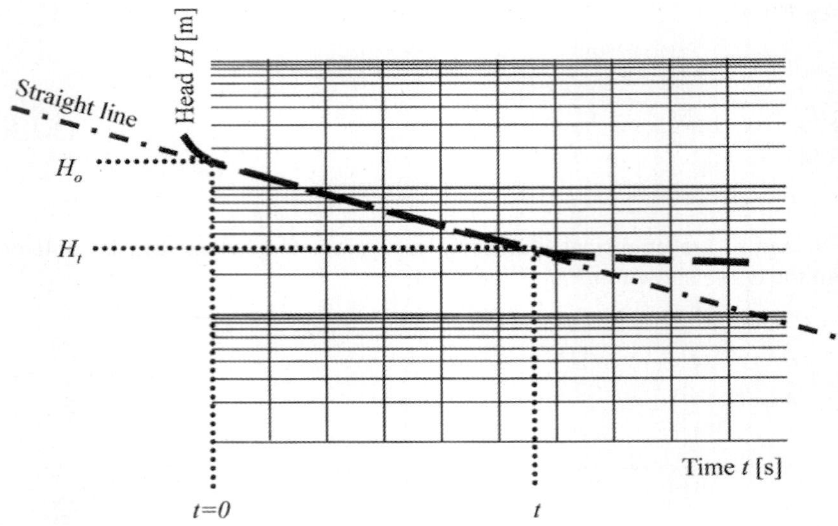

Fig. 5.2-67: Example of test data, SCHREINER & KREYSING (1998)

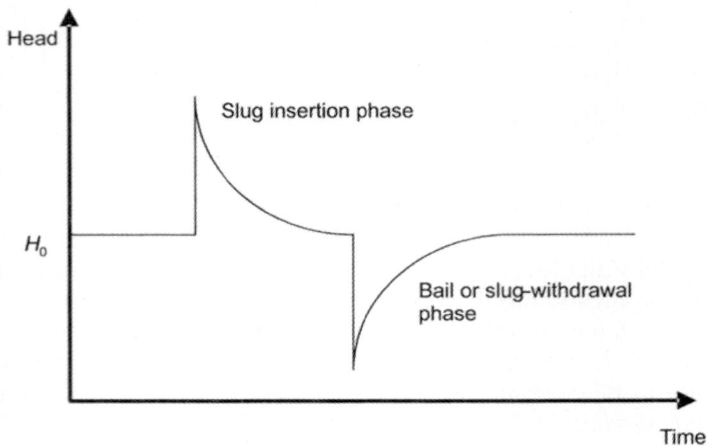

Fig. 5.2-68: Example of an overdamped response for an SB test

Another possible recovery pattern, which is observed mainly in highly conductive aquifers such as coarse sand and gravel, is when the water level oscillates about the initial water level with its magnitude of oscillation decreasing until the oscillations completely cease (*Fig. 5.2-69*). This kind of response is termed underdamped and is less often observed than an overdamped response. Different methods of analysis are used for these conditions. For overdamped response the evaluation method developed by BOUWER & RICE (1976) is recommended. For information about other analysis methods see FETTER (2001).

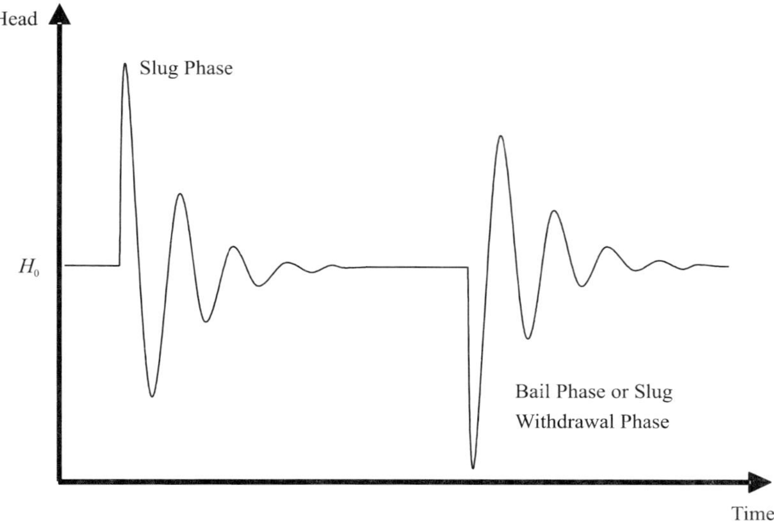

Fig. 5.2-69: Example of underdamped response for an SB test

A main advantage of SB tests is the extensive knowledge on this method. It is an easy and quick method which can be easily carried out by two persons. The volume for which the k_f value measured by this test is representative and comparable to that of a pumping test (see *Fig 5.2-46*). A large number of subsurface parameter values have to be collected for evaluation. The distance from the groundwater table to the aquiclude has to be known. The manner in which the water table recovers its initial state (underdamped and overdamped responses) may influence the final results.

Pulse test (PT)

The pulse test is a modified slug/bail test (SB test) for single boreholes and is highly suitable for determining the hydraulic conductivity of low permeability sediments ($k_f < 10^{-8}$ m s^{-1}) in the saturated zone. The equipment is the same as that for an SB test with packers. PT methods can also be used to determine transmissivity and skin effect. A short pressure pulse is generated in a test interval separated from other parts of the borehole by packers. The interval is then closed by a valve so that the decay of the pressure impulse (overpressure) is possible only through the rock surrounding the well. Despite the low hydraulic conductivity of the tested formation, the test lasts only 15 to 30 minutes.

Figure 5.2-70 shows a schematic of a test setup for pulse tests. If the formation consists of consolidated rock, then the test can be conducted without a casing in a test section closed by a single or double packer. A temporary casing is recommended for unconsolidated rock. The packers are filled with compressed air from a compressor or a gas cylinder. After installation of the test equipment (including inflating the packers) the valve is closed and the hydrostatic pressure of the groundwater in the test interval is measured. The installation in the borehole must be checked for water tightness. Then the casing above the closed valve is filled with water until the desired pressure head is reached. The pressure head must be less than that which would crack the rock. The required water is taken from a water tank using a pump if necessary. When the desired pressure head (overpressure) is reached, the valve is opened for a short time (< 10 s) and then closed. The pressure decay in the test interval is measured using a pressure transducer. A high sampling rate is necessary, particularly in the initial phase of pressure decay. *Figure 5.2-71* shows the course of a single pulse test run.

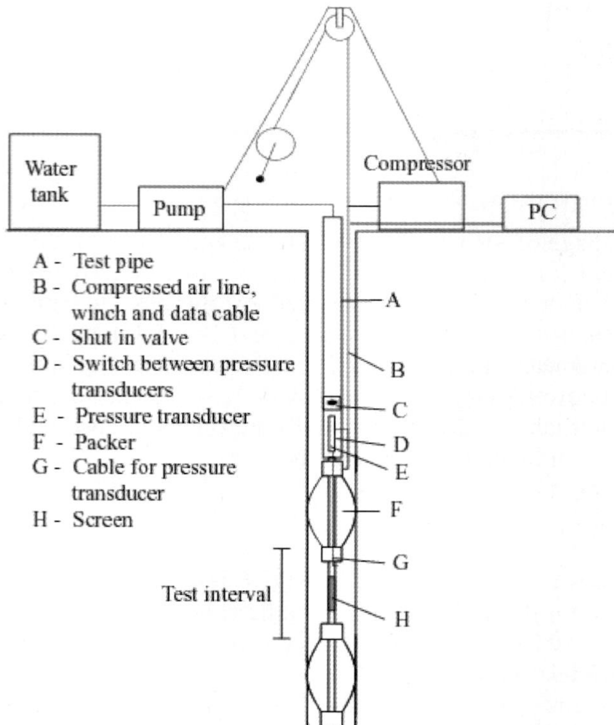

Fig. 5.2-70: Schematic diagram of a test setup for pulse tests, POIER (1998)

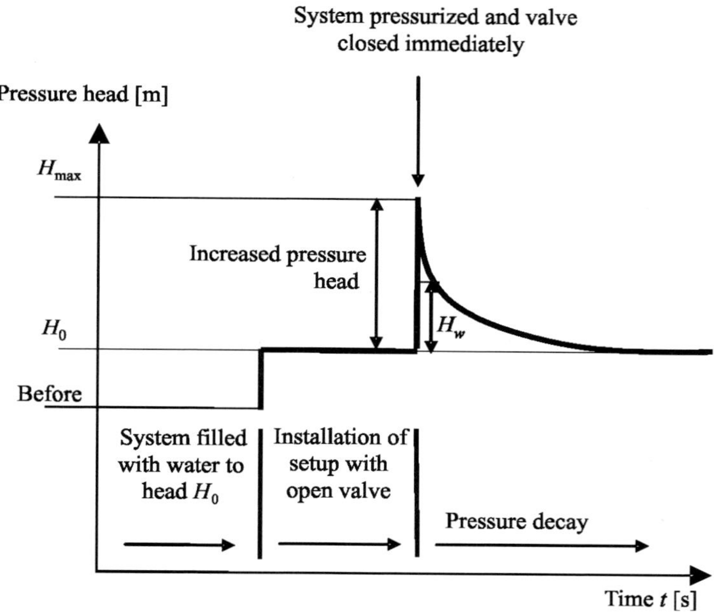

Fig. 5.2-71: Course of a single-pulse test run, modified from POIER (1998)

Several methods are available for the analysis of pulse test data (POIER, 1998). Owing to the consistency and reproducibility of the test results, the method developed by PERES et al. (1989) is one of the best ones available to analyze PT data. Peres' equation for determining hydraulic conductivity is

$$k_f = \frac{1.151\ \beta \rho V g}{2 \pi m L}, \tag{5.2.37}$$

where
- L length of test interval [m],
- V volume of test interval ($V = \pi\ r_w^2\ L$) [m³],
- r_w radius of well [m],
- β compressibility of water [$m^2\ N^{-1}$],
- ρ density of water [$kg\ m^{-3}$],
- g acceleration of gravity [$m\ s^{-2}$],
- m slope of the fit line [s per log-cycle], see below.

The value of m is determined graphically: First, values $I(dH)$ need to be calculated for plotting the graph:

$$I(\mathrm{d}H) = \left(\frac{(H_t - H_0) + (H_{t+1} - H_0)}{2(H_t - H_{max})} \right) (t_{t+1} - t_0), \tag{5.2.38}$$

where H_0 initial pressure head (water table) [m],
H_{max} maximum pressure head [m],
t_0 time at application of pressure (i.e. for H_{max}) [s],
H_t observed head at time t [s].

$I(\mathrm{d}H)$ versus time t is plotted on semi-logarithmic paper (example in *Fig. 5.2-72*), and a straight-line is fitted to the data. The slope of this line is the required value of m. For more details see PERES et al. (1989) and POIER (1998).

Pulse tests have been successfully used to determine the hydraulic conductivity of formations with k_f values as low as 10^{-16} m s^{-1} (BREDEHOEFT & PAPADOPULOS, 1980). One disadvantage associated with pulse tests is that it provides hydraulic parameter values only within a volume of 1 m around the well or borehole. This means the test has to be repeated at a large number of test points. The test should be repeated several times in each hole to ensure accuracy of the final results. Fracturing the rock surrounding the borehole is one factor which may affect the final results.

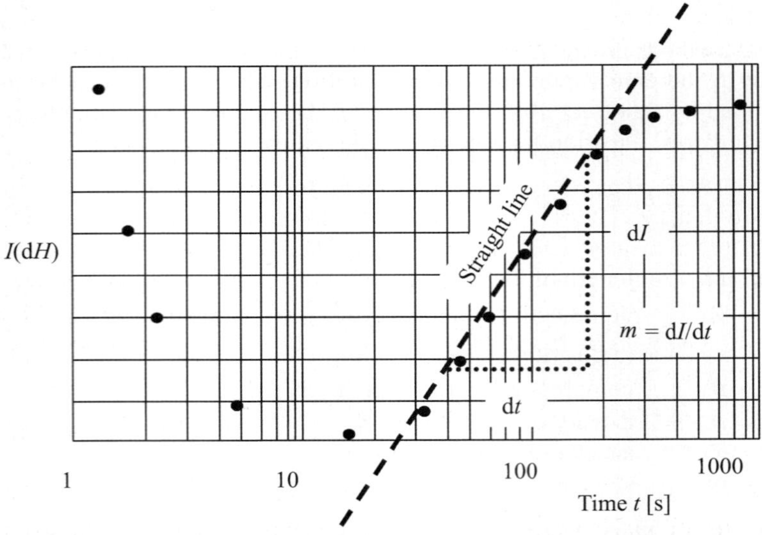

Fig. 5.2-72: Typical plot of $I(\mathrm{d}H)$ versus time t, modified from POIER (1998)

5.2.7.2 Pumping Tests

ULRICH BEIMS

Pumping tests are comparatively time-consuming controllable experiments to determine the hydraulic parameters of an aquifer and performance characteristics of wells. Pumping tests work on a very simple principle: Water is pumped from a well tapping the aquifer, called the pumped well, and the discharge from this well and the changes in the water level in it are observed. The changes in water levels in observation wells (also called piezometers) located at known distances from the pumped well are also observed. From the changes in the water levels in the pumped and observation wells, and from the quality and temperature of the groundwater, it is possible to obtain hydrogeological, hydrochemical, and operational information. Pumping tests are the most reliable methods for characterization of aquifers.

According to the objectives and the mode of implementation, the following types of pumping tests can be distinguished:

- *Well tests*: Well tests predominately serve to obtain the performance characteristics of a well. Moreover, the geohydraulic parameters as well as the quality of the groundwater can be estimated. The results of a well test form the basis for planning further pumping tests.
- *Aquifer tests*: Aquifer tests are used to determine the hydraulic conductivity and storage characteristics of an aquifer and the adjacent beds. They provide information about the structure of the aquifer, its hydrochemistry, as well as other information for constructing a groundwater model, e.g., geohydraulic boundary conditions or multiple-aquifer formations. In combination with other investigation methods (e.g., geophysical methods) the results can be interpreted in more detail and in combination with tracer tests, the groundwater flow parameters can be better understood.
- *Intermediate pumping tests*: Pumping tests during drilling help to decide about the well depth and well design.
- *Pumping tests for well development*: Pumping tests for well development are carried out to optimize well performance by improving the hydraulics in the well surroundings, minimizing skin effects, and ensuring that the well screen is free of any fine materials that could affect well performance.
- *Operational tests*: Before starting a pumping test or setting up a well for production, it is important to know the reaction of the water table to the pumping. This information is obtained in operational tests to optimize pump settings and later well operation.

- *Long-term pumping tests* are conducted to check the ecological effects of groundwater withdrawal and to obtain data on the sustained yield, long-term groundwater composition, and the origin of the water.

Figure 5.2-73 shows the stages of a well test with three pumping rates and a subsequent aquifer test. The pumping has to be interrupted for a recovery phase and can be carried out in various combinations.

The objectives of a pumping test, such as determination of geohydraulic parameters, information about aquifer structure, optimization of well performance, sustained yield and information on the groundwater composition, determine the type of pumping test, its planning, implementation and evaluation. Not all objectives can be achieved with one type of test. When several types of tests have to be carried out, practicability, order, and combination of the tests have to be considered. A pumping test involves the following phases:

- *Planning*: A tentative program has to be established prior to the start of a pumping test. It includes decisions about the pumping regime, well and piezometer design, the duration of the test, the test parameters and measuring intervals, the measuring instruments, required accuracy, etc. It is appropriate to model the pumping test with estimated parameter values in advance.

- *Implementation*: The implementation phase includes organization, supervision, and documentation of the test.

- *Evaluation*: The first step in the evaluation of a pumping test is to develop a hydraulic model on the basis of the obtained data. Subsequently, the required parameters can be determined and the initial model can be verified.

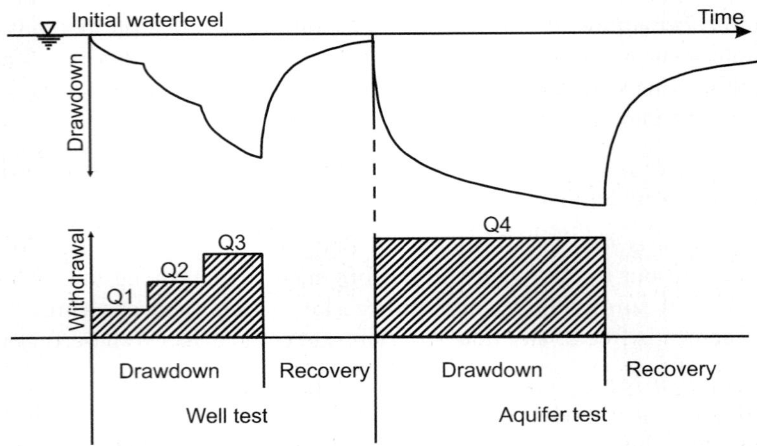

Fig. 5.2-73: Stages of a pumping test

Planning of pumping tests

The type of the test and the geological and hydrogeological conditions determine what needs to be done to prepare for a pumping test. The following items and their timing have to be considered:

- all regulations regarding groundwater extraction and infiltration of extracted water back into the ground,
- determination of the need for new observation wells,
- briefing everyone involved,
- safety and protection measures,
- briefing the field crew about the measurements and required accuracy,
- instruction about continuation/discontinuation of the test in case of emergencies,
- ensuring accurate recording of the drawdown and recovery data (data loggers are preferable to manual measurements in the case of frequent measurements),
- selection of a suitable pump and control devices for maintaining a constant pumping rate,
- function test of the groundwater observation wells by pumping, borehole geophysics methods or, if necessary, filling the well with water to observe its recovery behavior, and
- preparation for sample collection and analysis of the pumped water.

For aquifer tests and long-term tests, in addition to the above-mentioned items, attention must be paid to the following aspects:

- The drawdown, which immediately follows the withdrawal, is characterized by different phases. Therefore, the duration of the pumping test has to be long enough that the parameter values for the individual drawdown phases can be determined accurately, as well as additional information is ensured.
- The speed of reaction of the groundwater piezometric surface strongly depends on whether the aquifer is confined or unconfined. This has to be considered when deciding on the pumping duration.
- The degree and distribution of inhomogeneity and anisotropy of the subsurface must be considered when the observation points and measuring intervals are specified.

A detailed test plan must contain the following information about the pumping test:

- the pumping rate,
- probable duration of pumping and the time for the recovery measurements,
- measurement of the water level, including the initial water level, drawdown and recovery, measuring intervals, required accuracy, and tools,
- sample collection (chemical, physical, and microbiological), including sampling tool, sampling location, quantity, extraction time and extent of analysis and sample treatment,
- accompanying measurements of precipitation, air pressure, temperature as well as the water level and condition of surface water,
- special investigations such as pressure measurements in multiple aquifers, geophysical borehole measurements, chemical profile of indicator, and special parameters as well as isotope and tracer measurements, and
- evaluation of the test and calculation of the aquifer parameter values.

Implementation of pumping tests

Well tests are carried out in several stages and the drawdown relative to the initial water level is recorded. The duration of each pumping stage is usually 4 to 24 hours. The shorter time is suitable for small wells in high permeability aquifers, the longer time is suitable for large wells in low permeability aquifers. There should be three to five pumping stages with pumping rates Q_i, i = 1, 2, 3 (, 4, 5) uniformly distributed within the selected range of rates. The measuring intervals should be selected according to *Table 5.2-16*. This sequence is meant to cover the start of the pumping test, any changes in the withdrawal phase and in the recovery phase. If data loggers are used, measurements are usually taken more frequently. The recovery phase that follows the pumping should be measured for at least two thirds of the pumping time.

Table 5.2-16: Measuring intervals as a function of elapsed time since starting a pumping stage

Elapsed time since start of a pumping stage	Measuring interval
< 10 min	1 min
10 to 60 min	5 min
1 to 3 h	10 min
3 to 5 h	30 min
> 5 h	1 h

Accurate measurements of the water level at suitable intervals are essential in all pumping stages. Water level measurements in wells at a considerable distance (depends on the aquifer parameters) from the pumped well are usually not necessary. The drawdown s for each pumping stage Q_i is plotted versus the elapsed time t since the start of pumping with that rate. The relevant drawdown for the Q-s plot is the drawdown at the end of each pumping stage relative to the water level at the beginning of that stage. In the case of repetition of a well test in order to check the current yield, the later test has to be carried out, whenever possible, with the same specifications as the earlier one. Water samples should be taken for chemical analysis at the end of each pumping stage.

Aquifer tests provide comprehensive information about the well and the aquifer. The results of an aquifer test are the basis for all subsequent studies of groundwater quantity and quality as well as on watershed areas. Beyond this, aquifer tests provide well-specific data on well storage and skin effect.

For the aquifer under investigation, the pumping duration must be long enough that a geohydraulic model is recognizable and verifiable. Pumping between 200 and 400 hours at a constant pumping rate is needed for this purpose. The shorter duration is suitable for porous rocks, since the parameter values can be determined faster, whereas in hard-rock aquifers with distinct inhomogeneities or leakages the longer duration is needed. For remarks about measuring intervals see section on well tests. The recovery measurements should generally take 0.5 to 0.75 the duration of pumping. At least three water samples should be taken for chemical and/or isotope analysis during the test.

For the construction of observation wells for aquifer tests, the hydrological and hydrogeological conditions in the test field, as well as the distance from the pumped well are important. The observation wells should be located in different directions and at different distances from the pumped well in order to observe the drawdown in all directions. The observation well network has to be set up such a way that it takes into consideration local features of the investigation area. Examples of such local features are

- geological boundaries, e.g., faults or terrace edges,
- hydrological boundaries, such as lakes or rivers,
- competitive-use structures, like waterworks or ecologically sensitive areas, and
- preferential water flow paths, e.g., drainage ditches.

The thickness of the aquifer determines the minimum distance between the observation wells and their respective distances to the pumped well. In homogeneous aquifers the observation wells should be located in the drawdown area not too close to the pumped well and not very near the edges of the cone of depression. Generally, it can be stated that the following values can be used as guidance when positioning the observation wells:

confined groundwater $M < r < 20\,M$

unconfined groundwater $H < r < 10\,H$

where M thickness of a confined aquifer,
H thickness of an unconfined aquifer,
r distance of observation well from the pumped well.

For the installation of the observation well network, a rough drawdown calculation using estimated geohydraulic parameters can be used. An imperfect well has no influence on the evaluation if the observation well is at a distance $r > M$ or $r > H$. Large-scale inhomogeneities, as present in karst and jointed rock aquifers, are not always detectable with a dense observation net. To detect and understand leakage effects during an aquifer test, it is necessary to observe the behavior of the adjacent layers.

For aquifer tests, the groundwater levels should be measured with high accuracy, which depends on the expected drawdown and will be on the order of 0.1 % of the drawdown or 1 mm. For manual measurements, an accuracy of about 1 cm has to be achieved. With automatic data recording using data loggers, higher accuracies can be achieved even with short measuring intervals.

Depending on the objectives of the test, the following parameters have to be recorded:

- water levels in the well and in the annular fill,
- pumping rate,
- hydrochemical parameters, and
- sand content/turbidity.

Operational tests yield data important for the well specifications and for later well operations. Operational tests result in a better understanding of the typical behavior of a well during the start-up, i.e., the water level fluctuation immediately after the start of pumping. Operational tests also help in deciding the depth of pump placement and its hydraulic program.

Intermediate pumping tests are mainly carried out to test the characteristics of different layers during drilling. They can provide information on differences in yield, for example. They are also carried out in the case of sudden changes in hydraulic properties encountered during the drilling, for example,

- water level changes,
- separation of multiple aquifers,
- fracture zones, and
- changes in water quality.

In hard rocks intermediale pumping tests are carried out in open boreholes or boreholes with temporary casings, whereas in porous/unconsolidated rocks they are carried out in boreholes with temporary casings. Intermediate pumping tests are essential for the construction of technically perfect wells or adequately functioning observation wells.

Pumping tests for well development help to reduce skin effects, to remove any fine material that may be present in the well screens and to improve the well hydraulics. Objectives of well development are

- removal of undersized material in filter gravel,
- removal of drilling mud residues,
- removal of fine-grained material in the aquifer within a short distance of the well,
- construction of a stable grain-skeleton in fine-grained aquifers,
- washing out of all mobile particles in the joint system of a hard rock aquifer.

In general, well development immediately follows well construction, however in consolidated rock it can be done in open holes. Pumping tests for well development serve only this restricted purpose and are of limited use for further evaluation.

Long-term pumping tests are used under the following conditions and/or to achieve the following goals:

- in the case of difficult hydrogeological conditions where an analytical approach or modeling is not sufficient,
- recognition of long-term changes in the groundwater,
- investigation of quantitative and qualitative interactions between groundwater and surface water,
- to determine the appropriate pumping rate for a production well which will not significantly affect the ecology and water budget of the aquifer system, and
- to demonstrate the effects of groundwater withdrawal on the water budget and ecology.

A carefully considered measuring grid and a well coordinated investigation program are necessary for successful implementation of a long-term pumping test. Long-term pumping tests have an impact on the groundwater budget and should be properly justified.

The documentation of a pumping test should be done in a pumping test report and comprise the following parts:

- pumped well,
- observation wells,
- measured values of the pumped well, and
- measured values of the observation wells.

An example of a form for the Pumping Test Report – Pumped Well is shown in *Fig. 5.2-74.* It contains general data about the drilling, times of the start, implementation and end of pumping test, as well as data on well design, and turbidity and color of water. The report is supplemented by a sketch of the well design.

The "Pumping Test Report – Observation Wells" is similar to "Pumping Test Report – Pumped Well" and contains technical data about all of the observation wells involved in the test. A map is included in the report showing the locations of the observation wells, their location with respect to lithological and other boundaries, receiving streams, etc., and the design of the observation wells, particularly the position of the screen.

The "Pumping Test Report – Measured Values of the Pumped Well" contains all the parameter values recorded during the drawdown and recovery phases. An example showing the minimum requirements is shown in *Fig. 5.2-75.* The parameters are water levels in the wells, discharge rate (which is recorded by e.g., overfall weirs, water meters or inductive flowmeters), and time of measurement, as well as the properties of the groundwater. The "Pumping Test Report – Measured Values of the Observation Wells" is similar to the one for the pumped well.

Automatic systems are often used to record the data. These systems record water level, quantity pumped, as well as selected parameters on groundwater quality. Such systems generally consist of a probe for measuring the data, and a storage device. Many systems can transfer the data directly to a PC, which can evaluate the data with special programs. The accuracy of many devices for measuring water level is close to 1 mm, while the measuring frequency can be as low as 1 s.

Pumping Test Report – Pumped Well

Site: …………………….………….. Borehole / well no.: ………………………………

Client: …………………..………… Contract no.:

Drill foreman: ………….. Test supervisor: ……………. Pumping test no.: ……………

Map sheet no.: ………… Easting: …………………….. Northing: …………………....

Surface elevation: ………….. m above MSL

Measuring point is ……………………………………….. = …… m above / below surf.

Drainage pipes: …….. m Discharge into …………………………………..

Overfall width of flow rate box: ………………………… Slit shape rectangle / triangle

Water meter reading beginning: ………………….. end: …………………………..

Other discharge measurements:

Pumping test:	from: …….....	to: …………	o'clock	
1st pumping interval	from: …….....	to: …………	o'clock =	…… hrs
recovery interval	from: …….....	to: …………	o'clock =	…… hrs
2nd pumping interval	from: …….....	to: …………	o'clock =	…… hrs
recovery interval	from: …….....	to: …………	o'clock =	…… hrs
3rd pumping interval	from: …….....	to: …………	o'clock =	…… hrs
recovery interval	from: …….....	to: …………	o'clock =	…… hrs
Total time			pumping	…… hrs
			recovery	…… hrs

Drilling method: ……………………… Drilling mud additives: …………………………

Water samples: ……………………… (record on sheet "measured values")

Borehole depth: ……… m below surface Position of filter: …………… m below surface

Depth of pump: ……… m below surface Initial water level:………….. m below surface

Turbidity of water (according to DIN 38404)

0 = clear 1 = slightly turbid 2 = very turbid 3 = opaque

Color of water (according to DIN 38404)

0 = colorless 1 = slightly colored 2 = strongly colored (e.g., brownish)

Well design (borehole and casing diameters, screened sections and type of casing, seals) see back side

Fig. 5.2-74: Pumping Test Report – Pumped Well

Pumping Test Report – Measured Values Pumped Well

Site: Borehole / well no.: Contract no.:
Pumping test no.: page:

Time information			Water level		Amount of water		Properties							
date	time	time since pump start	water level below meas. point	draw-down	specific measure-ment value	discharge rate	electr. cond.	pH	temp.	sand content	turbidity	color	remarks
			m	m		$L\ s^{-1}$ $m^3\ h^{-1}$	$\mu S\ cm^{-1}$		°C	$cm^3\ L^{-1}$			

Fig. 5.2-75: Pumping Test Report – Measured Values of the Pumped Well

Evaluation of pumping tests

The evaluation of pumping tests is a classical "black-box" problem, since only the reaction of the aquifer (drawdown and recovery) to an action (withdrawal) is known, while its hydraulic properties and structure are not known.

One basic assumption made for the evaluation of the pumping tests is that the groundwater is confined. However, the evaluation methods discussed in the following sections can also be used for unconfined groundwater by introduction of "reduced drawdown", s_r, which corrects for the fact that in unconfined aquifers the cross-sectional area of groundwater flow changes with the water level. The variables s (drawdown) and M (aquifer thickness) in the formulas for confined conditions then have to be substituted by s_r and h_0, respectively.

$$s_r = s - \frac{s^2}{2h_0}, \tag{5.2.39}$$

where s_r reduced drawdown,
s measured drawdown,
h_0 initial (static) water level.

Aquifer tests and long-term tests are suitable for parameter value determination and model identification. The long-term pumping tests supply further information. There are three types of evaluation methods:

- graphical/analytical,
- computer-aided, and
- numerical methods.

Graphical/analytical evaluation

Non-steady-state flow of groundwater during a pumping test is described by the Theis equation, which is rather complex and not solvable analytically. Using "Theis type curves", pumping tests can be evaluated graphically by matching type curves to the measured pumping test data. Another method to simplify the Theis equation is the one developed by COOPER & JACOB (1946). The Cooper & Jacob solution of Theis' equation, however, is valid only if certain conditions are met (basically, this solution is applicable for data that is collected after a sufficient amount of time has passed following the beginning of pumping). The Cooper & Jacob approximation uses a semi-logarithmic plot of the measured data with a best-fit straight line.

The following sections introduce the different variants of the best-fit straight line procedure. The basic assumptions made are that the aquifer is quasi infinite in areal extent and does not receive any flow across its boundaries. This can be regarded as the standard case for pumping test implementation and evaluation.

Best-fit straight line method

The best-fit straight line method involves

- defining the simplest relationships between the dependent and independent parameters, e.g., $s = f(t)$, $s = f(r)$, $s = f(t/r^2)$, $s = f(t/t')$, or $s - s' = f(t')$, where $f(t)$ means function of t, for explanation of the other symbols see Equation (5.2.40),
- plotting the parameter values and finding the best-fit straight line,
- calculation of the required parameter values from the best-fit straight line, and
- checking of the validity of the approach.

In the case of pumping tests with a constant pumping rate (see *Fig. 5.2-76*, for an example) there are several possibilities for evaluating the data on the basis of Jacob's equation:

$$s = \frac{Q}{4\pi T}\ln\left(\frac{2.25Tt}{r^2 S}\right), \tag{5.2.40}$$

where (in SI units)

s	drawdown [m],
Q	pumping rate [$m^3\ s^{-1}$],
T	transmissivity [$m^2\ s^{-1}$],
t	time after start of pumping when s is measured [s],
r	distance of the drawdown observation well from the pumped well [m],
S	storage coefficient [dimensionless].

Different possibilities of evaluation are discussed in the following sections.

Evaluation with respect to time

The relationship between the measured parameters s and t is described by the straight line

$$s = 0.183\frac{Q}{T}\lg t + 0.183\frac{Q}{T}\lg\frac{2.25T}{r^2 S}. \tag{5.2.41}$$

When s and t are plotted semi-logarithmically with t on the log-axis, the result approximates a straight line. The transmissivity T can be determined from the best-fit straight line of the graph (*Fig. 5.2-72*) and from Equation (5.2.41):

$$T = 0.183\,Q\frac{\Delta(\lg t)}{\Delta s}. \tag{5.2.42}$$

Solving Equation (5.2.40) for S and setting $t = t_0$ and $s = 0$, we obtain

$$S = \frac{2.25\,t_0\,T}{r^2}. \tag{5.2.43}$$

The value of t_0 can be obtained by extending the straight line so that it intersects the t-axis (see *Fig. 5.2-77*). Then the storage coefficient S can be calculated using this equation. The validity of the Cooper & Jacob approximation must then be verified for the permissible margin of error:

$$t_G > 2.3\frac{S}{T}r^2 \quad \text{for } \varepsilon < 5\%, \tag{5.2.44a}$$

$$t_G > 5.0\frac{S}{T}r^2 \quad \text{for } \varepsilon < 2\%, \tag{5.2.44b}$$

$$t_G > 8.3\frac{S}{T}r^2 \quad \text{for } \varepsilon < 1\%, \tag{5.2.44c}$$

where

- t_G is the time when the measurements approach the straight line (*Fig 5.2-77*) and
- ε is the margin of error.

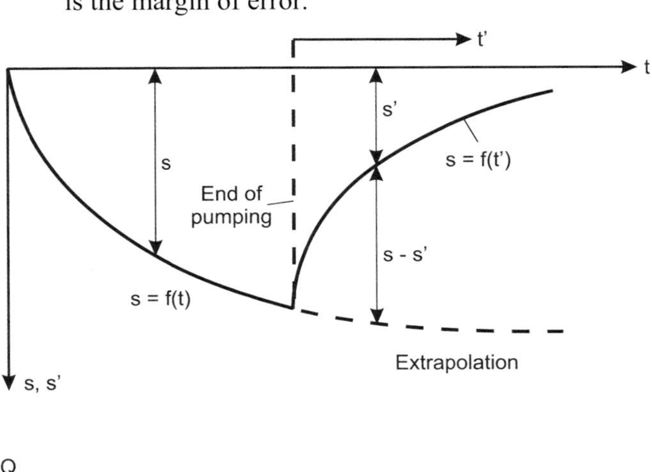

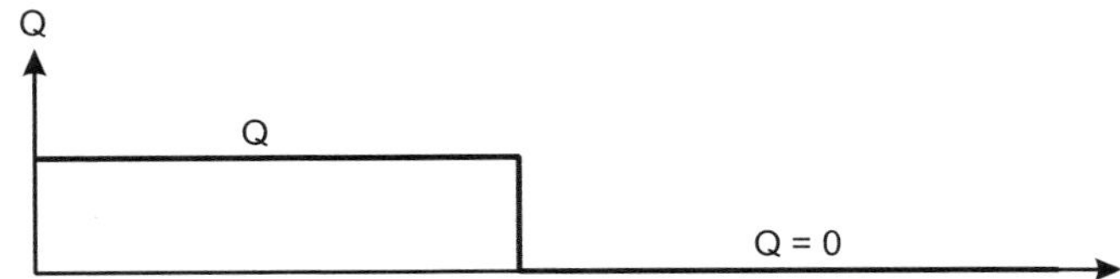

Fig.5.2-76: Drawdown and recovery of a pumping test

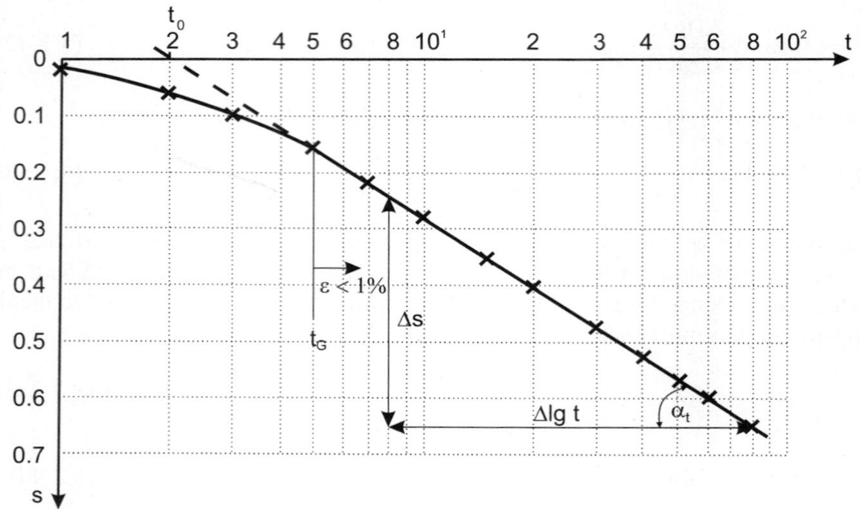

Fig. 5.2-77: Evaluation of drawdown with respect to time

Spatial Evaluation

The spatial evaluation is based on the following equation:

$$s = -0.366\frac{Q}{T}\lg r + C\,.$$

Figure 5.2-78 shows a plot of drawdown s versus distance r. Observation wells P_1 to P_4 are shown on the r axis at their respective distances from the pumped well. The drawdown distances in the different observation wells at several pumping time stages t_1, t_2,... are joined by straight lines. The figure also shows the radius R_t of the cone of depression, which is obtained by extending the straight line through the drawdown values for different wells at a single time until it intersects the r axis (i.e., at a point away from the pumped well where $s = 0$). The condition for extending such a line for a particular time (t_6 in this example) is that for the most distant well the condition in Equation (5.2.44c) has to be met (i.e., error margin less than 1 %). When the measured parameter values are plotted as shown in *Fig. 5.2-78*, transmissivity and storage coefficient can be calculated from the straight line for time t_i:

$$T = -0.366\,Q\frac{\Delta(\lg r)}{\Delta s} \quad \text{and} \tag{5.2.45}$$

$$S = \frac{2.25\,t_i\,T}{R_t^2}\,. \tag{5.2.46}$$

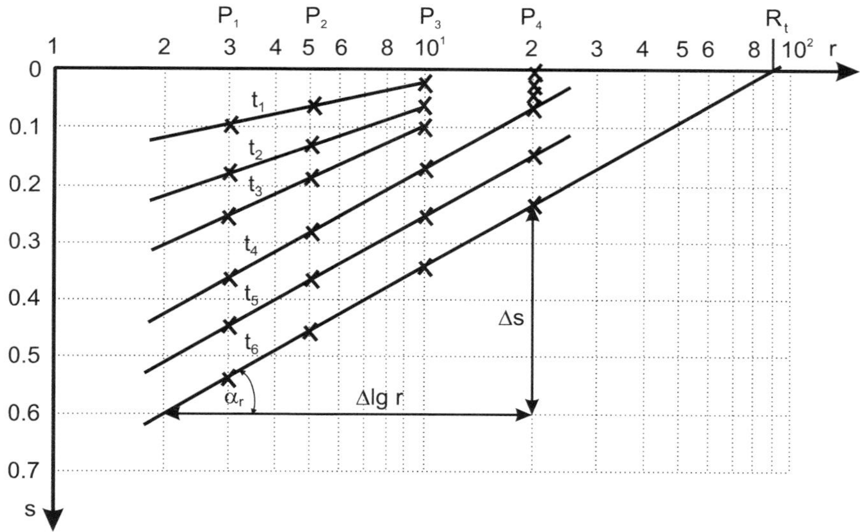

Fig. 5.2-78: Spatial evaluation of drawdown

Spatial Evaluation with respect to time

The spatial-temporal evaluation of pumping tests is based on the relationship of *s*, *t* and *r*, expressed as follows:

$$s = 0.183\frac{Q}{T}\lg\frac{t}{r^2} + C .$$

The plot in *Fig. 5.2-79* shows the drawdown in different wells plotted as a function of t/r^2. The data for three observation wells (P_1, P_2, and P_3) is shown in this example. Theoretically, all plotted values should lie on the same curve. If this is not the case, this indicates that at least one assumption of the underlying model is violated (e.g., no rotation-symmetric flow to the well). The values for transmissivity and storage coefficient can be obtained from this plot:

$$T = 0.183Q\frac{\Delta\left(\lg t/r^2\right)}{\Delta s} \quad \text{and} \qquad (5.2.47)$$

$$S = 2.25T\left(\frac{t}{r^2}\right)_0 . \qquad (5.2.48)$$

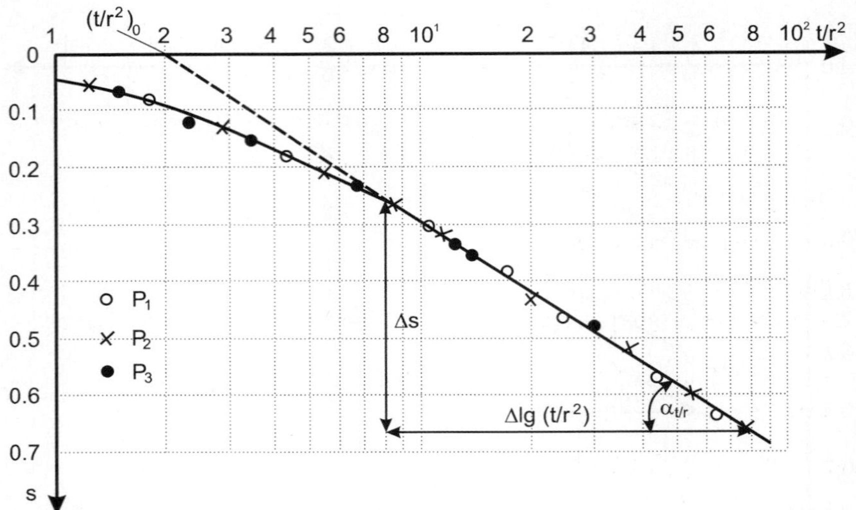

Fig. 5.2-79: Spatial evaluation with respect to time of a pumping test with three observation wells

Recovery analysis

Recovery analysis is based on the superposition principle, since the sudden change in the pumping rate when the pump is switched off, cannot be described mathematically. Instead, the assumption is made that the pump continues to withdraw water from the well. At the time when the pump is stopped, the pump in a second, "virtual" well is switched on into which water infiltrates with an identical rate Q. Hence, the sum of the flow rates in the two wells is zero (reflecting the real-world situation that the pump has been switched off), while mathematically a two-well system with constant Q exists (*Fig. 5.2-80*).

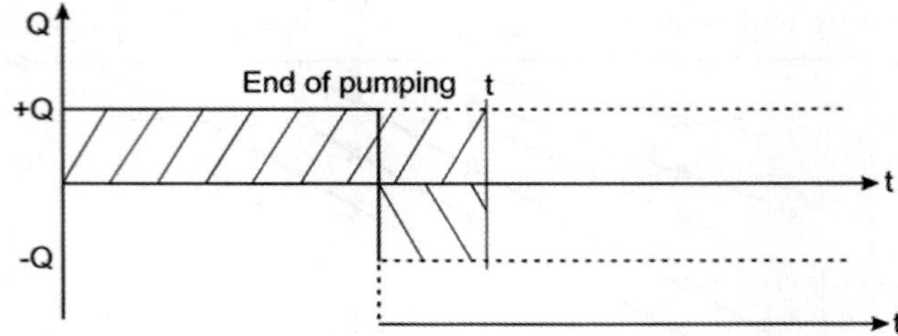

Fig. 5.2-80: Model for evaluating recovery

Methods for evaluation of recovery

For recovery analysis, two more variables need to be introduced:

- residual drawdown, which is the difference between the initial water level and those during recovery: $s' = s - h_0$ and
- time after pumping has stopped: t'.

There are two evaluation methods:

Method 1:
This method uses the residual drawdown s' directly, which is described by

$$s' = 0.183\frac{Q}{T}\left(\lg\frac{2.25Tt}{r^2S} - \lg\frac{2.25Tt'}{r^2S}\right), \tag{5.2.49}$$

which can be solved as follows:

$$s' = 0.183\frac{Q}{T}\lg\frac{t}{t'}. \tag{5.2.50}$$

Transmissivity can be calculated from the best-fit straight line drawn through the observed data, as shown in *Fig.5.2-81*:

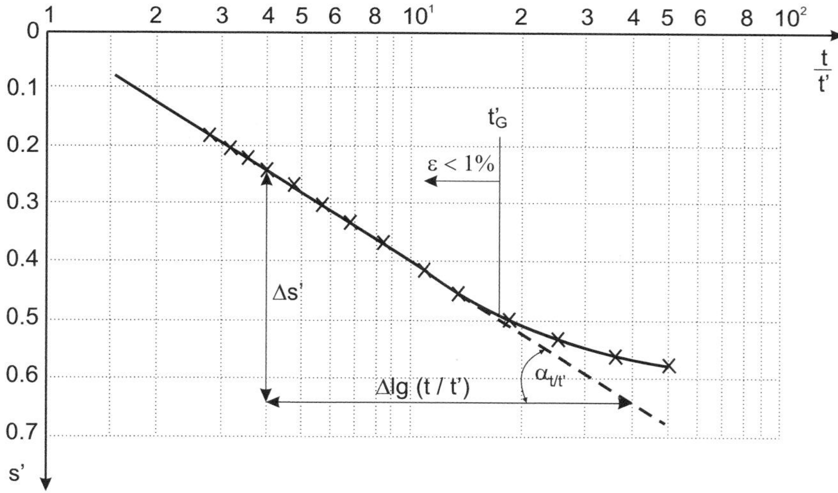

Fig. 5.2-81: Evaluation of recovery (method 1)

$$T = 0.183Q\frac{\Delta \lg \frac{t}{t'}}{\Delta s'}. \tag{5.2.51}$$

If the best-fit straight line intersects the t/t' axis at values > 1, groundwater inflow prevails; if it intersects at values < 1, groundwater outflow prevails.

A disadvantage of this method is that it is generally not possible to determine storativity S from well recovery.

Method 2:

This method uses the difference between the actual and residual drawdown for evaluation

$$s - s' = 0.183\frac{Q}{T}\left(\lg\frac{2.25Tt}{r^2S} - \left(\lg\frac{2.25Tt}{r^2S} - \lg\frac{2.25Tt'}{r^2S}\right)\right), \tag{5.2.52}$$

which is solved as follows:

$$s - s' = 0.183\frac{Q}{T}\lg\frac{2.25Tt'}{r^2S}. \tag{5.2.53}$$

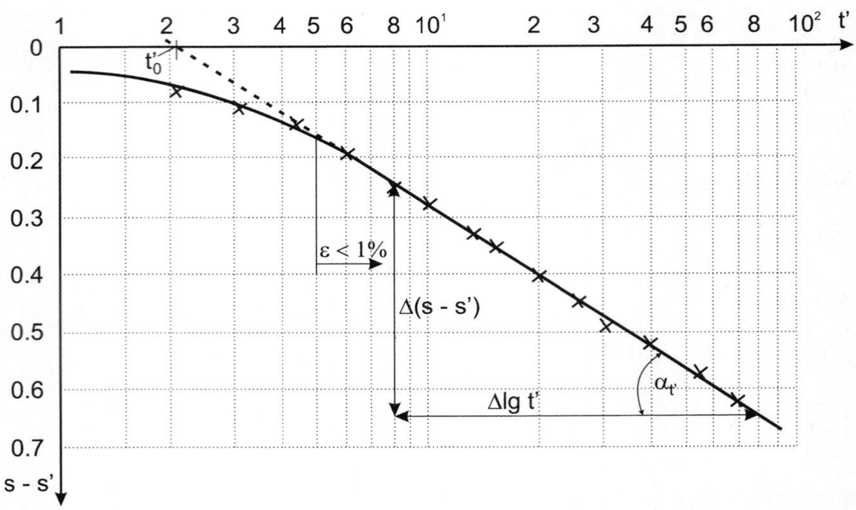

Fig. 5.2-82: Evaluation of recovery of a pumping test (method 2)

After plotting the data as shown in *Fig. 5.2-82*, transmissivity and storage coefficient can be obtained using

$$T = 0.183 Q \frac{\Delta(\lg t')}{\Delta(s - s')} \quad \text{and} \tag{5.2.54}$$

$$S = \frac{2.25 t'_0 T}{r^2}. \tag{5.2.55}$$

This method can be used to determine the skin effect by comparison with the drawdown values. The extrapolation of the drawdown phase on the semi-logarithmic plot can usually carried out without any problem.

Besides these evaluation methods for the normal pumping test, a number of further practical evaluation methods exist:

Pumping tests with constant drawdown

For a well where a constant drawdown s_c is maintained, the following equation can be used for times $t > 10^3\ S/T\ r_0^2$:

$$Q(t) = \frac{T s_c}{0.183 \lg\left(\frac{2.25 t T}{r_0^2 S}\right)}. \tag{5.2.56}$$

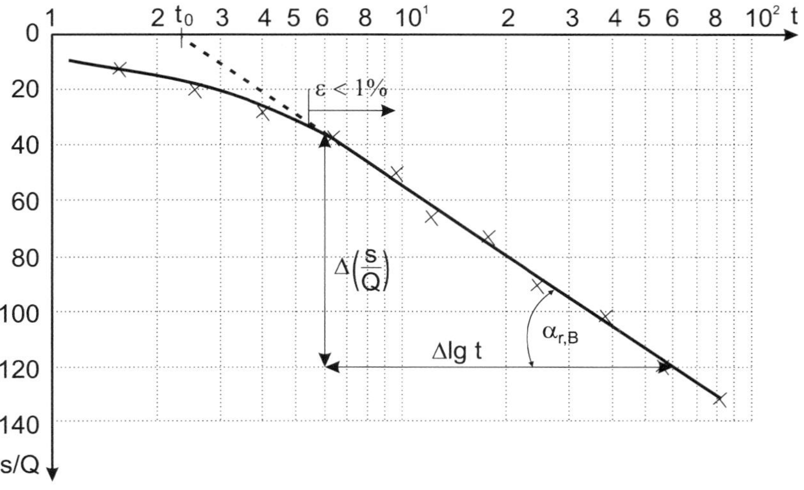

Fig. 5.2-83: Pumping test evaluation with constant drawdown s_c

When the observed parameter values are plotted as shown in *Fig. 5.2-83*, transmissivity and storage coefficient can be obtained from the following equations:

$$T = 0.183\frac{\Delta(\lg t)}{\Delta(s/Q)} \text{ and} \tag{5.2.57}$$

$$S = \frac{2.25 t_0 T}{r_0^2}. \tag{5.2.58}$$

Pumping test with a variable pumping rate

In the case of a pumping test with several pumping stages, each with a constant rate, Q_j, yielding a well hydrograph that can be approximated by a "step function", and when $(t_j - t_{j-1}) > 8.3\ S/T\,r^2$ holds, the following equation describes the drawdown:

$$s = 0.183\frac{1}{T}\sum_{j=1}^{m}\left((Q_j - Q_{j-1})\lg\frac{2.25(t - t_{j-1})T}{r^2 S}\right). \tag{5.2.59}$$

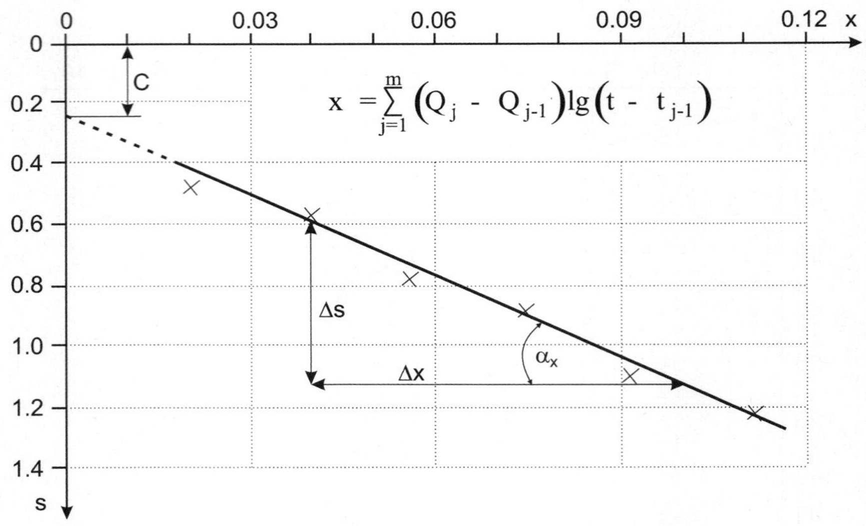

Fig. 5.2-84: Evaluation of pumping test with variable pumping rate

When the drawdown s is plotted as a function of the auxiliary variable x as shown in *Fig. 5.2-84*, transmissivity and storage coefficient can be calculated using the following equations:

$$T = 0.183/\alpha_x \text{ and} \tag{5.2.60}$$

$$\lg\frac{S}{T} = \lg\frac{2.25}{r^2} - \frac{C}{\alpha_x Q_m}, \tag{5.2.61}$$

where α_x is the slope of the best-fit straight line and
Q_m the pumping rate in the last stage.

Pumping test with groundwater inflow from adjacent layers

For a well with inflow from an adjacent semi-permeable layer, a formula that includes a *leakage factor B* can be used:

$$s = \frac{0.366Q}{T}\lg\left(1.12\frac{B}{r}\right). \tag{5.2.62}$$

When the measured drawdown values are plotted as shown in *Fig. 5.2-85*, the parameter values can be determined as follows:

$$T = -0.366Q\frac{\Delta(\lg r)}{\Delta s} \text{ and} \tag{5.2.63}$$

$$B = \frac{R_B}{1.12}. \tag{5.2.64}$$

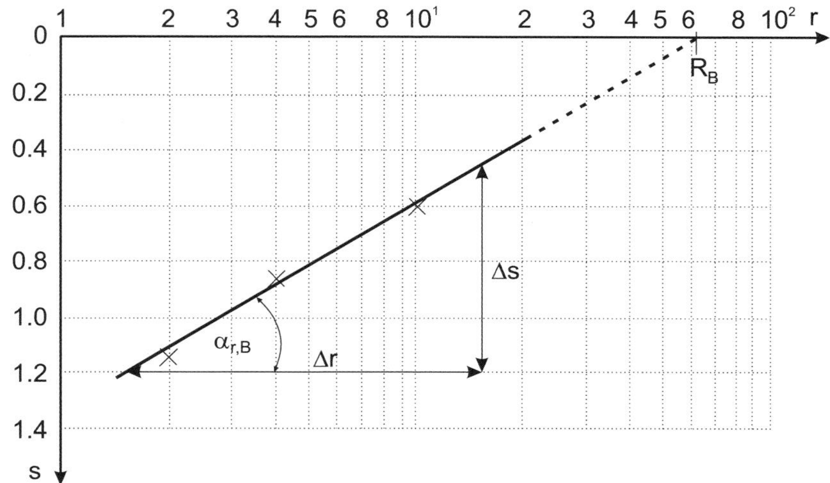

Fig. 5.2-85: Evaluation for inflow from adjacent layers

Pumping test in a two-layered aquifer

The evaluation of a two-layered aquifer system is shown in *Fig. 5.2-86*. In this case, pumping takes place in the lower layer of the aquifer (denoted in the equations by the index *l*). The drawdown analysis shows three phases. In the first phase, only the pumped lower layer reacts. In the second phase, inflow from the upper aquifer layer (index *u*) begins, while in the third phase, both aquifers react like a single aquifer, reflecting the drainage properties of the upper layer. From the data of phase I, transmissivity and storage coefficient can be calculated:

$$T = \frac{0.183Q}{\alpha_{\mathrm{I}}} \text{ and} \tag{5.2.65}$$

$$S_{\mathrm{l}} = \frac{2.25T\,t_{0,\mathrm{I}}}{r^2}. \tag{5.2.66}$$

Another set of equations is used for Phase III:

$$T = \frac{0.183Q}{\alpha_{\mathrm{III}}} \text{ and} \tag{5.2.67}$$

$$S_{\mathrm{u}} = \frac{2.25T\,t_{0,\mathrm{III}}}{r^2}. \tag{5.2.68}$$

For the evaluation of phase II, the auxiliary variable $t_{\mathrm{y},\beta}$ is introduced:

$$t_{\mathrm{y},\beta} = \frac{T\,t_\beta}{S_{\mathrm{u}} r^2}. \tag{5.2.69}$$

The required value of t_β is the time at the intersection of the best-fit straight line of phase II with the best-fit straight line of phase III. If phase III is not distinct enough for a best-fit straight line to be determined, then the auxiliary variable $t_{\mathrm{s},\beta}$ can be used instead:

$$t_{\mathrm{s},\beta} = \frac{T\,t_{\mathrm{s}}}{S_{\mathrm{l}} r^2}. \tag{5.2.70}$$

The value of t_{s} is obtained from the intersection of the best-fit straight line of phase II with the best-fit straight line of phase I. Next, a second variable β is defined for the conductivity values for vertical and horizontal flow. If

$4 \le t_{\mathrm{y},\beta} \le 100$ or $4 \le t_{\mathrm{s},\beta} \le 100$,

β can be determined by

$$\beta = \frac{0.195}{t_{\mathrm{y},\beta}^{1.105}} \text{ or } \beta = \frac{0.195}{t_{\mathrm{s},\beta}^{1.105}}. \tag{5.2.71}$$

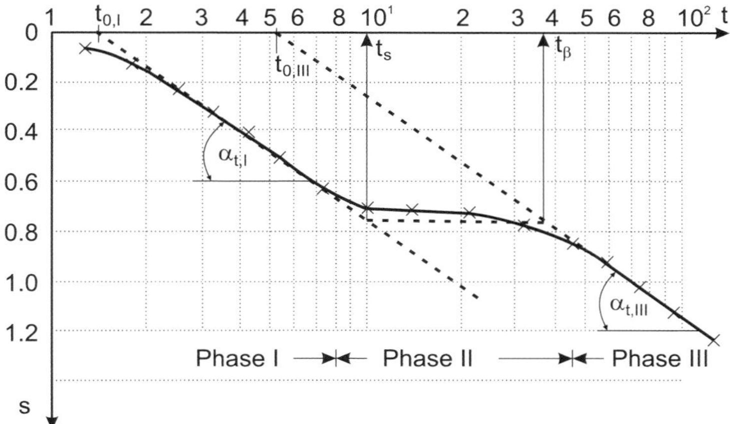

Fig. 5.2-86: Evaluation for delayed yield

The k value for vertical flow through the aquifer can then be obtained from

$$k_v = 3\frac{\beta m_u T}{r^2}, \tag{5.2.72}$$

where m_u is the thickness of the upper layer. The factor "3" in this equation was obtained hypothetically from numerical calculations and by comparison with the evaluation pattern of the mathematical model of HANTUSH (1964).

Pumping test with boundary conditions

If boundary conditions are given, the drawdown at the beginning is similar to the drawdown in an aquifer of infinite extent. When the cone of depression reaches the boundary conditions, the drawdown curve bends (see *Fig. 5.2-87*):

(1) boundary condition "river": upwards bend

(2) boundary condition "barrier": downwards bend

Boundary conditions can also be modeled by appropriate fictitious wells (superposition principle). In addition to the evaluation of phase I as usual, the value of T can be obtained from phase II of drawdown curve (1) as follows:

$$T = \frac{Q}{2\pi s}\ln\frac{\rho}{r}, \tag{5.2.73}$$

where ρ is the distance between the fictitious well and an observation well. Accordingly, from phase II of drawdown curve (2), T can be determined using the following equation:

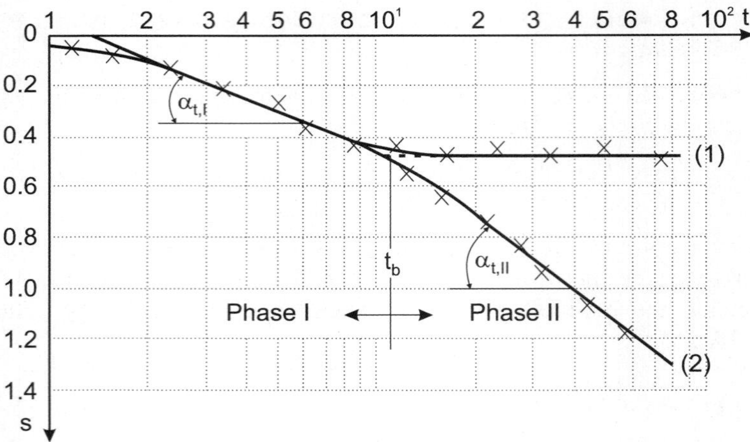

Fig. 5.2-87: Influence of boundary conditions

$$T = 0.366Q\frac{\Delta(\lg t)}{\Delta s}. \tag{5.2.74}$$

The effective distance λ^* of the boundary condition can be calculated from the time at which the curve bends (time t_b):

$$\lambda^* = 0.75\sqrt{\frac{t_b T}{S}}. \tag{5.2.75}$$

Procedure of type curves

When a type curve procedure is used to evaluate the data, the complete mathematical model of the pumping test is used instead of equations that define the relationships between the parameters. It is necessary to do so when the mathematical model is so complicated that the simplifying assumptions used for best-fit straight line methods are not valid, or there is no analytical solution. In this case, type curves have to be obtained by numerical methods.

Even in the case of simple pumping test procedures, it can occasionally be necessary to use type curves if the assumptions for the straight-line approach are not satisfied. If the curves have to be selected from a set of curves, it is sometimes possible to determine not only S and T but also additional independent parameters. If both procedures, best-fit straight line and type curves, can be applied, the one which is more suitable for the actual parameters should be used. The following steps are carried out when type curves are used:

- The well function is plotted on logarithmic graph paper to obtain a curve, called the type curve.
- The observed values are then plotted on a transparent sheet of paper of same scale as the type curve.
- The transparent sheet is then superimposed on the type curve so that both curves match as closely as possible.
- A match point is selected to obtain four values (two curves, each with two coordinates of the match point, see *Fig. 5.2-88*), which are then used to calculate the required parameter values as explained in the following sections.

Wells with constant delivery

The Theis well function, $y = W(\sigma)$, is plotted as a function of $1/\sigma$ on a sheet of graph paper, and the drawdown s as a function of time t is plotted on another sheet. The four values s_m, t_m, y_m, and $(1/\sigma)_m$ for the match point (*Fig. 5.2-88*) are used to calculate the transmissivity and storage coefficient for the aquifer:

$$T = \frac{Q}{4\pi}\frac{y_m}{s_m}, \qquad (5.2.76)$$

$$S = \frac{4T}{r^2}\frac{t_m}{(1/\sigma)_m}. \qquad (5.2.77)$$

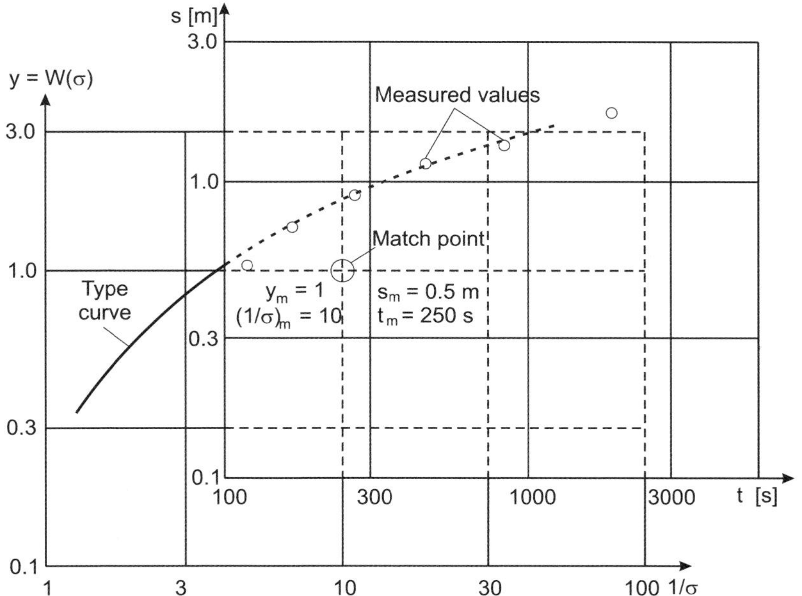

Fig. 5.2-88: Pumping test evaluation using the type curve $W(\sigma)$

Wells with constant delivery and inflow from adjacent layers

A set of curves of $y = W(\sigma)$ is plotted as a function of $1/\sigma$ and $\beta = r/B$ on logarithmic paper and drawdown s is plotted as a function of time t on another sheet of logarithmic paper. As shown in *Fig. 5.2-89*, again four values are obtained for the match point. Transmissivity, storage coefficient, and leakage factor can be calculated using the following set of equations:

$$T = \frac{Q}{4\pi}\frac{y_m}{s_m}, \tag{5.2.78}$$

$$S = \frac{4T}{r^2}\frac{t_m}{(1/\sigma)_m}, \tag{5.2.79}$$

$$B = \frac{r}{\beta_m}. \tag{5.2.80}$$

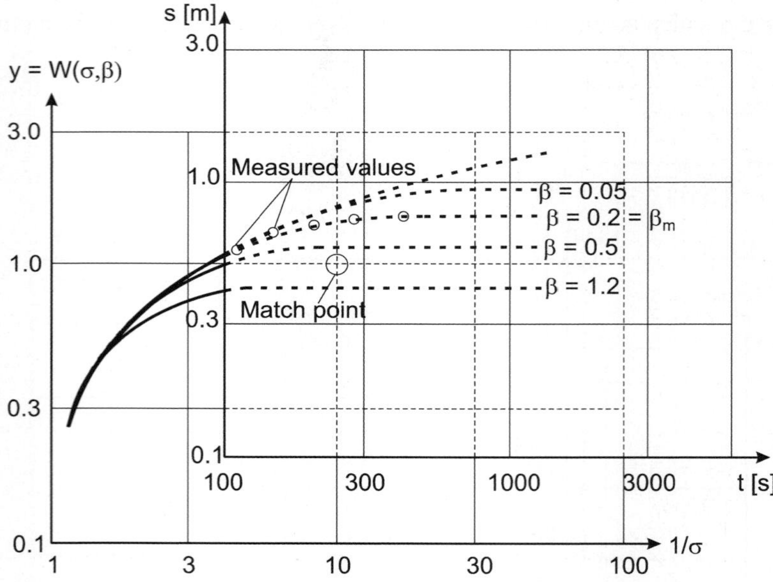

Fig. 5.2-89: Pumping test evaluation using a set of type curves $W(\sigma,\beta)$

Computer-aided procedures

The procedures described above – best-fit straight line and type curves – are in wide use because they are easy to use. However, they have some important shortcomings, as for example:

- Only some of the data, which has often been obtained with great effort, is used for evaluation.

- Often only a temporal or spatial evaluation is done and any possible difference is not analyzed.
- From the individual hydraulic values obtained from the spatial, temporal, drawdown, and recovery analyses, one representative value for the investigated part of the aquifer has to be selected (BUSCH et al., 1993).

These disadvantages are mostly avoided by numerical search techniques, because all measured values are taken into consideration simultaneously.

Method of evaluation

One characteristic of these techniques is the iterative solution of the problem. An initial value which is estimated or approximately determined with other methods is used to solve the problem. Then the values for the potential function *h*, *s* and/or for the volume flux *Q* are compared with the observed values. On the basis of this comparison a new set of geohydraulic parameters is chosen that will yield smaller deviations when the problem is solved again. This is done until the deviations do not become significantly smaller. The biggest difficulty is choosing the new set of parameter values on the basis of the deviation. *Fig. 5.2-90* shows a flow chart for the procedure.

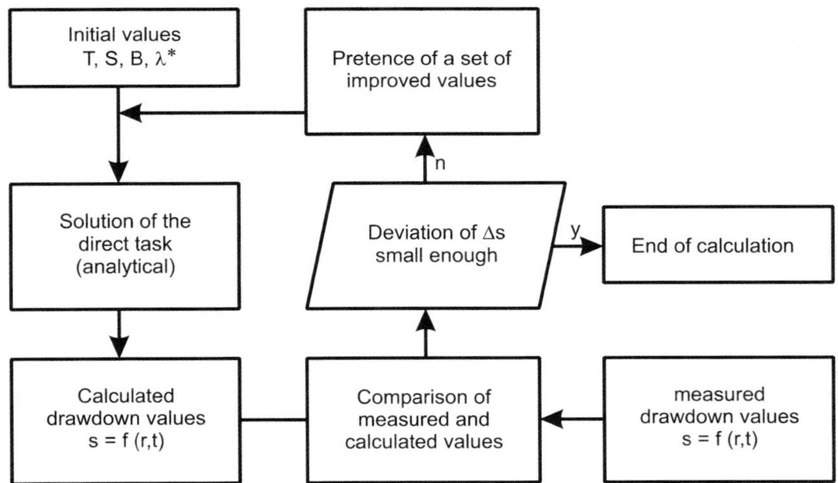

Fig. 5.2-90: Flow chart for computer-aided iterative determination of parameter values

The aim of this iterative procedure is to find a strategy for choosing a new set of parameter values.

The presently used strategies employ a quality function which is determined by the deviation between measured and calculated values. This quality function has to be minimized by the iteration procedure. For example, the sum of the squares of the differences can be used as a quality function:

$$G = \sum_i \alpha_i \left(h_{M,i} - h_{C,i}\right)^2,$$

where $h_{M,i}$ are measured values,
$h_{C,i}$ calculated values, and
α_i arbitrary weighting factors.

The quality function is a function of the parameters that are to be determined. The minimization of the quality function depends greatly on the number of investigated parameters and on the shape of the quality function "peak". It has to be checked whether the investigated parameters are independent of each other, since only then can an unambiguous set of parameters be determined. In the case of dependent variables (e.g., T, S and λ^*), the values are preferably determined in the sensitive areas (parts where there is a clearly established dependency), which often results in a stepwise iterative approach. *Figure 5.2-91* presents the result of the program PSU2G for the evaluation of a pumping test, showing the parameter values for each iteration.

Iteration	Value of quality function	Deviation	*T* value	Step regarding *T* value	*S* value	Step regarding *S* value
0	229.1206	0.000283	0.001	0.0002	0.001	0.0002
1	71.67189	0.000779	0.001764	0.0002	0.000846	0.0002
2	20.81663	0.000826	0.002482	0.0002	0.001253	0.0002
3	7.332969	0.000747	0.003223	0.0002	0.001352	0.0002
4	1.254234	0.001082	0.003898	0.0002	0.002198	0.0002
5	0.134818	0.000943	0.004837	0.0002	0.002283	0.0002
6	0.115335	0.000278	0.004993	0.0002	0.002053	0.0002
7	0.076056	0.000973	0.005633	0.0002	0.001319	0.0002
8	0.073032	0.000343	0.005888	0.0002	0.001089	0.00005
9	0.072904	0.000044	0.005915	0.00005	0.001054	0.0000125
10	0.072903	0.000002	0.005913	0.0000125	0.001055	0.0000031

The desired parameters have been determined!
T value: 0.00591288
S value: 0.00105504

Fig. 5.2-91: Output of the program PSU2G for a digital pumping test evaluation with three observation wells

Numerical procedures

All procedures described above are based on an analytical solution of the flow equation. This requires the aquifer-well system being simplified to a high degree. There is the danger that the real system is simplified to such an extent that the results of a graphical-analytical evaluation are dubious.

Pumping test simulator WELL89

For this reason the pumping test simulator WELL89 has been developed. It makes it possible to evaluate pumping tests in a complex aquifer and with a complex well design without inadequate approximations. *Figure 5.2-92* presents a discrete model for the program WELL89 for a two-dimensional rotation-symmetric numerical solution. It is a finite-volume model in which the hydraulic quantities are computed on the basis of Darcy's Law and the geometric dimensions of the segments of the flow field. With the pumping test simulator, the following characteristics of the aquifer can be taken into account:

- confined or unconfined,
- isotropy or anisotropy,
- homogeneous or layered, and
- capillary zone.

It also offers an opportunity to incorporate the well design characteristics, such as

- perfect or imperfect well,
- varying well diameter,
- multiple screened sections,
- capacity of the well,
- inflow through the bottom of the well, and
- screen losses.

The aquifer properties are discretized vertically on the basis of the geological structure and the well geometry and horizontally according to the following rule:

$$\rho_i = \rho_{i-1} 10^{0.25} . \qquad (5.2.81)$$

The node points are located in the center of gravity of the elements, i.e., they are shifted outwards from the geometric center by an amount of

$$\Delta r_i = \frac{(\rho_{i+1} - \rho_i)^2}{6(\rho_{i+1} + \rho_i)}. \tag{5.2.82}$$

The time steps are initially small and increased as follows:

$$\Delta t_i = \Delta t_{i-1} 10^{0.1}.$$

Both horizontal and vertical transmissivity and storage coefficient components are considered, as well as the specific elastic storage coefficient and the gravity storage coefficient. The model treats an unconfined groundwater table as a mobile boundary. Transmissivity factors are calculated according to the location of the unconfined groundwater table. The described program has been used successfully for the following four tasks of pumping tests:

- analysis of the importance of approximations and assumptions on which the different analytic solutions are based,
- determination of flow conditions close to the well,
- calculation of type curves for aquifer and well conditions for which no analytical solutions exist, and
- interactive pumping test evaluation by means of manual data manipulation.

Figure 5.2-93 shows some results of the program WELL89.

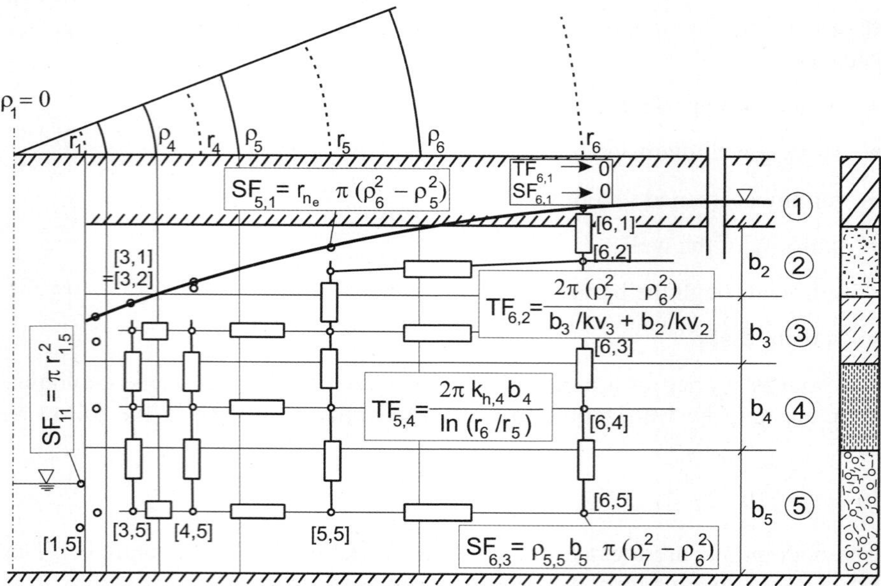

Fig. 5.2-92: Rotation-symmetric discretization with pumping test simulator WELL89

Simulation of drawdown after 2184.7 sec = 36.412 min
No. of time step: 16

Task: Comparison WELL – Lohmann
drawdown in the well (m): 2.000

Radius drawdown in (m)

	0.2	0.29	0.51	0.9	1.6	2.85	5.07	9.02	16.03
layer 1	2	0.01	0.01	0.01	0.01	0.01	0.01	0.01	0.01
layer 2	2	0.51	0.51	0.51	0.5	0.48	0.44	0.38	0.38
layer 3	2	1.9	1.73	1.56	1.39	1.22	1.06	0.89	0.72
layer 4	2	1.9	1.73	1.56	1.39	1.22	1.06	0.89	0.72
layer 5	2	1.9	1.73	1.56	1.39	1.22	1.06	0.89	0.72

Radius drawdown in (m)

	28.51	50.07	90.16	160.32	285.1	506.98	901.55	1603.21
layer 1	0.01	0	0	0	0	0	0	0
layer 2	0.23	0.15	0.09	0.04	0.01	0	0	0
layer 3	0.56	0.39	0.24	0.11	0.03	0	0	0
layer 4	0.56	0.39	0.24	0.11	0.03	0	0	0
layer 5	0.56	0.39	0.24	0.11	0.03	0	0	0

Well inflow layer 1:	0
Well inflow layer 2:	0
Well inflow layer 3:	0.005506
Well inflow layer 4:	0.007337
Well inflow layer 5:	0.005501
Inflow through bottom:	0

Fig. 5.2-93: Result of the program WELL89

5.2.7.3 Laboratory Methods

FLORIAN JENN, CLAUS NITSCHE & HANS-JUERGEN VOIGT

In addition to field tests, laboratory methods are used to determine physical parameters of soil, sediment, and rock. Mostly soil and unconsolidated sediment samples are tested in laboratory for petrophysical properties. Consolidated rocks are seldom used for such testing because of their heterogeneous distribution of cracks and joints. Methods for taking disturbed and undisturbed soil and unconsolidated sediment samples are described in Section 5.3.2.2.

Small soil/sediment samples collected in the field are tested in the laboratory under controlled conditions. The advantage of laboratory tests is that the experiments are carried out "under controlled conditions". Disadvantages are the problems of sample disturbance and representativeness of relatively small samples. Field and laboratory tests supplement and are a check on each other. Described here are laboratory methods for determining grain size distribution, porosity, and hydraulic conductivity of soil and sediments.

Grain size distribution

Grain size is the most fundamental physical property of soil, sediment and rock. It refers to the physical dimensions of individual particles. Grain-size analysis, also known as particle-size analysis or granulometric analysis, is the most basic sedimentological technique to characterize soil and sediment. Data from grain-size analysis is used to estimate parameters such as porosity and hydraulic conductivity, and to classify sediments.

The traditional method of determining the grain size distribution of soil and sediment samples is sieving for the coarse fractions and the pipette method, based on the "Stokes" sedimentation rates, for the fine fractions (sieve-pipette method). Grain-size analysis is now commonly done using high-tech laser instruments.

Granular material can range from very small colloidal particles (0.005 μm = 5 nm), to clay, silt, sand, and gravel, to boulders (in the meter range). Owing to the large range of grain sizes that are encountered (more than seven orders of magnitude), a logarithmic scale is used to represent the grain-size distribution. Different scales are used in the USA and in Europe. The European scale is based on particle size measured in mm or μm and plotted using base 10 logarithms. The U.S. scale is also based on measurements in mm or μm, but uses base 2 logarithms to obtain so-called ϕ (phi) values. The U.S. scale is related to the European scale as follows:

$$\text{phi} = \phi = -\log_2 d = -(\log_{10} d / \log_{10} 2), \tag{5.2.83}$$

where ϕ particle size in ϕ units and
d diameter of particle in mm.

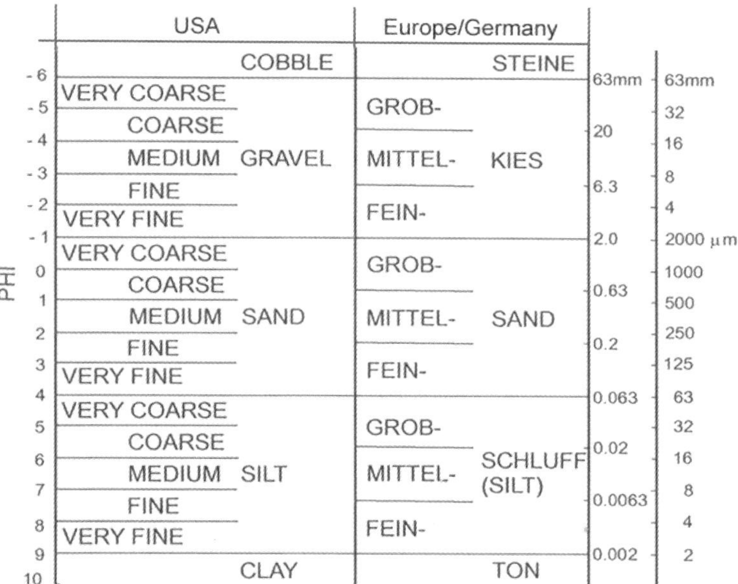

USA
Europe/Germany
PHI
-6
-5
-4
-3
-2
-1
0
1
2
3
4
5
6
7
8
9
10
COBBLE
VERY COARSE
COARSE
MEDIUM GRAVEL
FINE
VERY FINE
VERY COARSE
COARSE
MEDIUM SAND
FINE
VERY FINE
VERY COARSE
COARSE
MEDIUM SILT
FINE
VERY FINE
CLAY
STEINE
GROB-
MITTEL- KIES
FEIN-
GROB-
MITTEL- SAND
FEIN-
GROB-
MITTEL- SCHLUFF (SILT)
FEIN-
TON
63mm
20
6.3
2.0
0.63
0.2
0.063
0.02
0.0063
0.002
63mm
32
16
8
4
2000 μm
1000
500
250
125
63
32
16
8
4
2

Fig. 5.2-94: Comparison of grain-size classification according to the Atterberg scale of the European standard DIN EN ISO 14688 (right) with the Wentworth scale (left) used in USA (modified by DOEGLAS (1968), who moved the boundary between the silt and clay fractions from 4 μm to 2 μm), after FÜCHTBAUER (1988)

The grain-size classification according to the Atterberg scale of the ISO/European standard is compared with the Wentworth scale used in USA in *Fig. 5.2-94.*

For grain size analysis it is common practice to take soil samples from each meter of drill core or from each lithological unit. The samples can be split using either a micro-splitter, cone and quartering, or random bulk (KRUMBEIN & PETTIJOHN, 1938). The coarser the sediment is, the larger the volume of material that must be analyzed. The sample must contain enough material to give an adequate representation of the largest particle sizes. Approximately 500 g sample is considered to be appropriate for coarser sediment, while 50 g is sufficient for fine-grained sediment.

Standard sieving and sedimentation analysis includes

- sample preparation,
- drying of the sample in a convection oven (105 °C) and weighing,
- dispersion,
- wet sieving,
- sieving the coarse fraction,
- sedimentation of the fine fraction and analysis with the pipette technique.

Sample preparation comprises the removal of organic material and salts. The purpose of removing organic matter is that it does not form clastic grains and is, therefore, not to be considered in a grain-size analysis. Depending on the percentage of organic material in the sediment, sample preparation involves adding different volumes of water and 30 % hydrogen peroxide. Hydrogen peroxide is added until the dark color of the organic matter has largely disappeared. Then the sample is washed three times with NaOAc buffer solution (pH = 5) and once with methanol to remove the remaining released cations (JACKSON, 1956). If the sample contains mineralized water, the sample must be washed with distilled water to remove the salts.

Drying of the sample in a convection oven and weighing: Samples are dried in convection ovens at 105 °C (overnight is usually sufficient). The dry samples are then cooled in desiccators. The samples are weighed immediately after removal from the desiccators to minimize their exposure to the moisture in the ambient air (POPPE et al., 1998).

Dispersion: The purpose of this step is to break up the aggregates of clay-sized particles so that they will be analyzed individually and not as aggregates. This procedure consists of mixing the sediment with a dispersing agent such as sodium metaphosphate (PFANNKUCH & PAULSON, 2005).

Wet sieving is done to separate the soil sample into coarse (sand and gravel, which are retained on the sieve), and fine (clay and silt) fractions. The procedure involves placing a sample on a 40 µm (German standards DIN 18196 and DIN 4022) or 62 µm (U.S. standard ASTM D 2487) sieve. A funnel above a 1000 mL beaker is placed below the sieve to catch everything that passes through the sieve. If the fine fraction amounts to more than 5 % of the sample, then distilled water with a dispersing agent such as sodium metaphosphate is used during wet sieving. The procedure can be terminated when the wash water leaving the funnel is clear and apparently contains no fines. The fine fraction collected in the beaker is used for sedimentation analysis, while the material on the sieve is subjected to a sieve analysis.

Sieving the coarse fraction: The grain-size distribution of the coarse fraction is analyzed by sieving through a tower of sieves (*Fig. 5.3-95*). The size of mesh openings of the sieves is standardized in Europe as follows:

6, 3.15, 2.0, 1.0, 0.63, 0.315, 0.2, 0.1, 0.063 mm.

Table 5.2-17: U.S. standard sieve mesh, modified after PFANNKUCH & PAULSON (2005)

U.S. Standard sieve mesh No.	Millimeters (fractional)	Millimeters	phi (ϕ)	Wentworth size class
Use		4096	-12	
		1024	-10	boulder (-8 to -12 ϕ)
wire		256	-8	cobble (-5 to -8 ϕ)
		64	-6	
squares		16	-4	pebble (-2 to -5 ϕ)
5		4	-2	
6		3.36	-1.75	
7		2.83	-1.5	granule (-1 to -2 ϕ)
8		2.38	-1.25	
10		2	-1	
12		1.68	-0.75	
14		1.41	-0.5	very coarse sand (0 to -1 ϕ)
16		1.19	-0.25	
18		1	0	
20		0.84	0.25	
25		0.71	0.5	coarse sand (1 to 0 ϕ)
30		0.59	0.75	
35	1/2	0.5	1	
40		0.42	1.25	
45		0.35	1.5	medium sand (2 to 1 ϕ)
50		0.3	1.75	
60	1/4	0.25	2	
70		0.21	2.25	
80		0.177	2.5	fine sand (3 to 2 ϕ)
100		0.149	2.75	
120	1/8	0.125	3	
140		0.105	3.25	
170		0.088	3.5	very fine sand (4 to 3 ϕ)
200		0.074	3.75	
230	1/16	0.0625	4	
270		0.053	4.25	
325		0.044	4.5	coarse silt (5 to 4 ϕ)
Analyzed		0.037	4.75	
	1/32	0.031	5	
by	1/64	0.0156	6	medium silt (6 to 5 ϕ)
	1/128	0.0078	7	fine silt (7 to 6 ϕ)
pipette	1/256	0.0039	8	very fine silt (8 to 7 ϕ)
		0.002	9	clay (> 9 ϕ)
or		0.00098	10	
		0.00049	11	
hydrometer		0.00024	12	
		0.00012	13	

Fig. 5.2-95: Sieving shaker with a tower of sieves

The U.S. standard for sieve mesh size based on the phi scale is given in *Table 5.2-17*. Sieves are arranged in descending order of respective mesh size openings with the coarsest sieve at the top. The individual sand and gravel fractions retained on each sieve at the end of sieving are weighed and recorded. The sieve analysis data is normally directly entered into a computer and is analyzed using special software.

Sedimentation methods (hydrometer or pipette) to determine the fine-grained fraction: The fine-grained fraction consists of silt (particle size 2 - 62 μm) and clay (particle size less than 2 μm, particles smaller than 0.1 μm are colloidal clay). When the particles are too small for the dry sieving method (i.e. < 74 μm in the U.S. standard or < 63 μm in the European standard), a hydrometer or the pipette method is used to separate particles in this size range.

The sedimentation methods of particle-size analysis are based on Stokes' law, which relates the settling velocities of the spherical particles to the size (diameter) and density of the particles, and the viscosity of the fluid in which they are suspended. Stokes' law of settling velocities can be expressed as follows:

$$v = \frac{h}{t} = \frac{(\rho_1 - \rho_2) \mathrm{G} d^2}{18\eta} \qquad d = \sqrt{\frac{18\eta v}{\mathrm{G}(\rho_1 - \rho_2)}} \tag{5.2.84}$$

where v settling velocity [cm s^{-1}]
ρ_2 density of the fluid [g cm^{-3}]
h settling depth [cm]
G gravity constant [cm s^{-2}]
t settling time [s]
d equivalent diameter of the particle [cm]
ρ_1 density of the particles [g cm^{-3}]
η viscosity of the fluid (e.g., water) [g cm^{-1} s^{-1}].

Hydrometer and pipette methods involve preparing a homogeneous suspension of the fine fraction with a dispersing agent (usually 0.5 % sodium hexametaphosphate) in a 1000 mL graduated cylinder and allowing the particles to settle (KRUMBEIN & PETTIJOHN, 1938; FOLK, 1974; DIN 66115). *Table 5.2-18* gives the estimated times and corresponding depths to which a particle with certain grain size will settle for several suspension temperatures. Samples are withdrawn from the suspension and the *wt.* % are determined (in the pipette method) or measurements are made of the suspension (hydrometer analysis) at defined time intervals.

Table 5.2-18: Estimated times and corresponding depths to which a particle with a certain grain size will settle according to Stokes law of settling velocities (POPPE et al., 1998) at several suspension temperatures

ϕ	Grain size[µm]	at 20°C		at 24°C		at 32°C	
		Depth [cm]	Time	Depth [cm]	Time	Depth [cm]	Time
4	63	20	20 s	20	20 s	20	20 s
5	31	10	1 m 56 s	10	1 m 45 s	10	1 m 30 s
6	16	10	7 m 44 s	10	6 m 58 s	10	6 m
7	8	10	31 m	10	28 m	10	23 m
8	4	10	2 h 3 m	10	1 h 51 m	10	1 h 34 m
9	2	5	4 h 6 m	10	7 h 24 m	5	2 h 58 m
10	1	5	16 h 24 m	5	14 h 50 m	5	6 h 11 m
11	0.5	5	62 h 15 m	5	59 h 20 m	5	24 h 43 m
12	0.24			5	237 h 20 m		
13	0.12			5	949 h		

A hydrometer is a device used to measure the density or specific gravity of liquids. It consists of a weighted tube that floats vertically and determines the specific gravity of the liquid by comparison of the scale on the tube with the level it floats in the liquid. ASTM hydrometers are graduated by the manufacturer to read in either specific gravity or in grams per liter (g L^{-1}) and are calibrated for a standard temperature of 20 °C (68 °F).

After the initial shaking to homogenize the sediment, aliquots are drawn in the pipette method at a time that the last particles of a given size are just settling past the sampling depth. The aliquots are dried and weighed, which gives the weight of the material finer than the given size. Successive aliquots give the weights of silt and clay particles in each fraction. The weights of the fractions are recorded and the size distribution is determined from the weight of the sediment sample.

Conventional grain-size analyses of soil/sediment using a hydrometer or pipette are subject to a significant precision error (ca. 25 - 40 %); they can quantify only a limited number of grain-size classes and are time consuming (SPERAZZA et al., 2002). Laser particle size analyzers employ low-angle light scattering based on Mie theory to rapidly measure grain-size distributions. The light is emitted from a neon-helium laser source. The analyzer can determine a full range of particle sizes in as little as quarter-phi increments from 0.5 µm to nearly 4000 µm, i.e., from very fine clay to fine gravel. In conventional particle-size analysis (sieving and pipetting, for example) the most detailed analyses are commonly at one-phi intervals. Laser grain-size analysis takes typically less than 30 seconds per sample for a complete analysis. Time is required to remove cementing and flocculating agents prior to the laser analysis. In terms of laboratory efficiency, laser diffractometry is far superior to the other methods. This technology is perfect for compositionally homogeneous materials. However, the results of a laser diffractometer reading is influenced by the different optical properties and nonsphericity of the constituent particles of natural sediments (SPERAZZA et al., 2002). For example, the nonsphericity of the particles causes an increase in the observed mean diameter. The platy form of the clay particles causes a considerable difference (as much as eight size classes) between pipette and laser measurements. KONERT & VANDENBERGHE, (1997) found that the < 2 µm grain size determined by the pipette method corresponds to a grain size of 8 µm as determined by the Laser Particle Sizer. Because the different measurement techniques can provide different results, different methods should be used and the results compared for an evaluation of the material of the suspension.

The grain-size distribution of a soil or sediment is represented in a table or as grain-size distribution curve showing the cumulative weight percent of the different fractions. The data is plotted on semilogarithmic graph paper with particle diameter on the logarithmic x-axis and the percent weight of the fractions on *y*-axis. In Europe, curves of the cumulative percent weight retained by the sieves are generally plotted, which means the cumulative curve will start with the weight of the finest fraction, followed by the successive coarser fractions. In the USA curves of the cumulative weight percent passed by the sieve are normally used, in which each successive fraction is finer than the previous one. *Table 5.2-19* shows the results (in phi units) of a grain-size analysis together with the data for the two types of grain-size distribution curves (retained and passing curves) (*Fig. 5.2-96*). *Figure 5.2-97* shows some examples of grain-size distribution curves for different substrates.

Table 5.2-19: Example of grain-size analysis data from a sediment sample, PFANNKUCH & PAULSON (2005)

Grain size [mm]	Grain size (ϕ)	Weight of size fraction [g]	Weight percent	Cumulative weight % retained	Cumulative weight % passed
0.01	6.64	0.6	0.02	100	0
0.063	4	0.5	0.02	99.98	0.02
0.125	3	0.6	0.02	99.96	0.04
0.25	2	1.2	0.04	99.94	0.06
0.354	1.5	1.8	0.06	99.9	0.1
0.5	1	4.7	0.16	99.84	0.16
0.707	0.5	17.5	0.59	99.68	0.32
1	0	172.2	5.81	99.09	0.91
1.41	-0.5	2570.1	86.74	93.28	6.72
2	-1	152.7	5.15	6.54	93.46
2.83	-1.5	41.2	1.39	1.39	98.61
4	-2	0	0	0	100

Other statistical parameters can be derived from the determined grain-size distribution: The coefficient of uniformity (U) characterizes how well or poorly sorted the soil/sediment is:

$$U = d_{60} / d_{10} \,, \qquad (5.3.85)$$

where d_{60} and d_{10} are the grain diameters in mm for which 60 % and 10 % of the sample is finer (*Fig. 5.2-98*). Small values of U mean the soil or sediment is well sorted. Sorting is classified on the basis of the coefficient of uniformity according to different standards (*Table 5.2-20*). The coefficient of uniformity plays an important role in the estimation of porosity, hydraulic conductivity and effective grain-size diameter.

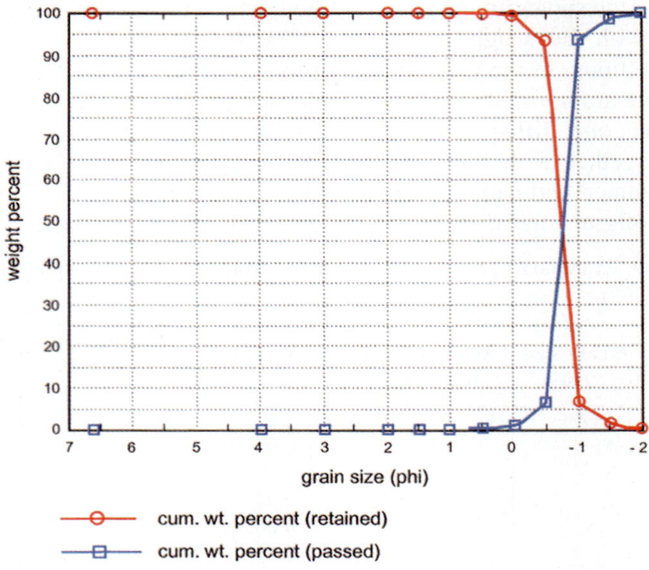

Fig. 5.2-96: The two types of cumulative weight percent curves, PFANNKUCH & PAULSON (2005)

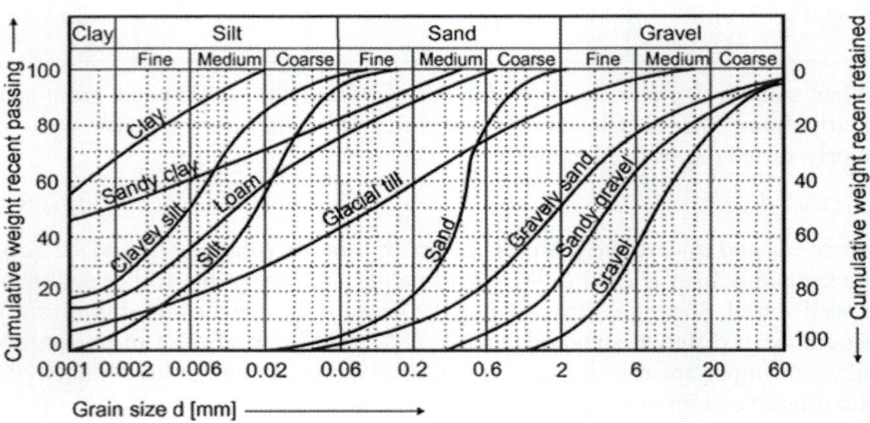

Fig 5.2-97: Grain size distribution curves of different substrates, FECKER & REIK (1996)

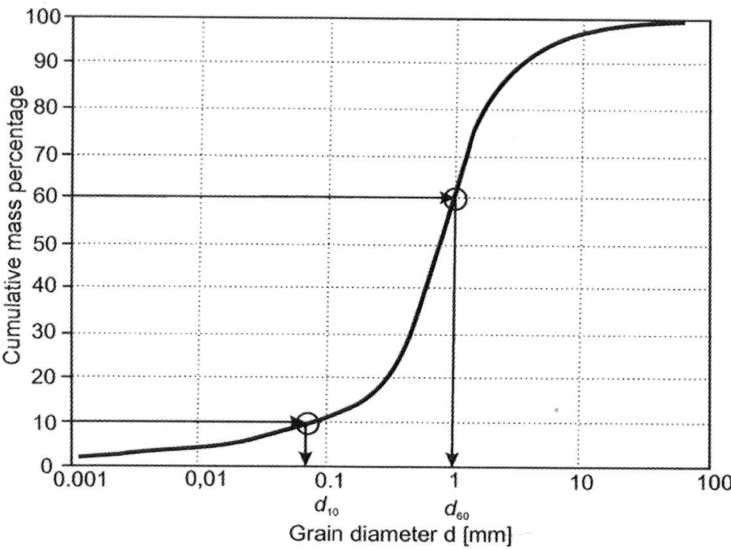

Fig. 5.2-98: Definition of the diameters d_{10} and d_{60} for the determination of the coefficient of uniformity

Table 5.2-20: Sorting classes of soil/sediment based on the coefficient of uniformity *U* according to different standards

	DIN 1054	DIN 18196	CASAGRANDE (1948)
well sorted	$U < 3$	$U < 6$	$U < 4$
moderately sorted			$4 < U < 6$
poorly sorted	$3 < U < 15$	$U > 6$	$U > 6$
very poorly sorted	$U > 15$		

According to KOZENY (1928), the effective grain-size diameter d_e of a mixture of grain sizes corresponds to the average equivalent sphere diameter of a material with a single grain size having the same specific surface as the mixture containing different grain sizes. In most cases d_e is equivalent to d_{10}. BEYER (1964) found a correlation between the effective grain-size diameter d_e, the coefficient of uniformity *U* and d_{10} (*Table 5.2-21*).

Table 5.2-21: Relationship between the effective grain-size diameter d_e, the coefficient of uniformity *U*, and d_{10} , BEYER (1964)

U	d_e / d_{10}
1 - 1.99	1.0 - 1.6
2 - 2.99	1.6 - 1.9
3 - 4.99	1.9 - 2.2
5 - 9.99	2.2 - 2.5
> 10	> 2.5

The results of grain-size analysis can be used to *classify soils and sediments* according to *Fig. 5.2-94* in the following manner: The soil or sediment is named by the fraction with the highest weight percent in the total sample. It can be further classified by denoting all fractions with a content of more than 30 % as "strongly" and all fractions with a content between 15 % and 30 % as "slightly". The remaining fractions are ignored for naming. Consider the example with following weight percent fractions according to grain-size analysis:

1.15 % gravel
22.8 % coarse sand
43.5 % medium sand
31.7 % fine sand
0.45 % silt
0.40 % clay

The sample name is "strongly fine-sandy, slightly coarse-sandy, medium sand".

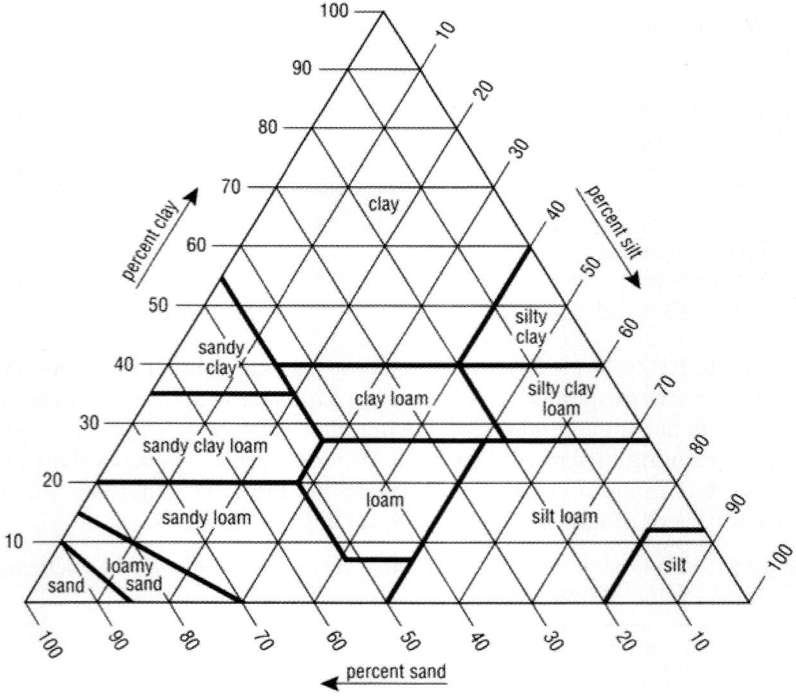

Fig. 5.2-99: U.S. Department of Agriculture method for naming soils (soil texture triangle), USDA (1993)

Another widely used classification system is the triangle of basic soil texture classes of the U.S. Department of Agriculture (*Fig. 5.2-99*). In this system the percentages of sand, silt and clay particles are used for the classification of the sample. The soil type is determined by plotting the percentages of each of the three soil particle classes in the soil sample on the soil texture triangle.

Porosity

Soil, sediment and rock are multi-phase systems consisting of a solid part (solid phase) and voids filled with gaseous and/or liquid phases. The primary voids (pores) can be modified by tectonic processes (development of cracks, joints and fissures), dissolution and cementation, to form secondary pore spaces. The primary minerals of the solid phase must be distinguished from the secondary cement minerals (e.g., quartz and carbonate). Therefore, porosity is a complicated parameter (*Figs. 5.2-100* and *5.2-101*).

Porosity depends on the structure and texture, uniformity of grain size (*Fig. 5.2-102*), grain shape (*Fig. 5.2-103*), packing density (*Fig. 5.2-104*) and degree of cementation of the pore and/or joint spaces (*Fig. 5.2-105*).

The porosity n is a fundamental property of soil/sediment/rock that controls the storage and the movement of water in the ground. Porosity is a dimensionless quantity and can be reported either as a decimal fraction or as a percentage. It can be measured in the laboratory.

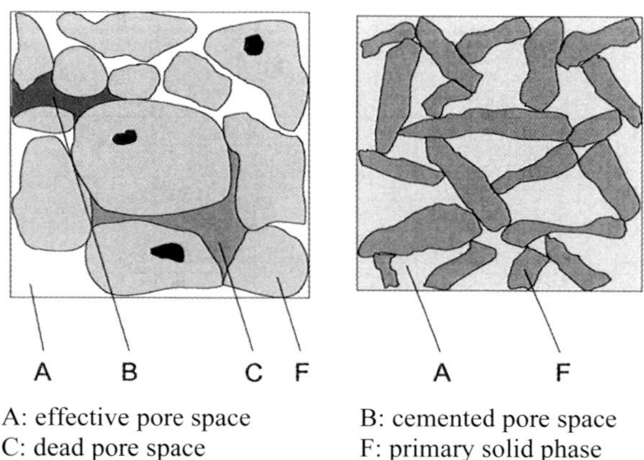

Fig. 5.2-100: Grain texture of a sediment with early cementation

Fig. 5.2-101: Texture of clay

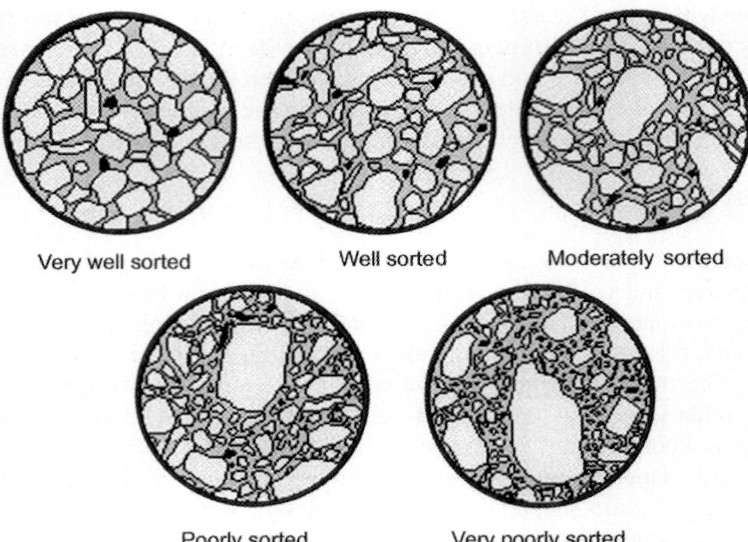

Fig. 5.2-102: Porosity is influenced by the degree of uniformity of grain size

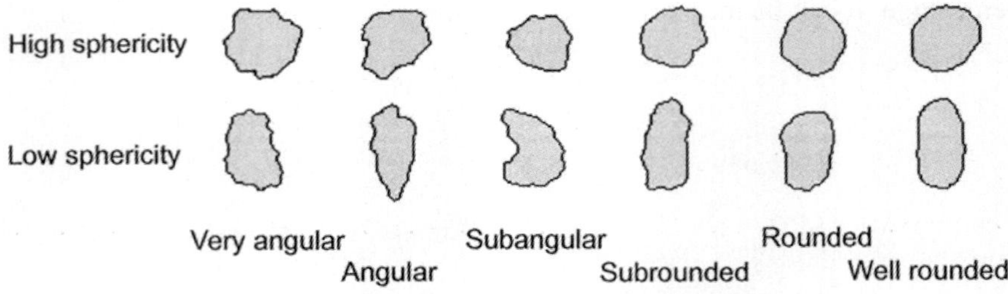

Fig. 5.2-103: Kinds of grain shape

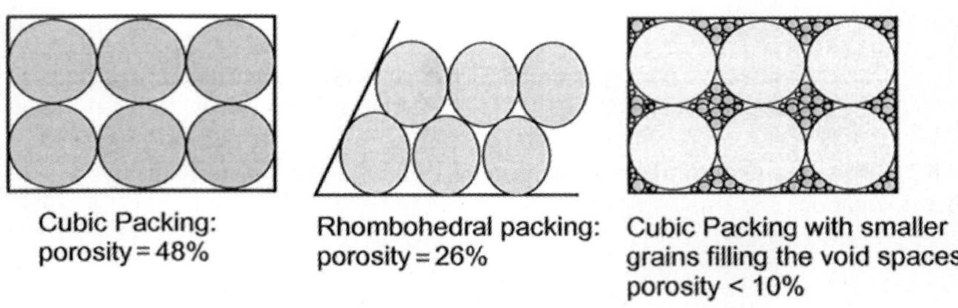

Fig. 5.2-104: Types of packing and their influence on porosity

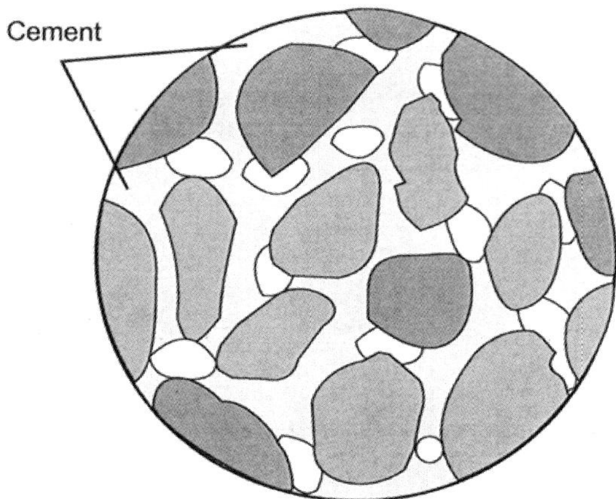

Fig. 5.2-105: Reduction of the pore space by cementation

The total porosity n of a porous medium is the ratio of the pore volume to the total volume of a representative sample. Assuming that the sample is composed of solid, liquid (water), and gas (air) phases, where V_s is the volume of the solid phase, V_l is the volume of the liquid phase, V_g is the volume of the gaseous phase, $V_p = V_l + V_g$ is the volume of the pores, and $V = V_s + V_l + V_g$ is the total volume of the sample, then the total porosity of the soil sample, n, is defined as

$$n = \frac{V_l + V_g}{V_s + V_l + V_g} = \frac{V_p}{V} . \tag{5.2.86}$$

The porosity of the soil/sediment/rock is expressed as total porosity, effective porosity, and specific yield (drainable porosity). Porosity ranges of common unconsolidated and consolidated rocks are given in *Table 5.2-22*. Only part of the water in a rock flows out when a saturated sample is drained under gravity. The volume of drainable pores is called specific yield or drainable porosity (n_d). The amount of water that a unit volume of a sample retains after drainage under gravity is called specific retention or field capacity. The sum of specific yield and specific retention is equal to the total porosity (*Fig. 5.2-106*). In unconfined aquifers the specific yield is equal to the storage coefficient. *Tables 5.2-23* and *5.2-24* compare the ranges of total porosity and specific yield for selected rocks.

Table 5.2-22: Porosity ranges of common unconsolidated and consolidated rocks, after DRISCOLL (1986)

Unconsolidated sediments	***n* [%]**	**Consolidated rocks**	***n* [%]**
clay	45 - 55	sandstone	5 - 30
silt	35 - 50	limestone/dolomite	
sand	25 - 40	(original & secondary porosity)	1 - 20
gravel	25 - 40	shale	0 - 10
sand & gravel mixture	10 - 35	fractured crystalline rock	0 - 10
glacial till	10 - 25	vesicular basalt	10 - 50
		dense, solid rock	< 1

Effective porosity, n_e, is the amount of pore space available for fluid flow. Effective porosity excludes pores occupied by water adsorbed onto the soil particle surfaces and water in dead-end pores. Effective porosity n_e is less than porosity n.

Table 5.2-23: Comparison of total porosity n and specific yield/drainable porosity n_d for selected unconsolidated sediments, DAVIS & DE WIEST (1966)

Material	**Porosity *n***	**Specific yield/drainable porosity n_d**
sandy gravel	0.25 - 0.35	0.20 - 0.25
gravelly sand	0.28 - 0.35	0.15 - 0.20
medium sand	0.30 - 0.38	0.10 - 0.15
silty sand	0.33 - 0.40	0.08 - 0.12
sandy silt	0.35 - 0.45	0.05 - 0.10
clayey silt	0.40 - 0.55	0.03 - 0.08
silty clay	0.45 - 0.65	0.02 - 0.05

Laboratory tests for effective porosity show that when groundwater velocity is low, effective porosity is normally equivalent to total porosity (SARA, 1993). In contrast to total porosity and specific yield, the effective porosity is not only a function of the grain size of the sediment, but also of the hydraulic boundary conditions.

Water in soils and in rock has often a dual character with regard to flow. The porous medium can consist of mobile and immobile regions (dual-porosity medium). Part of the water is mobile and flows along continuous joints, fractures, connected pores, while another part is immobile (*Fig. 5.2-107*). Exchange of water and solute between the two regions occurs by diffusion. Transfer of contaminants between the mobile and immobile regions is the main reason for the tail behind the contaminant front.

Table 5.2-24: Specific yield ranges for common sediments, after WALTON (1970)

Sediment	Specific yield [%]
clay	0.01 - 0.10
sand	0.10 - 0.30
gravel	0.15 - 0.30
sand and gravel	0.15 - 0.25
sandstone	0.05 - 0.15
shale	0.005 - 0.05
limestone	0.005 - 0.05

Total porosity

There are several ways to estimate the total porosity *n* of a given soil/sediment/rock sample.

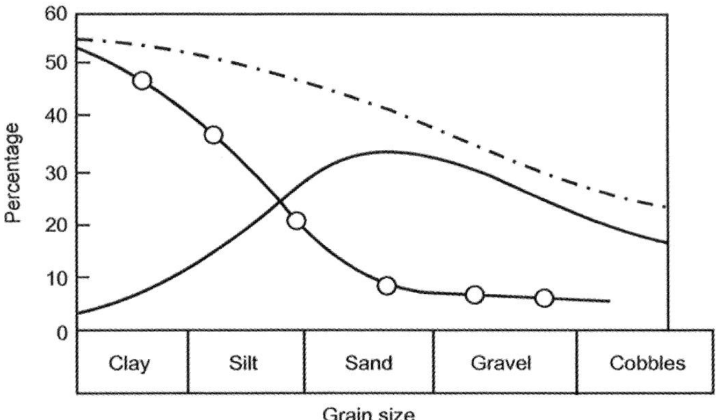

Fig. 5.2-106: Relationship between total porosity (dot-dashed), specific yield (solid line), specific retention (line with circles) and types of sediments, DAVIS & DE WIEST (1966)

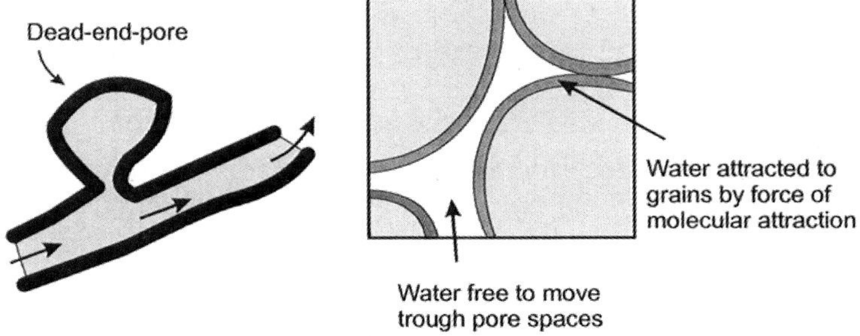

Fig. 5.2-107: Pore spaces with mobile and immobile water, modified from BEAR (2001)

Direct methods: According to the definition of total porosity in Equation (5.2.86), a sample can be evaluated for n by directly measuring the total volume (V) and the pore volume (V_p). The total volume is easily obtained by measuring the total volume of the sample with a ruler in the case of a consolidated rock sample and in case of unconsolidated material with a graduated cylinder. The pore volume can, in principle, be determined by measuring the volume of water needed to completely saturate the sample. In practice, however, it is always difficult to saturate the soil/sediment/rock sample completely and, therefore, the total porosity of the sample is rarely determined by this method.

Another possibility is to completely saturate the sample and weigh it. Then place the sample into an oven to dry it. Drying may take several days depending on the heat applied and the volume of the sample. Then the dried sample is weighed again. The difference between the weights of the saturated and the dried sample is equal to the volume of the water removed from the sample (measured in grams), which is equal to the pore volume (in cubic centimeters), since the density of water is 1 g cm^{-3}. The objections to this method are the same as for the previously described method.

Density method: The total porosity is usually determined indirectly by using the expression

$$n = \frac{V_p}{V} = 1 - \frac{\rho_b}{\rho_s}. \tag{5.2.87}$$

Equation (5.2.87) can be obtained by rearranging Equation (5.2.86). *Bulk* or *dry density* ρ_b is the ratio of the mass of the solid phase m_s of soil/sediment/rock to its total volume V. *Particle density* ρ_s of the soil/sediment/rock particles is the dry mass of the solid phase m_s divided by the solid (not bulk) volume V_s of the particles, in contrast to the bulk density. Units are for both densities kg m^{-3}. In this approach, the values of ρ_b and ρ_s are determined in the laboratory or in-situ and are then used to calculate the total porosity n.

Measurement of bulk density: The mass of the solid phase m_s is measured by weighing after drying the sample at 105 °C until a constant weight is reached. The drying temperature of 105 °C is to be taken only for samples which do not release water of crystallization at this temperature. For samples with a high swelling clay content drying temperature must not exceed 65 °C.

One method to determine the total volume V is to measure with a ruler a cylindrical sample taken with a steel pipe or a cubic sample and to calculate the volume. Another method is to coat the sample with a water repellant substance, such as paraffin, put the sample in a partly water filled graduated cylinder and determine the volume of the displaced water.

In the excavation method a quantity of soil/sediment is excavated, oven dried, and weighed. The volume of the hole left is measured by filling the hole with either sand or water and measuring the volume needed to fill the hole. In the water method, a plastic foil or back or balloon is placed in the hole and is filled with water from a graduated vessel.

Measurement of particle density: The sample is ground to a fine powder in order to open all pores and is then dried as described above. The particle density ρ_s can be determined using a pycnometer and obtaining the following four masses by weighing:

m_1 Mass of the empty pycnometer,
m_2 Mass of the pycnometer plus the dried sample,
m_3 Mass of the pycnometer plus the sample and filled with water up to a fixed volume, V_f, and
m_4 mass of the pycnometer filled with pure water up to the fixed volume V_f.

With the density ρ_w of water the particle density ρ_s is

$$\rho_s = \frac{(m_2 - m_1)}{(m_4 - m_1) - (m_3 - m_2)} \rho_w . \tag{5.2.88}$$

This method is precise, but it requires careful measuring of volumes and masses, and consideration of the effects of pressure and temperature conditions on water density. It is important to ensure that no air is entrapped in the sample or in the pycnometer during weighing.

Specific yield

The specific yield is mostly obtained from a water retention curve (pF-curve). Equipment for column tests can be used for sandy materials (Section 5.3.5.5). In this case the soil column will be slowly saturated from the bottom to the top over a time of 48 hours (*Fig. 5.2-108*).

After saturation it is determined how much water the soil column retains when a specified vacuum is applied to the bottom of the soil column (hanging water column). Once the outflow of the water stops, the column is weighed, and the specific yield can be calculated from the difference to the weight of the saturated column.

The water retention curve of the soil/sediment must be determined for detailed investigations. The water retention curve gives the relationship of volumetric water content θ in the sample [m^3 m^{-3} or vol. %] to the soil-water tension p_c (capillary pressure equal to the energy required to extract available water from the soil). The water retention curve determined for wetting a sample is different from that obtained while drying a sample due to hysteresis effects on the hydraulic properties of soil/sediment/rock (*Fig. 5.3-109* and Section 5.2.4).

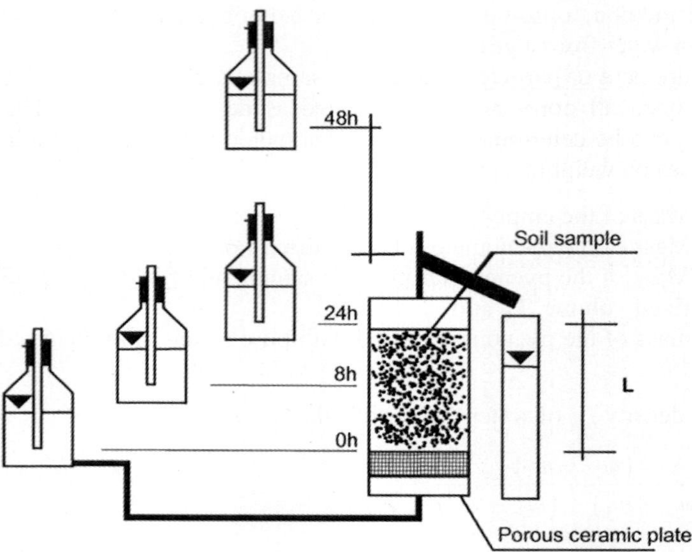

Fig. 5.2-108: A soil column is saturated with water by incrementally increasing the pressure head by raising the Mariotte bottle, L - length of soil column

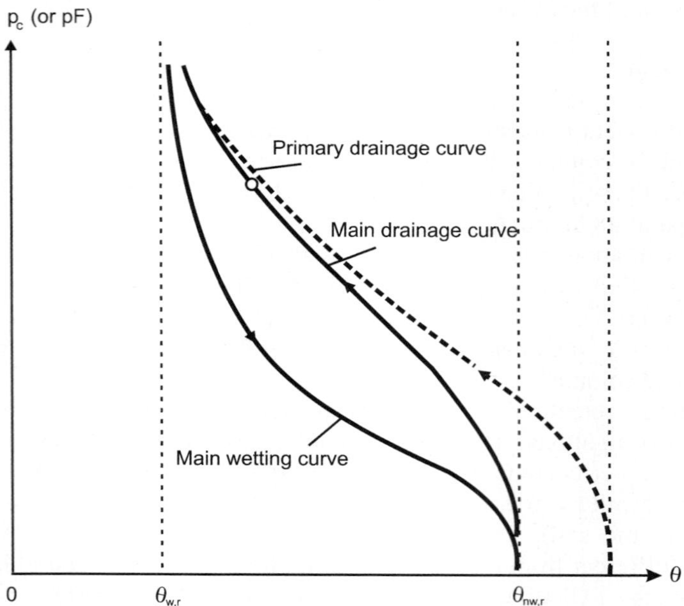

Fig. 5.3-109: Schematic fluid retention diagram for unconsolidated and consolidated rock; p_c: capillary pressure, θ_w and θ_{nw}: volumetric content of the wetting and nonwetting phases, respectively, $\theta_{w,r}$ and $\theta_{nw,r}$: residue of the wetting and nonwetting phases, respectively, *n* porosity

Submersible pressure cells are used for water retention measurements on small samples. Usually two methods are applied. For more information, see, for example, SU & BROOKS (1980) and ASTM D 4643.

A vacuum method or a pressure method is used to determine the hydraulic properties of an unsaturated soil or sediment. The main assets and drawbacks are the following:

- The vacuum method is in practice limited to pressures of less than 0.7 bar.
- The accuracy to which the equipment can be set to a desired pressure in the lower pressure range is much better with the vacuum method than with the pressure method when the state-of-the art instruments are used. The test setup as shown in *Fig. 5.2-110* has proven to work well for lower pressure settings.
- The pressure method is not limited with regard to pressure.

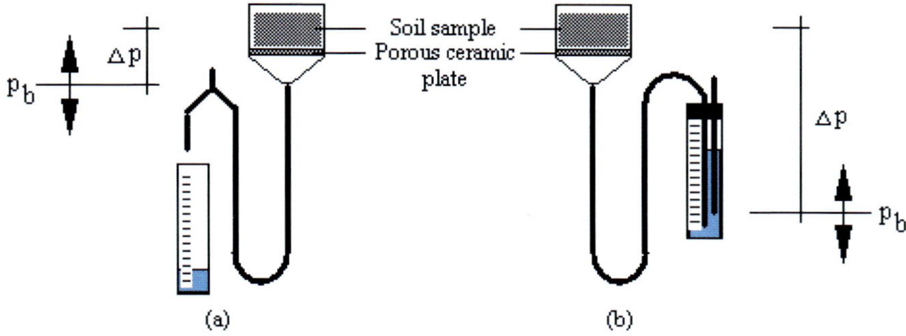

Fig. 5.2-110: Equipment for the vacuum method for the lower pressure range using a hanging water column. (a) for the drainage curve and (b) for the saturation curve of the water retention function using a Mariotte bottle; p_b: reference pressure; Δp: pressure difference

To overcome some of the major disadvantages and limitations of current methods for estimating the hydraulic properties of unsaturated soil, a new laboratory device called the AMHYP system (Automated Measurement of Soil Hydraulic Properties) has been developed and tested at the Dresden Groundwater Research Center in cooperation with the U.S. Salinity Laboratory in Riverside, California, the Department of Land, Air and Water Resources of the University of California, Davis, and the Soil and Groundwater Laboratory, Dresden. The system is now produced by the System Analysis and Automation Service Corp. Dresden. It is designed to collect the following data:

1. soil water retention data for the primary drainage curve starting at complete saturation, and for one of the scanning wetting curves,

2. volumetric soil-water outflow/inflow data for the same sample at selected capillary pressures (*Fig. 5.2-112*),
3. soil hydraulic conductivity data for the same sample at selected water contents,
4. identification of the unknown model parameters of the van Genuchten equation from the observed retention and outflow/inflow data using computerized optimization techniques, and
5. graphical representation of the final soil hydraulic functions.

The system comprises three levels: (i) a central or "master" level, (ii) a basic or "slave" level which executes commands received from the master level, and (iii) a process level consisting of the soil sample and instrumentation for the process control and data collection (*Fig. 5.2-111*). The software includes programs for test planning, automatic operation of the AMHYP system, and parameter identification.

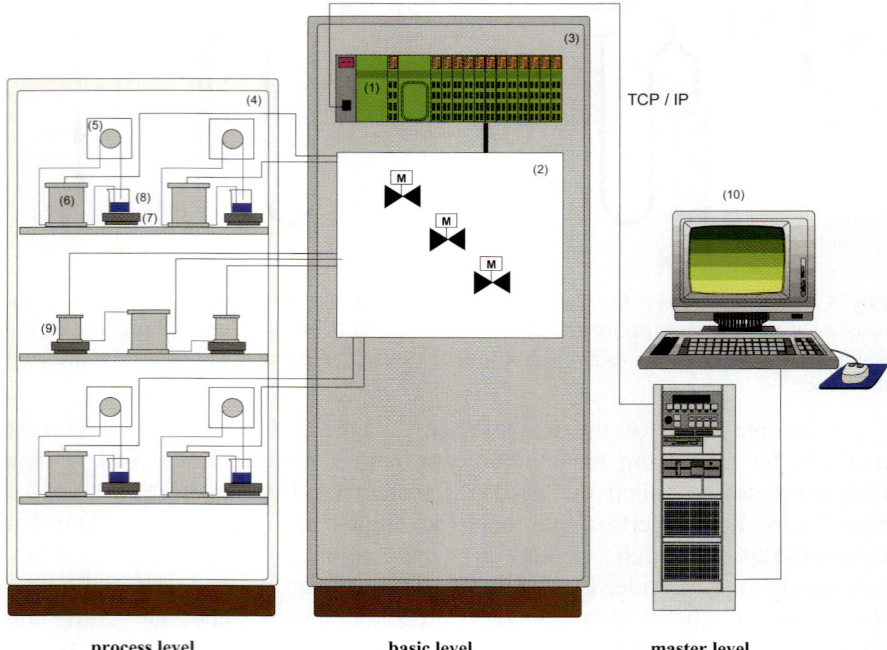

Fig. 5.2-111: Schematic diagram of the AMHYP system; 1: control unit, 2: pressure unit, 3: control box, 4: climate box, 5: peristaltic pump, 6: pressure membrane extractor with soil sample, 7: electronic balance, 8: beaker, 9: pressure tank for soil water, 10: computer

The AMHYP system is a dynamic batch test system which works on the principle of compressed air in a multistep inflow-outflow procedure. The tension pressure steps between the air-entry point and the residual saturation value θ_r must be planned prior to the beginning of the tests (*Fig. 5.2-112*). *Table 5.2-25* gives the recommended sample volume for determination of water retention curves in the AMHYP system.

Table 5.2-25: Recommended volumes of samples for determination of water retention curves using the AMHYP system

Soil type	Sample volume [cm³]
coarse sand, medium sand	550
fine sand, silty fine sand, silty medium sand	250

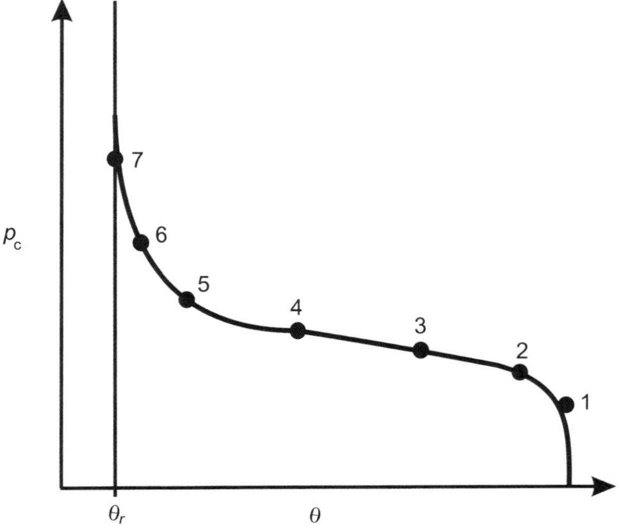

Fig. 5.2-112: An example of the relationship between planned tension pressure steps p_c and water content θ for determination of water retention function

Effective porosity

Effective porosity can be determined in the laboratory by tracer tests through relatively short columns with a length to diameter ratio of 3 (LUCKNER & SCHESTAKOW, 1991), see *Fig. 5.2-113*.

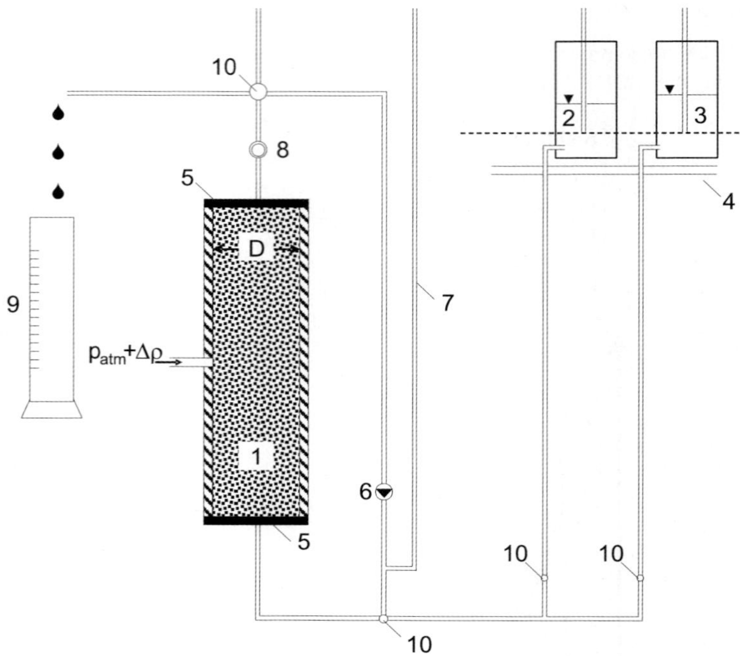

Fig. 5.2-113: Laboratory test device to determine effective porosity. 1-soil sample, 2-tracer-free water, 3-water containing the tracer substance, 4-elevation control mechanism, 5-porous ceramic plates, 6-peristaltic pump, 7-standpipe, 8-measuring cell, 9-burette, 10-two- or three-way valve

LUCKNER & NITSCHE (1984) describe in detail a method for estimating the effective porosity based on the breakthrough curve of the tracer (usually chloride or boron). The estimate is based only on the first measured values of the breakthrough curve in order to avoid influence of transphase exchange with the immobile fluid phase. The authors also recommend the tests be carried out with a relatively high flow rate, because the determined value of the effective porosity is related to the flow rate as shown in *Fig. 5.2-114*.

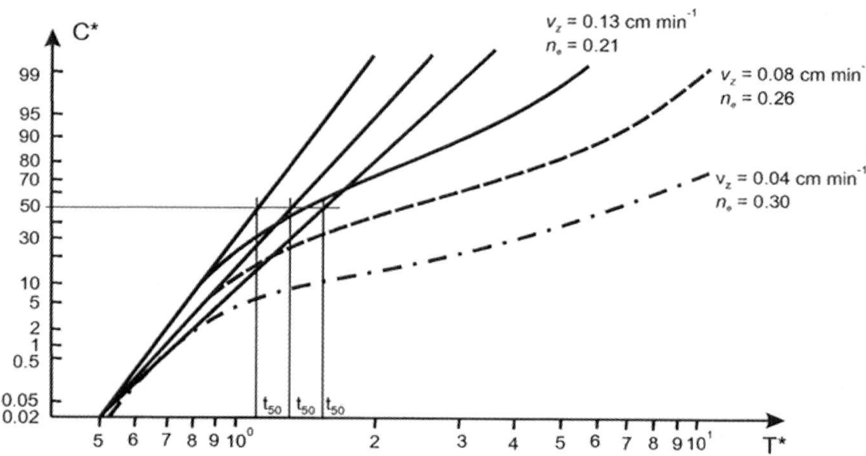

Fig. 5.2-114: Relationship between filtration velocity (v_z) and the traveltime (t_{50}) for 50 % of a tracer in a column of length z, where C* = C/C_0, T* = v_0 t/z, C_0 is the initial concentration of the tracer, C is the concentration measured at the end of the column after time t, and v_0 is the filtration velocity. In the example, the velocity ranges from 0.13 to 0.04 cm min^{-1} and the effective porosity n_e between 0.21 and 0.30.

Hydraulic conductivity

Analytical estimation: The Hazen approximation of hydraulic conductivity (HAZEN, 1911) is one of the oldest, but widely used methods to estimate hydraulic conductivity of sediments based on their grain-size distribution:

$$k_f = C d_{10}^2 \ [\mathrm{cm\ s^{-1}}], \tag{5.2.89}$$

where C is a coefficient whose values depend on the type of sediment (*Table 5.2-26*) and d_{10} has to be given in cm.

Table 5.2-26: Dependence of HAZEN's coefficient C on the type of sediment, sorting according to DIN 1054: well sorted $U < 3$, poorly sorted $U > 3$

Type of sediment	Coefficient C
very fine sand, poorly sorted	40 - 80
fine sand, well sorted	40 - 80
medium sand, well sorted	80 - 120
coarse sand, poorly sorted	80 - 120
coarse sand, well sorted	120 - 150

Hazen's method is applicable to sands with a d_{10} between 0.1 and 3.0 mm and with a coefficient of uniformity U of less than 5.

The Beyer approximation, a modified Hazen method, is the most commonly used method in Germany for estimating hydraulic conductivities based on grain-size distribution. This method has been validated for d_{10} between 0.06 and 0.6 mm, and for the coefficient of uniformity $U < 20$. BEYER (1967) developed a diagram (*Fig. 5.3-115*) for approximating the coefficient *C*. *C* can also be obtained from *Table 5.2-27*. Compared to Hazen's method (unit of hydraulic conductivity is cm s^{-1}), Beyer's method uses d_{10} in mm and yields hydraulic conductivity directly in m s^{-1}. Both methods were derived for water temperatures of about 10 °C.

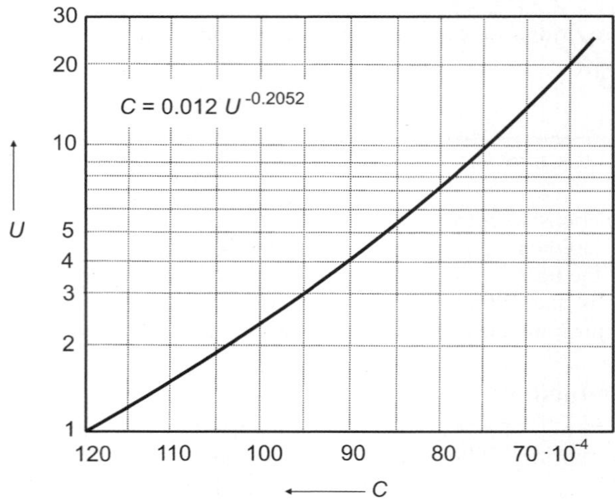

Fig. 5.2-115: BEYER diagram for determining the coefficient *C* as a function of the coefficient of uniformity *U*, BEYER (1967)

Table 5.2-27: Estimation of *C* in the BEYER approximation of hydraulic conductivity as a function of uniformity *U*, BEYER (1967)

Range of *U*	Range of *C*	Average values of *C*
1.0 - 1.9	(120 - 105) 10^{-4}	$110 \cdot 10^{-4}$
2.0 - 2.9	(105 - 95) 10^{-4}	$100 \cdot 10^{-4}$
3.0 - 4.9	(95 - 85) 10^{-4}	$90 \cdot 10^{-4}$
5.0 - 9.9	(85 - 75) 10^{-4}	$80 \cdot 10^{-4}$
10.0 - 19.9	(75 - 65) 10^{-4}	$70 \cdot 10^{-4}$
≥ 20.0	$< 65 \cdot 10^{-4}$	$60 \cdot 10^{-4}$

KASENOW (2001) mentions further methods for estimating hydraulic conductivity from grain size distribution.

Permeameter tests: Hydraulic conductivity of soil or unconsolidated sediment can be determined in the laboratory using permeameters. A permeameter has a cylindrical chamber in which the sample is placed. The sample can be a disturbed sample (for coarse sediments) or an undisturbed one taken in a liner or in a core sampler by various drilling or percussion methods. The following precautions must be taken during determination of hydraulic conductivities using permeameters in laboratory (JORDAN & WEDER, 1995):

- The diameter of the permeameter chamber must be at least 10times greater than the diameter of the largest grain in the sample.
- Laminar flow must be guaranteed through the sample.
- The hydraulic gradient should be ≤ 0.1.
- The test must be conducted under constant temperature conditions.
- The hydraulic conductivity of the porous plates on top and on bottom of the sample must be at least 100times higher than that of the sample.
- When disturbed samples are used, the sample must be filled in the chamber under water and by vibration.
- The walls of the chamber must be so constructed that flow between the walls and the sample is prevented.
- The sample must be saturated from the bottom to the top before beginning the test to avoid trapped air inclusions.

There are two types of permeameters: constant head permeameters (*Fig. 5.2-116a*) and falling head permeameters (*Fig. 5.2-116b*).

The hydraulic conductivity k_f of samples tested with constant head in a permeameter can be determined by rearranging Darcy's Equation (5.2.29) as follows:

$$k_f = \frac{\Delta V \, \Delta t \, L \, H}{A}, \qquad (5.2.90)$$

where ΔV volume of water discharging over a period of time Δt,
A cross-sectional area of the sample,
L length of the sample,
H hydraulic head.

The overflow chamber provides a constant head and a steady rate of flow through the sample. Constant head permeameters are usually used for coarse materials with a hydraulic conductivity greater than 10^{-5} m s^{-1}.

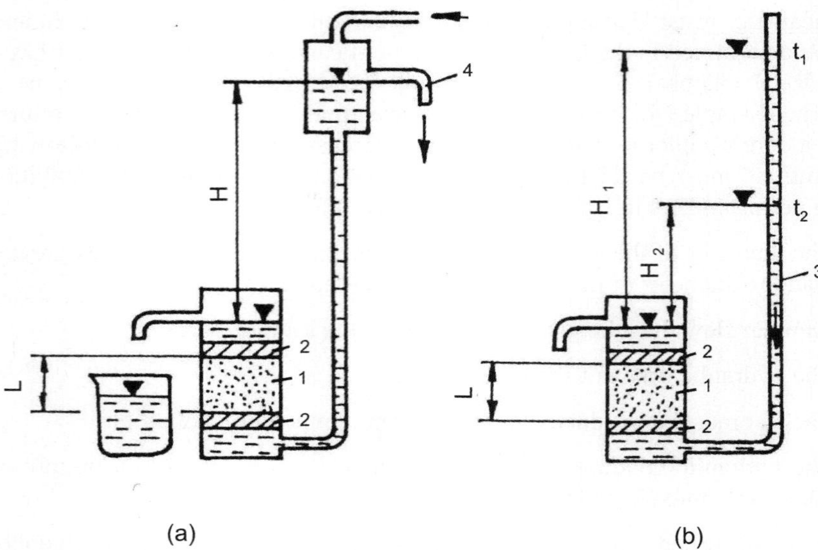

Fig. 5.2-116: Function of permeameters (a) with constant head and (b) with falling head, after BUSCH & LUCKNER (1973), 1 – permeameter chamber with the sediment sample, 2 – porous plates, 3 – falling head tube, 4 – overflow chamber

Falling head permeameters are used to determine the hydraulic conductivity of cohesive fine-grained materials. Hydraulic conductivity can be approximated by the following equation for a falling head permeameter:

$$k_f = \frac{a\,L}{A(t_2 - t_1)} \ln\left(\frac{H_1}{H_2}\right), \tag{5.2.91}$$

where a cross-sectional area of the falling head tube,
A cross-sectional area of the sample,
L length of the sample,
t_2 - t_1 time interval between two measurements,
H_1 initial head in the falling-head tube, and
H_2 final head in the falling-head tube.

Falling head permeameters similar to those shown in *Fig. 5.2-116* are used for materials with a hydraulic conductivity < 10^{-7} m s^{-1}. For sediments with lower k_f values (for example, clay) the test takes too long, so in this case compaction permeameter or triaxial cells are used to determine hydraulic conductivity (SARA, 1993; DAHMS & FRITZ, 1998). In addition to hydraulic conductivity, these devices can also be used to determine geomechanical properties of rocks, such as compressibility, strength, and swelling.

References and further reading

AARGAARD, P., BEDBUR, E., BIDOGLIO, G., CANDELA, L., NUETZMANN, G., TREVISAN, M., VANCLOOSTER, M. & VIOTI, P. (Eds.) (2004): COST Action 629. Proceedings of the International Workshop "Saturated and unsaturated zone: Integration of process knowledge into effective models". 5-7 May 2004, Rome.

ALLEN, R. G., PEREIRA, L. S., RAES, D. & SMITH, M. (1998): Crop evapotranspiration - Guidelines for computing crop water requirements - FAO Irrigation and drainage paper 56. FAO - Food and Agriculture Organization of the United Nations, Rome. www.fao.org/docrep/X0490E/x0490e00.htm

ASCE (1996): Hydrology Handbook. ASCE manuals and reports on engineering practice. American Society of Civil Engineers.

ASTM D 2487 (2000): Standard Classification of Soils for Engineering Purposes (Unified Soil Classification System). American Society for Testing and Materials, West Conshohocken, Pennsylvania.

ASTM D 4643 (2000): Standard Test Method for Determination of Water (Moisture) Content of Soil by the Microwave Oven Method. American Society for Testing and Materials, West Conshohocken, Pennsylvania.

BALKE K.-D., BEIMS U., HEERS F. W., HÖLTING B., HOMRIGHAUSEN R. & MATTHESS G. (2000): Lehrbuch der Hydrogeologie, **4**, Grundwassererschießung, Borntraeger, Berlin.

BARCZEWSKI, B., GRIMM-STREHLE, J. & BISCH, G. (1993): Überprüfung der Eignung von Grundwasserbeschaffenheitsmessstellen. Wasserwirtschaft **83**.

BAGARELLO, V. (1997): Influence of well preparation on field-saturated hydraulic conductivity measured with a Guelph permeameter. Geoderma, **80**, 169-180.

BAUMANN, G. & PETZOLD, H. (2005): Grundwassermessstellenuntersuchungen mit geophysikalischen Verfahren. In „Altlastenbearbeitung im Land Brandenburg, Teilthema Sickerwasserprognose 3", Fachbeiträge des Landesumweltamtes Brandenburg, Potsdam.

BAUMGARTNER, A. & REICHEL, E. (1975): Die Weltwasserbilanz, Niederschlag, Verdunstung und Abfluß über Land und Meer sowie auf der Erde im Jahrhundertdurchschnitt. Oldenbourg, München.

BEAR, J. (1972): Dynamics of Fluids in Porous Media. Dover.

BEAR, J. (2001): www.cmdlet.com/demos/mgfc-course/mgfcdarcy.html

BEYER, W. (1964): Zur Bestimmung der Wasserdurchlässigkeit von Kiesen und Sanden aus der Kornverteilungskurve. Wasserwirtschaft – Wassertechnik **14**, 165-168.

BEYER, W. (1967): Zur Analyse der Grundwasserfließbewegung. Wiss. Zeitschr. TU Dresden, **16**, 4.

BELL, F. G. (1998): Environmental Geology: Principles and Practice. Blackwell, Oxford.

BLACK, P. E. (1996): Watershed hydrology. 2nd ed., Ann Arbor Pr.

BMU (2003): Federal Ministry for the Environment, Nature Protection and Nuclear Safety BMU (2001) 3rd ed.: Hydrological Atlas of Germany, Berlin. Project co-ordination: Federal Institute of Hydrology, Koblenz (BfG) and Institute of Hydrology, University of Freiburg., http://had.bafg.de/

BOUWER, H. & RICE. R. C. (1976): A slug test for determining hydraulic conductivity of unconfined aquifers with completely or partially penetrating wells. Water Resources Research, **12**, 3, 423-428.

BREDEHOEFT, J. D. & PAPADOPULOS, I. S. (1980): A method of determining the hydraulic properties of tight formations. Water Resources Research, **16**, 1, 233-238.

BREDENKAMP, D. B., BOTHA, L. J., VAN TONDER, G. J. & VAN RENSBURG, H. J. (1995): Manual on quantitative estimation of groundwater recharge and aquifer storativity. Report TT 73/95, Water Research Commission, Pretoria, Rep. of South Africa.

BROUWER, C. & HEIBLOEM, M. (1986): Irrigation Water Management: Irrigation Water Needs. IRRIGATION WATER MANAGEMENT Training manual no. 3. FAO, Rome. www.fao.org/docrep/S2022E/s2022e00.htm#Contents

BRUCE, J. P. & CLARK, R. H. (1966): Introduction to hydrometeorology. Pergamon, Oxford.

BURKHAM, D. E. (1970): Depletion of streamflow by infiltration in the main channels of the Tucson Basin, southeastern Arizona. USGS Water-Supply Paper 1939-B, US Government Printing Office, Washington DC.

BUSCH, K. F. & LUCKNER, L. (1973): Geohydraulik. VEB Deutscher Verlag für Grundstoffindustrie, Leipzig.

BUSCH, K.-F., LUCKNER, L. & TIEMER, K. (1993): Geohydraulik. In: MATHESS, G. (Ed.): Lehrbuch der Hydrogeologie. Bornträger, Berlin.

BUTLER, J. J. (1997): The Design, Performance and Analysis of Slug tests. Lewis, Boca Raton, Florida.

CASAGRANDE, A. (1948): Classification and Identification of Soils. American Society of Civil Engineers, Transactions, **113**, 901-991.

CHOKNGAMWONG, R., CHIU, L. & VONGSAARD, J. (2005): Comparison of TRMM Rainfall and Daily Gauge Data in Thailand. American Geophysical Union, Spring Meeting 2005, abstract #H23A-10.

COOPER, H. H. jr. & JACOB, C. E. (1946): A generalized graphical method for evaluating formation constants and summarizing well field history. Am. Geophys. Union Trans., **27**, 526-534.

DAHMS, E. & FRITZ, L. (1998): Durchlässigkeit. In: HILTMANN, W. & STRIBRNY, B. (Eds.): Tonmineralogie und Bodenphysik. Handbuch zur Erkundung des Untergrundes von Deponien und Altlasten, **5**, Springer, Berlin, 131-148.

DAVIS, S. N. & DEWIEST, R. J. M. (1966): Hydrogeology. Wiley and Sons, New York.

DEHNERT, J., KUHN, K., GRISCHEK, T., LANKAU, R. & NESTLER, W. (2001): Eine Untersuchung zum Einfluss von voll verfilterten Messstellen auf die Grundwasserbeschaffenheit. Grundwasser, **4**, 6, 174-182.

Department of Meteorology, Thailand (1999): www.ldd.go.th/FAO/z_th/thmp131.htm

DIN 1054 (2005): Baugrund - Sicherheitsnachweise im Erd- und Grundbau. Subsoil - Verification of the safety of foundations. Beuth, Berlin.

DIN 4022: Subsoil and groundwater; classification and description of soil and rock. Beuth, Berlin.

DIN 18123 (1996): Soil, investigation and testing – Determination of grain-size distribution. Beuth, Berlin.

DIN 18196 (1988): Soil classification for civil engineering purposes. Note: To be replaced by DIN 18196 (2004-11) Beuth, Berlin.

DIN 66115 (1983): Particle size analysis; sedimentation analysis in the gravitational field; pipette method. Beuth, Berlin.

DOEGLAS, D. J. (1968): Grain-size indices, classification and environment. Sedimentology, **10**, 2, 83-100.

DÖRHÖFER, G. & JOSOPAIT, V. (1980): Eine Methode zur flächendifferenzierten Ermittlung der Grundwasserneubildungsrate. Geologisches Jahrbuch **C27**, 45-65.

DÖRHÖFER, G., LANGE, B. & VOIGT, H. (1991): Deponieüberwachungsplan Wasser – Niedersächsisches Landesamt für Bodenforschung und Niedersächsisches Landesamt für Wasser und Abfall. Richtlinienentwurf 2.1 für das Niedersächsische Umweltministerium, Hannover/Hildesheim.

DRACOS, T. (1980): Hydrologie – eine Einführung für Ingenieure. Springer, Wien.

DRISCOLL, F. D. (1986): Groundwater and Wells. 2nd edn. Johnson Screens, St. Paul, Minnesota.

DURNER, W. (2002): Skript „Bodenhydrologische Versuche im Praktikum Hydrologie (Laboratory notes on soil hydrology experiments). Institut für Geoökologie (Institute for Geoecology), Abteilung Bodenkunde und Bodenphysik, Technical University Braunschweig, Germany.

DVWK (1997): Tiefenorientierte Probenahme aus Grundwassermeßstellen (Depth-oriented sampling of groundwater monitoring wells). DVWK Merkblätter **245**. Deutscher Verband für Wasserwirtschaft und Kulturbau (German Association for Water Management and Land Improvement), Bonn.

DYCK, S. & PESCHKE, G. (1989): Grundlagen der Hydrologie. 2nd edn. Verlag für Bauwesen, Berlin.

EDMUNDS, W. M., DARLING, W. G. & KINNIBURGH, D. G. (1990): Solute profile techniques for recharge estimation in semi-arid and arid terrain. In: LERNER, D. N., ISSAR, A. & SIMMER, I.: Groundwater recharge: A guide to understanding and estimating natural recharge. International Contributions to Hydrogeology **8**. International Association of Hydrogeologists. Heise, Hannover, 257-270.

ELRICK, D. E., REYNOLDS, W. D., GEERING, H. R. & TAN, K.-A. (1990): Estimating Steady Infiltration Rate Times for Infiltrometers and Permeameters. Water Resources Research, **26**, 4, 759-769.

ELRICK, D. E. & REYNOLDS, W. D. (1992): Methods for analysing constant head well permeameter data. Soil Sciences Society of America Journal, **56**, 1, 320-323.

ENGLERT, A., HASHAGEN, U., JAEKEL, U., NITZSCHE, O., SCHWARZE, H. & VEREECKEN, H. (2000): Transport von gelösten Stoffen im Grundwasser – Untersuchungen am Testfeld Krauthausen. Grundwasser **3**, 5, 115-124.

FECKER, E. & REIK, G. (1996): Baugeologie. 2. Aufl., Enke, Stuttgart.

FETTER, C. W. (2001): Applied Hydrogeology. 4th edn. Prentice-Hall, Upper Saddle River, New Jersey.

FOLK, R. L. (1974): The petrology of sedimentary rocks. Hemphill Publishing Co., Austin, Texas.

FÜCHTBAUER, H. (Ed.) (1988): Sedimente und Sedimentgesteine. 4th edn., Schweizerbart, Stuttgart.

GARDNER, W. R. (1958): Some steady state solutions of the unsaturated moisture flow equation with application to evaporation from a water table. Soil Science, **85**, 4, 228-232.

GARTUNG, E. & NEFF, H. K. (1999): Empfehlungen des Arbeitskreises „Geotechnik der Deponiebauwerke" der Deutschen Gesellschaft für Geotechnik e.V. (DGGT) (Recommendations of the working group "Geotechnics of Landfills" of the German Society for Geotechnics). Bautechnik, **76**, 9, 723-746.

GERMANN, U. & JOSS, J. (2001): Mesobeta Profiles to Extrapolate Radar Precipitation Measurements above the Alps to the Ground Level. Journal of Applied Meteorology, **41**, 5, 542-557.

GIAKOUMAKIS, S. & TSAKIRIS, G. (1999): Quick Estimation of Hydraulic Conductivity in Unsaturated Sandy Loam Soil. Irrigation and Drainage Systems, **13**, 4, 349-359.

GLUGLA, G. & FÜRTIG, G. (1997): Dokumentation zur Anwendung des Rechenprogrammes ABIMO. Bundesanstalt für Gewässerkunde, Koblenz, Berlin.

HANTUSH, M. S. (1956): Analysis of data from pumping tests in leaky aquifers. Trans. Amer. Geophys. Union, **37**, 702-714.

HANTUSH, M. S. (1959): Analysis of data from pumping wells near a river. J. Geophys. Res., **94**, 1921-1932.

HANTUSH, M. S. (1960): Modification of the theory of leaky aquifers. J. Geophys. Res., **65**, 11, 3713-3725.

HANTUSH, M. S. (1961a): Drawdown around a partially penetrating well. Proc. Amer. Soc. Civil Eng., **87**, HY4, 83-98.

HANTUSH, M. S. (1961b): Aquifer tests on partially penetrating wells. Proc. Amer. Soc. Civil Engrs., **87**, HY5, 171-195.

HANTUSH, M. S. (1964): Hydraulics of wells. In: CHOW, V. T. (Ed.): Advances in hydroscience, **1**, 281-432, Academic Press, New York.

HANTUSH, M. S. (1967): Flow to wells separated by a semipervious layer. J. Geophys. Res., **72**, 6, 1709-1720.

HANTUSH, M. S. & JACOB, C. E. (1955): Non-steady radial flow in an infinite leaky aquifer. Trans. Amer. Geophys. Union, **36**, 1, 95-100.

HANTUSH, M. S. & THOMAS, R. G. (1966): A Method for analyzing a drawdown test in anisotropic aquifers. Water Resour. Res., **2**, 281-285.

HARDEGREE, S. P., VAN VACTOR, S. S., HEALY, K. R., ALONSO, C. V., BONTA, J. V., BOSCH, D. D., FISHER, D. S., GOODRICH, D. C., HARMEL, R. D., STEINER, J. L., VAN LIEW, M. W. (2003): Multi-Watershed Evaluation Of Wsr-88d (Nexrad) Radar-Precipitation Products. In: Proceedings Of The First Interagency Conference On Research In The Watersheds, October 28-30, 2003, Benson, Arizona, 486-491.

HAZEN, A. (1911): Discussion: dams on sand foundations. Transactions, American Society of Civil Engineers, **73**, 199.

HEITFELD, K.-H., OLZEM, R. & STOLPE, H. (1979): Erarbeitung von Kriterien und Untersuchungsmethoden für die Dichtigkeit von Böden zur Standortbestimmung von Sondermülldeponien (Development of criteria and investigation methods for the tightness of soils for location decisions for hazardous waste sites). Landesanstalt für Wasser und Abfall des Landes Nordrhein-Westfalen. Unpublished progress report.

HEITFELD, K.-H. & HEITFELD, M. (1992): Auswertung von WD-Tests in Gebirgsbereichen mit geringen Durchlässigkeiten (Evaluation of water pressure tests in low conductive rocks). Mitteilungen zur Ingenieur- und Hydrogeologie, **48**.

HEITFELD, M. (1998): Auffüll- und Absenkversuche in Rammkernsondierungen (Filling-up and drawdown tests in percussion core drillings). In: SCHREINER, M. & KREYSING, K. (Eds.) (1998): Handbuch zur Erkundung des Untergrundes von Deponien und Altlasten, **4**: Geotechnik, Hydrogeologie. Springer, Berlin, 397-425.

HEREFORD, R., WEBB, R. H. & GRAHAM, S. (2005): Precipitation History of the Colorado Plateau Region, 1900-2000. U.S. Geological Survey Fact Sheet 119-02. http://pubs.usgs.gov/fs/2002/fs119-02/

HERFORT, M., PTAK, T., HÜMMER, O., TEUTSCH, G. & DAHMKE, A. (1998): Testfeld Süd: Einrichtung der Testfeldinfrastruktur und Erkundung hydraulisch-hydrogeochemischer Parameter des Grundwasserleiters. Grundwasser **4**, 3, 159-166.

HISCOCK, K. M. (2005): Hydrogeology: principles and practice. Blackwell, Malden.

HÖLTING, B. (1992): Hydrogeologie. 4. Aufl., Enke, Stuttgart.

HORTON, R. E. (1940): An approach toward a physical interpretation of infiltration capacity. Soil Science Society of America, Proceedings **4**, 399-417.

HUDAK, P. F. (1998). A method for designing configurations of nested monitoring wells near landfills. Hydrogeology Journal, **6**, 341-348.

HUNTER, S. M. (1996): WSR-88D Radar Rainfall Estimation: Capabilities, Limitations and Potential Improvements. www.srh.noaa.gov/mrx/research/precip/PRECIP.htm.

HURTADO, E., ARTIGAO, M. M. & CASELLES, V. (1997): Models for determining evapotranspiration. www.fao.org/documents/show_cdr.asp?url_file=/docrep/W7320B/w7320b27.htm

HUDSON, N. W. (1993): Field measurement of soil erosion and runoff. Food and Agriculture Organization of the United Nations, Rome. www.fao.org/documents/show_cdr.asp?url_file=/docrep/T0848E/T0848E00.htm

JACKSON, M. L. (1956): Soil chemical analysis: advanced course. Published by the author, Madison, Wisconsin.

JORDAN, H. & WEDER, H.-J. (Hrsg.), (1995): Hydrogeologie – Grundlagen und Methoden. Enke, Stuttgart.

KALERIS, V. (1992): Strömungen zu Grundwassermessstellen mit langen Filterstrecken bei der Gewinnung durchflussgewichteter Mischproben. Wasserwirtschaft **82**.

KASENOW, M. (2001): Applied Ground-Water Hydrology and Well Hydraulics. 2nd edn., Water Resources Publications, Highlands Ranch, Colorado.

KONERT, M. & VANDENBERGHE, J. (1997): Comparison of laser grain size analysis with pipette and sieve analysis: a solution for the underestimation of the clay fraction. Sedimentology, **44**, 3, 523.

KOREIMANN, C., GRATH, J., WINKLER, G., NAGY, W. & VOGEL, W. R. (1996): Groundwater Monitoring in Europe. EEA Topic Report 14/96, Amsterdam.

KOZENY, J. A. (1928): Die Durchlässigkeit des Bodens. Kulturtechniker, **35**, 478-486

KRUMBEIN, W. C. & PETTIJOHN, F. J. (1938): Manual of Sedimentary Petrology. Appleton, Century, and Crofts, New York.

LAUTERBACH, D. & LUCKNER, L. (1999): Grundlagen und Konzepte des Grundwasser-monitoring aus der Sicht des Dresdner Grundwasserforschungszentrums. In: Proceedings des DGFZ e.V., **17**, 19-33, Dresden.

LAUTERBACH, D. & VOIGT, H.-J. (1993): Ausgewählte Aspekte zum Auftreten von Grundwasserverunreinigungen und deren Überwachung. In: Schriftenreihe des Hessischen Landesamtes für Umwelt **157,** Wiesbaden.

LAWA (1991): Länderarbeitsgemeinschaft Wasser - Pegelvorschrift, Anlage D: Richtlinie für das Messen und Ermitteln von Abflüssen und Durchflüssen.

LE BLANC, D. R., GARABEDIAN, S. P., HESS, K. M., GELHAR, L. W., QUADRI, R. D., STOLLENWERK, K. G. & WOOD, W. W. (1991): Large-scale natural gradient tracer test in sand and gravel, Cape Cod, Massachusetts, 1, Experimental design and observed tracer movement. Water Resources Research **27**, 5, 895-910.

LEGATES, D. R. (2004): Real-Time Calibration of Radar Precipitation Estimates. The Professional Geographer, **52**, 2, 235-246.

LERNER, D. N. & TEUTSCH, G. (1995): Recommendations for level-determined sampling in wells. Journal of Hydrology **171**, 3-4, 355-377.

LIEBSCHER, H. J. & KELLER, R. (1979): Abfluss. In: Deutsche Forschungsgemeinschaft (Hrsg). Hydrologischer Atlas der Bundesrepublik Deutschland. 90-159.

LUCKNER, L. & NITSCHE, C. (1984): Beitrag zur laborativen Ermittlung von Porositätsparametern. Wissenschaftliche Zeitschrift der TU Dresden, **33**, 3.

LUCKNER, L., NITSCHE, C. & EICHHORN, D. (1992): Das SGM-System – eine neue Technik und Technologie zur Boden- und Grundwasserüberwachung in Deutschland. Die Geowissenschaften **10**, 2, 37-44.

LUCKNER, L. & SCHESTAKOW, W. M. (1991): Migration processes in the soil and groundwater zone. Lewis, Michigan.

LVOVITCH, M. L. (1972): World water balance. In: Symposium of World Water Balance. IASH-UNESCO. Report Number 92.

MAF (2002): Soil water. Ministry of Agriculture and Forestry, Wellington, New Zealand. www.maf.govt.nz/mafnet/schools/activities/swi/swi-04.htm

MATTHES, G. & UBELL, K. (1983): Lehrbuch der Hydrogeologie, **1**, Allgemeine Hydrogeologie – Grundwasserhaushalt. Borntraeger, Berlin.

MCCAVE, I. N. & SYVITSKI, J. P. M., (1991): Principles and methods of particle size analysis. In: SYVITSKI, J. P. M. (Ed.): Principles, Methods, and Applications of Particle Size Analysis. Cambridge University Press, New York, 3-21.

MCLAREN, R. G & CAMERON, K. C. (1990): Soil Science. An Introduction to the Properties and Management of New Zealand Soils. Oxford University Press, Auckland, New Zealand.

MOORE, J. E., ZAPOROZEC, A. & MERCER, J. W. (1995): Groundwater: a primer. American Geolog. Inst., Alexandria.

MOORE, J. E. (2002): Field hydrogeology: a guide for site investigations and report preparation. Lewis, Boca Raton, Florida.

NATIONAL WEATHER SERVICE, ARKANSAS-RED BASIN RIVER FORECAST CENTER (2003): Point Precipitation Measurement, Areal Estimates And Relationships To Hydrologic Modeling. www.srh.noaa.gov/abrfc/map/map.shtml.

NILLERT, P. & THIERBACH, J. (1999): Exemplarische Darstellung von entscheidungs-orientiertem Grundwasser-Monitoring im innerstädtischen Bereich. Proceedings des DGFZ e.V., **17**, 129-140, Dresden.

PERES, A. M. M., ONUR, M. & REYNOLDS, A. C. (1989): A new analysis procedure for determining aquifer properties from slug test data. Water Resource Research, **25**, 7, 1591-1602.

PFANNKUCH, H. O. & PAULSON, R. (2005): Grain Size Distribution and Hydraulic Properties. http://web.cecs.pdx.edu/~ian/geology2.5.html.

PHILIP, J. R. (1957): The theory of infiltration. Soil Science, **83**, 5, 345-357.

PIDWIRNY, M. (2005): Fundamentals of Physical Geography. www.physicalgeography.net/home.html

POIER, V. (1998): Pulse Test. In: SCHREINER, M. & KREYSING, K. (Eds.) (1998): Handbuch zur Erkundung des Untergrundes von Deponien und Altlasten, **4**: Geotechnik, Hydrogeologie. Springer, Berlin. 361-376.

PONCE, V. M. (2005): Factors affecting precipitation. www.ponce.sdsu.edu/factors_affecting_precipitation.html

POPPE, L. J., ELIASON, A. H., FREDERICKS, J. J., RENDIGS, R. R., BLACKWOOD, D. & POLLONI, C. F. (1998): Grain-Size Analysis of Marine Sediments: Methology and Data Processing, US Geological Survey Open-File Report 00-358.

PRZYBILSKI, U., & RECHENBERG, B., (1999): Montanhydrogeologisches Monitoring als Grundlage der effektiven Führung der Tagebauentwässerung in der LAUBAG. Proceedings des DGFZ e.V. **17**, 107-115, Dresden.

REYNOLDS, W. D. & ELRICK, D. E. (1985): In situ measurement of field-saturated hydraulic conductivity, sorptivity and the α-parameter using the Guelph permeameter. Soil Science, **140**, 4, 292-302.

REYNOLDS, W. D. & ELRICK, D. E. (1986): A method for simultaneous in situ measurement in the vadose zone of field saturated hydraulic conductivity, sorptivity and the conductivity pressure head relationship. Ground Water Monitoring Review, **6**, 1, 84-95.

REYNOLDS, W. D., ELRICK, D. E. & TOPP, G. C. (1985): The constant head well permeameter: effect of unsaturated flow. Soil Science, **139**, 2, 172-180.

RENGER, M. & WESSOLEK, G. (1996): Jährliche Grundwasserneubildung in Abhängigkeit von Bodennutzung und Bodeneigenschaften. In: Ermittlung der Verdunstung von Land- und Wasserflächen. DVWK-Merkblätter zur Wasserwirtschaft **238**.

RICHTER, W. & LILLICH, W. (1975): Abriß der Hydrogeologie. Schweizerbart, Stuttgart.

RUSSELL, S. O. (2006): Notes on basic hydraulics. www.culvertbc.com/Text/class-basichydraulics.html

SARA, M. N. (1993): Standard Handbook for Solid and Hazardous Waste Facility Assessments. Lewis, Boca Raton, Florida.

SARA, M. N. (2003): Site Assessment and Remediation Handbook. 2nd edn. Lewis, Boca Raton, Florida.

SCHLEE, John (1966): A Modified Woods Hole Rapid Sediment Analyzer: Journal Sedimentary Research, **30**, 403-413.

SCHLINKER, K. (1969): Komplexmethodik der regionalen Grundwassererkundung im Großeinzugsgebiet Küste Warnow-Peene. Wiss. Z. Univ. Rostock, **18**, 7.

SCHREINER, M. & KREYSING, K. (Eds.) (1998): Handbuch zur Erkundung des Untergrundes von Deponien und Altlasten, **4**: Geotechnik, Hydrogeologie. Springer, Berlin.

SCHWAB, G. O., FREVERT, R. K., EDMINSTER, T. W. & BARNES, K. K. (1981): Soil and Water Conservation Engineering. 3rd edn., Wiley, New York.

SERRANO, S. E. (1997): Hydrology for engineers, geologists, and environmental professionals an integrated treatment of surface, subsurface and contaminant hydrology. HydroScience Inc., Lexington KY.

SIMMERS, I. (1997): Recharge of Phreatic Aquifers in (semi-)arid Areas. Balkema, Rotterdam.

SINGH, V. P. (Ed.) (1995): Environmental hydrology. Kluwer, Dordrecht.

SPERAZZA, M., MOORE, J. N. & HENDRIX, M. S. (2002): Methodology for sediment grain size analysis by laser diffraction: a high resolution application. http://gsa.confex.com/gsa/2002AM/finalprogram/abstract_44839.htm

STEPHENS, D. B. (1996): Vadose Zone Hydrology. Lewis, Boca Raton, Florida.

STONE, W. J. (1999): Hydrogeology in practice: a guide to characterizing ground-water systems. Prentice Hall, Upper Saddle River.

SU, C. & BROOKS, R. H. (1980): Water retention measurement for soils. Journal of the Irrigation and Drainage Division of the American Society of Civil Engineers, **106**, 2, 105-112.

THOMPSON, S. A. (1999): Hydrology for Water Management. Balkema, Rotterdam.

TORSTENSSON, B. A. (1984): A New System for Groundwater Monitoring. Ground Water Monitoring Review, **4**, 4, 131-138.

TRÖGER, U. & POIER, V. (2000): Abschlußbericht zum Forschungsvorhaben „Durchlässigkeitsverhalten natürlicher paläozoischer Untergrundabdichtungen und Schadstoffausbreitung entlang Trennflächen". Technical University Berlin, Germany.

USBR (1974): Earth Manual – A guide to the use of soils as foundations and as construction materials for hydraulic structures. US Bureau of Reclamation, Department of the Interior, Washington, D.C.

USDA (1993): Soil survey manual. Soil Conservation Service, US Department of Agriculture Handbook 18. Note: also available online: http://soils.usda.gov/technical/manual/

USDA-SCS (1964): Hydrology. Section 4, Part I, Watershed planning. In: National Engineering Handbook. US Department of Agriculture, Soil Conservation Service, Washington DC.

USDI (1975): Water Measurement Manual. US Department of the Interior, Bureau of Reclamation, Denver, Colorado.

VOIGT, H.-J. (1990) : Hydrogeochemie – Eine Einführung in die Beschaffenheitsentwicklung des Grundwassers. Springer, Berlin.

VOIGT, H.-J., ASBRAND, M. & BEIMS, U. (2000): Schematisierung der hydrogeologischen Verhältnisse als Grundlage für die Auswahl des numerischen Simulationsverfahrens. Tagungsband zum Kolloquium „Wasserbewirtschaftung – einzugsbezogen und integrativ", Teil II, 46 - 53. Cottbus.

WALTON, W. C. (1970): Groundwater Resource Evaluation. McGraw-Hill, New York.

WANG, Q. J., MCCONACHY, F. L. N., CHIEW, F. H. S., JAMES, R., DE HOEDt G. C. & WRIGHT, W. J. (2001): Climatic Atlas of Australia, Maps of Evapotranspiration. Bureau of Meteorology of Australia. www.bom.gov.au/climate/averages/climatology/evapotrans/et.shtml

WANIELISTA, M. P., KERSTEN, R. & EAGLIN, R. (1997): Hydrology: Water Quantity and Quality Control. 2nd edn., John Wiley & Sons, New York.

WATSON, I. & BURNETT, A. D. (1993): Hydrology: An Environmental Approach. Buchanan, Cambridge.

WECHMANN, A. (1964): Hydrologie. Oldenbourg, München.

WEIGHT, W. D. & SONDEREGGER, J. L. (2001): Manual of applied field hydrogeology. McGraw-Hill, New York.

WENDLAND, F., TETZLAFF, B., KUNKEL, R. & DÖRRHÖFER, G. (2001): GIS-basierte Grundwasserneubildung von Niedersachsen. Arbeitshefte Wasser, **1**, 47-53. Hannover.

WENTWORTH, C. K. (1929): Method of computing mechanical composition of sediments. Geol. Soc. of America Bull., **40**, 771-790.

WILSON, J. W. & BRANDES, E. A. (1979): Radar measurement of rainfall. Bull. Amer. Meteor. Soc., **60**, 1048-1058.

WINNEGGE, R. & MAURER, T. (2002): Water Resources Management, Country Profile Germany. A contribution to the Global Water Information Network WWW.GLOBWINET.ORG. Report No. 27, Global Runoff Data Centre (GRDC), Federal Institute of Hydrology (BfG), Koblenz, Germany.

YANG, Y. J. & GATES, T. M. (1997): Wellbore skin effect in slug-test data analysis for low-permeability geologic materials. Ground Water, **35**, 6, 931-937.

YOUNGS, E. G. (1987): Estimating hydraulic conductivity values from ring infiltrometer measurements. Journal of Soil Science, **38**, 4, 623-632.

5.3 Methods for Characterizing the Geochemical and Microbiological Conditions

SVEN ALTFELDER, MANFRED BIRKE, REINER DOHRMANN, HAGEN HILSE, FLORIAN JENN, STEPHAN KAUFHOLD, KLAUS KNÖDEL, CLAUS NITSCHE, KATHRIN R. SCHMIDT, ANDREAS THIEM & HANS-JÜRGEN VOIGT

A geochemical investigation in the study area should be primarily focused on characterizing the complex chemical inventory of the groundwater, surface water, soil, rock, stream and lacustrine sediments, and soil gas. Geochemical site characterization includes a determination of geogenic background values and anthropogenic input. This distinction is possible only if the size of the area and the number of sampling points is adequate for a statistical evaluation. A geochemical investigation usually takes place following the geological, hydrogeological and geophysical surveys, the results of which are used for a focused and representative sampling strategy. The approach and the scope of a geochemical site investigation depend on the following:

- the objective and phase of the investigation,
- the contamination potential,
- the compartment under consideration (e.g., the groundwater, soil, or air), and
- the natural environmental conditions.

In the case of planning and construction of a new landfill, the geogenic background and anthropogenic input from former land use have to be determined for comparison with later changes caused by the landfill (base line). Orientating investigations of sites suspected to be hazardous should enable an inventory of substances and determine contaminant concentrations in the various parts of the site. Detailed investigations of sites suspected to be hazardous must determine the mobile and/or potentially mobilized contaminant contents for the pathways groundwater, surface water, soil, and air. Natural retention and decomposition processes (natural attenuation) are to be included. Detailed site investigations provide the information for hazard assessment and decisions for action.

The samples must be properly collected in order to obtain useable analytical results and thus a correct assessment of the site. Errors made in collecting the samples cannot be recognized in the laboratory and hence cannot be corrected for afterwards. The sampling procedure depends on the geological composition of the subsurface (rock and soil), types and expected vertical and horizontal distribution of contaminants, affected pathways, objectives and phase of the investigation, size and land use of the area to be investigated. All of these conditions must be summarized in a "contamination hypothesis" following DIN ISO 10381-5 or in a process model (see *Fig. 5.2-31* in Section 5.2.6).

Sampling procedures, sample preparation, substances and parameters to be determined, and analysis methods are described in Sections 5.3.1 to 5.3.3 for groundwater, surface water, soil, stream and lacustrine sediments, and soil gas. Short descriptions of instruments and methods for chemical analysis are given in Section 5.3.4 and in the Glossary.

Groundwater

Leachate is the main carrier of contaminants from landfills and contaminated sites, entering initially the unsaturated zone and from there into the groundwater. Therefore, the chemical contents of the groundwater must be determined and monitored. A prerequisite for the investigation and assessment of the distribution of contaminants in the groundwater as a function of time is to have best possible knowledge of the geological and hydrogeological conditions in the investigation area and of the groundwater dynamics (Chapters 5.1 and 5.2). Groundwater monitoring zones (Sections 5.2.6) can be defined on the basis of this information, as well as the location and type of groundwater observation wells. The properties of the contaminants and their migration behavior must also be taken into account when the investigation strategy is prepared. In addition to convective transport of substances via the groundwater, dispersion, diffusion, matrix diffusion, sorption and decomposition result in retention of dissolved substances and a reduction in contaminant concentration in the water, as well as causing different migration velocities for the different contaminant substances. The migration behavior of specific substances has a great influence on the scope and the selection of parameters as well as on sample collection intervals. The more frequent analysis of fewer key parameters can provide more reliable results than a larger number of parameters with longer intervals between measurements.

The assessment of the extent of changes in the groundwater contents is based on the natural groundwater composition (geogenic background). Deviations from the geogenic background contents indicate groundwater contamination. The natural groundwater quality depends both on the lithological and geochemical composition of the rocks as well as on various hydrodynamic factors. A higher rate of groundwater flow results in considerably lower substance loads due to the shorter dwell times. In addition, chemical conditions (redox potential and pH) also play a role. The distribution of these parameters can change with time. VOIGT & WIPPERMANN (1998) have published an overview of geogenic quality patterns of groundwater taking into account flow components and classified according to hydrogeological structural units. These are generally based on an analysis of the distribution pattern of the main constituents in the water. The patterns of the geogenic groundwater composition for certain hydrogeological structural units and types of locations can be used as an initial estimate of geogenic contents and for distinguishing anthropogenic contamination from landfills and contaminated sites. However, a site-specific determination of geogenic background carried out in the groundwater inflow to the investigation site or in the area around it is necessary. To determine the natural hydrochemical conditions it is important to take both vertical and horizontal zoning into account. In cases where the groundwater is already subject to influx of unspecified anthropogenic substances, the local background must be taken as a reference.

Soil, rock, stream and lacustrine sediments

Geochemical site characterization includes the taking of soil, rock, stream and lacustrine sediments (summarized as "soil") samples to determine contaminant loads, background concentrations, and soil properties. Due to the inhomogeneity of the subsurface and spreading of contaminants via water and air, it is vital to take representative samples in order to have reliable investigation results.

Soil is formed from unconsolidated and consolidated rocks by weathering, new mineral formation and humification and frequently superimposed by anthropogenic influences. The natural geogenic constituents generally reflect the elemental and mineralogical content of the source rock (lithogenic component). The soil-forming processes begin with enrichment or removal of substances leading to the formation of typical soil horizons (pedogenic component). Therefore the specific properties of a soil type, the source rock of soil, chemical environmental parameters (pH, redox potential) and components capable of sorption (TOC, clay minerals) are significant for the migration behavior of contaminants and/or the retention potential of a soil.

Stream and lacustrine sediments provide information about the catchment area and are used for the geochemical characterization of large areas. Such sediments provide a good indication of the input of heavy metals and low-solubility organic substances to the site – mostly bound within the organic and the fine-grained fraction.

When the background is determined, it is necessary to not only take the geogenic background into consideration, but also the anthropogenic background resulting from the land use in the part and present. Knowledge of the background values for the soils and rock at the site is a prerequisite for identification of anthropogenic input.

Most organic contaminants do not occur naturally in the lithosphere. Other natural sources (for example, forest fires and migration of hydrocarbons from deep sources) can generally be considered negligible or are only of importance at certain locations. Therefore, it is only necessary to consider the anthropogenic background from the surrounding area. In contrast to organic contaminants, environmentally relevant heavy metals do occur naturally, derived from rock and soil formations. Hence, it is necessary to distinguish heavy metal contamination from the geogenic background. VINOGRADOV (1954), BOWEN (1979) and VOIGT & WIPPERMANN (1998) give global background concentrations (Clarke values) for inorganic trace elements and metals in soil and stream sediments as well as lacustrine sediments. These background values can be used as an initial estimate of geogenic concentrations. In the case of detailed investigations, the local background can be determined using statistical methods (Sections 5.4.1 and 6.2.2.3).

Soil gas

Soil gas (all gaseous substances formed in the soil and/or transported into the soil) consists, depending on the local conditions, of the following components:

- atmospheric gases,
- gases formed by bacterial metabolism as well as by aerobic or anaerobic decomposition of organic materials,
- geogenic gases (natural gas, petroleum gases, coal gases) formed by thermal processes from organic matter in the petroleum source rock deep underground, producing methane and other organic hydrocarbons, nitrogen, etc., which migrate to the Earth's surface,
- radiogenic gases (radon and helium) formed by the natural decay of radioactive elements in the Earth's crust which also migrate to near-surface zones from the deep underground,
- anthropogenic gases, which are primarily landfill gas, but also include gaseous and highly volatile organic contaminants from other sources.

Analysis of soil gas can be used to verify seepage from landfills and to monitor the spreading of contaminants from contaminated sites. The concentration, composition and distribution of soil gas in the pore space, in the soil water (dissolved gas phase) and in the soil matrix (adsorbed gas phase) depend on a multitude of factors. Consequently, soil gas concentrations and composition are subject to considerable variation and fluctuation. Such factors are the meteorologically influenced parameters of water content, air pressure and temperature, as well as the lithological and pedological conditions (e.g., permeability and adsorption properties). These factors must be taken into account when soil gas investigations are planned and conducted.

In the case of landfills in which organic waste has been deposited, landfill gas can be formed by microbial decomposition. Landfill gas can also contain any number of volatile trace substances of anthropogenic origin. There are two main pathways by which gases leave a landfill: direct escape into the atmosphere and migration into the subsurface or adjacent buildings.

Migration of gas from a landfill into the subsurface can be clearly detected by measuring methane as the key substance in soil gas. This provides an early indication of contaminant migration from the landfill.

If the physical conditions are suitable, it is possible to use soil gas analyses to investigate the horizontal and vertical extent of contamination and to identify the sources of gaseous and volatile contaminants (e.g., volatile halogenated hydrocarbons, BTEX) in the unsaturated zone. Extent and distribution of these contaminants are dependent on physical parameters of the soil. Porous and jointed rocks promote propagation of contaminants, whereas areas of low permeability with high water content and water saturated areas tend to inhibit it. Volatile components can collect below retaining interbeds. Contaminants with the appropriate density (e.g., volatile halogenated hydrocarbons) move downwards in the unsaturated zone and spread in the capillary zone or migrate into the groundwater. Determination of volatile substances in soil gas provides a rapid qualitative survey of contaminant loads. Owing to its qualitative character, soil gas analysis is suitable for orientating investigations and as a basis for decisions for detailed investigations based on soil and groundwater sampling. The use of inexpensive soil gas investigations can help minimize the need for expensive drilling.

Migration behavior

The geochemical investigation of groundwater, soil, and soil gas as described in the previous sections provides information about the distribution of geogenic and anthropogenic material concentrations in the various compartments groundwater, soil and air, for example. By analysis of the contaminant distribution over time and taking into account the geological, hydrogeological, and geochemical conditions at the site as well as the physical parameter values, it is possible to qualitatively estimate migration behavior. The migration behavior, as well as the eco-toxicological effect of a substance in the ground, depends not only on the concentration of the substance, but also to a decisive degree on the chemical form (species) and on how the substance is bound in the soil. In addition to species and type of bond it is also necessary to take into account the numerous processes by which these substances interact with mineral constituents, organic substances, soil water and groundwater, and soil gas. These processes include

- retention processes (filtering, ion exchange, sorption),
- dissolution and precipitation,
- microbial conversion,
- association and dissociation of dissolved particles,
- acid-alkaline reactions,

- oxidation-reduction, and
- transport (convection, hydrodynamic dispersion, diffusion).

Various laboratory techniques are available to obtain information on concentrations, species, type of bonding, chemical reactions, and transport of the substances of interest. The data obtained can be used together with conditions at the site to predict contaminant spreading using flow and transport models (Sections 5.4.3 and 5.4.4).

The species or chemical form in which an element is present is especially important in the case of trace elements. The oxidation state of some elements (e.g., Fe, Mn, and Cr) depends on pH and redox potential. Other elements (e.g., As, Hg, Pb, or Sn) can form metal-organic complexes. The different mobilization and eco-toxicological behavior of these species has to be taken into account. The speciation of heavy metals in leachate is often ignored when landfills are investigated, although this can be very significant in assessing the long-term mobility of an element. In the case of leachate from a municipal waste landfill, the main focus should be on sulfide precipitation and the formation of complexes by organic substances in connection with the chemical conditions in the soil and groundwater at the site during the various phases of the degradation of the contaminant. In the case of waste incineration slag, the ageing of the mineral phases is the main long-term mobility factor.

The mobilization of contaminants in soil, sediments and waste materials is basically determined by the type of bonding. The various types of bonds must be taken into account both for the inorganic elements as well as for organic substances. The main components of a soil or sediment controlling the element and substance bonding are

- clay minerals,
- carbonates,
- hydrated iron and manganese oxides, and
- organic matter.

In order to arrive at a distinguishing assessment of the various degrees of mobilization of different substances as a consequence of their type of bonding to the soil it is, therefore, necessary to assign the individual substances to the various solid body phases or to determine the fractions releasable under certain conditions. To achieve this, various methods of separation can be used, for example, single-stage and sequential extraction (elution). Batch or column tests can be used to assign substances to individual phases of a soil or sediment (soil air, soil water, or solid material) or to evaluate the leaching behavior. Such extractions, assuming careful selection of the experimental boundary conditions, are comparable with reactions under natural conditions. Since an investigation of the mobilization conditions and processes under natural conditions in the field would only be possible at great expense and consumption of time, it is preferred to use extraction investigations in the laboratory.

In contrast to single-stage elution tests, which only simulate a single mobilization aspect, the more complex, sequential extraction tests can provide information on

- the distribution of elements within the sample,
- alterations and phase changes in the material,
- changes of the chemical parameters in the leachate,
- the influence of pH and redox potential conditions, and
- changes in mobilization conditions over time.

According to the classification as presented by LAKE et al. (1984) and URE & DAVIDSON (1995), a method with sequential elution steps is used for assessment of the mobilization behavior of contaminated soils and sediments. Each step represents a different range of environmental conditions, and allows the various leached elements and substances to be assigned to the following fractions:

- water soluble,
- exchangeable,
- specially sorbed, carbonate bound,
- easily reducible substrates,
- easily extractable organics,
- moderately reducible oxides,
- oxidizable oxides and sulfides,
- crystalline Fe-oxides, and
- residual minerals.

The large number of in part very special, but not mutually balanced elution techniques means that the results are not comparable or cannot be evaluated together. At the European level, there are efforts being made to standardize the elution processes with respect to environmental assessment issues (QUEVAUVILLE, 1996).

A selection of elution methods, which have proven themselves for certain application areas, is summarized in *Table 5.3-1*. The summary of the application of single stage and sequential leachate investigations with respect to the mobilization of contaminants from soils, sediments, and waste materials is presented in VOIGT & WIPPERMANN (1998). Furthermore, REICHERT & RÖMER (1996) provide a detailed description and assessment of the various extraction methods available, whereby most methods have been developed to study heavy metals and water soluble substances. As far as organic substances are concerned, in particular poorly soluble substances, there are to date only a few test methods available (*Table 5.3-1*). The elution tests allow investigation of the behavior both in the saturated zone where the groundwater flows as well as in the unsaturated soil zone through which the leachate flows.

Table 5.3-1: Summary of selected elution methods

Elution method	**Area of application**
single-stage extraction	identification of components bound to certain mineral phases that will be released under certain environmental conditions, VOIGT & WIPPERMANN (1998)
multi-stage (sequential) extraction	determination of the types of bonding of the metallic elements present; identification of metallic elements that will be released under certain environmental conditions, VOIGT & WIPPERMANN (1998), TESSIER et al. (1979), ZEIEN & BRÜMMER (1991)
pH_{stat} method	estimate of long-term behavior of metals in mineral solids under specified pH conditions, OBERMANN & CREMER (1992)
cascade test	simulation of enrichment of percolating solutions, investigation of the release of substances under variable environmental conditions, REICHERT & RÖMER (1996)
column test (percolation and recirculating columns)	examination of an elution or mobilization process as a function of time, VOIGT & WIPPERMANN (1998), REICHERT & RÖMER (1996)
elution of organic contaminants	further development of the pH_{stat} method for organic substances, SCHRIEVER & HIRNER (1994)
EPA Method 1311	special elution process that is also suitable for organic contaminants
standardized methods: elution after DIN 38414-4 (DEV S4); multiple elution after DIN 38414-4	determination of water-soluble components of waste, soil, or sediment; determination of maximum amount of mobilizable water-soluble components

The various elution methods, in particular the sequential extraction method, represent an important instrument in analytical environmental chemistry. They investigate the distribution within a soil or sediment and the mobilization behavior of substances in contaminated soils and sediments under changing environmental conditions, as well as the leachate behavior of waste materials with respect to long-term contaminant migration behavior. These methods are particularly well suited in combination with other methods in that they are supported by, for example, mineral analyses, investigations of leachate and geochemical modeling or complementary to determination of contaminant concentration in solids.

In addition to determination of contaminant release behavior (elution studies) of soils, sediments and waste materials (changes in the composition of leachate with time), an important aspect within general assessment of sites is the analysis of the contaminant migration behavior based on the processes referred to above by which contaminants interact with the components in the ground. Depending on the required quality of the results, batch, reactor and column tests can be used to analyze storage, exchange and transformation processes. The complex, but for the most part realistic column tests offer not only the simulation of reactions occurring in soils and sediments, but also the possibility of investigating transport processes. Similar to the elution methods, the test conditions have a significant influence on the quality of the results. *Table 5.3-2* summarizes three of the most common methods, the main areas of application as well as the main advantages and disadvantages. The migration parameter values so determined can be used in the mathematical modeling of contaminant migration (Section 5.4.4).

Table 5.3-2: Methods to investigate migration processes

Method	Applications	Advantages and disadvantages
batch test	orientating investigation of storage (sorption) and chemical reactions in the soil (also microbial conversion), contaminant release (leaching) from wastes (see also elution methods) Section 5.3.5.4, VOIGT & WIPPERMANN (1998), HILTMANN & STRIBRNY (1998)	simple equipment and simple to carry out, rapid data acquisition, test conditions not realistic
reactor test (dynamic batch test)	investigation of storage and chemical reactions in the soil (also microbial conversion), investigation of non-steady-state processes in the unsaturated zone, Section 5.3.5.4, VOIGT & WIPPERMANN (1998)	complicated equipment, use of undisturbed samples possible, close to natural conditions, continuous process monitoring using electrodes possible
column test	investigation of transport and exchange processes, as well as storage and chemical reactions in the soil, lysimeter tests for aeration zone, aquifer simulation tests for saturation zone, Section 5.3.5.5, VOIGT & WIPPERMANN (1998), HILTMANN & STRIBRNY (1998)	complex equipment and execution, use of undisturbed samples necessary, realistic test conditions (and realistic fluid dynamics)

Geochemical modeling is an important additional instrument in assessing the mobilization and migration behavior of substances in the ground. Geochemical modeling allows the composition and concentration of species involved in groundwater and seepage water as well as in the matrix to be simulated under constant or variable chemical environmental conditions. Physical parameters controlling migration, e.g., porosity, are not taken into account in the purely chemical considerations. The modeling is based on the laws of conservation of mass and energy, the law of mass action and the Debye-Hückel theory. The required data is drawn from chemical analysis of groundwater and seepage water and, where appropriate, mineralogical analysis of the soil itself, analysis of species and type of bonds, as well as from elution tests. The models available today are suitable for describing the following processes in aqueous systems in particular:

- element speciation and distribution,
- behavior of elements under various pH and redox conditions,
- complexing of metal ions.

The most widely used and well proven geochemical model is PHREEQC. For details see VOIGT & WIPPERMANN (1998) as well as SCHULZ & KÖLLING (1992). APPELO & POSTMA (2005) give an introduction to geochemistry and modeling with PHREEQC.

In the case of geochemical modeling, there are, however, a number of limitations which have to be taken into account with respect to the quality of the results. These are, in particular, the ignoring of nonequilibrium reactions and the kinetics of chemical reactions. Newer programs, for example, the current vision of PHREEQC, start to overcome these limitations. Coupled program systems are also currently under development in which chemical reactions are modeled together with flow and substance transport. These programs allow much more comprehensive modeling to be done.

5.3.1 Sampling and Analysis of Groundwater and Surface Water

CLAUS NITSCHE & HANS-JÜRGEN VOIGT

There is a multitude of standards, guidelines, and recommendations for the collection and analysis of groundwater and surface water samples. Normally, groundwater samples are taken from an observation well. The quality of the groundwater is influenced by the drilling and constrution of the well. The sampling should be carried out according to DIN 38402-13 and ISO 5667-1. Preservation and handling of the samples are described in DIN EN

ISO 5667-3. Sampling locations and procedures must be carefully chosen on the basis of the objective. The particularities of each case are to be recorded in as much detail as possible. The documentation must include all the details of the sampling site, the area surrounding it, the sampling method, the equipment and materials used, the personnel, and the details of the sample collection. In the case of pumped groundwater samples, it is important to continually record changes in the well water during the sampling (pH, electrical conductivity, temperature, oxygen concentration, depth to the water table, and pumping rate) in order to determine the appropriate time for taking the sample and as documentation for the interpretation. The regulations dealing with how contaminated water is to be handled are to be observed. How samples are to be taken from surface water bodies is described in DIN 38402-12 for standing water bodies and in DIN 38402-15 for rivers. The procedures are to be adjusted according to the expected contaminant. The sampling of other types of water (e.g., water percolating through the soil or leachate from a landfill) depends on the objective and can be different from case to case. Water samples are prepared according to ISO 5567-3.

5.3.1.1 Planning and Preparation of Work

Sampling and analysis of groundwater is a main objective of a site investigation. The importance of the proper collection of groundwater samples is clearly highlighted given the fact that 85 percent of the errors in the chemical analysis of groundwater occur during sampling (*Fig. 5.3-1*).

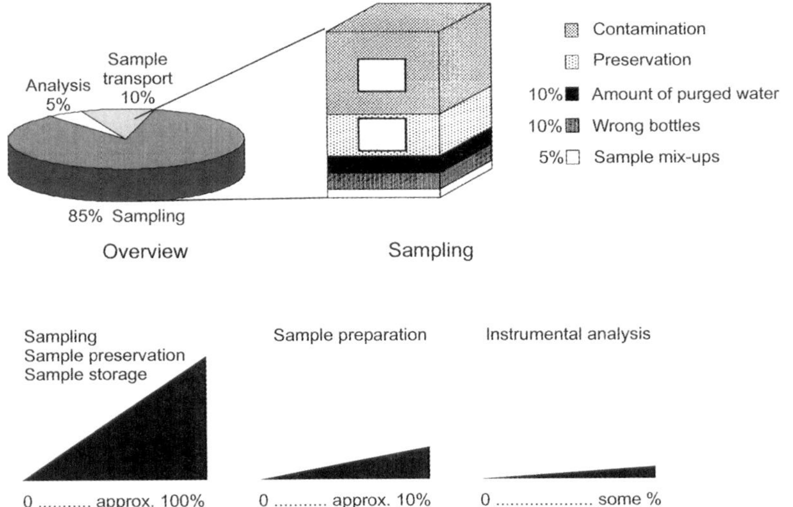

Fig. 5.3-1: Sources and margins of error that occur during groundwater sampling, after SÖHNGEN (1992) and VAN STRAATEN & HESSER (1998)

Given the state-of-art of the methods available for chemical analysis of samples, it may be assumed that the analytical error, expressed by, e.g., the error in the ion balance, is very low. Therefore, significant reduction of the total error can be attained only by reducing the sampling errors.

If representative groundwater samples are to be collected, it is necessary to know how the observation well was designed (Section 5.2.6). The following information is needed:

- type of well completion (fully screened, multiple screens at different depths, groups of wells with each well with a screen at a different depth, etc.),
- name and location of the well,
- casing and screen diameters,
- type of material the casing and screen are made of,
- depth and length of the screen within the aquifer,
- lithologic log,
- depth to the groundwater table, and
- borehole diameter.

Usually all this information should be part of a monitoring well identity (ID) sheet. The ID sheet of the well includes the name of the owner of the well and the owner of the investigated area. Before starting the sampling operation both owners must be informed about it (Chapter 2.3).

Based on the information about all wells to be sampled, the following preparatory steps must be done:

- preparation of a well location map showing the ID number of the wells,
- specification of sampling sequence of the wells,
- estimation of length of hose for the pump used for sampling,
- calculation of volume of water to be purged from each well,
- calculation of the pumping rate,
- calculation of the time needed for sampling,
- decision on the treatment needed for the contaminated water,
- selection of the type and material of sampling equipment,
- specification of health and safety measures.

Sequence of the wells

The groundwater observation wells should be sampled in a sequence such that the sampling equipment does not have to be cleaned in the field. This means that the sampling must be started in the least contaminated well. If it is suspected that a nonaqueous liquid phase is floating on the groundwater, the thickness of the nonaqueous phase must be determined first. The expected volume of the nonaqueous phase must be removed before the sampling equipment is placed in the borehole in order to avoid contamination of the sample or a protective casing must be installed (*Fig. 5.3-4*).

Length of the hose to the pump

The hose to the pump must be long enough to reach the pump installed 1 m above the screen.

Volume of water to be purged from the well

Purging is the process of removing a sufficient amount of water from a well so that no stagnant water remains. Proper purging ensures that representative samples of groundwater can be obtained. The volume V_F of water to be purged from a monitoring well is set to 1.5 times the screened volume (volume of screen and gravel pack, see *Fig. 5.3-2*) and can be estimated using Equation (5.3.1). The estimation is based on an installation depth of the pump 1 m above the screen.

$$V_F = 0.0015 \frac{\pi}{4} L_F \, d_{BH}^2 , \tag{5.3.1}$$

where L_F length of the gravel pack and counterfilter in cm (see *Fig 5.3-2*) and
d_{BH} diameter of the borehole in cm (see *Fig. 5.3-2*).

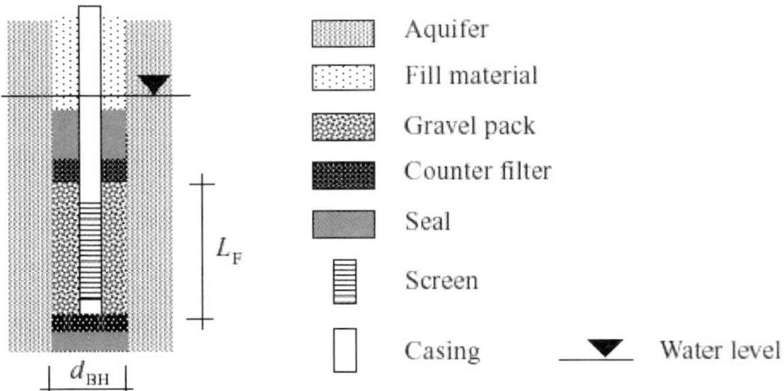

Fig. 5.3-2: Estimation of the volume of water to be purged from a monitoring well, after DVWK (1997)

Following this rule the sample will be collected from a distance of 3 to 4 times the radius of the well depending on the porosity of the aquifer. Other regulations recommend that a volume corresponding to 1 to 4 times the total volume of water in the well must be withdrawn for purging before sampling.

Estimating the pumping rate

The pumping for purging and during sampling must be such that the drawdown of the groundwater table is minimal. This is achieved by adjusting the pumping rate to the geohydraulic conditions of the aquifer. American regulations recommend that the lowering of groundwater table must not exceed 2 m, while in Germany a lowering down to 2 m above the screen is permitted. It must be taken into account that there will be considerable lowering of the groundwater table in an aquifer of fine-grained sand or silty fine-grained sand even at low pumping rates. The optimum pumping rate for a given drawdown can be calculated using the following semi-empirical equation developed by NITSCHE & LUCKNER (NITSCHE, 1998):

$$Q_\mathrm{p} = 20\, s\, k_f\, L_\mathrm{F}\, F_\mathrm{SA}\,, \tag{5.3.2}$$

where s drawdown of the groundwater table [cm],
Q_p pumping rate [L min^{-1}],
k_f hydraulic conductivity of the aquifer [m s^{-1}] (the guide values in *Table 5.3-3* can be used for orientation),
F_SA screen area factor, i.e. percentage of open area of the screen; provided by the manufacturer; guide values are given in (*Table 5.3-4*),
L_F length of the filter [cm].

The factor 20 is a constant that takes into account the distance in the aquifer and the geometry of the aquifer volume influenced by the well.

If a pumping rate Q_p of less than or equal to 0.1 L min^{-1} is desired or the well does not yield a higher rate, then the pump should be lowered to about 0.5 m above the sump or 0.5 m above the bottom of the well and the well pumped dry. This is to be repeated three times until the water within the screened volume has been withdrawn. It is best to repeat this pumping on three successive days. The sample should be taken on the fourth day at a pumping rate of about 0.1 L min^{-1} or with a bailer.

Table 5.3-3: Hydraulic conductivity for consolidated and unconsolidated rocks

Consolidated rock	Hydraulic conductivity k_f [m s^{-1}]	Unconsolidated rock	Hydraulic conductivity k_f [m s^{-1}]
dolomite	10^0 - 10^{-1}	sandy gravel	$3 \cdot 10^{-3}$ - $5 \cdot 10^{-4}$
limestone	10^{-1} - 10^{-2}	gravelly sand	$1 \cdot 10^{-3}$ - $2 \cdot 10^{-4}$
marly limestone	10^{-2} - 10^{-3}	medium sand	$4 \cdot 10^{-4}$ - $1 \cdot 10^{-4}$
calcareous sandstone	10^{-3} - 10^{-4}	fine sand	$2 \cdot 10^{-4}$ - $1 \cdot 10^{-5}$
quartzite	10^{-4} - 10^{-5}	sandy silt	$5 \cdot 10^{-5}$ - $1 \cdot 10^{-6}$
sandstone	10^{-5} - 10^{-6}	clayey silt	$5 \cdot 10^{-6}$ - $1 \cdot 10^{-8}$
siltstone	10^{-7} - 10^{-8}	silty clay	$< 10^{-8}$
claystone	10^{-8} - 10^{-9}		
granite, gneiss	$< 10^{-9}$		

Table 5.3-4: Screen area factors (F_{SA}) for different types of screens

Material	Type and size of screen holes	F_{SA}
PVC/HDPE	0.2 - 3.0 mm slots	0.08 - 0.13
special screen	0.1 - 0.6 mm mesh	0.07 - 0.23
steel	wire-wound continuous slot screen 0.1 - 3.0 mm	0.35 - 0.40
	bridge-slot screen 1 - 4 mm	up to 0.15
	slot hole screen 1 - 4 mm	up to 0.12
gravel pack screen	slot shape width depends on the type of screen	0.07 - 0.23

Calculation of the time needed for sampling

The time needed to collect a groundwater sample can be calculated using the following equation:

$$t_{\text{tot}} = \frac{V_{\text{F}}}{Q_{\text{p}}} + t_{\text{SA}} + t_{\text{IR}}, \tag{5.3.3}$$

where t_{tot} is the total time needed to collect the groundwater sample [min],
V_{F} volume of the screened section (in L) according to Equation (5.3.1),
Q_{p} pumping rate (in L min^{-1}) according to Equation (5.3.2),
t_{SA} time needed for on-site analysis (e.g., acid and base capacities), and for filling and treating the water samples [min],
t_{IR} is the time needed for installing and removing the sampling equipment [min]; on average 30 min.

The pumping time ($V_{\text{F}}/Q_{\text{p}}$) can be also estimated by a graphical method developed by NIELSEN (1991). An example showing the relationship between the required pumping (purging) time and the diameter of the well, the transmissivity of the aquifer and the pumping rate is given in *Fig. 5.3-3*.

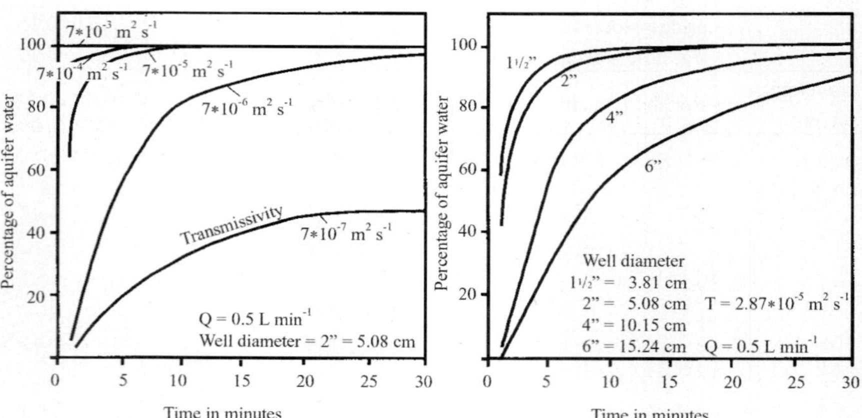

Fig. 5.3-3: Estimation of the purging time to obtain a good quality groundwater sample with respect to the diameter of the well, the transmissivity of the aquifer, and the pumping rate, NIELSEN (1991)

Planning the water treatment and disposal

Prior to the groundwater sampling, it must be agreed with the property owner and the governmental water management authority whether the withdrawn water is to seep back into the ground or pumped into the sewer lines, and whether it is to be disposed of directly or only after treatment. If treatment is necessary, a water treatment system must be installed that can handle the amount of water to be withdrawn and the type of contamination that may be encountered. The following systems are used:

- containers for separating phases;
- containers for stripping light-weight hydrocarbons, including an active carbon filter for the exhaust;
- containers for purifying the water with active carbon;
- containers for diluting contaminated water;
- containers for storage and disposal of contaminated water.

If the contamination cannot be sufficiently removed by one of these systems, then the contaminated water must be transported to a sewage plant or other facility that can decontaminate it. The container volume should be at least 1.5 times of the estimated volume to be withdrawn.

Equipment and other technical requirements

The equipment and other technical requirements are determined by the investigation objectives, the type of observation well, the geohydraulic conditions and the substances in the water to be analyzed (*Tables 5.3-5* and *5.3-6*).

A submersible pump together with a pressure-retaining bailer best fulfills the requirements for sampling. Particularly with respect to trace element analysis, this kind of bailer is the only practical way to obtain a representative groundwater sample. Other types of bailers are used only if the estimated pumping rate is smaller than 0.1 L min^{-1} and after three times the volume of the screened interval has been withdrawn.

Table 5.3-5: Geohydraulic properties and sampling equipment needed

Geohydraulic property	Sampling equipment				Other requirements
	Suction pump	**Submersible pump**	**Bladder pump (diaphragm pump)**	**Bailer (pressure retaining bailer)**	
well yield	$> 10\ L\ min^{-1}$	$> 1\ L\ min^{-1}$ $< 30\ L\ min^{-1}$	$< 3\ L\ min^{-1}$	depends on pump	measurement by filling several containers, inductive flow-through measuring system, or water meter
depth to groundwater table	< 7 m	< 80 m	< 70 m	water pressure at sampling depth: < 700 kPa	depth to water table < 50 m: electric sounder depth to water table > 50 m: pressure sensor

Table 5.3-6: Suitability of equipment depending on the parameter to be determined

Parameter to be analyzed	**Sampling equipment**			
	Suction pump	**Submersible pump**	**Diaphragm pump**	**Pressure retaining bailer**
physical and chemical parameters (e.g., T, pH, elect. conductivity, O_2, Eh)	0	2	2	2
mineralization, organic substances	1	2	2	2
pressure-sensitive parameters (e.g., dissolved gases, TIC, Eh, volatile and organic substances)	0	1	1	2
inorganic and organic substances present in traces (e.g., heavy metals and pesticides)	1	1	1	2
microbiological parameters (e.g., bacteria and viruses)	1	1	1	2

0 not suitable 1 limited suitability 2 suitable

Checking the equipment

The equipment must be checked and serviced according to the operation manual before it is used. The measuring instruments must be recalibrated in the laboratory before use and recorded in the record book for the instrument.

Cleaning of the sampling system

The sample collecting system (including the flow-through cell) must be cleaned after each use.

Preparation of sampling containers/Contact with the laboratory before sampling

Before the sampling takes place it is recommended to contact the laboratory that will carry out the chemical analyses. The laboratory usually prepares the sample bottles with the chemical substances required for preserving the samples. As a rule, glass bottles have to be used for the analysis of organic compounds, and PVC bottles for the analysis of inorganic compounds.

Safety and health plan

A safety and health plan must be prepared for working at contaminated sites, which includes for example,

- instructions for field personnel,
- protective clothes for field personnel,
- protective measures to reduce the contact with hazardous substances (including vapor phases), and
- containers for washing and cleaning of field personnel and equipment.

5.3.1.2 Groundwater Sampling

Checking the exterior condition of the well

The part above ground of any monitoring well must first be checked as described in Section 5.2.6.

Checking the interior condition of the well

The first step after a well is opened is to check whether there is contamination with volatile hydrocarbons. This can be done by either smelling or using gas detector tubes. If the initial check indicates the presence of any liquid hydrocarbons, then the thickness of the nonaqueous phase must be determined. Any floating phase of hydrocarbons (LNAPL), if detected, must be removed completely before sampling. This is necessary in order to avoid the contamination of the sampling equipment. In wells where a LNAPL phase is detected above the groundwater table, a protective casing is used to avoid contamination of the sampling equipment if the phase cannot be removed. The procedure of installation of protective casing is shown in *Fig. 5.3-4*.

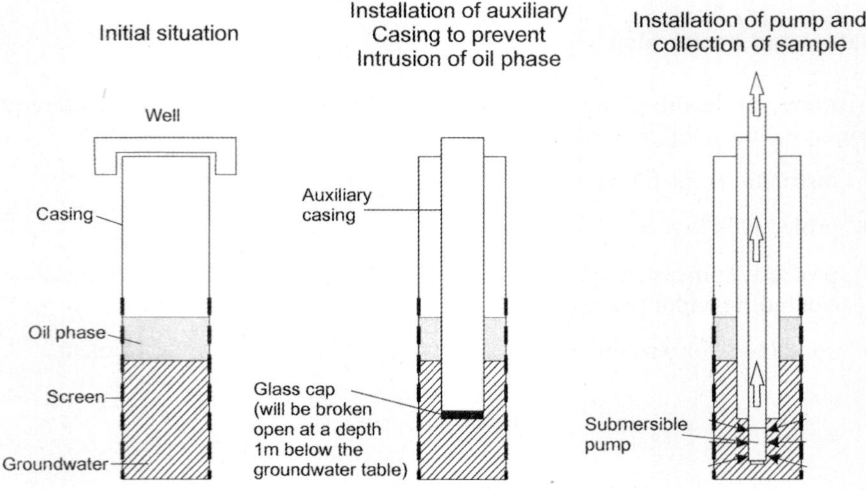

Fig. 5.3-4: Procedure of installation of a protective casing in a well with free hydrocarbon phase above groundwater

The depth to the groundwater table and the depth to the bottom of the well are then measured with special probes with an optical liquid sensor and conducting electrodes (interface meters, Section 5.2.6) and recorded (the thickness of a floating hydrocarbon phase must taken into consideration for the depth to the groundwater table). If the depth to the bottom of the well varies more than 10 cm from the well completion documentation, this should be noted in the sampling documentation. A reason for this could be sedimentation at the bottom of the well. The electrical measuring tapes (water level meter, Section 5.2.6) should be checked for the presence of sediment and, if necessary, cleaned to prevent false water level readings. Once the depth of the well and the depth to the groundwater table have been determined, the groundwater sampling system is tested before the first sample is collected. If the system does not function properly in the test, no sample is to be taken. This is recorded in the field book.

Installation of the sampling system

Normally the submersible pump should be placed 1 m above the top of the well screen. If the water table is within the screened interval, the position of the pump is given by the following equation:

$$Z_{pd} = Z_{wt} + Z_{ed} + 2\text{ m}, \tag{5.3.4}$$

where Z_{pd} is the depth of the pump in m below the top of the well,
Z_{wt} the depth of the water table in m below the top of the well, and
Z_{ed} is the estimated drawdown of the water table in m.

In any case, the pump should not be installed within 2 m of the bottom of the well to avoid collection of any sediment from the bottom. If the well is contaminated with volatile halogenated hydrocarbons, a pressure-retaining bailer is installed immediately above the pressure transducer for water level monitoring and the submersible pump. The vertical distance between the pump and the well screen(s) must be at least 0.5 m. In a layered aquifer all measures necessary to take samples from the different layers separately are to be taken.

Preparation for measurements

The electrodes used to measure oxygen content and redox potential must be calibrated using the standard procedures prior to inserting them into the flow-through cell. *Figure 5.3-5* shows a flow-through cell with electrodes in place.

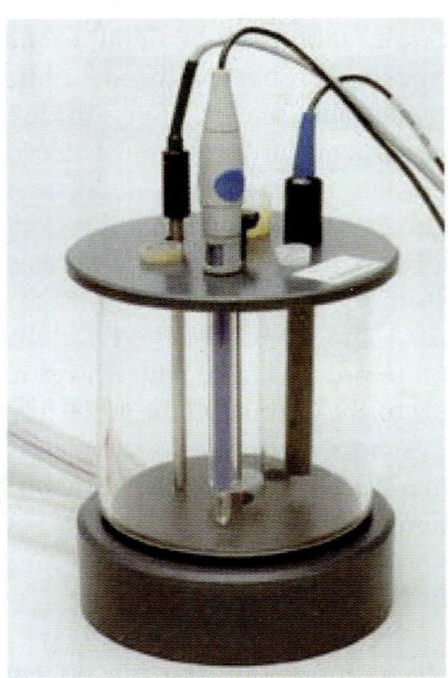

Fig. 5.3-5: Flow-through cell

Purging

Pumping is started at the planned pumping rate and the water level is monitored with a pressure transducer or an electric sounding light. The pumping rate has to be reduced if the expected drawdown of the water table is exceeded. The values of the following parameters are to be recorded every five minutes for pumping times less than 90 minutes and every ten minutes for pumping times longer than 90 minutes:

- depth to the water table in the well,
- pumping rate,
- temperature of the water,
- pH,
- electrical conductivity,
- oxygen content, and
- redox potential.

If the oxygen content exceeds 1 mg L^{-1} and the redox potential is negative, then one of the following sources of error must be assumed:

- a leak in the well casing and/or
- a leak in the pump hose/pipe.

Leaking equipment must be sealed. After sealing and reinstallation the purging would have to started anew.

When such leaks are present, there is a continually inflow of water above the pump and this would be included in the amount being pumped. In this case, the pumping rate is to be increased to lower the water level to about 1 m above the pump. The pumping rate is then decreased until a gradual rise in water level is observed. This is repeated several times, also during the collecting of samples. This is done to guarantee that the water sample is from the desired level in the aquifer. The encountered situation and the measures taken must be properly documented. It is necessary to check the joints between pump hose or pipe sections after removal. If leaks in the casing are indicated by the initial check of the well, then this will be confirmed in the data. If the measured drawdown in the well is more than 2 m than the predicted drawdown at the planned pumping rate, then the well must be regenerated. If a pumping rate of less than or equal to 0.1 L min^{-1} is necessary, the procedure described above in the Section "Estimating the pumping rate" or a bailer should be used. *Figure 5.3-6* shows different types of bailers - from a simple sampling tube to a pressure-retaining bailer.

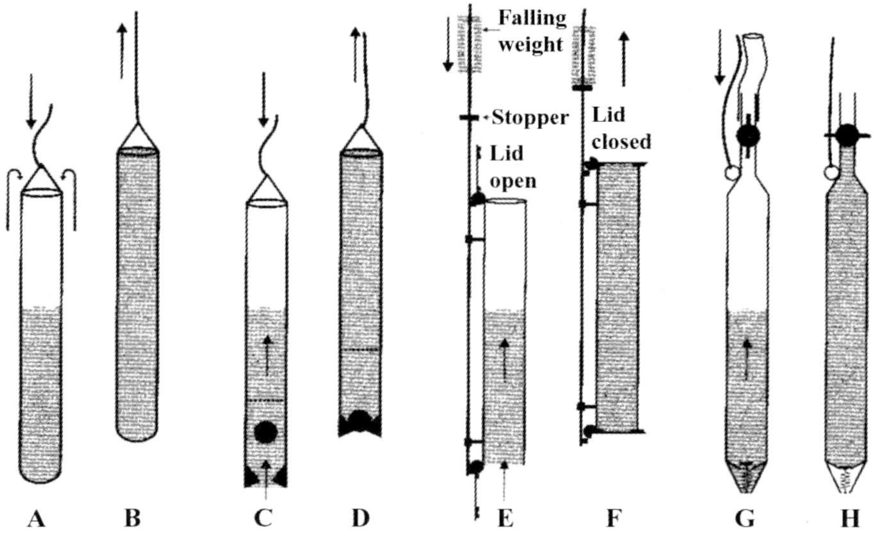

Fig. 5.3-6: Different types of bailers, BUWAL (2003), A and B - simple bailers (A - while filling, B - while being pulled up), C and D - bailer with ball valve (C - while filling, D - while pulled up), E and F - bailer with controlled lid (E - filling, F - closing by a falling weight), G and H - pressure-retaining bailer (G - filling, H - closed head space container)

Groundwater sampling

It is necessary to monitor the hydrochemical and physical parameters during the entire purging time (see above). The sampling is started after 1.5 times the screen volume has been purged. There are two basic conditions that must be satisfied during the groundwater sampling: The samples must be collected at the same pumping rate as the purging rate, and the sample must be collected when the control parameters do not fluctuate more than the following ranges during the purging operation:

- electrical conductivity ± 5 µS cm^{-1},
- pH ± 0.1,
- temperature ± 0.1 °C,
- oxygen content ± 0.1 mg L^{-1}, and
- redox potential ± 1 mV.

Redox and conductivity probes must not be switched on at the same time (DVWK, 1992). *Figure 5.3-7* shows examples of plots of the control parameters during purging.

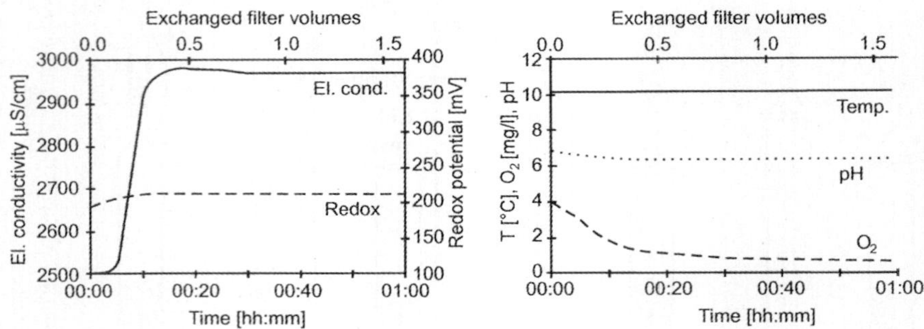

Fig. 5.3-7: Examples of measuring curves of the control parameters over the purging time, BWK (2004)

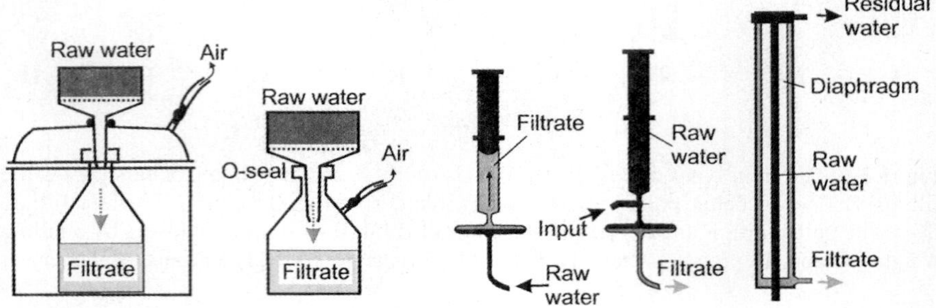

Fig. 5.3-8: Systems for on-site water filtration, BUWAL (2003)

Experience has shown that the following procedure is very useful during collection of groundwater samples: The 1.0 L and 2.5 L glass bottles used to transport the samples should be labeled and then rinsed three times with pumped groundwater before being filled with the groundwater sample. The bottles, for example for the TIC and sulfide analysis, must be completely filled with no air bubbles in the water. Bottles containing the appropriate preservatives are filled from the 2.5 L bottles. Groundwater samples that are filtered on-site (*Fig. 5.3-8*) are also taken from the 2.5 L bottles. In this case, the filter is first rinsed with the water from the bottle. If samples for analysis of volatile halogenated hydrocarbons are to be collected, a pressure-retaining bailer (e.g., BAT system, *Fig. 5.3-9*) should be used and the samples are collected in three headspace vials for GC analysis.

After the sampling equipment is removed from the well, the sample container of the pressure-retaining bailer is transported to the laboratory in a cool box. Besides the stable values of the physicochemical parameters (see above), the organoleptic observations of the sample together with any observations of gas bubbles or precipitation of any substances are recorded in the field book.

The groundwater samples must be filtered in the field if

- it is necessary to preserve the samples with acid to avoid precipitation of certain species (e.g., carbonate, sulfate, iron), or
- water conditions change occur, for example, sorption or ion exchange on fine-grained particles.

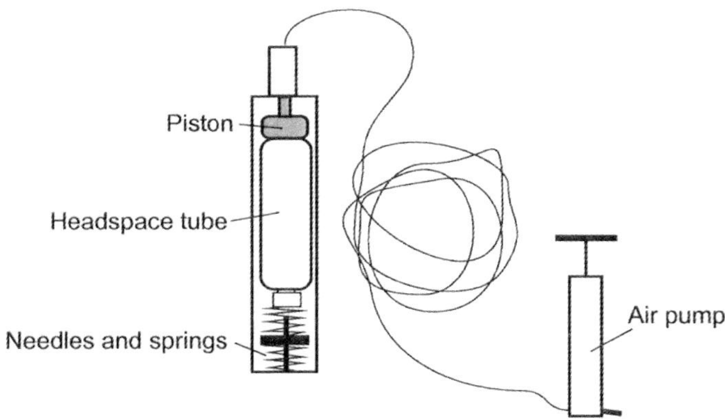

Fig. 5.3-9: BAT-system for pressure-retaining sampling

Preservation and on-site analysis of samples

Whether the samples should analyzed on site, as well as which sample preservation measures must be used, are decided by the laboratory staff on the basis of the objectives of the investigation. *Table 5.3-7* gives an overview of methods of preservation, their effects and purpose. Detailed information on preservation methods can be found in EN ISO 5667-3.

Table 5.3-7: Preservation methods

Preservative	**Used for samples for analysis of**	**Effect**
HNO_3 (nitric acid)	metals (e.g., Fe, Ca, Mg)	prevents precipitation
H_2SO_4 (sulfuric acid)	NH_4^+, amines	forms salts with organic bases
	organic substances (COD, org. C, oil)	bacterial inhibition
NaOH (caustic soda)	phenols, organic acids, cyanides	forms salts with volatile compounds
$HgCl_2$ (mercury chloride)	N and P compounds	bacterial inhibitor
cooling / freezing	acidity, alkalinity, org. C, org. N, org. P, organic substances	

All groundwater samples must be transported to the laboratory in a cool box as quickly as possible (usually in a period of 12 hours but not later than one week). During transport and storage, the samples must be kept at 2 to 4 °C in the dark.

Labeling of the samples

The bottles must have adhesive labels containing the following information:

- sample number,
- well name/number,
- aim of analysis,
- sampling date and time,
- company/laboratory taking the sample.

Cleaning the equipment

After sampling, the equipment must be properly cleaned.

Documentation of the sampling

The sampling operation is to be documented as follows:

- All information about the examination of the condition of the well must be recorded.
- The pumping rate, the water level in the well, and the amount of the pumped water resulting from the drawdown of the groundwater table are plotted as a function of time.
- The control parameters are plotted as a function of time.
- The collection of the samples is recorded in a groundwater collection sheet.

The groundwater collection sheet should contain

- number and kind of sampling points (monitoring well, water supply well, spring),
- coordinates of the sampling points,
- weather,
- well completion (casing material and diameter, screen depth and length),
- information on the sampling devices used (type of pump or bailer, depth of installation, material of pump housing and hose/pipe, maintenance data),
- depth to the groundwater table before and after sampling,
- date and time of the start and end of purging, time of sampling,
- list of samples with sample numbers, material of storage bottles, filtration and preservation information, analyzed parameters,
- smell, color, turbidity, bubbles, etc.,
- name of the company or laboratory that conducted the sampling,
- transportation conditions (cool box, temperature),
- date and time of delivery of the sample to the laboratory, stamp and signature of the laboratory.

Passive samplers

Passive groundwater samplers are being increasingly used in monitoring wells. Passive samplers must be used in combination with packer systems, which help isolate the well screen to avoid exchange of gas between the water in the aquifer and water in other parts of the well. It must be guaranteed that the well screen is hydraulically connected to the aquifer. The sampling cell is placed in the middle of the screen for a predetermined time (usually 10 to 30 days). Over this time period equilibrium is attained between the groundwater and the water inside the cell. There are two types of passive samplers (*Fig. 5.3-10*): "peeper" or diaphragm cells and diffusion cells.

The advantage of diaphragm cells is that fine particles and microorganisms have no influence on the water quality. They give good results for dissolved inorganic compounds in groundwater (STEINMANN & SHOTYK, 1996). Diffusion cells are recommended for measurement of volatile organic compounds in groundwater (BARBER & BRIEGEL, 1987).

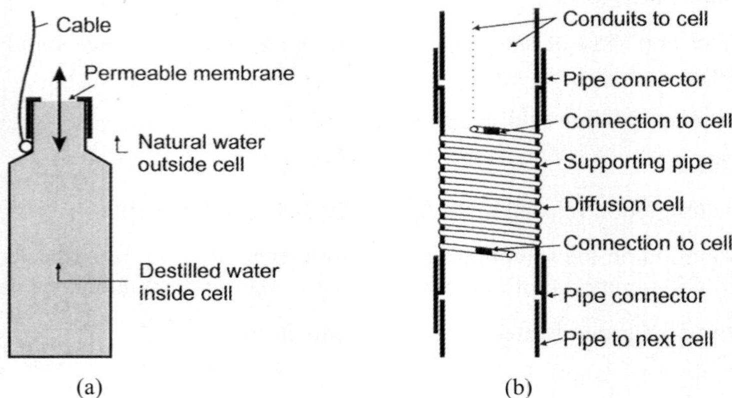

Fig. 5.3-10: Passive samplers, (a) - "peeper" or diaphragm cell, BUWAL (2003), (b) - diffusion cell, BARBER & BRIEGEL (1987)

Quality assurance

Some of the problems associated with groundwater sampling that can affect the quality of the samples are listed here:

- In the planning and preparation phase:
  - There is insufficient information about the site and the wells (e.g., no well ID sheets).
  - The selected equipment has not been proven reliable in hydrogeological conditions like that of the investigation site (JIWAN & GATES, 1992).
  - The materials to be used in the field have been insufficiently prepared.
- Before sampling:
  - The volume of purged stagnant water is too low relative to the estimated screened volume.
  - The pumping rate is too high for the hydraulic conditions of the aquifer.
  - The generator is not placed on the downwind side – exhaust fumes (containing, for example, BTEX) can change the water quality.
  - The equipment is not placed on a clean plastic surface but directly on the potentially contaminated soil surface.
  - Contamination from other wells occurs due to chemical residues in the pump or the well casing.
- In the sampling phase:
  - Sampling was not started with the least contaminated well.
  - Groundwater from other aquifers is included in the sample due to leakage in the casing.
  - Transported fines contaminate the water sample.
  - Gases are released during purging.
- In the bottle filling phase:
  - Contamination due to unclean bottles occurs.
  - Atmospheric oxygen enters the sample, resulting in oxidation of organics, sulfides, ferrous iron, ammonium and other compounds.
  - Exhaust fumes of the compressor contaminate the sample.
- During preservation and transportation of the samples:
  - Bottles are incorrectly labeled or labels are lost.

- If no preservation measures are taken (*Table 5.3-7*), chemical and biochemical reactions can occur.
- Samples are not cooled during transportation.

5.3.1.3 Groundwater Analysis

The following tables (*Tables 5.3-8* to *5.3-14*) give an overview of standard methods for determination of physico-chemical parameters and for chemical analysis of inorganic and organic compounds in water.

Table 5.3-8: Physico-chemical parameters for water analysis, BANNERT et al. (2001)

Parameter	**Method**	**Method according to**
color	photometry	DIN EN ISO 7887 (C1)
turbidity	photometry	DIN EN ISO 7027 (C2)
dry residue and ignition residue	gravimetry	DIN 38409-H1
pH		DIN 38404-C5
electrical conductivity		DIN EN 27888 (C8)
dissolved oxygen	electrochemical or iodometric determination	DIN EN 25814 (G22) DIN EN 25813 (G21)

Table 5.3-9: Analysis of anions in water, modified after BANNERT et al. (2001)

Parameter	**Method**	**Method according to**	**Detection limit**
fluoride, bromide, chloride, nitrate, o-phosphate, sulfate, nitrite	ion chromatography	DIN EN ISO 10304-1 (D19)	0.05 to 0.1 mg L^{-1}
fluoride	fluoride-sensitive electrode	DIN 38405-D4	0.2 mg L^{-1}
cyanide	flow analysis	DIN EN ISO 14403	0.005 mg L^{-1}
nitrate	UV/VIS photometry flow analysis (CFA and FIA)	DIN 38405-D9 DIN EN ISO 13395 (D28)	0.5 mg L^{-1} 0.2 mg L^{-1}
nitrite	UV/VIS photometry flow analysis (CFA and FIA)	DIN EN 26777 (D10) DIN EN ISO 13395 (D28)	0.01 mg L^{-1} 0.01 mg L^{-1}
sulfate	titration	DIN 38405-D5	20 mg L^{-1}
sulfide	UV/VIS photometry	DIN 38405-D26	0.04 mg L^{-1}

Table 5.3-10: Determination of elements in water, BANNERT et al. (2001)

Parameter	Method	Method according to	Detection limit
As, Cd, Cr, Co, Cu, Ni, Mo, Pb, Sb, Se, Sn, Tl, Zn, Ba, Ca, Fe, Mg, Mn, Na, K, Sr, etc.	ICP-AES or ICP-QMS; detection limits and disturbance by high concentrations must be taken into consideration	DIN EN ISO 11885 (E22) ISO 17294-2	0.05 to 1 mg L^{-1}, depends on element and spectral line used for the analysis
ammonium ion (NH^{4+})	UV/VIS photometry flow analysis (CFA and FIA)	DIN EN ISO 11885 (E22) ISO 17294-2	0.03 mg L^{-1} 0.1 mg L^{-1}
antimony (Sb)	AAS	DIN 38405-D32	0.01 mg L^{-1}
barium (Ba), strontium (Sr)	AAS	DIN 38406-E28; analogous for Sr	0.5 mg L^{-1}
lead (Pb)	AAS	DIN 38406-E6	0.5 to 0.005 mg L^{-1}
cadmium (Cd)	AAS	DIN EN ISO 5961 (E19)	0.05 to 0.001 mg L^{-1}
calcium (Ca), magnesium (Mg)	AAS	DIN EN ISO 7980	0.02 to 0.05 mg L^{-1}
total chromium (Cr)	AAS	DIN EN 1233 (E10)	0.5 to 0.005 mg L^{-1}
chromium(VI)	UV/VIS photometry ion chromatography	DIN 38405-D24 DIN EN ISO 10304-3 (D22)	0.05 mg L^{-1} 0.05 mg L^{-1}
iron (Fe)	AAS	DIN 38406-E32	5 to 0.002 mg L^{-1}
sodium (Na), potassium (K)	flame photometry AAS AAS	DIN ISO 9964-3 (E27) DIN 38406-E14 DIN 38406-E13	1 to 2 mg L^{-1}
cobalt (Co)	AAS	DIN 38406-E24	0.2 to 0.005 mg L^{-1}
copper (Cu)	AAS	DIN 38406-E7	0.1 to 0.002 mg L^{-1}
manganese (Mn)	AAS	DIN 38406-E33	0.2 to 0.001 mg L^{-1}
nickel (Ni)	AAS	DIN 38406-E11	0.5 to 0.001 mg L^{-1}
mercury (Hg)	AAS cold vapor method AAS amalgam method	DIN EN 1483 (E12) DIN EN 12338 (E31)	0.1 µg L^{-1} 0.01 µg L^{-1}
thallium (Tl)	ETAAS	DIN 38406-E26	0.002 mg L^{-1}
zinc (Zn)	AAS	DIN 38406-E8	0.05 to 0.0005 mg L^{-1}

Table 5.3-11: Determination of sum parameters in water, BANNERT et al. (2001)

Parameter	Method	Method according to	Detection limit
chemical oxygen demand (COD)	titration	DIN 38409-H41	15 mg L^{-1}
total phosphorus (TP)	oxidative decomposition, UV/VIS photometry	ISO 6878	10µg L^{-1}
total nitrogen (TN)	catalytic reduction, distillation, titration	DIN EN 25663 (H11)	1 mg L^{-1}

Table 5.3-12: Determination of the concentration of inorganic contaminants in eluates and leachate, BANNERT et al. (2001)

Parameter	Method	Method according to
As, Cd, Cr, Co, Cu, Mo, Ni, Pb, Sb, Se, Sn, Tl, Zn	ICP-AES (ICP-QMS possible)	on basis of DIN EN ISO 11885 *)
arsenic (As), antimony (Sb)	hydride-generation AAS	DIN EN ISO 11969
lead	AAS	DIN 38406-6
cadmium	AAS	DIN EN ISO 5961
total chromium	AAS	DIN EN 1233
chromium VI	spectrophotometry, ion chromatography	DIN 38405-24 DIN EN ISO 10304-3
cobalt	AAS	DIN 38406-24
copper	AAS	DIN 38406-7
nickel	AAS	DIN 38406-11
mercury	cold vapor AAS	DIN EN 1483
selenium	AAS	DIN 38405-23
zinc	AAS	DIN 38406-8
total cyanide	spectrophotometry	DIN 38405-13 DIN EN ISO 14403
easily liberatable cyanide	spectrophotometry	DIN 38405-13
fluoride	fluoride-sensitive electrode, ion chromatography	DIN 38405-4 DIN EN ISO 10304-1
cyanides	spectrophotometry	DIN 38405-D14

*) The detection limit has to be adapted to the investigation goal by suitable measures or suitable equipment.

Table 5.3-13: Determination of concentrations of organic contaminants in percolation water, BANNERT et al. (2001)

Parameter	Method	Method according to
benzene	GC-FID	DIN 38407-9 *)
BTEX	GC-FID, high concentrations must be taken into consideration	DIN 38407-9
volatile halogenated organic compounds (VHC)	GC-ECD	DIN EN ISO 10301
aldrin	GC-ECD (GC-MS possible)	DIN 38407-2
DDT	GC-ECD (GC-MS possible)	DIN 38407-2
phenols	GC-ECD	ISO 8165-2
chlorophenols	GC-ECD or GC-MS	DIN EN 12673 (F15)
chlorobenzenes	GC-ECD (GC-MS possible)	DIN 38407-2
total PCB	GC-ECD GC-ECD or GC-MS	DIN EN ISO 6468 DIN 51527-1 DIN 38407-3
total PAH	HPLC-F	DIN 38407-8
naphthalene	GC-FID or GC-MS	DIN 38407-9
petroleum hydrocarbons (PHC)	extraction with petroleum ether, quantification by gas chromatography	after ISO/TR 11046

*) adjustment of detection limit necessary

Table 5.3-14: Determination of organic contaminants in groundwater and surface water, modified after BANNERT et al. (2001)

Parameter	Method	Method according to	Detection limit
petroleum hydrocarbons (PHC)	extraction with acetone/petroleum ether, chromatographic cleaning, GC-FID	ISO 9377-2 DIN EN ISO 9377-2 (H53)	0.1 mg L^{-1}
volatile aromatic hydrocarbons: naphthalene, benzene, toluene, ethylbenzene, xylene, styrene, cumene, trimethylbenzene	1. headspace analysis, GC-MS or GC-FID	DIN 38407-F9-1	5 µg L^{-1}
	2. extraction with pentane, GC-MS or GC-FID	DIN 38407-F9-2	5 µg L^{-1}
six PAH listed in the German Drinking Water Ordinance	extraction with hexane, HPLC-UV/F, GC-MS	ISO 17993	0.01 to 0.1 µg L^{-1}
chlorobenzenes: chlorobenzene to hexachlorobenzene	1. extraction with pentane or hexane, if necessary chromatographic cleaning, GC-MS or GC-ECD	DIN 38407-F2	0.05 to 0.5 µg L^{-1}
	2. extraction with pentane or hexane or petroleum ether, if necessary chromatographic cleaning, GC-MS or GC-ECD	DIN EN ISO 6468 (F1)	0.05 to 5 µg L^{-1}
volatile halogenated organic compounds (VHC)	1. headspace analysis, GC-MS or GC-ECD	DIN EN ISO 10301 (F4)	0.01 to 5 µg L^{-1}
	2. extraction with pentane, GC-MS or GC-ECD	DIN EN ISO 10301 (F4)	0.01 to 5 µg L^{-1}
vinyl chloride	headspace analysis, GC-MS or GC-FID	DIN 38413-P2	5 µg L^{-1}
chlorophenols	extraction with hexane, derivatization with acetic anhydride, GC-MS or GC-ECD	DIN EN 12673 (F15)	0.1 µg L^{-1}

Parameter	Method	Method according to	Detection limit
phenols	1. extraction with hexane after acidification, derivatization with pentafluorobenzoyl chloride, GC-ECD	ISO 8165-2	0.1 μg L^{-1}
	2. extraction with hexane after acidification, derivatization with pentafluorobenzoyl chloride, GC-MS, GC-FID	DIN EN 12673 (F15)	1 μg L^{-1}
organic chlorinated pesticides, low volatile halogenated hydrocarbons	extraction with hexane, chromatographic cleaning, GC-ECD or GC-MS	DIN 38407-F2	0.01 μg L^{-1}
PCB (IUPAC numbers 28, 52, 101, 138, 153, 180)	extraction with hexane, if necessary chromatographic cleaning, GC-ECD or GC-MS	DIN 38407-F3 DIN 38407-F2	0.01 μg L^{-1} for each substance
phenol index	without distillation or dye extraction; with distillation and dye extraction; with distillation and without dye extraction; UV/VIS photometry	DIN 38409-H 16-1 DIN 38409-H 16-2 DIN 38409-H 16-3	0.01 mg L^{-1} 0.01 mg L^{-1} 0.01 mg L^{-1}
adsorbable organic halogens (AOX)	adsorption on active carbon, combustion to hydrogen halides, titration, coulometry	ISO 9562	0.01 mg L^{-1}
extractable organic halogens (EOX)	extraction with pentane, hexane or heptane, combustion to hydrogen halides, titration, coulometry	DIN 38409-H 8	0.02 mg L^{-1}
total organic carbon (TOC)	conversion of organically bound carbon to CO_2	DIN EN 1484 (H14) ISO 8245	0.1 mg L^{-1}

Parameter	Method	Method according to	Detection limit
plant protection products (containing N and P)	enrichment on RP 18, elution, GC-PND	ISO 10695 ISO 15913	0.05 µg L^{-1}
plant protection products	enrichment on RP 18, elution, HPLC-UV	ISO 11369	0.05 µg L^{-1}
phenoxyalkyl carbonic acids	after solid-liquid extraction and derivatization, GC-MS	DIN 38407-F14 ISO 15913	0.05 µg L^{-1}
Overview analysis of PCDD/F			
2,3,7,8 – tetrachlordibenzodioxin	HRGC-HRMS	EPA 1613	0.03 pg L^{-1}
1,2,3,7,8-pentachlordibenzodioxin	HRGC-HRMS	EPA 1613	0.05 pg L^{-1}
1,2,3,4,7,8-hexachlordibenzodioxin	HRGC-HRMS	EPA 1613	0.1 pg L^{-1}
1,2,3,6,7,8-hexachlordibenzodioxin	HRGC-HRMS	EPA 1613	0.1 pg L^{-1}
1,2,3,7,8,9-hexachlordibenzodioxin	HRGC-HRMS	EPA 1613	0.1 pg L^{-1}
1,2,3,4,6,7,8-heptachlordibenzodioxin	HRGC-HRMS	EPA 1613	0.03 pg L^{-1}
tetrachlordibenzodioxin	HRGC-HRMS	EPA 1613	0.17 pg L^{-1}
octachlorodibenzodioxin	HRGC-HRMS	EPA 1613	0.33 pg L^{-1}
2,3,7,8-tetrachlorodibenzofuran	HRGC-HRMS	EPA 1613	0.03 pg L^{-1}
1,2,3,7,8-pentachlorodibenzofuran	HRGC-HRMS	EPA 1613	0.05 pg L^{-1}
2,3,4,7,8-pentachlorodibenzofuran	HRGC-HRMS	EPA 1613	0.05 pg L^{-1}
1,2,3,4,7,8-hexachlorodibenzofuran	HRGC-HRMS	EPA 1613	0.1 pg L^{-1}
1,2,3,6,7,8-hexachlorodibenzofuran	HRGC-HRMS	EPA 1613	0.1 pg L^{-1}
2,3,4,6,7,8-hexachlorodibenzofuran	HRGC-HRMS	EPA 1613	0.1 pg L^{-1}

Parameter	Method	Method according to	Detection limit
1,2,3,7,8,9-hexachlorodibenzofuran	HRGC-HRMS	EPA 1613	0.1 pg L^{-1}
1,2,3,4,6,7,8-heptachlorodibenzofuran	HRGC-HRMS	EPA 1613	0.17 pg L^{-1}
1,2,3,4,7,8,9-heptachlorodibenzofuran	HRGC-HRMS	EPA 1613	0.17 pg L^{-1}
octachlorodibenzofuran	HRGC-HRMS	EPA 1613	0.33 pg L^{-1}
I-TEQ NATO/CCMS	HRGC-HRMS	EPA 1613	0.16 pg L^{-1}
TEQ WHO 1997	HRGC-HRMS	EPA 1613	0.19 pg L^{-1}

5.3.2 Sampling and Analysis of Soil, Rock, Stream and Lacustrine Sediments

MANFRED BIRKE, CLAUS NITSCHE & HANS-JÜRGEN VOIGT

Samples are taken for laboratory analysis to investigate

- texture, structure, and environmental parameter values,
- contents and spatial distribution of substances,
- physical, chemical, and biological properties,
- interaction between the groundwater and the sediments during migration,
- natural microbiological and geochemical attenuation processes.

The sampling methods are chosen according to the objectives of the investigation, the migration paths, size of the investigation area, the vertical and horizontal distribution of the contaminants, and the geology (pedology, lithology, hydrogeology, foreign materials) of the area, as well as according to the current land use, the legally allowed, and the earlier use. All of this is to be summarized in a "contamination hypothesis" following DIN ISO 10381-5.

The sample must be representative of a defined area and depth at a defined time. Determination of contaminant distribution in the ground is important because concentrations vary distinctly with depth. Proper sampling is necessary to obtain reliable analytical results for site assessment. Errors made in sample collection cannot be recognized in the laboratory and, therefore, not be corrected afterwards.

5.3.2.1 Planning and Preparation of Work

Sampling density, sample size, and sampling method depend on the substance(s) being investigated (type of contaminant, its chemistry, soil particle-size distribution, and retention capability of the soil for the contaminant), the purpose of the investigation (hazard evaluation, sanitation, disposal of contaminated material, securing evidence, determination of the degree of contamination), and what is known about the area (obtained from historical records dealing with the contamination, the subsoil and soil associations, as well as soil and geological maps). The following aspects have to be planned:

- node distribution of the sampling grid,
- number and spacing of the sampling sites,
- number and size of samples to be collected,
- equipment needed to guarantee representative samples, and
- time needed to collect the samples.

Structure of the sampling grid

The area of investigation should be subdivided into appropriate subareas for the sampling. This should be done on the basis of differences in estimated hazard, differences in land use, relief, soil type, or conspicuous differences, such as in vegetation, or on the basis of the inventory made in the preliminary investigation. If there are no differences that can be used as a criterion, the subdivision can be done on the basis of a grid that will permit sufficient coverage and reliability. If contaminant distribution is to be determined in the area, sampling is to be done on a grid. Grid size and sampling sites are to be selected on the basis of the "contamination hypothesis" or "process model" (see *Fig. 5.2-31* in Section 5.2.6) available for the site. The term "grid" is used here for a planned array of sampling sites, which does not necessarily have to be regularly spaced. The "contamination hypothesis" is to be documented as a basis for the sampling strategy. On-site analyses can also be done for selecting the sampling sites. Areas of suspected elevated contaminate concentrations are to be targeted for sampling. The number and distribution of the sampling sites must be sufficient to properly assess the hazard and to define the areal extent of the elevated contaminant concentrations. As far as possible, samples should be taken from each soil horizon.

Number and spacing of sampling sites

If the number and spacing of the sampling sites is not specified in advance, these are selected on the basis of previous studies (e.g., from geological, soil, and hydrogeological maps and cross-sections). In an investigation of contamination around a landfill, the soil samples should be taken with a high sampling density, so that geostatistical methods can be used and the results can be presented on a map without gaps. Depending on the land use and the impact paths, the samples must normally be taken in the depth range 0 to 35 cm and in agricultural areas down to 60 cm. The number of sampling sites is determined on the basis of the expected types and concentrations of contaminants. In soils, at least one sample must be taken from each soil horizon. For greater depths, samples are collected on the basis of CPT soundings. The samples should be collected from those sites with the lowest concentrations to those with higher concentrations in order to avoid spreading of contaminants from areas of high concentration to areas of low concentration during sampling.

Number and size of samples

Sample size is determined by grain size (DIN 18123) and must be sufficient for analysis and for reserve samples. The type of contamination must be taken into consideration when the sampling method is selected. The number and size of the samples depend on the number of analytical parameters to be determined. Normally, each soil horizon is to be individually sampled at each sampled location. If disturbed samples are to be taken from a man-made pile (e.g., drill cuttings, waste heaps, or even storage piles), the number of samples that should be taken is given in *Table 5.3-15*. Sample size in this case depends on the soil type and the kind of contaminant as shown in *Table 5.3-16*.

Table 5.3-15: Minimum number of samples to be taken from a pile of drill cuttings, a mine waste heap, or a pile of stored construction materials

Maximum grain size	Minimum number of samples to be taken from piles of the following sizes		
	< 50 t	50 t – 150 t	> 150 t
> 20 mm	5	1 per 10 t	15
< 20 mm	3	3 per 50 t	8

1 t = 1000 kg

Table 5.3-16: Minimum sample size (including duplicate analysis), modified after BANNERT et al. (2001)

Parameter	Minimum dry mass of the ≤ 2 mm fraction in g	Minimum dry mass for reserve samples of the ≤ 2 mm fraction in g
hydrocarbons	40	80 - 150
heavy metals	6	50
Hg	6	50
BTEX	40	80 - 150
PAH	40	80 - 150
lightly valatile halogenated hydrocarbons	40	80 - 150
phenol index	40	80 - 150
individual phenols	40	80 - 150
PCB	40	80 - 150
eluate (anions, CN^-, etc.)	200	400 - 800
dry weight	40	80 - 150
grain-size distribution	150	300 - 600
density	40	80 - 150
carbonate	10	20 - 40
loss on ignition	60	120 - 240
total carbon	2	50
pH	20	40 - 80
cation exchange capacity	5	50
electrical conductivity	40	80 - 150

Selection of the sampling method

The method of sample collection depends on the objectives of the laboratory analyses, on the geology of the survey area, and the substances to be analyzed (*Table 5.3-17*). Undisturbed samples are needed to determine the following parameters:

- structural parameters, e.g., density and porosity,
- hydraulic conductivity,
- capillary pressure vs. water retention function,
- phase ratios (e.g., hydrocarbons or volatile halogenated hydrocarbons).

Disturbed soil samples should be taken to determine values for the following parameters:

- structural parameters (e.g., grain-size distribution),
- cation exchange capacity,
- proportion of substances in the pore water and the sediment (batch tests),
- concentrations of inorganic and organic substances.

Mixed samples from different locations and/or soil horizons are not suitable for evaluating contaminant distribution or for a toxicological assessment of the hazard. To avoid loss of information due to uncontrolled dilution, mixed samples should be used only in exceptional cases, e.g., in the case of relatively thin interbedded layers. In such cases the mixed sample is to be handled like a single sample.

Table 5.3-17: Sampling method according to sampling depth

Sampling depth	Sampling method	Collection method	
		Disturbed samples	**Undisturbed samples**
0 - 0.2 m	none	sample collected in a glass or plastic container or paper or plastic bag	core sampling using a steel pipe
0 - 2 m	trench	sample collected in a glass or plastic container or paper or plastic bag	core sampling using a steel pipe
0 - 2 m	hand auger	sample collected in a glass or plastic container or paper or plastic bag	core sampling using a steel pipe
0 - 10 m	percussion core drilling	sample collected in a glass or plastic container or paper or plastic bag	core sampling using a liner in a steel pipe
> 10 m	borehole	sample collected in a glass or plastic container or paper or plastic bag; loss of fine-grained material must be avoided	core sampling using a liner in a steel pipe

Drilling fluid may not be used during drilling if the sample is to be chemically analyzed.

Preparation of the equipment

The sampling equipment and sample containers must be clean. At contaminated sites, gasoline-motor tools, painted tools, and percussion core drilling with a diameter of < 60 mm or a length > 1 m should not be used. Gasoline generators must always be placed upwind at least 10 m from the sampling point. The maintenance records must be checked to determine whether the equipment that is to be used has been maintained properly. If the scheduled maintenance interval has been exceeded, the equipment must be checked and serviced according to the operations manual before it can be used for sampling.

5.3.2.2 Sampling of Soil, Rock, Stream and Lacustrine Sediments

Sampling is always carried out by two persons, one of whom must be a geologist or have a sufficient field experience (more than one year). Appropriate health and safety measures are to be taken. General rules:

- The soil must be disturbed as little as possible, i.e. little or no changes in the bedding, loss of interstitial water, loss of any volatile substances.
- The coordinates of all sampling sites are to be determined by GPS.
- Every sample has to be documented on the sampling sheet (field record).
- Individual samples must be representative of an entire or part of a layer or horizon taken at a single point.
- Mixed samples are a mixture of several individual samples from a specified core interval.
- Sterile samples are collected with a steel pipe sterilized in the laboratory.

The ≤ 2 mm fraction should be separated in the laboratory. If this is not possible, the following steps should be carried out on site:

- weighing of the sample,
- sieving using a 2 mm sieve with round holes,
- weighing the ≤ 2 mm fraction.

Collection of disturbed samples

Disturbed soil samples are collected from soils near the surface as a mixed sample. At sampling sites on cropland or pasture, the samples are taken from the Ap or Aw horizon of the soil; at woodland sites, the samples are taken separately from the organic litter layer (0 - 10 cm) and the underlying humus-rich mineral soil (10 - 20 cm). To guarantee that the sample is representative, all samples must be taken in the same way. The sampling should be done as follows:

- A hole is dug about 30 cm deep.
- The soil profile is described in the field record according to soil mapping guidelines.
- The sample is taken from the side of the hole, taking care that the same size of sample is taken from each depth interval.
- Samples of the determined soil intervals are taken at the corners of a triangle 50 cm to a side. The three samples are mixed to obtain a composite sample.
- Large pieces of organic matter (e.g., roots), large pebbles (> 10 cm), and any recognizable manmade materials (> 10 mm) are removed.

Any horizon that is clearly of anthropogenic origin (e.g., slag, asphalt, gravel) is to be excluded from sampling. The humus litter on forest soil and the root zone of pasture or meadowland must be sampled separately, as these horizons contain a high proportion of organic matter that would impede comparison of samples.

From percussion core drillings

- The core is removed from the drill.
- The sediment material is pushed out of the core sampler.
- A folding meter stick is placed alongside the core.
- The core is photographed.
- The lithology of the core is recorded.
- Samples are taken layer-wise from the center of the core (to avoid any contamination that was dragged up the walls from other layers).
- The samples are placed in labeled containers.

From trenches

- Scrap the side of the trench with the steel spatula from top to bottom.
- A folding meter stick is placed vertically at the side of the trench.
- The soil profile is photographed.
- A lithological log is recorded.
- Samples are taken layer-wise from the bottom of the profile to the top using a steel spatula (to avoid contamination by falling material).
- The samples are placed in labeled containers.

From piles of soil

The number of samples to be taken from a pile of soil is given in *Table 5.3-15*.

- The volume of the pile is estimated.
- The weight of the pile is estimated.
- The pile is photographed with something that can serve for a comparison of scale, e.g., a person, a car, or a folding meter stick.
- The soil texture is documented.
- The sample is quartered for analysis in the laboratory as follows:
  - Four buckets made of high-grade steel are placed in a square with a quartering funnel above them.
  - The profile is sampled from top to bottom using a shovel. Each shovelfull is placed in the quartering funnel. In the case of a small heap, the entire heap is quartered.
- The samples are places in labeled containers.

Stream and lacustrine sediments

- Stream and lacustrine sediment samples are taken at the locations marked on a 1 : 25 000 or 1 : 10 000 map of the area to be sampled.
- A representative sample (sand, clay, organic matter) is taken. Sediments from lakes or ponds are to be taken from near the shore.
- The sediment samples are sieved with a 0.2 mm nylon mesh and also with a 2.0 mm nylon mesh. The < 0.2 and < 2.0 mm fractions are placed in sample containers (jar) using a funnel.
- In the case of lacustrine sediment samples, the water standing above the sediment is to be decanted after 24 hours.

- If the stream bed is dry, the sample should be taken from the lower part of the stream bank. The sample size must be sufficient to obtain about 100 g each of the < 0.2 mm and < 2.0 mm size fractions. The sample is collected in a paper or plastic bag.

Collecting of undisturbed samples

Using short core samplers

- Short core samplers are short steel pipes with a cutting edge screwed on one end and an anvil head on the other.
- A flat horizontal surface is prepared using a knife (grass roots must be removed).
- The sampler is driven entirely into the ground with a sledgehammer.
- The anvil head is removed and replaced with a screw-on lid.
- The soil around the sampler is removed so that the sample can be cut from the undisturbed soil below.
- The cutting edge is removed and also replaced with a lid.
- The sampler is then put in the sample box and fastened in place.
- The sampling site is photographed.

From liners

- The liner is taken from the drill pipe, closed at both ends, and placed on a table.
- About 10 cm is cut from each end of the liner and the ends are closed with a lid.
- The residue part of the liner is labeled with core number and depth interval.
- The sample is placed in a freezer box if necessary for its preservation.
- The sample is transported to the laboratory: vertically if unfrozen, horizontally with insulation if frozen.

Steps to be carried out in the laboratory:

- The liner is cut open along the length of the sample.
- A 2 cm thick layer is removed to avoid distortion of the stratigraphy due to dragging of sediment along the liner.
- A folding meter stick is placed alongside the core.
- The core is photographed.

- The lithology is recorded.
- Samples are taken layer-wise from the center of the core (to avoid any contamination that was dragged up the liner walls from other layers).
- The samples are placed in labeled containers.

Sampling sediments under water

The type of sampling apparatus to be used depends on the water depth and the type of sample (a mixed sample of the top 25 cm or a layered sample). The samples are taken as follows when taken from a boat:

- mixed sample – collected with a grab sampler pressed into the sediment,
- layered sample – collected using a steel pipe. Sediment thickness must be determined with a sounding rod beforehand, in order to assess the compaction by being pressed into the pipe.

Collecting samples for analysis of volatile organic compounds

Samples to be analyzed for volatile organic substances are taken from the ≤ 2 mm fraction of undisturbed soil (trench, percussion core, steel pipe). Analysis of volatile substances in soils with grain sizes > 2 mm is appropriate only as an exception. Generally, three samples are taken at each sampling site. The containers (vials) are cooled before the samples are taken. All preparations (baking out, weighing) are done in the laboratory before starting the field investigation. Glass containers with a gas-tight Teflon seal in the lid are to be used for collecting volatile substances. The samples are to be taken from selected layers in the core immediately after the percussion drill is removed from the ground or the liner is cut. A 5 mL disposable syringe from which the bottom has been cut off is pressed into the soil. The piston should be about 5 mm from the bottom of the syringe before being pressed into the soil and then carefully pulled back as the syringe is pressed into the soil. The samples are then to be immediately placed in a glass container and sealed. No material should be on the mouth of the container when it is sealed. The sample is to be covered in the glass container with a defined amount of a suitable solvent, which can be placed in the container either before or after sampling. The kind (usually methanol) and amount of solvent is to be agreed upon with the laboratory that conducts the analyses. The lithology of the core is recorded in detail after the samples have been taken. The containers are transported in cool boxes.

The sampling method should not lead to changes in the composition and concentration of the contaminant(s). The sampling equipment must be carefully cleaned after each sample is collected. The sampling tools (e.g., spoons, spatulas) must be made of stainless steel.

Preservation, storage and transport of the samples

The samples should be placed in containers for transport and storage that do not admit light. The sample containers (material, size, method of sealing, color in the case of glass) are to select on the basis of the objectives of the investigation and the kind of contamination. They must be made so that nothing can enter the container after it has been sealed. The container material itself should not emit anything into the sample and nothing should be adsorbed on the container surfaces. In general, glass or plastic containers are to be used for soil samples (*Table 5.3-18*). Appropriate materials are to be used to seal samples of moderately volatile to nonvolatile substances; for example, aluminum foil can be placed in the lid/cap. All samples are to be clearly labeled immediately after being sealed in the container.

Table 5.3-18: Material of the sample container depends on the substance to be analyzed, BANNERT et al. (2001)

Substance	**Container material**
heavy metals, arsenic	glass, plastic
organic mercury compounds	brown glass
petroleum hydrocarbons, PAH, PCB, phenols	brown glass
volatile organic substances	glass with a gas-tight seal

The types of sample preservation required for different survey objectives are given in *Table 5.3-19*.

Table 5.3-19: Samples for which preservation is necessary and type of preservation

Objective	**Type of preservation**
field analysis of pore water in soil samples from the saturated zone	Not necessary for core samples from boreholes drilled using drilling mud to which bentonite has been added.
laboratory analysis of pore water in soil samples from the saturated zone analysis of unaerated soil samples from the saturated zone	Drill cores are frozen immediately after removal from the coring equipment when the borehole was drilled using drilling mud to which bentonite has been added. If the sample will be not analyzed within four weeks, it should be stored under an inert gas.
analysis of undisturbed samples from the unsaturated zone	Ends of the core are to be sealed with a bentonite paste. If the sample will not analyzed within four weeks, it should be frozen and stored under an inert gas.
analysis of disturbed soil samples from the saturated and unsaturated zones	If the sample will be not analyzed within four weeks, it should be frozen and stored under an inert gas.
analysis of disturbed soil samples from the saturated and unsaturated zones for their content of volatile organic substances	About 10 mL of soil is collected with a disposable syringe with the needle and bottom cut off. The sample is pressed into a weighed headspace vial and then reweighed. The sample is covered with 10 mL of boiled distilled water, the vial is sealed, and cooled for transport.

The samples should be analyzed as soon as possible after they have been collected. This is particularly the case for biological parameters. During transport and storage, the samples should be kept at 2 to 4 °C in the dark.

Cleaning of the equipment

On-site cleaning of the equipment

- cleaning with a scraper, brush, and tissue paper,
- cleaning with tissue paper dipped in acetone with crucible tongs,
- drying with tissue paper,
- washing with a brush and water with detergent in it,
- rinsing with a brush and pure water (These two steps are done in separate containers), and
- drying with tissue.

Work to be done in the workshop after sampling campaign

- cleaning with a scraper, brush, and tissue,
- cleaning with tissue paper dipped in acetone with crucible tongs,
- drying with tissue paper,
- washing with a high-pressure washer with detergent in the water,
- rinsing in the washer using pure water,
- drying with compressed air, and
- storing of the equipment, labeling it with the date of cleaning and the name of the person doing the cleaning.

Sampling report

The sampling report should contain

- the on-site sample descriptions,
- descriptions of the methods and equipment used, including photographic documentation.

5.3.2.3 Analysis of Soil, Rock, Stream and Lacustrine Sediments

The samples must be delivered to the laboratory the day after sample collection at the latest. In order to obtain values that can be compared for a particular sampling method (values for different methods cannot be compared in any case), the samples must be analyzed within 24 hours after receipt. If physicochemical properties (*Table 5.3-20*) are to be determined or inorganic substances (*Table 5.3-21*) are to be analyzed, the samples are to be prepared (including drying) according to DIN ISO 11464. If heavy metals and other elements are to be analyzed, the air-dried sample is to be ground to a grain size of 150 µm. Preparation for analysis of organic substances is to be done according to DIN ISO 14507.

Sample preparation includes sample splitting, sieving, addition of a preservative, e.g., acidification, homogenization, drying, crushing, and grinding – any steps necessary to obtain a representative laboratory sample from the field sample. Further sample steps to prepare a sample for a specific analysis include digestion, extraction, enrichment, and cleaning. It must be kept in mind that the preparation methods affect the soil properties and the chemical contaminants. The choice of sample preparation procedures depends on the analysis that will be carried out. An overview is shown in the flow chart in *Fig. 5.3-11*. Depending on the largest grain size, the sample can be relatively large. Large samples can be split using suitable tools. In order to estimate the hazard emanating from the site, all of the mass in the grain-size fractions must be accounted for (dry mass basis). For this reason, the weight of all samples collected in the field and all splits made in the laboratory must be recorded. Volatile organic substances (e.g., benzene and its homologues, naphthalene, volatile halogenated hydrocarbons) are analyzed directly without sample preparation. The dry mass is determined on separate samples according to DIN ISO 11465 and taken into consideration in the analytical results.

Soils are to be analyzed according to the procedures given in *Tables 5.3-20* to *5.3-22*. If other procedures are used, the reasons must be given and it must be documented that the results are equivalent to those of the procedures in *Tables 5.3-20* to *5.3-22*. Contaminant concentrations are to be given relative to the dry mass (dried at 105 °C). Concentrations of inorganic substances in eluates and leachate are to be determined using the procedures given in *Table 5.3-23*; those of organic substances using the procedures in *Table 5.3-13*. If other procedures are used, the reasons must be given and it must be documented that the results are equivalent to those of the procedures in *Tables 5.3-23* and *5.3-13*. The analytical procedure used must be recorded for the analysis of any percolation water, groundwater, extract, eluate, or soil sample.

Table 5.3-20: Analysis of physicochemical properties, BANNERT et al. (2001)

Parameter	Remarks on method	Method according to
determination of dry weight	field fresh or air-dried sample	DIN ISO 11465
organic carbon and total carbon after dry combustion	air dried soil samples	DIN ISO 10694
pH value ($CaCl_2$)	suspension of the field fresh or air-dried soil samples in a 0.01 M $CaCl_2$ solution	ISO 10390
particle size distribution	1) “finger test” in the field *)	DIN 19682-2
	2) sieving, dispersion, pipette analysis*)	ISO 11277 ISO 11277
	3) sieving, dispersion, hydrometer method	DIN 18123 ISO 11277
bulk density	collection of sample while preserving the original volume, drying at 105 °C, reweighing	DIN ISO 11272 ISO 11272
electrical conductivity	air-dried soil sample	DIN ISO 11265:

*) recommended methods

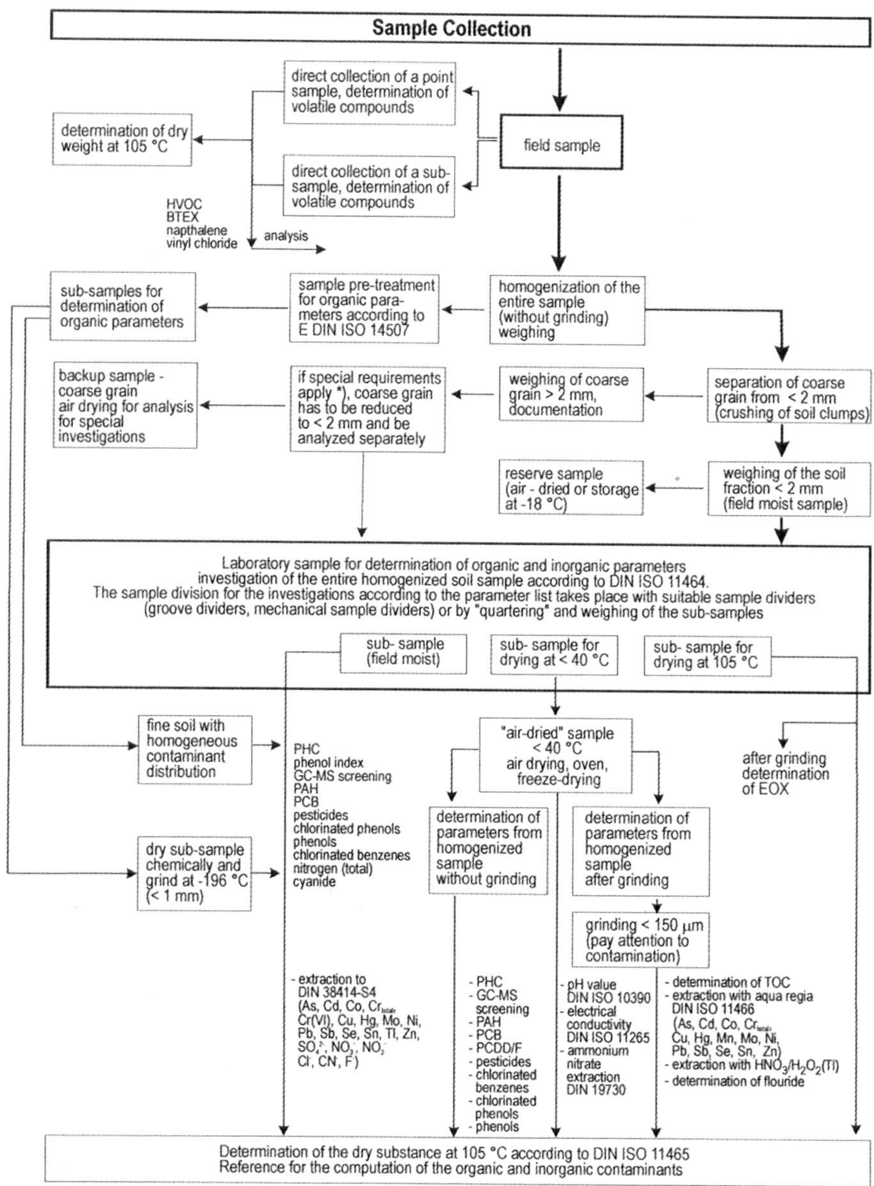

Fig. 5.3-11: Flowchart of treatment of soil samples, BANNERT et al. (2001)

Table 5.3-21: Analysis of inorganic contaminants, modified after BANNERT et al. (2001)

Parameter	Method	Method according to	Detection limit
Cd, Cr, Cu, Ni, Pb,Tl, Zn	AAS	ISO 11 074	0.1 mg kg^{-1}, Cd 0.05 mg kg^{-1}
As, Cd, Cr, Cu, Ni, Pb, Tl, Zn	ICP-AES (ICP-QMS possible), spectral disturbances must be taken into consideration when high concentrations are present in the solid part of the sample	DIN EN ISO 11885	0.005 to 0.01mg kg^{-1}, As, Zn 0.1 mg kg^{-1}
arsenic (As)	ET-AAS	in analogy to ISO 11 074	0.5 mg kg^{-1}
	Hydride AAS	DIN EN ISO 11969	0.5 mg kg^{-1}
mercury (Hg)	AAS cold vapor technology, during sample pretreatment, drying temperature must not exceed 40 °C	DIN EN 1483 reduction with tin(II) chloride or NaBH4	0.01 mg kg^{-1}
chromium (VI)	1) extraction with phosphate-buffered aluminium sulfate solution	spectrophotometry DIN 19734	0.01 mg kg^{-1}
	2) elution with water, separation of Cr(III), determination of soluble Cr(VI) in soils	DIN 38405 - 24	0.01 mg kg^{-1}
cyanide		DIN ISO 11262	0.05 mg kg^{-1}
determination of total element content	1. digestion with hydrofluoric and perchloric acid	ISO 14869-1	
	2. fusion decomposition for major constituents, lithium borate melt	ISO 14869-2	
extractable elements: Co, Mn, Mo, Sb, Se, Sn, Zn, etc.	after extraction with aqua regia according to DIN ISO 11466; the analysis is calibrated with aqua regia of the same concentration, Fe, Ca, Al, Mg matrix adjustment and measurement with internal standard, spectral disturbances and transportation problems have to be taken into consideration when high acid and high concentrations are present in the solid part of the sample; ICP-AES or ICP-QMS	DIN EN ISO 11885 (E22) ISO 17294-2	0.5 to 5 mg kg^{-1} < 0.5 mg kg^{-1}
As, Sb, Se	after extraction with aqua regia according to DIN ISO 11466; disturbances due to high matrix concentrations; ET-AAS and hydride AAS	(soil standard at present under development)	0.05 mg kg^{-1}

Parameter	Method	Method according to	Detection limit
mercury (Hg)	extraction with aqua regia according to DIN ISO 11466; drying temperature must not exceed 40 °C, AAS and AFS	ISO CD 16772 in preparation	0.03 mg kg^{-1}
thallium (Tl)	approx. 0.5 g soil with 2.5 mL HNO_3 (65 %) and 1.5 mL H_2O_2 (30 %) in pressure decomposition vessel at approx. 160 °C, > 80 bar, or at the return flow extract for 2 h. Zeeman ET-AAS with palladium nitrate/magnesium nitrate modifier, ET-AAS	DIN 38406-E26 (soil standard at present under development)	0.1 mg kg^{-1}
tin (Sn)	extraction with aqua regia according to DIN ISO 11466; disturbances due to high concentrations in the solid part of the sample; ET-AAS	in analog to ISO 11 074	0.1 mg kg^{-1}
easily liberatable and total cyanide	field-moist soil, < 2 mm, distillation, photometry, titration, flow analysis, after separation by distillation	edited standard ISO/DIS 11262:2001 E and DIN EN ISO 14403	0.1 to 1mg kg^{-1}
total fluoride	0.2 g soil ≤ 2 mm, grind to ≤ 100 μm, sinter 30 min at 950 °C with 0.3 g ZnO and 1 g Na_2CO_3, extract with water, adjust to pH = 5.8 with HNO_3, fill to 100 ml with water. Measure electrometrically with F-sensitive electrode using TISAB (1:1); sinter decomposition and ionometric determination	DIN 51084	20 mg kg^{-1}
total nitrogen	Kjeldahl determination with titanium dioxide as catalyst	DIN ISO 11261	0.5 to 1 mg kg^{-1}

Table 5.3-22: Analysis of organic contaminants, BANNERT et al. (2001)

Parameter	Method	Method according to	Detection limit
polycyclic aromatic hydrocarbons (PAH): 16 PAH (EPA), benzo(a)pyrene	Soxhlet extraction with toluene, chromatographic clean-up; analysis by GC-MS *)	Bulletin No. 1 of the LUA NRW, 1994 *)	
	extraction with tetrahydrofurane or acetonitrile; quantification by HPLC UV/DAD/F *)	Bulletin No. 1 of the LUA NRW, 1994 *)	
	extraction with acetone, addition of petroleum ether, removal of the acetone, chromatographic cleaning of the petroleum ether extract, absorption in acetonitrile; quantification by HPLC UV/DAD/F	DIN ISO 13877	
	extraction with a water/acetone/petroleum ether mixture in presence of NaCl; quantification by GC-MS or HPLC UV/DAD/F	VDLUFA Method Book, Vol. VII; Handbook of Contaminated Sites Vol. 7, LfU HE	
hexachlorobenzene	extraction with acetone/cyclohexane mixture or acetone/petroleum ether, if necessary chromatographic cleaning after removal of acetone; quantification by GC-ECD or GC-MS	DIN ISO 10382	
pentachlorophenol	Soxhlet extraction with heptane or acetone/heptane (50:50); derivatization with acetic anhydride; analysis by GC-ECD or GC-MS	DIN ISO 14154	
aldrin, DDT, HCH mixture	extraction with petrol ether or acetone/petroleum ether mixture, chromatographic cleaning; analysis by GC-ECD or GC-MS *)	DIN ISO 10382 *)	
	extraction with a water/acetone/petroleum ether mixture in presence of NaCl; analysis by GC-ECD or GC-MS	VDLUFA Method Book, Vol. VII	

Parameter	**Method**	**Method according to**	**Detection limit**
polychlorinated biphenols (PCB): 6 PCB congeners (nos. 28, 52, 101.138, 153, 180 after Ballschmiter)	extraction with heptane or acetone/petroleum ether, chromatographic cleaning; analysis by GC-ECD (GC-MS possible)	DIN ISO 10382	
	Soxhlet extraction with heptane, hexane or pentane, chromatographic cleaning on AgNO3/silica gel column; analysis by GC-ECD (GC-MS possible)	DIN 38414-20	
	extraction with a water/acetone/petroleum ether mixture in presence of NaCl; analysis by GC-ECD (GC-MS possible)	VDLUFA Method Book, Vol. VII	
polychlorinated dibenzodioxine and dibenzofurane	Soxhlet extraction of freeze-dried samples with toluene, chromatographic cleaning; analysis by GC-MS	German Sewage Sludge Ordinance taking into consideration DIN 38414-24. VDI Guideline 3499 Part 1	
GC-MS screening, qualitative and semiquantitative orientation analysis	extraction with cyclohexane/acetone (50:50) GC-MS	standard laboratory work instructions	
organic carbon and total carbon after dry combustion	air-dried soil samples, grain fraction ≤ 2 mm, ground to ≤ 150 µm (elementary analysis)	DIN ISO 10694	10 mg kg^{-1}
petroleum hydrocarbons (PHC)	1. dry with Na_2SO_4, extraction with 1,1,2-trichlorotrifluoroethane, chromatographic cleaning, IR spectrometry	ISO/TR 11046	20 mg kg^{-1}
	2. extraction with acetone/petrol ether (50:50), removal of acetone, GC-FID with suitable injector	ISO 16703 *)	100 mg kg^{-1}
highly volatile aromatic hydrocarbons benzene, toluene, ethylbenzene, xylene, styrene, cumene,	1. cover with methanol, transfer of an aliquot into water, headspace analysis, GC-MS or GC-FID or GC-PID	DIN 38407 F9-1 *)	1 to 10 mg kg^{-1}
	2. extraction with pentane, GC-MS or GC-FID or GC-PID	DIN 38407 F9-2	1 to 10 mg kg^{-1}

Parameter	Method	Method according to	Detection limit
trimethylbenzene, naphthalene	3. extraction with methanol, addition of an aliquot to water in a purge vessel, purge and trap method and thermodesorption, GC-MS or GC-FID or GC-PID	ISO 15009	0.1 mg kg^{-1}
polycyclic aromatic hydrocarbons (PAH) 16 PAH (EPA): acenapthene, acenaphthylene, anthracene, benzo(a)anthracene, benzo(b)fluoranthene, benzo(k)fluoranthene, benzo(ghi)perylene, benzo(a)pyrene, chrysene, dibenzo[a,h]anthracene, fluoranthene, acetone, indeno[1,2,3-cd]pyrene, naphthalene, phenanthrene, pyrene	1. Extraction with acetone/cyclohexane, chromatographic clean-up, GC-MS	Bulletin No. 1 of the LUA NRW, 1994 *)	0.05 mg kg^{-1} for each substance
	2. extraction with methanol HPLC-UV/F	Bulletin No. 1 of the LUA NRW, 1994 *)	0.05 mg kg^{-1} for each substance
	3. extraction with water/acetone/petroleum ether mixture in the presence of NaCl, GC-MS or HPLC-UV/F	LfU HE, Manual of Contaminated Sites, vol. 7, 1998 DIN 38414-pp23	0.1mg kg^{-1} for each substance 0.05 mg kg^{-1} for each substance
chlorobenzenes: trichlorobenzenes to hexachlorobenzene	1. extraction with acetone/cyclohexane or acetone/petroleum ether mixture, if necessary chromatographic cleaning after removal of acetone, GC-ECD or GC-MS	DIN ISO 10382 *)	0.01 to 0.1 mg kg^{-1}
	2. extraction with pentane or cyclohexane, GC-ECD	DIN 38407-F2	0.01 to 0.1 mg kg^{-1}
phenol index	slurry the sample with demineralized water, pH = 0.5, water vapor distillation (20 g sample and 150 ml water, distill off 100 ml), UV/VIS photometry	DIN 38409-H16-3	0.1 mg kg^{-1}
chlorophenols	extraction with acetone/hexane (50:50), removal of acetone, derivatization with acetic anhydride, GC-ECD or GC-MS	DIN ISO 14154	0.2 to 1 mg kg^{-1}

Parameter	Method	Method according to	Detection limit
phenol	extraction with acetone/hexane (50:50), removal of acetone, derivatization with acetic anhydride, GC-MS or GC-FID	DIN ISO 14154	1 mg kg^{-1}
volatile halogenated organic compounds (VHC)	1. cover with methanol, transfer an aliquot into water, headspace analysis, GC-MS or GC-ECD	DIN EN ISO 10301 (F4) *)	0.1 to 1 mg kg^{-1}
	2. extraction with pentane, GC-ECD or GC-MS	DIN EN ISO 10301 (F4)	0.1 to 1 mg kg^{-1}
	3. extraction with methanol, addition of an aliquot to water in a purge vessel, purge and trap method and thermodesorption, GC-ECD or GC-MS	ISO 15009	0.01 mg kg^{-1}
vinyl chloride	cover with methanol, transfer an aliquot into water, headspace analysis, GC-MS or GC-FID or GC-PID	DIN 38413-P2	5 mg kg^{-1}
chlorinated organic pesticides; low volatility, halogenated hydrocarbons without PCB	1. extraction with acetone/petroleum ether mixture (50:50), if necessary chromatographic cleaning after removal of acetone, GC-ECD or GC-MS	DIN ISO 10382 *)	0.01 mg kg^{-1}
	2. extraction with acetone/petroleum ether/NaCl mixture, if necessary chromatographic cleaning, GC-ECD or GC-MS	VDLUFA Method Book, Vol. VII	0.1 mg kg^{-1}
herbicides	extraction with acetone/petroleum ether/NaCl mixture, if necessary chromatographic cleaning, HPLC-UV	DIN ISO 11264	0.01 mg kg^{-1}
extractable organic halogens (EOX)	Soxhlet extraction with heptane, combustion to hydrogen halides and determination, titration, coulometry	DIN 38414-S17	0.2 to 1 mg kg^{-1}

*) recommended methods

Table 5.3-23: Determination of the concentration of inorganic contaminants in eluates and leachate, BANNERT et al. (2001)

Parameter	Method	Method according to
As, Cd, Cr, Co, Cu, Mo, Ni, Pb, Sb, Se, Sn, Tl, Zn	ICP-AES (ICP-QMS possible)	on basis of DIN EN ISO 11885 *)
arsenic (As), antimony (Sb)	hydride AAS	DIN EN ISO 11969
lead	AAS	DIN 38406-6
cadmium	AAS	DIN EN ISO 5961
total chromium	AAS	DIN EN 1233
chromium (VI)	spectrophotometry, ion chromatography	DIN 38405-24 DIN EN ISO 10304-3
cobalt	AAS	DIN 38406-24
copper	AAS	DIN 38406-7
nickel	AAS	DIN 38406-11
mercury	cold vapor AAS	DIN EN 1483
selenium	AAS	DIN 38405-23
zinc	AAS	DIN 38406-8
total cyanide	spectrophotometry	DIN 38405-13 DIN EN ISO 14403
easily liberatable cyanide	spectrophotometry	DIN 38405-13
fluoride	fluoride-sensitive electrode, ion chromatography	DIN 38405-4 DIN EN ISO 10304-1
cyanides	spectrophotometry	DIN 38405-D14

*) The detection limit has to be adapted to the investigation objectives by suitable measures or suitable equipment.

5.3.3 Sampling and Analysis of Soil Gas and Landfill Gas

HAGEN HILSE & HANS-JÜRGEN VOIGT

Soil gas analyses are suitable only for qualitative and orientating investigations, e.g., to determine contaminant sources and estimate contaminant distribution. Quantitative data about contamination can be obtained only by direct sampling of soil and groundwater.

The results of the analysis of soil gas samples taken with different sampling procedures are not comparable. Conclusions about the actual contaminant concentrations in soil or groundwater are not possible. For this reason, the values cannot be used to check whether concentration limits have been exceeded. They can be used, however, to determine whether the original objectives should be changed or supplemented. All of the different procedures have a potential for application, depending on the geological or technical conditions. To be able to conduct any analysis necessary, a laboratory must be able to carry out a number of soil gas sample collection methods.

Many organic substances classed as groundwater pollutants can be identified in the soil gas. These substances include volatile halogenated hydrocarbons, volatile aromatic hydrocarbons, other volatile hydrocarbons, and volatile organic compounds (esters, ethers, alcohols, ketones).

Landfill gas plays an important role at waste disposal sites and surrounding area. Landfill gas is generated from organic substances in the waste by microbes under aerobic and anaerobic conditions. Atmospheric air penetrates only into the upper parts of a landfill. Therefore, only near the landfill surface are biodegradable substances converted aerobically according to the equation:

$$\text{organic substances} + O_2 + \text{microbes} \rightarrow CO_2 + H_2O + \text{energy}. \qquad (5.3.5)$$

Anaerobic conditions exist in the deeper parts of a landfill. The anaerobic zone starts approximately three or four meters below the landfill surface, depending on the degree of compaction of the waste. Anaerobic degradation can be expressed as follows:

$$\text{organic substances} + \text{microbes} \rightarrow CH_4 + CO_2 + H_2O + \text{energy}. \qquad (5.3.6)$$

Biomass in a landfill is degraded to landfill gas mainly by anaerobic processes. On a smaller scale, gas is generated by the passing of dissolved, absorbed, liquid, or solid substances into the gaseous phase due to changes of physicochemical conditions in the landfill (e.g., rise in temperature). In addition, chemical reactions between waste materials or with leachate can be responsible for some of the landfill gas. The composition of the waste determines which processes occur. A considerable part of the hazard potential of landfill gas is attributed to these components due to their toxicological properties.

In general, landfill gas is characterized by a high proportion of methane. During the plateau stage of methane production in a landfill, methane concentration in the landfill gas will be above 50 % and that of carbon dioxide above 35 %. The equilibrium CH_4/CO_2 ratio ranges from 1.2 to 1.5 (BILITEWSKI et al., 2000). Nitrogen, oxygen and hydrogen are produced only during the initial phase of microbial degradation of the waste (two to six months) and decrease to zero during the plateau stage, which can last for several years (*Fig. 5.3-12*). Besides the main components, CH_4 and CO_2, further compounds such as hydrogen sulfide, ammonia and volatile organic compounds are always present (*Table 5.3-24*). In most cases they never make up more than 0.1 % of the landfill gas.

Depending on the type and concentration of biodegradable hydrocarbons and on the physicochemical conditions (moisture content, temperature, etc.), 120 to 400 m^3 landfill gas is generated per 1000 kg of domestic waste.

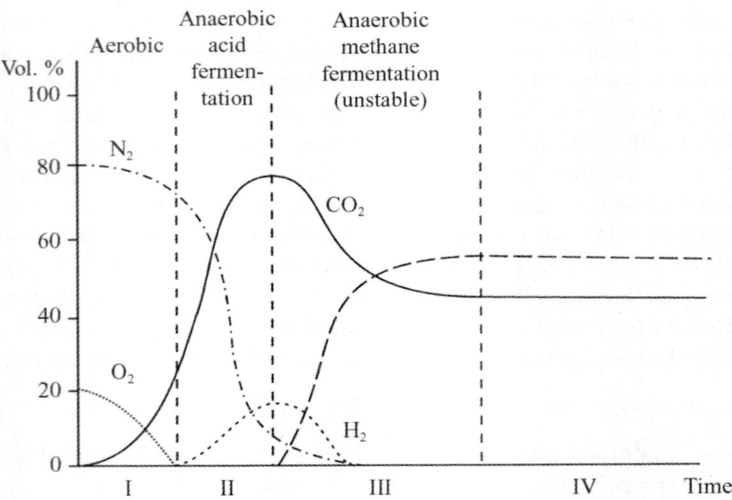

Fig. 5.3-12: Composition of landfill gas in phases I to IV of landfill development, BILITEWSKI et al. (2000)

Table 5.3-24: Components of landfill gas in the plateau phase of methane production

Compound	Volumetric content
methane	55 - 65 %
carbon dioxide	35 - 45 %
hydrogen	$\ll$ 1 %
nitrogen	$\ll$ 1 %
carbon monoxide	$\ll$ 1 %
ammonia	0 - 100 ppm
H_2S	0 - 100 ppm
other VOC per component	0 - 150 ppm

A mixture of methane with air in proportions between 5 and 15 vol. % is explosive. There is a risk of explosion both on the landfill itself and in the surrounding area. The risk of suffocation from landfill gas depends on the ratio of CH_4 to CO_2. If the percentage of methane (lighter than air) by volume is relatively small compared to carbon dioxide (heavier than air), as is the case in the early phase of landfill gas production, the landfill gas remains on the surface or in the drainage pipes and depressions. The risk of suffocation is high in these cases. There is no separation of the CH_4 and CO_2. Besides suffocation by carbon dioxide and carbon monoxide, toxic (e.g., H_2S) and carcinogenic trace components (e.g., VOCs) also represent a risk to human health.

Released methane, carbon dioxide and halogenated compounds affect the environment due to their contribution to the greenhouse effect and ozone layer destruction, respectively. Furthermore, methane in combination with NO_x can cause the formation of surface-near ozone (photochemical smog) (RETTENBERGER & METZGER, 1992).

Soil and landfill gas analyses can be used for the following purposes:

- determination of the kind of contamination (substances; order of magnitude of their concentration),
- localization of the sources or center of concentration,
- general determination of the lateral and vertical distribution,
- mapping of groundwater contamination.

A large range of concentrations must be analyzed for the above-named purposes, from traces to saturation concentrations. Thus, several sample collection procedures are necessary, since concentrations can be in the range between $< 100\ \mu g\ m^{-3}$ and $>100\ 000\ 000\ \mu g\ m^{-3}$. Methods not involving enrichment are preferable, e.g., direct measurement using gas collection vessels (e.g., test tube closed with a septum or a pasteur pipette). Even trace concentrations of less than $100\ \mu g\ m^{-3}$ must also be reliably measured. Methods involving enrichment are necessary for such concentrations, e.g., adsorption on activated charcoal or an adsorbent resin. Enrichment methods should be used, as a rule, for orientating investigations, since the concentrations at the site are not known and can vary strongly over short distances. The depth at which the samples should be taken depends on the objective of the investigation and the geology. Depths of 8 - 12 m can be necessary for investigations of groundwater contamination or of the depth to which the contaminant(s) has migrated. To determine the depth of contaminant migration, the soil gas must be taken from known depths. The opening of the sampler must be correspondingly small (10 cm maximum). Integrated sample collection is not acceptable, because the spreading of the contaminant within the borehole cannot be quantified. Especially when the ground is heterogeneous, consisting of layers of very different permeabilities with respect to gases, the results of integrated sample collection cannot be interpreted.

5.3.3.1 Planning and Preparation of Work

The following information is needed for planning the distribution and the depth of gas sampling points:

- geological or, if possible, soil maps at a scale greater than 1 : 25 000,
- depth to the groundwater table,
- lithology of the unsaturated zone, and
- network of sewage and water pipes, and underground electric power lines.

The sampling sites are selected on the basis of site history. Potential input locations or centers of elevated concentrations are to be determined in the orientating investigations. A detailed knowledge of the underground structures is needed to determine the sampling depth and for interpretation of the analytical results. If the subsurface structure is not known, it should be determined with a small percussion drill or similar method. The boreholes can then be used to collect the soil gas samples (as soon as possible after drilling).

The depth of soil gas sampling, especially for the determination of CO_2, must normally be deeper than the highly organic soil cover to exclude the influence of natural soil gas on the measurement. The sampling depth for soil gas mapping is usually 1.5 m below the Earth's surface. For the detection of volatile compounds above the groundwater table, the sampling depth should be 1 m above the capillary fringe. If there is a clay-rich bed in the unsaturated zone, the gas samples must be taken beneath this bed. When mapping gas contamination in the subsurface, the sampling points should be no more than 30 m apart. Samples are taken from either temporary drill holes or from permanently installed "wells".

In both types, the sample intake must be protected from entry of atmospheric air. A pump whose flow rate can be regulated is used to collect the soil gas sample. The flow rate is between 0.1 and a maximum of 1 L min^{-1}, depending on the permeability of the ground. If the pump does not have pressure and flow gauges, external instruments must be used and continually monitored. In this case the manometer is installed at the top of the probe. The entire pumping system must be gas tight. The tubing connecting the probe to the pump should be as short as possible and frequently replaced. It must be made of a nonadsorbing material (e.g., HDPE). The gas volume which is pumped out must be at least 1.5 times the amount in the casing and in the pipes or hoses.

Before the field investigations are started, the sampling and percussion equipment, as well as the sample containers must be clean. In the field, the sampling sequence has to start at the least contaminated site. If the sampling operation takes longer than one day, the first sampling point of the next day must be in the same place as the last one on the day before. A control sample of the atmospheric air 20 cm above the surface must be collected each day before the soil gas sampling is started. Gas sampling is usually carried out together with cone penetration testing.

5.3.3.2 Sampling and Analysis of Soil Gas and Landfill Gas

The techniques for detection and measurement of soil gas components can be divided into direct measurement and sampling with analysis in the laboratory.

Landfill gas monitors are instruments for estimating the concentrations of the typical components of landfill gas. They are designed for both stationary and manual temporary use and are suitable for use in areas of the risk of explosion. The gas is pumped into an instrument equipped with special sensors. The main components (CH_4 and CO_2) are measured in integrated infrared-absorption cells. Oxygen is measured in electrochemical cells. The data is recorded for immediate use or over a period of time with a data logger, which collects and stores data and makes it available to a computer via an interface. The precision of these instruments varies between ± 0.5 to 3 vol. %. It is often possible to include electrochemical cells to measure hydrogen, hydrogen sulfide, carbon monoxide, HCN, NO_2, SO_2, etc. Some instruments can also estimate and store other parameters, e.g., gas flow, temperature, air pressure, date and time.

Indicator tubes: Also used are glass tubes containing chemicals which turn a characteristic color in the presence of a specific gas. First the tubes are opened using special cutting tools, then a defined volume of the gas is pumped through the test tube by a vacuum pump. If the expected substance is present in the air, a color indication of defined length and intensity is observed in the tube. The concentration of the gas can be estimated either by the scale on the tube (in ppm or vol. %) or by means of a reference scale. Indicator tubes are useful for both direct testing and long-term measurements. Indicator tubes are cost-effective, robust, and even under difficult conditions a reliable technique for gas measurements. Furthermore, they are available for a wide range of gases and trace substances. However, when these tubes are used, possible cross sensitivities of the measured substances must be taken into account. Ambient temperature, humidity, and atmospheric pressure conditions may influence the results. The volume of gas passed through the tube also has to be taken into consideration.

Sorption tubes contain substances on which the detected gas components are adsorbed and analyzed later in the laboratory. A defined volume of gas is passed through the sorbent when the tube is opened until it is closed. Activated carbon and silica gel are common sorbents. The concentration of the substance is estimated from the mass of the added gas and the volume of the tube. Sorption tubes often contain the adsorption layer itself and a second inert layer. A separate analysis can be carried out to determine whether the total amount of the component to be analyzed was adsorbed. If this is not the case, the sampling must be repeated. Sorption tubes are well suited for the sorption of the organic compounds that are characteristic of landfill gas. The laboratory analysis should follow the sampling immediately to minimize losses caused by leakage of the tubes.

Special gasometers can be used if large gas volumes are required or gases have to be analyzed which cannot be adsorbed and, therefore, cannot be measured using sorption or test tubes. These are closed after filling and sent to the laboratory for analysis.

Gas collectors are glass cylinders with tapered ends equipped with valves. Atmospheric air is pumped from the cylinder until a pressure of about 200 Pa is reached. During sampling, both valves are open to allow entry of the gas to be analyzed. Both valves are closed when the gas cylinder is completely filled. The cylinders should be transported immediately to the laboratory to minimize gas losses. Gas collectors are often used to estimate high gas concentrations.

Gas bags have commonly volumes up to 1 L and are filled with the gas through a valve. Before sampling the bags should be purged twice with pure nitrogen and completely emptied afterwards. However, about 5 mL gas often remains in the bag, which can cause inaccuracies in the measurements. Further inaccuracies may be caused by CFCs which are part of the bag material itself. Gas bags lose some gas components very quickly, for instance hydrogen sulfide, carbon dioxide, and VOCs. On the other hand, oxygen and argon can diffuse into the bag. If it is necessary to transport the samples by aircraft after sampling in remote areas, the bags should not be completely filled, owing to the lower air pressure at high altitudes. In general, transport to the laboratory needs to be very fast (*Fig. 5.3-13*).

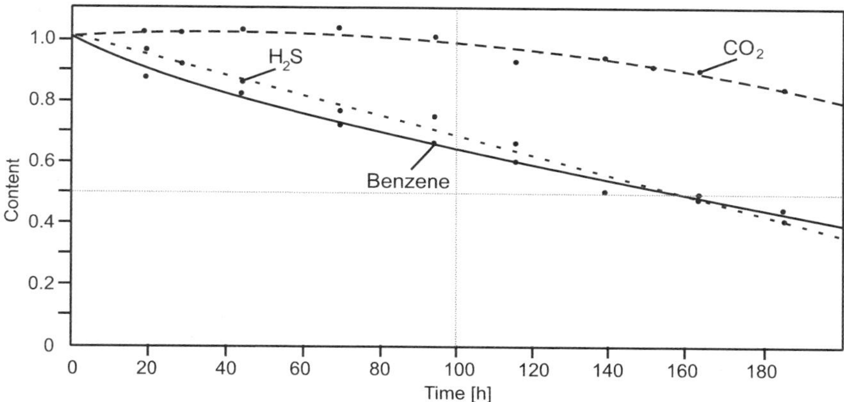

Fig 5.3-13: Decline of the gas concentration in Tedlar bags

Gas soundings are made at various points on a landfill or contaminated site to take samples of soil, air, or gas. Gas samples can be taken from holes made using a sounding rod or by drilling. Penetration of atmospheric air into the upper layers of the landfill and its influence on the landfill gas is to be taken into account, especially for uncovered landfills. Gas concentration gradients in the landfill depend on the rate of gas production and the ambient air pressure. Therefore, soundings should be deeper than 1.5 m in the case of uncompacted waste. To obtain a representative gas concentration for the landfill, there should be a closely spaced monitoring network of measuring points all over the landfill. Boreholes of the required depth are drilled into the landfill using a sounding rod. The boreholes should be protected with perforated polyethylene tubes, especially in noncohesive soils. Measuring points and support tubes are to be sealed to prevent the penetration of atmospheric air. Special cements are available for sealing the edges of the boreholes and plastic caps are available for the tubes. The measuring points should be left at least for several hours to be filled again by landfill gas. After that, a comparatively small amount of soil gas can be sampled and analyzed, which will provide a good value for the soil gas composition near the measuring point. Soundings are simple and cost-effective for sampling and analysis of soil and landfill gas. *Figure 5.3-14* shows an example of a landfill gas-depth profile.

Soil air pumping is used to collect larger gas volumes than for gas soundings. The appropriate gas collectors should be positioned centrally in the sampling area. A gas collector is placed in the landfill after drilling. The landfill gas can then be continually pumped with a vacuum pump or radial fan. The pumping rate depends on the diameter of the borehole, the filter length and the pump capacity. Guide values for the pumping rate is 25 to 30 $m^3 h^{-1}$ and for the radius around the borehole is 25 - 30 m. Pumping measurements cover a large monitoring area. Fewer boreholes are necessary for the monitoring network when this method is used than with the sounding method. On the other hand, it is technically more complicated.

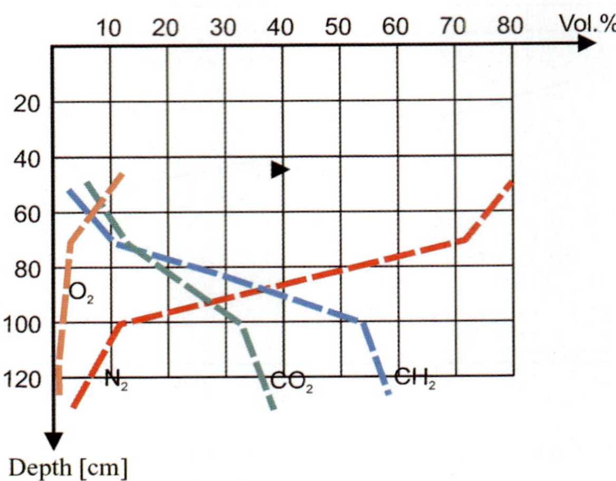

Fig. 5.3-14: Example of a landfill gas depth profile, RETTENBERGER & METZGER (1992)

Gas emissions from landfills are measured with "gas boxes" and FID or PID instruments. Plastic buckets or modified barrels can be used as gas boxes. The plastic should not be PVC; PE-HD is especially suitable. The box is installed in a shallow trench which is sealed with cement to prevent penetration of ambient air. The upper part of the box is connected to a flow-meter (e.g., rotameter or gas bag with a defined volume). Gas boxes are simple methods for measuring landfill gas. However, they measure only one point. Therefore, a closely spaced monitoring network should be used. Flame ionization detectors (FID) are suitable to detect even the smallest emissions on abandoned landfills. A mixture of N_2/H_2 and the gas sample is burned in a hydrogen flame, which normally has a low electrical conductivity. If hydrocarbons or other combustible components are present in the sample at a concentration of more than 1 ppm, the electrical conductivity of the mixture increases considerably and can be detected. The conductivity is measured by the FID, yielding an estimated methane value. FIDs are suitable for a rapid screening of emission sources. But it is to be taken into account that they are potential causes of ignition and should be used only in combination with explosion warning devices. Photo-ionization detectors (PID) are suitable for volatile alcohols, chlorinated hydrocarbons, esters, aliphatic and aromatic hydrocarbons, hydrogen sulfide, and ammonia. The detected values are aggregate parameters and cannot provide information on the individual components. However, PIDs can often be calibrated for a special substance to provide, for example, a toluene equivalent value. A PID is used analogously to an FID; they can only be used in addition to an FID due to the fact that they cannot detect methane. A PID can provide only qualitative data on trace substances.

Depending on how the gas chromatograph is equipped, the samples should be collected in 10 or 20 mL vials sealed with a septum. Pasteur pipettes can also be used. The septums should be coated with Teflon; other materials adsorb gases. Self-sealing septums (made of butyl rubber) should be used. Depending on the equipment used, aluminum caps are preferable. The materials used should be recorded. The gas collection vessels and septums should be heated before use. The vials should be careful closed with the septum: The lips should be absolutely clean, the septum should be taut, the septum tongs should be properly used.

A field protocol must be recorded for each sampling operation. This record should contain the following information:

- number and kind of the sampling point (temporary or permanent),
- coordinates of the sampling point,
- weather conditions,
- design of the well or gas sampler (casing material and diameter, depth and length of the filter),
- sampling devices and kind of field measurement,
- date and time of sampling,
- volume of the pumped gas and the pumping rate,
- volume of the gas sample,
- storage and transport conditions,
- name of the company and the operator,
- date and time of delivery of the samples to the laboratory.

Samples in glass collectors must be stored in a dark container without cooling. Sorption tubes must be preserved in a dark cool box at a temperature below 4 °C.

5.3.4 Methods for Chemical Analysis used in Geochemical Investigations

FLORIAN JENN

A wide spectrum of chemical analysis methods is available for the determination of the chemical composition of soil, water, and gas samples. Analytical chemistry methods are either qualitative and quantitative. Most of the methods used in environmental geochemistry are quantitative. In some cases not only the total concentration of an element is determined but also the amount of that element in a specific chemical species.

Decomposition and extraction

The samples must normally be prepared for chemical analysis by separating the complex mixtures of substances they contain. This is done in some cases by physical breakdown of the sample into smaller particles, or by chemical decomposition. This preparation step is often used in conjunction with extraction. Suitable solvents are used to extract specific components of a solid or liquid sample. In a solid-liquid extraction, the solvent containing the extracted material is filtered and then evaporated. In a liquid-liquid extraction, the extracting liquid must be immiscible with the sample liquid.

Chromatographic Separation

Chromatography is a common method for separating the complex mixtures of substances in a sample.

Chromatography systems separate substances by taking advantage of differences in the strength of binding to a stationary phase as the fluid phase containing the sample is passed over it. Different substances in the mobile phase have different affinities to the stationary phase. Substances with a low affinity to the stationary phase (i.e. only weakly adsorbed) travel almost as quick as the carrier fluid, whereas substances with a high affinity to the stationary phase are considerably retarded, resulting in a separation of the sample components. The stationary phase typically consists of small particles packed in a glass or metal column or is sometimes a film on the inside of a long tube. The mobile phase is a fluid. The material to be separated is placed at the top of the column in dissolved form and is carried by the mobile phase as it moves through the column. The individual substances can be captured or detected at the bottom of the column. Chromatographic methods include thin layer chromatography (TLC), liquid chromatography (LC), gas chromatography (GC), gas-liquid chromatography (GLC), and high performance liquid chromatography (HPLC).

Gas chromatography is an important chromatographic method with which gas-phase mixtures can be separated. In this method, an inert carrier gas is used as the mobile phase. A column is packed or lined with a substrate that forms the stationary phase. Particularly, good separation can be achieved by employing a long (up to 100 m) capillary column with an inside diameter of 0.2 - 1 mm without a substrate fill; separation takes place owing to differences in the retention of the different components to the capillary surface. Gas chromatography coupled with a mass spectrometer as detector (GC-MS) is an efficient system that makes the identification and determination of complex mixtures of organic substances possible. Other conventional detectors suitable for different classes of substances include flame ionization detector (GC-FID), photo-ionization detector (GC-PID), and electron capture detector (GC-ECD).

Analytical Methods

Various analytical methods based on the interaction of material with a physical field (e.g., electromagnetic, electric or magnetic field) are used to identify and quantify substances in a sample. This interaction creates a signal that is detected (see detectors below in this section) and measured. The following types of analytical methods are commonly used:

Spectrometry (or spectroscopy) is a general term for methods in which information about a sample is obtained from the interaction of electromagnetic radiation (e.g., visible or ultraviolet light, radio waves) with matter (atoms, ions, molecules). The following interactions can take place: absorption (attenuation of the intensity of radiation by the sample), emission (emission of radiation by the sample), scattering (with or without transfer of energy between the radiation and sample), reflection, refraction, and change in the polarization of the radiation. The elements or compounds in the sample and their concentration can be determined from the recorded spectra. The method name is often associated with the name of the method used to prepare the substance for recording the spectrum, e.g., graphite furnace AAS (GF-AAS).

Atomic absorption spectrometry (AAS) is a reliable analytical technique for determination of the concentration of metallic elements. About 70 metals can be analyzed by this technique; however, each one must be analyzed separately. The method uses liquid samples. However, techniques have been developed in which finely dispersed solid samples can be used directly. The sample is evaporated and atomized, e.g., by injecting into a hot flame or furnace. The vaporized atoms of the element absorb light according to its electron configuration. Monochromatic light of the appropriate wavelength (usually from a cathode tube containing the same metal as the one to be analyzed) is passed through the vapor or flame and its extinction (the ratio of the intensity of incident and transmitted light) is measured. The extinction is proportional to the concentration of the free atoms in the vapor according to the Beer-Lambert law. AAS can be classified into two types on the basis of the method of vaporization: flame AAS (F-AAS) and flameless AAS. Typical flameless methods are graphite furnace AAS (GF-AAS, ETA-AAS), cold vapor AAS (CV-AAS), as well as hydride generation AAS (HG-AAS).

Atomic emission spectroscopy (AES) or optical emission spectrometry (OES) is used to analyze sample composition based on the principle that electrons when excited (i.e. heated to a high temperature) release light of a particular wavelength. The presence or absence of various elements is established by examining the spectral lines characteristic of those elements. This method generally gives an accuracy of only 25 % and has been superseded by ICPS (inductively coupled plasma emission spectrometry).

In *flame atomic emission spectrometry (FAES)*, also called *flame photometry*, a sample solution (usually aqueous) is evaporated and atomized in a hot flame. The thermally excited atoms emit radiation in the visible and ultraviolet spectra at wavelengths characteristic of the emitting element. The intensity of radiation is proportional to the concentration of the element in the sample. This method is predominantly used to analyze for alkali and alkaline-earth metals. It has been replaced by AAS in many fields of application (WELZ & SPERLING, 1997).

Atomic fluorescence spectrometry (AFS) is based on the observation that thermally produced atoms in the gaseous state can be moved to a higher energy state by electromagnetic radiation, from which they return to the ground state by emitting energy in the form of fluorescent radiation. The sensitivity of AFS is proportional to the concentration in the analyte and the intensity of the primary radiation. The detection limits of AFS are comparable to those of GF-AAS. This method has not found widespread application.

Infrared spectroscopy (IR spectroscopy) is based on absorption within the infrared spectrum (typical of organic compounds). The infrared absorption spectrum of a substance is called its molecular fingerprint. It is a technique for determining the molecular species present in a material and measuring their concentrations by measuring the intensity of the wavelengths the analyzed material characteristically absorbs infrared radiation. A typical application is the analysis of hydrocarbons.

X-ray fluorescence spectroscopy (XRF) is used for determination of the total major and trace elements by measuring secondary X-ray emissions after a solid sample has been bombarded with a primary X-ray beam. The wavelength of the secondary X-ray emission of an element is characteristical of that element; the intensity of the emission is a function of the concentration. Two types of x-ray fluorescence spectroscopy are distinguished: wavelength dispersive X-ray spectroscopy (WDX) and energy dispersive X-ray spectroscopy (EDXRF).

Electrochemical methods involve chemical reactions that take place at electrodes in an electrochemical cell. Certain parameters can be measured that are proportional to concentration or are specific to particular substances. These parameters include voltage (potential difference), current, electric charge, conductivity. Common electrochemical methods are amperometry, coulometry, conductometric analysis, polarography, potentiometry, voltametry.

Amperometry or amperometric titration is based on keeping voltage constant during a titration and measuring the current. The current flowing between two electrodes in a solution depends upon the concentration of the materials dissolved in the solution. Oxidized and reduced substances in solution can be detected by amperometric titration. Amperometric titration is based on the fact that during titration of a solution in an electrochemical cell, polarization occurs at the electrodes while voltage is kept constant. As a reagent solution (titrant) is added to the titration cell, the reaction can be observed by the changes in the current in the electrochemical cell. When all of the substance being measured has reacted, the current changes abruptly. The concentration of the substance being analyzed is determined from the volume of the titrant and the measured current. A classical example of an amperometric titration is the determination of water content after Karl Fischer, in which sulfur dioxide is oxidized in the presence of water and iodine: $SO_2 + I_2 + 2H_2O \rightarrow H_2SO_4 + 2HI$. At the titration end point I^- is present in excess, causing depolarization of the electrode and a abrupt rise in the current flowing through the cell.

Coulometry is based on the number of coulombs necessary to release a substance during electrolysis. The analysis can be carried out holding either voltage constant (potentiostatic coulometry, measurement of the electric charge versus time) or current constant (amperostatic coulometry or coulometric titration, measurement of the electric potential versus time). In a coulometric titration, the reagent necessary for titration is produced electrochemically with current kept constant. The endpoint of titration can be detected, for example, with a colored indicator or by potentiometry. The concentration is calculated from the product of current, time and equivalent weight.

Conductometry is based on the determination of the electrical conductivity of an aqueous solution, which is a function of the concentration of free ions, using a low-frequency alternating current. Most common uses are conductometric titration and as a detection method in chromatography. These techniques are especially employed with highly diluted or strongly colored solutions, where the use of color indicators is limited or impossible.

In *polarography*, also called *voltammetry*, current is plotted versus voltage (polarogram), most commonly using a dropping mercury electrode (a capillary filled with mercury dropping at a slow but steady rate) as an indicator electrode (microelectrode) against a much larger nonpolar reference electrode.

Polarographic measurements are based on the principle of electrolysis of substances at a mercury electrode. When voltage is applied between the indicator electrode (Hg electrode) and the reference electrode, current flows in the presence of electrochemically active materials. The cell is connected in series with a galvanometer (for measuring the current) in an electrical circuit that contains a battery or other source of direct current and a device for varying the voltage applied to the electrodes. With the dropping mercury electrode connected to the negative side of the polarizing voltage, the voltage is increased in small increments and the corresponding current is observed on the galvanometer. The current is very small until the applied voltage is increased to a value large enough to cause reduction of the analyzed substance at the dropping mercury electrode. The current increases rapidly at first as the applied voltage is increased above this critical value but gradually attains a limiting value and remains more or less constant as the voltage is increased further. The critical voltage required to cause the rapid increase in current is characteristic of, and also serves to identify the substance that is being reduced (qualitative analysis). Under proper conditions the constant limiting current is governed by the rates of diffusion of the reducible substance up to the surface of the drops of mercury and its magnitude is a measure of the concentration of the reducible substance (quantitative analysis). Limiting currents also result from the oxidation of certain oxidizable substances when the dropping electrode is used as the anode. Polarographic analysis can be used to identify most chemical elements and is especially useful for the analysis of alloys and various inorganic compounds. Polarography is also used to identify numerous types of organic compounds and to study chemical equilibrium and reaction rates in solution.

Two solid electrodes in an electrolytic cell can also be used and the change in voltage measured with respect to time. In water and waste water analysis, this form of voltammetry is used to analyze for lead, cadmium, cobalt, copper, nickel, thallium and zinc.

Potentiometry is a branch of analytical chemistry in which measurements of electrical potential are used to determine the concentration of some component of the analyte solution. The potential difference between a measuring electrode (e.g., an ion-selective electrode) and a reference electrode is measured in a galvanic cell by keeping voltage constant with no flow of current. The ion concentration in the solution can be determined directly from the measured potential difference.

<u>Mass spectrometry (MS)</u> is used to separate compounds based on differences in their mass, as well as qualitative and quantitative analysis and to determine the structure of molecules. The substance to be analyzed is vaporized, ionized, and then the ions are accelerated through a magnetic field to separate the ions by mass. Mass spectrometry can result in the exact identification of a compound, and is a very powerful analytical method, especially when combined with chromatography. It can be used to identify and quantify

extremely small amounts of substances on the basis of their mass-fragment spectrum. Common types of mass spectrometry include FAB (fast atom bombardment), ESI (electrospray ionization), MALDI (matrix assisted laser desorption), TOF (time of flight), FT (Fourier transform), and ICP (inductively coupled plasma). Mass spectrometry is often combined with other analytical techniques, such as gas chromatography (GC) and liquid chromatography (LC).

In *mass spectrometry with inductively coupled plasma (ICP-MS)*, the sample is ionized in an ICP and the resulting ions are analyzed in a mass spectrometer. A wide range of elements can be determined with this method, which has a low detection limit. It is being increasingly used instead of GF-AAS or ICP-AES, owing to its higher throughput, superior detection capabilities, and the ability to obtain isotopic information. On the other hand, analysis of trace elements in samples with high level of chloride is difficult. Ions generated in inductively coupled quadrupole plasma mass spectrometry (ICP-QMS) are extracted electrostatically from the ionization chamber and introduced into a quadrupole mass filter.

Thermal analysis: *Calorimetry* is based on the measurement of heat generated (exothermic process) or consumed (endothermic process) by chemical reactions or physical changes. The measurements are made using a calorimeter. A simple calorimeter may consist of a thermometer in an insulated container. Several types of calorimeters are used, for example, constant-volume calorimeters, constant-pressure calorimeters, differential scanning calorimeters, and isothermal titration calorimeters.

Thermogravimetry is based on changes in weight which are measured as a function of change in temperature, pressure or gas composition.

Other methods: *Titration* is a chemical method for determining the concentration of a substance in solution by adding a standard reagent of known concentration in carefully measured amounts until the reactant has completely reacted as shown by a color change or by electrical measurement. The concentration of the titrated substance is then calculated from the amount of reagent used. Titration is used, for example, to determine the acidity of a solution.

Photometry is a method of determining the concentration of a substance dissolved in water by adding one or more suitable reagents that form a colored product with that substance. The color intensity is a function of the concentration of the substance. A photometer is required to measure the intensity accurately. The less accurate visual assessment of the color intensity is called “colorimetry”.

Fluoride-sensitive electrodes are used to determine the content of fluoride in nonvolatile organofluoro compounds using an ion-selective electrode.

Common detectors

An *electron capture detector (ECD)* is used for gas chromatography. The carrier gas is ionized by a radioactive source; a voltage is applied, creating a small current (caused by electrons released by the ionization). When an electronegative substance enters the detector, it captures electrons, leading to a detectable drop in current amplitude. This reduction is a measure of concentration. An ECD is typically used for organic compounds containing halogen, sulfur, nitrogen, phosphorus, or heavy metals (e.g., pesticides).

The *flame ionization detector (FID)* is a nearly universal gas chromatographic detector. It responds to almost all organic compounds. An FID does not respond to nitrogen, hydrogen, helium, oxygen, carbon monoxide or water. This detector ionizes compounds as they reach the end of the chromatographic column by burning them in an air/hydrogen flame. As the compounds pass through the flame, the conductivity of the flame changes, generating a signal.

Photo-ionization detectors (PID) use ultraviolet light to ionize an analyte. The ions produced by this process are collected by electrodes. The current generated is a measure of the analyte concentration. PIDs are used to detect volatile organic compounds (VOCs), such as benzene, chlorinated hydrocarbons, ethyl benzene, toluene, and xylene (BETX). Offering rapid response and high sensitivity, photo ionization is the preferred method for detecting VOCs. Since PIDs are normally calibrated for benzene, the directly readable output is valid only for benzene contamination.

Mass spectrometers (MS) are frequently used as detectors for gas chromatography (GC-MS) as mentioned above.

The results of chemical analysis rely on quality assurance, i.e. scrupulous attention must be paid to cleanliness, sample preparation, compliance with standards and detection limits.

Analytical chemistry research is conducted to improve sensitivity, selectivity, robustness, linear range, accuracy, precision, and speed, as well as to reduce cost of purchase, operation, training, and space. Therefore, much effort is being expended on shrinking the analysis methods to chip size.

More information about analytical chemistry methods can be found in the corresponding entries in the glossary of this handbook.

Sections 5.3.1.3 and 5.3.2.3 contain a tabular overview of standard methods for various parameters, and their detection limits.

5.3.5 Laboratory Methods for the Determination of Migration Parameters

SVEN ALTFELDER with a contribution by CLAUS NITSCHE

Dissolved substances migrate with the groundwater through permeable soil, sediment or rock in the unsaturated and saturated zones. Investigation of the fate of these contaminant substances during their migration within these zones is crucial for a risk assessment of landfills, mines and industrial plants. Adsorption/desorption and biodegradation are the most important processes that reduce the mass, toxicity, mobility, volume or concentration of contaminants in soil/sediment/rock and groundwater (natural attenuation) during migration.

It is important to keep in mind that the migration zone (extends from the soil into the lithosphere) varies spatially with time and is an anisotropic multiphase system (JURY et al., 1987; VAN DER ZEE & VAN RIEMSDIJK, 1987; STRECK & RICHTER, 1999). Many of the processes governing contaminant migration are of a nonequilibrium and nonlinear nature (SCHEIDEGGER et al., 1994; STRECK et al., 1995; TOTSCHE, 2001). Experimental investigation and mathematical description of these processes are made difficult by the complexity of the system. Experiments in the field and laboratory are conducted to investigate this complex system, using mathematical models to analyze the data.

Multiple processes and aspects have to be considered when investigating contaminant mobility in the unsaturated and saturated zones. These include sorption, biodegradation, chemical reactions, water flow, preferential flow, scale problems, variability and randomness. The following sections will deal with experimental methods for studying sorption and flow-related processes in the laboratory. It should be keep in mind, however, that these processes and aspects affect each other, so the investigation of a single process such as sorption requires complicated preparation. The interpretation of the experimental data and the application of the results to a problem require careful consideration of all relevant processes.

Solute migration in soil and sediment can be investigated on the molecular scale; for a risk assessment this is rather impractical. An alternative approach is to average the molecules of interest over a certain volume and to regard the average as a homogeneous phase. The migration and interaction of the phases of the substances being considered are described on a macroscopic scale. By averaging over a certain volume of the porous medium, macroscopic variables are defined that allow treatment of the medium as a continuum. These variables include porosity, bulk density, matrix potential and dispersivity. The application of volume-averaged variables implies that a *representative elementary volume* (REV) exists. The REV is defined as the minimum volume which – when further increased – does not lead to a change in the averaged variable (BEAR, 1972). This is a necessary condition for the use of local statistical averages of variables describing geometry, kinematics and kinetics, as well as physical, chemical, and biological processes. On the other hand, the REV must be small compared to the dimensions of the site, so that the local statistical averages may be regarded as continuous subsurface parameter and variable values. Migration processes may then be mathematically described by differential equations. The scale of interest is called the continuum scale. As the size of a REV is not known a priori, any change in scale – for instance from the laboratory to the field scale – may be accompanied by errors because the REV can be different on the two scales.

Depending on the focus of the investigation, batch and column tests are commonly used to determine migration parameter values. Occasionally, batch reactor (dynamic batch) tests are used.

5.3.5.1 Basic Theory of Sorption

In general, sorption can be subdivided into absorption and adsorption. Absorption is the assimilation of a liquid or gas into a solid or a gas into a liquid. Adsorption is the adhesion of the molecules of gases, liquids, or dissolved substances (adsorbate) to the surface of a solid or liquid particle (adsorbent). The adsorbent may either immobile (e.g., the soil) or may itself be mobile, e.g., the dispersed particles of a colloid or dissolved organic material (KÖGEL-KNABNER & TOTSCHE, 1998). Since the latter case is rather complex it is not further addressed in the following treatment of sorption.

Adsorption mechanisms are generally categorized as either physical, chemical, or electrostatic. Weak molecular forces, such as van der Waals forces, provide the driving force for physical adsorption. There is no significant redistribution of electron density in either the adsorbed molecule or the adsorbent surface. In chemical adsorption, a chemical bond is formed between the adsorbed molecule and the surface of the solid, involving substantial rearrangement of electron density between the adsorbate and adsorbent. Electrostatic adsorption involves Coulomb forces, and is normally referred to as ion exchange. In liquids, interactions between solute and solvent also play an important role in establishing the degree of adsorption.

Desorption is the opposite of adsorption and absorption, generally referred to together as sorption. Experimental evidence shows that sorption is often nonlinear and kinetically controlled.

Equilibrium controlled sorption

In order to describe sorption we must derive a relationship between the concentration S of the adsorbed substance and its dissolved concentration (solute concentration) C. When equilibration is rapid compared to transport, we can consider these two phases to be in equilibrium and the relationship between the two phases can be described by a sorption isotherm. *Figure 5.3-15* shows the three most common types of isotherms.

The simplest sorption model is the *linear sorption isotherm*, which may be expressed by

$$S = K_d\, C, \tag{5.3.5}$$

where K_d, the slope of the isotherm, is called the distribution coefficient, sorption coefficient, or partition coefficient. This isotherm is commonly applied to the partitioning of organic compounds in two-phase systems containing water and an organic liquid. Partitioning has also been claimed to be analogous to the sorption of organic contaminants on natural organic matter (CHIOU et al., 1979; CHIOU et al., 1983). In certain situations the linear isotherm may also be used to approximate the *Freundlich isotherm*:

$$S = K_f\, C^{1/n}. \tag{5.3.6}$$

The Freundlich isotherm describes situations in which the number of adsorption sites is large relative to the number of contaminant molecules, K_f is the Freundlich coefficient and n is a substance-specific coefficient. Values of n near 1.0 are valid for some organic contaminants (partitioning) but lower values (0.4 - 0.6) have to be used, for example, for heavy metals.

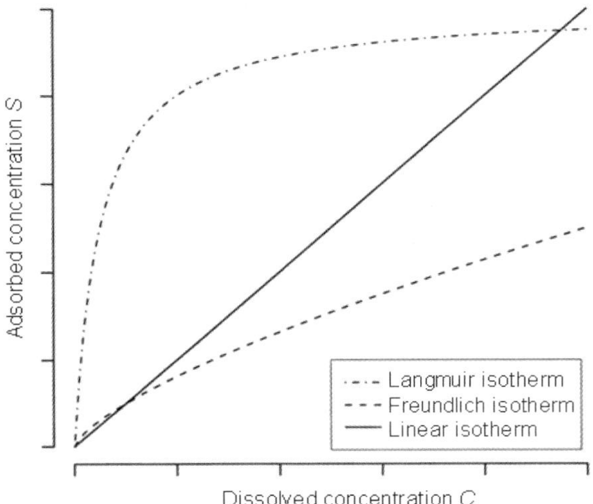

Fig. 5.3-15: Types of frequently used adsorption isotherms

The *Langmuir adsorption isotherm* describes the formation of a layer of molecules on an adsorbent surface as a function of the concentration of the dissolved contaminant in the liquid. In the Langmuir model, adsorption increases almost linearly with increasing solute concentration at low C values and approaches a constant value at high concentrations. The adsorbed concentration approaches a constant value because there is a limited number of adsorption sites on the soil particles. The Langmuir isotherm can be described mathematically as follows:

$$S = \frac{Q_0 \, b C}{1 + bC}, \tag{5.3.7}$$

where b is the equilibrium constant for the adsorption and Q_0 is the number of sorption sites (maximum amount of adsorbed contaminant). Other types of adsorption isotherms exist – an example is the BET isotherm, which was developed for situations in which the adsorbate forms multiple layers. A detailed description is beyond the scope of this section.

The retardation factor or retardation coefficient R can be calculated from the distribution coefficient K_d as follows:

$$R = 1 + \frac{\rho_b K_d}{n}, \tag{5.3.8}$$

where ρ_b is the bulk density and n is the porosity. The retardation coefficient expresses how much slower a contaminant moves than the water itself does. If v_w is the average linear groundwater flow velocity and v_c is the velocity of an adsorbate substance (contaminant), then

$$v_c = \frac{v_w}{R}. \tag{5.3.9}$$

The value of v_w can be determined by column tests with a nonadsorbent tracer.

A retardation factor calculator is available as online tool for site assessment from U.S. EPA at www.epa.gov/Athens/learn2model/part-two/onsite/retard.htm. The retardation factor for any specific substance is strongly affected by temperature, pH, redox potential, salinity, natural organic content, and concentrations of other chemical species. Therefore, each chemical species has its own retardation factor which changes as the chemical composition changes. This complexity has resulted in unsuccessful predictions of retardation factors from general principles. R must be measured for each species under each condition for each system using batch tests and/or column tests. *Table 5.3-25* gives examples of calculated retardation factors.

Table 5.3-25: Examples of retardation factors calculated with the EPA online tool assuming a particle density $\rho_s = 2.65$ g cm^{-3}

Chemical	Porosity	Fraction organic carbon	K_{oc} [L kg^{-1}]	ρ_b [g cm^{-3}]	K_d [L kg^{-1}]	Retardation factor R
toluene	0.25	0.0001	300	1.99	0.03	1.2
ethylbenzene	0.22	0.0002	1100	2.07	0.22	3.1
benzene	0.22	0.0002	83	2.07	0.0166	1.2

A general expression (ZHENG & BENNET, 1995) for the retardation factor is

$$R = 1 + \frac{\rho_b}{n}\frac{\delta S}{\delta C}, \tag{5.3.10}$$

where S is the adsorbed concentration and C is the dissolved concentration. For the linear isotherm, Equation (5.3.10) reduces to Equation (5.3.8). Equation (5.3.10) can be used to derive retardation factors for the nonlinear Freundlich and Langmuir isotherms.

Kinetically controlled sorption

Two common models for kinetically controlled sorption under nonequilibrium conditions are described here. Both models divide the sorbent into two domains. In the exterior domain, the solid phase concentration is assumed to be in equilibrium with the solution phase concentration. In the interior domain, nonequilibrium conditions and kinetically controlled sorption are assumed. Sorption in the equilibrium domain can be described by the Freundlich isotherm given in Equation (5.3.6). Sorption in the nonequilibrium domain is modeled either as first-order mass-transfer or as spherical diffusion, as explained below. Neglecting degradation, the mass balance in each sorption or desorption step is

$$\frac{dC_t}{dt} = 0, \tag{5.3.11}$$

where C_t is the total concentration in the batch container, and t is time. For a given solution-phase concentration, both models predict the same solid-phase concentration at equilibrium:

$$S = K_f\, C^{1/n} \quad t \to \infty. \tag{5.3.12}$$

First-order model

In this modeling approach described by STRECK et al., (1995), sorption in the nonequilibrium domain is described by

$$(1-f)\frac{dS_2}{dt} = \alpha\left(K_f C^n - S_2\right), \tag{5.3.13}$$

where S_2 is the solid-phase concentration in the nonequilibrium domain, α is the sorption rate coefficient, and f the fraction of equilibrium sites. S_1 is defined in *Fig. 5.3-16*. The total solid-phase concentration of sorbed solute, S is given by

$$S = f S_1 + (1 - f) S_2. \tag{5.3.14}$$

The linear form ($n = 1$) of this model is mathematically equivalent to the model of BRUSSEAU & RAO, (1989). In this case the notation of K_f changes to K_d, the distribution coefficient. *Figure 5.3-16* displays two sorbent geometries that may be described by the model.

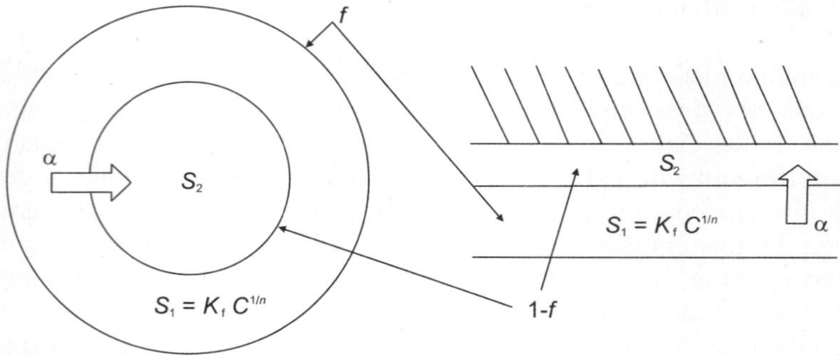

Fig. 5.3-16: Illustration of two geometries involving exterior and interior sorption sites that may be described by the first-order model

Diffusion model

In this approach, sorption in the nonequilibrium domain is modeled as diffusion into and out of uniform spheres (ALTFELDER & STRECK, 2006). Spherical diffusion is described by

$$\frac{\partial S_2^*}{\partial t} = D^* \left(\frac{\partial^2 S_2^*}{\partial r^{*2}} + \frac{2}{r^*} \frac{\partial S_2^*}{\partial r^*} \right), \qquad (5.3.15)$$

where S_2^* stands for the local sorbate concentration in the sphere, and $r^* = r / a$ denotes the radial coordinate normalized with respect to the radius a of the sphere. $D^* = D_a / a^2$ is the diffusion rate constant calculated from D_a the apparent diffusion coefficient, a definition which is in line with that of BRUSSEAU et al. (1991).

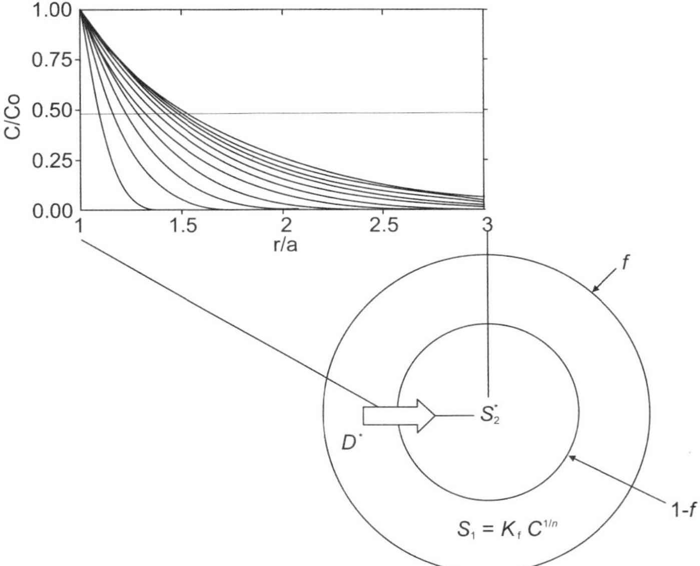

Fig. 5.3-17: Schematic diagram of the spherical diffusion model – the inset illustrates the development of concentration in the sphere with respect to time as a function of radius

While the solute flux at $r^* = 0$ vanishes for symmetry reasons, the sorbate concentration at $r^* = 1$, i.e., at the interface between the equilibrium and nonequilibrium domains, is given by $S^*_{2(r^*=1)} = S_1$, where S_1 is a nonlinear function of the solution phase concentration given by Equation (5.3.6). The change in the solid-phase concentration in the nonequilibrium domain can be calculated from the solute flux across the surface of the sphere using the solution of Equation (5.3.15):

$$(1 - f)\frac{\partial S_2}{\partial t} = -3J_0^*, \qquad (5.3.16)$$

where the normalized flux density J_0^* is given by

$$J_0^* = -D^* \left.\frac{\partial S_2^*}{\partial r^*}\right|_{r^*=1} . \qquad (5.3.17)$$

Figure 5.3-17 displays a sorbent geometries that may be described by the model.

5.3.5.2 Basic Theory of Transport

Bulk contaminant flow is commonly described by the convection-dispersion solute transport equation. Taking a single dimension into consideration, we start by defining the local mass balance for a unit element of soil:

$$\frac{\partial C_t}{\partial t} = -\frac{\partial J_s}{\partial z} + \sum_i r_i , \qquad (5.3.18)$$

where C_t and J_s denote total concentration and solute flux density, respectively, t and z are time and space coordinates, and r_i are source or sink terms, accounting, for instance, for degradation. For contaminants that are not volatile, the total concentration is

$$C_t = \theta C + \rho S, \qquad (5.3.19)$$

where S is adsorbed concentration, C is the dissolved concentration, ρ is the bulk density, and θ the volumetric water content. If the solute transport is truly convective-dispersive, the solute flux density is

$$J_s = -D_s^w \frac{\partial C}{\partial z} + J_w C, \qquad (5.3.20)$$

where D_s^w and J_w are the apparent dispersion coefficient and the volumetric water flux density, respectively. The relationship between D_s^w and J_w is given by $D_s^w = \lambda J_w / \theta$. The parameter λ is the dispersivity. These equations yield

$$\frac{\partial \theta C}{\partial t} + \frac{\partial \rho S}{\partial t} = \frac{\partial}{\partial z}\left(D_{\mathrm{s}}^{\mathrm{w}} \frac{\partial C}{\partial z} \right) - \frac{\partial J_{\mathrm{w}} C}{\partial z} + \sum_{\mathrm{i}} r_{\mathrm{i}} . \tag{5.3.21}$$

To solve this equation, the water content θ and water flux density J_{w} must be known. They are readily available from steady-state column experiments. Experiments with non-steady-state flow require solution of the equation describing the water regime. A detailed treatment of this case is beyond the scope of this book. Source and sink terms should also be known, for instance, from degradation experiments (see Section 5.3.6). Moreover, the relationship between sorbed-phase and solution-phase concentrations as described in the previous paragraphs needs to be known. This relationship may be described by either a sorption isotherm (equilibrium sorption) or a kinetic term (nonequilibrium sorption).

5.3.5.3 Sampling and Preparation of Soil or Sediment for the Determination of Migration Parameter Values

To ensure the reproducibility of the experimental results, detailed documentation of the following sampling and sample preparation steps is necessary:

- sample collection via drilling, digging, etc.,
- preservation,
- transport,
- partitioning, and
- installation in the test facility.

Literature or manuals with detailed information on these steps are rare. As a consequence, these steps are implemented in quite different ways in the various scientific studies. In addition, sampling schemes depend on the contaminant to be investigated. A good strategy is to first define the purpose of the study and then scan the existing literature for similar problems. A suitable sampling scheme may be developed on the basis of the results.

An example of the difficulties associated with sampling is the risk of cross contamination when samples are collected by drilling or digging. If the difference in certain properties is large between soil or sediment layers, small quantities of soil or sediment from another layer may change the values obtained for that sample. When an auger is used for sample collection, removal of the outer part of the sample is one way to avoid cross contamination (STRECK & RICHTER, 1997).

The importance of proper sample preparation is illustrated by the following example: A common technique for preserving samples is to dry them by either air-drying or drying at 105 °C. Especially for organic contaminants with an affinity to natural organic matter this treatment leads to an increase in the measured sorption when the content of natural organic matter (NOM) in the investigated soil or substrate is relatively high. The rearrangement of NOM upon dehydration – a process that is not readily reversible after rewetting – leads to additional sorption sites for organic contaminants (ALTFELDER et al., 1999). A possible solution to this problem could be to freeze the field-fresh sample. Besides drying of the sample, homogenization and sieving to certain size fractions are commonly used methods for preparing a sample for analysis. Possible consequences of these two methods are mineralization of organic matter and changes in the cation exchange properties.

Samples are quite often collected under anaerobic conditions. To avoid misinterpretation in the case of such samples, it may be necessary to carry out all preparation steps in an anaerobic atmosphere. WINKLER (1989) describes an appropriate procedure using a glove box containing an inert gas.

5.3.5.4 Batch Tests

An important advantage of batch tests is the relatively easy and quick setup, the ability to study processes that influence migration without having to cope with possible effects of contaminant transport complicating data interpretation (as in column studies) and the ability to investigate large sample quantities in parallel experiments. The equipment is cheap and the handling of many parallel samples is simple.

Common processes investigated with batch tests are sorption, degradation, dissolution and precipitation. Measures to inhibit other processes are necessary if the focus is on sorption. These measures include sterilization of the sample and creating chemical conditions that inhibit precipitation or abiotic degradation (e.g., by hydrolysis). Typical applications of batch tests are determination of contaminant retention in a soil, sediment or man-made clay landfill seal. Typical contaminants are heavy metals, radionuclides, and organic compounds (FRANK, 1991; KLOTZ & FOLIV, 1983).

In a batch test, a certain amount of soil or sediment is shaken with an aqueous solution with a known concentration of the test substance for a certain time period. The aqueous suspension is then separated by centrifugation or filtration and the concentration of the test substance in the aqueous solution and possibly in the soil or sediment is determined either by direct analysis or after extraction from the sediment (e.g., with an organic solvent). If the test substance concentration is measured only in the aqueous solution, the concentration in the solid phase may be calculated from this value. The contaminant must be stable for the length of time of the test. Briefly, batch tests are carried out as follows:

1. A known amount of dissolved solute is added to a test tube/flask.
2. A known mass of soil/sediment is added to the test tube/flask (minimum 2 g).
3. Two blanks are prepared: (a) one without soil/sediment, (b) one without solute.
4. The test tubes are shaken for a certain time period at a constant temperature.
5. The concentration of the solute is measured.
6. How much mass was removed from the solution is calculated.
7. The aqueous concentration of the test substance is plotted versus the adsorbed mass and a model (isotherm or kinetic model) is fitted to the data.

Agitation is not standardized, an overhead shaker or equivalent equipment should keep the soil in suspension during shaking. It is recommended that all experiments be performed in duplicate. Usually the test substance is dissolved in a solution of 0.01 M $CaCl_2$. The common practice of using distilled or deionized water is not recommended because of unwanted cation exchange and difficulties in centrifugation. The solute concentration of the stock solution should be approximately three orders of magnitude higher than the detection limit of the analytical method used. This threshold safeguards accurate measurements. The solution should be prepared just before shaking with the soil samples. An appropriate solubilizing agent may be used for poorly soluble substances. The agent should not behave as a surfactant or react with the test chemical. The solubilizing agent should be miscible with water and should not exceed 0.1 % of the total volume of the solution added to the soil or sediment sample. For a detailed review on possible pitfalls of spiking poorly soluble substances into a soil or sediment see the study of NORTHCOTT & JONES (2000). The two blank samples (a) and (b) are used to detect sorption to the vessel and interfering substances or contaminated soils or sediments, respectively. The size of the test tube/flask is determined by the required ratio R of soil to solution, in extreme cases the ratio may be up to 1:100. For the minimum amount of soil or sediment (2 g), a 250 mL flask would be appropriate. Probes that allow constant monitoring of the solution may be inserted into the solution to monitor the changes that occur.

When contaminant concentration in the solid phase is calculated from the decrease in the aqueous concentration, it is important to select an appropriate soil/solution ratio for the experiment. Otherwise the error introduced in the calculation is unacceptable. More than 50 % of the initial test substance should be adsorbed and the remaining concentration should be above the detection limit. *Figure 5.3-18* gives appropriate soil/solution ratios R based on the distribution coefficient K_d.

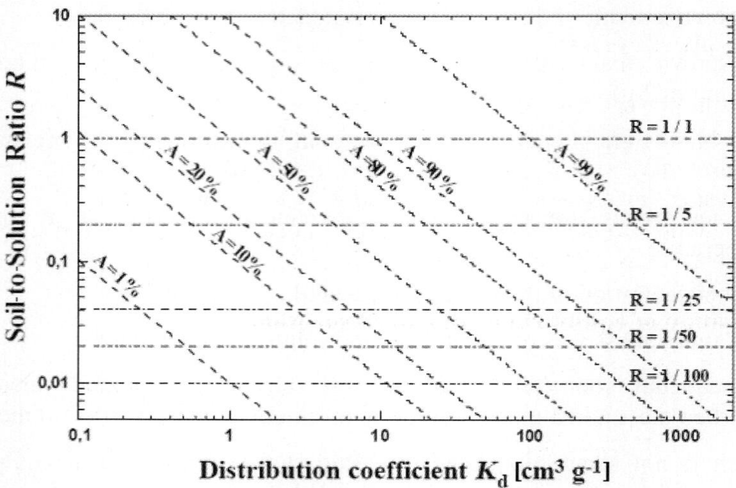

Fig. 5.3-18: Relationship between the ratio R of soil to solution and K_d at various percentages of adsorbed test substance A, OECD (2000)

The approximate distribution coefficient K_d may be derived from preliminary studies or may be approximated from empirical data. A good example for estimating K_d is the KOC method (EPA, 2006b). This method is based on the assumption that sorption of organic contaminants on naturally occurring organic materials depends on the properties of the contaminant. K_d is calculated from the fraction of organic carbon, f_{oc} (e.g., in an aquifer) and the organic carbon partition coefficient, K_{oc} of the chemical:

$$K_d = K_{oc} f_{oc}. \tag{5.3.22}$$

Some K_{oc} values are given on the "chemical properties page" of EPA (2006b) in *Table 5.3-26*.

Table 5.3-26: Partition coefficients K_{oc} for organic carbon (EPA, 2006b)

Chemical	K_{oc}
MTBE	11
benzene	83
toluene	300
ethylbenzene	1100
o-xylene	830
p-xylene	870
m-xylene	982
total xylenes	894

Possible pitfalls of batch tests

Batch tests are carried out with relatively large soil-to-solution ratios, most commonly 1:2 to 1:3. These ratios are far from natural conditions. Shaking the suspensions may lead to particle abrasion and increased sorption due to new sorption sites and a delay in equilibration while new sorption sites appear during agitation. When transport and sorption occur at the same time in nature, batch experiments may not yield a realistic picture.

Investigation of equilibrium controlled sorption

Studying equilibrium controlled sorption with batch experiments is fairly straightforward. A prerequisite is that the flasks are agitated long enough to guarantee that equilibrium between the dissolved and the adsorbed concentrations is reached. A plot of the measured solute concentrations in a simple x,y-plot provides a first impression of the sorption isotherm. An example of equilibrium controlled sorption data is given in *Fig. 5.3-19*. The data appear to be linear.

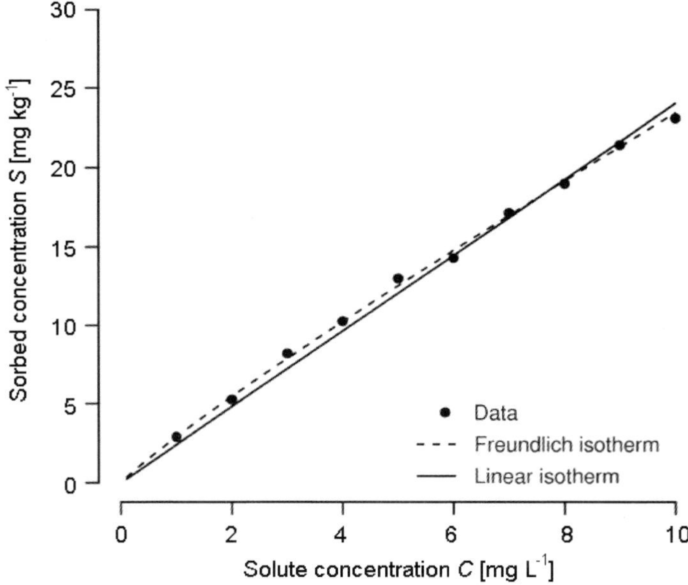

Fig. 5.3-19: Fit of a linear and a Freundlich isotherm to a measured set of data

An example of isotherm parameter values obtained by fitting a linear and a Freundlich isotherm is given in *Table 5.3-27*. The values for the fitted curves reveal that the measured data is actually slightly nonlinear, as indicated by the fact that the root mean square error (RMSE) of the Freundlich isotherm is smaller than that of the linear isotherm. However, *Fig. 5.3-19* also illustrates that a linear isotherm is a good approximation of sorption behavior between $C = 1$ and 10 mg L^{-1}. In this special case it may be more convenient to use a linear isotherm without introducing a significant error in the description of sorption.

Table 5.3-27: Parameters of the linear and Freundlich model for the fit shown in *Fig. 5.3-19*

Linear model		**Freundlich model**		
K_d	RMSE	K_f	n	RMSE
2.41	0.352	2.90	0.908	0.215

Fitting a linear isotherm to measured data is straightforward. The same holds for the Freundlich isotherm. Even though the equation is nonlinear, a simple way to estimate its parameters is the rearrangement of Equation (5.3.6) to

$$\log S = \log K_f + 1/n \log C. \qquad (5.3.23)$$

K_f and $1/n$ may be estimated by plotting log S versus log C or by simple linear regression.

Investigation of kinetically controlled sorption

The simplest and most common method to study kinetically controlled sorption in batch experiments is the *parallel method*. In this method n samples are prepared, where n is the number of time intervals at which measurements are made. After each time interval, a sample is either centrifuged or filtered and an aliquot of the aqueous phase is analyzed. The time intervals used for sampling should be increased on a log scale because rapid sorption at the beginning of the experiment masks the slow uptake of the test substance later, which makes it appear that equilibrium has been attained even though nonequilibrium conditions are present. *Figure 5.3-20* illustrates data obtained with the parallel method. The investigated contaminant is slowly sorbed with an equilibration time in the range of several months.

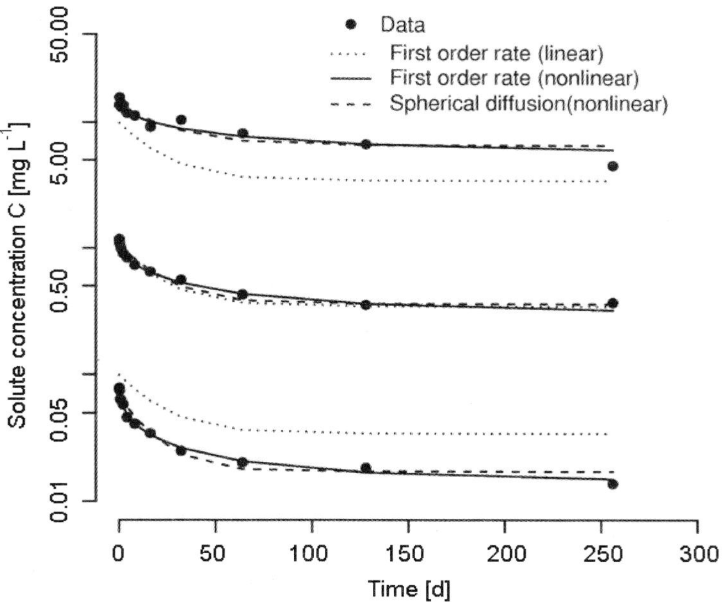

Fig. 5.3-20: Kinetic sorption data measured with the parallel method for three initial concentrations of the solute. Three models for kinetically controlled sorption were fitted to the data, the first-order rate model (linear and nonlinear) and the nonlinear spherical diffusion model.

Figure 5.3-20 illustrates that when sorption is nonlinear it is important to carry out experiments over a range of initial concentrations (several orders of magnitude). *Figure 5.3-20* also shows the results of fitting the linear and nonlinear first-order rate model to the data (Equations (5.3.14) and (5.3.13)). The two models are almost indistinguishable at intermediate concentrations. However, the linear model ($n = 1$) is not able to adequately describe sorption at the higher and the lower concentrations. Applying the parallel method at only one concentration is, therefore, insufficient to detect sorption nonlinearity. A linear model cannot be recommended for the concentration range illustrated in *Fig. 5.3-20*.

Comparison of the fit of the nonlinear first-order model with the fit of the spherical diffusion model (Equation 5.3.15) reveals that the first-order model predicts sorption equilibrium too early compared to the measured data, while the diffusion model is able to portray ongoing uptake beyond 100 days. The deviations are small, however, and the nonlinear first-order model may be adequate to describe contaminant uptake in this special case. The estimated parameter values obtained with the different models are listed in *Table 5.3-28*. It can be seen that nonlinearity is pronounced, with m equal to about 0.7. The FITHYST program was used to fit the models to the measured data (STRECK et al., 1995).

Table 5.3-28: Parameter values for the different kinetic sorption models

First-order model (linear)		
α	[d^{-1}]	0.0144
f	[-]	0.212
K_d	[L kg^{-1}]	9.62
First-order model (nonlinear)		
α	[d^{-1}]	0.0164
f	[-]	0.270
K_f	[mg$^{(1-m)}$L^{m}kg^{-1}]	6.87
m	[-]	0.723
Spherical diffusion (nonlinear)		
D^*	[d^{-1}]	0.000769
f	[-]	0.183
K_f	[mg$^{(1-m)}$L^{m}kg^{-1}]	7.85
m	[-]	0.722

5.3.5.5 Column Experiments

During solute migration in natural systems consisting of soil or unconsolidated or consolidated rock, the amount of material adsorbed within a sample or a layer depends on the location in the sample/layer relative to the contaminant source and the time since the start of migration. As contaminated fluid enters the sample or layer, it comes into contact with adsorbing material. Solute is adsorbed, filling some of the available sites. Soon, the adsorbent near the input end of the column is saturated and the contaminant penetrates further into the sample or layer. The area in which the contaminant is sorbed progresses through the sample/layer with time. Fluid emerging from the sample/layer contains little or no solute until the bulk of the sample/layer is saturated. The "breakpoint" occurs when the concentration of contaminant in the fluid leaving the sample/layer increases. This process is described and analyzed by column experiments. The investigation of solute transport with column experiments has the advantage that contaminant flow and interaction with a soil or sediment may be studied simultaneously. The experimental design is closer to the transport processes observed in natural soil or sediment systems than those of batch experiments. In a column experiment the investigated solute is added to the inlet of a soil column. The concentration of the substance at the column outlet is recorded as the *breakthrough curve* (BTC). A breakthrough curve is commonly plotted as the relative concentration of a given substance in solution versus time, where relative concentration is defined as the ratio of the actual concentration to the source concentration. Breakthrough curves may be analyzed graphically, analytically or with numerical models. In recent years, breakthrough curves are usually analyzed numerically.

Simple experimental design

The most common form of column experiment uses saturated steady-state flow at relatively high flow velocities with a pulse- or step-type boundary condition at the column inlet. The design is aimed at the determination of transport as well as sorption parameter values. It is still fairly easy to implement and cost efficient. The breakthrough curves are either s- or bell-shaped depending on the chosen boundary condition. Transport or sorption-related parameter values are determined from the deviations of the measured BTC from the BTC of an ideal tracer. *Figure 5.3-21* shows a simple column experiment setup that is similar to the setup described in DIN V 19736 (1998). A solution of the test substance enters the bottom of the column via, for example, a peristaltic pump; the solution passes through the soil in the column, with which it may or may not interact; the concentration of the test substance is analyzed from time to time as the solution leaves the top of the column.

Figure 5.3-22 shows details of the column itself and may serve as an example for the design of a column experiment. It is very important that all materials involved in the experiment setup are inert with regard to the test substance. A nonpolar organic compound may require the use of a piston pump with stainless steel tubing because the compound may sorb to the PVC tubing of a peristaltic pump.

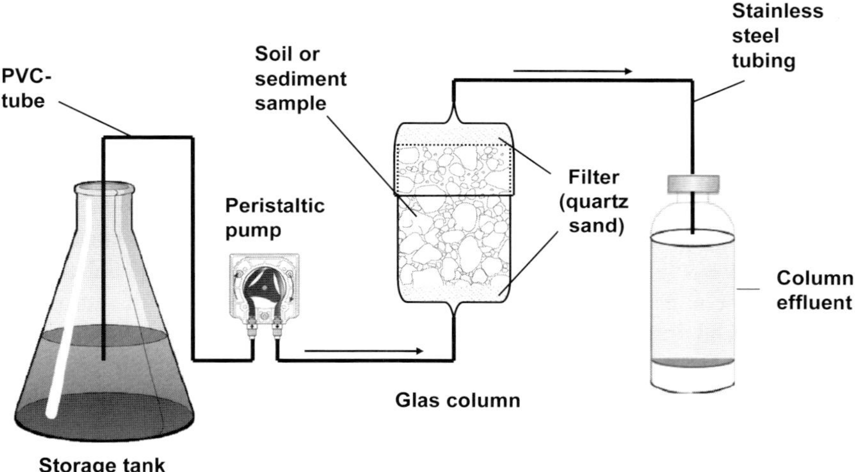

Fig. 5.3-21: Schematic setup of a simple saturated column experiment

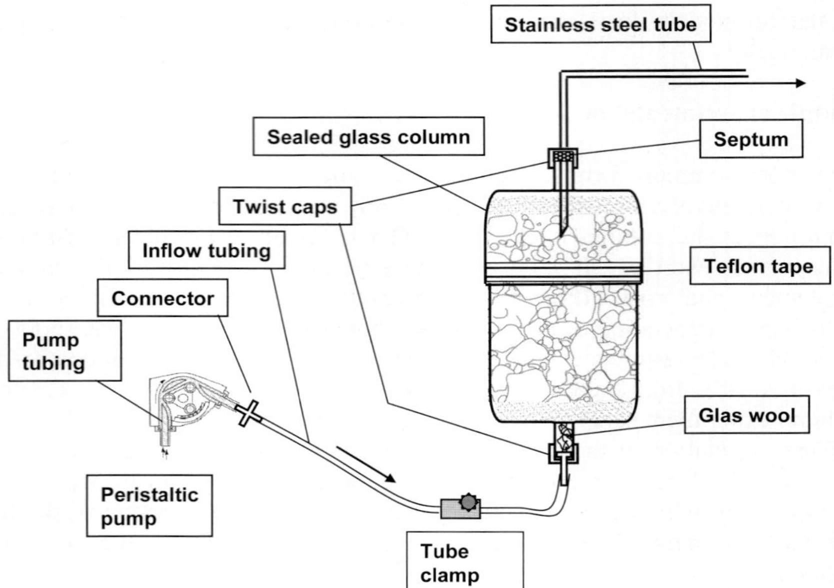

Fig. 5.3-22: Construction details of the column vessel in the simple saturated column experiment

Sophisticated experimental design

More sophisticated setups have been developed that allow better control of the flow conditions than the simple experimental setup. An example of a soil or sediment column designed for this purpose is shown in *Fig. 5.3-23*. A perforated Teflon disc is used to spread the test solution on the top of the column material. In addition to testing saturated solutions, this setup may also be used to investigate unsaturated solutions. The porous plate at the outlet end of the column can be used to apply a predetermined suction pressure, allowing a more natural flow regime.

The design of the process control of an experiment for which these columns are used is shown in *Fig. 5.3-24*. The design allows both steady-state and non-steady-state transport experiments. The percolate may be collected in sample vessels that have to be changed manually or with automated equipment. The entire system is connected to a PC with software for feedback control of the experiment.

The feedback control system allows the experimental conditions to be adjusted so that parameter values can also be determined – for instance by inverse modeling – even in the presence of concurrent processes influencing solute migration.

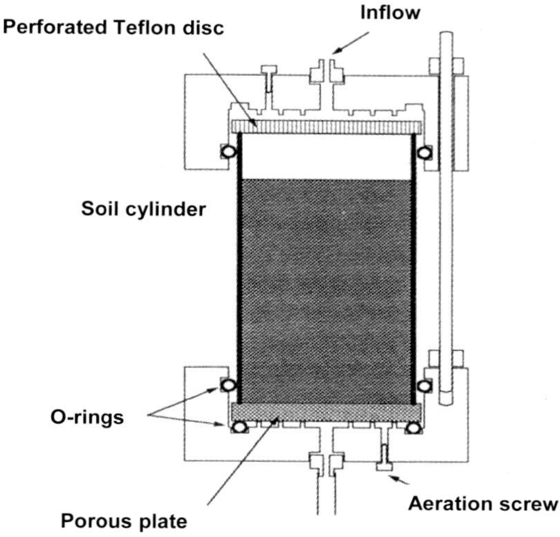

Fig. 5.3-23: Construction details of the column vessel for a saturated/unsaturated column experiment with variable boundary conditions, ZURMÜHL (1994)

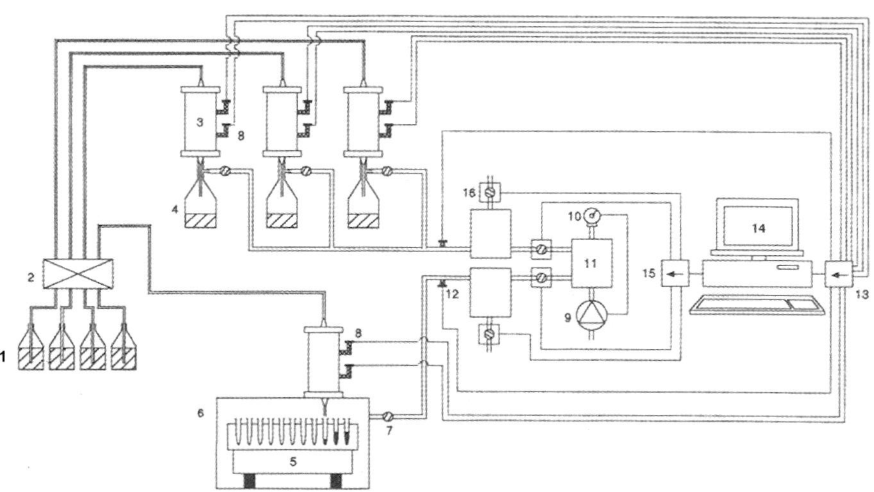

1: vessels with solutions, 2: piston stroke pump, 3: soil column, 4: vessels for percolate collection, 5: fraction collector, 6: vacuum chamber, 7: gate valve, 8: tensiometers with pressure sensors, 9: vacuum pump, 10: manometer, 11: vacuum vessel for pressure balance, 12: pressure sensor to record suction pressure in the column, 13: A/D-card to record pressure values, 14: computer with software for digital control of the experiment, 15: digital I/O card for adjustment of magnetic valves, 16: magnetic valves for pressure regulation

Figure 5.3-24: Example of an experimental setup for steady state and non-steady state transport experiments, ZURMÜHL (1994)

Possible pitfalls of column experiments

The setup and handling of column experiments is complicated and requires skilled experimenters and extensive experience with the system. The advantage of column experiments over batch experiments, namely their closeness to natural conditions in which transport and interaction between contaminant and soil take place simultaneously is also a considerable disadvantage. The concurrent processes have very similar effects on the breakthrough curves in column studies, leading to ambiguity in the interpretation of the experimental results. This is demonstrated below.

Process identification with column experiments

The nonlinearity of the processes involved in solute migration and their concurrent influence on transport complicates the unique determination of the corresponding parameter values. Especially the simple experimental design leads to results that are sometimes difficult to interpret. An example is given in *Fig. 5.3-25* in which breakthrough curves corresponding to a step input at the columns inlet are displayed. *Table 5.3-29* lists the synthetic data sets used to calculate the curves. Equation (5.3.21) was solved forthe calculation of the curves by direct modeling using CXTFIT – a program, which is part of STANMOD (Studio of Analytical Models, SIMUNEK et al., 1999). STANMOD is freely available on the web (www.pc-progress.cz/Fr_STANMOD.htm). The program CXTTFIT may be used for transport experiments with appropriate boundary conditions that allow the analytical solution of Equation (5.3.21).

Table 5.3-29: Parameter values for the calculation of the synthetic breakthrough curves in Fig. 5.3-25

Column length [l]	J_w [$l\ t^{-1}$]	**Input conc.** C_{inp} [$m\ l^{-3}$]	θ [-]	ρ [$m\ l^{-3}$]	K_d [$m^{-1}\ l^3$]
10	0.1	1	0.4	1.5	1.07

The graph on the left shows breakthrough curves for a substance subject to linear equilibrium sorption and a dispersion coefficient D_s^w of different magnitudes. Forward modeling is done using Equation (5.3.21) with Equation (5.3.6) assuming $n = 1$. The graph on the right shows a substance subject to dispersion with a constant dispersion coefficient of 0.05 and linear nonequilibrium sorption according to the model for kinetic sorption described in Section 5.3.5.1. Forward modeling is done using Equation (5.3.21) with Equations (5.3.13) and (5.3.14) assuming $n = 1$ and $f = 0$. The distribution coefficient is the same as in the left side of the figure, however only rate limited sites ($f = 0$) are available for sorption with varying values for the rate coefficient α.

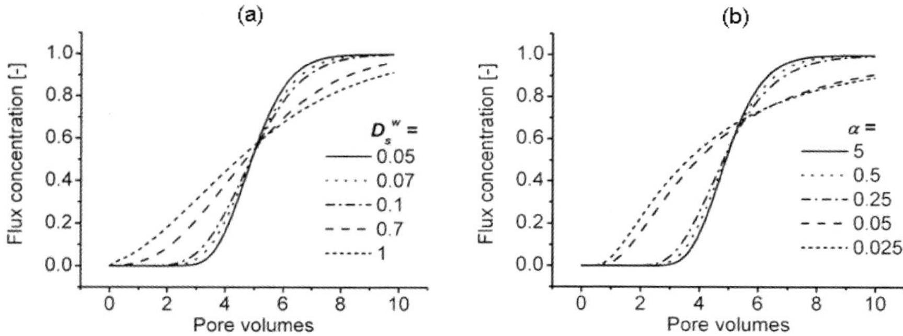

Figure 5.3-25: Breakthrough curves for a sorbing test solute with stepwise input at the column inlet, (a) Influence of an increasing dispersion coefficient D_s^w, (b) influence of a decreasing rate coefficient α

It can be seen that two completely different processes have a very similar effect on the shape of a breakthrough curve in response to stepwise input in a steady-state column experiment. An attempt to determine migration parameter values from such data (for instance by inverse modeling using CXTFIT) is doomed to fail. It is the responsibility of the experimenter to avoid these situations and to plan column experiments that yield unambiguous results. For the case described above, a possible workaround would be an independent estimate of the dispersion coefficient in an accompanying experiment using an ideal tracer.

However, if the mass transfer process is not related to sorption but to the presence of mobile and immobile pore water, (a process that is mathematically equivalent to rate limited sorption and leads to the same effect in breakthrough curves) this solution is not feasible because an ideal tracer will also be affected. This is where a more sophisticated experimental design allows a better control of the experiment.

A column experiment with the boundary conditions given in *Table 5.3-30* is presented to illustrate how an experimenter may be able to identify concurrent processes.

Table 5.3-30: Experimental parameter values for a ten-day column experiment

L [cm]	J_w [cm day^{-1}]	C_{inp} [mg L^{-1}]	θ [-]	ρ [g cm^{-3}]
10	6	10	0.43	1.5

The whole experiment had an overall duration of ten days. In the initial phase of the experiment, a test solution with a concentration of 10 mg L^{-1} was pumped through the column at a rate of 6 cm day^{-1}. After two days the flow was stopped. It was resumed seven days later on day 9 of the experiment. The measured breakthrough curve is shown in *Fig. 5.3-26*. The curve shows two interesting features that illustrate the complexity of column experiments.

Within the initial phase of the experiment (up to around three pore volumes), the breakthrough curve appears to behave as expected for an ideal tracer. The solute breakthrough occurs at approximately one pore volume and the breakthrough curve appears to reach the input concentration soon afterwards. This initial phase of the experiment suggests that the solute travels through the column without any interaction. However, when inflow was stopped for seven days the concentration in the liquid phase in the column apparently declined to approximately 6 to 7 mg L^{-1} as illustrated by the low solute concentrations in the effluent when flow was resumed on day 9. This behavior is unexpected as an ideal tracer would not be affected by flow interruption. Apparently the investigated solute is subject to a rate-limited retention within the column that is very slow. Because the flow rate in the initial phase of the experiment was rather high, this retention was impossible to detect because the transport velocity of the solute is so fast that almost none of the solute is retained within the column.

One consequence is that the termination of the experiment on day 2 would have led an experimenter to conclude that the investigated solute is not retained at all by the substrate and may be subject to easy leaching at a disposal site. This conclusion is, however, only an artifact of the relatively high flow rate in the initial phase of the column experiment. It would have been completely incorrect as shown by the remaining part of the experiment in which a retention process is identified.

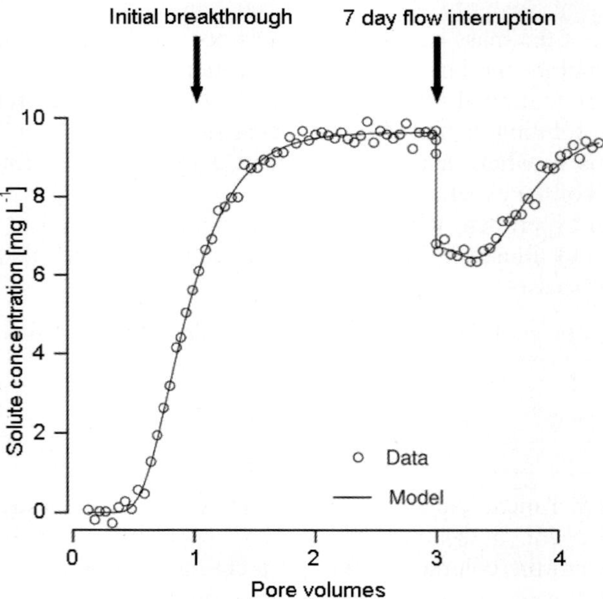

Fig. 5.3-26: Breakthrough curves of a sorbing test solute in a ten-day column experiment with a seven-day flow interruption on day 2

Assuming that the rate-limited retention process can be explained by the linear rate-limited sorption model in which only rate-limited sites are available for sorption ($f = 0$ and $n = 1$), Equation (5.3.21) together with Equations (5.3.13) and (5.3.14) was solved by inverse modeling using HYDRUS1D (SIMUNEK et al., 1998). Different from CXTFIT, the program HYDRUS1D does the calculations for a numerical model that allows modeling of experiments with arbitrary boundary conditions (such as flow interruption) that exclude analytical solutions. The estimated parameters are the sorption coefficient, the mass transfer coefficient and the dispersion coefficient. HYDRUS1D is also freely available on the web (www.pc-progress.cz/Pg_Hydrus_1D.htm). *Table 5.3-31* shows the fitted parameter values.

Table 5.3-31: Results of fitting a numerical model to the observed breakthrough data of *Fig. 5.3-26*

K_f [$L^3\ kg^{-1}$]	α [day^{-1}]	λ [cm]
3	0.005	1

Fitting these parameters simultaneously and obtaining unique estimates is possible because the concurrent influence of the two processes in the flow phase of the experiment is stopped during interruption of the flow. In this phase of the experiment, only the retention process is active while dispersion has stopped – the breakthrough curve in this part of the experiment is shaped solely by the retention process. However, even this experimental technique has shortcomings, as we are unable to identify the rate-limited retention process. As stated before the physical nonequilibrium model of mobile/immobile water is mathematically equivalent to certain forms of the chemical nonequilibrum model describing rate-limited sorption. Concurrent batch experiments may clarify this, as structures responsible for mobile/immobile regions are destroyed in such experiments. Additionally, as only a single concentration level was investigated it is impossible to detect sorption nonlinearity.

Tips for quality assurance of batch and column tests are given in VOIGT & WIPPERMANN (1998).

Example

Contamination by heavy metals was detected in an industrial area. Migration tests were carried out to determine parameter values relevant for assessing contaminant spreading. The following processes in uncontaminated sediment influencing the retardation behavior of the metal ions of interest (arsenic, zinc, copper, lead, and aluminum) had to be considered for planning column tests:

- buffering of hydrogen ions by sediment constituents (mainly by carbonates);
- exchange of ions (calcium, magnesium, sodium and potassium) between the sorption layer and pore water of the sediment and the intruding ions; this includes competitive sorption of the intruding ions among each other;
- precipitation and solution that are controlled mainly by pH, which in turn depends on the two processes described above. If precipitation takes place, the precipitates cannot be completely redissolved within a foreseeable period of time due to aging processes (e.g., condensation) and changes in hydraulic properties (accessibility).

Classical batch tests are carried out to determine the relationship between pH and added water volume (cumulative) as shown in *Fig. 5.3-27*. Such tests were carried out with groundwater from the contaminated area.

With these tests it was possible to estimate that column tests with material in the original liners from drilling (length 100 cm, diameter 10 cm) would have taken about two years to exhaust the sediment's buffer capacity. Therefore, pooled samples were used for the column test shown in *Fig. 5.3-28*. The parameter values for the column test are given in *Table 5.3-32*. To prevent washing out of the fine-grain fraction, the Darcy velocity must be chosen for a hydraulic gradient of I = 1, i.e. a Darcy velocity equal to the hydraulic conductivity (k_f). The flow rate (mL h^{-1}) is set via the pumping rate.

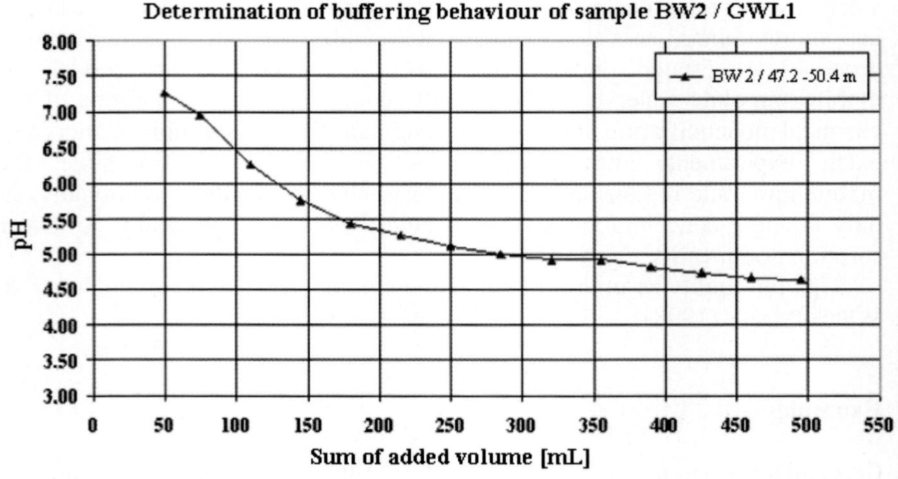

Fig. 5.3-27: pH as a function of added groundwater in a cumulative batch test

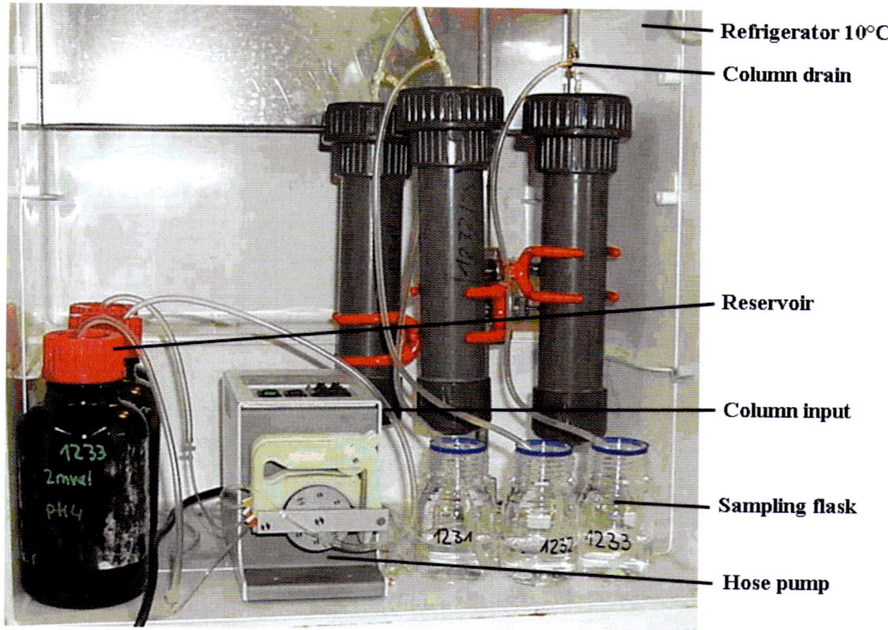

Fig. 5.3-28: Column test equipment (in the open cabinet)

Table 5.3-32: Parameter values of a selected column test

Parameter	Units	Value
sample diameter	cm	4.6
sample length	cm	23
sample volume	cm^3	382
sample mass	g	635
density	g cm^3	1.66
k_f	m s^{-1}	3.47 10^{-7}
porosity*	-	0.37
pore space*	mL	141
Darcy velocity	cm d^{-1}	3
flow rate	mL h^{-1}	2
time for exchanging	d	2.9
pore volume once	h	70.5

* determined by a column test

Figure 5.3-29 shows the pH of the percolate during the test as a function of the exchanged pore volume. The measured breakthrough curves are plotted in *Fig. 5.3-30* as the ratio of the concentration in the percolate to that in the column input (normalized concentration) versus the exchanged pore volume. Normalized concentrations above 1.0 (percolate concentration is higher than input concentration), for example for zinc, are typical of systems in which strongly adsorbing substances displace those more weakly adsorbed. *Figure 5.3-30* shows the different sorption and mobilization behavior of the heavy metals tested.

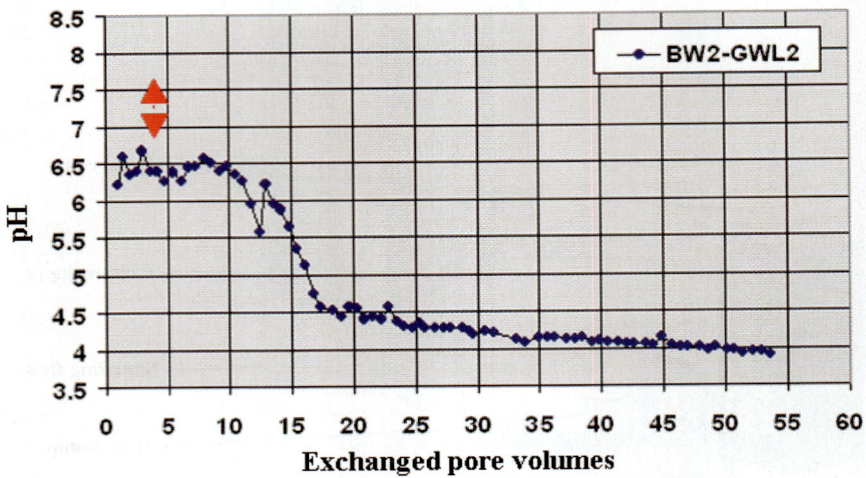

Fig. 5.3-29: Column test with groundwater from observation well BW2-GWL2: pH in the percolate as a function of exchanged pore volume

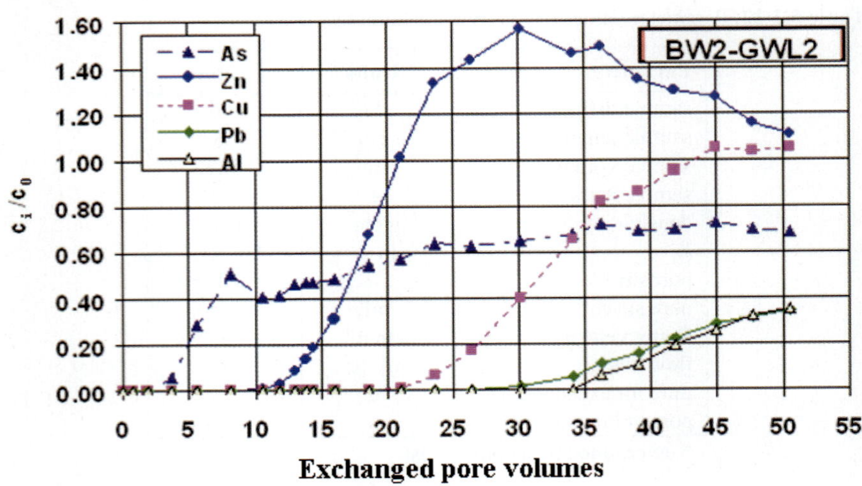

Fig. 5.3-30: Column test with groundwater from observation well BW2-GWL2: breakthrough curves

5.3.5.6 Clays and Clay Minerals

REINER DOHRMANN

The most abundant components of the continental crust are feldspars, and the most frequent mineral found in sediments is quartz. However, clay minerals are the most abundant components of fine-grained near-surface rocks. The difference in behavior of the various types of clays is determined by the fascinating properties of the clay minerals, mainly phyllosilicates and iron oxides. Clays are studied in a wide interdisciplinary field of scientific research for applications in the petroleum, chemical, steel and food industries, farming, civil engineering, environmental protection, geotechnics, ceramics, for radioactive waste repositories, and site investigations. Clays and clay minerals play an important role in everyday life – from sanitary ceramics, paper, and paints to wine and cosmetics.

The field of clays and clay minerals requires the work of mineralogists, geologists, mechanical and civil engineers, chemists, physicists, and soil scientists to be fully understood. A single discipline is not capable of describing all aspects of the fine-grained flaky minerals. Interdisciplinary research is needed for a sufficient understanding and a variety of tools for characterization of clay minerals and clays are required. Clay minerals have their fine-grained nature in common, but they have completely different physicochemical properties, such as swelling and sorption, and this is due to their crystal chemical properties. Their micro-scale properties influence their macroscopic behavior. This is illustrated in *Fig. 5.3-31* by a comparison of the scales of investigation.

The clays and clay minerals community is organized in national and international societies with the aipea (Association internationale pour l'etude des argyles, International Association for the Study of Clays) as umbrella organization. In 1990, within the OECD, the Nuclear Energy Agency (NEA) established a Working Group known informally as the "Clay Club". The Clay Club examines the various argillaceous rocks that are being considered for the deep underground disposal of radioactive waste, ranging from soft clays to indurated shale. Intensive research is being conducted within the framework of the European Union program for research and technological development to understand the processes in clays that are relevant to their function as barriers.

The most important earth science journals are Clays and Clay Minerals, Clay Minerals, and Applied Clay Science. However, many articles are published in other scientific journals as well. Valuable textbooks focus on methods such as XRD (MOORE & REYNOLDS, 1997) and IR spectroscopy (FARMER, 1974) or give an overview of methods designed for clay-rich materials such as soils (AMONETTE & ZELAZNY, 1994). Other textbooks summarize the current knowledge on clays (MEUNIER, 2005), on selected clay minerals such as illite (MEUNIER & VELDE, 2004), on the impact of properties such as microstructure on engineering performance (PUSCH & YONG, 2006), or on the whole range of structures, properties, applications, and analysis (BERGAYA et al., 2006).

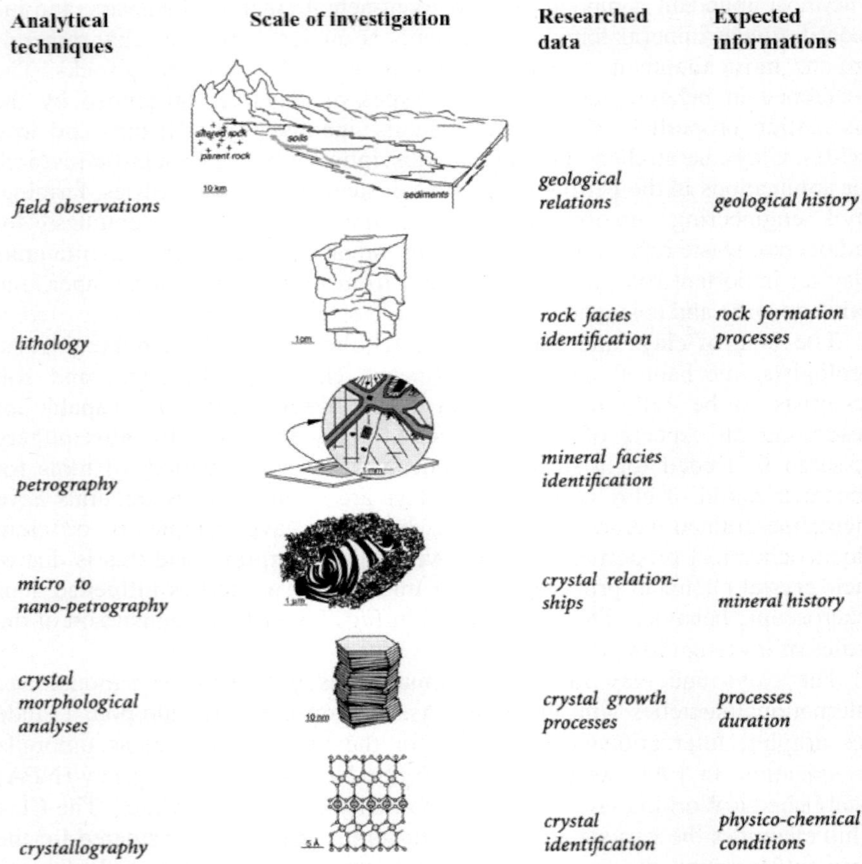

Analytical techniques
Scale of investigation
Researched data
Expected informations
altered rock
soils
parent rock
10 km
sediments
field observations
geological relations
geological history
1 cm
lithology
rock facies identification
rock formation processes
1 mm
petrography
mineral facies identification
1 µm
micro to nano-petrography
crystal relation-ships
mineral history
crystal morphological analyses
10 nm
crystal growth processes
processes duration
5 Å
crystallography
crystal identification
physico-chemical conditions

Fig. 5.3-31: Sketch showing the relationships between the different observation scales in geological studies of clay-bearing rocks, MEUNIER (2005)

Crystal structures

Clay minerals are hydrous aluminum silicates of < 2 μm size (equivalent sphere diameter) with commonly platy morphology and perfect cleavage. They show considerable variation in chemical and physical properties. Much of what we understand about the structural or chemical details of these minerals has been extrapolated from XRD studies of their macroscopic counterparts (MOORE & REYNOLDS, 1997).

Clay minerals are built of fundamental structural units, tetrahedra and octahedra. Like all phyllosilicates, clay minerals are characterized by two-dimensional sheets of corner-sharing tetrahedra, predominantly SiO_4, AlO_4, and occasionally $Fe^{III}O_4$.

One triangular face of each tetrahedron is parallel to the phyllosilicate sheet and shares three of its oxygen atoms with the three neighboring tetrahedra. The fourth vertex is not shared with another tetrahedron and all of the tetrahedra "point" in the same direction. The tetrahedra form a quasi infinitely extending plane.

In clays, the tetrahedral sheets are always bonded to octahedral sheets containing small cations, such as aluminum or magnesium, each coordinated with six oxygen atoms. The octahedral sheets can be thought of as two planes of closest-packed oxygen ions with cations occupying the resulting octahedral sites between the two planes. An octahedron is obtained when the six oxygen atoms around the cation are connected. Sharing of neighboring oxygen ions forms a sheet of edge-linked octahedra, extending quasi-infinitely in two dimensions.

The oxygen atoms in the tetrahedral sheet that are not shared with the other tetrahedral are shared with the octahedra of the octahedral sheet. In the center of each ring of six tetrahedral is an oxygen atom that is not shared with the tetrahedral sheet. This oxygen atom is bonded to a hydrogen atom to form an OH group in the clay structure.

Clay minerals are classified on the basis of the proportion of sites occupied by metal ions in the octahedral sheet: In trioctahedral minerals, 3 of 3 octahedral sites are occupied by Me^{2+} cations, the cation/anion ratio is 1:2. In dioctahedral minerals, 2 of 3 sites are occupied by Me^{3+}, the cation/anion ratio is 1:3. If the vacant site in dioctahedral minerals lies on a pseudomirror plane, the structure is called "transvacant".

Clays are categorized according to the ratio of tetrahedral and octahedral sheets in the layers (*Fig. 5.3-32*). A clay mineral of the 1:1 layer type consists of layers with only one tetrahedral and one octahedral sheet (called a T-O layer type); a clay mineral of the 2:1 layer type consists of layers with two tetrahedral sheets (called a T-O-T clay mineral) with the unshared vertexes of each sheet pointing towards each other, forming each side of the octahedral sheet. Clay minerals of the 1:1 layer type have many more hydroxyl groups.

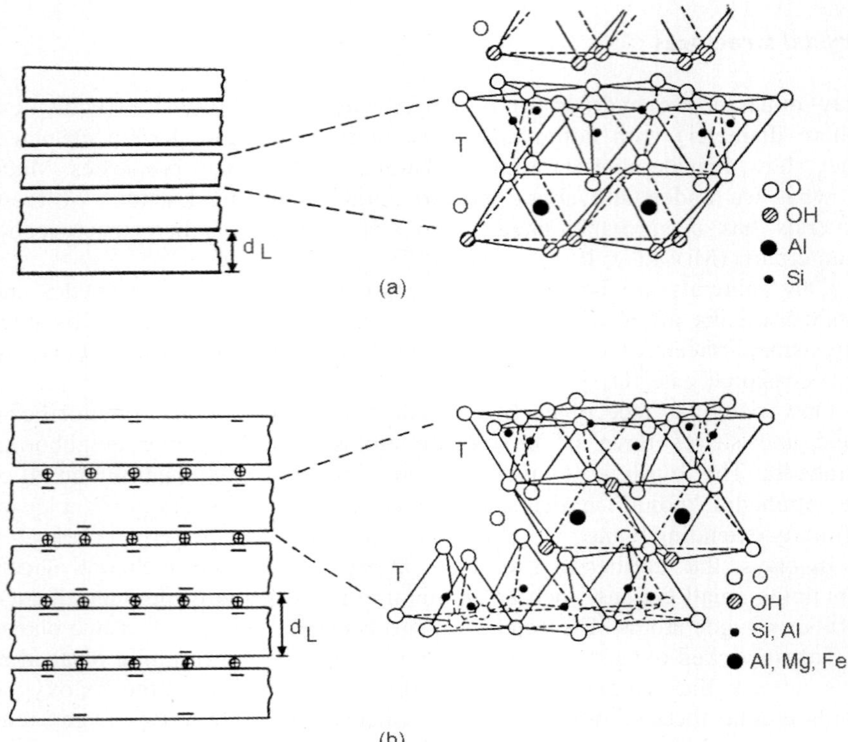

Fig. 5.3-32: Structures of clay minerals, (a) 1:1 clay mineral, (b) 2:1 clay mineral, d_L = distance between the layers, T = tetrahedral layer, O = octahedral layer, LAGALY & KÖSTER (1993)

An important parameter is the layer charge density. In contrast to other minerals the structures of clay minerals can have a permanent net negative charge, which is balanced by cations in outer sphere or inner sphere complexes. If the layer charge density is low, like in smectites, hydrated cations can be intercalated between 2:1 layer silicates. These cations can be readily replaced by competing cations. This can cause swelling or shrinkage. If the layer charge density increases, then the forces between the layers will not allow cation exchange and the cations are no longer hydrated. This is the case for illites, in which cations can be exchanged only on the outer surface of the crystals. In natural clays, Na^+, K^+, Ca^{2+} and Mg^{2+} usually form the exchange population if the pH of the surrounding solution is near neutral.

The 2:1 and 1:1 layers can be stacked on each other to form crystals. It is also possible to stack layers of different types. If the stacking of layers of different types is regular, superstructures are formed. The minerals are called mixed-layer minerals. If ordering of the stacking is poor, irregular mixed-layer minerals are present. The interlayer regions of these minerals often are swellable and, therefore, of smectitic character.

Classification (following Bailey, 1980)

The phyllosilicates are divided into groups (*Table 5.3-33*), each containing dioctahedral and trioctahedral subgroups. Each subgroup is divided into mineral species.

Table 5.3-33: Classification scheme for phyllosilicates related to clay minerals

Layer type	Group (x = charge per formula unit)	Subgroup	Species*
1:1	kaolinite-serpentine	kaolinite	kaolinite, dickite, halloysite
	x ~ 0	serpentine	chrysotile, lizardite, amesite
2:1	pyrophyllite-talc	pyrophyllite	pyrophyllite
	x ~ 0	talc	talc
	smectite	dioctahedral smectite	montmorillonite, beidellite
	x ~ 0.2-0.6	trioctahedral smectite	saponite, hectorite, sauconite
	vermiculite	dioctahedral vermiculite	dioctahedral vermiculite
	x ~ 0.6-0.9	trioctahedral vermiculite	trioctahedral vermiculite
	mica**	dioctahedral mica	muscovite, paragonite
	x ~ 1	trioctahedral mica	phlogopite, biotite, lepidolite
	brittle mica	dioctahedral brittle mica	margarite
	x ~ 2	trioctahedral brittle mica	clintonite, anandite
	chlorite	dioctahedral chlorite	donbassite
	x variable	di- and trioctahedral chlorite	cookeite, sudoite
		trioctahedral chlorite	clinochlore, chamosite, nimite

* Only a few examples are given.
** The status of illite (or hydromica), sericite, etc. must be left open at present, because it is not clear whether or at what level they would enter the table: many materials so designated may be interstratified.

5.3.5.7 Cation Exchange Capacity

REINER DOHRMANN

The cation exchange capacity (CEC) is an important property of clay minerals. "The CEC is defined as a measure of the ability of a clay or a soil to adsorb cations in such a form that they can be readily desorbed by competing ions" (BACHE, 1976). Cation exchange is reversible and the CEC value is a sum parameter.

The CEC results from an excess negative charge of the alumosilicate lattice caused by isomorphic substitution of either Al(VI) (= six-fold coordinated Al) by Mg(VI) within the octahedral sheet or Si(IV) by Al(IV) within the tetrahedral sheet. Both substitutions cause an excess negative charge which is balanced by the exchangeable cations between the T-O-T layers.

The common exchangeable cations in clay minerals are Ca^{2+}, Mg^{2+}, Na^{+}, and K^{+}. Illite/smectite mixed-layer minerals which formed in the vicinity of petroleum reservoirs can in some instances contain NH^{4+}. Clay minerals in soils might contain exchangeable Al^{3+} or Fe^{3+}, as they are products of weathering.

The excess negative charge caused by isomorphic substitution is a permanent charge since it is independent of the ambient pH. Depending on the ambient pH, smectites have a variable charge, which results from the differing degrees of surface protonation.

Tetrahedral charge, in contrast to octahedral charge, is located near the surface of the clay mineral, which results in a relatively localized charge. Therefore, a localized permanent charge can be distinguished from the more delocalized charge in the octahedral sheet (*Fig. 5.3-33*). CEC values represent the total negative charge.

It is difficult to correctly determine the CEC and the exchangeable cations of common clays and soils for a number of reasons. With the exception of some special conditions, such as an acidic pH < 5 or in saline and alkaline (electrolyte-rich) clays, the correct value of the CEC is more or less independent of pressure, temperature, solution/solid ratio, electrolyte composition, and concentration (BACHE, 1976).

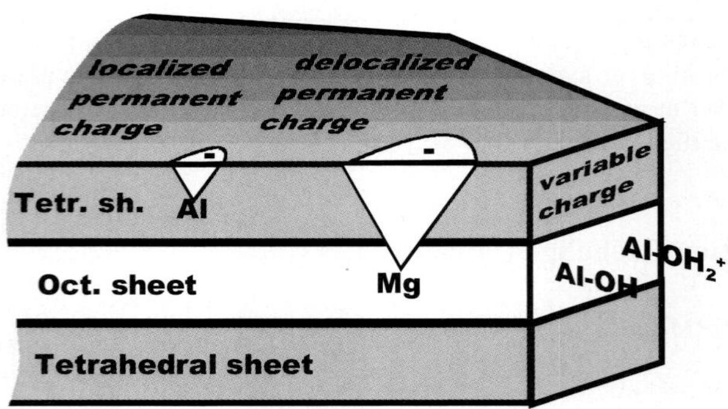

Fig. 5.3-33: Types of charges of swellable clay minerals

The cation exchange behavior of highly charged smectites, vermiculites, and zeolites is different from that of a common clay mineral. Highly charged smectites and vermiculites have such a high layer charge density that the interlayer cations cannot be readily desorbed quantitatively. In case of vermiculites, the rate of cation transport between the 2:1 layers is reduced by the high layer charge (WALKER, 1959; GRAF VON REICHENBACH, 1966; MALCOLM & KENNEDY, 1969). Complete cation exchange, therefore, is possible only at elevated temperatures or in long-term experiments. Zeolites, on the other hand, show a pronounced selectivity for certain cations; in this case, the overall CEC cannot be determined for unspecified cations as required by the definition.

CEC data is frequently used for the characterization and quantification of adsorbed species in clays and soils. For many scientific investigations it is important to know which exchangeable cations are present on the exchanger surfaces. In soil science, exchangeable nutrients are essential for plant growth, and in geotechnical engineering the stability of a building foundation may depend on the type of exchangeable cations in the soil. In the host rocks of radioactive waste disposal sites, the exchangeable cations and the cations in the pore water have to be determined exactly in order to characterize fluid flow pathways through the barrier (NAGRA, 2002).

CEC and exchangeable cations of soils and clays have been determined since the early work of WAY (1852). Numerous publications about various methods discuss the range of validity of the results of CEC procedures for the wide variety of natural materials. A number of methods for determining the CEC of a clay mineral have been developed during the last several decades: ammonium acetate (LEWIS, 1949), triethanolamine-buffered barium chloride (MEHLICH, 1948; BASCOMB, 1964), radioactive tracers (^{133}Ba, BACHE, 1970; ^{85}Sr, FRANCIS & GRIGAL, 1971), nephelometry (ADAMS & EVANS, 1979), and several other methods. However, common CEC methods like ammonium acetate (LEWIS, 1949) or barium chloride (MEHLICH, 1948) are very time consuming and for natural materials the results are often poor (DOHRMANN, 2006a). A disadvantage of these procedures is that the resulting CEC and exchangeable cation values are often inaccurate to a large extent (KICK, 1956; DELLER, 1981). Problems usually do not occur for pure clay fractions (< 2 μm equivalent diameter), but there are problems if the sample contains minerals unfavorable for the method, such as carbonates. If CEC and exchangeable cation values are determined for calcareous clayey materials, these minerals are partially dissolved during the exchange experiments. This is caused by a decrease in the stability of the minerals when they interact with electrolyte-rich solutions. DOHRMANN (2006c) reports on a new method in which these problems have been solved.

Other approaches use the higher selectivity of metal-organic complexes, such as the one-step silver thiourea method (CHHABRA et al., 1975), Cu(II) ethylenediamine (BERGAYA & VAYER, 1997), or Cu(II) triethylenetetramine

(MEIER & KAHR, 1999). Less time is required for analysis than for the common methods.

In general, problems in CEC determination are caused by interaction of the components of soils and clays with the exchange solutions used. The main sources of error are soluble mineral phases (usually Ca minerals), and in the faster modern methods, hydrophobic interaction can cause unrealistic CEC values (DOHRMANN, 2006b).

The interaction of cationic organic compounds with clay minerals is generally well understood. It is the subject of several reviews (MORTLAND, 1970; THENG, 1974; JOHNSTON, 1996). In clay chemistry, excess adsorption of cationic organic compounds on clay minerals is a well known phenomenon. When organic cations react with kaolinite, adsorption exceeds the CEC due to van der Waals forces (GRIM, 1962). DÉKÀNY et al. (1978) report on the adsorption behavior of organophilic montmorillonite and kaolinite in alcohol-benzene mixtures. They find that excess adsorption is dependent on the nature of the exchange population of the clay minerals. NARINE & GUY (1981) report that the extent of adsorption of six organic compounds is strongly dependent on the electrolyte concentration of the solution.

For scientific and industrial purposes the determination of the smectite content is of considerable interest. The methylene-blue method (MB method) is commonly used to estimate smectite content. This molecule is adsorbed at cation exchange sites. This method has to be used carefully because dimers and trimers are formed in an MB solution with a concentration higher than 10^{-5} mole L^{-1} (BREEN & ROCK, 1991). BUJDÁK & KOMADEL (1997) observed dimer formation even below this concentration (at 2.5×10^{-6} mole L^{-1}). Adsorption of dimers or trimers on one negative charge of the smectite makes it appear that two or three charges are present instead of the one which is actually present. This causes the CEC to be overestimated. KAUFHOLD & DOHRMANN (2003) summarize problems of this method, and discuss the potential of the Cu triethylenetetramine method for quantification of smectite in bentonites.

Common ion exchange analyses are done with a high salt concentration (e.g., ammonium acetate, barium chloride). Samples have to be treated several times because otherwise exchange competition allows only partial saturation of the exchanger surfaces by the "index cation" used. MCBRIDE (1979) describes these procedures in detail. It is possible to conduct column or batch tests with the samples. For acidic samples, the effective CEC (= CEC_{eff}) is usually determined using unbuffered exchange solutions. For nonacidic samples, buffered solutions are used to obtain the potential CEC (= CEC_{pot}) of the sample. If an acidic sample is treated with a buffered CEC exchange solution, the resulting CEC is usually higher than it should be because due to the higher pH of the solution, variable charges can be generated, giving a potentially higher CEC.

The interaction of silver thiourea complexes (AgTU) with montmorillonite clay was described by CREMERS & PLEYSIER (1973a, b). This observation was then used to develop a relatively fast one-step determination of CEC and exchangeable cations. This is possible because of the much higher selectivity of this cation complex to clays than that of naturally occurring cations. The selectivity is explained mainly by the higher polarizability of the silver complex than that of the hydrated cations. As shown for smectites from three different locations, when the silver-thiourea complex is intercalated between the 2:1 layers of smectite, the silver: thiourea ratio in the complex equilibrates to a 1:2 ratio (CREMERS & PLEYSIER, 1973b). Therefore, the adsorbed complex can be described as $[Ag(TU)_2]^+$. MEIER & KAHR (1999) and BERGAYA & VAYER (1997) use cation complexes such as Cu(II) triethylenetetramine and Cu(II) ethylenediamine. CEC data obtained with these last two index cations correspond well with ammonium acetate CEC values of standard clays. Polymer adsorption on clay minerals commonly appears to be irreversible (LAGALY, 1985), and consequently, polymers are not used as index cations in CEC methods. The adsorption of silver thiourea complex on smectite is reversible and, therefore, the method can be used for CEC analysis (PLEYSIER & CREMERS, 1975).

DOHRMANN (2006d) has developed a model that allows the recognition of flawed CEC and exchangeable cation data. CEC is determined using two different masses of clay and the results are plotted in a graph. As described above, the CEC of common clays should be independent of the mass used for the analysis. The correctness can be evaluated by a visual checking of the location of the data in the graph (ideally the values should lie on the $y = x$ line). This model is simple to use.

CEC methods

New silver-thiourea method ($AgTU_{mod}$) for noncalcareous clays

The silver-thiourea method $AgTU_{mod}$ (DOHRMANN, 2006c) should be used for noncalcareous clay materials.

Thiourea (15.2 g) is dissolved in deionized water (about 1400 mL) with a few minutes ultrasonic treatment. Subsequently, 3.397 g $AgNO_3$ (dissolved in about 300 mL deionized water) are added very slowly (2 minutes) while stirring the thiourea solution. After that, 200 mL of 1 M ammonium acetate (no need to do this slowly) are added and the resulting solution is brought to 2000 mL with deionized water. This AgTU exchange solution has to be freshly prepared and should not be stored for longer than 48 hours.

Generally, 50 mL of AgTU exchange solution are added to a clayey sample (< 2 mm) and the suspension is shaken (batch) for two hours in an 80 - 100 mL centrifuge tube. The ratio of solution to solid needed for an optimum CEC value depends on the expected CEC. For samples with a CEC up to 30 meq/100 g, a 0.5 - 1 g sample is a good choice, while 0.2 - 0.3 g is feasible for bentonites or other samples rich in pure smectites with a CEC up to 130 meq/100 g. For kaolin and other low-CEC materials, up to 10 g can be used. 25 - 30 % of the initial silver content should be present in solution after the completed ion exchange to be sure that all exchange sites are occupied by $[Ag(TU)_2]^+$ cations (CHHABRA et al., 1975). Two different solution/solid ratios are recommended (e.g., 0.2 g and 0.3 g for bentonites) for checking CEC values using the *Carbonate and Sulphate Field Model* (CSF model; DOHRMANN, 2006d). Differences in the CEC results usually indicate interactions of the exchange solution with the mineral phases or other sources of error. A blank is run each time as a control of the initial silver content. Required precision of the solid mass is $\pm$ 0.0005 g and for the amount of solution $\pm$ 0.1 mL. The suspension is shaken in an end-over-end shaker for usually 2 h or until the exchange reaction is finished.

After the exchange is completed, the clay suspension is centrifuged and the clear supernatant is decanted through a wet cellulose acetate filter paper or a fiber glass filter into a 250 mL volumetric flask to which 25 mL of 0.5 M HNO_3 has been added for chemical stabilization. The filter has to be washed out immediately after contact with the solution to prevent reaction with AgTU. Fifty mL of deionized water are added to the sample and after shaking the reaction run for another 10 minutes. The suspension is then centrifuged and filtered into the first supernatant. This is repeated with 25 mL deionized water but shaking need be done only briefly. Fill the flask to the volumetric mark and proceed with chemical analysis.

Na^+, K^+, Ca^{2+}, Mg^{2+} (exchangeable cations) and Ag^+ (for the CEC value) are determined by elemental analysis (atomic absorption or induced coupled plasma). CEC is determined by the difference between the initial and the remaining silver content.

Calcite-saturated silver-thiourea exchange solution, the AgTUcalcite method for calcareous clays

The $AgTU_{calcite}$ method should be used for calcareous clays. When a $AgTU_{calcite}$ exchange solution is used, calcite in the sample can no longer be dissolved but exchangeable Ca^{2+} is desorbed quantitatively. The case for dolomite is similar, because dissolution of dolomite is minimized. However, reasonable exchangeable Ca^{2+} values for samples containing gypsum cannot be determined because gypsum is soluble in the exchange solution.

The AgTU exchange solution contains 0.01 M Ag^+, 0.1 M thiourea, and 0.1 M ammonium acetate. Thiourea (15.2 g) is dissolved in 1400 mL of deionized water with a few minutes ultrasonic treatment. Then about 300 mL of aqueous $AgNO_3$ containing 3.397 g $AgNO_3$ is added slowly (2 minutes) while stirring the thiourea solution. Finally, 200 mL of 1 M ammonium acetate solution are added (no need to do this slowly) and the resulting solution is brought to 2000 mL with deionized water in a volumetric flask. The solution is transferred into a 2000 mL beaker, 1 g of fine-grained calcite (e.g., Merck) is added, and the solution is stirred vigorously for 2 h. After a few hours of settling (e.g., overnight) the supernatant can be filtered using a quick filtering paper to remove any suspended calcite grains. The filter may become brownish due to traces of AgS_2. This $AgTU_{calcite}$ solution should be freshly prepared each time and should not be stored longer than 48 - 72 hours to prevent precipitation of Ag_2S.

The sample (0.2 - 2 ± 0.0005 g) is placed in a 80 - 100 mL centrifuge tube, 50 mL (± 0.1 mL) of the $AgTU_{calcite}$ solution are added and the sample is equilibrated for 2 h in a shaking machine. The mass of the samples depends on the nature and quantity of the clay minerals as described for the $AgTU_{mod}$ method. After the exchange procedure is completed, the clay suspension is centrifuged and 100 µL are transferred into a 10 mL volumetric flask containing 1 mL of 0.5 M HNO_3 (for chemical stabilization). The flask is made up to volume by adding deionized water and the concentrations of Ca^{2+} and the other exchangeable cations are determined by elemental analysis (atomic absorption or induced coupled plasma). This is done to make sure that no further calcite dissolution (during the following washing steps) affects the determination of the correct exchangeable Ca^{2+} value. Exchangeable Ca^{2+} is determined from the difference between the measured Ca^{2+} content and the Ca^{2+} concentration of the blank. To proceed with the CEC index cation determination, 25 mL of 0.5 M HNO_3 are added to a 250 mL volumetric flask for chemical stabilization. The rest (99.8 %) of the supernatant from the centrifuged suspension is added into the 250 mL volumetric flask using a wet cellulose acetate filter paper. It is necessary to immediately wash the filter after contact with the solution to prevent reaction with $Ag(TU)_n^+$ cations. 50 mL of deionized water are added to the sample and shaken for another 10 minutes. If the slurry becomes very dark then this step should be done without the 10 minutes in the shaking machine. In this case there might be a loss of Ag due to precipitation of Ag_2S, which can be avoided by shortening the reaction time. The suspension is centrifuged and the filtrates are combined. This is repeated with 25 mL of deionized water; the slurry is dispersed quickly and centrifuged immediately to prevent a pH increase above pH 8. The flask is filled and the solution is analyzed for silver and the CEC is calculated from the difference between the initial and the remaining silver content after the experiment. The exchangeable Ca^{2+} cannot be determined from the measured Ca^{2+} content of the collected filtrates because additional calcite may have been dissolved during the washing steps (see below).

Cu-triene Method

A fast and easy method for determining CEC values with the blue-colored triethylenetetramine Cu(II) complex (Cu-triene) was presented by MEIER & KAHR (1999). It is suitable for common clays and especially if only the CEC is required. A UV-VIS spectrometer is sufficient for the analysis of the complex.

The Cu-triene solution is produced by adding 1.596 g $CuSO_4$ (dried) to 100 mL deionized water, then 1.463 g triethylenetetramine (triene), and finally water is added up to 1.0 L. The solution can be stored for months. For the CEC determination, samples (e.g., in the case of bentonites: 0.080 g and 0.120 g) are weighed in sealable 80 - 100 mL centrifuge tubes. Then 10 mL of the 0.01 M Cu-triene solution and 50 mL water are added (precision ± 0.1 mL). Before centrifugation, the suspension is kept for 2 h in an end-over-end shaker. The supernatant is analyzed with respect to VIS absorption at 578 nm against water. The VIS absorption of the pure Cu-triene solution diluted 1:5 (= 0.002 M) is also recorded. After reaction, the solution should contain at least 25 % of the initial complex. Otherwise the selectivity decreases and the experiment has to be repeated using a smaller mass of clay.

The CEC is calculated from the difference between the absorption of the 0.002 M Cu-triene solution and the absorption of the supernatant of the sample. The concentration is calculated from this and related to the dry mass of the sample.

Triethanolamine-buffered $BaCl_2$ solution for noncalcareous and non-smectitic clays

This $BaCl_2$ method is suitable for many clays but not for calcareous, gypsiferous, and smectitic clays (DOHRMANN, 1997). First, there are interactions with calcite and gypsum which result in incorrect exchangeable calcium values. Secondly, barium carbonate will precipitate during contact with calcite and redissolve during the re-exchange procedure, which can cause incorrect elevated CEC values. Thirdly, the triethanolamine can be protonated and exchanged into the interlayer spaces of swellable clay minerals (smectites), which lowers the CEC because Ba^{2+} cations cannot replace them. Fourthly, gypsum may not be dissolved completely during the first five steps if the crystals are large or the content of gypsum in the sample is high. Then during the re-exchange all desorbed Ba^{2+} cations could be trapped as barite and CEC can be lowered to zero. Despite of the problems discussed, the method is widely used and accepted.

This method is a batch procedure with triethanolamine-buffered 0.1 M $BaCl_2$ followed by re-exchange with aqueous 0.1 M $MgCl_2$. All analyses should be performed in duplicate with two different solution/solid ratios. This can be done by varying the solid content (e.g., 0.5 g and 2.0 g, ± 0.0005 g) with a fixed amount of solution. For each experiment, a 250 mL volumetric flask should be used containing 25 mL 0.5 M HNO_3 for chemical stabilization.

Exchange solutions: Exchange solution (A) is prepared by dissolving 50 g $BaCl_2$ dihydrate in about 1 L of deionized water in a 2 L volumetric flask; 45 mL triethanolamine is added and after that the pH is adjusted to 8.1 using dilute HCl. The flask is then filled to the 2000 mL mark. For solution (B), 25 g of $BaCl_2$ dihydrate is dissolved in deionized water in a 1 L volumetric flask and then filled to the 1000 mL mark. Solution (C) is prepared by dissolving 20 g of $MgCl_2$ hexahydrate in deionized water in a 1 L volumetric flask, which is then brought to the 1000 mL mark.

Exchange of the naturally adsorbed cations by Ba^{2+}: The naturally adsorbed cations are desorbed in five steps with different shaking times and exchange solutions. The exchange solutions are added in sequence to an oven-dried clayey sample (< 2 mm, dried at a maximum of 60 °C) that has been weighed in a sealable 80 – 100 mL centrifuge tube. Each suspension is shaken in an end-over-end shaker and then centrifuged for five minutes. The clear supernatant is decanted through a filter into a 250 mL volumetric flask.

Step 1: 50 mL of solution (A) added to the sample, 60 minutes shaking time.

Step 2: 50 mL of solution (A) added to the residue from step 1, 60 minutes shaking time.

Step 3: 30 mL of buffer-free solution (B) added to the residue from step 2, 30 minutes shaking time.

Step 4: 30 mL of deionized water added to the residue from step 3, 10 minutes shaking time.

Step 5: 30 mL of deionized water added to the residue from step 4, 10 minutes shaking time.

During the second washing step (step 5) it cannot be avoided that sometimes the supernatant is cloudy, resulting in loss of a small portion of the sample, resulting in a slight decrease in the CEC value obtained. Repeated centrifugation helps in this case. However, all filtrates are placed together in the volumetric flask, which is then filled to exactly 250 mL with deionized water. The exchangeable cations Na^+, K^+, Mg^{2+} and Ca^{2+} in this solution are determined by elemental analysis (atomic absorption or induced coupled plasma).

Re-exchange of adsorbed Ba^{2+} with Mg^{2+}: The adsorbed Ba^{2+} is re-exchanged by saturating the clay five times with solution C. At each step the suspension is shaken with 0.1 M $MgCl_2$ for 20 minutes. The suspension is then centrifuged and the clear supernatant is decanted into a 250 mL volumetric flask, containing 25 mL of 0.5 M HNO_3 for stabilization. All filtrates are placed in the flask together. The resulting solution is analyzed for Ba^{2+} in order to calculate the CEC value.

Steps 6-10: 40 mL of solution (C) added to the residue of the previous step, 20 minutes shaking time.

5.3.5.8 Carbonates

STEPHAN KAUFHOLD

The most abundant carbonates are calcite ($CaCO_3$, trigonal), aragonite ($CaCO_3$, rhombical), dolomite ($CaMg(CO_3)_2$), ankerite ($CaFe(CO_3)_2$), and siderite ($FeCO_3$). In contrast to most of the other major constituents of soils and rocks, carbonates are partially soluble in water (e.g., calcite about 12 mg L^{-1}).

$CaCO_3 + H_2O \leftrightarrow Ca^{2+} + HCO_3^- + OH^-$ reaction 1

The equilibrium of this reaction, and hence the carbonate solubility, is significantly affected by CO_2 owing to the following reaction:

$H_2O + CO_2 \leftrightarrow HCO_3^- + H^+$ reaction 2

Due to the above reactions and the comparably high solubility, carbonates significantly influence the geochemical milieu. In view of these reactions it is evident that the amount of CO_2 dissolved in a water determines the solubility of any carbonate in contact with it and hence the pH of the solution. In this context it is important to note that the distilled water (degassed and deionized) commonly used in the laboratory is undersaturated with respect to CO_2. If carbonate-containing solids are placed in distilled water, the pH may rise to even higher than 10. Subsequent introduction of CO_2 (e.g., from air) decreases the pH owing to the production of HCO_3^- by reaction 2, which also shifts the equilibrium of reaction 1 to the left, decreasing OH^- concentration. In addition, the electrolyte concentration has to be considered, which also affects carbonate solubility (LIPPMANN, 1973; BUHMANN & DREYBRODT, 1987).

Carbonates occur in almost every grain-size fraction. Macroscopically visible carbonate particles can be detritus (e.g., shell fragments) and precipitated carbonate might form the mineral cement (macropore and mesopore filling).

Determination of carbonates

Like other important soil or rock-forming minerals, carbonates can be determined either by selective extraction methods or by the common bulk mineralogical methods as X-ray diffraction (XRD). Grinding prior to analysis is usually required. In this context, it is worth mentioning that carbonates are mechanically relatively unstable. Extensive mechanical forces lead to the formation of highly energetic surface sites which, in the case of carbonates, tend to release CO_2 (KÖSTER, 1979). Accordingly, careful grinding, e.g., by a hand mortar or addition of ethanol or other alcoholic solutions during grinding, is recommended.

XRD is the most important method for the mineralogical analysis of soils and rocks. The detection limit is generally between 1 and 5 wt. %. However, the XRD intensity depends on structural parameters such as mineral crystallinity: Amorphous phases (e.g., gels) cannot be detected by XRD. Short-range order minerals, such as allophane or 2-L ferrihydrite, only show very broad peaks often termed bands. The XRD detection limit of short-range order minerals is higher than 5 %. Therefore, the qualitative mineralogical analysis still is a challenge and hence the use of complementary methods as infrared (IR) spectroscopy or thermal analysis (DTA) is recommended. An example of carbonate detection by XRD and IR is given in *Fig. 5.3-34*. The sample was prepared by mixing standard materials of 100 % purity.

Carbonates readily react with acids to produce CO_2. This reaction is frequently used for the qualitative analysis of carbonates (described in detail by the German Industry Norm DIN 4022-1, 1987). However, this reaction also can be used for the quantitative analysis of carbonate (known as the *Scheibler method*, DIN 18129, 1996): The sample is dried, ground to < 60 µm, and placed in a "gasometer". After addition of dilute hydrochloric acid, either the volume of the produced CO_2 is measured or the pressure generated by it. KLOSA (1994) developed an automatic apparatus in which the gas pressure is measured and recorded by a computer, which plots the desired curves. The advantages of this technique over the common Scheibler-method are (1) 60 - 70 % less time required, and (2) the results are immediately available. The measured CO_2 volume or pressure (Scheibler method or KLOSA, 1994) is used to determine total carbonate. However, carbonates that are only slightly soluble, such as siderite, ankerite, and dolomite, might withstand HCl treatment, which in turn results in a too low carbonate content being determined. The third advantage of KLOSA's method is the possibility to interpret the pressure curve with respect to kinetic aspects. Calcite, dolomite, and siderite can be differentiated on the basis of the slope of the curve. Accordingly, reaction time can be increased in order to make sure that all carbonates have reacted. Generally, the methods based on HCl treatment provide a value for the total amount of carbonate.

According to the GDA recommendation (GDA, 1993) calcite, dolomite, and siderite can be distinguished on the basis of their different chemical stabilities. Calcite, for example, is extracted by 0.1 M ethylenediamminetetraacetic acid (EDTA) at pH 4.5 or alternatively by an acetate buffer solution (TRIBUTH & LAGALY, 1986a, b). Dolomite can be selectively dissolved by 0.1 M EDTA at pH 8.0 and siderite by 0.1 M HCl solution.

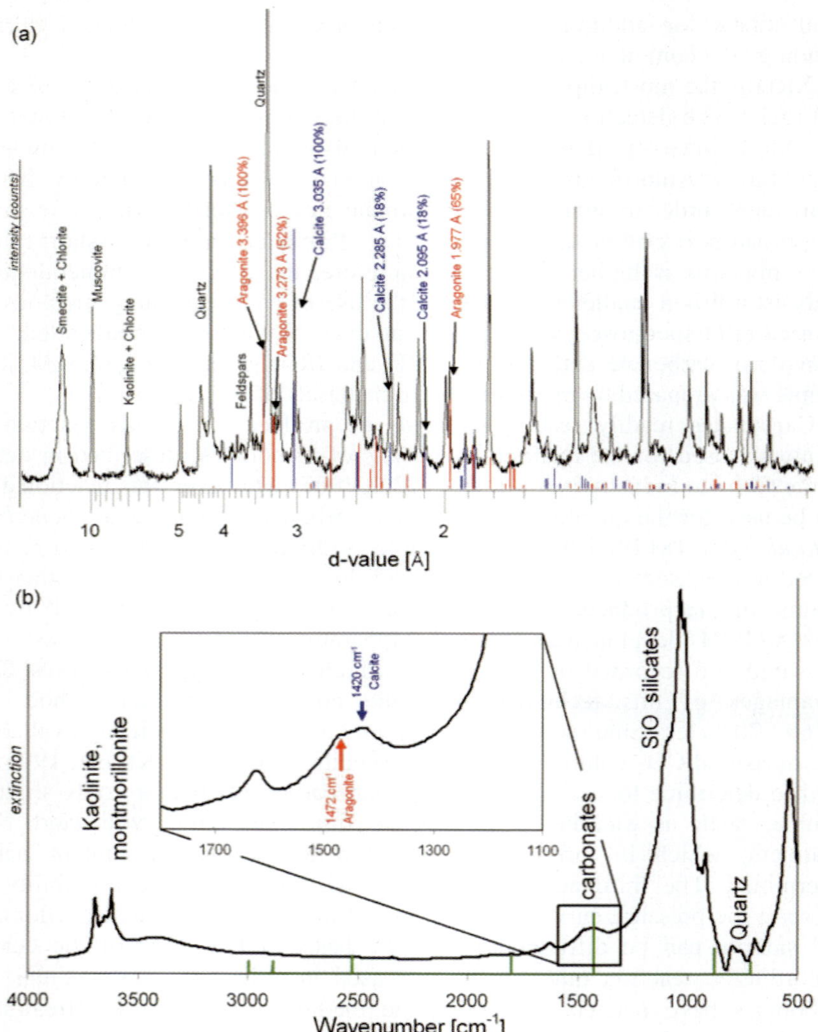

Fig. 5.3-34: Example of (a) XRD and (b) IR detection of carbonates in a synthetic mixture (swelling clay) containing calcite (5 %) and aragonite (2 %) in a mixture of smectite (ca. 34 %), quartz (ca. 30 %), kaolinite (ca. 10 %), muscovite (ca. 10 %), feldspars (ca. 6 %), chlorite (ca. 2 %), and rutile (ca. 1 %).

An alternative method for the quantitative analysis of total carbonate is coulometry: The sample is heated to about 1200 °C under nitrogen. Under these conditions all CO_2 is degassed from carbonates. Due to the absence of oxygen, organic carbon is preserved. The CO_2 is pumped through a barium perchlorate solution (pH 10.1) and witherite ($BaCO_3$) is formed. The amount of OH^- consumed by the carbonate formation is compensated for by electrolysis and the current required for the OH^- compensation is recorded. This current is proportional to the amount of CO_2 which is degassed at 1200 °C and hence to the total amount of carbonate in the sample (ISO 10694, 1995; HEINRICHS & HERMANN, 1990).

An additional method for the determination of the total carbonate is the use of a LECO analyzer to determine the amount of total inorganic carbon (LECO CS-444-Analysator, produced by Leco Instruments, a supplier of analytical and laboratory instruments). Like for the Scheibler method, information about the carbonate species (e.g., from XRD) is required for the calculation of the carbonate minerals. The calcite content of ten different Opalinus clay samples determined using the modified Scheibler method (KLOSA, 1994) and the LECO method are compared in *Fig. 5.3-35* (EU Heater Project, 2006).

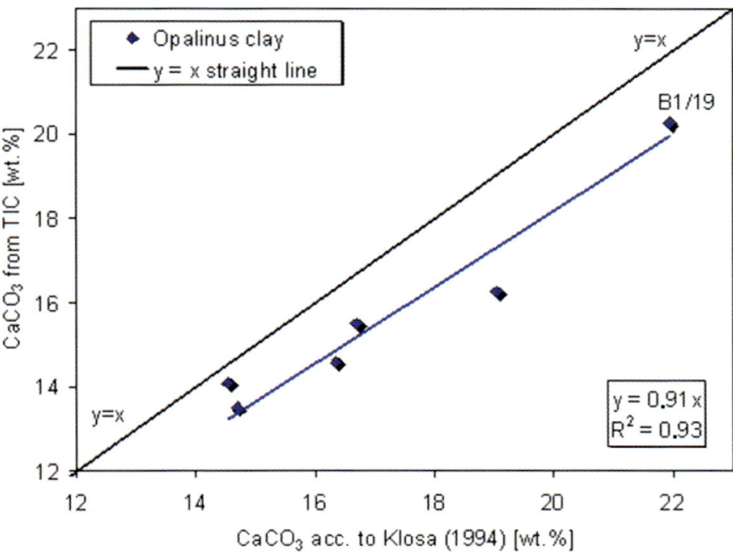

Fig. 5.3-35: Comparison of the $CaCO_3$ content obtained by two different methods, EU Heater Project (2006)

As can be seen in *Fig. 5.3-35*, the $CaCO_3$ content determined by the Scheibler method (modified by KLOSA, 1994) is ca. 10 % higher than the value derived from the total inorganic carbon content (TIC). This can be explained by the presence of sulfides (pyrite). If sulfides react with the acid they form H_2S, which contributes to the total pressure from which the $CaCO_3$ content is calculated. On the basis of sample B1/19 (which does not contain gypsum), 1.1 wt. % S is from pyrite (≈ 2 wt. %). Using an approximation following the ideal gas law, it can be seen that 1 wt. % S from sulfide leads to overestimation of $CaCO_3$ by 3 wt. %. Therefore, it can be concluded that pyrite does not dissolve completely in HCl. Otherwise, an even higher $CaCO_3$ content would have been determined by the Scheibler method.

Recommendation

Extraction methods, as well as the Scheibler and coulometric methods, are frequently used because they provide reproducible and comparably robust quantitative information about the carbonate content. However, quantitative XRD measurements have advanced in recent years. These improved methods are either based on peak-fitting models or on the *Rietveld refinement*[1]. The results of the Reynolds Cup, an international mineral quantification contest, proved that these methods are more successful than extraction or XRD single-line methods (MCCARTY, 2002; KLEEBERG, 2005). The strengths and the weaknesses of both methods are well known and can be compensated for by complementary methods. For the evaluation of the results of these two methods, it is valuable to know the chemical composition of a sample, which can be accurately determined by XRF. It is recommended to additionally determine the C and S contents, e.g., by LECO analysis. The mineralogical composition, which can be obtained, for example, by Rietveld refinement of XRD diagrams, has to fit with the chemical composition: Using the mineral formula (chemical composition of each phase) a theoretical chemical composition can be calculated which has to fit with the experimental data. Misfits clearly indicate weaknesses of the Rietveld refinement which can be corrected by the chemical composition. It is worth mentioning that even the qualitative determination of all phases of a sample remains an underestimated challenge, which is proven by the Reynolds Cup contest.

Due to the limitations of each of the presented methods, the combination of complementary methods is required in order to ensure the quality of the results (quality control). It is recommended to combine accurate chemical analysis

[1] Powder samples containing up to 10 or even more minerals are frequently used in XRD analysis. This technique, using material in the form of very small crystallites, yields XRD patterns with mineral peaks that partly overlap, thereby preventing proper determination of the structure. The Rietveld method, also called Rietveld refinement, Rietveld analysis, Rietveld technique, Rietveld calculation, creates an effective separation of these overlapping peaks, thereby allowing an accurate determination of the mineral abundances.

(e.g., XRF and LECO-S/C) with results obtained by the XRD Rietveld method, which has been improved in recent years.

5.3.5.9 Iron and Manganese Oxides

STEPHAN KAUFHOLD

The term oxide encompasses all metal oxides, oxyhydroxides (or oxohydroxides) and hydroxides, e.g., goethite (FeOOH), ferrihydrite (hydrous ferric oxide, more than one ideal formula) or gibbsite ($Al(OH)_3$). Some of them are characterized by a relatively high specific surface area which, even at low mineral concentrations, can dominate the specific surface area of the soil or rock they are in. They additionally possess a varying amount of hydroxyl groups which are able to interact with the surrounding fluid by ligand exchange and adsorption/desorption of protons. Both properties are meant when a high "reactive surface area" is mentioned. These properties play an important role in the retention of specific pollutants and the supply of plant nutrients.

Iron oxides

Iron oxides are characterized by their intense color, which determines the color of many soils and rocks. The color of the different iron oxyhydroxide minerals is classified according to MUNSELL (e.g., goethite: 7.5YR - 2.5Y or hematite: 5R - 2.5YR; JASMUND & LAGALY, 1993) and gives valuable information for soil classification. Iron oxides are generally more abundant than Al or Mn oxyhydroxides (KRAUSKOPF & BIRD, 1995). Therefore, they often determine the contaminant retention capacity of soils and rocks. Various metals are incorporated into iron oxides as they precipitate. The iron oxide surfaces display a strong affinity to arsenate, phosphate, and carboxylic acids. However, the displacement of sorbed arsenate by carbonate, a ubiquitous species in almost all geological environments, is overlooked in many studies (e.g., APPELO et al., 2002). On the crystallographic scale, most iron oxides consist of Fe oxide/hydroxide octahedra (Fe + 6 O^{2-} or + (3 O^{2-} + $3OH^-$)). The minerals differ in the way these octahedra are linked (JASMUND & LAGALY, 1993). As an example, goethite consists of double chains of octahedra linked along their edges. An appreciable amount of the iron in goethite can be replaced by Al, which results in a change of the unit cell dimensions and hence can be detected by XRD (JASMUND & LAGALY, 1993).

However, the structure of ferrihydrite is still not fully understood. Particularly the high amount of Si which has been reported in it (up to 9 wt. %) is subject to discussion (CHILDS, 1992).

Iron oxides are industrially synthesized and used as pigments and as adsorbant, e.g., for the removal of arsenate from drinking water.

The most common iron oxides in soils are ferrihydrite and goethite. The anhydrous minerals hematite and magnetite are more abundant in rocks than in soils due to dehydration during diagenesis. However, the various iron minerals can be transformed into each other by ripening, hydration or dehydration. Transformation schemes have been presented, for example, by JASMUND & LAGALY (1993), HOUBEN & TRESKATIS (2006): see *Fig. 5.3-36*, and CORNELL & SCHWERTMANN (2003), who provide a fairly complete compilation of iron oxide structures, properties, reactions, occurrences, and applications.

The most important reaction relevant to pollutant retention in common soils and sediments is the transformation from ferrihydrite to goethite. Recent studies on the transformation mechanisms are provided, for example, by KUKKADAPU et al. (2003). HOUBEN (2003) showed that anions, such as arsenate or phosphate, which are adsorbed on the surface of ferrihydrite are released during the transformation.

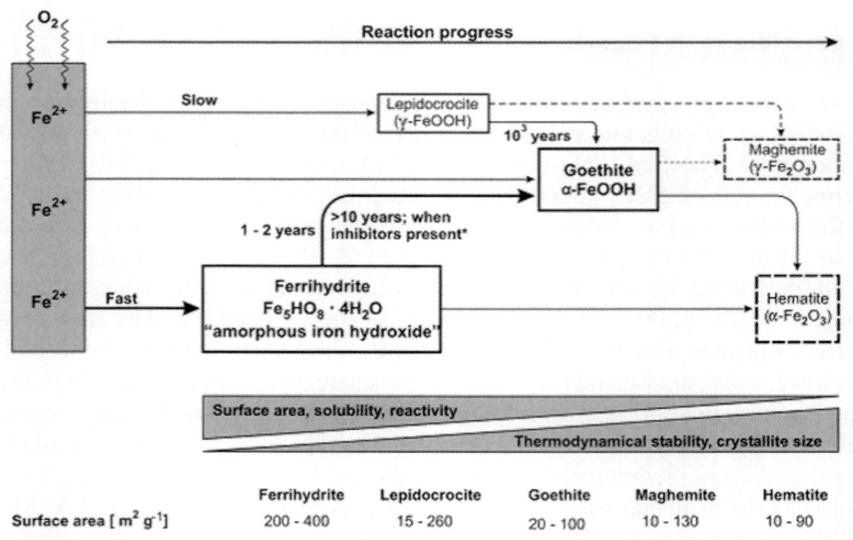

Fig. 5.3-36: Ageing sequence ("Ostwald ripening") of iron oxides and its effects on mineral content and mineral properties, HOUBEN & TRESKATIS (2006)

Iron oxides are the most abundant incrustations in water wells, pipelines and drainage systems (HOUBEN, 2003). These incrustations have to be removed and this is by far easier if ferrihydrite is present instead of goethite. This is relevant, since ferrihydrite is transformed to goethite within a few years (SCHWERTMANN & MURAD, 1983).

The XRD detection limit for ferrihydrite is 5 - 10 % (CHILDS, 1992). However, the actual detection limit strongly depends on its crystallinity: By XRD it is possible to distinguish "two-line ferrihydrites" (2-L) from "six-line ferrihydrites" (6-L) simply by the number of XRD peaks. The broad ferrihydrite XRD peaks are often referred to as bands because of their comparatively large halfwidths.

In soil science, extraction techniques are commonly used to quantify goethite and amorphous iron hydroxides (including ferrihydrite). The total amount of iron oxyhydroxides is determined by dithionite extraction according to MEHRA & JACKSON (1960). The content of XRD "amorphous" iron oxide (including ferrihydrite) is measured by determining the amount of iron that can be extracted by ammonium oxalate solution (pH 3, 2 hours). The acid oxalate method can be used as approximate value for the quantification of ferrihydrite but not for its identification (CHILDS, 1992). XRD or infrared spectroscopy is recommended for the identification of the different iron oxides. A detailed compilation of IR data is given, for example, by CORNELL & SCHWERTMANN (2003).

The specific surface area (SSA) of iron oxides, particularly ferrihydrite, is very high. Therefore, even minor amounts can govern the SSA of a soil or rock (CHILDS, 1992). The SSA of 2-L ferrihydrites as determined by the widely used N_2 adsorption method (applying the BET equation[2]) ranges from 300 to 340 $m^2 g^{-1}$. Unpublished analyses by the author show that the specific surface area decreases continuously during transformation of ferrihydrite to goethite to a minimum of 20 $m^2 g^{-1}$. (This value for pure goethite was verified by IR spectroscopy.)

Infrared spectra of ferrihydrite and goethite are shown in *Fig. 5.3-37* together with their BET SSAs. The ratio of OH-stretching vibrations of ferrihydrite and goethite can be correlated with the specific surface area.

In mineralogical practice today, XRD (particularly the Rietveld method) or spectroscopic methods are preferred over extraction methods. However, for the quantification of minor amounts (< 5 wt. %) of iron oxyhydroxides, particularly ferrihydrite, extraction techniques – despite their well known weaknesses – still lack of alternatives. LOAN et al. (2002) reasonably propose to use XRD, MÖßBAUER spectroscopy, transmission electron microscopy, and extraction techniques.

[2] The BET theory, developed by BRUNAUER, EMMETT & TELLER (1938), can be used to calculate surface areas of solids by physical adsorption of gas molecules (e.g., nitrogen).

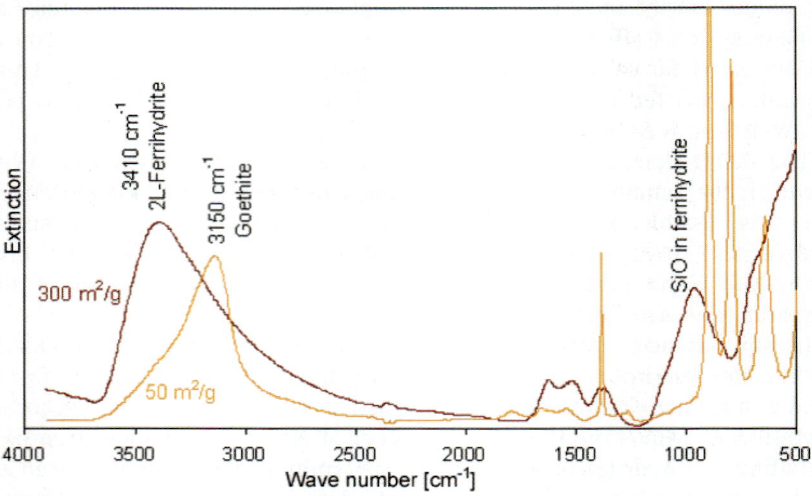

Fig. 5.3-37: Infrared (IR) spectra as well as BET surface area (N2 Adsorption) of 2L-ferrihydrite (including some Si) and goethite (synthetic)

Manganese oxyhydroxides

In contrast to Fe, Mn not only occurs as Mn^{2+} and Mn^{3+} but also as Mn^{4+}. Therefore, there are more Mn oxyhydroxides than Fe oxyhydroxides. Interestingly, a group of Mn minerals exists that are structurally related to clay minerals. Their basic units are layers consisting of MnO_6 octahedrons (layered manganates or phyllomanganates). Depending on their hydration state, they have a basal spacing of 10 Å (hydrated) or 7 Å (dehydrated).

A negative charge deficiency is known for some Mn oxyhydroxides which is similar to that observed for swelling clay minerals like montmorillonite, the charge deficiency being balanced by exchangeable cations.

The amount of manganese oxyhydroxides in soils and rocks is generally less than that of iron oxyhydroxides. Nevertheless, some scenarios were investigated which demonstrated the relevance of Mn oxyhydroxides (e.g., poorly ordered oxidic Mn minerals) as most significant sink for toxic heavy metals in floodplain sediments and river channels (COTTER-HOWELLS et al., 2000).

5.3.5.10 Organic Carbon

KLAUS KNÖDEL

Besides clay minerals, carbonates, hydrated iron and manganese oxides, the content of organic substances in soil and sediment is a key factor for the sorption of organic contaminants. Reversibility of adsorption is observed to be considerably lower in materials with a high concentration of organic substances than in materials with low concentrations. Organic matter provides the largest number of binding sites because it has an extremely large surface area and is very reactive chemically. Organo-clay complexes are formed by the inclusion of natural organic substances like humins, lignins, etc. in clay minerals (THENG, 1979, TADJERPISHEH & ZIECHMANN, 1986). They have a substantially higher retention capacity with respect to polar organic compounds and heavy metal ions than the clay minerals alone (KOHLER, 1985). Reactive organic materials are the most important geochemical controlling factor for microbial processes and formation of redox zones and they influence the interaction of the microbes with the dissolved contaminants. Organic matter content in soil and sediment depends on climate, vegetation, location in the landscape, soil and sediment texture, and farming practices. The natural organic fraction of soil and sediment includes plant and animal residues at various stages of decomposition, cells and tissue of soil organisms, and substances produced by the soil population. Soil organic matter is frequently subdivided in humic substances, such as humic acids[3], fulvic acids[4] and humins[5], and nonhumic substances, as carbohydrates, lipids and amino acids.

Therefore, the content of organic matter in soil and sediment as well as in groundwater and surface water must be determined. The parameter determined is called total organic carbon (TOC). TOC includes all the carbon atoms covalently bonded in organic molecules. It is a measure of the sum of all organic carbon compounds determined in units of mg L^{-1} or mg kg^{-1}.

Soil, sediment, groundwater and surface water normally also contains inorganic forms of carbon (i.e., carbonates and elemental carbon). The total inorganic carbon content (TIC) is also reported in units of mg L^{-1} or mg kg^{-1}. Total carbon (TC) is the sum of organically bound carbon (TOC) and inorganically bound carbon (TIC) present in soil, sediment, and water.

---

[3] Humic acids are soluble by alkaline extraction and can be precipitated by strong acids. They have a higher molecular weigth and less functional groups than fulvic acids.

[4] Fulvic acids are soluble by alkaline extraction and cannot be precipilated by strong acids.

[5] Humins are humic substances that are not soluble by alkaline extraction.

Analytical methods for the determination of TOC, TIC, and TC

All quantitative analysis methods for determining total organic carbon (TOC) are based on the conversion of organic matter present in the soil, sediment or water to CO_2. This is done either chemically or by heating to a high temperature. The CO_2 is measured directly or indirectly and converted to TOC or TC content, depending on the presence or absence of inorganic carbon compounds. If TOC is to be determined, any carbonates (see Section 5.3.5.8) present have to be removed. In all cases a careful sample preparation is necessary.

Preparation of soil or sediment samples starts with the removal of inorganic particles greater than 2 mm in diameter (gravel, pebbles and rocks) and sample homogenization. This changes the particle-size distribution of the sample and is recorded for completeness of sample characterization. Removal of large organic particles, such as twigs, roots, wood chips, branches, etc., will affect sample TOC concentrations and should also be recorded for completeness of sample characterization. Samples should be stored at 4 °C until analysis.

Material that would interfere with the analysis (e.g., carbonates in the case of TOC determination or water in the case of the dry combustion technique) is removed. This is not done in the case of a TC determination, and the carbonate minerals will be decomposed along with the organic matter. A simple test can be performed to determine if carbonates are present: A few drops of 1 N to 4 N HCl is added to the sample and it is observed whether the sample effervesces. Care must be made when dolomite is present, since it does not as rapidly effervesce like calcite. Alternatively, the pH of the soil or sediment may be determined and if the pH is 7.8 to 8.2, then calcium carbonates are indicated in the sample (MCLEAN, 1982). If carbonates are present, the most common method for their removal is the addition of an acid or a combination of acids (HCl and H_2SO_4, alternatively H_2SO_4 and $FeSO_4$ may be used to minimize oxidation and decarboxylation of organic matter by the H_2SO_4). Other sample components that interfere with wet chemistry techniques include Fe^{2+} and Cr, which lead to overestimation of TOC content, and MnO_2, which leads to underestimation of TOC content (SCHUMACHER et al., 1995). For a TOC determination by combustion, the sample must first be dried, e.g., by air drying or oven-drying the sample at 105 °C overnight. One problem with this is the loss of volatile organic compounds.

Several factors for both sample preparation and sample analysis must be considered when the method for a TOC determination is chosen. These factors include the availability of equipment, ease of use, health and safety concerns, cost, sample throughput, and comparability to standard reference methods.

Wet chemical method

A content of organic matter greater than 0.2 wt. % in a sample can be determined by oxidation with strongly oxidizing solutions (e.g., hydrogen peroxide (H_2O_2) or potassium dichromate ($K_2Cr_2O_7$)). Rapid oxidation of organic matter with dichromate is the standard wet chemistry method. The Walkley-Black method is the best known of the rapid dichromate oxidation methods and has been the reference method for other methods in numerous studies. It is widely used because it is simple, rapid, and has minimal equipment requirements.

Between 0.5 g and 1.0 g of soil or sediment sample, depending on organic carbon content, is transferred to a 250 mL Erlenmeyer flask containing 10 mL of 1 N potassium dichromate ($K_2Cr_2O_7$) and 10 mL of concentrated H_2SO_4. Because the reaction is exothermic when potassium dichromate and sulfuric acid are mixed with the sample, the suspension is swirled and must be cooled prior to adding water to halt the reaction. After 30 minutes, 50 mL of deionized water, 3 mL of concentrated H_3PO_4 and 0.5 mL of 1 % diphenylamine indicator are added. The resulting solution is then titrated slowly with 1 N $FeSO_4$ up to the brilliant green color end point. The sulfuric acid used in this procedure is a corrosive, strong oxidant and should be handled with caution.

However, this procedure has been shown to lead to the incomplete oxidation of organic substances and is particularly poor for digesting elemental C forms. Studies have shown that the recovery of organic C using the Walkley-Black procedure range from 60 to 86 % with a mean recovery of 76 % (WALKLEY & BLACK, 1934). Owing to the incomplete oxidation, a correction factor of 1.33 is commonly applied to the results. To overcome the concern of incomplete digestion of the organic matter, the Walkley-Black procedure has been modified to include strong heating of the sample during sample digestion. Due to health, safety, and disposal concerns the Walkley-Black method is often replaced by other methods.

Dry combustion method

Several hundred milligrams of a soil or sediment sample are heated to temperatures between 1350 and 1500 °C in an oven. Temperatures greater than or equal to 1350 °C are necessary so that the inorganic carbon compounds are also decomposed. To ensure complete oxidation of the sample, a pure oxygen stream is used as well as various catalysts or accelerators (e.g., vanadium pentoxide, Cu, CuO, and aluminum oxide, LECO, 1996). The produced CO_2 is quantified by titrimetric, gravimetric, manometric, spectrophotometric, or gas chromatographic techniques and the TC or TOC content calculated. The dry combustion method is often used in order to avoid the health, safety, and disposal concerns of wet chemical methods.

Coulometric method

After freeze-drying and grinding, the soil or sediment samples are analyzed for organic and inorganic carbon compounds (TC or TOC depending on sample treatment) in a Ströhlein Coulomat 702. This method can be used to analyze both very low and very high carbon concentrations (range 0.001 - 100 % C). The method is based on combustion of 10 mg to 1 g samples in oxygen at a temperature up to 1350 °C and coulometric detection of released CO_2 (pH change). Coulometric carbon analysis involves no calibration and can be used for checking other analytical methods. The resolution of this method is 0.1 ppm and the precision is about 0.5 %. Analysis time amounts 120 to 300 s per sample.

TOC analyzers

TOC analyzers from several manufacturers are available for the analysis of solid samples and water samples. The Shimadzu gas sampling unit can measure TOC in gas samples (SHIMADZU, 2006). An auto-sampler and PC control including data export facility can be used for high-throughput automation. Online TOC analyzers are designed to operate in remote locations without continuous surveillance by an operator. For a comparison of TOC analyzers see http://cfpub.epa.gov/safewater/watersecurity/guide/productguide.cfm?page=chemicalsensortotalorganiccarbonanalyzer. The primary differences between TOC analyzers are in the methods used for oxidation and CO_2 quantification. Oxidation methods are, for example, combustion, combustion catalytic oxidation, and wet chemistry/UV oxidation. The oxidation step can be carried out at high or low temperature. In general, high temperature (combustion) analyzers achieve more complete oxidation of carbon compounds than low temperature (wet chemistry/UV) analyzers do. There are three main methods used for detection and quantification of carbon dioxide produced in the oxidation step:

- nondispersive infrared (NDIR) detector,
- colorimetric methods, and
- aqueous conductivity methods.

The most common method is the use of a nondispersive infrared detector. The detection limit for organic carbon depends on the methods used for oxidation and quantification and the type of the analyzer. For high-temperature methods

(680 °C and higher) the detection limit is 1 mg L^{-1} carbon, and for low-temperature methods (below 100 °C) the detection limit is 0.2 mg L^{-1} carbon. High-temperature methods can efficiently oxidize hard-to-decompose organic compounds, whereas low-temperature methods have a limited oxidation potential. The response time of a TOC analyzer may vary depending on the manufacturer's specifications, but it usually takes from 1 to 15 minutes for a trustworthy reading.

European standard for the determination of total organic carbon (TOC) and dissolved organic carbon (DOC) in water is DIN EN 1484 and for the determination of organic carbon and total carbon after dry combustion of air-dried soil samples DIN ISO 10694. American standards for the determination of total organic carbon (TOC) in water are EPA Method 415.1 (Combustion or Oxidation), EPA Method 415.2 (UV Promoted, Persulfate Oxidation) and SW-846 Method 9060.

Screening or indicator tests

TOC, TIC and TC are determined as a screening or indicator test in site investigations. The results indicate that a class of substances is present without identification of the individual compounds. In many cases a costly chemical analysis can be avoided by first screening the samples in this way (JIMÉNEZ et al., 2002).

Other frequently used screening or indicator tests, sometimes also called group parameter analysis, include the determination of

- biochemical oxygen demand (BOD),
- chemical oxygen demand (COD),
- dissolved organic carbon (DOC),
- absorbable organic halogens (AOX) – e.g., LANIEWSKI et al. (1999),
- extractable organohalogens (EOX),
- total organic halides (TOX),
- total petroleum hydrocarbons (TPH),
- phenol index (PI).

Analysis of individual organic compounds

The analysis of individual organic compounds requires much more sophisticated instrumentation and a well trained and experienced operator. Typical analytical techniques comprise chromatographic (gas or liquid chromatography), spectrometric (UV, IR, MS) or microbiological methods. Typical compound classes analyzed are

- benzene, toluene, ethylbenzene, xylene (BTEX),
- volatile organic compound (VOC),
- polycyclic aromatic hydrocarbons (PAH),
- polychlorinated biphenyl (PCB),
- polychlorinated dibenzo-p-dioxins and dibenzofurans (PCDD/F).

For European and ISO[6] standards see Sections 5.3.1 and 5.3.2.

5.3.6 Methods to Evaluate Biodegradation at Contaminated Sites

ANDREAS TIEHM & KATHRIN R. SCHMIDT

Biodegradation is the decomposition of organic substances by microorganisms such as bacteria, yeast, and fungi or their enzymes in the environment. Decomposition is the breaking down of a substance into smaller molecules, by which its original physical and chemical properties are changed. This can be caused by abiotic (chemical, photochemical) and microbial reactions or processes. Decomposition usually takes place in several stages. Products of biotic processes are called metabolites. An example of a decomposition series is the dechlorination of chlorinated ethenes. The chemicals perchloroethene and trichloroethene, used as fat-dissolving cleaning agents, degrade in a strictly anaerobic environment via *cis*-1,2-dichloroethene into vinyl chloride and finally ethane.

This section focuses on the evaluation and assessment of microbial degradation processes. Biodegradation processes vary greatly depending on the substance, microorganism, and the hydrogeological and geochemical conditions at the site. Biodegradation is a key aspect of natural attenuation (NA) as well as for stimulated bioremediation (e.g., enhanced natural attenuation, ENA) of environmental pollution.

Natural attenuation is defined as the sum of processes which cause a decrease in contamination without human intervention. A situation in which contamination from various sources is naturally attenuated in an aquifer is shown schematically in *Fig. 5.3-38*. In contrast to abiotic attenuation – dilution, dispersion and adsorption – biodegradation results in a net loss of

---

[6] International Organization for Standardization (ISO), Geneva, (www.iso.ch), an international organization with responsibility for international technical standards working with the United Nations. ISO is a worldwide federation of national standards bodies from some 140 countries, one from each country.

contaminant mass within the contaminant plume. The use of these natural processes for site remediation is being increasingly considered as a sound alternative to technical measures. However, it has to be considered that using NA for site remediation is not a "do-nothing option". The site-specific degradation processes have to be characterized in detail, quantified and then evaluated to determine whether they are sufficiently efficient. When natural attenuation is used for site remediation it has to be adequately monitored (SUAREZ & RIFAI, 1999; RÖLING & VAN VERSEVELD, 2002; TIEHM & SCHULZE, 2003; SCHULZE & TIEHM, 2004).

The identification, quantification, and assessment of natural biodegradation in the field are a prerequisite for the acceptance of monitored NA or ENA approaches as an alternative to more costly technical remediation measures. This section is intended to give an introduction into microbial degradation processes (Section 5.3.6.1), to introduce the methods currently available for assessment of microbial degradation (Section 5.3.6.2), and to highlight some recent results of investigations at two German sites contaminated with tar oil and chlorinated ethenes (Section 5.3.6.3).

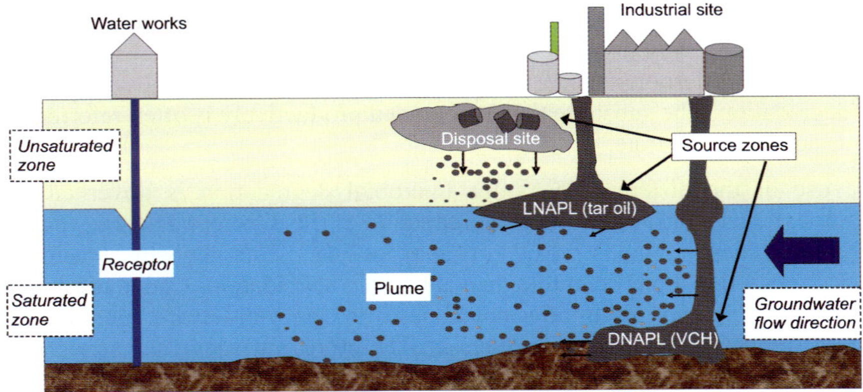

Fig. 5.3-38: Schematic view of natural attenuation of contaminants in groundwater (LNAPL = light nonaqueous phase liquid, DNAPL = dense nonaqueous phase liquid, VCH = volatile chlorinated hydrocarbons)

5.3.6.1 Microbial Processes in the Subsurface

Microorganisms obtain energy for cell production and maintenance by catalyzing the transfer of electrons from electron donors to electron acceptors (*oxidation-reduction*). Electron donors are compounds that are in a reduced state, such as natural organic material and most contaminants (e.g., aromatic hydrocarbons), and are oxidized during degradation. Compounds serving as *terminal electron acceptors* (TEA) in microbial metabolism are in an oxidized state and are reduced during microbial metabolism. Naturally occurring TEAs in the subsurface include dissolved oxygen, nitrate, sulfate, iron(III) and manganese(IV) minerals, and carbon dioxide. Anthropogenic halogenated compounds can also serve as TEAs.

Microbial degradation processes are grouped as follows (WIEDEMEIER et al., 1999):

- Productive degradation: The organic compound (natural or anthropogenic) is used as primary growth substrate.
  - Oxidative degradation: The organic compound acts as electron donor and an electron acceptor is required for degradation.
  - Reductive degradation: The organic compound acts as electron acceptor and an electron donor (frequently referred to as auxiliary substrate) is required for degradation.
  - Fermentation: The organic compound serves as both electron donor and acceptor and no further substrate is needed for degradation (proceeds under anaerobic conditions).
- Co-metabolic degradation: The organic compound is transformed by enzymes normally used for the productive degradation of a primary growth substrate. Microorganisms do not have any benefit from co-metabolic degradation. The substrate that is degraded productively is often referred to as auxiliary substrate.

The use of the different TEAs for microbial degradation is governed by thermodynamic constraints since the reduction of the different TEAs results in different energy gains $\Delta G°$ (*Table 5.3-34*). Energetically, aerobic biodegradation is more favorable than anaerobic degradation. Under anaerobic conditions, reduction of nitrate, manganese or iron yield more energy than sulfate reduction or methanogenesis. Therefore, in general, sequential utilization of TEAs is observed: Aerobic degradation is followed by reduction of nitrate, manganese, iron, sulfate, and methanogenesis (CHRISTENSEN et al., 2001).

The oxidation of different organic substrates (natural or anthropogenic) also yields different energy gains (*Table 5.3-35*). Under normal conditions, only when there is a net energy gain for the complete redox reaction (electron consuming reactions in *Table 5.3-34* and electron donating reactions in *Table 5.3-35*) will degradation proceed. Hence, reduced contaminants such as aromatic hydrocarbons and reduced halogenated substances are rather degraded via oxidative mechanisms, whereas oxidized pollutants such as oxidized halogenated substances are rather degraded reductively.

Table 5.3-34: Standard free energy changes for various microbial respiration processes, McFarland & Sims (1991) and Wiedemeier et al. (1999), modified, (TEA = terminal electron acceptor, PCE = perchloroethene, TCE = trichloroethene, cDCE = *cis*-1,2-dichloroethene, VC = vinyl chloride)

Electron-consuming reaction (reduction of TEA)		ΔG° [kcal/mol e⁻]
aerobic respiration	$4\ e^- + 4\ H^+ + O_2 \rightarrow 2\ H_2O$	- 18.5
nitrate reduction	$5\ e^- + 6\ H^+ + NO_3^- \rightarrow 0.5\ N_2 + 3\ H_2O$	- 16.9
manganese reduction[1]	$2\ e^- + 4\ H^+ + MnO_2 \rightarrow Mn^{2+} + 2\ H_2O$	- 8.6
iron reduction[1]	$e^- + Fe^{3+} \rightarrow Fe^{2+}$	- 17.8
	$e^- + 2\ H^+ + FeOOH + HCO_3^- \rightarrow FeCO_3 + 2\ H_2O$	+ 1.1
sulfate reduction	$8\ e^- + 9.5\ H^+ + SO_4^{2-} \rightarrow 0.5\ HS^- + 0.5\ H_2S + 4\ H_2O$	+ 5.3
methanogenesis	$8\ e^- + 8\ H^+ + CO_2 \rightarrow CH_4 + 2\ H_2O$	+ 5.9
reductive dechlorination		
of PCE	$C_2Cl_4 + H^+ + 2\ e^- \rightarrow C_2HCl_3 + Cl^-$	- 9.9
of TCE	$C_2HCl_3 + H^+ + 2\ e^- \rightarrow C_2H_2Cl_2 + Cl^-$	- 9.6
of cDCE	$C_2H_2Cl_2 + H^+ + 2\ e^- \rightarrow C_2H_3Cl + Cl^-$	- 7.2
of VC	$C_2H_2Cl + H^+ + 2\ e^- \rightarrow C_2H_4 + Cl^-$	- 8.8

[1] Reduction of manganese and iron results in different energy gains depending on the type of minerals occurring in the aquifer (an example is given for the reduction of iron).

Table 5.3-35: Standard free energy changes for the oxidation of various pollutants, Wiedemeier et al. (1999), modified, (cDCE = *cis*-1,2-dichloroethene, VC = vinyl chloride)

Electron-donating reaction (oxidation of substrate)		ΔG° [kcal/mol e⁻]
benzene oxidation	$12\ H_2O + C_6H_6 \rightarrow 6\ CO_2 + 30\ H^+ + 30\ e^-$	- 7.0
toluene oxidation	$14\ H_2O + C_6H_5CH_3 \rightarrow 7\ CO_2 + 36\ H^+ + 36\ e^-$	- 6.9
ethylbenzene oxidation	$16\ H_2O + C_6H_5C_2H_5 \rightarrow 8\ CO_2 + 42\ H^+ + 42\ e^-$	- 6.9
m-Xylene oxidation	$16\ H_2O + C_6H_4(CH_3)_2 \rightarrow 8\ CO_2 + 42\ H^+ + 42\ e^-$	- 6.8
naphthalene oxidation	$20\ H_2O + C_{10}H_8 \rightarrow 10\ CO_2 + 48\ H^+ + 48\ e^-$	- 6.9
cDCE oxidation	$4\ H_2O + C_2H_2Cl_2 \rightarrow 2\ CO_2 + 10\ H^+ + 8\ e^- + 2\ Cl^-$	- 16.1
VC oxidation	$4\ H_2O + C_2H_3Cl \rightarrow 2\ CO_2 + 11\ H^+ + 10\ e^- + Cl^-$	- 11.4

Biodegradation of aromatic hydrocarbons

Abandoned coal gasworks sites and landfills containing residues from gasworks are major emitters of tar oil pollutants, including BTEX (benzene, toluene, ethylbenzene, o-, m-, p-xylene), PAH (polycyclic aromatic hydrocarbons), and heterocyclic aromatic hydrocarbons (heterocycles). Chemical structures of these compounds are shown in *Fig. 5.3-39*. At present, all BTEX except the xylenes and 16 of the PAH belong to the "priority pollutants" selected by the U.S. Environmental Protection Agency (EPA) (EPA, 2006) and are of concern due to their harmful, toxic and/or carcinogenic properties (WHO, 1998). Heterocycles are included in this discussion since recent studies demonstrate that these compounds are also toxic and mutagenic and occur in groundwater plumes at even larger distances from the source than the BTEX and PAH compounds (SAGNER & TIEHM, 2005; SAGNER et al., 2006).

In the presence of oxygen, the BTEX, most PAH with up to five rings and most heterocycles are biodegradable (CERNIGLIA, 1992; FRITSCHE & HOFRICHTER, 2000; KÄSTNER, 2000; TIEHM & SCHULZE, 2003; SAGNER & TIEHM, 2005). Biodegradation rates of PAH decrease with increasing molecular weight (TIEHM & FRITSCHE, 1995; TIEHM et al., 1997; TIEHM & STIEBER, 2001) and additional growth substrates might be necessary as was shown for co-metabolic degradation of benzo[a]pyrene (JUHASZ & NAIDU, 2000). Under aerobic conditions, oxygen is not only used as terminal electron acceptor, but also as reactant for the enzymatic oxygenase reactions that cleave the aromatic ring structure (FRITSCHE & HOFRICHTER, 2000; KÄSTNER, 2000).

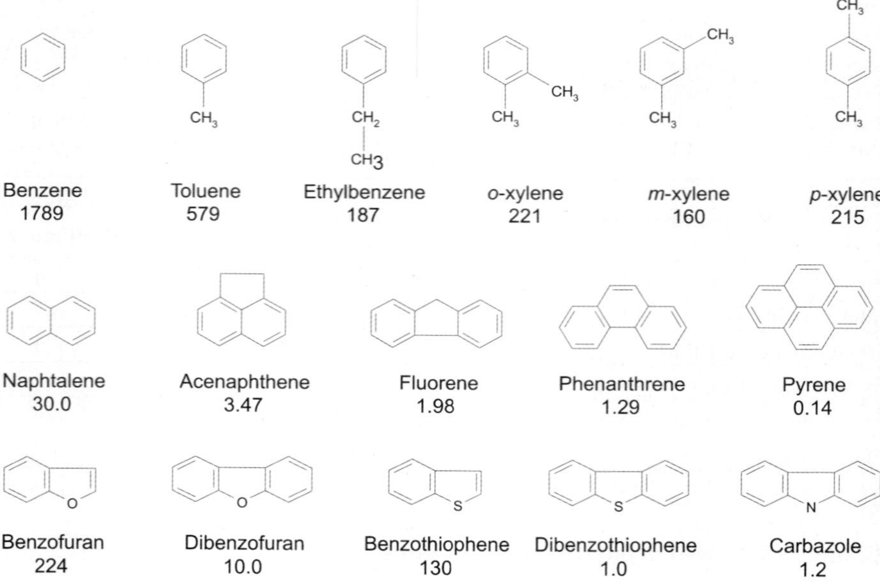

Fig. 5.3-39: Chemical structure and water solubility (mg L^{-1}) of BTEX (top) and selected polycyclic (middle) and heterocyclic (bottom) hydrocarbons, data from MILLER et al. (1985); SIMS & OVERCASH (1983), and JOHANSEN et al. (1996)

Under anaerobic conditions the biodegradability of aromatic hydrocarbons varies significantly. With nitrate as electron acceptor, biodegradation has been observed for all BTEX (HUTCHINS, 1991; LOVLEY, 2000; SCHINK, 2000), and has been reported for PAH in a few cases (LEDUC et al., 1992; MCNALLY et al., 1998; MILHELCIC & LUTHY, 1988; ROCKNE et al., 2000). Field studies indicate that degradation of BTEX and PAH is also possible in the presence of iron(III) (LYNKILDE & CHRISTENSEN, 1992; RAMSAY et al., 2001). Utilization of manganese(IV) as electron acceptor has been observed for the biodegradation of naphthalene (LANGENHOFF et al., 1996). Benzene (LOVLEY et al., 1995) and naphthalene (MECKENSTOCK et al., 2000) have been shown to be degraded in the presence of sulfate. Biodegradation of benzene and toluene has been observed under methanogenic conditions (GRBIC-GALIC & VOGEL, 1987; GRBIC-GALIC, 1990). Some heterocycles also are degradable under anaerobic redox conditions (LICHT et al., 1996; DYREBORG et al., 1997; SAGNER & TIEHM, 2005).

Not only the availability of suitable electron acceptors, but also the lack of nutrients can limit microbial natural attenuation (ALTHOFF et al., 2001; SAGNER & TIEHM, 2005). Moreover, the presence of easily degradable compounds may enable co-metabolic degradation pathways and, therefore, affects the degradation of tar oil contaminants (PROVIDENTI et al., 1993; DYREBORG et al., 1997). For example, anaerobic co-metabolism of polycyclic and heterocyclic hydrocarbons has been demonstrated during degradation of naphthalene (SAFINOWSKI et al., 2006).

Biodegradation of chloroethenes

Perchloroethene (PCE) and trichloroethene (TCE) are widely used as dry cleaning solvents and as degreasing agents (MIDDELDORP et al., 1999). Due to accidents and improper disposal practices, PCE and TCE are among the contaminants most frequently found in groundwater systems (FETZNER, 1998). Both are U.S. EPA priority pollutants (EPA, 2006) and have an EPA maximum contaminant level for drinking water of 5 µg L^{-1} (BRADLEY, 2003).

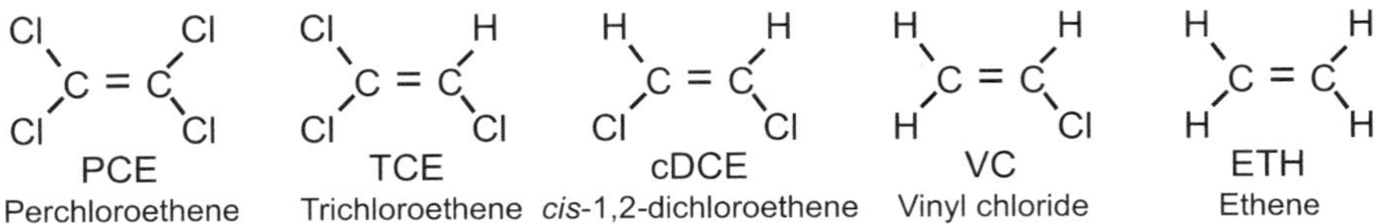

Fig. 5.3-40: Chemical structures of chloroethenes and ethene

Complete dehalogenation of chloroethenes (see *Fig. 5.3-40* for chemical structures) can be achieved by reductive dechlorination to ethene or by sequential reductive/oxidative degradation (*Fig. 5.3-41*) (TIEHM et al., 2002a; BRADLEY, 2003). During complete reductive dechlorination, PCE and TCE are anaerobically dechlorinated via *cis*-1,2-dichloroethene (cDCE) and vinyl chloride (VC) to the dehalogenated degradation end-products ethene and ethane (EL FANTROUSSI et al., 1998). In the 1980s, only co-metabolic processes were considered for reductive microbial dechlorination. However, in recent years halorespiration, i.e. the use of halogenated compounds as TEA, has been demonstrated to be the most important reductive degradation process at many sites contaminated with PCE or TCE. Reductive dechlorination of the higher chlorinated ethenes PCE and TCE is often faster than reductive dechlorination of cDCE and VC. Therefore, an accumulation of the lower chlorinated ethenes is frequently observed at contaminated sites (MIDDELDORP et al., 1999), which is of concern due to their toxic (cDCE and VC) and carcinogenic (VC) properties (VERCE et al., 2002). The monochlorinated compound VC is among the U.S. EPA priority pollutants (EPA, 2006). At many sites, the metabolites of reductive dechlorination cDCE and VC form long plumes, often reaching oxidative, aerobic zones of groundwater.

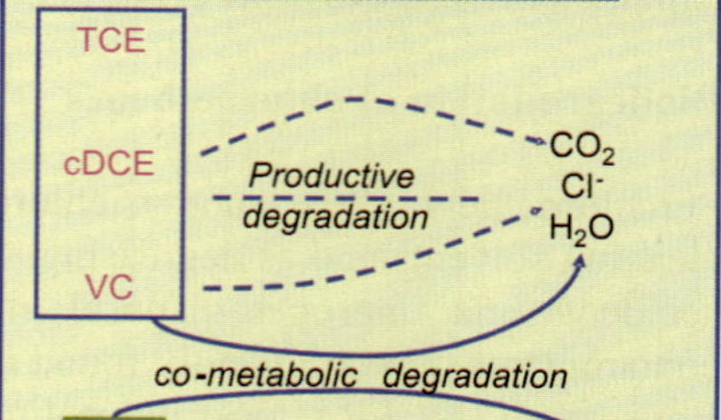

Fig. 5.3-41: Microbial degradation of chloroethenes (PCE = perchloroethene, TCE = trichloroethene, cDCE = *cis*-1,2-dichloroethene, VC = vinyl chloride)

The less chlorinated compounds cDCE and VC, as well as TCE, are co-metabolically degradable under aerobic conditions with auxiliary substrates such as ammonia, methane, or ethene (BIELEFELDT & STENSEL, 1999; BRAR & GUPTA, 2000; HOURBRON et al., 2000; TIEHM et al., 2002a; TAKEUCHI et al., 2005). Recent findings have also demonstrated productive degradation of VC (COLEMAN et al., 2002b; DANKO et al., 2004; SINGH et al., 2004; MATTES et al., 2005; TIEHM et al., 2006a and 2006b) as well as co-metabolic dechlorination of cDCE with VC as primary substrate (VERCE et al., 2002; SINGH et al., 2004; FATHEPURE et al., 2005; TIEHM et al., 2006a and 2006b). Furthermore, the use of cDCE as sole carbon source has been reported by COLEMAN et al. (2002a).

5.3.6.2 Assessment Methods

Regulatory acceptance of monitored natural attenuation approaches and active remediation techniques is dependent on credible evidence that an observed decrease in contaminant concentration is truly due to biodegradation of these compounds. Different methods have been developed to assess in situ bioremediation under field conditions. In many cases, several techniques are applied simultaneously in order to demonstrate microbial degradation of pollutants by multiple lines of evidence (CHAPELLE et al. 1995; CHO et al. 1997; LEVINE et al. 1997; CHRISTENSEN et al. 2001; BRADLEY, 2003; HIRSCHORN et al., 2004; TIEHM et al., 2006a).

In order to assess intrinsic bioremediation at contaminated sites multiple lines of evidence approaches are recommended. Depending on the conditions at the site, different parameters should be analyzed to obtain an understanding of the site-specific microbial degradation processes. The following analytical methods proved to be suitable for the assessment of microbiological natural attenuation:

- *Analysis of pollutant profiles and metabolites*: Microbial degradation is indicated by a more rapid decline of less hydrophobic compounds and the formation of metabolites.
- *Hydrochemical and geochemical analysis*: Redox zones and availability of auxiliary substrates and electron acceptors show which microbial degradation processes are possible from a hydrochemical point of view.
- *Microbial survey with "most-probable-number (MPN)" techniques*: Detection of microbial groups with different physiological and contaminant degrading capabilities indicates which microbial degradation processes are predominant.

- *Detection of specific degraders with polymerase chain reaction (16S-PCR)*: Detection of degrading microorganisms/enzymes shows the presence/activity of specific pollutant-degrading microorganisms.
- *Analysis of isotopic signatures in the field*: Changing isotopic signatures indicate microbial degradation processes in the field and enable differentiation from abiotic NA processes.
- *Determination of isotope enrichment factors in the laboratory*: Knowledge of the extent of microbial fractionation enables quantification of microbial degradation in the field.
- *Microcosm studies*: Identification of microbial degradation processes and analysis of specific redox processes lead to an extensive knowledge of microbial degradation processes occurring in the field.

Analysis of pollutant profiles and metabolites

In the field, assessment of pollutant biodegradation by monitoring a single compound is not possible since abiotic processes such as dilution, dispersion and sorption also have to be considered. However, the spatial distribution of contaminants provides information about the occurrence of biodegradation in the field. Conclusions can be drawn from changing ratios of different pollutants with different physicochemical properties and/or the detection of specific metabolic products of biodegradation along the flow path. A more rapid decline of a less hydrophobic and, therefore, less retarded compound indicates that biodegradation occurs at the site.

The water solubility of the BTEX (benzene, toluene, ethylbenzene, o-, m-, p-xylene) decreases as follows: benzene > toluene > o-xylene > p-xylene > ethylbenzene > m-xylene (*Fig. 5.3-39*). At a BTEX-polluted site, toluene and o-xylene concentrations were observed to decrease more rapidly in the anaerobic zone than ethylbenzene (BORDEN et al., 1995). In a gasoline plume, toluene and ethylbenzene declined most rapidly with distance from the source, followed by m-xylene, p-xylene, o-xylene and benzene (BORDEN et al., 1997). These examples illustrate that the horizontal and vertical distribution of pollutants provides evidence for biodegradation.

Another approach is to use distinctive metabolites for documenting in-situ biodegradation. Such metabolites should have the following characteristics: (i) an unequivocal and unique biochemical relationship to their parent compound, (ii) no commercial or industrial use, and (iii) sufficient biological and chemical stability (BELLER et al., 1995). In the case of metabolites that are subject to further degradation, their absence does not indicate that in situ biodegradation does not occur.

For chlorinated ethenes, the detection of the metabolites of reductive dechlorination *cis*-1,2-dichloroethene (cDCE), vinyl chloride (VC) and ethene (*Fig. 5.3-41*) along the flow path shows that microbiological reductive degradation occurs in the field (SCHMIDT et al., 2006a; TIEHM et al., 2006a). FENNELL et al. (2001) investigated a trichloroethene-(TCE)-contaminated aquifer and observed a significant decrease of TCE and an increase of cDCE and VC in parts of the plume. In contrast, aerobic oxidative degradation of the less chlorinated ethenes cannot be proven by contaminant profiles, since the oxidation leads to complete degradation to carbonate and chloride and not to the formation of stable intermediate metabolites. In these cases, changing isotopic signatures of the contaminants in the field indicate that microbial degradation processes occur on site (see below for further explanation).

The formation of metabolites can also be observed in hydrocarbon plumes. Under anaerobic conditions, metabolic by-products include benzoic acid, mono- to tri-methyl benzoic acids and other aromatic acids that are structurally related to alkyl benzene precursors, alicyclic acids and low molecular weight aliphatic acids (COZZARELLI et al., 1994 and 1995; BELLER et al., 1995; LEVINE et al., 1997). It has also been shown that succinate derivatives of toluene, xylene and n-alkanes, for example, are metabolites that indicate anaerobic degradation (BELLER, 2002; GIEG & SUFLITA, 2002; GRIEBLER et al., 2004). Because methylation can be the first step in the degradation of some polycyclic aromatic hydrocarbons (PAH), the corresponding methyl succinic acids can be detectable metabolites, too (SAFINOWSKI et al., 2006).

Since pollutant concentration data are available at most investigated sites, the contaminant profiles represent readily available evidence of in-situ biodegradation.

Hydrochemical and geochemical analysis

The biodegradability of groundwater pollutants differs significantly in different redox environments (*Tables 5.3-34* and *5.3-35*), which makes it necessary to understand the redox conditions in contaminated aquifers in order to predict in situ biodegradation. Redox conditions are deduced from measurements of redox-sensitive hydrochemical/geochemical parameters such as the redox potential (oxidation-reduction potential, ORP) and concentrations of oxygen, nitrate-nitrite, sulfate-sulfide, dissolved iron(II) and manganese(II), methane and hydrogen (CHAPELLE, 2001). Microbial activity affects the redox regime by consumption of terminal electron acceptors (TEAs) and formation of reduced degradation products. At contaminated sites with high levels of biodegradable material, most of the TEAs present upgradient in the plume are rapidly consumed at or near the source, resulting in a typical sequence of redox zones (*Fig. 5.3-42*).

The availability of the different TEAs significantly affects microbial utilization of specific pollutants. Depletion of oxygen, nitrate, and sulfate within a plume relative to their concentrations outside the plume and elevated concentrations of iron(II) and methane produced within the plume can be used to roughly estimate the consumption of electron acceptors in groundwater (BORDEN et al., 1995). However, calculation of the mass of degraded contaminant on the basis of electron acceptor consumption and by-product production is hampered by precipitation and abiotic reactions of compounds in the subsurface. For instance, elevated concentrations of iron(II) only reflect reduction of iron(III) if there is minimal sulfate reduction, because generation of sulfide results in the precipitation of iron sulfide. Iron(II) also may precipitate as ferrous carbonate (HERON & CHRISTENSEN, 1995). The actual distribution of iron(II) will be controlled by the chemical content of the groundwater and the sediment type. The sulfide and carbonate precipitates increase the reducing capacity of the sediment. This will dramatically increase the demand for oxygen if stimulation of microbial degradation by the addition of oxygen is considered (HERON & CHRISTENSEN, 1995; HARTOG et al., 2002).

Analysis of hydrochemical/geochemical site data provides rough information which microbial degradation processes are possible considering the prevailing redox zonation and the availability of auxiliary substrates and electron acceptors.

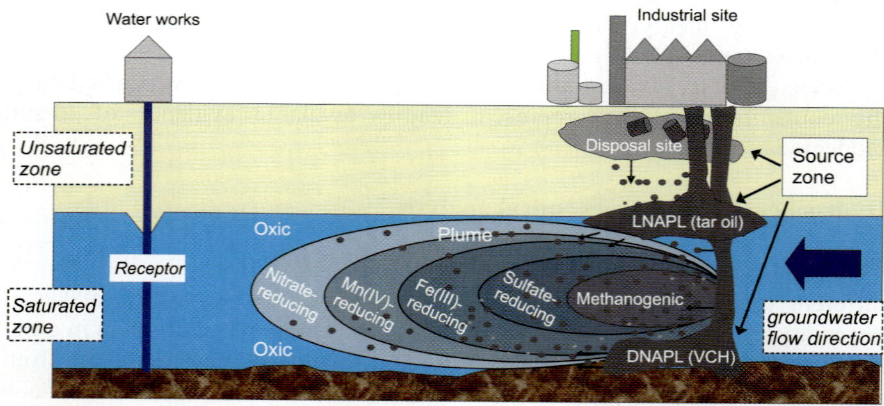

Fig. 5.3-42: Characteristic sequence of redox zones in an aquifer contaminated with biodegradable organics (LNAPL = light nonaqueous phase liquid, DNAPL = dense nonaqueous phase liquid, VCH = volatile chlorinated hydrocarbons)

Isotope fractionation

Isotopes are atoms of the same element with different numbers of neutrons, resulting in different atomic weights. Stable, nonradioactive isotopes behave slightly different in chemical and biological reactions owing to this difference in mass. Natural as well as synthetically produced compounds consist of a mixture of heavy (at a low percentage) and light (the majority) isotopes. The ratio of heavy to light isotopes is expressed by the isotopic signature δ, which can be measured by gas chromatography – isotope ratio mass spectrometry (GC-IRMS). The isotopic signatures of chemicals depend on differences in the chemical feedstock and the production process and, therefore, differ for different production periods and different manufacturers. Thus, isotopic signatures can be used to assign the responsibility for contamination, since isotopic signatures are hardly influenced by abiotic processes in the ground, such as sorption (SCHUTH et al., 2003).

Molecules consisting of light isotopes (e.g., ^{12}C) are more rapidly biologically degraded than molecules containing heavy isotopes (e.g., ^{13}C). Thus, microbial degradation causes enrichment of the heavy isotope in the remaining substrate (*Fig. 5.3-43*). Isotope fractionation of carbon ($^{13}C/^{12}C$), hydrogen ($^{2}H/^{1}H$) and chlorine ($^{37}Cl/^{35}Cl$) atoms during contaminant degradation has been observed for a wide range of contaminants (MORASCH et al., 2001; PELZ et al., 2001; SLATER et al., 2001; BARTH et al., 2002; HUNKELER et al., 2002; NUMATA et al., 2002; MANCINI et al., 2003; CHU et al., 2004; NIJENHUIS et al., 2005) and was reviewed by MECKENSTOCK et al. (2004). Determination that isotope fractionation takes place in the aquifer gives strong evidence that microbial degradation occurs and makes it possible to distinguish biodegradation from abiotic natural attenuation (POND et al., 2002; SHERWOOD-LOLLAR et al., 2001; SONG et al., 2002; CHARTRAND et al., 2005). In the case of complete degradation of contaminants (without formation of metabolites) detection of changing isotopic signatures in the field is the only way to verify microbial degradation analytically.

The extent of microbial fractionation is expressed by an isotope enrichment factor. These factors differ for different compounds as well as for different reaction conditions (*Fig. 5.3-44*), enzymes and bacterial species (BLOOM et al., 2000). Quantification of microbial degradation in the field is possible if appropriate enrichment factors are known (GRIEBLER et al., 2004). Specific enrichment factors can be determined in laboratory degradation studies (e.g., using microcosms) under suitable conditions (HIRSCHORN et al., 2004). The wide range of possible enrichment factors in the literature demonstrates that site-specific enrichment factors should be determined in order to realistically quantify microbial natural attenuation in the field.

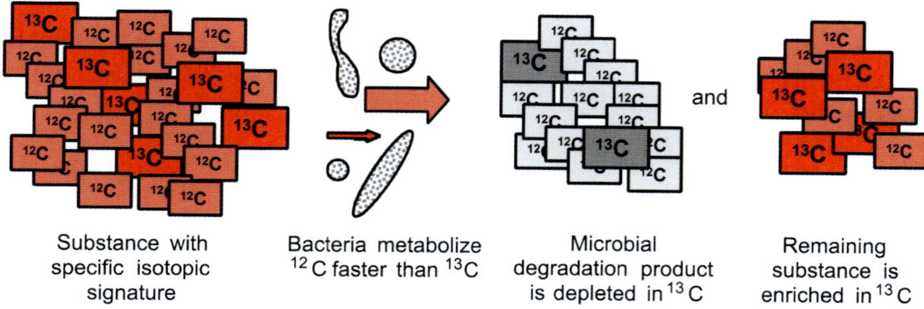
and
Substance with specific isotopic signature
Bacteria metabolize ^{12}C faster than ^{13}C
Microbial degradation product is depleted in ^{13}C
Remaining substance is enriched in ^{13}C

Fig. 5.3-43: Principle of isotope fractionation

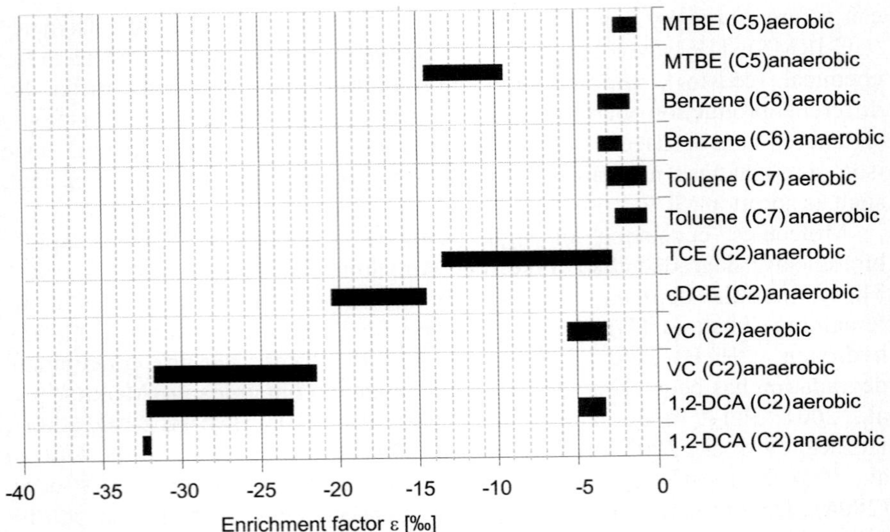

Fig. 5.3-44: Range of fractionation factors for degradation of different compounds under different conditions (data from MORASCH et al., 2005; MTBE = methyl tertiary butyl ether, TCE = trichloroethene; 1,2-DCA = 1,2-dichloroethane; cDCE = *cis*-1,2-dichloroethene; VC = vinyl chloride)

Isotope fractionation enables the detection of microbial degradation processes by measurement of isotopic signatures in field samples. Furthermore, isotope fractionation can be used to quantify microbial degradation processes in the field.

Microbial community analysis

Most-probable-number (MPN) techniques are used to analyze microbial community composition in terms of metabolically active contaminant degraders or to determine the microorganisms that use different electron acceptors. Metabolically active microorganisms are counted in groundwater samples or soil eluates. Samples are diluted in the wells of microplates (*Fig. 5.3-45, left*). After incubation, the turbid wells, i.e. wells in which microorganisms grew, are counted (*Fig. 5.3-45, right*). The most probable numbers are determined using appropriate tables or software based on a Poisson distribution.

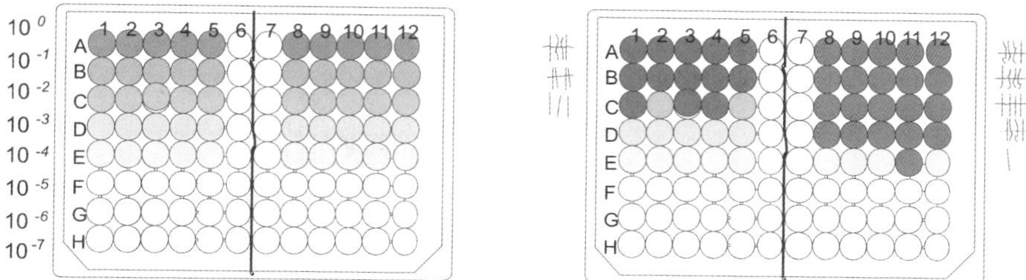

Fig. 5.3-45: Most-probable-number determination; *left*: dilution series; *right*: counting of the turbid or colored wells

Microorganisms with different metabolic physiologies can be detected with selective media, making it possible to distinguish between active organisms in different redox zones. Total aerobic heterotrophs as well as methanotrophic, ammonia-oxidizing and nitrite-oxidizing bacteria can be detected separately. Anaerobic organisms can be classed as nitrate-reducing, nitrate-ammonifying, iron-reducing, sulfate-reducing or methanogenic microorganisms. Specific contaminant-degrading bacteria can be assessed by adding the relevant contaminants as the sole carbon source (STIEBER et al., 1994). This method can easily be adapted to new compounds, even if the biodegrading microorganisms have not been previously characterized.

Analysis of the microbial community present at the site provides an estimate of the number of metabolically active contaminant degraders or microorganisms using different electron acceptors. In combination with hydrochemical/geochemical analyses, redox zones relevant to degradation can be identified. In addition, changes in the microbial community composition during active bioremediation can be monitored.

Molecular biological approaches

Molecular techniques such as 16S-PCR (polymerase chain reaction, see *Fig. 5.3-46* for a general scheme of a PCR reaction) can be used for fast detection of microbial groups, single species, or specific enzymes. The detection of specific contaminant degraders is a promising tool for assessing the degradation potential of a site (FENNELL et al., 2001; HOHNSTOCK-ASHE et al., 2001). During a 16S-PCR reaction, specific parts of 16S-rDNA are amplified, allowing their subsequent detection with agarose-gel electrophoresis. A nested PCR approach is used to increase sensitivity: In the first PCR reaction, universal bacterial 16S-rDNA is amplified, augmenting the DNA material serving as template for the second PCR reaction, the amplification of species-specific rDNA regions.

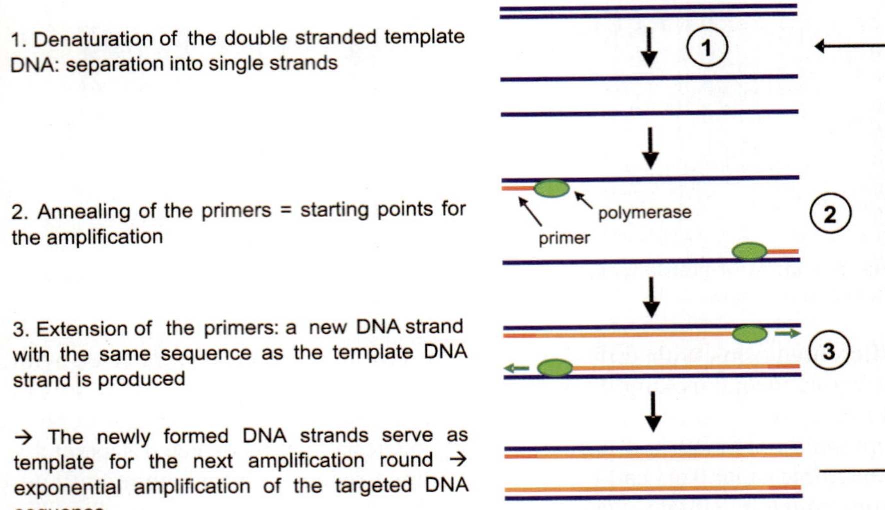

1. Denaturation of the double stranded template DNA: separation into single strands
2. Annealing of the primers = starting points for the amplification
3. Extension of the primers: a new DNA strand with the same sequence as the template DNA strand is produced
→ The newly formed DNA strands serve as template for the next amplification round → exponential amplification of the targeted DNA sequence
1
polymerase
primer
2
3

Fig. 5.3-46: Principle of a polymerase chain reaction (PCR)

PCR techniques allow a more specific site assessment than MPN methods, since specific degraders can be detected. Moreover, PCR is a very fast technique and, therefore, allows analysis of numerous samples for site assessment. On the other hand, 16S-PCR only provides information about the presence but not about the activity of specific microorganisms. Analysis of the mRNA encoding for degrading enzymes can be a means to compensate for this drawback in the future.

Microcosm studies

Materials placed in laboratory vessels to measure microbial activity are referred to as microcosms (*Fig. 5.3-47*). A microcosm is defined as a community or other small unit that is representative of a larger unit. The reasoning for microcosms is that by understanding the activity in a small portion of an aquifer, much can be learned about the aquifer as a whole (TIEHM & SCHULZE, 2003).

Microbial degradation studies using microcosms containing the autochthonous, site-specific microflora allow an assessment of the degradation processes occurring in the field. Laboratory (HUNT et al., 1997; BJERG et al., 1999; ALTHOFF et al., 2001; FENNELL et al., 2001; AULENTA et al., 2002; TIEHM & SCHULZE, 2003; TIEHM et al., 2006a) and in-situ microcosms (GILLHAM et al., 1990a and 1990b; ACTON & BARKER, 1992; BJERG et al., 1996; BJERG et al., 1999; GEYER et al., 2005) have frequently been used to demonstrate biodegradation of contaminants by indigenous microorganisms under conditions closely resembling ambient conditions.

Redox conditions are maintained by appropriate techniques for sampling (*Fig. 5.3-47*, *left*), transport, handling (*Fig. 5.3-47*, *bottom*) and storage (*Fig. 5.3-47*, *right*) of the aquifer material. It is also possible to intentionally change conditions in the laboratory in order to understand the effect of changing, for example, redox conditions and availability of specific electron acceptors or auxiliary substrates in the field (HE et al., 2002; LYEW et al., 2002; SCHULZE & TIEHM, 2004; AULENTA et al., 2005; SAGNER & TIEHM, 2005). This is of special interest for understanding the impact of technical remediation or enhanced natural attenuation measures on groundwater chemistry.

With microcosm studies, substantiated information is gathered about degradable pollutants and metabolites formed by their degradation under different redox conditions. Whereas field techniques, such as measuring contaminant profiles or isotope fractionation, represent sum parameters for assessing in situ biodegradation, microcosm studies are an appropriate tool for identifying degradation processes.

Fig. 5.3-47: Anaerobic treatment of groundwater microcosms; *top left*: filling of 2-L bottles with groundwater in the field; *top middle*: microcosm in the laboratory; *top right*: anaerobic incubation in an anaerobic vessel equipped with an oxygen-consuming reagent; *bottom*: anaerobic chamber for the anaerobic handling of microcosms

5.3.6.3 Case Studies

Natural attenuation approaches have been applied for virtually all classes of pollutants (landfill leachates, BTEX (benzene, toluene, ethylbenzene, o-, m-, p-xylene), polycyclic aromatic hydrocarbons (PAH), heterocycles, solvents, pharmaceuticals, pesticides, heavy metals, arsenic and a wide range of halogenated and nonhalogenated organic substances). The most widespread pollutant classes BTEX/PAH and chlorinated solvents are taken as examples to illustrate the methods for site assessment and two contaminated sites are presented as detailed case studies (TIEHM et al., 2002b; SCHULZE et al., 2003; TIEHM & SCHULZE, 2003; SCHULZE & TIEHM, 2004; SAGNER & TIEHM, 2005; TIEHM et al., 2005; SCHMIDT et al., 2006a and 2006b; MÜLLER et al., 2006; TIEHM et al., 2006a and 2006b).

Tar oil polluted landfill

Site description: The Stürmlinger sand pit (*Fig. 5.3-48*) is located in Karlsruhe, Germany. The aquifer consists of porous sand and gravel with a composition typical of the upper Rhine valley. The groundwater flow rate averages 0.8 m d^{-1} and the hydraulic conductivity (k_f) is about $2 \cdot 10^{-3}$ m s^{-1}. Between 1925 and 1956, the former sand pit was filled with municipal and industrial waste, including slag, combustion, and gasworks residues. The disposal site covers an area of about 1500 m^2. The bottom of the landfill extends to the groundwater table at about 8 m depth. In the northern part of the dump, tar oil was found as nonaqueous-phase liquid in the groundwater fluctuation zone.

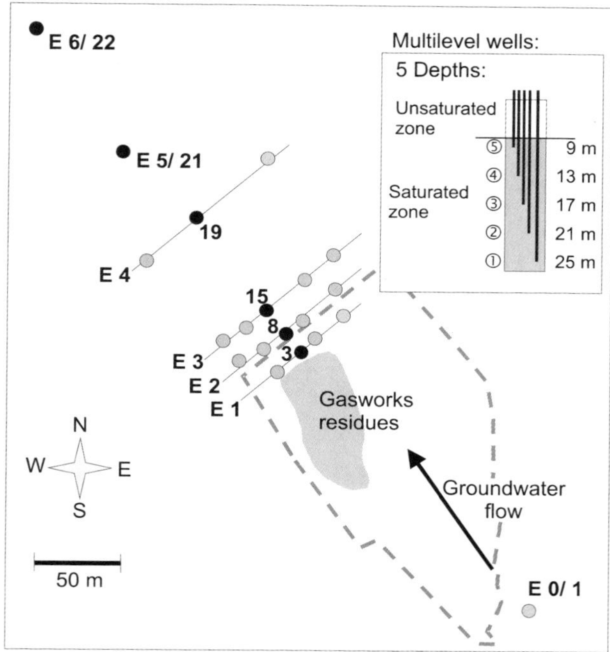

Fig. 5.3-48: Stürmlinger sand pit site, locations of multilevel sampling wells

Contaminant profile: After passing through the landfill, the groundwater is heavily contaminated with BTEX (benzene, toluene, ethylbenzene, o-, m-, p-xylene), PAH (polycyclic aromatic hydrocarbons) and heterocycles. Adjacent to the landfill, the maximum concentration of dissolved BTEX, PAH and heterocycles (concentration given for benzofuran exemplarily) reached values of 50 000 μg L^{-1}, 17 000 μg L^{-1} and 1300 μg L^{-1}, respectively. The maximum contamination was found at depths of 9 m near the landfill and at 13 m in most of the multilevel sampling wells further downgradient.

Concentrations of BTEX and PAH decline along the plume (*Fig. 5.3-49*). In *Fig. 5.3-49*, pollutant concentrations are plotted as average concentrations over all five sampling depths of the multilevel wells to eliminate flow path deviations and to increase statistical certainty. All contaminants decrease significantly with increasing distance from the disposal site. The xylenes are no longer detectable (detection limit: 1 μg L^{-1}) at a distance of 170 m and at 260 m only benzene is still measured. At this distance, the concentration of naphthalene is below 1 μg L^{-1}, despite the high initial aqueous phase concentration of about 5000 μg L^{-1}.

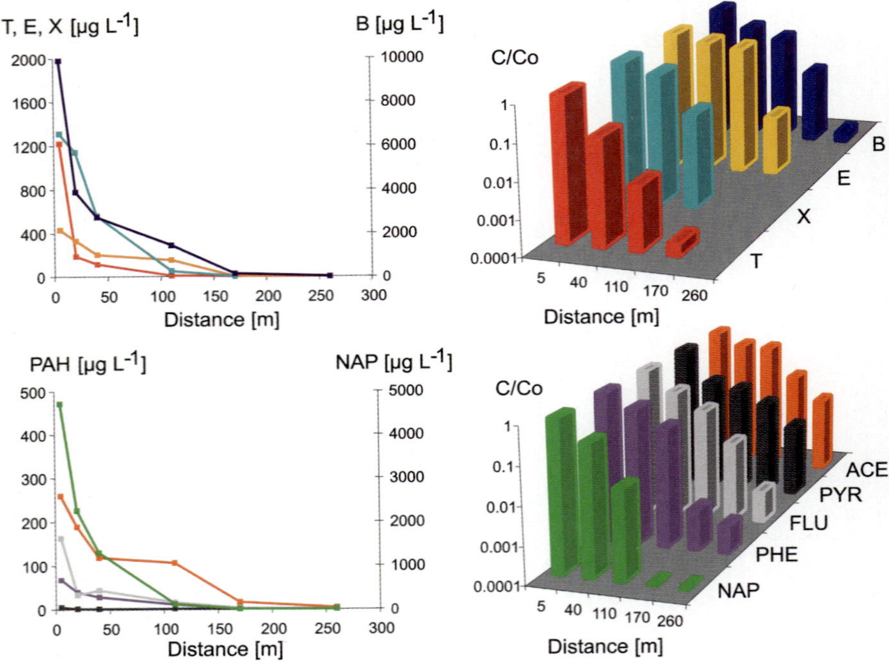

Fig. 5.3-49: Stürmlinger sand pit site, benzene, toluene, ethylbenzene, o-, m-, p-xylene (BTEX, top) and polycyclic aromatic hydrocarbons (PAH, bottom) contamination (average concentrations over the entire depth of the well), absolute (left side) and relative C/C_0 (right side, semilogarithmic plot) concentrations are shown (NAP = naphthalene, PHE = phenanthrene, FLU = fluorene, PYR = pyrene, ACE = acenaphthene)

Adjacent to the landfill, 55 % of the total mass of the BTEX and PAH was attributed to benzene, 26 % to naphthalene, and 16 % to toluene, ethylbenzene, and the xylenes (*Fig. 5.3-50*). At 260 m downgradient, 34 % of the contamination was still attributed to benzene, but naphthalene and the alkyl benzenes represented only 2 % of the remaining contaminants. The contribution of acenaphthene increased from 1 % at a distance of 5 m to 57 % after 260 m. Obviously, both the retardation and biodegradation of acenaphthene in the groundwater plume are poor, resulting in an increasing predominance of this compound. Therefore, acenaphthene might be a suitable analytical substance to monitor maximum plume length. The concentration of naphthalene in the aqueous phase decreased more rapidly (four orders of magnitude within 170 m) than the other, more hydrophobic and more sorptive PAH (two to three orders of magnitude), and toluene was degraded faster than the other alkyl benzenes. This finding is one line of evidence of active biodegradation in the anaerobic zones of the plume.

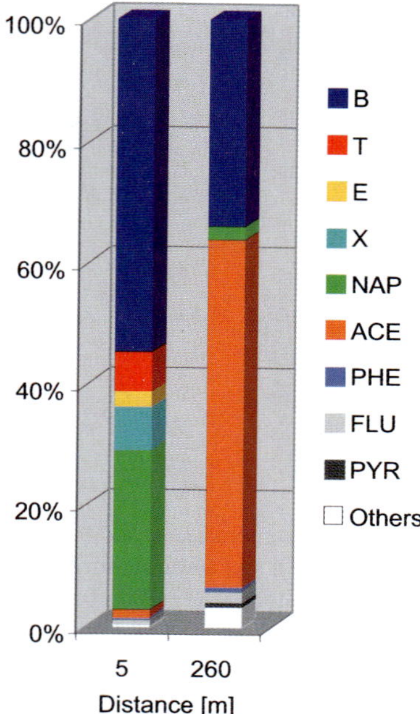

Fig. 5.3-50: Stürmlinger sand pit site: percentage composition of the sum of benzene, toluene, ethylbenzene, and xylenes (BTEX) and polycyclic aromatic hydrocarbons (PAH) in the contaminant plume at 5 m and at 260 m distance from the landfill source (NAP = naphthalene, PHE = phenanthrene, FLU = fluorene, PYR = pyrene, ACE = acenaphthene)

Redox zonation: Microbial degradation of dissolved contaminants from the landfill is reflected in the concentrations of electron acceptors and donors in the groundwater. Several redox zones were detected in the plume in which reduced concentrations of electron acceptors were observed with the simultaneous appearance of the corresponding degradation products (*Fig. 5.3-51*).

Oxygen and nitrate were detected only in the upper part of the aquifer close to the depth of the water table (9 m); 260 m downgradient, these electron acceptors were also present in the deeper horizons (*Fig. 5.3-51 A*). Elevated iron(II) concentrations in a large area of the plume indicated reduction of iron (*Fig. 5.3-51 B*). Sulfate reduction was the predominant redox process in the highly contaminated areas of the plume near the disposal site (*Fig. 5.3-51 C*). Due to the spatial proximity of iron- and sulfate-reducing processes resulting in precipitation of iron sulfide, only low concentrations of iron(II) and sulfide were found. Elevated methane concentrations were observed only in a small area near the landfill (*Fig. 5.3-51 D*). The redox zones did not have sharp boundaries and transition zones were observed, indicating that the groundwater samples might have represented composite samples of different microenvironments.

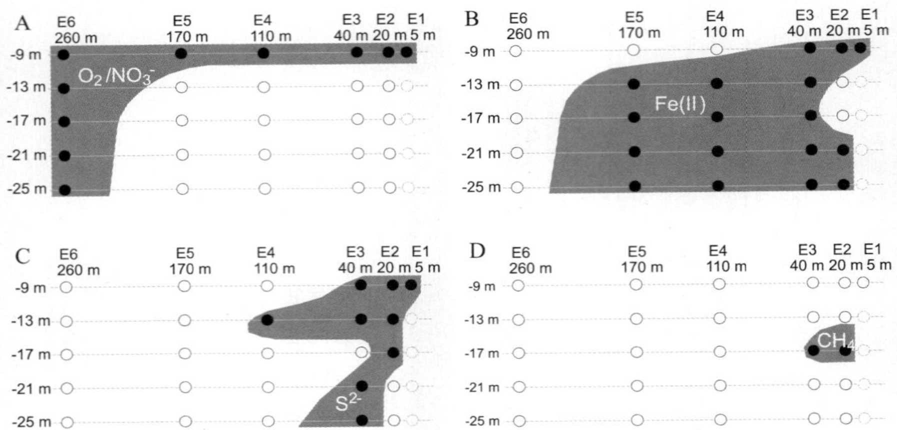

Fig. 5.3-51: Stürmlinger sand pit site, schematic redox zonation in the longitudinal section (groundwater flow is from right to left): (A) aerobic nitrate-reducing zone ($O_2 \geq 0.5$ mg L^{-1} and $NO_3^- \geq 1$ mg L^{-1}); (B) iron-reducing zone (Fe(II) ≥ 1 mg L^{-1}); (C) sulfate-reducing zone ($S^{2-} \geq 100$ µg L^{-1}); (D) methanogenic zone ($CH_4 \geq 100$ µg L^{-1})

Microbial numbers: The numbers of bacteria were determined in sediment and water samples (SCHULZE & TIEHM, 2004). The spatial distribution of bacteria in groundwater from four wells along the centerline of the plume is shown in *Fig. 5.3-52* (see *Fig. 5.3-48* for well location). Three samples were taken from each well at different depths. High numbers of total heterotrophic and of contaminant-degrading bacteria (up to 10^6 g^{-1} dry sediment) were found. Bacteria that use oxygen, nitrate, iron(III), or sulfate as terminal electron acceptors were detected in most of the samples. Most-probable-number (MPN) results indicated that iron(III) and sulfate reduction were occurring in the same areas and are not sharply separated in the field. In addition, higher microbial numbers and elevated ratios of PAH- and BTEX-degrading bacteria to total heterotrophic bacteria in the highly contaminated areas indicate active biodegradation.

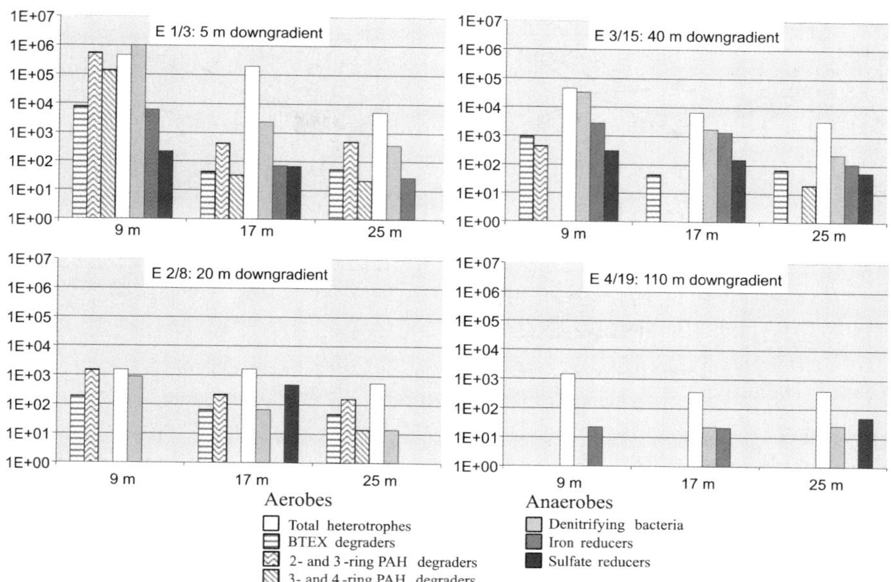

Fig. 5.3-52: Stürmlinger sand pit site, aerobic and anaerobic microbial numbers (per g of sediment dry substance) of four wells in the center line at three different depths (BTEX = benzene, toluene, ethylbenzene, o-, m-, p-xylene, PAH = polycyclic aromatic hydrocarbons)

Degradation processes: Results from microcosm studies show different degradation patterns for BTEX, PAH and heterocycles under different redox conditions. Under sulfate-reducing conditions, toluene and ethylbenzene were degraded (*Fig. 5.3-53, left*), whereas benzene and the PAH were not metabolized during 100 days of incubation. In the presence of iron(III), stimulated biodegradation of the model compounds was observed (*Fig. 5.3-53, right*). Toluene and ethylbenzene were transformed with faster kinetics than under sulfate-reducing conditions. Benzene and naphthalene were also degraded. Of the heterocycles, only benzofuran was degraded under sulfate-reducing conditions. Iron-reducing conditions led to the degradation of benzothiophene and dibenzothiophene, whereas the latter was only degraded when additional nutrients were available. Under nitrate-reducing conditions degradation of dibenzofuran was also observed (*Table 5.3-36*). The fastest degradation was observed in the presence of oxygen. After the addition of air, all BTEX, PAH and heterocycles were degraded to concentrations below the detection limit within 14 days (*Table 5.3-36*).

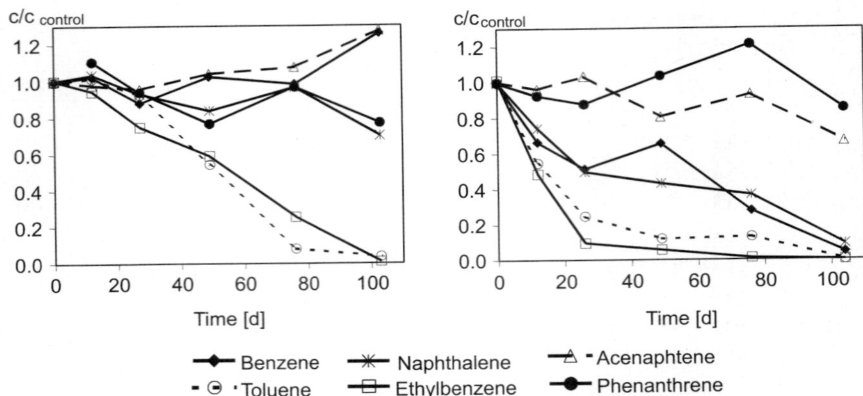

Fig. 5.3-53: Stürmlinger sand pit site, degradation of benzene, toluene, ethylbenzene and polycyclic aromatic hydrocarbons (PAH) under sulfate-reducing (*left*) and iron-reducing (*right*) conditions

Table 5.3-36: Stürmlinger sand pit site, degradation of benzene, toluene, ethylbenzene, PAH (polycyclic aromatic hydrocarbons) and heterocycles in groundwater microcosms under different redox conditions: - = no degradation; (+) = slight degradation; + = degradation of > 50 % of the sterile control

Distance to the pit	**20 m**	**110 m**	**110 m**	
Depth below water table	**10 m**	**10 m**	**2 m**	
predominant redox conditions in the field	sulfate-reducing	iron-reducing	nitrate-reducing/aerobic	
redox conditions in the microcosms	sulfate-reducing	iron-reducing	nitrate-reducing	aerobic
addition of	-	Fe(II)/Fe(III)[1]	-	oxygen
degradation of				
benzene	-	+[2]	-	+
toluene	+	+	+	+
ethylbenzene	(+)	+	+	+
naphthalene	+	+[2]	+[2]	+
acenaphthene	-	-	+[2]	+
phenanthrene	-	+[2]	-	+
pyrene	-	-	-	+
benzofuran	+	n.i.	n.i.	n.i.
dibenzofuran	-	-	+[2]	+
benzothiophene	-	+[2,3]	(+)[2]	+
dibenzothiophene	-	+[2]	+[2,3]	+
carbazole	-	-	-	+

[1] Fe(III) was available for the first 200 days, later sulfate was consumed.

[2] Only degraded in microcosms with additional nutrient (ammonia, phosphate, microelements).

[3] Slight degradation was observed without nutrients.

n.i. = not investigated

The microcosm experiments demonstrated that anaerobic biodegradation contributes significantly to contaminant elimination in the field. However, the efficiency of the biodegradation varied, depending on which electron acceptors were available (*Table 5.3-36*). It is clear that sulfate-reducing and especially iron-reducing microbial communities contribute significantly to the elimination of aromatic hydrocarbons in the anaerobic zones of the plume and can even degrade benzene, a persistent priority pollutant. All model contaminants were degraded rapidly in the presence of oxygen. This indicates complete biodegradation at the plume fringes and adjacent to the groundwater fluctuation zone, where nitrate and oxygen are available.

Investigation of the Stürmlinger sand pit site showed that in-situ biodegradation in the plume originating from the landfill for municipal waste and gasworks residues resulted in the formation of different redox zones. Three-dimensional hydrogeochemical analysis and pollutant profiles indicated that both anaerobic and aerobic processes contribute to the degradation of BTEX, PAH and heterocycles. However, the reduction of aromatic hydrocarbons could not be related to specific redox processes on the basis of the field data alone. Corresponding microcosm experiments demonstrated that toluene, ethylbenzene, naphthalene, and benzofuran were degradable under sulfate-reducing conditions. Benzene, phenanthrene, benzothiophene, and dibenzothiophene were degraded in the presence of iron(III). In the presence of oxygen, the anaerobically resistant contaminants were also degraded. Due to low retardation and slow biodegradation, acenaphthene, dibenzofuran, and dibenzothiophene were suitable compounds for monitoring the maximum plume length at this site.

Chloroethene polluted site

Site description: The Killisfeld site (*Fig. 5.3-54*) is in an industrial area in Karlsruhe, Germany, in the Rhine river valley (MÜLLER et al., 2006). The highly permeable sand and gravel aquifer in the area (hydraulic conductivity k_f = 1.5 - 3.5 10^{-3} m s^{-1}; groundwater velocity: 0.4 - 0.9 m d^{-1}; water table: 2 - 3 m below surface; aquifer thickness: 12 - 16 m) is interspersed with less permeable areas rich in humic acids. There is a chloroethene plume in the aquifer originating from at least three different source areas. There are also several domestic waste landfills downgradient from the chloroethene source areas from which organic substances infiltrate into the groundwater.

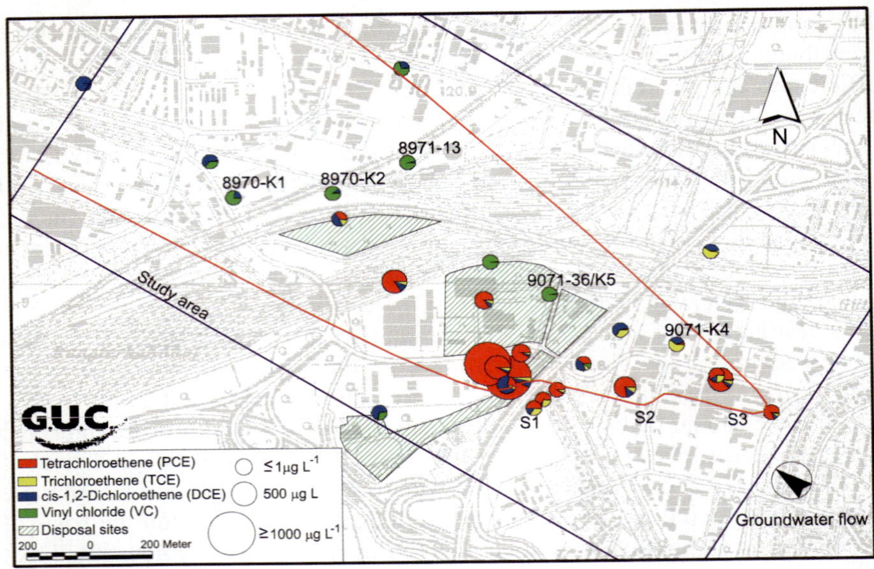

Fig. 5.3-54: Killisfeld site, chloroethene contamination (average vertical concentrations) originating from three source zones (S1-S3), locations of domestic waste disposal sites and investigated wells (with permission of G.U.C. GEO UMWELT CONSULT GmbH, Karlsruhe, Germany)

Contaminant profiles: The groundwater is contaminated with the chlorinated ethenes perchloroethene (PCE) and trichloroethene (TCE) from at least three source zones (S1-S3 in *Fig. 5.3-54*), resulting in three parallel plumes close to the sources. Further downgradient the three plumes merge and in total reach a plume length of at least 2.5 km. Immediately downgradient from the source zones, chloroethene concentrations of > 2000 μg L^{-1} have been determined. For approximately the first 600 m of the flow path the contamination mainly consists of PCE and to a lesser extent TCE and *cis*-1,2-dichloroethene (cDCE). Vinyl chloride (VC) is not detected until the plumes pass the disposal sites further downgradient, where it is the predominating contaminant. Ethene and ethane, dechlorinated end-products of microbial reductive degradation, are also detected downgradient from the disposal sites (see *Fig. 5.3-55* for the northern part of the plume). Since the chloroethene contamination decreases along the flow path and the typical microbial degradation products cDCE, VC, ethene, and ethane are observed downgradient from the disposal sites, microbial degradation processes can be assumed to occur at this site.

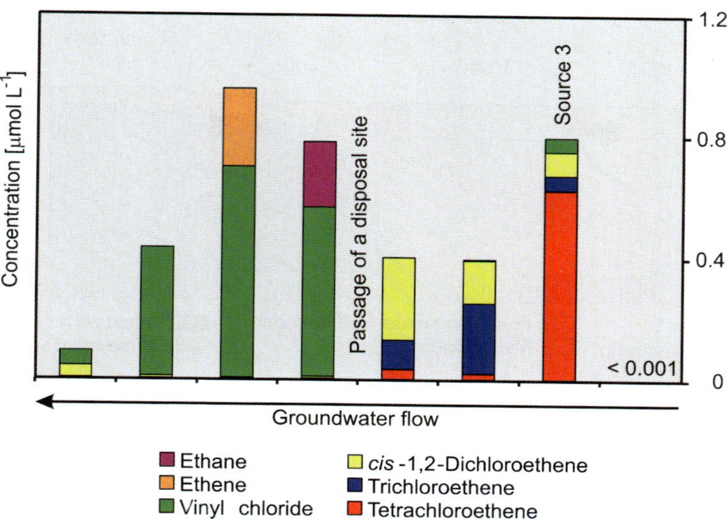

Fig. 5.3-55: Killisfeld site, distribution of chloroethenes, ethene and ethane (average vertical concentrations) along the flow path of the northern plume

Redox zonation: A distinct redox zonation is observed in all three plumes downgradient from the source zones (see *Fig. 5.3-56*). Nitrate and nitrite are detected close to the source zones along with a relatively high redox potential of 288 - 343 mV, indicating nitrate-reducing conditions. Iron-/manganese-reducing conditions can be deduced from the detection of increased concentrations of the reduced, soluble iron and manganese forms iron(II) and manganese(II) 700 - 1000 m from the source. The soluble iron and manganese ions are formed by reduction of iron(III) and manganese(IV) minerals in the aquifer. Dissolved sulfide was not detected in the groundwater samples. However, iron(II) sulfide was found in the sediment. Since iron(III) as well as sulfate-reducing microorganisms were detected in the plume (*Fig. 5.3-57*), both processes are most probably occurring in a mixed iron(III)-/sulfate-reducing area. Organic substances (elevated total organic carbon) from the disposal sites create locally more strongly reducing methanogenic conditions (detection of up to 2000 $\mu g\ L^{-1}$ methane).

The strongly reducing regions of the aquifer are favorable for the reductive dechlorination of the chloroethenes. Therefore, VC is mainly detected in methanogenic zones. The prevailing redox conditions indicate that aerobic degradation is possible only at the plume fringes, i.e. further downgradient and in the groundwater fluctuation zones, where oxygen is available.

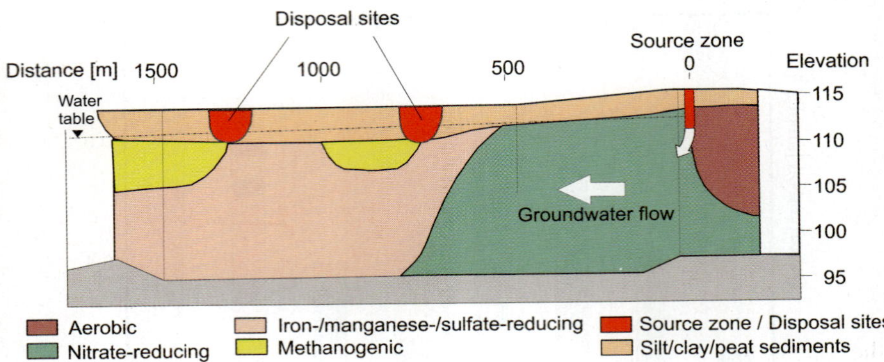

Fig. 5.3-56: Killisfeld site, schematic representation of the redox zones along the northern plume, aerobic zone: $O_2 \geq 0.5$ mg L^{-1}; nitrate-reducing zone: $NO_3^- \geq 1$ mg L^{-1}; iron-/manganese-/sulfate-reducing zone: Fe(II) ≥ 1 mg L^{-1}, Mn(II) > 0.2 mg L^{-1}; methanogenic zone: $CH_4 \geq 0.5$ mg L^{-1}

Microbial numbers: A microbial survey was conducted using several most-probable-number (MPN) tests on sediment samples taken at different depths during construction of four groundwater wells (see *Fig. 5.3-54* for well locations):

- 9071-K4: downgradient close to the chloroethene source; moderately reducing conditions; detection of PCE, TCE and cDCE.
- 9071-K5: downgradient close to one of the disposal sites; strongly reducing conditions; detection of VC and ethane.
- 8970-K2: further downgradient; reducing conditions; cDCE and VC predominate.
- 8970-K1: further downgradient; reducing conditions; cDCE and VC predominate.

The microbial numbers of total aerobic heterotrophs, BTEX (benzene, toluene, ethylbenzene, o-, m-, p-xylene) degraders, ammonia- and nitrite-oxidizing bacteria, and methanotrophic bacteria under aerobic conditions were determined. Anaerobic nitrate-/iron(III)-/sulfate-reducing bacteria and methanogenic bacteria were investigated (see *Fig. 5.3-57* for results of two wells).

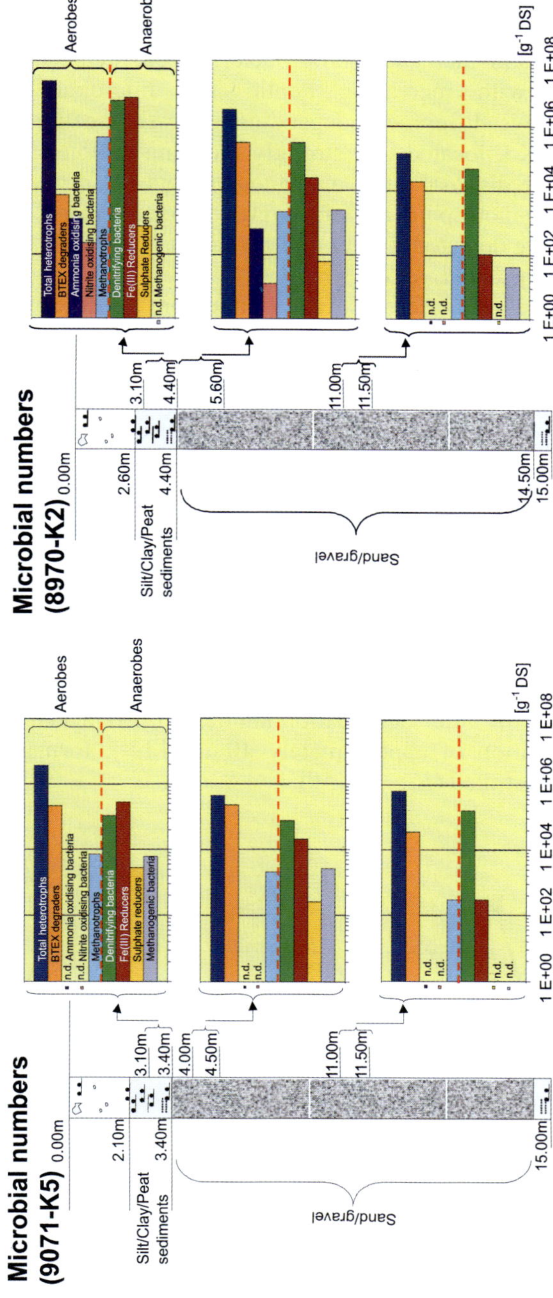

Fig. 5.3-57: Killisfeld site, aerobic and anaerobic microbial numbers (per g of sediment dry substance (DS)) of two wells at three different depths (n.d. = not detected, BTEX = benzene, toluene, ethylbenzene, o-, m-, p-xylene)

All four wells investigated showed very high microbial numbers in regions where silt/clay/peat sediments predominate. High numbers of aerobic, BTEX-degrading, nitrifying and methanotrophic bacteria were detected. Microbial numbers decreased with increasing depth. The microbial community composition also shows differences depending on well location within the plume. In well 9071-K5, located in a strongly reducing region of the aquifer downgradient from the disposal sites, methanogenic but no nitrifying bacteria and only a few methanotrophic bacteria were detected. Large numbers of aerobic organisms were detected further downgradient in well 8970-K2, where less reducing conditions occurred.

Degradation processes: Aerobic and anaerobic dechlorination was observed with groundwater from different parts of the plume. Results are shown for the following two wells:

- 9071-36: downgradient close to a disposal site; strongly reducing conditions; detection of VC and ethane.
- 8971-13: further downgradient; reducing conditions; VC predominates.

Perchloroethene (9 µmol L^{-1}) was completely dechlorinated during anaerobic incubation with acetate and hydrogen as electron donors (*Fig. 5.3-58*). The intermediary metabolites TCE and cDCE were detected, but not VC and ethene. It is assumed that VC and ethene were formed in concentrations below the detection limit (306 µg L^{-1} = 4.9 µmol L^{-1} for VC and 133 µg L^{-1} = 4.7 µmol L^{-1} for ethene) due to the relatively low initial concentration of PCE. In other degradation studies with repeated addition of PCE (cumulative addition of approximately 40 µmol L^{-1}), formation of VC and ethene was demonstrated (*Fig. 5.3-59*).

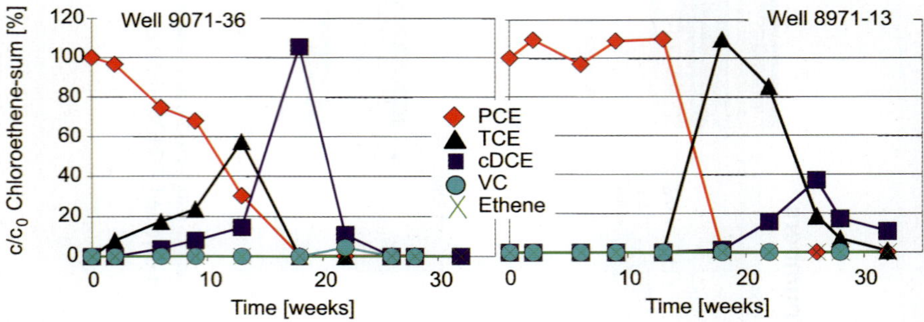

Fig. 5.3-58: Killisfeld site, reductive dechlorination in groundwater from different zones of the plume (PCE = perchloroethene, TCE = trichloroethene, cDCE = *cis*-1,2-dichloroethene, VC = vinyl chloride)

Groundwater from well 9071-36 directly downgradient from a disposal site in a strongly reducing zone of the plume showed faster dechlorination than groundwater from well 8971-13 further downgradient. A lag phase of about 12 weeks occurred with groundwater from well 8971-13 before the onset of dechlorination (*Fig. 5.3-58*). Addition of yeast extract, trace elements and vitamins did not result in an increase in the dechlorination rate (data not shown). Hence, groundwater from the investigated wells contained all nutrients required by dechlorinating microorganisms.

The lower chlorinated metabolites of reductive dechlorination were also degradable under aerobic conditions (*Fig. 5.3-60*). Again, degradation of cDCE (4.1 µmol L-1) and VC (19.2 µmol L-1) was faster with groundwater from well 9071-36 close to the source zones. The increasing degradation rate of VC after repeated spiking with VC indicates productive degradation of this compound. Cis-1,2-dichloroethene was degraded as long as VC was available in the microcosm, suggesting co-metabolic degradation of cDCE with VC as auxiliary substrate.

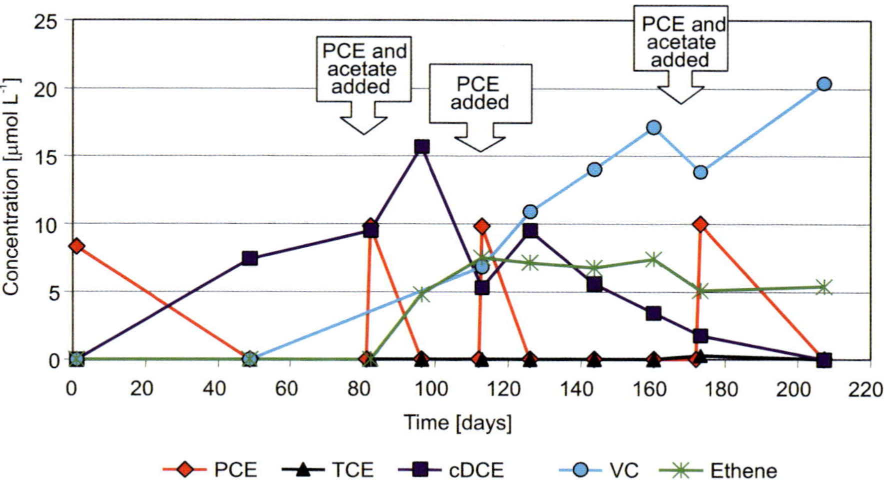

Fig. 5.3-59: Killisfeld site, formation of vinyl chloride (VC) and ethene during reductive dechlorination of perchloroethene (PCE) (TCE = trichloroethene, cDCE = *cis*-1,2-dichloroethene)

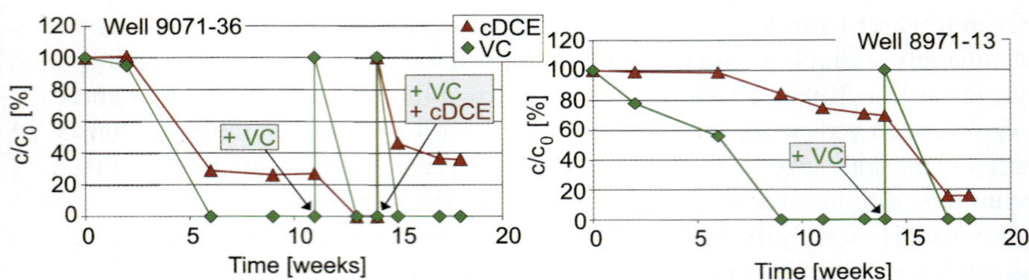

Fig. 5.3-60: Killisfeld site, aerobic dechlorination of *cis*-1,2-dichloroethene (cDCE) and vinyl chloride (VC) in groundwater from different zones of the plume

The results are consistent with site data. The groundwater downgradient from the disposal sites contains VC, ethene and ethane – products of microbial reductive dechlorination. Organic substances from the disposal sites evidently create strongly reducing conditions, and provide auxiliary substrates required for reductive dehalogenation. Aerobic degradation is possible at the plume fringes, i.e. further downgradient and in the groundwater fluctuation zones, where oxygen is available.

Analysis of specific degrading bacteria: Groundwater samples were assayed for the halorespiring microorganisms *Dehalobacter* sp., *Desulfuromonas* sp., *Desulfomonile tiedjei*, *Desulfitobacterium* sp., and *Dehalococcoides* sp. (BRADLEY, 2003) with nested polymerase chain reaction (PCR). All sampling wells are in the plume downgradient from the source areas. DNA sequences from *Desulfuromonas* sp., *Desulfomonile tiedjei* and *Dehalococcoides* sp. were detected in groundwater samples (*Fig. 5.3-61*). Bacteria belonging to the *Dehalococcoides* cluster are the only microorganisms reported so far to catalyze the complete reductive dechlorination of PCE to ethene (MAYMÓ-GATELL et al., 1997; FETZNER, 1998; MIDDELDORP et al., 1999; BRADLEY, 2003). Their presence is, therefore, of particular interest for sites contaminated with chloroethenes, since they are often detected at sites where complete reductive dechlorination occurred (HENDRICKSON et al., 2002; SCHMIDT et al., 2006a).

At the Killisfeld site *Dehalococcoides* were even found in areas within the plume where no VC or ethene were detected and, therefore, the contaminant distribution did not indicate the presence of *Dehalococcoides*. There are two possible explanations for this: (i) Since 16S-PCR detection of *Dehalococcoides* does not differentiate between the various strains, it is possible that incompletely dechlorinating strains were also detected (LÖFFLER & RITALAHTI, 2005) and (ii) another explanation might be that the detected *Dehalococcoides* sequences reflect previous site conditions more favorable for the growth of *Dehalococcoides* (HENDRICKSON et al., 2002).

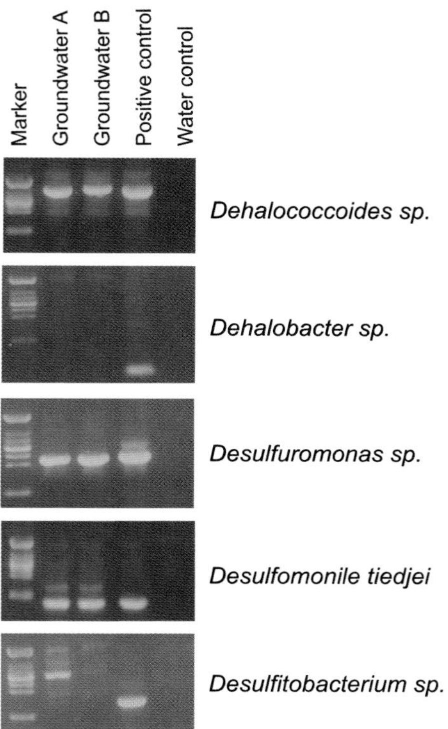

Fig. 5.3-61: Killisfeld site, polymerase-chain-reaction (PCR) detection of halorespiring bacteria in groundwater

Moreover, PCR allowed growth of *Dehalococcoides* during dechlorination of PCE to be monitored in a microcosm study. The microcosm was assayed for concentrations of chloroethenes and auxiliary substrates, total cell numbers (TCN), protein content and specific *Dehalococcoides* sequences (*Fig. 5.3-62*) at different points in time during degradation. *Figure 5.3-62A* shows the complete degradation of PCE by reductive dechlorination consuming the electron donors acetate and hydrogen (*Fig. 5.3-62B*). VC and ethene were detected in only trace amounts below the limit of quantification and are, therefore, not shown on the graph. Growth of microorganisms is demonstrated by increasing total cell number (TCN) and protein content (*Fig. 5.3-62C*). The increase in the amount of the *Dehalococcoides*-specific 16S-PCR product on day 16 of incubation (*Fig. 5.3-62D*) demonstrates that the number of *Dehalococcoides* increased during PCE degradation.

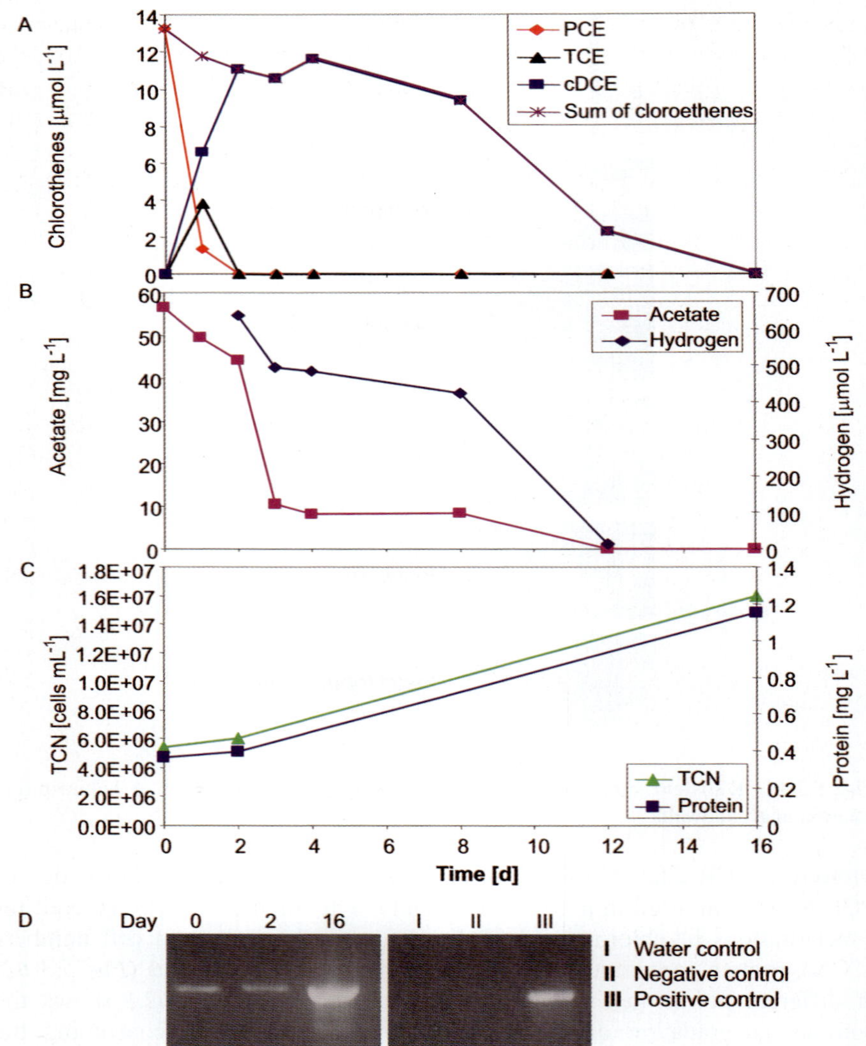

Fig. 5.3-62: Killisfeld site, degradation of perchloroethene (PCE) by a *Dehalococcoides* containing a mixed enrichment culture, A: chloroethene concentrations; B: acetate and hydrogen concentrations; C: total cell number (TCN) and protein content; D: direct 16S-PCR (polymerase chain reaction) (TCE = trichloroethene, cDCE = *cis*-1,2-dichloroethene)

In order to prove the assumption that *Dehalococcoides* bacteria only occur when complete reductive dechlorination takes place, incomplete dechlorinating microcosms from another chloroethene contaminated site (Frankenthal, Germany) were compared with Killisfeld microcosms. Six reductively dechlorinating groundwater microcosms from the Killisfeld site and nine from the Frankenthal site were analyzed for rDNA sequences of *Dehalococcoides* with nested 16S-PCR. All microcosms originating from the Killisfeld site showed complete anaerobic degradation of PCE, TCE and cDCE (see *Fig. 5.3-63*, *left*). In contrast, the Frankenthal microcosms only dechlorinated PCE via TCE to cDCE, which accumulated (see *Fig. 5.3-63*, *right*). The 16S-PCR results correspond to the observed dechlorinating capabilities of the microcosms. Organisms of the *Dehalococcoides* cluster were detected only in completely dechlorinating microcosms (*Fig. 5.3-63*, *bottom*). These results confirm previous reports that only *Dehalococcoides* are capable of degrading PCE completely to ethene (MAYMÓ-GATELL et al., 1997; FETZNER, 1998; MIDDELDORP et al., 1999; BRADLEY, 2003).

Further comparison of the two sites Killisfeld and Frankenthal has shown that contaminant profiles, hydrochemical site characteristics, dechlorination activity in microcosms, and the presence of *Dehalococcoides* correlated well (*Table 5.3-37*). Bacteria belonging to the *Dehalococcoides* cluster were only detected in microcosms where complete reductive dechlorination was observed. Other dechlorinating microorganisms were detected at both sites. 16S-PCR detection of specific contaminant-degrading bacteria may complement analyses of contaminant profiles at contaminated sites and may be used instead of rather long degradation studies in the laboratory.

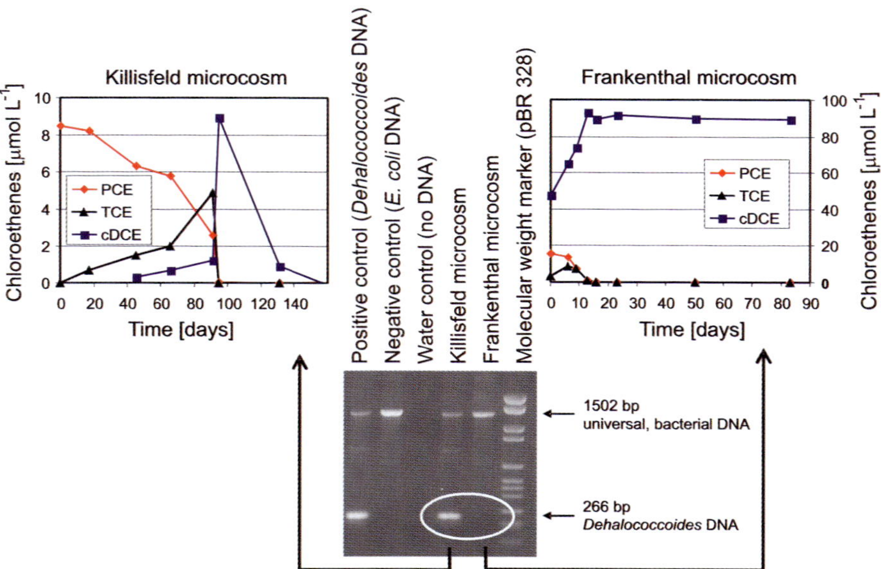

Fig. 5.3-63: Killisfeld and Frankenthal sites, dechlorination activity and 16S-PCR (polymerase chain reaction) results of groundwater microcosms (PCE = perchloroethene, TCE = trichloroethene, cDCE = *cis*-1,2-dichloroethene)

Table 5.3-37: Killisfeld and Frankenthal sites, site characteristics and polymerase chain reaction (PCR) results (PCE = perchloroethene, TCE = trichloroethene, cDCE = *cis*-1,2-dichloroethene, ORP = oxidation-reduction-potential)

	Killisfeld	**Frankenthal**
primary contaminants on site	PCE and TCE	PCE and TCE
redox conditions on site (ORP)	47 - 469 mV	140 - 555 mV
degradation products detected on site	cDCE, VC, ethene	mainly cDCE
dechlorination in groundwater microcosms	PCE → → → → ethene	PCE → → cDCE
16S-PCR detection of Dehalococcoides (PCE → ethene) in vertically averaged groundwater samples and microcosms	positive	negative[1]
16S-PCR detection of other halorespirers (PCE → cDCE) in groundwater samples and microcosms	positive	positive

1 *Dehalococcoides* were only found in one of 17 investigated wells when multilevel sampling techniques were used. Only in this area was vinyl chloride (VC) also detected.

Isotope fractionation: Selected aerobic microcosms from the Killisfeld site were assayed for isotope fractionation. Significant fractionation was observed during aerobic degradation of cDCE with VC (*Fig. 5.3-64*). Analyses for the Frankenthal site also showed isotope fractionation during anaerobic degradation of PCE and TCE (data not shown). The enrichment factors can be used for conservative quantification of biodegradation in the field (MARTIN et al., 2006).

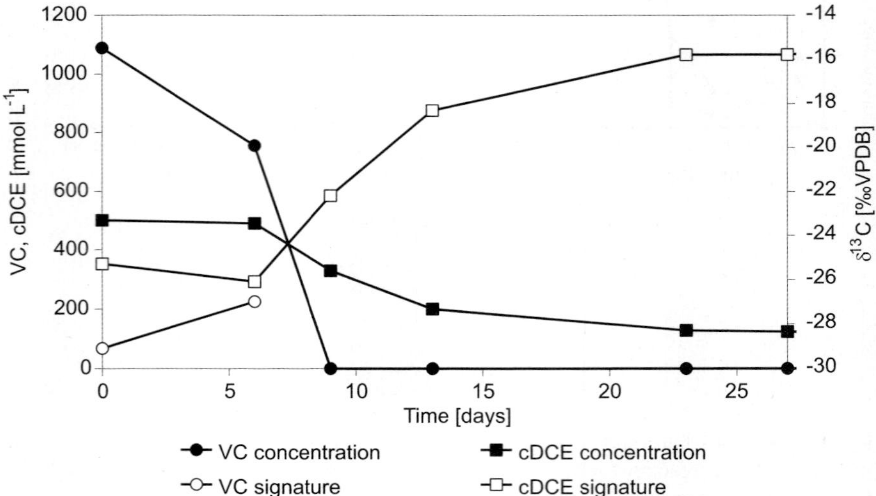

Fig. 5.3-64: Killisfeld site, isotope fractionation during aerobic degradation of *cis*-1,2-dichloroethene (cDCE) and vinyl chloride (VC)

At the Killisfeld site, the analysis of the hydrochemical data showed the occurrence of strongly reducing conditions along with elevated total organic carbon (TOC) (possible auxiliary substances), indicating site conditions favorable for reductive dechlorination of the contaminating chloroethenes. The detection of specific metabolites of microbial degradation (cDCE, VC and ethene) then demonstrated that microbial dechlorination occurs at this site. Investigation of microbial numbers and detection of specific halorespiring microorganisms further confirmed that microorganisms with the metabolic traits needed for degradation are present and viable in the field. Finally, microcosm studies proved that the autochthonous microflora is able to reductively degrade PCE completely to ethene under anaerobic conditions and to degrade cDCE and VC under aerobic conditions. The applied methods complemented each other. The results led to the conclusion that reductive dechlorination of PCE to TCE, cDCE, VC, ethene, and ethane occurred in the strongly anaerobic parts of the plume and that the lower chlorinated metabolites are partly degraded in regions where oxygen is available (plume fringes).

Summary of case studies: The results obtained at the Stürmlinger sand pit and Killisfeld sites demonstrate that the different assessment methods complement each other and allow a well-founded evaluation of the degradation processes occurring in the field. The different techniques can be combined depending on the specific site conditions. Similar strategies are recommended for other classes of biodegradable contaminants.

References and further reading

ACTON, D. W. & BARKER, J. F. (1992): In-situ biodegradation potential of aromatic hydrocarbons in anaerobic groundwaters. J. Contamin. Hydrol., **9**, 325-352.

ADAMS, J. M. & EVANS, S. (1979): Determination of the cation-exchange capacity (layer charge) of small quantities of clay minerals by nephelometry. Clays Clay Minerals, **27**, 137-139.

AG BODEN (1996): Bodenkundliche Kartieranleitung, 4. Aufl. (AG Boden ("Soil" Working Group of the German Federal and State Geological Surveys) (1996). Soil mapping manual, 4th edn.).

ALLABY, M. (1977): A Dictionary of the Environment. Van Nostrand Reinhold, New York.

ALTES, J. (1976): Die Grenztiefe bei Setzungsberechnungen. Bauingenieur, **51**, 93-96.

ALTFELDER S., STRECK T. & RICHTER J. J. (1999): Effect of air-drying on sorption kinetics of the herbicide Chlorotoluron in soil. J. Environ. Qual., **28**, 1154-1161.

ALTFELDER, S. & STRECK, T. (2006): Capability and limitations of first-order and diffusion approaches to describe long-term sorption of chlortoluron in soil. J. Contam. Hydrol., **86**, 279-298.

ALTHOFF, K., MUNDT, M., EISENTRAEGER, A., DOTT, W. & HOLLENDER, J. (2001): Microcosm-experiments to assess the potential for natural attenuation of contaminated groundwater. Water Res., **35**, 3, 720-728.

AMONETTE, J. E. & ZELAZNY, L. W. (Eds.) (1994): Quantitative Methods in Soil Mineralogy. Miscellaneous Publication, Soil Science Society of America, Madison.

ANDREWS, J. E., BRIMBLECOMBE, P., JICKELLS, T., LISS, P. & REID, B. (2004): An Introduction to Environmental Chemistry. Blackwell.

APPELO, C.A., VAN DER WEIDEN, M. J., TOURNASSAT, C. & CHARLET, L. (2002): Surface complexation of ferrous iron and carbonate on ferrihydrite and the mobilization of arsenic. Environ. Sci. Technol., **36**, 14, 3096-3103.

APPELO, C. A. J.; POSTMA, D. (2005): Geochemistry, groundwater and pollution. 2nd edn., A. A. Balkema Publishers

AULENTA, F., Bianchi, A., Majone, M., Petrangeli Papini, M., Potalivo M. & Tandoi V. (2005): Assessment of natural or enhanced in-situ bioremediation at a chlorinated solvent-contaminated aquifer in Italy: a microcosm study. Environ Int., **31**, 2, 185-190.

Aulenta, F., MAJONE, M. VERBO, P. & TANDOI, V. (2002): Complete dechlorination of tetrachloroethene to ethene in presence of methanogenesis and acetogenesis by an anaerobic sediment microcosm. Biodegradation, **13**, 411-424.

BACHE, B. W. (1970): Barium isotope method for measuring cation-exchange capacity of soils and clays. J. Sci. Food and Agric., **21**, 169-171.

BACHE, B. W. (1976): The measurement of cation exchange capacity of soils. J. Sci. Food Agric., **27**, 273-280.

BAILEY, S. W. (1980): Summary of recommendations of AIPEA nomenclature committee on clay minerals. American Mineralogist, **65**, 1-7.

BANNERT, M., BERGER, W., FISCHER, H., HORCHLER, D., KEESE, K., LEHNIK-HABRINK, P., LÜCK, D., PRITZKOW, J. & WIN, T. (2001): Anforderungen an Probennahme Probenvorbehandlung und chemische Untersuchungsmethoden auf Bundesliegenschaften. (Eds.), Bundesanstalt für Materialforschung und -prüfung (BAM) Berlin, Amts- und Mitteilungsblatt der BAM, Sonderheft 2/2001. www.bam.de/de/service/publikationen/publikationen_medien/probennahme.pdf

BARBER, C. & BRIEGEL, D. (1987): A method for the in-situ determination of dissolved methane in groundwater in shallow aquifers. J. of Contaminant Hydrology, **2**, 51- 60.

BARTH, J. A. C., SLATER, G., SCHÜTH, C., BILL, M., DOWNEY, A., LARKIN, M. & KALIN, R. M. (2002): Carbon isotope fractionation during aerobic biodegradation of trichloroethene by Burkholderia cepacia G4: a tool to map degradation mechanisms. Appl. Environ. Microbiol., **68**, 4, 1728-1734.

BASCOMB, C. L. (1964): Rapid method for the determination of the cation exchange capacity of calcareous and non-calcareous soils. J. Sci. Food and Agric., **15**, 821-823.

BEAR, J. (1972): Dynamics of fluids in porous media. Elsevier, New York.

BEHRENS, W. & FEISER, J. (1995): Anmerkungen zur Berechnung der Setzungen von Deponiebauwerken. AbfallwirtschaftsJournal, **7**, 9, 545-549.

BELLER, H. R. (2002): Analysis of benzylsuccinates in groundwater by liquid chromatography/tandem mass spectrometry and its use for monitoring in situ BTEX biodegradation. Environ. Sci. Technol., **36**, 2724-2728.

BELLER, H. R., DING, W.-H. & REINHARD, M. (1995): Byproducts of anaerobic alkylbenzene metabolism useful as indicators of in situ bioremediation. Environ. Sci. Technol., **29**, 2864-2870.

BERGAYA, F., THENG, B. K. G. & LAGALY, G. (Eds.) (2006): Handbook of Clay Science. Elsevier.

BERGAYA, F. & VAYER, M. (1997): CEC of clays; measurement by adsorption of a copper ethylenediamine complex. Appl. Clay Sci., **12**, 275-280.

BIELEFELDT A. R. & STENSEL H. D. (1999): Biodegradation of aromatic compounds and TCE by a filamentous bacteria-dominated consortium. Biodegradation, **10**, 1-13.

BILITEWSKI, B., HÄRDTLE, G. & MAREK, K. (2000): Abfallwirtschaft (Waste management). 3rd edn. Springer, Berlin.

BILITEWSKI, B. (2001): Praktikumsskript „Altlasten“ (Laboratory notes for lecture on hazardous sites). Dresden-Pirna.

BJERG, P. L., BRUN, A., NIELSEN, P. H. & CHRISTENSEN, T. H. (1996): Application of a model accounting for kinetic sorption and degradation to in situ microcosm observations on the fate of aromatic hydrocarbons in an anaerobic aquifer. Water Resources Research, **32**, 6, 1831-1841.

BJERG, P. L., RÜGGE, K., CORTSEN, J., NIELSEN, P. H. & CHRISTENSEN, T. H. (1999): Degradation of aromatic and chlorinated aliphatic hydrocarbons in the anaerobic part of the Grindsted landfill leachate plume: In situ microcosm and laboratory batch experiments. Ground Water, **37**, 1, 113-121.

BLOOM Y., ARAVENA R., HUNKELER D., EDWARDS E. & FRAPE S. K. (2000): Carbon isotope fractionation during microbial degradation of trichloroethene, cis-1,2-dichloroethene, and vinyl chloride: Implications for assessment of natural attenuation. Environ. Sci. Technol., **34**, 13, 2768-2772.

BORDEN R. C., DANIEL R. A., LEBRUN IV L. E. & DAVIS C. W. (1997): Intrinsic biodegradation of MTBE and BTEX in a gasoline-contaminated aquifer. Water Resources Research, **33**, 5, 1105-1115.

BORDEN R. C., GOMEZ C. A. & BECKER M. T. (1995): Geochemical indicators of intrinsic bioremediation. Ground Water, **33**, 2, 180-189.

BOWEN, H. J. M. (1979): Environmental chemistry of the elements. Academy Press, London.

BRADLEY, P. M. (2003): History and ecology of chloroethene biodegradation: A review. Bioremediation Journal, **7**, 2, 81-109.

BRAR, S. K. & GUPTA, S. K. (2000): Biodegradation of trichloroethylene in a rotating biological contactor. Water Res., **34**, 17, 4207-4214.

BREEN, C. & ROCK, B. J. (1991): The competitive adsorption of dyes on clays. 7th Euroclay Conference, Dresden.

BRUNAUER, S., EMMETT, P. H. & TELLER, E. (1938): Adsorption of gases in multimolecular layers. J. Amer. Chem. Soc., **60**, 309-319.

BRUSSEAU, M. L. & RAO P. S. C. (1989): Sorption nonideality during organic contaminant transport in porous media. Crit. Rev. Environ. Control, **19**, 33-99.

BRUSSEAU, M. L., JESSUP, R. E. & RAO, P. S. C. (1991): Nonequilibrium Sorption of Organic Chemicals: Elucidation of Rate-Limiting Processes. Environ. Sci. and Technol., **25**, 134-142.

BUHMANN, D. & DREYBRODT, W. (1987): Calcite dissolution kinetics in the system H2O-CO2-CaCO3 with participation of foreign ions. Chem. Geology, **64**, 89-102.

BUJDÁK J. & KOMADEL P. (1997): Interaction of Methylene Blue with reduced charge montmorillonite. J. Phys. Chem. B, **101**, 9065.

BUWAL (2003): Praxishilfe Grundwasserprobenahme. Bundesamt für Umwelt, Wald und Landschaft, Bern, Switzerland.

BWK (2004): Erarbeitung von Leistungsbeschreibungen und Leistungsverzeichnissen zur Grundwasserprobenahme bei Altlasten im Lockergestein. BWK-Merkblatt 5. Gelbdruck. Bund der Ingenieure für Wasserwirtschaft, Abfallwirtschaft und Kulturbau, Pfullingen, Germany.

CERNIGLIA, C. E. (1992): Biodegradation of polycyclic aromatic hydrocarbons. Biodegradation, **3**, 351-368.

CHAPELLE F. H., MCMAHON P. B., DUBROWSKY N. M., FUJII, R. F., OAKSFORD E. T. & VROBLESKY D. A. (1995): Deducing the distribution of terminal electron-accepting processes in hydrologically diverse groundwater systems. Water Resources Research, **31**, 2, 359-371.

CHAPELLE, F. H. (2001): Ground-water Microbiology and Geochemistry. Wiley & Sons, New York.

CHARTRAND, M. M. G., WALLER, A., MATTES, T. E., ELSNER, M., LACRAMPE-COULOUME, G., GOSSETT, J.M., EDWARDS, E. A. & SHERWOOD LOLLAR, B. (2005): Carbon isotopic fractionation during aerobic vinyl chloride degradation. Environ. Sci. Technol., **39**, 4, 1064-1070.

CHHABRA, R., PLEYSIER, J. & CREMERS, A. (1975): The measurement of the cation exchange capacity and exchangeable cations in soils: A new method. Proc. Int. Clay Conf. 1975, Wilmette, Illinois, 439-449.

CHILDS, C. W. (1992): Ferrihydrite: A review of structure, properties and occurrence in relation to soils. Z. Pflanzenernähr. Bodenk., **155**, 441-448.

CHIOU, C. T., PETERS L. J., & FREED, V. H. (1979): A Physical Concept of Soil-Water Equilibria for Non-Ionic Compounds Science, **206**, 831-832.

CHIOU, C. T., PORTER P. E. & SCHMEDDLING D. W. (1983): Partition equilibria of nonionic organic compounds between soil organic matter and water. Environ. Sci. Technol., **17**, 227-231.

CHO, J. S., WILSON, J. T., DIGIULIO, D. C., VARDY, J. A. & CHOI, W. (1997): Implementation of natural attenuation at a JP-4 jet fuel release after active remediation. Biodegradation, **8**, 265-273.

CHRISTENSEN, T. H., BJERG, P. L. & KJELDSEN, P. (2001): Natural attenuation as an approach to remediation of groundwater pollution at landfills. In: Treatment of Contaminated Soil, STEGMANN, R., BRUNNER, G., CALMANO, W. & MATZ, G. (Eds.), Springer, Berlin, 587-602.

CHU, K.-H., MAHENDRA, S., SONG, D. L., CONRAD, M.E. & ALVAREZ-COHEN, L. (2004): Stable carbon isotope fractionation during aerobic biodegradation of chlorinated ethenes. Environ. Sci. Technol., **38**, 11, 3126-3130.

COHEN, R. M. & MERCER, J. W. (1993): DNAPL site evaluation. Smoley, Boca Raton, FL.

COLEMAN, N. V., MATTES, T. E., GOSSETT, J. M. & SPAIN, J. C. (2002a): Biodegradation of cis-dichloroethene as the sole carbon source by a β-proteobacterium. Appl. Environ. Microbiol., **68**, 6, 2726-2730.

COLEMAN, N. V., MATTES, T. E., GOSSETT, J. M. & SPAIN, J. C. (2002b): Phylogenetic and kinetic diversity of aerobic vinyl chloride-assimilating bacteria from contaminated sites. Appl. Environ. Microbiol., **68**, 12, 6162-6171.

CORNELL, R. M. & SCHWERTMANN, U. (2003): The iron oxides: structure, properties, reactions, occurrences and uses. Wiley-VCH, Weinheim.

Cotter-Howells, J. D. Campbell, L. S., Valsami-Jones, E. & Batchelder, M. (2000): Environmental Mineralogy: Microbial Interactions, Anthropogenic Influences, Contaminated Land and Waste Management. Mineralogical Society Book Series Vol. 9.

COZZARELLI, I. M., BAEDECKER, M. J., EGANHOUSE, R. P. & GOERLITZ, D. F. (1994): The geochemical evolution of low-molecular-weight organic acids derived from the degradation of petroleum contaminants in groundwater. Geochim. Cosmochim. Acta, **58**, 2, 863-877.

COZZARELLI, I. M., HERMAN, J. S. & BAEDECKER, M. J. (1995): Fate of microbial metabolites of hydrocarbons in a coastal plain aquifer: The role of electron acceptors. Environ. Sci. Technol., **29**, 458-469.

CREMERS, A. & PLEYSIER, J. (1973a): Adsorption of the silver-thiourea complex in montmorillonite clay. Nature Phys. Sci., **243**, 86-87.

CREMERS, A. & PLEYSIER, J. (1973b): Coordination of silver in silver-thiourea montmorillonite. Nature Phys. Sci., **244**, 93.

DANKO, A. S., LUO, M., BAGWELL, C. E., BRIGMON, R. L. & FREEDMAN, D. L. (2004): Involvement of linear plasmids in aerobic biodegradation of vinyl chloride. Appl. Environ. Microbiol., **70**, 10, 6092-6097.

DAVIS, S. N. & DE WIEST, R. J. M. (1966): Hydrogeology. Wiley & Sons, New York.

DÉKÀNY, I., SZÁNTÓ, F. & NAGY, L. G. (1978): Selective adsorption of liquid mixtures on organophilic clay minerals. Progress in Colloid and Polymer Science, **65**, 125-132.

DELLER, B. (1981): Determination of exchangeable acidity, carbonate ions and change of buffer in triethanolamine-buffered solutions percolated through soil samples containing carbonates. Commun. Soil Sci. Plant Analysis, **12**, 161-177.

DIN 1319-3 Fundamentals of metrology – Part 3: Evaluation of measurements of a single measurand, measurement uncertainty. Beuth, Berlin.

DIN 1319-4 Fundamentals of metrology – Part 4: Evaluation of measurements; uncertainty of measurement. Beuth, Berlin.

DIN 4019-1, Subsoil; Settlement Calculations for Perpendicular Central Loading. Beuth, Berlin.

DIN 4021 Soil; exploration by excavation and borings; sampling. Beuth, Berlin.

DIN 4022-1 Subsoil and groundwater; classification and description of soil and rock; borehole logging of soil and rock not involving continuous core sample recovery. Beuth, Berlin.

DIN 4022-2 Subsoil and groundwater; Designation and description of soil types and rock; Stratigraphic representation for borings in rock. Beuth, Berlin.

DIN 4022-3 Subsoil and groundwater; Designation and description of soil types and rock; Borehole log for boring in soil (loose rock) by continuous extraction of cores. Beuth, Berlin.

DIN 4023 Borehole logging; graphical representation of the results. Beuth, Berlin.

DIN 4030-1 Assessment of water, soil and gases for their aggressiveness to concrete; principles and limiting values. Beuth, Berlin.

DIN 4030-2 Assessment of water, soil and gases for their aggressiveness to concrete; collection and examination of water and soil samples. Beuth, Berlin.

DIN 4049-1 Hydrology; basic terms. Beuth, Berlin.

DIN 4049-2 Hydrology; terms relating to quality of waters. Beuth, Berlin.

DIN 4049-3 Hydrology – Part 3: Terms for the quantitative hydrology. Beuth, Berlin.

DIN 4084 Subsoil; Calculations of terrain rupture and slope rupture. Beuth, Berlin.

DIN 4124 Excavations and trenches – Slopes, planking and strutting, breadths of working spaces. Beuth, Berlin.

DIN 18123 Soil, investigation and testing – Determination of grain-size distribution. Beuth, Berlin.

DIN 18129 Soil, investigation and testing – Determination of lime content. Beuth, Berlin.

DIN 19682-2 Methods of soil investigations for agricultural water engineering – Field tests – Part 2: Determination of soil texture. Beuth, Berlin.

DIN 19684-3 Methods of soil investigations for agricultural water engineering – Chemical laboratory tests – Part 3: Determination of the loss on ignition and the residue of soil after ignition. Beuth, Berlin.

DIN 19711 Graphical and letter symbols for hydrogeology. Beuth, Berlin.

DIN 19730 Soil quality – Extraction of trace elements with ammonium nitrate solution. Beuth, Berlin.

DIN 19731 Soil quality – Utilization of soil material. Beuth, Berlin.

DIN 19734 Soil quality – Determination of chromium(VI) in phosphate extract. Beuth, Berlin.

DIN 19738 Soil quality – Absorption availability of organic and inorganic pollutants from contaminated soil material. Beuth, Berlin.

DIN 32645 Chemical analysis; decision limit; detection limit and determination limit; estimation in case of repeatability; terms, methods, evaluation. Beuth, Berlin.

DIN 38402-12 German standard methods for the examination of water, waste water and sludge; general information (group A); sampling from barrages and lakes (A 12). Beuth, Berlin.

DIN 38402-13 German standard methods for the examination of water, waste water and sludge; general information (group A); sampling from aquifers (A 13). Beuth, Berlin.

DIN 38402-15 German standard methods for the examination of water, waste water and sludge; general information (group A); sampling of flowing waters (A 15). Beuth, Berlin.

DIN 38404-5 German standard methods for examination of water, waste water and sludge; physical and physico-chemical characteristics (group C); determination of pH value (C5). Beuth, Berlin.

DIN 38405-1 German standard methods for the examination of water, waste water and sludge; anions (group D); determination of chloride ions (D 1). Beuth, Berlin.

DIN 38405-4 German standard methods for the examination of water, waste water and sludge; anions (group D); determination of fluoride (D 4). Beuth, Berlin.

DIN 38405-5 German standard methods for the examination of water, waste water and sludge; anions (group D); determination of sulfate ions (D 5). Beuth, Berlin.

DIN 38405-9 German standard methods for examination of water, waste water and sludge; anions (group D), determination of nitrate ion (D9). Beuth, Berlin.

DIN 38405-13 German Standard Methods for the Analysis of Water, Waste Water and Sludge; Anions (Group D); Determination of Cyanides (D 13). Beuth, Berlin.

DIN 38405-14 German standard methods for the examination of water, waste water and sludge; anions (group D); determination of cyanides in drinking water, and in groundwater and surface water with low pollution levels (D 14).

DIN 38405-23 German standard methods for the examination of water, waste water and sludge – Anions (Group D) – Part 23: Determination of selenium by atomic absorption spectrometry (D 23). Beuth, Berlin.

DIN 38405-24 German standard methods for the examination of water, waste water and sludge; anions (group D); photometric determination of chromium(VI) using 1,5-diphenylcarbonohydrazide (D 24). Beuth, Berlin.

DIN 38405-26 German standard methods for the examination of water, waste water and sludge; anions (group D); determination of dissolved sulfide by spectrometry (D 26). Beuth, Berlin.

DIN 38405-32 German standard methods for the examination of water, waste water and sludge – Anions (group D) – Part 32: Determination of antimony by atomic absorption spectrometry (D 32). Beuth, Berlin.

DIN 38406-5 German standard methods for the examination of water, waste water and sludge; cations (group E); determination of ammonia-nitrogen (E 5). Beuth, Berlin.

DIN 38406-6 German standard methods for the examination of water, waste water and sludge – Cations (group E) – Part 6: Determination of lead by atomic absorption spectrometry (AAS) (E 6). Beuth, Berlin.

DIN 38406-7 German standard methods for the examination of water, waste water and sludge; cations (group E); determination of copper by atomic absorption spectrometry (AAS) (E 7). Beuth, Berlin.

DIN 38406-E7 German standard methods for the examination of water, waste water and sludge; cations (group E); determination of copper by atomic absorption spectrometry (AAS) (E 7). Beuth, Berlin.

DIN 38406-8 German standard methods for examination of water, waste water and sludge – Cations (group E) – Part 8: Determination of zinc – Method by atomic absorption spectrometry (AAS) using an air-ethine flame (E 8). Beuth, Berlin.

DIN 38406-11 German standard methods for the examination of water, waste water and sludge; cations (group E); determination of nickel by atomic absorption spectrometry (AAS) (E 11). Beuth, Berlin.

DIN 38406-13 German standard methods for the examination of water, waste water and sludge; cations (group E); determination of potassium by atomic absorption spectrometry (AAS) using an air-acetylene flame (E 13). Beuth, Berlin.

DIN 38406-14 German standard methods for the examination of water, waste water and sludge; cations (group E); determination of sodium by atomic absorption spectrometry (ASS) using an air-acetylene flame (E 14). Beuth, Berlin.

DIN 38406-24 German standard methods for the examination of water, waste water and sludge; cations (group E); determination of cobalt by atomic absorption spectrometry (AAS) (E 24). Beuth, Berlin.

DIN 38406-26 German standard methods for the examination of water, waste water and sludge – Cations (group E) – Part 26: Determination of thallium by atomic absorption spectrometry (AAS) using electrothermal atomisation (E 26). Beuth, Berlin.

DIN 38406-28 German standard methods for the examination of water, waste water and sludge – Cations (group E) – Part 28: Determination of dissolved barium by atomic absorption spectrometry (E 28). Beuth, Berlin.

DIN 38406-32 German standard methods for the examination of water, waste water and sludge – Cations (group E) – Part 32: Determination of iron by atomic absorption spectrometry (E 32). Beuth, Berlin.

DIN 38406-33 German standard methods for the examination of water, waste water and sludge – Cations (group E) – Part 33: Determination of manganese by atomic absorption spectrometry (E 33). Beuth, Berlin.

DIN 38407-2 German standard methods for the determination of water, waste water and sludge; jointly determinable substances (group F); determination of low volatile halogenated hydrocarbons by gas chromatography (F 2). Beuth, Berlin.

DIN 38407-3 German standard methods for the determination of water, waste water and sludge – Jointly determinable substances (group F) – Part 3: Determination of polychlorinated biphenyls (F 3). Beuth, Berlin.

DIN 38407-8 German standard methods for the examination of water, waste water and sludge – Jointly determinable substances (group F) – Part 8: Determination of 6 polynuclear aromatic hydrocarbons (PAH) in water by high performance liquid chromatography (HPLC) with fluorescence detection (F 8). Beuth, Berlin.

DIN 38407-9 German standard methods for the examination of water, waste water and sludge; substance group analysis (group F); determination of benzene and some of its derivatives by gas chromatography (F 9). Beuth, Berlin.

DIN 38407-14 German standard methods for the examination of water, waste water, and sludge – Jointly determinable substances (group F) – Part 14: Determination of phenoxyalkyl carbonic acids by gas chromatography and mass-spectrometric detection after solid-liquid-extraction and derivatization (F 14). Beuth, Berlin.

DIN 38409-1 German standard methods for the examination of water, waste water and sludge; parameters characterizing effects and substances (group H); determination of total dry residue, filtrate dry residue and residue on ignition (H 1). Beuth, Berlin.

DIN 38409-8 German standard methods for the examination of water, waste water and sludge; summary indices of actions and substances (group H); determination of extractable organically bonded halogens (EOX) (H 8). Beuth, Berlin.

DIN 38409-16 German standard methods for the examination of water, waste water and sludge; general measures of effects and substances (group H); determination of the phenol index (H 16). Beuth, Berlin.

DIN 38409-41 German Standard Methods for Examination of Water, Waste Water and Sludge; Summary Action and Material Characteristic Parameters (Group H); Determination of the Chemical Oxygen Demand (COD) in the Range over 15 mg/l (H41). Beuth, Berlin.

DIN 38413-2 German standard methods for the examination of water, waste water and sludge; individual constituents (group P); determination of vinyl chloride by headspace gas chromatography (P 2). Beuth, Berlin.

DIN 38414-4 German standard methods for the examination of water, waste water and sludge; sludge and sediments (group S); determination of leachability by water (S 4). Beuth, Berlin.

DIN 38414-17 German standard methods for the examination of water, waste water and sludge; sludge and sediments (group S); determination of strippable and extractable organically bound halogens (S 17). Beuth, Berlin.

DIN 38414-18 German standard methods for the examination of water, waste water and sludge; sludge and sediments (group S); determination of adsorbed organically bound halogens (AOX) (S 18). Beuth, Berlin.

DIN 38414-20 German standard methods for the examination of water, waste water and sludge – Sludge and sediments (group S) – Part 20: Determination of 6 polychlorinated biphenyls (PCB) (S 20). Beuth, Berlin.

DIN 38414-22 German standard methods for the examination of water, waste water and sludge – Sludge and sediments (group S) – Part 22: Determination of dry residue by freezing and preparation of the freeze dried mass of sludge (S 22). Beuth, Berlin.

DIN 38414-23 German standard methods for the examination of water, waste water and sludge – Sludge and sediments (group S) – Part 23: Determination of 15 polycyclic aromatic hydrocarbons (PAH) by high performance liquid chromatography (HPLC) and fluorescence detection (S 23). Beuth, Berlin.

DIN 38414-24 German standard methods for the examination of water, waste water and sludge – Sludge and sediments (group S) – Part 24: Determination of polychlorinated dibenzodioxins (PCDD) and polychlorinated dibenzofuranes (PCDF) (S 24). Beuth, Berlin.

DIN 51084 Testing of oxidic raw materials for ceramic, glass and glazes; determination of fluoride content. Beuth, Berlin.

DIN 51527-1 Testing of petroleum products; determination of polychlorinated biphenyls (PCB); preseparation by liquid chromatography and determination of six selected PCB compounds by gas chromatography using an electron capture detector

DIN EN 932-1 Test for general properties of aggregates – Part 1: Methods for sampling; German version EN 932-1:1996. Beuth, Berlin.

DIN EN 1233 Water quality – Determination of chromium – Atomic absorption spectrometric methods. Beuth, Berlin.

DIN EN 1483 Water quality – Determination of mercury. Beuth, Berlin.

DIN EN 1484 (H14) Water analysis – Guidelines for the determination of total organic carbon (TOC) and dissolved organic carbon (DOC). Beuth, Berlin.

DIN EN 12338 Water quality – Determination of mercury – Methods after enrichment by amalgamation. Beuth, Berlin.

DIN EN 12673 Water quality – Gas chromatographic determination of some selected chlorophenols in water. Beuth, Berlin.

DIN EN 25663 Water quality; determination of Kjeldahl nitrogen; method after mineralization with selenium. Beuth, Berlin.

DIN EN 25667-2 Water quality; sampling; part 2: guidance on sampling techniques. Beuth, Berlin.

DIN EN 25813 Water quality; determination of dissolved oxygen; iodometric method. Beuth, Berlin.

DIN EN 25814 Water quality; determination of dissolved oxygen by the electrochemical probe method. Beuth, Berlin.

DIN EN 26777 Water quality; determination of nitrite; molecular absorption spectrometric method. Beuth, Berlin.

DIN EN 27888 Water quality; determination of electrical conductivity. Beuth, Berlin.

DIN EN ISO 5667-3 Water quality – Sampling – Part 3: Guidance on the preservation and handling of water samples (ISO 5667-3); German version EN ISO 5667-3. Beuth, Berlin.

DIN EN ISO 5961 Water quality – Determination of cadmium by atomic absorption spectrometry. Beuth, Berlin.

DIN EN ISO 6468 Water quality – Determination of certain organochlorine insecticides, polychlorinated biphenyls and chlorobenzenes – Gas-chromatographic method after liquid-liquid extraction. Beuth, Berlin.

DIN EN ISO 7027 Water quality – Determination of turbidity. Beuth, Berlin.

DIN EN ISO 7887 Water quality – Examination and determination of colour. Beuth, Berlin.

DIN EN ISO 7980 Water quality – Determination of calcium and magnesium – Atomic absorption spectrometric method. Beuth, Berlin.

DIN EN ISO 9377-2 Water quality – Determination of hydrocarbon oil index – Part 2: Method using solvent extraction and gas chromatography. Beuth, Berlin.

DIN EN ISO 10301 Water quality – Determination of highly volatile halogenated hydrocarbons – Gas-chromatographic methods. Beuth, Berlin.

DIN EN ISO 10304-1 Water quality – Determination of dissolved fluoride, chloride, nitrite, orthophosphate, bromide, nitrate and sulfate ions, using liquid chromatography of ions – Part 1: Method for water with low contamination. Beuth, Berlin.

DIN EN ISO 10304-3 Water quality – Determination of dissolved anions by liquid chromatography of ions – Part 3: Determination of chromate, iodide, sulfite, thiocyanate and thiosulfate. Beuth, Berlin.

DIN EN ISO 11732 Water quality – Determination of ammonium nitrogen – Method by flow analysis (CFA and FIA) and spectrometric detection. Beuth, Berlin.

DIN EN ISO 11885 Water quality – Determination of 33 elements by inductively coupled plasma atomic emission spectroscopy. Beuth, Berlin.

DIN EN ISO 11969 Water quality – Determination of arsenic – Atomic absorption spectrometric method (hydride technique). Beuth, Berlin.

DIN EN ISO 10695 Water quality – Determination of selected organic nitrogen and phosphorus compounds – Gas chromatographic methods. Beuth, Berlin.

DIN EN ISO 13395 Water quality – Determination of nitrite nitrogen and nitrate nitrogen and the sum of both by flow analysis (CFA and FIA) and spectrometric detection. Beuth, Berlin.

DIN EN ISO 14403 Water quality – Determination of total cyanide and free cyanide by continuous flow analysis. Beuth, Berlin.

DIN EN ISO 15913 Water quality – Determination of selected phenoxyalkanoic herbicides, including bentazones and hydroxybenzonitriles by gas chromatography and mass spectrometry after solid phase extraction and derivatization. Beuth, Berlin.

DIN EN ISO 17294-2 Water quality – Application of inductively coupled plasma mass spectrometry (ICP-MS) – Part 2: Determination of 62 elements (ISO 17294-2:2003); German version EN ISO 17294-2:2004. Beuth, Berlin.

DIN ISO 9964-3 Water quality – Determination of sodium and potassium – Part 3: Determination of sodium and potassium by flame emission spectrometry. Beuth, Berlin.

DIN ISO 10381-1 Soil quality – Sampling – Part 1: Guidance on the design of sampling programmes. Beuth, Berlin.

DIN ISO 10381-2 Soil quality – Sampling – Part 2: Guidance on sampling techniques. Beuth, Berlin.

DIN ISO 10381-3 Soil quality – Sampling – Part 3: Guidance on safety. Beuth, Berlin.

DIN ISO 10381-4 Soil quality – Sampling – Part 4: Guidance on the procedure for investigation of natural, near-natural and cultivated sites. Beuth, Berlin.

DIN ISO 10381-5 Soil quality – Sampling – Part 5: Guidance on the procedure for the investigation of urban and industrial sites with regard to soil contamination. Beuth, Berlin.

DIN ISO 10382 Soil quality – Determination of organochlorine pesticides and polychlorinated biphenyls – Gas-chromatographic method with electron capture detection. Beuth, Berlin.

DIN ISO 10694 Soil quality – Determination of organic and total carbon after dry combustion (elementary analysis). Beuth, Berlin.

DIN ISO 11047 Soil quality – Determination of cadmium, chromium, cobalt, copper, lead, manganese, nickel and zinc in aqua regia extracts of soil – Flame and electrothermal atomic absorption spectrometric methods. Beuth, Berlin.

DIN ISO 11074-1 Soil quality – Vocabulary – Part 1: Terms and definitions relating to the protection and pollution of the soil. Beuth, Berlin.

DIN ISO 11261 Soil quality – Determination of total nitrogen – Modified Kjeldahl method. Beuth, Berlin.

DIN ISO 11262 Soil quality – Determination of cyanide. Beuth, Berlin.

DIN ISO 11264 Soil quality – Determination of herbicides using high performance liquid chromatography with UV-detection. Beuth, Berlin.

DIN ISO 11265 Soil quality – Determination of the specific electrical conductivity. Beuth, Berlin.

DIN ISO 11272 Soil quality – Determination of dry bulk density. Beuth, Berlin.

DIN ISO 11277 Soil quality – Determination of particle size distribution in mineral soil material – Method by sieving and sedimentation. Beuth, Berlin.

DIN ISO 11464 Soil quality – Pretreatment of samples for physico-chemical analysis. Beuth, Berlin.

DIN ISO 11465 Soil quality – Determination of dry matter and water content on a mass basis – Gravimetric method. Beuth, Berlin.

DIN ISO 11466 Soil quality – Extraction of trace elements soluble in aqua regia. Beuth, Berlin.

DIN ISO 13877 Soil quality – Determination of polynuclear aromatic hydrocarbons – Method using high-performance liquid chromatographic. Beuth, Berlin.

DIN ISO 14154 Soil quality – Determination of some selected chlorophenols in soils – Gaschromatographic method. Beuth, Berlin.

DIN ISO 14507 Soil quality – Pretreatment of samples for determination of organic contaminants. Beuth, Berlin.

DIN V 4019-100 Soil – Analysis of settlement – Part 100: Analysis in accordance with partial safety factor concept. Pre-standard. Beuth, Berlin.

DIN V 19736 (1998): Soil quality – Derivation of concentrations of organic pollutants in soil water (Vornorm). Beuth, Berlin.

DOHRMANN, R. (1997): Kationenaustauschkapazität von Tonen – Bewertung bisheriger Analysenverfahren und Vorstellung einer neuen und exakten Silber-Thioharnstoff-Methode. PhD thesis RWTH Aachen, AGB-Verlag No. 26.

DOHRMANN, R. (2006a): Problems in CEC determination of calcareous clayey sediments using the ammonium acetate method. J. Plant Nutr. Soil Sci., **169**, 330 – 334.

DOHRMANN, R. (2006b): Cation Exchange Capacity Methodology II: proposal for a modified silver-thiourea method. Appl. Clay Sci., **34**, 38-46.

DOHRMANN, R. (2006c): Cation Exchange Capacity Methodology III: Correct exchangeable calcium determination of calcareous clays using a new silver-thiourea method. Appl. Clay Sci., **34**, 47-57.

DOHRMANN, R. (2006d): Cation Exchange Capacity Methodology I: An Efficient Model for the Detection of Incorrect Cation Exchange Capacity and Exchangeable Cation Results. Appl. Clay Sci., **34**, 31-37.

DRÄGER (2001): Dräger-Röhrchen- / CMS-Handbuch (Manual for indicating test tubes / CMS). Dräger Sicherheitstechnik, Lübeck, Germany.

DRESCHER, J. (1997): Deponiebau. Ernst, Berlin.

DVWK (1992): Entnahme und Untersuchungsumfang von Grundwasserproben. DVWK-Regeln 128.

DVWK (1997): Tiefenorientierte Probenahme aus Grundwassermessstellen. DVWK-Merkblätter 245, Bonn.

DYREBORG, S., ARVIN, E. & BROHOLM, K. (1997): Biodegradation of NSO compounds under different redox conditions. J. Cont. Hydrol., **25**, 177-197.

EL FANTROUSSI, S., NAVEAU, H. & AGATHOS, S. N. (1998): Anaerobic dechlorinating bacteria. Biotechnol. Prog., **14**, 167-188.

EN ISO 5667-3 (2003): Water quality – Sampling – Part 3: Guidance on the preservation and handling of water samples. International Organization for Standardization, Geneva.

EPA (2006a): http://oaspub.epa.gov/wqsdatabase/wqsi_epa_criteria.rep_parameter, downloaded: 13/11/2006.

EPA (2006b): EPA On-line Tools for Site Assessment Calculation. www.epa.gov/athens/learn2model/part-two/onsite/ard_onsite.htm

EU Heater project (2006): Heater Experiment: Rock and bentonite thermo-hydro-mechanical (THM) processes in the near field of a thermal source for development of deep underground high level radioactive waste repositories. – Final technical publishable report project: FISS-2001-00024, Contract No. FIKW-CT-2001-00132.

FANG, Z. (1995): Flow injection atomic absorption spectrometry. Wiley & Sons, New York.

FARMER, V. C. (1974): The infrared Spectra of Minerals. Mineralogical Society Monograph **4**, London.

FATHEPURE, B. Z., ELANGO, V. K., SINGH, H. & BRUNER, M. A. (2005): Bioaugmentation potential of a vinyl chloride-assimilating Mycobacterium sp., isolated from a chloroethene-contaminated aquifer. FEMS Microbiol. Lett., **248**, 227-234.

FENNELL, D., CARROl, A., GOSSETT, J. & ZINDER, S. (2001): Assessment of indigenous reductive dechlorinating potential at a TCE-contaminated site using micorocosms, polymerase chain reaction analysis and site data. Environ. Sci. Technol., **35**, 1, 1830-1839.

FETTER, C. W. (1994): Applied hydrogeology. MacMillan College Publishing, Co.

FETZNER, S. (1998): Bacterial dehalogenation. Appl. Microbiol. Biotechnol., **50**, 633-657.

FRANCIS, C. W. & GRIGAL, D. F. (1971): A rapid and simple procedure using 85Sr for determining cation exchange capacities of soils and clays. Soil Sci., **112**, 17-21.

FRANK, K. (1991): Tongesteine. Retention von Schwermetallen und die Einflußnahme künstlicher Komplexbildner. Schriftenreihe Angew. Geologie, **11**.

FREEZE, R. A. & CHERRY, J. A. (1979): Groundwater. Prentice-Hall, Englewood Cliffs, N. J.

FRITSCHE, W. & HOFRICHTER, M. (2000): Aerobic degradation by microorganisms, in Biotechnology, Rehm, H.-J. & Reed, G. (Eds.), Vol. 11b Environmental Processes II, Wiley-VCH, Weinheim.

GDA (1993): Empfehlungen des Arbeitskreises „Geotechnik der Deponien und Altlasten" GDA. 2. Aufl., Deutsche Gesellschaft für Erd- und Grundbau e.V., GDA E 3-3 3; Ernst, Berlin.

GEYER, R., PEACOCK, A. D., MILTNER, A., RICHNOW, H. H., WHITE, D. C., SUBLETTE, K. L. & KÄSTNER, M. (2005): In situ assessment of biodegradation potential using biotraps amended with ^{13}C-labeled benzene or toluene. Environ. Sci. Technol., **39**, 13, 4983-4989.

GIEG, L. M. & SUFLITA, J. M. (2002): Detection of anaerobic metabolites of saturated and aromatic hydrocarbons in petroleum-contaminated aquifers. Environ. Sci. Technol., **36**, 17, 3755-3762.

GILLHAM, R. W. & O'HANNASIN, S. F. (1994): Enhanced degradation of halogenated aliphatics by zero-valent iron. Groundwater, **32**, 6, 958-967.

GILLHAM, R. W., ROBIN, M. J. L. & PTACEK, C. J. (1990a): A device for in situ determination of geochemical transport parameters 1. Retardation. Ground Water, **28**, 5, 666-672.

GILLHAM, R. W., STARR, R. C. & MILLER, D. J. (1990b): A device for in situ determination of geochemical transport parameters 2. Biochemical Reactions. Ground Water, **28**, 5, 858-862.

GRAF VON REICHENBACH, H. (1966): Anomalien des Kationenaustausches bei Vermiculiten. Z. Pflanzenernährung Bodenkunde, **113**, 203.

GRBIC-GALIC, D. (1990): Methanogenic transformation of aromatic hydrocarbons and phenols in groundwater aquifers. Geomicrobiol. J., **8**, 167-200.

GRBIC-GALIC & D. VOGEL, T. M. (1987): Transformation of toluene and benzene by mixed methanogenic cultures. Appl. Environ. Microbiol., **53**, 2, 254-260.

GRIEBLER, C., SAFINOWSKI, M., VIETH, A., RICHNOW, H. H. & MECKENSTOCK, R. U. (2004): Combined application of stable carbon isotope fractionation for assessing in situ degradation of aromatic hydrocarbons in a tar oil-contaminated aquifer. Environ. Sci. Technol., **38**, 2, 617-631.

GRIM, R. E. (1962): Applied Clay Mineralogy. McGraw-Hill, New York.

HANG, P. T. & BRINDLEY, G. W. (1970): Methylene blue adsorption by clay minerals. Determination of surface areas and cation exchange capacities. Clays Clay Minerals, **10**, 203-212.

HARRISON, R. M. (Ed.) (1999): Understanding Our Environment. An Introduction to Environmental Chemistry and Pollution. Third Edition. Royal Society of Chemistry.

HARTOG, N., GRIFFIOEN, J. & VAN DER WEIJDEN, C. H. (2002): Distribution and reactivity of O_2-reducing components in sediments from a layered aquifer. Environ. Sci. Technol., **36**, 2338-2344.

HE, J., SUNG, Y., DOLLHOPF, M. E., FATHEPURE, B. Z., TIEDJE, J. M. & LÖFFLER, F. E. (2002): Acetate versus hydrogen as direct electron donors to stimulate the microbial reductive dechlorination process at chloroethene-contaminated sites. Environ. Sci. Technol., **36**, 18, 3945-3952.

HEINRICHS, H. & HERRMANN, A. G. (1990): Praktikum der analytischen Geochemie. Springer, Berlin.

HENDRICKSON, E., PAYNE, J., YOUNG, R., STARR, M., PERRY, M., FAHNESTOCK, S., ELLIS D. & EBERSOLE, C. (2002): Molecular analysis of Dehalococcoides 16S ribosomal DNA from chloroethene-contaminated sites throughout North America and Europe. Appl. Environ. Microbiol., **68**, 2, 485-495.

HERON, G. & CHRISTENSEN, T. H. (1995): Impact of sediment-bound iron on redox buffering in a landfill leachate polluted aquifer. Environ. Sci. Technol., **29**, 187-192.

HILTMANN, W. & STRIBRNY, B. (1998): Handbuch zur Erkundung des Untergrundes von Deponien und Altlasten, Bd. 5, Tonmineralogie und Bodenphysik, Springer, Berlin.

HIRSCHORN, S. K., DINGLASAN, M. J., ELSNER, M., MANCINI, S. A., LACRAMPE-COULOUME, G., EDWARDS, E. A., SHERWOOD LOLLAR, B. (2004): Pathway dependent isotopic fractionation during aerobic biodegradation of 1,2-dichloroethane. Environ. Sci. Technol., **3**, 18, 4775-4781.

HOHNSTOCK-ASHE, A. M., PLUMMER, S. M., YAGER, R. M., BAVEYE, P. & MADSEN, E. L. (2001): Further biogeochemical characterization of a trichloroethene-contaminated fractured dolomite aquifer: electron source and microbial communities involved in reductive dechlorination. Environ. Sci. Technol., **35**, 22, 4449-4456.

HOPMANS, J. W. & DANE, J. H. (1986): Temperature Dependence of Soil Hydraulic Properties. Soil Sci. Soc. Am. Journal, **50**, 1.

HOUBEN, G. J. 2003: Iron oxide incrustations in wells. Part 1: genesis, mineralogy and geochemistry. – Applied Geochemistry, **18**, 927-939.

HOUBEN, G. & TRESKATIS, C. (2006): Rehabilitation of water wells. McGraw Hill, New York.

HOURBRON, E., ESCOFFIER, S. & CAPDEVILLE, B. (2000): Trichloroethylene elimination assay by natural consortia of heterotrophic and methanotrophic bacteria. Water Sci. Technol., **42**, 5-6, 395-402.

HUNKELER, D., ARAVENA, R. & COX, E. (2002): Carbon isotopes as a tool to evaluate the origin and fate of vinyl chloride: laboratory experiments and modeling of isotope evolution. Environ. Sci. Technol., **36**, 3378-3384.

HUNT, M. J., SHAFER, M. B., BARLAZ, M. A. & BORDEN, R. C. (1997): Anaerobic biodegradation of alkylbenzenes in laboratory microcosms representing ambient conditions. Bioremediation J., **1**, 1, 53-64.

HUTCHINS, S. R. (1991): Biodegradation of monoaromatic hydrocarbons by aquifer microorganisms using oxygen, nitrate, or nitrous oxide as terminal electron acceptor. Appl. Environ. Microbiol., **57**, 8, 2403-2407.

ISO 5667-1 Water quality; Sampling; Part 1: guidance on the design of sampling programmes and sampling techniques. International Organization for Standardization, Geneva.

ISO 5667-3 Water quality; Sampling; Part 3: Guidance on the preservation and handling of water samples. International Organization for Standardization, Geneva.

ISO 5667-4 Water quality; Sampling; Part 4: guidance on sampling from lakes, natural and man-made. International Organization for Standardization, Geneva.

ISO 5667-6 Water quality; sampling; Part 6: guidance on sampling of rivers and streams. International Organization for Standardization, Geneva.

ISO 5667-11 Water quality; sampling; Part 11: guidance on sampling of groundwaters. International Organization for Standardization, Geneva.

ISO 5667-18 Water quality – Sampling – Part 18: guidance on sampling of groundwater at contaminated sites. International Organization for Standardization, Geneva.

ISO 6878 Water quality – Determination of phosphorus – Ammonium molybdate spectrometric method. International Organization for Standardization, Geneva.

ISO 8165-2 Water quality – Determination of selected monovalent phenols – Part 2: Method by derivatization and gas chromatography. International Organization for Standardization, Geneva.

ISO 8245 Water quality – Guidelines for the determination of total organic carbon (TOC) and dissolved organic carbon (DOC). International Organization for Standardization, Geneva.

ISO 9377-2 Water quality – Determination of hydrocarbon oil index – Part 2: Method using solvent extraction and gas chromatography. International Organization for Standardization, Geneva.

ISO 9562 Water quality – Determination of adsorbable organically bound halogens (AOX). International Organization for Standardization, Geneva.

ISO 10390 Soil quality – Determination of pH. International Organization for Standardization, Geneva.

ISO 10694 Soil quality – Determination of organic and total carbon after dry combustion (elementary analysis). International Organization for Standardization, Geneva.

ISO 10695 Water quality – Determination of selected organic nitrogen and phosphorus compounds – Gas chromatographic methods. International Organization for Standardization, Geneva.

ISO 11074-2 Soil quality – Vocabulary – Part 2: Terms and definitions relating to sampling. International Organization for Standardization, Geneva.

ISO 11262 Soil quality – Determination of cyanide. International Organization for Standardization, Geneva.

ISO 11272 Soil quality – Determination of dry bulk density. International Organization for Standardization, Geneva.

ISO 11277 Soil quality – Determination of particle size distribution in mineral soil material – Method by sieving and sedimentation. International Organization for Standardization, Geneva.

ISO 11369 Water quality – Determination of selected plant treatment agents – Method using high performance liquid chromatography with UV detection after solid-liquid extraction. International Organization for Standardization, Geneva.

ISO 14869-1 Soil quality – Dissolution for the determination of total element content – Part 1: Dissolution with hydrofluoric and perchloric acids. International Organization for Standardization, Geneva.

ISO 14869-2 Soil quality – Dissolution for the determination of total element content – Part 2: Dissolution by alkaline fusion. International Organization for Standardization, Geneva.

ISO 15009 Soil quality – Gas chromatographic determination of the content of volatile aromatic hydrocarbons, naphthalene and volatile halogenated hydrocarbons – Purge-and-trap method with thermal desorption. International Organization for Standardization, Geneva.

ISO 15913 Water quality – Determination of selected phenoxyalkanoic herbicides, including bentazones and hydroxybenzonitriles by gas chromatography and mass spectrometry after solid phase extraction and derivatization. International Organization for Standardization, Geneva.

ISO 16703 Soil quality – Determination of content of hydrocarbon in the range C_{10} to C_{40} by gas chromatography. International Organization for Standardization, Geneva.

ISO 17294-2 Water quality – Application of inductively coupled plasma mass spectrometry (ICP-MS) – Part 2: Determination of 62 elements. International Organization for Standardization, Geneva.

ISO 17993 Water quality – Determination of 15 polycyclic aromatic hydrocarbons (PAH) in water by HPLC with fluorescence detection after liquid-liquid extraction. International Organization for Standardization, Geneva.

ISO CD 16772 Soil quality – Determination of mercury in aqua regia soil extracts with cold-vapour atomic spectrometry or cold-vapour atomic fluorescence spectrometry. International Organization for Standardization, Geneva.

ISO/DIS 5667-1 Water quality – Sampling – Part 1: Guidance on the design of sampling programmes and sampling techniques. International Organization for Standardization, Geneva.

ISO/DIS 9377-4 Water quality – Determination of hydrocarbon oil index – Part 4: Method using solvent extraction and gas chromatography. International Organization for Standardization, Geneva.

ISO/IEC 17025 General requirements for the competence of testing and calibration laboratories. International Organization for Standardization, Geneva.

ISO/ TR 11046 Soil quality – Determination of mineral oil content – Methods by infrared spectrometry and gas chromatographic method. International Organization for Standardization, Geneva.

JASMUND, K. & LAGALY, G. (1993): Tonminerale und Tone. Steinkopff, Darmstadt.

JIMÉNEZ, L., ALZAGA, R. & BAYONA, J. M. (2002): Determination of organic contaminants in landfill leachates: a review. Int. J. Environ. Anal. Chem. **82**, 7, 415 – 430.

JIWAN, J. & GATES, G. (1992): A Practical Guide to Groundwater Sampling, 1st ed. New South Wales Department of Water Resources, Technical Services Division, TS 92 080.

JOHANSEN, S. S., HANSEN, A. B., MOSBAEK, H. & ARVIN, E. (1996): Method development for trace analysis of heteroaromatic comounds in contaminated groundwater. J. Chromatogr. A., **738**, 295-304.

JOHNSTON, C. (1996): Sorption of organic compounds on clay minerals: A surface functional group approach. In: SAWHNEY, B. (Ed.), Organic pollutants in the environment, CMS Workshop Lectures, Vol. 8, The Clay Minerals Society, Boulder, Colorado, 1-44.

JUHASZ, A. L. & NAIDU, R. (2000): Bioremediation of high molecular weight polycyclic aromatic hydrocarbons: a review of the microbial degradation of benzo[a]pyrene. Int. Biodeterior. Biodegrad., **45**, 1-2, 57-88.

JURY W. A., RUSSO D. & SPOSITO, G. (1987): The spatial variability of water and solute transport properties in unsaturated soil: II. Scaling models of water transport. Hilgardia **55**, 33-56.

KANY, M. (1974): Berechnung von Flächengründungen. 2. Aufl. Ernst, Berlin.

KÄSTNER, M. (2000): Degradation of aromatic and polyaromatic compounds, in Biotechnology, REHM, H.-J. & REED, G. (Eds.), Vol 11b Environmental Processes II, Wiley-VCH, Weinheim.

KAUFHOLD, S. & DOHRMANN, R. (2003): Beyond the Methylene Blue method: determination of the smectite content using the Cu-triene method. Zs. Angew. Geol., **49**, 13-17.

KICK, H. (1956): Bemerkungen zur T- und S-Wert-Bestimmung nach Mehlich in Böden mit höheren Gehalten an $CaCO_3$. Z. Pflanzenernährung Düngung Bodenkunde, **75**, 67-69.

KLEEBERG, R. (2005) Results of the second Reynolds Cup contest in quantitative mineral analysis. IUCr CPD Newsletter, **30**, 22-26.

KLOSA, D. (1994): Eine rechnergestützte Methode zur Bestimmung des Gesamtkarbonatgehaltes in Sedimenten und Böden. – Z. Angew. Geol., **40**, 1, 18-21.

KLOTZ, D. & FOLIV, F. (1983): Eine einfache Methode zur Bestimmung der Verteilungskoeffizienten von Radionukliden im Grundwasser. GWF-Wasser/Abwasser **124**, 139-141.

KÖGEL-KNABNER, I. & TOTSCHE, K. U. (1998): Influence of dissolved and colloidal phase humic substances on the transport of hydrophobic organic contaminants in soils. Phys. Chem. Earth, **23**, 2, 179-185.

KOHLER, E. E. (1985): Mineralogische Veränderungen von Tonen und Tonmineralen durch organische Lösungen. – In: MESECK, H. (Hrsg.): Abdichten von Deponien, Altlasten und kontaminierten Standorten. – Mitt. Inst. Grundbau u. Bodenmechanik TU Braunschweig **20**, 87-94.

KÖSTER, H. M. (1979): Die chemische Silikatanalyse. Springer, Berlin.

KRAUSKOPF, K. B. & BIRD, D. K. (1995): Introduction to Geochemistry. 3rd edn., McGraw-Hill, New York.

KUKKADAPU, R. K., ZACHARA, J. M., FREDRICKSON, J. K., SMITH, S. C., DOHNALKOVA, A. C. & RUSSELL, C. K. (2003): Transformation of 2-line ferrihydrite to 6-line ferrihydrite under oxic and anoxic conditions. – Am. Mineralogist, **88**, 11-12, 1903-1914.

LAGALY, G. & KÖSTER, H. M. (1993): Tone und Tonminerale. In JASMUND, K. & LAGALY, G. (Ed.): Tonminerale und Tone. Steinkopff, Darmstadt.

LAGALY, G. (1985): Clay organic interactions: Problems and results. In: SCHULTZ, L. G., VAN OLPHEN, H. & MUMPTON, F.A. (Eds.): Proc. Int. Clay Conference, Denver, 1985. The Clay Min. Soc., Bloomington, Indiana, 343-351.

LAKE, D. L., KIRK, W. W. & LESTER, J. N. (1984): Fractionation, characterization, and speciation of heavy metals in sewage sludge and sludge-amended soils: A Review.

LANG, H-J., HUDER, J. & AMANN, P. (1996): Bodenmechanik und Grundbau. 6. Aufl. Springer, Berlin.

LANGENHOFF, A. A. M., ZEHNDER, A. J. B. & SCHRAA, G. (1996): Behaviour of toluene, benzene and naphthalene under anaerobic conditions in sediment columns. Biodegradation, **7**, 267-274.

LANIEWSKI, A. K., DAHLEN, J., BOREN, H. & GRIMVALL, A. (1999): Determination of group parameters for organically bound chlorine, bromine and iodine in precipitation. Chemosphere, **38**, 4, 771-782.

LECO (1996): CNS-2000 Elemental Analyzer – Instruction Manual. LECO Corp., St. Joseph, MI.

LEDUC, R., SAMSON, R., AL-BASHIR, B., AL-HAWARI, J. & CSEH, T. (1992): Biotic and abiotic disappearance of four PAH compounds from flooded soil under various redox conditions. Wat. Sci. Technol., **26**, 1-2, 51-60.

LEVINE, A. D., LIBELO, E. L., BUGNA, G., SHELLY, T., MAYFIELD, H. & STAUFFER, T. B. (1997): Biogeochemical assessment of natural attenuation of JP-4-contaminated ground water in the presence of fluorinated surfactants. Sci. Total Environ., **208**, 179-195.

LEWIS, D. R. (1949): Analytical data on reference clay materials. Sect. 3, Base-exchange data. 1950, American Petroleum Institute Project 49 Clay Mineral Standards, Preliminary Report No. 7.

LICHT, D., AHRING, B. K. & ARVIN, E. (1996): Effects of electron acceptors, reducing agents, and toxic metabolites on anaerobic degradation of heterocyclic compounds. Biodegradation, **7**, 83-90.

LIPPMANN, F. (1973): Sedimentary carbonate minerals. Springer, Berlin.

LOAN, M., PIERRE, T. G. S., PARKINSON, G. M., NEWMAN, O. G. M. & FARROW, J. B. (2002). Identifying nanoscale ferrihydrite in hydrometallurgical residues. JOM, **54**, 12, 40.

LÖFFLER, F. E. & RITALAHTI, K. M. (2005): Bioaugmentation: Insights from the field and application of molecular diagnostic tools. In: DECHEMA, Tagungshandbuch, Perspektiven molekularer und isotopischer Methoden zum Nachweis des natürlichen Schadstoffabbaus in Böden, 29-30 September 2005, Braunschweig, Germany, 51-57.

LOVLEY, D. R. (2000): Anaerobic benzene degradation. Biodegradation, **11**, 107-116.

LOVLEY, D. R., COATES, J. D., WOODWARD, J. C. & PHILLLIPS, E. J. P. (1995): Benzene oxidation coupled to sulfate reduction. Appl. Environ. Microbiol., **61**, 3, 953-958.

LUA NRW (1994): Determination of PAH in soil samples. Bulletins of the Environmental Agency of the State of Northrhine-Westphalia, 1. Essen, Germany.

LUCKNER, L., VAN GENUCHTEN, M. Th. & NIELSEN, D. R. (1989): A Consistent Set of Parametric Models for two-phase flow of immiscible fluids in the subsurface. Water Resources Research, **25**, 10, 2187-2193.

LUCKNER, L. & SCHESTAKOW, W. M. (1991): Migration Processes in the Soil and Groundwater Zone. Lewis, Boca Raton, FL.

LYEW, D., TARTAKOVSKY, B., MANUEL, M.-F. & GUIOT, S. R. (2002): A microcosm test for potential mineralization of chlorinated compounds under coupled aerobic/anaerobic conditions. Chemosphere, **47**, 695-699.

LYNKILDE, J. & CHRISTENSEN, T. H. (1992): Fate of organic contaminants in the redox zones of a landfill leachate pollution plume (Vejen, Denmark). J. Contamin. Hydrol., **10**, 291-307.

MALCOLM, R. L. & KENNEDY, V. C. (1969): Rate of cation exchange on clay minerals as determined by specific ion electrode techniques. Soil Sci. Soc Amer. Proc., **34**, 247-253.

MANAHAN, S. E. (2004): Environmental Chemistry. CRC Press.

MANCINI, S. A., ULRICH, A. C., LACRAMPE-COULOUME, G., SLEEP, B., EDWARDS, E. A., & SHERWOOD LOLLAR, B. (2003): Carbon and hydrogen isotopic fractionation during anaerobic biodegradation of benzene. Appl. Environ. Microbiol., **69**, 1, 191-198.

MARTIN, H., HEIDINGER, M., ERTL, S., EICHINGER, L., TIEHM, A., SCHMIDT, K., KARCH, U. & LEVE, J. (2006): 13C-Isotopenuntersuchungen zur Bestimmung von Natural Attenuation – Abgrenzung und Charakterisierung eines CKW-Schadens am Standort Frankenthal. TerraTech., **3-4**, 14-17.

MATTES, T. E., COLEMAN, N. V., SPAIN, J. C. & GOSSETT, J. M. (2005): Physiological and molecular genetic analyses of vinyl chloride and ethene biodegradation in Nocardioides sp. strain JS614. Arch. Microbiol., **183**, 95-106.

MAYMÓ-GATELL, X., CHIEN, Y.-T., GOSSETT, J. M. & ZINDER, S. H. (1997): Isolation of a bacterium that reductively dechlorinates tetrachloroethene to ethene. Science, **276**, 1568-1571.

McBride, M. B. (1979): An interpretation of cation selectivity variations in M+-M+ exchange on clays. Clays Clay Minerals, **27**, 417.

McCarty, D. K. (2002): Quantitative mineral analysis of clay bearing mixtures: The Reynolds Cup contest. IUCr CPD Newsletter, **27**, 12-16.

McFarland, M. J. & Sims, R. C. (1991): Thermodynamic framework for evaluating PAH degradation in the subsurface. Ground Water, **29**, 6, 885-896.

McLean, E. O. (1982): Soil pH and lime requirement. In: Methods of Soil Analysis, Part 2, 2nd edn., Page A. L. et al., Ed., Agronomy., **9**, 199-224. Am. Soc. of Agron., Inc., Madison, WI.

McNally, D. L., Mihelcic, J. R. & Lueking, D. R. (1998): Biodegradation of three- and four-ring polycyclic aromatic hydrocarbons under aerobic and denitrifying conditions. Environ. Sci. Technol., **32**, 2633-2639.

McNaught, A. D. & Wilkinson, A. (1997): IUPAC Compendium of Chemical Terminology – The Gold Book. 2nd edn. Blackwell Science. Also available online: www.iupac.org/publications/compendium/index.html

Meckenstock, R. U., Morasch, B., Griebler, C. & Richnow, H. H. (2004): Stable isotope fractionation analysis as a tool to monitor biodegradation in contaminated aquifers. J. Contam. Hydrol., **75**, 215-255.

Meckenstock, R. U., Annweiler, E., Michaelis, W., Richnow, H. H. & Schink, B. (2000): Anaerobic naphthalene degradation by a sulfate-reducing enrichment culture. Appl. Environ. Microbiol., **66**, 7, 2743-2747.

Mehlich, A. (1948): Determination of cation- and anion-exchange properties of soils. Soil Science, **66**, 429-445.

Mehra, O. P, & Jackson, M. L. (1960): Iron oxide removal from soils and clays by a dithionite-citrate systems buffered with sodium bicarbonate. Clays Clay Minerals, **7**, 317-327.

Meier, L. P. & Kahr, G. (1999): Determination of the Cation Exchange Capacity (CEC) of Clay Minerals using the Complexes of Copper (II) Ion with Triethylenetetramine and Tetraethylenepentamine. Clays Clay Minerals, **47**, 386-388.

Meunier, A. (2005): Clays. Springer, Berlin.

Meunier, A. & Velde, B. (2004): Illite. Origins, Evolution and Metamorphism. Springer, Berlin.

Middeldorp, P., Luijten, M., Van De Pas, B., Van Eekert, M., Kengen, S., Schraa, G. & Stams, A. (1999): Anaerobic microbial reductive dehalogenation of chlorinated ethenes. Bioremediation Journal, **3**, 3, 151-169.

Mihelcic, J. R. & Luthy, R. G. (1988): Degradation of polycyclic aromatic hydrocarbon compounds under various redox conditions in soil-water systems. Appl. Environ. Microbiol. **54**, 5, 1182-1187.

MILLER, M. M., WASIK, S. P., HUANG, G.-L., SHIU, W.-Y. & MACKAY, D. (1985): Relationships between octanol-water partition coefficient and aqueous solubility. Environ. Sci. Technol., **19**, 6, 522-529.

MOORE, D. M., & REYNOLDS, R. C. (1997): X-ray Diffraction and the Identification and Analysis of Clay Minerals, 2nd edn., Oxford University Press, New York.

MORASCH, B. & HUNKELER, D. (2005): Isotopenfraktionierung zur Bestimmung des natürlichen Abbaus von chlorierten und nicht chlorierten Kohlenwasserstoffen. In: DECHEMA, Tagungshandbuch, Perspektiven molekularer und isotopischer Methoden zum Nachweis des natürlichen Schadstoffabbaus in Böden, 29.-30. September 2005, Braunschweig, Germany, 31-38.

MORASCH, B., RICHNOW, H. H., Schink, B. & Meckenstock, R. U. (2001): Stable hydrogen and carbon isotope fractionation during microbial toluene degradation: Mechanistic and environmental aspects. Appl. Environ. Microbiol., **10**, 4842-4849.

MORTLAND, M. M. (1970): Clay organic complexes and interactions. Advances in Agronomy, **22**, 75-117.

MULL, R., NORDMEYER, H., BOOCHS, P. W. & LIETH, H. (1994): Pflanzenschutzmittel im Grundwasser. Springer, Berlin.

MÜLLER, A., SCHÄFER, W., WICKERT, F., TIEHM, A. (2006): Nachweis und Identifikation von Natural Attenuation Prozessen in einer LCKW-Fahne. altlasten spektrum 6/2006: in press.

NAGRA (2002): Projekt Opalinuston – Synthese der geowissenschaftlichen Untersuchungsergebnisse – Entsorgungsnachweis für abgebrannte Brennelemente, verglaste hochaktive sowie langlebige mittelaktive Abfälle. Technischer Bericht (NTB), 02-03.

NARINE, D. R. & GUY, R. D. (1981): Interactions of some large organic cations with bentonite in dilute aqueous solutions. Clays Clay Minerals, **29**, 205-212.

NAS-NRC: National Academy of Science, National Research Council (1994): Alternatives for groundwater cleanup. National Academy Press, Washington, D.C..

Nielsen, D. M. (1991): Practical Handbook of Ground-Water Monitoring. Lewis Publishers, Boca Raton, Fl..

NIJENHUIS, I., ANDERT, J., BECK, K., KÄSTNER, M., DIEKERT, G. & RICHNOW, H. H. (2005): Stable isotope fractionation of tetrachloroethene during reductive dechlorination by *Sulfurospirillum multivorans and Desulfitobacterium* sp. strain PCE-S and abiotic reactions with cyanocobalamin. Appl. Environ. Microbiol., **71**, 7, 3413-3419.

NITSCHE, C. (1998): Grundwasserprobennahme im Umfeld von Tagebaurestlöchern und – seen. Proceedings des DGFZ e.V., Heft 13, 239-251.

NORTHCOTT, G. L. & JONES, K. C. (2000): Spiking hydrophobic organic compounds into soil and sediment: A review and critique of adopted procedures. Environ. Toxicol. and Chem., **19**, 2418-2430.

NUMATA, M., NAKAMURA, N., KOSHIKAWA, H. & TERASHIMA, Y. (2002): Chlorine isotope fractionation during reductive dechlorination of chlorinated ethenes by anaerobic bacteria. Environ. Sci. Technol., **36**, 20, 4389-3494.

OBERMANN, P. & CREMER, S. (1992): Entwicklung eines Routinetests zur Elution von Schwermetallen aus Abfällen und belasteten Böden. Abschlußbericht Landesamt für Wasser und Abfall NRW von der Ruhruniversität Bochum.

OECD (2000): Adsorption – Desorption Using a Batch Equilibrium Method. Organisation for Economic Co-operation and Development (OECD). OECD guideline for the testing of chemicals no. 106.

PELZ, O., CHATZINOTAS, A., ANDERSEN, N., BERNASCONI, S. M., HESSE, C., ABRAHAM, W.-R. & ZEYER, J. (2001): Use of isotopic and molecular techniques to link toluene degradation in denitrifying aquifer microcosms to specific microbial populations. Arch. Microbiol., **175**, 270-281.

PLEYSIER, J. & CREMERS, A. (1975): Stability of silver-thiourea complexes in montmorillonite clay. Journal of the Chemical Society-Faraday Transactions I, 256 - 264.

PLEYSIER, J. & JUO, A. S. R. (1980): A single-extraction method using silver-thiourea for measuring exchangeable cations and effective CEC in soils with variable charges. Soil Sci., **129**, 205-211.

POND, K. L., HUANG, Y., WANG, Y. & KULPA, C. F. (2002): Hydrogen isotopic composition of individual n-alkanes as an intrinsic tracer for bioremediation and source identification of petroleum contamination. Environ. Sci. Technol., **36**, 724-728.

PROVIDENTI, M. A., LEE, H., & TREVORS, J. T. (1993): Selected factors limiting the microbial degradation of recalcitrant compounds. J. Ind. Microbiol., **12**, 379-395.

PUSCH, R. & YONG, R. N. (2006): Microstructure of Smectite Clays and Engineering Performance. Taylor & Francis, London.

QUEVAUVILLE, P. (1996): Harmonization of leaching/extraction tests for environmental risk assessment. Sci. Tot. Environ. **178**, 1-132.

RAMSAY, J., ROBERTSON, K., MEYLAN, S., LUU, Y.-S., LEE, P. & RAMSAY, B. (2001): Naphthalene and phenanthrene mineralization coupled to Fe(III)-reduction and mechanisms of accessing insoluble Fe(III), in Natural Attenuation of Environmental Contaminants: Proceedings of the 6th Int. In Situ and On Site Bioremediation Symposium, LEESON, A., KELLEY, M. E., RIFAI, H. S. & MAGAR, V. S. (Eds.), vol. 6, 2, Battelle Press, Columbus, OH.

REICHERT, J.-K. & ROEMER, M. (1996): Eluatuntersuchungen. In: Fachgruppe Wasserchemie in der GDCh (Hrsg.), Chemie und Biologie der Altlasten. VCH, Weinheim.

RETTENBERGER, G & METZGER, H. (1992): Der Deponiegashaushalt in Altablagerungen (Landfill Gas Budget of Abandoned Waste Sites). In: Handbuch Altlasten (Manual for Hazardous Sites). Landesanstalt für Umweltschutz Baden-Württemberg (State Institute of Environmental Protection, Baden Württemberg), Germany.

ROCKNE, K. J., CHEE-SANFORD, J. C., SANFORD, R. A., HEDLUND, B. R., STALEY, J. T. & STRAND, S. E. (2000): Anaerobic naphthalene degradation by microbial pure cultures under nitrate-reducing conditions. Appl. Environ. Microbiol., **66**, 4, 1595-1601.

RÖLING W. F. M. & VAN VERSEVELD H. W. (2002): Natural attenuation: What does the subsurface have in store? Biodegradation, **13**, 53-64.

RUMP, H.-H. & SCHOLZ, B. (1995): Untersuchung von Abfällen, Reststoffen und Altlasten (Investigation of wastes, residues, and hazardous sites). Wiley-VCH, Weinheim.

SAFINOWSKI, M., GRIEBLER, C. & MECKENSTOCK, R. U. (2006): Anaerobic cometabolic transformation of polycyclic and heterocyclic aromatic hydrocarbons: evidence from laboratory and field studies. Environ. Sci. Technol., **40**, 13, 4165-4173.

SAGNER, A., BRINKMANN, C., EISENTRÄGER, A., HINGER, G., HOLLERT, H. & TIEHM, A. (2006): Vorkommen und Ökotozität von heterozyklischen Kohlenwasserstoffen (NSO-HET). In: Tagungshandbuch des BMBF-Workshops zum KORA Themenverbund 2: "Gaswerke, Kokereien, Teerverarbeitung": MNA bei der Altlastenbehandlung, Duisburg 12. Juni 2006 und Dresden 19. Juni 2006, 37-41.

SAGNER, A. & TIEHM, A. (2005): Enhanced natural attenuation of heterocyclic hydrocarbons: biodegradation under anaerobic conditions and in the presence of H_2O_2. In: UHLMANN, O. & ANNOKKÉE, F. (Eds.): Proceedings (CD) of the 5th international FZK/TNO conference on soil-water systems, 03-07 October, Bordeaux, France, 1629-1636.

SCHEIDEGGER, A., BÜRGISSER, C. S., BORKOVEC, M., STICHER, H. MEEUSSEN & VAN RIEMSDIKK, W. H. (1994): Convective Transport of acids and bases in porous media. Water Resour. Res., **30**, 2937-2944.

SCHINK, B. (2000): Principles of anaerobic degradation of organic compounds, in Biotechnology, REHM, H.-J. & REED, G. (Eds.), vol. 11b Environmental Processes II, Wiley-VCH, Weinheim.

SCHMIDT, K.R., STOLL, C. & TIEHM, A. (2006a): Evaluation of 16S-PCR detection of Dehalococcoides at two chloroethene-contaminated sites. Water Sci. Technol., in press.

SCHMIDT, K. R., STOLL, C. & TIEHM, A. (2006b): 16S-PCR detection of halorespiring bacteria at two chloroethene-contaminated sites in Germany. In: UFZ Centre for Environmental Research Leipzig-Halle, Book of Abstracts, International Conference on Environmental Biotechnology, 09-14 July, Leipzig, Germany, 233.

SCHREINER, M., & KREYSING, K. (1998): Handbuch zur Erkundung des Untergrundes von Deponien und Altlasten, Bd. 4, Geotechnik Hydrogeologie, Springer, Berlin.

SCHRIEVER, M. & HIRNER, A. (1994): Entwicklung von Routinetests zur Elution von organischen Komponenten aus Abfällen und belasteten Böden. Abschlußbericht Landesamt für Wasser und Abfall NRW von der Universität GH Essen.

SCHULZ, H.-D. & KÖLLING, M. (1992): Grundlagen und Anwendungsmöglichkeiten hydrogeochemischer Modelle. In: DVWK (Hrsg.): Anwendung hydrogeochemischer Modelle.- DVWK-Schriften 100, Parey, Berlin.

SCHULZE, S. & TIEHM, A. (2004): Assessment of microbial natural attenuation in groundwater polluted with gasworks residues. Water Sci. Technol., **50**, 5, 347-353.

SCHULZE, S., BIRKLE, M., EBNER, R., BARCZEWSKI, B. & TIEHM, A. (2003): Natural microbial degradation in a BTEX/PAH polluted groundwater plume. In: Proceedings of ConSoil 2003 (CD), 8th Int. FZK/TNO Conf. Contamin. Soil, 12.-16.05.2003 in Gent, Belgium; Theme C, 2212-2218.

SCHUMACHER, B. A., NEARY, A. J., PALMER, C. J. et al. (1995): Laboratory Methods for Soil and Foliar Analysis in Long-Term Environmental Monitoring Programs. EPA/600/R-95/077. U.S. EPA, Las Vegas.

SCHUMACHER, B. A. (2002): Methods for the determination of total organic carbon (TOC) in soil and sediments. United States Environmental Protection Agency, Environmental Sciences Division National, Exposure Research Laboratory Las Vegas, NCEA-C- 1282, www.epa.gov/esd/cmb/research/papers/bs116.pdf.

SCHUTH, C., TAUBALD, H., BOLANO, N. & MACIEJCZYK, K. (2003): Carbon and hydrogen isotope effects during sorption of organic contaminants on carbonaceous materials. J. Contam. Hydrol., **64**, 3-4, 269-281.

SCHWERTMAN, U. & MURAD, E. (1983): Effect of pH on the formation of goethite and hematite from ferrihydrite. Clays and clay minerals, **31**, 277-284.

SEARLE, P. L. (1986): The measurement of soil cation exchange properties using the single extraction, silver-thiourea method: An evaluation using a range of New Zealand soils. Australian J. of Soil Res., **24**, 193-200.

SEEL, F. (1979): Grundlagen der analytischen Chemie unter besonderer Berücksichtigung der Chemie in wäßrigen Systemen. (Basics of analytical chemistry with a particular focus on chemistry of aqueous systems). 7th edn. VCH, Weinheim.

SHERWOOD LOLLAR, B., SLATER, G. F., SLEEP, B., WITT, M., KLECKA, G. M., HARKNESS, M. & SPIVACK, J. (2001): Stable carbon isotope evidence for intrinsic bioremediation of tetrachloroethene and trichloroethene at Area 6, Dover Air Force Base. Environ. Sci. Technol., **35**, 2, 261-269.

SHIMADZU (2006): Shimadzu's TOC-V Series. www.ssi.shimadzu.com/products/product.cfm?product=visionarytoc

SIMS, R. C. & OVERCASH, M. R. (1983): Fate of polynuclear aromatic compounds (PNAs) in soil plant systems. Residue Rev., **88**, 1-68.

SIMUNEK, J., SEJNA, M.. & VAN GENUCHTEN, M. T. (1998): The HYDRUS-1D software package for simulating the one-dimensional movement of water, heat, and multiple solutes in variably-saturated media. US Salinity Laboratory, Agricultural Research Service, US Department of Agriculture, Riverside, CA.

SIMUNEK, J., VAN GENUCHTEN, M. T., SEJNA, M., TORIDE, N. & LEIJ, F. J. (1999): The STANMOD Computer Software for Evaluating Solute Transport in Porous Media Using Analytical Solutions of the Convection-Dispersion Equation, U.S. Salinity Laboratory, USDA/ARS, Riverside, CA.

SINGH, H., LÖFFLER, F. E. & FATHEPURE, B. Z. (2004): Aerobic biodegradation of vinyl chloride by a highly enriched mixed culture. Biodegradation, **15**, 197-204.

SLATER, G. F., SHERWOOD LOLLAR, B., SLEEP, B. E. & EDWARDS, E. A. (2001): Variability in carbon isotopic fractionation during biodegradation of chlorinated ethenes: Implications for field applications. Environ. Sci. Technol., **35**, 5, 901-907.

SMALL, H. (1989): Ion Chromatography. Plenum Press, New York.

SÖHNGEN, K. (1992): Strategie einer sach- und fachgerechten Entnahme von Grundwasserproben. eretec. Institut für chemische Analytik und Umwelttechnik, Gummersbach, Germany.

SONG, D. L., CONRAD, M. E., SORENSON, K. E. & ALVAREZ-COHEN, L. (2002): Stable carbon isotope fractionation during enhanced in situ bioremediation of trichloroethene. Environ. Sci. Technol., **36**, 2262-2268.

STANGER, G. (1994): Dictionary of hydrology and water resources. Lochnan, Adelaide.

STEINMANN, P. & SHOTYK, W. (1996): Sampling anoxic pore waters in peatlands using "peepers" for in-situ filtration. Fresenius' Journal of Analytical Chemistry, **354**, 709-713.

STIEBER, M., HAESELER, F., WERNER, P. & FRIMMEL, F. H. (1994): A rapid screening method for micro-organisms degrading polycyclic aromatic hydrocarbons in microplates. Appl. Microbiol. Biotechnol., **4**, 753-755.

VAN STRAATEN, L. & HESSER, F. (1998): Grundwasserprobennahme. In: VOIGT, H.-J. & WIPPERMANN, Th., Geochemie, Bd. 6 des Handbuch zur Erkundung des Untergrundes von Deponien und Altlasten, Springer, Berlin.

STRECK, T., POLETIKA, N., JURY, W. A. & FARMER, W. J. (1995): Description of simazine transport with rate – limited, two-stage, linear and nonlinear sorption. Water Resources Research, **31**, 811-822.

STRECK, T. & RICHTER, J. (1997): Heavy metal displacement in a sandy soil at the field scale: I. Measurements and parameterization of sorption. J. of Environ. Qual., **26**, 49-56.

STRECK, T. & RICHTER, J. (1999): Field-scale study of chlortoluron movement in a sandy soil over winter: II. Modeling. J. Environ. Qual., **28**, 1824-1831.

SUAREZ, M. P. & RIFAI, H. S. (1999): Biodegradation rates for fuel hydrocarbons and chlorinated solvents in groundwater. Bioremediation Journal, **3**, 4, 337-362.

TADJERPISHEH, N. & ZIECHMANN, W. (1986): Genese und Analyse von Ton-Huminstoff-Komplexen. – Mitt. Dtsch. Bodenkdl. Gesellsch. **45**, 155-160.

TAKEUCHI, M., NANBA, K., IWAMOTO, H., NIREI, H., KUSUDA, T., KAZAOKA, O., OWAKI, M. & FURUYA, K. (2005): In situ bioremediation of a cis-dichloroethylene-contaminated aquifer utilizing methane-rich groundwater from an uncontaminated aquifer. Water Res., **39**, 2438-2444.

TESSIER, A., CAMPBELL, P. G. C. & BISSON, M. (1979): Sequential extraction procedure for the speciation of particulate trace metals. Anal. Chem., **51**, 844-851.

THENG, B. K. G. (1974): The chemistry of clay organic reactions. Hilger, London/ Wiley, New York.

THENG, B. K. G. (1979): Formation and properties of clay-polymer complexes. Elsevier, Amsterdam.

TIEHM, A. & SCHULZE, S. (2003): Intrinsic aromatic hydrocarbon biodegradation for groundwater remediation. Oil & Gas Science and Technology – Rev. IFP, **58**, 4, 449-462.

TIEHM, A., GOZAN, M., MÜLLER, A., SCHELL, H., LORBEER, H. & WERNER, P. (2002a): Sequential anaerobic/aerobic biodegradation of chlorinated hydrocarbons in activated carbon barriers. Water Sci. Technol.: Water Supply, **2**, 2, 51-58.

TIEHM, A., SCHMIDT, K. R., STOLL, C., MÜLLER, A., LOHNER, S., HEIDINGER, M., WICKERT, F. & KARCH, U. (2006a): Assessment of natural microbial dechlorination. Ital. J. of Engin. Geol. and Envir., in press.

TIEHM, A., SCHMIDT, K., STOLL, C., MÜLLER, A. & LOHNER, S. (2005): Natürlicher mikrobieller Abbau (Natural Attenuation) von CKW: Fallbeispiele, Abbaumechanismen und Nachweismethoden. In: Ressourcen- und Grundwasserschutz, Veröffentlichungen aus dem Technologiezentrum Wasser (ISSN 1434-5765), Bd. **28**. 53-73.

TIEHM, A., SCHMIDT, K. R., MARTIN, H. & HEIDINGER, M. (2006b): Stable isotope fractionation during PCE halorespiration and aerobic cisDCE and VC biodegradation. In: UFZ Centre for Environmental Research Leipzig-Halle, Book of Abstracts, International Conference on Environmental Biotechnology, 09-14 July, Leipzig, Germany, 408.

TIEHM, A. & FRITZSCHE, C. (1995): Utilization of solubilized and crystalline mixtures of polycyclic aromatic hydrocarbons by a Mycobacterium sp. Appl. Microbiol. Biotechnol., **42**, 964-968.

TIEHM, A. & STIEBER, M. (2001): Strategies to improve PAH bioavailability: Addition of surfactants, ozonation and application of ultrasound. In: STEGMANN, R., BRUNNER, G., CALMANO, W. & MATZ, G. (Eds.): Treatment of Contaminated Soil, Springer, Berlin.

TIEHM, A., SCHULZE, S. & MÜLLER, A. (2002b): Handlungsoption „Natural Attenuation" – Natürlicher Abbau von Schadstoffen im Grundwasser. In: Aktuelle Themen bei der Trinkwassergewinnung, Veröffentlichungen aus dem Technologiezentrum Wasser (ISSN 1434-5765), Bd. **18**, 65-78.

TIEHM, A., STIEBER, M., WERNER, P. & FRIMMEL, F. H. (1997): Surfactant-enhanced mobilization and biodegradation of polycyclic hydrocarbons in manufactured gas plant soil. Environ. Sci. Technol., **31**, 2570-2576.

TOLMAN, C. F. (1937): Groundwater. McGraw-Hill, New York.

TORIDE, N., LEIJ, F. J. & VAN GENUCHTEN, M. T. (1995): The CXTFIT Code for Estimating Transport Parameters from Laboratory or Field Tracer Experiments, Version 2.0. Research Report No. 137, U.S. Department of Agriculture, Riverside, CA.

TORSTENSSON, B. A. & PETSONK, A. M. (1988): A hermetically isolated sampling method for groundwater investigations. ASTM American Special Technical Publication **963**, 274-289.

TOTSCHE, K. U. (2001): Reaktiver Stofftransport in Böden: Optimierte Experimentdesigns zur Prozeßidentifikation. Bayreuther Bodenkundl. Ber., **75**.

TRIBUTH, H. & LAGALY, G. (1986a): Aufbereitung und Identifizierung von Boden- und Lagerstättentonen I. – GIT Fachz. Lab., **6**, 524-529.

TRIBUTH, H. & LAGALY, G. (1986b): Aufbereitung und Identifizierung von Boden- und Lagerstättentonen II. – GIT Fachz. Lab., **8**, 771-776.

UIT (2000): Druckhaltende Schöpfer. Umwelt- und Ingenieurtechnik GmbH, Dresden, Germany.

URE, A. M. & DAVIDSON, C. M. (1995): Chemical speciation in the environment. Blackie, London.

USEPA (1994): Terms of environment, glossary, abbreviations, and acronyms. U.S. Environmental Protection Agency EPA 175-B 94-015.

VERCE, M. F., GUNSCH, C. K., DANKO, A. S. & FREEDMAN, D. L. (2002): Cometabolism of *cis*-1,2-dichloroethene by aerobic cultures grown on vinyl chloride as the primary substrate. Environ. Sci. Technol., **36**, 10, 2171-2177.

VINOGRADOV, A. P. (1954): Geochemie seltener und nur in Spuren vorhandener chemischer Elemente in Böden. Akademie Verlag, Berlin.

VOIGT, H.-J. & WIPPERMANN, T. (1998): Handbuch zur Erkundung des Untergrundes von Deponien und Altlasten, Bd. 6, Geochemie. Springer, Berlin.

WALKER, G. F. (1959): Diffusion of exchangeable cations in vermiculite. Nature, **184**, 1392-1393.

WALKLEY, A. & BLACK, I. A. (1934): An Examination of Degtjareff Method for Determining Soil Organic Matter and a Proposed Modification of the Chromic Acid Titration Method. Soil Sci., **37**, 29-37.

WEISS, J. (1991): Ionenchromatographie. 2. erw. Aufl., VCH, Weinheim.

WELZ, B. & SPERLING, M. (1997): Atomabsorptionsspektrometrie (Atomic absorption spectrometry). 4th edn. Wiley-VCH, Weinheim.

WHO (1998): International Programme on Chemical Safety (IPCS); Environmental Health Criteria 202: Selected Non-heterocyclic Aromatic Hydro-carbons. World Health Organization. Wissenschaftliche Verlagsgesellschaft, Stuttgart.

WIEDEMEIER, T. H., RIFAI, H. S., NEWELL, C. J. & WILSON, J. T. (1999): Natural Attenuation of Fuels and Chlorinated Solvents in the Subsurface. Wiley & Sons, New York.

WILLIAMS, I. (2001): Environmental Chemistry: A Modular Approach. Wiley & Sons, New York.

WILSON, W. E. & MOORE, J. E. (Eds.) (2001): Glossary of Hydrology. American Geological Institute.

WINKLER, A. (1989) Untersuchungen zur Mobilität von Technetium (und Selen) in norddeutschen Grundwasserleitern und Technetium im Kontakt mit natürlich vorkommenden Mineralien., Berliner geowiss. Abh., **117**.

VAN DER ZEE, S. E. A. T. M. & VAN RIEMSDIJK, W. H. (1987): Transport of Reactive Solute in Spatially Variable Soil Systems. Water Resources Research, **23**, 11, 2059-2069.

ZEIEN, H. & BRÜMMER, G. W. (1991): Chemische Extraktionen zur Bestimmung von Schwermetallbindungsformen in Böden. Mitt. Dt. Bodenk. Ges. 59/1.

ZHENG, C. & BENNETT, G. D. (1995): Applied contaminant transport modeling: theory and practice. Van Nostrand Reinhold, New York.

ZURMÜHL, T. (1994): Validierung konvektiv-dispersiver Modelle zur Berechnung des instationären Stofftransports in ungestörten Bodensäulen. Bayreuther Bodenkundliche Berichte 36.

5.4 Interpretation of Geological, Hydrogeological, and Geochemical Results

FLORIAN JENN, CLAUS KOFAHL, MIKE MÜLLER, JENS RADSCHINSKI & HANS-JÜRGEN VOIGT

5.4.1 Statistical Methods

FLORIAN JENN & HANS-JÜRGEN VOIGT

Almost all data obtained in geoscientific investigations need some kind of statistical treatment for interpretation, as well as for the assessment of reliability and errors. The following statistical methods are frequently used to analyze data sets:

- univariate analysis, i.e., analysis of an individual parameter (e.g., concentration of a substance, pH, or electrical conductivity at several locations),
- multivariate analysis, i.e., analysis of several parameters together, to determine the relationship between parameters (e.g., the relationship between concentrations of different substances and/or of several environmental parameters), and
- time series analysis (e.g., analysis of a parameter as a function of time, for example, monitoring of water level or the concentration of a substance in a groundwater observation well).

Geostatistics, interpolation of spatial data, and tests specific for hydrogeochemical data are used to evaluate geological, hydrogeological and geochemical data. Commonly used statistical software are, for example, Origin, SPSS, S-Plus and R. Simple statistical evaluations can also be performed using spreadsheet software (e.g., Microsoft Excel, OpenOffice Calc). Various aspects and examples of interpretation using statistical methods are described in Part 6 of this handbook.

Measurements are usually made to determine or estimate the typical (representative, average) values and the range of variation of a parameter (e.g., hydraulic conductivity) of a geological unit. All values of a parameter, not only those actually observed, but those that are potentially observable, are called a statistical population (SWAN & SANDILANDS, 1995). The population is, therefore, established by the choice of the investigation target.

It is, of course, not feasible to carry out all those possible measurements. Instead, a limited set of measurements is carried out. In statistics, the individual measurements are commonly called *observations*. Taken together, they comprise a subset of the population, the *statistical sample*, which is the data available for statistical analysis.

The statistical and the geological meanings of the word "sample" must not be confused. A sample in geological terms is, for example, a piece of rock or a quantity of water drawn from a well, and yields a single value for any one property. Several measurements of that property comprise a statistical sample, which can be analyzed to estimate other properties, for example the hydraulic conductivity of a geological unit (e.g., an aquifer) or the concentration of a substance in the aquifer.

5.4.1.1 Univariate Statistics

Statistical analysis is commonly carried out stepwise. The first step in the statistical treatment of a data set is a univariate analysis of the data, i.e. each variable (e.g., water level, concentration of substances, layer thickness) in a data set is analyzed separately. The aim of this kind of analysis is to describe the properties of the population by analyzing the sampled data. In the following, the selected variable is denoted X, with the observed values $x_1,\ldots,x_n$, where n is the *sample size*.

Histogram and frequency distribution

Histograms are used to visualize the *frequency distribution* of the given values x_i. The range of observed values is divided into (usually equally sized) classes. A rectangle is drawn for each class with a width from the lower to the upper boundary of the class and with a height corresponding to the number or percentage of values x_i observed within the class. An example is given in *Figure 5.4-1*. When the number of classes n_c is chosen, a compromise has to be found between too few classes, which may suppress information, and too many classes, which causes the variation in the individual values to have too much influence. As a starting point, the statistical software may suggest a number of classes, or the following rules of thumb may be used.

Number of classes n_c according to DVKW (1990):

$$\begin{aligned} n_c &= \sqrt{n} \quad \text{for } n < 1000 \\ n_c &= 10\lg n \quad \text{for } n > 1000. \end{aligned} \tag{5.4.1a}$$

Number of classes n_c according to Sturges' Rule (EVERITT, 1998):

$$n_c = \log_2 n + 1. \tag{5.4.1b}$$

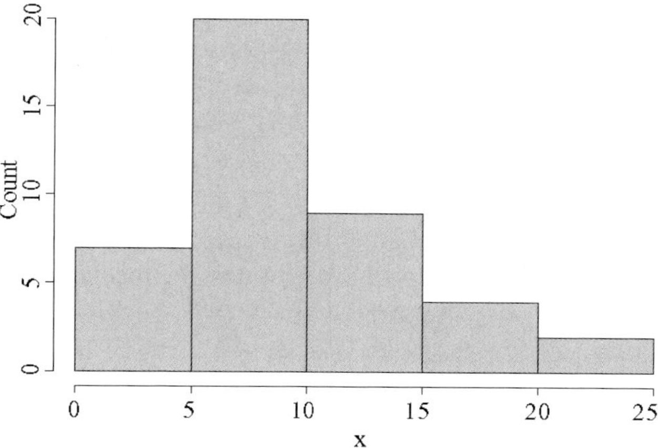

Fig. 5.4-1: Example of a histogram

Between 5 and 20 classes are usually used (DVKW, 1990). A histogram shows the frequency distribution of the statistical sample, which approximates the frequency distribution of the population. The latter is often drawn as a smooth curve instead of a histogram, and sometimes described in an idealized way by mathematical probability functions (described in more detail below).

The frequency distribution normally shows one or more distinct peaks, called modes. These indicate values for x that have been observed most often. Depending on the number of modes, the distribution is called *unimodal* (one mode), *bimodal* (two modes) or *multimodal* (more than two modes). If observations are taken from a single population (e.g., chloride concentration in groundwater samples from various parts of an aquifer) the resulting distribution is usually unimodal. Several modes often indicate that several populations are superimposed (e.g., chloride concentration in samples from an uncontaminated and a contaminated region of an aquifer).

Summary statistics and properties of the population

There are several parameters to describe the properties of a statistical sample and the corresponding population. Although they could be computed for any data set, they usually only make sense for unimodal data sets. The most important statistical parameters are those for average or typical values for the sample or the population. The *sample mean* is defined as

$$\bar{x} = \frac{1}{n}\sum_{i=1}^{n} x_i \qquad (5.4.2)$$

and is an estimate of the *population mean* μ. The *median* is the value in an ordered set of values for which there is an equal number of values larger and smaller than it. The *mode* is the most frequently occurring value in a data set. In a symmetrical, unimodal distribution, the values for these three parameters are identical. Asymmetric, or *skewed*, distributions are often found (*Fig. 5.4-1*). This may be caused by a general characteristic of the population (natural concentrations of trace substances are usually steep on the left side and have a longer tail on the right side) or by *outliers* (individual very high or low values). If the histogram is steep on the left-hand side gradual slopes to the right, it is called *positively skewed*; if it is steep on the right-hand side, it is called *negatively skewed*. In skewed distributions, the median and mean move in the direction of the tail with respect to the mode:

$$\text{mode} < \text{median} < \text{mean, if positively skewed} \tag{5.4.3a}$$

$$\text{mean} < \text{median} < \text{mode, if negatively skewed.} \tag{5.4.3b}$$

This also shows that the mean is more strongly affected by asymmetry and outliers than the median. On the other hand, the mean is easier to deal with mathematically, and many statistical methods and tools are based on it.

Another group of parameters describes the *dispersion*, or spread, of a distribution, i.e. the amount by which a set of observations deviate from their average value. The most simple is the *range*, the difference between the maximum and minimum values observed. This, however, is very sensitive to outliers. A more robust approach involves *quartiles* or *percentiles*: The *first quartile* is the value below which the lowest quarter of the sorted set of observations can be found. The *third quartile* is the value above which is the highest quarter of the observations. The second quartile is the median. The range between first and third quartile is called the *interquartile range* (*IQR*) and comprises half of the samples – the ones that are closest to the center.

Percentiles are defined in an analogous way. For example, the 5th percentile is the value below which the lowest 5 percent of observations are found. The range between 5th and 95th percentiles is often used as a range of representative values. The observations outside of this range are considered outliers.

Another way to characterize spread is to quantify, on average, how far each observation is from the center. The *sample variance* s^2 is thus defined as the average of the squared deviations of each observation x_i from the mean $\overline{x}$:

$$s^2 = \frac{1}{n-1}\sum_{i=1}^{n}\left(x_i - \overline{x}\right)^2 \tag{5.4.4}$$

and the *standard deviation s* is the square root of the variance.

These parameters can be used to roughly describe a set of observations, e.g., by giving mean and standard deviation, or the so-called five-number summary (minimum, 1st quartile, median, 3rd quartile, maximum). The latter is the basis for box-and-whisker plots (also called boxplots for short), of which several variations are in use. The basic principle is to draw a box extending from the 1st to 3rd quartile (representing the IQR), intersected by a line at the median. Two lines ("whiskers") extend from the box to one of the following:

- the minimum and maximum values,
- selected percentiles (e.g., 5th and 95th), or
- the furthest observation within a distance of 1.5 IQR from the box (considering anything further away at least a "mild" outlier).

This design can be extended to include additional information: outliers can be indicated separately, the mean shown by an additional dot, or notches to indicate medians in other distributions that are significantly different. *Figure 5.4-2* shows a sample distribution and the resulting basic boxplot. The boxplot is an alternative to the histogram and is a condensed view of the sample distribution. It quickly shows the median, possibly the mean, and the range of typical values (50 % of the observations are within the box), and indicates the spread of the distribution by the length of the box and the length of the whiskers. Skewness can be seen by off-center placement of the median line as well as different lengths of the "whiskers". However, it is not possible to discern the modality of the distribution, i.e. whether multiple independent populations are superimposed in the sample (*Fig. 5.4-3*).

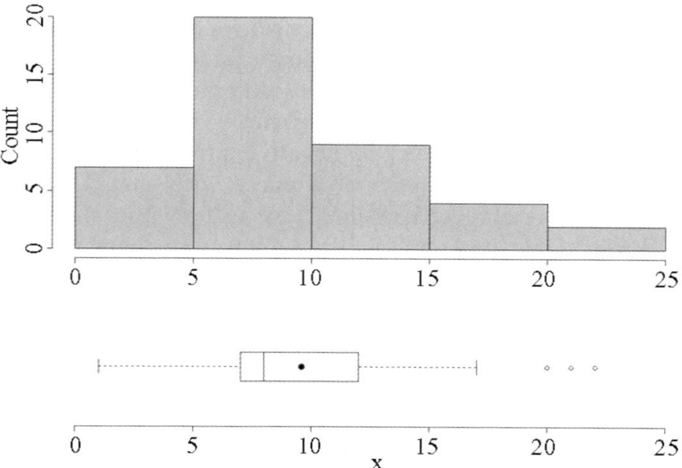

Fig. 5.4-2: Histogram of the frequency distribution and corresponding boxplot containing the line for the median. Circles indicate outliers, the black dot is the mean. "Whiskers" stretch to the furthest observations no more than 1.5·IQR away.

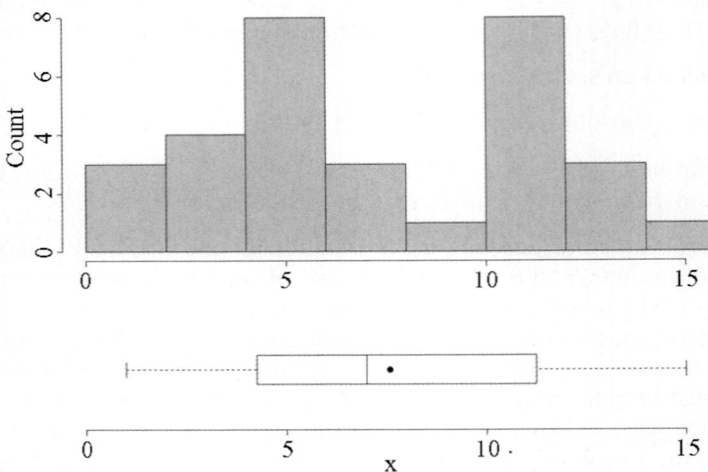

Fig. 5.4-3: Histogram of a multimodal distribution and corresponding boxplot (same design as in *Fig. 5.4-2*)

Probability distributions

The *probability density function* $f(x)$ of a continuous random variable X is a statistical function that shows how the density of possible observations in a population is distributed. It is the mathematical ideal corresponding to the empirical frequency distribution of samples described above. The probability density function is the derivative of the *cumulative distribution function* $F(x)$ if $F(x)$ is differentiable. It can never be negative and its total integral is unity. Geometrically, $f(x)$ is the ordinate of a curve such that $f(x)\,dx$ yields the probability that the variable will assume some value within the range dx of x. For each differentiable probability density function, there is a cumulative probability distribution

$$F(x) = \int_{-\infty}^{x} f(x)dx, \tag{5.4.5}$$

which describes the probability of $X \leq x$. Values of $F(x)$ for different distributions are available as tables or from statistical software. The cumulative probability distribution can be used to calculate the probability that X falls between two values x_1 and x_2:

$$P(x_1 \leq X \leq x_2) = F(x_2) - F(x_1) \tag{5.4.6}$$

There are several mathematical functions that describe such probability distributions. The most important is the *normal distribution*, which is a symmetrical, bell-shaped curve (*Fig. 5.4-4*). Given the distribution mean μ and the *standard deviation* σ, the distribution is described by

$$f_N(x) = \frac{1}{\sigma\sqrt{2\pi}} e^{-\frac{(x-\mu)^2}{2\sigma^2}}. \tag{5.4.7}$$

The normal distribution has been found to describe remarkably well the errors of observation in physics; many variables measured in environmental investigations are (approximately) normally distributed (WEBSTER & OLIVER, 2001). Also, many statistical methods are based on the assumption of data with a normal distribution.

If $\mu = 0$ and $\sigma = 1$, the result is called the *standard normal distribution*, which is the only normal distribution that is usually tabulated. If a variable X has a normal distribution, the corresponding *standardized* variable

$$Z = \frac{X - \mu}{\sigma} \tag{5.4.8}$$

has a standard normal distribution. Conversely, values of any normal distribution can be calculated from the standard normal distribution by rearranging Equation (5.4.8)

$$X = \sigma Z + \mu. \tag{5.4.9}$$

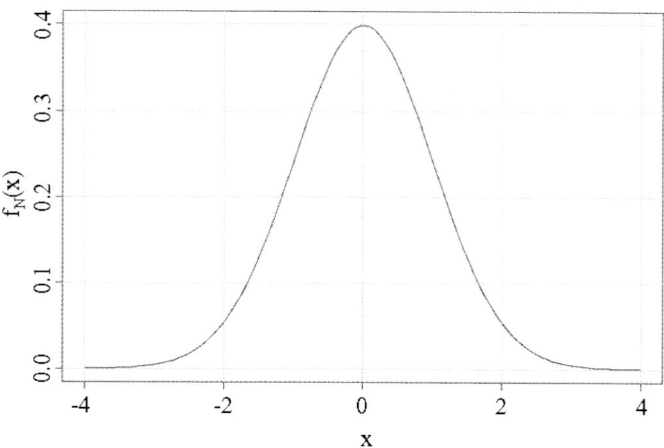

Fig. 5.4-4: Probability density of the standardized normal distribution

Another important distribution is the *log-normal distribution*. This is the distribution of a variable whose logarithm has a normal distribution. Given the probability density function of the normal distribution f_N, the log-normal distribution is defined as:

$$f_{LN}(x) = \frac{1}{x\sigma\sqrt{2\pi}} e^{-\frac{(\ln x-\mu)^2}{2\sigma^2}} . \qquad (5.4.10)$$

This distribution is not symmetrical, but positively skewed. A typical example of a variable with a log-normal distribution is the concentration of a trace substance.

There are many more distributions for various situations and purposes, e.g. Student's *t* distribution, *F* distribution, and χ^2 distribution for statistical inference (see below), which are not described here.

Example: Determining natural groundwater conditions by component separation analysis

The frequency distribution of concentrations in groundwater was used to determine the natural groundwater quality in aquifers in Germany (WENDLAND et al., 2005). They assume that the histograms can be modeled by a log-normal distribution for the natural concentration distribution and a superimposed normal distribution reflecting local or anthropogenic influences (*Fig. 5.4-5a*). Using the software package Origin, the frequency distributions of the observed data were separated. Then the 10th and 90th percentiles (i.e. comprising 80 % of values) were taken as the range of natural groundwater conditions (*Fig. 5.4-5b*). *Table 5.4-1* shows the natural groundwater conditions determined by this technique for the main hydrostratigraphic units of major interest to water management.

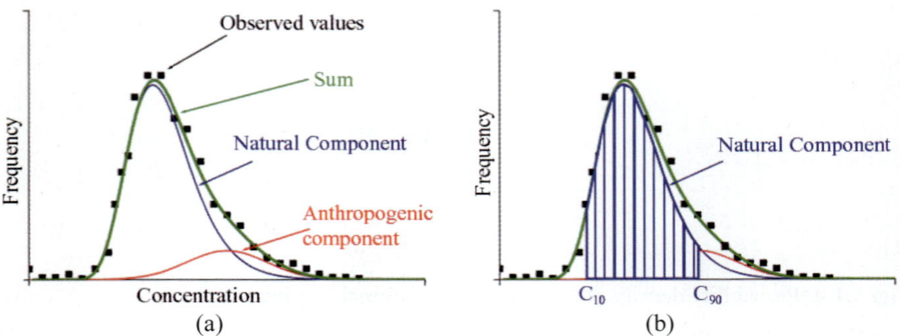

Fig. 5.4-5: Component separation analysis of groundwater data from Germany, WENDLAND et al. (2005). (a) Assumption of two superimposed distributions, (b) percentiles as the range of natural groundwater conditions

Determination of natural groundwater quality is not only a basis for assessing the influence of various contamination sources, but also for decisions on remediation goals. In this case, however, the possible influence of groundwater upstream from a contamination source, for example a landfill, has to be kept in mind.

Table 5.4-1: Results of assessment of natural groundwater conditions in Germany using component separation analysis, WENDLAND et al. (2005)

		Sands and gravel of aquifer F2 (Saale glaciaton)		**Jurassic limestone (Malm)**		**Triassic limestone (Muschelkalk)**		**Triassic sandstone (Buntsandstein)**	
		from	to	from	to	from	to	from	to
conductivity	$\mu S\ cm^{-1}$	186	521	387	704	637	939	50	256
O_2	$mg\ L^{-1}$	0.2	4.6	6	11	3	10	5	11
pH	–	6.0	7.8	7.1	7.7	7.0	7.5	6.8	7.7
DOC	$mg\ L^{-1}$	0.8	5.0	0.3	1.3	0.4	1.2	0.3	1.6
Ca	$mg\ L^{-1}$	29	143	69	126	99	154	7	29
Mg	$mg\ L^{-1}$	3	30	4	37	17	50	2	23
Na	$mg\ L^{-1}$	6	24	1.3	6.3	3.0	9.2	2	16
K	$mg\ L^{-1}$	0.8	4.0	0.3	1.9	0.6	2.1	1.3	3.6
NH_4	$mg\ L^{-1}$	<0.01	0.5	<0.01	<0.01	<0.01	<0.01	<0.01	<0.01
Fe	$mg\ L^{-1}$	0.1	5.0	<0.01	0.15	<0.01	0.1	<0.01	0.1
Mn	$mg\ L^{-1}$	0.04	0.64	<0.01	<0.01	<0.01	<0.01	<0.01	<0.01
HCO_3	$mg\ L^{-1}$	150	426	278	380	287	446	6	96
Cl	$mg\ L^{-1}$	9	43	5	37	9	49	4	17
SO_4	$mg\ L^{-1}$	4	68	13	32	30	147	5	58
NO_3	$mg\ L^{-1}$	<0.01	0.1						

Statistical inference

Statistical inference methods are used to draw conclusions about a population using data collected from a sample, for example:

- Calculation of the 95 % confidence interval for the mean chloride concentration in an aquifer.
- Testing whether a variable has a normal (or log-normal, etc.) distribution. This is often important, as many interpretation methods assume a normal distribution.

- Testing the statistical hypothesis, based on measurements of chloride concentrations in groundwater, that the investigated aquifer has a mean chloride concentration greater than 250 mg L^{-1}.
- Testing whether two statistical samples are derived from the same population.

Confidence interval for the mean of a normal distribution

A confidence interval is a measure of the precision of an estimated value. The interval represents the range of values that is believed to encompass the "true" value with a high probability (usually 95 %). The confidence interval is expressed in the same units as the estimate. Wider intervals indicate lower precision and narrow intervals indicate greater precision. For example, given a sample with a normal distribution, with a size n and a sample variance s, the lower and upper 95 % confidence limits for the mean μ_L and μ_U of a normal distribution can be calculated according to SWAN & SANDILANDS (1995) by

$$
\begin{aligned}
\mu_L &= \overline{x} - t_{2.5\%;n-1}\sqrt{s^2/n}\,, \\
\mu_U &= \overline{x} - t_{2.5\%;n-1}\sqrt{s^2/n}
\end{aligned}
\tag{5.4.11}
$$

where $t_{2.5\%;n-1}$ denotes the 2.5-percentile of a Student's t distribution with n - 1 degrees of freedom (these values can be found in statistical tables). The 2.5-percentile is chosen because a 95 % confidence interval leaves out 5 % for error, half of which (2.5 %) on each side. For 90 % confidence, one would use the 5th percentile of the Student's t distribution.

Hypothesis testing

A result is called significant if it is unlikely to have occurred by chance. In hypothesis testing, the significance level α is the criterion used for rejecting the null hypothesis. The null hypothesis is a hypothesis about a population parameter. The purpose of hypothesis testing is to test whether the null hypothesis is correct in the light of the experimental data. Depending on the data, the null hypothesis either will or will not be rejected as a useable possibility. The null hypothesis is often the reverse of what the experimenter actually believes. It is assumed to allow the data to contradict it. By convention, the probability of the observed value t of the test statistic (see below) is called the p value. If the p value is less or equal the significance level α, the null hypothesis is rejected. This is just another way of stating that the null hypothesis is rejected if the observed t is greater than a critical value t_α (i.e., calculated for a specified significance level) of the corresponding distribution. There are several misunderstandings when dealing with p values. In particular, it is not the probability of the null hypothesis being true, not the probability of the observation just being a random coincidence, and the significance level is not determined by p. The significance level is a value that has to be agreed upon *before* the investigation. The lower the significance level, the more the data must diverge from the null hypothesis to be significant. Therefore, the 0.01 level is more conservative than the 0.05 level.

The procedure of hypothesis testing is demonstrated here using an example about the mean concentration μ of chloride in an aquifer. First, two hypotheses have to be stated, the null hypothesis H_0 and the alternative hypothesis H_1, which is usually the one whose truth the analyzer is trying to establish. In the example, the hypotheses would be

$$\begin{aligned} H_0&: \mu = \mu_0 = 250 \\ H_1&: \mu > \mu_0 = 250. \end{aligned} \qquad (5.4.12)$$

Because the null hypothesis is stating that the observed value and the hypothetical value are equal, the null hypothesis is sometimes called the "no difference" hypothesis. Depending on the parameter(s) to be tested, a mathematical function, called the *test statistic*, has to be chosen (see, for example, SWAN & SANDILANDS (1995) or ISO 2854) and calculated for the measured data. For testing the equivalence of the population mean to a hypothetical mean, an appropriate test statistic is the Student's t distribution (other distributions apply with other test statistics):

$$t = \frac{\overline{x} - \mu_0}{\sqrt{s^2 / n}}. \qquad (5.4.13)$$

Then the *critical value* $t_{\alpha;n-1}$ of the Student's t distribution is looked up in the tables. If the calculated test statistic t exceeds the critical value, the null hypothesis is rejected and the alternative hypothesis accepted.

In the example, there were $n = 7$ measurements with a mean of 258 mg L^{-1} and a variance of 127 $mg^2\,L^{-2}$ (corresponding to a standard deviation of 11.3 mg L^{-1}). The value of the test statistic then is 1.878. On the other hand, for $\alpha = 5$ % the critical value $t_{5\%;6} = 1.943$. The calculated t does not exceed the critical value. Thus, the null hypothesis that the chloride concentration is 250 mg L^{-1} (as opposed to the alternative that it is greater than 250 mg L^{-1}) cannot statistically be rejected. This means that it is not safe to assume that the concentration of chloride in the groundwater exceeds 250 mg L^{-1}.

Additional methods of inference and statistical tests for other parameters and purposes are described in the literature and included in common software. Examples can be found in CHIANG (2003), SWAN & SANDILANDS (1995), and ISO 2854, ISO 2602 and ISO 3301.

5.4.1.2 Multivariate Statistics

Usually, more than one parameter is determined in an investigation. This gives rise to the question if and how those parameters are related, and how those data sets can be characterized and interpreted. First, treatment of two variables (*bivariate statistics*) is described here, then followed by analysis of more than two variables (*multivariate statistics*) and methods of classification (i.e. finding groups of related observations or measures that highlight similarities or differences).

Bivariate scatter diagram and correlation coefficient

A set of measurements of two variables X and Y (e.g., sodium and chloride concentrations in water samples) can be visualized on a coordinate grid by plotting a point for each measurement. An example of the resulting *scatter diagram* (or *scatter plot*) is shown in *Fig: 5.4-6.*

The shape of the "cloud" of test points indicates the general relationship of the two variables. If they are completely unrelated, the points are about evenly spread over the plot. If there is a relationship, for example sodium and chloride concentrations in concentrated rock salt solutions, the scatter approximates a linear shape: If one solution has twice as much salt in it than another, sodium and chloride concentrations are also both twice as high.

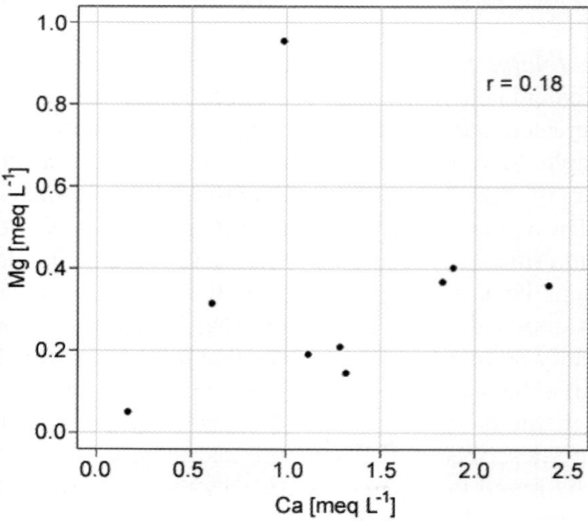

Fig. 5.4-6: Scatter plot of calcium and magnesium concentrations in groundwater from several wells. Their correlation coefficient r is low, which is caused by the single very high magnesium concentration. Without this outlier, r would be 0.73[1].

---

[1] The abbreviation meq stands for milliequivalent and is used instead of mol when it is important to take the valency of the dissolved substance into consideration.

The extent of the statistical relationship between two variables is represented by the *correlation coefficient* r, which is calculated as follows:

$$r_{xy} = \frac{\sum_{i=1}^{n}(x_i - \bar{x})(y_i - \bar{y})}{(n-1)s_x s_y}, \qquad (5.4.14)$$

where $\bar{x}$ and $\bar{y}$ are the mean values and s_x and s_y are the standard deviations of x and y, respectively. If $r = 0$, the variables do not correlate and are statistically unrelated; if $r = \pm 1$, they correlate perfectly. $r = 1$ means that the observations lie on a straight line with a positive slope of +1, and $r = -1$ means a negative slope of -1.

Multivariate scatter diagram and correlations

If more than two variables $X^1, \ldots, X^m$ (m is the number of parameters) are determined, an array of scatter diagrams can be plotted for each pair of variables. An example of this is shown in *Fig: 5.4-7*. Likewise, correlation coefficients r_{ij} can be calculated for each pair (analog to Equation 5.4.14) and included in the diagram.

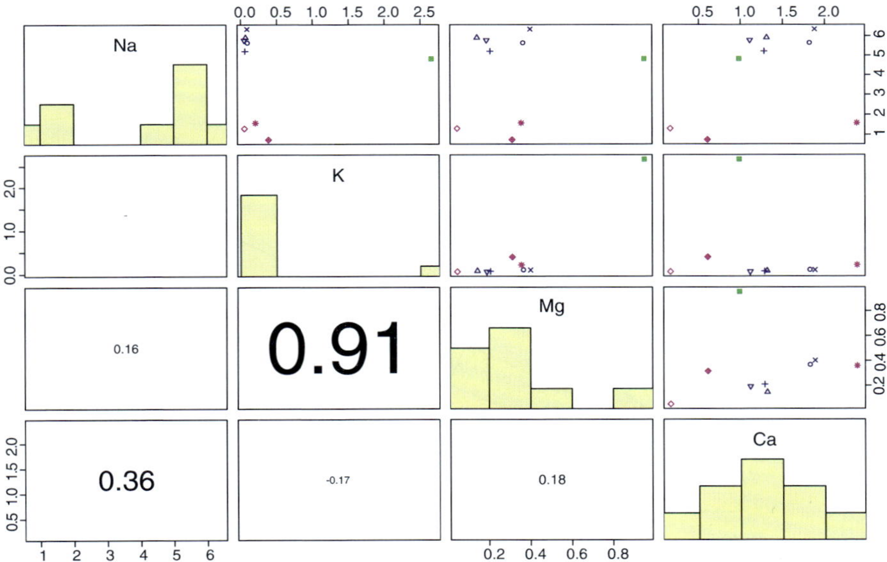

Fig. 5.4-7: Multivariate scatter for four cations in analyses of samples from several groundwater monitoring wells. The names of the ions and their (univariate) histogram is shown on the diagonal. Above the diagonal, the bivariate scatter plots are shown. The corresponding correlation coefficients are shown below the diagonal, with font sizes proportional to the degree of correlation. Colors indicate different types or depths of wells. The K-Mg correlation is high due to the very high concentrations in one well (green symbol).

In some cases, visual inspection of the multivariate scatter plot already shows observations or variables that are related and form groups. But often this is not clearly visible. Classification methods help establish groups of variables or observations. Two of these methods are described here:

Principal component analysis

Using principal component analysis (PCA), the measured variables are analyzed and a new set of variables is related to the original variables by a linear equation. These are called *principal components* and are chosen such that the first principal component accounts for most of the variation in the data set, and subsequent components for decreasing amounts of variation. Geometrically, this can be thought of as a new coordinate system. *Figure 5.4-8* illustrates this concept for the bivariate case. However, PCA usually exhibits its strength with a higher number of variables: A large set of variables is reduced to a few components. Higher-order principal components are commonly ignored because they are often dominated by statistical noise. The components can often be interpreted as physical processes; the position of an observation along this component's axis then reflects the status of the process at this point. The correlations of the original variables with the principal components are called *loadings*. Variables with high loadings indicate the factors that are crucial to the principal component and help in interpretation.

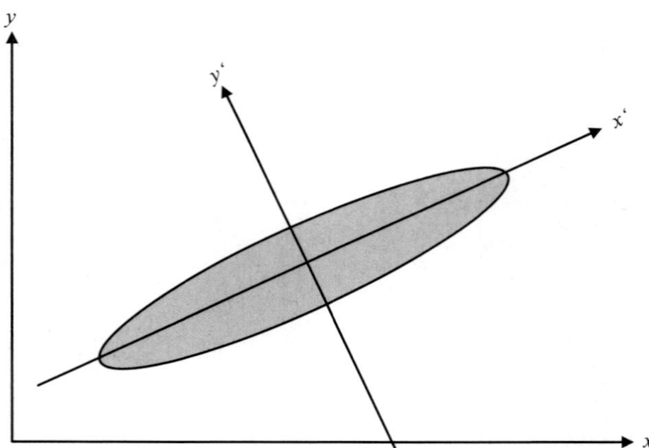

Fig. 5.4-8: Illustration of the concept of principal component analysis (PCA) using only two variables x and y. The ellipse symbolizes the variation within the data set. PCA yields a new set of axes x' and y', which are rotated and offset from the original x and y axes in such a way that x' encompasses the most variation in the data; y' is perpendicular to x' and accounts for the rest of the variation.

Table 5.4-2 shows the results of a PCA of groundwater samples in which the first three principal components have been interpreted as “nitrate reduction”, “calcite dissolution”, and “phosphate dissolution”. For example, the first component (“nitrate reduction”) has highly negative loadings of both nitrite + nitrate and dissolved oxygen; this suggests reduction of nitrate under anaerobic conditions. When oxygen concentrations are too low, microorganisms use nitrate as an electron acceptor. Thus, as one progresses along the first component's axis, increasingly denitrified water is found. The positive loadings of dissolved iron and manganese indicate that under increasingly reducing conditions (after nitrate has already been reduced), iron and manganese oxides are reduced and thereby dissolved.

Table 5.4-2: Example of PCA results of groundwater analysis from studies in southeastern USA. Only the first three components are shown. High loadings are indicated in bold. (NOLAN, 1999)
http://water.usgs.gov/nawqa/nutrients/pubs/jeq_v28_no5/jeq_v28_no5.pdf,Table 2

Water-quality variable *	**Component 1** “nitrate reduction”	**Component 2** “calcite dissolution”	**Component 3** “phosphate dissolution”
dissolved oxygen	**-0.8902**	-0.0566	-0.0007
manganese, μg L^{-1}	**0.8108**	-0.3396	-0.0539
nitrite-plus-nitrate as N	**-0.7861**	0.1267	-0.1443
ammonium as N	**0.7684**	0.1352	0.0581
iron, μg L^{-1}	**0.7630**	-0.4048	0.1154
bromide	0.6249	0.0042	0.0642
ammonia and organic nitrogen as N	0.6001	0.2606	0.0211
sodium	0.5925	0.1336	0.1667
silica as SiO_2	0.5496	-0.2365	0.3294
calcium	-0.0646	**0.8497**	-0.1197
alkalinity as $CaCO_3$	-0.0767	**0.8395**	0.0266
pH, standard units	-0.0567	**0.7879**	0.1292
specific conductance, μS cm^{-1}	0.1159	**0.7619**	0.0347
dissolved solids	0.1848	**0.7435**	0.0515
phosphorous	0.1716	0.0084	**0.8107**
orthophosphorus as P	0.1009	-0.0503	**0.7930**
temperature,°C	0.1034	-0.1746	-0.0435
magnesium	0.0548	0.2894	0.0674
potassium	0.0024	0.0526	-0.0527
nitrite as N	-0.1632	-0.0656	0.4130
chloride	0.0740	0.0323	-0.0347
dissolved organic carbon, as C	0.3794	0.0574	0.0689
fluoride	-0.2052	0.1567	**0.4667**
sulfate as $SO_4^{=}$	0.4209	0.3744	0.0684
percent variance explained:	23	18	9

* Units of the original variables are in mg L^{-1} except as noted. Principal components analysis was performed on the sample correlations of the rank-transformed variables. Components were rotated obliquely by the Direct Oblimin method.

Cluster analysis

This method allows groups ("clusters") of similar observations to be found in multivariate data sets, if there are any. The most common clustering methods link the most similar pairs of observations or clusters step by step until all points are grouped in a hierarchy of similarity. This can be visualized by a *dendrogram*, which is a tree-like structure with ever smaller clusters as one proceeds along the branches. *Figure 5.4-9* shows an example in which groundwater samples from several wells have been clustered according to the concentration of major ions.

There are several methods of measuring similarity (the Euclidian distance being the most common), and several methods of linking points or clusters. In all cases, the two most similar observations or clusters are linked together. However, there are other methods of establishing what "most similar" is (SWAN & SANDILANDS, 1995):

- *Nearest-neighbor linkage* (or *single linkage*): The similarity between observations and a new cluster is defined as equal to the similarity between that point and the most similar point in the cluster. This method tends to produce elongated clusters ("chaining"), which is often not desired.
- *Furthest-neighbor linkage* (or *complete linkage*): In contrast to single linkage, the similarity is defined as the least of all similarities in pairs. This leads to spherical clusters, and even slightly ellipsoidal clusters are likely to be broken up.
- *Average linkage*: In this case, the similarity between clusters is computed as the average of the similarities of their respective elements. This is an intermediate between the previous two methods and usually results in a decently structured hierarchy.
- *Ward's method* uses a different approach: Clusters are linked so that the increase in the sum of squared deviations from the cluster means is as small as possible. This method produces "good looking" cluster structures and is often the preferred method of clustering.

When performing cluster analysis, one should try different methods. If different methods yield very different results, it may be possible that a clear clustering of the data is not present at all.

When working with variables (parameters) of different kinds or different orders of magnitude, data should usually be standardized (e.g., subtracting the mean and dividing by the standard deviation), so that values that are very large or small just by virtue of their corresponding physical units do not dominate the clustering results.

The axis of the dendrogram represents a dissimilarity coefficient. By drawing a line through the dendrogram at a chosen dissimilarity value, clusters will be isolated from one another, and groups of clusters will be formed. The elements of the clusters will have a similarity that is greater than this value, and individual clusters are less similar than this value. This line is called the *phenon line*. Choosing the phenon line is always a matter of interpretation. Sometimes, a certain number of clusters is expected, or large gaps in the hierarchy suggest a "natural" choice of the phenon line (SWAN & SANDILANDS, 1995). *Figure 5.4-10* shows an interpretation of the example dendrogram.

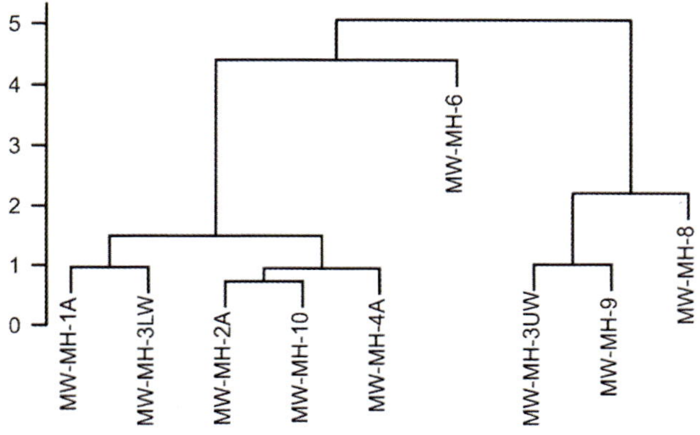

Fig. 5.4-9: Example of a dendrogram for cluster analysis of groundwater samples from a number of wells based on major ion concentrations

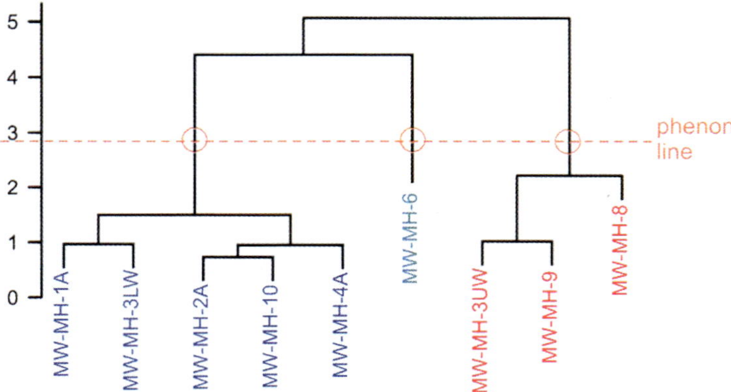

Fig. 5.4-10: Interpretation of the dendrogram in *Fig. 5.4-9*, showing the phenon line (orange), which separates three clusters (red, green, and blue) from each other

5.4.1.3 Time Series Analysis

Groundwater monitoring yields measurements over time at a single monitoring point, e.g., water level fluctuations over one or more years. The determination of trends, cycles, amplitudes and other characteristics describing the behavior is part of a time series analysis. The underlying mathematical theory is not presented here as it is mostly rather complex and the calculations are generally done with statistical software.

Data pretreatment

Often it is not possible to collect data at regular time intervals, although some interpretation methods require evenly spaced data. Moreover, there are sometimes singular outliers or peaks which cannot be assumed to be representative. Therefore, time series data should be interpolated and smoothed/filtered before further analysis. When smoothing data, one has to take care that not too much detail is smoothed out.

Trends

Time series data may exhibit long-term trends (e.g., rising chloride concentration or decreasing water level) onto which short-term fluctuations are imposed. Using regression analysis, a best-fit straight line can be found which describes the trend. Care must be taken in the interpretation. It is usually not possible to use a trend line for extrapolation into the future. This can be dangerous, especially with data over a short time period: If water levels are measured only during dry season, one might find a steadily decreasing trend, although this would be reversed when wet season begins. Not all pitfalls are that obvious, however.

Amplitudes and frequency analysis

If the data show cyclic behavior (e.g., water level changes over the year according to the season), the *amplitude* can be determined as the difference between maximum and minimum values. Also, the mean or median can be used to estimate a “long-term average value” of the parameter. This is, for example, useful to learn whether the water table may rise above the bottom of a landfill. For a more detailed investigation, frequency analysis tools can be used to obtain the *power spectrum* or the *periodogram*. Thus, the presence and amplitude of, for example, annual cycles can be established.

Correlation of time series

Given two time series, one can calculate their correlation coefficient analog to Equation (5.4.14). This can show whether the two parameters are synchronous in general. It is also possible to compute the correlation coefficient with the second time series offset by a certain amount (*time lag*). If this is done for all possible time lags (i.e. no lag, 1 unit lag, 2 unit lag and so on; also with negative lags), and the coefficients are plotted against time lag, a cross-correlation diagram of the two time series results. A peak in the diagram indicates that the second parameter follows the first with some delay behind (or vice versa). *Figure 5.4-11* shows an example in which the delayed correlation between water balance and water level changes in wells is shown.

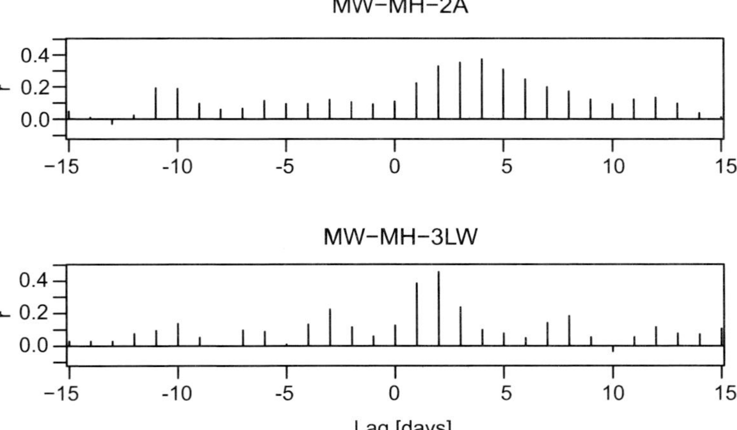

Fig. 5.4-11: Cross correlation diagrams showing the correlation between water level changes and climatic water balance (precipitation minus evaporation) for two monitoring wells. It can be seen that the water level in both wells follows the meteorological conditions with only a few days delay. In well MW-MH-3LW this behavior is more pronounced, however.

5.4.1.4 Geostatistics and Interpolation of Spatial Data

Geostatistics uses methods to analyze the spatial distribution of variables. The key tool is the *variogram*, which describes the degree of correlation over distance. Measurements that are made near each other are usually strongly correlated, whereas measurements made far from each other normally show little correlation. Besides the insights drawn from the variogram directly, it is also fundamental to the *kriging* method for interpolation of spatial data. There are also other interpolation methods besides those used in geostatistics, such as simple linear interpolation, splines, triangulation. Generally, care has to be taken when using any interpolation technique for geohydraulic purposes. Especially hydraulic boundaries (rivers, lakes, geological boundaries, etc.) have a strong influence on groundwater levels and flow. Some interpolation methods allow for including additional information, e.g., breaklines or auxiliary points, to model those features. The resulting isolines often have to be corrected manually afterwards (e.g., using the tools in a GIS). For geostatistical methods and interpolation of spatial data, see SWAN & SANDILANDS (1995) or WEBSTER & OLIVER (2001), for example. See also Part 6 of this handbook.

5.4.1.5 Specific Tests for Hydrogeochemical Data

Treatment of values below the detection limit

Most, if not all, statistical methods do not allow consideration of "not detected" or "below detection limit" observations. Setting these to a zero value is not recommended, however, as this may lead to strong distortions of the statistical results. Therefore, a certain fraction of the detection limit (e.g., one half) is commonly substituted. There are more sophisticated methods based on modeling the distribution of values as a basis for choosing the substitution value (DVWK, 1990).

Plausibility checks for chemical analysis results

There are several ways to check whether laboratory results are plausible or whether errors have to be suspected. For example, some ion may not be found at elevated concentrations if another ion is also present. *Table 5.4-3* shows several such plausibility checks.

If the analysis comprises all major ions, an *ion balance* (IB) can be calculated. This is the sum of the equivalent concentrations c_{eq} (i.e. concentration in mol L^{-1} times the charge z of the ion) of all ion species. As a sample is not electrically charged, the balance has to be zero. The error in the ion balance should be no higher than 5 %; errors of up to 10 % could be tolerated. Samples with higher balance errors should be checked for laboratory or sampling errors.

$$\Delta IB = \frac{\sum\limits_{\text{cations}} c_{\text{eq,i}} - \sum\limits_{\text{anions}} c_{\text{eq,i}}}{\sum\limits_{\text{all ions}} c_{\text{eq,i}}} \cdot 100\,, \qquad (5.4.15)$$

where ΔIB is in %.

If the electrical conductivity EC of the water sample is known, it can be compared to the ionic strength I (a measure of the content of dissolved solids) of the solution. It is also possible to compare ionic strength with total dissolved solids (*TDS*). Ionic strength is calculated from the molal concentrations m_i and the ion charge z_i as follows:

$$I = \frac{1}{2} \sum_{\text{all ions}} m_i z_i^2 \tag{5.4.16}$$

and is ideally related to electrical conductivity in µS cm^{-1} as described by MAIER & GROHMANN (1977):

$$I = \frac{EC}{54.5} \tag{5.4.17}$$

LANGLIER (in HÖLTING, 1996) gives a similar relation to the total dissolved solids TDS in mg L^{-1}:

$$I = 2.5 \cdot 10^{-2} \cdot TDS \tag{5.4.18}$$

If values for *I* calculated from concentrations differ significantly from those calculated from *EC* or *TDS*, a sampling or analysis error can be suspected. Additionally, scatter diagrams of *I* (from concentrations) versus *EC* or *TDS* can be drawn, including the straight lines described by Equations (5.4.17) and (5.4.18), respectively. This allows a quick assessment if a large number of samples are analyzed. *Figure 5.4-12* shows an example for conductivity.

Table 5.4-3: Checks for plausibility of water analyses, modified from HÖLTING (1996)

	Plausibility check
ion balance	ideally < 5 %, 10 % maximum; see Equation (5.4.15)
electrical conductivity (EC)	calculate EC from chemical analysis data (ROSSUM, 1975); compare with ionic strength (Equation 5.4.17); multiply TDS (in mg L^{-1}) by 0.7, should be same order of magnitude as EC in µS cm^{-1}
total dissolved solids (TDS)	compare with ionic strength (Equation 5.4.18)

If concentrations listed in this column are observed		**then the following concentrations are not plausible**	
O_2	> 5 mg L^{-1}	Fe^{2+} Mn^{2+} NO_2^- NH_4^+ H_2S	> 0.05 mg L^{-1} > 0.05 mg L^{-1} > 0.05 mg L^{-1} > 0.1 mg L^{-1} > 0.01 mg L^{-1}
Fe^{2+}	> 0.2 mg L^{-1}	NO_3^-	> 2.0 mg L^{-1}
Fe^{2+}	> 1.0 mg L^{-1}	H_2S	> 0.1 mg L^{-1}
Mn^{2+}	> 0.2 mg L^{-1}	NO_3^- H_2S	> 2.0 mg L^{-1} > 0.1 mg L^{-1}
H_2S	> 0.1 mg L^{-1}	NO_3^-	> 1.0 mg L^{-1}
pH	> 8.0 or < 5.5	$Ca^{2+} + Mg^{2+}$ > 1.0 mg L^{-1}	

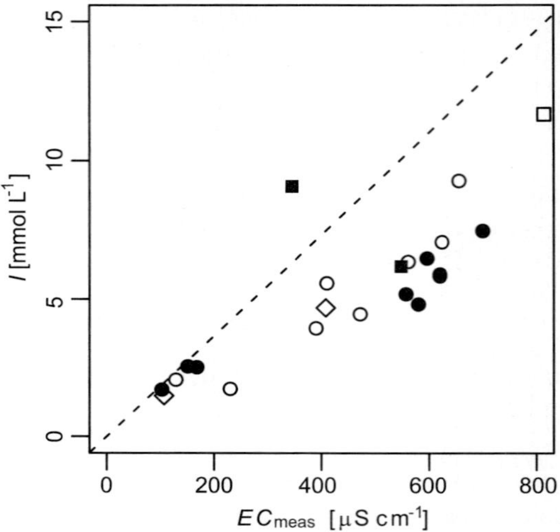

Fig. 5.4-12: Scatter plot of ionic strength I versus electrical conductivity EC_{meas}. The dashed line shows the ideal relationship described by Equation (5.4.17). Most samples, especially those with elevated mineral content, fall short of the I value they should have according to EC measurements, indicating that the measured concentrations of some ions might be too low or that some ion was not analyzed at all.

There are more sophisticated procedures for calculating the electrical conductivity from chemical analysis data that could be employed in plausibility testing. In most cases, the method of ROSSUM (1975) works very well.

5.4.2 Conceptual Model

HANS-JÜRGEN VOIGT & JENS RADSCHINSKI

A conceptual model is a mental image of an object, system, or process. It consists normally of a simplified, schematic written description and a visual representation. *Conceptual hydrogeological models* have been used for several decades to describe and understand hydrogeological systems. Such conceptual models are more concerned with the physical than the chemical aspects of the natural system. During the last few decades *conceptual models of investigated sites* have come into frequent use to assess ecosystem features and processes (including biological, physical, chemical and geomorphic components) of an environment (e.g., new or abandoned landfills, industrial site or mining site).

A conceptual model is developed as part of the site investigation and reflects the progress of site characterization. Therefore, the conceptual model is the basis as well as the final result of site characterization. The conceptual model is not a fixed rigid idea, but a flexible concept (ORESKES & BELITZ, 2001) which is upgraded iteratively on the basis of the information obtained during the field and laboratory investigations.

The first step in a site investigation is to define together with the client the objectives, the boundaries of the site, and the scale. A preliminary conceptual model is prepared on the basis of the orientating investigation of the site. This preliminary conceptual model is the basis for planning the geophysical, geological, hydrogeological, geochemical and microbiological investigations. The product of the site investigation is a verified conceptual model. This is a sound basis for site assessment and all remediation measures. Experience in Germany shows that remedial measures that are not based on a conceptual model are not effective or failed completely in almost 80 % of the cases.

The objectives and the sequence of site investigations are outlined in Part 1 of this handbook. Some aspects must be repeated in this section for the discussion of the development of a conceptual model. Site investigations usually take place in the following progression:

- orientating site investigation,
- preliminary conceptual model,
- field survey for a detailed site investigation,
- refining of the conceptual model,
- verified or final conceptual model, and
- risk assessment.

The refined conceptual model can be the basis for numerical modeling of groundwater flow and contaminant transport. Numerical modeling is one way to verify the conceptual model. *Table 5.4-4* gives elements of a conceptual site model.

The hydrogeological model of a site is the nucleus of every conceptual site model. Therefore, the development of the conceptual model should start with an analysis of the *hydrogeological system*:

- the spatial distribution of hydrostratigraphic units and their lithological and petrophysical properties,
- groundwater flow, and
- hydraulic boundary conditions.

Table 5.4-4: Elements of a conceptual site model

Topic	Information	Illustrations	Example figures/tables
main objective	characterization of a site for a new landfill, industrial site or mine site (e.g., spoil/slag heap, ore treatment plant)		
	inspection of an operating landfill, industrial site or mine site (e.g., spoil/slag heap, ore treatment plant)		
	characterization of a site suspected to be hazardous		
classification of sites suspected to be hazardous	waste deposit (landfill, sewage sludge deposition, small site with buried and/or heaped waste and litter)		
	industrial site (e.g., galvanizing plant, metallurgical workshop, mechanical workshop, motor vehicle garage, gas/petrol station, tannery, workshop in which wood is impregnated or worked)		
	mining site (e.g., spoil/slag heap, ore treatment plant)		
base data	- locally used site name; - owner/user of the site; - geographical coordinates (e.g., determined by GPS); - sheet number of topographical map; - site plan; - size of the site (in m^2); - classification of the site suspected to be hazardous (see above); - use of the site in the past and at present, e.g., type of industrial operation; - operation period for industrial sites, landfills or mining sites; - contamination expected on the basis of the type of industrial operation; - in the case of a landfill, type of waste deposition (e.g., stockpiling, heap on a slope, backfill of a quarry, backfill of a gravel pit), type of material deposited (e.g., uncontaminated excavated soil, natural rock, minerals, municipal waste, industrial waste), type of contaminants expected (e.g., water-soluble substances, gaseous and volatile substances, inorganic substances, organic substances, toxic, carcinogenic, mutagenic substances, warfare agents, explosives, radioactive materials); and - person(s) responsible for releases of hazardous substances at the site		*Fig. 5.4-17*
	topography, land use and vegetation, settlements, roads and railways	topographic map, map of land use	*Fig. 5.4-19*

	climate: precipitation, temperature, evapotranspiration, direction and velocity of the wind	long-term average climatic water balance graph	*Fig. 5.4-18*
	geogenic hazards: active faults, karst, earthquakes, subsidence, landslides, flooding	map of natural hazards	
	human activities: mining damage, buildings, quarries, gravel pits, clay pits, etc.		
	geotechnical stability		
geomorphology	relief, major landforms	topographic map, aerial photographs, ground-level photographs	
geological setting	major geological structures, stratigraphy and lithology, thickness and lateral extent of geological units, stratigraphy, lithology, homogeneity and heterogeneity, bedding conditions and tectonic structures, fractures, impact of weathering, soil	geologic column(s), well log(s), geologic map showing dip, strike, folds, faults, structure map(s), cross-section(s)	*Figs. 5.4-20, 5.4-21, 5.4-22, 5.4-23, 5.4-24, 5.4-25, 5.4-26, 5.4-27*
hydrologic and hydrogeological conditions	streams, lakes and ponds, springs, wells, use and quality of surface water and groundwater, runoff, water balance	map of streams, lakes and other surface water, hydrograph(s),	
	aquifer(s)/aquiclude(s), aquifer/aquiclude properties, groundwater table, soil water content, hydraulic heads, direction and rate of groundwater flow, groundwater recharge and discharge (area, type, rate), importance of the aquifer(s)	water well records (tables), well-construction diagrams, groundwater table map(s) (contours/flow directions for various aquifers, if different), water quality table(s), water quality map(s) (water type, concentration isopleths)	*Figs. 5.4-28, 5.4-29, 5.4-30*
mineralogical, geochemical and microbiological conditions	composition and properties of soil, rock and groundwater, estimation of contaminant retention, microbial activity, natural attenuation	map of contaminant sources, maps of contaminant distribution and environmental parameters, tables of mineralogical, physical and geochemical analyses	*Tables 5.4-6, 5.4-7, 5.4-8, 5.4-9*
integrated interpretation and assessment of the model elements			

Hydrostratigraphic units are the layers of rocks with similar hydrogeological properties (FH-DGG, 1999). Geological cross-sections are constructed on the basis of the results of geophysical measurements, remote sensing and drilling investigations. These provide an impression of the spatial distribution of geological units. These geological sections must be transformed into hydrostratigraphic sections using hydrogeological properties and stratigraphic information for correlation. Schemes for the classification of hydrostratigraphic units have been developed for the main geological units by a working group of the German State Geological Surveys. As an example, the classification of the unconsolidated Quaternary and Tertiary rocks of the North German Basin is given in *Table 5.4-5*. Layers are grouped into hydrostratigraphic units according to they are considered to be aquifers or aquicludes.

The next step in the development of a conceptual model of a site is to map the distribution of aquifers and aquicludes. Knowledge about hydrogeological connections (windows) between aquifers is especially important, because in these places contaminants can spread to deeper aquifer(s). Such connections can be caused by changes in lithology or tectonic activity (see for example *Figs. 6.2-25, 6.2-5, 6.2-9*, and *6.2-10*). Geophysical measurements can provide helpful information about hydrogeological windows (see Parts 4 and 6). Not only the spatial distribution of the hydrostratigraphic units, but also the representative properties of these units must be determined. Rock genesis, lithofacies distribution, tectonic evolution of the structures must be taken into account for quantification of the properties of hydrostratigraphic units.

In some cases (for example in colluvial sediments), it may be impossible to correlate aquifers and aquicludes over a large area owing to their heterogeneous lithology. Heterogeneous aquifers can be modeled using stochastic methods which use statistical values and probability laws to generate possible parameter distributions.

An initial evaluation of the *groundwater dynamics* is done by preparing groundwater equipotential contour maps derived from groundwater levels in observation wells over the period of a hydrological year (including both dry and wet seasons). It is important for the screen depths of the observation wells to correspond to the hydrostratigraphic units. Construction of groundwater isohypses using piezometric head data from different aquifers is a common error. Other errors result from not knowing or incorrectly interpreting the interaction between groundwater and surface water flow.

Table 5.4-5: Hydrostratigraphic units of the unconsolidated Quaternary and Tertiary rocks of the North German Basin, after MANHENKE et al. (2001); F – aquifers, C – aquicludes

Hydrostratigraphic unit	Lithology (predominant)	Lithogenetic/lithostratigraphic units	Stratigraphy
C1	peat, silt, loam	raised bog and fen peat, turfy moulder, meadow loam, sapropel	Holocene
F1	sand, gravel	flood-plain sand, dune sand, blown sand, river gravel, lowland sand, melt-water deposits, calcareous tufa	Holocene, Pleistocene (Vistulian)
C2	till, loess	Vistulian ground moraines, periglacial deposits	Pleistocene (Vistulian)
F2	sand, gravel	melt-water deposits of the Saalian detrital subsequent stage to foreset phase of the Vistulian, including calcareous tufa	Pleistocene (Saalian to Vistulian)
C3	tell, silt, clay	Saalian ground moraines, glacial-lake deposits	Pleistocene (Saalian)
F3	sand, gravel	melt-water deposits of the Elsterian detrital subsequent stage to Saalian foreset phase, river gravel in middle-terrace deposits, calcareous tufa	Pleistocene (Elsterian to Saalian)
C4	silt, clay, till	Lauenburg Clay Elsterian ground moraines, glacial-lake deposits	Pleistocene (Elsterian)
F4.1	sand, gravel	melt-water deposits, river gravel, middle-terrace deposits, younger and older main terrace deposits, upper-terrace deposits decomposed coarse gravel	Pleistocene (Lower Pleistocene to Elsterian)
F4.2	sand, gravel	melt-water deposits in deep channels	Pleistocene (Elsterian)
F4.3	sand	kaolin sands, micaceous fine sands of the Pliocene and Upper Miocene, sands of the Rauno strata	Tertiary (Pliocene, Miocene)
C5	clay, silt, lignite	Upper Micaceous Clay, 1st and 2nd Lusatian lignite horizons, silts of the Rauno strata	Tertiary (Miocene)
F5	sand	Upper Lignite Sands/marine sands, sands of the Oxlund strata, Lower Brieske Sands	Tertiary (Miocene)

Under semiarid climatic conditions, in which distinct wet and dry periods occur, the hydraulic interaction of surface water bodies and groundwater changes during the year. The water level in a stream rises during periods of intense precipitation. In this case the stream loses water, which means that surface water infiltrates into the aquifer. These conditions are called influent (losing stream) conditions. In periods with normal or low stream water levels, groundwater will discharge into the river, forming the base runoff of the stream. These conditions are called effluent (gaining stream). *Figure 5.4-13* shows the influent and effluent conditions of a stream in an area with a high groundwater table.

But over the course of a year, not only does the hydraulic interaction of surface water and groundwater change, but the groundwater table also varies as a result of changing groundwater recharge conditions. The changes in the groundwater levels are registered in hydrographs of observation wells. The groundwater table rises during recharge periods and declines at other times. The rise of the groundwater table can result in the leaching of waste or of other contaminated materials from landfills and other contaminated areas. *Figure 5.4-14* shows a groundwater hydrograph of an observation well near a landfill in Thailand. During the dry period the groundwater table is below the bottom of the landfill, in the wet season the groundwater table rises to a level higher than the base of landfill. The result is leaching of substances from the waste.

Withdrawal of groundwater from water wells, different recharge conditions in different years, human activity (irrigation, drainage) and other artificial and natural factors must also be taken into account when the groundwater dynamics is analyzed. Influenced by these factors, not only can the direction of groundwater flow change, but also the pressure gradients and, therefore, the recharge – discharge conditions between the aquifers.

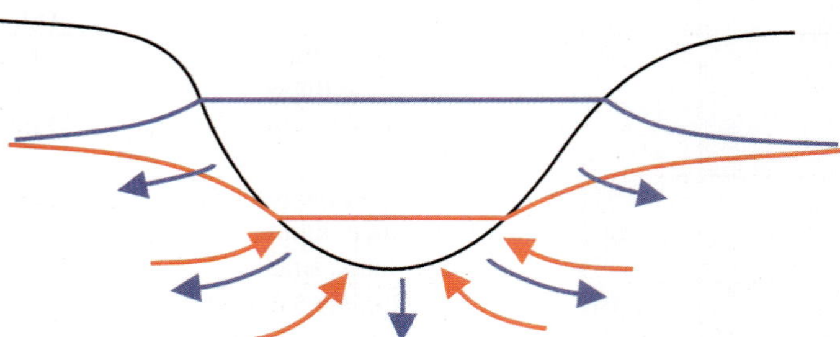

Fig. 5.4-13: Influent (losing stream, blue) and effluent (gaining stream, orange) conditions of a stream in an area with a high groundwater table

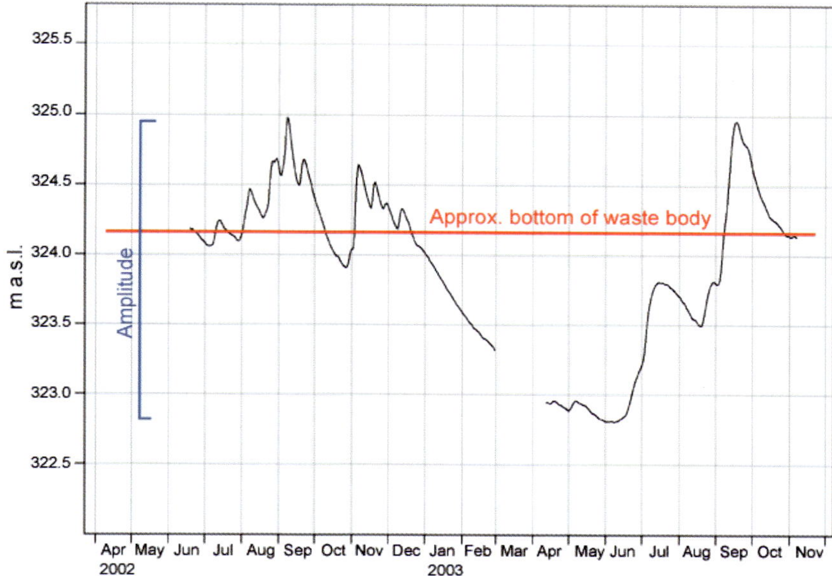

Fig. 5.4-14: Groundwater hydrograph of an observation well near the Mae Hia landfill of Chiang Mai, Thailand

The *boundary conditions* are derived from hydrogeological data. Boundaries can be faults, facies changes or changes in hydraulic conditions (for example, flow lines, groundwater divides). Determination of the hydraulic conditions at the margins of the model and features within the model area which affect the hydraulic regime is required for the modeling of groundwater flow and contaminant transport. The reliability and the accuracy of the model largely depend on the definition of the boundary conditions. Three types of boundaries are commonly used (see also Section 5.4.3 and, for example, CASTANY & MARGAT, 1977):

- a specified head or constant head boundary,
- a specified flux boundary, and
- head dependent boundary.

A boundary with a measured piezometric head is called a *specified head boundary*. Lakes, ponds, streams, and ditches are typical specific head boundaries of a conceptual model. For groundwater modeling, specified head boundaries are more desirable than specified flux boundaries because head data can be measured much easier than flux data (USACE, 1999). A specified head boundary in a flow model represents an unlimited supply of water at this location. This also means that it has to be checked whether the flux passing through the boundary in the flow model fits the flux of water available in reality.

Specified flux boundaries are boundary conditions which are characterized by known, measured or estimated flux across the boundary. Pumping and injection wells, springs, surface infiltration, and leakage from a confining layer are typical elements. A special case of the specified flux boundary is a "no-flow" boundary, where no flux occurs. A "no-flow" boundary can be an impermeable barrier or model hydraulic boundaries such as groundwater divides or flow lines ("parallel flow" boundaries). When groundwater divides and flow lines are used as the outer boundaries of a model, care must be taken because these lines can move with the seasons or when the aquifer is under stress, e.g., during pumping.

A *head dependent boundary* is present when the flux through the boundary is determined by a hydraulic conductivity term. An example of a head dependent boundary is a losing stream. The hydraulic head in the aquifer below the stream is dependent on the hydraulic conductivity of the stream bed, its thickness, and the head difference between the stream and the aquifer (ANDERSON & WOESSNER, 1992).

The *rate of groundwater recharge*, together with the inflow and the outflow of the aquifer, is a governing factor in the groundwater balance. In the same way as the aquifer properties, the boundary conditions and the elements of the *groundwater balance* must be quantified in space coordinates.

Before a numerical model can be prepared (Section 5.4.3), the refined conceptual hydrogeological model has to be transformed into a final conceptual or "information adequate" model. This involves a generalization of the model and assigning parameter values to the individual hydrogeological model units. How detailed an "information adequate" model needs to be depends on the objectives of investigation and on the quality of the data. *Figure 5.4-15* schematically shows two possibilities for generalizing a hydrogeological model. In the one case (case A), a rough generalization is made, which can be sufficient to estimate, for example, the potential withdrawal of wells from two main aquifers. This is not sufficient, however, to assess the risk emanating from a contaminated area or landfill. In this case, the units and parameters of the hydrogeological model must be defined as precisely as possible, as done in case B. The procedures for discretization, calibration, and verification of a numerical model are described in Section 5.4.3.

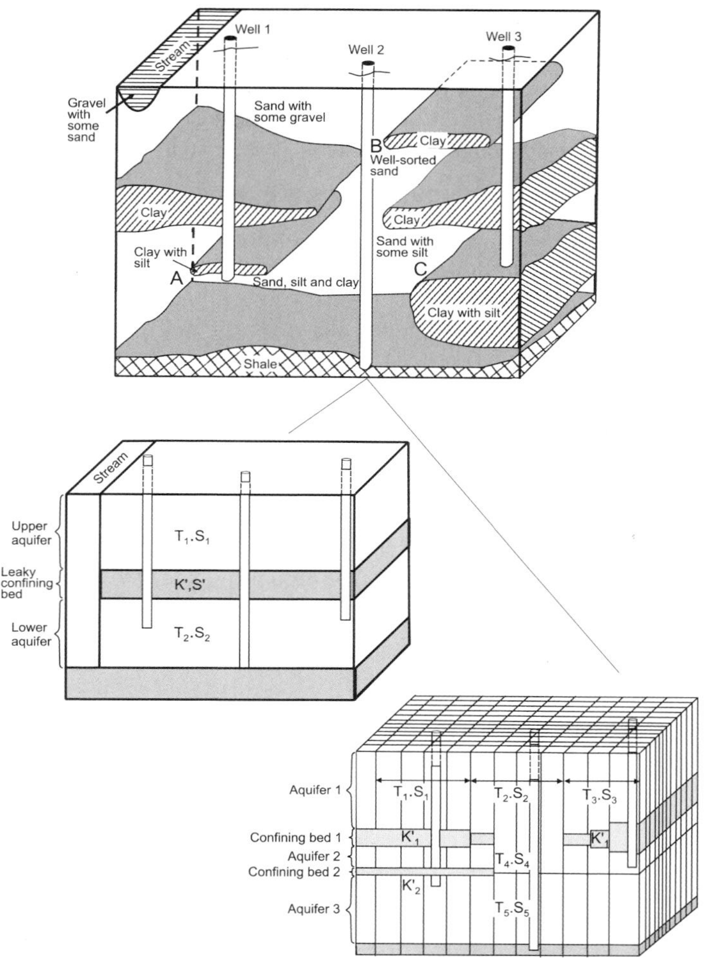

Fig. 5.4-15: Possibilities for the transformation of a hydrogeological model into an "information adequate" model, example from PETERS (1972)

For sites suspected to be hazardous, chemical data must be included in the conceptual hydrogeological model. A hazard ranking system (e.g., Hazard Ranking System (HRS) for uncontrolled waste sites from the U.S. EPA, http://www.epa.gov/superfund/programs/npl_hrs/hrsint.htm) uses information from screening level investigations to assess the relative potential of a site to pose a threat to human health and/or the environment. The HRS uses a structured analysis approach to assess a site. The computer program Quickscore http://www.epa.gov/superfund/programs/npl_hrs/quickscore.htm can be used to develop a conceptual model (collection of exposure pathways) for site assessment, to calculate "scores", and to identify gaps in the data. Scores are calculated for each pathway separately and are then combined using a root-mean-square equation to determine the overall score for the site. This ensures that the site score can be relatively high even if only one pathway score is high. Conservative assumptions are used for screening-level ecological risk assessment.

Example: Conceptional model for the abandoned Nong Harn municipal waste disposal site near Chiang Mai, Thailand[2]

The abandoned Nong Harn municipal waste disposal site is about 18 km north of the city center of Chiang Mai in northern Thailand. The site served the Chiang Mai Municipality and nearby sanitary districts as a central waste disposal facility from 1995 - 1998. The site is a former borrow pit for road construction material (clay, sand and gravel) with a lateral area of about 150 × 150 m and rather steep walls down to a depth of 40 - 50 m. The standards of that time were used for the planning and construction of the landfill, including HDPE liners, leachate collection and treatment systems as well as compactors (see *Fig. 3.1-1*). Consequently, the site was considered suitable for the disposal of a wide variety of waste types, including industrial waste. To the southwest of the Nong Harn landfill there is another smaller, older landfill, also in a former borrow pit.

Fig. 5.4-16: Nong Harn landfill in 2002

[2] Condensed version of the report by GRISSEMANN et al. (2006).

But, the information about how completely the HDPE liner covers the Nong Harn landfill is contradictory. The drainage system for the seepage water does not function properly. The originally constructed leachate collection system has been broken by the settling of the waste.

According to the available information, mainly municipal household waste has been disposed of. The typical composition of municipal solid waste in Chiang Mai in 1995 is shown in *Fig. 5.4-17*. It should be noted that the main component is organic waste, which can amount to 70 % of the total waste. The general moisture content was as high as 54 %. At that time, 250 tons of waste per day were generated in Chiang Mai, but only 200 tons per day were collected. A low degree of compaction is assumed. To avoid odor emission, the waste was covered with layers of clay. According to the Pollution Control Department (PCD) in Bangkok, this covering was not done on a daily basis nor was it done over the entire surface. The amount of cover material is not known but roughly estimated to amount up to one-third of the volume of the landfill. The volume of the landfill was calculated to be 530 375 m^3.

Due to settlement of the waste, the central part is lower than the margins of the landfill and a small pond several meters deep formed, covering approximately one-fifth of the landfill (*Fig. 5.4-16*). The pond overflows in the SW corner of the landfill during rainy season.

Intense degassing occurs from the surface of the pond, along fracture zones at the margins of the landfill, and from degassing wells on the site. There are twenty degassing wells in the landfill.

The *main objective* of this investigation was a risk assessment based on a comprehensive geological and hydrogeological model. The following work was carried out for the investigation:

- compilation of information from available geoscientific maps, reports, satellite images, air photos, field observations, archive material and interviews, preparation of a base map,
- implementation of geophysical field surveys for an assessment of the geological structure and the delineation of more detailed follow-up investigations,
- drilling and completion of observation wells; well logging and sampling of cores for laboratory investigations, hydraulic well tests,
- monitoring of the wells and groundwater sampling for chemical analysis,
- integrated data interpretation, assessment of the risk potential, recommendations for follow-up investigations and pollution prevention measures.

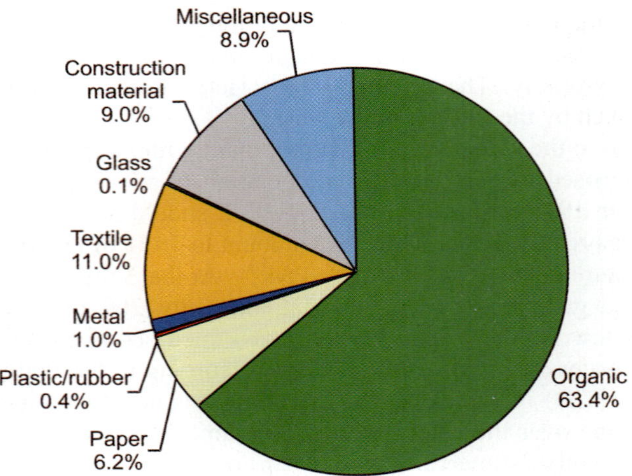

Fig. 5.4-17: Typical waste composition for Chiang Mai, TATONG (1997)

Climate: Annual rainfall in the area ranges from 800 mm in the lowland to more than 1500 mm in the surrounding mountains. The main period of precipitation is the rainy season between May and October with maximum values of more than 200 mm in August. Temperatures vary between a minimum of about 13 °C in January and a maximum of about 36 °C in April. The annual mean temperature lies around 25 °C (1951 - 2000). Potential evaporation exceeds rainfall, except between July and September (*Fig. 5.4-18*). This is the main period of groundwater recharge (MARGANE et al., 1999a).

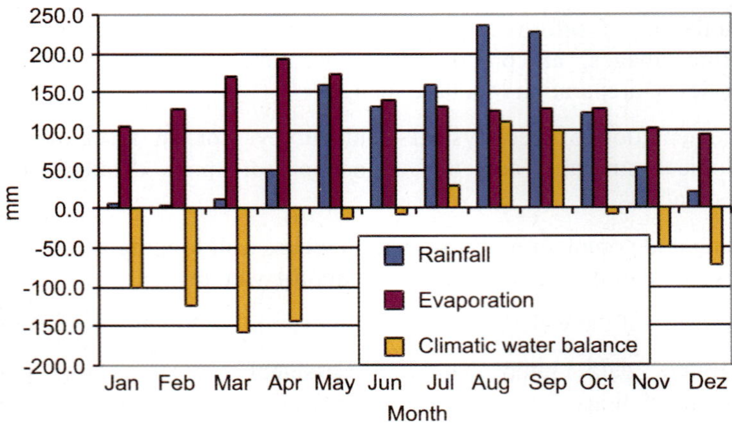

Fig. 5.4-18: Long-term average (1961 - 90) climatic water balance of the area around the city of Chiang Mai according to data provided by the Chiang Mai Airport weather station

Geological setting: The Nong Harn landfill is at the eastern rim of the Chiang Mai-Lamphun Basin, which is interpreted as a fault-controlled west-dipping half-graben system. Tectonic subsidence of the intramontane Chiang Mai-Lamphun Basin is believed to have started in the late Cretaceous/early Tertiary (TANTIWANIT & DORN, 1999; DORN & TANTIWANIT 2002). The surrounding mountains consist of Precambrian metamorphic and plutonic rocks as well as strongly consolidated Paleozoic sediments with interbedded volcanics. The basin fill, up to 2000 m in its deepest parts (WATTANANIKORN et al. 1995), consists of predominantly clastic rocks of fluviatile, alluvial and colluvial origin of Quaternary to probably also upper Tertiary age. Colluvial sediments occur on the mountain slopes and make up the foothills at the basin margins. These slope-derived sediments of Quaternary age are well exposed in numerous borrow pits as clay rich, poorly sorted material of mainly reddish color with intercalated silty-clayey sand and gravel beds. Gray-colored and more sandy horizons with erosional washout structures are also visible. Towards the basin center, the slope-wash deposits grade into alluvial and fluviatile sediments of the Ping River system.

Though the colluvial sediments are young in age they are remarkably well consolidated. This widespread phenomena is best documented in borrow pits where the material has been excavated partly down to 50 m depth along nearly vertical walls. It is assumed that the colluvium is underlain by deeply weathered basement rocks of Paleozoic age which show a fault controlled dip towards the basin center. Volcanic rocks are shown to be present at depth in the Nong Harn area by the presence of outcropping basalts in the southeastern vicinity of the site.

Various types of alluvial deposits occur between the landfill and the Ping River to the west. Besides flood plain deposits and unspecified alluvial deposits, upper terrace gravel beds of the Mae Taeng Group are part of these sediments. The terraces consist mainly of sandstone, quartz and quartzite gravel and pebbles, which are generally well rounded. Coarse gravel horizons are frequently interbedded with medium to fine-grained sands, silts or silty clays.

General groundwater situation: The local drainage pattern in the area of Nong Harn is directed towards the Ping River in the west, which acts as the central drainage channel of the Chiang Mai-Lamphun Basin.

The water table in the area around the landfill is at a depth of several tens of meters. For this reason, there are no dug wells, which are widely used in other parts of the basin. Water supply from groundwater is limited to deep wells below 50 m, and numerous rain-filled pits serve as reservoirs.

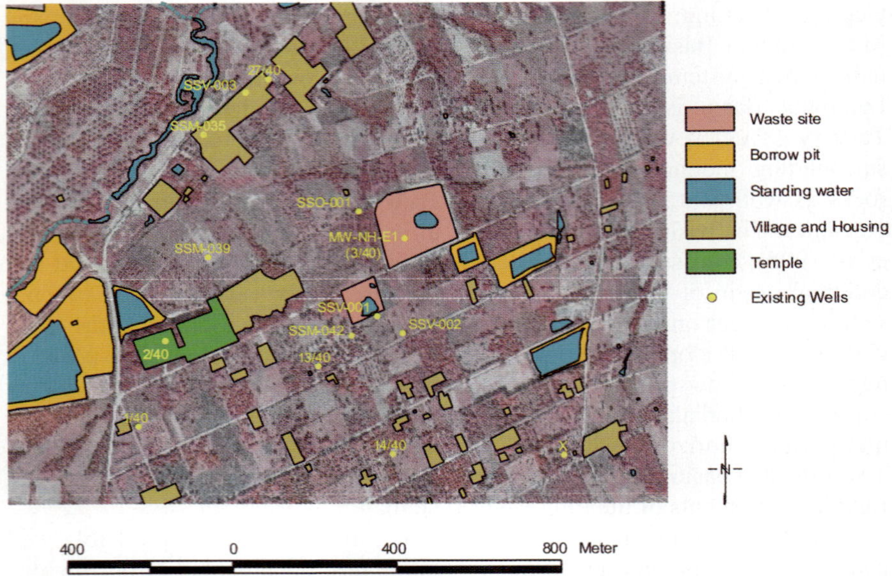

Fig. 5.4-19: Aerial photograph of the area around the Nong Harn landfill, together with information on geographic features

Results of the site investigation

An *interpretation of aerial photographs* is shown in *Fig. 5.4-20*. The map shows a set of several lineaments in the area around the landfill site and three main directions of linear features and geophysical indications of faults crossing the survey area.

At the site a *combined ground geophysical survey* was carried out with reflection/refraction seismic, gravity, magnetic, electromagnetic and dc resistivity methods (Geophysik GGD 2002).

The geological structures in the area around the Nong Harn landfill are best documented in the seismic survey. For reasons of clarity, the seismic section in *Fig. 5.4-28* does not show interpreted tectonic elements but only focus on the seismic sequences that have characteristic reflection patterns. Four seismic sequences can be distinguished within the investigation area. Sequences A and B show a slightly prograding reflection pattern to SSW, separated by an unconformity (red line). The lithologies of sequences A, B and D are documented in three wells close to the landfill (MW-NH1, MW-NH2, MW-NH3), which were completed as observation wells.

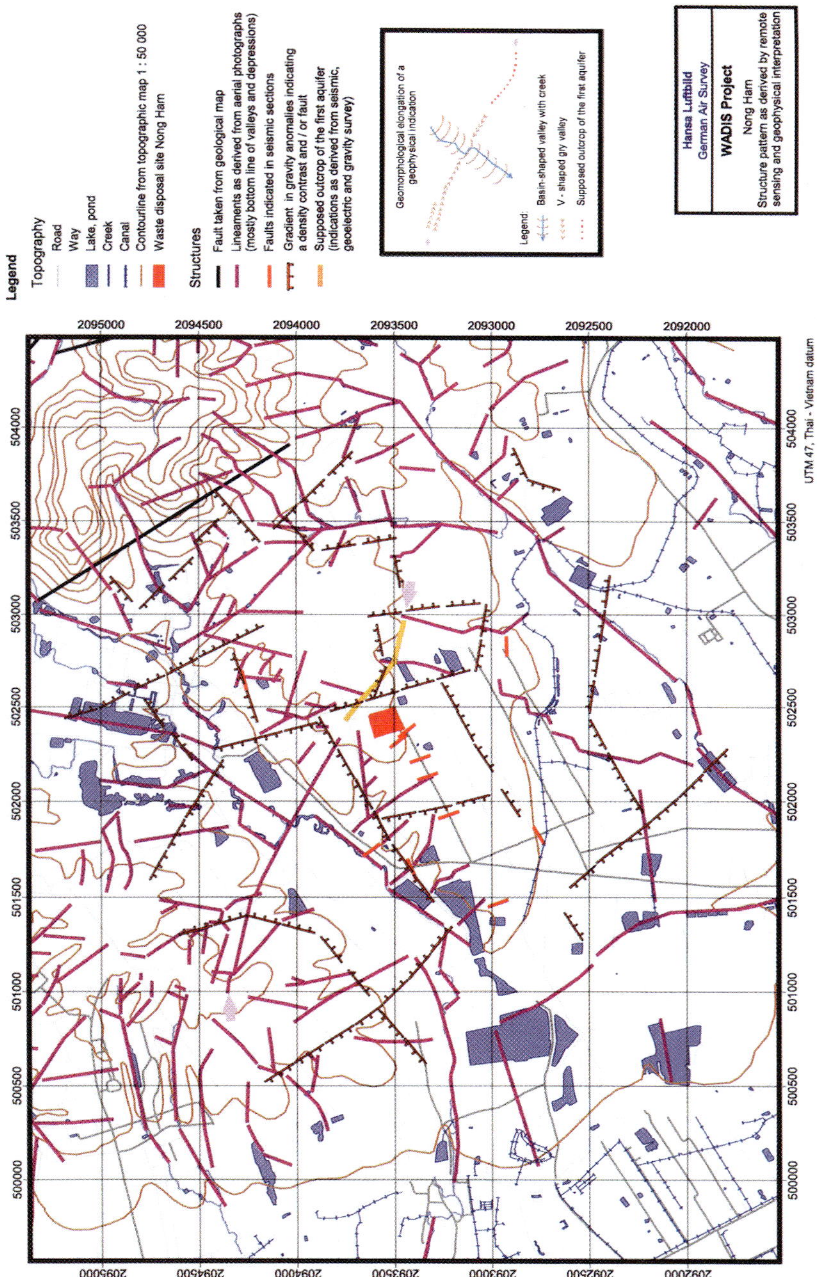

Fig. 5.4-20: Structure map as derived from interpretation of aerial photographs and geophysical data

Well logging and cuttings indicate sequence A to be predominantly clay-rich with interbedded sand horizons while sequence B is more sand-rich with clayey intercalations and less consolidated. Sequence D is mainly made up of red clays which become progressively harder (low drilling rate) down to 98 m. They can be considered as "basement rocks" with respect to the overlying and less consolidated colluvial and alluvial sediments.

Sequence C does not seem to exist directly below the landfill. Line NH0201 (*Fig. 5.4-28*) indicates that this sequence wedges out against sequence D, which rises steeply towards the east and might reach the surface at some distance east of the site. Interpreted fault patterns along this line suggest local uplift of sequence D beneath the landfill. Genetically, sequence C appears to consist of slope-derived colluvial sediments from the nearby mountains.

In addition to the seismic survey, the gravity and magnetic surveys provide further information about the geology of the area around the Nong Harn landfill, in particular the structure of the basement. Combining the results of all three methods, it was possible to produce a structure map of the investigation area (*Fig. 5.4-21*). According to this map, the geology can be expected to be fairly complicated. A number of local gravity highs (red-brown) and lows (green) are observed. The gravity gradients which mark faults and structure boundaries or material changes trend in very different directions and show a number of intersections. The main regional trend is NW-SE but other directions (e.g., E-W or NE–SW) can be found too. This means that there has been considerable tectonic activity and erosion in the area.

The highest elevation in the area is northwest of the landfill. The landfill is at the north edge of a local gravity high, which possibly marks a block in the ground. The almost E-W-trending southern gradient of this gravity high can be related to a fault indicated in the seismic data. A small graben structure found on seismic profile NH0201 southwest of the landfill cannot be resolved by the gravity measurements, probably due to the large spacing between stations.

The local gravity high directly south of the landfill also seems to be relevant with regard to the groundwater flow. Considering the surface topography, groundwater movement towards the south would be more likely than towards the north. But, the groundwater investigations show that the groundwater flows to the northwest, which correlates with the observed gravity gradient. Assuming a bulk density of 2.1 g cm^{-3} for the unconsolidated sediments above the siltstone found in boreholes NH01 and NH02 and a bulk density of 2.4 g cm^{-3} for the siltstone, the gravity gradient northwest of the landfill can be modeled as the dipping surface of the siltstone. The depth of the siltstone calculated by gravity modeling in the eastern part of the seismic profile NH0101 correlates well with the depth of the basement determined by the seismic survey (about 160 m below ground surface).

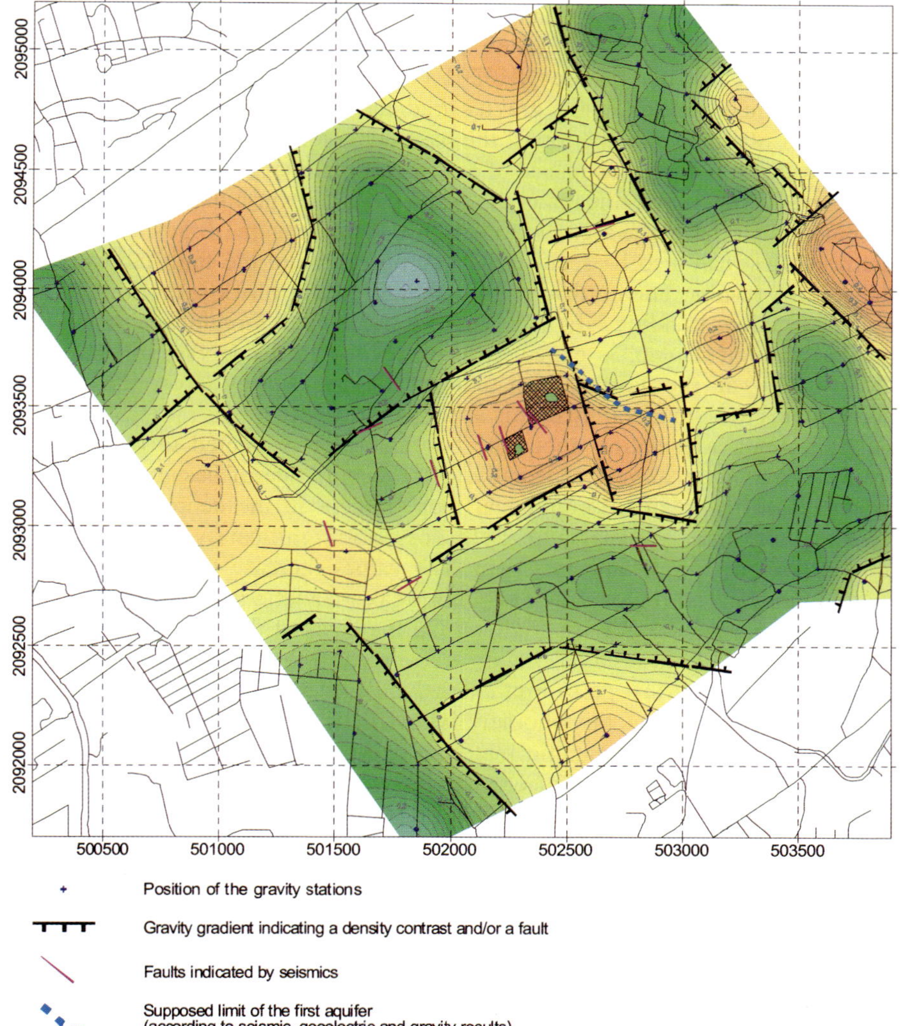

Fig. 5.4-21: Structure map of the Nong Harn area based on geophysical data: residual gravity field from high-pass wavelength filtering with a cut-off wavelength of 2 km

The central gravity high continues to the northeast of the landfill, but its amplitude is lower than it is south of the landfill. Comparing this gravity anomaly with seismic profile NH0201 (*Fig. 5.4-28*), it can be seen that a facies change correlates with a gravity gradient. Northeast of this gradient, the seismic results indicates the top of the basement is at about the same depth as southwest of it but that it consists of a different material (facies II). According to the gravity anomaly, this material should have a lower bulk density than the basement material in the area of the landfill. The extension of facies II towards the northwest can also be seen in the gravity field and the fault indicated by the seismic results. The northwestern boundary of this block is confirmed by a gravity gradient trending NE-SW. Another larger fault structure, also marked on the geological map, is expected to be related to the NNW-SSE-trending gravity low at the northeastern margin of the survey area. This zone corresponds to a small valley in this area.

Additional information about the basement was expected to be obtained from the magnetic survey, especially as the occurrence of basaltic rocks was indicated on the regional geological map. Except in the central part of the investigation area, the magnetic field does not significantly change – only about 20 nT over a distance of 3 km. The values increase from the valley to the mountains. The regional trend of this weak magnetic gradient corresponds to that of the regional gravity gradient. On the basis of this agreement, it may be assumed that both regional trends are related to the basement topography. Due to the fact that just part of an extensive magnetic anomaly was surveyed and that the gradient of this anomaly is low, it is assumed that the magnetic anomaly is caused by magnetic material at greater depth (probably basaltic rocks). The other possible cause, material with low magnetization near the surface, is less probable as the residual gravity field shows local changes and not a smooth picture like the magnetic data.

The geophysical data also provide essential information about the landfill itself. In particular, information about the influence of the landfill on the surrounding area was obtained from the 2-D resistivity survey. Gravity and magnetic modeling indicate the shape of the landfill and composition of the material in it. Electromagnetic measurements proved their usefulness to quickly determine the lateral boundaries of a landfill.

The profiles of the 2-D resistivity survey were placed in such a way that they could provide information about the landfills and about possible contamination plumes – some profiles cross the landfills, some run close to both sites and some are at some distance to see if any influence of the landfill can be detected.

Profile P2 (*Fig. 5.4-22*) crosses the center of the Nong Harn landfill from west to east. The location of the waste pit is clearly marked in the resulting section by very low resistivities. Different parts of the landfill show different resistivities. Slightly higher resistivities near the surface are related to the cover material and there are two regions with very low resistivities in the interior. The boundaries of the pit are steeply dipping with no indication of significant lateral flux of water. Only at the eastern wall can leakage from the landfill be expected as the resistivity gradient in profile 2 is not inclined inwards as would be expected from the construction of the pit. Corresponding to the gravity data, the western part of the pit is either not as deep as the eastern or higher resistivity material is in this part of the pit (see *Fig. 3.1-1*). The depth of the pit (40 to 50 m) could not be determined with these measurements as the depth of penetration of the method (about 40 m) was too small.

Profile 1 (*Fig.5.4-23*) extends west of the main landfill and east of an old landfill (see *Fig. 5.4-19* for orientation). Low resistivities can be observed at the surface of the main landfill. This could be caused by conductive material at the surface either due to either waste disposal in shallow pits or depressions or water overflowing from the depression in the middle of the landfill or seeping from the sides during the rainy season. Down to a depth of about 20 - 25 m there is no indication of leakage from the pits, but below this depth, the decrease in electrical resistivity seems to be related to the landfills. It is assumed that seepage water is leaving the pits in this area. This assumption is supported by measurements in borehole NH3a, where groundwater conductivities of more than 1900 µS cm^{-1} were determined at the northern end of the profile. However, it can be seen that the influence of the landfill decreases rapidly, i.e. no strong contamination plume can be found in the direction of groundwater flow at the depth range investigated.

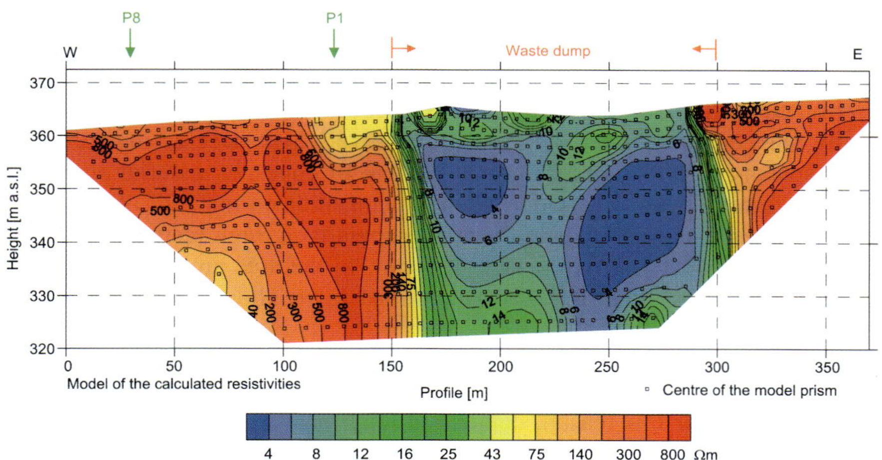

Fig. 5.4-22: Profile 2, 2-D resistivity survey across the Nong Harn landfill

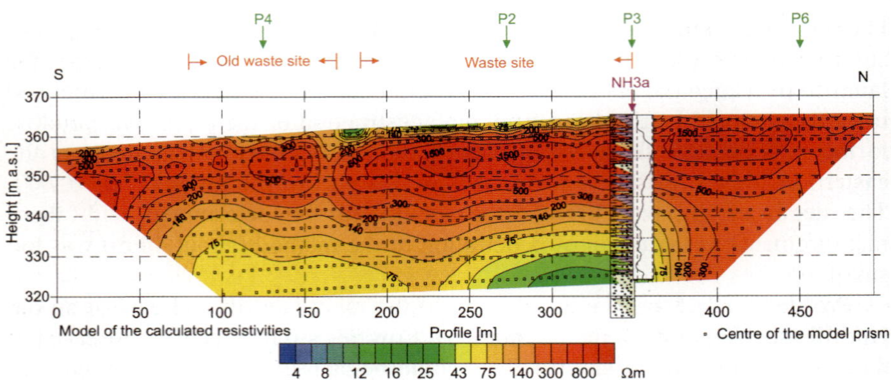

Fig. 5.4-23: Profile 1, 2-D resistivity survey at the western margin of the Nong Harn landfill

As the depth of penetration of the 2-D resistivity survey was limited for technical reasons to about 40 m, resistivity soundings (VES) were also carried out to obtain information about the local groundwater system and to fill the gap between results of the near-surface 2D-resistivity survey and those of the seismic survey. Except for the VES in the vicinity of the landfill, almost all of the sounding curves are similar – lower resistivities at the beginning of the curve, high resistivities in the middle part and decreasing values for the large spacings, i.e. at greater depths. The low resistivities of the layers near the surface could be caused by weathered lateritic material. Below that there is a zone which is also above the groundwater level, but apparently not weathered, thus, causing high resistivities. Although a high content of cohesive material was found in the area around the landfill, the very high resistivities partly determined for this layer (up to more than 2000 Ω m), indicate that, at least locally, the content of sand and gravel is relatively high. Decreasing values at larger spacings are related to a higher content of cohesive material and/or to increasing groundwater saturation of the unconsolidated material observed in the drill holes. The main reason for the comparatively high resistivity values, even for clayey material, is the lack of water. In the area around the landfill, the depth of the groundwater was determined to be about 40 m, which was confirmed by the boreholes.

As the depth of the Nong Harn waste pit was roughly known, it was possible to estimate the average bulk density of the waste from the gravity measurements. For this reason, detailed gravity measurements were made on profiles across the Nong Harn landfill and the older landfill southwest of it in order to obtain better data for modeling. A residual field was used for comparison with the model gravity (see upper left picture in Fig. *5.4-24*). This field was derived from the BOUGUER gravity by removing a linear trend. In this residual field, the gravity lows caused by the waste material, which has a lower bulk density than the surrounding rock, can clearly be seen.

The only input information for profile 1 across the Nong Harn landfill is that the pit was expected to be 40 - 50 m deep. Assuming a depth of 45 m, the calculated average density of the waste is about 1.70 g cm^{-3}. This is a value significantly higher than for normal domestic waste (~1.5 g cm^{-3}) or organic material (about 1.3 g cm^{-3}). Whether this higher value is caused by the clay layers which were alternately deposited or denser waste material was also disposed of cannot be determined on the basis of the gravity data. From the shape of the gravity anomaly, it must be assumed that the eastern wall of the pit is much steeper than the western one. This result corresponds to the information that there was an access ramp on the western side. The same density for the waste in the old landfill and a depth of about 20 m for that pit was calculated from the data from profile 2 (*Fig. 5.4-24*). This density is in quite good agreement with the result of the 2-D resistivity measurements along two profiles, where the depth of the older pit was estimated to be 20 - 25 m.

As strong magnetic anomalies were found in the area of the waste disposal sites, a more closely spaced grid was measured to define these anomalies more precisely and to possibly find an explanation for the elevated values, as it was not expected that a significant amount of iron was present in the waste. In Nong Harn there is a single clear anomaly with a minimum in the center and two maxima north and south. A similar anomaly is present in the area of the old landfill. The shape of the anomalies with its large E-W trend is typical of the Earth's magnetic field at the low latitude of northern Thailand. According to model calculations, such anomalies can be caused by the presence of iron material in the waste. The measured field is shown in the *upper left* of *Fig. 5.4-25*. A model magnetic body which would fit the measured anomaly on a N-S profile is shown on the *upper right*. The calculated anomaly caused by such a body is shown on the *lower left*. It can be seen that the shape of this model anomaly corresponds to the shape of the measured anomaly.

As this rough estimate demonstrates, a body at a depth of about 30 m and a thickness of about 15 m with a susceptibility of 0.075 (cgs) would fit the measured anomaly. The amplitude of the anomaly is more than 2000 nT. An iron content of 0.1 to 1 % in the model body would be sufficient to produce such an anomaly.

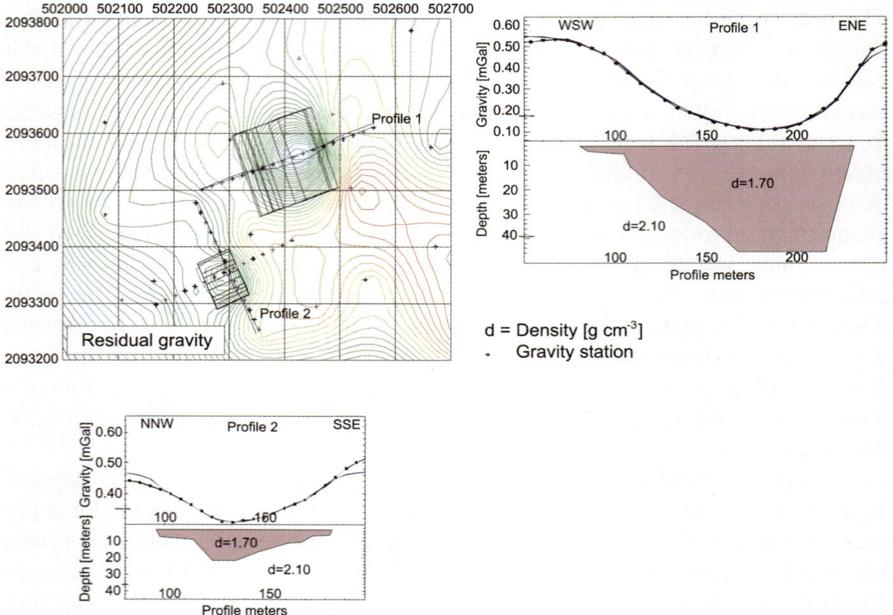

Fig. 5.4-24: Gravity method – modeling of the waste pits

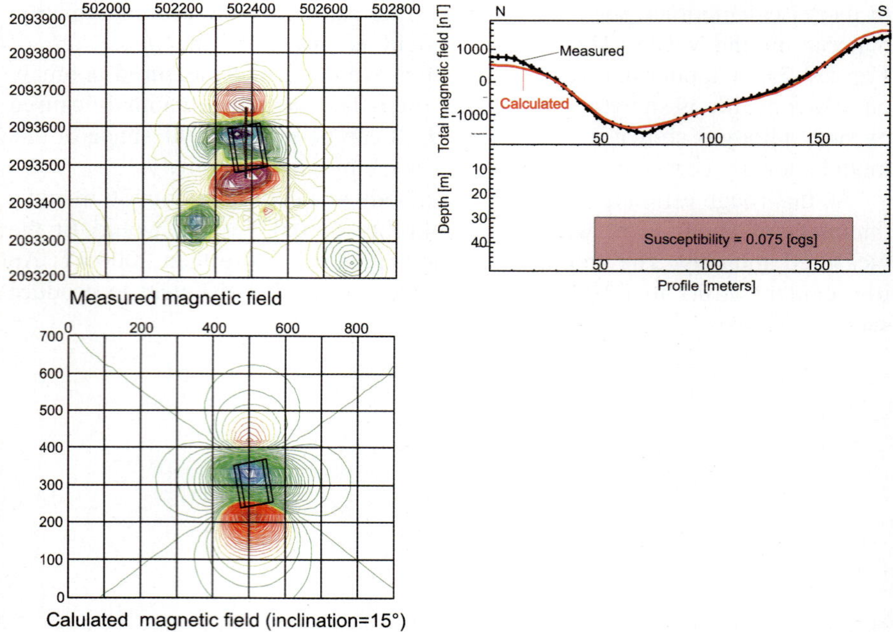

Fig. 5.4-25: Magnetic method – modeling of the anomaly related to the Nong Harn landfill

Three boreholes were drilled and geophysical borehole logging conducted to obtain information about the *geology and the hydrogeology* of the area. The material was recovered from one of the boreholes to obtain an accurate lithological description of the sediments, to correlate them with the geophysical well logs and for laboratory investigations. Three groundwater observation wells were installed in the area around the site to detect possible groundwater contamination and to determine the direction of groundwater flow. Eight shallow percussion core boreholes with a maximum depth of 4 m were drilled to obtain soil gas samples in order to detect near-surface gas migration. Groundwater samples were collected from the observation wells. The samples were filtered in the field, conserved in prepared bottles, saved in ice-cooled boxes and analyzed in laboratory. After well completion, pumping tests were carried out in each observation well to determine its hydrodynamic aquifer parameter values.

Well MW-NH-1 was drilled to a depth of 98.60 m. The lithological profile and well logs are shown in *Fig. 5.4-26*. The lithological profile shows clay-rich sediments in the top 33 m, with only little sandy intercalation. Sand and gravel predominates between 33 and 65 m. Clay is present from 65 to 88.5 m, followed by siltstone. Drilling progress slowed noticeably below 65 m and below 85 m penetration rates became increasingly slower.

The lithological sequences found in the seismic survey are well represented by the results from this borehole. The sequences A (0 - 33 m b.g.l), B (33 - 65 m b.g.l.) and D (below 65 m b.g.l) in *Fig. 5.4-28* were penetrated. In the interval from 13 to 19 m drilling mud was lost, indicating potential groundwater migration pathways in this depth range.

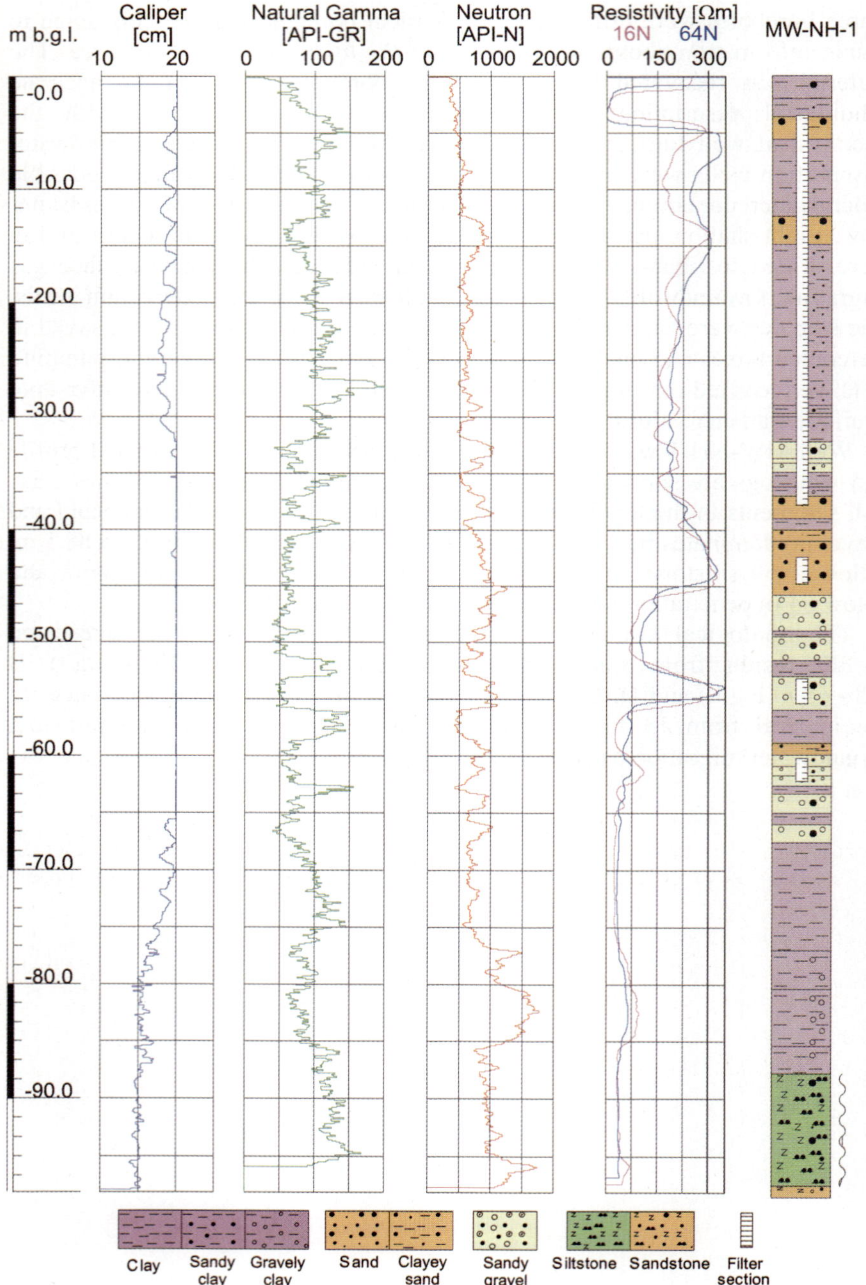

Fig. 5.4-26: Geophysical well logs in borehole MW-NH-1 and the lithological interpretation (*right*) based on the drill cuttings. Caliper measurements are limited to a diameter of 20 cm (AZTEC Engineering, Lampang, commissioned by WADIS-project).

The investigation area seems to be characterized by very frequent facial changes. Correlation of wells MW-NH-1 and MW-NH-2a shows that similar sediment sequences are found over short distances, despite the fact that colluvial sediments generally change properties within short distances.

The results of lithological interpretation of boreholes MW-NH-1 and MW-NH-2a are shown in the cross-section in *Fig. 5.4-27*. In the top 30 m, the bedding is approximately horizontal and the layer thicknesses are similar. The lower depths are characterized by layers ascending to the east. Thickness of the sediments decreases in the same direction and the hard material (siltstone) is present at a depth of 60 m. This is in good accordance with the results from reflection seismic survey.

In the area of the landfill, the seismic section NH0201 extends NE–SW. If boreholes MW-NH-1 and MW-NH-2a are projected onto this section, it can be seen that a fault crossing the profile causes uplift of the eastern part (or downslip of the western part) of the top of the basement. If this fault extends to the north, crossing the landfill, an alternative correlation pattern in the lower part of the profile (*Fig. 5.4-27*) has to be assumed. However, the attempt to connect fault indications on seismic profiles results in a tectonic pattern that shows this fault running NNW–SSE, not crossing the profile line connecting the two boreholes.

Prior to drilling, two soil samples were taken from the unfilled pit just southeast of the landfill and three samples from the "big pit" approximately 2 km SW of the landfill and analyzed for *sediment properties and mineral composition*. Quartz is the predominant mineral in all of the samples. The proportion of material of grain size < 63 μm is especially high in the two samples from the unfilled pit (49 % and 65 %) and one sample from the "big pit" (59 %) (*Table 5.4-6*). These three samples show a very uniform grain-size distribution. This fact and the generally high proportion of silt and clay causes a low effective porosity and thus to practically impermeable sediment layers. Characteristic of all five samples is a medium effective cation exchange capacity (CEC_{eff}) of 1.0 - 2.8 $cmol^+/kg$. They are little mineralized, but the eluate shows high nitrate and ammonium concentrations. All of the samples yield a strongly to moderately acidic solution when stirred with deionized water and none of the samples contain carbonate. All of the carbon is in organic carbon material.

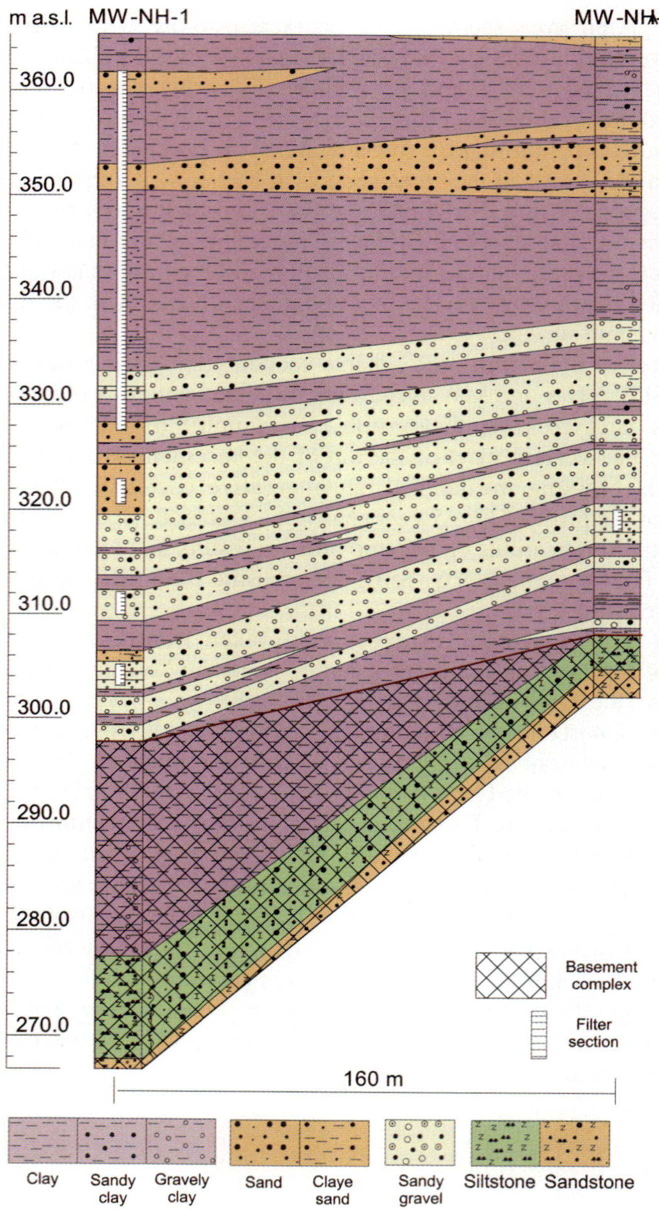

Fig. 5.4-27: Schematic geological cross-section between the two boreholes MW-NH-1 and MW-NH-2a (vertical exaggeration)

Table 5.4-6: Soil parameters of samples from the area around the Nong Harn landfill; NH: samples from the unfilled pit to the southeast of the landfill; BP: samples from the "big pit" approximately 2 km southwest of the landfill

	Units	NH1/01	NH2/01	BP1/01	BP2/01	BP3/01
Soil						
proportion < 2 mm	%	98	100	97	44	100
proportion < 63 µm	%	65	49	27	14	59
soil type		silt, fine to medium sand	silt, fine to medium sand	silty sand	silty gravel, fine to medium sand	silt, fine to medium sand
pH	-	5.11	4.84	4.93	5.13	4.88
electrical conductivity	$\mu S\ cm^{-1}$	24.1	41.2	54	21.2	15.9
total carbon (TC)	%	0.071	0.042	0.021	0.025	0.028
total organic carbon (TOC)	%	0.069	0.039	0.020	0.029	0.026
total inorganic carbon (TIC)	%	< 0.005	< 0.005	< 0.005	< 0.005	< 0.005
Effective CEC	$cmol^+/kg$	2.4	2.8	1.3	1.0	2.9
pH (after adding $BaCl_2$ solution)	-	4.3	4.4	4.8	4.8	4.5
sum of exchangeable cations	-	1.4	1.9	0.83	0.51	2.1
calcium	$cmol^+/kg$	0.31	0.06	0.55	0.34	1.2
magnesium	$cmol^+/kg$	0.08	0.02	0.23	0.12	0.48
sodium	$cmol^+/kg$	< 0.02	< 0.02	< 0.02	< 0.02	< 0.02
potassium	$cmol^+/kg$	0.06	0.06	0.05	0.05	0.09
iron	$cmol^+/kg$	< 0.02	< 0.02	< 0.02	< 0.02	< 0.02
aluminium	$cmol^+/kg$	0.99	1.8	< 0.02	< 0.02	0.30
Eluate after DIN 38414-4 (S4)						
ammonium	$mg\ L^{-1}$	0.63	1.1	0.13	0.18	0.15
nitrate	$mg\ L^{-1}$	1.3	5.3	0.53	0.58	0.89
o-phosphate	$mg\ L^{-1}$	0.003	< 0.002	0.003	0.003	< 0.002

Table 5.4-7: Chemical parameters of sediments from liner core samples from the drilling of MW-NH-3a

	Units	NH3-3	NH3-6	NH3-10	NH3-11
Sediment Investigation					
proportion of > 2 mm fraction	%	31	20	23	30
proportion of the < 2 mm fraction of the sample without cobbles	%	42	62	56	46
pH	-	8.43	6.65	6.96	8.14
electrical conductivity	$\mu S\ cm^{-1}$	52.9	30.5	30.9	34.8
total carbon (TC)	%	0.10	0.047	0.081	0.085
total organic carbon (TOC)	%	0.090	0.046	0.081	0.068
total inorganic carbon (TIC)	%	< 0.005	< 0.005	< 0.005	< 0.005
Eluate after DIN 38414-4 (S4)					
pH	-	7.83	6.94	6.98	7.46
electrical conductivity	$\mu S\ cm^{-1}$	13.5	< 10	< 10	11.7
dissolved organic carbon (DOC)	$mg\ L^{-1}$	1.5	1.8	1.6	1.1
SAK254	m^{-1}	n.m.*	< 0.1	< 0.1	< 0.1
ammonium-N	$mg\ L^{-1}$	< 0.05	< 0.05	< 0.05	< 0.05
nitrate-N	$mg\ L^{-1}$	0.16	0.11	0.09	< 0.05
nitrite-N	$mg\ L^{-1}$	< 0.02	< 0.02	< 0.02	< 0.02
o-phosphate-P	$mg\ L^{-1}$	0.015	0.003	0.003	< 0.002
sulfate	$mg\ L^{-1}$	1.58	0.64	< 0.5	0.50
iron	$mg\ L^{-1}$	0.203	0.011	0.007	0.020
manganese	$mg\ L^{-1}$	0.054	0.013	0.075	0.028
zinc	$mg\ L^{-1}$	0.015	0.003	< 0.002	< 0.002
calcium	$mg\ L^{-1}$	0.570	0.134	0.307	1.25
magnesium	$mg\ L^{-1}$	0.103	< 0.042	0.077	0.229
sodium	$mg\ L^{-1}$	6.35	< 0.029	1.35	2.41
potassium	$mg\ L^{-1}$	< 0.06	0.337	0.804	1.43

* not measured

Further soil analyses were carried out on core samples recovered in a liner from borehole MW-NH-3a (*Table 5.4-7*). These samples were taken from a depth of 43 - 49 m. As geophysical well logging in borehole MW-NH-1 indicated contamination at a depth of 46 – 48 m, this interval was sampled in the nearby located drilling MW-NH-3a. Samples NH3-10 and NH3-11 were taken from this interval, while the upper two samples NH3-3 and NH3-6 come from an uninfluenced part of the aquifer. The analysis of these samples revealed no significant differences between the upper and the lower two samples (*Table 5.4-7*).

The *emission of landfill* gas was determined on the landfill and the immediate surroundings. The gas was analyzed to determine its composition, the rate of seepage was determined, and the distribution of emission.

Municipal waste contains a large amount of organic bound carbon in the form of food and garden wastes. Within a landfill, the organic carbon is transformed into gaseous substances by microorganisms. The resulting gas generally consists of up to 99 % methane (CH_4) and carbon dioxide (CO_2). The increasing gas pressure within the landfill results in diffuse migration of the gas leading to uncontrolled release of gas to the atmosphere if no measures are taken to collect the gas.

Landfill gas poses a potential risk to health in different ways. The primary danger is the potential for the methane to form an explosive mixture with air. Mixtures containing 4.5 - 15 vol. % methane are explosive. Mixtures with a higher proportion of methane are flammable. The risk of explosion is present on the landfill itself as well as where gas is emitted in the area around the site.

The gas emitted from the Nong Harn landfill (measured at points where it has not mixed with atmospheric air) consists of 57 - 61 % methane and 35 - 37 % carbon dioxide. This composition corresponds to steady-state anaerobic fermentation of waste with an exceptionally high organic content.

The compounds H_2S and NH_3 were detected in landfill gas in trace concentrations of up to a maximum of 94 ppm (H_2S) and 180 ppm (NH_3). The wide range of organic substances observed in traces in the landfill gas was very much as expected. Remarkable is the composition of the gas in well W9, about 160 m NE of the landfill – 58 % methane and 28 % carbon dioxide, which is comparable to that of the gas seeping from the landfill. A very high rate of flow was measured at some of the main seepage points, which indicates a high gas production rate in the landfill. At well W9, the flow rate of 5 L gas per minute was measured.

Hydrogeological investigations show that the geology of the area around the Nong Harn landfill correlates well with the occurrence of groundwater in the area. Water wells at and close to Wat Wiwek Waharam (a temple about 600 m west of the site), together with the evidence from the observation wells at the site and from water supply wells in the broad valley to the north, indicate that aquiferous horizons are present only in seismic sequence B. Wells which penetrated only sequence A remained dry. This situation is shown in *Fig. 5.4-28*, which displays an expanded part of seismic line NH0201 in the area around the landfill with projections of nearby wells.

The 80 m deep borehole about 160 m NE of the site (see above) most likely reached sequence D, but did not find any groundwater. It only emitted methane gas, which obviously originated from the landfill. The aquifer presumably wedges out east of the landfill against the rather impermeable sequence D. Within the landfill the depth of groundwater table is believed to be near the bottom of the former borrow pit and therefore within its natural fluctuation interval. No wells are documented in the area, which penetrate sequence C.

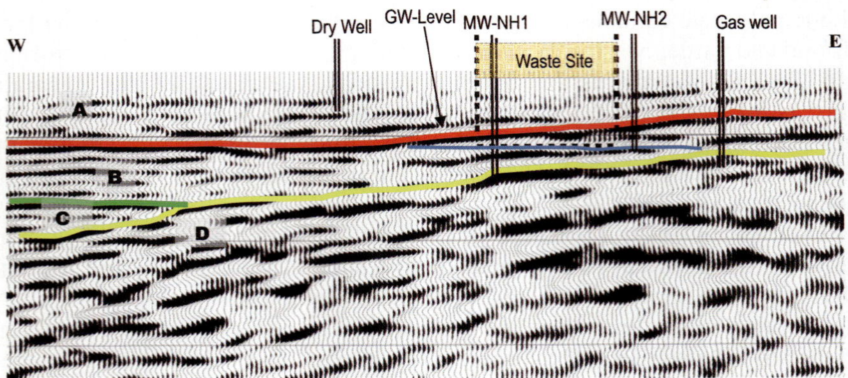

Fig. 5.4-28: Expanded part of seismic line NH0201, length: 700 m

The water level in the unfilled pit southeast of the landfill is about 29 m below the surrounding ground level. The groundwater table in the area is at a depth of around 40 m, approximately the depth of the bottom of the landfill, varying several meters over the year (*Fig. 5.4-29*). Therefore, water in the unfilled pit, with a maximum depth of approximately 4 m, is well above the groundwater table and is derived solely from precipitation.

Only wells MW-NH-2a, MW-NH-3a and MW-NH-E1 could be used to determine the direction of local groundwater flow. These wells have screens at depths between 45 - 49 m (MW-NH-2a and 3a), and a not exactly known section starting at approximately 30 m (MW-NH-E1). NW-NH-1 could not be used for this determination. The general groundwater flow direction between the wells at the edges of the landfill is to the NNW, as shown in *Fig. 5.4-30*. The difference in elevation of the groundwater table of 1.25 m over a distance of 106 m represents a groundwater gradient of $I = 0.0118$.

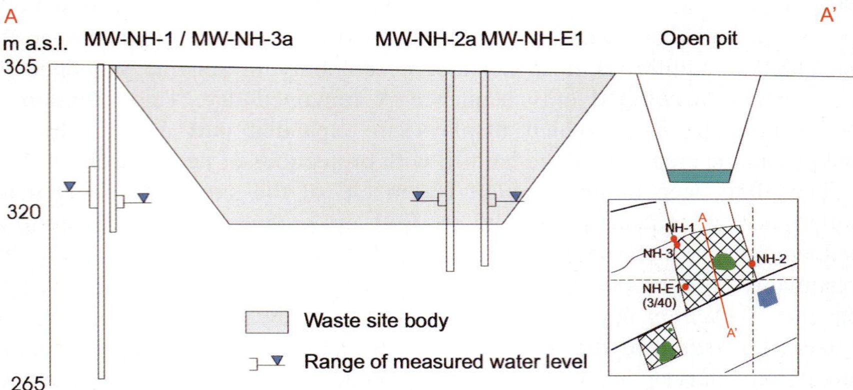

Fig. 5.4-29: Cross-section across the Nong Harn landfill site; wells and unfilled pit are projected onto the profile line. The depth of the bottom of the landfill is not exactly known and is shown here with an assumed depth of 45 m.

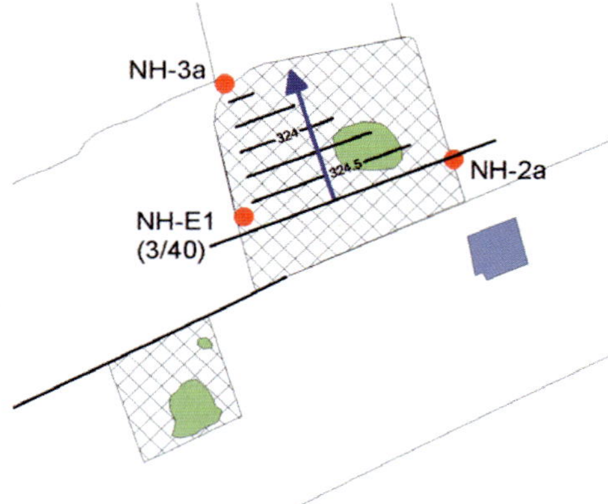

Fig. 5.4-30: Local groundwater gradient between the observation wells at the Nong Harn waste disposal site on April 8, 2003. The groundwater gradient is indicated by the blue arrow.

Since the landfill is sealed with an HDPE liner, although incompletely, there is some limited hydraulic connection to the groundwater flow regime. Repeated water level measurements show that the direction of the gradient is rather consistent over all climatic seasons.

The groundwater flow direction to the NNW seems to be in contrast to the surface topography of the area, which slopes slightly to the south. The gravity survey data showing a gravity gradient corresponding to the groundwater gradient might explain the direction of groundwater flow.

The groundwater table in the area immediately surrounding the Nong Harn landfill is at a depth of approximately 40 m. The results of drilling show that the top 33 m consist of very clay-rich material with only thin layers of sand (*Figs. 5.4-26* and *5.4-27*). The main aquiferous horizons are below 33 m depth. Nevertheless, there are observations which indicate that perched water occurs in the upper layers during the rainy season.

Pumping tests to determine hydraulic properties could be conducted only in well MW-NH-1. Wells MW-NH-2a and MW-NH-3a did not yield enough water to obtain a steady-state water level. Since well MW-NH-1 has screens at different depths, the determined k_f value represents an average of all the screens. In a pumping test, the hydraulic conductivity of the aquifers was determined to be $k_f = 6.6 \cdot 10^{-7}$ m s^{-1}. However, while parts of the 20.8-m-thick aquifer have a low hydraulic conductivity, other sections will have a higher conductivity than the average value.

The objective of the *hydrochemical investigation* was to determine whether landfill leachate from the Nong Harn landfill is present in the groundwater or not.

Since the leachate drainage system for the landfill is not functioning properly, it was not possible to collect a water sample from inside the landfill. Due to the high volume of gas production, drilling directly on the site was excluded. Samples were taken in concrete rings on the site that were part of the drainage system early in the operation of the landfill. Originally extending to the bottom of the landfill, but they have now collapsed due to settlement of the waste, the depths of these pipes are not exactly known. Leachate from inside the landfill reaches the surface of the water in these pipes as a result of convection driven by gas rising in the water and seen emerging at the water surface. Even though the water also contains rain water, thus changing concentrations of its contents, it can provide an indication of the original composition and of the contaminant contents.

This water has a high electrical conductivity (1660 $\mu S\ cm^{-1}$) relative to the background values and water from the observation wells. The water in the lake on the landfill and from well MW-NH-3a have similar conductivities. When these values are compared with those for leachate from other landfills, it is apparent that dilution is occurring. At the Mae Hia landfill, which was used for waste from Chiang Mai before Nong Harn, conductivities of up to 25 600 $\mu S\ cm^{-1}$ were measured. Water taken from the concrete rings showed high chloride (282 $mg\ L^{-1}$) and ammonium (50 $mg\ L^{-1}$) contents, as well as organic compounds in low concentrations. Volatile organic compounds (VOCs) such as benzene, toluene, xylene and ethylbenzene were found in concentrations that make closer investigation recommendable, but should be at least monitored.

Water samples were taken from the small lake on the landfill (NH-lake) and the unfilled pit (NH-open pit) southeast of the site. Besides the main ions, organic compounds were analyzed, especially with respect to toxicity and biodegradability.

The water samples are alkaline (NH-lake) or slightly acidic (NH-open pit) and differed in their content of diluted and undiluted substances (*Table 5.4-8*). The NH-lake sample was more concentrated than the NH-open pit sample. Analysis for metals (As, Cd, Cr, Co, Cu, Ni, Pb, Zn) showed concentrations below the 0.1 $mg\ L^{-1}$ (high limit because of the small samples).

Fluorescence measurements showed that the humic contribution to the DOC is low. The NH-lake sample contained traces of resorcin and one other alkylphenol. Other investigations have shown that polycyclic aromatic hydrocarbons (PAH) are unlikely to be present. Furthermore, herbal protein, typical of phytoplankton bloom, was found in this sample. The NH-lake sample was rich in bacteria with a high diversity of species and a high abundance of plankton in, typical of eutrophic surface water. A luminescent bacteria test revealed no acute toxicity.

Table 5.4-8: Selected analytical results for surface water samples from Nong Harn

Sample ID	Units	NH-lake	NH-open pit
sample location		small lake on the landfill	unfilled pit south of the landfill
sampling date		17.04.01	17.04.01
temperature	°C	34	33
pH (sampling)		8.5	5 - 5.5
electric conductivity	$\mu S\ cm^{-1}$	1940	85.2
chloride	$mg\ L^{-1}$	501	4.16
total organic content (TOC)	$mg\ L^{-1}$	135	2.6
dissolved organic content (DOC)	$mg\ L^{-1}$	52	2.5
biological oxygen demand (BOD)	$mg\ L^{-1}$	10.8	5.2
luminescence bacteria test		not toxic	not toxic

Water samples were taken from the new observation wells MW-NH-1, MW-NH-3a, the existing observation well MW-NH-E1 and the production well PW-NH-2/40 at the nearby temple. The observation wells are near the landfill, the "temple" well is about 600 m west of the landfill. Before sampling, the observation wells had to be considered possibly affected by seepage water, whereas the temple well (PW-NH-2/40) represents natural background water.

The samples showed pH values between 4.5 and 6 and a broad range of electric conductivities. While the water from the temple well and well MW-NH-1 have a low mineral content, higher concentrations were found in wells MW-NH-E1 and MW-NH-3a (*Table 5.4-9*). The sample from well MW-NH-3a shows the highest concentration of all main ions, with only chloride being exceptionally high (271 $mg\ L^{-1}$). The iron and manganese concentrations are very high in the observation wells. Of the heavy metals, only cadmium in MW-NH-1 (not measured in all wells) was detected in amounts above the guideline values for drinking water (Thai Ministry of Industry). No bacteria were found in the analyzed samples. Phenolic substances were present in significant concentrations. While the concentrations in the well at the temple (PW-NH-2/40) are similar to background values in other parts of the Chiang Mai/Lamphun Basin and the Nong Harn area, concentrations in wells MW-NH-1 and MW-NH-E1 are clearly elevated, and especially so in well MW-NH-3a.

Multiple sensor probe measurements were carried out continuously in well MW-NH-3 to determine groundwater parameter values. Several groundwater parameters, including water table fluctuations, were monitored in this way and found to correlate to some extent with other hydrogeological and hydrochemical data.

Table 5.4-9: Selected parameters from analyses of groundwater samples from Nong Harn

Well number		**MW-NH-1**	**MW-NH-E1**	**MW-NH-3a**	**PW-NH-2/40**
Sampling date	**Units**	24 Apr. 2002 21 Oct. 2002	22 Oct. 2002	24 Apr. 2003	24 Apr. 2003
temperature	°C	27.3 - 29.1	29.7	30.4	-
pH		4.5 - 4.9	4.91	5.98	4.99
EC	$\mu S\ cm^{-1}$	65 - 194	259	1359	31.5
O_2	$mg\ L^{-1}$	0.1 - 1.5	0.1	0.1	3.3
Eh	mV	205 - 265	27	106	312
total solids (TS)	$mg\ L^{-1}$	128 - 216	202	742	40
suspended solids (SS)	$mg\ L^{-1}$	2 - 12	28	11	< 2
biochemical oxygen demand	$mg\ L^{-1}$	< 0.60	2.14	< 0.60	< 0.60
chemical oxygen demand	$mg\ L^{-1}$	< 4.0 - 4.0	28	16	4
total organic carbon	$mg\ L^{-1}$	0.46 - 0.88	5.6	2.9	0.5
total inorganic carbon	$mg\ L^{-1}$	18.9 - 54	61	28	5.5
dissolved organic carbon	$mg\ L^{-1}$	0.56	5.6	3.1	0.59
phosphate phosphorus	$mg\ L^{-1}$	0.01	0.01	0.04	0.03
nitrite nitrogen (NO_2-N)	$mg\ L^{-1}$	0.05	0.03	< 0.01	< 0.01
nitrate nitrogen (NO_3-N)	$mg\ L^{-1}$	0.14 - 0.28	0.12	0.02	0.03
ammonia nitrogen (NH_3-N)	$mg\ L^{-1}$	0.02 - 0.6	0.42	9.51	0.02
chloride (Cl^-)	$mg\ L^{-1}$	45.5 - 47.9	53.4	271	6.75
sulfate (SO_4^{2-})	$mg\ L^{-1}$	8.6 - 15.1	8.93	12	8.94
magnesium (Mg^{2+})	$mg\ L^{-1}$	0.86 - 4.6	3.94	15.3	2.59
calcium (Ca^{2+})	$mg\ L^{-1}$	3.7 - 14.9	13.2	55.3	6.5
sodium (Na^+)	$mg\ L^{-1}$	5.6 - 35.4	21.9	116	4.23
potassium (K^+)	$mg\ L^{-1}$	0.5 - 2.84	1.2	10.8	5.88
boron (B)	$\mu g\ L^{-1}$	< 100	< 50	31	5.3
iron (Fe)	$\mu g\ L^{-1}$	100 - 1365	9190	1517	434
manganese (Mn)	$\mu g\ L^{-1}$	1 - 753	1442	8570	37
zinc (Zn)	$\mu g\ L^{-1}$	< 20 - 147	52	50	< 20
aluminum (Al)	$\mu g\ L^{-1}$	< 20	28	33	< 20
arsenic (As)	$\mu g\ L^{-1}$	< 5.0	7.8	< 5.0	< 5.0
cadmium (Cd)	$\mu g\ L^{-1}$	0.11 - 17	< 0.50	-	-
chromium total (Cr)	$\mu g\ L^{-1}$	2 -2.6	8.5	7.3	< 2.0
cobalt (Co)	$\mu g\ L^{-1}$	0.2 - 6.2	7.3	-	-
copper (Cu)	$\mu g\ L^{-1}$	1.2 - 2	3.6	< 2.0	2.3
fluoride (F^-)	$\mu g\ L^{-1}$	< 0.01	< 0.01	-	-
molybdenum (Mo)	$\mu g\ L^{-1}$	0.34	< 5.0	< 5.0	< 5.0
nickel (Ni)	$\mu g\ L^{-1}$	6.2	7	< 5.0	< 5.0
lead (Pb)	$\mu g\ L^{-1}$	1.4	< 5.0	-	-
total coliform bacteria	MPN/100 mL	< 2	< 2	-	-
fecal coliform bacteria	MPN/100 mL	< 2	< 2	-	-
phenol	($mg\ L^{-1}$)	0.013 - 0.044	0.011	-	-

MPN = most probable number

To estimate migration parameter values, *column tests* were carried out using leachate from the landfill and sediment from the area around the landfill. The column tests were carried out with undisturbed core samples with the water entering the column at the bottom using the same groundwater velocity as in the in-situ situation, and at a temperature of 20 °C. The soil columns were 5.3 cm in diameter and 30 cm long. Due to the lack of availability of undiluted leachate from the landfill, groundwater from the observation well MW-NH-3 was used for the tests. An electrical conductivity of 1.1 mS cm^{-1} was measured in the groundwater sample, a value of 3.1 mg L^{-1} was determined for DOC. The analyses were limited to pH, electrical conductivity (EC), total and dissolved organic carbon contents (TOC/DOC) of the water after passing through the column. The measured pH in the samples at the outlet of the column was between 7.0 and 8.0, increasing slightly during the experiment. The conductivity in the samples from experiments was almost the same after the second exchange of the pore volume. With an average of 1.2 mS cm^{-1}, the conductivity of the samples was slightly higher than that of the test water (1.1 mS cm^{-1}). This shows that all of the ions in the test water were transported through the column without retardation. The estimated retardation coefficient is about 1.

The DOC concentrations show a typical elution behavior: The DOC content in the soil samples was much higher than in the water. After about six pore volume exchanges, the DOC in the water after passing through the column is nearly constant and is about the same as the DOC of the test water. The source of the high DOC content in this sample is not known. Spectrometric and/or chromatographic methods would be necessary to obtain more information about this. One possible explanation of the nonexistent retardation capacity and the elution of DOC from the sediment is that the sediment was in contact with water containing leachate from the landfill before it was sampled during the drilling of the wells.

Risk Assessment

Two aspects have to be taken into consideration when the risk of emission of landfill gas is assessed:

- the direct risk to human health by poisoning and
- risk of gas accumulation and explosion in the area around the landfill.

There is no risk of the spreading of toxic gases outside of the Nong Harn landfill area. The concentration of toxic substances in the landfill gas is so low that threshold values are only reached immediately around the outlets. There are also no significant odor problems. Nevertheless, in the direct vicinity of the gas outlets, an explosive atmosphere may exist due to the high concentration of methane. But at a first glance there is no possibility of accumulation of explosive atmosphere, because no oxygen is available under the surface seal (perhaps in gas bubbles) and the gas, due to its low density, escapes into the air immediately after leaving the landfill.

Nevertheless, there is a certain risk for persons who are on the landfill surface, owing to the toxicity of the gases, as well as the danger of fire and explosion. Admission to unauthorized persons should strictly be prohibited (e.g., playing children). With respect to the environment, the largest problem is the emission of methane, which promotes the greenhouse effect. This can be avoided only by collecting and burning the landfill gas.

It is difficult to assess the dispersion of the gas underground. The radial distance of the migration is not known, neither the depth nor the areas where migration and accumulation is taking place. A dry borehole approximately 160 m NNE of the landfill, originally drilled as water well, was found to yield gas originating from the landfill. Thus, the migration of considerable amounts of landfill gas is without dispute. In general, risk for the population that lives in the surrounding area cannot be excluded. However, a risk of poisoning is not likely because the relevant substances (especially H_2S and NH_3) are naturally attenuated by either absorption in the ground or by biodegradation.

Important for the assessment of the risk of *groundwater contamination* from the Nong Harn landfill are the potential for spreading of the leachate, the effectiveness of the technical and geological barriers, the hydraulic conditions, and the type and location of drinking water wells and population that needs to be safeguarded.

It is difficult to calculate the amount of leachate produced in the Nong Harn landfill, because many input parameter values are unknown. First of all, there is no information about the actual amount of cover material used during operation of the site nor of the physical properties of the waste. The density after dumping and compaction of the waste can only be roughly estimated. At the beginning of operation, leachate was pumped off and transported to a waste water treatment plant. Information about how much leachate was pumped off was not available. Two more uncertain variables are the exact depth of the landfill and the possible discharge through incomplete and damaged HDPE liner. The bottom of the landfill is at about the depth of the groundwater table. Whether groundwater flows through landfill waste depends on whether the bottom of the landfill is above or below the groundwater table.

Generation of leachate by rain should be rigorously limited. When the landfill was closed, the top of the waste was higher than the surrounding area, enabling rapid runoff of rainwater. With time, the surface formed a depression in which rainwater accumulates, forming a small lake. The climatic water balance in the area is negative overall. Evaporation is especially high from a free water surface and the final landfill cover seems to be of rather low permeability. Most of leachate may be expected to originate from the high initial water content of the waste, which was rich in organic material, and from rain during the operation phase. Most of this water will be retained within the landfill owing to the high field capacity of the waste. Part of the water in the waste may be trapped between intermediate layers of cover material and the intact parts of the landfill liner.

Since no pure leachate could be obtained from inside the landfill, the risk had to be assessed on the basis of samples from the small lake on the landfill surface and water from the remains of the former leachate collection system. However, this water is diluted by rain and affected by atmospheric deposition and does not represent the deeper part of the landfill. Therefore, it was assumed the leachate has a composition similar to that in the old Mae Hia landfill.

The typical leachate of the Mae Hia landfill, where samples could be collected directly from the center of the landfill, has a high concentration of chloride (max. 3400 mg L^{-1}), sodium (1900 mg L^{-1}), potassium (2930 mg L^{-1}), iron (35.7 mg L^{-1}), manganese (2.1 mg L^{-1}), zinc (0.56 mg L^{-1}), and aluminum (3.6 mg L^{-1}), which is reflected by high values for total solids (TS): 13 to 15 g L^{-1}. The values for total organic carbon (1800 mg L^{-1}), inorganic carbon (2140 mg L^{-1}), and nitrogen compounds (max. ammonium-nitrogen 974 mg L^{-1}) are high. This leachate contains very little PAH, pesticides and other organic components, which were reported to be lower than 0.2 µg L^{-1}, and mostly even below the detection limit. The main risk to human health found in this leachate is posed by the high nitrogen content.

Concentrations of the inorganic substances in the samples from the concrete rings and the small lake on top of the Nong Harn landfill show that much higher concentrations have to be expected in the waste itself. The water from Nong Harn showed only small noticeable concentrations of volatile organic components (VOC). They might be health threatening if found in higher concentrations in undiluted leachate. The occurrence of these substances in the samples is an indication of contamination in the deeper parts of the landfill.

Magnetic measurements indicate the presence of barrels of unknown content within the landfill. Barrels are a typical sign of industrial waste. These barrels pose a risk of contamination when they have corroded and begin leaking. This possibility cannot be investigated without additional drilling and sampling inside the waste. But drilling within the landfill is considered dangerous owing to the considerable gas production.

The top 30 m of sediment in the area around the landfill are rich in clay and silt. A medium CEC was determined for all samples from the area around the landfill. This indicates a good retention capacity for possible contaminants. Loss of drilling mud and lateral gas migration indicate potential pathways for migration in parts of this depth interval. The predominantly fine-grained material suggests migration along secondary pathways, such as shrinkage cracks, rather than in pore spaces. Such shrinkage cracks can be seen in the lower part of the open pit wall. Migration is also possible in the sandy intercalations of not more than a few meters thickness observed in borehole MW-NH-1 (*Figs. 5.4-26* and *5.4-27*). Sandy material and aquiferous layers are present at the depth of the landfill bottom. The porosity of the sediments at this depth was determined to be $n = 0.33$. Column tests with this material showed that dissolved components in the test water are not retarded.

It may be concluded that continuous discharge of leachate from the landfill is possible only if there is a hydraulic connection of the bottom part of the landfill with the groundwater. Considering that the base of the landfill is at a depth of 40 - 50 m and the depth to the groundwater table averages 39 - 40 m in the directly neighboring observation wells, contact of the waste to aquiferous sediments is likely. Whether leachate enters the groundwater through the bottom of the landfill then depends on the condition of the liner. Reconstruction of a hilltop relief on the landfill and the filling of the fissures at the edges of the landfill should lead to reduction of leachate production.

In the area immediately around the landfill the groundwater table has a gradient of $I = 0.0118$ to the NNW, as determined between the wells MW-NH-2a, MW-NH-3a and MW-NH-E1. On the basis of the average hydraulic conductivity ($k_f \sim 1 \cdot 10^{-6}$ m s^{-1}) determined from a pumping test in MW-NH-1 and a porosity of $n = 0.33$, the groundwater velocity was calculated to be approximately $v = 1.1$ m a^{-1}. Assuming a maximum hydraulic conductivity of $k_f \sim 1 \cdot 10^{-5}$ m s^{-1} for the part of the aquifer with the highest conductivity, water from the landfill may travel approximately 11 m a^{-1}.

A small village is present about 370 m downstream from the landfill. The closest known deep well in this direction is about 400 m NNW of the landfill. The nearby temple area, with a well approximately 600 m west of the landfill is not in the possible direction of groundwater flow.

MW-NH-3a and MW-NH-E1 show clear indications of landfill leachate. MW-NH-1 shows much lower concentrations than in wells MW-NH-3a and MW-NH-E1. The reason is either the longer distance from the landfill or dilution with water from screens at shallower depth than in the other wells. The results of a 2-D resistivity survey show that the influence of the landfill decreases rapidly to the north, while a small resistivity minimum is visible at the depth of the screen in well MW-NH-3a (*Fig. 5.4-23*). Contamination is probably present only in the narrow interval of 46 - 48 m depth at well MW-NH-3a. Lateral migration of seepage water is obviously limited (*Fig. 5.4-22*). None of the profiles not directly bordering the landfill shows an impact from the site.

Conclusions

It can be concluded from the geology of the area around the landfill that the geological barrier properties of the sediments are not perfect. Even though the top 33 m consist of clays, which provide a good seal, and geophysical results show a sharp vertical boundary at the edges of the landfill, indicating there is little or no lateral spreading of contamination from the upper 40 m of the landfill. The base of the landfill is near or within the groundwater and hydraulically conductive layers. Therefore, there is a risk of groundwater contamination by seepage from the site. The possibility of industrial waste within the landfill and the occurrence of volatile organic components (VOC) in water samples from the site indicate a potential for contamination in the deeper part of the landfill. This potential must be kept in mind in the future.

Limited leachate production and the barrier of the HDPE liner have retarded the spreading of contamination from the site up to now. Five years after closure, concentrations of contaminants outside the site, even within only a few meters, are relatively low and the small hydraulic gradient of the area and medium hydraulic conductivity of sediments contribute to slow groundwater movement and consequently a slow spreading of contamination.

More examples for conceptual models are given in Chapter 6.2.

5.4.3 Groundwater Flow Modeling

CLAUS KOHFAHL

Groundwater modeling has become an important tool for management of water supply in the last several decades. Consumption and contamination of groundwater have increased worldwide due to population growth, and a sustainable management of groundwater supply with regard to its quality and quantity is indispensable. The aims of groundwater modeling can be summarized as follows:

- to describe the behavior of subsurface systems,
- to construct theories or hypotheses that account for the observed behavior, and
- to use these theories to predict future behavior.

In principle, groundwater modeling can be divided into flow and transport modeling. In general, flow modeling treats the movement of water, whereas transport modeling deals with the migration of particles in the water and therefore also has to take into account dispersion and chemical reactions. A transport model is usually based on a flow model, i.e. a flow model has to be constructed when a transport model is developed. The most important applications of flow modeling are water supply management, supervision of hydraulic works, and simulation of natural hazards. Transport modeling is a helpful tool for problems concerning groundwater quality. As the computer programs for such modeling depend on a larger number of variables than flow modeling and substance migration is mostly transient, transport models are far more difficult and less common than flow models. An illustrative introduction to groundwater modeling is given by ANDERSON & WOESSNER (1992).

5.4.3.1 Fundamentals of Groundwater Flow Modeling

Like many processes in natural science, groundwater flow involves rates, which are often written as derivatives. Therefore, groundwater flow can be described by a partial differential equation involving spatial coordinates and time. This equation can be derived using the continuity equation and Darcy's law. To take the three-dimensionality of groundwater flow into account, the spatial variables are written as tensors and vectors. Gradients are written using the nabla operator to simplify mathematical notation.

Darcy's law shows that the flow velocity (v in m s^{-1}) in a saturated porous medium is proportional to the gradient of the piezometric head ∇h, assuming constant density of the water:

$$v = -K \nabla h, \tag{5.4.19}$$

where v is the vector of Darcy velocity,
K the tensor of hydraulic conductivity[3], and
h is the hydraulic potential (head).

The proportionality constant, K (in m s^{-1}), in the equation is referred to as hydraulic conductivity. It has the units of velocity and differs significantly depending on the geological formation involved. The values range from 10^{-13} m s^{-1} in unfractured metamorphic rocks to 10^{-1} m s^{-1} in highly permeable gravel (Section 5.2.7.3 and FREEZE & CHERRY, 1979).

Application of the law of mass conservation leads to the continuity equation. Assuming constant density, the *continuity equation* is

$$\nabla v = \frac{\partial h}{\partial t} S_s + w, \tag{5.4.20}$$

where v is the vector of the Darcy velocity,
h hydraulic potential (head),
t time,
S_s specific storage coefficient, and
w is sources/sinks.

The terms of the equation represent the difference between the amounts of flow into and out of the elementary volume called control volume. This difference equals the change in water storage in the porous medium corrected by sink and source terms (w), e.g., pumping and infiltration.

---

[3] In numerical modeling the tensor of conduvtivity is usually denoted K instead of k_f.

Combining Equations (5.4.19) and (5.4.20) yields a differential equation for flow with constant density:

$$\nabla(K\nabla h) - w = \frac{\partial h}{\partial t} S_s . \qquad (5.4.21)$$

This *groundwater flow equation* is a linear second-order partial differential equation and describes groundwater flow in a saturated porous medium assuming constant density. Under steady-state conditions, the term on the right side equals zero, because there is no change of hydraulic head with time. Therefore, no storage coefficients have to be defined for steady-state calculations. The mathematical description of density-driven flow leads to nonlinear differential equations, which require special numerical solution techniques. Most of them are described by HOLZBECHER (1998).

Linear partial differential equations can be solved analytically or numerically. Analytical solutions consist of direct integration of the differential equations, for which boundary conditions must be defined. Analytical solutions are used more for problems with spatially independent parameters than for spatially dependent ones.

In numerical solutions, space and time are discretized and the differential quotients of the differential equations are approximated by difference quotients. Important for the categorization of computer programs is the method for the discretization of space. The most common options for spatial discretization of flow problems are

- finite differences, and
- finite elements.

Finite differences (FD)

In the method of finite differences, a rectangular grid is used for discretization (*Fig. 5.4-31*). The term on the left side of the flow equation (5.4.21) can also be written in the following form:

$$\begin{aligned}\nabla(K\nabla h) &= \frac{\partial}{\partial x}\left(K_x \frac{\partial h}{\partial x}\right) + \frac{\partial}{\partial y}\left(K_y \frac{\partial h}{\partial y}\right) + \frac{\partial}{\partial z}\left(K_z \frac{\partial h}{\partial z}\right) \\ &= K_x \frac{\partial^2 h}{\partial x^2} + K_y \frac{\partial^2 h}{\partial y^2} + K_z \frac{\partial^2 h}{\partial z^2} ,\end{aligned} \qquad (5.4.22)$$

where K is the tensor of hydraulic conductivity (with the spatial components K_x, K_y, K_z),

h is the hydraulic potential, and

x, y, z are distances in the x ,y, and z directions.

These differential quotients in this equation can be approximated by difference quotients. For example in the x direction, where the indices represent the number of rows and columns of nodes or block center points, the following equation can be used:

$$\frac{\partial^2 h}{\partial x^2} = \frac{\partial\left(\dfrac{\partial h}{\partial x}\right)}{\partial x} \approx \frac{\dfrac{h_{i+1,j} - h_{i,j}}{\Delta x} - \dfrac{h_{i,j} - h_{i-1,j}}{\Delta x}}{\Delta x} = \frac{h_{i-1,j} - 2h_{i,j} + h_{i+1,j}}{(\Delta x)^2}. \tag{5.4.23}$$

Insertion of these difference quotients in flow equation (5.4.22) and application of this equation to each node in the grid leads to a set of linear equations that can be solved for hydraulic head at each node by direct or iterative linear solvers. Direct solvers, e.g., the Gauss and Cholesky methods, can normally be used for grids up to several hundred nodes. For a larger number of nodes, iterative solvers are preferred because storage requirements become a problem. There is no alternative at present to iterative solvers. For simple models, the simplest methods perform well, e.g., Jacobi, Gauss-Seidel, or the overrelaxation methods. In professional programs, the conjugate gradient method has become very commonly used and leads to good results.

For nonlinear problems, such as unconfined aquifers, density-driven flow, or unsaturated flow, nonlinear solvers, for example, Picard iterations, quasi-Newton approach, the incomplete Newton method, or the Newton-Raphson method, have to be used. For a more detailed discussion, see HOLZBECHER (2002).

Experience shows that FD programs give good results, in principle, but rectangular grids are difficult to adapt to more complicated shapes, e.g., rivers and lakes. Local grid refinement is also difficult in most of the common programs. FD grids normally allow only refinement of entire columns or rows.

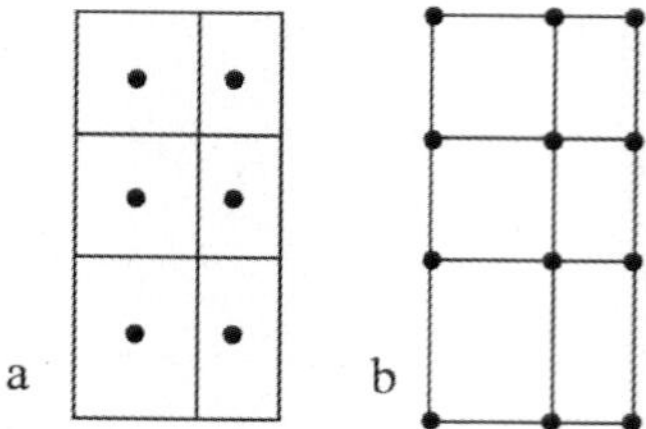

Fig. 5.4-31: (a) Block-centered and (b) node-centered grids

Finite elements (FE)

The concept of finite elements is to subdivide the model region into geometric elements, each of which is approximated by a relatively simple function. The type of function depends on the number of nodes that form the element. For example in 2-dimensions, elements can be triangles and/or quadrangles, etc, each with a different type of function. The functions are usually polynomials. A detailed description of the method is given by ISTOK (1989). Examples of FE grids are shown in *Fig. 5.4-32*. The greatest advantage of FE grids is their ability to reproduce complex shapes, e.g., geological boundaries and rivers. Local grid refinements can also be performed easily.

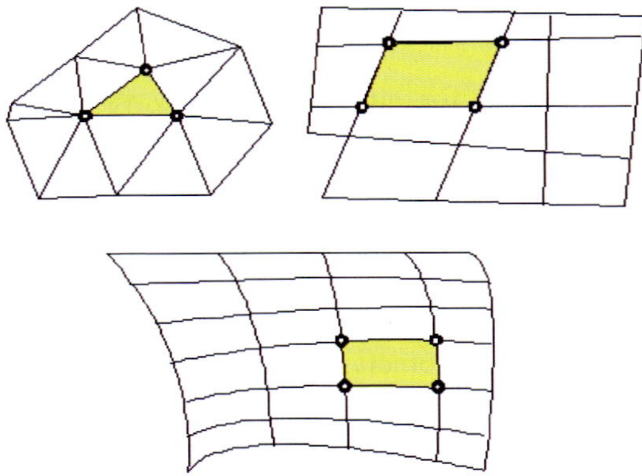

Fig. 5.4-32: Finite element examples, HOLZBECHER (2002)

Flow boundary conditions and initial conditions

During the computer simulation, values for a variable, mostly for piezometric head, are calculated for the modeled area. For both analytical and numerical solutions, flow boundary conditions have to be defined at the model boundaries. Three kinds of boundary conditions are distinguished:

Dirichlet (1st kind) boundary conditions: Defined piezometric head h:

$h = const.$

Examples are the measured hydraulic head in an observation well or the level of surface water with direct hydraulic coupling to groundwater.

Neumann (2nd kind) boundary conditions: Defined flux or velocity v:

$v = const.$

An example is the lateral groundwater flux from a mountain range into the aquifer of a valley.

Cauchy (3rd kind) boundary conditions: Potential driven flux v and taking the leakage factor $k_{f,l}$ into consideration:

$$v = -k_{f,l}\left(\frac{\partial h}{\partial x}\right).$$

An example is surface water with poor hydraulic contact to groundwater.

As boundary conditions have to be defined and are not calculated by the program, they should be chosen at locations where sufficient information is available. Boundaries should already be considered when defining the model shape. The initial values for piezometric head have to be defined for all nodes before the computer program is started.

Particle tracking

Particle tracking is helpful for simulating pathways and traveltimes of a conservative particle, i.e. disregarding diffusion, dispersion and chemical reactions. After the piezometric heads have been calculated for the flow model, one or more particles are started and their pathways and traveltimes are calculated using the velocity field that surrounds the particle. The porosity has to be defined by the user in order to calculate the average linear velocities from the known piezometric heads. Important applications are estimation of the extent needed for a water protection area, initial simulation of contaminant transport and optimization of hydraulic works. An example is given in *Fig. 5.4-33.*

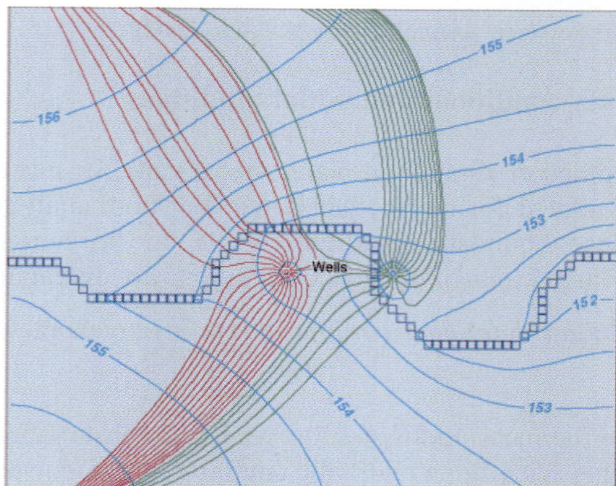

Fig. 5.4-33: Flow paths of particles discharging at two wells. Pathlines were calculated by MODPATH and are denoted by red and green lines. Blue squares are cells representing a river and light blue lines are groundwater-level contours in feet above sea level, computed by MODFLOW, POLLOCK (1994)

Calibration and inverse modeling

To prove the plausibility of a groundwater model, it has to be tested using a measured hydrodynamic reference state. Therefore, model results, normally piezometric heads, are fitted to observed field data by systematic adjustment of parameter values within reasonable magnitudes. Various parameters may be used as input data for this calibration of the model, e.g., recharge, flow boundaries, leakage coefficients, pumping, infiltration, and layer thickness. To reduce the number of parameters that are varied during calibration, the modeler normally uses parameters for which less information is available, e.g., permeability, leakage or storage coefficients, for adjustment while the better known parameters, which are mostly natural recharge and boundary conditions, remain constant. Experience shows that model calibration, especially for complex distributions of the parameters, is often the most time-consuming step of setting up a model.

To reduce the time and effort required for calibration, parameter estimation techniques that automatically adjust calibration parameters systematically to fit model results to field data can be used in many cases. Common programs are PEST (DOHERTY et al., 1994) and UCODE (POETER & HILL, 1998), which are integrated in various groundwater simulators. They are able to "take control" of a model, executing it as many times as needed, adjusting its parameters until the discrepancies between selected model output values of selected model parameters and a complementary set of field or laboratory measurements are reduced to a minimum using the weighted-least-squares method. Depending on the problem, PEST can calibrate models with as many as hundreds of parameters on the basis of thousands of observations.

Flow in fractured-porous media

Fractured subsurface systems have received increasing attention in several geoscientific and geotechnical disciplines, such as hydrology, reservoir engineering, and environmental engineering. Problems of practical interest include groundwater production from karstified aquifers, exploitation of petroleum reservoirs, utilization of geothermal energy (both hydrothermal and hot dry rock reservoirs), safe deposition of waste, and numerous geotechnical applications, for example, tunnel and cavern construction, and stability analysis of manmade structures.

Modeling of flow in fractured-porous media requires special techniques for spatial discretization. The finite element program ROCKFLOW (KRÖHN & LEGE, 1994, ISEB) has been developed to simulate flow and transport of one or more fluid phases and provides special finite element techniques for discretization of fractured-porous media.

Transport of dissolved substances

Analogous to flow Equation (5.4.21), a *differential equation for the transport* of nonreactive particles can be derived from the mass balance of called a control volume:

$$\frac{\partial (cV_{\mathrm{w}})}{\partial t} = -\nabla j_{\mathrm{a}} + \nabla j_{\mathrm{d}} + \sigma V_{\mathrm{G}}, \tag{5.4.24}$$

where v is the vector of the Darcy velocity,
t time,
c solute concentration,
j_{a} advection term,
j_{d} dispersion term
σ external source/sink term,
V_{G} total water volume, and
V_{w} is the effective transported water volume.

The changes in mass within the control volume per time step (*left side*) equals the sum of influx and out-flux of particles due to advection, dispersion and source/sink terms (right side).

The advection term describes the migration of particles due to water movement and depends on the flow velocity, which normally is known if the transport model is based on a flow model. The dispersion term describes the dilution of the particles resulting from the fact that flow velocities in porous media are different at the microscopic scale than at the macroscopic scale. Sources and sinks can, in principle, be local or diffusive. Contaminant spills are typical point sources, whereas diffusive sources could be agricultural areas or precipitation.

For reactive transport, a reaction term has to be added to the equation. This term takes into consideration changes in concentration resulting from chemical reactions. Sometimes simple processes like sorption, decay or degradation are included for one or several species in commercial transport programs. For more complicated reactions, the transport model has to be coupled with a geochemical model. In this case, the transport model calculates concentrations resulting from advection, dispersion and source/sink terms for a time step. On the basis of these concentrations, concentrations resulting from chemical reaction are calculated and returned to the transport model.

Similar to the differential flow equation, the transport equation can be solved either analytically or numerically. For a more detailed description of transport algorithms, see Section 5.4.4 and ZHENG & BENNET (1995).

A large number of simulation programs are multipurpose programs that allow a flow model to be set up as a first step and a transport model, based on the flow model, in a second step. This procedure is permissible only if transport has no influence on flow. This is not the case in density-driven flow, which is described in the following section.

Density-driven flow

Various cases of density-driven flow, e.g., seawater intrusion, saltwater rise, interaction between saltwater lakes and groundwater, or mixing of NAPLs with water, are of practical interest. In all these cases groundwater flow is controlled mainly by large differences in fluid density.

For density-driven flow, the differential flow Equation (5.4.21) has to be written in a general form in terms of pressure because flow can no longer be expressed as a function of piezometric head alone. Densities also have to be considered. Thus, the general form of Darcy's law has to be used to derive the differential equation for density-driven flow, which leads to

$$\frac{\partial(\varphi\rho)}{\partial t} = -\nabla\rho\frac{k}{\mu}(\nabla p - \rho g), \tag{5.4.25}$$

where t is the time,
ρ density,
φ porosity,
k tensor of permeabilities,
μ dynamic viscosity,
g acceleration of gravity, and
p is the pressure.

In this equation, flow influences transport and transport also influences flow. Hence, flow and transport cannot be treated separately – flow first and transport second, as described above. Special computer programs have to be used to manage the problem. Details cannot be given here for all of these cases and their appropriate modeling approaches. Most of them are described by HOLZBECHER (1998).

5.4.3.2 Programs

A large number of modeling programs have been compared and assessed by PANDIT et al. (1993) and VAN DER HEIJDE et al. (1985). Therefore, the comments in this section are focused on the following commonly used programs:

- MODFLOW (HARBAUGH et. al., 2000, USGS WRD),
- PROCESSING MODFLOW (CHIANG et al.,1998, PMWIN),
- VISUAL MODFLOW (Visual MODFLOW Pro),

- GMS (Groundwater Modeling System from EMS-I),
- SUTRA (VOSS, 1984, USGS WRD), and
- FEFLOW (DIERSCH, 1996, WASY GmbH).

Owing to the rapid development of software, it is recommended to obtain detailed and up-to-date information from the corresponding internet web pages of the programs.

MODFLOW is a modular three-dimensional finite-difference groundwater model of the U.S. Geological Survey for calculation of groundwater flow. PROCESSING MODFLOW and VISUAL MODFLOW are graphical user interfaces which couple versions of MODFLOW with various tools for calculating transport (MT3D, MT3DMS, MOC, RT3D), particle tracking (PMPATH), and estimating parameter values (PEST, UCODE). GMS also is a graphical user interface, which allows the coupling of various flow simulators with various transport tools and offers parameter estimation by UCODE and PEST. SUTRA and FEFLOW are multipurpose programs. The large number of programs can be evaluated and compared using the following aspects:

- numerical method implemented,
- types of application,
- practicability and price.

Numerical method

In principle, groundwater models are subdivided into finite difference (FD) and finite element (FE) models. The most obvious difference between these methods is the generation of the grids/meshes (see above). FD grids normally allow only refinement of entire columns or rows and adaptation of the grid to boundary shapes can be done by the defining "inactive" cells (*Fig. 5.4-34*). FE meshes are generally easy to adapt to irregular geographical elements, e.g., geological boundaries, rivers, lakes, etc. An FE mesh can be refined for individual cells or groups of cells, which is often necessary in areas with a high hydraulic gradient, for example in the vicinity of a well (*Fig. 5.4-35*).

MODFLOW is a very popular finite difference program, for which the graphical user interfaces VISUAL MODFLOW and PROCESSING MODFLOW have been developed. The program GMS provides both the FE and FD methods for calculating flow.

Frequently used finite element programs are FEFLOW, SUTRA and ROCKFLOW. In FEFLOW, various options allow efficient generation of complex model grids. Triangular and quadrilateral elements can be used, whereas SUTRA is based on quadrilateral elements only. Both methods provide acceptable results.

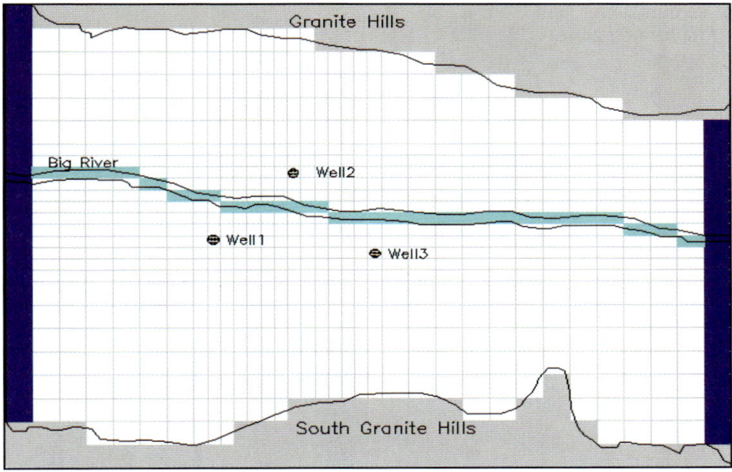

Fig. 5.4-34: FD grid generated by PROCESSING MODFLOW, flow boundaries are indicated by dark blue cells, "inactive" cells are grey

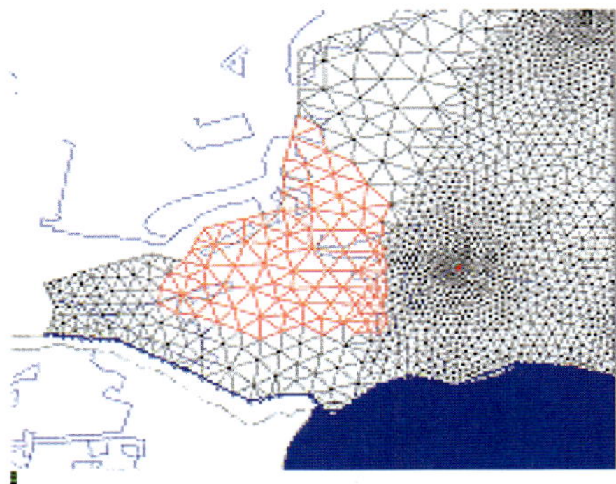

Fig. 5.4-35: Part of a FE mesh generated by FEFLOW

Types of application

Besides the numerical method, the types of applications for which a program can be used are also of great importance. *Figure 5.4-36* gives an overview of the different program types and their frequencies of use.

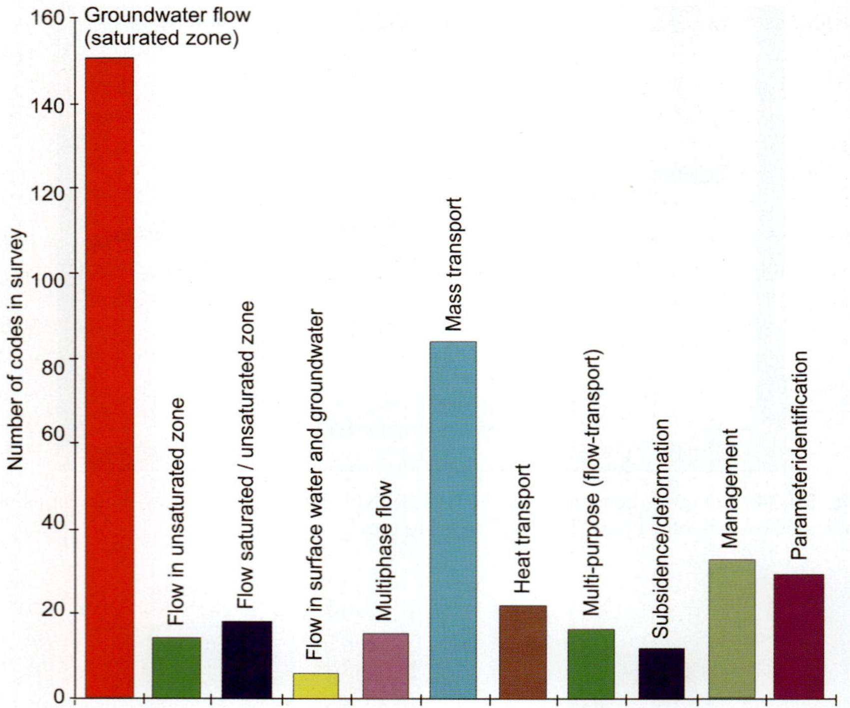

Fig. 5.4-36: Types of groundwater modeling programs and their frequency of utilization, VAN DER HEIJDE et al. (1985)

Most programs are multipurpose. When these programs are used in the two-step procedure mentioned above, flow and transport modeling are usually based on the same spatial and temporal discretization. This has the advantage that both grid construction and time stepping have to be done only once – one task less for the modeler, thus saving time. SUTRA and FEFLOW include modeling of both flow and transport in one program. Some graphical user interfaces, e.g., PROCESSING MODFLOW, VISUAL MODFLOW and GMS, combine one or more transport models with one or more flow models. Saturated flow and transport modeling in confined and unconfined aquifers for steady-state or transient flow, for example, can be done by all of the programs mentioned above.

Most transport modeling programs allow calculations for particle tracking, advection, diffusion, dispersion, linear and nonlinear retardation, and zero-order and first-order processes, e.g., radioactive decay and biochemical degradation. The differences between popular transport simulators are their flexibility for defining chemical reactions and their ability to handle more difficult aspects, such as density-driven flow and heat transport.

Often used transport programs are MT3D, MT3DMS and RT3D. MT3D can model 3-D contaminant transport with simulation of advection, dispersion, sources/sinks, and processes such as radioactive decay, biochemical degradation, and linear and nonlinear sorption. MT3DMS is an expanded form of MT3D for multiple species, RT3D is a modified version of MT3DMS that utilizes alternate chemical reaction modules. Numerous possible reactions are available, including aerobic degradation, BTEX degradation with numerous electron acceptors, sequential anaerobic degradation of PCE/TCE, and combined aerobic/anaerobic degradation of PCE/TCE. For special cases, an option is provided for creating user-defined reactions. GMS, VISUAL MODFLOW and PMWIN (PROCESSING MODFLOW for Windows) contain interfaces for MT3D, MT3DMS and RT3D and, therefore, provide great flexibility with respect to reactive transport modeling. FEFLOW allows multiple species transport simulation, including advection, diffusion, dispersion and both linear and nonlinear retardation. Calculations can be carried out for zero-order and first-order reactions of single species. To simulate chemical reactions of greater complexity, transport programs can be coupled with geochemical simulators such as PHREEQC (PARKHURST, 1995, USGS WRD).

There are large differences between common simulators for special applications such as density-driven flow and unsaturated flow. There is no general numerical approach for density-driven flow. The approach and the appropriate software always depend on the specific case and its boundary conditions. Most of them are described by HOLZBECHER (1998). MODFLOW provides for calculation of density-driven flow with densities held constant, while GMS, SUTRA and FEFLOW also solve problems involving changing densities.

The unsaturated zone can be modeled with FEFLOW, SUTRA and GMS but not with MODFLOW. FEFLOW also provides for simulation of heat transport.

Usability and price

Another criterion for modeling software is its practicability, which can be characterized by the type of interface, handling of input and output data, visualization of model results and organization of the processing steps. Graphical user interfaces are now standard for simulation software and they generally provide a more efficient way for the user to treat a problem. A very important point for efficiency is the organization of data handling, both for preprocessing (input data) as well as for post-processing (output data). Especially when groundwater in large areas is modeled a large amount of data for various time and spatial parameters has to be managed. Geographical information systems (GIS) are efficient instruments for organizing all the data that is needed for a groundwater model. An interface of the simulator software to GIS-based data can be very helpful and time saving. GMS, FEFLOW and SUTRA, for example, offer, in addition to ASCII and SURFER (.GRD) formats, an interface module for ARCVIEW (shape format) and, therefore, provide powerful tools for assigning data to grid elements and for data visualization. PROCESSING MODFLOW and VISUAL MODFLOW allow import and export of ASCII (*x*, *y*, *z*) files and SURFER (.GRD) files, but direct compatibility with GIS data formats is not provided, so the data handling in general is less efficient. The prices of modeling software vary over a wide range and differ from free programs like PROCESSING MODFLOW up to EUR 10 490 for a complete FEFLOW network version license (Dec. 2006 price).

5.4.3.3 Guide for Construction and Use of a Groundwater Model

The general steps required for constructing and use of a groundwater model can be described as follows (see also *Fig. 5.4-37*): The first three steps correspond to the development of the conceptional model (see *Fig. 5.2-31*, Section 5.2.6 and Section 5.4.2). Together with the impact model, numerical modeling yields the process model, i.e. the understanding and description of the processes at the investigated site.

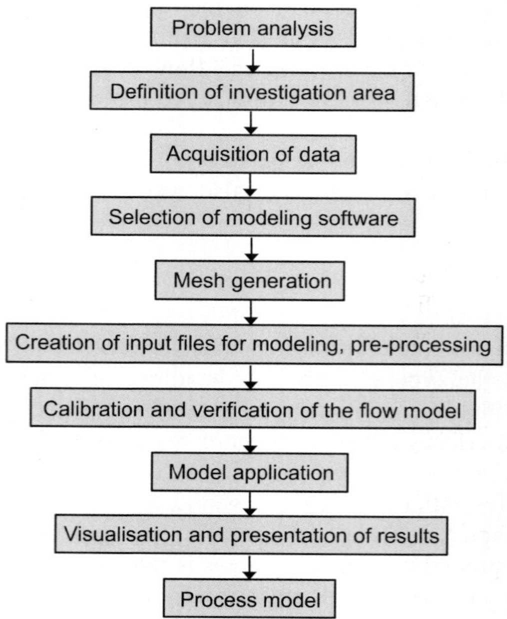

Fig. 5.4-37: Steps for setting up a model

Problem analysis

The first step is a detailed analysis of the problem. Therefore, all available information, both literature and base data have to be taken into consideration. The best modeling approach can be determined on the basis of this information.

Definition of the investigation area

The size of the investigation area depends on the problem to be solved. In general, it should exceed the area covered by the groundwater model that is to be constructed. In principal, the investigation area should provide all the information required for the groundwater model.

Data Acquisition

All available information about aquifer geometry, flow parameters, and boundary conditions, which characterize the hydrogeology and hydrodynamics of the investigation area, is important. Information can sometimes be obtained directly from reports of previous field investigations but often base data like drilling protocols or measured piezometric heads have to be evaluated. If the available information is not sufficient, further field studies need to be done. The most important information is

topographic data: morphology, surface water bodies, built-up areas, land use;
hydrology: natural recharge, infiltration and exfiltration of surface water, flow rates;
geology: lithology, stratigraphy, tectonics;
hydrogeology: hydrogeologic units and their geometries, hydraulic conductivities, storage coefficients, porosities, leakage coefficients;
hydrodynamics: piezometric heads, time series of piezometric heads and maps of isolines at different times, surface water levels, pumping rates, artificial recharge.

The collected data are then interpreted and used to develop the conceptional model (Section 5.4.2).

Selection of modeling software

The modeling software can be selected on the basis of detailed knowledge about the problem and the hydrogeologic setting of the investigation area. The characteristics of the different methods and programs mentioned above can be evaluated to select the optimum tool for the modeling approach. The mathematical operations that have to be done by the program are the most important aspect to be taken into consideration. If transport in the unsaturated zone or density-driven flow with variable densities has to be modeled, there are significantly fewer adequate programs available. A further aspect is the heterogeneity of the hydrogeological conditions. Complicated model shapes can be more easily generated with FE models, and complex data sets of flow parameters are easier to handle with modeling software that has the required GIS interface module(s). In general, these programs are more expensive. Problems with less complex parameter distributions and model shapes can be handled very well by FD models and do not necessarily need coupling with GIS data. Small-scale problems with spatially independent variables can often be solved by analytical approaches.

Grid/Mesh generation

The first step in the generation of a grid/mesh is the definition of the lateral model boundaries. Therefore, the following points have to be taken into consideration:

- Boundary conditions should be independent of the problem. For example, if a pumping well for treating contaminated groundwater (as a pump-and-treat remediation measure) has to be planned, the lateral model boundaries should not be influenced by the simulated pumping.
- Model boundaries should be defined in areas with sufficient information about groundwater flow to permit the definition of realistic boundary conditions. Therefore, boundaries are often orientated with respect to wells or surface water bodies, e.g., rivers and lakes, where water level data is available.
- To obtain better water balance results, the axes of the model should be parallel and vertical to groundwater flow. Model elements that are orientated parallel to the groundwater flow direction do not represent flow boundaries (Neumann type, 2nd kind). Model elements that are orientated vertical to the flow direction can be defined parallel to isolines by Dirichlet type (1st kind) boundaries. Information about groundwater flow can be obtained from isoline maps of piezometric head and should be studied for different hydrologic periods to ascertain whether the direction of groundwater flow is constant with time.

The grid/mesh is generated when the topographic boundary of the model has been defined. In principle, the number of mesh elements should be kept as small as possible to save computing resources. The number should be large enough, however, to answer all questions with satisfactory exactness and numerical stability. The following recommendations are given here for the horizontal discretization of the model:

- To facilitate assignment of flow parameter values to the grid cells, the grid/mesh should account for all important hydrogeologic features such as surface water bodies, position of wells, differences in land use, sharp geologic boundaries, etc.
- No general values can be given for the appropriate size of grid cells. The size always depends on the desired precision in the corresponding zone and its hydraulic conditions. Grid/mesh refinement is generally done in zones with large hydraulic gradients, for example, near pumping wells, and small elements are often used to simulate the shape of surface water bodies, such as rivers and lakes. In peripheral zones of less interest, cell size may increase gradually towards the boundaries of the model. To prevent numerical instabilities large differences between the sizes of neighboring cells should be avoided. In zones where transport has to be simulated, elements should keep small to fulfill the stability criteria.

The structure of the aquifer and its hydrogeologic units are taken into consideration for the vertical discretization. Hydrogeologic units are normally represented by model layers. For most programs the topology data for the upper and lower surfaces of these layers can be in ASCII (x, y, z coordinates) or SURFER (.GRD) format. As vertical flow between two layers is always calculated using the mean value of their hydraulic conductivities, an aquiclude has to be represented by at least two model layers to simulate its very low permeability. Very thick layers should be avoided in order to more precisely calculate the vertical flow component. For this reason, thick hydrogeologic units should be subdivided into several model layers with equal hydraulic parameter values.

Creation of input files for modeling (preprocessing)

Once the horizontal and vertical topology of the model has been defined, flow parameter values and boundary conditions have to be assigned to the grid cells. As the data base normally only provides values for defined points, e.g., boreholes or observation wells, generalization has to be done by any of several methods. This is commonly done by defining polygons representing areas of equal values and applying statistical methods. For generalization, see also remarks in Sections 5.4.1.4 and 6.1.1.

The spatial distribution of natural recharge is, for example, often represented by polygons generated by GIS-based intersection of different parameter layers, e.g., land use, rainfall, etc. For large areas, more than 10 000 polygons can be necessary. In these cases, it is indispensable for the modeling program to have a GIS interface. Hydraulic conductivity distributions can be handled as polygons similar to those for groundwater recharge and are often generated using interpolation techniques or stochastic approaches. The disadvantages of polygon definition are large differences in the hydraulic conductivities of neighboring polygons which lead to irregular pathlines in particle tracking and transport simulation. On the other hand, if the boundaries of the geologic units are well known, their hydraulic conductivities can be assigned easily to the polygon elements representing these geologic units, giving a more realistic distribution of parameter values.

Parametrization is commonly done using geostatistical interpolation techniques such as *kriging* offered by many modeling programs. *Stochastic modeling* is done for hydraulic computer simulations with various, but statistically equivalent distributions of parameter values, which are evaluated on the basis of the measured and calculated hydraulic heads. A popular method of stochastic modeling is the Monte Carlo method. For a detailed description of geostatistical methods and techniques, see WACKERNAGEL (1995).

Calibration and verification of the flow model

When all the required input parameters and boundary conditions have been defined, the flow model has to be optimized on the basis of measured field data and all other available information. Calibration means the fitting of model results to a measured hydrodynamic reference state, which is normally the hydraulic head of the groundwater. Verification of the calibrated model should be based on another hydrodynamic reference state. In principle, both calibration and verification can be done in a steady-state or transient manner. Normally, a steady-state calibration is carried out first. Because more data is required, transient calibrations are less common than steady-state calibrations.

For a steady-state calibration, a reference state has to be documented first, which normally is in the form of a map with contour lines of hydraulic heads and values at the observation wells. Either measured or averaged hydraulic head values can be employed for the reference state used for a steady-state calibration. When measured values are used, a stable hydrodynamic period should be selected which provides a sufficient number of measured hydraulic heads. Occasional lack of data can be improved by using statistical methods, such as covariance analysis of time series from selected observation wells. Sometimes averaged values for hydraulic heads can help to obtain a better data base for calibration. The correct proceeding and technique for data compilation always depends on the specific case. When all the needed data has been collected, a map of hydraulic head isolines has to be constructed. Besides the hydraulic head data, all other available information has to be taken into consideration for the interpolation of the isolines. Often the relationship between surface water and groundwater is known only from chemical data or flow rate analysis. Flux from adjacent mountains into the modeled area is sometimes understood determining other parameter values than hydraulic head. So in principle, the interpolation should be done manually by an experienced person and not by a computer program. Interpolation programs should be used only if all available information is used in addition to the hydraulic heads and coordinates. Popular software packets are SURFER and ARCGRID, which offer various algorithms and options for interpolation (see also Section 6.1.1). The resulting isoline map of flow velocities and flow directions represents the reference state for calibration. Since all further work and all later model-based conclusions depend on this reference state, it should be documented with great exactitude and plausibility.

Transient data, e.g., pumping tests or flood data, is also useful for calibration. The dynamic characteristics of the model can be optimized on the basis of measured changes in hydraulic head at selected locations together with storage coefficients.

Calibration consists of the fitting of model results to the reference state by iteration. In principle, a flow model should be calibrated using the parameters hydraulic head, isoline contours, flow direction, flow velocity, chemical data and water budget. Experience shows that hydraulic heads are often measured with pedantic exactness, but little attention is paid to the water budget. Influx at the model boundaries have to be in agreement with the recharge in the catchment area outside the modeled area; infiltration and outflow rates of surface water bodies should correspond to flow rates of surface water and chemical data, etc.

Depending on the problem, calibration and verification of a model often requires more time than the construction of the model. In some cases, automatic calibration provides fast and good results. The programs FEFLOW, PROCESSING MODFLOW, VISUAL MODFLOW and GMS mentioned above provide interfaces to common tools for inverse modeling, e.g., UCODE or PEST.

After calibration, the flow model should be verified using another reference state with its corresponding boundary conditions. The degree of conformity of the measured and calculated hydraulic heads indicates the quality of the model and its ability to predict conditions for future remediation measures and/or groundwater/surface water exploitation.

Model application

Common applications of groundwater modeling are prediction of flow and transport, determination of protection areas for waterworks, simulation and optimization of hydraulic works, counter measures for contamination, etc. The results and predictions always depend on the exactness of the calibration and verification, and on the choice of boundary conditions.

Visualization and presentation of the results

Analogous to preprocessing, the visualization of model results also involves with spatial data, e.g., hydraulic conductivity distribution, piezometric head, pathline contours, which are normally presented in the form of maps. Especially for problems with complex hydrogeologic structures and parameter distributions, interface modules of the modeling software to GIS are efficient tools for presentation of model results.

5.4.4 Contaminant Transport Modeling

MIKE MÜLLER

Groundwater quality is important for long-term environmental development. Contaminants in groundwater can affect aquifers and their usage for decades or centuries. Transport of these contaminates by the flowing groundwater is a complex process that is determined by several interacting mechanisms. Models are needed to describe and evaluate these transport phenomena. The models can differ considerably with regard to their degree of sophistication. Building on a conceptual model, the mathematical model that is used to quantify the processes may range from back-of-the-envelope calculations to analytical solutions to complex three-dimensional numerical models. Which method or combination of methods is appropriate depends on the task at hand. For instance, analytical solutions can be found relatively quickly but provide only limited answers, because numerous conditions such as homogeneity of the aquifer, a point or linear contaminant source and restricted dimensionality must be fulfilled. For more realistic insight into the problem and especially for predictions of water quality, numerical models are required.

A transport model is always based on a flow model that has to be available before any transport can be investigated. The basics of flow modeling are described in Section 5.4.3.1. As explained there, a flow model that is designed only for modeling the groundwater flow can be very different from a flow model that is to serve as a basis for transport modeling. While a flow-only model may yield good results with a simplified approach, such as two dimensions and spatially averaged hydraulic conductivity, flow models that serve as a basis for transport modeling typically need to incorporate more details. In the flow-only model, the actual pathway of the individual water particle does not matter. Only the pressure head and water balance need to be correct. On the other hand, a flow model that provides flow velocities for transport models needs to model the flow pathways of the groundwater as closely as possible. While different pathways may not change the amount of water, the concentration distribution can be considerably different. Therefore, before using an existing flow model as a basis for transport modeling this model must be evaluated to determine whether it meets all requirements to provide useful groundwater velocity fields.

This section covers only transport of dissolved contaminants. Other contaminations that can undergo transport in groundwater, such as NAPLs, are not addressed. Furthermore, density-dependent flow resulting from high contaminant concentrations is not included. For all models it is assumed that concentrations are low and thus concentration differences have no effect on groundwater flow. The distinction between reactive and nonreactive processes is not very sharp. In the following discussion, "nonreactive" means there is no interaction between different chemical species and thus only source/sink terms in the transport equation are included, while "reactive" includes terms for these interactions, mainly chemical reactions and microbiological processes.

5.4.4.1 Fundamentals of Transport Modeling

Transport equation

The equation that describes the transport of a nonreactive substance can be derived from the mass balance of a control volume. Because of mass conservation, the fluxes through the boundaries of the control volume can change the storage S in the control volume. Other terms that may change the storage parameter value are for sinks and sources. Using effective porosity n_e, which describes the portion of the porous medium that is used for groundwater flow, the equation describing the transport can be formulated as follows:

$$S = \frac{\partial(c n_e)}{\partial t} = -\nabla j_{adv} + \nabla j_{diff} - \nabla j_{disp} + \sigma n_e, \tag{5.4.26}$$

where S is storage in the control volume per unit time,
c fluid phase concentration,
n_e effective porosity,
t time,
j_{adv} advection flux,
j_{diff} diffusive flux,
j_{disp} dispersive flux, and
σ is for source(s)/sink(s).

The fluxes through the boundaries of the control volume are caused by advection, molecular diffusion, and dispersion. These transport mechanisms will be explained in more detail in the following sections. There are several processes, such as chemical and biological reactions, withdrawal from pumping wells or input from contamination sources, that can act as sources or sinks during transport. Some of the most important chemical reactions will be presented below.

Advection

Movement of solute with the average velocity of the pore fluid is called *advective transport*. The advective flux can be expressed as

$$j_{adv} = u n_e c, \tag{5.4.27}$$

where u is the average pore fluid velocity ($u = v/n_e$, calculated from the Darcy velocity v and the effective porosity n_e).

The transported substance travels with the average velocity of the groundwater through the effective pore space. A solute that travels only by advection would result in a sharp concentration peak that has the same magnitude at all observation points along the flow path. Mass added to a "stream tube" will remain only in that stream tube. If the flow regime is steady-state, the mass will spread along the flow paths. Even though advection is often the predominant process, advection-only transport is not observed in the field. Other processes such as molecular diffusion and to a much greater extent, hydrodynamic dispersion overlay advection, leading to a different overall behavior of plume spreading. The term *convection* is sometimes used synonymously with advection to describe water movement due to density or temperature gradients.

Molecular diffusion

Brownian motion is the main cause for molecular diffusion. Other causes such as osmotic forces, thermal diffusion, or electroosmosis are less common in groundwater systems. Brownian motion means that water molecules move randomly on a small-scale along irregular pathways. This causes movement of transported substances toward regions of low concentration, because more dispersed transported substances move from places of high concentration to low concentration than in the other direction. For example, if there are two volume elements with the same volume, one with no transported substances and the other with a concentration c of transported substance, both elements will have a concentration of about $c/2$ after a long period of time relative to the speed of Brownian motion and size of volume element. Using Fick's first law this can be described from a macroscopic point of view:

$$j_{\mathrm{diff}} = -D_{\mathrm{m}} \nabla c \,, \qquad (5.4.28)$$

where D_{m} is the molecular diffusion coefficient.

The diffusive flux is proportional to the concentration gradient. The molecular diffusion coefficient is the proportionality factor. The diffusive flux is independent of direction of groundwater flow and groundwater velocity. Therefore, diffusion can lead to spreading of solute in all directions towards lower concentration zones, which is indicated in the equation by the negative sign before D_{m}. The typical range of diffusion velocities is on the order of decimeters and, therefore, two to three orders of magnitude smaller than typical advection velocities. Due to this and the fact that dispersion in field investigations is usually much greater than diffusion, the diffusive flux is often neglected or incorporated into the dispersive flux.

Dispersion

Leaving the pore scale, where only advection and diffusion are present, different conditions and processes influence transport, which are included in the "dispersion" term. At the local scale, i.e. at the scale of pore channels, heterogeneity of permeability and storage capacity cause four dispersion mechanisms as shown in *Fig. 5.4-38*:

a) The flow velocity varies across the pore channels. The highest velocity occurs in the middle of the pore channels, decreasing towards the fringe where the velocity approaches zero.

b) The cross-sectional areas of the flow channels differ, which leads to different flow velocities in these channels.

c) The flow velocities along different flow paths through the pores that end at the same point are different. Therefore, particles starting at the same location and time, will arrive at the same destination at different times.

d) Splitting and rejoining of flow channels results in transverse propagation of transported substances.

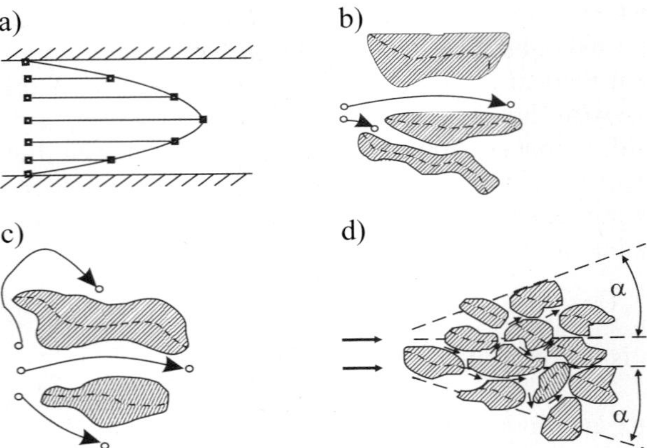

Fig. 5.4-38: Different causes of dispersion (see text for explanation)

Moving up to the macroscale, heterogeneities can be explicitly accounted for in numerical models by choosing appropriate parameter values for all model elements. However, in most cases there is not enough data to allow such a detailed approach. Moreover, analytical solutions, as described in a later section, very often assume homogenous parameter distributions. The effects of macro-scale heterogeneities on transport can be represented by macro dispersion. Macro dispersion can be further divided into small-scale macro dispersion, on the order of 10 m, and large-scale macro dispersion, up to about 1000 m (KINZELBACH & RAUSCH, 1995). Due to dispersion, some of the transported substances travel at velocities greater than the average groundwater velocity, while others travel at less than the average groundwater velocity. The larger the dispersion, the more pronounced the spread of transported substances in all directions along the travel path. Dispersion along the flow path (longitudinal) is different from that perpendicular to the flow path (transversal). Fick's first law can also be applied to dispersion:

$$j_{\mathrm{L}} = n_{\mathrm{e}} D_{\mathrm{L}} \frac{\partial c}{\partial s_{\mathrm{L}}}, \qquad (5.4.29)$$

$$j_{\mathrm{T}} = n_{\mathrm{e}} D_{\mathrm{T}} \frac{\partial c}{\partial s_{\mathrm{T}}}, \qquad (5.4.30)$$

$$j_{\mathrm{disp}} = j_{\mathrm{L}} + j_{\mathrm{T}}, \qquad (5.4.31)$$

where j_{L} is the dispersive flux in flow direction,
j_{T} dispersive flux perpendicular to flow direction,
n_{e} effective porosity,
D_{L} longitudinal dispersion coefficient,
D_{T} transversal dispersion coefficient,
$\frac{\partial c}{\partial s_{\mathrm{L}}}$ concentration gradient in longitudinal direction, and
$\frac{\partial c}{\partial s_{\mathrm{T}}}$ is the concentration gradient in transversal direction.

Dispersion coefficients can be expressed as a tensor. There are four components in two dimensions, and nine components in three dimensions. Since the dispersion is also dependent on the magnitude of groundwater velocities, dispersivities can be derived by

$$D_{\mathrm{L}} = \alpha_{\mathrm{L}} u \qquad (5.4.32)$$

$$D_{\mathrm{T}} = \alpha_{\mathrm{T}} u, \qquad (5.4.33)$$

where α_{L} is the longitudinal dispersivity,
α_{T} transversal dispersivity, and
u is the average pore fluid velocity.

Dispersivity can be described as mixing length. This length can be measured in the field. BEIMS (1983) and GELHAR & AXNESS (1983) compiled numerous values from field experiments that show a systematic relationship between mixing length and dispersivity. HÄFNER et al. (1985) derived the following empirical formula for longitudinal dispersivity:

$$D_{\mathrm{L}} = 0.017\, x, \qquad (5.4.34)$$

where x is distance in the longitudinal direction.

Lateral transversal dispersivity is about one order of magnitude smaller than the longitudinal dispersivity. It is even smaller vertically. Dispersivity values may span orders of magnitude. Care has to be taken in selecting values for field applications. At small scale, tracer experiments can be used to determine dispersivity. At greater scale, environmental tracers such as tritium or atmospheric trace gases can be used.

Dispersion is an important process in transport. A more detailed coverage of this topic can be found in DOMENICO & SCHWARTZ (1990) and BEAR (1988), who gives an extensive mathematical background.

Final transport equation

On the basis of the above discussion of advection, diffusion, and dispersion, we can rewrite Equation (5.4.26). Assuming the effective porosity n_e as constant and combining diffusion with dispersion ($\mathbf{D} = D_m + D$), we obtain

$$\frac{\partial c}{\partial t} = -u\ \nabla c + \nabla\left(\mathbf{D}\nabla c\right) + \sigma - \frac{q}{n_e}\left(c - c_{in}\right), \tag{5.4.35}$$

where c_{in} and c are the input concentrations of the source and sink, respectively.

If groundwater velocity u, dispersion $\mathbf{D}$, external source/sink term σ, external inflow or outflow q with concentration c_{in} are known, the partial differential equation can be solved for c.

Transport of reactive substances

Solutes in groundwater can be altered by different kinds of reactions. If reaction rates are very small compared to the time scale of transport, reaction may be neglected. Solutes such as sodium chloride act as tracers and can be used to quantify fluxes of nonreactive substances. There is a multitude of reactions that can be classified as chemical, biochemical, or physicochemical. Other criteria are reversible vs. irreversible processes, fast vs. slow reactions, and reactions between water/dissolved substances and solid material vs. reactions between dissolved species.

Sorption processes

Sorption (Section 5.3.5.1) is a reversible process between solutes and the rock. Both transported substance and rock material remain chemically unchanged. *Adsorption* is the binding of solute to the interstitial surfaces of the rock. On the other hand, *desorption* is the release of the sorbed substance. There are several types of bonds, such as van der Waals attraction, Coulombic attraction, and chemical bonding. The solid material of the rock is called the ion exchanger and can be characterized by its ion exchange capacity. Depending on whether anions or cations are exchanged, it is commonly referred to as anion exchange capacity (AEC) or cation exchange capacity (CEC) (Section 5.3.5.7).

Dissolution and precipitation processes

Solutes may precipitate and form solids that cover the interstitial surfaces of the rock or formerly precipitated material may dissolve. These processes are usually reversible. Depending on the physicochemical conditions, precipitated material may redissolve. Dissolution processes are typically kinetically controlled processes. Reaction rates vary greatly depending on the species involved. Precipitation rates are often fast compared to the time scale of interest, so once saturation or supersaturation occurs, precipitation can be considered instantaneous.

Redox reactions

An important type of reaction in the groundwater is the redox reaction. In redox reactions, electrons are exchanged between the reaction partners. Reductants are electron donors and oxidants are electron acceptors. An example of a redox reaction is the reaction of iron with oxygen in water:

$O_2 + 4H^+ + 4e^- \rightarrow 2H_2O$ reduction

$4Fe^{2+} \rightarrow 4Fe^{3+} + 4e^-$ oxidation

$O_2 + 4Fe^{2+} + 4H^+ \rightarrow 4Fe^{3+} + 2H_2O$ complete redox reaction.

Redox reactions are kinetically controlled reactions. Reaction rates may be slow compared to time scale of interest. Slow reactions cannot be treated as instantaneous reactions. Different redox reactions occur in a sequence. The order depends on the redox potential, which describes the ability to change to a reduced or oxidized state. The pE parameter expresses the electron activity, which may be described as the relative tendency of a solution to accept or transfer electrons (STUMM & MORGAN, 1996). *Table 5.4-10* shows the sequence of redox reactions. These reactions are mediated by microbes. The redox range for different reactions may have significant overlap. This may be explained by the adaptation of microorganisms to certain environmental conditions. Still some reactions have less energetic yield and are therefore less likely to occur than others. Microbial mediation supports these reactions (LENSING, 1995; SCHÜRING et al., 2000) (Section 5.3.6).

Table 5.4-10: Sequence of microbially mediated redox reactions, after STUMM & MORGAN (1996)

Reaction	Starting at pE
reduction toward lower pE	
O_2 reduction	+14
denitrification	+13
Mn(IV)oxide → Mn(II)	+10
NO_3 reduction	+7
Fe(III)oxide → Fe(II)	+1
reduction of organic material	-1
SO_4^{2-} reduction	-2
CH_4 fermentation	-3
$N_2 \rightarrow NH_4^+$	-4
H_2 formation	-6
oxidation toward higher pE	
oxidation of organic material	-6
sulfide → SO_4^{2-}	-4
Fe(II) oxidation	0
$NH_4^+ \rightarrow NO_3^-$	+6
Mn(II) oxidation	+10
$N_2 \rightarrow NO_3^-$	+12
O_2 formation	+14

Data for the solution of the transport equation

Application of a model to the real world requires a solid data foundation. Unfortunately data collection is time-consuming and expensive. Therefore, in nearly all cases where modeling is done, data are scarce and approximations have to be made for missing data. Moreover, the available data may not all be useable for the project due to inconsistency in the data or lack of tools to evaluate the quality of the data. To alleviate this problem, modeling should not be treated as an end in itself that only consumes previously collected data. Data collecting and modeling should rather be treated as mutually supplementing methods to accomplish a common goal to shed light on the processes at the site. The collected data can be used to set up an initial model, even though educated guesses have to be used for some of model input. Preliminary model results can provide a basis for planning new measurements that in turn will provide more and – even more important – more significant data. This approach is inherently dynamic in nature. There is no final model but only a model at a certain stage that continuously needs to be updated to the current state of affairs. Even though this process seems to be rather intuitive, it is not always carried out that way in practice.

The data needed to set up an adequate transport model depend considerably on the problem at hand. So no detailed guidelines can be given.

The solution of the transport problem is always based on the solution of the flow problem, therefore, everything discussed in Section 5.4.3 applies here also. Adding transport and possibly reactions increases the complexity of the problem considerably.

In order to solve the transport equation, first, the values of the transport parameters are needed, and the initial values and boundary values must be selected.

Parameters

The values of the parameters that appear in the transport equation need to be determined. These parameters include groundwater velocity u, effective porosity n_e, dispersivity in different directions, source-sink terms and concentration of transported substances in the inflowing water. Determining these parameter values can be a major effort. Groundwater velocity is typically taken from the results of flow modeling, which can by itself be an effort comparable to transport modeling (see Section 5.4.3). Other parameter values can be obtained from the literature, laboratory and/or field experiments, as well as boreholes. The main problem is the small sample size compared to the area of interest in the subsurface. Heterogeneities in the field can only be approximated. Therefore, the range of possible parameter values can be rather large. This may result in a high uncertainty of the modeling results, because other combinations of parameter values may fit the observed values as well, but predictions with these different sets of values may yield different results. This nonuniqueness is one of the main difficulties in groundwater flow and transport modeling. Since transport modeling is heavily based on the results of flow modeling, the nonuniqueness is compounded for transport.

Determining source/sink terms can be a very difficult task. Very often the amount and place where contamination took place are only vaguely known. Examination of historical documents, asking contemporary witnesses and drawing conclusions from other soft factors, such as size of production facilities or age of landfill sites, are widespread methods for estimating source-sink terms. The effort necessary for this task can be considerable. Despite these efforts the result is often no more than an educated guess.

Initial conditions

At the beginning of the simulation, initial concentrations must be given for all of the aquifer represented in the model. As with the determination of parameter values, considerable effort may be involved. Even if there is a rather intensive investigation program, measured concentration values are sparsely distributed compared with the area of interest. Interpolation is used to obtain values for areas where there are no measured values. Different interpolation methods may result in different concentrations. Since the initial concentration strongly affects the simulation, this concentration is an important part of modeling. The combined knowledge of experts such as hydrogeologists, chemists, and biologists is needed to obtain a reasonable distribution of initial concentrations.

Boundary conditions

Because a model is an approximation of a part of reality, boundaries need to be defined. Even if in reality there are no boundaries, but rather a continuous medium, with either an aquifer or the atmosphere at the upper boundary, the model boundaries need to be defined. But transport does not stop at these boundaries. Therefore, conditions have to be defined how transport is to continue across these boundaries. Analog to groundwater flow modeling (see Section 5.4.3.1), there are three types of boundary conditions, the Dirichlet, Neumann, and Cauchy boundary conditions. In transport modeling there is an additional boundary condition that is applied only to an outflow boundary. It is called the transmission boundary condition or Shamir boundary condition (SHAMIR & HARLEMAN, 1967).

Dirichlet boundary condition (1st kind)

The concentration is fixed at a Dirichlet boundary. The concentration may change over time as specified by the model user. This boundary condition can be used to represent the background concentration at a boundary with inflowing water. If the substance of interest is not in the inflowing water, a concentration of zero would be used. If there is little advection, this boundary condition should not be used. Special care has to be taken when this boundary condition is used if a source of contamination is present. In reality, the source concentration might change due to exhaustion of the source. This boundary condition will keep the source concentration at the same value independent of the amount of contaminant leaving the source. Therefore, this would provide an indefinite amount of contaminant. A modification of the Dirichlet boundary condition that would make it dynamic would enable changes in the concentration depending on a specified amount of the substance at the source and the amount that has been transported away.

Neumann boundary condition (2nd kind)

Diffusive and dispersive flux at the boundary are specified at a Neumann boundary. Although there is no advection across the boundary, diffusive and dispersive fluxes may occur. Diffusive and dispersive fluxes are zero if the boundary is a natural boundary of the aquifer or there is no concentration gradient across this boundary. In many models the dispersive and diffusive fluxes are incorrectly assumed to be zero. Therefore, a plume should not approach the boundary unless it is a natural boundary. If there is significant advection across the boundary, this type of boundary should not be used.

Cauchy boundary condition (3rd kind)

This boundary condition is a linear combination of the two boundary conditions described above. Both advection and diffusion/dispersion are considered in this boundary condition. This is especially useful for inflow boundaries because typically all these processes occur there.

Shamir boundary condition (transmission)

If there is advection across an outflow boundary, the diffusive and dispersive fluxes across this boundary are not known because the concentration gradient cannot be calculated due to the lack of concentration values outside the model. Extrapolation of concentration across the boundary solves this problem.

Analytical methods

In simple cases and under particular boundary conditions it is possible to employ analytical formulas. Analytical methods are sufficient when the geometry of the model area is simple, when parameters are constant, and special boundary conditions can be assumed. Analytical methods can be used, for example, to treat the following cases:

- Case 1: Maximum expected substance concentration in a pumping well.
- Case 2: What is the travel time to a pumping well?
- Case 3: As case 2, but taking dispersion into account.
- Case 4: As case 1, but taking decomposition into account.
- Case 5: Range of a specific substance concentration in an aquifer.

The last example marks the limits of a pocket calculator. Sixteen programs for a programmable pocket calculator that analytically solve one-, two- and three-dimensional transport problems are presented in LUCKNER et al. (1987). Numerical modeling methods are used for more complex underground conditions and complicated boundary conditions.

Use of analytical solutions has decreased considerably in the last decade because of readily available numerical programs with graphical user interfaces that facilitate the setting up and operating of numerical models. Numerical models can solve a much wider range of problems than analytical methods. However, analytical solutions are frequently used to test whether newly developed numerical programs work correctly. Compilations of analytical solutions can be found in LUCKNER & SCHESTAKOW (1991), HÄFNER et al. (1992), and FETTER (1992).

Numerical methods

Over the last 10 to 20 years numerical methods have become the method of choice to solve transport problems with models. Unlike analytical solutions, numerical methods can be applied to complex problems involving a high degree of heterogeneity, complicated geometries, and a variety of technical interventions, such as pumping, drainage or injection of water and solutes. Powerful desktop computers provide the CPU power to solve sizeable problems within an acceptable time frame. Graphical user interfaces greatly simplify and expedite preprocessing and post-processing tasks. However, this makes the use of numerical models available to users who do not have a strong mathematical or numerical background. But a basic knowledge of the underlying numerical methods is required to apply the models correctly. There is a variety of methods in use that have different advantages and disadvantages. Depending on the specific problem at hand, the application of some methods may be appropriate while the use of others may lead to incorrect results.

Time discretization

In order to solve the transport Equation (5.4.35) numerically, it has to be discretized in time. There are three methods: the explicit Euler method, the implicit Euler method, and the Crank-Nicolson method. The explicit Euler method uses the transport equation for the current time step to calculate the change in concentration for the next time step. This yields an algebraic equation that can be solved directly. Therefore, this method can be easily implemented and is computationally effective. Unfortunately this method is not stable. If the time steps are too large, the method starts to oscillate with growing amplitudes resulting in unreasonable, physically implausible model outputs. Small time steps lead to a long simulation time. Criteria for optimizing the time step in combination with the space discretization will be explained below.

The implicit Euler method uses the transport equation for the time step after the current one. This yields a system of equations that needs to be solved simultaneously. CPU and memory requirements are considerably higher than for the explicit method. On the other hand, the time steps can be much larger without resulting in unreasonable model output. Thus, the runtime of the model may be less than with the explicit method.

The Crank-Nicolson method combines the explicit and implicit Euler methods. This leads to a fairly stable solution in which oscillations do occur but do not grow with increasing size of the time steps. One main advantage is the behavior of the discretization error. While the error that is caused by discretization grows linearly with the increase in the grid size for the explicit and implicit Euler methods, it grows only to the square root of the grid size for the Crank-Nicolson method.

Even though the explicit time discretization scheme is rarely used for flow models, it is used in several transport models, including the widely used MT3D (ZHENG, 1990). There are several reasons for this, one being that an explicit time discretization can be used in mixed Eulerian-Lagrangian space discretization (see below). Moreover, the additional accuracy required for transport simulations puts constraints on the size of the time step even when implicit or Crank-Nicolson schemes are used. Finally, an explicit solution requires less computer memory than the other schemes. Considering the fast development of computer technology, this is of practical importance only for very large models.

Space discretization

Discretization of space can be achieved by different methods. Generally, there are two main types of numerical solutions for transport in groundwater, the Eulerian and the Lagrangian approaches. The Eulerian method is a grid-based method in which the problem is viewed from a fixed point in space. The groundwater with its dissolved substances flows by the viewpoint. The Lagrangian method is a particle-based approach in which the problem is viewed from the moving particle. This moving particle can be thought of as part of the groundwater that is moving through the aquifer. Both types of methods have several advantages and disadvantages. Moreover, both types can be implemented with different methods. By combining these two methods into a mixed Eulerian-Lagrangian approach, the advantages of both methods can be utilized.

Eulerian methods

The three Eulerian methods – finite difference, finite volume, and finite elements – which are also used for groundwater flow modeling, are the most known methods. A method called total variation diminishing (ZHENG & BENNETT, 2002) has also been recently applied. All three of these methods have advantages and disadvantages. Describing these methods in detail is beyond the scope of this section. There are several textbooks that cover this field in great detail (ZHENG & BENNETT, 2002; RAUSCH et al., 2002; LUCKNER & SCHESTAKOW, 1992; HÄFNER et al. 1992).

Finite difference method

The finite-difference method discretizes space in rectangular elements with nodes at the corners or at the center of these elements. A partial differential equation is written for each node. Section 5.4.3 explains the use of this method for flow modeling. The same principles apply to transport modeling. As in flow modeling, there are node-centered grids or block-centered grids and differential quotients are replaced by difference quotients.

When the finite difference method is used, numerical instabilities may occur that can considerably distort the model results or lead to no solutions at all. These instabilities include oscillation and numerical dispersion. If numerical dispersion is present, the model output shows large dispersivities, even though the dispersion in the model is small. This occurs because the transport equation has a hyperbolic part for advection and a parabolic part for dispersion. Since all Eulerian methods are mainly concerned with solving the parabolic part, the hyperbolic part leads to effects that is commonly called numerical dispersion. The *Peclet criterion* or *Peclet number* describes the ratio between advection and dispersion. Applied to one-dimensional space discretization, it can be expressed as

$$Pe = \frac{\Delta x}{\alpha_{\mathrm{L}}} \leq 2, \tag{5.4.36}$$

where Δx is the grid spacing, and
α_{L} is the longitudinal dispersivity.

If this condition is satisfied, it can be assumed that numerical dispersion is small compared to physical dispersion.

Several criteria must be met to guarantee numerical stability of the method. All of them restrict the size of discretization in time and/or space. Small discretization steps will prevent numerical instabilities. On the other hand, this might increase computation times beyond reasonable limits. To find the coarsest discretization that still prevents numerical instabilities, several criteria can be used. These criteria are only applicable for time discretization in the explicit Euler method. Nevertheless, the other time discretization methods also have limits on the time and space discretization if sufficient accuracy is to be achieved. The following criteria are presented for the one-dimensional case for simplicity reasons. For higher dimensions, see ZHENG & BENNETT (2002). The principle is the same as in the one-dimensional case.

- The *Courant criterion* or *Courant number Cr* can be used as a measure for whether oscillation will occur. It addresses the advection part in the transport equation. It requires that no particle is able to travel more than one cell size within one time step.

$$Cr = \left|\frac{\Delta t\, u}{\Delta x}\right| \leq 1, \tag{5.4.37}$$

where Δt is the time step,
u groundwater velocity, and
Δx is the grid spacing.

If this criterion is fulfilled, no oscillation and no physically impossible concentrations are to be expected.

- The *Neuman criterion* states that the concentration gradient cannot be reversed by dispersion alone.

$$\frac{2D}{\Delta x^2} \leq 1, \qquad (5.4.38)$$

where D is the dispersion coefficient, and
Δx is the grid spacing.

- The *Courant-Neuman criterion* is the combination of the two previous criteria, defining a criterion that covers both advective and dispersive transport.

- The *well criterion* requires that at a withdrawal node the maximum mass that can be extracted per time step must be smaller than or equal to the mass at this node at the beginning of the time step:

$$\frac{\Delta c}{c} \leq 1, \qquad (5.4.39)$$

where c is the concentration at beginning of time step and
Δc is the change of concentration caused by extraction.

Finite volume method (FV)

The finite volume method is also known as control volume method. Unlike the finite difference method that approximates the partial differential equation at the nodes, the finite volume method applies the approximation to the volume surrounding the node. This means that the differential equations are integrated. The balance over all flow into and out of the control volume is calculated. This method allows a much more flexible space discretization using not only rectangular but also triangular cells of different sizes. The method is numerically more complex than those commonly used in finite difference approaches.

Finite element method

The finite element method as described in Section 5.4.3 for groundwater flow can also be used for transport modeling. Finite element discretization approximates the solution of the partial differential equation by a linear equation for each element. Unlike the finite volume method, which integrates over each control volume, the finite element method integrates over the whole model. Detailed descriptions of this method can be found in SUN (1996) or HUYAKORN & PINDER (1983).

Total variation diminishing methods

Total variation diminishing methods (TVD) are higher-order finite difference or finite volume methods that are mass conservative. These methods are characterized by a decrease in the sum of concentration differences of adjacent cells over successive time steps. This helps to avoid oscillations during calculations. These methods, which use higher order differential equations, help prevent numerical dispersion. Typically, a TVD method uses a flux limiter, to be able to represent sharp concentration fronts. The main limitations of TVD methods are that they are much more computational intensive and more difficult to implement. Numerical dispersion is still somewhat larger than with Lagrangian approaches.

Lagrangian methods

In contrast to Eulerian methods, which view the water and dissolved substances as they flow past a particular standpoint, Lagrangian approaches tackle the transport problem from the viewpoint of the moving particles. Lagrangian methods do not solve the transport equation directly but rather approximate it using a large number of particles. These methods do not exhibit numerical dispersion and can solve advection-only problems.

Particle tracking method

If only advective transport is to be considered, the pathlines of dissolved substances coincide with the pathlines of groundwater flow. Applying different interpolation algorithms for different kinds of flow models, i.e. finite difference or finite volume flow models, groundwater flow path lines can be delineated. Along these lines particles can be tracked that represent the concentration of a contaminant. This tracking can be achieved by semi-analytical or numerical methods. Every particle has a fixed mass and, therefore, a large number of particles are necessary to account for large differences in concentrations.

By laying a grid over the modeled space and counting the number of particles in each cell, the concentration in these cells can be calculated from the number of particles times the mass per particle. At sources and sinks, particles have to be generated and destroyed, respectively. Boundaries that are nonpermeable "reflect" the particles back into the modeled space. In cells with a prescribed concentration, particles have to be added or withdrawn to keep the concentration at the defined level. Bookkeeping of cell locations and the numbers of particles is a large part of the implementation of the method.

Random-walk method

Based on the particle-tracking method, the random-walk method applies a method from statistical physics which approximates diffusion phenomena to account for dispersion. The theoretical background can be traced back to 1940 (CHANDASEKHAR, 1943). Dispersion and diffusion are represented by random movement of the particles, hence the name "random walk". This random walk is superimposed on advective movement modeled by particle tracking as described in the previous section. The statistical properties of this random movement coincide with the properties of dispersive/diffusive transport. By using a large number of particles a reasonable concentration distribution can be achieved.

Mixed Eulerian-Lagrangian methods

Eulerian methods are best suited for dispersion-dominated problems, while Lagrangian methods are better applied to advection-dominated problems. To overcome the disadvantages and to use the advantages of both methods, the mixed Eulerian-Lagrangian approach was developed. Advection is simulated by a Lagrangian method while dispersion is accounted for by an Eulerian approach. Depending on the method used for Lagrangian advective transport, the following methods can be distinguished:

- method of characteristics (forward tracking),
- modified method of characteristics (backward tracking),
- hybrid method of characteristics (combination of the two above), and
- Eulerian-Lagrangian localized adjoint method.

Method of characteristics

The method of characteristics (MOC) calculates the advective term using the particle tracking method (GARDER et al., 1964). Particles are distributed regularly or randomly in the model cells. Unlike the random-walk approach, particles do not represent mass but rather are assigned the concentration of the cell they are present in. The moving particles may be described by either a uniform or a dynamic approach. In the uniform approach, the number of particles is the same in all cells of the model. In contrast, the dynamic approach assigns particles only to cells where the concentration is greater than a specified value. The number of particles per cell depends on the concentration in this cell. Particles are added or removed from the model according to the changes in concentration.

The dynamic approach is especially useful for cases where the plume is present in only a small portion of the modeling domain. Since this is quite common, use of the dynamic approach can reduce the number of particles considerably. The dynamic approach results in a significant increase in the efficiency of the method compared to the uniform approach.

Each particle is assigned spatial coordinates and a concentration. The particle-tracking method tracks the particles forward in time. The concentration of each cell is calculated from the concentrations of the particles in this cell. The advective part of the transport can be accounted for with this algorithm. The dispersive part is calculated using an Eulerian method. The average concentration, which can be calculated from the concentration before and after the advection step, is used to calculate the change in concentration due to dispersion. This can be achieved by applying a weighting factor, typically between 0.5 and 1, to this newly calculated concentration and to the old concentration:

$$c_{\text{new}}\left(x,t+\Delta t\right)=\omega c_{\text{av}}\left(x,t+\Delta t\right)+\left(1-\omega\right)c\left(x,t\right), \tag{5.4.40}$$

where $c_{new}(x, t+\Delta t)$ is the concentration at new time step,
$c_{av}(x, t+\Delta t)$ average concentration of the pre- and post-advection steps,
$c(x,t)$ initial concentration, and
ω is the weighting factor in the interval between 0.5 and 1.

The MOC is nearly free of numerical dispersion. The computational requirements in terms of memory and bookkeeping are high. Moreover, the local mass balance is not guaranteed for every time step.

Modified method of characteristics

The modified method of characteristics (MMOC) was developed in order to overcome some of the disadvantages of the MOC (DOUGLAS & RUSSEL, 1982). The MMOC employs backward tracking in time. One particle is placed at a nodal point of the grid for each new time step. The position at the previous time step is found by tracking the particle backwards. The advective part of the transport can be calculated with this procedure. The dispersive term is calculated with the same algorithm as in the MOC.

Since there is only one particle per nodal point as opposed to several for the MOC, the MMOC is computationally more efficient. Also less memory is needed because much less bookkeeping about positions of particles is necessary. For sharp concentration fronts numerical dispersion may be significant. Application of higher order interpolation schemes can solve this problem but at the cost of greater computational effort.

Hybrid method of characteristics

The hybrid method of characteristics (HMOC) is a combination of the MOC and the MMOC methods (ZHENG, 1990). The MOC is used in areas with sharp concentration fronts, while the HMOC is used in areas with smooth concentration gradients. The switching between the two methods is done automatically by the simulation software according to a specified criterion. Again, advantages of two methods are combined while trying to avoid their disadvantages.

Eulerian-Lagrangian localized adjoint method

The Eulerian-Lagrangian localized adjoint method (ELLAM) was developed in the late 1980s (RUSSEL, 1990, CELIA et al., 1990) and implemented in the MODFLOW package by HEBERTON et al. (2000). ELLAM uses an implicit method for the dispersion calculation, allowing for large time steps without stability constraints. Again, the advection problem is solved by particle tracking, followed by solution of the dispersion problem using a fixed grid. This method solves integral equations, leading to local and global mass conservation. Unlike the other MOC approaches, mass is tracked as fluid volumes, which provides the foundation for mass conservation at all times and locations.

Selection of transport method

Numerous methods were presented in the previous section. All of them have certain advantages and disadvantages. The advantages and disadvantages of the different methods are given in *Table 5.4-11* after RAUSCH et al. (2002). Guidelines for their selection can be derived from these advantages and disadvantages (*Table 5.4-12*).

Since different solution methods are implemented by modern simulation software, e.g., MT3D (ZHENG, 1990), several methods for a problem can be applied with relatively little effort. Comparison of the results of different methods yields a deeper insight into the problem at hand and the fitness of the methods used.

Table 5.4-11: Comparison of numerical methods for solute transport, after RAUSCH et al. (2002)

Method	Advantages	Disadvantages
explicit time discretization	- small memory requirement - simple implementation	- small time step constraint by Courant criterion
implicit time discretization	- large time steps possible - therefore less computational effort	- very large systems of equations for a large number of nodes
finite difference method	- easily comprehensible method - mass conservative	- confined by a fixed grid - limited local refinement - difficult adaptive refinement - numerical dispersion and oscillation
finite volume method	- very flexible spatial discretization - local refinement and adaptive refinement possible - locally and globally mass conservative	- more effort than with the finite difference method - numerical dispersion or oscillation
finite element method	- very flexible spatial discretization - local refinement and adaptive refinement possible	- more effort than with the finite difference method - numerical dispersion or oscillation - only globally mass conservative
TVD	- much less numerical dispersion than with other Eulerian methods - no oscillation	- computationally demanding - more numerical dispersion than Lagrangian methods
random walk	- no numerical dispersion - mass conservative	- particle movement partially decoupled from water movement, which may lead to unreasonable concentrations, especially at the fringe of plumes
MOC	- no numerical dispersion - large grid Peclet number and therefore large grid spacing possible	- no mass conservation - memory requirements and bookkeeping may be enormous
MMOC	- much less memory needed than for MOC - large grid Peclet numbers	- numerical dispersion possible - no mass conservation
HMOC	- less memory usage than MOC - large grid Peclet number	- no mass conservation - more complicated to implement
ELLAM	- globally and locally mass conservative - large time steps - no oscillation	- difficult to implement for higher dimensions

Table 5.4-12: Guidelines for method selection for some model features, after RAUSCH et al. (2002)

Model features	**Recommended method**
high spatial concentration gradient (e.g., plume with a point source, laboratory experiments with a discontinuous source)	FD, FE, FV with higher order interpolation methods, no "upwind" weighting; MOC, random walk, TVD, ELLAM
low concentration gradient from a dispersed source (e.g., non-point nitrate contamination by agriculture)	FD, FE, FV, TVD, ELLAM
coupling of transport with nonlinear reaction terms (no oscillation at all, exact mass balance)	FD, FV, FE with "upwind" weighting or "upwind" stabilization
large time step required (e.g., long distance transport for long periods of time)	FD, FV, FE with implicit time discretization, TVD, MMOC, random walk, ELLAM
advection-dominated transport with large grid Peclet numbers (small dispersivities or large grid spacing)	MOC, MMOC, random walk, ELLAM

Numerical solution of reactive transport

Coupling of transport and reactions can be done with any of several methods, such as explicit and implicit coupling. CIRPKA (1997) gives an overview of the available methods and suggests that the explicit method (also called "operator splitting") is the most suitable method, especially for large problems with several ten thousand nodes. The operator splitting method solves the transport model first. Then the resulting solute concentrations are applied to the reactions that take place. Since the model error is unknown, the time step needs to be small. The only way to estimate the model error is to use different time steps for the same simulation and compare the results. If there is no significant deviation between the two time steps, a larger one can be used for this problem. Reactive transport modeling is still used more extensively in research projects than in applied studies. This is not only due to the considerably increased computation time, but also due to the immense increase in complexity that places high demands on staff and time, which is currently typically only available in academia.

5.4.4.2 Model Application

Even though all numerical methods seem to be complex, application of the model to the real situation is a much greater challenge due to the high degree of uncertainty encountered in nature. The selection of an appropriate model is crucial for modeling success. This section will briefly introduce some models and the appropriate fields of application. Calibrating a model is an important task in modeling. Therefore, some guidelines for model calibration are given as well.

Available modeling programs

There are many modeling programs available which are more or less capable of simulating solute transport in groundwater. Many of them are research programs with certain restrictions for users other than the model developers. Besides providing the correct numerical solution of the transport problem using one or several of the above methods, some features of the modeling programs are of considerable practical importance. The following criteria seem to be crucial if a model is to be applicable:

- model must in be actively maintained,
- there must be comprehensive documentation,
- some kind of preprocessor and/or postprocessor is necessary, and
- public domain or open source code is preferable.

Only modeling programs that are kept up to date should be considered for modeling. This means one or several model developers should be continually assigned to work on the model. This work should include bug fixing as well as improvements in the model, such as new algorithms or data import and export facilities.

Documentation is crucial for successful application of a model. Only if all relevant information is available in a comprehensive written form does the modeler have a realistic chance to master the model. Mastery is a prerequisite for dealing with all aspects of modeling, such as inadequate data or communication with other scientists or authorities.

The trend is towards bigger and more complex models. To efficiently work with these large models that include a large amount of data, effective preprocessing software has to be available. This software can ideally provide a tool that frees the modeler from most of the tedious work. This means the modeler is enabled to work towards the solution of the problem rather than on administrating the data.

Since programming a model for numerical transport in groundwater is a complex task, the resulting modeling programs are complex. Naturally, the software will contain some errors as all software does to a certain extent. If the source code is public domain or open source it can be scrutinized by peers. This leads to improvement of the code and information exchange between developers. Especially for universities that would like to improve, extend, or couple a model, availability of the source code is often invaluable.

Only a few models meet the above criteria. Because of the popularity of MODFLOW (MCDONALD & HARBAUGH, 1988), transport models that are compatible with this flow model are probably the most widely used. This group of models includes the various versions of MT3D (ZHENG, 1990, ZHENG, 1996, ZHENG & WANG, 1998), RT3D (CLEMENT, 1997), and MOC3D (KONIKOW et al., 1996). All of these models have thorough documentation. They are maintained by the developers. Generally, the source code is available to the public. The choice of pre- and post-processors for MODFLOW is large. Currently there are at least six programs (VISUAL MODFLOW, GMS, PMWIN, GROUNDWATER VISTAS, ARGUS ONE, and CADSHELL – an internet search will retrieve the current version of each program). These pre- and post-processors can, for the most part, also be used to process input and output data for the mentioned transport models. With these models only those problems can be solved for which MODFLOW can provide a suitable solution of the flow problem. For some problems MODFLOW cannot produce reliable results. This includes rising groundwater where water enters previously dry model cells or local grid refinements. Two other models will be briefly presented that can solve problems the current version of the MODFLOW-based models are not capable of yet.

The finite element FEFLOW (DIERSCH, 1991; DIERSCH, 2002) is a rather sophisticated three-dimensional modeling program for saturated and unsaturated groundwater flow, solute and heat transport, as well as density-driven flow. A mature graphical user interface facilitates work with large data sets and model outputs. This model is not as widely used as MODFLOW for several reasons, including the rather high retail price, the much higher hardware requirements, and the closed source code.

Another model that is also not as widely used is PCGEOFIM (SAMES & BOY, 1999). This model solves the groundwater flow equations using the finite volume method. It offers a variety of modified boundary conditions that can be used to incorporate elements such as the influence of mining. Also the interaction of aquifer(s) with lakes and rivers can be represented in a useful manner by the model. The transport can be modeled with the random-walk method or with a finite volume method employing a flux limiter that suppresses numerical dispersion. Even though this model is used primarily for mining areas it is general enough to be useful for other applications as well. The user interface is only partially graphical. The use of database files for input allows the manipulation of input with a few SQL-like statements. It is closed source, which does not allow for implementation of the user's own modifications.

Model calibration

In nearly all cases, models do not produce satisfying results with the initial input parameter values because measured values can never reflect reality in all details. The adjustment of input parameters according to the outcome of a comparison of measured and calculated values is commonly called *calibration*. A related task is *verification* or *validation*. ZHENG & BENNETT

(2002) define verification as the process in which the model shows good correspondence between calculated values and measured values not used for calibration. Validation can establish that the model can produce reasonable future predictions.

Calibration of a transport model is a complicated task. Since a variety of processes are involved, frequently no single reason for an effect can be found. This nonuniqueness is compounded if reactions are involved in the transport process. In order to calibrate a transport model, the expertise of a team is necessary that may include geologists, chemists, biologists, and people experienced in numerical analysis. Again, because of the uniqueness of each site no general strategy can be recommended that can be applied to all cases. Usually a combination of manual, automatic, and semi-automatic techniques with different weighting for different sites seems most promising.

A *trial and error method* that involves changing parameter values and reexecuting the model on the basis of the previous results is the conceptually simplest method. This method incorporates the knowledge and intuition of the modeling team at every iteration. This can be of great value for the solution of the problem. On the other hand, this method might lead to solutions that correspond to the expected results. Unconventional scenarios might not be considered at all. Also the number of scenarios is rather small because data handling has to be done manually even though a graphical user interface is used.

Automatic procedures such as gradient search methods may lead to physically unreasonable results because the numerical algorithms commonly cannot take into account all the conditions that influence the problem like a human team could. *Semi-automatic procedures* can be a good compromise. A large number of scenarios can be tried in automatically executed scenarios that are set up by customized programming. This programming usually involves only a very few lines of computer code that vary parameter values within a predefined range with a specified step size. Model results also have to be evaluated automatically, for instance by ranking deviations of measured and calculated results. Even though calculations are done automatically, humans define useful parameter ranges that avoid scenarios which are not possible.

Model calibration is a complex topic and often controversially discussed in the literature. As found by ZIMMERMANN et al. (1998) by comparison of seven approaches for model calibration, the expertise and experience of the modeler with the applied methods is at least as important as the choice of method itself. This leads to the conclusion that sophisticated algorithms and their implementation in complex software packages requires even more knowledge from the modeler, rather than making him obsolete.

Usability of transport models for risk assessment

There is a general argument on whether modeling is to be used in risk assessment. Since the problem at hand is usually rather complex and knowledge of the parameter values is rarely plentiful, the argument against modeling is that a model cannot cope with the situation and results are rather dubious. While it is true that problems are complex and the available data is sparse, modeling still can help to improve the understanding of the situation. The inherent nature of the complex problems can be rarely comprehended by human analysis alone. Models can help to uncover the results of nonlinear cause-effect chains. The stress has to be on the use of adequate models. Neither no modeling at all, nor unconditional belief in modeling results are viable options. Models have a lot to offer to solve problems. Used as a tool that extends human abilities, they can be of great value. On the other hand, if used wrongly, the damage can be enormous.

References and further reading

ANDERSON, M. P. & WOESSNER, W. W. (1992): Applied Groundwater Modeling – Simulation of Flow and Advective Transport. Academic Press, San Diego, CA.

BEAR, J. (1988): Dynamics of Fluids in Porous Media. Dover, New York.

BEIMS, U. (1983): Planung, Durchführung und Auswertung von Gütepumpversuchen, Geohydraulische Erkundung. Zs. f. angew. Geologie, **29**, 10, 484-492.

BREDEHOEFT, J. (2005): The conceptualization model problem – surprise. Hydrogeology Journal, **13**, 1, 37-46.

CASTANY, G. & MARGAT, J. (1977): Dictionaire francais d'hydrogeologie. BRGM, Orleans, France.

CELIA, M. A., RUSSELL, T. F., HERRERA, I. & EWING, R. E. (1990): An Eulerian-Lagrangian localized adjoint method for the advective-diffussion equation. Water Resources, **13**, 4, 187-206.

CHANDRASEKHAR, S. (1943): Stochastic problems in physics and astronomy. Rev. Mod. Phys., **15**, 1, 1-89.

CHIANG, C. L. (2003): Statistical methods of analysis. World Scientific Publishing.

CHIANG, W.-H., KINZELBACH, W. & RAUSCH, R. (1998): Aquifer Simulation Model for Windows – Groundwater flow and transport modeling, an integrated program. Borntraeger, Berlin.

CIRPKA, O. (1997): Numerical simulation of reductive dechlorination of tetrachloroethene in a sandbox model. In: XXVII IAHR Congress, San Francisco.

CLEMENT, T. P. (1997): RT3D, A Modular Computer Code for Simulating Reactive Multispecies Transport in 3-Dimensional Groundwater Aquifers. Pacific Northwest National Laboratory, Richland, WA.

DIERSCH, H.-J. G. (1991): FEFLOW – An interactive, graphics-based finite-element simulation system for modeling groundwater contamination processes, User's Manual, Version 3.0, February 1991. WASY GmbH, Berlin.

DIERSCH, H.-J. G (1996): FEFLOW – Physikalische Modellgrundlagen. WASY GmbH, Berlin.

DIERSCH, H.-J. G. (2002): FEFLOW – A Finite Element Subsurface Flow and Transport Simulation System, Reference Manual. WASY GmbH, Berlin.

DIN 38414-4 (S4): Deutsche Einheitsverfahren zur Wasser-, Abwasser- und Schlammuntersuchung; Schlamm und Sedimente (Gruppe S); Bestimmung der Eluierbarkeit mit Wasser (S 4). German standard methods for the examination of water, waste water and sludge; sludge and sediments (group S); determination of leachability by water (S 4)

DOHERTY, J., BREBBER, L. & WHYTE, P. (1994): PEST – Model-independent parameter estimation, User's manual. Watermark Computing Australia.

DOMENICO, P. A. & SCHWARTZ, F. W. (1990): Physical and Chemical Hydrogeology. Wiley & Sons, New York.

DORN, M. & TANTIWANIT, W. (2002): New methods for the delineation of geological barrier rocks for waste disposal sites in northern Thailand. Episodes **25** (4), 240-247.

DOUGLAS JR. J. & RUSSEL, T. F. (1982): Numerical methods for convection-dominated diffusion problems based on combining the method of characteristics with finite element or finite difference procedures. Society for Industrial and Applied Mathematics, Journal on Numerical Analysis, **19**, 871-885.

DVKW (1990): Methodensammlung zur Auswertung und Darstellung von Grundwasserbeschaffenheitsdaten (Collection of methods for evaluation and presentation of groundwater quality data). DVKW Schriften, **89**. Deutscher Verband für Wasserwirtschaft und Kulturbau (German Association for Water Management and Land Improvement), Bonn.

EMS-i: Environmental modeling systems Inc., South Jordan, UT. http://www.ems-i.com/

EVERITT, B. S. (1998): The Cambridge Dictionary of Statistics. Cambridge University Press, Cambridge.

FETTER, C. W. (1992): Contaminant Hydrogeology. Macmillan, New York.

FH-DGG (1999): Hydrogeologische Modelle – Ein Leitfaden für Auftraggeber, Ingenieurbüros und Fachbehörden (Hydrogeological models – a guideline for clients, engineering consultants and agencies). Hydrogeologische Beiträge der Fachsektion Hydrogeologie in der Schriftenreihe der Deutschen Geologischen Gesellschaft **10** (Hydrogeological contributions of the Hydrogeology Section, in Publications of the German Geological Society 10).

FREEZE R. A. & CHERRY J. A. (1979): Groundwater. Prentice Hall, Upper Saddle River, New Jersey.

GARDER, A. O., PEACEMAN, D. W. & POSSI, A. L. J. (1964): Numerical calculation of multidimensional miscible displacement by the method of characteristics. Society of Petroleum Engineers Journal, **6**, 2, 175-182.

GELHAR, L. W. & AXNESS, C. L. (1983): Three-dimensional stochastic analysis of macrodispersion in aquifers. Water Resources Research, **19**, 1, 161-180.

Geophysik GGD (2002): SCHUCK, A. & SEIDEL, K.: Geophysical survey for the exploration of a waste disposal site in Nong Harn /Thailand. Report Thai-German Project WADIS. (WADIS-project funded by BMBF/Germany, FKZ 0261218).

GRISSEMANN, C., RADSCHINSKI, J., JENN, F., LIETZ, J., VOIGT, H.-J., KNÖDEL, K., LANGE, G., SEIDEL, K., SCHUCK, A. (2006): Practice of Waste Disposal Site Investigations in Thailand - Three Case Studies. Report of the Thai-German Research Project "Investigation of abandoned landfills and proposed areas for new waste disposal sites in Thailand (WADIS)", funded by the Ministry of Education and Research (BMBF) of the Federal Republic of Germany (FKZ 0261218), Bangkok, Berlin, Hannover, Cottbus.

HÄFNER, F., SAMES, D. & VOIGT, H. D. (1992): Wärme- und Stofftransport – Mathematische Methoden. Springer, Berlin.

HÄFNER, F., VOIGT, H. D., BAMBERG, H. F. & LAUTERBACH, H. (1985): Geohydrodynamische Erkundung von Erdöl-, Erdgas- und Grundwasserlagerstätten. Wissenschaftlich-Technischer Informationsdienst des Zentralen Geologischen Instituts, **26**, 1-232.

HARBAUGH, A. W., BANTA, E. R., HILL, M. C. & MCDONALD, M. G. (2000): MODFLOW-2000, the U.S. Geological Survey modular ground-water model – User guide to modularization concepts and the Ground-Water Flow Process. U.S. Geological Survey Open-File Report 00-92.

HEBERTON, C. L., RUSSEL, T. F., KONIKOW, L. F. & HORNBERGER, G. Z. (2000): A Three-Dimensional Finite-Volume Eulerian-Lagrangian Localized Adjoint Method (ELLAM) for Solute-Transport Modeling. U.S. Geological Survey Open-File Report 00-4087.

HÖLTING, B. (1996): Hydrogeologie (Hydrogeology). 5th edn. Enke, Stuttgart.

HOLZBECHER, E. (1998): Modeling density-driven flow in porous media. Springer, Berlin.

HOLZBECHER, E. (2002): Groundwater Modeling – Computer Simulation of Groundwater Flow and Pollution. FiatLux Electronic Book Series, Fremont, CA.

HUYAKORN, P. S. & PINDER, G. F. (1983): Computational Methods in Subsurface Flow. Academic Press, San Diego, CA.

ISEB: Institut für Strömungsmechanik und Elektronisches Rechnen im Bauwesen, Arbeitsgruppen Geohydrologie und Geotechnik (Institute for Fluid Dynamics and Computer Applications in Civil Engineering), University of Hannover. http://www.hydromech.uni-hannover.de/Projekte/Grundwasser/software_pack.html

ISO 2854 (1994): Statistical interpretation of data. Techniques of estimation and tests relating to means and variances, International Organization for Standardization, Geneva.

ISO 2602 (1980): Statistical interpretation of test results; estimation of the mean; confidence interval, International Organization for Standardization, Geneva.

ISO 3301 (1975): Statistical interpretation of data; Comparison of two means in the case of paired observations, International Organization for Standardization, Geneva.

ISTOK, J. (1989): Groundwater Modeling by the Finite Element Method. American Geophysical Union, Water Resources Monograph **13**.

KINZELBACH, W. & RAUSCH, R. (1995): Grundwassermodellierung – Eine Einführung mit Übungen. Borntraeger, Berlin.

KÖNIG, C. (1994): Operator Split for Three Dimensional Mass Transport Equation. Computational Methods in Water Resources **X**. Heidelberg, London.

KONIKOW, L. F., GOODE, D. J. & HORNBERGER, G. Z. (1996): A Three-Dimensional Method-of-Characteristics Solute-Transport Model (MOC3D). U.S. Geological Survey Open-File Report 96-4267.

KRÖHN, K. P. & LEGE, T. (1994): ROCKFLOW Teil 2: TM2: Transportmodell für nichtkompressible Fluide, Theorie und Benutzeranleitung. Bericht des Inst. Für Strömungsmechanik und Elektronisches Rechnen im Bauwesen (ISEB). Universität Hannover

LENSING, H. J. (1995): Numerische Modellierung mikrobieller Abbauprozesse im Grundwasser. Ph.D. Thesis, University of Karlsruhe, Germany.

LUCKNER, L., ECKHARTD, A. & MANSEL, H. (1987): Handbuch zur Lösung von Migrationsproblemen im Boden und Grundwasserbereich mit dem Taschenrechner SHARP PC-1401. Water Section of the Technical University of Dresden.

LUCKNER, L. & SCHESTAKOW, W. M. (1991): Migration processes in the soil and groundwater zone. Lewis, Boca Raton, FL.

MAIER, D. & GROHMANN, A. (1977): Bestimmung der Ionenstärke natürlicher Wässer aus deren elektrischer Leitfähigkeit. Zeitschrift für Wasser- und Abwasser-Forschung, **10**, 1, 9-12.

MANHENKE, V., REUTER, E., LIMBERG, A., LÜCKSTÄDT, M., NOMMENSEN, B., SCHLIMM, W., TAUGS, R. & VOIGT, H.-J. (2001): Hydrostratigraphische Einheiten des norddeutschen Lockergesteinsgebietes (Hydrostratigraphic units of the northern German unconsolidated sediments region). Materialien der 68. Tagung der AG Nordwestdeutscher Geologen in Bremerhaven, 5.–8.6.2001 (Proceedings of the 68th conference of the Working Group of Northwestern German Geologists, Bremerhaven, 5 - 8 June 2001).

MARGANE, A & TATONG, T. (1999a): Aspects of the hydrogeology of the Chiang Mai-Lamphun basin, Thailand that are important for groundwater management. Zs. f. angew. Geologie, **45**, 4, 188-197.

McDonald, M. G. & Harbaugh, A. W. (1988): MODFLOW, A modular three-dimensional finite difference ground-water flow model. U.S. Geological Survey Open-File Report 83-875, Chapter A1.

Nolan, B. T. (1999): Nitrate behavior in ground waters of the southeastern USA. Journal of Environmental Quality, **28**, 5.

Oreskes, N. & Belitz, K. (2001): Philosophical issues in model assessment. Chapter 3, In: Anderson, M. G. & Bates, P. D. (Eds). Model validation, Perspectives in hydrological sciences. Wiley & Sons, New York, 23-41.

Pandit, A., Panigrahi, B. K., Peyton, L. & Sayed, S. M. (1993): Strengths and Limitations of Commonly Used Ground Water Models. Advances in Hydroscience and Engineering, Vol. 1, Proceedings of the International Conference on Hydroscience and Engineering, Washington, D.C., June 7 - 11, 1993.

Parkhurst D. L. (1995): PHREEQC – a computer program for speciation, reaction-path, advective transport, and inverse geochemical calculations. U.S. Geological Survey, Water-Resources Investigations Report 95-4227.

Peters, J. G. (1972): Description and comparison of selected models for hydrologic analysis of ground-water flow, St. Joseph River Basin, IN. U.S. Geological Survey Water-Resources Investigations Report 86-4199.

PMWIN: http://www.pmwin.net/

Poeter E. P. & Hill, M. C. (1998): Documentation of UCODE, a computer code for universal inverse modeling. U.S. Geological Survey Water-Resources Investigations Report 98-4080.

Pollock, D. W. (1994): User's guide for MODPATH/MODPATH-PLOT, version 3: a particle tracking post-processing package for MODFLOW, the U.S. Geological Survey finite-difference ground-water flow model. U.S. Geological Survey Open-File Report 94-464.

Rausch, R., Schäfer, W. & Wagner, C. (2002): Einführung in die Transportmodellierung im Grundwasser. Borntraeger, Berlin.

Rossum, J. R. (1975): Checking the accuracy of water analyses through the use of conductivity. Journal of the American Water Works Association, **67**, 4, 204-205.

Russel, T. F. (1990): Eulerian-Lagrangian Localized Adjoint Methods for Advection-Dominated Problems. In: Griffiths, D. F. & Watson, G. A. (editors): Numerical Analysis 1989, Proceedings of the 13th Dundee Conference, June 1989. Pitman Research Notes in Mathematics Series 228. Longman Scientific & Technical, Harlow.

Sames, D. & Boy, S. (1999): PCGEOFIM – Anwenderdokumentation. Ingenieurbüro für Grundwasser GmbH, Leipzig, Germany.

Schüring, J., Schulz, H. D., Fischer, W. R., Böttcher, J. & Duijnisveld, W. H. M. (2000): Redox – Fundamentals, Processes and Applications. Springer, Berlin.

Shamir, U. Y. & Harleman, D. R. F. (1967): Numerical solutions for dispersion in porous media. Water Resources Research, **2**, 557-581.

STUMM, W. & MORGAN, J. J. (1996): Aquatic Chemistry – Chemical Equilibria and Rates in Natural Waters. 3rd ed., Wiley & Sons, New York.

SUN, N.-Z. (1996): Mathematical Modeling of Groundwater Pollution. Springer, Berlin.

SURFER, Golden Software Inc., Golden, CO.

SWAN, A. R. H. & SANDILANDS, M. (1995): Introduction to Geological Data Analysis. Blackwell Science Ltd., Oxford.

TANTIWANIT, W. & DORN, M. (1999): The Site Searching Process for Waste Disposal Sites in the Chiang Mai - Lamphun Basin. Delineation and Presentation of the Geological Barrier. In: KHANTAPRAB, CH. (Ed.): Proceedings of the Symposium on Mineral, Energy and Water Resources of Thailand: Towards the year 2000. Bangkok, 386-396

TATONG, T. (1997): Country report about solid waste management and night soil treatment in Thailand, Course No.: J-97-00111, Environmental Geology Section, Geological Survey Division, DMR, Thailand.

USACE (1999): Engineering and Design – Groundwater Hydrology. U.S. Army Corps of Engineers, Publication Number EM 1110-2-1421. February 1999. Also available online:
http://www.usace.army.mil/inet/usace-docs/eng-manuals/em1110-2-1421/toc.htm

USGS-WRD: US Geological Survey, Water Resources Division.
http://water.usgs.gov/nrp/gwsoftware/

VAN DER HEIJDE, P., BACHMAT, Y., BREDEHOEFT, J., ANDREWS, B., HOLTZ, D. & SEBASTIAN, S. (1985): Groundwater management: the use of numerical models, American Geophysical Union, Water Resources Monograph **5**. Washington, D.C..

Visual MODFLOW Pro: A Division of Scientific Software Group. Sandy, UT. http://www.visual-modflow.com/

VOSS, C. I. (1984): SUTRA: A FE simulation model for saturated-unsaturated, Fluid-Density-Dependant Groundwater Flow with Energy Transport or Chemically-Reactive Single-Species Solute Transport. U.S. Geological Survey, Water Resources Investigations Report 84-4369.

WACKERNAGEL, H. (1995): Multivariate Geostatistics. Springer, Berlin.

WASY GmbH: Institute for Water Resources Planning and Systems Research. Berlin-Bohnsdorf. http://www.wasy.de/

WATTANANIKORN, K., BESHIR, J. A. & NOCHAIWONG, A. (1995): Gravity Interpretation of Chiang Mai Basin, Northern Thailand: concentrating on the Ban Siep Area. Journal of SE Asian Earth Sciences, **12**, 53-6

WEBSTER, R. & OLIVER, M. A. (2001): Geostatistics for Environmental Scientists. Wiley & Sons, Chichester.

WENDLAND, F., HANNAPPEL, S. KUNKEL, R., SCHENK, R., VOIGT, H.-J. & WOLTER, R. (2005): A procedure to define natural groundwater conditions of groundwater bodies in Germany. Water Science and Technology, **51**, 3-4, 249-257.

ZHENG, C. (1990): MT3D: A Modular Three-Dimensional Transport Model for Simulation of Advection, Dispersion and Chemical Reactions of Contaminants in Groundwater Systems. Report to the U.S. Environmental Protection Agency, Ada, OK.

ZHENG, C. (1996): MT3D Version DOD 1.5, a modular three-dimensional transport model. Hydrogeology Group, University of Alabama, Tuscaloosa, AL.

ZHENG, C. & BENNETT, G. D. (1995): Applied contaminant transport modeling. Van Nostrand Reinhold, New York.

ZHENG, C. & BENNETT, G. D. (2002): Applied Contaminant Transport Modeling. 2nd edn. Wiley & Sons, New York.

ZHENG, C. & Wang, P. P. (1998): MT3DMS, A modular three-dimensional multispecies transport model for simulation of advection, dispersion and chemical reactions of contaminants in groundwater systems. Documentation and user's guide. Department of Geology and Mathematics, University of Alabama, Tuscaloosa, AL.

ZIMMERMAN, D.A., DE MARSILY, G., GOTOWAY, C. A., MARIETTA, M. G., AXNESS, C. L., BEAUHEIM, R. L., BRAS, R. L., CARRERA, J., DAGAN, G., DAVIES, P. B., GALLEGOS, D. P., GALLI, A., GOMEZ-HERNANDEZ, J., GRINDROD, P., GUTJAHR, A. L., KITANIDIS, P. K., LAVENUE, A. M., MCLAUGHLIN, D., NEUMAN, S. P., RAMARAO, B. S., RAVENNE C. & RUBIN Y. (1998): A comparison of seven geostatistically based inverse approaches to estimate transmissivities for modeling advective transport by groundwater flow. Water Resources Research, **34**, 6, 1373-1414.

6 Integration of Investigation Results

THEKLA ABEL, MANFRED BIRKE, ANTJE BOHN, KLAUS KNÖDEL, GERHARD LANGE, ALEJANDRA TEJEDO, MARKUS TOLOCZYKI & UGUR YARAMANCI

The main objective of a site investigation is to develop a model of the site. A model is a (simplified) representation of the reality that cannot be observed directly or that is difficult to observe directly. A comprehensive model of a site contains information about the structure (geometry), the materials present (minerals, rocks, chemical substances), as well as the past and ongoing processes. Models can be confined to aspects relevant to an assessment of the situation or to solve a problem. Elementary models map only one or a few aspects of the reality (e.g., 1-D model derived from dc resistivity sounding, thickness of unconsolidated rocks derived from boreholes, groundwater table derived from measurements in observation wells). Elementary models can be integrated to a comprehensive model.

There are conceptual, descriptive, physical and mathematical models. Examples: The "geological barrier" is a conceptual model. A geological map is a descriptive model of an area. The description of the propagation of seismic waves in a bedded half-space is a physical model. The representation of the subsurface by a sequence of coplanar strata and the representation of an intercalation in a rock massive by a sphere in a homogeneous half-space are mathematical models.

Models can be derived from data of various investigation methods by several procedures: interpretation of remote sensing images and geological and hydrogeological observations, processing and interpretation of results from geophysical measurements, interpretation of geochemical results, etc. Mathematical models are the most frequently used in the processing and interpretation of geoscientific data. There are models of different dimensionality depending on the number of coordinates (geometric variables) (one-dimensional model, two-dimensional model, etc.).

Mathematical models assuming the distribution of a physical parameter (e.g., seismic longitudinal wave velocity) are used to calculate the propagation of seismic waves using the seismic wave equation (forward modeling). The results of the model calculations can then be compared with the measured values (e.g., seismograms). Automated derivation of a model from measured data is called inversion. Beginning with an "initial guess" (starting model), one or more model parameters are changed iteratively – normally with a computer program – until the calculated data corresponds within certain limits to the measured data. The program modifies the model parameters using an algorithm. Individual parameters can be set to fixed values using a-priori information. This can improve the convergence behavior and reliability of the models.

Because of the lack of uniqueness in geoscientific interpretation the coincidence between observations and results derived from the model does not "prove" that the model represents the actual situation. Integration of the results obtained with independent investigation methods can improve the reliability of "site models". A useful tool for integrating investigation results is a geographic information system (GIS). In this part of the book, methods for the development of "site models" by integration of investigation results and examples of this are discussed.

6.1 Data Fusion

KLAUS KNÖDEL, MARKUS TOLOCZYKI, ANTJE BOHN, THEKLA ABEL, GERHARD LANGE & ALEJANDRA TEJEDO

Data fusion is a term recently offered for the use of information from various sources for interpretation. The data may be independent or redundant. The data can be obtained using different methods (remote sensing, geophysics, hydrogeology, etc.), different sensors (e.g., spectral scanners, magnetometers, groundwater leveling tools) at the same time or at different times, as well as a-priori information (e.g., state-of-the-art knowledge, information from outcrops and boreholes, control points). The suitable combination of investigation methods can substantially increase the profit of information in comparison with added up profit of information from single methods. The terms "data fusion" and "information fusion" can be used synonymously. Instead of fusion the terms "combination", "integration", "assimilation", "merging", "synergy", and 'interaction' are used. There are different levels of data fusion in use, depending on the degree of knowledge extraction and representation. In pixel-level fusion, the data (e.g., images) from different sensors are combined at the pixel level. In attribute-level fusion, the features and objects are extracted from different representations (e.g., images, maps, geophysical and/or geochemical results) and combined. Decision-level fusion is done, for example, with conclusions or actions. In all cases an improvement in the accuracy of the interpretation is intended.

usually represent square areas of the ground, but other shapes may be used. The resolution of a raster data set is its cell width on the Earth's surface.

An attribute is the characteristic or property of an object, such as weight, size, color, or any detail that serves to qualify, identify, classify, quantify, or express the state of an object. These characteristics can be quantitative and/or qualitative in nature. Attribute data are linked to spatial data that define the location. In vector data, attributes are assigned to the objects; in raster data, attributes are assigned to cells (pixels).

There are advantages and disadvantages in using raster or vector data to represent the real world. Raster data sets contain values for all points of an area while vector data has values only where needed. Therefore, vector data usually require less memory than data in a raster format. Overlay operations can be carried out easier with raster data than with vector data. Raster data are represented as images. If raster data are of low spatial resolution (pixel size is too large in a given scale), object boundaries appear in the form of a step function. Vector data can be displayed as vector graphics on conventional maps.

6.1.2.2 Hardware, Networks, Software, and Manpower

Owing to the complexity of a GIS (*Fig. 6.1-2*), specialists are needed for different aspects of its development and maintenance: on the one hand, to design the database, and on the other hand, to handle the spatial and attribute input of the data. The input of coordinate data and the selection of appropriate cartographic representation for the visualization of the data require cartographic training. Good system management and regular system maintenance are prerequisite for the long-term functioning of an information system. Specialists are needed for the system management, the scientific processing, development of the database, programming of the GIS tools and control mechanisms, and for data input and presentation. Normally, at least one scientist and one IT assistant are needed. The task of the scientist is not only to develop the GIS system and its database, but also to function as mediator between the GIS users and the data sources and interpreters.

The most expensive investment for a GIS system is not the hardware or software, but the training of the personnel, the developed methods, and the data. The data can be collected by the GIS team themselves, in exchange with others, or bought. The cost of a GIS is not a one-time expense, the database must be continually updated, which must be taken into consideration during budget planning.

The more complex the software is, the more intensively the personnel who are to use it must be trained. The setup of an information system requires profound knowledge of the structuring and programming of GIS applications. Training appropriate for the tasks to be carried out, as well as money for the necessary software upgrades, must be included in the budget. The IT personnel should have a basic understanding of the data, so that they will be able to recommend possibilities for utilization.

The decisions on hardware and software should take into consideration the medium-term and long-term development of the GIS. Seldom, only a single GIS program should be used. Instead, different applications of the data will require different software. The different software selected must be able to use the same database without data conversion.

Is the GIS to be used only for digital production of information? Is the data to be processed and analyzed? The answers to these questions may require large investments. For this reason, the decision on the system to be used should be made well-founded. Especially the long-term use of the system should be carefully considered. The hardware must be appropriate for the software. Some tasks require high-performance hardware; for other tasks standard PCs are sufficient. The trained personnel, the continuing training measures, the acquisition of data, the programming of applications, and making the data available to the user can be very expensive, depending on the size of the information system.

CAD programs and graphics software are more economical than GIS systems for the production of maps and graphics. Data can be presented with such software within a very short time. However, these systems are not capable of searching a database for georeferenced data. Overlays of spatial and attribute data are also not possible. Therefore, in the long-term the use of GIS software is superior.

GIS software is offered in versions from low-end to high-tech. Before any software is bought, the following aspects are an example of what should be taken into consideration:

- What is the long-term objective?
- Who will be using the data?
- What data for which purpose must be available?
- In what form must the data be made available?
- Which methods will be used for data processing?
- Where will the data come from?
- How and how often the data will be updated?

Only when all of these questions have been answered can planning of the information system start.

- Good digitizing tools must be available if you are going to collect your own data.
- Import functions are important if data from other people is to be used.
- Which data formats will be needed?
- What requirements must the database fulfill?
- How are the data to be visualized: as map on the screen or as a high-quality printed map?

Software is available from several sources, both as commercial products and free open-source software. Inexpensive software for the visualization and limited processing of spatial and attribute data is available. More demanding tasks – depending on the task at hand and the working conditions – require, however, the use of high-quality software. The cost of such high-performance software is correspondingly higher. For a relatively large working group, a "floating license" to be adapted to the task at hand is recommended. Such a license allows the use of a "light" version for simple tasks and a "professional" version for more complicated ones. The database is a fundamental part of a GIS system and thus the choice of the database management system, its structure and development must be considered right at the beginning. In addition to the license for the GIS software, a license is necessary for the software needed if the data is to be made available in the internet or distribution on CD.

The most commonly used software is listed in *Table 6.1-2*. It is not meant to be comprehensive.

Normally, a software license must be renewed annually. This must be included in the annual budget. It usually amounts to about 18 % of the initial cost.

The number of appropriately equipped computers should be sufficient for the work to be done. The computers should be linked in a local area network (LAN) adequate to handle the large amounts of data[5]. Large amounts of data and highly sophisticated, complicated software require high-performance computers. Like for the software, maintenance and updating of the hardware must be included in the budget.

[5] Between 0.3 and 3 GB of data is needed for even a medium-sized GIS project to make geoscientific data available for regional planning, depending on the size and number of satellite images and digital elevation models (DEM) used.

Table 6.1-2: Selected GIS software

Software	**Company or institution**	**Web adress**
ArcGIS 9.x (sold in three license versions with upscaling functionality: ArcView, ArcEditor, & ArcInfo)	ESRI, Redlands, CA (USA)	http://www.esri.com
ArcView 3.x older version of GIS software, still in use	ESRI, Redlands, CA (USA)	http://www.esri.com
ArcIMS (for servers)	ESRI, Redlands, CA (USA)	http://www.esri.com
ArcPad	ESRI, Redlands, CA (USA)	http://www.esri.com
MapInfo	MapInfo Corporate Headquarters, Troy, New York (USA)	http://www.mapinfo.co.uk
IDRISI	Clark Labs, Clark University, Worcester, MA (USA)	http://www.clarklabs.org/Home.asp
ERDAS Imagine	Leica Geosystems Geospatial Imaging, LLC, Norcross, GA (USA)	http://gis.leica-geosystems.com
PCI	PCI Geomatics, Richmond Hill, Ontario, Canada, L4B	http://www.pcigeomatics.com
TNT Mips	NPA Group, Edenbridge, Kent, (UK)	http://www.npagroup.com/gisit/mipsinfo.htm
FreeGIS	FreeGIS Project	http://freegis.org
GRASS	(freeware/open source)	http://grass.itc.it
MapServer 4.6.1	(University of Minnesota, MN, USA)	http://mapserver.gis.umn.edu

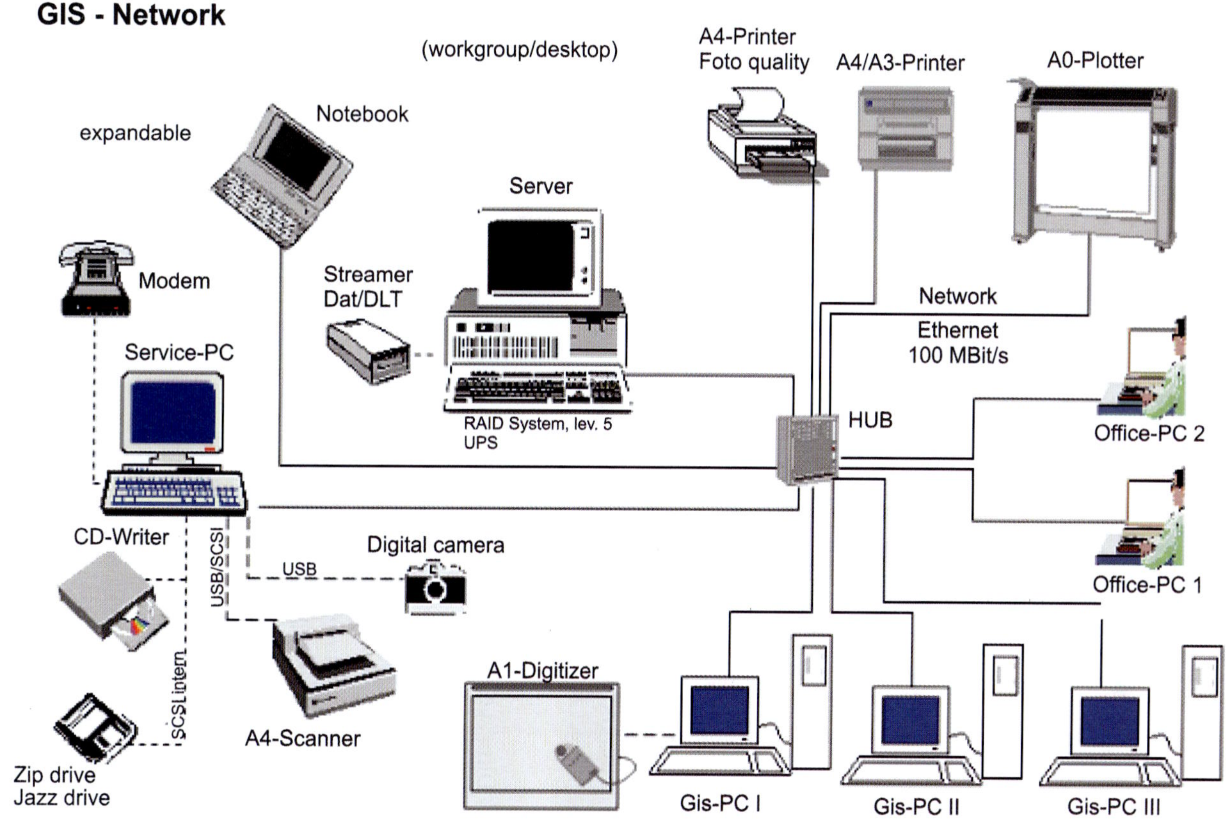

Fig. 6.1-6: Local area network for a GIS project in Ghana

Well equipped, commercially available PCs are adequate for GIS work. There must be sufficient hard-disk storage capacity with fast retrieval times for the large amounts of data. The graphics card must be fast and support high screen resolution. Memory (RAM) should be larger than required by the software producer, since only the minimum requirements are usually given in the specifications, which are not sufficient for a moderately large GIS project.

Because work will normally be done at more than one computer, the computers should be linked in a LAN. It is recommended to store the data on a central server with file sharing. All data for a GIS should be stored in a main directory, even if this means several files, such as for logos and some graphics, are stored in more than one place. Owing to the large amounts of data, data transfer must be fast and efficient (rates of at least 100 kbit/s) in all parts of the network (LAN card, cables, switches, etc.). A backup program must backup the data at regular intervals. Periphery hardware should include printers and scanners for both large- and small-size paper (e.g., A0 and A3/A4). The plotter for printing maps should have adequate memory capacity. Depending on the location, an independent power supply is recommended. The emergency power supply should be sufficient to properly store the data in memory, and shut down the system in the case of a power failure. Access to the internet is also needed.

The example of a GIS network shown in *Fig. 6.1-6* was for an urban geology project in Ghana.

6.1.2.3 Data Acquisition and Analysis

Some of the data can be bought or acquired in exchange with other projects. Data acquired by the project itself must be vectorized after scanning of the printed information or entered manually into the database. The budget must include cost for acquiring data from other parties and personnel cost for input the project's own data into the database.

A significant aspect of the storing of GIS data is that the form in which it is stored must be compatible with the data of other GIS users. The use of georeferenced data, as well as the use of the commonly used topographic reference system are basic requirements for a GIS project. A topographic reference system unique to the project should never be used unless it is absolutely necessary, and only after discussion with other users. Data stored using such a reference system cannot be exchanged and cannot be used for further work without great effort and expense.

There are three main methods for entering data into the database:

- vectorization
  - by digitizing analog maps on a digitizing table, or
  - from scanned maps or satellite images interactively on the screen;
- direct input from
  - another database,
  - GPS records, or
  - remote-sensing data records;
- importing and conversion of data in another digital format.

When analog maps are digitized, all of the cartographic reference data for the map projection must be included: type of map projection, units of measure, geodetic datum, map scale, publisher, publication year and location, as well as the right of usufruct.

Data can be input into the database directly from a GPS receiver with a corresponding interface on a computer. The Global Positioning System is operated by the government of the USA (Chapter 2.5). It consists of 24 satellites that orbit the Earth twice a day, transmitting information to GPS receivers on the Earth. From this information, a GPS receiver can calculate the coordinates of its location in the parameters of the selected geographic projection.

The information can be stored as point, line, and areal data in ASCII or other text format or in shape format together with detailed attribute data. The most used formats for GIS data are listed in *Table 6.1-3*.

Tools for data management and for “geoprocessing” are an integral characteristic of GIS software. “Geoprocessing” is the generation of a new georeferenced data set from available data sets according to specified rules and tools. The data are first analyzed using an overlay process to compare the different thematic information. Data for the new data set is selected according to defined criteria.

Important functions of GIS software for data analysis include

- data selection (with respect to spatial and attribute properties),
- intersection of areas,
- generation of buffer zones,
- aggregation and subtraction of overlays.

The following example of a search for a suitable landfill site illustrates the use of these analytical tools.

Table 6.1-3: The most used formats for GIS data

Data format	Description	File extensions
Shapefile (ArcGIS, ArcView)	ArcView shapefiles use a simple, non-topological format for storing the geometric location and attribute information of geographic features. Shapefiles define the geometry and attributes of georeferenced features in as many as five files with different file extensions that should be stored in the same projectworkspace.	.shp for the feature geometry .shx for the index of the feature geometry .dbf dBASE file containing the attribute information of the feature, and others
Coverage (ArcInfo workstation)	ArcInfo coverage is a topological data structure for geographic features. In an ArcInfo coverage, features are stored as vector data and their attributes are stored in tables known as feature attribute tables. Each class of features stored in a coverage has its own attribute table. Feature attribute tables are INFO data files containing a number of predefined items and additional user-defined attributes for each feature. A coverage comprises several data files. If you look inside a coverage directory, for example, you may find files with the names arc.adf, tic.adf, aat.adf, etc. Most ArcInfo data files have an .adf extension, which stands for 'Arc Data File'. Note: The coverage format will not be supported in the future by ESRI	*Feature attribute tables:* .PAT point attribute table .AAT arc attribute table .NAT node attribute table .PAT polygon attribute table *Data files (selected):* .tic tic file .bnd boundary coordinate table .txt annotation text file .pat polygon attribute table .aat arc attribute table .prj projection parameter file .grd grid file
Personal Geodatabase (ArcGIS)	This geodatabase format is based on the MS ACCESS relational database management system. Single-user system.	.mdb
SDE (Spatial Data Engine) Geodatabase (ArcGIS)	ArcGIS connect to an external database system, like SQL Server or ORACLE using the ArcSDE gateway. It is a multi-user system.	
TIN	triangulated irregular network	
Raster	image processing	.tiff .img
CAD drawings (AutoCAD, Microstation)	Computer-aided design (CAD) drawings typically have many layers, each of them represents a different type of geographic feature.	.dwg AutoCAD drawing file .dxf industry standard interchange file format .dgn microstation design file

6.1.2.4 Examples

Search for a sanitary landfill site in the area of Osorno, Chile

In a Technical Cooperation environmental geology project between Chile (SERNAGEOMIN) and Germany (BGR), a search for potential sites for a landfill for domestic waste was carried out in the area around the town of Osorno in the Los Lagos Region. Within the scope of the project, data on the geology, hydrogeology, mineral deposits, contamination sources, natural hazards, etc. in the area was geoprocessed (*Figs. 6.1-7* and *6.1-8*). First, all areas that cannot be used for a landfill owing to their present use were classified as "exclusion area I", for example, residential areas, roads, and forests. Buffer zones were defined around each of these areas depending on the present land use. The buffer zone around an urban area was made wider to provide space for growth. Areas of natural hazards (e.g., landsliding and flooding) were classified as "exclusion area II", each with a buffer zone. An analysis of the geological map was then carried out to determine which geological units have the lithological characteristics and sufficient thickness to function as a barrier between the landfill and the groundwater and with a sufficient load-capacity for the landfill body.

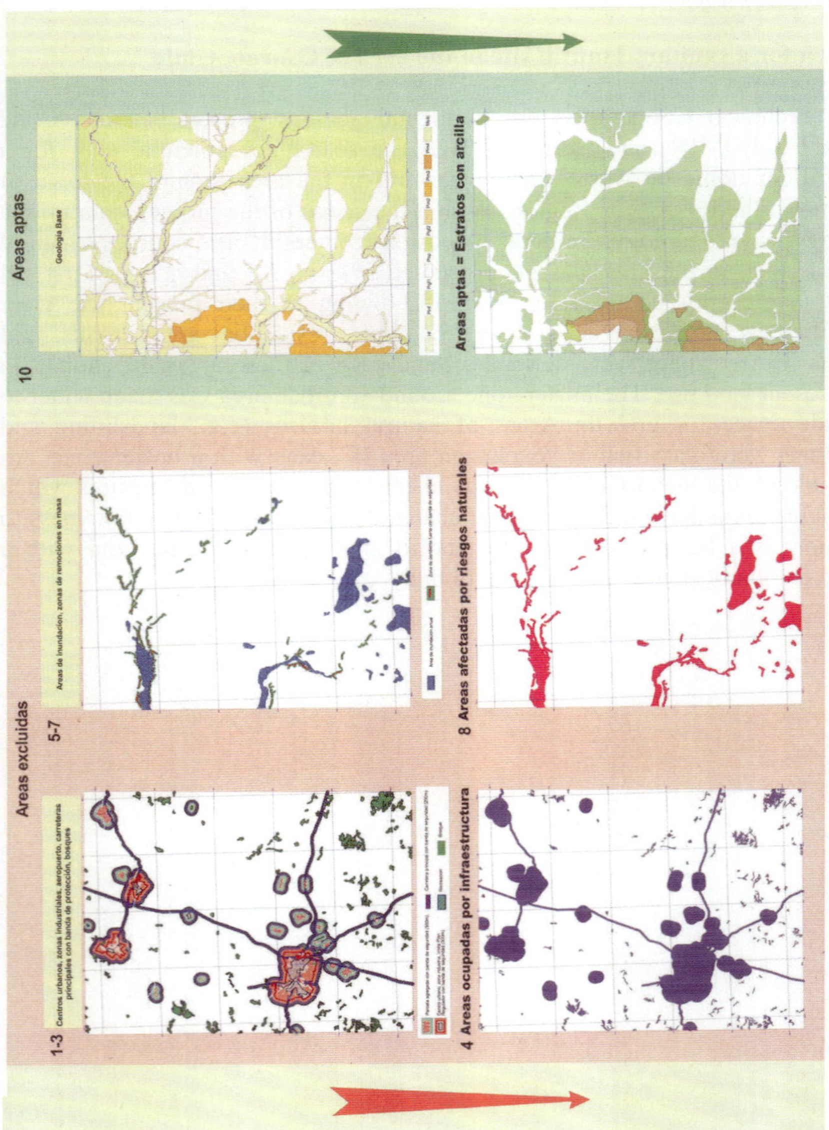

Fig. 6.1-7: Preliminary selection of suitable areas on the basis of land use, natural hazards (exclusion areas), and lithology (suitable areas)

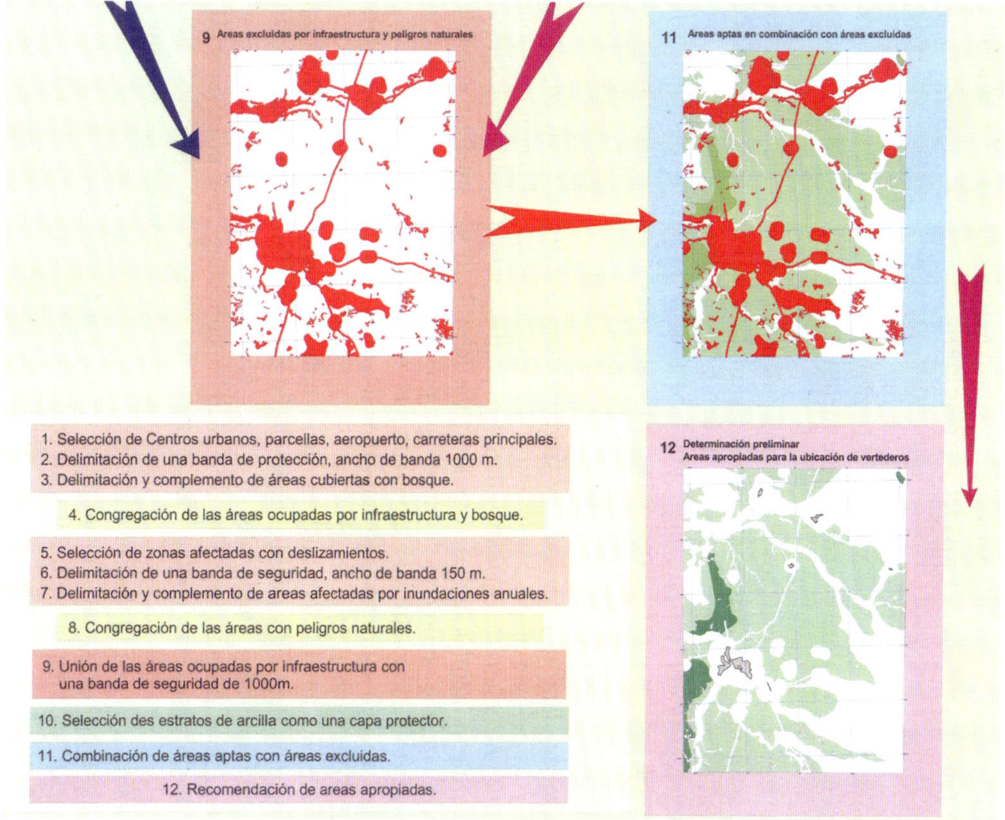

9 Areas excluidas por infraestructura y peligros naturales
11 Areas aptas en combinación con áreas excluidas
12 Determinación preliminar
Areas apropiadas para la ubicación de vertederos
1. Selección de Centros urbanos, parcellas, aeropuerto, carreteras principales.
2. Delimitación de una banda de protección, ancho de banda 1000 m.
3. Delimitación y complemento de áreas cubiertas con bosque.
4. Congregación de las áreas ocupadas por infraestructura y bosque.
5. Selección de zonas afectadas con deslizamientos.
6. Delimitación de una banda de seguridad, ancho de banda 150 m.
7. Delimitación y complemento de areas afectadas por inundaciones anuales.
8. Congregación de las áreas con peligros naturales.
9. Unión de las áreas ocupadas por infraestructura con una banda de seguridad de 1000m.
10. Selección des estratos de arcilla como una capa protector.
11. Combinación de áreas aptas con áreas excluidas.
12. Recomendación de areas apropiadas.

Fig. 6.1-8: Preliminary selection of suitable areas on the basis of land use, natural hazards (exclusion areas), and lithology (suitable areas), final assessment

Identification of a suitable area for the construction of a sanitary landfill in San Ignacio, Misiones Province (Argentina)

Within the framework of the urban planning project between the geological surveys of Argentina (SEGEMAR) and Germany (BGR), suitable areas for the construction of a sanitary landfill were identified in 2005 for the city of San Ignacio (Misiones Province).

San Ignacio is a small town with a population of about 6500 in Misiones Province in northern Argentina, about 60 km east of Posadas, the province capital. The town lies on the east bank of the Paraná River in a hilly region. Agriculture and forestry are the main land use in the area. The climate is subtropical to tropical. Triassic to Jurassic sandstone and Jurassic to Cretaceous basalt are dominant. The sandstone is part of the widespread Guaraní aquifer system.

A sanitary landfill is defined by the American Society of Civil Engineers (ASCE) as a man-made construction for the disposal of waste without damaging the environment and without impairing public health and safety. Engineering principles are used to limit the landfill to the smallest possible area, reducing the volume to the minimum practicable. The waste deposited in this way is to be frequently covered with a layer of soil.

There are two approaches to determine the most suitable area for a sanitary landfill. One method is a detailed site investigation of preselected locations. This means the landfill will be constructed only at a preselected site. The sites are ranked by detailed studies, which will be the basis for further planning. In the second method, suitable sites in a large area are identified on the basis of thematic map information. The first step is collecting and processing all available information. Further information is obtained in repeated field surveys and groundchecks. The result is a recommendation for decision-makers for areas that fit the search criteria.

To find the best suitable place in the area around San Ignacio and to avoid damage to the environment, the main environmental aspects were analyzed, like topography, geomorphology, slope inclination, geology, vegetation, groundwater and surface water, soil, land use, and geohazards.

In the process of site selection the availability of the following infrastructure and planning aspects were considered:

- access to the road network,
- sufficient size of land for a long-term use (more than ten years),
- land in concordance with the long-term urban development,
- suitable geological units which will function as a geological barrier to avoid groundwater contamination,

- cover material, like clay or loam, in the surrounding area for the construction of the sanitary landfill, e.g., the liner and drainage system of the landfill and the intermediate layers covering the waste deposited within a certain time interval, and
- a flat area without significant slopes.

All evaluated characteristics support the prevention of contamination of surface water and groundwater and impairment of the landscape. The results are adequate for regional planning according to governmental regulations.

The collected data was stored and evaluated in a GIS. In addition to the thematic data a numerical field "suitability" was added to each attribute table. Values from "0" for unsuitable to "1" for highly suitable were entered in this field. The values in this field are used for the suitability analysis of all relevant thematic layers.

For a better understanding of this step in the evaluation of spatial units, this was done for each thematic layer and the result was represented in the thematic maps. Shades of red are used on the maps for "low suitability" and green for "high suitability".

Not all of the data needed for the site searching were available. For example, no depth to the groundwater table was available. This initial result was supplemented by field surveys and groundchecks.

Except for the flooding that will result from enlarging the Yacivetá dam, there are no natural hazards in the area surrounding San Ignacio, for example, low-load-bearing foundation ground, area prone to landslides or rockfall, or other hazards. Therefore, no geological hazards were considered in the evaluation step. Tourism aspects and conservation and protection of the environment were considered, but these are covered and included on other thematic maps, for example for vegetation and municipal land. The different themes were weighted according to the local needs (*Table 6.1-11*).

Geology

The lithological characteristics of the geological units are important. The aim is to find a geological unit which will function as a natural barrier to protect the groundwater against water seeping from the future sanitary landfill and having sufficiently high bearing capacity. Only the basaltic lava flows satisfactorily fulfill these criteria (*Table 6.1-4* and *Fig. 6.1-9*).

Table 6.1-4: Suitability on the basis of lithological characteristics

Lithology	Suitability
gravel, sand, silt	0
silty sand	0
sandy silt	0
sand	0
sand with organic material	0
heterogeneous mixture of iron-rich nodules and sand	0
cemented sandstone	0
basalt dikes, basaltic lava flows	1
sandstone	0

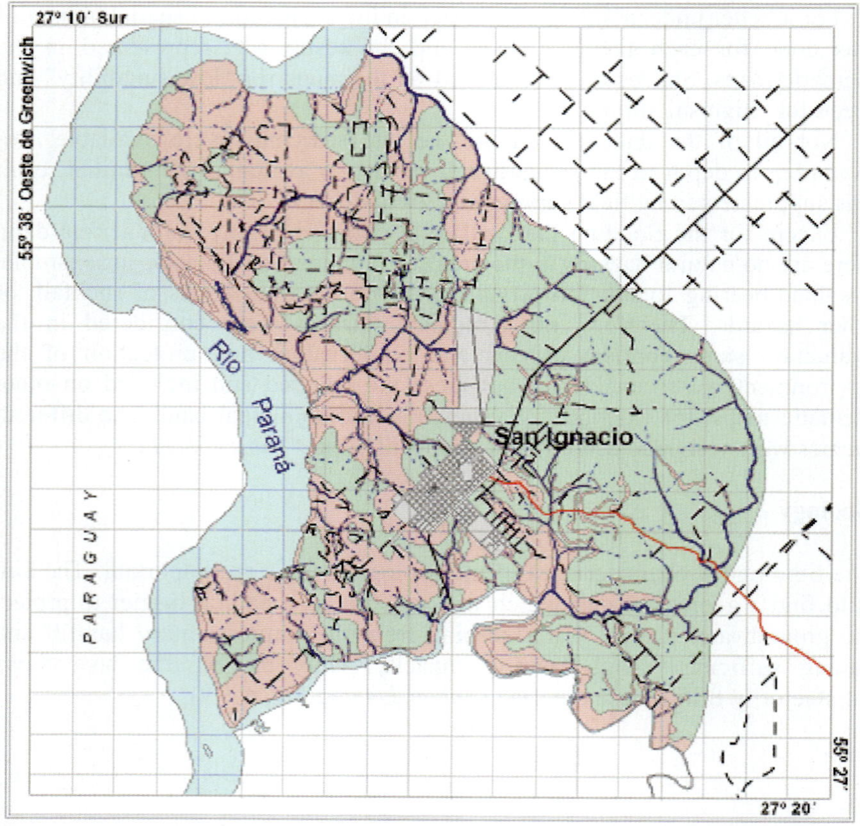

Fig. 6.1-9: Suitability Map: Lithology (red = not suitable, green = suitable)

Slope

For the construction of a reliable liner with a collection system for seeping waste water the slopes should not be more than 2° (*Table 6.1-5* and *Fig. 6.1-10*). A higher slope angle might also result in mass movement within the waste.

Table 6.1-5: Suitability on the basis of slope angle

Slope angle [degrees]	Suitability
0 – 2	1
> 2 – 70	0

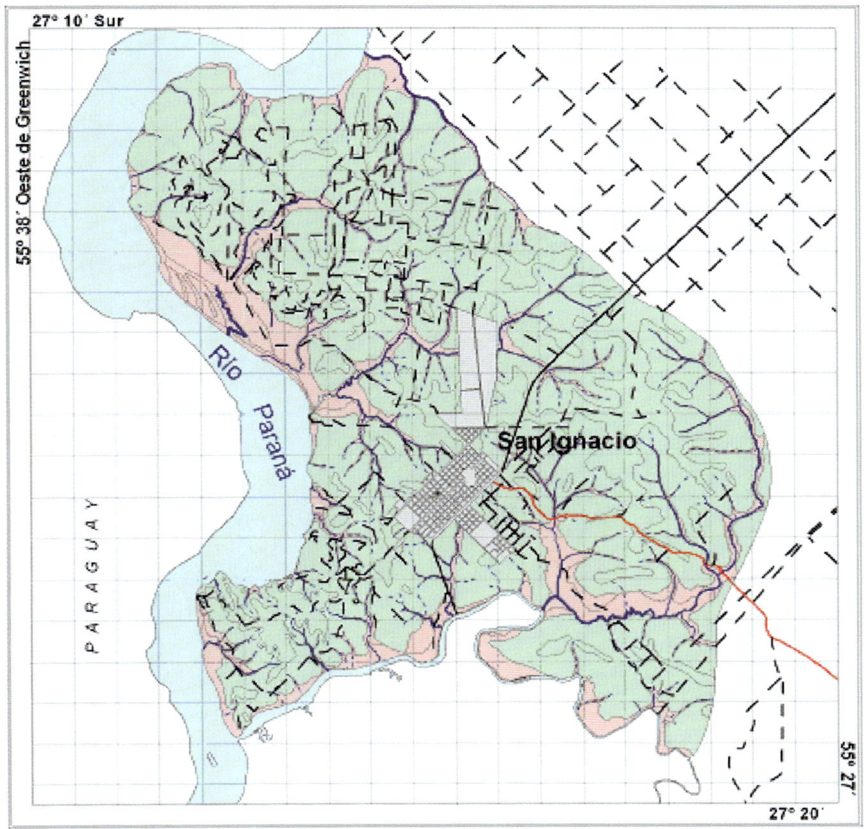

Fig. 6.1-10: Suitability Map: Slopes (red = not suitable, green = suitable)

Geomorphology

In addition to the lithologic descriptions and slope inclination, the geomorphological units were also assigned a suitability value (*Table 6.1-6* and *Fig. 6.1-11*). Not suitable are present and abandoned riverbeds with unconsolidated sediments that are part of the upper, unprotected aquifer system. Suitable sites are found at the top of hills, where the water seeping from the waste can be collected more easily than in depressions, and on smooth slopes over basaltic lava flows.

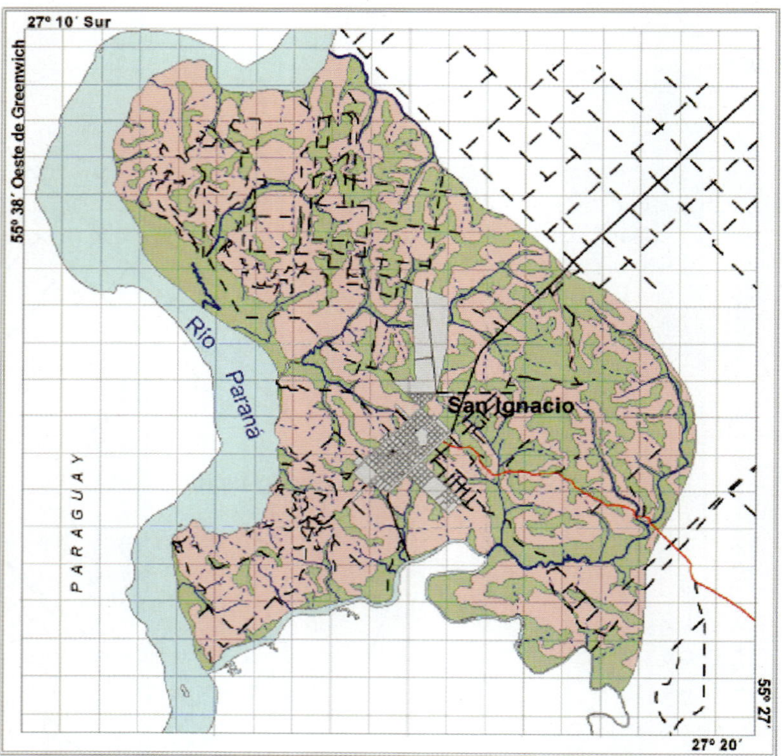

Fig. 6.1-11: Suitability Map: Geomorphology (red = not suitable, green = suitable)

Table 6.1-6: Suitability by geomorphologic units

Geomorphological units	Suitability
alluvial fans and colluvium	0
coast packsaddle of the Yabebiry River	0
valley floor of tributary streams	0
abandoned channels of the Paraná River	0
gentle slopes over sandstone and basalts	1
alluvial plain of the tributary streams of the Paraná River	0
alluvial terrace of the tributary streams of the Paraná River	0
alluvial terrace of the Paraná River (level I)	0
alluvial terrace of the Paraná River (level II)	0
hill tops	1

Vegetation

There is still some isolated virgin jungle (jungle I and II) in Misiones Province. These natural units should not be destroyed by the construction of a sanitary landfill. Jungles and other forests along the rivers, fertile agriculturally used land, and reforestation areas should also be protected (*Table 6.1-7* and *Fig. 6.1-12*).

Table 6.1-7: Suitability by vegetation cover

Land units vegetation cover	Suitability
jungle I	0
jungle II	0
jungle III	1
jungle IV	1
jungle and riverside moisture-loving shrubs	0
wetlands	0
grassland	0
wet grasslands	0
clear-cut areas and fallow land	1
clear-cut areas and farm land	0
clear-cut areas and reforestation areas	0
reforestation areas	0

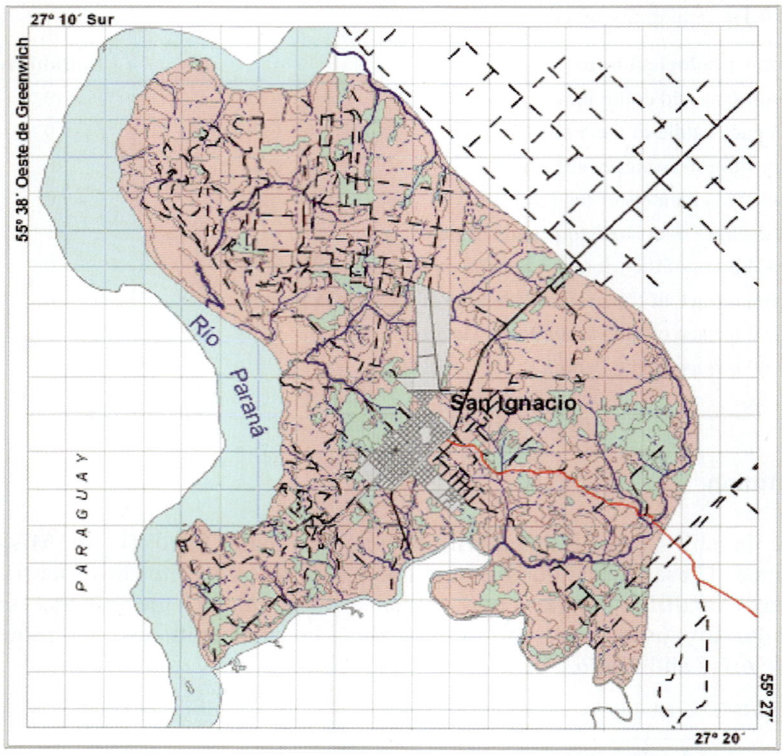

Fig. 6.1-12: Suitability Map: Vegetation (red = not suitable, green = suitable)

Accessibility to roads

The site for a sanitary landfill has to be located close to the existing roads and to the center of the area from which waste will be delivered in order to avoid additional costs for road construction and transport (*Table 6.1-8* and *Fig. 6.1-13*). To avoid problems with wind-blown plastic and other waste material, the landfill should be placed at least 100 m away from any road. This buffer zone between the road and sanitary landfill also serves to hide the site by a "green belt".

Table 6.1-8: Suitability on the basis of distance to roads

Distance from the road	Suitability
<100 m	0
>100 m	1

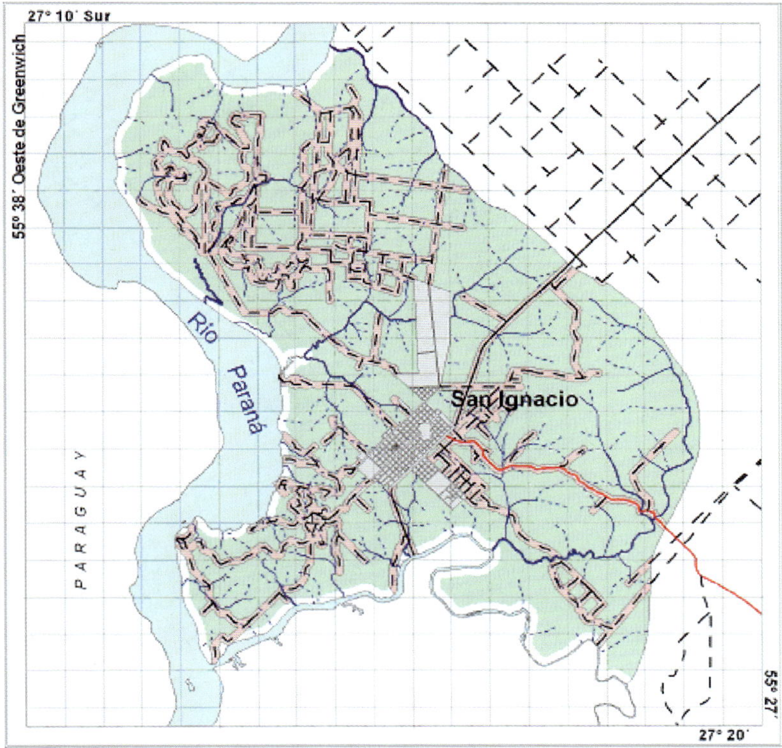

Fig. 6.1-13: Suitability Map: Buffer zone along roads (red = not suitable, green = suitable)

Rivers and streams

To keep waste and contaminants from entering surface water, a buffer zone between a stream or other water body and a sanitary landfill was defined (*Table 6.1-9* and *Fig 6.1-14*). A buffer distance of 200 m was defined for the Paraná River and its tributaries near their confluence with the Paraná. These rivers are assigned to the highest protection category. A buffer distance of 100 m was defined for the intermediate category streams. For the small streams associated with the lowest category a buffer zone of 50 m was defined.

Table 6.1-9: Suitability on the basis of distance to rivers and streams

Distance from rivers and streams	Suitability
<200, <100 or <50 m (for the three protection categories, respectively)	0
>200, >100 or >50 m (for the three protection categories, respectively)	1

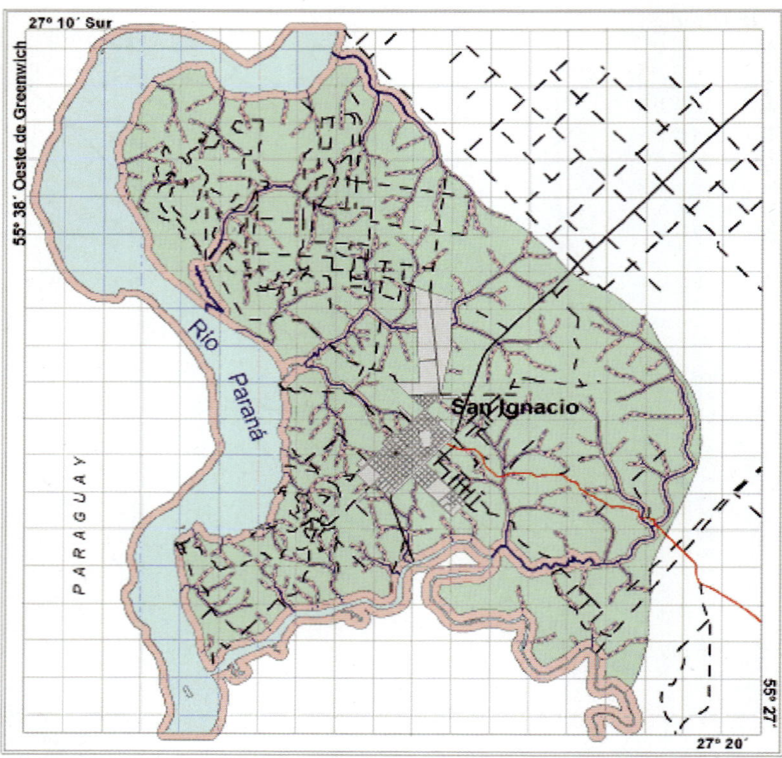

Fig. 6.1-14: Suitability Map: Protection zone along rivers and streams (red = not suitable, green = suitable)

Built-up areas and public land

Construction of a sanitary landfill has to be in accordance with the legislation for urban regions on environmental protection and preservation of the natural resources. This legislation gives the distances that have to be present between a populated center and the location of a landfill. Space for expanding the landfill to take care of future development of the urban population, and industry of the area must also be taken into consideration. The transport costs from the areas of waste generation will have to be calculated as well as the cost of acquisition or leasing of land for the landfill. The cost in some cases can be very low if the land is public property. On the basis of the above-described analysis, a buffer zone of 2.3 km around the city of San Ignacio was defined (*Table 6.1-10* and *Fig. 6.1-15*).

Table 6.1-10: Suitability on the basis of built-up areas

Built-up areas and public land	Suitability
urban public land and sphere of influence (buffer = 2300 m)	0
area beyond the buffer area	1

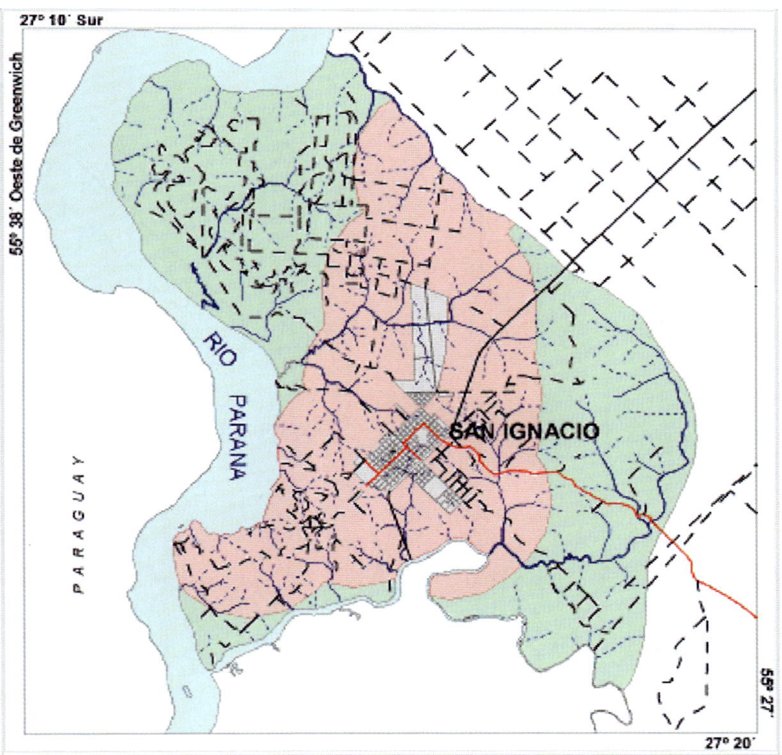

Fig. 6.1-15: Suitability Map: Availability of public urban land (red = not suitable, green = suitable)

Areas potentially flooded by the Yaciretė reservoir (on the Paraná River)

The areas expected to be flooded by raising the water level of the Yaciretė reservoir up to an estimated elevation of 85 m above MSL will not be suitable for a sanitary landfill (*Fig. 6.1-16*). Therefore, a "suitability" value of 0 was assigned to these areas, while those areas higher than 85 m above MSL were assigned a value of 1 (*Table 6.1-11*).

Table 6.1-11: Suitability on the basis of the risk of flooding

Risk of flooding	Suitability
areas along the Paraná River that will be flooded by increasing the water level of the Yaciretá reservoir up to 85 m above MSL	0
areas higher than 85 m above MSL	1

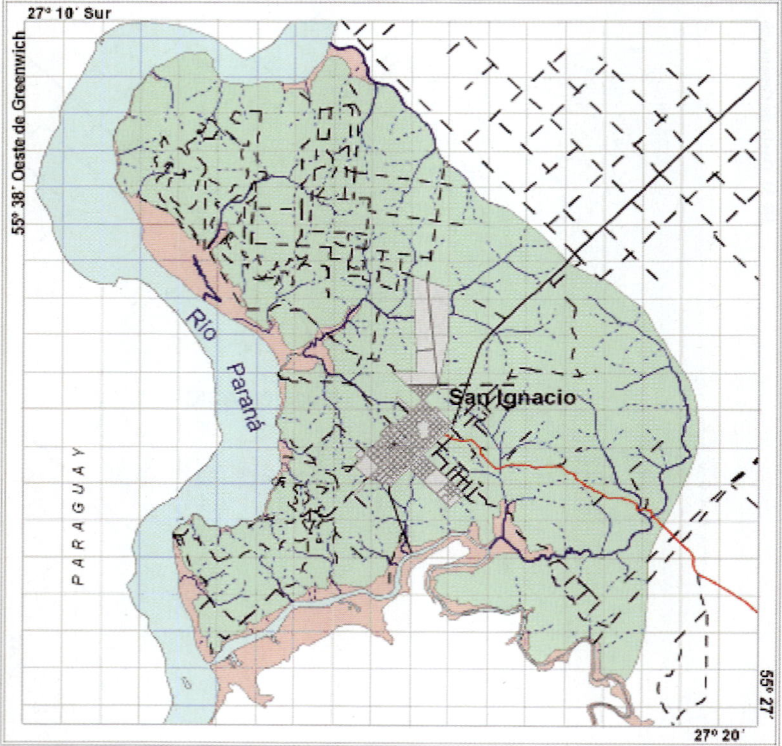

Fig. 6.1-16: Suitability Map: Areas potentially flooded by raising the water level of the Yaciretá reservoir (Paraná River) (red = not suitable, green = suitable)

Integration of the information

All related vector data were converted to a matrix of 614 rows and 505 columns and a cell size (pixel) of 30 m. The cells were assigned to values of 1 (suitable) and 0 (not suitable).
The following formula is used for the data in raster format:

$$Aa = \sum_{i=1}^{m} (P_i a) C_i a$$

where Aa the final suitability for a sanitary landfill (raster cell/area),
i number of spatial unit for the activity a,
P_ia the weighting factor assigned to each map,
C_ia the partial suitability, and
m the number of maps generated by the project.

The data for each map was also assigned a "weight" (*Table 6.1-12*) indicating their importance in the integration of the information to determine the suitability of a planned landfill site using a calculation wizard (*Fig. 6.1-17*).

Table 6.1-12: Weighting factor for the thematic map data

Thematic map	Weighting factor
lithology	16
geomorphology	14
rivers	14
flood zones	16
vegetation	9
slopes	10
roads (accessibility)	10
urban public land	11

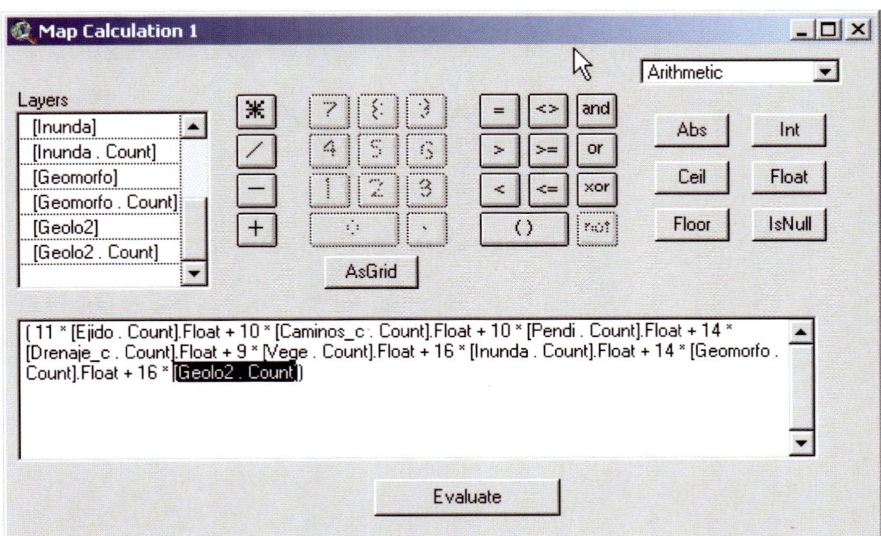

Fig. 6.1-17: Calculation of the suitability of a site for a sanitary landfill on the basis of thematic maps using a calculation wizard

The results of this evaluation were used to prepare a map of suitability (*Fig. 6.1-18*). The final suitability of the areas on the map ranged between 0 and 9. The lower numbers indicate areas with a low suitability. The intermediate values indicate moderate suitability. Areas with these values would require some technical measures to improve some of the thematic factors whose initial suitability is 0. Higher values indicate a very good suitability. The color green was used for “suitable”, yellow for “moderately suitable” and red for “not suitable”.

The most important thematic factors that were evaluated for San Ignacio were geology and geomorphology, followed by vegetation, drainage network, urban public land, roads, and paths. Therefore, if the geological and geomorphological suitability values are 0, the final value for them is also 0. The final decision is left to the authorities and decision-makers who deal with this kind of information. It is possible to undertake technical measures to improve the suitability and make construction feasible.

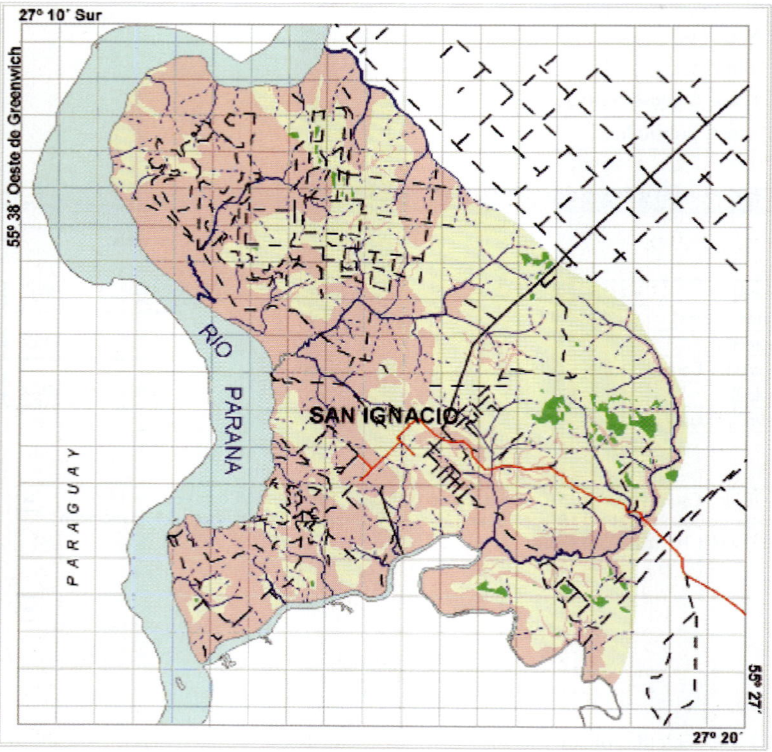

Fig. 6.1-18: Map of suitability for the installation of a sanitary landfill (red = not suitable, yellow = moderately suitable, green = suitable)

6.1.3 Other Data Fusion Examples

KLAUS KNÖDEL & ANTJE BOHN

Digital elevation models

A digital elevation model (DEM) is a three-dimensional representation (x, y and z coordinates) of the Earth's surface derived, e.g., from stereoscopic aerial photographs and satellite images as well as from laser scanner images, mostly in connection with ground control points (Chapter 3.2). If cloud cover impedes imaging with optical systems, radar interferometry (Chapter 3.3) can be used to complete and to improve the DEM. The relief model can be plotted as contour map and/or as shaded relief map.

Classification of images

Image classification (also called pattern recognition) is the process used for creating thematic maps from satellite images. A thematic map shows the spatial distribution of a particular feature, e.g., type of rock, soil, vegetation, land use, flood prone areas, faults, etc. The image classification process can be summarized in three steps:

- feature extraction,
- "training of the pixels", and
- labeling.

Feature extraction is optional in the classification process, serving only to reduce the amount of data and/or enhance the multispectral features of the image using spatial filters or spectral transforms. A multispectral image is transformed into a feature image in the feature extraction step. In the second step, pixels from the image are extracted to "train" the classification program of pattern recognition that helps to differentiate the classes. Based on these patterns, the program creates discriminant functions to assign each pixel to a class. The "training of the pixels" can be either supervised or unsupervised. In supervised training knowledge about the spatial distribution of the classes in the image is used as the starting point. The training points for each site are selected prior to the application of discriminant functions. Classes are labeled before clustering. In unsupervised training classes are separated using parametric and non-parametric classification algorithms without any previous assumptions about their spatial distribution on the image. Various image classification tools are provided by image processing programs.

Landsat 7 data from seven multi-spectral bands were used for mapping flood prone areas in the Dan Khun Thot region/Nakhon Ratchasima (Thailand) at a scale of 1 : 50 000 (*Fig. 3.3-15*). In order to classify selected land cover types in the area, a combined approach of unsupervised and supervised classification was carried out. This classification divided the pixels into statistically defined classes for certain types of relevant land cover: standing water, swamps, elevated areas, plantations, soil salinization areas, and settlements.

Map of risks for a land use

The following conditions represent risks for the sustainable sociological and ecological development of Stassfurt, Saxony-Anhalt, Germany:

- uplift and subsidence due to halokinetic processes and subrosion;
- subsidence and sinkholes as a result of potash, rock salt and subsurface lignite mining below and around the city;
- pumping of groundwater to lower the groundwater table since the beginning of the twentieth century (about 850 m^3/day with about 28 t/day of salt) to avoid water-logged surface soils;
- 28 waste heaps in the central part of the subsidence area containing residues from potash and chlorine chemical plants, from coal-fired power plants, from metallurgical and electronic equipment plants, and from weapons production during World War II.

The risks must be estimated in as much detail as possible for the development of land-use plans in urban industrial areas. For this purpose, a map of mechanically instable areas, water-logged and flood-prone areas, and areas of geogenic and anthropogenic contamination in and around Stassfurt was prepared (*Fig. 6.1-19*). The risks are partly synergetic, increasing the risk of each other.

Subsidence areas are mechanically unstable. Subsidence in the Stassfurt area is due to convergence of mine and subrosion cavities in the salt rock. The hazard emanating from the subsidence can be characterized by the amount of subsidence and the occurrence of collapse sinkholes in the past, the current subsidence rates, and the expected future subsidence. In the last ten years, buildings have been damaged due to tilting caused by subsidence. Further subsidence of 0.5 - 1.0 m is predicted. Mechanical instabilities are also expected in the area of latent risk of collapse. Where sinkholes have been filled with mining and industrial waste, further settling of the underlying loose rock can be expected. An analysis of the present conditions leads to the conclusion that there is a latent risk of collapse in about 70 ha of the mine subsidence area (*Fig. 6.1-19*). Seismic monitoring, however, does not indicate any immediate danger of collapse sinkholes in this part of the city, but cannot exclude it.

Subsidence has lowered central parts of the city to below the groundwater table and is kept dry by pumping. If the pumping system fails, if there is high water in the Bode River with infiltration into the groundwater, there is the risk of water-logging of the soil and flooding. On the map, the currently water-logged areas are indicated by the depth to the water table in March 1977. The groundwater table was observed at very shallow depth in the southwestern part of the city. The continuing subsidence will require to extend pumping to further areas. Certain areas of the city will be allowed to become waterlogged when the pumping is carried out decentralized and others will be allowed to become covered with water as subsidence continues.

The areas near the Bode River are in danger of flooding during times of high water in the river, as well as the areas of the city that have subsided to below the mean water level of the river. The extent of flooding in April 1994 (*Fig. 6.1-19*) gives an idea of the extent of the flood-endangered area. With further subsidence, especially near the river, the size of the flood-endangered area will increase.

In Stassfurt, there are a number of organic and inorganic contaminants in the soil from both geogenic and anthropogenic sources. The hazards emanating from these contaminants are increased by the risks described at the beginning of this section. As an example of organic contamination, the concentration of benzo(a)pyrene after EPA has been selected as representative of polycyclic aromatic hydrocarbons (PAH). The distribution of benzo(a)pyrene in Stassfurt corresponds to a large extent to the PAH distribution. The legal limits for benzo(a)pyrene are exceeded in several areas. These areas are also contaminated with other organic substances.

Besides ascending groundwater from depth with its high salt content, water percolating from abandoned industrial sites is also corrosive to concrete. When the groundwater table is at shallow depth, building foundations can be damaged by direct contact with the groundwater or by capillary action. The water in the near-surface aquifers was analyzed: Elevated corrosion of concrete, mortar, and iron (*Fig. 6.2-19*) can be expected with magnesium concentrations of more than 1000 mg L^{-1}, chloride concentrations of more than 200 mg L^{-1}, and sulfate concentrations of more than 600 mg L^{-1}.

Contamination of groundwater and soil by organic and inorganic substances is marked on the map by colored dots. Old waste sites are shown on the map as not evaluated because the hazard assessment of the waste heaps and landfills is based partly on their location in subsidence areas and in areas of water-logging and potential flooding, which would give them a double weighting. The groundwater flowing into the Bode river from the city area contains contamination from contact with the waste heaps and landfills. Pumping in the city area also delivers a salt load of 28 t per day into the Bode river.

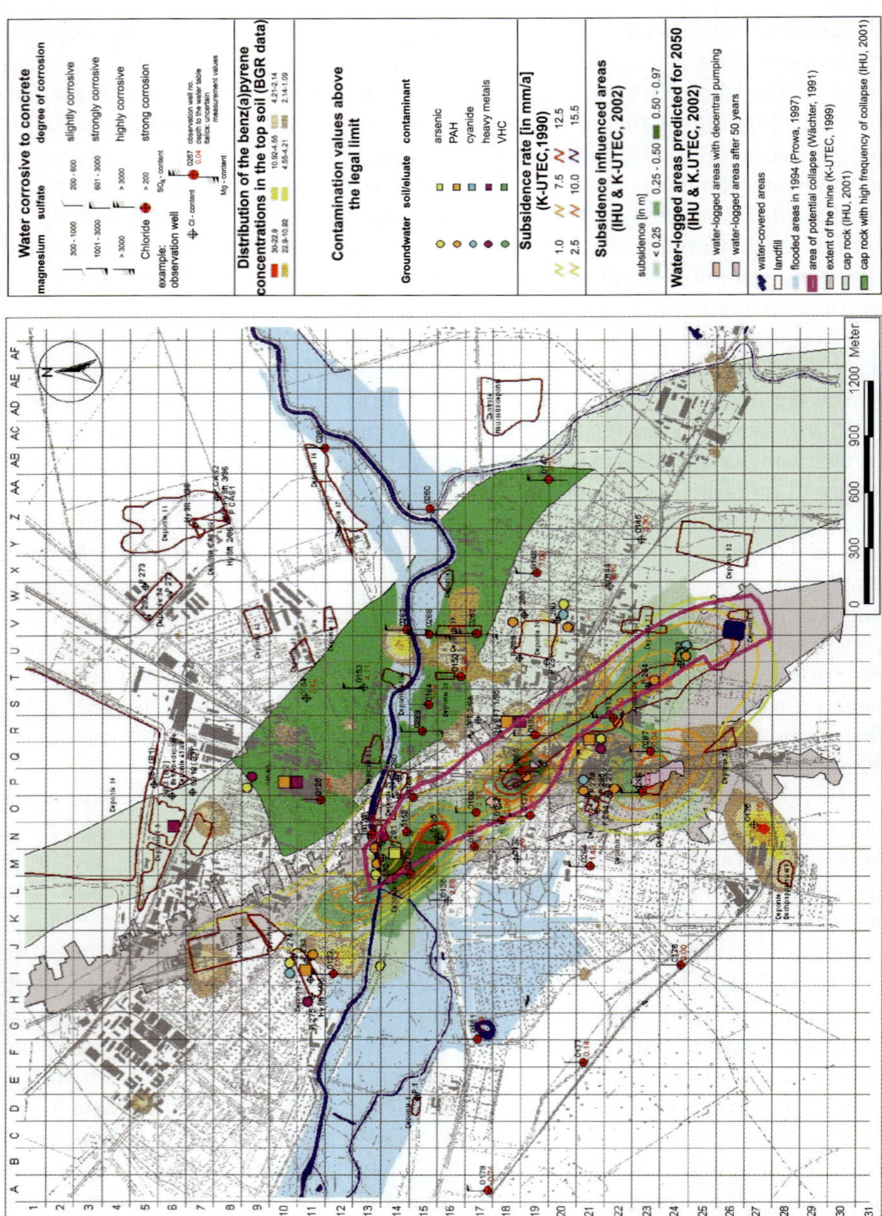

Fig. 6.1-19: Map of the risks for land use in Stassfurt, Saxony-Anhalt, Germany

Analysis of the current and potential hazard areas yields the following results:

- areas with multiple synergic hazards (southern part of the city, map areas O-Q 21-23; eastern part, U 14),
- areas with single hazards (western part of the city, G-K 17-19; eastern part Y-Z 17-18),
- areas with no recognizable hazards (northwestern part of the city, A-F 7-10; southeastern part, X-Z 20-22).

The analysis thus shows areas in which measures must be taken and areas in which land use must be restricted. Recommendations can be made for land use on the basis of the analysis.

References and further reading

ARAI, K. (1992): A supervised Thematic Mapper classification with a purification of training samples. International Journal of Remote Sensing, **13**, 11, 2039-2049.

ARNDT, C., HEIN, A., LOFFELD, O. (1998): Information theory in data fusion. In: Proceedings of the second conference "Fusion of Earth data: merging point measurements, raster maps and remotely sensed images", Sophia Antipolis, France, January 28-30, 1998, T. RANCHIN and L. WALD (Eds.), published by SEE/URISCA, Nice, France, 85-90.

ASCH, K. (Ed.) (1999): GIS in Geowissenschaften und Umwelt. Springer, Berlin.

ATKINSON, P. M. & TATE, N. J. (1999): Advances in Remote Sensing and GIS Analysis. Wiley & Sons, Chichester.

BARTELME, N. (2005): Geoinformatik. 4. Aufl., Springer, Berlin.

BÄHR, H.-P. & VÖGTLE, T. (1999): GIS for Environmental Monitoring. Schweizerbart, Stuttgart.

BEHR, F.-J. (2000): Strategisches GIS-Management. Grundlagen und Schritte zur Systemeinführung. 2. Aufl., Wichmann, Heidelberg.

BERRY, J. K. (1993): Beyond Mapping: Concepts, Algorithms and Issues in GIS. Fort Collins, CO: GIS World Books.

BILL, R. & ZEHNER, M. (2001): Lexikon der Geoinformatik. Wichmann, Heidelberg.

BILL, R. (1999a): Grundlagen der Geo-Informationssysteme. Bd 1: Hardware, Software und Daten. Wichmann, Heidelberg.

BILL, R. (1999b): Grundlagen der Geo-Informationssysteme. Bd 2: Analysen, Anwendungen und neue Entwicklungen. Wichmann, Heidelberg.

BURROUGH, P. A. & MCDONELL, R. (1998): Principles of Geographical Information Systems. Oxford University Press, Oxford.

CHEN, K. S., TZENG, Y. C., CHEN, C. F. & KAO, W. L. (1995): Land-cover classification of multispectral imagery using a dynamic learning neural network. Photogrammetric Engineering and Remote Sensing, **61**, 4, 403-408.

CHEN, Y., TAKARA, K., CLUCKIE, I. & De SMEDT, F.H. (Eds.) (2004): GIS and Remote Sensing in Hydrology, Water Resources and Environment. IAHS Publication 289, Wallingford.

CIVCO, D. L. (1993): Artificial neural networks for land-cover classification and mapping. International Journal of Geographical Information Systems, **7**, 2, 173-186.

COPPOCK, J. T. & RHIND, D. W. (1997): The History of GIS. In: MAGUIRE, D. J. & DASARATHY, B. V. (Eds.) (1997): Proceedings of the SPIE conference "Sensor Fusion: Architectures, Algorithms, and Applications", Orlando, Florida, 24-25April, 1997, SPIE vol. 3067.

DUCROT, D., SERY, F., SASSIER, H., GOZE S., PLANÈS, J. G. (1998): Classification and fusion of optical and radar satellite data for land use extraction. In: T. RANCHIN & L. WALD (Eds.): Proceedings of the second conference "Fusion of Earth data: merging point measurements, raster maps and remotely sensed images", Sophia Antipolis, France, January 28-30, 1998, published by SEE/URISCA, Nice, France, 45-50.

EVANS, D. (1998): Data fusion applied to geologic mapping and natural hazards. In: T. RANCHIN & L. WALD (Eds.): Proceedings of the second conference "Fusion of Earth data: merging point measurements, raster maps and remotely sensed images", Sophia Antipolis, France, January 28-30, 1998, published by SEE/URISCA, Nice, France, 117-122.

GOLDEN SOFTWARE (2002): Surfer8 users guide. Golden, Co.

GOODCHILD, M. F. & RHIND, D. J. (Eds.) (2005): Geographical Information Systems. 21-43. Wiley & Sons, New York.

GURNELL, A. M. & MONTGOMERY, D. R. (Eds.)(1999): Hydrological Applications of GIS. Wiley & Sons, Chichester.

HALL, D. L. (1992): Mathematical techniques in multi-sensor data fusion. Artech House, Norwood, Massachusetts.

HARDIN, P. J. (1994): Parametric and nearest-neighbor methods for hybrid classification: a comparison of pixel assignment accuracy. Photogrammetric Engineering and Remote Sensing, **60**, 12, 1439-1448.

HEYWOOD, I., CORNELIUS, S., & CARVER, S. (2002): An Introduction to Geographical Information Systems. 2nd edition. Andison Wesley Longman. 2nd edn.

LI, X. & GÖTZE, H. J. (1999): Comparison of some gridding methods. The Leading Edge, 898-900.

LONGLEY, P. A., GOODCHILD, M.F. & MAGUIRE, D. J. (1999): Geographical Information Systems: Principles, techniques, applications and management (vols. 1 & 2). Wiley & Sons, New York.

LONGLEY, P. A., GOODCHILD, M. F., MAGUIRE, D. J. & RHIND, D. W. (2005): Geographic Information Systems and Science. 2nd edn., Wiley & Sons, Chichester.

MINGHELLI, A., POLIDORI, L., MANGOLINI, M. (1998): Image fusion for the simulation of hyperspectral satellite images. In: T. RANCHIN & L. WALD (Eds.): Proceedings of the second conference "Fusion of Earth data: merging point measurements, raster maps and remotely sensed images", Sophia Antipolis, France, January 28-30, 1998, published by SEE/URISCA, Nice, France, 97-102.

NACEUR, M. S., ALBUISSON, M., BOUSSEMA, M. R. (1998): Extraction of textural information by fusion of cartographic and remotely sensed data. In: RANCHIN, T. & WALD, L. (Eds.): Proceedings of the second conference "Fusion of Earth data: merging point measurements, raster maps and remotely sensed images", Sophia Antipolis, France, January 28-30, 1998, SEE/URISCA, Nice, France, 141-148.

NCGIA (1992): A Glossary of GIS Terminology, compiled by G. Padmanabhan and Jeawan Yoon, North Dakota State University, and Mark Leipnik, University of California Santa Barbara. Technical report of the National Center for Geographic Information and Analysis, Santa Barbara, CA. http://www.ncgia.ucsb.edu/publications/tech_reports/92/92-13.pdf

POLIDORI, L. & MANGOLINI, M. (1996): Potentialities and limitations of multisensor data fusion. In: RANCHIN, T. & WALD, L. (Eds.): Proceedings of the conference "Fusion of Earth data: merging point measurements, raster maps and remotely sensed images", Cannes, France, February 6-8, 1996, SEE/URISCA, Nice, France, 13-20.

DEL RÍO, M., BÓ, M. A., BERNASCONI, B., OSTERRIETH, F. & FERRARO, L. y T. (1993): Planificación territorial sobre criterios geoambientales en la cuenca del arroyo y laguna De Los Padres; in XII Congreso Geológico Argentino y II Congreso de Exploración de Hidrocarburos. Actas T VI (293-302); Argentina.

RANCHIN, T. & WALD, L. (Eds.) (1996): Proceedings of the conference "Fusion of Earth data: merging point measurements, raster maps and remotely sensed images", Cannes, France, 6-8 February, 1996, published by SEE/URISCA, Nice, France.

RANCHIN, T. & WALD, L. (Eds.) (1998): Proceedings of the second conference "Fusion of Earth data: merging point measurements, raster maps and remotely sensed images", Sophia Antipolis, France, 28-30 January, 1998, SEE/URISCA, Nice, France.

RANCHIN, T., WALD, L. & MANGOLINI, M. (1998): Improving spatial resolution of images by means of sensor fusion. A general solution: the ARSIS method. In: DONNAY, J.-P. & BARNSLEY, M. (Eds.): Remote Sensing and Urban Analysis. Taylor & Francis.

SCHOWENGERDT, R. A. (1997): Remote Sensing, Models and Methods for Image Processing. 2nd edn., Academic Press, San Diego, CA.

SHEFFIELD, T. M., MEYER, D., LEES, J., KAHLE, G., PAYNE, B. & ZEITLIN, M.J. (1999): Geovolume visualization interpretation: Color in 3-D volumes. The Leading Edge, **18**, 668 - 674

THE OPEN GIS CONSORTIUM (2000): Geospatial fusion services testbed. Wayland, MA.

THURSTON, J., POIKER, T. K. & MOORE, J. P. (2003): Integrated Geospatial Technologies: A Guide to GPS, GIS, and Data Logging. Wiley & Sons, Hoboken, New Jersey.

U.S.P.H.S. Bureau of Solid Waste Management (1968): National Solid Waste Survey.

VIEUX, B. E. (2001): Distributed Hydrologic Modeling Using GIS. Kluwer, Dordrecht.

WALD, L., RANCHIN, T. & MANGOLINI, M. (1997): Fusion of satellite images of different spatial resolutions: assessing the quality of resulting images. Photogrammetric Engineering & Remote Sensing, **63**, 6, 691-699.

WALD, L. (1998): A European proposal for terms of reference in data fusion. International Archives of Photogrammetry and Remote Sensing, XXXII, part 7, 651-654.

WALD, L. (1999): Some terms of reference in data fusion. IEEE Transactions on Geosciences and Remote Sensing, **37**, 3, 1190-1193.

WALD, L. (2002): Data Fusion, Definitions and Architectures, Fusion of Images of Different spatial Resolutions. Les Presses de l'Ecole des Mines, Paris.

WISE, S. (2002): GIS Basics. Taylor & Francis, London.

WORBOYS, M. & DUCKHAM, M. (2004): GIS: a computing perspective. CRC Press, Boca Raton, FL.

6.2 Joint Interpretation

MANFRED BIRKE, KLAUS KNÖDEL, GERHARD LANGE & UGUR YARAMANCI

6.2.1 Qualitative and Semiquantitative Approach

KLAUS KNÖDEL & GERHARD LANGE

Search for a new waste disposal site near Luederitz, Namibia

Geophysical and photogeological surveys of four preselected potential disposal sites for waste from the town and the harbor were carried out in the vicinity of the town of Luederitz on the coast of Namibia. The area under consideration in the Namib desert is geologically part of the 1200 Ma old Namaqualand Metamorphic Belt, consisting essentially of highly deformed, high-grade schist and granitic gneiss intruded by granite and granodiorite. At least four different stages of faulting were identified. Aquifers containing significant amounts of water are unknown from the area. The survey objective was to determine the location of any fractures in the bedrock that might serve as paths for contaminants from a planned landfill. Since the bedrock is largely covered by sand dunes and weathering debris, geophysical methods were required to map fractures hidden by unconsolidated sediments. Using photogeology together with geophysical methods made it possible to focus the geophysical ground surveys on the sand-covered areas.

Aerial photographs were taken at two different altitudes. The flight lines were oriented approximately E-W. Enlargements were made at scales of 1 : 10 000 and 1 : 40 000. Before the survey flights, ground control points (GCP) were laid out using differential GPS to achieve the exact positioning required by the envisaged small-scale analysis of the photos and to assign them to the local coordinate system. Tectonic features in and around the preselected sites were mapped with a small positional error of mostly no more than two meters. Evaluation of the aerial photographs was also done in the vicinity of each of the four pre-selected sites to enable identification of structures of more regional extent. The geological interpretation of the present geomorphological forms was based on the tectonic and orogenic history of the region. The photographs were analyzed stereoscopically to determine trend directions of mountain ranges and valleys produced by differential erosion of steeply dipping strata resulting from orogenesis. A ground truth check was made of selected tectonic features.

Weathering processes are able to penetrate fractured rock farther than rocks not exposed to stress. As a consequence, fractures often contain moisture, clay, silt, etc., indicated by conductivity anomalies. Therefore, an electromagnetic mapping survey was conducted to reveal any faults. Refraction seismics and dc resistivity depth soundings (VES) were used to determine the thickness of the sand and weathering zone covering the basement rock. This information is useful for further planning, firstly to know about the nature and bedding of the sand; secondly, to obtain an estimate of the volume of material needed for the bottom seal of the landfill and to cover the waste.

The results will be discussed taking the site "Luederitz North", which is closest to Luederitz, as example. This site is located in a N-S-trending valley with smooth slopes to the west and east, where granitoid bedrock crops out or is overlain by a thin sand cover. To the west and south it is sheltered from the prevailing wind by low ridges of basement rocks. A dune belt is present to the east and north. The site itself is largely covered by sand mined for diamonds in the past.

Fig. 6.2-1: Luederitz North, Namibia, view from the southeast corner of the geophysical ground survey area (*Fig. 6.2-2*) towards the fractured granitic gneiss at the west side of the shallow valley

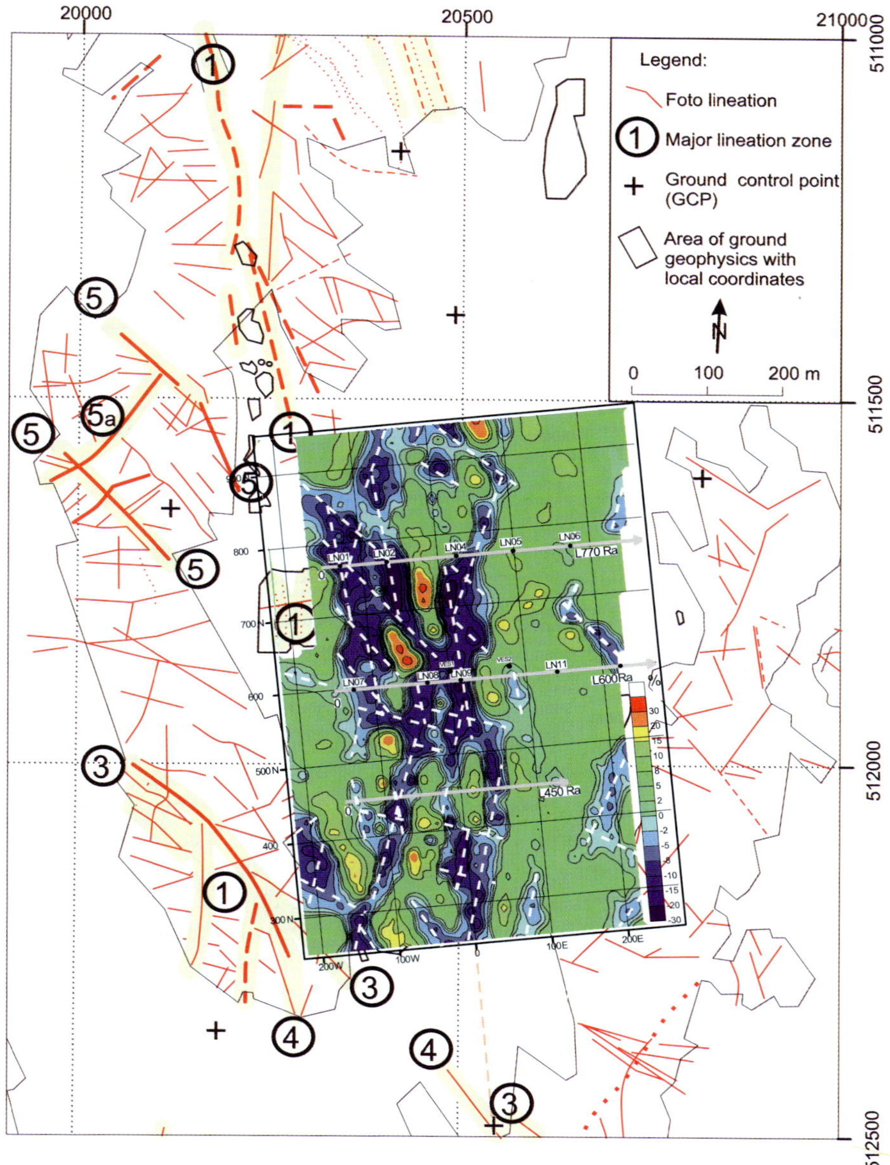

Fig. 6.2-2: Luederitz North, Namibia, lineaments identified from detailed aerial photographs and the ground conductivity of horizontal loop electromagnetic (HLEM) data (7040 Hz) showing the fault indications (dashed lines), dc resistivity sounding points (LN01 to LN11) and refraction seismic profiles (L450Ra, L600Ra and L770Ra)

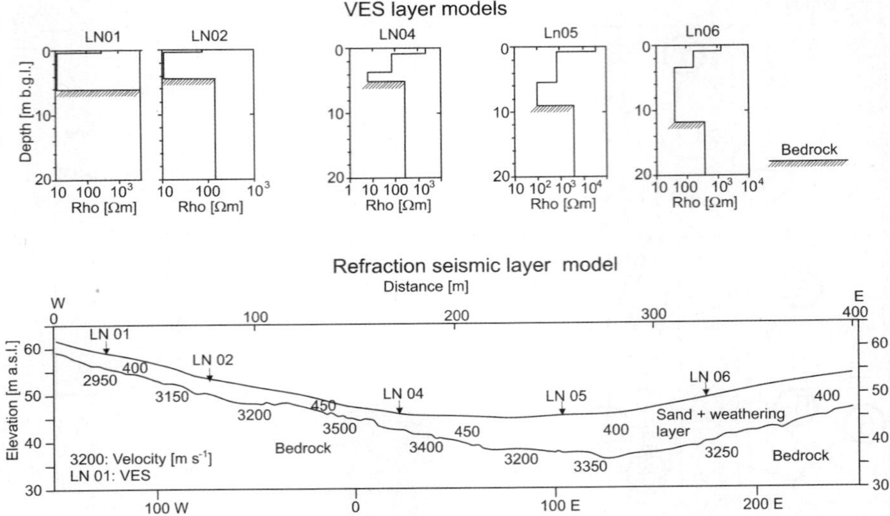

Fig. 6.2-3: Luederitz North, Namibia, layer models derived from dc resistivity sounding data (VES), refraction seismic layer model of the profile L770Ra

Five major lineation zones were recognized in the aerial photographs of the area (*Fig. 6.2-2*). Their tectonic relevance was assessed in the field.

Electromagnetic data were collected with an Apex MaxMin system with 40 m separation between the transmitter and receiver coils (HLEM) at frequencies of 3520, 7040 and 14 080 Hz. The stations along the profiles were spaced 10 m apart with a profile spacing of 20 m. The out-of-phase maps were interpreted assuming that increased conductivity (indicated by negative out-of-phase values) reflects fractured bedrock due to enhanced weathering. Linear trends are clearly reflected by the 7040 Hz out-of-phase data (*Fig. 6.2-2*).

Refraction seismic measurements were made on three E–W survey lines (L450N, L600N and L770N) across the N-S-trending valley (*Fig. 6.2-2*). The results best fit a two-layered model with a low-velocity (400 - 650 m s^{-1}) layer (unconsolidated, dry sediments) overlying high-velocity (3000 - 4000 m s^{-1}) unweathered bedrock. Velocities lower than the 3000 - 4000 m s^{-1} range in the western part of lines 600N and 770N may indicate fractured bedrock. The upper layer has an average thickness of just 5 m, which increases to about 15 m towards the northern and eastern parts of the valley (line 770N, *Fig. 6.2-3*), where sand dunes have accumulated by the continuous winds.

To verify the two-layer model resulting from the refraction seismic survey, eleven Schlumberger VES were conducted on lines 600N and 770N. Horizontal layer interpretation of the sounding curves produced a simple model with a high resistive basement overlain by conductive material of varying thickness and resistivity in all cases. The results obtained on line 770N from five geoelectrical depth soundings are compared with the results of the refraction seismic survey in *Fig. 6.2-3*. Both methods show that the depth to bedrock and the thickness of the cover sediments increase from west to east. The increase in resistivity values from west to east reflects either an increasing volume/thickness of dune sands and/or a wedging out of the conductive bottom of the cover layer.

The shallow valley hosting the Luederitz North site is likely to be part of a major tectonic feature. The cover sediments are fairly thin (only 5 m on average) with no indication of any soil moisture or groundwater. According to observations made in nearby abandoned diamond prospecting trenches, the low-resistivity bottom part of the cover sediment consists of coarse sand and/or reddish silty/calcrete (ferrocrete) material, and might prove to be largely impermeable to leachate from a landfill. Samples of these calcareous silty sands were analyzed, revealing an inorganic weathering product consisting of a high proportion of ferric oxides, salt contents of 13.8 cmol/100 g, and about 20 % $CaCO_3$. The high buffering capacity of carbonates in conjunction with a low coefficient of hydraulic conductivity ($\leq 10^{-6}$ m s^{-1}) make this material promising as a geological barrier. The bedrock does not seem to be deeply weathered. Several fault or fracture trends were identified. The fault pattern derived by evaluation of the aerial photographs is in good agreement with that derived from the ground electromagnetic survey data (*Fig. 6.2-2*). There are three main trends: N-S, NNE-SSW and NNW-SSE. The granitic gneiss on the west side of the valley is strongly fractured. The east side of the valley, in contrast, has almost no fractures. Where the bedrock is covered by sand the use of suitable geophysical methods proved to be highly successful. Only by the combination of the results of geophysical and aerial photograph interpretation was it possible to produce a complete fracture pattern of the site under consideration.

Abandoned Eulenberg Landfill

The site of the abandoned Eulenberg landfill near Arnstadt, Thuringia, Germany, consists of consolidated Mesozoic rocks with a complicated structure. This unofficial landfill was begun in the 1960s, and deposition continued until 1979. Because of its position within the outer protection zone of the Arnstadt waterworks, the site was investigated as a possible source of contamination using a combination of geoscientific methods.

The landfill is in a former quarry on a slope in the Upper Muschelkalk formation. The sides of the quarry vary between 10 and 26 m in height. The maximum thickness of the waste is 26 m. The top of the landfill is terraced with steep slopes between the terraces 4 - 12 m high extending beyond the quarry pit. The total height of the slopes of 33 m increases the problems of surveying the site. Because the slope angles are often more than 45°, the slope stability is vulnerable. The total volume of the landfill is estimated to be 430 000 m^3. The waste material deposited here included building rubble, ash from home furnaces, slaughterhouse and tannery wastes, galvanic sludge, sewage sludge from the polishing of leaded glass, cyanide-bearing sludge, and phenol-bearing wastes from asphalt production.

In the 1940s, the natural geological conditions were disturbed by the construction of subterranean chambers. To understand the effect of this extensive underground excavation on the functioning of the rock as a geological barrier to the spreading of contaminants, the investigations had to include an analysis of the site history. Moreover, the geology below the landfill had to be investigated through waste up to 26 m thick. The two aspects – manmade changes and the natural geological conditions below the waste – largely affect the pollution potential of the landfill.

Simultaneously with the determination and analysis of the history of the site, a preliminary geological/hydrogeological model was prepared basing on the regional geology, previously drilled boreholes, shallow boreholes at selected points, and groundwater analyses. This was followed by the preparation of a map showing tectonic features and the location of borehole and trenching sites.

Objective of the remote-sensing investigation was to reconstruct the history of deposition and to determine the presence, condition, and distribution of impermeable rocks, as well as possible migration pathways for contaminated water flowing off the landfill (*Fig. 6.2-5*).

At the beginning of the investigation, the location of the underground chambers was unknown – aerial photographs and blueprints from the time before the deposition of waste were not yet available. Therefore, gravity measurements were made to obtain information about their location. A distinct gravity minimum was expected above the waste-filled former quarry owing to the difference between the density of the rocks of the Keuper and the Muschelkalk and that of the waste. The objective of the magnetic measurements was to determine the boundaries of the landfill and the location of large magnetic bodies in the landfill and steel-reinforced concrete in the underground chambers (*Fig. 6.2-6*).

Direct current geoelectric measurements were made to investigate the geological structures and to detect contaminant plumes originating from the landfill. A resistivity map was prepared using the Schlumberger configuration with an electrode spacing of AB/2 = 50 m. 30 Schlumberger depth soundings (including six cross arrays to estimate the effect of lateral conductivity changes) were carried out with AB/2 electrode spacings up to 420 m. Whereas the resistivity mapping provided information to a depth of about 15 m, the depth soundings yielded information down to a depth of about 100 m (*Fig. 6.2-7*).

A standard refraction seismic survey was carried out, as well as refraction tomography and high-resolution 2-D reflection seismics – sometimes in combination. The objective was to determine the bedding below and around the landfill, to map the fracture zones and inhomogeneities (tectonically stressed areas), to determine local and regional fault systems, locate the boundary between the landfill and undisturbed rock, determination of the thickness and the relief of the bottom of the waste as well as structures within the waste body and the location of manmade cavities (*Figs. 6.2-8* to *6.2-12*).

Geophysical cone penetration tests were conducted to determine zones within the landfill with different types of waste, to determine the depth of the base of the waste, and the nature of the rock just below and in the vicinity of the landfill. On the basis of the results, soil samples were taken from selected depths. Percolation tests were made directly below the landfill to obtain supplementary hydrogeological information (*Fig. 6.2-12*).

Geophysical borehole logging was conducted using the following methods and objectives: gamma logging for clay content, gamma-gamma logging for soil density, resistivity logging for lithology, neutron-neutron logging for porosity and water content, caliper logging for borehole diameter, inclinometer logging for borehole inclination, dipmeter logging for the dip angle of inclined beds, flowmeter logging for flow conditions, borehole TV and televiewer for the condition of the borehole wall and determination of fracture zones. Packer tests were carried out in two boreholes to determine the hydraulic conductivity and storage capacity of rock material. The tests included pulse tests, slug tests, injection tests, pumping tests, and WD tests.

Electrical conductivity, temperature, and pH of the groundwater were determined as a function of depth using multi-parameter probes. This was done four times three months apart, each time followed by taking water samples from different depths selected on the basis of the multiparameter probe results (*Fig. 6.2-13*). Depth to groundwater table and runoff flow rates were measured to estimate the groundwater recharge in the area. The water pumped by the Schönbrunn waterworks was analyzed for boron and AOX. A spatial and temporal model was derived from the results, describing the contamination emanating from the landfill.

The following investigation results show how a geological model of the site can be derived from the results of the individual methods. The detailed geological map in *Fig. 5.1-2* (SCHMIDT, 1993) shows the fault block structure in the area of the abandoned Eulenberg landfill. *Figure 5.1-11* shows the results of the detailed investigation of a fault along the north side of the abandoned Eulenberg landfill indicated by the reflection-seismic section EU04. Reconstruction of the fault genesis on the basis of trench observations shows that the fault is not open and, therefore, spreading of contaminants via this fault is unlikely.

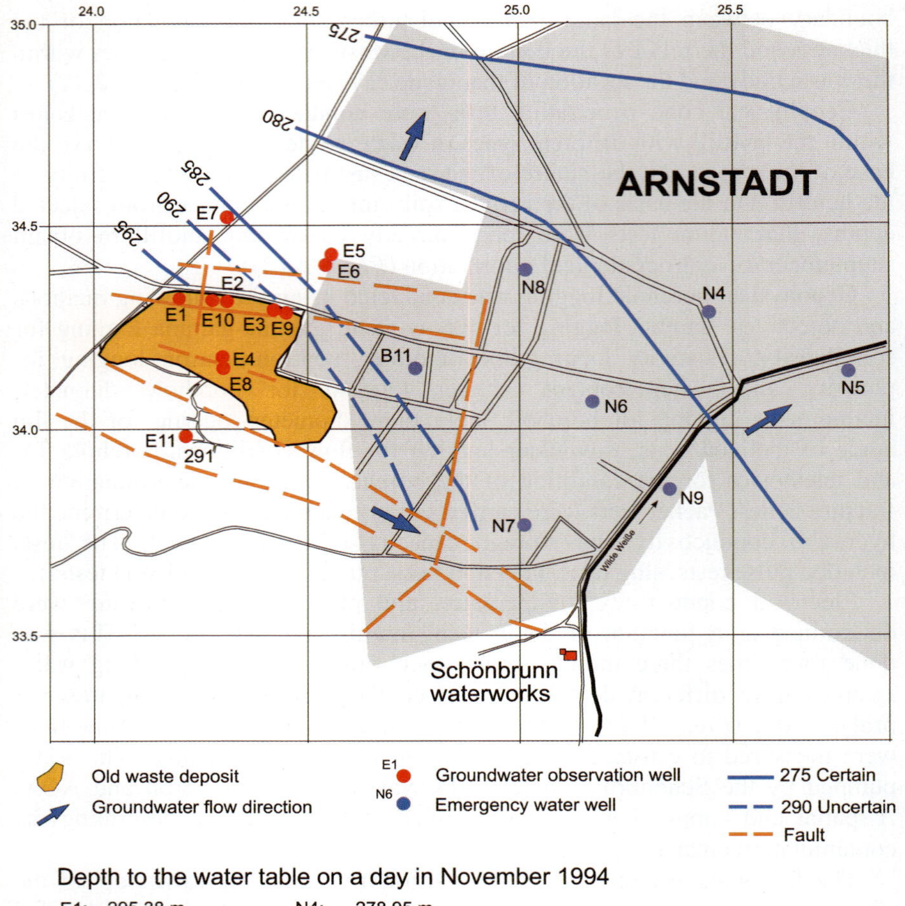

Depth to the water table on a day in November 1994

E1:	295.38 m	N4:	278.95 m
E2:	294.76	N5:	276.40
E3:	289.42	N6:	280.56
E4:	312.46	N7:	283.14
E5:	280.79	N8:	280.75
E6:	281.27	N9:	281.62
E7:	285.20		
E8:	307.63		
E9:	293.67		
E10:	295.89		
E11:	290.76		

Fig. 6.2-4: Abandoned Eulenberg landfill near Arnstadt, Thuringia, Germany: Groundwater isohypses derived from measurements on a day in November 1994. GFE Halle 1995, commissioned by BGR

Figure 6.2-4 shows, that the groundwater table does not form a continuous surface. In some places the aquifer is confined and modified by fracture and joint zones. The highest point of the groundwater table is within the landfill (borehole E4), sloping steeply downwards to the northeast and southwest.

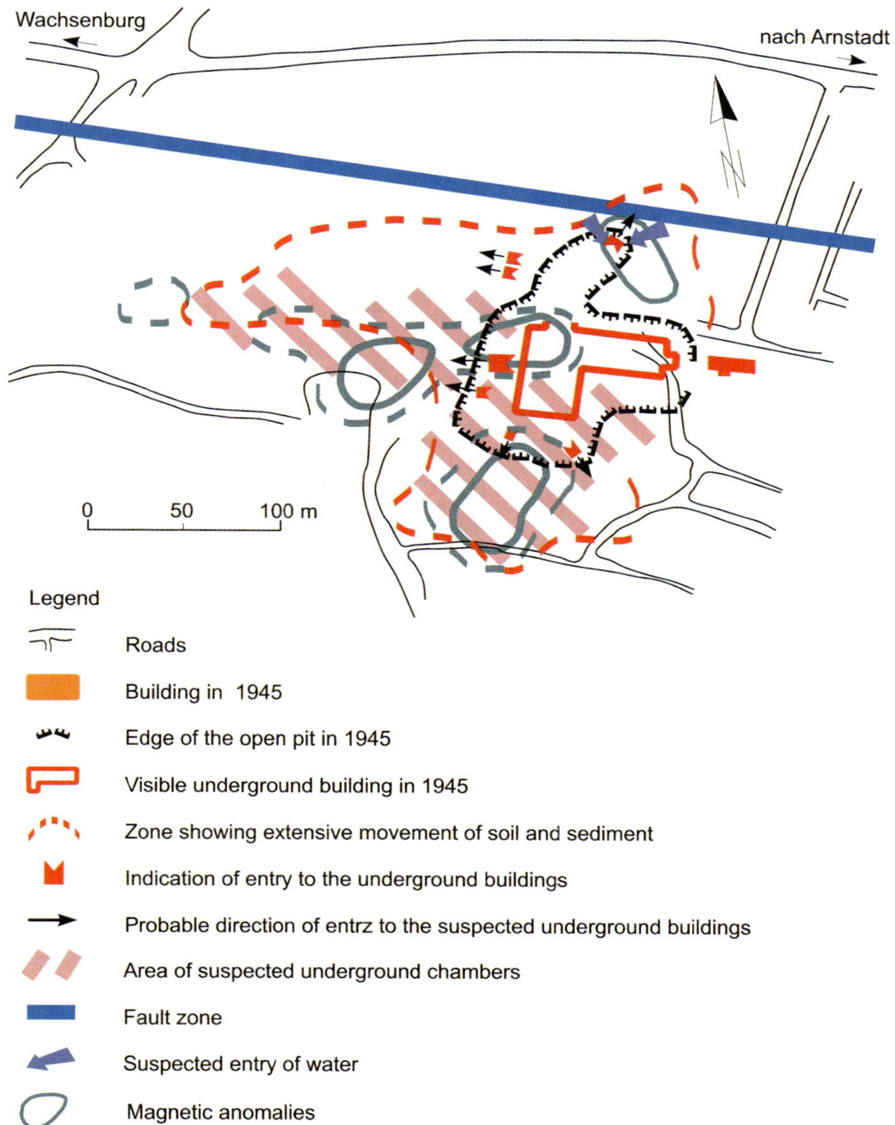

Fig. 6.2-5: Abandoned Eulenberg landfill near Arnstadt, Thuringia, Germany: Results of the analysis of archived aerial photographs (KÜHN & HÖRIG, 1995)

The aerial photograph shows a deep pit with an L-shaped, ca. 55 m × 70 m, apparently uncompleted underground manmade structure. It is not clear in the photograph whether there is the roof of an underground chamber or a foundation in the long eastern part of the pit. Geophysical cone penetration tests (CPT) show that there is a large chamber below a concrete layer.

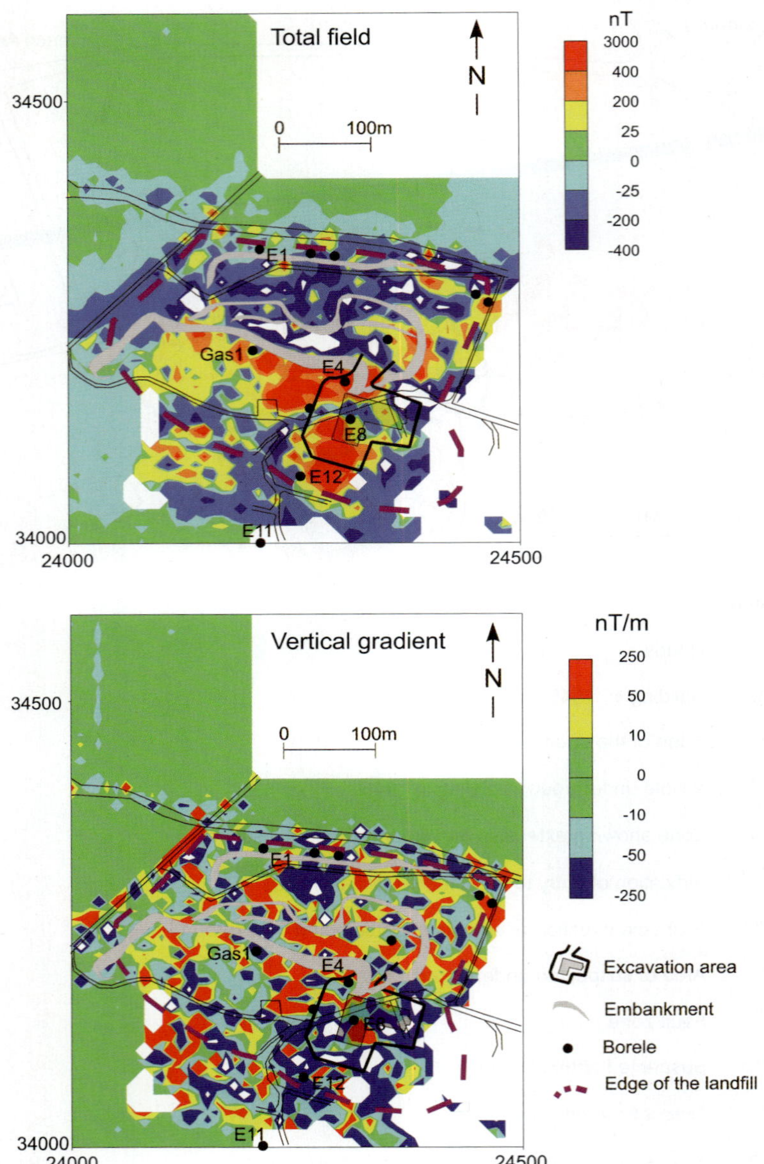

Fig. 6.2-6: Abandoned Eulenberg landfill near Arnstadt, Thuringia, Germany: Boundaries of the landfill, anomalies of total magnetic intensity 2.7 m above the ground (*top*) and the vertical gradient as derived from the measured total intensity at 1.3 and 2.7 m above the ground (*bottom*); Geophysik GGD 1992, commissioned by BGR

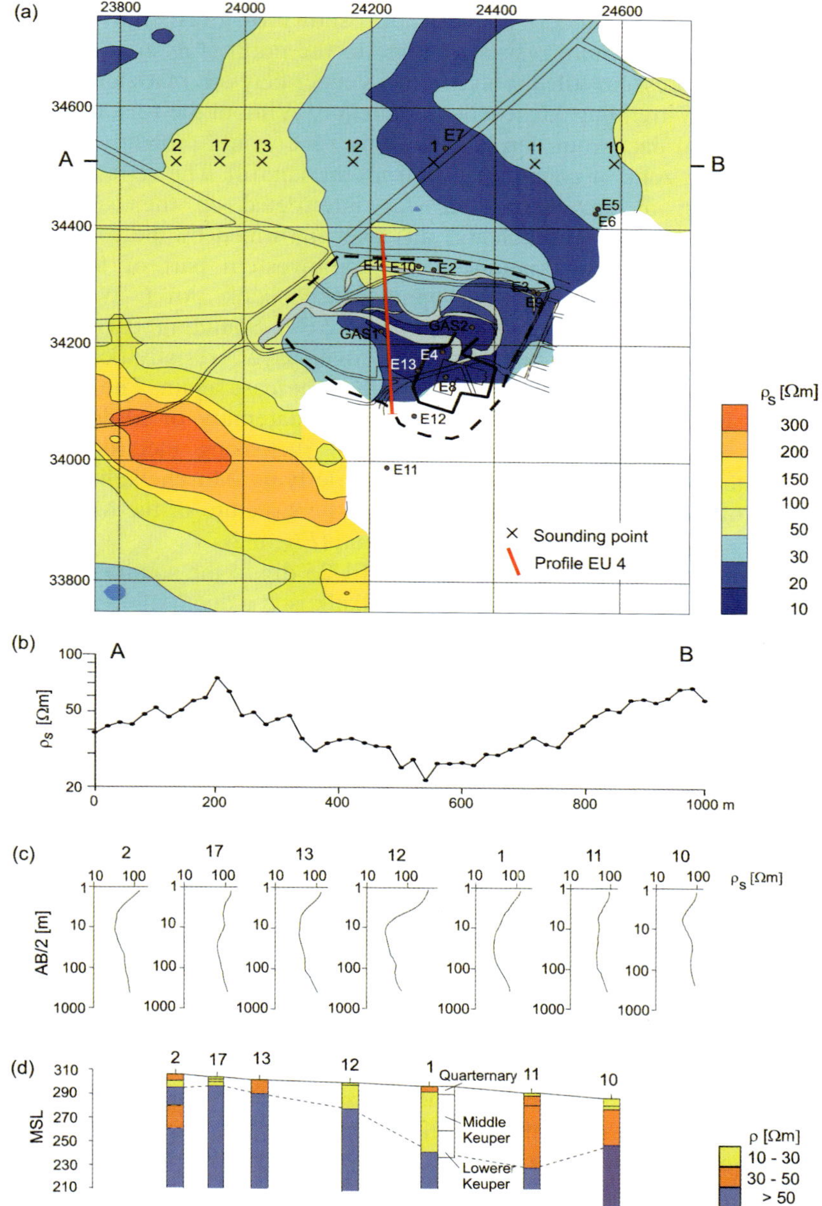

Fig. 6.2-7: Abandoned Eulenberg landfill near Arnstadt, Thuringia, Germany: dc resistivity measurements: (a) results of the resistivity mapping using the Schlumberger method; (b) apparent resistivity along profile A—B; (c) sounding curves from the Schlumberger soundings; (d) geoelectric profile after inversion of the curves in (c)

Besides the distinct boundary of the landfill shown by a resistivity minimum, there is another minimum resistivity zone to the north (*Fig. 6.2-7a*). This is caused by a depression filled with Middle Keuper clay-silt marlstone bounded by a NNE-SSW-trending Muschelkalk anticline. Without the seismic data and hydrogeological data from borehole E6, this zone would probably have been interpreted as a zone of contamination from the landfill. The zone of elevated resistivity west of the depression was interpreted on the basis of the geoelectric data to be Muschelkalk; this was confirmed by the seismic data. The strong resistivity maximum in the southwestern part of the area of investigation corresponds to outcropping Muschelkalk. An E–W resistivity profile across the area is shown in *Fig. 6.2-7b*. Sounding curves for the marked points show that below a thin cover layer, resistivities of about 100 Ωm initially decrease with depth and then increase again. The decreasing resistivity is due to low-resistivity, cohesive Quaternary or Middle Keuper sediments. The increase in resistivity with increasing electrode spacing indicates the increasing influence of the underlying Muschelkalk. The 1-D inversion of the individual soundings shows the structure of the Muschelkalk below the Keuper in the area of the depression (*Fig. 6.2-7d*).

The strata on reflection seismic profile EU25 dip to the south (*Fig. 6.2-9*). At the beginning of the profile, the Muschelkalk is at a depth of approximately 20 m. This interpretation is based on the results of borehole E6 connected over the profile EU30 (*Fig. 6.2-10*). The reflection at the beginning of profile EU29 indicates a relatively rapid decrease in depth to the south. Phase offsets, changing reflection dip, and diffracted waves indicate the presence of numerous fracture zones, which makes interpretation of the phases nearly impossible. The interpretation is based on the geological map and the results of the neighboring profiles.

The dominating structure in the area of investigation is a NW–SE depression. The lowest parts are between profiles EU27 and EU30 and in the area where profiles EU24/EU25/EU29 and EU24/EU26 cross (see *Figs. 6.2-8* and *6.2-11*). Features on profile EU30 running in the eastern part parallel to profile EU27 can be interpreted with relative high certainty supported by the connection to data from borehole E6. The fault seen at AP100 on this seismic line may be assumed to be identical with the one observed in the borehole E6 15 m south of the profile. The half graben located between AP85 and AP100 corresponds to a low in the base of the Cenozoic, which may be related to the fault. Profile EU24 east of the landfill shows a general ascent of the beds to the southwest, interrupted by faults with only a small offset. The interpretation of the Keuper/Muschelkalk boundary in the southwestern part of the profiles (AP60 – 107) is confirmed by the data from crossing profile EU02. The low in the surface of the Muschelkalk at the beginning of the profiles (AP7 – 60) is confirmed by the correlation to profiles EU25/EU29 and EU26. The thrust dipping to the southwest identified on profiles EU30 (AP45 – 65) and EU25 (AP95 – 105) could end between profiles EU25 and EU26 at the end of the southwestern fault zone or displace it. It may disappear gradually to northwest or at the northeast fault of the half graben.

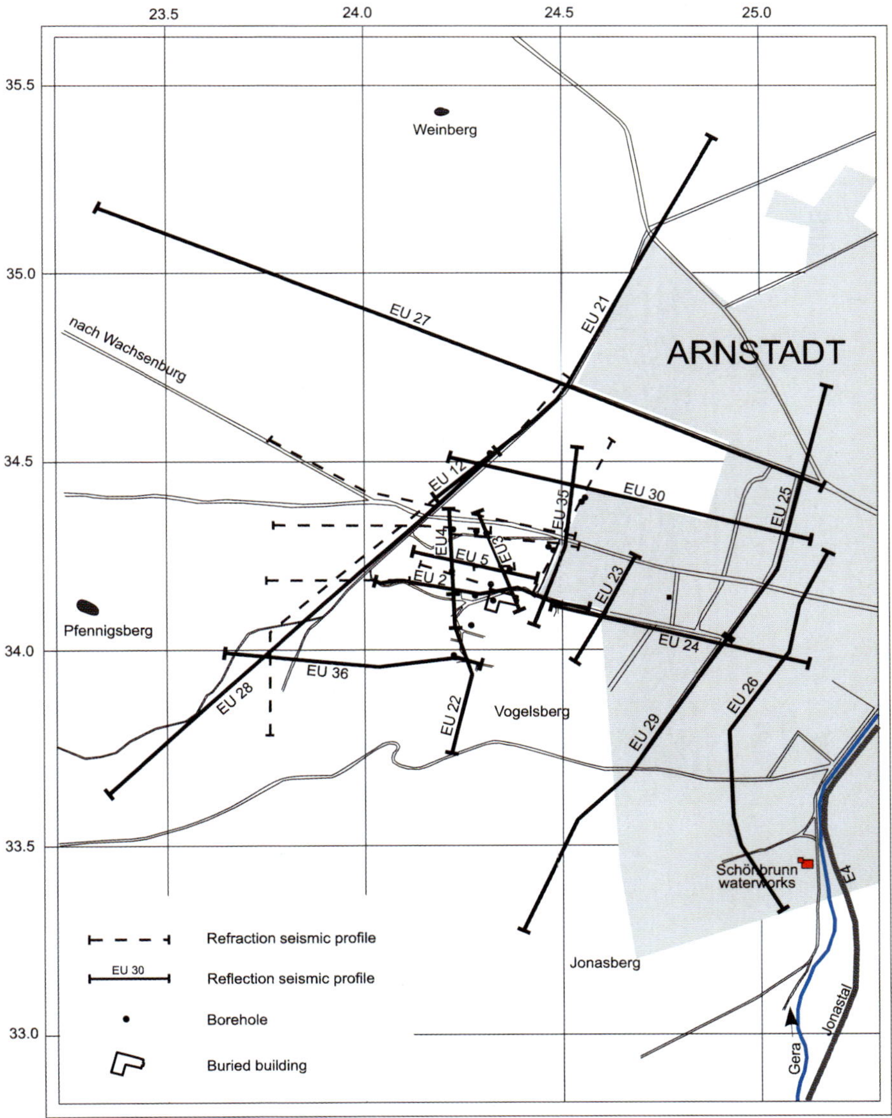

Fig. 6.2-8: Abandoned Eulenberg landfill near Arnstadt, Thuringia, Germany: Locations of the seismic profiles

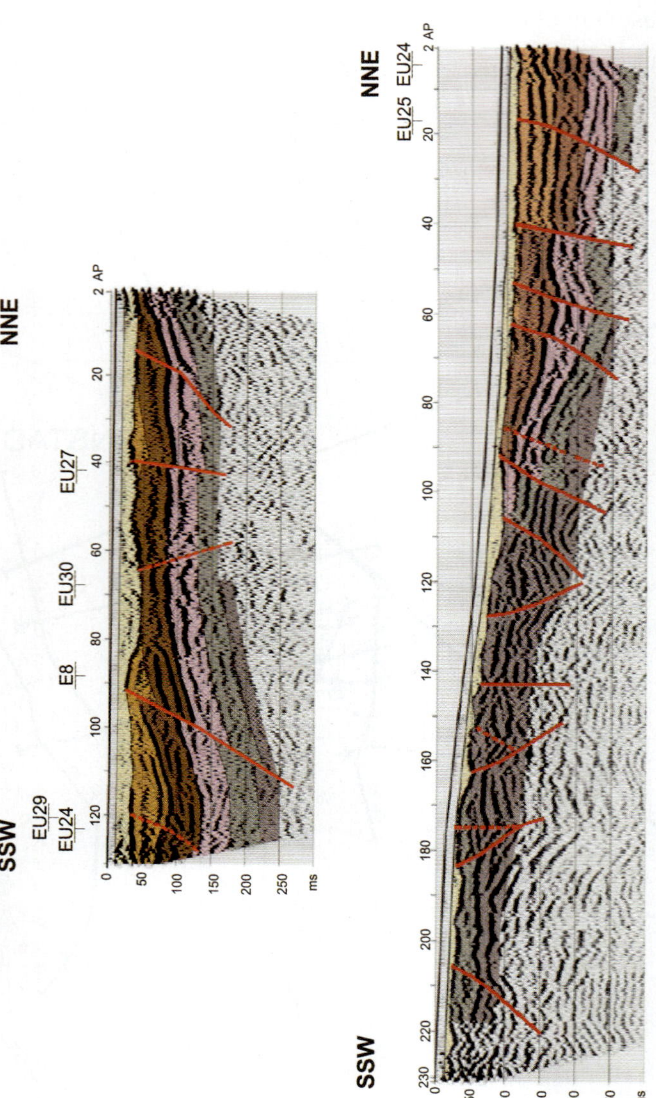

Fig. 6.2-9: Abandoned Eulenberg landfill near Arnstadt, Thuringia, Germany: Reflection seismic sections of profiles EU25 (*top*) and EU29 (*bottom*); Geophysik GGD 1993, commissioned by BGR (Legend see *Fig. 6.2-10*)

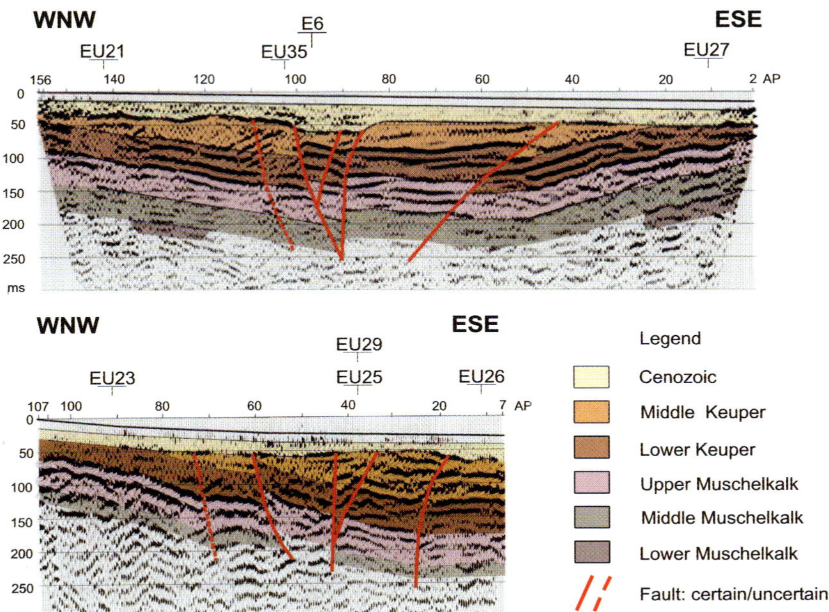

Fig. 6.2-10: Abandoned Eulenberg landfill near Arnstadt, Thuringia, Germany: Reflection seismic sections for profile EU30 (*top*) and EU24 (*bottom*); Geophysik GGD 1993, commissioned by BGR

The map in *Fig. 6.2-11* showing fracture zones and blocks results from the interpretation of 17 reflection seismic profiles, which provided the locations, dip directions, and character of the faults in the investigation area, as well as the dip of the beds underlying the Cenozoic or the weathering layer. Lithological/stratigraphic interpretation of the reflections was difficult, owing to the complicated reflection patterns between the blocks. A plausible interpretation was achieved using borehole data, geological mapping, and the geoelectric data.

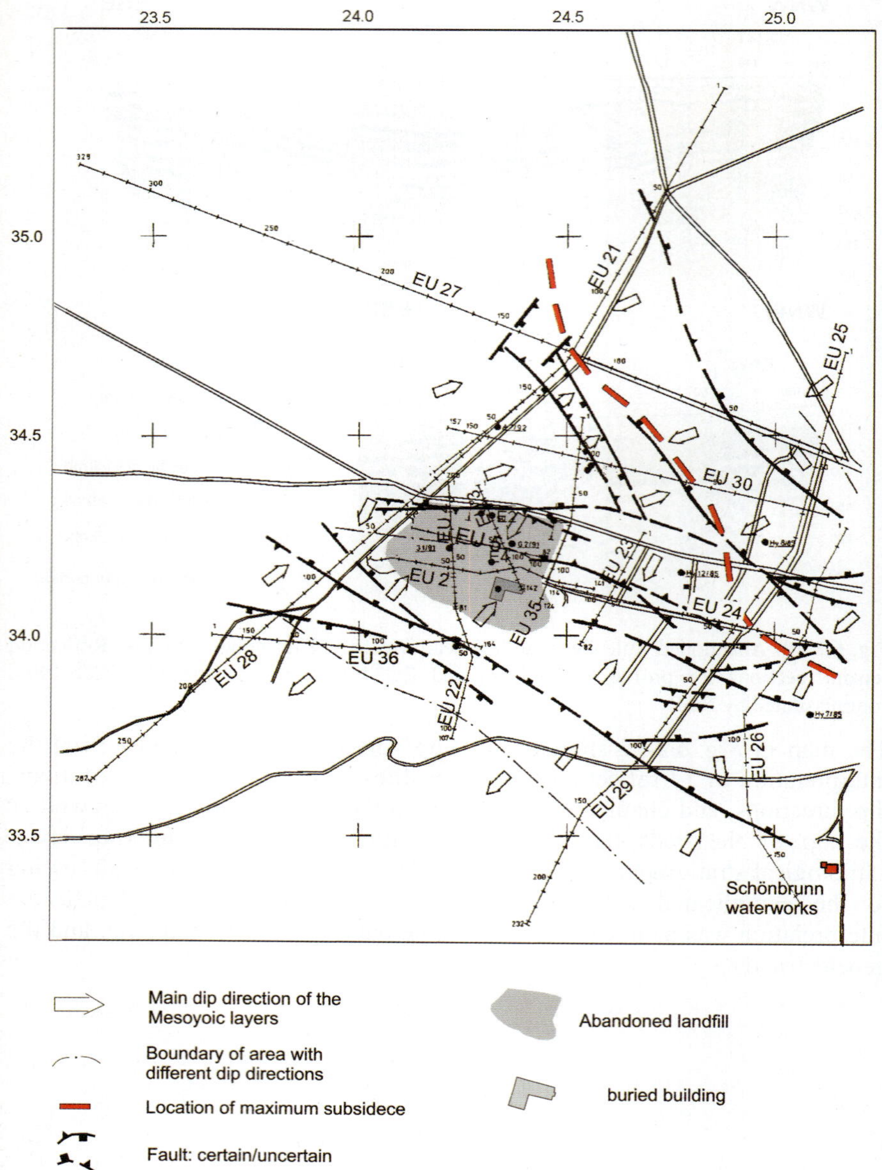

Fig. 6.2-11: Abandoned Eulenberg landfill near Arnstadt, Thuringia, Germany: Fracture zones and block boundaries as determined by reflection seismics, Geophysik GGD 1993, commissioned by BGR

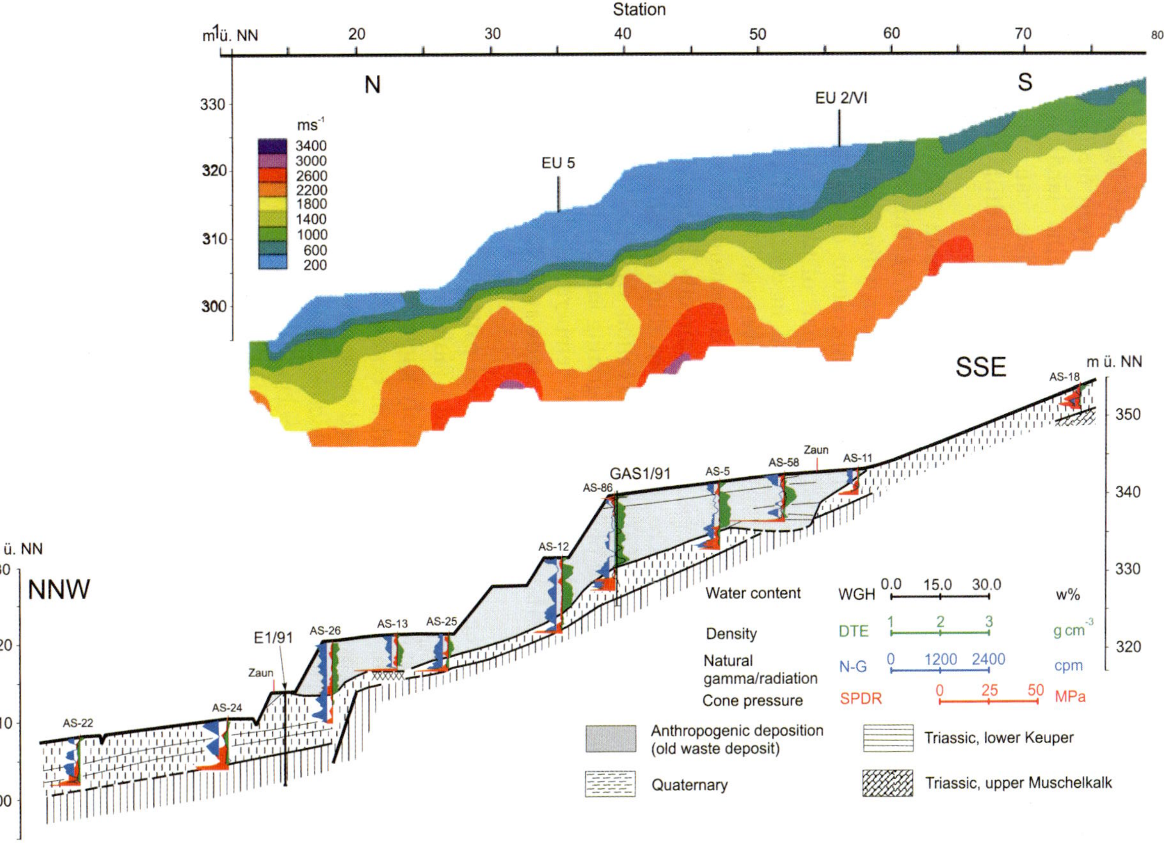

Fig. 6.2-12: Abandoned Eulenberg landfill near Arnstadt, Thuringia, Germany: Comparison of cone penetration testing profile IV with refraction tomography profile EU4

Cone penetration testing profile IV and refraction tomography profile EU4 were measured over the same traverse N–S across the landfill. The comparison in *Fig. 6.2-12* shows excellent correlation between the tomogram and the lithological conditions. The transition from the landfill to the surrounding rock is distinct. The thickness of the waste can be exactly determined. The offset fault zone at the north side of the landfill can be seen (*Fig. 5.1-11*).

None of the parameter values measured in borehole E8, except for temperature, change significantly with depth (*Fig. 6.2-13*). This can be viewed as an indication of vertical flow of percolation water within the waste body. This is underlined by the consistently high temperature of 19 - 20 °C. The drop in temperature by about 4 °C at the bottom part of the profile is due to inflow of freshwater. There was a large increase in electrical conductivity from 8000 $\mu S\ cm^{-1}$ in March 1993 (blue curve) to 18 000 $\mu S\ cm^{-1}$ in November 1994 (black curve), indicating increased contaminant mobilization by groundwater recharge in the landfill. The indirectly indicated inflow of water by the drop in temperature in the bottom third of the profile is not observable in the conductivity profile. Observation well E6 was drilled to monitor whether a contaminant plume was migrating in the originally assumed flow direction north of the landfill. The pH, temperature, and conductivity values measured in this borehole indicate groundwater with no contaminant load.

The investigation area shows a complicated pattern of blocks formed by faults of the Eichenberg – Gotha – Saalfeld – fault zone and various faults perpendicular to them. The area is extensively karstified. Karst cavities with a diameter of more than 1 m were observed in borehole E11. Several aquifers that are confined in parts form a complicated hydraulic system.

Except for thin Quaternary aquifers, which become significant north of the Eulenberg landfill, the aquifers below the landfill in the Muschelkalk and Keuper formation are connected in a complicated way via local fracture zones and karst cavities. The base of this aquifer system is the relatively impermeable claystone of the Upper Buntsandstein (Röt).

The groundwater table does not form a smooth surface. It is morphologically structured, in some places confined and strongly influenced by fracture and joint zones. The highest point of the groundwater table is within the landfill (borehole E4), sloping steeply to the northeast and southwest as shown in *Fig. 6.2-4*. It can be assumed that groundwater recharge within the landfill is higher than in the surrounding rock, owing to the higher hydraulic conductivity of the waste. The pressure compensation takes place by the relatively slow groundwater flow from the landfill into the surrounding.

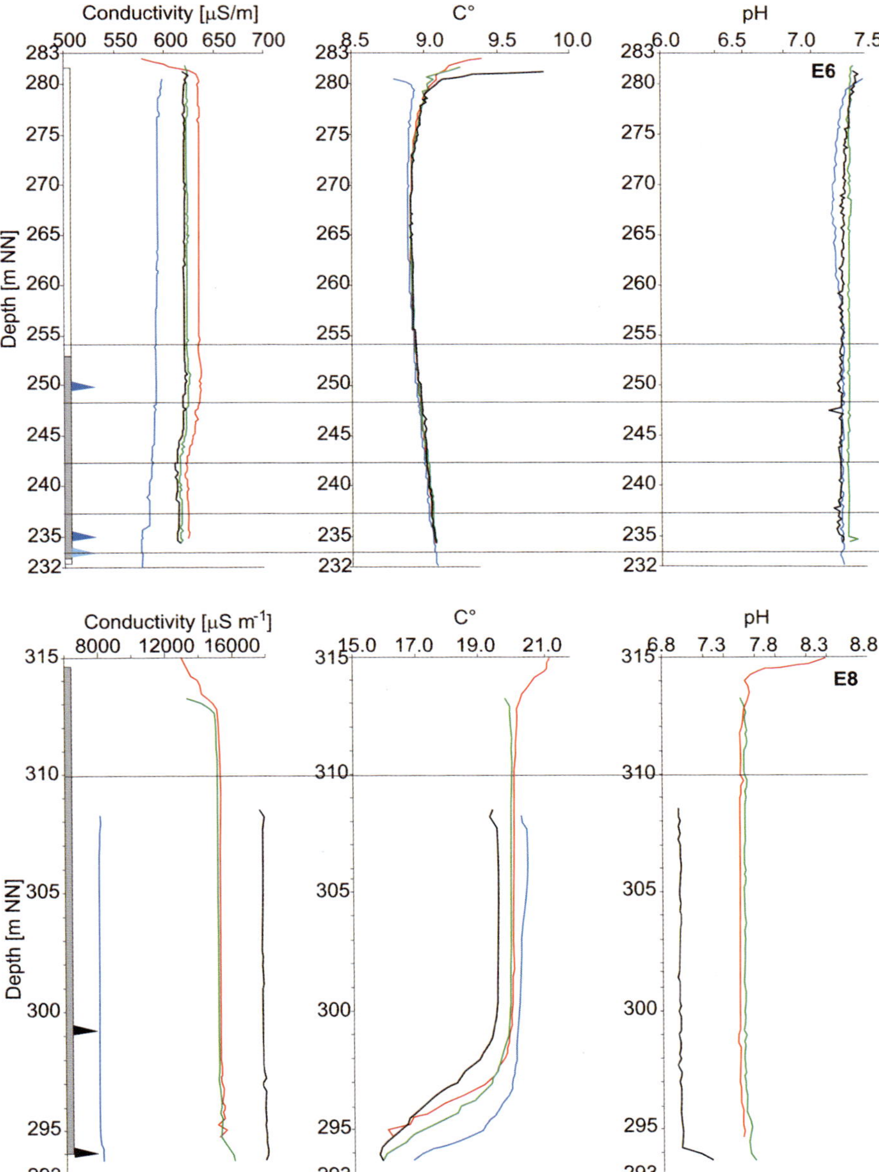

Fig. 6.2-13: Abandoned Eulenberg landfill near Arnstadt, Thuringia, Germany: Multi-parameter probe profile at the wells E8 (*bottom*), in the center of the landfill, and E6 (*top*), in the Keuper depression north of the landfill, location see *Fig. 6.2-4*, blue: March 1993, red: June 1994, green: September 1994, black: November 1994

The results of the hydrogeological investigations support the conceptual model, which predicts groundwater and percolation water flow below the landfill to greater depth and at the southern edge of the landfill along a fault towards the southeast. The NE–SW striking faults are significant, because they provide a path for water flowing from the landfill to the regional fault system. The highest flow rates are in the karstified Middle Muschelkalk in the NW-SE striking fault zones. The time between an accident in the neighboring town Bittstädt and the arrival of contaminants in the Schönbrunn waterworks shows that displacement velocities of 200 m h^{-1} can be expected.

Planned landfill Rabenstein

In a comparison of ten preselected sites, the Galgenberg near Rabenstein west of Chemnitz in Saxony, Germany, was chosen as the preferred site for a new landfill. The area consists of a sequence of metamorphic rocks. At the prospective landfill site, there is a NE-SW to N-S-striking sequence of phyllite, mica-schist, and black shale with a 20-50° SE to E dip, in places with only a thin cover layer. These rocks contain intercalated amphibolites, quartzites, and metacarbonates. The black shale has a lens structure, owing to sandy interbeds, as well as layers with a high percentage of pyrite and marcasite. It is platy and contains numerous joints filled with calcite and quartz. The amphibolites often contain volcanogenic material. In the metacarbonates, which are mined nearby, subrosion has created cavities and fissures since the Late Cretaceous. The corrosiveness of the subrosion solutions is attributed to the high pyrite content of the black shale. Rotliegend sediments were widely deposited due to subsidence during the Late Carboniferous and Early Permian, but there occur only local remnants in the Rabenstein area. In the immediate vicinity of the proposed landfill site, there are Rotliegend conglomerates and clay beds containing interbedded pyroclastics and other volcanic rocks.

The present structural and geomorphological conditions were formed during Variscan, Permian, Saxonian, and Cenozoic tectogenesis. In the area of investigation, the rock has been deeply weathered and karstified – related to uplift during the Late Cretaceous and Paleogene. The Rabenstein R3 borehole shows weathering products to a depth of more than 60 m. Aggressive solutions have deeply weathered the phyllite, mica-schist, and black shale to clayey loam containing no carbonate, which was overlain with loess during the Weichselian. Indications of periglacial landslides during the Pleistocene are observed in shallow boreholes.

The lineations observed in the aerial photographs have three predominant directions, indicating the existence of several fault zones: NNW-SSE, NNE-SSW, and E-W. The pattern of fracturing and the good cleavage observed in individual outcrops indicates that the hydraulic conductivity is anisotropic. Since these fractures are steeply dipping, it can be expected that percolation water from a landfill would rapidly enter the joint aquifers in the phyllite, amphibolite, marble basement rocks, at least where they are covered with only a thin weathering or talus layer.

Several methods were used to investigate the geological barrier at the Rabenstein site. First, geological mapping of the study area was conducted based on clitter, accompanied by an aerial photograph interpretation and fault analysis. This was followed by magnetic, geoelectric, and electromagnetic mapping as well as geoelectrical soundings to investigate the structural conditions and lithology in the area. New boreholes were drilled to investigate the basement rocks in the Galgenberg area and earlier boreholes drilled in uranium exploration surveys by the SDAG Wismut were evaluated.

In addition to shallow boreholes, 1 to 3 observation wells for hydraulic tests were drilled around boreholes R1, R3, and R5. Around the 100 m deep R5 borehole at the east edge of the proposed landfill site, three 35 m deep boreholes were drilled 7 m from it. This group of wells is located in an area with thin unconsolidated cover rocks. All boreholes were cored and geophysical well logs were made.

Packer tests were conducted in the uncased boreholes R1 and R2 to determine hydraulic parameter values and the initial formation pressure. Following the analysis of the drill cores, the permeability of the rocks was investigated using a TV camera to determine the rock fabric. Hydraulic tests were then conducted in borehole R5 and surrounding boreholes, beginning with pumping tests, slug tests, and bailing tests. The drawdown and recovery phases in the pumped well and the observation wells show a quasi-homogeneous picture of the penetrated rocks. The transmissivity values determined in the pumping tests are between $T = 5.1 \cdot 10^{-5}$ and $2.9 \cdot 10^{-4}$ m^2 s^{-1}. Five slug- and bail-tests were conducted in each borehole. The mean values from these tests are between $T = 2.1 \cdot 10^{-5}$ and $1.0 \cdot 10^{-4}$ m^2 s^{-1}, close to the values from the pumping tests. This shows that the slug and bailing tests are well suited for permeability determinations; their disadvantage is that they characterize only a small area around the borehole.

Borehole R5 was divided by packers into 18 intervals of 2.6 m length each. Slug and bail tests, hydraulic pressure tests, pulse tests, and drill-stem tests were then conducted. The slug and bail tests provided the best results. The hydraulic pressure tests also proved to be useful. The results are very close together, providing information about a ground volume up to 10 m around the borehole. On the basis of the results, the drill-stem test cannot be recommended in hard rock. The pulse test is the only suitable method to determine hydraulic conductivity values $k_f < 10^{-7}$ m s^{-1}. This test requires considerable technical efforts. The test itself is very simple and yields reliable results, as shown by repeated measurements. Interference tests were conducted in two of the tested intervals in borehole R5. This allowed a detailed picture of the water flow within the test area. Water flows here mainly along bedding planes and fractures, where the highest flow rates with the highest tracer concentrations were observed.

Fig. 6.2-14: Planned waste disposal site at Rabenstein near Chemnitz, Germany: Lineaments derived from b/w aerial photographs from 1988 (TU Bergakademie Freiberg, 1993, commissioned by BGR)

The interference tests were followed by a tracer test in each of the packer intervals to simulate the spread of contaminants. The packer intervals with the highest tracer inflow rates correspond to those indicated by the interference tests. The highest tracer concentrations did not occur in the intervals with the highest flow rates but in the intervals with moderate flow rates of the interference tests. Thus, contaminant transport will occur mainly along the planes which are hydraulically the most effective. The rate of contaminant bedding transport is greatest in the primary flow direction, as also shown by the interference tests. The highest tracer concentrations, however, were observed in the direction of moderate flow rates. This is because in the primary flow direction there is more water available for dilution. Thus, in the primary flow direction the contaminant front is considerably further moved than the maximum concentration front. Nothing can be said, however, about long-term behavior, considering the short time of the tests (seven hours). Nevertheless, the tests show the main hydraulic flow directions.

The site under investigation is characterized by numerous NNW-SSE, NNE-SSW, and several ENE-WSW to E-W striking lineaments in the aerial photographs. Two NNE-SSW lineaments of about one kilometer length associated with a buried channel structure can be seen in *Fig. 6.2-14*.

The magnetic, geoelectric, and geomagnetic survey data support the results of the geological survey with respect to the structures in the basement and cover rocks, and the regional and local groundwater system. Because it was known that some of the rocks in the area under investigation are magnetic (phyllite, hornblende shale, and amphibolite), it was expected that the magnetic survey (*Figs. 4.1-7* and *4.1-8*) would provide information about the basement lithology and structures. The electromagnetic data (*Fig. 4.4-11*) provided information about the thickness and material of the cover and about fracture zones in the basement which is of primary importance for the assessment of the geological barrier. On the basis of the geological map, thick (>10 m) Holocene sediments were expected only near streams in the valleys. In the center of the investigated site, only a thin loess loam cover and weathering layer was expected on the steeply sloping surface.

The model of the boundaries of different lithotypes and the fracture zones (*Fig. 6.2-15*) is based mainly on the dc geoelectric depth soundings and mapping. The interpretation also used the geological and magnetic survey data. The distribution and thickness of the unconsolidated cover layer as derived from geoelectric depth soundings and mapping (*Fig. 6.2-16*) was confirmed by data from the Wismut boreholes, the BGR shallow boreholes FB1-13, and the cored boreholes R1-5, and has been converted into a thickness map of the unconsolidated cover rocks (*Fig. 6.2-17*).

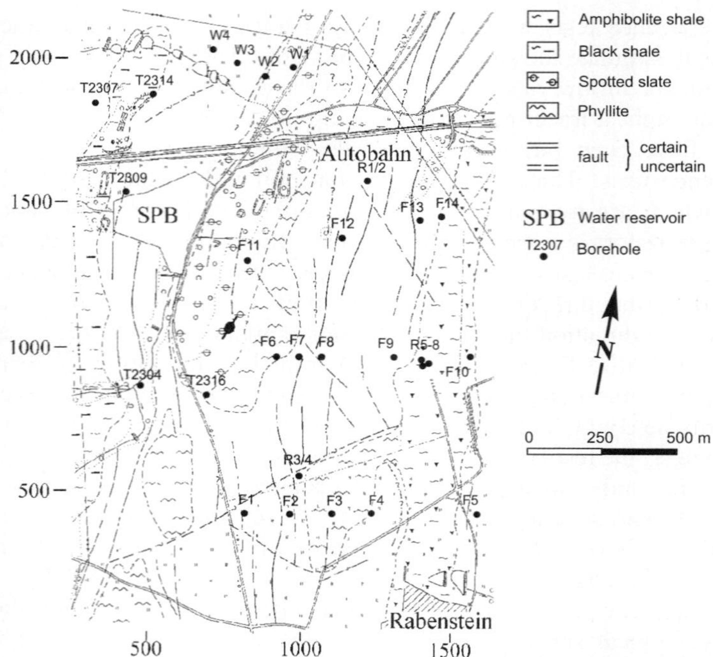

Fig. 6.2-15: Planned waste disposal site at Rabenstein near Chemnitz, Germany: Lithotectonic map prepared from geological mapping based on clitter, as well as geoelectric, and geomagnetic survey data, Geophysik GGD 1993, commissioned by BGR

The lithological evaluation of the geophysical well logs is a typical example of joint interpretation. The lithologs were used with other data in the next joint interpretation step. The well logs in the 100 m deep R5 borehole and the lithological logs from the cores are shown in *Figs. 6.2-18* and *6.2-19*. The natural gamma log (GR) shows no variation in the upper part of the borehole. In the lower part there are prominent maxima up to six times the amplitude of the curve for the overlying phyllites and amphibolites (*Fig. 6.2-18*). This elevated gamma activity is from the radioactive nuclides of the U-Ra series in the black shale. The correlation with low resistivities ($< 10\ \Omega$m) in the electric well logs (FEL, KN, GN) and anomalies in the self-potential log (SP) are due to the high electronic conductivity of the graphite in the black shale (*Fig. 6.2-19, left*).

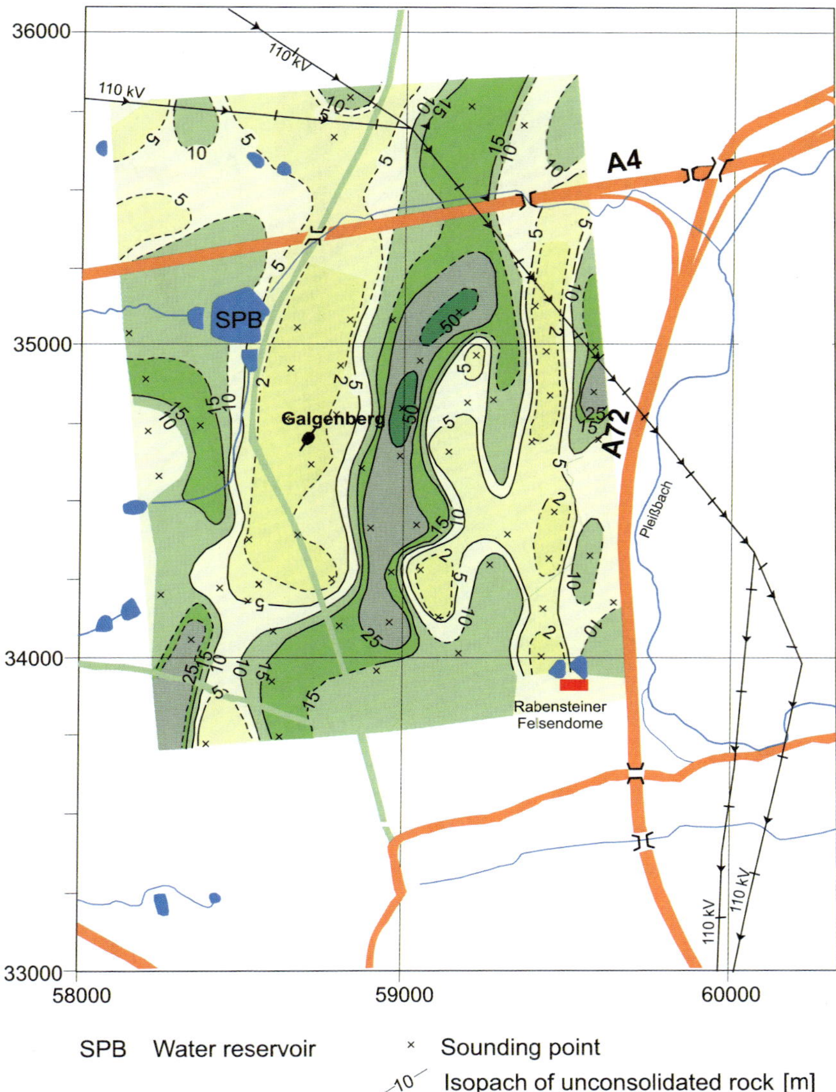

Fig. 6.2-16: Planned waste disposal site at Rabenstein near Chemnitz, Germany: Thickness of the unconsolidated cover rocks according to the results of the geoelectric depth soundings (VES) and mapping, Geophysik GGD 1993, commissioned by BGR

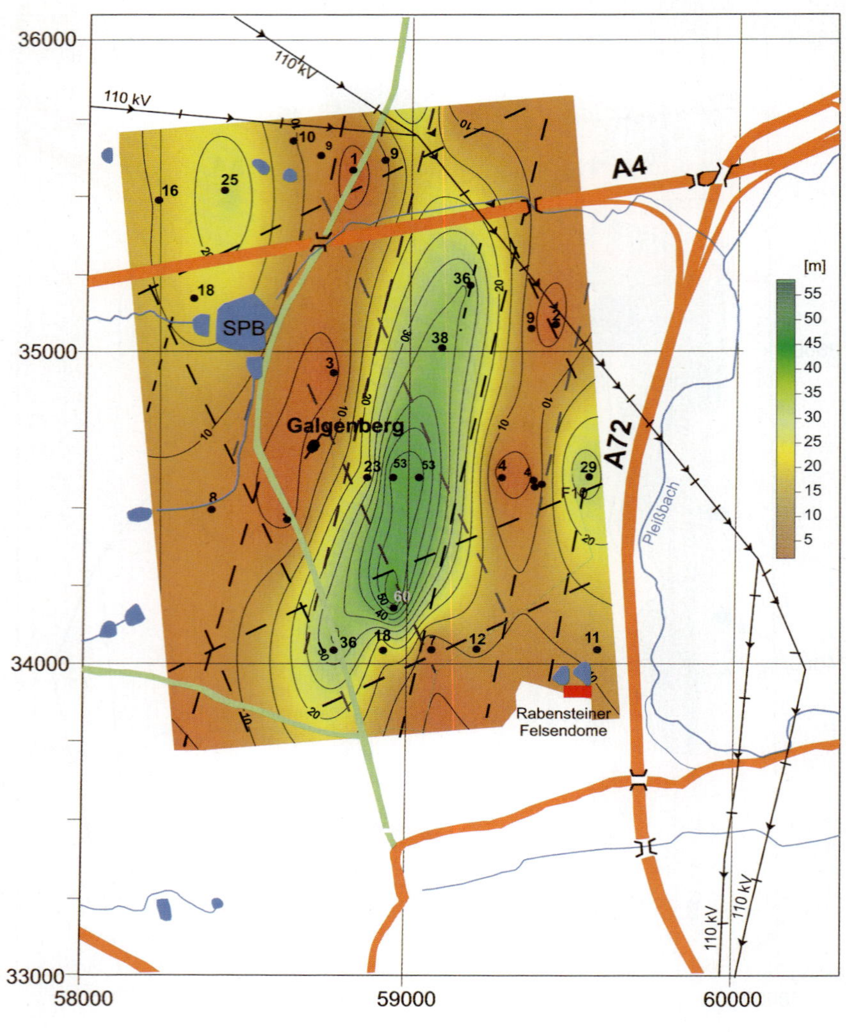

SPB Water reservoir

52.5 Borehole / thickness of the weathering layer [m]

Fault according to geophysical data and aerial photo interpretation

Fig. 6.2-17: Planned waste disposal site at Rabenstein near Chemnitz, Germany: Thickness of the unconsolidated cover rocks according to the borehole data

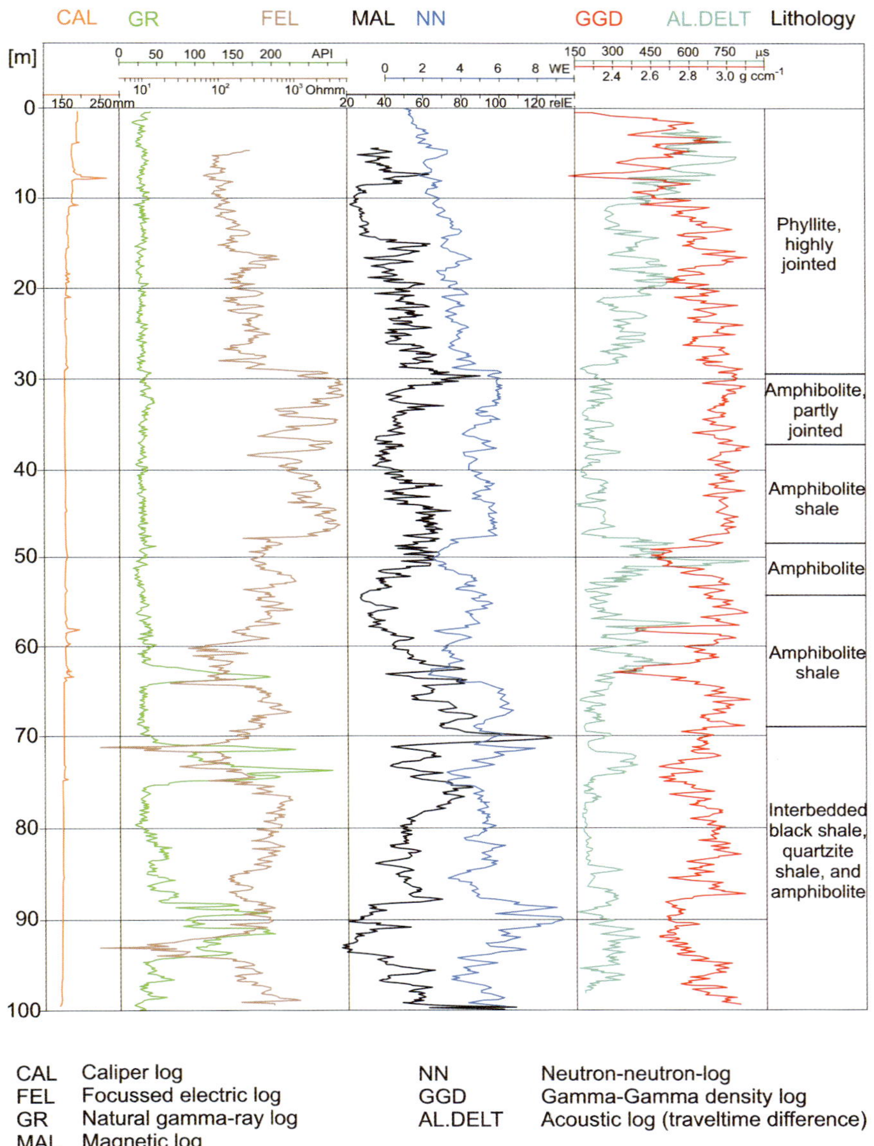

Fig. 6.2-18: Planned waste disposal site at Rabenstein near Chemnitz, Germany: Geophysical well logs in borehole R5, WE = water unit (WU), BLM 1994, commissioned by BGR

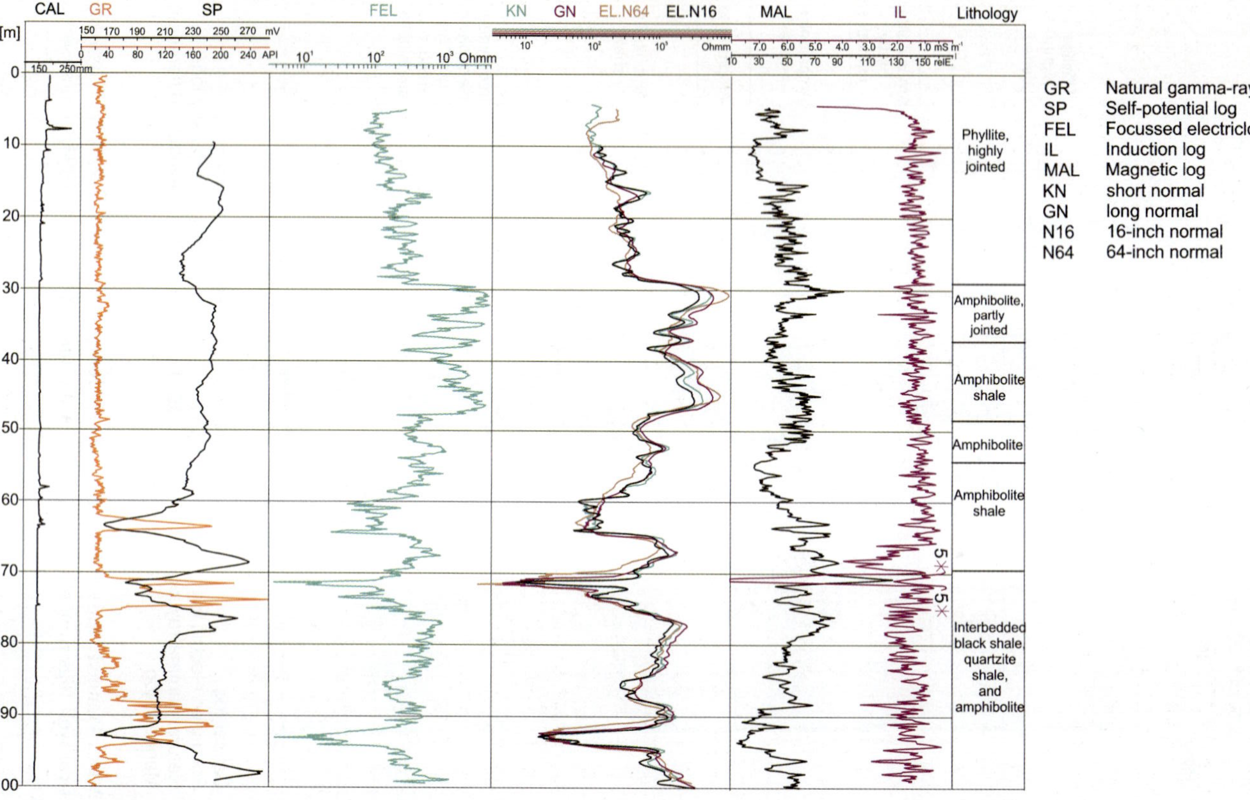

Fig. 6.2-19: Planned waste disposal site at Rabenstein near Chemnitz, Germany: Further geophysical well logs in borehole R5, BLM 1994, commissioned by BGR

A lithological subdivision can be made using the density log (GGD), and zones of different degrees of consolidation, fracturing and weathering can be distinguished. The increase in bulk density within the phyllite complex from ~2.7 to 2.9 g cm^{-3} is due to the increasing consolidation of the phyllites with depth. Above a depth of 8 m, minima indicate weathered layers within the phyllite. Below 45 m, strongly fractured amphibolite shale is indicated by minima in the GGD and NN logs and maxima in the acoustic traveltimes AL.DELT (*Fig. 6.2-18 right*).

The neutron log was, besides the lithological interpretation, used to evaluate the degree of consolidation and extent of fracturing. The values in the NN log increase in the phyllite complex from < 2 WU to about 4 WU. In the upper part of the borehole, the individual weathering horizons can be identified by the distinct increase in the WU values. The base of the phyllite complex is marked by an increase from 4 WU to 6 WU at a depth of about 29 m. In general, fractured water-bearing zones appear as minima and highly consolidated zones as maxima. The black shale has distinctly higher maxima than the phyllite and amphibolite, due to its lower water content (*Fig 6.2-18 middle*).

The values of the conventional resistivity logs (EL.GN/KN and FEL) were also used for the lithological subdivision and estimating the degree of consolidation. Layer boundaries were often identified on the basis of abrupt increases or decreases in resistivity. Zones of increasing weathering and loosening of the rock are indicated by decreasing resistivity. The distinct increasing resistivity trend recognizable in the FEL and SP logs down to about 29 m indicates the increasing compactness of the phyllite complex with depth, as also shown by the GGD log. Local maxima in all of the resistivity logs within the phyllite are caused by the intercalated phyllitic shale with higher resistivity. The abrupt increase in resistivity at 29 m indicates the phyllite/amphibolite boundary. The average values in the FEL log increase here from 200 to 2000 Ωm, and in the EL logs from 400 to 3000 Ωm. The fluctuations below 29 m indicate an alternation between zones of higher and lower schistosity. Distinct minima (< 20 Ωm) at 72 m and between 90 and 94 m depth are caused by high-conductivity black shale layers. The magnetic log (MAL) indicates ferromagnetic and paramagnetic material in the rocks or filling fissures (e.g., magnetite and pyrite) and indicates variations in the lithology.

Because unconsolidated and fractured rock shows longer compression-wave traveltimes than compact rock, the acoustic log (AL) could be used to identify fracture zones. In the uppermost part of the phyllite layer (above 10 m), highly weathered zones are indicated by a distinct increase in traveltimes in the AL.DELT log. Black shale beds and slightly weathered phyllite with little fracturing are indicated by short traveltimes. Particularly noticeable is the decrease in mean traveltime from 350 μs in the phyllite to 220 μs in the amphibolite between about 29 and 47 m, abruptly increasing again at 47 and 57 m. The black shale and the intercalated black shale/quartzite/amphibolite below 74 m depth show constant traveltimes of about 200 μs (*Fig. 6.2-18*).

The combined evaluation of fluid logs, flowmeter logs, temperature logs and redox potential logs permitted the determination of inflow and outflow intervals and preparation of a "flow profile". The fluid and flowmeter logs, and the conductivity, redox potential, pH, and temperature logs made with the multi-parameter probe are shown in *Fig. 6.2-20*. The arrows indicate the inflow and outflow zones and the direction of vertical flow. As expected, the zones indicated by the electric and NN logs to be fractured are also hydraulically active. The main inflow zones are indicated in the flowmeter log by increasing values and in the fluid log by changes with time. The highest inflow rate (3.5 L min^{-1}) was observed between 21 and 26 m depth. This strong inflow was shown by using a salt tracer moving upwards in the well. The highest outflow rate (1.2 L min^{-1}) was measured between 7.4 and 8 m. The temperature, redox potential and conductivity logs clearly indicate the inflow zones. The drilling logs also document highly permeable zones via drilling mud losses.

Of particular value for the characterization of the groundwater is the pH. The pH shows an average value of 9.0 with little variation. The slight decrease in pH between 54 and 64 m depth confirms the zone of inflow shown in the "flow profile". The relatively high pH of 9.0 indicates the groundwater in the well has flowen out of the carbonate rocks. A further indication that this is deep groundwater is the very low nitrate concentration in the groundwater, despite intensive agriculture in the area. The cores showed the first carbonaceous rock at a depth of about 80 m. Below 64 m depth the flow in the well is relatively weakly downwards. It may be assumed that vertical flow occurs around the well through joints between layers. Above 64 m depth, the flow is mainly upwards in the well (*Fig. 6.2-20 middle*).

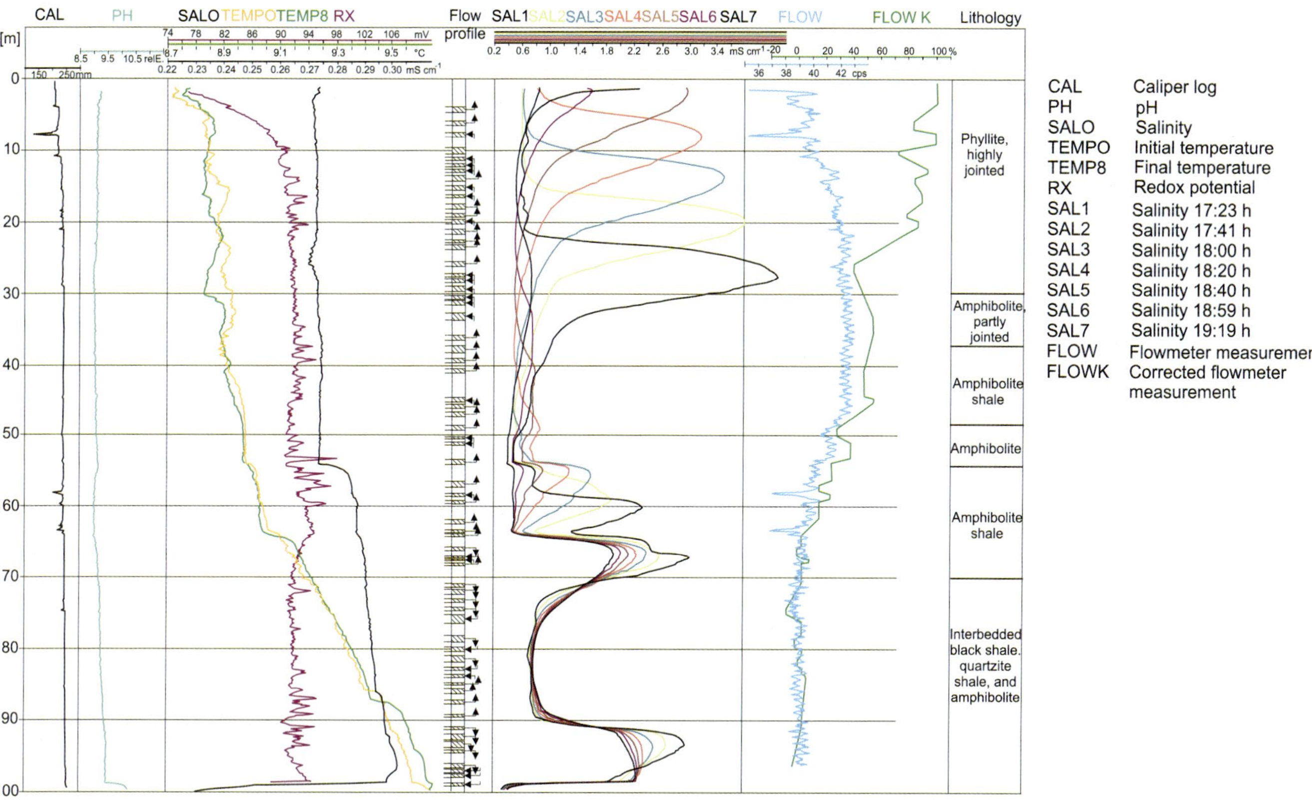

Fig. 6.2-20: Planned waste disposal site at Rabenstein near Chemnitz, Germany: Fluid logs and flowmeter logs to identify hydraulically active zones and to prepare a "flow profile". The results of the multi-parameter probes are also shown (*left*). BLM 1994, commissioned by BGR

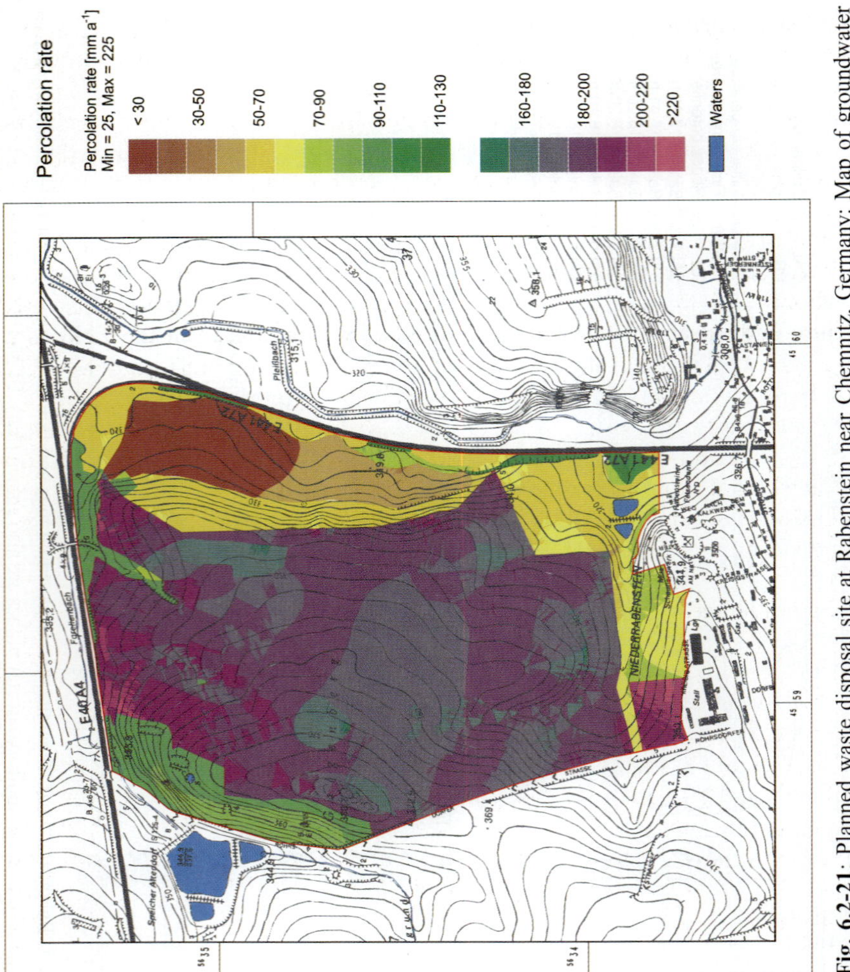

Fig. 6.2-21: Planned waste disposal site at Rabenstein near Chemnitz, Germany: Map of groundwater recharge rates, FUCHS (2003)

The groundwater recharge rate was determined using soil data from the area of the proposed landfill base and its immediate vicinity using the regression approach of RENGER & WESSOLEK (1996). This value is needed to prepare a groundwater flow model of the site. In addition to the soil data, the mean precipitation over a 30-year period is used in the calculations: at the Chemnitz weather station, 273 mm for the winter half year (October 1 - March 31) and 427 mm for the summer half-year (April 1 - Sept. 30). The map in *Fig. 6.2-21* shows the areal variation in groundwater recharge. The total amount of water that infiltrates into the ground in the 168 ha investigation area (1.68 km^2) has been calculated to about 250 000 m^3 annually.

The most important result for the assessment of the geological barrier at the Rabenstein site is the distribution and thickness of unconsolidated rock, derived from the dc electric and electromagnetic mapping and resistivity soundings (*Fig. 6.2-16*). This model was confirmed by the shallow boreholes drilled at a later investigation stage (*Fig. 6.2-17*). The previous knowledge of the unconsolidated cover rock and thus, the geological barrier had to be revised.

East of the Galgenberg, the area of investigation is crossed by a NNE-SSW-trending buried channel structure 400 m wide, up to 60 m deep, and about 1500 m long. In the northern part of the investigation area this structure ends at an E–W-trending fault zone, which is seen in the field as a topographical bench and as a lineation in the aerial photographs. It is not clear in the borehole and mapping data whether this structure was formed before or after the karstification of the metacarbonate rocks. The fill mainly consists of clayey weathering products, presumably with several aquifer levels that are not connected. For this reason, confined groundwater occurs locally. From the groundwater divide, which extends eastwards from the Galgenberg, the groundwater flows to the ENE and SSW.

The Schöneiche-Mittenwalde area

Schöneiche-Mittenwalde southeast of Berlin is an area of unconsolidated rock with the pollution of nearly a hundred years of industry, waste disposal, intensive agricultural use, especially intensive animal husbandry.

The bedding conditions of the Quaternary sediments are complicated. Aquifers and aquicludes are not persistent, due to glacial compression and erosion, and the aquifers are hydraulically connected. The underlying Tertiary sediments – silty fine-grained sand; coarse and medium-grained sand containing little silt; coaly silt and lignite – are normally undisturbed. The results of the study can be applied to numerous other sites in northern Germany.

The study area (*Fig. 6.2-22*), with the operating Schöneiche (domestic waste) and the Schöneicher Plan (mixed waste) landfills in the center, was sized especially large in order to study the groundwater system and the influence of other sources of contaminants. The results show that a smaller area would have led to incorrect conclusions. Remote sensing, airborne geophysics, geochemistry, and to a certain degree geoelectric methods can be used to facilitate the study of such a large area.

A geological model was prepared on the basis of archive documentation, geological mapping 1 : 50 000, revision of the sheet Königs Wusterhausen, reports of geophysical surveys, coring and trenching from soil surveys, and new boreholes. This model shows the lithology, facies, and stratigraphy of the area and was used to prepare a hydraulic model of the area.

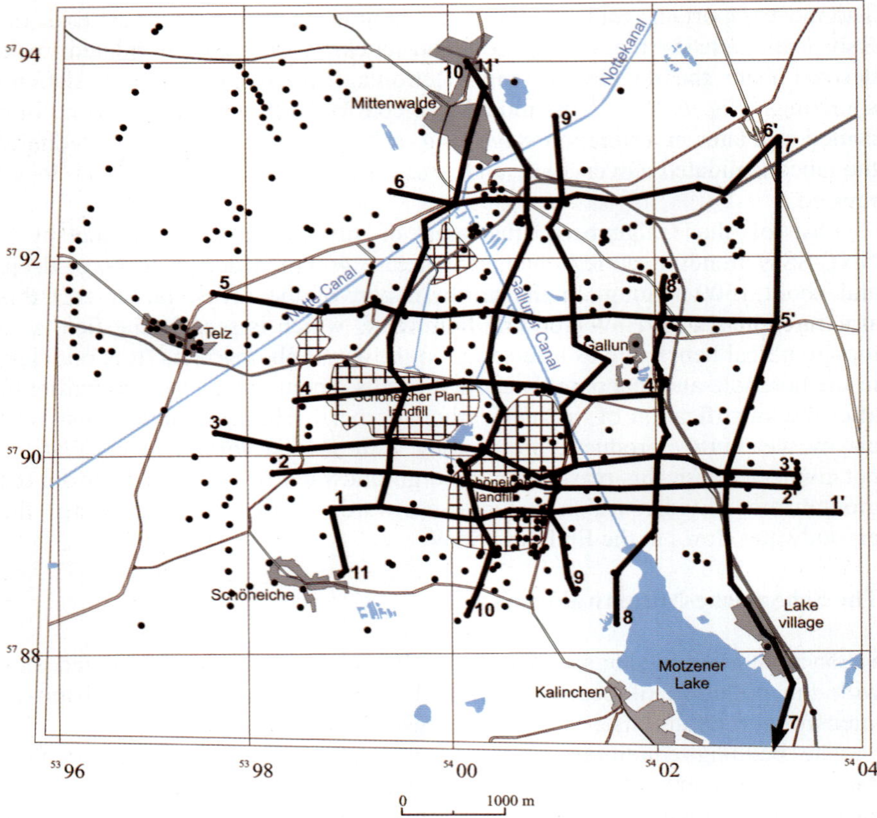

Fig. 6.2-22: Schöneiche-Mittenwalde area: Map showing boreholes and geological cross-section lines (see *Fig. 6.2-25*)

In addition to the multitemporal analysis of satellite images and aerial photographs, the area was surveyed with a thermal scanner, new CIR aerial photographs and spectroradiometer images were analyzed. The last was done to evaluate the extent of stress caused by pollutants in plants typical of the area.

Information about underground structures, groundwater flow, and presence of contaminants was provided by geophysical measurements. Hidden contamination was detected by an airborne geophysical survey (electromagnetics, magnetics, and gamma-ray spectroscopy), which also provided information about the geology and hydrogeology of the area. dc geoelectric and induced polarization surveys provided information to a maximum depth of 100 m about the distribution of cohesive sediments that could hinder entry of contaminants from the landfills into the groundwater. Both electrical resistivity and chargeability were measured using the IP method, the latter should provide a better distinction between cohesive and noncohesive sediments, and between contaminated and uncontaminated areas. It was found, however, that the IP effect was very slight in the study area. A seismic survey of the Schöneicher Plan landfill was made with high coverage. Refraction-seismic, refraction-tomographic, and reflection-seismic evalution methods were used to analyze the data. Cone penetration tests (CPT) were carried out to a maximum depth of 30 m, in combination with geophysical logging. At certain depths water and soil samples were taken.

Multi-parameter probes were used at five different times to measure electrical conductivity, temperature, pH, and redox potential of groundwater as a function of depth in ten observation wells.

An analysis of historical land use in the area was made to determine possible sources of contamination. Archived aerial photographs are useful for this purpose. Other sources of information are archived documents and interviews with persons living or working in the area in the past.

A geochemical analysis was carried out of the complex distribution of contaminant chemicals in the soil as well as stream and lacustrine sediments in the study area to determine background values and anthropogenic input. An inventory was made of the chemicals in the environment emitted from the Schöneiche industrial plants (chemical, metal processing, metal painting, and galvanic plants) and landfills (domestic, mixed, and industrial wastes), as well as from intensive agricultural land use. Anthropogenic enrichment and the geogenic background of heavy metals can be distinguished on the basis of areal distribution trends, so that hazardous areas can be identified. The geochemical data was analyzed using univariate and multivariate statistical methods. Maps of the distribution of individual elements were prepared, as well as geochemical anomaly maps (factor values and cluster points), maps of the geochemical contaminant load in the top soil, and maps of the differences between the measured element concentrations and limiting values given in the Berlin List. For the first time, inorganic and organic contaminants were analyzed together using multivariate methods, as factor and cluster analyses.

The following selected results show how a holistic model of an area with a multitude of contaminants and a complicated geology can be prepared from the individual results.

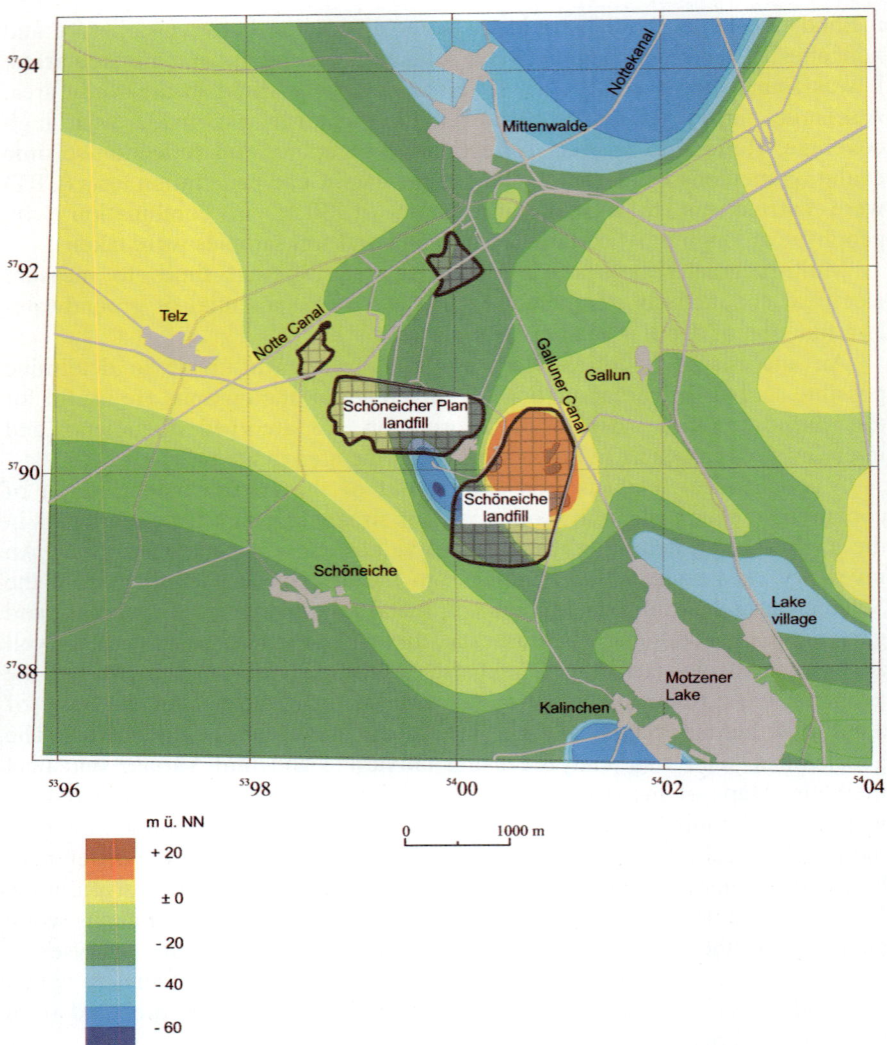

Fig. 6.2-23: Schöneiche-Mittenwalde area: Map of the base of the Quaternary

A geological model (*Fig. 6.2-23* and *Figs. 6.2-24* and *6.2-25*) was derived for the Quaternary in the investigation area using new and old drilling data.

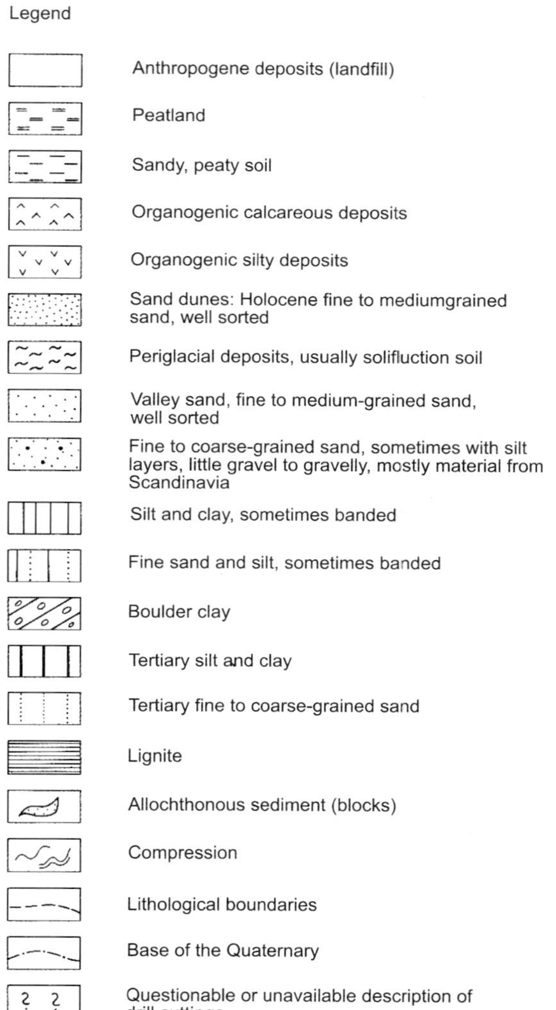

Fig. 6.2-24: Schöneiche-Mittenwalde area: Legend for the geological cross-sections

The geological cross-sections show gaps in the aquicludes. In some areas, permeable sand extends to the base of the Quaternary and into the Tertiary. The corresponding remote sensing results are presented in Chapters 3.1 and 3.3.

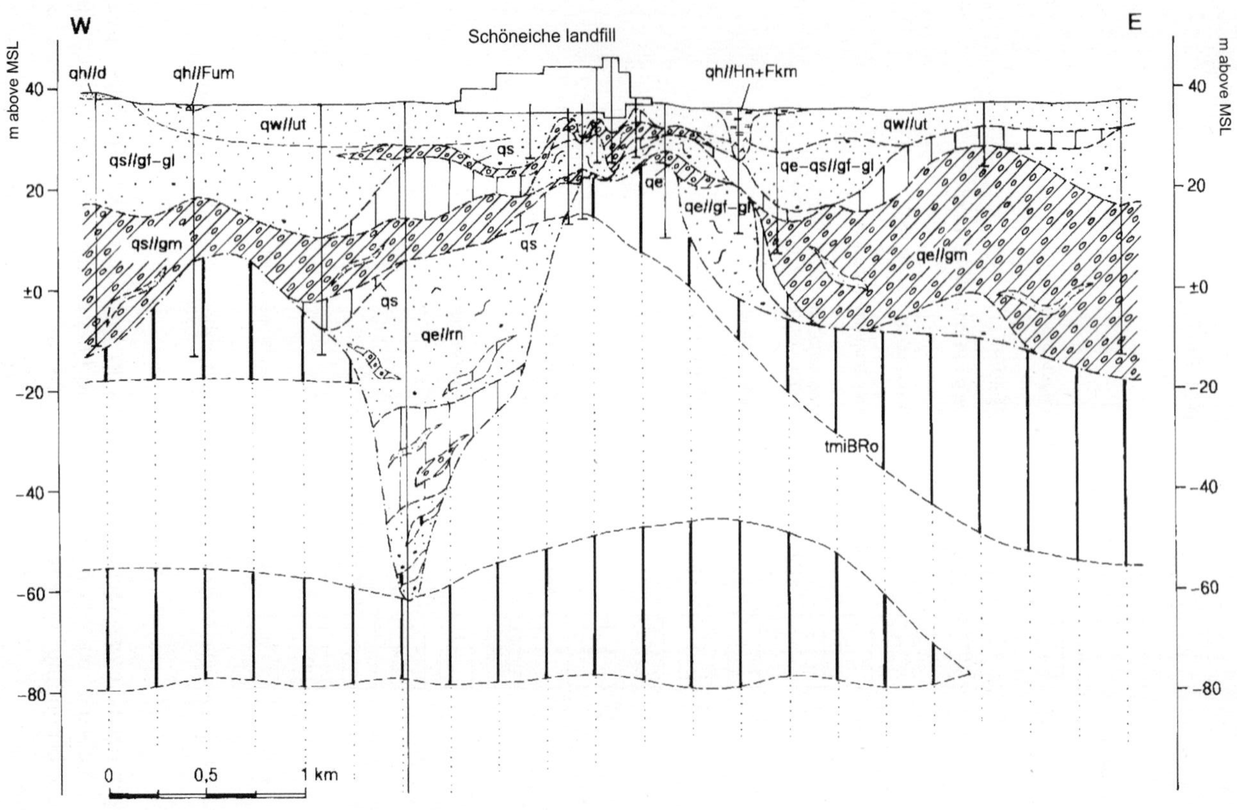

Fig. 6.2-25: Schöneiche-Mittenwalde area: Geological cross-section 2—2'

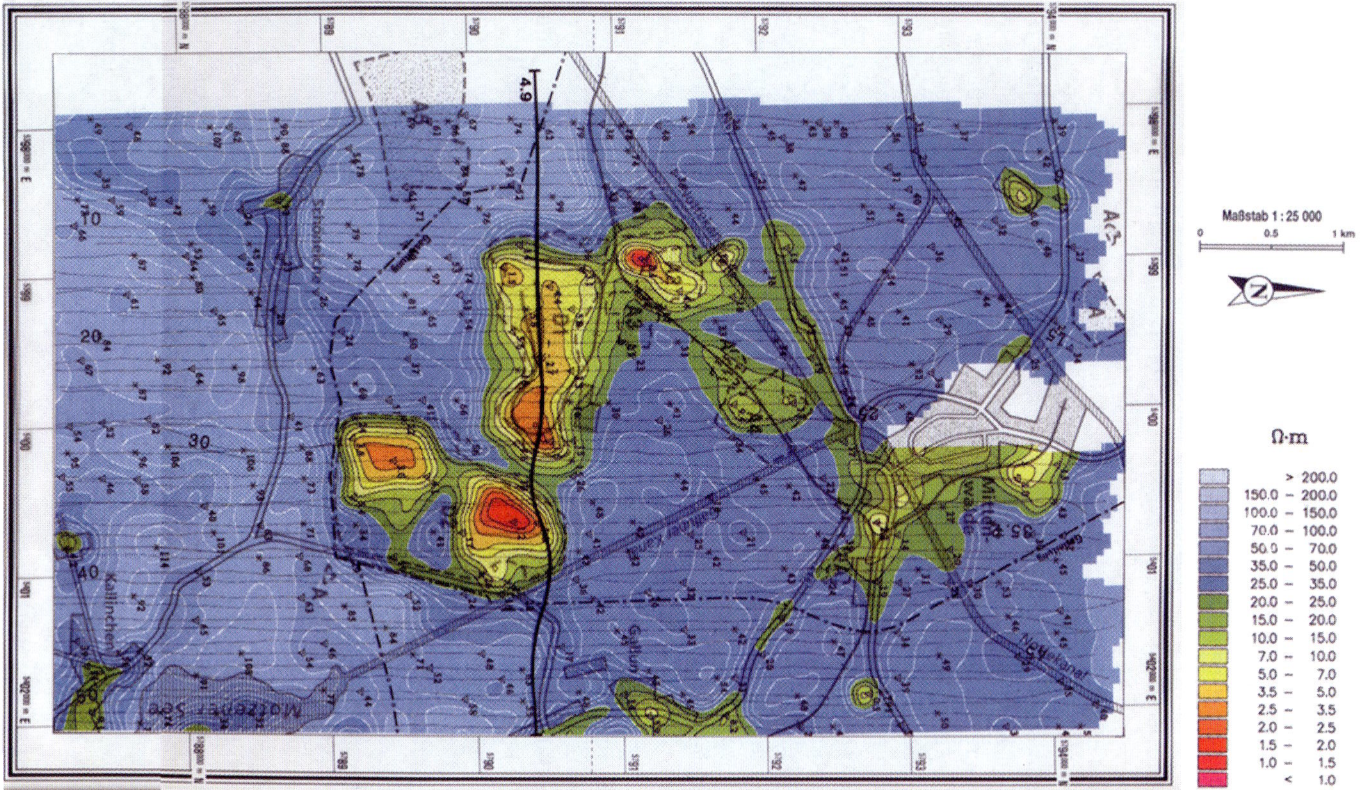

Fig. 6.2-26: Schöneiche-Mittenwalde area: Map of apparent resistivity ρ_a [Ωm] at 3640 Hz derived from airborne electromagnetic data

The distribution of apparent resistivity ρ_a at 3640 Hz, derived from the airborne geophysical data, is shown in *Fig. 6.2-26*. The Schöneicher Plan (D1) and Schöneiche (D2) landfills, and the old abandoned landfills A1 – A4 have distinctly lower apparent resistivities ($\rho_a < 25\ \Omega$m) than the surrounding areas. Only the area of landfill A5 (mainly sewage) does not show a similarly low apparent resistivity. Outside the areas of the old abandoned landfills the apparent resistivities are between 30 and 100 Ωm, except for the Motzener lake area ($\rho_a < 30\ \Omega$m), probably due to the clayey sediments, and in built-up areas, and along the road B246, which is probably due to anthropogenic causes.

The radiometric data (equivalent concentrations for the natural radioactive isotopes of K, U, and Th) provide information about the surficial geology and areas where material has been dumped. Terrestrial gamma radiation is completely absorbed by about 50 cm of water; thus gamma radiation measured above surface water with a depth of more than this is only from the surrounding area. The lake Motzener See and the water-logged areas can be seen to have very low equivalent concentrations. The maximum Th equivalent concentrations (*Fig. 6.2-27*) at the large landfills D1 (together with A1 and A2) and D2 (8 ppm and 6 ppm, respectively) are significantly higher than in the surrounding areas. The highest equivalent concentrations (12 ppm) are at the old abandoned landfill A4 (called “Alte Asche”, i.e., “Old Ash”). The other old landfills (designated with an A) have equivalent concentrations of 1 - 3 ppm, which is normal for areas outside landfills. Other elevated Th equivalent concentrations are observed in built-up areas. The distributions of U and K equivalent concentration are similar to those for Th: Landfills and built-up areas show significantly higher values. The highest U values were observed for the old landfill A4 (6 ppm); the highest K values are for landfills D1 (11 ‰) and D2 (12 ‰), with an abrupt drop at the edge of the landfills.

The distribution of apparent resistivity (dc resistivity mapping with AB/2 = 24 m) is shown in *Fig. 6.2-28*. The old landfill between the Notte and Gallun canals is characterized by low values. This can be due to the leaching of electrolytes from the waste or a crest in the underlying cohesive sediments. Low values around this landfill indicate leachate plumes to the west and north. Leachate plumes are also recognizable extending from the unofficial waste dumps around an industrial site (to the WSW on the west side, to the N on the north side, and to the NNE on the east side). Northeast of the Schöneicher Plan landfill (D1), there is a relatively wide zone of low resistivity. In this zone, these values can be due to spreading of leachate and/or a crest in the underlying cohesive sediments. The possible migration pathway for contaminants north of the Schöneiche landfill is less noticeable than for shallower depths of investigation. In contrast, the area east of this landfill shows very low values.

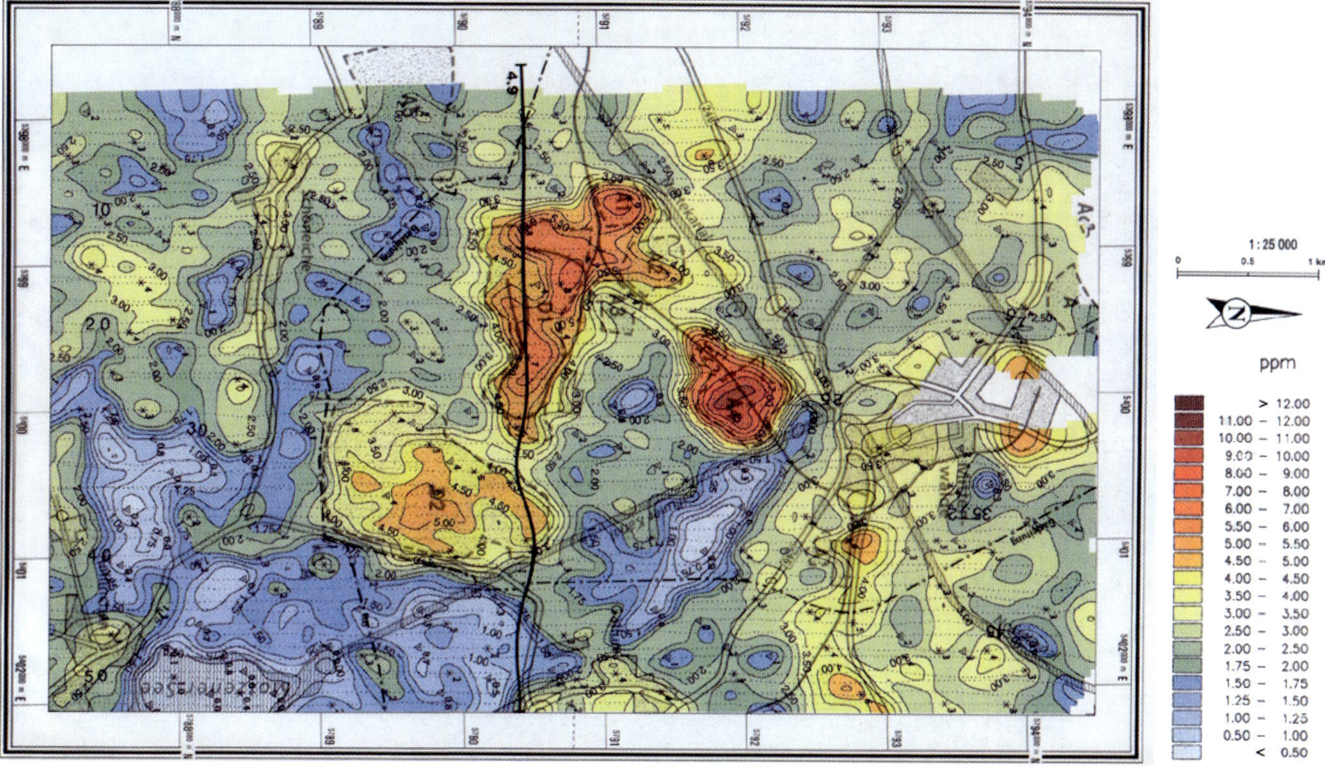

Fig. 6.2-27: Schöneiche-Mittenwalde area: Map of the radiometrically determined equivalent thorium concentrations [ppm] derived from airborne geophysical data

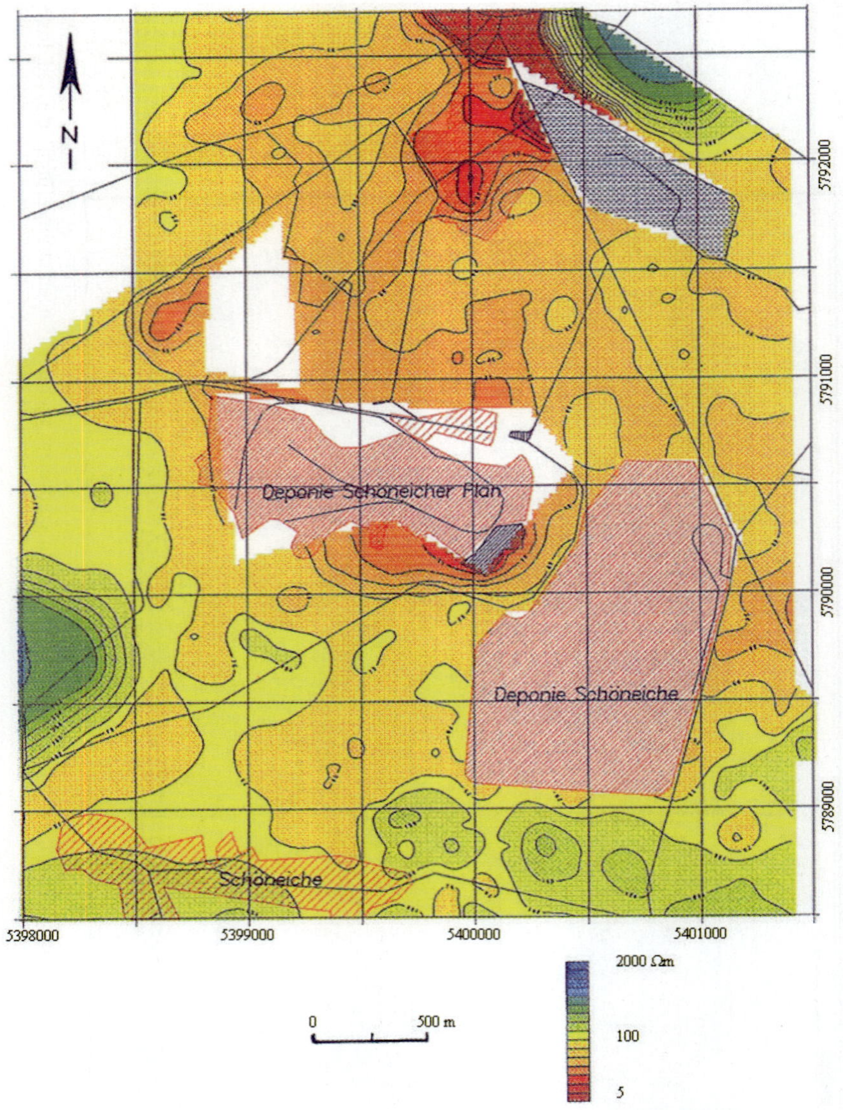

Fig. 6.2-28: Schöneiche-Mittenwalde area: Isoline map of apparent resistivity ρ_a [Ωm] derived from the dc resistivity mapping (AB/2 = 24 m)

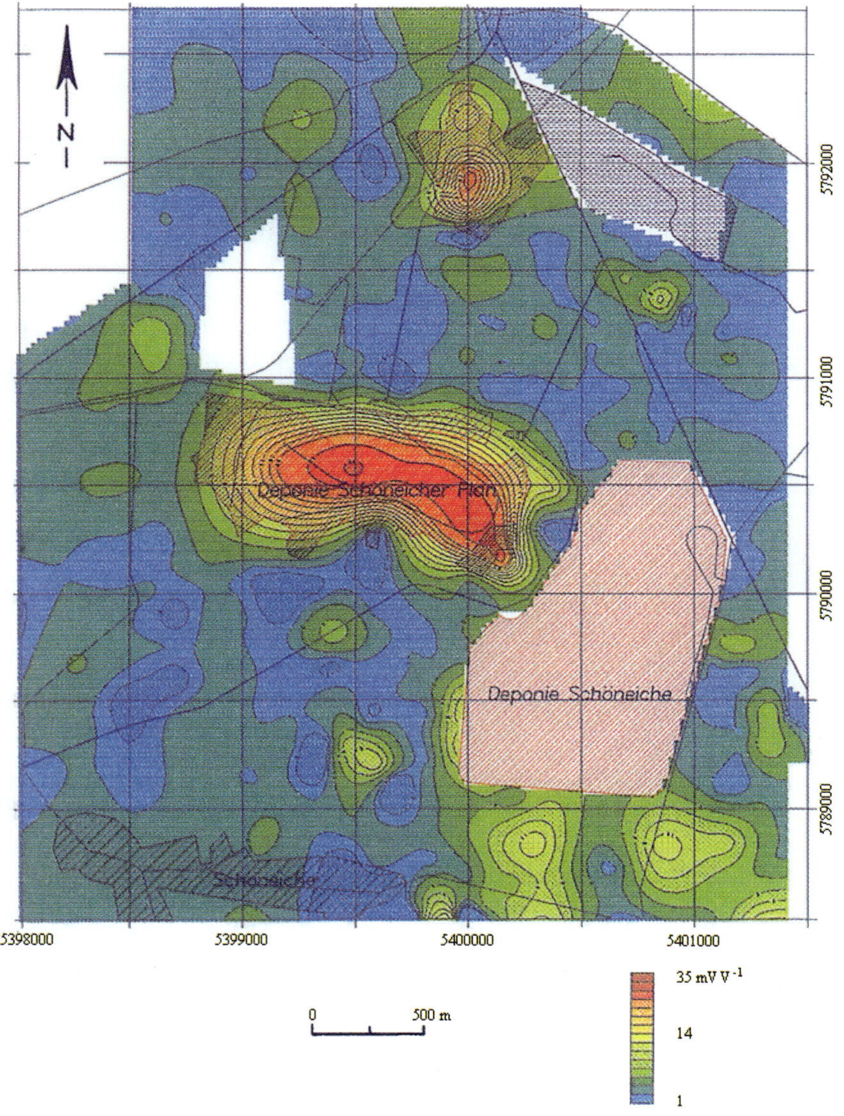

Fig. 6.2-29: Schöneiche-Mittenwalde area: Isoline map of apparent chargeability [mV V^{-1}] derived from the IP mapping data (AB/2 = 24 m)

The distribution of apparent chargeability derived from the IP mapping data (AB/2 = 24 m) is shown in *Fig. 6.2-29*. The values are, in general, very low. The values at the Schöneicher Plan and "Alte Asche" landfills are distinctly higher than the surrounding areas. The values to the south and west of the Schöneiche landfill are also generally higher.

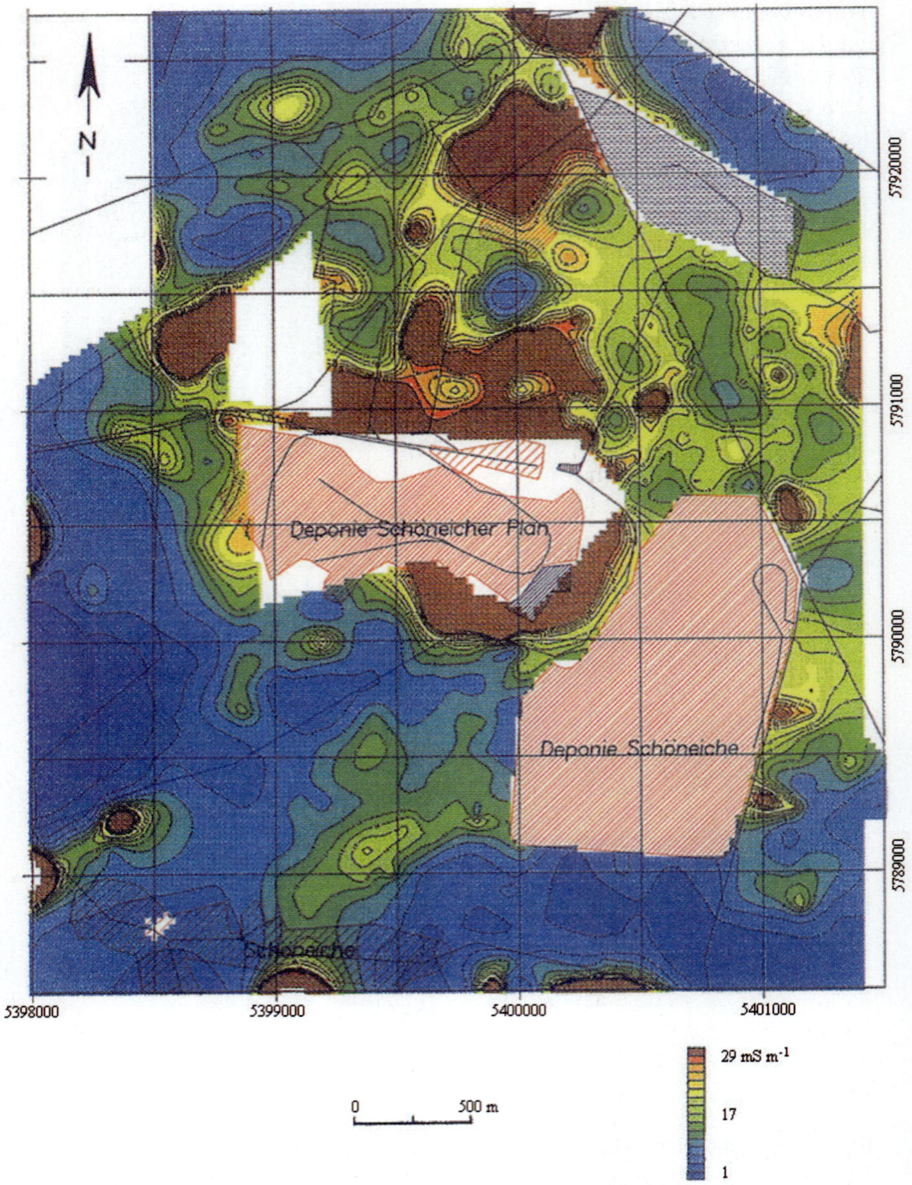

Fig. 6.2-30: Schöneiche-Mittenwalde area: Distribution of a proxy-parameter K as an indicator of anthropogenic contamination of the uppermost aquifer

The low apparent resistivities (higher conductivities) in the near-surface layers can be caused by cohesive materials (clay, silt, boulder clay) or contaminated groundwater. A proxy-parameter K (with dimension of electrical conductivity) derived from the apparent resistivity and the depth to the first cohesive layer indicated by the geoelectric soundings was tested (*Fig. 6.2-30*). Normalization with respect to the depth to the first cohesive layer minimizes the influence of this layer on the apparent resistivity (AB/2 = 13 m). Contaminated water causes significant anomalies. The areas marked brown and yellow have higher contaminant concentrations than the green and blue areas. The areas around the Schöneicher Plan landfill and the Alte Asche landfill, an area northwest of the Schöneicher Plan landfill, an area north of the road B246 from Telz to Mittenwalde, and areas east and northeast of the Schöneiche landfill show significant anomalies.

Conventional refraction seismic analysis of the data yields a two-layer model. Such a model is shown at the top of *Fig. 6.2-31*, together with the velocity distribution. The layer model shows variations within a relatively large range with respect to both the depth of the refractor and seismic velocity. Topography is here the main reason for the variation in depth. Local velocity anomalies are observed mainly within the landfill. The first layer is a cover layer (landfill waste or dry Weichselian sand); the underlying second layer is also unconsolidated material (water-saturated Weichselian sand and/or Saalian clay or boulder clay).

A refraction tomographic analysis of the data (*Fig. 6.2-31, middle*) confirms the results of the conventional refraction seismic analysis in general. Complex structures are observed below the refractor down to a depth of 60 m. Relatively high seismic velocities are observed outside the landfill and relatively low velocities in the central part. A depression is observed below the landfill. Seismic velocities of 1600 - 1700 m s^{-1} are obtained for depths considerably greater than those resolved by the conventional analysis. The depression is probably due to the excavation of clay before the landfill was established.

Reflection seismic analysis of the data yields time and depth sections. The depth sections were derived using the stacking velocities for time-to-depth conversion. The seismic time sections show correlatable reflection horizons down to two-way traveltimes > 800 ms. In the northern and southern parts of the profile (*Fig. 6.2-31, bottom*), the first significant reflector is at two-way traveltimes of 75 - 100 ms. Reflection bands between 100 and 700 ms dip to the north. Only scattered reflections are observed below the area of the landfill. A strong reflector is observed south of the landfill at traveltimes of about 850 ms. The associated layer boundary dips to the south, in contrast to the overlying layers, which indicates an unconformity.

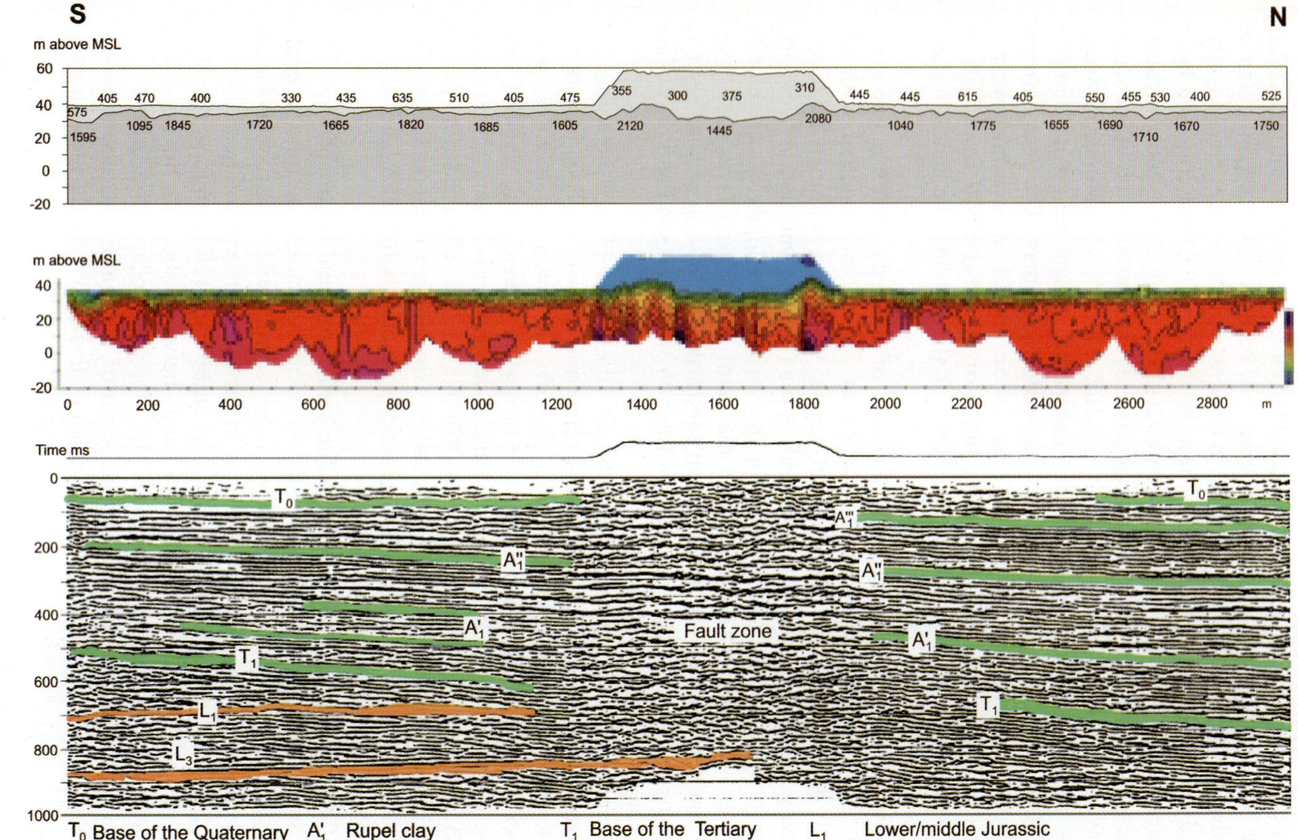

Fig. 6.2-31: Schöneiche-Mittenwalde area: N-S seismic profile across the Schöneicher Plan landfill: (*top*) refraction seismic section, (middle) refraction tomography section, (*bottom*) reflection seismic section, THOR, commissioned by BGR

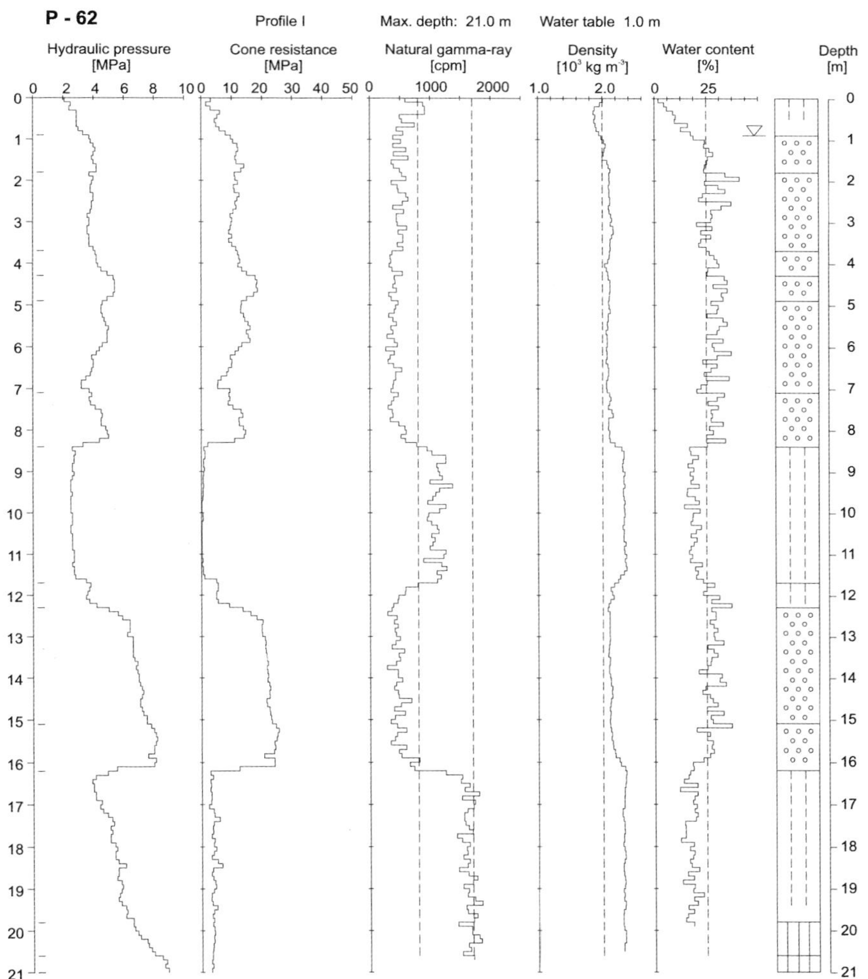

Fig. 6.2-32: Schöneiche-Mittenwalde area: Results of cone penetration test and geophysical logging at the location P-62, ELGI, commissioned by BGR

Cone penetration tests with geophysical logging were conducted in the study area (*Figs. 6.2-32* and *6.2-33*). In addition water and soil samples were collected at selected depths. The natural gamma activity was measured, as well as gamma-gamma density logs and neutron-neutron logs. The clay content can be derived from the natural gamma activity in unconsolidated sediments. Gamma activity is an important parameter for lithological interpretation. The density of the penetrated material can be determined from the backscatter from a small gamma-ray source. Water content (i.e., hydrogen content) can be determined from the neutron-neutron logs.

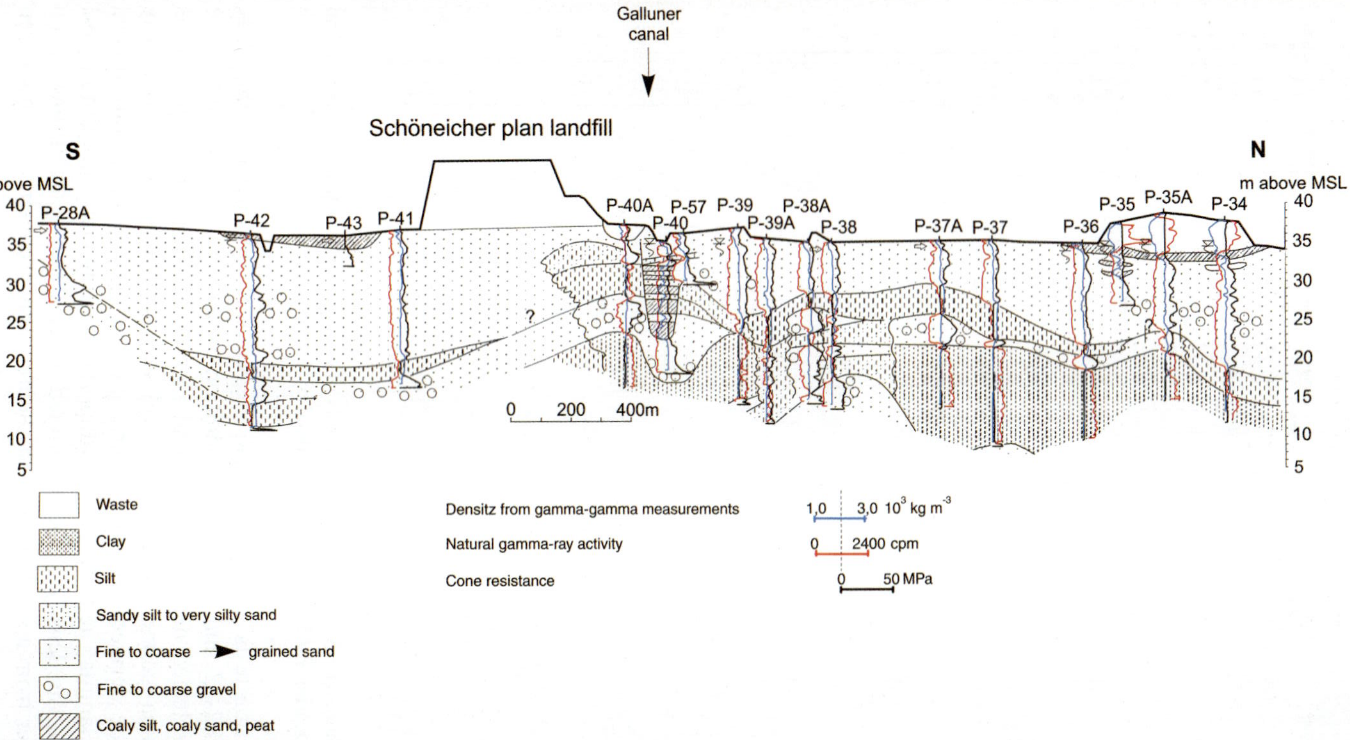

Fig. 6.2-33: Schöneiche-Mittenwalde area: Geophysical-geological profile III as derived from the cone penetration tests with geophysical logging, ELGI, commissioned by BGR

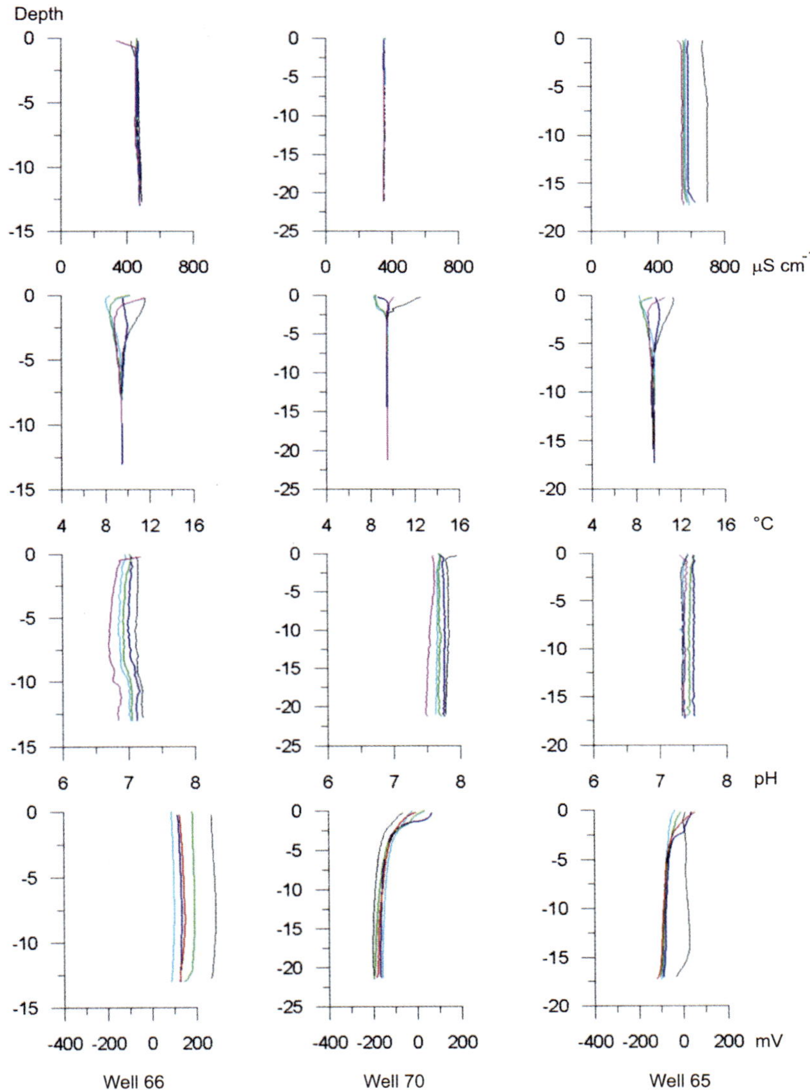

Fig. 6.2-34: Schöneiche-Mittenwalde area: Multi-parameter profiles in observation wells P65, P66, and P70 measured between October 1994 and June 1995

Selected results from the multi-parameter probe measurements are shown in *Figs. 6.2-34* and *6.2-35*. The reproducibility of the parameters is very good in all the observation wells. The main direction of groundwater flow is to the NNE. The groundwater table lies between 1.6 and 8.0 m below the well head. The fully screened observation wells are in the topmost aquifer. The wells, mostly in sandy to silty layers, end at thick, nonpersistent layers of boulder clay.

The multi-parameter values for groundwater observation wells P65, P66 and P70 (*Fig. 6.2-34*) reflect the geogenic content of groundwater of drinking water quality. The electrical conductivity values are between 350 and 600 µS cm^{-1}. The pH is neutral to weakly basic. The water temperature in the deeper parts of the wells is 9.2 - 9.5 °C, corresponding to the long-term annual mean of the air temperature in the region. Annual temperature fluctuation is observable in the upper part of the wells showing highest temperatures measured in October, when the summer maximum reached the deeper parts of the wells. The redox potential was in the range between +300 and -300 mV: more strongly reducing in P65 and P70 and weakly reducing in P66. The redox potential decreases with increasing depth, indicating an increasingly reducing environment.

The multi-parameter values for groundwater observation wells P62, P63 and P71 (*Fig. 6.2-35*) indicate anthropogenic contamination. Water temperature in P63 is higher than normal (15 °C, which is more than 5 °C above the regional value) indicating exothermic reactions. The pH is neutral to weakly acidic, and redox potentials are between +200 and -300 mV. Like for the uncontaminated water, the water in these wells is weakly to strongly reducing. Decreasing redox potential with increasing depth indicates an increasingly reducing environment. The most important parameter for anthropogenic contamination is the electrical conductivity. From 0 to 6 m in P62 (at the southern edge of the Schöneicher Plan landfill) the mean conductivity is 1300 µS cm^{-1}; between 6 and 7.5 m the conductivity increases to more than 4000 µS cm^{-1}. Contaminated water apparently collects at the surface of the boulder clay aquiclude. The high conductivities are due to leachate from both the Schöneicher Plan landfill and an old abandoned landfill. Observation well P63 (within the old ash landfill "Alte Asche") has conductivities of more than 2000 µS cm^{-1}, indicating contamination by the high salt content of the ash. Contaminated water apparently also accumulates at the boulder clay aquiclude. Seasonal differences in conductivity in the top 5 m can be attributed to differences in amount of precipitation, i.e., groundwater recharge. The pH is neutral, which is probably due to the high ash content of the landfill. The electrical conductivity in P71 (downstream from the Schöneicher Plan landfill) increases with depth from about 1000 µS cm^{-1} to about 3000 µS cm^{-1}. At shallow depth, conductivity, pH, and redox potential show the effect of groundwater recharge, i.e., influx of water with little or no contamination. In this well too, contamination is due to leachate from both the Schöneicher Plan landfill and an old abandoned landfill.

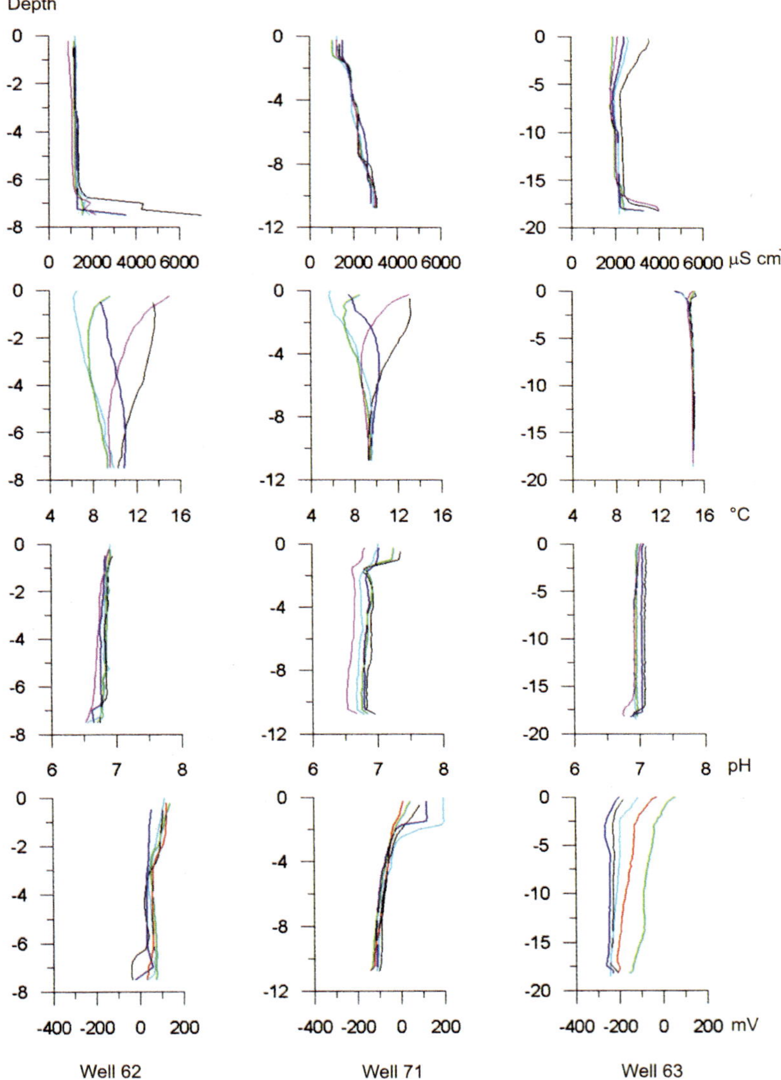

Fig. 6.2-35: Schöneiche-Mittenwalde area: Multi-parameter profiles in observation wells P62, P63, and P71 measured between October 1994 and June 1995

The results of geochemical investigations in the Schöneiche-Mittenwalde area are described in Section 6.2.2.3.

6.2.2 Quantitative Approach

MANFRED BIRKE, KLAUS KNÖDEL, GERHARD LANGE & UGUR YARAMANCI

6.2.2.1 Joint Quantitative Interpretation of Several Geophysical Measurements and Core Analysis Results

UGUR YARAMANCI & GERHARD LANGE

The most important objectives in site characterization are the investigation of the geology and of the groundwater conditions. Therefore, the structure of the subsurface, lithology, water table, and hydraulic properties must be investigated. These were the objectives for investigation at a site near Nauen, northwest of Berlin, Germany. The geophysical methods surface nuclear magnetic resonance (SNMR), ground penetrating radar (GPR), 2-D geoelectric sections, 1-D spectral induced polarization (SIP) soundings, and refraction seismics were used, together with core drilling and laboratory measurements of porosity and grain-size distribution carried out on core samples (YARAMANCI et al., 2002).

The geology of the Nauen site is quite typical for large parts of northern Germany. It consists of Quaternary sediments deposited during the last glacial stage (Weichselian) and the underlying Tertiary clays. In the area of investigation, these sediments mainly consist of fluvial sands bordered by glacial till. The topography is characterized by low hills consisting of till and plains of glaciofluvial sand and gravel. There is an unconfined shallow aquifer consisting of fine- to medium-grained sand underlain by an aquiclude of marly to clayey glacial till. North of the site, the glacial till crops out at the surface with a nearly E-W strike.

SNMR soundings were measured at five locations (B2, B5 - B8) 25 m apart on the N-S running main profile using figure-eight antennas 50 m in diameter. The axis of the antennas was directed E-W, parallel to the strike of subsurface structures. The results of the inversion of the SNMR soundings are shown in *Fig. 6.2-39*. A least squares smooth inversion (matrix inversion) using a regularization parameter was carried out to determine the distribution of water content with depth (LEGCHENKO & SHUSHAKOV, 1998). In a smooth inversion it is not possible to detect the sharp boundaries or changes in water content with depth as would be expected for aquifers in medium to coarse-grained sand (which have a relatively small capillary fringe) or in the presence of a distinct aquiclude, as is the case here.

The most important feature of SNMR is that it allows direct determination of water content. The SNMR data indicate the content of mobile (free) water in the pores of the aquifer as about 25 - 30 % (*Fig. 6.2-39*). This value is an average for at least three locations to the south – B8, B7 and B2. Although the sharp boundary of the water table is not well determined by the smooth inversion used, the vadose zone, with a mean water content of 10 - 20 %, is distinguishable to some degree. This water corresponds to percolation water, which needs some time to reach the aquifer. Below the aquifer, material with a water content of 5 - 10 % is indicated, which should be sandy glacial till. This, in fact, is in agreement with the free water content that can still be accommodated in the glacial till. The higher water contents found for the deeper regions, suggesting a second aquifer, have not yet been confirmed by another method.

The average decay times from the SNMR can be used to estimate the hydraulic conductivity with an equation proposed earlier by YARAMANCI et al. (1999). This empirical equation, based on SNMR decay times and mean grain size made available by SCHIROV et al. (1991), is

$$k_f \approx T^4, \tag{6.2.1}$$

where k_f is the hydraulic conductivity in m s^{-1} and T is the decay time in s. An average decay time of 150 ms yields a hydraulic conductivity of $5 \cdot 10^{-4}$ m s^{-1}, which fits quite well with the hydraulic conductivities determined on the core material (*Fig. 6.2-40*).

An extensive GPR survey was carried out at the site. To achieve optimum penetration and resolution, 200 MHz antennas were used for the measurements. The profiles were orientated N-S and E-W each 25 m apart (*Fig. 6.2-36*). As an example, a N-S profile, perpendicular to the strike, is shown in *Fig. 6.2-37b*. CMP measurements were also carried out at selected locations (*Fig. 6.2-36*) in order to derive proper velocities for time-to-depth conversion. GPR can be used to map the water table if there is a sharp discontinuity and not a broad transition zone due to the capillary fringe. Usually, an aquifer will not be penetrated properly to see reflections from its bottom due to the high absorption, i.e. energy loss, in the aquifer. In Nauen favorable situations are met: There is a sharp water table due to the high hydraulic conductivity of the coarse sand, and therefore a good reflection, and there is a low electrical conductivity of the water, and because of low attenuation good bottom reflections, i.e. reflections from the top of the glacial till arise due to contrast in the dielectric properties. According to the GPR measurements, the water table is at about 2 m depth (*Fig. 6.2-38a*), depending on the slowly varying topography. The bottom of the aquifer, i.e. the top of the glacial till, is about at 15 m depth in the southern part of the investigation area (*Fig. 6.2-38b*) and gradually ascending to surface in the northern direction. North of this, there is no radar signal due to the high energy absorption of the electrical highly conductive till.

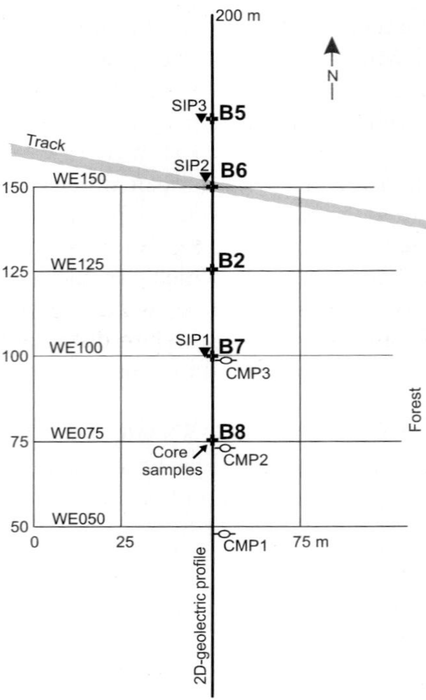

Fig. 6.2-36: Location map of measurements at the Nauen/Berlin site: Location of profiles for GPR sections and GPR CMP soundings, of sounding points for SIP, of the 2D-geoelectric profile and of core sampling locations

Owing to the clear indication of a two-dimensional structure in the GPR measurements 2-D-geoelectric measurements were made in order to obtain more detailed information about the resistivity structures. In situations comparable with Nauen, where sharp boundaries occur, block inversion should be used (*Fig. 6.2-37a*). With this kind of inversion, distinct resistivities are found for individual blocks, i.e. formations. The following resistivities were determined: 3000 Ωm for the vadose zone, 280 Ωm for the aquifer, and 90 Ωm for the sandy glacial till. These values are in good agreement with those known from other geoelectrical surveys for Quaternary aquifers in the Berlin region. The comparatively low resistivity of 35 Ωm for near-surface sections within the glacial till in the northern part of the profile indicates an internal structure, probably with an elevated clay content.

Initially, 1-D geoelectrical soundings in Schlumberger configuration were performed to see the main features of the resistivity structures in the area. They were carried out with equipment for spectral induced polarization (SIP) and, therefore, the phase information of the complex resistivity was also available. The soundings were centered at three locations, SIP1, SIP2 and SIP3, along profile SN050 (*Fig. 6.2-36*) with E–W layouts. The results of 1-D soundings for greater depths were not that reliable, as the structure is basically 2-D and side effects may dominate for larger layouts.

Supplementary seismic refraction measurements were carried out along line WE075. Shotpoints were located at B8 and at a location 52 m west of B8. Geophone spacing of 1 m was used up to 12 m distance from the shot points, and spacing of 2 m at greater offsets. The water level can be clearly identified in the resulting data at a depth of 2 m near point B8, with velocities of 400 ± 15 m s^{-1} in the vadose zone and 1360 ± 80 m s^{-1} in the aquifer. The refraction from surface of the glacial till was not very distinct, so that the determined velocity of 3700 ± 700 m s^{-1} and a corresponding depth of approximately 19 m are not reliable.

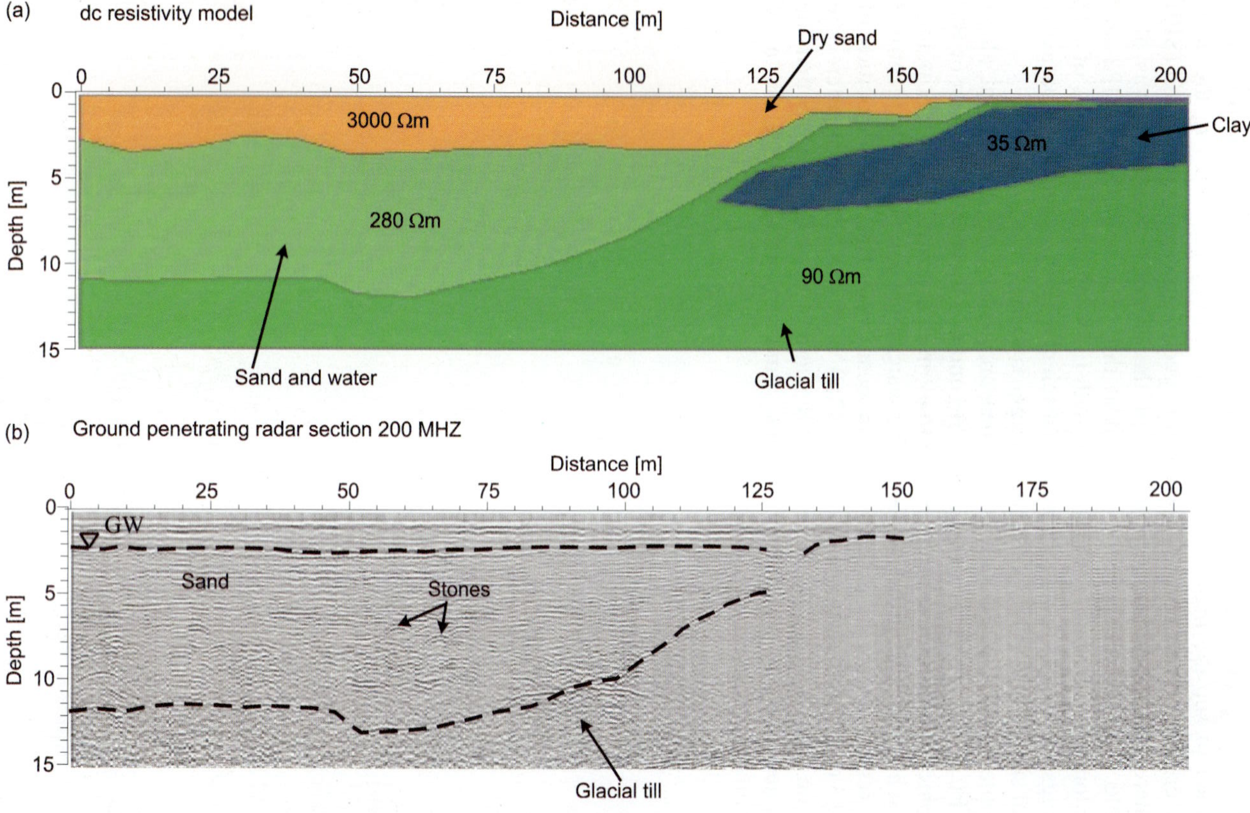

Fig. 6.2-37: (a) dc resistivity model along profile SN050 derived by block inversion, (b) GPR section along profile SN050, 200 MHz antennas

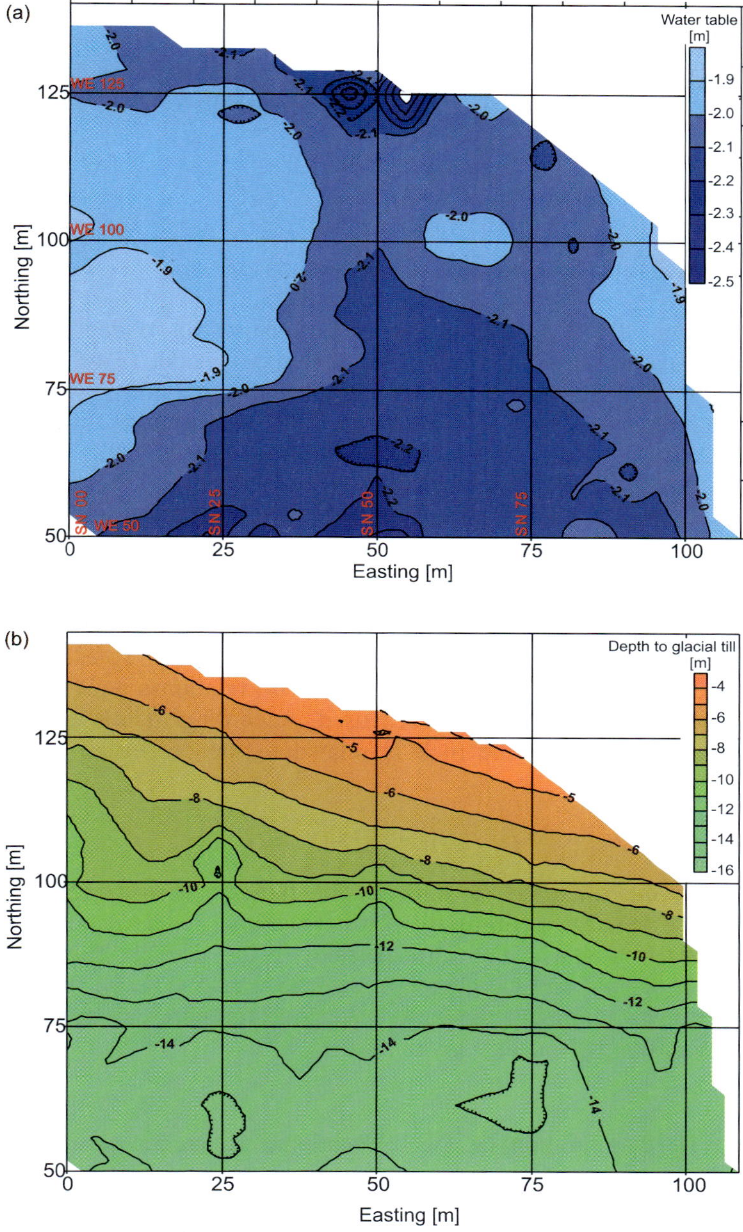

Fig. 6.2-38: (a) Depth to the water table and (b) depth of glacial till as determined by GPR measurements

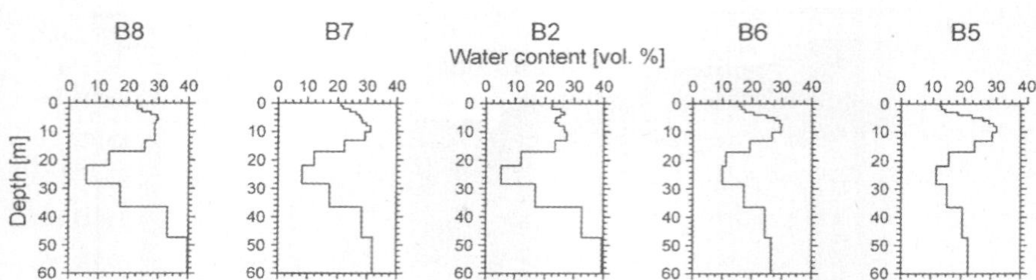

Fig. 6.2-39: Water content as derived from inversion of SNMR data

A combination of 2-D geoelectric survey data, dc resistivity depth soundings, and GPR data was used to derive the geological model (*Figs. 6.2-37* and *6.2-38b*), and to delineate the groundwater table (*Fig. 6.2-38a*).

The estimation of water content from GPR data is quite straightforward and widely used (GREAVES et al., 1996; HUBBARD et al., 1997; DANNOWSKI & YARAMANCI, 1999). The dielectric permittivity ε of the rock can be determined from the velocity v of electromagnetic waves measured with radar using the following relationship:

$$\varepsilon = \left(c_0 / v\right)^2, \tag{6.2.2}$$

with c_0 being the velocity of electromagnetic waves in air. To estimate water content from CMP velocity data, the usual CRIM relationship (WILSON et al., 1995; MAVKO et al., 1998) has been used. According to this equation, the dielectric permittivity of the rock is related to the properties of the rock components as follow:

$$\sqrt{\varepsilon} = (1 - \phi)\sqrt{\varepsilon_m} + S\phi\sqrt{\varepsilon_w} + (1 - S)\phi\sqrt{\varepsilon_a}, \tag{6.2.3}$$

where ε, is the dielectric permittivity of rock, ϕ porosity, S degree of saturation, ε_w dielectric permittivity of water ~ 80, ε_m dielectric permittivity of the rock matrix $\approx$ 4.5, ε_a dielectric permittivity of air = 1. The porosity is defined as $\phi = V_p/V$ and the degree of saturation as $S = V_w/V_p$, where V, V_w, and V_p are the volumes of the rock, water and the pores, respectively. The water content $G = V_w/V$ is then given by

$$G = S\,\phi. \tag{6.2.4}$$

In case of 100 % saturation, i.e. $S = 1$, Equation (6.2.3) reduces to

$$\sqrt{\varepsilon} = (1 - \phi)\sqrt{\varepsilon_m} + \phi\sqrt{\varepsilon_w}, \tag{6.2.5}$$

and the water content is equal to porosity, $G = \phi$. The porosity determined using Equation (6.2.5) is then used in Equation (6.2.3) to estimate the water content in the vadose zone with the dielectric permittivity of the vadose zone.

In order to interpret the resistivity and its local variations, the physical cause of resistivity and the influencing factors must be well understood. A model widely used for a variety of rocks is the well-known Archie's law (ARCHIE, 1942) which is widely used for interpreting resistivity well logs. The electrical conductivity σ of rocks can be modeled by the sum of two types of conductivity in parallel, (SCHOPPER, 1982; GUEGUEN & PALICIAUSKAS, 1994; MAVKO et al., 1998):

$$\sigma = \sigma_v + \sigma_q, \tag{6.2.6}$$

where σ_v, is the volume conductivity caused by the ionic conductivity of the free electrolyte in the pores, and σ_q is the capacitive interface (excess) conductivity due to water adsorbed on the pore surfaces. The conductivity σ_q is, in contrast to σ_v, strongly frequency dependent, being very small for zero frequency and becoming increasingly large with increasing frequency. For rocks with a large internal surface (for example, those containing a high proportion of clay as shaley sands), σ_q can be very high and it is, therefore, called the "clay term". For aquifers in more sandy formations and at low frequencies, σ_v is much higher than σ_q, and thus $\sigma \cong \sigma_v$. Going back to the more familiar expression in terms of resistivity, with $\rho = 1/\sigma$, the ohmic resistivity of rock is

$$\rho = \rho_w \phi^{-m} S^{-n} = \rho_w FI, \tag{6.2.7}$$

where ρ_w is the resistivity of water, ϕ the porosity, m the Archie exponent (or cementation factor), S the degree of saturation, and n, the saturation exponent. The dependence of the resistivity on the pores is expressed by the formation factor $F = \phi^{-m}$ and saturation index $I = S^{-n}$. For a saturated rock (i.e. $S = 1$), the resistivity in Equation (6.2.7) becomes

$$\rho_o = \rho_w \phi^{-m}, \tag{6.2.8}$$

where the index 'o' stands for saturated rock.

Equation (6.2.8) is used first to estimate the water content in the aquifer, and the porosity determined in this way is then used to estimate the porosity in the vadose zone. Very often it can be assumed that m-n $\cong$ 0, and consequently, $S^{m-n} \cong 1$. Even though the effect of S is not that high, it should not be neglected a priori, especially in such cases when values are available for m and n. The parameters used for the estimation of porosity and water content at the Nauen site and the corresponding results are shown in *Table 6.2-1*.

The estimates were first made for the aquifer. Then some of the estimated parameter values were used to estimate values for the vadose zone. Note that the SNMR estimates are for free water and estimates from GPR and geoelectric data are for the total water content (like for neutron logs in well logging).

Table 6.2-1: Estimation of water content with different methods at the Nauen site

	GPR	Geoelectric sounding	Geoelectric 2D section	SNMR
aquifer (2-13 m)	20 % using $v = 0.86$ m/ns, $\varepsilon = 12.2$, $\varepsilon_w = 80$, $\varepsilon_m = 4.5$	28 % using $\rho_o = 170\ \Omega$m, $\rho_w = 33\ \Omega$ m, $m = 1.3$	20 % using $\rho_o = 280\ \Omega$m, $\rho_w = 33\ \Omega$ m, $m = 1.3$	25-30 %
vadose zone (0-2 m)	5 % using $v = 0.130$ m ns^{-1}, $\varepsilon = 5.3$, $\phi = 20$ %, $\varepsilon_w = 80$, $\varepsilon_m = 4.5$, $\varepsilon_a = 1$	4 % using $\rho_o = 8 \cdot 10^3\ \Omega$m, $\phi = 28$ %, $\rho_w = 33\ \Omega$m, $m = 1.3$	5 % using $\rho_o = 3 \cdot 10^3\ \Omega$m, $\phi = 20$ %, $\rho_w = 33\ \Omega$m, $m = 1.3$	15-20 %

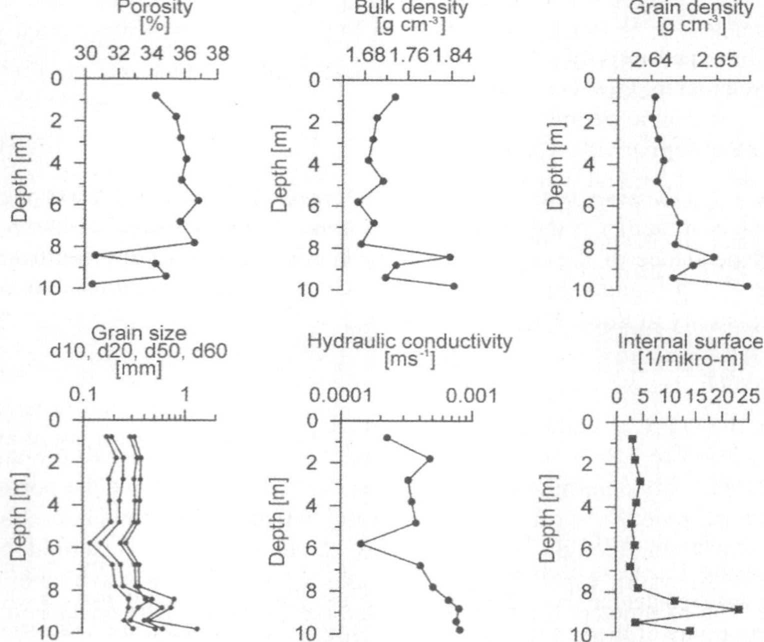

Fig. 6.2-40: Measurements on core samples at location B8

A number of laboratory measurements were carried out to determine petrophysical parameter values on core samples (*Fig. 6.2-40*). Cores were available down to 10 m depth, well representative for the aquifer material. Down to 8 m, the material was quite uniform, consisting of sands with an average grain size of approximately 0.25 mm. Below 8 m depth, there was some variation in the sandy material with large grain sizes, broader grain-size distributions and, consequently, slightly higher hydraulic conductivities. The most important information derived from the laboratory measurements is porosity values determined from the bulk density and grain density. The porosity is about 35 % down to a depth of 8 m, and slightly less below that. This porosity is quite typical of Quaternary sands in northern Germany. From other sites with similar sandy material, it is known that about 4 - 5 % adhesive water can be held in these sands (YARAMANCI et al., 1999). This means that the content of free water is about 30 %.

6.2.2.2 Joint Inversion

KLAUS KNÖDEL, UGUR YARAMANCI & GERHARD LANGE

For almost all of the geophysical methods there is a linear relationship between the parameters of a model $\boldsymbol{m}$ and the observed data $\boldsymbol{d}^{(o)}$ expressed by a parameter $\boldsymbol{G}$ governing the specific measurement (e.g., dc resistivity sounding, magnetic measurement). In some cases the relationship between model and data is nonlinear and a linearization has to be made. Then the relationship between model and data is developed in an expansion (e.g., Taylor series expansion) and terms of higher order are neglected:

$$\boldsymbol{d} = \boldsymbol{G}\,\boldsymbol{m} \qquad (6.2.9)$$

where $\boldsymbol{d} = [d_i]$, i = 1, 2,..., N; $\boldsymbol{m} = [m_j]$, j = 1, 2,..., M; and $\boldsymbol{G} = [G_{ij}]$; $\boldsymbol{d}$ and $\boldsymbol{m}$ are vectors and $\boldsymbol{G}$ is a matrix. The matrix element G_{ij} describes how the observation d_i is related to the model parameter m_j. If $\boldsymbol{m}$ and $\boldsymbol{G}$ are known and $\boldsymbol{d}$ is calculated using Equation (6.2.9), the procedure is referred to *forward modeling* with $\boldsymbol{d}$ being denoted by $\boldsymbol{d}^{(c)}$.

Forward modeling can be used to study how changes in model parameters affect the data. The elements $\partial d_i/\partial m_j$ of the Jacobian matrix $\boldsymbol{J} = [\partial d_i/\partial m_j]$ were calculated for this study. At the one extreme, a model parameter will not influence a particular measured value; in another extreme case, a measured value depends only on one model parameter. But in the normal case, a measured value is influenced by many model parameters to different degree. Determination of which parts of the subsurface actually contribute to the measurement and to which extent is possible by the use of such a "sensitivity matrix". Design of the set-up for survey measurements can be aided by the use of sensitivity matrices (example in Chapter 4.3).

Inversion is the (automated) determination of the parameter values of a model from observed data (LINES & TREITEL, 1984; HJELT, 1992; RUBIN & HUBBARD, 2005). Beginning with an "initial guess" (starting model), the model parameter values $\boldsymbol{m}$ are changed iteratively until the calculated data $\boldsymbol{d}^{(c)}$ corresponds within certain limits to the observed (measured) data $\boldsymbol{d}^{(o)}$. The deviation between the observed data and the calculated model data, i.e. the error $\boldsymbol{e}$, is defined as

$$\boldsymbol{e} = \boldsymbol{d}^{(o)} - \boldsymbol{d}^{(c)} \text{ and using } \boldsymbol{d}^{(c)} = \boldsymbol{G}\,\boldsymbol{m} \text{ yields } \boldsymbol{e} = \boldsymbol{d}^{(o)} - \boldsymbol{G}\,\boldsymbol{m}. \tag{6.2.10}$$

An appropriate measure of the fit is the least squares error (error function) E

$$E = \boldsymbol{e}^{\mathrm{T}}\,\boldsymbol{e} = (\boldsymbol{d}^{(o)} - \boldsymbol{G}\,\boldsymbol{m})^{\mathrm{T}}\,(\boldsymbol{d}^{(o)} - \boldsymbol{G}\,\boldsymbol{m}), \tag{6.2.11}$$

or in terms of the elements,

$$E = \Sigma e_{\mathrm{i}}^2 = \Sigma\left(d_{\mathrm{i}}^{(o)} - \Sigma G_{\mathrm{ij}} m_{\mathrm{j}}\right)^2. \tag{6.2.12}$$

The smaller E is, the better the model, i.e. the model parameter values $\boldsymbol{m}$ have to be changed so that E becomes a minimum:

$E \Rightarrow$ Min!

Minimization of E is achieved taking the derivative of E with respect to the parameters m_{j} and setting these derivatives equal to 0:

$$\partial E / \partial m_{\mathrm{j}} = 0, \quad \mathrm{j} = 1, 2, \dots, \mathrm{M}. \tag{6.2.13}$$

This leads to a linear equation system for the model parameters:

$$\boldsymbol{m} = [\boldsymbol{G}^{\mathrm{T}}\boldsymbol{G}]^{-1}\,\boldsymbol{G}^{\mathrm{T}}\,\boldsymbol{d}^{(o)} \tag{6.2.14}$$

and consequently to a set of equations to be solved, commonly expressed in matrix and vector form. This equation stands for matrix inversion, which is the classical least-squares solution of the inversion problem.

Many different algorithms exist for inversion. There are also algorithms available that change the model parameters randomly at each iteration. If this random change does not improve the fit, this approach is then abandoned. Such schemes, called random search algorithms, e.g., Monte-Carlo, simulated annealing, genetic algorithms, neuronal networks, are very often used. One advantage of this kind of inversion is that there is no need of a formal analytical minimization of the error E. A physical model relating observed data to model parameters, i.e. forward model, is sufficient. Thus, it is very favorable for complex problems and, in particular, also for non-linear problems. Another advantage of random search algorithms is that they are able to find the absolute (true) minimum of the error function E and not terminate when a local minimum is found.

More than one method is usually used for a site investigation in order to improve the reliability of the "site model". Data from the individual methods are separately processed and jointly interpreted. The joint interpretation can be improved by using joint inversion if certain prerequisites are fulfilled. The data obtained with two or more methods and used for joint inversion must be affected by the same object(s), i.e. they must contain information at least partly from the same depth and/or volume range and about at least one conjoint rock parameter.

Table 6.2-2 shows the relationships between geophysical parameters and rock parameters. From this table one can conclude that electrical (dc) resistivity and SNMR-relaxation amplitude data can be used, in principle, to determine porosity by joint inversion. Other combinations of geophysical parameters in joint inversion are possible.

Table 6.2-2: Relationships between geophysical parameters and rock parameters (structure, texture, geometry, composition). (xxx = strong influence, xx = moderate influence, x = weak influence, blank = no influence)

Geophysical parameter	Rock parameter							
	chemical composition of minerals	geometric arrangement of minerals	porosity	pore size – (capillarity)	internal surface	pore connectivity	pore fluid content	chemical composition of pore fluid
velocity of seismic waves	xx	x	xxx				xx	
attenuation of seismic waves	x	x	xx	x			xx	
bulk density	xxx		xxx					
magnetic susceptibility	xxx							
electrical (dc) resistivity	x	x	xxx			xxx	xxx	xx
phase of electrical resistivity				xxx	xxx		xx	xx
velocity of electromagnetic waves	xx		xx				xx	
attenuation of electromagnetic waves	x		xxx					xx
SNMR-relaxation amplitude			xxx				xxx	
SNMR-relaxation time				xxx	xx		xx	xx

Joint inversion is based on the strategy of optimizing a model $\boldsymbol{m}$ for data sets $\boldsymbol{d}^{(o1)}$ and $\boldsymbol{d}^{(o2)}$ from measurements with two geophysical methods (e.g., seismic refraction and geoelectric measurements). The physical relationships connecting the model parameters of $\boldsymbol{m}$ to the observations are the same as in the individual inversions, $\boldsymbol{G}^{(1)}$ and $\boldsymbol{G}^{(2)}$. The corresponding calculated data is expressed as follows:

$$\boldsymbol{d}^{(c1)} = \boldsymbol{G}^{(1)}\,\boldsymbol{m} \text{ and } \boldsymbol{d}^{(c2)} = \boldsymbol{G}^{(2)}\,\boldsymbol{m}. \tag{6.2.15}$$

The error or misfit is defined as

$$\boldsymbol{e}^{(1)} = \boldsymbol{d}^{(o1)} - \boldsymbol{d}^{(c1)} \text{ and } \boldsymbol{e}^{(2)} = \boldsymbol{d}^{(o2)} - \boldsymbol{d}^{(c2)}. \tag{6.2.16}$$

The individual errors for the sets of data,

$$E^{(1)} = \boldsymbol{e}^{(1)\mathrm{T}}\,\boldsymbol{e}^{(1)} \quad \text{and} \quad E^{(2)} = \boldsymbol{e}^{(2)\mathrm{T}}\,\boldsymbol{e}^{(2)}, \tag{6.2.17}$$

are combined to a joint error which is to be minimized with respect to the model parameters

$$E = w_1\,E^{(1)} + w_2\,E^{(2)} = \text{Min!} \tag{6.2.18}$$

The weighting factors w_1 and w_2 allow different weights to be assigned to the error functions, corresponding to the significance of the data sets for the model.

The joint inversion is quite difficult in least-squares inversion schemes because it is hardly possible to determine the matrix of partial derivatives of two different linear operators simultaneously. But, the implementation of a guided random search algorithm such as those mentioned above, where only the forward problem has to be solved, provides a powerful tool for model calculations.

HERTRICH & YARAMANCI (2000) have described an algorithm to jointly invert data of dc geoelectric and surface nuclear magnetic resonance soundings in order to obtain a model of the subsurface that best fits the data of both measurements. The basic assumption is that both methods depend on water content. For SNMR this condition is fulfilled, as only water protons contribute to the measured signal. The connection between water content and electrical conductivity measured by dc methods is drawn by Archie's Law and other petrophysical relationships. In SNMR measurements, only the amount of free water in the pore space contributes to the signal, whereas the total amount of water, including the water bound to the grain surfaces, is responsible for the electrical conductivity of rocks. The numerical description of the subsurface is given by a finite number of layers with a certain thickness, each containing values for the amount of free and adhesive water. The joint inversion scheme using the simulated-annealing algorithm is based on forward modeling of both sounding curves. An extended characterization of an aquifer includes the ratio of mobile to adhesive water.

Joint inversion of ground geophysical data together with data from geophysical borehole logging and core analysis is the subject of extensive research. Improved determination of aquifer parameters can be expected from this technique in the near future.

6.2.2.3 Joint Interpretation Using Statistical Methods

MANFRED BIRKE

A joint interpretation of the results of the geochemical investigation in the Schöneiche-Mittenwalde area was carried out using statistical methods. Other results of the interdisciplinary investigation are described in Section 6.2.1, a case study of the Schöneiche-Mittenwalde area.

A geochemical analysis was made of the complex distribution of contaminant chemicals in the soil as well as stream and lacustrine sediments in the study area to determine background values and anthropogenic input. An inventory was made of the chemicals in the environment emitted from the Schöneiche industrial plants (chemical, metal processing, metal painting, and galvanic plants) and landfills (domestic, mixed, and industrial wastes), as well as from intensive agricultural land use. Anthropogenic and geogenic enrichments of heavy metals can be distinguished on the basis of areal distribution trends, so that hazardous areas can be identified. The geochemical data was analyzed using univariate and multivariate statistical methods. Maps of the distribution of the individual elements were prepared, as well as geochemical anomaly maps (factor values and cluster points), maps of the geochemical contaminant load in the topsoil, and maps of the differences between the measured element concentrations and limiting values given in the Berlin List. For the first time, inorganic and organic contaminants were analyzed together using multivariate methods, as well as factor and cluster analyses. The main information about the historical industrial land use in the study area and the distribution of areas of suspected contamination are shown on the map in *Fig. 6.2-41*.

A total of 56 inorganic parameters (11 main elements, 42 trace elements, TOC, pH in H_2O and $CaCl_2$, and electrical conductivity) were determined in the < 2.0 mm fraction of 170 soil samples. The organic contaminants AOX, EOX, cyanide, hydrocarbons, phenol, BTEX, volatile chlorinated hydrocarbons, PCB, and PAH were analyzed in 160 soil samples. PCDF and PCDD were analyzed in 90 soil samples.

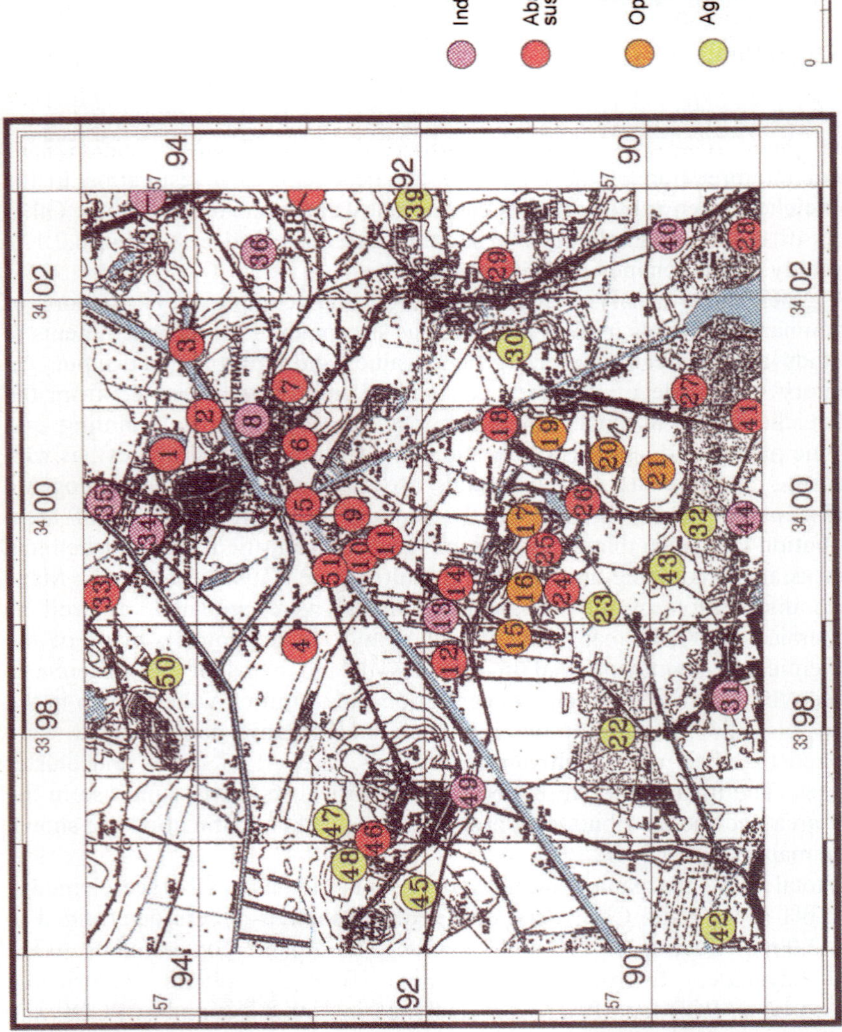

Fig. 6.2-41: Distribution of contaminant sources in the Schöneiche-Mittenwalde area

The first step in the statistical treatment of a data set is a univariate analysis of the data, determining the mean, minimum, maximum, standard deviation, median, and mode values for each of the individual geochemical parameters (see also Section 5.4.1). This initial information about the study area must be viewed critically, however. This is particularly clear for the mean value: Assume the study area has areas with nearly no contamination and other areas that are highly contaminated; assume the samples are collected on a regular grid. If the contaminated areas make up about 10 % of the study area and the contaminant concentration is about three times higher than in the uncontaminated areas, the mean for the study area as a whole would be 20 % higher than otherwise. Assume further that the highest contaminant concentrations are about ten times higher than in the uncontaminated areas, making up, e.g., about 2 % of the total area, the mean for the study area as a whole would be 35 % higher than the mean for the uncontaminated area alone. If the sampling density is doubled in addition in the contaminated areas in order to better characterize and delimit them, the mean values for the total area would be raised by 50 % and 78 % in the two cases. Neither the uncontaminated areas nor the contaminated ones can be characterized by such values.

As the above example shows, the mean value in the case of a varying sample density is not dependent only on the measured values but also on the sample distribution. Similar problems are present with other statistical parameters, e.g., percentiles.

Assuming more than 50 % of the samples are taken in the areas that have little or no contamination, the median and mode values provide important information about the study area. The distribution of the data is important to know in the case of the mode values. For this reason, a histogram analysis must be carried out for each parameter. Histogram analysis is of particular importance in the case of multimodal distributions and for computer-aided map preparation.

It is important to realize that nearly all commercial software for statistics calculates the smallest, not the highest mode value. For many trace elements, this value corresponds to the detection limit for that element. For this reason, a special iteration procedure must be used to determine the highest mode value.

Whereas in a univariate statistical evaluation, each parameter is treated separately, the purpose of a multivariate analysis is to determine the relationships between the parameters. Two such multivariate analyses are factor analysis and cluster analysis.

The objective of factor analysis is to reduce the number of parameters, i.e., the results should contain fewer parameters than the original data. This is done by combining parameters that have the same data distribution to a single parameter, called a factor. Each of these factors is a linear combination of the parameters in the original data set; the coefficients of the individual original parameters in the factor are called "factor values". These factor values are normalized and lie between -1 and 1. High absolute values of the "factor values" indicate a large influence of the corresponding parameter on the factor, low values indicate little influence. The sign of the factor value indicate the kind of influence: A positive sign indicates a positive correlation, a negative sign indicates an inverse correlation. Factor values near zero indicate the parameter has little or no influence on the factor value.

The factor values for each element in the individual samples can be calculated from the original data using various mathematical procedures. The factor values are normalized values, i.e., for the data set as a whole they have a mean of 0 and a standard deviation of 1.

Factor analysis is of great importance for the geochemical evaluation of large data sets. Element concentrations and other parameters that have the same or very similar or opposite distributions appear in the factors with high absolute factor values. In this way, geogenic and anthropogenic influences are indicated by the factor values. The contribution (in percent) of the individual influences for each sampling site can be obtained by interpretation of the factor values. Thus factor analysis provides a means of reducing the number of parameters that need to be taken into consideration, as well as a means of identifying the source of contamination.

The quality of the factor analysis results depends on the size of the data set, the data precision, the number of gaps in the data, and ability to control the order of the statistical calculations as needed by the data set. The "SPSS for Windows, v. 6.0" was used for the investigation of the Schöneiche region. Since there were few gaps in the data set, it was possible to exclude samples that lacked a value for an element from the calculations for that element. A varimax rotation was carried out to improve interpretability. The factor values were determined by regression.

Owing to the usually large number of influences, their complexity and interrelationships in anthropogenic influenced areas, there is a certain amount of uncertainty in the analysis of the results that must be interpreted according to the experience of the analyzing geochemist.

While the objective of a factor analysis is to identify samples with similar element relationships, the aim of cluster analysis (Q-mode) is to group samples with similar element concentrations.

Prerequisite for significant results is a normalization of the parameter values in the original data set. This makes the influence of the parameter values all the same. There are two types of cluster analysis: a partitioning procedure and a hierarchical classification. In the partitioning procedure, the samples are divided among a set number of classes. In the hierarchical procedure, similar samples are assigned to classes which are then grouped in classes with less similarity, forming larger classes, which are then grouped in further classes until all of the samples form a single cluster. Both procedures have their advantages and disadvantages.

The partitioning classification procedure is a simple, fast method that requires little computer time and, hence, is advantageous for large data sets. By adjusting the locations of the class boundaries to optimize the locations of the class centers and reclassifying the samples several times, an unambiguous classification of every sample can be achieved. A considerable disadvantage is that the number of classes must be fixed before the classification. Moreover, the relationships between the classes can be determined only with difficulty or not at all.

The hierarchical classification procedure, in contrast, requires a large RAM memory and is, therefore, suitable only for smaller data sets. The main advantage of the method is that the dendrogram of the result provides the optimum number of classes and the relationships of these classes to one another. Especially in the case of complicated geochemical relationships, the results can be interpreted less easily than the results of the partitioning procedure, owing to the hierarchical order. This is because when a sample is assigned to a particular class at one stage of the procedure, there is no later checking whether this sample can be more optimally assigned.

To take benefit of the advantages of both methods, a combination of the partitioning and hierarchical classification procedures can be used. First, the data is divided into a set number of classes. The locations of the class centers are then shifted and the samples are reclassified until there is no further shifting of the samples between the classes. The centers of the resulting classes are then used for a hierarchical classification in order to investigate the relationships between the classes and to determine the optimum number of classes.

The results of a cluster analysis are interpreted in two ways: The areal distribution of the classes is determined, and each class is subjected to a univariate evaluation.

Cluster analysis is also of considerable importance for a geochemical evaluation. The efficiency of the evaluation is increased, because classes of samples are taken into consideration instead of the individual samples. And the quality of the results is also increased by determination of element background values and the identification of multi-element anomalies, i.e., areas characterized by the element/parameter associations observed within them.

Maps of the distribution of the individual elements/parameters and maps of the distribution of the factor values (multi-element maps, e.g., *Fig. 6.2-46*) enable successful interpretation of geochemical data. Cluster maps (*Fig. 6.2-47*) are prepared by hand and then digitized and edited using the capabilities of the computer.

The selected results of a statistical treatment of the analysis results of soil samples are discussed in the remainder of this section.

Anomalous electrical conductivity values (*Fig. 6.2-42*) mark the location of the Schöneicher Plan landfill (which shows the highest values:

2850 µS cm^{-1}, see *Table 6.2-3*), the abandoned "Alte Asche" landfill next to the Notte canal, and large areas containing old waste dumps east, northeast, west, and northwest of Mittenwalde. Distinctly elevated values > 500 µS cm^{-1} extend eastwards 500 - 600 m from the waste incinerator plant, marking its emissions plume. Slightly elevated values up to 200 µS cm^{-1} (background = 21 µS cm^{-1}) characterize the agriculturally used areas (mostly pasture and fallow land) on the southeast side of the study area.

Table 6.2-3: Statistical values for the parameters pH, electrical conductivity (EC), and total organic carbon (TOC) in the topsoil of the Schöneiche area

Parameter	Units	Statistical values (N=170)				
		Mean	Minimum	Maximum	Mode	Median
pH in dist. H_2O		6.6	3.9	8.6	7.8	6.9
pH in $CaCl_2$		6.2	3.6	8.3	7.4	6.4
EC	µS cm^{-1}	165	9	2850	21	57
TOC	%	2.3	0.1	27.0	0.5	1.2

The soil pH depends on the chemical composition of the soil; in addition to the clay and humus content, pH is the main parameter affecting the mobility of elements in the soil. pH (measured after leaching the soil with a $CaCl_2$ solution or with distilled water) is a measure of the acidity/alkalinity of the soil. Soil can become acidified by anthropogenic emissions (mostly industrial, e.g., SO_2, NO_x, SO_4^{2-}, NO_3^-, HCl) and by natural processes (e.g., decomposition of biomass, leaching by high amounts of precipitation, metabolism of soil organisms).

The relatively small wooded areas, wetlands, and grasslands in the Schöneiche study area are characterized by soil pH values < 4.0. The soil pH values show a typical bimodal frequency distribution with peaks between 4.2 and 5.2 and between 6.5 and 8.5. The maximum in the acid range is due to the dissolving of Ca, Mg, and K compounds; pH values < 4.2 can dissolve Al (potentially toxic) from clay minerals. The peak in the neutral range is due to buffering by amphoteric ions.

Examination of the distribution of pH values in the ranges of buffers shows that in most soil samples at pH > 5, acids are neutralized with the release of alkali and alkaline earth ions – at pH > 6.2, for example, by the dissolving of calcium carbonate – and only small amounts of plant nutrients are leached from the soil. Locally, pH values < 5 result in a decrease in cation exchange capacity and the acids are buffered by aluminum compounds (clay minerals) with the release of Al ions (pH < 4.2) and by iron compounds with the dissolving of iron oxides (pH < 3.2). These two buffer ranges occur only in wooded areas. The soil pH is increased to values > 8 by building rubble, providing good buffering of acid input.

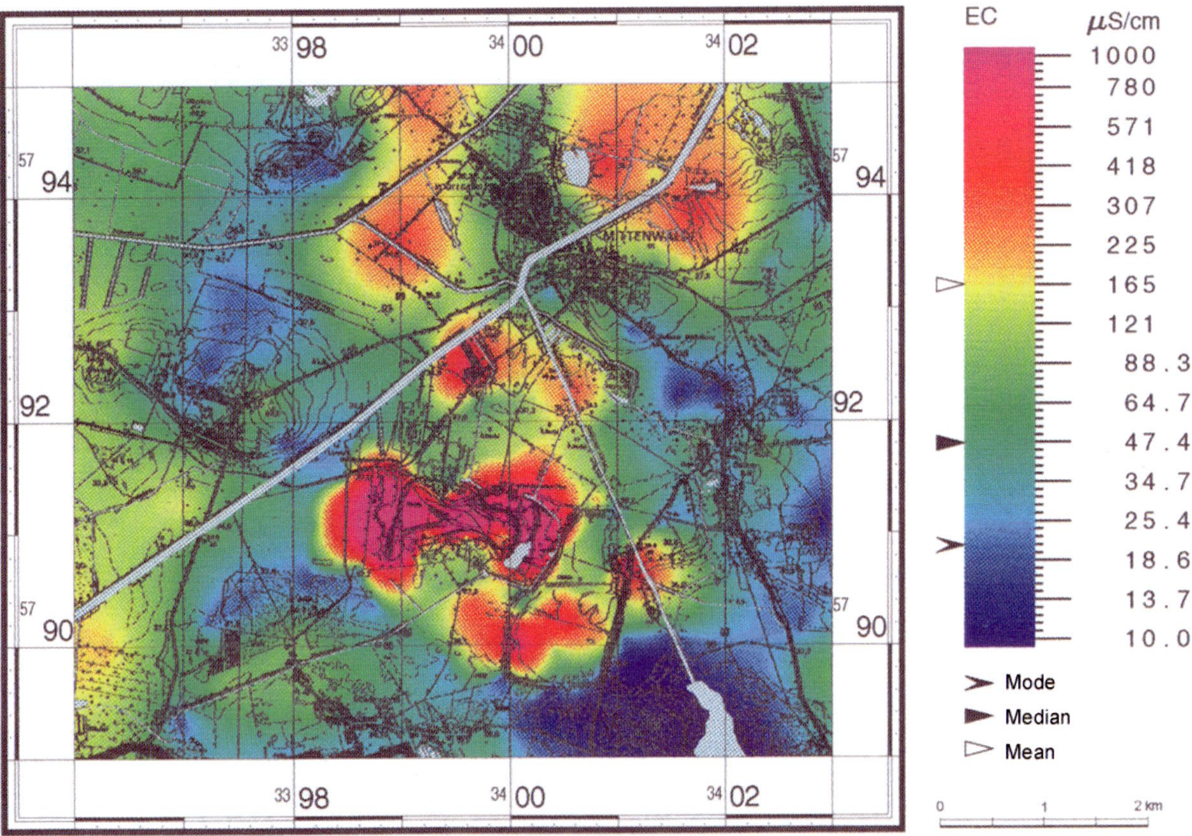

Fig. 6.2-42: Distribution of electrical conductivity in the topsoil of the study area

Because pH alone is not a sufficient indicator of acid input, additional information is needed about the basic ions in the form of cation exchange capacity, degree of base saturation, and the acid/base ratio. Acid input is always associated with a loss of basic ions and buffer capacity and an increase in Al and Fe ion concentrations in the soil water and the leaching of these ions into the groundwater. The K_1 = Al/(Ca+Mg+K) value can be used as a proxy for the buffering capacity of the soil. In areas with a ratio < 2.5, the soil has a good buffering capacity ($K_1 < 1$). The distribution of the concentration coefficient maxima (*Fig. 6.2-43*) is the same as that for the pH minima. Only two very small problem areas of acidification and lack of buffer capacity ($K_1 > 2.5$) were observed in the study area (southeast of Mittenwalde).

The humus content of the soil can be calculated from the TOC value via multiplying by 1.724 (a factor based on the mean carbon content of 58 % in organic compounds). The regional TOC background value of 1.2 % indicates a relatively low humus content. Corresponding to the range of TOC values (0.1 - 27.0 %), the soils in the study area can be classified as very weakly humic to peaty (> 30 % humus). TOC values > 3 % occur mainly in areas of humic gley, fen, and peaty gley. The TOC maxima near the Notte canal, in the southwestern part of the study area, and in the western part of the Schöneicher Plan landfill are often accompanied by elevated concentrations of organic contaminant parameters (PAH, PCDD, CN, phenols, hydrocarbons).

The main elements in the soil include the plant nutrients P, K, Ca, Na, Mg, N, and S. Although Al, in contrast to Fe and Mn, is not a plant nutrient, it is included together with Si and Ti in the main element analysis.

Silicon is an essential element for all life. The mean SiO_2 content of 84.6 % and a geochemical background value of 88.4 % (*Table 6.2-4*) in the Schöneiche study area shows that the sand and gravel in the region has a high silica and silicate content. SiO_2 minima (< 60 %) are observed only in the areas of complex multi-element anomalies (*Fig. 6.2-47*) southeast of Mittenwalde (the “Alte Asche” landfill, anomaly no. 30) and northeast of Mittenwalde (anomaly no. 15), the eastern, central and western parts of the Schöneicher Plan landfill, and the southwest side of the study area (anomaly no. 68). Local low Si contents are observed in the areas of known abandoned landfills just north of the Schöneiche landfill and in the entrance area to this landfill. The Si minima occur in the same areas as TOC and P maxima. Low pH values cause a decrease in mobility of Si and application of P fertilizers an increase. This is confirmed by the high inverse correlations between SiO_2 and TOC and between SiO_2 and P_2O_5. Weak positive correlation coefficients are observed only for Cs and Zr.

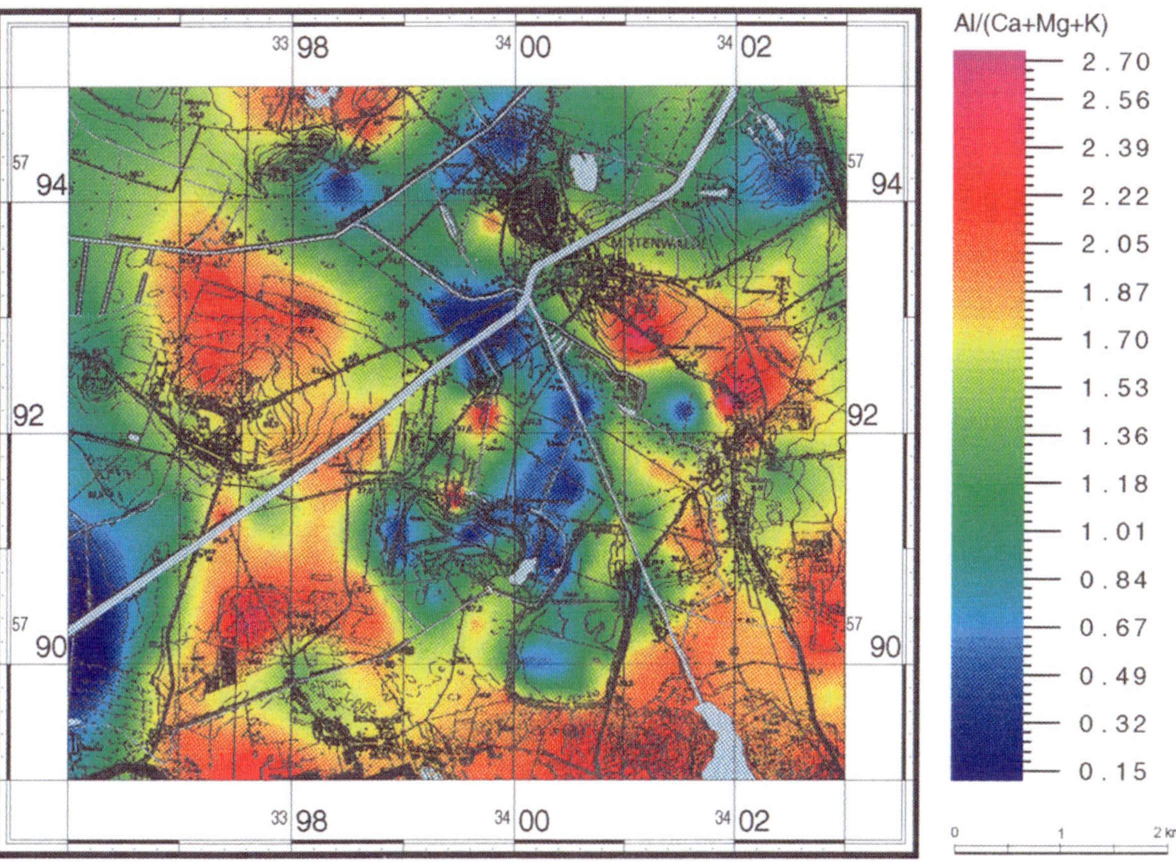

Fig. 6.2-43: Distribution of the K_1 = Al/(Ca+Mg+K) value in the topsoil of the Schöneiche area

Table 6.2-4: Statistical values [in mg kg^{-1}] for the main elements in the topsoil of the Schöneiche study area

Element	Statistical values, N=162					Global values for soils (after Bowen, 1979)		
	Mean	Minimum	Maximum	Mode	Median	Median (Clarke value)	range Min.	range Max.
Si	395 181	121 892	439 729	428 422	412 911	330 000	250 000	410 000
Ti	1 091	432	3 699	935	935	5 000	150	25 000
Al	20 575	7 726	78 110	15 982	18 866	71 000	10 000	300 000
Fe	8 910	1 679	68 961	3 077	5 490	40 000	2 000	550 000
Mn	248	62	1 100	132	201	1 000	20	10 000
Mg	1 170	181	9 048	483	724	5 000	400	9 000
Ca	10 956	929	160 093	1 286	3 359	15 000	700	500 000
Na	2 931	742	8 829	2 448	2 745	5 000	150	25 000
K	7 555	2 740	15 774	6 808	6 932	14 000	80	37 000
P	441	22	3 885	218	327	800	35	5 300

In contrast to most organic contaminants, heavy metals occur naturally in rocks and ore minerals. For this reason, sediments and soils contain a geogenic (natural) concentration of these elements. The global background values (Clarke values) are given in *Table 6.2-5* together with the background values for the Schöneiche region. Both the local and regional background concentrations of heavy metals must be determined in order to ascertain whether the soil being investigated has values significantly higher than the regional value. Thus the natural geochemical background values provide a solid basis for estimating hazards and assessment of the contaminant load.

Relative to the regional background values, the most important heavy metals – Cd, As, Cu, Pb, Zn, Hg, Cr, Mo, Sn, Ni and Sb – are highly enriched ($K \geq 10$; K = ratio of the sample concentration of the element to the regional background) in the areas covered by the two large landfills, in the contaminated areas SW and SE of Mittenwalde (*Fig. 6.2-41*), and just north of the Schöneiche landfill (Cu, Zn, Cd, Pb, Be, Tl, Sn, Bi, Sb, As, and Cr), W and NW of Telz (Sb, Bi, F, and As), NNW of Mittenwalde (V, U, and As), E of Mittenwalde (As and Ni), and on the SE, SW, and NW sides of the study area.

Table 6.2-5: Statistical values [in mg kg^{-1}] for the trace and heavy-metal elements in the topsoil in the study area

Element	Statistical values (N=170)					Global values for soils (after Bowen 1979)		
	Mean	Range		Mode	Median	Median (Clarke value)	Range	
		Min.	Max.				Min.	Max.
Ag	0.5	0.1	25.6	0.1	0.1	0.05	0.01	8
As	3.0	0.51	44.8	1.0	1.6	6.0	0.1	40
B	17.4	6.3	114.0	11.0	12.0	20	2	270
Ba	266.5	139.0	2 590.0	166.0	211.0	500	100	3000
Be	1.2	0.7	14.0	0.9	1.0	0.3	0.01	40
Bi	0.4	0.2	9.7	0.2	0.2	0.2	0.1	13
Br	6.4	0.4	69.5	1.4	3.1	10	1	110
Cd	0.52	0.01	30.9	0.07	0.19	0.35	0.01	2
Ce	25.8	9.3	170.0	20.2	22.8	50	3.0	100
Co	2.3	0.2	75.8	0.2	0.2	8.0	0.05	65
Cr	21.3	1.2	301.0	8.8	14.2	70	5	1500
Cs	3.2	0.1	10.9	0.1	3.3	4.0	0.3	20
Cu	31.5	2.2	711.0	4.6	8.8	30	2.0	250
F	400	100	26 300	300	300	200	20	700
Ga	4.6	2.0	20.8	4.2	4.2	20	2	100
Ge	1.0	0.2	3.0	1.0	1.0	1	0.1	50
Hf	12.2	10.0	32.0	10.0	10.0	6	0.5	34
Hg	0.25	0.01	10.6	0.02	0.06	0.06	0.01	0.5
In	0.04	0.03	0.25	0.03	0.03	1.0	0.7	3
I	1.5	0.1	9.6	0.1	0.9	5.0	0.1	25
La	15.8	7.4	53.5	12.3	14.1	40	2	180
Mo	4.6	0.4	52.0	4.2	4.3	1.2	0.1	40
Nb	4.8	2.4	16.8	3.7	4.4	10	6	300
Ni	6.04	0.03	112.0	0.03	1.08	50	2	750
Pb	59.3	11.6	1 330.0	16.2	26.2	35	2	300
Rb	32.1	12.2	88.0	27.4	29.5	150	20	1000
Sb	2.37	0.05	23.9	2.2	2.2	1.0	0.2	10
Sc	2.5	1.0	25.0	1.0	2.0	7.0	0.5	55
Se	0.1	0.1	1.4	0.1	0.1	0.4	0.01	12
Sn	9.7	0.4	631.0	1.3	2.1	4.0	1	200
Sr	76.9	24.7	767.0	36.0	52.9	250	4	2000
Ta	0.05	0.03	1.2	0.03	0.03	2	0.4	6
Te	0.5	0.3	2.0	0.3	0.3			
Th	3.4	0.4	17.5	2.8	3.2	9	1	35
Tl	0.3	0.1	2.7	0.1	0.1	0.2	0.1	0.8
U	1.0	0.03	12.1	0.03	0.5	2.0	0.7	9.0
V	11.6	0.7	206.0	2.7	4.8	90	3.0	500
W	1.0	0.2	38.0	0.8	0.8	1.5	0.5	83
Y	9.7	3.6	231.0	5.2	7.3	40	10	250
Zn	95.4	10.7	1 510.0	26.3	34.2	90	1	900
Zr	172.9	50.3	281.0	175.0	172.0	400	60	2000

Arsenic concentrations in the Schöneiche region range from 0.5 to 44.8 mg As kg^{-1}. The values increase with the intensity of anthropogenic land use. The map of As distribution (*Fig. 6.2-44*) shows high technogenic concentrations (more than 10 times the regional background) in the central, western, and eastern parts of the Schöneicher Plan landfill, the old landfill just north of the Schöneiche landfill, the south heap of the Schöneiche landfill, and the entry to the waste incinerator plant. The anthropogenic anomaly on the former brickworks site east of Mittenwalde also exceeds the limits (10 mg As kg^{-1}) of the Brandenburg and Berlin lists. Other distinct anomalous concentrations (>3 mg As kg^{-1}) which can be attributed to anthropogenic sources are present northwest of Vogelsang (NW of Mittenwalde), west of Telz, and the southwestern part of the study area. Further As contamination occurs in the area of the "Alte Asche" landfill and along the Gallun canal in the west. Local As maxima do not correlate with pH minima except around the old landfills along the Notte canal. The pH at which As begins to be mobilized is not reached, however. The As concentrations in the area of the old landfills along the Notte canal are accompanied by elevated concentrations in the surface water and stream sediments. An assessment of trace element concentrations is given in the section about factor and cluster analyses.

Organic contaminants of anthropogenic origin (e.g., industrial plants and storage and transport, waste disposal, landfills, waste incineration, sewage and sewage sludge, and pesticides) are ubiquitous in all parts of the biosphere. With respect to distribution in soils, toxicity, persistence, and enrichment mechanisms, organic compounds – e.g., polycyclic aromatic hydrocarbons (PAH), polychlorinated biphenyls (PCB), chlorinated pesticides, and polychlorinated dibenzo-p-dioxins and furans (PCDD/F) – are becoming increasingly a problem.

The soil samples were analyzed for the main organic parameters using standard digestion and extraction procedures in accredited laboratories. No volatile halogenated or aromatic hydrocarbons were detected.

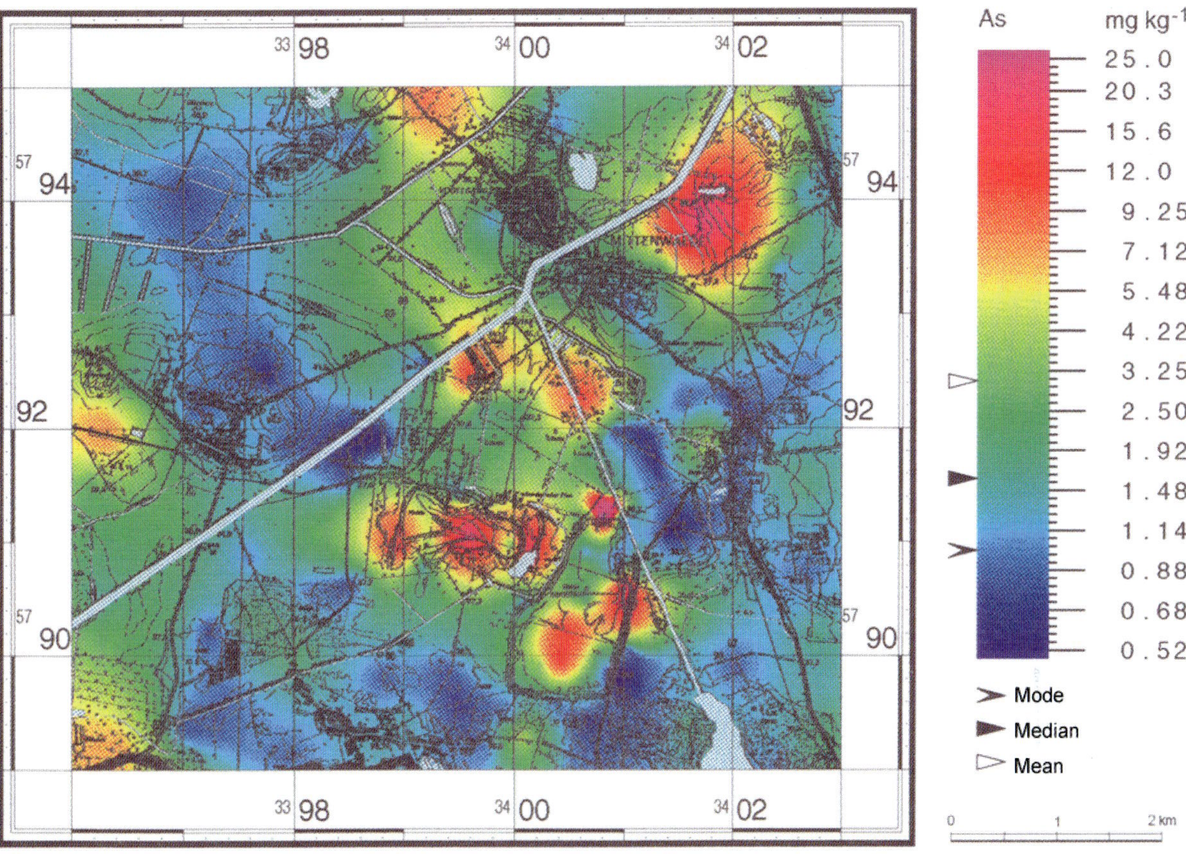

Fig. 6.2-44: Distribution of arsenic in the topsoil in the Schöneiche–Mittenwalde area

Table 6.2-6: Statistical values [in mg/kg dry substance] for selected organic contaminants in the topsoil in the Schöneiche area together with the background and limits of the Berlin and Netherlands lists

Parameter	Schöneiche				Berlin List (1996)[6] sand, gravel, marl		Netherlands List (1998)[6] soil, sediment	
	Number of samples	Mean	Median	Max.	Background value	Intervention limit	B-value (Level 1 limit)	C-value (Intervention value)
EOX	116	3.2	1.8	17	1[1)]	3[2)]	10[3)]	15[4)]
AOX	25	14	12	32		500[5)]		
petroleum hydrocarbons	119	43	13	1500	10	20	50	5000
phenol index	4	1.4	0.4	4.6	1.0[6)]	10.0[7)] (30)[8)]	0.05	40
$cyanide_{tot}$	28	4.7	0.2	50	1	2 (25)[9)]	10	100
PCB_6[10)]	14	0.008	< 0.002	0.27	0.05	0.1	0.02[11)]	1.0[11)]
PAH	116	2.0	0.3	63	0.4	0.8	1	40
PCDD/F (ng TE/kg)[12)]	89	4.7	1.2	111	10[13)]	100[14)]	2[15)]	5[16)]

1) Class 0 value – for unlimited use (LAGA, 1994)
2) Class 1.1 value – for unlimited, uncovered use (LAGA, 1994)
3) Class 1.2 value – for unlimited, uncovered use (LAGA, 1994)
4) Class 2 value – for use under certain restrictions
5) limit for sewage sludge
6) Level 1 limit (Bayern,1991)
7) Level 2 limit (intervention value) (Bayern, 1991)
8) limit for playgrounds (Berlin List, 1996)
9) intervention value for humus soils (Berlin List, 1996)
10) after BALLSCHMITTER & BACHER (1996)
11) sum of seven PCBs: PCB28, 52, 101, 118, 138, 153 and 180
12) TE = toxicity equivalent according to the German Federal Health Office (BGA)
13) Class I soil value, geogenic background, various uses of the soil possible (EIKMANN & KLOKE, 1993)
14) limit for sewage sludge
15) background value (Baden-Württemberg 1993)
16) limit in Baden-Württemberg

6 Also referred to as “Berliner Liste”, “Brandenburger Liste”, and “Niederländische Liste”, after its origin (see references)

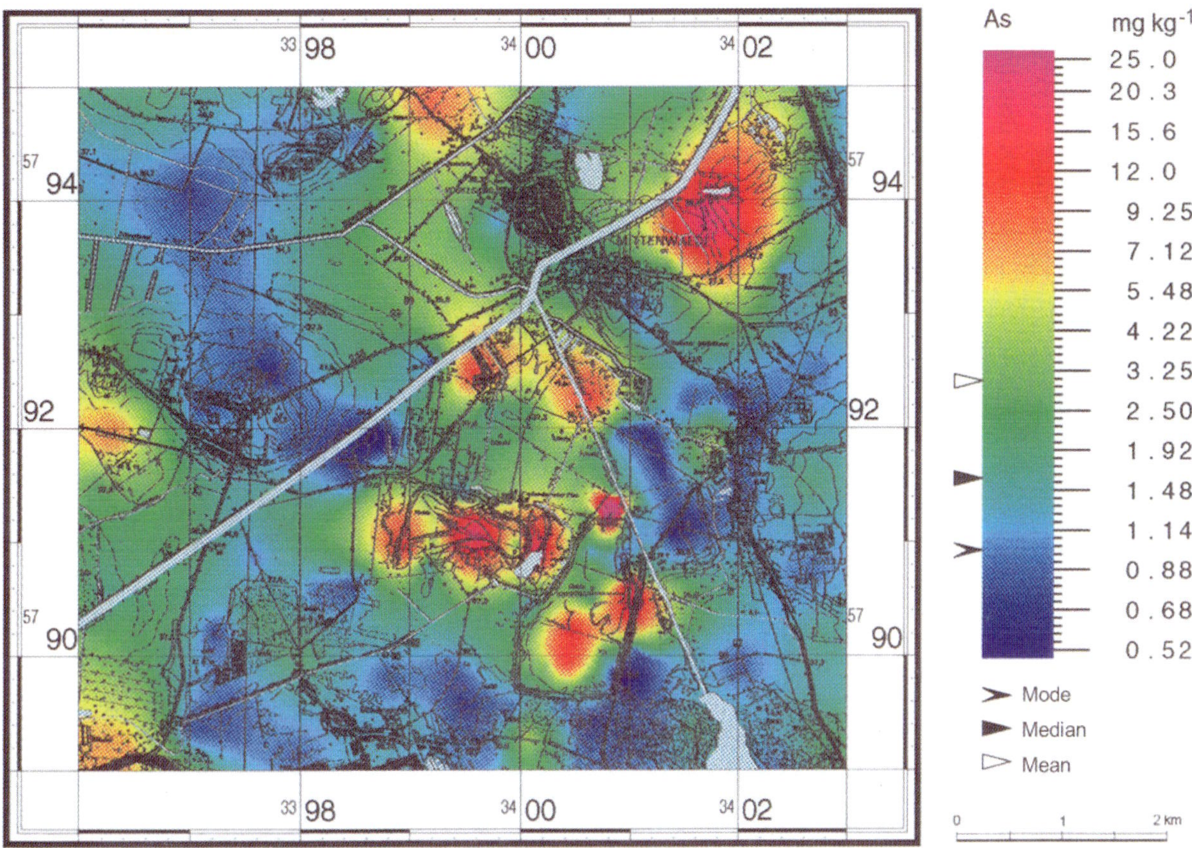

Fig. 6.2- 45: Schöneiche-Mittenwalde area: EOX distribution in the topsoil

The chemical parameters HC, AOX, EOX, and the phenol index are important for an overview of organic contamination. Chemicals that are of particular importance from environmental and toxicological points of view are summed in the parameters AOX (adsorbable organic halogen compounds) and EOX (extractable organic halogen compounds). The AOX and EOX parameter values are a basis for decisions on further analytical procedures. AOX were identified only in 15 % of the samples; EOX in 73 %. The EOX distribution varies considerably over very short distances (*Fig. 6.2-45*). Values above the regional background (1.8 mg EOX kg^{-1}) are mostly local, represented by only a single sample. Frequently, high concentrations (>10 mg EOX kg^{-1}) were measured in the immediate vicinity to sampling sites where no EOX was detected. Elevated concentrations over large areas were not observed. EOX values >20 mg EOX kg^{-1} are reasons for further analysis for PCB, volatile halogenated hydrocarbons, and chlorinated pesticides, e.g., chlorinated phenol and benzene. The distribution of EOX values is seldom similar to that of other organic compounds. Further analyses are required when the threshold value of 8 mg EOX kg^{-1} of the Netherlands List (1988) is exceeded, which in the study area exceeded by up to twice this value. The areas where this threshold is exceeded are all in agricultural land (E, SE, W of Mittenwalde, and W of Gallun) and industrial areas (N of Mittenwalde and along the autobahn SE and E of Mittenwalde), except for the west side and east and southeast edges of the Schöneicher Plan landfill and the “Alte Asche” landfill. The wooded areas have considerably lower values and thus input in the entire area via the atmosphere can be neglected.

Factor analysis (R-mode) was used to investigate the linear correlations between the many inorganic and organic soil parameters. Twelve factors were obtained with factor analysis, which took both organic and inorganic contaminants into consideration. The classification of the factors shown in *Table 6.2-7* shows that the results help considerably to recognize element and contaminant migration in the soil as well as the factors influencing element distribution in the study area. Factor analysis permits the examination of the complex causes of anomalies and their influence on element enrichment and migration. The classification of the factors (*Table 6.2-7*) reveals three main causes of element migration in the study area: anthropogenic, geogenic, and environmental influences.

Despite many varied anthropogenic influences, the geogenic, i.e., natural, factors still explain 41 % of the total variance. The geogenic factors characterize the near-surface geology. Elevated Sc-V-Be-Al-Ga-Th-Ge-U-Nb-Co-Mg-Rb-Ti-La-Ni-Sr-Fe factor values correlate with the distribution of boulder clay in the ground moraines east and northwest of Mittenwalde and west of Telz. Anomalous Zr-K-Rb-Na-Ti-Nb factor values occur in the areas of glaciofluvial sands and ground moraines (marl and clay) north and northeast of Telz and east and southeast of Mittenwalde. The elevated factor values in the landfill areas relate to the sources of the cover material used.

Zr-K-Rb-Na-Ti-Nb factor value maxima also occur in the old landfill areas north of the Schöneiche and Schöneicher Plan landfills.

Table 6.2-7: Classification of the multi-element factors

Contaminant source	**Element association**
1. Anthropogenic contamination factors	
landfill factor: (domestic waste and abandoned landfills, industrial residues)	W-Mo-HC-Sb-Sn-Cu-Cr-Pb-Zn-Ba-(As-Fe-Ti-Ni)
sewage and composting factors: sewage sludge and liquid manure as fertilizer industrial residues, irrigation with liquid manure, sewage sludge treatment plants	PI-Cd-Ag-Hg-Bi-PCB-P-PCDD PAH-(PCB-AOX-pH)
industrial emissions factor: ash, waste incineration industrial residues, abandoned landfill, industrial and trades area, area around operating and abandoned landfills (former chemical storage)	PCDF-PCDD-(pH-Zn) CN-Se-Hf-(As-TOC-PCDD) Ta-(B-Hf-EOX-EC), In-(EC-S-TOC-B-Sn-Hg-PCDD), minus Na, minus AOX
change in land use	minus Te-[minus Cs]-[minus I]-[minus La]-(pH-Ca-Mn-Tl)
factor for building rubble and/or construction materials (also former chemical and construction materials storage, chemical industry)	F-Y-Ce-S-EC-Sr-La-Pb-Ca-Ni-B-(As-Tl-Fe-Mg-Zn-Ba-TOC-Hg-Co)
2. Natural, and geogenic factors with anthropogenic contributions	
boulder clay in ground moraines (clay minerals) glaciofluvial sand (alkaline earth factor, brown soils and leached earth)	Sc-V-Be-Al-Ga-Th-Ge-U-Nb-Co-Mg-Rb-Ti-La-Ni-Sr-Fe Zr-K-Rb-(Na-Ti-Nb)
3. Environmental factors	
geochemical barriers (humus-rich soils, e.g., fen and gley soils)	Br-Mn-Ca-TOC-(Sr-U-EC), minus Si

PI: phenol index, CN: cyanide, EC: electrical conductivity,
TOC: total organic carbon

The geogenic enrichment of trace elements is increased (factor values ≥ 1) by anthropogenic influence in the areas of the large landfills. The alkaline earth factor Zr-K-Rb-(Na-Ti-Nb) explains only 2.4 % of the total variance and is therefore of subordinate significance relative to the clay minerals factor. The environmental factor Br-Mn-Ca-TOC-(Sr-U-EC) has elevated values mainly in grassland and wetlands areas (NW, W, SW. and NE of Mittenwalde and the SW edge of the study area), as well as croplands with humus-rich soils south and north of the Schöneicher Plan landfill, west of Telz, and WNW of Gallun. These are areas of marl and some Holocene fen peat (on either side of the Notte canal north of the Schöneicher Plan landfill and SSE of Mittenwalde).

The slightly positive factor values (especially for the component EC) and elevated factor values for anthropogenic contamination, as well as the cluster analysis (Q-mode) results, point very definitely to the influence of the application of lime and fertilizers in the agricultural areas.

The anthropogenic factors characterize the various technogenic influences in the study area and are an important aid for delimiting areas with little or no contamination from those containing contamination to various degree and complexity.

The landfill factor W-Mo-HC-Sb-Sn-Cu-Cr-Pb-Zn-Ba-(As-Fe-Ti-Ni) is elevated in the areas of both of the large landfills, the roads leading from the towns and villages, old landfills, and disposal areas for industrial residues. The Schöneicher Plan landfill, the central part of the Schöneiche landfill, and the downwind area from the waste incinerator plant all have factor value maxima > 2. Slightly elevated factor values are observed in areas of humus-poor soils north and southeast of Gallun, south of the Schöneiche landfill, and northwest of Telz. This factor explains 11.6 % of the total variance, underscoring the influence of the landfills on the contaminant distribution in the study area. It also confirms its ability to explain the influences in effect and the causes of element migration, even in the areas of weakly elevated factor values.

The building rubble factor F-Y-Ce-S-EC-Sr-La-Pb-Ca-Ni-B shows elevated values in the parts of the landfills that contain nearly only building rubble (central and west side of Schöneicher Plan, southwestern part of Schöneiche, and Telz). Elevated values are observed in the industrial area at the south edge of Mittenwalde (storage of chemicals and construction materials) and in the new trades area southeast of Gallun. The slightly elevated values east of Grossmachnow at the NW edge of the study area and east of Mittenwalde are due to the influence of relatively recent industrial areas and old landfills.

Industrial influence is sometimes of very complex nature. Elevated values for the factor PCDF-PCDD-(pH-Zn) (*Fig. 6.2-46*) characterize the locations of industrial residues (ash and slag from the chemical industry and waste incineration, residues from large and small fires). Elevated values mark the town of Mittenwalde and immediate surroundings (industrial plants, old landfills, fertilizer storage), the old "Alte Asche" landfill southwest of

Mittenwalde, the central and southern parts of the Schöneiche landfill, the downwind area east of the waste incineration plant, and old contaminated areas at the northern and southern edges of the Schöneicher Plan landfill. Slightly elevated values are observed in the village of Schöneiche and the area west of Telz (former chemical storage). To investigate the sources of the PCDD/F contamination, samples were analyzed for the distribution of homologous halogen compounds of all degrees of halogenation, the total contents of PCDD/F, and the percentage of specific congener isomers, comparing the results with reference samples with known contamination sources. The main sources of dioxins and furans in the Schöneiche study area were determined to be the use of sewage sludge as fertilizer and residues from low-temperature combustion.

In contrast to factor analysis (combination of parameters with similar distributions to a new parameter), Q-mode cluster analysis divides the data into different classes on the basis of similarities in their geochemical relationships, representing the regional and sometimes local multi-element background and geochemical anomalies. A table of statistical parameters for the different classes and the ratio with respect to the background values was prepared and a map of the multi-element geochemical anomalies was made (*Fig. 6.2-47*).

Application of Q-mode cluster analysis to the main element and organic contaminant parameters permits to investigate the main causes of element migration (enrichment and/or depletion) for each sampling site. The use of both cluster analysis and factor analysis distinctly improves the interpretation of the anomaly and background classes. None of the 93 geochemical multi-element anomalies can be unambiguously assigned to a geogenic origin. There are, in contrast, many very different anthropogenic influences, mostly from the two large operating landfills and the old, abandoned landfills. Characteristic of the complex types of anomalies is the large number of different out of the ordinary combinations which point to an anthropogenic origin.

The data set included 170 soil samples, divided into forty classes. The element associations in the geochemical anomalies were defined on the basis of the ratio of the class mean of the element concentrations in that class to those of the background classes. Ratios of 1.8 - 1.9 were described as "elevated", those of 2.0 - 2.9 as "distinctly elevated", those of 3.0 - 9.9 as "anomalous", and those > 10 as "very anomalous".

A total of 93 geochemical multi-element anomalies were mapped (*Fig. 6.2-47*). The main sources were industrial sites and landfills (27 anomalies), old landfills and built-up areas (19 anomalies). Agricultural activities (fertilizers and spreading of liquid manure and sewage sludge) were the source of 16 anomalies. Most of the anomalies (31) have multiple sources.

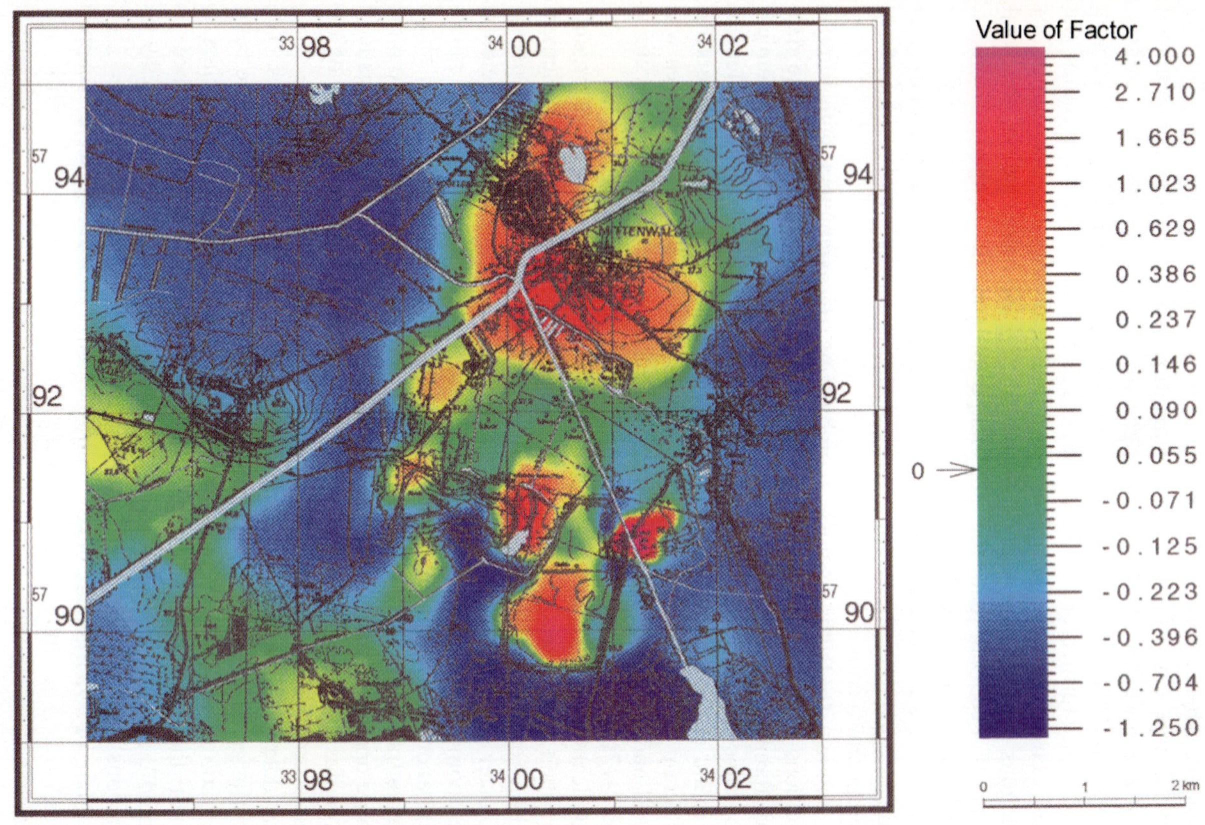

Fig. 6.2-46: Value of Factor 9: PCDF-PCDD-(pH-Zn)

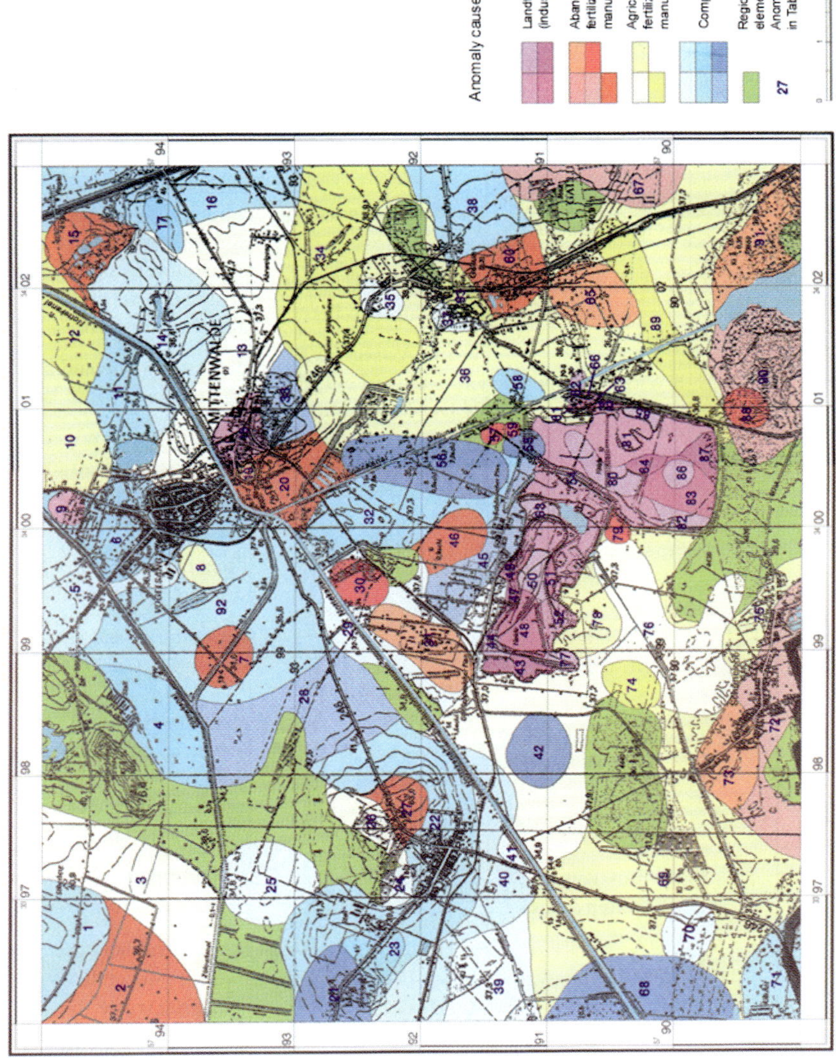

Fig. 6.2-47: Multi-element geochemical anomalies in the Schöneiche-Mittenwalde area

Table 6.2-8: Characterization of selected geochemical multi-element anomalies in the Schöneiche study area (see *Fig. 6.2-47*)

Anomaly number	Class number	Locality	Anomalous element association	Causes/remarks
1	4	ca. 1 km E of Grossmachnow	Ni-Hg-CN-PAH-U-Ca-(Mn-P-EC-Cd-Cr-Br-Ag-Mg-Zn)	industrial waste, (old landfills), irrigation with liquid manure, agricultural land use
2	33	SE of Grossmachnow	Ni-Ca-PAH-TOC-Br-EC-S-Hg-Fe-Sn-Cd-Se-Cu-Ta-As	illegal burning of waste
3	35	N–S anomaly E of Grossmachnow	PAH-Ni-Se-PCDD-Br	use of sewage sludge in agriculture, other agricultural land use
4	38	S of Großmachnow Weinberg	Ni-EOX-Ca-U-Hg-(Br-S-Cd-Sn-P-Cr-PI)	agricultural land use (animal husbandry, fertilizers), abandoned landfill
5	37	NW of Mittenwalde	Ni-EC-As-Hg-Ca-Br-Co-S-U-Cd-Fe-TOC-PCDD-Mg-P-Mn-Cr	domestic waste and building rubble, agricultural land use
6	38	within Mittenwalde	Ni-EOX-Ca-U-Hg-(Br-S-Cd-Sn-P-Cr-PI)	complex anthropogenic influences (industrial and trades area), agricultural land use
7	33	W of Mittenwalde	Ni-Ca-PAH-TOC-Br-EC-S-Hg-PCDD-Fe-Sn-Cd-Se-Cu-Ta-As	illegal dumping of waste or sewage, agricultural land use
8	28	W edge of Mittenwalde	Ni-PCDD-Ca-(Hg-U)	agricultural land use (sewage sludge as fertilizer)
9	25	N of Mittenwalde	Ta-Ni-Ca-Hg-EOX-PCDD/F-S-Fe-Zn-TOC-U-(Cd-Mn-Br-P-AOX-EC-Mg-Cr)	abandoned landfill, industrial and trades area
10	28	NE of Mittenwalde	Ni- PCDD-Ca-(Hg-U)	agricultural land use, (abandoned landfill)
11	8	E of Tonsee, Mittenwalde	Ni-Ca-EC-Br-Hg-TOC-S-PCDD-PAH-Fe-P-Cd-Mn-Ta-As-Sn-Zn	agricultural land use, former brickworks, residential areas
12	4	NE of Mittenwalde	Ni-Hg-CN-PAH-U-Ca-(Mn-P-EC-Cd-Cr-Br-Ag-Mg-Zn)	agricultural land use (liquid manure), secondarily industrial waste
13	4	E of Mittenwalde	Ni-Hg-CN-PAH-U-Ca-(Mn-P-EC-Cd-Cr-Br-Ag-Mg-Zn)	agricultural land use, industrial and trades area
14	37	E of Mittenwalde	Ni-EC-As-Hg-Ca-Br-Co-S-U-Cd-Fe-TOC-PCDD-Mg-P-Mn-Cr	agricultural land use, illegal dumping of waste, former brickworks

Technogenic/complex anthropogenic influences are present in all of the geochemical multi-element anomalies (*Table 6.2-8* and *Fig. 6.2-47*). Geogenic, i.e., purely natural anomalies, cannot be unambiguously identified in the study area on the basis of the geochemical multivariate analysis. A geogenic influence from the marl in the ground moraines can be identified in only a few of the anomaly classes. The high Ca concentrations of the anomalies is locally increased by the marly constituents of the ground moraines.

Areas of geogenic/pedogenic influence (e.g., alkaline-earth factor) can be delineated only by comparison of the cluster analysis result with the factor analysis result and the maps of the distribution of the individual parameters. The cluster analysis reveals a strong technogenic influence overlying the geogenic components.

Environmental enrichments at organic material occur in many of the anomalies in addition to the overriding anthropogenic input. TOC concentrations in some of the anomaly classes indicate the presence of a large variety of organic substances, especially in the areas of both large operating landfills and agricultural areas that have been treated with liquid manure and sewage sludge.

The mode for each element determined in the topsoil in the Greater Berlin area and (in some cases the detection limit) was used as the geogenic background value for the study of the Schöneiche landfill and surrounding area (*Table 6.2-9*).

The limiting values provide a basis for comparison to evaluate the contaminant load of the soil. If the limiting value is exceeded, this does not necessarily mean a deterioration of the multi-functionality of the soil. Limiting values are determined according to legal regulations and are the highest acceptable values. Guideline values, on the other hand, serve as orientation for substances for which there is no legally binding limiting value. Limiting values and guideline values do not represent toxicological threshold values. However, if they are exceeded, the cause must be clarified in order to estimate the risk represented.

The soil solids contents in the Berlin List (1996) are primarily for estimating potential hazards. The background values, depending on the geology and the depth to the groundwater table, represent the natural background and the anthropogenic content generally present in the soil. The intervention limits in the Berlin List represent contaminant concentrations above which groundwater contamination can be expected in the Berlin area.

Table 6.2-9: Comparison of the regional and global background values (mode) and threshold values (in mg kg^{-1})

Element	Regional background for the area around Berlin	Regional background for the Schöneiche region	Global mean soil value, after Vinogradov (1954)	Global median soil value, after Bowen (1979)	Guideline value for agricultural soils, after Kloke (1980)
As	2.2	1.0	5	6.0	20
B	13.0	11.0	10	20	25
Ba	201	166	500	500	
Be	1.0	0.9		0.3	10
Cd	0.02	0.07	0.5	0.35	1.5 [1)] – 3
Co	4.0	4.0	8	8.0	50
Cr	13	8.8	200	70	100
Cu	6	4.6	20	30	100
F	200	300	200	200	200
Hg	0.04	0.02	0.05	0.06	2
Mn	240	132	850		
Mo	1.0	1.0		1.2	
Ni	1.0	1.0	40	50	50
Pb	22	16	10	35	100
Sb	2.2	1.0		1.0	
Se	0.45	0.45		0.4	
Sn	1.6	1.3		4.0	50
Th	2.0	2.8		9	
Ti	1061	935	4600		5000
U	1.0	1.0		2.0	
Cs	3.0	2.7		4.0	
I	2.0	1.5		5.0	
V	9.0	5.4	100	90	50
Zn	24.0	26	50	90	300
Zr	231	175	300	400	300

[1)] for light soils

The Kloke List (KLOKE, 1980) contains guideline values for the tolerable concentrations of heavy metals in agriculturally used soils. The values do not take the paths for intake by humans and other animals into account. A hazard assessment should take into consideration the land use of the area.

The PAH anomalies (*Fig. 6.2-48*) are clearly from technogenic sources (old landfills S and SE of Mittenwalde and S of Gallun, the Schöneicher Plan landfill with a maximum in the central part and the NW edge). The limit (*Fig. 6.2-6*) is exceeded by more than 50 in the southwest part of the Schöneiche landfill and by more than 5 south of Telz, at the NW edge and N of Mittenwalde (industrial and trades area).

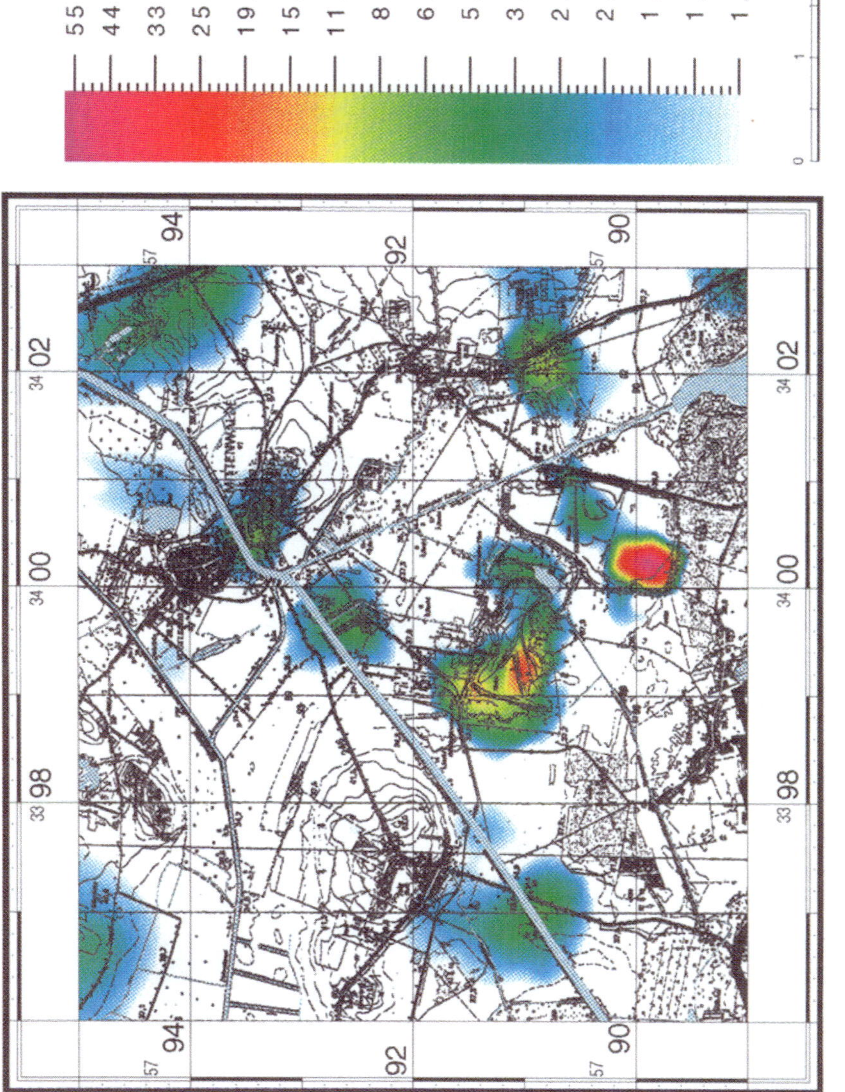

Fig. 6.2-48: Schöneiche-Mittenwalde area: Difference between the PAH concentrations and the limit in the Berlin List (1996) (0.8 mg kg^{-1} for water protection areas II and IIIA, sand, gravel, marl)

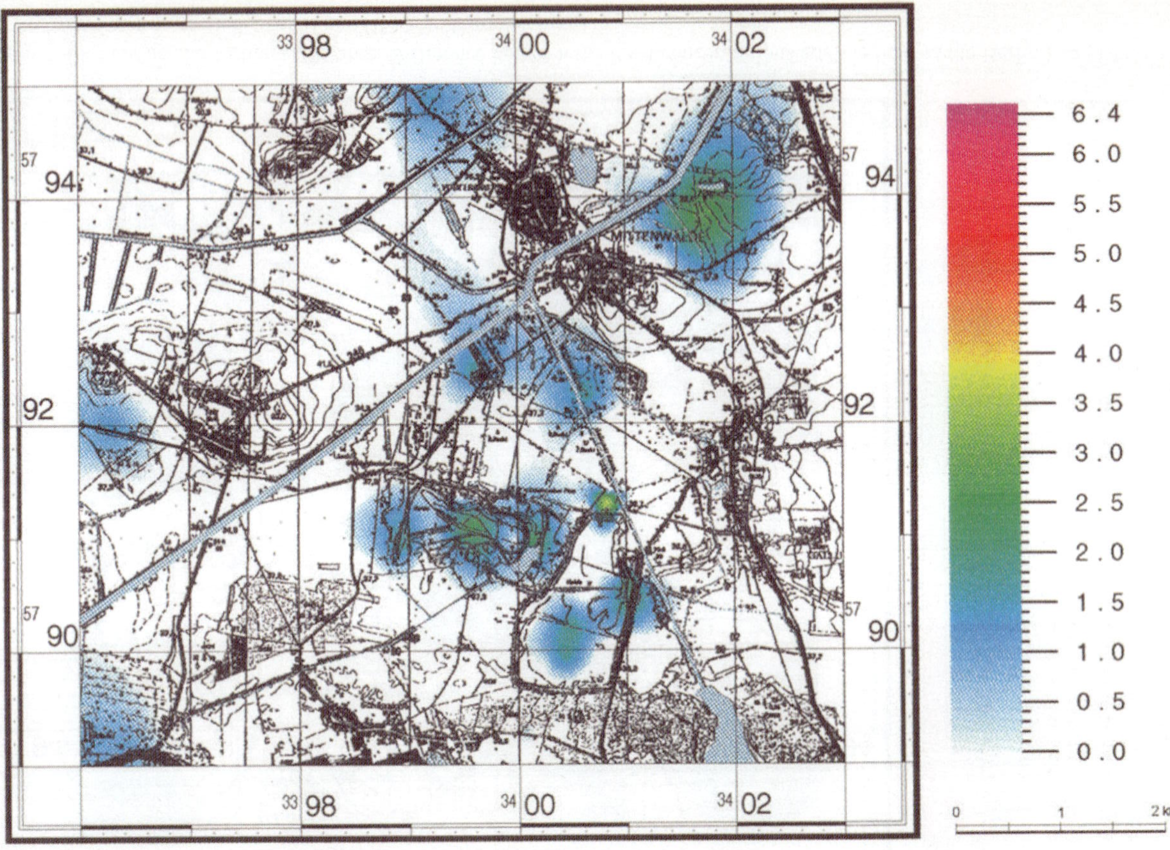

Fig. 6.2-49: Schöneiche-Mittenwalde area: Geochemical contamination index $I_{geochem}$ for arsenic based on the geogenic background of the Berlin region (2.2 mg kg^{-1})

To evaluate heavy-metal contamination, a geochemical contamination index ($I_{geochem} = \log_2$ (C/1.5 B), where C = concentration in the sample and B = geogenic background of the Berlin region) based on the geogenic background of the Berlin region (*Table 6.2-9*) was calculated. To minimize variations due to the slight influence of anthropogenic background or changes in the geological source rocks, the background concentration of each element was multiplied by 1.5, setting the lower limit of contamination class I. This takes large differences in soil properties (pH, clay and humus content) into consideration. The highest contamination index (contamination index IV) for As (*Fig. 6.2-49*) was in the areas of the old landfills east of the Notte canal NE of Mittenwalde and at the northern edge of the Schöneiche landfill. In the moderately contaminated areas along the Notte canal ("Alte Asche" landfill), $I_{geochem}$ values > 1.5 indicate technogenic influence (presumably old landfills). The landfills (western, central, and eastern parts of the Schöneicher Plan landfill; the south heap and eastern edge of the Schöneiche landfill, waste incinerator) are moderately to strongly contaminated (Igeochem III). The geochemical index $I_{geochem}$ for PAH has local maxima > 5 in the anthropogenic contaminated parts of the study area (*Fig. 6.2-50*).

A comparison of the maximum concentrations of toxic inorganic elements in the study area with the Level II limits of the Brandenburg List (1993) shows slightly elevated values for As, Pb, Sn, and Cu; values for Hg and Cd are considerably above the Level II values. Determination of measures for action must take the pH distribution and land use into consideration. The maximum concentrations of organic contaminants (cyanides, petroleum hydrocarbons, PAH, PCB, PCDD/F, phenol index) exceed, in some cases considerably, the intervention values of the Brandenburg and Berlin lists (*Fig. 6.2-6*) and mark the location of contaminated areas on the anomaly maps. Outside of the areas of the landfills, there are 16 areas in which the values of the Berlin list are exceeded, some of them previously unrecognized as contaminated areas.

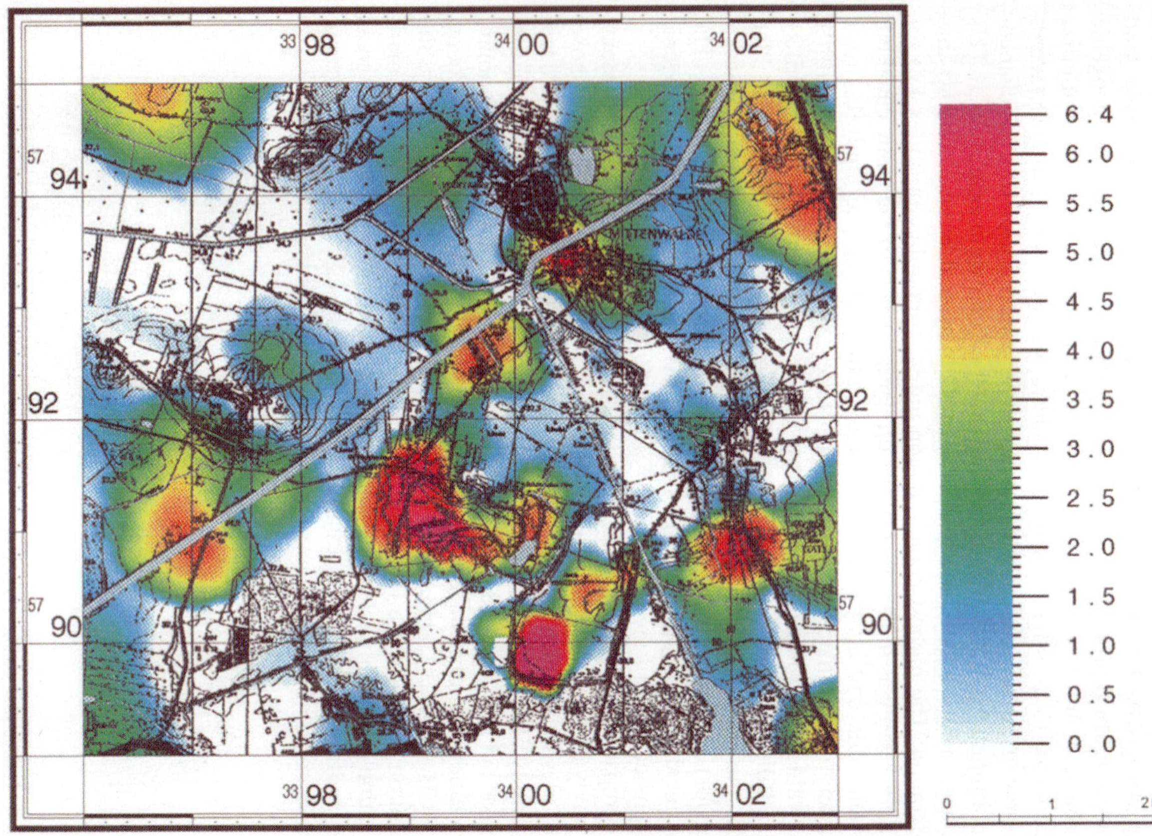

Fig. 6.2-50: Geochemical contamination index $I_{geochem}$ for PAH based on the geogenic background of the Schöneiche area (0.1 mg kg^{-1})

References and further reading

ARCHIE, G. E. (1942): The electrical resistivity as an aid in determining some reservoir characteristics. Transactions of the American Institute of Mining, Metallurgical and Petroleum Engineers, **146**, 54-62.

AVSETH; P., MUKERIJ, T. & MAVKO, G. (2005): Quantitative Seismic Interpretation. EAGE Bookshop.

BADEN-WÜRTTEMBERG (1993): Gbl. v. 30. 11. 1993, Nr. 33, 1115-1123: Orientierungswerte für die Bearbeitung von Altlasten und Schadensfällen.

BALLSCHMITTER, K.-H. & BACHER, R. (1996): Dioxine. VCH Verlagsgesellschaft, Weinheim.

BAYERN (1991): Bayerisches Landesamt für Wasserwirtschaft (Hrsg.): Hinweise zur wasserwirtschaftlichen Bewertung von Untersuchungsbefunden über Grundwasser- und Bodenbelastungen, München.

Berliner Liste (1996): Amtsblatt Nr. 15, 46. Jhg., 20. 03. 1996, A 1262A. Bewertungskriterien für die Beurteilung stofflicher Belastungen von Böden und Grundwasser in Berlin, 957-984.

BIRKE, M. & RAUCH, U. (1997): Ergebnisse der Umweltgeochemischen Bestandsaufnahme Teststandort Schöneiche. Abschlussbericht, BGR, Hannover, Archiv-Nr. 114580, unpubl., 233p.

BIRKE, M., RAUCH, U. & KEILERT, B. (2003): Kapitel5. Geochemische Untersuchungen. In: LANGE, G. & KNÖDEL, K. (Eds.): Handbuch zur Erkundung des Untergrundes von Deponien und Altlasten, Erkundungspraxis, Fallbeispiel Schöneiche-Mittenwalde, Eulenberg/Arnstadt und Rabenstein/Chemnitz. Springer, Berlin, 209-427.

BOWEN, H. J. M. (1979): Environmental chemistry of the elements. Academy Press, New York, 316.

Brandenburger Liste 1993: Teil 1 - Eingreifwerte zur Sanierung kontaminierter Standorte.

BUTLER, D. L. (Ed.) (2005): Near Surface Geophysics. Part 1: Concepts and fundamentals; Part 2: Applications and case histories. SEG Investigations in Geophysics, no.13.

CARMICHAEL, R. S. (Ed.) (1982): Handbook of physical properties of rocks. 3 vols., CRC Press, Boca Raton, FL.

CIRIA C562; McDOWELL, P. W. et al. (2002): Geophysics in Engineering Investigations. Geological Society Engineering Geology Special Publication 19. Construction Industry Research and Information Association (CIRIA), London

DANNOWSKI, G. & YARAMANCI, U. (1999): Estimation of water content and porosity using combined radar and geoelectrical measurements. European Journal of Environmental and Engineering Geophysics, **4**, 71-85.

EIKMANN, T. & KLOKE, A. (1993): Nutzungs- und schutzgutbezogene Orientierungswerte für (Schad-) Stoffe in Böden. In: ROSENKRANTZ, D., EISELE, G. & HARRESS, G.: Bodenschutz. Ergänzbares Handbuch der Maßnahmen und Empfehlungen für Schutz, Pflege und Sanierung von Böden, Landschaft und Grundwasser. I. Bd.: Grundlagen, Informationen, Bodenbelastung, Berlin 1993, Lfg. X/93, Kennzahl 3590.

FUCHS, M. (2003): Fallbeispiel Rabenstein/Chemnitz – Ermittlung der Grundwasserneubildungsrate aus bodenkundlichen Daten. In: LANGE, G. & KNÖDEL, K.: Handbuch zur Erkundung des Untergrundes von Deponien und Altlasten, Bd. 8, Erkundungspraxis. Springer, Berlin, 911-924.

GREAVES, R. J., LESMES, D. P., LEE, J. M. & TOKSÖZ, M. N. (1996): Velocity variations and water content estimated from multi-offset, ground penetrating radar. Geophysics, **61**, 683-695.

GUEGUEN, Y. & PALICIAUSKAS, V. (1994): Physics of Rocks. Princeton University Press, Princeton.

HERTRICH, M. & YARAMANCI, U. (2000): Improvement on aquifer characterisation by joint inversion of surface-NMR and geoelectrics. Proceedings of 6th Meeting of Environmental and Engineering Geophysics, Bochum Germany. EEGS-ES, Lausanne.

HJELT, S. E. (1992): Pragmatic Inversion of Geophysical Data. In: Lecture Notes in Earth Sciences, **39**, Springer, Berlin.

HUBBARD, S. S., PETERSON Jr., J. E., MAJER, E. L., ZAWISLANSKI, P.T., WILIAMS, K. H., ROBERTS, J. & WOBBER, F. (1997): Estimation of permeable pathways and water content using tomographic radar data. The Leading Edge, **16**, 11, 1623-1628.

KLOKE, A. (1980): Richtwerte '80: Orientierungsdaten für tolerierbare Gesamtgehalte einiger Elemente in Kulturböden. Mitt. VDLUFA, 1-3, 9-11.

KOWALSKY, M. B., CHEN, J. & HUBBARD, S. S. (2006): Joint inversion of geophysical and hydrological data for improved subsurface characterization. The Leading Edge, **25**, 730-734.

KÜHN, F. & HÖRIG, B. (1995): Geofernerkundung,. Handbuch zur Erkundung des Untergrundes von Deponien und Altlasten, Bd. 1, Springer, Berlin.

LAGA (1994): Vereinheitlichung der Untersuchung und Bewertung von Rohstoffen. Länderarbeitsgemeinschaft Abfall (LAGA).

LANGE, G. & KNÖDEL, K. (2003): Handbuch zur Erkundung des Untergrundes von Deponien und Altlasten, Bd. 8, Erkundungspraxis. Springer, Berlin.

LEGCHENKO, A. V. & SHUSHAKOV, O. A. (1998): Inversion of Surface NMR data. Geophysics, **63**, 75-84.

LINES, L. R. & TREITEL, S. (1984): Tutorial: A review of least-squares inversion and its application to geophysical problems. Geophys. Prosp, **32**, 159-186

MAVKO, G., MUKREJI, T. & DVORKIN, J. (1998): The Rock Physics Handbook. Cambridge Univ. Press, Cambridge.

NIEDERLÄNDISCHE LISTE (1988): Leidroad bodensanering afl. 4. November 1988, Staatssiutgeverig's - Gravenhage.

NIEDERLÄNDISCHE LISTE (1994): Interventions- (I-Werte) und Referenzwerte (S-Werte) für Böden und Grundwasser. In: Bodenschutz. Schmidt Verlag, 18. Lfg., 1995, 8936.

RENGER, M. & WESSOLEK, G. (1996): Jährliche Grundwasserneubildung in Abhängigkeit von Bodennutzung und Bodeneigenschaften. In: Ermittlung der Verdunstung von Land- und Wasserflächen. DVWK-Merkblätter zur Wasserwirtschaft **238**.

RUBIN, Y. & HUBBARD; S. S. (2005): Stochastic forward and inverse modeling: The „Hydrogeophysical" Challenge. In: RUBIN, Y. & HUBBARD; S. S. (Eds.): Hydrogeophysics. Water Science and Technology Library vol. **50**. Springer, Dordrecht, 487-511.

RUBIN, Y. & HUBBARD, S. S. (Eds.) (2005): Hydrogeophysics. Water science and technology, **50**, Springer, Dordrecht.

SCHIROV, M., LEGCHENKO & A., CREER, G. (1991): A new direct noninvasive groundwater detection technology for Australia. Exploration Geophysics, **22**, 333-338.

SCHMIDT, J. (1993): Der Untergrund der Altlast „Eulenberg" bei Arnstadt – ein tektonisches Modell und Beiträge zur Umweltgeologie. Bergakademie Freiberg, Diplomarbeit.

SCHOPPER, J. R. (1982): Electrical conductivity of rocks containing electrolytes. In: Hellwege, K.-H. (Ed.), Landolt-Börnstein Numerical Data and Functional Relationships in Science and Technology, Group V, Physical Properties of Rocks, vol. lb. Springer-Verlag, Berlin, 276 -291.

Special Section Hydrogeophysics (2006): The Leading Edge, **25**, 713-740

TARANTOLA, A. (1987): Inverse problem theory. Methods for data fitting and model parameter estimation. Elsevier

WILSON, L. G., EVERETt, L. G. & CULLEN, St. J. (Eds.) (1995): Handbook of Vadose Zone Characterization and Monitoring. Lewis, Boca Raton, FL.

WUNDERLICH, J. (1992): Ergebnisbericht über die Resultate der Schürfe an der ehemaligen Deponie Eulenberg, Landkreis Arnstadt. GEOS Ingenieurbüro GmbH Jena im Auftrag der BGR.

YARAMANCI, U., LANGE, G. & KNÖDEL, K. (1999): Surface NMR within a geophysical study of an aquifer at Haldensleben (Germany). Geophysical Prospecting, **47**, 923-943.

YARAMANCI, U., LANGE, G. & HERTRICH, M. (2002): Aquifer characterisation using Surface NMR jointly with other geophysical techniques at the Nauen/Berlin test site. Journal of Applied Geophysics, **50**, 47-65.

Glossary

abandoned waste disposal site (abandoned landfill)
Abandoned or shut down site for the disposal of domestic, industrial or mining waste

abandoned industrial site
Sites with closed down plants which pose a (potential) threat to the water, soil and air, and thus for human health, as a result of the previous operation of the plants or the handling of environmentally hazardous materials.

absorption
1. Absorption, in chemistry, is a physical or chemical process in which atoms, molecules, or ions (absorbate) enter some bulk phase – liquid or solid material (absorbent). This is a different process from adsorption, since the atoms, molecules, or ions are taken up by the volume, not attached to the surface. A more general term is sorption, which covers both adsorption and absorption. 2. The process by which electromagnetic radiation is converted into other forms of energy when it enters a substance, usually causing a rise in temperature.

accuracy
Accuracy is the closeness to the truth or the true value. Accuracy relates to the quality of a result, and is distinguished from precision, which relates to the quality of the operation by which the result is obtained.

adhesive water
→ pellicular water

adsorbable organic halogens (AOX)
The sum of adsorbable organic halogen compounds in the groundwater determined by chemical analysis after adsorption on activated carbon.

adsorption
Adhesion of the molecules of gases, liquids, or dissolved substances (adsorbate) to the surface of a solid (adsorbent). It is different from absorption, in which a substance diffuses into a liquid or solid. cf. absorption. Adsorption is usually described by isotherms, i.e., functions that relate the amount of adsorbate on the adsorbent with respect to its partial pressure (if gas) or concentration (if liquid). Adsorption mechanisms are generally categorized as either physical adsorption, chemisorption, or electrostatic adsorption. Weak molecular forces, such as van der Waals' forces, provide the driving force for physical adsorption. There is no significant redistribution of electron density in either the molecule or at the substrate surface. In chemisorption chemical reaction forms a chemical bond between the compound and the surface of the solid, involving substantial rearrangement of electron density between the

adsorbate and substrate. Electrostatic adsorption involves the Coulombic forces between ions and the substrate surface. Ion exchange involves ions electrostaticly adsorbed on a substrate. In liquids, interactions between the solute and the solvent also play an important role in establishing the degree of adsorption.

(ad)sorption isotherm

The graphic representation of the mass of a substance (e.g., a contaminant) adsorbed per unit dry mass of soil versus the concentration of the substance when an instantaneous equilibrium is reached between the sorbent and the sorbate. (Ad)sorption isotherms are used to analyze the sorption behavior of a substance.

aeration zone

→ unsaturated zone

aerial photography

Despite technical progress in digital imaging, standard aerial photographic film remains an important remote-sensing tool. Aerial photography can be used for a multitude of geoscientific tasks. The cost of aerial photography is relatively low, the data are informative and easy to manage, and the film does not require special image processing systems for analysis. Cameras for aerial photography, also referred to as metric cameras and frame reconnaissance cameras, are analogous to normal film cameras in that they use lenses, shutters, and film. Panchromatic (black and white), color, and color-infrared (CIR) film are commonly utilized for aerial photography. It is expected that in the near future digital cameras will be used for most topographic mapping surveys because it is easier to develop digital elevation models from digital camera images than from aerial photographs. Vertical aerial photographs provide a three-dimensional impression of an area if stereo-pair photographs (requires an overlap of 60–90 % with the next image along the flight line) are evaluated. Aerial photographs can be analyzed to identify rock formations and types of soil, as well as typical relief forms, man-made features, distinctive vegetation types, drainage patterns and specific types of land use. In general, as the incidence of man-made features increases in an area, it becomes more difficult to extract geologic information from aerial photographs. Even older aerial photographs have a proper spatial resolution and they often document changes in an area of interest over a period of several decades. They make also it possible to evaluate the geological and initial environmental situation. Archives of aerial photographs are maintained worldwide, e.g., in governmental survey offices, and can provide aerial photographs at low cost. Thematic interpretation of a site using aerial photographs is carried out at scales between 1:2000 and 1:10 000. Regional analyses (e.g., lineament analyses) are carried out at scales between 1:25 000 and 1:100 000.

aerobic

1. Conditions where free oxygen is present. 2. Organisms that depend on such conditions. Opposite of anaerobic.

A-horizon

The top horizon of a soil that contains humified organic matter (humus) interspersed with roots. It is the agriculturally used soil horizon. Soil from this horizon is sometimes removed and spread at another location.

altitude

Vertical distance above or below a reference level (usually mean sea level). It is often used synonymously with elevation when the datum is mean sea level. → elevation

amperometry

Amperometry or amperometric titration is an electrochemical analysis method based on keeping voltage constant during a titration and measuring current. The current flowing between two electrodes dipped in a solution depends upon the concentration of the materials dissolved in the solution. Oxidized and reduced substances in solution can be detected by amperometric titration. Amperometric titration is based on the fact that during titration of a solution in an electrochemical cell, polarization effects occur at the electrodes while voltage is kept constant. As a reagent solution (titrant) is added to the titration cell, the reaction can be observed by the changes in the current in the electrochemical cell. At the end of the reaction, the current changes abruptly. The concentration of the substance to be analyzed is determined from volume of titrant and the measured current. A classical example of an amperometric titration is determination of the water content after Karl Fischer, in which sulfur dioxide is oxidized in the presence of water and iodine: $SO_2 + I_2 + 2H_2O \rightarrow H_2SO_4 + 2HI$. At the titration end point, the reversible redox system I/I^- occurs in the solution and leads to depolarization of the electrode and an abrupt increase in the current flowing through the cell.

amplitude

The maximum deviation from the average, a zero or equilibrium value of any repeatedly changing quantity, such as the position of a vibrating object, a wave, voltage, current, etc.

anaerobic

1. Conditions where no free oxygen is present. 2. Organisms that are able to live under such conditions. Opposite of aerobic.

analytical model

A mathematical model in which the relationships among parameters can be solved using analytical mathematical tools, such as equations. → numerical model, → model

angular field of view (FOV)
The area visible to an imaging system, whether a camera or a nonphotographic system, expressed as the angle from the lens to the edges of the image.

anion
A negatively charged ion.

anisotropy
Variation of a physical property at one point with the direction in which it is measured caused by layering or fabric. In contrast heterogeneity involves variation from point to point. Both anisotropy and heterogeneity are depending on scale. → isotropy

anomaly
A deviation of physical parameters of a material (e.g., density, magnetic susceptibility or electrical conductivity) from homogeneity is called an anomaly. An anomaly of physical parameters leads to an anomalous field distribution (e.g., gravity anomaly, magnetic anomaly, anomalous current distribution) that can be measured on the Earth's surface or calculated from measured field values. The difference between measured field values and a theoretical (global or regional) field distribution (normal field, e.g., in gravity method or magnetic method) is also called anomaly.

antenna
Device that transmits and receives electromagnetic energy.

anthropogenic
Said of conditions that are caused by human activity (e.g., contamination of groundwater from a landfill).

aperture
Opening in a remote sensing system that controls the amount of electromagnetic radiation reaching the film or detector. Syn. opening angle.

apparent resistivity
In dc resistivity measurements the resistivity ρ of a homogeneous, isotropic ground is determined from the Ohm's law ratio of measured voltage ΔV to applied current I, multiplied by a geometric parameter k which depends on the electrode array: $\rho = k\,\Delta V/I$. In the case of inhomogeneous ground, the same formula is used and gives the resistivity of an equivalent homogeneous half-space. The term "apparent resistivity" ρ_a has been introduced for this value.

apparent velocity

The apparent velocity or Darcy velocity of groundwater is determined using Darcy's law from the ratio of the rate of the groundwater flow to the cross-sectional area perpendicular to the direction of flow. This is not the velocity with which groundwater moves. The term "specific discharge" has been introduced to avoid this misunderstanding.

aquiclude

A geologic unit that is incapable of transmitting significant quantities of groundwater under ordinary hydraulic gradients; such geologic units have a hydraulic conductivity $< 10^{-9}$ m s^{-1}.

aquifer

A permeable geological stratum or formation that can both store and transmit water in significant quantities.

aquitard

A geologic unit that has a low hydraulic conductivity relative to the units above and below it. See also leaky aquifer.

arid

Said of a climate characterized by dryness; less than 25 cm of annual rainfall; or a higher evaporation rate than precipitation rate. It is used synonymously with dry.

aromatic hydrocarbon

A hydrocarbon that contains one or more benzene rings. The name of the class comes from the fact that many of them have strong, pungent aromas. Many solvents and pesticides are aromatic hydrocarbons.

array

An array, also called a configuration, is a linear or 2-dimensional assembly of electrodes in geoelectrics, of antennae in electromagnetics and of seismic sources and/or recorders in seismics.

artesian aquifer

→ confined aquifer

artesian well

A well in an aquifer where the groundwater is confined under pressure and the water level stands above the top of the confined aquifer, but does not necessarily reach the land surface. The term is often used only for a flowing out artesian well.

atomic absorption spectrometry (AAS)

A reliable analytical technique for determination of the concentration of metallic elements. About 70 metals can be analyzed by this technique; however, only one can be measured in a single analysis. The method normally uses liquid samples. However, with modern techniques finely dispersed solid samples can be used directly. The sample is evaporated and atomized, e.g., by injecting into a hot flame or in a furnace. The vaporized atoms of the element absorb light according to its electron configuration. Monochromatic light of the appropriate wavelength (usually from a cathode tube containing the same metal as the one to be analyzed) is passed through the vapor or flame. Its extinction (the ratio of the intensities of incident and transmitted light) is measured. The extinction is proportional to the concentration of the free atoms in the vapor according to the Beer-Lambert law. Depending on the method of vaporization, AAS can be classified into two types: flame AAS (F-AAS) and flameless AAS. Typical flameless methods are graphite furnace AAS (GF-AAS, ETA-AAS), cold vapor AAS (CV-AAS), as well as hydride generation AAS (HG-AAS).

atomic emission spectroscopy (AES)
→ optical emission spectrometry (OES)

atomic fluorescence spectrometry (AFS)
Atomic fluorescence spectrometry is based on the observation that thermally produced atoms in the gaseous state can be moved to a higher energy state by electromagnetic radiation, from which they return to the ground state by emitting energy in form of fluorescent radiation. The sensitivity of AFS is proportional to analyte concentration and the intensity of the primary radiation. The detection limits of AFS are comparable to that of GF-AAS. This method has not found widespread application.

attenuation
1. The diminishing of a parameter value, e.g., contaminant concentration in groundwater or energy. Attenuation can occur in the environment as a result of filtering, biodegradation, dilution, sorption, etc. → natural attenuation 2. Attenuation also occurs in wave propagation and in measuring systems.

attitude
The orientation of an object relative to an external reference system: For a planar rock body, it is expressed as strike and dip; for an imaging system it is expressed as the angle with respect to the external reference system, e.g., as azimuth, tilt, yaw, or pitch.

audio frequency range
Frequencies that can be normally heard by a human, i.e., 15 Hz to 20 kHz.

audio-magnetotelluric method (AMT)
Geophysical method involving measurement of natural electromagnetic signals in the 10 to 10^4 Hz range to determine subsurface electrical resistivity.

auger

Rotating drilling tool with a helical shaft that carries the cuttings to the top of the borehole. Can also be built as a hollow-stem auger to collect sediment core samples while drilling.

azimuth

Direction of a horizontal line from the observer to an object or of a flight direction given as the angle (in degrees) measured clockwise from north (in the US from south).

bacteria

A large group of single-cell microorganisms. Bacteria have a simple cell structure with no distinct nucleus and reproduce by subdivision. Bacteria live in soil, water, plants, organic matter, or the bodies of animals or people, some can cause infections and disease in animals and humans.

bailer

1. A device used to withdraw a water sample from a small-diameter well. A bailer typically is a piece of pipe attached to a wire and having a check valve in the bottom. 2. A device used in cable tool drilling to remove drill cuttings from a well. It consists of a simple tube suspended on a cable, open at the top, with a foot-valve at the bottom. The foot-valve opens when the bailer touches the bottom of the drilled hole, permitting water with drill cuttings in suspension to enter the tube. When the bailer is raised to be emptied, the foot-valve closes instantly as it loses contact with the bottom of the hole and retains the water and drill cuttings.

bandwidth

The width of the band of frequencies measured.

bank infiltration

Process of surface water seeping from the bank or bed of a river or lake to production wells. During the water's passage through the ground, its quality changes due to microbial, chemical and physical processes, and due to mixing with groundwater. → influent flow

barrier

→ geological barrier

baseflow

The quantity of streamflow that results from groundwater discharge. Groundwater flows underground until the water table intersects the land surface and the flowing water becomes surface water in the form of springs, streams, lakes and wetlands. Baseflow is an important source of streamflow between rainstorms. It is estimated by employing a hydrograph separation method whereby long-term streamflow records are reconstructed so that most of the stormwater flow is subtracted to theoretically yield the groundwater contribution to the stream. This is also referred to as groundwater runoff or dry-weather flow.

base map
A large-scale map showing sufficient topography for geographic reference for all field operations of a site investigation, as well as for the subsequent data documentation and interpretation. If large-scale topographic maps are not available, base maps have to be prepared from aerial photographs or high-resolution satellite images.

basement
The igneous or metamorphic rock underlying sedimentary rocks.

base station
A reference station that may be used to normalize data from other stations (e.g., in gravity or magnetic measurements).

bathometer
A device for measuring depth in water.

beam
A collection of nearly parallel rays (e.g., in laser scanning) or stream of particles.

bed
A layer of sedimentary rock which is distinguishable from the strata above and below it. → layer

bedrock
Solid rock that underlies soil or other unconsolidated rock, or is exposed at the Earth's surface.

belt of soil water
The upper part of the unsaturated zone. The soil water in this zone enables plant growth. Syn. soil water zone.

benchmark
A permanently fixed metal tablet or other marker embedded on the Earth's surface showing the precisely determined elevation and coordinates of that point and used as a reference in topographic surveys.

bentonite
A colloidal clay consisting mostly of montmorillonite formed from volcanic ash which can absorb large amounts of water and expands to many times its normal volume. Used in drilling mud.

benzene, toluene, ethylbenzene, xylene (BTEX)
A group of toxic, volatile organic compounds present in coal tar, gasoline and other petroleum products. They are commonly found as groundwater contaminants near gas stations and industrial facilities.

biochemical oxygen demand (BOD)
The amount of oxygen needed for the biological decomposition of organic material by microorganisms in aquatic environments, expressed in parts per million. It is an indicator for the amount of biodegradable organic matter in a sample.

biochemical process
Chemical reaction caused by the metabolism of organisms.

biodegradation
Decomposition of matter, normally organic, by microorganisms.

bioremediation
The use of microorganisms to reduce contamination.

biotic
Said of ecological parameters that refer to the living environment, e.g., the reproduction potential of microorganisms.

bladder pump
A device used to sample groundwater. The pump uses a bag made of fluorocarbon material in order to prevent contamination of the sample and loss of volatile components.

borehole
A circular hole of small diameter made by boring. It is used synonymously with drillhole, boring, bored well. Boreholes yield direct information about the structure of the ground below the surface. Boreholes can be outfitted to serve as groundwater observation wells, this is called borehole completion.

borehole logging
Borehole logging (also called downhole logging or well logging) provides in-situ information in addition to the direct geological information from drill cuttings or cores. In all environmental investigations, geophysical borehole logging is required in all boreholes. These logs allow the rock structure to be investigated with considerable certainty. Cores can be used for comparison and calibration of borehole logs. Time and cost can be saved by borehole logging because the well does not necessarily need to be cored, the well logs provide sufficient data to allow correlation of lithological profiles in several boreholes. In many cases, recovered cores are broken during the drilling process and as a result of stress release and do not provide a basis for reliable information about the rock jointing. Numerous logging methods are available. Which one is used depends on

the objectives of the investigation and the conditions in the borehole (with or without casing, dry, filled with water or drilling mud, etc.). In principle, the various methods of geophysical borehole logging are used for the following tasks:

- preparation and review of lithological profiles,
- inspection of the completion and technical status of wells,
- quantitative determination of hydrogeological parameters (e.g., density, clay content, porosity).

borehole storage effect
The amount of water in a well before pumping is begun influences the change in water level in the well during the initial pumping.

boundary conditions
1. The conditions that are specified at the boundaries of hydrogeological models. Boundary conditions are necessary to define how the model interacts with the groundwater system in the surrounding area. Model boundaries correspond to the actual physical boundaries and hydraulic conditions. Water levels and flow, faults, facies changes, groundwater divides, solute concentrations, stage/discharge relationships, etc., are typical boundary conditions. 2. Boundary conditions are also used in geophysical modeling.

bound water
Water present in rock, soil and substances which cannot be removed without changing the structure or composition of the material and cannot react as free water does.

brackish water
Water with a content of solved solids between 1000 and 10 000 mg L^{-1}. Water in the transition zone between freshwater and saltwater.

break
Any sudden change in topography.

breakthrough curve
The graphic representation of the relative concentration of a given substance in a solution versus time where relative concentration is defined as the ratio of the actual concentration to the source concentration. Breakthrough curves are determined, for example, in tracer and column experiments.

bulk (dry) density
The bulk or dry density is the ratio of the mass of the solid phase of soil/rock (i.e., oven-dried soil/rock) to its total volume (solids and pores). Units are kg m^{-3} or g cm^{-3}.

buffer capacity

Capacity of an aqueous solution or soil to resist changes of its pH by added acidic or basic substances.

calibration

1. Process of comparing instrument readings with a standard. 2. Mathematical models, e.g., a groundwater flow model, must be calibrated with measured data. → model calibration

calorimetry

Measurement of quantities of heat generated (exothermic process) or consumed (endothermic process) by chemical reactions or physical changes using a calorimeter. A simple calorimeter may consist of a thermometer in an insulated container. Sophisticated calorimeters are used in chemical analysis: for example, constant-volume calorimeters, constant-pressure calorimeters, differential scanning calorimeters and isothermal titration calorimeter.

capacity

The maximum amount of water that a soil can contain; the maximum amount of water that a pump, well, or reservoir can yield; the maximum amount of sediment that a stream or wind can transport.

capillarity

The action by which the surface of a liquid is elevated or depressed where it is in contact with a solid. Water rises in soil against the pull of gravity due to capillary action in the soil pores.

capillary action

The movement of liquid through tiny spaces in material (e.g., spaces between soil particles) due to molecular forces against gravity. Capillary action is essential to the transport of nutrients to plant roots.

capillary electrophoresis (CE)

Electrophoresis is a separation technique in which charged particles are separated in a thin silica capillary tube filled with an electrolytic solution to which an electrical field (up to 30 kV, 100 mA) is applied. Similar to chromatographic methods (differences in retention), substances are characterized in electrophoresis by their electrophoretic mobility. Owing to the sample volume of only several nanoliters, only very small sample quantities are needed for analysis. The separated substances are identified using spectroscopic methods. Due to their high separation capability, high analysis speeds and very good reproducibility, capillary electrophoresis is increasingly replacing chromatographic techniques.

capillary fringe

That part of the unsaturated zone immediately above the water table in which the interstices are filled with water to a certain extent due to capillarity. It is used synonymously with capillary space or zone of capillarity. The height above the free water level to which water will rise as a result of capillarity depends mainly on the size of interstices (grain size) of the soil or rock.

capillary rise

→ capillary fringe

casing

Steel or plastic pipes used in boreholes to keep them from caving in and to shut off water entering the borehole through the sides of the hole usually made in pieces of three meters (ten feet) length which are screwed together. The casing is sealed to the borehole wall near the land surface with the annular seal to protect groundwater.

catalyst

A catalyst is a substance that accelerates the rate of a chemical reaction but without itself being consumed by the reaction.

catchment area

The area of land that drains water to a river, lake, sea, ocean or reservoir. Also called drainage basin or watershed.

cation

A positively charged ion.

cation-exchange capacity (CEC)

Cation-exchange capacity is defined as the degree to which a soil can adsorb and exchange cations with a solution. Cations are positively charged ions (NH^{4+}, K^+, Ca^{2+}, Fe^{2+}, etc.). Soil particles and organic matter have negative charges on their surfaces, which are, in general, permanent lattice charges with pH-dependent variable charges at functional groups. Cations can be adsorbed on the negative surfaces. CEC is expressed as the equivalent amount of maximum exchangeable cations per unit of soil mass (the SI unit is cmol(+)/kg but in the literature meq/100 g is still frequently used). It is highly dependent on soil texture and organic matter content. In general, the more clay and organic matter in the soil, the higher the CEC. Clay content is important because these small particles have a high ratio of surface area to volume. Different types of clays have different CEC values. Smectites have the highest CEC (70 - 130 meq/100 g), followed by illites (15 - 40 meq/100 g) and kaolinites (3 - 15 meq/100 g). In general, the CEC of most soils increases with an increase in soil pH.

chemical oxygen demand (COD)

The amount of oxygen required for the oxidation of all oxidizable matter in a specific amount of water. Usually expressed as the corresponding amount of $KMnO_4$ (or $K_2Cr_2O_7$) needed for complete oxidation (mol ($KMnO_4$)/L or mol ($K_2Cr_2O_7$)/L).

chemical species

Chemically identical atomic or molecular structural units.

chemical transformation

Used when chemical reaction specifics are not relevant or not available.

chlorinated solvent

Chlorinated solvents are common groundwater contaminants. They are used to dissolve substances that do not dissolve easily in water, for example, in manufacturing, degreasing, and dry cleaning. A chlorinated solvent contains at least one chlorine atom in its structure.

chlorofluorocarbon (CFC)

→ fluorochlorinated hydrocarbon

chromatography

Several related techniques for separating a mixture of compounds between two phases. The separation of substances takes place via mass transfer and exchange processes in a selective two-phase system consisting of a stationary (an adsorbing material) and a mobile phase (e.g., a solution). Different substances in the mobile phase have different affinities to the stationary phase. Substances with low affinity to the stationary phase (i.e. only weakly adsorbed) travel almost as fast as the mobile phase, whereas substances with high affinity to the stationary phase are considerably retarded. Thus, a mixture is separated into peaks. Additionally, these peaks tend to broaden due to diffusion. Chromatographic separation is typically carried out using a column of glass or metal containing either a packed bed (consisting of small particles) or a narrow-diameter tube lined with a thin film of stationary phase and hollow at the center. The stationary phase is usually a highly viscous liquid covering the surface of the packed bed particles or the tube. These surfaces themselves could also constitute the stationary phase. The mobile phase is a fluid. The analyte is introduced at the top of the column and separated as the mobile phase moves through the column. At the bottom, the individual substances can be extracted or detected. Chromatographic methods include thin layer chromatography (TLC), liquid chromatography (LC), gas chromatography (GC), (sometimes called gas liquid chromatography or GLC) and high performance liquid chromatography (HPLC).

cistern

An artificial underground reservoir for storing water.

Clarke values

In contrast to most organic contaminants, heavy metals occur naturally in rocks and ore minerals. For this reason, sediments and soils contain a geogenic (natural) concentration of these elements. Global background values of inorganic elements in the Earth's crust are called Clarke values.

classification

Process of grouping things into categories (classes). Taxonomic classification is the placing of an object or concept into a set of categories. Almost anything (e.g., objects and events) may be classified according to some taxonomic scheme. Statistical classification is another commonly used type of classification. The classification of images is an example: Image classification (also called pattern recognition) is the process used for creating thematic maps from satellite images. A thematic map shows the spatial distribution of a particular feature, e.g., type of rock, soil, vegetation, land use, flood-prone areas, faults.

clay

Clay is fine-grained, earthy material with a particle size less than 0.002 mm in diameter that becomes plastic when wet. They consist mainly of clay minerals. During lithification, compacted clay layers are transformed into shale. Under intense heat and pressure, the shale can be metamorphosed into slate and then phyllite and schist. Properties of clays include plasticity, shrinkage when air dried or fired. Clays are divided into various classes or groups. The purest clays are china clays and kaolins. Bentonites are clays composed of very fine particles derived usually from volcanic ash. They are composed chiefly of the hydrous magnesium-calcium-aluminum silicate called montmorillonite. Clays often form colloidal suspensions in water; in saline water the clay particles flocculate (clump) and quickly settle. The three principal particle sizes in soils are clay, silt, and sand. A certain amount of clay is a desirable constituent of soil, since it binds other kinds of particles together and makes it retain water. Clays are divided into two groups according to origin: residual clay, found where it was formed, and transported clay, also known as sedimentary clay, removed from the place of origin by erosion and deposited in a new and possibly distant location. Residual clays are most commonly formed by surface weathering, which gives rise to clay in three ways-by the chemical decomposition of rocks, such as granite, which contain silica and alumina; by the dissolution of rocks, such as limestone, containing clayey impurities, which, being insoluble, are deposited as clay; and by the disintegration and dissolution of shale.

clay minerals

Clay minerals are hydrous aluminum silicates of < 0.002 mm in size (equivalent sphere diameter), commonly with platy morphology and perfect cleavage. They show considerable variation in chemical and physical properties. Clay minerals consist of the fundamental structural units tetrahedra and octahedra. Like all phyllosilicates, clay minerals are characterized by two-dimensional sheets of corner-sharing tetrahedra, predominantly SiO_4. The tetrahedral sheets are always bonded to octahedral sheets containing small cations, such as aluminum or magnesium, each coordinated with six oxygen atoms. Some oxygen atoms form an OH group in the clay structure. Clay minerals are classified on the basis of the proportion of sites occupied by metal ions in the sheets and according to the ratio of tetrahedral and octahedral sheets in the layers. A clay mineral of the 1:1 layer type consists of layers with only one tetrahedral and one octahedral sheet; kaolinite is the most common mineral of this type. A clay mineral of the 2:1 layer type, such as illite, smectite or vermiculite, consists of layers with two tetrahedral sheets and one octahedral sheet. Clay minerals of the 1:1 layer type have many more hydroxyl groups. The 2:1 and 1:1 layers can be stacked on each other to form crystals. It is possible to stack layers of different types, such minerals are called mixed-layer minerals.

cleanup

→ remediation

climate

General pattern of weather conditions in a region over a long period time (usually 30 years), taking into account temperature, precipitation, humidity, wind, etc. The major influence governing the climate of a region is its latitude. The influence of latitude on climate is modified by one or more secondary influences, including position relative to land and water masses, altitude, topography, prevailing winds, ocean currents, and prevalence of cyclonic storms.

climate zone

A broad latitudinal division of the Earth's surface into climatic zones based on global winds includes the equatorial zone, characterized by high temperatures with small seasonal and diurnal change and heavy rainfall; the subtropical zone, including the trade-wind belts and the horse latitudes; a dry region with uniformly mild temperatures and little wind; the intermediate zone, the region of the prevailing westerly winds that, because of several secondary influences, displays wide temperature ranges and marked changeability of weather; and the polar zone, a region of short summers and long winters, where the ground is generally permanently frozen (permafrost). The transitional climate zone between the subtropical and intermediate zones, known as the Mediterranean type, is found in areas bordering the Mediterranean Sea and on the west coasts of continents. It is characterized by mild temperatures with moderate winter rainfall under the influence of the moisture-laden prevailing westerly winds and dry summers under the influence of the horse latitudes or the trade winds.

cluster analysis

Method of analytical statistics for investigation of multidimensional random variables. Clustering consists of partitioning a data set into subsets (clusters), so that the data in each subset share some common trait.

coefficient of uniformity

A measure of the uniformity of the particle size in a soil or sediment as given by the slope of the grain size distribution curve. It is defined as $U = d_{60}/d_{10}$, where d_{60} and d_{10} represent the grain diameter in mm, for which, 60 % and 10 % of the sample, respectively, are finer than.

cold vapor AAS (CV-AAS)

CV-AAS is an atomic absorption spectrometric technique used for direct determination of mercury in soil and sediment samples as well as in aqueous solutions. This method is similar to conventional atomic absorption spectrometry, except that – because mercury can exist in a gaseous state at ambient temperatures – no flame or furnace is needed. Solid samples are quickly heated (approx. 1000 °C) and the released elemental mercury is spectroscopically determined. In liquids, mercury ions are first reduced to elemental mercury with stannous chloride ($SnCl_2$) or sodium tetrahydrobromate ($NaBH_4$). If necessary, the mercury is concentrated by amalgamation. See also atomic absorption spectrometry.

coliform bacteria

→ fecal coliform bacteria

colloid

A mixture in which one substance is divided into minute particles (called colloidal particles) and dispersed throughout a second substance. The mixture is also called a colloidal system, colloidal solution, or colloidal dispersion. Colloidal particles are usually on the order of 10^{-7} to 10^{-5} cm in size. They are larger than molecules but too small to be observed directly with a microscope; however, their shape and size can be determined by electron microscopy. One way of classifying colloids is to group them according to the phase (solid, liquid, or gas) of the dispersed substance and of the medium of dispersion. There are two basic methods of forming a colloid: reduction of larger particles to colloidal size, and condensation of smaller particles (e.g., molecules) into colloidal size particles. Condensation of smaller particles to form a colloid usually involves chemical reactions, typically displacement, hydrolysis, or oxidation and reduction. The particles of a colloid selectively absorb ions and acquire an electric charge. All particles of a given colloid take on the same charge (either positive or negative) and thus are repelled by one another. If an electric potential is applied to a colloid, the charged colloidal particles move toward the oppositely charged electrode; this migration is called electrophoresis. If the charge on the particles is neutralized, they may precipitate out of the suspension. A colloid may be precipitated by adding another colloid with oppositely charged particles; the particles are attracted to one another, coagulate, and precipitate out. Addition of soluble ions may precipitate a colloid; the ions in seawater precipitate the colloidal clay in river water, forming a delta.

colluvium

Colluvium or hillwash is an unconsolidated, unsorted, heterogeneous mixture of Earth material that has moved down a slope by direct gravitational action and by local, unconcentrated runoff and settled at the base of the slope. Colluvium normally forms fan or wedge-shaped deposits that seal former ground surfaces. This build-up process is called colluviation.

compaction

The bulk volume and pore space within unconsolidated material is reduced due to the increasing weight of overlying material that is continually being deposited or due to the pressures resulting from Earth movements. Unconsolidated material is converted by compaction to consolidated rock. Compaction is often accompanied by cementation of the particles. The combined process of compaction and cementation is called consolidation.

compartment

The different parts of a complex system are sometimes called compartments. The Earth, for example, can be subdivided into five compartments: atmosphere, pedosphere, lithosphere, hydrosphere, and biosphere. In the case of a site suspected to be hazardous, the environmental media (compartments) that can be affected by contamination are groundwater, surface water, air, and soil. In groundwater modeling, compartments are delimited areas of a complex system, in which equilibrium predominates with respect to an analyzed variable.

composting

Composting is used in waste management to reduce the amount of waste to be deposited with the help of aerobic degradation of organic waste by microorganisms. The result of the composting process is compost mainly composed of partly decayed organic material that is used to fertilize the soil and to increase its humus content. Compost is usually made from plant materials (e.g., grass clippings, vegetable tops, garden weeds, tree leaves, sawdust, and peat) together with manure and some soil.

concentration

The ratio of the quantity of a substance in a sample of a given volume or weight to the volume or weight of the sample. Typical concentration units: molar (e.g., 0.5 M) = mol L^{-1}; normal (e.g., 1.0 N) = eq L^{-1}; molal (e.g., 1.0 m) = mol kg^{-1}; ppm.

conceptual model

A mental image of an object, system, or process. It consists normally of a simplified, schematic written description and a visual representation. Conceptual hydrogeological models have been used for several decades to describe and understand hydrogeological systems. Such conceptual models are more concerned with the physical than the chemical aspects of the natural system. During the last few decades conceptual models of investigated sites have come into frequent use to assess ecosystem features and processes (including biological, physical, chemical and geomorphic components) of an environment (e.g., new or abandoned landfills, industrial sites or mining sites). A conceptual model is developed as part of the site investigation and reflects the progress of site characterization. Therefore, the conceptual model is the basis as well as the final result of site characterization.

condensation

Change of a substance from the gaseous to the liquid state. Condensation is the reverse of evaporation/vaporization, i.e. the change from liquid to gas. The condensation of water vapor into clouds and precipitation as the temperature drops or the atmospheric pressure rises is a basic part in the hydrologic cycle.

conductometry

Term used for analytic determination procedures in which the electrical conductivity of aqueous solutions, which is a function of the concentration of free ions, is determined using a low-frequency alternating current. Most common uses are conductometric titration and as a detection method in chromatography. These techniques are especially employed with highly diluted or strongly colored solutions, where the use of color indicators is limited or impossible.

conductor

A body/layer within which an electrical current can flow readily. Often, the "target" of an electromagnetic survey. The conductors are subdivided into metallic and electrolytic conductors. The conductivity depends in the first case on the mobility of electrons and in the second case on concentration and mobility of ions.

cone of depression

Depression in the groundwater surface in the shape of an inverted cone around a well from which water is being withdrawn. It defines the area of influence of a well. The size of the cone depends on the duration of pumping, the pumping rate, and the hydraulic conductivity, specific yield and thickness of the aquifer.

cone penetration test (CPT)

→ direct push testing

configuration

1. → array. 2. The setup of an individual computer which consists of hardware and software, and how it is put together.

confined aquifer

Aquifer covered by an aquiclude or aquitard (confining bed) and whose piezometric surface is above the base of the confining layer. Syn. artesian aquifer.

confined groundwater

Groundwater in a confined aquifer.

confining bed

Geologic unit which is relatively impermeable and does not yield usable quantities of water. Confining beds are also referred to as aquitards or aquicludes, depending on their hydraulic conductivity.

consolidation

Consolidation is the process by which unconsolidated material is converted into a solid mass by compaction and cementation. In the course of compaction and cementation coarse sediments become conglomerates, sands become sandstone, and mud is converted to shale.

contaminant

A substance or material that makes air, water, or rocks unfit for use. It is often used more or less synonymously with pollutant.

contaminant loading

The amount (volume and concentration) of a contaminant discharged to soil or groundwater.

contaminant retention

The delayed transport of contaminants compared to the movement of water only.

contaminant retention capacity

Geochemical and mineralogical properties which influence contaminant retention capacity are

- clay mineral composition,
- organic carbon content (a key factor for the sorption of organic contaminants),
- presence of iron compounds and carbonates as buffers and (particularly for heavy metals) for sorption,
- cation exchange capacity, ion coating and exchangeable cations (which affect the retention capacity for organic and heavy metal ions),
- main ions on clay minerals,
- pH and acid and alkali capacity (which affect the long-term acid and base buffering capacity).

In addition, hydraulic conductivity, dry bulk density, grain density, water content, and porosity must be determined for the assessment of the contaminant retention capacity.

contaminant retention potential

→ contaminant retention capacity

contaminant transport

Spreading of contaminants in a fluid by convection, diffusion, dispersion, and sorption, and is affected by degradation.

contamination

The presence of substances that prevent or reduce the usability of water, soil and air.

contour (line)

A line connecting points of equal value on a map. The smallest interval between contours should be at least twice the uncertainty in the measured values.

contour map

A map that portrays the Earth's surface or the results of geophysical measurements by means of contour lines drawn at regular intervals (e.g., topographic map, magnetic anomaly map).

contrast

The difference between parameter values (e.g., density contrast in gravity method or a color difference in an image).

controlled-source electromagnetics

Any electromagnetic method which uses artificially generated fields.

control point

A reference point that can be accurately located on a photograph and for which its coordinates and/or elevation on the Earth's surface (e.g., from GPS measurements) is available.

Coordinated Universal Time (UTC)

Official world time reference for civil and scientific purposes. Coordinated Universal Time is determined from six standard atomic clocks at the International Bureau of Weights and Measures (BIPM) in Paris, France. Implemented in 1964.

core

A cylindrical rock sample cut from a borehole.

coulometry

Coulometry is a quantitative analysis method based on the amount of charge necessary to release a substance during electrolysis. The analysis can be carried out either holding potential constant (potentiostatic coulometry, measurement of the electric charge versus time) or current constant (amperostatic coulometry or coulometric titration, measurement of the electric potential versus time). In a coulometric titration, the reagent necessary for titration is produced electrochemically with current kept constant. The endpoint of titration can be detected, for example, with a colored indicator or by potentiometry. The concentration is calculated from the product of current, time and equivalent weight.

crosshole method

Method for investigating the area between two or more boreholes by measuring traveltimes and amplitudes of seismic or electromagnetic (radar) waves from a source in one borehole to geophones or antennae in other borehole(s).

crosshole tomography

Mathematical reconstruction of geological structures from measurements with crosshole methods. Traveltime tomography is based on arrival times, attenuation tomography on amplitude, and diffraction tomography on the scattered wave field.

crossplot

A graph used to show the relationship between two parameters, e.g., the values obtained with two different methods.

cross section

A vertical plane through a geologic formation or structure. A diagram representing this can show the geology, the spatial relationship of physical or other parameter values, or seismic reflections as a function of depth, in the case of seismic and ground penetrating radar as a function time, or in geoelectrical profiling as a function of electrode spacing.

crystalline rock
A term for an igneous or metamorphic rock, as opposed to a sedimentary rock.

cumulative runoff
The total volume of runoff over a specified period of time (usually one year). It is often visualized in a cumulative curve for comparison with the other parameter distributions.

current
1. An electric current is the flow of electric charge. 2. The movement or flow of fluids, especially water, is also described as current.

current meter
In hydrology a device for measuring the speed and direction or speed only of flowing water. Current meters can be mechanical, electric, electromagnetic, acoustic, or a combination thereof.

cuttings
Material obtained from drill holes, especially with rotary drilling. Cuttings collected at closely spaced intervals are used to provide a record of the strata penetrated. But due to changes caused by the drilling process, care in interpretation is advisable.

cycle
An interval during which a recurring sequence of events occurs.

Darcian flow
Darcy's law is a linear function of hydraulic gradient and discharge. Flow of groundwater that follows this law is called Darcian flow.

darcy
Permeability is expressed in darcies: equivalent to the passage of one cubic centimeter of fluid of one centipoise viscosity flowing in one second under a pressure differential of one atmosphere through a porous medium having a cross-sectional area of one square centimeter and a length of one centimeter. The commonly used unit is the millidarcy.

Darcy's law
Darcy's law states that the discharge Q of a fluid through a porous medium is proportional to the product of hydraulic conductivity k_f, the cross-sectional area of flow A and the hydraulic gradient J normal to that area:

$$Q = k_f A J$$

It is assumed that the flow is laminar and stationary. The hydraulic conductivity k_f depends on the internal structure of the medium the fluid flows through and the viscosity of the fluid.

Darcy velocity

The Darcy velocity q is determined by Darcy's law from the ratio of the rate of flow Q to the cross-section A passed by the groundwater flow perpendicular to the direction of flow: $q = Q/A$. The term is used synonymously with apparent velocity or specific discharge.

database

A collection of information stored in a computer and organized in such a way that a computer program can quickly select desired pieces of data. Traditional databases are organized by fields, records, and files. A field is a single piece of information; a record is one complete set of fields; and a file is a collection of records stored in the database. To access information from a database, a database management system (DBMS) is necessary. This is a collection of programs that enables the user to enter, organize, and select data in a database.

data logger

Electronic instrument that records measurements (temperature, pressure, precipitation, water level, voltage, etc.) over time. Data loggers are typically small, battery-powered devices that are equipped with a microprocessor, data storage and sensor. Most data loggers utilize turn-key software on a personal computer to initialize the logger and view the collected data. Data loggers are often used for monitoring in remote areas with telemetric units. The term is used synonymously with data recorder.

datum

The reference value to which other measurements are referred, e.g., mean sea level as a base for elevation. Geodetic datums define the size and shape of the Earth and the origin and orientation of the coordinate systems used to map the Earth. Different nations and agencies use different datums as the basis for coordinate systems. The diversity of datums in use today and the technological advancements that have made possible global positioning measurements with accuracies of less than a meter requires careful datum selection and careful conversion between coordinates referenced to a different datum.

DDT

Dichlorodiphenyltrichloroethane: an insecticide commonly used in the mid-1900s to destroy disease-carrying, crop-eating insects. It is very soluble in fats and most organic solvents and practically insoluble in water. DDT is persistent in the environment, with a reported half-life of between 26 days in river water to 15 years, and is immobile in most soils. It was banned by many countries in the 1970s as a pollutant which is a probable human carcinogen and causes damage to internal organs.

decomposition

1. Chemical or physical breakdown of material into smaller particles, compounds or into the component chemical elements, usually as a sample preparation technique. This sample preparation step is often used in conjunction with extraction. 2. Breakdown of organic matter into smaller parts or inorganic constituents by decomposing organisms. The terms disintegration and digestion are used synonymously with decomposition.

denser-than-water nonaqueous-phase liquid (DNAPL)

A liquid that is not soluble in water and has a density greater than that of water.

density

1. Density (of matter) is the mass per unit volume, normally expressed in kg per m^3 or grams per cubic centimeter (g cm^{-3} or g/cc). 2. Current density is the amperage per unit of length, area or volume.

deoxyribonucleic acid (DNA)

DNA carries the genetic information that encodes proteins and enables cells to reproduce and perform their functions.

depletion curve

The curve showing the decrease in water in a stream channel, surface soil, or groundwater in an aquifer; it can be for baseflow, direct runoff, or total runoff.

deposit

Material that has accumulated after being dropped by water, wind, ice, volcanoes, etc.

deposition

1. Laying down of sediment transported by wind, water, or ice. 2. The process of depositing something (e.g., rubble, waste, mining debris).

depth

The vertical distance from the terrain surface to a point in the subsurface. Other reference points for depth are mean sea level (MSL) and well head.

desorption

Opposite of adsorption, absorption or more general sorption. The result is a decrease in the amount of the adsorbed/absorbed/sorbed substance.

detectability

1. Size of the smallest object that can be discerned on an image. 2. Size of the smallest object/structure in the ground that can be recognized with a geophysical method. 3. The lowest concentration of a substance that can be detected by a chemical analysis method.

detection limit

1.The minimum concentration of a substance that in a given material and with a specific method has a 99 % probability of being identified, qualitatively or quantitatively measured, and reported to be greater than zero. 2. The lowest concentration of a substance that can be measured by a method. Also referred to as method detection limit (MDL). See also limit of quantitation (LoQ).

detector

An electronic device that responds in a predictable way to a physical phenomenon (e.g., electromagnetic radiation, seismic waves, particle radiation) and can produce a response signal suitable for measurement, recording, and analysis. Examples are charge-coupled devices (CCD) in remote sensing systems, geophones and Geiger counters.

deterministic

A process, model, simulation or variable whose outcome, result, or value does not have random or probabilistic aspect but proceeds in a fixed predictable fashion, governed by and predictable in terms of definite laws. A function or algorithm is deterministic if the output is uniquely determined by the input. The opposite of probabilistic, random or stochastic.

diagenesis

Chemical, physical, and biological changes in sediments after initial deposition and burial that convert a sediment to consolidated rock. Such changes might result from compaction, cementation, recrystallization or replacement, but exclude metamorphism and fracturing resulting from tectonic stress.

dielectric constant

→ permittivity

differential global positioning system (DGPS)

A method of improving GPS accuracy using measurements at a known location to improve measurements made by other GPS receivers within the same general area. Using FM radio transmissions, the system transmits error corrections to those receivers. Differential corrections can be applied in either real time or by post-processing.

differential thermal analysis (DTA)

Thermal analysis is a group of methods in which physical properties of a sample are measured as a function of temperature. In DTA, a sample and an inert reference are heated (or cooled) under identical conditions. The temperature difference between sample and reference is plotted against time or temperature. Changes in the difference indicate physical or chemical changes in the sample that produce or consume heat (exothermic or endothermic reactions, phase transitions).

differential scanning calorimetry (DSC)
A technique in which the difference in energy input into a substance and a reference material is measured as a function of temperature while the substance and reference material are subjected to a controlled temperature regime.

diffusion
1. Spontaneous spreading of something such as particles, heat, or momentum. The phenomenon is readily observed when a drop of colored water is added to clear water. 2. Diffusion is the result of random motion of atoms and/or molecules in solids, liquids, and gases from a more concentrated to a less concentrated area.

digestion
1. The process of decomposing organic matter (as in sewage) by bacteria, chemical action or heat. Aerobic digestion takes place in the presence of oxygen and anaerobic digestion takes place with the absence of oxygen. 2. Synonym of decomposition.

digital elevation model (DEM)
A digital elevation model is a three-dimensional representation (in x, y, and z coordinates) of the Earth's surface on the basis of a (square) grid in the x-y plane. The term digital terrain model (DTM) is used synonymously. DEMs can be derived, for example, from stereoscopic aerial photographs, maps and nonphotographic remote sensing images.

digital terrain model (DTM)
→ digital elevation model

dilution
1. Adding a solvent (e.g., water) to a solution to lower the concentration of the solute. 2. The reduction of the concentration of a contaminant in groundwater due to dispersion.

diode array detector (DAD)
A photo diode array detector used, for example, in HPLC.

dip
One of the directional properties of a geologic structure such as a layer or a fault. Dip is the inclination angle of the formation as measured at right angles to the strike.

direct-current (dc) resistivity methods
Direct-current resistivity methods use an artificially produced current. It is introduced into the ground through point electrodes (C_1, C_2) and the potential field is measured using two other electrodes (the potential

electrodes P_1 and P_2). The source current can be direct current or low-frequency (0.1 - 30 Hz) alternating current. The aim of generating and measuring the electrical potential field is to determine the spatial resistivity distribution (or its reciprocal – conductivity) in the ground. As the potential between P_1 and P_2, the current introduced between C_1 and C_2, and the electrode configuration are known, the resistivity of the ground can be determined; this is referred to as the "apparent resistivity". Resistivity measurements may be made at the Earth's surface, between boreholes or between a borehole and the surface. With special cables, measurements can be made underwater in lakes, rivers and coastal areas. The following basic modes of operation can be used:

- profiling (mapping),
- vertical electrical sounding (VES),
- combined sounding and profiling (two-dimensional resistivity imaging),
- three-dimensional resistivity survey (3-D resistivity imaging), and
- electrical resistivity tomography (ERT).

Profiling methods use fixed electrode spacing to detect lateral resistivity changes along a profile down to a more or less constant investigation depth, which is governed by the electrode spacing. The results are normally interpreted qualitatively. The main aim of sounding methods is to determine the vertical resistivity distribution in the ground. Several soundings in an area will provide information about structures. Soundings for investigation depths up to several hundred meters may be interpreted both qualitatively and quantitatively. Sounding and profiling can be combined in 2-D resistivity imaging to investigate complicated geological structures with strong lateral resistivity changes. Three-dimensional (3-D) resistivity surveys and ERT measurements provide information about complex structures.

direct push testing
Direct-push testing, cone-penetration tests (CPT), and percussion drilling can provide a highly detailed, 3-dimensional picture of the subsurface in less time than needed by traditional methods (e.g., drilling). Different tools are used to identify lithology, stratigraphy, depth to the groundwater table and the capillary fringe, as well as to determine geotechnical parameter values and to detect the presence of contaminants. There are also tools for collecting soil, water, and soil gas samples. Logging methods (e.g., gamma measurements, neutron-neutron measurements) for small diameter holes provide very detailed information about the strata. Special tools are available for detecting types of contaminants, e.g., petroleum hydrocarbons, polycyclic aromatic hydrocarbons (PAH), phenols, volatile

halogenated hydrocarbons (VHC), and benzene, toluene, ethylbenzene, xylene (BTEX). Individual compounds cannot be identified and the results are only semiquantitative. X-ray fluorescence (XRF) and laser-induced breakdown spectroscopy (LIBS) are used in direct push testing to identify inorganic substances (e.g., heavy metals). In addition to the above-described tools, video tools are used to recognize very thin soil and contamination layers. Prerequisite for the use of direct-push testing is unconsolidated rock. Owing to the small diameter of the tools, the effect on the soil ecosystem is small. To avoid migration of any contaminants, the resulting hole is filled with bentonite when the testing is completed. Because there is no drilling, there are no drill cuttings to be disposed of, which would be expensive in the case of contaminants.

discharge

1. Rate of fluid flow passing a given point or area within a given period of time, expressed as volume per unit of time. It can be applied to describe the flow of water from a pipe, a groundwater flow system, or from a drainage basin. 2. The conversion of the chemical energy of a battery into electrical energy, and the withdrawal of the electrical energy into a load.

disk-tension infiltrometer

A constant-head permeameter for determining hydraulic parameter values for unsaturated soil from infiltration at negative pressure, i.e. under positive moisture suction.

disk-tension pemeameter

→ disk-tension infiltrometer

dispersion

1. Hydrodynamic dispersion is the process whereby a contaminant dissolved in groundwater spreads out in the direction of and perpendicular to groundwater flow, causing the contaminant to become diluted. The sum of the effects of mechanical mixing and molecular diffusion on a dissolved contaminant results in differences in flow path length and velocity for different molecules. 2. The dependence of the velocity of an electromagnetic or seismic wave on the frequency of the wave. A medium in which waves of different frequencies propagate at different speeds is said to be dispersive. 3. In optics, dispersion is a phenomenon that causes the separation of a wave into spectral components with different frequencies, due to dependence of the wave velocity on its frequency. 4. Suspension of minute particles in a suitable medium.

dissolved organic carbon (DOC)

All organic carbon dissolved in a given volume of water at a particular temperature and pressure. DOC comprises soluble carbohydrates, amino acids, and other acids whose sources include photosynthesis, leaching from plant leaves and roots, and soil organic matter. Unit is mg L^{-1}.

dissolved oxygen (DO)
The amount of oxygen dissolved in water, expressed in ppm (mg kg^{-1}) or in mg L^{-1}. It is critical to aquatic life and biodegradation of contaminants in water.

dissociation
The breaking up of a compound into its simpler components such as molecules, atoms, or ions. Results from the action of some form of energy on gases and from the action of solvents on substances in solution.

dissolution
The process of dissolving and dispersing a substance in a liquid.

dissolved solids
The solid matter remaining when a solution is evaporated.

distortion
Any change in shape and position of objects with respect to their true shape and position on an image.

domestic waste
Waste generated by daily household activities. Domestic waste may contain a significant amount of hazardous waste.

downgradient
The direction in which groundwater flows; the term is analogous to downstream for surface water.

downhole measurements
Measurements made in a borehole.

downhole logging
→ borehole logging

drain
A natural or artificial channel that drains excess water from an area, e.g., so that land can be farmed or built on.

drainable porosity
→ specific yield, → permeability

drainage
Natural or artificial removal of surface and groundwater from a given area. Many agricultural soils need drainage to improve production or to manage water supplies.

drainage basin
→ catchment area

drainage divide

→ watershed

drawdown

The distance that the water level in the well is lowered by pumping. It is the difference between the static water level and the pumping level.

drift

1. Drift is the slow long-term variation of an attribute or value of a system or device. Drift is usually undesirable and unidirectional, but may be bidirectional, or cyclic. It usually refers to instrument drift that can be due to aging, temperature effects, sensor contamination, etc. It is usually modeled as an exponential, linear or quadratic function of time. 2. The angle between the heading of an aircraft and its flight path over the ground as affected by winds. 3. A horizontal (or nearly horizontal) passageway in a mine. 4. Surficial, unconsolidated rock debris transported from one place and deposited in another.

drill

A device used for making holes in rock. Such devices include rotary drills, percussion drills, augers, and water or air jets.

drillhole

→ borehole

drilling

The process of making a circular hole with a drill or other cutting tool. The main criteria for selection of the most suitable drilling method are reliable investigation of the subsurface down to the required depth, collection of a good quality samples, and cost efficiency. When selecting the drilling method it is also advisable to take into consideration the possible subsequent installation of groundwater observation wells. The chosen method should not have any long-term influence on the characteristics of the groundwater. At least one core drilling is recommended at new landfill sites for each hectare of landfill area. This recommendation, however, neither reflects the difference between unconsolidated and consolidated rock nor the very different geological and hydrogeological conditions at different locations and also cannot be applied to abandoned landfill sites. A large number of cored holes would only be deemed necessary at locations with unconsolidated rock (in general, sites with Quaternary and/or Tertiary rock). The wells must be backfilled prior to construction of a new landfill; this would involve expensive drilling with a larger diameter bit, since the gravel fill of the borehole required for a water well would also need to be removed. Drilling wells at locations with hard rock entails more work and expense than in unconsolidated rock and, hence, is more expensive. This is the reason why the number of wells is often kept to a minimum. From a geological point of view this is justified since it can be assumed that the rocks will – as a consequence of their geological genesis – not have any significant lateral inhomogeneities in an area the size of a landfill that are likely to remain undetected. A further factor which greatly influences cost is the depth of drilling. Boreholes drilled into unconsolidated rock should have depths of 10 - 15 m, with some drilled to greater depths depending upon the geological conditions. In hard rock greater depths should be drilled up to 20 - 30 m, in individual cases also to greater depths, for example when investigating deeper aquifers. In the case of unconsolidated rock, direct push testing can also be a suitable method to provide information about the subsurface. Geophysical methods can be used to provide information between the boreholes in order to reduce the drilling cost.

drilling fluid
→ drilling mud

drilling mud
A mixture of clay, water, chemical additives, and high-density materials that flushes rock cuttings from a well, lubricates and cools the drill bit, maintains the required pressure at the bottom of the well, stabilizes the walls of the borehole, and prevents other fluids from entering the well.

drilling rig
The derrick, power supply, draw works, and other surface equipment needed to drill a well/borehole/drillhole.

drill pipe
In rotary drilling, the pipe that is rotated so that the bit at the bottom can bore into the rock and through which the drilling fluid is conducted to cool the bit. It is made up of 3 - 10 m sections.

driven well
A normally small diameter (3 - 10 cm), shallow well made by forcing pipe by a percussion method into unconsolidated sediment to reach groundwater.

dry density
→ bulk (dry) density

dual-porosity medium
Water in structured soils and in rock has often a dual character with regard to flow. The porous medium is divided into the mobile and immobile regions. Part of the water is mobile and flows along the continuous joints, fractures, connected pores, while another part remains immobile. Exchange of water and solutes between the two parts occurs by diffusion.

Contaminant mass transfer between the mobile and immobile regions is the main reason for the tailing effect after contaminant breakthrough.

dug well
A large diameter well for individual domestic water supplies, mostly excavated by hand, often lined with concrete or hand-laid bricks. Such wells typically reach less than 15 m (~50 feet) in depth and are very susceptible to contamination.

dump (site)
A site used to dispose of solid waste in a disorderly or haphazard fashion without regard to protecting the environment.

dwell time
Time required for a detector to sweep across a ground resolution cell of a nonphotographic remote-sensing system.

dynamic viscosity
A measure of the resistance to flow of a fluid under an applied force. The greater the resistance to flow, the greater the dynamic viscosity. The dynamic viscosity depends on the temperature. It increases for gases and decreases for liquids with increasing temperature. The SI unit of dynamic viscosity is the pascal-second (Pa s), which is equal to $1\ \mathrm{kg\ m^{-1}\ s^{-1}}$. Coefficient of viscosity and absolute viscosity are used synonymously. See also kinematic viscosity.

earthquake
A sudden motion or trembling in the Earth caused by the abrupt release of accumulated strain on a fault or by magmatic activity.

ecology
The study of how organisms interact with each other and their physical environment. The study of the structure and function of ecosystems.

ecosystem
A community of plants, animals, and microorganisms that are linked by energy and nutrient flows and that interact with each other and with the physical environment. An ecosystem can be of any size – a pond, field, forest, or the Earth's biosphere – but it always functions as a whole unit.

effective permeability
A parameter indicating the ability of a single-phase fluid to flow through a rock when the pore spaces of the rock are not completely filled or saturated with fluid. → intrinsic permeability, → relative permeability, → absolute permeability

effective pore volume
The pore space in a rock in which water can freely move; this does not include spaces containing trapped air or pellicular water.

effective porosity

The portion of pore space in a saturated permeable material where movement of water takes place.

effective velocity

1. The actual velocity of groundwater percolating through water-bearing material. It is the volume of groundwater passing through a unit cross-sectional area per time unit. 2. The flow of groundwater divided by the effective porosity.

effluent

Any material that flows outward from something, as in a stream flowing from a lake, or wastewater, treated or untreated, that flows from a water treatment plant, sewer, or factory.

effluent flow

Flow of water from an aquifer to a receiving stream due to the hydraulic gradient.

electrical conductivity

1. The physical property of material enabling it to conduct an electrical current. It is expressed in siemens per meter ($S\ m^{-1}$). For water samples, it depends on the concentration and type of dissolved ions in the water and temperature. 2. The reciprocal of electrical resistivity.

electrical resistivity

Electrical resistivity (also known as specific electrical resistance) is a parameter indicating how strongly a material opposes the flow of electric current. It is the reciprocal of electrical conductivity. A material that readily allows the movement of electrons or ions has a low resistivity. The SI unit for electrical resistivity is ohm meter (Ωm).

electrical profiling

Measuring the resistivity changes along profiles by dc resistivity, induced polarization or electromagnetic methods utilizing a fixed spacing of electrodes or antennas moved progressively along profile lines.

electrical sounding

Increasing the electrode spacing in a dc resistivity or induced polarization survey yields data for increasingly greater depth; horizontal layering is assumed for this method.

electrical survey

Measurements to evaluate the distribution of electrical resistivity or conductivity at shallow depth by dc resistivity, induced polarization or electromagnetic methods.

electrochemical methods

A group of methods in which chemical reactions take place at or between electrodes. Certain quantities can be measured that are proportional to concentration or are specific to particular substances. Among these are voltage (potential difference), current, electric charge, conductivity, etc. Common electrochemical methods are amperometry, coulometry, conductometric analysis, polarography, potentiometry, voltametry, and others.

electrode

A conductor (usually metal, porous pot or graphite) through which electrical current enters or leaves liquids, gases, or solid bodies. A nonpolarizable electrode consists of a porous pot filled with electrolyte in which a metal rod is inserted (e.g., copper sulfate and copper).

electrolyte

A substance that dissociates fully or partially into ions when dissolved in a solvent, producing a solution that conducts electricity by movement of ions between two electrodes. Acids, bases, and salts are common electrolytes. In geoelectric measurements, soil water functions as an electrolyte.

electromagnetic methods

Electromagnetic (EM) inductive methods are in geophysics an excellent means to obtain information about electrical conductivities of the subsurface. They can be classified as natural field methods and controlled-source methods. Systems with transmitter and receiver coils have been commercially available since the 1970s. The controlled-source methods can be divided into frequency-domain methods (FEM) and time-domain or transient electromagnetic methods (TEM). In the most commonly used FEM systems, a transmitter coil, continuously energized with a sinusoidal audio-frequency current, forms a magnetic dipole. Its primary magnetic field induces very weak eddy currents in the conductive ground. These eddy currents in turn generate a secondary magnetic field that is of the same frequency but with a different phase and much lower amplitude than the primary field. The primary and secondary magnetic fields are detected by the receiving coil. To obtain the ground conductivity information, the weak secondary field has to be separated from the primary field. The secondary magnetic field strength depends on the electrical conductivity of the ground, on the transmitter-receiver coil separation, and on the operating frequency. Like FEM techniques, TEM methods use a dipole magnetic source field. The waveform of this primary field is a modified square wave. By abruptly turning off a constant current in the ungrounded transmitter loop, a transient electromagnetic field is produced. This rapid change induces eddy currents in nearly horizontal circles in the ground below the loop, which in turn create a secondary magnetic field. With increasing time after transmitter current turn-off, the induced fields have penetrated deeper into the Earth by "diffusion". The decay of the secondary magnetic field is recorded during the transmitter's off-time, i.e., in absence of the primary field. The decay rate is a function of the ground conductivity and the time after the transmitter current has been switched off; it determines the investigation depth. Investigation depths of commercially available equipment range from about 5 to 3000 m. FEM and TEM methods are available for land-based and airborne surveys and for borehole logging. There are two kinds of measurements: mapping/profiling and sounding. Profiling/mapping provides information about the variation in lateral conductivity within a certain depth range. This depth range is directly proportional to the transmitter-receiver coil separation and is inversely proportional to the operating frequency and ground conductivity. FEM methods are mainly used for profiling as they are not as time consuming as TEM methods. Interpretation of profiling data yields mainly qualitative results. FEM soundings are done by changing the frequency and/or the coil separation (parametric/geometric sounding). TEM techniques generally provide better sounding data than FEM techniques. The very low frequency (VLF), the VLF-resistivity (VLF-R), and the radiomagnetotelluric (RMT) methods are passive methods which employ the electromagnetic fields of remote radio transmitters. EM surveys are less time consuming and require less manpower than dc resistivity surveys. As the transmitter-receiver array necessary to investigate a certain depth range is much smaller than the equivalent dc array. EM methods provide a better lateral resolution and a higher survey efficiency. The disadvantage of EM methods is their lower vertical resolution and higher sensitivity to electromagnetic noise.

electromagnetic monitoring system

Electrical conductivity is a sensitive parameter for monitoring contaminants from landfills or the position of a fresh water/salt water interface. The real and imaginary components of electrical conductivity are functions of the frequency of the transmitted electromagnetic waves. The water and electrolyte content can be derived from the conductivity function. This function is determined for selected frequencies from the measured signal attenuation and phase shift between the transmitter and receiver antennas.

electron capture detector (ECD)

A radioactive source is used to ionize the carrier gas as it leaves a gas chromatograph. A voltage is applied, creating a current. Any electronegative substance carried by the gas from the column captures electrons, causing a decrease in the current. The size of the decrease is a function of concentration. An ECD is typically used for organic compounds containing halogen, sulfur, nitrogen, phosphorus, or heavy metal groups (e.g., pesticides).

electron conductor

Electron conductors consist of metal atoms. The electrons from the outer shell of the atoms are released. The atoms become positively charged ions and form a lattice in which the free electrons move like a cloud. When a voltage is applied to the conductor, the cloud of electrons (electron current) moves from the negative pole to the positive pole. Unlike ion conductors the electron current in electron conductors does not change the metal. The term is used synonymously with metallic conductor.

electrothermal AAS (ET-AAS)

Atomic absorption spectrometry with the sample being vaporized by electrothermal atomization.

elevation

Vertical distance from a reference level (usually mean sea level) to a point on the Earth's surface. Although the terms elevation and altitude are often used synonymously, the term altitude is normally used for points above the Earth's surface, and elevation is used for points on the surface.

elution

The process of removing adsorbed materials from the surface of an adsorbent such as soil or activated charcoal by washing with a liquid. The liquid in this process is called the eluent. The solution consisting of the eluted sorbate dissolved in the eluent is called eluate. The various elution methods, in particular the sequential extraction method, are important tools in analytical environmental chemistry. They are used to investigate the substances within a sample and the mobilization behavior of substances in contaminated soils and sediments under changing environmental conditions, as well as the leachate behavior of waste materials with respect to long-term contaminant migration behavior. Depending on the requirement to the quality of results, batch, reactor and column tests can be used for the analysis of the reaction processes (storage, exchange and transformation processes) as well as to determine their characterizing parameters.

emission

1. The release or discharge of a substance into the environment. Generally refers to the release of gases or particulates into the air. 2. The process by which a body emits electromagnetic radiation.

emissivity

The ratio of electromagnetic radiation emitted by an object to that emitted by an ideal blackbody at the same temperature per unit area.

endothermic

A chemical reaction in which heat is absorbed. Opposite of an exothermic reaction, which releases heat.

energy

The capacity of a physical system to do work. The units of energy are joule (J) or erg. Energy includes thermal, mechanical, electrical, and chemical forms. Energy may be transformed from one form into another.

enhancement

Filtering or removal of noise improves the quality of the data obtained in a geophysical survey.

equivalent concentration

Equivalent concentration is the molar concentration times the charge of the dissolved ion. The equivalent charge of an ion is the number of moles multiplied by its charge. For singly charged ions, equivalent concentration therefore equals molar concentration, for doubly charged ions, it is twice the molar concentration, and so on. Equivalent concentration is usually given in milliequivalents per liter (meq L^{-1}). Equivalent concentrations are used, for example, for ion balances. Example: The concentration of sulfate (SO_4^{2-}) in a water sample is 9 mmol L^{-1}. As sulfate is doubly charged, the equivalent concentration is 2×9 mmol L^{-1} = 18 meq L^{-1}.

error

A discrepancy between measured or computed value or condition and the correct value or condition.

evaporation

1. The conversion of a liquid to a vapor at a temperature below its boiling point. The opposite of condensation. 2. The quantity of water that is evaporated.

evapotranspiration

Water discharged to the atmosphere by evaporation from the soil and surface-water bodies and by plant transpiration. The term is also used for the volume of water lost by evapotranspiration.

exaggeration

1. The act of making something more noticeable than usual. 2. The factor by which something has been exaggerated. See also vertical exaggeration.

exothermic

A chemical reaction that releases heat. Opposite of an endothermic reaction, which absorbs heat.

extractable organohalogens (EOX)

Extractable organohalogens as chlorides (µg L^{-1}) are used for assessment of contamination. EOX gives a better indication of the amount of organic halogens susceptible to lipophilic absorption than AOX. Usually, EOX is significantly lower than AOX (≤ 10 % of AOX).

extraction

Separation processes in which components of solid or liquid samples are removed with the help of suitable solvents. In a solid-liquid extraction, the solvent containing the extracted material is filtered and then evaporated. In a liquid-liquid extraction, the extracting liquid must be immiscible in the sample liquid.

fabric

The spatial arrangement of the solid particles and associated voids of a soil; the spatial orientation of the particles of a sedimentary rock.

factor analysis

A branch of multivariate statistical analysis based on correlation coefficients, used to examine the interrelations among a set of variables, or items, in order to identify an underlying structure to those items and seek to reduce the number of variables.

fecal coliform bacteria

A class of bacteria that is unique to the intestinal tract of warm-blooded animals (including birds and humans), one of the main species is Escherichia coli (usually abbreviated to E. coli). The presence of fecal coliform bacteria in groundwater is a common indicator of pollution and possible contamination by pathogens (harmful viruses, bacteria or parasites). Fecal coliform bacteria are normally measured by filtration and culture on disk media (refered to as [MPN/100 mL]; most probable number).

far-field

In electromagnetic wave propagation the immediate vicinity of the transmitting antenna is called "near-field". The "far-field" is further away from the transmitting antenna depending on the wavelength of the transmitted electromagnetic wave, the antenna parameters, and the electromagnetic properties of the ground. The distance is greater than one wavelength. In the far-field, spherical waves can be treated to a good approximation in the same way as plane waves.

fault

1. A surface or zone of rock fracture along which there has been vertical or lateral displacement at a scale of a few centimeters to a few kilometers. Compression mainly causes folding, and faulting, e.g., normal and strike-slip faults, and thrusts. Stretching leads to faulting (ruptures), flexures and

fissures. Faults are often imaged in the results of geophysical measurements. It is, for example, an indication of a fault if in a seismic profile correlatable reflections are offset or disrupted or diffraction waves can be found. Likewise, strong gradients in gravimetric, magnetic or geoelectric results indicate faults. Abrupt movement of tectonic units (e.g., tectonic plates) along a fault are the cause of most earthquakes.

field

1. In physics, it is the space in which a particular type of force occurs and can be measured. Depending on the force, one can thus speak of a gravity field, magnetic field, electric field, electromagnetic, and seismic wave field. The laws of physics suggest that fields represent more than a possibility of force being observed, but that they can also transmit energy and momentum. 2. In applied geophysics, the terrain where measurements are made.

field capacity

The amount of water soil can hold against the force of gravity after saturation and runoff usually 1 to 3 days after rainfall. This water is used by plants during the growing season. In soil science, the term field capacity is used, whereas in hydrogeology the same property is called specific retention. → specific yield

field intensity

A term used to describe the strength of a gravity field, magnetic field, electric field, electromagnetic or seismic wave field.

filter

1. Filters are components of an electronic or mathematical system which are used to separate a signal (measured values, field distribution, functions) into its components. There are analog filters (hardware filters, electronic filters) and digital filters (software filters, numeric filters). Differences in frequency, wavelength, amplitude, phase, velocity, coherence, etc. are used to separate the signal components. Analog and digital filters are widely used in geophysical measuring and interpretation methods. Examples: low-pass filters to eliminate high-frequency components, high-pass filters to eliminate low-frequency components, notch filters to eliminate components within a very narrow frequency band (e.g., interfering signals of power lines or of electrified railway lines, (in Germany: 50 Hz or 16⅔ Hz electromagnetic fields, respectively)). 2. A membrane or other porous device which removes undissolved, suspended particles from a liquid or gas. 3. A device (optical filter) used to absorb electromagnetic radiation, particularly light, of specific wavelengths.

filter gravel
Gravel composed of sub-angular, hard, durable, and dense grains of predominately siliceous material. The gravel is washed, kiln dried, and screened to meet specifications. The gravel is used for as gravel packing in the annulus of groundwater wells to form a barrier between the screens and the natural deposit of the soil formation in order to protect the well from silting up.

filtration
The process of passing a fluid through one or more permeable membranes or media of limited diameter to physically remove particulate matter.

finite-difference model
A numerical model that utilizes a grid of rectangular cells. Spatial and temporal derivatives of the parameters are approximated using Taylor series expansions. → numerical model, → model design

finite-element model
A numerical model where the modeled space is described by a mesh of polygonal cells. Spatial derivatives are approximated using polynomials that are functions of spatial coordinates. Finite-element models can generally be better adapted to complex geometries than finite-difference models. → numerical model, → model design

fissure
A fracture or crack in rock along which there is a distinct separation. It may be partially or totally filled with soil or weathering material or, if open, it can act as a conduit for water.

fixed groundwater
Water in the saturated zone that is held to the pore walls or is in pores so small that it is unavailable for pumping.

fixed moisture
Moisture held in the soil below the hygroscopic limit.

flame atomic absorption spectrometry (F-AAS)
F-AAS is an atomic absorption spectrometric technique in which a hot flame is used for evaporation and atomization of the sample. Most commonly, an air/acetylene flame is used, because the attainable temperatures are high enough for the atomization of most elements and at the same time are low enough that disturbances due to ionization remain small. For elements that form very stable compounds in flames (e.g., refractory oxides of V, Ti, Zr, etc.), a N_2O/acetylene flame is employed. Additionally, air/propane or air/butane flames are used with F-AAS. Chemical disturbances (formation of stable bonds with the components of the flame gases or the sample), ionization disturbances (decrease in

sensitivity), and spectral disturbances (nonspecific background absorption by molecules) can occur in the flame. These must be considered and corrected for in the measured data.

flame atomic emission spectrometry (FAES)

In flame emission spectrometry, a sample solution (usually aqueous) is evaporated and atomized in a hot flame. The thermally excited atoms emit radiation in the visible and ultraviolet range at wavelengths characteristic of the emitting element. The intensity of the radiation is proportional to the concentration of the element in the sample. This technique is predominantly used to analyze for alkali and alkaline-earth metals. It has been replaced by AAS in many fields of application. Also called flame photometry.

flame ionization detector (FID)

A nearly universal gas chromatographic detector. It responds to almost all organic compounds. An FID does not respond to nitrogen, hydrogen, helium, oxygen, carbon monoxide or water. This detector ionizes compounds as they reach the end of the chromatographic column by burning them in an air/hydrogen flame. As the compounds pass through the flame, the conductivity of the flame changes, generating a signal.

flameless AAS

→ atomic absorption spectrometry

flame photometry

→ flame atomic emission spectrometry

flow injection analysis (FIA)

Together with a carrier liquid, a liquid sample is passed through a detection system (commonly a spectrophotometer, AAS or OES); reagents can be added to react with the substance of interest to yield compounds that can be detected by the detection system used.

fluid

1. adj. Characteristic of a fluid; capable of flowing and easily changing shape. 2. n. Substance without any definite shape, including both gases and liquids. Some solids (e.g., rock salt) can also exhibit fluid behavior under pressure on long time scales.

fluorescence spectroscopy

Spectroscopic methods that measure the fluorescence of gaseous, liquid, or solid substances. They are used, for example, to analyze for metals, pesticides, or PAHs. Two kinds of spectra can be measured: excitation and emission spectra. Excitation spectra are obtained by irradiating the sample with ultraviolet light of different wavelengths and measuring the intensity of fluorescence at a single wavelength that is characteristic of the analyzed element or compound. Emission spectra are obtained by irradiating the sample with monochromatic UV light and measuring the intensity of the emitted fluorescence over its whole spectral range.

fluorochlorinated hydrocarbon (HCFC)
A hydrocarbon with some of the hydrogens replaced by fluorine and chlorine. If all of the hydrogens are replaced, it is called a chlorofluorocarbon (CFC). CFCs have been generally prohibited by the Montreal Protocol owing to their destructive effects on the ozone layer. HCFCs are now used as CFC substitutes.

fluviatile
A term for the results of river action (e.g., fluviatile sands). The term is used by some as a synonym of fluvial.

formation
A rock unit (e.g., layer) identified by lithological characteristics and/or stratigraphic position and is mappable at the Earth's surface or traceable in the subsurface. It is not necessarily a time unit. Formations have formal names, generally derived from the geographic localities where the unit was first recognized and described and the dominant rock type (e.g., Leine Rock Salt in the Upper Permian).

formation water
Water that is naturally present in an underground formation, in contrast to water that has been injected into the ground.

forward modeling
Geophysical field values (e.g., magnetic field strength) or other quantities (apparent resistivities) are calculated for a given model. The effects of simply shaped bodies on measurements can often be calculated with analytical methods. For complicated subsurface structures, numerical methods must be applied. Another example of forward modeling is the calculation of sounding graphs for a layered subsurface in geoelectrics.

fracture
A break in a rock formation due to structural stresses, e.g, faults, fissures cracks, and joints.

fracture permeability
Permeability due to fracturing of the rock, as opposed to permeability due to pores.

fracture porosity
A type of secondary porosity produced by the fracturing of rock. Fractures typically do not have much volume, but by joining preexisting pores, they enhance permeability significantly.

free water

Water in soil or rock that can be removed by physical or biological processes without changing the structure or composition of soil or rock.

free water surface

The surface of a body of water that is open to the atmosphere, both surface water bodies and the water table. It is often desirable to know the amount of evaporation that occurs from free water surfaces. This can be calculated from inflow and outflow to and from the water body provided the surface area of the impoundment is known and provided the level of the surface does not change during the calculation period. Even when a water impoundment is shrinking in size, the free water surface evaporation for a period can be calculated if the surface elevations and areas of the impoundment are known for the beginning and ending of the period.

frequency

1. The number of times something happens within a certain period of time (usually one second), in the case of a wave, the number of complete cycles per second. Frequency is expressed in hertz [Hz]. 2. Angular frequency ω is expressed in radians per second, is related to frequency f as follows: $\omega = 2\pi f$.

frequency domain (FD)

Representation of a process in which frequency is the independent variable.

frequency domain method

Any geophysical survey method in which parameters of interest are estimated from characteristics of amplitude and phase spectra.

fresh water

Water that contains less than 1000 milligrams per liter (mg L^{-1}) of dissolved solids, especially sodium chloride. Water that contains more than 500 mg L^{-1} is generally undesirable for drinking and many industrial uses. All freshwater ultimately comes from precipitation of atmospheric water vapor, reaching inland lakes, rivers, and groundwater directly or after the melting of snow or ice. → brackish water, → saline water

fully penetrating well

A well in which the screen covers the entire aquifer.

gauge

1. An instrument for measuring a quantity such as the thickness of wire or the amount of rain (rain gauge), also spelled gage. 2. Device for registering water level, discharge, velocity, pressure, etc.

gas chromatography (GC)

Gas chromatography is an important variant of the chromatographic methods with which gases and volatile materials can be separated. In this method, an inert carrier gas is used as the mobile phase. A column is packed or lined with a substrate that forms the stationary phase. The substances to be separated are present in gaseous or vapor state. Particularly good separation results can be achieved by employing long (up to 100 m) capillary columns with an inside diameter of 0.2 - 1 mm without a substrate fill, in which the separation takes place by distribution between the gas phase and the capillary surface. The use of a mass spectrometer as a detector (GC-MS) results in an efficient system that makes the identification and determination of complex mixtures of organic substances possible. Other conventional detectors suitable for different classes of substances include flame ionization detector (GC-FID), photo ionization detector (GC-PID), and electron capture detector (GC-ECD).

geochemistry

The study of the properties, distribution, circulation, and interactions among the chemical elements and compounds found in the minerals, ores, rocks, soils, water and atmosphere of the Earth.

geocode

A geographical code (e.g., coordinates) assigned to a geographic feature (e.g., street address), photograph, or anything else with a geographic component.

geogenic

Said of conditions (e.g., heavy metal concentration in soil, sediment or groundwater which originate in natural processes (e.g., elution of soil, weathering of rock), as opposed to those of anthropogenic origin.

geographic information system (GIS)

A computer software system and associated geo-referenced database designed to efficiently capture, store, update, manipulate, analyze, retrieve and display all forms of geographically referenced information. The system generally can utilize a variety of data types, such as imagery, maps, tables, etc. Usually the data are organized in "layers" representing different aspects such as hydrology, remote-sensing and geophysical results, land use, topography, etc. A geographic information system permits information from different layers to be easily integrated and analyzed.

geological barrier

The geological barrier is understood to be the natural substratum beneath and surrounding a landfill which, on the basis of its properties and dimensions, extensively prevents the spread of pollution. The geological barrier in all cases comprises naturally arranged, slightly permeable, unconsolidated or consolidated rock of several meters thickness and exhibiting a high pollution retention capacity extending beyond the area of the dumping site. The geological barrier must be as homogeneous as possible beneath the dumping site (TASi, 1993). A key criterion of the geological barrier is low permeability (hydraulic conductivity $k_f \leq 10^{-7}$ m s^{-1}), which impedes migration of contaminants by convection and diffusion. A further prerequisite for a geological barrier is a high contaminant retention capacity. The < 2 µm grain-size fraction should make up at least 20 wt. %, and the clay mineral content at least 10 wt. %. This assumes that clay minerals represent the main sorbents besides organic matter (humin) and crystalline and amorphous oxides and hydroxides. The low-permeability, contaminant-retarding rock of must be "several meters" thick (minimum thickness 5 m). A sufficient thickness is an important prerequisite in order to delay the gradual seepage of contaminants over time and also improve the filter effect. The thicker the barrier rock layer, the larger the reaction volume, the longer the period of contact of the leachate with the barrier rock and, the greater the effect of physical and chemical reactions and the contribution to retardation and/or fixing of contaminants. The barrier rock must extend beyond the direct limits of the planned landfill in order to effectively prevent lateral movement of contaminants over the long term. How large the area must be depends on the geological structures and the quality of the barrier rock. Previous recommendations of "50 m beyond the landfill" have been shown to be inadequate in a number of cases. The geological barrier should be as homogenous as possible under the planned landfill and the surroundings. Structural heterogeneities in the form of faults, joints, inclusions, zones of loosening and facies transitions all represent weaknesses in the overall function of the geological barrier, leading to water pathways.

geologic map

A geological map shows the distribution of rock units according to kind and age, as well as structures (folds, faults, joints), mineral deposits, and sites of fossil occurrences. Structural information may be indicated by outcrop locations, symbols showing strike and dip, or by contour lines.

geological and hydrogeological methods

Methods to be used to investigate lithological structures, to determine the homogeneity of the rock, to locate fractures, to determine the permeability of the rock with respect to water, gases and various contaminants, to assess the mechanical stability of the ground, and to obtain data on the groundwater system. Flow and transport models must be developed to estimate groundwater recharge and the potential for groundwater contamination. The main tasks of geological and hydrogeological surveys are to gain information directly by examining outcrops, digging trenches and drilling boreholes, conducting hydraulic tests (e.g., pumping tests and tracer tests) in wells to determine hydraulic properties in-situ. This work is augmented by geological mapping, examination of drill cores, construction and expansion of a network of groundwater observation wells. Rock, soil and groundwater samples are taken to determine physical, chemical, petrographic and mineralogical parameters. Special laboratory experiments can be carried out to estimate migration parameters and the texture of rock and soil samples. Data from cone penetration tests and other field and laboratory methods are used to assess the stability of the ground.

geology
The science and study of the Earth, its composition, structure, history, and properties together with the forces and processes to produce change within and on the Earth.

geometric correction
Procedure in image processing that corrects spatial distortions in an image.

geomorphology
The study of the Earth's surface features (landforms), the processes that form them, and the geologic structures beneath.

geophysical borehole logging
→ borehole logging

geophysical downhole logging
→ borehole logging

geophysical methods
Geophysical methods are used to develop a geological model of subsurface of a site, to locate fracture zones, to investigate the groundwater system, to detect and delineate abandoned landfills and contamination plumes, as well as to obtain information on the lithology and physical parameters of the ground. A broad spectrum of geophysical methods is available for the investigation of the subsurface: magnetic methods, gravity methods, dc resistivity, electromagnetics, ground penetrating radar, refraction seismics, reflection seismics, surface nuclear magnetic resonance (SNMR), geophysical borehole logging, and geophysical monitoring methods. A necessary condition for a meaningful use of geophysical methods is the existence of differences in the physical parameter values (magnetization, susceptibility, density, electrical resistivity, seismic velocities, etc.) of different geological structures. It must be first determined which parameters can be expected to have sufficiently different values in the different structures at the site before a geophysical survey is conducted. Geophysical methods supplement each other because they detect different physical properties. Seismic methods are used to investigate structures and lithology. Electrical and electromagnetic methods are very sensitive to changes in electrolyte concentrations in pore water. Ground penetrating radar can be used in areas with dry, low-conductivity rocks. Both magnetic

and electromagnetic mapping have proved useful for locating and determining the edges of concealed landfills. Both methods are fast and easy to conduct, enabling large areas to be investigated in a short time. Seismic, dc resistivity, electromagnetic and gravity methods are used to investigate groundwater systems on a regional scale. Geophysical surveys help to find suitable locations for drilling groundwater observation wells and provide information between boreholes and for areas where it is impossible to drill. Borehole logging is absolutely necessary. Logging data are not only necessary for processing and interpretation of surface geophysical data but also as a bridge between geophysical surveys and hydrogeological modeling.

geophysical monitoring
→ monitoring

geophysics
The study of the Earth, especially its interior by physical methods.

georeferencing
The assignment of coordinates of a geographic reference system to data points, geographic features or other field attributes with specific locations on the Earth's surface either in real-time or by post-processing. It is part of the rectification of aerial photographs.

geotechnics
The use of engineering principles to understand how Earth materials, such as soils and rocks, behave. This knowledge is applied to better design engineering structures (e.g., buildings, landfills, dams).

glacial
1. adj. Pertaining to features and materials resulting from the action of glaciers, e.g., a glacial deposit — sediments remaining after the ice recedes. Such deposits consist of an unlayered mixture of material ranging widely in size, shape, and composition. 2. n. A cold episode of the Pleistocene, in contrast to a warmer interglacial period.

global positioning system (GPS)
A worldwide radio-navigation system developed by the US Department of Defense consisting of a series of 24 geosynchronous satellites that continually transmit their position and receiving devices used to compute positions on the Earth (longitude, latitude, elevation). GPS is used in air, land and sea navigation, mapping, surveying and other applications where precise positioning is necessary. The system is unaffected by weather and provides a worldwide common grid reference system.

glow discharge mass spectrometry (GD-MS)

A mass spectrometric method in which an electrical discharge is created between an electrically conductive sample and a counter electrode in a vacuum or in low-pressure inert-gas atmosphere. This discharge causes material to be released from the sample surface, atomized and ionized. The resulting ions are detected by a mass spectrometer.

glow discharge optical emission spectrometry (GD-OES)

Similar to GD-MS, a discharge is created between an electrically conductive sample and a counter electrode. This releases material from the sample, which is subsequently atomized and ionized with the emission of light. This emission is detected by a spectrometer. The emitted wavelength is element-specific and its intensity a measure of concentration.

gradient

1. A spatial rate of change of a physical quantity with respect to distance, such as temperature or pressure, in the direction of maximum change. 2. Vector of first partial derivatives of a function (assumed to be differentiable at least once) denoted grad $f(x)$, where f is the function and x is a point in its domain. 3. The steepness of a slope.

grain size analysis

The diameter distribution of the grains of an unconsolidated rock or soil is determined, for example, microscopically on an etched plane surface of the metal as described in ASTM Standard Method E112 and or by sieve and sedimentation analysis.

graphite furnace AAS (GF-AAS)

→ atomic absorption spectrometry

gravimetric analysis

A type of chemical quantitative analysis in which a dissolved substance is determined by weighing a precipitate which consists of a stable, defined compound or an element.

gravimetry

Method for the measurement of gravity. → gravity method

gravity method

Gravity is defined as the force of mutual attraction between two bodies; it is a function of their masses and the distance between them and is described by Newton's law of universal gravitation. The gravity field at each location on Earth consists of a global field which is superimposed by a local anomaly field. In a gravity survey, measurements are made of the local differences in the gravity field resulting from density variations in the subsurface. The effects of small-scale masses are very small compared with the effects of the global part of the Earth's gravity field (often on the order of 1 part in 10^6 to 10^7). Highly sensitive gravimeters are necessary to accurately measure such variations in gravitational attraction. Special data processing and interpretation techniques are used to interpret the shape and amplitude of the anomalies in terms of subsurface geological or anthropogenic structures. Gravity measurements can be performed on land, at sea and in the air. For environmental problems, land measurements are generally made. A necessary condition for the application of this method is the existence of density contrasts.

Greenwich Mean Time (GMT)

Mean solar time of the meridian at Greenwich, England (zero degrees longitude), used as the basis for standard time throughout most of the world. This is the old term for Coordinated Universal Time (UTC). Also referred to by the military as Zulu (Z).

grid

1. A two-dimensional network consisting of a set of equally spaced parallel lines superimposed upon another set of equally spaced parallel lines so that the lines of one set are perpendicular to the lines of the other, thereby forming square areas. The intersections of the lines provide the basis for an incremental location system. 2. A pattern of lines on a chart or map, such as those representing latitude and longitude, which helps determine absolute location. 3. An electric grid is a system of interconnected power lines and generators. A grid can also refer to the layout of a gas distribution system in which pipes are laid in both directions and connected at intersections.

ground penetrating radar (GPR)

Geophysical method increasingly used for geological, engineering, environmental, and archaeological investigations since the 1980s. Pulses of electromagnetic waves are transmitted into the ground and the traveltime and amplitudes of the electrical field strength of reflections and scattering from layer boundaries and buried objects are recorded. Electromagnetic waves are reflected and diffracted at boundaries between rock strata and objects that have large differences in their electrical properties (particularly permittivity and electrical conductivity). Broadband dipole antennas are normally employed for transmission and reception of the signals. Frequencies between 10 and 1000 MHz are used for geological and engineering investigations. For noninvasive testing, frequencies higher than 1000 MHz are also used. High frequencies, i.e., short wavelengths, provide a higher resolution; on the other hand, owing to absorption and scattering, higher frequencies have a lower penetration depth than lower frequencies. GPR is particularly suitable for materials with higher resistivities, such as dry sand with a low clay content or for consolidated rocks. In these cases GPR is the geophysical method with the highest resolution (achieving the centimeter range) for subsurface imaging. A high pulse rate enables measurements to be made quasi-continually by pulling the antennas along a profile. Several kilometers can be surveyed per day. When the underground conditions are suitable, the advantages of the method are that it is noninvasive and provides very high horizontal and vertical resolution, as well as yielding results in real time in the form of radargrams on a monitor or plotter. In many cases, a preliminary interpretation is possible in the field.

ground resolution
The minimum distance on the ground between two closely located objects at which they are distinguishable as separate objects. The ground resolution is determined by the instantaneous field of view (IFOV).

ground resolution cell
The area on the terrain that is covered by the instantaneous field of view of a detector. The size of the ground resolution cell is determined by the altitude of the remote sensing system and the instantaneous field of view of the detector.

ground swath
The strip of ground viewed by an imaging system.

groundwater
Water beneath the Earth's surface that saturates the pores and fractures of sand, gravel, and rock formations. Groundwater is a major source of water for agricultural and industrial purposes and is an important source of drinking water. This water moves under the influence of gravity. A velocity of meters per day is considered to be very fast.

groundwater budget
→ groundwater equation

groundwater discharge
The amount of water leaving a defined volume of an aquifer.

groundwater flow
Movement of water in the saturated zone of an aquifer. It is considerably slower than surface runoff.

groundwater flow model
A model of groundwater flow in which the aquifer is described commonly by numerical equations (numerical model), with specified values for boundary conditions that are solved on a digital computer. Groundwater flow modeling starts with a conceptual model (conceptualization of aquifer-aquitard systems, specification of boundary conditions, and hydrological stresses), followed by the design of the numerical model, the preparation of model inputs (initial conditions, boundary conditions, hydrogeological parameters, and hydrological stresses), the model

calibration and validation, and the simulation of scenarios. Numerical models of groundwater flow are well suited to exploring scenarios, because they are based on approximate representations of the physical principles, but they are relatively time consuming and expensive to calibrate, require much information to characterize the system, and the uncertainty of the model predictions is hard to quantify. Numerical models generate a spatially complete hydrologic prediction, but it is inevitable that assumptions are made in parameterizing the model, which renders error analysis difficult.

groundwater equation

Most important for characterizing the hydrologic conditions of an area is the groundwater equation (groundwater balance equation): $P = ET + R_d + R_b + \Delta S$, where P is precipitation, ET is evapotranspiration, R_d is direct runoff (overland flow and interflow), R_b is baseflow and ΔS is change in groundwater storage due to lateral in/outflow.

groundwater hydraulics

The study of the natural or induced movement of water through permeable rock formations (LOHMANN et al., 1972).

groundwater movement

→ groundwater flow

groundwater observation well

A well equipped to monitor groundwater level fluctuations and groundwater quality. In many areas there are observation well networks. Sometimes automatic monitoring systems are used.

groundwater pollution

Groundwater pollution occurs when substances change the chemical or biological characteristics of the water and degrade water quality so that animals, plants or human uses of the water are affected. Pollutants include plant nutrients, bacteria, viruses, pesticides, herbicides, hydrocarbons (including gasoline and oil), heavy metals and other toxic chemicals. Permissible limits for pollutants are regulated by law. Shallow groundwater is often affected by land use. Groundwater in deeper (confined) aquifers beneath layers that do not let water through (aquitards) has better protection from pollution because it is not directly connected to the surface environment. The most severe pollution often results from local spills of chemicals, or where contaminated water is disposed of in soak wells or unlined pits. The contaminant moves with the groundwater, in some cases reacting with it and the soil, spreading out to form a plume in the same direction as the groundwater flow. The resulting groundwater contamination plume may extend several hundred meters or even further from the source of pollution. Groundwater can also be contaminated by diffuse sources over a wide area, for instance widespread use of fertilizers. Diffuse contamination may have greater environmental impacts than contamination from point sources because a much larger volume of water is affected.

groundwater protection zones

Zones delineated according to specific aquifer categories or source protection areas and the associated vulnerability ratings. The zones are classified, for example, as outer source area with a high vulnerability, as regionally important aquifer with an extreme vulnerability, etc. The activities that are permitted in each zone are narrowly defined according to the degree of vulnerability of the aquifer, as well as the response to any exceeding of the legal limit of a contaminant.

groundwater runoff

That part of runoff which has passed into the ground as precipitation or snowmelt, has become groundwater, and has been discharged into a stream channel as spring or seepage water.

groundwater storage

The quantity of water in the saturated zone and the amount of recoverable water in the unsaturated zone of an aquifer.

groundwater vulnerability

→ vulnerability

groundwater withdrawal

1. Water removed from groundwater for use. 2. The quantity of water removed from groundwater.

gully

A channel (deep ditch) resulting from erosion and caused by the concentrated but intermittent flow of water usually during and immediately following heavy rains.

half-life

1. The time it takes for a radioactive substance to lose half of its radioactivity from decay. At the end of one half-life, 50 % of the original radionuclide remains. Radioactive isotopes have half-lives ranging from less than a second to thousands of years. 2. The period of time necessary for one half of a substance introduced to a living system or ecosystem to be eliminated or disintegrated by natural processes or to be converted to another substance(s).

half-space

A model used in physics and or mathematics to simplify reality: either of the two parts into which a plane divides a space, e.g. in simple geophysical models the Earth's surface separates the air half-space from the ground half-space.

half-width of a geophysical anomaly

If the anomaly of a geophysical field parameter (e.g., gravity anomaly) is depicted along a profile, the half-width of the anomaly (half the horizontal distance across an anomaly peak) is taken where the amplitude of the anomaly is half the maximum value. The half-width of an anomaly is used in the magnetic and gravity methods to estimate the depth of the anomaly source. The half-width of a peak is used in many other fields of physics.

hazard

In this context, a source of pollution.

HD

Abbreviation for "high density" used, for example, in PE-HD for high-density PE plastics.

head

A measure of water pressure expressed in terms of an equivalent weight or pressure exerted by a column of water. The height of the equivalent column of water is the head.

herbicide

A class of pesticides used to destroy or inhibit the growth of plants such as weeds. Selective herbicides kill certain targets while leaving the desired crop relatively unharmed. Herbicides used to clear waste ground are nonselective and kill every plant with which they come into contact.

heterocyclic compounds

Organic compounds that contain a ring structure containing other atoms in addition to carbon, such as sulfur, oxygen or nitrogen, as part of the ring. Examples of this class of organic substances are found, for example, among the nucleic acids, vitamins, hormones, pigments, pharmaceuticals, and pesticides.

heterogeneity

A characteristic of a medium in which material properties vary from point to point.

hexachlorocyclohexane (HCH)

A persistent insecticide. It does not break down easily in the environment, and can be transported great distances in the atmosphere.

high-performance thin-layer chromatography (HPTLC)

Chromatographic method in which the TLC plates are developed in special chambers. Development and analysis of the plates is largely automatic and, therefore, highly reproducible. Unlike other chromatographic techniques, HPTLC allows for the simultaneous analysis of multiple samples in one operation, which facilitates the processing of large numbers of sample. HPTLC is used, for example, in the analysis of PAHs in water and sewage. See also thin-layer chromatography (TLC).

high-performance liquid-chromatography or high-pressure liquid-chromatography (HPLC)

This is a liquid chromatographic method in which the stationary phase is packed in thin pressure-resistant columns (diameter 2 - 4 mm, length 10 - 20 cm, pressure approximately 300 bar) and the mobile phase is transported through the column with a constant flow by a high-pressure pump. Both polar materials (silica gel, ion exchanger or polymer based) and nonpolar materials ("reverse phase" (RP) silica gel modified by adding groups of hydrocarbons, e.g., C-18) can be used as the stationary phase. Good separation can be obtained with moderate pressures applied to substrates with defined particle size within a range of 3 - 10 μm. At the time of analysis several microliters of the analyte is brought into the stream of the mobile phase by means of a valve. The different substances in the sample are then separated chromatographically in the column and detected in the stream leaving the column by spectroscopic or electrochemical methods. The signal intensity is a function of the amount of substance and is evaluated after calibration of the system. HPLC is useful in water and wastewater analysis. There are some variations of HPLC: HPLC with fluorescence detector (HPLC-F), HPLC with UV detector (HPLC-UV), and HPLC with diode array detector (HPLC-DAD). See also liquid chromatography (LC) and thin layer chromatography (TLC).

HLEM, HEM

Horizontal loop electromagnetic method, a configuration in which both transmitter and receiver antennae are in the horizontal plane. Also called horizontal coplanar.

homogeneity

A characteristic of a medium in which its material properties are identical everywhere (LOHMAN et al., 1972).

homogeneous

1. Refers to anything which displays a uniform or consistent composition.
2. A characteristic of a soil/rock mass in which physical properties (e.g., density, hydraulic conductivity) are independent of location within the mass.

horizon

1. A distinct layer of soil encountered in a vertical section. 2. A particular stratigraphic level or time interval, definable geologically or by the fauna or artifacts in it. 3. The line where the sky and the ground seem to meet.

HRGC-HRMS

High-resolution gas chromatography coupled with high-resolution mass spectrometry. → gas chromatography, → mass spectrometry

humic substance

When plant or animal material decomposes, there are numerous steps in the breakdown of the organic matter. The macromolecular matter of the intermediate steps is called humus, and is an important part of a fertile soil.

humidity

The quantity of moisture (vapor) in the air. It can be expressed as absolute humidity, which is the mass of water in a specific mass of air or, more commonly, as relative humidity, which is the percent of moisture relative to the actual amount which air at any given temperature can retain without precipitation.

humidification

The process of adding water vapor to air.

humification

1. The process of decomposition whereby organic material is transformed and converted to humic substances through biochemical and abiotic processes and becomes humus. 2. The degree of decomposition of organic matter.

hydraulic conductivity

Volume of groundwater that will move within a unit of time under a unit hydraulic gradient through a unit cross-sectional area that is perpendicular to the direction of flow. Hydraulic conductivity is preferred to the term permeability. Hydraulic conductivity is the permeability to water only, whereas permeability is a general property applicable to different fluids. For a particular soil or rock layer, the hydraulic conductivity may not be the same in the horizontal direction as in the vertical direction.

hydraulic gradient

1. Rate and direction of change in the total head per unit distance of groundwater flow. 2. The difference in hydraulic head between two measuring points within a porous medium, divided by the distance between the two points. It is the driving force of fluid flow in a porous medium.

hydraulic head

The sum of the elevation head and the pressure head. (It is the same as "total head" in the case of groundwater.) Same as potentiometric/piezometric head. The pressure caused by the weight of an equivalent column of liquid upon a unit area expressed by the height or distance of the liquid above the point at which the pressure is measured. Although head is expressed as distance or height, it refers to the pressure resulting from the weight of a liquid since the weight is directly proportional to the height.

hydride generation AAS (HG-AAS)

→ atomic absorption spectrometry

hydrocarbons (HC)

A class of chemical substances made of hydrogen and carbon. Petroleum consists primarily of hydrocarbons. Some hydrocarbon compounds are major air pollutants. Chlorinated hydrocarbons are artificially made and generally toxic. See also petroleum hydrocarbons.

hydrogeology

The study of groundwater and its relationship to the geologic environment.

hydrograph

1. A graphical representation showing for a given point on a stream the discharge, depth (stage), velocity, or other property of water as a function of time. 2. A device which records the depth of water, as in a well, or flow, as in a stream. 3. An instrument that provides a continuous recording of relative humidity with time.

hydrologic cycle

The continuous circulation of water between the Earth and the atmosphere, through condensation, precipitation, runoff, percolation, evaporation, transpiration, groundwater storage and seepage.

hydrologic year

Same as water year.

hydrology

The science of the behavior of water in the atmosphere, on the Earth's surface, and in the subsurface.

hydrometer

A device used to measure density or specific gravity of liquids. It consists of a tube that weighted so that it floats vertically. The specific gravity of the liquid is shown by the level of the liquid on the scale on the tube. ASTM hydrometers are graduated by the manufacturer to read in either specific gravity or in grams per liter of a suspension and are calibrated at a standard temperature of 20 °C (68 °F). The device is used to determine, for example, the grain size of the suspended fine fraction of soil or sediment samples or the salinity of water.

hyperspectral scanner (in remote sensing)

Hyperspectral scanners are airborne imaging spectrometers. The term hyperspectral is used to indicate the large number of contiguous spectral bands. Whereas opto-mechanical and opto-electronical scanners commonly used for multispectral sensing have 4 - 10 spectral channels with bandwidths of 100 - 200 nm, hyperspectral scanners record images of the Earth's surface in 50 - 300 narrow bands with a bandwidths between 1 - 20 nm in the visible and near-infrared parts of the electromagnetic spectrum. Thus, a quasi-continuous spectral signature for each pixel or ground resolution cell can be obtained. Radiation-splitting optics, CCD linear, and two- and multi-dimensional detector arrays are used to measure the radiation intensity. Imaging spectrometry is based on reflectance spectroscopy – the study of the reflection, absorption and scattering of the solar radiation using laboratory and/or field spectrometers. The spectral features of minerals, rocks, vegetation types, manmade materials, water and dissolved organic compounds and environmental contaminants are recorded in digital spectral libraries. If these materials have unique and identifiable absorption features, they can be mapped by analysis of the imaging spectroscopy data. The wavelength region of interest is the portion of the electromagnetic spectrum in solar reflection band from 0.4 to 2.5 μm. Airborne imaging spectrometers provide data with a ground resolution of 0.5 - 20 m.

image

Used as a general term for a pictorial representation of a scene or object, typically produced by an optical or electronic device. The term is often restricted to representations acquired by nonphotographic methods. Common examples include remotely sensed data such as satellite data, scanned data, and photographs. An image is stored as a raster data set of binary or integer values that represent the intensity of reflected light, heat, or another range of values on the electromagnetic spectrum.

incineration (of waste)

Solid, liquid, and gaseous wastes can be disposed of by incineration – controlled burning at high temperatures. Hazardous waste is converted in this way to carbon dioxide, water and ash. Burning destroys organic substances, reduces the volume of waste, and vaporizes water and other liquids the wastes may contain. The ash must be considered as potentially hazardous waste.

indicator test

→ screening test

induced polarization (IP)

A method of geophysical surveying in which the slow decay of voltage is measured in the ground following the cessation of an excitation dc current pulse (time-domain method) or the change in the impedance of the ground is measured as a function of the frequency of a low-frequency (below 100 Hz) ac current (frequency-domain method).

inductively-coupled plasma (ICP)

A high-temperature conductive gaseous mixture, contained and energized by a radio-frequency electromagnetic field.

inductively-coupled plasma AAS (ICP-AAS)

→ atomic absorption spectrometry, → inductively-coupled plasma

inductively-coupled-plasma atomic-emission spectrometry (ICP-OES, ICP-AES)

Atomic emission spectrometry is similar to the atomic absorption spectrometry, a method for obtaining information about components and their concentrations in a sample. In ICP-OES, an aqueous sample is converted into aerosol by an atomizer and then is passed through hot plasma where it is vaporized and atomized. Under the plasma conditions, the atoms are brought into an excited state from which they revert to the ground state by emitting photons of element-specific wavelengths. The intensity of the emitted light is proportional to the concentration of the element to be analyzed in the sample. For the analysis of dissolved materials, inductively coupled argon plasma is generally preferred over other plasma types (e.g., microwave induced plasma, MIP). Generally, the plasma technique is much more applicable than flame AAS, since many elements can be tested at same time and with high precision (multi-element analysis with simultaneous and/or sequential spectrometers). Atomic emission spectrometry with inductively coupled plasma is suitable for qualitative and quantitative analysis of most of the elements. The analysis of dissolved materials can be accomplished simultaneously or sequentially, whereby the simultaneous element determination means substantial savings in measuring time. Owing to its lower detection limits, plasma spectrometry is being increasingly preferred to optical emission spectrometry.

inductively-coupled-plasma emission spectrometry (ICPS)

Based on the same basic principles as optical emission spectrometry (OES), but the generation of much higher temperatures reduces problems of interference and produces more accurate results.

infiltration

Movement of atmospheric precipitation into the soil. Infiltration depends on surface characteristics such as slope, vegetation cover, land use and soil properties such as porosity, soil water content etc. Secondary porosity (e.g., shrinkage cracks, pathways left by decayed roots, earthworm tunnels) is also of high importance.

infiltration capacity
The maximum rate at which infiltration can occur under given conditions. It is a function of soil moisture content. The unit is depth of water per unit time, for example mm h^{-1} or cm s^{-1}.

infiltration rate
The quantity of water that enters the soil in a specified time interval. It is expressed as depth of water per time unit, for example mm h^{-1} or cm s^{-1}.

infiltrometer
An instrument for measuring the rate of infiltration of water into soil.

inflow
1. Flow of water into a stream, lake, reservoir, container, basin, aquifer system, etc. 2. The process of flowing into something.

influent flow
Flow of water from a body of surface water (e.g., river, lake) to an aquifer. → bank infiltration

influent stream
A stream flowing above the water table that contributes water to the saturated zone. Syn. losing stream.

infrared (IR)
The part of the electromagnetic spectrum whose wavelength range is above that of the visible spectrum, but below that of microwaves extending from 0.78 to 1000 μm. It is the form of radiation used for making noncontact temperature measurements.

infrared spectroscopy (IR spectroscopy)
An analytical method based on the absorption by chemical compounds (typically organic compounds) within the infrared spectrum. The infrared absorption spectrum of a substance is called its molecular fingerprint. It is a technique for determining the molecular species present in a material and measuring their concentrations by measuring the intensity of the characteristic wavelengths at which the analyzed material absorbs infrared radiation. A typical application is the analysis of hydrocarbons.

inhomogeneity
A characteristic of a medium in which material properties are not identical everywhere. Opposite of homogeneity.

inhomogeneous
Opposite of homogeneous.

inorganic
Any chemical compound that does not contain the element carbon; carbon dioxide, carbon monoxide and carbonate compounds considered inorganic.

insecticide
A class of pesticides (substances or mixture of substances) intended to kill insects.

in-situ
Being in the natural or original position. Many experiments/methods are conducted in-situ, i.e., in the field.

instantaneous field of view (IFOV)
The cone angle within the incident energy is focused on the detector. All energy propagating toward the detector within the IFOV contributes to the detector response at any instant.

interaction
1. Generally, an interaction is a kind of action which occurs as two or more objects have an effect upon one another. 2. In physics, the transfer of energy between elementary particles or between an elementary particle and a field or between fields. 3. The effect of two or more substances acting on each other. 4. In a statistical model in which the effect of two, or more, variables is not simply additive.

interactive
Referring to computer programs or applications that respond directly to the user, taking instructions and giving feedback.

interferogram
In remote-sensing radar, an image that displays interference patterns created by superposing images acquired by two antennas that are separated by a short distance.

interferometry
A method for increasing resolving power by the use of interference of electromagnetic waves with each other.

interflow
That portion of rainfall that infiltrates into the soil and moves laterally through the unsaturated zone during or immediately after a precipitation event and discharges into a stream or other body of water. This usually takes longer to reach stream channels than runoff.

interpolation
Mathematical procedure for estimating unknown values from neighboring known data.

interpretation

Derivation of a plausible geological model that is compatible with observed data and other available information.

intrinsic permeability

The relative ease with which a porous or jointed medium can transmit a fluid under a hydraulic or potential gradient. It is a property of the medium and is independent of the nature of the fluid or the potential field.

inventory

1. A detailed list of all items with selected characteristics. In the context of environmental investigations, a list of sites suspected to be hazardous and/or a list of chemicals produced at a site. 2. The process of preparing such a list.

inversion

Inversion is the derivation of a model from measured data. This is normally done with a computer program. Beginning with an initial guess (starting model) the model is changed by the program iteratively until the calculated data corresponds within certain limits to the measured data. The program modifies the model parameter values according to an algorithm. Individual parameters can be "fixed". The use of a-priori information (fixed parameters) can improve the convergence behavior and the reliability of the models.

iodometry

A method of volumetric chemical analysis commonly used to determine the concentration of chlorine in water.

ion balance

Because solutions of ions are electrically neutral as a whole, there must be as many positive charges as there are negative ones. This also means that the sum of the equivalent concentrations of all cations must equal the sum of the equivalent concentrations of the anions. This calculation is called an ion balance. Usually, because of sampling and analysis errors and because not all, but only the more important species are analyzed, the ion balance turns out to be different from zero. However, the ion balance error IBE (or charge balance error CBE) *IBE* = *CBE* = (*sum of cations* – *sum of anions*) / (*sum of cations* + *sum of anions*) should be < 5 % to accept an analysis as valid (general practice, FREEZE & CHERRY, 1979). Otherwise, the analysis should not be trusted as is, and reasons for the error should be investigated.

ion chromatography (IC)

Ion chromatography is a special case of high performance liquid chromatography in which ionic substances can be separated into ions and quantified. The IC separation columns are usually filled with polymer materials (resins) capable of ion exchange (anion exchangers: e.g., quaternary ammonium bases NR_4^+; cation exchangers: e.g., sulfonic acids RSO_3^-). The mobile phase normally contains a buffer system (e.g., carbonate, phosphate, or carbonic acids). Electrochemical or spectroscopic techniques are used to identify the separated substances. Ion chromatography is particularly used for water and waste water analysis to determine concentrations of fluoride, chloride, nitrate, nitrite, sulfate and phosphate ions using conductivity detectors (SMALL, 1989; WEISS, 1991).

ion exchange
A chemical process in which ions (anions or cations) are selectively removed from solution and replaced by other ions by interaction with a stationary substrate. Water softening is a common application of ion exchange.

ion conductor
→ conductor → electrolyte

isanomalies
Line on a map connecting points of equal anomaly values. → isolines

isoline
A line in a map which connects points of equal value (e.g., line of equal gravity anomaly values – isogal; line of equal magnetic field intensities – isogam; line of equal (apparent) resistivity – isoohm). For some other kinds of isolines, there are also special terms:

- isobar: line of equal barometric pressure,
- isobath: line of equal depth to the water table,
- isochore: line of equal interval between two beds or two seismic events,
- isochron: line of equal reflection time, equal time difference, or equal delay-time in seismic,
- isohypse: line of equal elevation (e.g., above mean sea level),
- isopach: line of equal thickness of a rock type, formation, group of formations, etc., and for equal distances between two seismic reflectors,
- isopleth: line of equal geochemical value (e.g., contaminant content).

isotherm
→ adsorption isotherm

isotope
Atoms of an element having the same number of protons (atomic number), but different numbers of neutrons. Isotopes of the same element have the same atomic number but have different atomic masses. Isotopes have the same chemical and gross physical properties, but can have greatly different nuclear properties.

isotropic

Having physical properties or field values (e.g., intensity of electromagnetic radiation) that do not vary with direction.

iteration

Repeated execution of the same sequence of instructions in a calculation, each time varying the value of parameter(s) by a certain amount in an attempt to improve the accuracy of the result.

joint

A fracture, crack or parting in rock without any apparent displacement. Often parallel joints form a joint set. See also jointing.

joint aquifer

An aquifer consisting of solid rock with predominantly fracture permeability.

jointing

Fractures in rock along which there has been no movement.

joint inversion

1. Simultaneous inversion of two or more data sets of related measurements with different methods. 2. In site investigations, usually more than one method is used in the same area in order to improve the reliability of the "site model". Data from individual methods are commonly processed separately and jointly interpreted. The joint interpretation can be improved by using joint inversion if certain prerequisites are fulfilled. The data observed by two or more methods and used for joint inversion must be affected by the same object(s), i.e. they must contain information at least partly from the same depth and/or volume range and at least one rock parameter in common.

karst

A type of topography formed on limestone, gypsum, and other rocks, primarily by dissolution, and that is characterized by sinkholes, caves, and underground drainage. Karst formations are not only present in sediments but also in metamorphic rocks (e.g., marble). Karstification may have occurred in the past (paleokarst) or may be still in progress. Karstification is often associated with anthropogenic causes, such as mine pumping, flooding of mines, brine production, etc. Karst regions are particularly vulnerable to contamination because contaminants seep quickly into the groundwater.

kinematic viscosity

A coefficient defined as the ratio of the dynamic viscosity of a fluid to its density. The SI unit for kinematic viscosity is $m^2\ s^{-1}$.

kriging

An interpolation method for obtaining statistically unbiased estimates for field attributes (values of geophysical anomalies, groundwater levels, elevation) from a set of neighboring points. Named after the South African mining engineer D. G. Krige.

laminar flow

A type of smooth, viscosity-controlled flow at relatively low velocity in which the surface, bed, and internal flow vectors are all parallel to one another, thus there is no mixing. In contrast to turbulent flow.

landfill

A site for disposing of solid waste on land for an unlimited time. State-of-the-art landfills are constructed according to the multibarrier safety concept so that it will reduce the risk to public health and safety. There are three main kinds of landfills: (1) Sanitary landfills are controlled disposal sites for nonhazardous solid wastes at which the waste is spread in layers, compacted to the smallest practical volume, and covered with material at the end of each operating day. (2) Secure chemical landfills are disposal sites for hazardous waste. They are designed to minimize the chance of release of hazardous substances into the environment. (3) Old landfills were built without modern day protections. They may contain hazardous waste.

landfill gas

Gas that is produced when organic waste materials naturally decompose in a municipal solid waste landfill. Landfill gas is approximately 55 % methane, the primary component of natural gas, and 45 % carbon dioxide. Besides methane and carbon dioxide, landfill gas contains ammonia, carbon monoxide, hydrogen, hydrogen sulfide, nitrogen and oxygen. Landfill gas can be collected and used as a fuel for heating or generating electricity.

Landsat

The Landsat program, operated by the US Earth Observation Satellite Company (EOSAT), is the longest operating enterprise for acquisition of imagery of Earth from space. The first Landsat satellite was launched in 1972 and the most recent, Landsat 7, was launched 1999. The images, archived in the United States and at Landsat receiving stations around the world, are a unique resource for applications in geology, agriculture, forestry, regional planning.

landslide

1. The downslope movement of soil and/or rock under the influence of gravity. It is often indicated by a distinct surface rupture or zone of weakness which separates the slide material from underlying more stable material. 2. It also describes the landform that results.

land use

The range of uses of the Earth's surface by humans. Land use classes include urban, rural, agricultural, forested, with more classes and subclasses for specific purposes. It affects the amount and character of runoff, groundwater vulnerability, and erosion.

laser scanning (in remote sensing)

Airborne laser scanning is an active remote sensing system. A pulsed laser beam scans the Earth's surface in a strip across the flight line. The pulse frequency and the speed of the aircraft determine the measured point density and the resulting mesh size of the data grid. An opto-mechanical scanner with rotating mirror guides the laser beam across the flight path. The path of the reflection points is a zigzag line due to the forward movement of the aircraft and the sideward oscillation of the scanner's mirror. Complicated data processing derives the coordinates (position and elevation) of each terrain point. Data collection is possible even during the night and under cloudy conditions. The current standard for image accuracy is about 15 cm horizontally and 5 - 50 cm vertically for a flying altitude of 1000 meters above ground during data acquisition.

layer

A body of rock of relatively greater horizontal than vertical extent that is characterized by properties different from those of overlying and underlying materials, usually sedimentary or volcanic. Sometimes used synonymously with bed for a sedimentary rock.

leachate

Water that has percolated through solid material, e.g., the waste in a landfill, and has dissolved substances from it. Chemicals such as fertilizer are leached from the soil by percolating rainwater.

leaching

The process by which soluble organic and inorganic substances in the soil, such as salts, nutrients, pesticide chemicals or contaminants, are washed into a lower layer of soil or are dissolved and carried away by water.

leak

A hole or crack that allows something, such as gas or liquid, to escape its boundaries (e.g., in a landfill seal system).

leakage

The act of leaking, e.g., the flow of water from one hydrogeologic unit to another. The leakage may be natural, as through a semi-impervious confining layer, or man-made, as through the walls of an uncased well. Another example is the flow of water from a landfill or sewer to an aquifer. In the case of geophysical measurements it is the unintended passage of electric current that causes noise in the measurements.

leaky aquifer
An aquifer that loses or gains significant quantities of water through semipermeable confining beds.

less-dense-than-water nonaqueous-phase liquid (LNAPL)
A liquid that is not soluble in water and has a density less than that of water.

limit of detection (LoD)
→ detection limit

limit of determination
→ detection limit

limit of quantitation (LoQ)
Lowest concentration of analyte that can be determined with an acceptable degree of accuracy and precision. It is usually the lowest point on a calibration curve under exclusion of a blank experiment.

lineament
A linear topographic feature that may be due to geological structures in the crust.

linear
1. Relating to line or lines. 2. Straight line-like features on the terrain, on images or photographs. 3. A process is linear if a plot of input versus output is a straight line.

lineation
Any linear arrangement of features found in a rock.

liner
1. A relatively impermeable artificial barrier designed to avoid groundwater recharge and keep leachate inside the landfill. Liner materials include geo-textiles and dense clay. 2. A string of pipe used to case an open well hole below existing casing. 3. A cylinder within the drillstem just behind the bit used to recover undisturbed cores from boreholes in unconsolidated rock.

lining
Covering a landfill with a relatively impermeable barrier designed to avoid groundwater recharge and keep leachate inside the landfill.

liquid chromatography (LC)
This is a chromatographic method using a liquid mobile phase. The modern variant is high-pressure/performance liquid chromatography (HPLC).

lithology

Mineralogical composition, color, grain size, texture, and other physical properties of soil, sediment, or rock.

lithostratigraphy

The classification of rock layers on the basis of their lithology and stratigraphic position.

log

1. Well log, a continuous record of a geophysical measurement as a function of depth in a borehole or well. 2. To carry out borehole logging. See also borehole logging. 3. Listing that contains a record of events. 4. Abbreviation and symbol for logarithm to the base 10.

lysimeter

1. An instrument to investigate the infiltration of water in the unsaturated zone by measuring the quantity or rate of downward water movement through a block of soil usually undisturbed. 2. Device to collect soil water in the unsaturated zone for analysis.

magnetic method

Everywhere on Earth there is a natural magnetic field which moves a horizontally free-moving magnetic needle (magnetic compass) to magnetic north. The magnetic field is a vector field, i.e., it is described by its magnitude and direction. The magnetic field consists of three parts: the main field, a fluctuating field, and a local anomaly field. The main field, whose origin is within the Earth, varies very slowly with time (years to decades) and is superimposed by a rapidly varying (in fractions of seconds to days) field component, whose origin is outside the Earth (external field, time-varying field). In addition to these geomagnetic field components, there is an almost constant local anomaly field, which results from the magnetization of material in the upper crust. Not only do geological structures, consisting of, for example, basalt, metamorphic rocks, and some ore deposits, cause local magnetic anomalies, but also metal objects at waste disposal sites. The magnetic flux density has values up to 50 000 nT for the main field, 0.1 to 100 nT for the fluctuating field, and up to several 1000 nT for the local anomaly field. The local anomaly field is estimated from magnetic measurements and conclusions are made about the sources. To obtain the local anomaly field, the main field and the time-varying field must be eliminated from the measured total field. The local anomaly field is represented on an isoline map, (pseudo 3-D plot) or along profiles. The sources of the magnetic anomalies are interpreted from these maps and, in some cases, by two and/or three-dimensional modeling.

magnetic permeability

The ratio of the magnetic flux density in a substance to the external field strength. In SI units, permeability is expressed in henrys per meter ($H\ m^{-1}$).

magnetic susceptibility

The ratio of the induced magnetic field to the strength of the magnetic field causing the magnetization. In SI units, magnetic susceptibility is dimensionless.

magnitude (of earthquake)

Magnitude (abbreviated M) is a measure of the energy released by an earthquake. The term was introduced by C. F. Richter in 1935. Magnitude scales, like the Richter scale, measure the strength of an earthquake at its source. Like all later defined scales, the Richter scale is logarithmic: A difference of 1 means an energy difference of thirty. The strongest recorded earthquake occurred in Alaska with a magnitude of 8.8.

map

1. n. A diagram, drawing, or other graphic representation of selected physical features (e.g., gravity anomalies) of a part or the whole of the surface of the Earth or any desired surface or subsurface area, by means of signs and symbols, so that the relative position and size of each feature on the map corresponds to its correct geographic location according to a definite and established scale and projection. If the top of the map does not show true north, an arrow must indicate its direction. 2. v. The transfer of geoscientific data (e.g., gravity anomaly values) into a graphic representation (e.g., gravity anomaly map).

mapping

1. The term mapping is often used synonymously for the preparation of a two-dimensional representation of a physical parameter (e.g., elevation of the terrain, apparent electrical resistivity, total intensity of the magnetic field). 2. In geophysical surveying, → profiling.

map projection

A mathematical model for converting locations on the Earth's surface from spherical to planar coordinates, allowing flat maps to depict three-dimensional features. Some map projections preserve the integrity of shape, others preserve accuracy of area, distance, or direction.

mass spectrometry (MS)

An analytical technique used to separate compounds based on differences in their molecular mass, as well as to identify unknown compounds, quantify known compounds, and determine the structure of molecules. The substance to be analyzed is vaporized, ionized and then the ions are accelerated through a magnetic field to separate the ions by molecular weight. Mass spectrometry can result in the exact identification of an unknown compound, and is a very powerful analytical technique, especially when combined with chromatography. This technique can identify and quantify extremely small amounts of compounds on the basis of their mass-fragment spectrum. Common types of mass spectrometry include FAB (fast atom bombardment), ESI (electrospray ionization), MALDI (matrix assisted laser desorption), TOF (time of flight), FT (Fourier transform), and ICP (inductively coupled plasma). Mass spectrometry is also often combined with other analytical methods, such as gas chromatography (GC) and liquid chromatography (LC).

mass spectrometry with inductively coupled plasma (ICP-MS)
In ICP-MS, the sample is ionized by an ICP and the resulting ions are analyzed in a mass spectrometer. This method allows for the determination of a wide range of elements, and has a low detection limit. This method is increasingly used as a replacement of GF-AAS or ICP-AES owing to its higher throughput, superior detection capabilities, and the ability to obtain isotopic information. On the other hand, analysis of trace elements in samples with high level of chloride is difficult. In inductively coupled quadrupole plasma mass spectrometry (ICP-QMS) ions are electrostatically extracted from the ionization chamber and introduced into a quadrupole mass filter.

matric potential
A water potential component, always negative, resulting from capillary and adsorptive forces. Matric potential is another term for pressure potential or pressure head. Also called matrix potential.

matrix
1. The solid material in a porous rock or soil. 2. A tabular representation of a set of data. It is characterized by its dimensionality measured by the number of rows and columns. Matrices in equations are normally symbolized by bold capital letters.

mean sea level (MSL)
The average height of the sea surface, based upon hourly observation of the tide height over a long period of time on the open coast or in adjacent waters that have free access to the sea. In the United States, it is defined as the average height of the sea surface for all tidal stages over a nineteen year period. It is usually the datum considered zero elevation used in the surveying of land. Elevations above zero elevation are positive and those below zero elevation are negative.

messenger RNA (mRNA)
RNA that serves as a template for protein synthesis.

metabolism
The chemical changes that take place in living organisms by which substances are transformed to obtain energy or converted to other substances.

metabolites
Substances produced by the metabolism of other substances.

metallic conductor

→ electronic conductor

method detection limit (MDL)

→ limit of detection

methyl tertiary butyl ether (MTBE)

A gasoline additive which increases octane rating. Has been required to be phased out.

microwave

Electromagnetic radiation with a wavelength between 1 mm and 30 cm.

migration

The movement of fluids and dissolved substances through permeable rock.

microorganisms

Single-celled organisms, also known as microbes, which are so small that they can only be seen under the microscope, including bacteria, simple fungi, protozoans, yeasts, viruses, and algae. Most are beneficial but some produce disease. Some are involved in composting and sewage treatment. When alive in a suitable environment, they grow rapidly and may divide, i.e., reproduce every 10 to 30 minutes, reaching large populations very quickly.

mineralogy

The study of the composition, structure, appearance, stability, occurrence, associations, and physical and chemical properties of minerals.

mobile analysis

Usually during investigation of contaminants and their concentrations in the environment, samples are taken in field and transported to the laboratory. Transport, storage and sampling conditions introduce changes in physical and chemical composition of the samples. In order to avoid some of these error sources the use of mobile analysis methods (Syn. on-site analysis, field screening) has increased in the recent times, which give quick on-site results. As compared to conventional laboratory-supported analysis methods, the methods of mobile analysis are generally characterized by simpler sample preparation and measuring techniques, which result in shorter analysis times and, therefore, analysis of more samples. Methods for mobile analysis range from simple tests to highly-sophisticated measuring techniques.

mobilization

1. The release and migration of substances that are bound to soil or rock as a result of physical, chemical and/or biological processes. See also desorption. 2. In geophysical field work all operations conducted to

prepare for field work and move staff and equipment to the investigation area.

model

A (simplified) representation of reality is called a model. There are, for example, conceptual, physical, and mathematical models. A conceptual model is a mental image of an object, system, or process. It consists normally of a simplified, schematic written description and a visual representation. A physical model is used in various contexts to mean a physical representation of some thing. The representation of the propagation of seismic waves in a bedded half-space is a physical model. The representation of the subsurface by a sequence of coplanar strata and the representation of the embedding a sphere in a half-space are mathematical models. In mathematical models, the distribution of parameter values (e.g., seismic longitudinal wave velocity) is assumed, e.g., in seismic model calculations in order to calculate the propagation of seismic waves using the wave equation. The results of the model calculations can then be compared with the measured values (e.g., seismograms). Simple models are often constructed of bodies with a simple shape in a homogeneous environment. More complicated models of geological reality are developed by assigning parameter values (e.g., electrical conductivity or seismic wave velocity) to points of a grid. Models are made with different dimensionality (one-dimensional model, two-dimensional model, etc.). → model with simple shape, → numerical model, → forward modeling

model calculation

→ model, → model with simple shape, → numerical model, → forward modeling, →inversion.

model calibration

The procedure that changes model parameters so, that the model matches field conditions within some acceptable criteria.

model design

Model design is the process of translating a conceptual model into a numerical model by defining grids, establishing boundaries, and providing model parameter values for each computational cell. Grid design differs greatly between finite-difference and finite-element models, and is handled differently by different software.

modeling

→ model, → model with simple shape, → forward modeling, → numerical model, → inversion.

model validation
The process by which a model is compared to known analytical solutions or to measured values.

model with simple shape
The simplest model of the geological underground is the homogeneous half-space. Deviations from this homogeneous underground are often represented in models as simply shaped bodies (sphere, cylinder, prism, etc.). Geological bodies, however, can have much more complex forms. They cause local differences in the values of geophysical fields (and thus anomalies). In many methods, the simple shape model concept has become obsolete because of the highly developed computer programs that allow the use of more complicated models.

moisture
Water dispersed or condensed in air, soil, etc. in very small amounts.

moisture content (in soil or rock)
The amount of water in a soil or rock, expressed as a percentage of the dry weight (weight of a soil sample after drying it to a constant weight).

moisture meter
A device used to measure moisture content, usually by measuring electrical resistivity.

moisture potential
The tension on the pore water in the unsaturated zone due to the attraction of the soil-water interface.

molality
The number of moles of solute dissolved in one kilogram of solvent. There are two key differences between molarity and molality. Molality uses mass rather than volume and uses solvent instead of solution. Unlike molarity, molality is independent of temperature because mass does not change with temperature.

molarity (M)
The number of moles of solute dissolved in one liter of solution. For instance: 4.0 liters of solution containing 2.0 moles of a dissolved substance is a 0.5 molar solution, written 0.5 M. Working with moles can be highly advantageous, as they enable measurement of the absolute number of particles in a solution, irrespective of their weight and volume or ionic charge.

mole

SI unit of the amount of substance, the symbol is mol. One mole contains as many elementary entities (e.g., atoms, molecules) as there are atoms in 12 grams of Carbon-12 (^{12}C). It is always necessary to specify the compound when giving a mole value (e.g., 5 moles of hydrochloric acid).

monitoring

Periodic or continuous observation of a parameter, to determine the level of compliance with statutory requirements, to determine pollutant levels in soil, air or water, or to determine changes in water level, streamflow, or other parameters. The area around both vulnerable sites (e.g., waterworks) and hazardous sites (e.g., landfills and industrial plants) has to be monitored for contaminants. The integrity of contaminant barriers and the success of rehabilitation of contaminated sites must also be monitored. The spread of contaminants occurs via three paths: water, soil, and air. Of these, groundwater and surface water are the most important. Current practice is to take water samples from observation wells and soil and gas samples at vulnerable and hazardous sites at regular intervals for chemical analysis on site and/or in the laboratory. It is also current practice to use multi-parameter probes to monitor several environmental parameters in wells, such as temperature, electrical conductivity, pH, redox potential and oxygen concentration in the water. Repeated visits to observation wells are personnel intensive and hence expensive. Further disadvantages are the uncertainties deriving from the sample collection, transport, and the chemical analysis, in addition to the masses of data produced. The advantages are that numerous substances can be analyzed and that the chemical analyses can be used in court. All of these procedures yield point data. To adequately monitor a site, numerous sampling points are necessary. A new technological development permits continual in-situ monitoring over long periods of time. The concept is based on a combination of measurements made with multi-parameter probes and an optical sensor system to obtain point data together with an electromagnetic (EM) system to obtain spatial data.

monitoring well

1. A well used to measure water quality or groundwater levels continuously or periodically. 2. A well drilled at a hazardous-waste management facility or Superfund site to collect groundwater samples for physical, chemical, or biological analysis to determine the amounts, types, and distribution of contaminants in the groundwater beneath the site (USEPA, 1994).

mosaic

A composite image or map made from parts from more than one source covering adjacent areas. See also photomosaic.

mud cake

The layer of concentrated solids from the drilling mud or cement slurry that forms on the walls of the borehole opposite permeable formations; also called filter cake or wall cake.

multi-barrier concept

This concept for planning, construction, operation and follow-up measures is based on the use of several largely independent barriers which are supposed to make a landfill safe and environmentally compatible. The barriers include the landfill body, the landfill seal and drainage system (engineered barrier), and the landfill site (geological barrier). The multi-barrier concept for landfills was introduced in Germany in 1993 with the "Technical Instructions on Wastes from Human Settlements" (TASi).

multi-parameter probe

An instrument for measuring several environmental parameters at the same time, such as electrical conductivity, temperature, pH, redox potential, and oxygen concentration. The information about the quality of groundwater and surface water can be derived from environmental parameter values.

multispectral scanner (MSS)

A remote-sensing device carried in aircraft or satellites that simultaneously acquires images of the same scene in several wavelength bands.

municipal solid waste (MSW)

Waste generated in households, commercial establishments, institutions, and businesses. MSW includes used paper, discarded cans and bottles, food scraps, yard trimmings, and other items. Industrial process wastes, agricultural wastes, mining waste, and sewage sludges are not MSW (USEPA, online glossary).

nadir

1. The point directly below an observer. 2. The point on the ground directly in line with the remote sensing system and the center of the Earth. Opposite of zenith.

natural attenuation (NA)

The term generally refers to physical, chemical, or biological processes that under favorable conditions lead without human intervention to the reduction of mass, toxicity, mobility, volume or concentration of contaminants in soil and/or groundwater. The reduction takes place as a result of processes such as biological or chemical degradation, sorption and others (USEPA, 1999).

near-field

For explanation see far-field.

noise

A disturbance that affects a signal distorting the information carried by that signal, e.g., geophysical and remote-sensing data. Examples: The movement of trees in the wind or traffic causes interference in seismic registrations. Electrical noise fields caused by electric trains, power lines and other current consumers affect electrical, electromagnetic and seismic measurements. Electronic noise from measuring instruments affects geophysical and remote-sensing data. "Geological noise" is caused by small-scale changes in geological conditions than cannot be resolved or registered by geophysical measurements or considered in a model.

nonphotographic imaging systems

Nonphotographic imaging systems detect the incoming electromagnetic (EM) radiation with semiconductor detectors or special antennas instead of using light-sensitive film. Whereas photography is limited to the spectral range 0.3 to 0.9 μm, the range of a multispectral scanner (MSS) covers wavelengths from 0.3 to approximately 14 μm. An MSS can use very narrow spectral bands. Nonphotographic imaging is done with either passive or active methods to obtain geometric, geological, and environmental information from the Earth's surface. Passive methods detect the natural electromagnetic radiation reflected, scattered, or transmitted at the Earth's surface. In active methods, EM radiation is transmitted (e.g., laser beam or radar wave) and the response of the ground is detected. The most important nonphotographic imaging systems for geoscientific site investigations are optical-mechanical line scanners, often called whisk broom scanners; optical-electronic line scanners, often referred to as charge-coupled devices (CCD); linear scanners, called push broom scanners, digital cameras (CCD arrays); imaging spectrometers; microwave, mainly radar-based systems; and laser scanners. Most of them are operated from aerial platforms as well as from space-borne platforms. Space-based imaging systems operate in almost the same manner as airborne systems, except that there is a greater distance between the sensor and the object being evaluated and the methods of data transfer are different. There are some limitations on the use of space-borne data. For geologic applications, data acquisition is sometimes restricted to a particular season and the study area must be free of clouds. Thus, the chances of adequate data capture are somewhat reduced, because the Landsat Thematic Mapper, for example, covers the same site on the Earth only every 16 days, which may not coincide with optimum weather conditions. Bad weather might result in delaying data acquisition or requires the use of archival material. Processing and interpretation of nonphotographic data require special sophisticated software.

nuclear magnetic resonance (NMR)

NMR is a physical phenomenon involving the interaction of atomic nuclei placed in an external static magnetic field with an applied electromagnetic field oscillating at a particular frequency. The conditions within a material are measured by monitoring the radiation absorbed and emitted by the atomic nuclei. NMR methods are applied in the laboratory.

numerical model

The behavior of a system, e.g., groundwater flow in aquifers, is commonly described by differential equations. In numerical modeling the differential equations are approximated by simpler mathematical forms (e.g., difference equations) suitable for solution on a digital computer. The continuous parameters (length, breadth, height, time, physical properties, etc.) are replaced by a finite number of discrete space and time steps and discrete parameter values. The subsurface area to be modeled is replaced by a two- or three-dimensional cellular structure. The model is improved by changing parameter values iteratively and solving the equations of the model anew. Finite-difference and finite-element models are frequently used for numerical modeling. See also finite-difference model, finite-element model, model design.

oblique photograph

Aerial photography may be divided into two major types, the vertical and the oblique. Each type depends upon the attitude of the camera with respect to the Earth's surface when the photograph is taken. An oblique photograph is taken with the camera inclined about 30° to 60° from the vertical.

observation well

A well drilled into an aquifer for the purpose of obtaining water level, water temperature, or water quality data.

one-dimensional model (1-D model)

In this model there are changes of parameters in only one coordinate direction (e.g., the electrical conductivity is only a function of depth below the observation point). Often the subsurface is approximated by a horizontal layer model (layers coplanar with the Earth's surface).

open hole

Any well/borehole or part of it in which casing has not been set. → cased hole

optical axis

An imaginary line that runs through the focus and center of the lens of a camera.

optical emission spectrometry (OES)

A method used to analyze sample composition based on the principle that electrons when excited (i.e. heated to a high temperature) release light of a particular wavelength. The presence or absence of various elements is established by examining the spectral lines characteristic of those elements. This method generally gives an accuracy of only 25 % and has been superseded by ICPS (inductively-coupled-plasma emission spectrometry).

optical square

A double prism used in land surveying for setting out survey lines at right angles. It allows to view an object straight ahead and another perpendicular to it.

optosensor system

Such sensor systems permit simultaneous measurement of transmission/absorption, scattering, fluorescence, and refraction. (a) Transmission: Molecules absorb radiation energy and convert it to another form (e.g., heat or radiation at another frequency). The amount of absorbed energy depends on the concentration of the substance. The concentration of substances such as phenols, BETX, dyes, explosives, and nitrate can be determined in this way. Humic acids also absorb light and can serve as an indicator of a pulse of seepage water from a landfill. (b) Radiation can also be scattered by molecules and particles. The extent of scattering depends on concentration and particle size and provides a measure of turbidity and macromolecule content. (c) Fluorescence is the emission of radiation after excitation of a molecule by absorption of radiation energy. The intensity of fluorescence is proportional to the concentration of the fluorescing substances. Because few substances fluoresce, e.g., PAH, this parameter is selective for certain groups of substances. (d) When radiation enters a liquid, it changes its velocity and direction. This is called refraction. The index of refraction depends on the liquid and the substances in it. Refraction measurements can be used to determine substances that do not absorb in the UV/VIS/NIR spectral range, e.g., salts and fatty acids.

orbit

The path of a satellite, planet, or heavenly body around another, larger, body in space under the influence of gravity. For example, the Earth is in orbit around the Sun. The Moon is in orbit around the Earth. Orbits differ in their eccentricities (deviation from circular).

ordnance

A general term for military equipment of all kinds including, bombs, artillery projectiles, rockets and other munitions; military chemicals, bulk explosives, chemical warfare agents, pyrotechnics; explosive waste, etc. See also UXO.

organic
1. Of or related to a substance that contains carbon atoms linked together by carbon-carbon bonds. 2. Of or relating to or derived from living organisms, e.g., organic soil matter.

organoleptic
Relating to the senses (taste, color, odor, feel). Testing techniques are considered organoleptic if investigators perform a variety of procedures such as visually examining, feeling, and smelling.

orthogonal
1. Pertaining to or involving right angles (90 degree angles). Two curves are orthogonal at a point P where they cross if the tangents at P are perpendicular to each other. 2. The property of an experimental design which ensures that different measured parameters are statistically unrelated.

orthophotograph
An aerial photograph that has all the distortions due to camera tilt and perspective as well as surface relief removed. An orthophotograph has the advantages of a photograph (all visible features displayed) and the constant scale and accuracy of a map.

outcrop
That part of a rock unit that is exposed at the surface of the Earth, includes bedrock covered only by surficial deposits such as alluvium.

out-of-phase
Not in the same phase, or not synchronized or coordinated with each other. Two or more periodic signals (e.g., electromagnetic signals) with identical periods (or equivalently, identical frequencies) are out-of-phase if their periods are not synchronized.

overburden
1. Soil or rock overlying a valuable mineral deposit that must be removed before the mineral can be removed in open pit or strip mining. 2. Unconsolidated material overlying bedrock.

overland flow
The flow of rainwater or snowmelt over the land surface in a down-gradient direction toward stream channels which does not infiltrate into the soil.

oxidation
Any process in which electrons are removed from an atom or ion (for example, adding oxygen or removing hydrogen). It always occurs accompanied by reduction of the oxidizing agent. Oxidation occurs at the anode in an electrochemical reaction. The opposite of reduction.

oxidation-reduction potential (ORP)
Ability of water to oxidize or reduce. The ORP is measured in millivolts (mV), with positive values indicating oxidizing behavior and negative values indicating reducing behavior. Reference point is the "standard hydrogen electrode", by definition having an ORP of zero. Also known as → redox potential.

packer
An inflatable "sleeve", used to create a temporary seal in a borehole or well. It is made of rubber or other nonreactive materials. Single and double packers are used. Two rubber cuffs are connected by a steel rod which can be equipped with pressure transducers and valves. The sleeves are filled with air until a proper seal is attained.

packer test
An aquifer test performed in an open borehole; the interval of the aquifer to be tested is sealed off by packers above and below the interval of interest in the borehole.

panchromatic film
Photographic black-and-white film sensitive to all visible wavelengths that records all colors in shades of about the same relative brightness as the human eye sees in the original scene.

parallax
1. The apparent displacement of an object as seen from two different points which are not on a line with the object. 2. The angle between the directions in which an object is seen from two different positions. The parallax of an object seen with the left and right eye enables depth perception. 3. Apparent difference in the position of a target on two consecutive aerial photographs.

partially penetrating well
A well in which the screen covers only part of an aquifer. See also fully penetrating well.

particle density
The density of the soil/sediment/rock particles. The dry mass of the particles divided by the solid (not bulk) volume of the particles, in contrast to bulk density, which includes the pore volume. Used in the determination of porosity and degree of saturation of soil/rock. Units are kg m^{-3} or g cm^{-3}.

pass point
A point whose horizontal and/or vertical position is determined from photographs by photogrammetric methods and which is intended for use as a control point for the orientation of other photographs. → control point

PCB

Polychlorinated biphenyl, a synthetic organic chemical once widely used for manufacturing plastics, in electrical equipment, in specialized hydraulic systems, heat transfer systems, and other industrial products. Highly toxic and persistent environmental pollutant and tends to accumulate in animal tissues. Sale was banned in the US in 1978.

PCDD/F

Polychlorinated dibenzo-p-dioxins and dibenzo-furans, some of these chlorinated hydrocarbons are extremely toxic.

pellicular water

Water adhering to soil particles in the unsaturated zone in layers more than 1 or 2 molecules thick and cannot be removed by gravity but is accessible to absorption by roots.

perched aquifer

A permeable layer containing unconfined groundwater separated from the main aquifer below it by a layer of impermeable material.

perched groundwater

→ perched aquifer

percolation

In hydrologic terms, the downward movement of water, under hydrostatic pressure, through the interstices of soil, sediment or rock. Percolation is expressed in terms of permeability by distance per unit time (for example $cm\ a^{-1}$).

percolation zone

→ vadose zone

percussion drilling

1. → direct push testing. 2. The term percussion drilling is also used for drilling with cable tools, i.e. the lifting and dropping of a heavy cutting tool will chip and excavate material from a hole.

permeability

A measure of the ability of a fluid (such as water, gas or oil) to flow through a rock formation when the formation is saturated with the fluid. The permeability of a rock filled with a single fluid is different from the permeability of the same rock filled with two or more fluids. Usually measured in millidarcies (mD) or darcies (D). (1 darcy = 10^{-12} m^2; 1 millidarcy = 10^{-3} darcy). → hydraulic conductivity, → effective permeability, → intrinsic permeability, → relative permeability, → fracture permeability, → darcy; magnetic permeability

permeameter

An instrument for determining permeability of soil, sediment and rock by measuring the discharge through the material when a known hydraulic head is applied.

permittivity

Describes the dielectric behavior of a material. It indicates the ability of a material to store electrical energy when a voltage is applied to it. It is the ratio of the electric flux density produced by an electric field in a medium to that produced in a vacuum by the same field. Also referred to as dielectric constant.

pesticide

A substance or mixture of substances intended to kill any species designated a "pest", including weeds, insects, rodents, fungi, bacteria, or other organisms. The term pesticide includes herbicides, insecticides, rodenticides, fungicides, and bactericides.

petrography

The description of rocks and their textures.

petroleum hydrocarbons (PHC)

The hydrocarbons found in or derived from crude oil (petroleum as taken from the well). Petroleum hydrocarbons are found in gasoline, oil, diesel, and jet fuels. Includes hundreds of chemical compounds.

pF value

Water tension Ψ is also reported as *pF* value, defined as $pF = \log_{10}\Psi$, where $\Psi = \Psi_\sigma + \Psi_m$ is the total potential of the unsaturated zone and Ψ_σ is the gravitational potential and Ψ_m is the matrix potential. Ψ can be given in [cm] or [Pa] or [mbar] units. The *pF* curve or water retention curve is a graphical representation of water tension Ψ of a soil as a function of water content θ.

pH

Abbreviation of p(otential of) H(ydrogen) also referred as Latin pondus hydrogenii, the negative logarithm of the hydrogen ion concentration in a solution: $pH = -\log_{10}[H+]$. If the hydrogen ion concentration of a solution increases, the pH will decrease, and vice versa. The value for pure distilled water is defined as neutral; solutions with pH values of 0 to 7 are acidic, those with values of 7 to 14 are alkaline or basic.

pH electrode

In the electrochemical determination of pH, a pair of electrodes is used: a reference electrode and a measuring electrode (the measuring electrode is usually a glass electrode). The determination of pH is based on the measurement of the electrical potential between a glass electrode and a reference electrode, which is proportional to the pH of the solution. Glass and reference electrodes are combined into a single electrode measuring cell in a test setup (usually with an integrated temperature compensator).

phenol

A white, crystalline compound (C_6H_5OH) derived from benzene. It is a by-product of petroleum refining, tanning, and textile, dye, and resin manufacturing. Used in the manufacture of weed killers, plastics, and disinfectants, as well as in petroleum refining. Phenol is toxic.

phenol index (PI)

Parameter for the assessment of summed concentration of phenol and its derivatives in a sample. The contribution to the phenol index of humic acids produced by biochemical decomposition of leaves and wood in soil is difficult to take into account.

photogrammetry

Reliable measurements can be obtained from photographic images using photogrammetry techniques. Not only size, shape, and position, but also color, texture, and patterns can be measured. The main objective of photogrammetry is to relate the pixel coordinates on a photograph as exactly as possible to the geographic coordinates (longitude, latitude, elevation) of terrain points and to remove the distortion caused by the data acquisition system (aerial camera or scanner), perspective, and motion of the aircraft or space-borne platform. This processing, called rectification of the image, also requires the specifications of the camera and the coordinates of ground control points. The coordinates of the ground control points have to be surveyed on the ground if they cannot be extracted accurately from topographic maps. This can be carried out quickly and inexpensively using GPS technology. Photogrammetry techniques are used to prepare maps and digital elevation models (DEM) from remote-sensing data. Other tasks are the preparation of orthophotographs, photomosaics, terrain cross sections, and vector data (e.g., the elevation difference between the top and bottom of a terrain slope, length of a stream, volume of a landfill).

photo ionization detector (PID)

PIDs use ultraviolet light to ionize an analyte. The ions produced by this process migrate to a pair of electrodes. The current generated is a measure of the analyte concentration. PIDs are used for detection of volatile organic compounds (VOCs), such as benzene, chlorinated hydrocarbons, ethylbenzene, toluene, and xylene (BETX). With its fast response and high sensitivity, photo ionization is the preferred method for the detection of VOCs. Since PIDs are normally calibrated for benzene, the directly readable output is valid only for benzene contamination.

photometry

A method of determining the concentration of a substance dissolved in water by adding one or more suitable reagents that form a colored product with that substance. The color intensity is a function of the concentration of the substance. A photometer is used to measure the color intensity accurately. The less accurate visual assessment of the color intensity is called “colorimetry”.

photomosaic

An assemblage of several overlapping and/or adjacent aerial photographs, each of which shows part of a region, put together in such a way that each point in the region appears once and only once in the assemblage and scale variation is minimized.

phreatic aquifer

→ unconfined aquifer

piezometer

1. A device (tube or pipe) that allows the hydraulic head in an aquifer to be determined at a given point. Some types of piezometers can also be used for collecting water samples. As a result, wells designed specifically for collecting water samples are often referred to, incorrectly, as piezometers.
2. An instrument to measure pore water pressure.

piezometric head

The elevation to which water rises in a well. It is measured with a water level meter. It is the elevation above a datum plus the pressure head. In an unconfined aquifer it is at the elevation of the water table. In a confined aquifer, the piezometric head is typically at a higher elevation than the top of the aquifer due to the confining pressure. The preferred term is potentiometric head.

piezometric surface

The water level surface that can be obtained from the mapping of water level elevations in wells drilled into an aquifer.

pixel

Term used for “picture element”, the smallest hardware element that a device (e.g., digital photograph, scanner) can address. The quality of a digital image depends on the number of pixels per inch that make up the image. A number of different shades or colors can be represented by each pixel.

plume

An area of chemicals in a medium, such as air or groundwater, moving away from its source in a long band or column. A plume can be smoke from a chimney or chemicals moving with groundwater. It defines an area where exposure would be dangerous. In groundwater a plume can extend for some distance, depending on groundwater flow and the chemical substances in the water. The highest concentration is at the source, with concentration decreasing with increasing distance from the source.

polarization

1. The restriction of the transverse oscillations of an electromagnetic wave to a single plane. Polarization is typically referred to as being horizontal or vertical, but the actual polarization can be at any angle. Circular polarization is also possible. Reception of a horizontally polarized signal with an antenna oriented for vertically polarized signals, or vice versa, will reduce the amount of signal received. 2. In electrostatics, the polarization is the vector field that results from permanent or induced electric dipole moments in a dielectric material. 3. The inability of an electrode, for example, a glass pH electrode, to reproduce a reading after a small electrical current has been passed through its membrane owing to a static layer of charged particles on the electrode.

polarography

Polarography is a special case of the voltametry in which current-voltage curves (polarograms) are plotted using a dropping mercury electrode (a capillary filled with mercury dropping at a slow but steady rate) as an indicator electrode against a nonpolar reference electrode. Polarographic measurements are based on the principle of electrolysis of substances at a mercury electrode. When voltage is applied between the indicator electrode (Hg electrode) and the reference electrode, current flows in the presence of electrochemically active materials. The cell is connected in series with a galvanometer (for measuring the flow of current) in an electrical circuit that contains a battery or other source of direct current and a device for varying the voltage applied to the electrodes. With the dropping mercury electrode connected to the negative side of the polarizing voltage, the voltage is increased in small increments and the corresponding current is observed on the galvanometer. The current is very small until the applied voltage is increased to a value large enough to cause reduction of the analyzed substance at the dropping mercury electrode. The current increases rapidly at first as the applied voltage is increased above this critical value but gradually attains a limiting value and remains more or less constant as the voltage is increased further. The critical voltage required to cause the rapid increase in current is characteristic of, and also serves to identify the substance that is being reduced (qualitative analysis). Under proper conditions the constant limiting current is governed by the rates of diffusion of the reducible substance up to the surface of the drops of mercury and its magnitude is a measure of the concentration of the reducible substance (quantitative analysis). Limiting currents also result from the oxidation of certain oxidizable substances when the dropping electrode is used as anode. Polarographic analysis can be used to identify most chemical elements and especially for the analysis of alloys and various inorganic compounds. Polarography is also used to identify numerous types of organic compounds and to study chemical equilibrium and rates of reactions in solution.

pollutant
→ contaminant

pollution
→ contamination

polychlorinated biphenyl
→ PCB

polycyclic aromatic hydrocarbons (PAH)
A class of organic compounds with two or more joined benzene rings in various, more or less clustered forms; they are present in petroleum and related materials and used in the manufacture of materials such as dyes, insecticides, and solvents. Some higher molecular weight PAHs (e.g., fluoranthene, benzo(a)pyrene) are products of combustion. Lower molecular weight PAHs (e.g., naphthalene, fluorene) are generally obtained from unburned petroleum. In general, the lower molecular weight PAHs are more soluble, more volatile, and more acutely toxic than the higher molecular weight PAHs, whereas the higher molecular weight PAHs have a stronger tendency to accumulate in the biosphere. PAHs are considered to be cancerogenic to various degrees.

polyethylene (PE)
A thermoplastic material composed of ethylene polymers. It is normally a translucent, tough solid which is unaffected by water and/or by a large range of chemicals.

polyvinyl chloride (PVC)
A thermoplastic material composed of vinyl chloride polymers. PVC is a colorless solid that is highly resistant to water, alcohols, and concentrated acids and alkalis. PVC is used for cold water pipes (outside/underground), as well as for sewers, drains, and venting pipes and fittings. Used instead of rubber as insulation of electric cables.

pond
A body of water smaller than a lake, often artificially formed, sometimes ephemeral or seasonal.

pore
Any opening or passageway admitting passage of a fluid (liquid or gas) in a soil, rock or other material.

pore aquifer

An aquifer consisting mostly of unconsolidated rock with groundwater predominantly present and flowing in the pores.

pore volume

Volume of the open interstitial spaces in soil or rock; preferred term: pore space.

porosity

The percentage of open pore space in soil, rock or other materials. It is the ratio of the pore volume to the total volume of the material. The porosity is determined from cores or from borehole logs.

porous

A material containing pores capable of being filled (permeated) by a fluid, e.g., water, oil or gas.

positioning

Determining the location and elevation, usually with respect to a geodetic datum but sometimes with respect to bench marks whose geodetic locations may not be known. The use of satellite supported positioning (GPS, DGPS) for geoscientific purposes is a relatively new development. A precision of about 1 cm to several millimeters is attainable by post-processing. The coordinates and altimetry data determined by GPS must be converted to the national or local frame of reference.

potential

1. In vector calculus, any vector field of a certain type has an associated scalar field called the potential. Potentials find broad applications in physics, e.g., the electric potential is the difference in electrical charge between two points expressed in volts. Electric, magnetic, and gravitational fields have potential fields that obey the Laplace equation. 2. A measure of the driving force behind an electrochemical reaction that is reported in units of volts. 3. Possible but not yet in existence.

potential evaporation

The amount of evaporation that would occur if sufficient water were available. Surface and air temperatures, insolation, and wind all affect this. The amount of evaporation that takes place from open water.

potential evapotranspiration

The amount of water that could be evaporated or transpired if it were available.

potential gradient

1. → hydraulic gradient. 2. Gradient of potential fields. → potential

potentiometry

A branch of analytical chemistry in which measurements of electrical potential are used to determine the concentration of some component of the analyte solution. The potential difference between a measuring electrode (e.g., an ion-selective electrode) and a reference electrode is measured in a galvanic cell by keeping voltage constant with no flow of current. The ion concentration in the solution can be determined directly from the measured potential difference.

precession

The slow change in the direction of the rotation axis of a spinning object caused by some external influence.

precipitation

1. The falling to Earth of any form of water (rain, snow, hail, sleet or mist) as part of the weather. Precipitation is a major component of the hydrologic cycle. 2. The quantity of water falling to Earth at a specific place within a specified period of time. 3. The formation of a solid from a solution during a chemical reaction or as a result of a lowering of the temperature.

precision

Precision is the repeatability of an instrument measurement as expressed by the mean deviation of a set of measurements from the average value. Precision is different from accuracy, which relates to the deviation of the measurement from the true value.

pressure gradient

The amount of pressure change occurring over a given distance in a given direction.

pressure head

The height of a column of fluid, normally water, that will produce a given pressure. It is equal to the force per unit area divided by the product of the density of the fluid and the acceleration due to gravity.

pressure transducer

Device that converts pressure into an analog electrical signal. There are various types of pressure transducers, one of the most common is a strain-gauge base transducer. The conversion of pressure into an electrical signal is achieved by the physical deformation of strain gauges affixed on the diaphragm of the pressure transducer and wired in a wheatstone bridge configuration. Pressure applied to the pressure transducer produces a deflection of the diaphragm which introduces strain to the gauges. The strain will produce an electrical resistance change proportional to the change in pressure.

principal point

1. The center point of an aerial photograph. 2. The point on the Earth where a satellite sensor is focused at any time during its orbit.

principle of equivalence

Potential methods, e.g., magnetics, gravity method, geoelectrics etc., basically do not allow a model (e.g., density distribution in the ground) to be derived unambiguously from the measured data (e.g., gravity anomalies), which means for a measured field distribution there is theoretically an indefinite number of models (distribution of the corresponding physical parameters). Previous geological knowledge, plausibility considerations and/or comparison with the results of other geophysical methods have to be used in order to reduce the diversity of the possible models. The principle of equivalence in the 1-D case of dc resistivity sounding means that it is impossible to arrive at a unique solution for the layer parameters (thickness and resistivity). For conductive layers, only the thickness/resistivity ratio can be determined (S equivalence), whereas for high resistivity layers only the product of thickness times resistivity can be determined (T equivalence). This means that when resistivity and thickness of a layer vary within certain limits, their product remaining constant, no differences can be seen in the sounding curve. Thickness and resistivity are coupled in both cases of equivalence and cannot be independently determined.

probabilistic

Of or relating to or based on probability. → stochastic

probability

A measure of how likely it is that some event will occur. Probabilities are expressed as numbers between 0 and 1. The probability of an impossible event is 0, while an event that is certain to occur has a probability of 1.

processing

1. Carrying out any set of operations on data, including organizing, adapting, altering, retrieving, aligning, combining, etc. 2. Geophysical processing operations include applying corrections, rearranging the data, filtering it, combining data elements, transforming, migrating, measuring attributes, etc.

profile

1. A line representing the Earth's crust in a vertical section; a profile is a line on which is marked the locations of the ground surface, the boundaries of the underlying strata, the water table, etc. 2. A graphical representation of geophysical data as a function of distance. 3. A line along which geophysical measurements are made or samples taken.

profile assembly

Representation of sections (e.g., geological cross sections, seismic sections) according to their location in the surveyed area. Parallel sections are represented below one another in a figure. More complicated programs allow the (pseudo)-spatial representation of user-defined profiles (fence diagrams).

profile line

The line along which observations and/or measurements are made. → profile

profiling

This term is used for a geophysical survey in which the measuring system is moved along lines called profiles to determine lateral variations in the parameters measured. Syn. mapping.

pseudosection

A cross section showing the distribution of a geophysical property, such as apparent resistivity as a function of position and electrode separation, from which a two-dimensional model of the geological structure can be derived by inversion.

pump

A mechanical device using suction or pressure to raise or move fluids.

pumpage

The total quantity of liquid pumped in a given interval, usually a day, month, or year.

pumping test

A test made by pumping a well for a period of time and observing the change of water level in the aquifer. A pumping test may be used to determine the capacity of a well and the hydraulic characteristics (the transmissivity and storativity) of the aquifer. Also called an aquifer test.

pycnometer

Vessel (e.g., flask with a close fitting stopper with a fine-diameter hole) with a precisely known volume used to measure the density of solids or liquids. Sometimes called specific gravity bottle or density bottle.

pyrohydrolysis

Decomposition by the combined action of heat and water vapor. It is used for the separation of organic halogens from samples for AOX/TOX determination.

qualitative

1. Relating to quality. 2. Referring only to the characteristics of something being described, rather than the amount. 3. Indicative only of relative sizes or magnitudes, rather than their numerical values. A qualitative comparison would say whether one thing is larger, smaller, or equal to another, without specifying the size of any difference. As opposed to quantitative. 4. Chemical analysis which merely determines the constituents of a substance without any regard to the quantity of each ingredient, in contrast to quantitative analysis.

quality assurance

All of the actions taken to provide adequate confidence that the quality of work is within defined tolerances and a product or service is of the type and quality needed and expected by the customer. Also referred to as quality control. Quality assurance is part of the field work, processing, interpretation and reporting. It must be documented by field protocols and, in condensed form, in the final report. Aspects of quality assurance are

- completeness of the report and comprehensibility for nonexperts: The report must contain all specifications and representations necessary for understanding. The path from measured data to the interpretation must be comprehensible to the expert;
- convincing representation of the results;
- delivery of the measured data to the client in machine-readable, well-documented form for preservation of evidence and subsequent reinterpretation in connection with the results of other methods.

quality control

Work undertaken by contractor, client's representative or, possibly a quality assurance company/agency to ensure the quality of materials and on-site work is in accordance with the agreed specifications and accepted standards. Also referred to as quality assurance.

quantitative

1. Refers to a measured numerical value, e.g., amount (either absolute or concentration), rate, velocity, or depth. 2. Analysis which determines the amount of each component of a substance or mixture by weight or volume; in contrast to qualitative analysis.

radar methods (in remote sensing)

Radar (acronym for Radio Detection and Ranging) illuminates an area with its self-produced coherent radiation in the cm wavelength range. Radar radiation passes through clouds. Due to the oblique illumination angle of the transmitted radar signal to the Earth's surface, radar images are especially useful for detecting slight changes in topography due to the radar shadowing effect. This property is useful for mapping topographically expressed faults. Radar measurements are also sensitive to differences in soil moisture content. Common disadvantages of radar data are the complexity of data acquisition, geometric distortions, usually lower spatial resolution than optical systems and challenges in data interpretation. In side-looking airborne radar (SLAR) systems, a transmitter and receiver antenna is mounted on the side of the fuselage with a radiation direction perpendicular to the azimuth direction (flight direction). In common radar systems the same antenna is alternately operated as transmitter and receiver antenna. The ground resolution of SLAR images is different in the range direction and the flight direction (azimuth resolution). The range resolution is related to the duration of the transmitted radar pulse and to the look angle. Short pulses and a smaller look angle improve the range resolution. The resolution in the direction of flight depends on the antenna characteristics and, therefore, on antenna length and on the wavelength of the transmitted radar wave. Shorter wavelengths and a longer antenna improve the resolution in the direction of flight. The possible antenna length is limited. With respect to azimuth resolution, radar systems are subdivided into real-aperture radar (RAR) and synthetic-aperture radar (SAR). In RAR systems, the azimuth resolution is determined mainly by the physical length of the antenna. In SAR systems, the azimuth resolution is determined by processing Doppler shift data for multiple return pulses from the same object to create synthetically longer antenna. Consecutive antenna positions are treated as if they were elements of one large antenna. Current aerial and space-borne systems use SAR technology. Whereas conventional SAR imaging usually only uses the amplitudes of the radar signals, interferometric SAR (InSAR) and polarimetric SAR (PolSAR) utilize the phase measurements as well. Detailed digital elevation models (DEM) can be derived from the phase differences of the interferograms. Differential interferometric SAR uses interferograms acquired over the same area at different times to detect displacements of the Earth's surface in the centimeter range due to earthquakes, volcanic events, landslides or land subsidence.

radiation

1. Transmission of energy from a source into the surrounding medium in the form of waves or particles. Radio waves, microwaves, light (infrared, visible or ultraviolet), x-rays and gamma rays, as well as α and β radiation, are types of radiation. 2. Energy that is radiated or transmitted.

radiometer

Device that detects and measures electromagnetic radiation.

radiometric resolution

1. The sensitivity of a sensor for recording variations in the electromagnetic spectrum. High resolution means that more subtle changes in the image can be detected. 2. The number of levels into which the spectral range of the sensor is subdivided is the radiometric resolution. In general, the greater the number of levels, the greater the detail. The number of levels is normally expressed as the number of binary digits needed to store the value of the maximum level, for example, an 8-bit radiometric resolution would be 256 levels.

radius of influence

1. The maximum distance from an extraction or injection well where lowering or increase in pressure, respectively (soil gas or groundwater movement) occurs. 2. The radial distance from the center of a well to the point where there is no lowering of the water table or potentiometric surface (the edge of the cone of depression). Area of influence is also used.

rainfall

The amount of water that falls in a given period of time as rain. Not synonymous with precipitation.

rain gauge

A device, usually a cylindrical container, for measuring precipitation (melted snow, sleet, or hail as well as rain). Also spelled: rain gage.

Raman spectroscopy

Spectroscopy is an analytical method used to identify the composition and determine the structure of chemical species by studying the electromagnetic radiation they emit, transmit, or absorb. Raman spectroscopy relies on the scattering of light from a gas, liquid or solid with a shift in wavelength from that of the usually monochromatic incident radiation.

range resolution

In radar images, the least radial separation between two targets in the same direction from a radar device that allows them to be distinguished. This separation equals one-half the transmitted pulse length. Targets closer together than this distance are not resolved and appear as a single target on the display.

ranging pole

Surveying instrument consisting of a straight rod painted in bands of alternate red and white each 50 cm wide; used by surveyors for sighting points on lines or for marking the position of a ground point. Syn.: range rod, range pole, lining pole, line rod, sight rod.

receiving stream

A stream into which runoff or effluent is discharged.

recharge

Water added to an aquifer or the process of adding water to an aquifer. Groundwater recharge occurs naturally as the net gain from precipitation or runoff. Injection of water into an aquifer through wells is a form of artificial recharge. The land area where recharge occurs is called the recharge area. This area is particularly vulnerable to pollution. Any paving of the area lessens the amount of water that can enter the aquifer.

recombinant DNA (rDNA)

Recombinant DNA is DNA that has been created artificially. DNA from two or more sources is incorporated into a single recombinant molecule.

reconnaissance

A preliminary survey to gain general information of an area. It is often followed by a detailed survey.

rectification (of a photograph)

The process of eliminating scale variations within a photograph and relief displacement from an aerial photograph taken at an oblique angle or when the plane has a slight tilt. The angular relationship between the photograph and the reference (e.g., Earth's surface) is determined at ground control points. In digital image processing, it also refers to correcting for geometric distortions, radiometric calibrations, and noise removal.

redox potential (Eh)

1. The potential of a reversible oxidation-reduction electrode measured with respect to a reference electrode, corrected to the hydrogen electrode, in a given electrolyte. 2. A measure of the relative tendency of a solution to accept or transfer electrons. 3. The reduction-oxidation potential (redox potential) is a measure of the availability and concentration of oxygen to enter into chemical reactions.

redox process

Chemical reactions involving loss of one or more electrons by one molecule (oxidation) and simultaneous gain by another (reduction).

reduction

1. Any process in which electrons are added to an atom or ion (as by removing oxygen or adding hydrogen). It always occurs accompanied by oxidation of the reducing agent. Reduction occurs at the cathode in an electrochemical reaction. 2. The transformation of "raw" data to a more usable form. In geophysics, this often refers to the conversion of an observed value to the value that it theoretically would have at some selected or standard level, e.g., mean sea level.

regional

Related or limited to a particular region, e.g., "regional geology". A region can be any area that has some unifying feature. Refers to areas ranging from few to hundreds of kilometers across; it is always larger than an area.

registration

In image processing, the process of adjusting two images so that equivalent geographic points coincide.

relative permeability

1. The ratio of the effective permeability to a given fluid at partial saturation to the permeability to that fluid at 100 % saturation. It ranges in value from zero at low saturation to 1.0 at 100 % saturation of that fluid. In the presence of other fluid phases, it is the ratio of the amount of the fluid in question that will flow at a given percentage saturation to the amount that would flow at a saturation of 100 %, other factors remaining the same. Since different fluid phases inhibit the flow of each other, the sum of the relative permeabilities of all phases is always less than unity. 2. Magnetic permeability of a material relative to the permeability of free space.

relief

Vertical irregularities of a surface (e.g., Earth's surface).

relief displacement

Geometric distortion in vertical aerial photographs in which features at higher elevations are displaced away from the center point of the photograph. Relief displacement aids in the calculation of heights of features on an aerial photograph, but also results in the image not having a uniform scale like a planimetric map, unless it is converted to an orthophoto.

remediation

The removal of hazardous materials from a contaminated site or aquifer.

remobilization

Release and renewed migration of fixed (by immobilization) or detained (by retention) substances in a soil or rock.

remote sensing

Obtaining data about the environment and the surface of the Earth from a distance without having the sensor in direct contact with the object, for example, from an aircraft or satellite. Remote sensing is restricted to methods that record electromagnetic radiation (ultraviolet, visible, infrared, microwave, or radar) reflected or radiated from objects, which excludes magnetic and gravity surveys that record force fields.

remote-sensing methods

Remote-sensing methods can provide geoscientific data for large areas in a relatively very short time. They are not limited by extremes in terrain or hazardous conditions that may be encountered during an on-site appraisal. In many cases aerial photographs and satellite images must be used to prepare a base map of the investigation area. Remote sensing methods can enable a preliminary assessment and site characterization of an area prior to the use of more costly and time-consuming methods, such as field mapping, geophysical surveys and drilling. The data obtained from satellite-based remote-sensing systems is best suited for regional studies as well as for detecting and monitoring large-scale environmental problems. Mapping scales of 1 : 10 000 or larger are required for a detailed geoenvironmental assessment of landfills, mining sites or industrial sites. High-resolution aerial photographs, airborne scanners, and some satellite-based remote-sensing systems provide data at the required spatial resolution (e.g., 70 cm or better). Aerial photographs made at different times can reveal the changes at sites suspected to be hazardous. Photographic images can be evaluated not only with respect to size, shape, and position of objects, but also to color, hue, texture, and patterns of distribution of these aspects with regard to their geoscientific information. Photogrammetry is a method to obtain reliable measurements from photographic images. The main tasks of photogrammetry are to prepare maps and digital elevation models (DEM), orthophotographs, photomosaics, terrain cross sections, and obtain vector data (e.g., height of a terrain slope, length of a creek, volume of a landfill) from remote-sensing data. Nonphotographic imaging systems detect electromagnetic (EM) radiation with semiconductor detectors or special antennas instead of using light-sensitive film. While photography is limited to the spectral range 0.3 to 0.9 μm the range of sensing for a multispectral scanner (MSS) covers wavelengths from 0.3 to approximately 14 μm. MSS can operate within very narrow spectral bands. Both passive and active methods are used for nonphotographic imaging. Passive methods sense the natural electromagnetic radiation reflected, scattered, and transmitted at the Earth's surface. In active methods, EM radiation (e.g., laser beam or radar wave) is transmitted and the response of the ground is detected.

resolution

The smallest distance at which two objects can be distinguished in geophysical survey measurement or in remote sensing data.

resolve

To separate into parts.

retained water

→ adhesive water

retardation

The movement of a solute through a geologic formation at a velocity less than that of the groundwater. Preferential retention of contaminants in the subsurface by one or more physical, chemical, or biological factors.

retardation coefficient

The retardation coefficient expresses how much slower a contaminant moves than does the water itself.

retention

1. → contaminant retention. 2. The amount of water from precipitation that has infiltrated into the ground and not escaped from the area as runoff or through evapotranspiration.

ribonucleic acid (RNA)

A single-stranded nucleic acid (similar to the double-stranded nucleic acid DNA) that plays an important role in protein synthesis and other chemical activities of the cell. The four main types of RNA are heterogeneous nuclear RNA (hRNA), messenger RNA (mRNA), transfer RNA (tRNA) and ribosomal RNA (rRNA).

risk assessment

A qualitative and quantitative evaluation of the risk posed to human health and/or the environment by the actual or potential presence and/or use of specific contaminants (USEPA, 1994).

rock

1. A naturally occurring aggregate of minerals. Rocks are classified by mineral and chemical composition, the texture of the constituent particles, and also by the processes that formed them. Rocks are formed by igneous, sedimentary, or metamorphic processes. 2. Sometimes used synonymously with "stone".

roentgen fluorescence analysis (RFA)

→ x-ray fluorescence spectrometry

rotary drilling

A drilling method in which a hole is made by a rotating bit to which a downward force is applied. The bit is fastened to and rotated by the drill stem, which also provides a passageway through which the drilling fluid is circulated in order to carry the rock cuttings out of the hole.

root zone

The zone of soil in which most (about 90 %) of the roots of plants occur.

runoff

That part of precipitation that flows toward streams on the surface of the ground or within the ground. Runoff includes both groundwater runoff and surface runoff.

salinity

The total quantity of dissolved salts (usually sodium chloride) in a given volume of water. It is normally reported in parts per thousand (ppt). Normal seawater has a salinity of about 35 ppt = 35 000 ppm (parts per million).

saline water

Water containing more than 1000 parts per million (ppm = mg L^{-1}) of dissolved solids of any type (e.g., dissolved salts). The US Geological Survey (USGS) classifies the degree of salinity as follows: [1] slightly saline: 1000 - 3000 ppm; [2] moderately saline 3000 - 10 000 ppm; [3] very saline 10 000 - 35 000 ppm; and [4] brine: more than 35 000 ppm.

sample

1. A small part of something that is assumed to be representative of the whole (such as water sample or soil sample). 2. In statistics, a group of observations selected from a statistical population by a set procedure in order to analyze the population of interest. Samples may be selected at random or systematically.

sample preparation

Before a sample can be chemically analyzed, various treatment steps are necessary depending on the substance to be analyzed and the method to be used. Sample preparation steps include grinding, digestion, extraction, etc.

sampling

1. The removal of a representative portion (sample) of a material for examination or analysis. 2. Sampling, or analog-to-digital conversion, is the process of converting an analog signal to a series of digital samples (numbers). 3. The process of statistically testing a data set for the likelihood of relevant information.

sanitary landfill

→ landfill

satellite

An object that orbits the Earth or other planet. For example, smaller bodies orbiting around planets are called those planets' satellites (or occasionally, "moons"). Probes launched into orbit around the Earth are called "artificial satellites".

saturated zone

The zone below the water table where every available space (pores and fractures) is filled with water and the fluid pressure is greater than atmospheric.

scatter

The act and result of scattering.

scattering

1. The physical process in which particles are deflected haphazardly as a result of collisions with other particles. 2. The process by which electromagnetic radiation interacts with and is redirected by the molecules of the atmosphere, ocean, or land surface.

scene

Area on the ground that is covered by an image or photograph.

schistosity

The type of foliation that characterizes schist, resulting from the parallel arrangement of coarse-grained platy minerals, such as mica, chlorite, and talc.

screen

A device to separate particles according to size: Smaller particles pass through the holes in the screen, the larger ones are held back. The term is also used for the section of a well in which groundwater enters the well but which prevents sediment to enter the well.

screening

1. Use of screens to separate material according to size of its particles. 2. A test to identify potential contamination.

screening test

A fast and inexpensive test or combination of tests intended to detect possible contamination. Examples are TOX, TOC, and VOA. A positive TOX test indicates the presence of man-made organic chemicals containing chlorine and bromine. TOC is the total amount of organic carbon in the water sample. VOA is a testing procedure for volatile organic chemicals. Screening test is used synonymously with indicator test.

sealing

1. Covering the ground with concrete, asphalt, pavement, etc. 2. Covering a landfill with a relatively impermeable barrier designed to avoid groundwater recharge and keep leachate inside the landfill.

section

A graphical representation showing the geologic layers that would be exposed in a vertical cut. It is often called a cross section. The results of geophysical surveys (conductivity, seismic wave velocity, apparent resistivity, the seismic wave field, etc.) are also represented in sections. Horizontal sections through the ground are used as well (e.g., time sections or depth sections in seismic). → pseudosection

sediment

Material that has been held in suspension by flowing water or wind and deposited as the flow rate decreased; material in or on ice that was deposited when the ice melted; material that has precipitated from solution or secreted by living organisms. Typical sedimentary material is sand, gravel, silt, mud, till, loess, and alluvium. Sediment can be either unconsolidated or consolidated; typical consolidated sediments are sandstone, conglomerate, siltstone, claystone.

sedimentary rock

Rock formed of sediment, especially from mechanical, chemical, or organic processes, and specifically: clastic rock, such as conglomerate, sandstone, and shale, formed of fragments of other rock transported from their sources and deposited in water, and rocks formed by precipitation from solution, as rock salt and gypsum, or from secretions of organisms, such as most limestone. Many sedimentary rocks show distinct layering, which is the result of different types of sediment being deposited in succession. Sedimentary rock is one of the three main rock classes (along with igneous and metamorphic rocks).

seepage

1. Movement of water through a porous medium, often used in the context of water movement from a groundwater system to surface water, or vice versa. 2. The loss of water by infiltration into the soil from a river, ditch, or other watercourse, reservoir, or other body of water.

seismic methods

Geophysical methods using the controlled generation of elastic (seismic) waves by a seismic source (e.g., explosion, vibrator, weight drop, sledgehammer) to investigate the subsurface. During their propagation within the subsurface, seismic waves are reflected, refracted or diffracted at boundaries between rock masses/layers with different properties (seismic velocity and/or bulk density) or at man-made structures. The seismic response is simultaneously detected by a number of receivers (geophones, seismometers) positioned along straight profile lines (2-D seismics) or over an area in 3-D seismic surveying. The signal is then sent to a recorder via cables. The recordings of individual geophones or groups of geophones by a seismograph are collected, processed and displayed in seismic sections to image the ground structure. The recording of seismic waves returning from the ground to the surface (seismic events) provides information about structures and lithological composition of the subsurface. A geological model of the subsurface can be derived by measuring the traveltimes of seismic waves and determining their material-specific velocities. The classical seismic methods are seismic refraction and seismic reflection. Seismic methods based on surface waves have been increasingly used in the last several years. Besides application from the

Earth's surface, seismic methods are used on inshore waters or in boreholes. In the seismic refraction method, only the traveltimes of first arrivals on recorded seismograms are mostly evaluated. The first arrivals have to be identified, selected (picked) and presented in traveltime curves (plots of arrival times vs. source-receiver distance). The main objective of refraction surveys is to determine the depths to rock layers, the refractor topography and layer velocities. Sophisticated methods are available for extracting this information from traveltime curves. The seismic refraction method requires an increase in velocity with depth. If a layer of lower velocity lies beneath a layer of higher velocity, then that layer cannot be detected. Such conditions may cause ambiguity in the interpretation. The seismic reflection method yields, for example, geological cross sections. It is the most important method for oil and gas prospecting at great depth. Reflection seismic has become increasingly used since the 1980s for depths of less than 50 m. There are special problems for shallow seismic surveys because unprocessed field records are characterized by a poor signal-to-noise ratio. Owing to the short traveltimes, source generated noise, surface waves, first arrivals of direct waves and refracted waves dominate and genuine reflections are masked. Thus, both acquisition and standard processing of conventional seismic methods cannot be transferred in the sense of a "scaling down" from deep reflection seismics. The surface wave method is based on analyzing the variation of seismic velocities with frequency (dispersion). Dispersion of surface waves is high when near-surface velocity layering occurs. In the late 1990s the method was successfully used, for example, for underground cavity detection and delineation of abandoned waste sites. Seismic methods are also referred to as seismics.

seismic waves

Elastic waves generated, for example, by an earthquake or a artificial seismic source (e.g., explosion, vibrator, weight drop, sledgehammer). Seismic waves from this impulse may travel either along or near the Earth's surface (surface waves) or through the Earth (body waves). There are two types of body waves:

- P-waves (primary, longitudinal or compressional waves) with particle motion parallel to the direction of propagation.
- S-waves (secondary, shear or transverse waves) with particle motion perpendicular to the direction of propagation.

The velocities of seismic waves depend on the elastic properties and the bulk density of the media and vary with mineral content, lithology, porosity, pore fluid saturation and degree of compaction. P-waves have a higher velocity than S-waves. S-waves cannot propagate in fluids. Surface waves, often considered to be a source of noise, contain valuable information about the properties of shallow ground. But although their use has considerably increased in engineering studies, body waves are of prime interest in seismic investigations.

semiarid
Refers to land that is characterized by little precipitation (25 - 50 cm a^{-1}) and sparse, but still present, vegetation.

semiconfined aquifer
→ leaky aquifer

semiquantitative
A term for numerical results from an chemical analysis technology with relatively high error limits and which are only an approximation of the true value.

sensitivity
1. A measure of the minimum change in an input signal that an instrument can detect. 2. Degree of responsiveness of a film to light.

settlement
1. A village. 2. Lowering of the ground resulting from compaction of the grain lattice in the soil. Settlement takes place in unconsolidated sediments. The amounts of settlement are measured by leveling surveys as well as using high-precision differential global positioning system (DGPS) measurements.

sewage
Wastewater generated by commercial or domestic use of water and discharged into sewers. Sewage typically contains everything from soap to solid waste, and must be purified before it can be safely returned to the ecosystem. The preferred term is wastewater.

sewage sludge
A solid or semi-solid residue generated during the treatment of domestic sewage in a sewage treatment plant.

sewer
A system of underground pipes that carries away sewage and/or direct runoff to treatment facilities (or receiving stream).

sheet flow
Flow that occurs overland during intense local rainfall in places where there are no defined channels. The flood water spreads out over a large area at a uniform depth.

sieve analysis

Determination of the particle-size distribution of soil or unconsolidated sediment by measuring the percentage of the particles that will pass through standard sieves of various sizes.

signal

1. Any carrier of information, as opposed to noise. 2. In geophysics, physical parameter values (e.g., voltage or current or field strength) whose modulation represents coded information about the source from which it comes.

signal-to-noise ratio (S/N or SNR)

The ratio between the amplitude of a signal (S) being measured and that of the noise (N), e.g., the amplitude of the seismic signal and the amplitude of noise caused by seismic unrest and/or the seismic instruments. Sometimes the denominator is the total energy, i.e., S/(S + N). The signal-to-noise ratio is measured in decibels (dB). Larger numbers are better. A component with a low signal-to-noise ratio is noisy.

signature

Set of characteristics by which a material or an object may be identified in an image or photograph or on a geophysical map.

significance

1. The quality of being significant. 2. A result is significant when it equals or exceeds certain probability criteria. A result is significant when there is a low probability that the results of an experiment occurred by chance alone. 3. The probability that a stated statistical hypothesis will be accepted or rejected; referred to as the level of significance.

sinkhole

Sink holes are a feature of karst landscapes, for example in areas of limestone bedrock. The result is a depression in the surface topography. This may range from a small, gentle earth-lined depression, to a cliff-lined chasm. Also known as sinks, dolines, and cenotes. A sinkhole found above a buried pipe indicates pipe failure.

SI system

Système International d'Unités (International System of Units): a metric system of units. It is used in almost all countries of the world except the United States, Liberia and Myanmar, and it is almost globally used in scientific and engineering work. In 1960, a subset of the meter-kilogram-second system of units (MKS) was selected and designated as the SI system. The fundamental quantities are length (meter), mass (kilogram), time (second), electric current (ampere), temperature (kelvin), amount of matter (mole), and luminous intensity (candela).

site investigation

An investigation to determine the properties, distribution, and geology of the surface and subsurface materials at a site being considered for some type of use. The information is needed for planned engineered structures and/or for determining the type and extent of contamination present, and an estimate of its impact on human health and the environment.

site suspected to be hazardous

Abandoned landfills, abandoned industrial and mining sites, and other sites may represent a hazard to human health, water, soil, air, as well as flora and fauna. Sites suspected to be hazardous are classified as follows:

- industrial site (e.g., galvanizing plant, metallurgical workshop, machinist workshop, motor vehicle garage, gas/petrol station, tannery, workshop in which wood is impregnated or worked),
- waste deposit (landfill, sewage sludge deposit, small site with buried and/or heaped waste and litter), and
- mining site (e.g., spoil/slag heap, ore treatment plant).

skin effect

1. Installation of a well affects the permeability of the rock next to the well screen; if permeability is reduced, there is a positive skin effect, if permeability is increased, the skin effect is negative. 2. The skin effect is what happens when an alternating electric current distributes itself within a conductor so that the current density near the surface of the conductor is greater than that at its core. It causes the effective resistance of the conductor to increase with the frequency of the current.

sludge

Any muddy waste generated from municipal, commercial, or industrial wastewater treatment plant, water supply treatment plant, or air pollution control facility or any other such waste having similar characteristics, e.g., as produced by metallurgical processes. It can be a hazardous waste.

slug test

A single well aquifer test conducted to determine the in-situ hydraulic conductivity of low to moderate hydraulic conductivity formations by the instantaneous addition or removal of a known volume of water or solid object to or from a well. The subsequent well recovery is measured.

soil

1. The top layer of the Earth's surface in thickness from centimeters to a meter or more. A complex system consisting of weathered rock, organic matter, air and small living organisms. It contains water and nutrients used by plants. Soil is made up of distinct, more or less horizontal layers. These layers are called horizons. They range from the upper layers rich in organic

matter (topsoil) to underlying rocky layers (subsoil, regolith). Soil-forming processes involve the interaction between climate, living organisms, and relief acting on soil and soil parent material. 2. In engineering, all unconsolidated material above bedrock.

soil gas

Air and other gases found in the pore spaces of soils in the unsaturated zone. These gases may include vapor of hazardous chemicals as well as water vapor.

soil horizon

A layer of soil or soil material nearly parallel to the land surface and differing from adjacent genetically related layers in physical, chemical, and biological properties or characteristics such as color, structure, texture, consistency, kinds and number of organisms present, degree of acidity or alkalinity, etc. Soil horizons are generally designated by a capital letter, with or without numerical subdivision, e.g., A horizon, A2 horizon.

soil moisture

→ soil water

soil moisture potential

→ moisture potential

soil type

System of classification that refers to the different sizes of mineral particles in soil. Soil is made up in part of rock particles, grouped according to size as sand, silt, and clay. Each size plays a different role. For example, the largest particles, sand, determine aeration and drainage characteristics, while the smallest, clay particles, are chemically active, binding with water and plant nutrients. The ratio of these constituents determines soil type: clay, loam, clay-loam, silt-loam, and so on. The FAO developed in the early 1970s a supra-national classification (World Soil Classification) which offers useful generalizations about soil pedogenesis with respect to the interactions with the main soil-forming factors. Since 1998 this system has been replaced by a much-improved classification system (World Reference Base for Soil Resources).

soil water

Water contained in the soil above the water table, including water vapor in the soil pores. In some cases this term refers strictly to the moisture contained in the root zone of plants. Syn. soil moisture.

soil water zone

→ belt of soil water

solid

One of the three phases of matter. A solid is characterized by a definite volume and a definite shape in which the molecules are very close together and cannot move around.

solid phase extraction

A sample purification method based on the affinity of either the desired or undesired components of a reaction mixture for a solid material and subsequent filtration of the solid material from the reaction.

solid rock

British term for bedrock.

solubility

1. The amount of a substance that can be dissolved in a given amount of water or other liquid under specified conditions. Usually expressed in weight per unit volume (e.g., milligrams per liter) or percent. 2. The capacity of a solid to dissolve in a liquid.

sorption

A general term used for physical and chemical absorption and adsorption. The mechanism of sorption is not specified. It is often used when the specific mechanism is not known.

sorption coefficient

The ratio of contaminant concentration associated with the solid to the contaminant concentration in the surrounding aqueous solution when the system is in equilibrium. The sorption coefficient is also referred to as the distribution coefficient.

sorption isotherm

→ adsorption isotherm

sounding

1. Measurement of physical parameters or geophysical field values or calculation from measured values as a function of depth. Often, the distance between source (transmitter) and receiver or the frequency of a signal is changed in order to obtain information from different depths. In contrast to profiling. 2. Measuring the depth of water.

sounding rod

Tool to take samples from soil and unconsolidated rock to a maximum depth of 2 m.

source

1. Any place or object from which contaminants are released. A source can be a power plant, factory, dry cleaning business, gas station, car repair workshop or farm. 2. A device or process that releases energy into a system (e.g., seismic sources, electromagnetic transmitters). 3. The

beginning of a river. 4. A document (or organization) from which information is obtained.

source protection area (SPA)
The catchment area around a groundwater source which contributes water to that source (zone of contribution), divided into two areas: The inner protection area is designed to protect the source from human activities that may have an immediate effect on the source, in particular microbiological pollution. It is defined by a 100-day time of travel (TOT) from any point below the water table to the source. The outer protection area is the remainder of the catchment of the groundwater source.

soxhlet extractor
A device for discontinuous solid-liquid extraction at boiling temperature, especially suited for hard-to-extract substances (e.g., EOX).

spark emission spectrometry
In spark emission spectrometry, the sample is used as one of the two electrodes. A high voltage is applied between the electrodes, the sample is atomized by the resulting sparks. The atoms emit the absorbed energy in form of light, which is detected with a spectrometer. Its intensity is a function of the element concentration in the sample.

spatial resolution
→ ground resolution

species
→ chemical species

specific retention
→ field capacity

specific storage
The volume of water released from storage per unit volume of a saturated porous medium per unit decrease in the hydraulic head. Specific storage is the primary expression for the release of water from confined aquifers. It has units of m^{-1}.

specific yield
The ratio of the volume of water which will drain from a porous medium (drainable porosity) by gravity to the volume of the porous medium. See also effective porosity.

spectral resolution
The ability of a sensing system to differentiate between electromagnetic radiation of different wavelengths. The term also refers to the width of the spectral bands that, for example, a remote sensing system can detect.

spectrophotometry
The quantitative study of spectra by measurement of the relative amounts of radiant flux at each wavelength of the spectrum.

spectrometer
1. An instrument for measuring the intensity of radiation as a function of wavelength. 2. An instrument which separates radiation into energy bands (or in the case of a mass spectrometer, particles into mass groups) and indicates the relative intensities in each band or group.

spectrometry (or spectroscopy)
Spectrometry (or spectroscopy) is a general term for analytical chemical procedures in which information about a sample is gathered from the interactions of electromagnetic radiation (e.g., light of the visible or ultraviolet range, radio waves) with matter (atoms, ions, molecules). The following interactions can take place: absorption (attenuation of radiation intensity by the sample), emission (emission of radiation by the sample), scattering (with or without transfer of energy between the radiation and sample), reflection, refraction and change in the polarization of the radiation. The elements or compounds contained in the sample and their concentration can be determined from the recorded spectra. The method names are often linked to the name of the method used for preparing the substance for measuring the spectrum, e.g., graphite furnace AAS (GF-AAS).

spectroscopic procedures
→ spectrometry (or spectroscopy)

spectrum
1. A curve showing amplitude, phase and/or intensity as a function of frequency/wavelength or period. 2. A band of colors which forms when visible light passes through a prism. The band ranges in color from violet (shorter wavelength) to red (longer wavelength).

SPOT
Systeme Pour l'Observation de la Terre (French Earth observation system). Polar-orbiting Earth observation satellites capture either panchromatic (resolution 10 m) or 3-band multispectral imagery (resolution 10 m). SPOT images are available commercially and are intended for such purposes as environmental research and monitoring.

stage meter
Device used to measure the height of a water surface above an arbitrarily established datum plane.

stereo pair
Two overlapping images or photographs that give a 3-D impression when viewed in a stereoscope, in which simultaneously each eye sees only one of the pair of images.

stereoscopic vision
Binocular vision which enables the observer to view an object simultaneously from two different positions to obtain the mental impression of a three-dimensional model.

stochastic
1. Applied to processes that have random characteristics. Synonym for random. 2. Adjective applied to any phenomenon obeying the laws of probability. As opposed to deterministic.

stochastic model
Any mathematical model whose inputs are expressed as random variables, and whose output is a distribution of possible results. In contrast to a deterministic model. In many cases, stochastic models are used to simulate deterministic systems that include small-scale phenomena that cannot be accurately observed or modeled.

storage coefficient
→ storativity

storativity
The volume of water that an aquifer releases from or takes into storage per unit surface area of the aquifer per unit change in head. Same as storage coefficient. It is equal to the product of specific storage and aquifer thickness. Specific storage is the primary expression for the release of water from confined aquifers. In an unconfined aquifer, the storativity is equal to the specific yield.

storm water
Water from heavy rainfall or melting snow which runs off the land and does not infiltrate into the soil, frequently containing contaminants. This untreated water is collected by (ditches) drains and discharges further to creeks, wetlands, rivers, and the ocean. Syn. direct runoff.

strain
Deformation of a physical body caused by the action of stress on the body, e.g., small changes in shape and volume associated with deformation of the Earth by tectonic stresses or by the passage of seismic waves.

stratigraphic column
Chronological sequence of sedimentary or volcanic rock layers. Syn. geologic column.

stratigraphy

1. The branch of geology that deals with the origin, composition, distribution and succession of strata. 2. The study of sedimentary and volcanic rock layers sequentially deposited over time.

streamflow

The volume of water passing a given point per unit time in a stream channel usually expressed in cubic meters per second ($m^3\ s^{-1}$). Streamflow is a function of depth, width, and velocity of the water in a channel.

stream line

In fluid dynamics, a path that an imaginary particle without mass would make if it followed the flow of the fluid in which it is embedded.

stress

Force per unit area that produces strain in a physical body. Stress is a tensor quantity with nine terms, but which can be described fully by six terms due to symmetry: three normal components and three shear components. Simplifying assumptions are often used to represent stress as a vector for engineering calculations.

strike

The direction of a horizontal line drawn upon an inclined plane. It lies at right angles to the direction of dip. The general direction of other structures (e. g., faults, trenches and horsts, mountains, rivers) is called trend.

structure

1. The position and arrangement of rock layers in the subsurface is called geological structure if it differs from homogeneity (homogeneous or horizontally rock layers). Examples: graben, horst, intrusion, fault. The term structure is also used for physical parameters (density, conductivity, etc.) and anomalies of geophysical fields (gravity, magnetic field, traveltime of seismic waves) if they differ from the homogeneous (or horizontally layered) case. Examples: density structures, conductivity structures, high position of a seismic reflector or refractor. 2. Anything constructed or erected with a fixed location on the ground, or attached to something having a fixed location on the ground including, but not limited to, buildings, walls, fences, radio towers, bridges, etc., but not including roads, driveways, sidewalks, the sole purpose of which is to provide access to a structure. 3. In petrology, the megascopic features of a rock mass as seen best in outcrops. For smaller scale features of rocks, see texture.

submersible pump

A water pump with the motor and pump assembly located entirely below the water surface.

subrosion
Dissolution of carbonate, sulfate and chloride rocks creating voids is referred to as subrosion. The relative dissolution rates of limestone, gypsum and rock salt are 1, 100, 10 000. The loss of mass caused by subrosion can result in subsidence or collapse, e.g., sinkholes and caving to surface.

subsidence
The process of sinking or settling of a land surface due to natural or artificial causes (e.g., tectonic processes, compaction, karstification, eruption of lava or withdrawal of large amounts of groundwater, oil, or other underground material).

subsoil
The material between the topsoil and the bedrock.

subsurface water
All water that occurs below the Earth's surface in the liquid, solid or gaseous state.

suction lysimeter
A device consisting of a porous ceramic cup to withdraw soil pore water using a vacuum.

superfund
The USEPA program that funds and carries out programs for remediation of contaminated water resources, solid-waste emergency activities, and long-term contaminant removal. Activities include establishing the National Priorities List, investigating sites for inclusion on the list, determining their priority, and conducting and/or supervising cleanup and other remedial actions (USEPA, 1994).

supervised classification
A type of digital classification of imagery, whereby pixels of unknown identity are classified using samples of known identity (i.e., pixels already assigned to informational classes by ground truthing or registration with known land cover) as training data.

surface nuclear magnetic resonance (SNMR, MRS)
SNMR is a new non-invasive geophysical tool for direct determination of water content of an aquifer and other aquifer parameter values. In contrast to SNMR, other geophysical techniques, e.g., dc resistivity, electromagnetics, seismics and groundpenetrating radar, estimate the lithological parameters of aquifers and aquicludes indirectly. The SNMR method utilizes the nuclear properties of water. Because hydrogen nuclei possess a magnetic moment (spin), they are aligned with the Earth's magnetic field. It is possible to excite them with an external magnetic field and to measure the signal response resulting from precession of the protons

after the external magnetic field decays. Since the amplitude of the SNMR signal is related to the number of excited hydrogen atoms, the method can be used to estimate the water content of rocks and soils. Moreover, information about other hydraulic properties, such as pore size and hydraulic conductivity can be derived from this signal. In contrast to NMR methods applied in the laboratory, which use two strong orthogonal magnetic fields generated by the NMR instrument, SNMR uses the Earth's magnetic field and an external magnetic field generated in a large loop on the Earth's surface. Extensive testing of SNMR worldwide under different geological conditions began in the early 1990s.

surface runoff

The portion of precipitation, snow melt, or irrigation water in excess of what can infiltrate into the soil or temporarily stored in small surface depressions. It is a major transporter of non-point source pollutants in rivers, streams, and lakes. Rain that falls on the stream channel is often lumped with this quantity. → overland flow → streamflow → sheet flow

surface tension

The attraction of molecules to each other on a liquid's surface. Thus, a barrier is created between the air and the liquid tending to minimize the area of the surface. Surface tension results from the force of cohesion between liquid molecules.

surface water

All water on the surface of the Earth naturally open to the atmosphere (e.g., rivers, lakes, reservoirs, ponds, streams, seas and oceans).

survey

1. The systematic collection, analysis and interpretation of information about a site or an area, including subsurface characteristics. 2. The determination of the size, location and physical description of a land area.

surveying

Accurately determining the position of points and the distances between them. This is done by measuring the relative horizontal and vertical position of points on the ground using a theodolite or a similar optical instrument. The points are usually, but not exclusively, associated with positions on the surface of the Earth.

swath

In remote-sensing, the strip of Earth's surface imaged by an orbiting sensor at one time or during one sweep during a single satellite overpass.

system

1. In geosciences, the system is the object under consideration. Together with the surroundings (everything not part of the system, is separated from it by a real or imaginary boundary). 2. A fundamental geologic time-rock unit of worldwide significance. The strata of a system are those deposited during a period of geologic time (for example, rocks formed during the Tertiary period are included in the Tertiary system). 3. A combination of interacting or interdependent components, assembled to carry out one or more functions (e.g., a seismic system includes the seismic source, geophones, cables, filters, amplifiers, recording, processing, and final presentations).

tachymeter

A surveying instrument (theodolite) designed for rapid measurements; in addition to direction and elevation difference, a tachymeter can also measure distance. Often called total station.

target

A specific object or site as the subject of an investigation.

telemetry

Measuring a quantity, transmitting the measured value wireless or by a landline system to a distant station, and there, recording and/or interpreting the quantity measured. Telemetry is typically used to gather data, e.g., groundwater level, precipitation, from distant, inaccessible locations, or when data collection would be dangerous or difficult for a variety of reasons. Telemetry can also mean radio signals from a spacecraft used to transmit data to a ground station and vice versa.

tensiometer

A device used to determine soil moisture tension, an indirect measure of soil moisture content. A porous, permeable ceramic cup is connected through a water-filled tube to a pressure-measuring device.

TEQ

Toxicity equivalence, the international method of relating the toxicity of various dioxin/furan congeners to the toxicity of 2,3,7,8-tetrachlorodibenzo-p-dioxin.

terrain

The surface of an area of land.

texture

1. The overall appearance of a rock in hand specimen or under the microscope reflected by the type, size, shape, and relative abundance of the constituents of the rock, their mutual spatial and size relationships and, in some cases, mutual orientation. 2. In a photographic image the frequency of change and arrangement of tones.

theodolite

Precision surveying instrument for measuring horizontal and vertical angles. In America called optical transit. See also tachymeter.

thermal analysis

Measurement of changes in physical or chemical properties of materials as a function of temperature, usually heating or cooling at a uniform rate. Differential thermal analysis (DTA), measures the temperature difference between a sample and reference material. Differential scanning calorimetry (DSC), measures the differential heat flow between a sample and reference material. Thermogravimetry (TG), measures weight changes of a sample.

thermogravimetry

A chemical analysis method in which changes in weight are measured as a function of changes in temperature, pressure and gas composition.

thin layer chromatography (TLC)

In thin layer chromatography a stationary phase with a thickness of 0.5 - 0.25 µm is immobilized on a sheet of glass, aluminum or plastic. This method is useful for separating organic compounds. Coated plates are commercially available. Silica gel or aluminum oxide is often used as the stationary phase. The mobile phase is usually a mixture of organic solvents. Which mobile phase is selected depends on the substances to be separated. The materials can be separated both by adsorption on the sorption layer and by distribution between the mobile and stationary phases. Samples to be separated are dissolved and are applied at one end of the TLC plate, which is then placed in a chamber containing the mobile phase and filled with its vapor. The mobile phase gradually rises by capillary forces in the layer and transports the individual substances up the TLC plate. Constituents of the sample are separated by distribution between stationary and mobile phase. The ratio of the distance traveled by the individual substance to the distance traveled by the mobile phase is called the Rf value and is substance specific. The separated substances can be detected using appropriate chemical or physical detection methods. TLC is used, for example, for water and waste water analysis, e.g., for PAH or pesticides.

three-dimensional model (3-D model)

A model in which the model parameters are a function of all three of the 3-dimensional coordinates.

tide

The periodic rise and fall of the Earth's surface, especially the oceans, occurring twice a day over of the Earth. It results primarily from the gravitational attraction of the moon and secondarily of the sun acting upon the rotating Earth.

tie line
A survey line that connects a point to bench mark(s) or other surveyed line(s).

time of travel (TOT)
The time required for a contaminant to move in the saturated zone from a source point to a well in according to the average linear velocity of flowing groundwater using Darcy's law.

titration
A chemical method for determining the concentration of a substance in solution by adding a standard reagent of known concentration in carefully measured amounts until the reaction is completed as shown by a color change or electrical measurement. The concentration of the titrated substance is then calculated from the amount reagent used. Titration is used, for example, to determine the acidity of a solution.

tomography
A method for the delineation of subsurface structures (e.g., conductivity distribution, distribution of the seismic wave velocities and geological structures) from geophysical measurements. A subsurface area is scanned with electromagnetic or other waves between a large number of transmitter and receiver positions. A model of this part of subsurface is derived from the measured data. Example: Seismic transmission tomography is used to calculate the distribution of the seismic wave velocity from traveltime measurements for a large number of source and receiver positions between two or more boreholes or the Earth's surface and a borehole. Tomographic methods are also used in geoelectromagnetic and in ground penetrating radar surveys. A variant of tomography uses measurements at the Earth's surface for tomographic interpretation (dc-resistivity geoelectrics, seismic refraction tomography).

topography
1. The physiogeographic characteristics of land in terms of elevation, slope, and orientation. 2. The representation of a portion of the Earth's surface showing natural and man-made features of a given locality, such as rivers, streams, ditches, lakes, roads, buildings and most importantly, ground elevations.

topsoil
Uppermost layer of the soil, also called the A-horizon, usually the most fertile.

tortuosity
The crookedness of the pore pattern. The ratio of the distance between two points by way of the connected pores to the straight-line distance.

total inorganic carbon (TIC)
The carbon containing compounds in soil, sediment and/or water which are not organic compounds, e.g., carbonate and bicarbonate species.

total organic carbon (TOC)
Sum of all organic carbon compounds in mg L^{-1} or mg kg^{-1}. TOC is a monitoring parameter in environmental water programs (screening test). It can influence the concentration of other compounds. Organic matter plays a major role in aquatic systems. It affects biogeochemical processes, nutrient cycling, biological availability, chemical transport and interactions. It also has direct implications in the planning of wastewater treatment and drinking water treatment.

total organic halides (TOX)
The sum of organic chemicals containing chlorine and bromine. The determination of TOX is used as a screening test for contamination.

total petroleum hydrocarbons (TPH)
Screening or indicator tests to evaluate the extent of petroleum product contamination. Petroleum hydrocarbons are a large family of several hundred chemical compounds that originally come from crude oil. It is not practical to measure each one separately. However, it is useful to measure the total amount of TPH at a site. The unit is mg kg^{-1} or ppm hydrocarbon.

total station
→ tachymeter

toxic
The quality of being poisonous or harmful.

toxicity
1. The potential of a substance to exert a harmful effect on humans or animals and a description of the effect and the conditions or concentration under which the effect takes place. 2. The quality of being poisonous or harmful.

trace element
A chemical element that appears in very small amounts in soil, rock, or water with concentrations of less than 1 % (often less than 0.001 %) in soil and rock and less than 1 mg L^{-1} in water. Included are arsenic, cadmium, chromium, copper, lead, mercury, nickel, selenium, and zinc. Organisms need certain trace elements to survive.

tracer
A substance that can be readily identified, such as an isotope or a dye, used to trace the path of a chemical or process or the movement of a fluid or determine parameters that cannot be directly measured (e.g., hydraulic conductivity of an aquifer).

trade-off

The losing of one quality or aspect of something in return for gaining another quality or aspect. An exchange that occurs as a compromise.

transmissivity

A measure of the capability of an aquifer to transmit water. Also known as coefficient of transmissivity. Transmissivity T is related to hydraulic conductivity K by the relationship $T = K\,b$, where b is the thickness of the saturated part of aquifer. Transmissivity is expressed in units of $m^2\ s^{-1}$. The transmissivity changes with the saturated thickness of an unconfined aquifer.

transpiration

The process by which water in plants, absorbed through the roots from the soil, is transferred into the atmosphere from the plant surface, principally from the leaves.

transport

Any process in which movement of matter and/or energy from one part of a system to another occurs.

transport model

Mathematical/physical model of the spreading of substances (e.g., contaminants) in a fluid (e.g., groundwater).

transverse dispersion

The spreading of a dissolved substance in groundwater in the direction normal (i.e., transverse) to that of the groundwater flow.

trench

An exact record of the vertical soil profile can be obtained from observation of the walls of a trench in unconsolidated and consolidated rocks in the unsaturated zone down to a depth of several meters. Trenches allow the taking of disturbed and undisturbed soil samples. The location of the trenches in the area under investigation can be selected relatively arbitrarily in areas of unconsolidated rock in order, for example, to investigate the depth and the structures formed by the loamification of a boulder clay or to determine the bedding conditions in an area of sand cover. Such trenches are easy to construct. In hard rock areas the walls of the trench are generally stable without the need for support measures. The problems in hard rock locations are associated with excavating the trench and depend upon the strength of the rock. The construction of trenches in consolidated rock can also be useful not only for investigating the weathering layers and zones of loosening but also for mapping the contact between adjacent beds and areas where beds thin out, as well as for investigating thin interbeds. Attention should also be paid to the presence of wet zones and layers semi-impervious to water.

trenching
Method of investigating the stratigraphy of the top several meters of soil and or rock.

tsunami
A giant long-wavelength sea wave with periods of 60 min or longer caused by a strong shallow earthquake, volcanic eruption, large landslide on the seafloor or meteorite impact. However, the most common cause is an undersea earthquake. An earthquake which is too small to create a tsunami by itself may trigger an undersea landslide quite capable of generating a tsunami. The effects of a tsunami can range from unnoticeable in the open ocean to devastating when it reaches the shore.

two-dimensional model (2-D model)
The model parameters are a function of two of the three 3-dimensional coordinates. A model based on the assumption that the subsurface structure perpendicular to the profile for which the calculation is made extends an infinite distance.

2.5-dimensional model (2.5-D model)
Perpendicular to the profile for which the calculation is made, the subsurface structure extends the same finite distance on both sides of the profile.

2.75-dimensional model (2.75-D model)
The subsurface structure extends different finite distances on the two sides of the profile for which the calculation is made. The strike direction of the subsurface structure does not necessarily have to be perpendicular to the profile.

uncertainty
1. A statistically defined discrepancy between a measured quantity and the true value of that quantity that cannot be corrected by calculation or calibration. The standard deviation of a sufficiently large number of measurements of the same quantity by the same instrument or method. The uncertainty of an instrument is caused by the unpredictable effects upon its performance of such factors as friction and electronic noise. 2. A measure of the lack of knowledge about certain factors in a study or the amount of doubt or distrust with which the data should be used.

unconfined aquifer
An aquifer in which the water is at atmospheric pressure. The water level in a well tapping an unconfined aquifer will rise only to the groundwater surface within the aquifer. Syn.: phreatic aquifer.

unconfined groundwater
Groundwater in an unconfined aquifer.

unconformity

A break or gap in the geologic sequence where a rock unit is overlain by another that is not next in stratigraphic succession, such as an interruption in continuity of a depositional sequence of sedimentary rocks or a break between eroded igneous rocks and younger sedimentary strata. It results from a change that caused deposition to cease for a considerable time and it normally implies uplift and erosion with loss of previously formed layer(s). An unconformity is often a good seismic reflector and can be recognized even where the layers above and below the unconformity are parallel.

unconsolidated material

Loose sedimentary material in which the particles have not been cemented together, also called unconsolidated rock.

Universal Transverse Mercator (UTM) coordinate system

The UTM coordinate system is a metric coordinate system commonly used for mapping at scales of 1:50 000 or larger and is popular with GPS users. Satellite imagery is usually supplied in UTM coordinates because it provides georeferencing at high levels of accuracy for the entire grid. The UTM divides the Earth into 60 north-south trending zones of 6° longitude between the latitudes 84 °N and 80 °S. Each zone extends 3° east and west from its central meridian and are numbered consecutively west to east from the 180° meridian. Transverse Mercator projections may then be applied to each zone.

unsaturated zone

The zone between the land surface and the top of the saturated zone, i.e., the water table. Some of the pore spaces in this zone are filled with air and some are filled with water. Used synonymously with vadose zone.

unsupervised classification

Classification of digital data by a computer program without interaction with the user.

UXO

Unexploded ordnance (or UXOs) are explosive weapons (bombs, shells, grenades, etc.) which have been primed, fused, armed or otherwise prepared for use or used, that did not explode when they were employed, and still pose a risk of detonation. Unexploded ordnance from as far back as World War I still poses a hazard in parts of the world. Cluster bombs are especially problematic in this regard.

vadose zone

→ unsaturated zone

vegetation

The plant cover of an area.

vegetation anomaly
Deviation from the normal distribution or properties of vegetation that may be caused by faults, anomalies in soil water content, contamination, or other factors.

vertical electrical sounding (VES)
Measurement of the apparent electrical resistivity with successively greater electrode spacing while maintaining a fixed center point of the array. VES gives the resistivity as a function of depth in the case of or assuming horizontal layering.

vertical exaggeration
The ratio of the vertical scale to the horizontal scale in two-dimensional or three-dimensional representation of data, for example in an elevation model in order to more clearly represent the landscape relief. Also used in the representation of geological sections.

viscosity
→ dynamic viscosity

volatile organic analysis (VOA)
Testing procedure for volatile organic chemicals; also called volatiles scan, volatiles screen, or referred to by specific EPA method number, EPA 601 or EPA 602.

volatile organic compound (VOC)
Any organic compound that evaporates readily at room temperature. These compounds are used as solvents, degreasers, paints, thinners and fuels. Due to their low water solubility, environmental persistence and widespread industrial use, they are commonly found in soil and groundwater. VOCs contribute significantly to photochemical smog production and certain health problems.

voltametry
Type of electro-chemical method in which current flows between solid electrodes in an electrolytic cell and the change in voltage is measured with respect to time. Polarography is a special case of voltametry. In water and waste water analysis, voltametry is used for determination of lead, cadmium, cobalt, copper, nickel, thallium and zinc.

volumetric water content
The soil-water content expressed as the volume of water per unit bulk volume (i.e. volume of solids and pores) of soil.

vulnerability

The potential to which a system may be harmed by something external to that system. The vulnerability (sensitivity) of groundwater resources (aquifer systems) to contamination depends on the contaminant retention capacity and homogeneity of the overlaying geological strata.

waste

Any materials remaining after use and rejected as worthless or unwanted from households, medical activities, and/or manufacturing process. Domestic waste is produced by household consumption and processed by traditional collection and separation methods. Municipal waste is generated in households and businesses. Industrial waste can be either nonhazardous or hazardous (toxic). Toxic industrial waste requires special handling during processing to protect the environment. Medical waste arises in doctor's offices and hospitals. See also municipal waste and domestic waste.

wastewater

Water that has been used and is no longer clean. Water that carries wastes from homes, businesses, and industries; a mixture of water and dissolved or suspended solids. See also sewage.

wastewater treatment

The physical, chemical, and/or biological processes of removing waste and contaminants from water that has been used. There are different stages of treatment. Primary sewage treatment involves screening the water to remove the largest particles and then letting the water sit in settling tanks so that the smaller particles sink to the bottom, forming a sludge. Secondary treatment involves another stage in which microbes are added to the wastewater to consume the biodegradable contaminants, or the wastewater is filtered again. The treated water is then disinfected and released back into the environment. The more steps included in the treatment, the more expensive the process. Also called as sewage treatment.

water capacity

The maximum amount of water that a rock or soil can hold against gravity.

water level

1. The elevation of the surface of a body of water (e.g., a lake or river). The elevation is measured with respect to a datum (e.g., mean sea level). 2. Water table, the underground surface below which the ground is saturated with water.

water level meter
A device used to measure the depth to water in a water well. Frequently made of a measuring tape with a probe on the end of the tape, which completes an electric circuit when it contacts water.

water quality
The chemical, physical, and biological characteristics of water, usually with respect to its suitability for a particular purpose. Water quality is affected by natural processes and human activities. Water quality is based upon an evaluation of measured quantities and parameters, which are then compared to water quality standards, objectives or criteria.

watershed
1. A term used in British English for a drainage divide: a line that divides two adjacent river systems or lakes with respect to the flow of water by natural channels into them; the natural boundary of a basin. 2. In the US and a number of international agencies it refers to the area drained by a river system.

water table
The level of the water at the top of the zone of saturation in an aquifer; at this level the pressure is atmospheric.

water year
A continuous 12-month period for which hydrologic records are compiled and summarized. Generally, 1 October to 30 September in the Northern Hemisphere, 1 July to 30 June in the Southern Hemisphere. The annual cycle that is associated with the natural progression of the hydrologic seasons. It commences with the start of the season of soil moisture recharge, includes the season of maximum runoff (or season of maximum groundwater recharge), if any, and concludes with the completion of the season of maximum evapotranspiration (or season of maximum soil moisture utilization).

wavelength
1. The distance between successive points of equal amplitude and phase on a wave of a single frequency (for example, crest to crest or trough to trough). 2. Distance of one period (wave repeat) of a single frequency wave.

weathering
The natural near-surface processes by which the actions of atmospheric and other environmental agents, such as wind, rain, temperature changes, plants, and microorganisms, result in the physical disintegration and chemical decomposition of rocks and Earth materials in place with little or no transport of the loosened or altered material. Weathering is one of the major factors in soil formation.

weir

A low dam built to raise the level of the stream behind it, often to divert flow into another channel, e.g., to a mill. Flow rate (discharge) can be measured by directing the overflow through a spillway designed for this purpose. 2. A fence or enclosure built across a stream to catch or retain fish.

well

1. A drilled or dug hole, often with a casing, whose depth is greater than the largest surface dimension and whose purpose is to tap an underground supply of water. 2. A hole drilled or bored into the Earth, usually cased with metal pipe, for the production of gas or oil. 3. A hole for injection of water or gas under pressure into a subsurface rock formation.

well capacity

The maximum well yield under a stipulated set of conditions, such as a given drawdown, pump and motor, or engine size.

well casing

A steel or plastic pipe which serves as the lining of a well, preventing it from collapsing and protecting groundwater from contamination by surface water. Well casing should extend to at least 30 cm above ground or above the 100-year flood level (the flood level that will occur with a 1 % probability).

well cuttings

→ cuttings

well development

Cleaning of a well by purging (e.g., pumping, airlifting), thereby removing fine-grained material from the formation surrounding the borehole as a process that can enhance well yield.

well filter

→ well screen

well hydrograph

A graphical representation of the fluctuations of the water level in a well, plotted as ordinate, against time, plotted as abscissa.

well logging

→ borehole logging

well screen

A tubular device with either slots or holes used as part of well casing to complete a well. The water enters the well through the well screen and the sediment is kept from entering the well by the filter gravel in the annular ring.

well whistle

A device to determine the depth of the water level in a well.

well yield

The volume of water discharged from a well in cubic meters per day.

WGS-84

→ World Geodetic System 1984 (WGS-84)

World Geodetic System 1984 (WGS-84)

The WGS-84 is the standard coordinate system (mathematical ellipsoid) used as a geodetic datum by GPS since January 1987. It will be valid up to about 2010. WGS84 coordinates are usually expressed as latitude, longitude and ellipsoid height as a measure of altitude.

x-ray diffraction (XRD)

A method of analyzing the structure of a mineral by examining the pattern created when x-rays are passed through a crystal of the mineral or a powdered sample in which crystals have varied and random orientation.

x-ray fluorescence analysis (XRF)

→ x-ray fluorescence spectroscopy

x-ray fluorescence spectroscopy (XRF)

An analytical method for the total determination of major and trace elements by measuring secondary x-ray emissions after a solid sample is bombarded with a primary x-ray beam. The energy level of the secondary x-ray emission of an element is characteristic of that element; the intensity of the emission is a function of the concentration. Two types of x-ray fluorescence spectroscopy are distinguished: wavelength dispersive x-ray spectroscopy (WDX) and energy dispersive x-ray spectroscopy (EDXRF).

zone of contribution (ZOC)

The area surrounding a pumped well that encompasses all areas or features that supply groundwater recharge to the well. It is defined as the area required to support abstraction from long-term groundwater recharge.

References and further reading

FANG, Z. (1995): Flow injection atomic absorption spectrometry. Wiley & Sons, New York.

FETTER, C. W. (1994): Applied hydrogeology. Macmillan College Publishing, Co.

FREEZE, R. A. & CHERRY, J. A. (1979): Groundwater. Prentice-Hall, Englewood Cliffs, N.J.

JACKSON, J. A. (Ed.) (1997): Glossary of Geology. 4th edn. American Geological Institute, Alexandria, Virginia.

LOHMANN, S. W. et al. (1972): Definitions of selected ground-water terms – revisions and conceptual refinements. US Geological Survey. Water-Supply Paper 1988.

MCNAUGHT, A. D. & WILKINSON, A. (1997). IUPAC Compendium of Chemical Terminology – The Gold Book. 2nd edition. Blackwell Science. Also available online: http://www.iupac.org/publications/compendium/index.html

NAS-NRC (1994): Alternatives for groundwater cleanup. National Academy of Science, National Research Council, National Academy Press, Washington, DC.

SABINS, F. F. (1997): Remote Sensing: Principles and Interpretation. 3rd edn. Freeman & Co., New York.

SEEL, F. (1979): Grundlagen der analytischen Chemie unter besonderer Berücksichtigung der Chemie in wäßrigen Systemen. (Basics of analytical chemistry with a particular focus on chemistry of aqueous systems). 7th edn. VCH, Weinheim.

SHERIFF, R. E. (1991): Encyclopedic Dictionary of Exploration Geophysics, 3rd edn. Society of Exploration Geophysicists.

SMALL, H. (1989): Ion Chromatography. Plenum Press, New York.

STANGER, G. (1994): Dictionary of hydrology and water resources. Lochnan, Adelaide, South Australia.

TASi (1993): Technical Instructions on Waste from Human Settlements (TA Siedlungsabfall), Germany.

USCE (1991): Glossary of hydrologic engineering terms. U.S. Army corps of Engineers, Hydrologic Engineering Center, Davis, CA.

USEPA (1994): Terms of environment, glossary, abbreviations, and acronyms. US Environmental Protection Agency EPA 175-B 94-015.

USEPA (1999): OSWER Directive 9200.4-17P. Use of monitored natural attenuation at Superfund, RCRA corrective action, and underground storage tank sites. US Environmental Protection Agency, April 21, 1999, Office of Solid Waste and Emergency Response. http://www.epa.gov/swerust1/directiv/d9200417.htm

WEISS, J. (1991): Ionenchromatographie. 2. erw. Aufl., VCH, Weinheim.

WELZ, B. & SPERLING, M. (1997): Atomabsorptionsspektrometrie (Atomic absorption spectrometry). 4th edn. Wiley-VCH, Weinheim.

WILSON, W. E. & MOORE, J. E. (eds.) (2001): Glossary of Hydrology. American Geological Institute.

Abbreviations

AAS	atomic absorption spectrometry
ABIMO	ABflussBIldungsMOdell (runoff generation model)
ac	alternating current
ACE	acenaphthene
A/D	analog to digital
ADC	analog-digital converter
AES	atomic emission spectroscopy (optical emission spectroscopy - OES)
AFS	atomic fluorescence spectrometry
AGC	automatic gain control (maintains a constant output signal level virtually constant for varying input signal levels)
AIS	Airborne Imaging System
AISA	Airborne Imaging Spectrometer for Applications
AL	acoustic log (sonic log)
AMT	audio-magnetotelluric method
AOX	adsorbed organic halogens
ASAS	Advanced Solid-State Array Spectrometer
ASCE	American Society of Civil Engineers
ASCII	American Standard Code for Information Interchange (standard code used in computers to represent letters, numbers and other characters)
ASGG.D	annular space gamma-gamma density scanner
ASTM	American Society for Testing Materials
AVIRIS	Airborne Visible Infrared Imaging Spectrometer
BGR	Bundesanstalt für Geowissenschaften und Rohstoffe (German Federal Institute for Geosciences and Natural Resources)
BHTV	borehole televiewer
BLM	Bohrlochmessung GmbH (German well logging company)
BMBF	Bundes Ministerium für Bildung und Forschung (German Federal Ministry for Education and Research)
BOD	biological (biochemical) oxygen demand

BTX	benzene, toluene, xylene
BTEX	sum of benzene, toluene, ethylbenzene, and xylene
BUWAL	Bundesamt für Umwelt, Wald und Landschaft (Swiss Federal Agency for the Environment, Forests, and Landscape)
CAD	computer aided design
CAL	caliper log
CASI	Compact Airborne Spectrographic Iimager
CBL	cement bond log
CCD	charge-coupled device (image sensor)
cDCE	*cis*-1,2-dichloroethene
CDP	common depth point
CE	capillary electrophoresis
CEC	cation exchange capacity
CET	Central European (standard) Time (1 hours ahead of GMT)
cf.	compare (Latin confer)
CFC	chlorofluorocarbon, fluorochlorinated hydrocarbon
CIR	color-infrared
CIS	Chinese Imaging Spectrometer
CMP	common midpoint (in reflection seismics and GPR)
CN	cyanide
COD	chemical oxygen demand
CPT	cone penetration testing
CPTU	piezocone test – cone penetration test with pore water pressure measurement
CRIM (formula)	complex refractive index method (empirical relation used in GPR analog to Wyllie's time average equation for seismic waves)
CSAMT	controlled-source audio-magnetotellurics
CSA/CCRS	Canadian Space Agency/Canada Centre for Remote Sensing
CV-AAS	cold-vapor AAS
DAD	diode array detector
DAIS	Digital Airborne Imaging Spectrometer
DBMS	database management system
dc	direct current

DDT	dichloro-diphenyl-trichlorethane
DEM	digital elevation model, → DTM
DEV	deviation log (borehole geometry)
DGPS	differential global positioning system
DGZfP	Deutsche Gesellschaft für Zerstörungsfreie Prüfung (German Society for Non-Destructive Testing)
DIN	Deutsches Institut für Normung (German Institute for Standardization)
DIN EN ISO	International, European, and German standard
DIP	dipmeter tool in borehole logging
DLL	dual (spacing) laterolog (provides two resistivity measurements with different investigation depths)
DLLd	dual laterolog, deep
DLLs	dual laterolog, shallow
DLR	Deutsches Zentrum für Luft- und Raumfahrt e.V. (German Aerospace Center)
DMC	Digital Modular Camera
DMO	dip moveout → NMO
DNA	deoxyribonucleic acid
DNAPL	denser-than-water nonaqueous-phase liquid
DO	dissolved oxygen
DOC	dissolved organic carbon
DRI	double ring infiltrometer
DSC	differential scanning calorimetry
DSO	digital storage oscilloscopes
DTA	differential thermal analysis
DTM	digital terrain model → DEM
DVWK	Deutscher Verband für Wasserwirtschaft und Kulturbau (German Association for Water, Wastewater and Waste)
EBS	elastically backscattered neutron spectrum (used in neutron-gamma spectrometry)
EC	electrical conductivity
ECD	electron capture detector
EDTA	ethylenediaminetetraacetic acid
EDXRF	x-ray fluorescence spectroscopy
EEA	European Environmental Agency

Eh	redox potential
EL	electric log
ELGI	Eötvös Loránd Geophysical Institute, Budapest, Hungary
EM	electromagnetics
EMR	electromagnetic reflection
ENA	enhanced natural attenuation
ENVI	Environment for Visualizing Images (software used for remote sensing)
EOX	extractable organohalogens
EPA	Environmental Protection Agency
ERDAS	software for image processing
ERT	electrical resistivity tomography
ESA	European Space Agency
ESRI	Environmental Systems Research Institute
ET	Evapotranspiration
ET-AAS	electrothermal AAS
ETM	Landsat Enhanced Thematic Mapper
EU	European Union
EWG	elastic wave generator (a seismic source)
FAAS	flame atomic absorption spectrometry
FAES	flame atomic emission spectrometry
FAO	Food and Agriculture Organization of the United Nations
FD	1. finite difference, 2. frequency domain
FDTD	finite difference time domain
FE	1. finite element, 2. frequency effect (used in induced polarization, IP)
FEL	focused electrolog
FEM	1. frequency domain electromagnetic method, 2. finite element method used for solving differential equations
FDEM	frequency domain electromagnetic method
FFT	fast Fourier transform (an algorithm for computing the Fourier transform of a set of discrete data values)
FIA	flow injection analysis
FID	flame ionization detector
FIR	far infrared

f-k-filter	frequency-wavenumber filter (velocity filter)
FLOW	flowmeter log
FMC	forward motion compensation
FOV	angular field of view (angle of view)
FUGRO	geotechnical, survey and geosciences company
GC	gas chromatography
GC-ECD	gas chromatography-electron capture detector
GC-FID	gas chromatography-flame ionization detector
GC-MS	gas chromatography-mass spectrometry
GC-PID	gas chromatography-photoionization detection
GC-PND	gas chromatography with phosphorus-nitrogen detection
GCM	ground conductivity meter (electromagnetic instrument operating under low induction number (LIN) conditions)
GD-MS	glow discharge mass spectrometry
GD-OES	glow discharge optical emission spectrometry
Geophysik GGD	Gesellschaft für Geowissenschaftliche Dienste mbH (International Geophysical Services GmbH)
GF-AAS	graphite furnace AAS
GG	gamma-gamma log
GGD/GG.D	gamma-gamma density log
GIS	geographic information system
GMT	Greenwich Mean Time
GP	Guelph permeameter
GPR	ground penetrating radar
GPS	Global Positioning System
GR	gamma-ray log, also called natural gamma ray log
GRI	gamma-ray index (a "clayiness" index)
GRM	generalized reciprocal method (a seismic analysis method)
GROWA	groundwater balance model
GRS	spectral natural gamma radiation
GRS 80	Geodetic Reference System 1980, adopted by the International Association of Geodesy (IAG)
GSMS	Geological Survey of Malaysia, Sarawak
GW, gw	groundwater

HC	hydrocarbons
HCH	hexachlorocyclohexane
HD	high density
HDPE	high-density polyethylene
HEM	horizontal-loop electromagnetics
HG-AAS	hydride generation AAS
HLEM	horizontal-loop electromagnetics
HMD	horizontal magnetic dipole
HPLC	high-performance liquid-chromatography or high-pressure liquid-chromatography
HPLC-F	HPLC with fluorescence detector
HPLC-DAD	HPLC with diode array detector
HPLC-UV	HPLC with UV detector
HPTLC	high-performance thin-layer chromatography
HRGC-HRMS	high-resolution gas chromatography coupled with high-resolution mass spectrometry
HRSC	High Resolution Stereoscopic Camera
HRV	high resolution visible infrared
HyMap	Airborne Hyperspectral Mapping System
HYDICE	Hyperspectral Digital Imagery Collection Experiment
IAEA	International Atomic Energy Agency
ICP	inductively-coupled plasma
ICP-AAS	inductively-coupled-plasma atomic-absorption spectrometry
ICP-AES	inductively-coupled-plasma atomic-emission spectrometry
ICP-MS	inductively-coupled-plasma mass spectrometry
ICP-OES	inductively-coupled plasma atomic-emission spectrometry
ICP-QMS	inductively-coupled-plasma quadrupole mass spectrometry
ICPS	inductively-coupled-plasma emission spectrometry
IDRISI	raster based GIS software package
IEEE	Institute of Electrical and Electronics Engineers
IFFT	inverse fast Fourier transformation
IFOV	instantaneous field of view
IGIS	International Geographic Information Systems

IGRF	International Geomagnetic Reference Field
IL	induction log
IMC	image motion compensation
ING.W	impulse neutron gamma log
INN	pulsed neutron lifetime log
INS	inertial navigation system
InSAR	interferometric synthetic-aperture radar
INTERPEX	US software company producing software for geophysical methods
IR	1. infrared, 2. irrigation water
IRIS	Infrared Imaging Spectroradiometer
IP	induced polarization
ISO	International Standards Organization
ISPRS	International Society for Photogrammetry and Remote Sensing
IT	information technology
KTB	German Continental Deep Drilling Program
LAGA	Länderarbeitsgemeinschaft Abfall (German Federal States Working Group on Waste)
LAN	local area network
LC	liquid chromatography
LIBS	laser-induced breakdown spectroscopy
LIN	low induction number
LNAPL	less-dense-than-water nonaqueous-phase liquid
LoD	limit of detection
LoQ	limit of quantification
lp	(visual) line pair
M	molarity
MAF	New Zealand Ministry of Agriculture and Forestry
MAL	magnetic log
MAS	MODIS airborne simulator
MASW	multichannel analysis of surface waves
MAPX	radar based mean areal precipitation
MDL	method detection limit

MIL	multi-parameter probe to measure local environmental parameters (pH, oxygen concentration, redox potential of the borehole fluid)
MIP	membrane interface probe
MIR	mid-infrared
MIVIS	Multispectral Infrared and Visible Imaging Spectrometer
MKSC	parameter indicating the proportions of the clay minerals muscovite, kaolinite, smectite and chlorite
ML	microlog, minilog
MLL	microlaterolog
MODIS	Moderate Resolution Imaging Spectrometer
MOS	modular optoelectronic scanner
MPN	most probable number
mRNA	messenger RNA
MRS	magnetic resonance sounding (→ SNMR)
MS	mass spectrometry
MSK	Medvedev-Sponheuer-Karnik macroseismic intensity scale
MSL	mean sea level
MSS	Landsat Multispectral Scanner
MSW	municipal solid waste
MTBE	methyl tertiary butyl ether
MW	microwave (radar)
NA	natural attenuation
NAP	naphthalene
NAPL	non-aqueous phase liquid
NASA	National Aeronautics and Space Administration (USA)
NATO/CCMS	North Atlantic Treaty Organisation, Committee on the Challenges of the Modern Society
NAVSTAR	Navigation Signal Timing and Ranging (GPS)
NCGIA	National Center for Geographic Information and Analysis (USA)
NDVI	normalized difference vegetation index
NEMO	Navy Earth Map Observer (imaging satellite program of the U.S. Naval Research Laboratory)

NEXRAD	acronym for Next Generation Weather Radar
NIR	near infrared
NMO	normal moveout
NMR	nuclear magnetic resonance
NN	neutron-neutron log
NOAA	National Oceanic and Atmospheric Administration of the U.S. Department of Commerce
NOAA-AVHRR	NOAA-Advanced Very High Resolution Radiometer
NUV	near-ultraviolet
OES	optical emission spectrometry
ODP	Ocean Drilling Program
ORP	oxidation reduction potential
OPT	video image of the borehole wall
OSHA	Occupational Safety & Health Administration (USA)
PAH	polycyclic aromatic hydrocarbons
PC	personal computer
PCB	polychlorinated biphenyl
PCDD	polychlorinated dibenzo-p-dioxin
PCDF	polychlorinated dibenzo-furans
PCDD/F	polychlorinated dibenzo-p-dioxin and furans
PCE	perchloroethylene (primarily used as a dry-cleaning agent)
PCR	polymerase chain reaction
PDOP	position dilution of precision (in GPS)
PE	polyethylene
pF	water retention
PHC	petroleum hydrocarbons
PHE	phenanthrene
PI	phenol index
PID	photo ionization detector
PMT	1. photo-multiplier tube, 2. photometric turbidity measurement
PMR	proton magnetic resonance (see also NMR, MRS)
ppb	parts per billion
ppm	parts per million
ppt	parts per thousand

PT	pulse test
PVC	polyvinyl chloride
PYR	pyrene
Q-mode	Q-mode factor analysis in multivariate statistics
RAM	random access memory (read and write memory)
RAR	real aperture radar
rDNA	recombinant DNA
RES	radio echo sounding
RFA	roentgen fluorescence analysis, x-ray fluorescence analysis
RMT	radiomagnetotellurics
RNA	ribonucleic acid
RS-232	serial communications standard, most commonly used for low-end professional and consumer equipment
RS-485	standard interface for professional equipment, allows up to 32 devices to share the same connection
Rx	receiver (loop) in electromagnetics
SAL	salinity log (electrical resistivity of the drilling fluid)
SAMP	sample collecting tool
SAR	synthetic aperture radar
SB test	slug and bail test
SDAG Wismut	Soviet-German corporation for Uran mining in the former GDR
SEG	Society of Exploration Geophysicists (Tulsa, Oklahoma)
SEG 2; SEG D; SEG Y	standard seismic data formats recommended by the Technical Standards Committee
SEGEMAR	Servicio Geologico Minero Argentino Geological Survey of Argentina)
SERNAGEOMIN	Servicio Nacional de Geología y Minería (Geological Survey of Chile)
SFR	stepped-frequency radar
SGL	segmented gamma-ray log
SGM	soil and groundwater measurement system
SI	Système International d'Unités (International System of Units)
SIP	spectral induced polarization

SIR	Subsurface Interface Radar
SIRT	simultaneous iterative reconstruction technique
SLAR	side-looking airborne radar
SNMR	surface nuclear magnetic resonance (see also MRS, PMR)
S/N	signal-to-noise ratio
SOR	solar reflection region
SP	self-potential, spontaneous potential
SPA	source protection area
SPOT	Systeme Pour l'Observation de la Terre (French Earth observation system)
SPSS	Statistical Package for the Social Sciences (used for univariate and multivariate statistical analyses)
SRTM	Shuttle Radar Topography Mission
SUSZ	magnetic susceptibility log
Syn.	Synonym
SWIR	short-wave infrared
TCE	trichloroethene
TCP	triangulated (ground) control point
TDS	thermal-desorption sampler
TDS	total dissolved solids
TE	Thornwaite's temperature efficiency index
TEA	terminal electron acceptor
TEM	transient electromagnetic method
TEMP	temperature log
TEQ	toxicity equivalence
TIC	total inorganic carbon
TIN	triangulated irregular networks
TIR	thermal infrared
TLC	thin layer chromatography
TM	Landsat Thematic Mapper
TOC	total organic carbon
TOT	time of travel
TOX	total organic halide
TPH	total petroleum hydrocarbons
TRWIS III	TRW Imaging Spectrometer

TWT	two-way travel time
Tx	transmitter (loop) in electromagnetics
U	coefficient of uniformity
UHF	ultra high frequency
UTC	Coordinated Universal Time
USGS	U.S. Geological Survey
US PHS	U.S. Public Health Service
USDA SCS	U.S. Department of Agriculture, Soil Conservation Service
USDI	U.S. Department of the Interior
UTM	Universal Transverse Mercator System
UV	ultraviolet light
UXO	unexploded ordnance
VC	vinyl chloride
VCH	volatile chlorinated hydrocarbons
VCP	vertical coplanar
VDI	Verein Deutscher Ingenieure (Organization of German Engineers)
VES	vertical electrical sounding
VETEM	very early time domain EM
VHC	volatile halogenated hydrocarbons
VHF	very high frequency
VIMS-V	Visual and Infrared Mapping Spectrometer
VIS	visible part of spectrum
VLEM	vertical loop electromagnetic method
VLF	1. very low frequency band (30 - 300 kHz) 2. electromagnetic survey method using radio broadcast transmitters
VLF-R	VLF resistivity
VMD	vertical magnetic dipole
VOA	volatile organic analysis
VOC	volatile halogenated hydrocarbons
VSP	vertical seismic profiling
WARR	wide-angle reflection and refraction
WD test	water pressure test to determine hydraulic properties of the ground

WDX	wavelength dispersive x-ray fluorescence spectroscopy
WHO	World Health Organization
WE	→ WU
WGS 84	World Geodetic System 1984
WIB GmbH	Weltraum-Institut Berlin GmbH (Berlin Space Institute Inc.)
WSR 88-D	Weather Surveillance Radar 88-Doppler
WU	water unit
XRD	x-ray diffraction
XRF	x-ray fluorescence
XSD	halogen-specific detector
ZOC	zone of contribution
4WD	four-wheel drive
1-D, 2-D, 3-D	one-dimensional, two-dimensional, three-dimensional

Units of Measure

The SI system, Système International d'Unités (International System of Units), a complete metric system of units is used in this book. In some countries and in older publications other unit systems are used. Therefore, some of such units are listed in the table "Non SI Units" and conversion factors are given. The units in these tables are restricted to those used in this book. For the preparation of these tables the following website was used: www.physics.nist.gov/cuu/Units/.

Symbol	Name	Quantity	Expression in terms of other SI units	Expression in terms of SI base units
m	meter	length		base unit
kg	kilogram	mass		base unit
s	second	time		base unit
A	ampere	electric current		base unit
K	kelvin	thermodynamic temperature		base unit
mol	mole	amount of substance		base unit
cd	candela	luminous intensity		base unit
m^2	square meter	area		SI derived unit
m^3	cubic meter	volume		SI derived unit
$m\ s^{-1}$	meter per second	speed, velocity		SI derived unit
$m\ s^{-2}$	meter per second squared	acceleration		SI derived unit
m^{-1}	per meter	wave number		SI derived unit
$kg\ m^{-3}$	kilogram per cubic meter	mass density		SI derived unit
$m^3\ kg^{-1}$	cubic meter per kilogram	specific volume		SI derived unit
$A\ m^{-2}$	ampere per square meter	current density		SI derived unit
$A\ m^{-1}$	ampere per meter	magnetic field strength		SI derived unit
$mol\ m^{-3}$	mole per cubic meter	concentration of a substance		SI derived unit
$cd\ m^{-2}$	candela per square meter	luminance		SI derived unit
$mg\ kg^{-1}$	milligram per kilogram	mass fraction		SI derived unit
rad	radian	plane angle		$m\ m^{-1}$

Symbol	Name	Quantity	Expression in terms of other SI units	Expression in terms of SI base units
Hz	hertz	frequency		s^{-1}
N	newton	force		$m\ kg\ s^{-2}$
Pa	pascal	pressure, stress	$N\ m^{-2}$	$m^{-1}\ kg\ s^{-2}$
J	joule	energy, work, quantity of heat	N m	$m^2\ kg\ s^{-2}$
W	watt	power, radiant flux	$J\ s^{-1}$	$m^2\ kg\ s^{-3}$
C	coulomb	electric charge, quantity of electricity		s A
V	volt	electric potential, electromotive force	$W\ A^{-1}$	$m^2\ kg\ s^{-3}\ A^{-1}$
F	farad	capacitance	$C\ V^{-1}$	$m^{-2}\ kg^{-1}\ s^4\ A^2$
Ω	ohm	electric resistance	$V\ A^{-1}$	$m^2\ kg\ s^{-3}\ A^{-2}$
S	siemens	electric conductance	$A\ V^{-1}$	$m^{-2}\ kg^{-1}\ s^3\ A^2$
Wb	weber	magnetic flux	V s	$m^2\ kg\ s^{-2}\ A^{-1}$
T	tesla	magnetic flux density	$Wb\ m^{-2}$	$kg\ s^{-2}\ A^{-1}$
H	henry	inductance	$Wb\ A^{-1}$	$m^2\ kg\ s^{-2}\ A^{-2}$
°C	degree Celsius	temperature		K, $T_{°Celsius} = T_{kelvin} - 273.15$
Bq	becquerel	activity (of a radionuclide)		s^{-1}
Sv	sievert	dose equivalent	$J\ kg^{-1}$	$m^2\ s^{-2}$
Pa s	pascal second	dynamic viscosity		
N m	newton meter	moment of force		
$N\ m^{-1}$	newton per meter	surface tension		
$W\ m^{-2}$	watt per square meter	heat flux density, irradiance		
$J\ K^{-1}$	joule per kelvin	heat capacity, entropy		
$J\ (kg\ K)^{-1}$	joule per kilogram kelvin	specific heat capacity, specific entropy		
$J\ kg^{-1}$	joule per kilogram	specific energy		
$W\ (m\ K)^{-1}$	watt per meter kelvin	thermal conductivity		
$J\ m^{-3}$	joule per cubic meter	energy density		
$V\ m^{-1}$	volt per meter	electric field strength		
$C\ m^{-3}$	coulomb per cubic meter	electric charge density		
$C\ m^{-2}$	coulomb per square meter	electric flux density		

Symbol	Name	Quantity	Expression in terms of other SI units	Expression in terms of SI base units
$F\ m^{-1}$	farad per meter	permittivity, dielectric constant		
$H\ m^{-1}$	henry per meter	magnetic permeability		
$J\ mol^{-1}$	joule per mole	molar energy		
cmol+/kg or cmol(+)/kg	centimoles of positive charge per kilogram	moles of charge per mass		

SI Prefixes

Factor	Name	Symbol
10^{12}	tera	T
10^{9}	giga	G
10^{6}	mega	M
10^{3}	kilo	k
10^{2}	hecto	h
10^{1}	deka	da

Factor	Name	Symbol
10^{-1}	deci	d
10^{-2}	centi	c
10^{-3}	milli	m
10^{-6}	micro	µ
10^{-9}	nano	n
10^{-12}	pico	p

Non SI Units

Symbol	Name	Explanation and/or expression in terms of SI units
a	year	equal to 365 or 366 (leap year) days (d)
API (unit)	American Petroleum Institute unit(s)	unit of counting rate for gamma-ray logs and neutron logs
Å	ångström	1 Å = 0.1 nm = 10^{-10} m
B	bel	1 B = (1/2) ln 10 Np
bar	bar	1 bar = 1000 hPa = 10^{5} Pa
bit	bit	smallest unit of information in a binary system, either 0 or 1
byte	byte	1 byte consists of 8 bits
cal	calorie	1 cal = 4.1868 J
cps	counts per second	unit of counting rate for gamma-ray log and/or neutron log
d	day	1 d = 24 h = 86 400 s
dB	decibel	logarithmic unit R of the linear ratio r of the actual signal amplitude to a reference amplitude. R (in dB) = 10 $\log_{10}(r)$. A decrease of 70 % in the signal amplitude (50 % in energy) corresponds to -3 dB.

Symbol	Name	Explanation and/or expression in terms of SI units
ft or '	foot	1 ft = 12 in = 0.3048 m
gal	gallon	3.785 41 L (US); 4.546 L (British)
Gal	Gal	10^{-2} m s^{-2}
g.u.	gravity unit	1 g.u. = 0.1 mGal =10^{-6} m s^{-2}
h	hour	1 h = 60 min = 3600 s
ha	hectare	1 ha = 10^4 m^2
in or "	inch	$2.54 \cdot 10^{-2}$ m
L	liter	1 L = 1 dm^3 = 10^{-3} m^3
meq	milliequivalent	1 mmole ≡ 1 milliequivalent of an ion with a charge of one
mile	mile	$1.609344 \cdot 10^3$ m
min	minute (time)	1 min = 60 s
mGal	milligal	unit of acceleration used with gravity measurements, 10^{-3} Gal = 1 mGal = 10 gravity units (g. u.)
Np	neper	1 Np = 1= 8.686 dB
pH	potential of hydrogen	the logarithm of the reciprocal of hydrogen ion concentration in an aqueous solution
t	metric ton	1 t = 10^3 kg
WU or WE	water unit	unit used in well logging for the quantity of formation water present in a unit volume of rock. The product of water saturation and porosity.
°	degree (angle)	1° = (π/180) rad
'	minute (angle)	1' = (1/60)° = (π/10 800) rad
"	second (angle)	1" = (1/60)' = (π/648 000) rad

Physical constants

Symbol	Name	Explanation and/or expression in SI units
c_0	velocity of light in vacuum	$2.998 \cdot 10^8$ m s^{-1} = 0.2998 m ns^{-1}
G	gravitational constant	$6.672 \cdot 10^{-11}$ m^3 kg^{-1} s^{-2}
k	Boltzmann constant	$1.38065 \cdot 10^{-23}$A s K^{-1}
γ	gyromagnetic ratio of hydrogen ^1H	0.267 518 Hz nT^{-1}
ε_0	permittivity of free space	$8.85419 \cdot 10^{-12}$ A s V^{-1} m^{-1}
μ_0	magnetic permeability of free space	$4\pi \cdot 10^{-7}$ V s $A^{-1}m^{-1}$
σ	Stefan-Boltzmann constant	$5.6703 \cdot 10^{-8}$ W m^{-2} K^{-4}

Mathematical Symbols and Constants

Symbol	Meaning	Explanation
a = b	a is equal to b	
a ≠ b	a is not equal to b	
a ≈ b or a ~ b	a is approximately equal to b	
a ≡ b	a is the same as b	
a < b	a is less than b	
a > b	a is greater than b	
a ≤ b	a is less than or equal b	
a ≥ b	a is greater than or equal b	
a ≪ b	a is much less than b	
a ≫ b	a is much greater than b	
$\sqrt{a^2 + b^2}$	square root of $a^2 + b^2$	
$x \perp y$	x is perpendicular to y	
$x \parallel y$	x is parallel to y	
$\boldsymbol{u}$	vector $\boldsymbol{u}$	
$\boldsymbol{u} \cdot \boldsymbol{v}$ or $(\boldsymbol{u},\boldsymbol{v})$	dot product of the vectors $\boldsymbol{u}$ and $\boldsymbol{v}$	the scalar product of two vectors
$\boldsymbol{u} \times \boldsymbol{v}$ or $[\boldsymbol{u},\boldsymbol{v}]$	cross product of the vectors $\boldsymbol{u}$ and $\boldsymbol{v}$	the vector product of two vectors
∂	partial derivative	
∇ , del	nabla operator	$\nabla = \boldsymbol{i}\frac{\partial}{\partial x} + \boldsymbol{j}\frac{\partial}{\partial y} + \boldsymbol{k}\frac{\partial}{\partial z}$
Δ	Laplace operator or operator	$\Delta = \frac{\partial^2}{\partial x^2} + \frac{\partial^2}{\partial y^2} + \frac{\partial^2}{\partial z^2} = \nabla \cdot \nabla = \nabla^2$
grad f	gradient, derivative of a scalar field f	$\text{grad}\, f = \frac{\partial f}{\partial x}\boldsymbol{i} + \frac{\partial f}{\partial y}\boldsymbol{j} + \frac{\partial f}{\partial z}\boldsymbol{k} = \nabla f$

Symbol	Meaning	Explanation
div $\boldsymbol{v}(x,y,z)$	divergence of a vector field $\boldsymbol{v}(x,y,z) = v_x \boldsymbol{i} + v_y \boldsymbol{j} + v_z \boldsymbol{k}$	$\text{div}\,\boldsymbol{v} = \frac{\partial v_x}{\partial x} + \frac{\partial v_y}{\partial y} + \frac{\partial v_z}{\partial z} = \nabla \cdot \boldsymbol{v}$
curl $\boldsymbol{v}(x,y,z)$	curl of a vector field $\boldsymbol{v}(x,y,z) = v_x \boldsymbol{i} + v_y \boldsymbol{j} + v_z \boldsymbol{k}$	curl $\boldsymbol{v}$ $= \left(\frac{\partial v_z}{\partial y} - \frac{\partial v_y}{\partial z}\right)\boldsymbol{i} + \left(\frac{\partial v_x}{\partial z} - \frac{\partial v_z}{\partial x}\right)\boldsymbol{j} + \left(\frac{\partial v_y}{\partial x} - \frac{\partial v_x}{\partial y}\right)\boldsymbol{k}$ $= \nabla \times \boldsymbol{v}$
$\sum_{k=1}^{n} a_k$	sum of a_k from 1 to n	$a_1 + a_2 + \ldots + a_n$
$\prod_{k=1}^{n} a_k$	product of a_k from 1 to n	$a_1\, a_2 \cdots a_n$
$\int$	indefinite integral	
$\int_a^b$	definite integral between a and b	
∞	indefinite	
π	pi	Its value is 3.14159265....
e	Euler's number, base of the natural logarithm and of the exponential function e^x	Its value is 2.71828….
log	logarithm, the inverse of an exponent	
$\log_b a$	logarithm of a to the base b	
$\ln a = \log_e a$	natural logarithm, the logarithm to the base e	
$\lg a = \log_{10} a$	logarithm to base 10 (common logarithm). On calculators denoted as lg or log. Sometimes "log" is used for the natural logarithm ln.	
$\exp(x) = e^x$	exponential function with base e	

Subject Index

abandoned industrial site → industrial site
abandoned landfill → landfill
abandoned mining site → mining site
absorption 25-29, 104, 109, 128-133, 148-150, 284-297, 301-303, 329, 343, 476, 481-484, 493, 499-502, 552, 802, 811, 817-818, 824, 858-860, 998, 1151
acceleration of gravity 679
accessibility 11, 170
acid-alkaline reaction 753
acoustic log 436, 453, 465-467, 1127
acoustic tomography 539
across-track scanner 97
adhesive water 406-407, 416-417, 1159, 1162
adsorbable organic halogens 480, 625, 783, 875
adsorption 752, 783, 810-812, 823-826, 856-876, 1026
adsorption-desorption isotherm → isotherm
advection 1008-1013, 1022-1041
aeration zone → unsaturated zone
aerial photography 4, 12-13, 23-67, 73-91, 98-99, 108, 125-130, 140-147, 274, 965, 976-977, 1055-1056, 1091, 1099-1107, 1118-1121, 1131-1133
aerobic condition 807, 832, 878-885, 902-911, 955
agriculture 607, 1062, 1078, 1128, 1184
A-horizon → soil horizon
airborne system 4, 12-13, 23-67, 73-91, 97-99, 73, 97-115, 121-136, 140-150, 274, 965, 976-977, 1055-1056, 1091, 1099-1107, 1118-1121, 1131-1133
aliasing 303, 362, 374, 375
alkali vapor magnetometer 169
along-track scanner 97, 106-107
altimetry 18-20
ambiguity 191, 199, 212, 225, 320, 338, 374, 842, 1055
ammunition site 15, 162, 169, 172-173, 180-182, 242-243, 258-264, 285
anaerobic (condition) 142, 752, 807, 832, 876-911, 955, 991, 1013
angular field of view 100-103, 116, 121-122
anion 778, 788, 851, 868, 1026
anion exchange capacity 1026
anisotropy 642, 683, 709
annular space gamma-gamma density scanner 635-636
annulus (of a borehole) → casing
antenna 97, 112, 113,257, 258, 284, 291-307, 313-329, 410, 476, 484-493, 1150, 1151, 1154
antenna characteristics 112
anthropogenic background → chemical background
anthropogenic emission 749-751, 1133, 1163, 1185
aperture 100, 108, 112, 127
apparent resistivity 205-209, 214, 221-228, 233-235, 255-257, 266, 270-277, 439, 1109, 1137-1143
apparent velocity 364
a-priori information 219, 222, 224, 413, 1054
aquiclude 207, 403-404, 410, 420, 423, 444, 627, 643-644, 677, 965-967, 1017, 1131-1150
aquifer 207, 223, 228, 231, 285, 307, 403-425, 456, 524, 536-538, 560, 612-617, 624-642, 643-646, 651-652, 671-676, 681-687, 691, 702-710, 725, 757-777, 834, 876-904, 942-951, 965-970, 985-1020, 1029-1045, 1078-1082, 1093, 1099, 1106, 1116-1119, 1131, 1142, 1147-1163
aquifer test 492, 638, 647, 677, 681-711, 985, 993, 1000, 1019, 1105, 1119
aquitard 456, 613, 617, 643-644
Archie's equation 1157, 1162
archival search 31-39, 54
array, dc geoelectrical 102, 107, 109, 118, 123, 129, 205-234, 243, 270-271, 786, 1102-1109, 1119, 1152, 1153
artesian aquifer → confined aquifer

ash 129, 188, 495, 569, 1138, 1148, 1179-1180, 1103
assessment → risk assessment
Atterberg scale→ grain size analysis
auger drilling → drilling
azimuth resolution 112, 113

background → chemical background
bacteria 752, 766, 774, 876, 887-909, 994-996
bailer 527, 529, 530, 762-775
bailing test → slug/bail test
bandwidth 101, 109, 122, 295-297, 326-327, 353
barrier → geological barrier
base flow 568, 590-594, 600-601, 613, 617
base map 4, 12, 24, 48, 86-91, 130, 514, 1055, 1056
base station 19, 100, 169-178, 194, 198
baseload → chemical background
batch test 531, 733, 757, 789, 826-846, 856, 835-845
bedding 507-510, 519, 539, 965, 987, 1100, 1104, 1119, 1121, 1131
bedrock 231, 254, 274, 340, 364, 382, 511, 517, 608, 1099, 1100, 1102, 1103
bentonite 467, 536, 542, 559, 562, 626, 627, 653, 796, 856, 858, 860
benzene, toluene, ethylbenzene, xylene 476, 490, 502, 503, 541, 551, 552, 541, 551, 753, 777-804, 822-827, 834, 856, 876-904, 994, 1013, 1163, 1178
Berlin List 1133, 1163, 1174, 1176, 1185, 1187, 1189
biochemical oxygen demand 500, 754-759, 875, 911, 968, 983, 995-996, 1015-1019, 1062, 1075, 1085, 1093, 1104-1105, 1116
biodegradable substance 807, 911
biodegradation 823, 876-887, 890-899, 910, 998
biological investigation 507, 963
biological process 824, 886, 1021, 1022
bioremediation 876, 883, 889
biotope 35-46, 130
Blaney-Criddle equation 588
borehole 209, 211, 298, 310-311, 316-325, 476, 484-498, 508-511, 517-546, 625-641, 649-665, 671-680, 683-690, 760-761, 796, 810-813, 1053-1054, 1104-1132, 1163
borehole deviation log 436, 446, 448, 465
borehole logging 12, 223, 241, 320, 325, 416-419, 431-470, 492, 494, 508-510, 537-543, 633, 635, 973, 978, 985, 990, 1105, 1157, 1163
borehole permeameter test 647, 649, 659-671
borehole seismic method 384, 385, 388-389
borehole televiewer 436, 441, 446, 448, 466
borehole TV 436, 1105
Bouguer anomaly 195, 196, 199-203, 982
Bouguer correction → corrections
boundary condition → modeling
Bouwer & Rice method 673-676
box-and-whisker plot 944-946
brackish water 243, 276, 469
Brandenburg List 1189
breakthrough curve 734, 838-848
brine 213, 286, 392, 498
Brownian motion 1023
buffer capacity 846, 1103, 1170
building rubble 162, 176, 1103, 1168, 1179-1184
bulk (dry) density 341, 344, 346, 347, 349, 351, 395-396, 463, 464, 728, 798, 823-830, 978-982, 1127, 1159-1161

cable tool drilling → drilling
calibration 81, 98, 125-126, 132-136, 250, 253, 265, 267, 361, 382, 409, 436-438, 448-467, 492-494, 509, 546, 547, 874, 970, 1007, 1018-1020, 1042-1044
caliper log 446, 448
calorimetry 821
capillarity 582, 603, 606, 644, 729, 730, 732, 788, 810, 1093, 1161

capillary fringe 541, 644, 810, 1150, 1151
carbonate 538, 754-755, 773, 788, 846, 855-886, 987, 1103, 1118, 1128, 1131, 1168
casing 385, 394, 415, 432, 436-468, 510, 525, 527, 535, 536, 539, 540, 558, 624-636, 651, 658, 663-665, 671-673, 677-678, 681, 687-689, 760-777, 811, 815
catalyst 801, 873
catchment area 577-581, 584, 590-594, 600-603, 614-615, 620, 751, 1019
cation 714, 832-833, 851-861 870, 941, 952-954, 964, 966, 969-970, 976, 980-981, 987, 989, 994-1013, 1018-1032, 1036-1045
cation exchange capacity 788-789, 853-861, 987, 989, 999, 1026, 1168, 1170
caving 40, 92, 447, 448, 463, 465, 507, 520, 526, 531, 533
cavity 306, 313, 340, 463
CCD array 97, 100, 106-109
cement bond log 436, 441
cementation 244, 432, 441, 443, 445, 723, 725, 1157
change analysis 126
channel flow 599
channel structure 1121, 1131
chargeability 214, 215, 266, 1132, 1141
chemical analysis 116, 129, 135, 136, 475, 476, 556, 622, 625, 685, 750-760, 778-806, 816-822, 833, 842, 858-885, 887, 899, 960-973, 1031, 1133, 1163
chemical background 258, 622, 624, 749-752, 994-995, 1026-1037, 1163-1190
chemical bonding 1026
chemical composition 750, 816, 826, 866, 1161, 1168
chemical factory 32
chemical oxygen demand 500, 625, 774, 780, 875, 890, 996, 1007-1010, 1019, 1042-1045
chemical reaction 31, 480-481, 499, 754-758, 778, 807, 819-823, 1001-1013, 1021-1022
chemical species 816, 826, 1021
chloride-budget method 618
chronological analysis 24, 99, 1132
cistern 975
Clarke value → chemical background
classification of sediments 549
clay 188, 213, 214, 228, 284-288, 406, 422, 423, 432-439, 444-451, 459-467, 495, 519, 527-538, 549-562, 592, 602-603, 606-607, 617, 634-635, 645, 659, 712-718, 722-726, 751, 754, 763, 792, 810, 832, 849-871, 904, 911, 965, 967, 972-975, 978, 982-987, 993, 999-1000, 1079, 1100-1110, 1116-1118, 1131, 1138, 1143-1152, 1157, 1168, 1178-1180, 1189
clay content 245, 728, 1105, 1145, 1152
clay minerals 461, 463, 751, 754, 849-871, 911, 1168, 1179, 1180
client → commissioning
climate 3, 567, 568, 570, 574, 581-586, 589, 591, 593, 613, 618, 732, 871, 965, 968, 974, 1078
climatic water balance 959-974, 998
clogging 613, 632, 636
cluster analysis 956-957, 1165, 1166, 1167, 1180, 1181, 1185
coefficient of uniformity 719-721, 735-736
coliform bacteria 996
colloid 481, 620, 712, 716, 824
colluvium 975, 1083
color infrared film 24-44, 52-64, 88, 89, 127-130, 144, 491
column test 729, 754-757, 824-826, 831-845-848, 997-1000
combustion 569, 783, 798-805, 872-875, 892, 1181
commissioning 11-22, 1055
common -midpoint 300-308, 352, 353, 365, 369, 374, 375, 377, 378, 379, 390, 391, 393,1151, 1152, 1156
common-midpoint section 352, 374
common-midpoint stacking 377, 378, 379, 391, 393
compaction 31, 91, 337, 794, 807, 973, 998
component separation analysis 948, 949

composting 1179
Compton effect 435, 437
conceptual model 511, 617, 622, 962-963, 966, 969-971, 1001, 1020, 1053, 1118
conductivity
-, electrical 205, 214, 235, 240-252, 255-265, 270-277, 283-290, 295, 301, 307, 329, 406, 410-415, 433, 435, 436, 440, 446, 469, 475-479, 485-498, 503, 548, 622, 625, 631, 638, 759, 770-798, 814-822, 941, 949, 960-962, 989-997, 1105, 1116, 1133, 1143, 1148-1151, 1157, 1162-1169, 1179
-, hydraulic 390, 403-404, 416-417, 426, 519, 556-562, 605, 607-608, 643-647, 649-680, 735-738, 762-763, 788, 846, 892, 899, 941-942, 970, 993, 1000-1003, 1103-1105, 1116-1121, 1151
cone (tip) resistance 543-550
cone of depression 685, 694, 703
cone penetration test → direct push sounding
confined groundwater 418, 423, 644, 646, 673, 686, 691, 725, 1004, 1012, 1131
confining bed 613, 644
conservative particle 618, 1006
constant head test 660, 665
contaminant 11, 30, 32, 35, 38, 40-41, 66, 100, 110, 132-136, 148-150, 233, 309, 314, 315, 338, 355-359, 369, 372, 386, 431, 433, 475-481, 484, 488-503, 507-508, 520-526, 540-543, 551-562, 590, 591, 603, 612, 620, 624-633, 642, 726, 749-759, 765-845, 855-910, 963-969, 994-1021, 1030-1036, 1061-1064, 1085, 1093, 1099-1105, 1116-1121, 1131-1133, 1138, 1143, 1163-1189
contaminant profile 885, 891, 893, 900, 909
contaminant retention 3, 4, 465, 507-508, 646, 725-733, 749-753, 786, 788, 817, 832, 844, 845, 867-871, 965, 999
contaminant transport 591, 753-757, 810, 823, 832, 963, 969, 1006-1020, 1121, 1178
contamination 28, 35-42, 52-66, 125, 133, 140-149, 206-207, 218, 222, 228, 263, 285, 470, 507-509, 525-526, 539-560, 569, 614, 620-633, 876, 949, 964, 973, 980-1001, 1020-1030, 1041, 1063
continuity equation 1002
contour map 162, 172, 195, 196, 202, 206, 221, 223, 228, 264-266, 270-272, 1060, 1061, 1091
contractor → commissioning
contribution matrix 413, 414, 1150, 1160
controlled source electromagnetics 239, 242, 255-257, 262, 269
convection 630, 713-714, 754, 830, 994, 1022
core sampling 524-532, 737, 789, 791, 793, 1152
coring 416, 431, 434, 455, 463, 466, 508-517, 524-526, 713, 737, 789-796, 971-973, 985-997,1063, 1119-1122, 1128, 1131, 1150-1152, 1158-1163, 1180
corrections 171-172, 195-199, 201, 224, 262-264, 267, 270, 448
crosshole methods 339, 384, 387
crossplot 449
customer → commissioning
cuttings → drill cuttings

Darcy velocity 846, 847, 1002, 1008, 1022
Darcy's equation 608, 644, 645, 652, 655, 709, 737, 1002, 1009
data fusion 1054-1055, 1091-1095
Debye-Hückel theory 758
decay process 142, 258, 266, 406, 407, 412-414, 417, 420, 1151
decomposition 749-752, 780, 797-801, 816, 871-876, 1168, 1174
decontamination 539, 540
deconvolution 302, 354, 379, 380, 387
degradation 754, 807-808, 823-832, 876-911, 998, 1008, 1012-1013

demagnetization factor 164
denser-than-water nonaqueous-phase liquid 315, 642, 877, 886
density 105, 185-193, 196-203, 348, 478-479, 537-550, 633, 645, 679, 716-718, 728-729, 753, 786-798, 823-831, 847, 852-855, 946-948, 978-983, 998-1016, 1021-1022, 1056, 1060, 1104-1105, 1127, 1145, 1159-1165
density log 188, 395, 435, 451-452, 458, 537, 633, 1127, 1145
density-driven flow 1003-1021, 1043
depletion curve 600-601
desorption 557, 804-805, 821-827, 867, 1026
destabilization 25, 40, 46, 56, 99
detection limit 190, 490, 552-556, 778-782, 800-806, 818-833, 863-875, 893-904, 960, 999, 1165, 1185
diagenesis 455, 868
diaphragm cell 776
dielectric permittivity 243, 432, 461, 479, 1156
differential global positioning system 19, 28, 44, 85-88, 115, 124-128, 134, 169, 1099
differential interferometric SAR 113, 133, 151, 152
diffraction 283-291, 302-308, 314-315, 347, 349, 350, 380, 391, 862
diffusion 240, 251-252, 276, 287, 726, 750-754, 776, 820-829, 837-838, 911, 1006, 1012-1013, 1022-1037
diffusion cell 776
diffusion coefficient 829, 1023
digestion/decomposition 749-752, 780, 797-801, 816, 871-876, 1174
digital camera 73, 97, 108, 115, 125-136, 149
digital elevation model 73-74, 81-86, 91-92, 99, 108, 113, 120, 127-130, 136, 149-150, 1061, 1091
dip 173, 338, 363-385, 389, 391, 432-433, 435-436, 440, 444-446, 449-451, 1099, 1105, 1110, 1113, 1118, 1143
dip moveout 370, 378, 381
dipmeter 435, 440, 466, 1105
dipol array detector 802
dipole-dipole array → array, dc geoelectrical
direct current (dc) resistivity methods 205-234
direct push sounding 541, 554-556, 631, 649, 787, 811, 1005, 1107, 1115-1116, 1133, 1145-1146, 1151
discharge 298, 594-597, 600-601, 612-617, 641-645, 649, 681, 688-690, 965-968, 998-1000
dispersion 286-288, 340, 351, 633, 713, 714, 750-754, 798, 823, 830, 842-845, 876-884, 944, 998-1043
dissolved organic carbon 492, 499, 625, 759, 769-775, 786-798, 831, 875-884, 949, 955, 990, 994-997
dissolved oxygen 480, 489, 778, 878, 955
dissolved solids 557, 955, 960-961
distortion 73-83, 98-104, 112, 126-127, 260, 321, 354, 360, 362, 380, 793, 960
domestic waste → waste
double ring infiltrometer 649-656
drainable porosity → specific yield
drainage 12, 16, 99, 124-129, 146-147, 587, 593, 610-613, 685, 702, 725, 731, 809, 869, 968-975, 994-1032, 1079, 1090
drainage basin 24, 36-46, 54-58, 62, 593, 973, 975, 994,1079
drainage divide 577, 593, 613, 685
drawdown 597, 638, 674, 675, 683-711, 762-775, 1119
drill cuttings 467, 524, 533, 542, 524-542, 787, 978, 986
drillhole → borehole
drilling 414-415, 423-424, 508-517, 524-542, 551-558, 626, 630-632, 651, 656, 659-664, 681, 686-689, 737, 753-758, 789-796, 810-813, 831, 846, 966, 973-999, 1015, 1128, 1134, 1150
drilling fluid log 436
drill-stem test 1119
drinking water regulation 478
dry-weather flow 592, 600, 601

dual-porosity medium 726
dug well 975
dump (site) → landfill
dwell time 101, 106-107, 750
dynamic correction 378
dynamic viscosity 1009

earthquake risk 3, 113, 507, 965, 1063
ecosystem 542, 962
effective grain size diameter → grain size analysis
effective permeability → permeability
effective porosity → porosity
effective rainfall → rainfall
effluent flow 968
elastic bulk modulus 340
elastic wave → seismic wave
electric log 435, 439, 452, 453, 636, 1122
electrical conductivity → conductivity
electrical resistivity 205-234, 241-249, 254-266, 270-277, 296-301, 308-309, 301, 308, 403, 406, 414-420, 432-470, 631, 633-636, 1132, 1053, 1100-1110, 1127-1143, 1152-1161
electrical resistivity tomography 206, 539
electrical sounding 206, 212, 216-225, 240, 287, 412, 415, 416, 1119, 1153
electrolyte content 475
electrolytic conductivity 244
electromagnetic (EM) monitoring system 475, 484, 490-491, 586, 622, 626
electromagnetic methods 218, 239-281, 406, 435, 440, 467, 478
electron acceptor 877-891, 895-896, 899, 955, 1013, 1027
electron capture detector 781-783, 802-805, 817-822
eluate 780, 788, 798, 806, 888, 987-990
elution 754-758, 784, 800, 997
emission 476, 552, 568, 620, 814-818, 973, 990-998, 1168, 1179
energy dispersive x-ray spectroscopy 818
environment log 436
environmental parameter 105, 475, 478, 495, 497, 751, 785, 941, 965
ephemeral stream 594
equivalence analysis → equivalence principle
equivalence principle 165, 223, 420
Euler deconvolution 174
evaporation 124, 567, 569, 576, 581-589, 614, 615, 653, 959, 974, 998
evapotranspiration 568, 581-593, 606, 610, 612-614, 618, 965
excavation 180, 286, 508, 519, 728, 1104, 1143
exothermic reaction 479, 1148
extractable organic halogens 1163, 1176, 1177, 1178, 1179, 1184
extraction 671, 683, 684, 1035, 1054-1055, 1091, 1174

factor analysis 1165, 1166, 1178, 1181, 1185
FAO Penman-Monteith method 585, 589-590
far-field 248, 289-294, 486-490
fault 25-46, 100, 112, 128, 129, 173, 186, 201-203, 207, 212, 242-243, 254, 260-264, 274 308, 340, 349-350, 368, 382, 390-392, 404, 422-425, 433, 441, 444-451, 458-459, 465-466, 507-520, 613, 644, 645, 686, 726, 965, 969, 973-980, 987, 1002, 1007, 1056, 1091, 1099-1121, 1127-1128, 1131
field capacity 646, 725, 998
field continuation 172
figure-8 loop 410, 415
filter (Darcy) velocity 846, 847, 1002, 1008, 1022
finite difference method → numerical model
finite element method → numerical model
finite volume method → numerical model
flame ionization detector 781-783, 803-805, 814, 817-822, 949, 950, 1028
flight parameter distortion 101
floating phase 768
flooding 13, 138, 139, 606, 965, 1063, 1075, 1079, 1088-1093

flow analysis 217, 436, 442, 445, 535, 567, 568, 590-618, 642-646, 691-692, 709-711, 734-738, 750-758, 766-801, 811-855, 884, 892-901, 965, 992-993, 1016, 1019, 1025
flow equation 709, 1003-1009, 1043
flow modeling → groundwater flow modeling
flowmeter 436, 442, 465, 636, 688, 1105, 1128, 1129
flow-through cell 767-770
fluorescence 476, 483-493, 502, 541, 552-556, 994
fluorescence spectrometry 818
Fluxgate magnetometer 168, 172
focused electric log 435, 439, 444, 446, 452-453, 461-463, 439, 467, 635-636, 1122, 1127
fouling 482, 489, 492, 500
fracture 25-46, 100, 112, 128, 129, 173, 186, 201-203, 207, 212, 242-243, 254, 260-264, 274 308, 340, 349-350, 368, 382, 390-392, 404, 422-425, 433, 441, 444-451, 458-459, 465-466, 507-520, 613, 644, 645, 686, 726, 965, 969, 973-980, 987, 1002, 1007, 1056, 1091, 1099-1121, 1127-1128, 1131
free air correction → corrections
frequency effect 214
frequency-domain electromagnetic method 214, 239-245, 248-270
Fresnel zone 295-296, 352-353, 375

gamma-gamma log 1122, 1145
gamma-ray log 435-437, 446, 447, 450-455, 457-469, 537, 543-550, 633, 636, 1122, 1145
gas analysis 25, 31, 539-541, 551-558, 568- 569, 620- 622, 671, 677, 723-725, 753, 812-813, 964, 973, 985-999, 1025
gas coking plant 100, 502-503
gas emission 100, 124, 142, 356, 568, 814
gas phase 480, 752
gas sampling 810, 811, 874
gasometer 812, 863
generalized reciprocal method 367-368, 389-390
geochemical analysis 3, 749-751, 883, 885, 899, 1133, 1163
geochemical modeling 757-758, 1008
geochemical simulator 1013
geocoding 1055-1056, 1065-1066
geogenic background → chemical background
geographical information system 13, 39, 91, 99, 126, 1013-1018, 1054, 1055, 1061-1074, 1079, 1086
geologic map 508-514, 980, 1053, 1075, 1105, 1110, 1113, 1119, 1121, 1122, 1131
geological barrier 1-4, 178, 214, 316, 325, 392, 432, 438, 467, 510, 570, 577-578, 703, 849, 855, 970, 998, 1000-1001, 1053, 1075-1079, 1103-1104, 1119-1121, 1131, 1179
geological cross-section 310-311, 511-518, 966, 988, 1132, 1135, 1136
geological mapping 5, 243, 1113, 1119, 1122, 1131
geological model 5, 174, 223, 382, 395, 456, 461, 624, 962-963, 970-973, 1104, 1105, 1131, 1134, 1156
geological setting 3, 4, 35, 161, 169-172, 185, 189, 191, 197, 201, 203, 206, 212, 217, 222, 231, 260, 410, 414, 422, 510-511, 519, 538, 709
geological site investigation 508-510
geological structure 3, 4, 35, 161, 169-172, 185, 189, 191,197, 201, 203, 206, 212, 217, 222, 231, 260, 510-511, 519, 538, 709
geophysical borehole logging → borehole logging
geophysical well logging → borehole logging
georeferencing 1063, 1066
geotechnical stability 3, 965
global positioning system 12, 19-21, 23, 27, 28, 34, 44, 74, 78-89, 115, 124-126, 128, 134, 169, 193, 790, 1063, 1066, 1073, 1099
grain size analysis 464, 712-723, 736, 790, 797, 798, 832

grain texture 723
gravimeter 186, 191-198, 203
gravimetry 778, 821
gravity 185-191, 196, 202-203, 603, 610, 646, 679, 709, 710, 717, 718, 725, 976-984, 993, 1009, 1060
gravity method 18, 185-203, 984
grid/mesh generation → numerical model
grid/mesh refinement → numerical model
gridding 196, 1056-1059
ground check 12, 34-46, 74, 82, 87, 135, 145, 1078, 1079, 1099
ground control point 73-74, 78-86, 99, 114, 1091, 1099
ground penetrating radar 213, 233, 243, 258, 283-329, 403, 1150-1158
ground resolution 100-113, 122, 128, 138, 149
ground roll 351, 362, 371-381
ground swath 101
groundwater 231, 233, 259, 276, 404-406, 420-425, 475-481, 490-503, 507-508, 514-530, 536, 541, 552-560, 567-571, 582-586, 590-591, 600-603, 612-645, 651-652, 660, 665-673, 677-691, 698, 701, 710, 726, 731, 749-785, 798, 807-810, 823-826, 846-848, 871-910, 941-1042, 1053-1054, 1063-1064, 1075-1079, 1092-1093, 1103-1106, 1116-1121, 1128-1133, 1143-1156, 1170, 1185
groundwater budget 615, 687, 970
groundwater contamination 39, 52, 56, 142, 143, 390, 614, 750, 809, 810, 985, 998, 1001, 1078, 1185
groundwater discharge 613, 615, 617
groundwater dynamics 499, 750, 966, 968
groundwater flow 207, 217, 436, 442, 451, 465, 479, 495, 498, 567, 617, 642, 643, 691, 750, 756, 826, 892, 896, 963-1010, 1016-1043, 1093, 1116, 1130-1132, 1147
groundwater flow modeling 436, 442, 445, 446, 465, 594, 595-698, 811, 826, 844, 846-847, 855, 884, 892-893, 900-901, 969-970, 991, 1001-1024, 1029-1042, 1064, 1079-1082, 1093, 1104-1105, 1116-1121, 1128-1132, 1147
groundwater hydrograph 614-617, 968, 969
groundwater monitoring 525, 560, 619-642, 683, 750, 761, 941, 953, 958, 985
groundwater observation well 5, 18, 432, 467, 468, 495, 498, 502, 503, 525, 538, 560, 561, 619, 622-640, 642, 683, 941, 953, 985, 1148, 1149
groundwater quality 502, 530, 619-622, 634, 688, 750, 948, 949, 1001, 1020
groundwater quantity 620, 685
groundwater recharge 3-5, 421, 479, 498, 568-571, 584, 600, 605, 610-619, 641, 965-974, 1007, 1015-1019, 1105, 1116, 1130, 1148
groundwater sampling 624, 628, 642, 753-777, 973, 560
groundwater storage 568, 615, 617
groundwater system 4-5, 390, 432, 507, 508, 881
groundwater table 225, 301, 307, 309, 519, 524, 541, 556, 582, 583, 591, 615-620, 632, 642, 644, 652, 660, 665-669, 673, 677, 710, 960, 966, 968, 982
groundwater withdrawal 614, 682, 687
grouting 525, 562, 630, 633, 636-637, 663
Guelph permeameter 647-650, 656-659, 1063, 1066, 1073, 1099, 1150, 1151, 1152, 1154, 1155, 1156, 1157, 1158
gyromagnetic ratio 404, 426
gyroscopic platform 115

Hales' method 367
half-width of a geophysical anomaly 172
halogen sensitive detector 552
hardpan 612
Hazard Ranking System 971
hazardous site → site suspected to be hazardous
hazardous substance → contaminant
Hazen approximation 735
health and safety measure 760, 790

heat capacity 129
heavy metal 541, 554, 555, 556, 751-756, 766, 788, 795-797, 822-825, 832, 845-848, 870-871, 892, 995, 1133, 1163, 1172, 1186
heterogeneity 83, 456, 459, 507, 508, 509, 657, 711, 965-966, 1016, 1023, 1032
history of the site 14, 1104
homogeneity 207, 208-209, 211, 341-343, 350-351, 377, 389, 507-509, 652, 683, 685, 709, 717, 718, 965, 1020
Hooke's law 341-343
horizontal coplanar loop configuration 248-250, 267, 1101, 1102
Hortonian overland flow 591, 606
humic matter 476, 492, 499, 871, 899
humification 751
humus content 463, 618, 1168-1170, 1189
Huygens' principle 347-349, 363
hydrated iron 754, 871
hydraulic conditions 433, 445, 567, 607, 628-629, 643-651, 680-682, 686, 707, 762-765, 777
hydraulic conductivity → conductivity
hydraulic gradient 618, 644, 645, 655, 737, 1001, 1010, 1017
hydraulic head 646, 649, 737, 965, 970, 1003-1005, 1018-1019
hydraulic model 633, 682, 685, 1131
hydraulic test 524, 646-680, 1119
hydrodynamic dispersion 633, 754, 1022
hydrogeological model 5, 456, 511, 624, 962-973, 1104
hydrogeological triangle 642, 643
hydrograph analysis 591, 594, 600, 601, 614-617, 700, 965-969
hydrologic cycle 567, 569, 584
hydrologic year 613
hydrometer 715-718, 798
hydrostratigraphic unit 948, 963-967
hydrotop 614, 615
hysteresis 608, 729
identification sheet 638
image classification 126, 1091
image map 1060, 1061
impulse neutron gamma log 417
incineration 754, 1174, 1179-1181
inclination 165, 166, 167, 174, 183, 1078, 1082, 1105
indicator test 811-812, 875
indirect recharge 613
induced magnetization 163, 165, 173
induced polarization 213-215, 1132, 1141, 1150-1153
induction log 435, 440, 452
industrial site 1-6, 24, 38, 66, 105, 962-972, 999-1001, 1093, 1138, 1181
infiltration capacity 594, 605, 606
infiltration rate 567, 591, 600, 603-619, 612, 633, 634, 638, 646, 652-658, 683, 970, 1002-1007, 1015-1019, 1093
infiltration test 633, 634, 638, 649-656
infiltrometer 647-656
inflow zone 1128
influent (losing stream) condition 613
information adequate model → conceptual model
information fusion 1054
infrared film 29-30
infrared spectroscopy 818, 849, 869
inhomogen 186, 187, 207, 208, 209, 220, 240, 245, 267, 343, 351, 389
inhomogeneity 606, 617, 683-686
initial condition → modeling
initial guess → inversion
injection test 1105
in-situ groundwater monitoring, geophysical 475, 625
instantaneous field of view 100-102, 116, 122
intercept-time method 365, 366, 367
interface conductivity 244
interfacemeter 642
interference test 1119, 1121
interferometric SAR → radar methods (in remote sensing)
interflow 568, 590, 591, 594, 613
International Geomagnetic Reference Field 171
interpolation 73, 81-84, 318, 941, 959-960, 1014, 1017-1041, 1057, 1059

inversion 172, 206-208, 212, 219-225, 231-234, 242, 255, 263-266, 276, 310, 351, 367, 368, 409, 412-425, 840, 843, 845, 1007, 1019, 1053-1056, 1109-1110, 1150-1163, 1192-1195,
-, block inversion 212, 225, 413, 1152, 1154,
-, smooth inversion 225, 420, 1150, 1151,
-, 1-D inversion 212, 219, 224, 231, 242, 264-266, 276, 424, 425, 1110,
-, 2-D inversion 206, 208, 212, 224, 225
ion balance 760, 960, 961
ion exchange capacity 788, 789, 853, 1168, 1170
iron oxide 849, 867-869, 1168
irrigation 40, 54, 56, 124, 143, 146, 588, 590, 613, 619, 968, 1179, 1184
isohyetal analysis 578, 580, 581
isotherm 127, 821-836
isotope 436, 437, 610, 619, 684, 685, 884-891, 910, 1138
isotope fractionation 886-891, 910
isotopic signature 884-888

Jacobian matrix 1159
joint interpretation 345, 451, 494, 1099-1190
joint inversion 413, 1159, 1161-1163

karst 129, 201-203, 404, 406, 412, 433, 507, 686, 965, 1007, 1116, 1118, 1131
k_f value → hydraulic conductivity
Kirchhoff depth migration 389
Kloke List 1186
kriging 959, 1018

Lamé parameter 340
land sliding 3, 40, 43, 113, 507, 1063, 1079, 1118
land subsidence 113-114, 133, 151-152
land surveying 15-20, 151-152, 193, 196, 198, 218, 220, 268, 1054
land use mapping 42-43, 64
landfill 1-14, 25, 30-66, 73-74, 86-91, 99-110, 128-149, 161-178, 186, 188, 192, 199-200, 206-207, 214-218, 222-230, 243, 259-263, 270-275, 388, 390, 392, 475-479, 495-499, 507-523, 538, 568, 620-625, 749-759, 787-814, 823, 832, 879, 892-899, 949, 958-1001, 1029, 1063-1064, 1073-1148, 1163-1189
landfill gas 568, 752, 807-814, 990-998
Larmor frequency 404-405, 409-411, 426
laser-induced breakdown spectroscopy → dirct push sounding
laterolog 439, 451, 454, 469
leachate 38-40, 49-58, 99, 145, 146, 243, 568, 750-759, 780, 798, 806-807, 892, 968-973, 993-1001, 1103, 1138, 1148
leaching 968, 1138, 1168-1170
leakage 217, 225, 445, 613, 633, 665, 685, 686, 701, 706, 777, 812, 970, 981, 1006-1007, 1015
leakage coefficient 701, 706, 1006-1007, 1015
less-dense-than-water nonaqueous-phase liquid LNAPL 642, 768, 877, 886
leveling → land surveying
lignite mining 149, 477, 1092
lineament analysis 24, 25, 42, 973-977, 1101, 1120-1121
lineation 973, 1102, 1118, 1131
liner coring 29, 57, 61, 526, 556, 737, 789, 793, 794, 846, 972, 973, 990, 993, 998-1001, 1079-1081
lithological classification 431, 433, 440, 444-445, 451, 548, 549
lithological log 416, 438, 451, 456, 792, 1122
lithology 3-5, 178, 213, 214, 222, 337, 340, 345, 419, 507-517, 527, 541-548, 556, 785, 791, 794, 810, 965-967, 1015, 1076-1080, 1089, 1105, 1119-1122, 1127-1131, 1150
lithostratigraphy 508
low-induction-number 240, 247, 249-250, 261
lysimeter 587, 588, 612, 614, 757

magnetic field 161-174, 180-183, 239-245, 249-259, 277, 286-287, 403-415, 418-426, 817, 820, 980, 983, 1060
magnetic gradient 173, 980
magnetic induction 183, 239
magnetic method 161-183, 435, 440, 467, 980-984, 1132
magnetic moment 165-167, 183, 252, 277, 403-406
magnetic permeability 163, 183, 243, 246, 251, 277, 283, 286, 288, 485
magnetic susceptibility 163-166, 172-173, 183, 435, 446, 451, 454, 461, 467-468, 539, 1060-1161
magnetization 162-173, 180-183, 980
magnetometer 162-163, 168-178, 404, 411, 416, 1054
magnetotellurics 239-242, 254-257
manganese oxide/manganese oxyhydroxide 754, 867, 871, 955
Manning equation 599
Manning's roughness coefficient 600
mapping 4-5, 25-30, 35-46, 63-64, 73-75, 82, 91, 100-112, 123-130, 136-138, 148-151, 206-207, 215-222, 227-229, 241-243, 255-263, 269-271, 340, 350, 390, 391, 392, 508-519, 791, 809, 810, 911, 1065, 1092, 1100-1113, 1119-1123, 1131, 1138-1141
matrix diffusion 750
matrix potential 823
mechanical stability 5
membrane interface probe 519, 551
messenger RNA -mRNA 890
metabolism 752, 876-891, 904, 905, 1168
metallic conductor → electron conductor
meteorological conditions 105, 268
methane 129, 142, 752, 753, 808-814, 883-901, 991, 997-998
microbial process 753, 757, 871, 877
microbial survey 883, 902
microbiological method 10, 875
microlog 439
microorganism 481, 499, 776, 807, 871, 876-878, 884, 885, 888-890, 901, 905-911, 955, 965, 991, 1027
microwave 25, 26, 97-98, 111-113, 120, 577
microwave remote-sensing method 111
migration 302, 316, 353, 380, 381, 389, 393, 527, 536, 542, 556, 562, 750-758, 785, 810, 823-845, 985, 991-1001, 1008, 1104, 1138, 1178, 1180, 1181
migration behavior 750, 751, 753, 756, 757, 758
migration test 527, 556, 845
mineral composition 438, 463, 464, 987
mineralization of groundwater 219, 223
mining damage 965
mining site 1-6, 11-15, 24, 38-40, 46, 92, 99-109, 130-132, 140-149, 162, 186, 507, 948, 962-965, 1019, 1029, 1062-1063, 1092, 1100
mobility 754, 823, 1168-1170
modeling 126, 127, 162-163, 172-176, 189-199, 212, 217-218, 222-226, 242, 263-264, 303, 347, 349, 364, 367-368, 389-390, 410, 413, 418, 420, 456, 458, 510, 511, 577, 606, 613-614, 617-620, 681, 687, 703-704, 754, 757-758, 828, 839-845, 960-970, 978-984, 1001-1044, 1053, 1056, 1064, 1159-1162,
-, analytical model 612, 842,
-, descriptive model 1053,
-, forward modeling 172, 189, 217, 224, 367, 368, 410, 413, 418, 1053, 1056, 1159, 1162,
-, inverse modeling → inversion
-, numerical model 617, 622, 839, 845, 963, 970, 1014, 1016-1017, 1020, 1024, 1031-1043,
-, stochastic model 1018,
-, 1-D modeling 264, 1053,
-, 2-D modeling 197, 218,
-, 2.75-D modeling 197,
-, 3-D modeling 162, 197
moisture content 24-29, 35-41, 46, 54-55, 99, 112-113, 124-133, 145-146, 218, 285, 292, 361, 568, 570, 571,

587, 588, 593, 594, 606, 612, 615, 714, 808, 1083, 1100-1103
molecular diffusion 1022, 1023
monitoring 99, 114, 128, 141, 207, 285, 314, 445, 475-503, 524, 530, 541, 558, 560, 586, 619-643, 750-776, 813-814, 818, 833, 848-857, 884, 899, 941, 953-959, 973, 1063, 1092
monitoring well 432, 467-468, 475, 492-503, 524, 530, 541, 619, 622-625, 626-642, 664, 673, 681-688, 692-696, 760-768, 775-776, 941, 966-976, 985-1005, 1017-1018, 1053, 1116-1119, 1133, 1147-1149
morphology 507, 510, 851, 1078, 1082, 1089, 1090
most probable number 883-902
multi-electrode system 208, 220
multi-parameter probe 442, 475-476, 488-498, 622, 625, 1105, 1117, 1128, 1129, 1133, 1147
multi-temporal analysis 99, 1132
multivariate analysis 470, 941, 952, 1133, 1163, 1165, 1185
municipal solid waste → waste

natural attenuation 749, 823, 876, 877, 881, 883, 887, 891, 892, 965, 996
natural gamma log → gamma log
navigation system 34, 115, 125, 127, 134
near-field 295, 486-488
neutron log 435-438, 451-460, 467-468, 537, 550, 588, 633, 1105, 1127, 1145, 1157
neutron-gamma log 417, 436
neutron-gamma spectrometry 435, 438
nitrate 470, 476, 490, 620, 778, 801, 878-902, 955, 987-996, 1041, 1128
non-aqueous-phase liquid 315, 626, 761, 877, 886, 892
nonphotographic imaging system 82, 97, 100, 115, 128
non-reactive transport → transport modeling
normal moveout 370
nuclear magnetic resonance 465, 470, 1150-1162
numerical dispersion → modeling
Nyquist frequency 362

oblique aerial photograph 33
operational safety 11-12
optical absorption magnetometer 168, 169
optosensor 475-476, 488-494, 499-502
organic carbon 129, 492, 499-501, 625, 714, 752-754, 791-792, 825, 832, 871-873, 783, 798-803, 827, 834, 865-875, 901, 911, 955, 987, 989, 990, 991, 996, 997, 999, 1168, 1179
organo-clay complex 871
organoleptic parameter 773
orthophotograph 38, 73-74, 81-88
orthorectification 74, 99, 126
outcrop 178, 276, 508, 511, 517, 975, 1054, 1110, 1118
overburden 201-203, 392
Overhauser effect 168
overland flow 568, 590, 591, 604, 605
oxidation-reduction potential → redox potential
oxygen concentration 442, 446, 475, 480-481, 488-503, 619, 622, 625, 638, 642, 759, 769-780, 808-812, 822, 851, 865-891, 895-911, 955, 995, 996, 997, 1027

packer 628-634, 660-665, 671-673, 677-678, 776, 1105, 1119-1121
packer test 1105, 1119
panchromatic film 28
parallax 38, 78-80
parameter resolution matrix 223, 419, 420
particle density 728, 729, 827
particle tracking 633, 1006-1012, 1018, 1036-1039
passive groundwater sampler 776
pathway 38-40, 51, 207, 508, 510, 520, 612, 645, 749, 752, 855, 881, 971-972, 985, 999-1006, 1021-1023, 1104, 1138
Peclet criterion → modeling
penetration depth 113, 247-248, 252-263, 283-288, 297-300, 307-308,

325-327, 356, 369, 439, 519, 527, 540-560, 981-985
percolation 315, 556, 582, 586, 588, 600, 612, 613, 617, 632, 756, 781, 798, 1105, 1116-1119, 1151
percolation test 556, 1105
percussion core drilling → drilling
percussion drilling → drilling
Peres' equation 679
permeability 39-43, 53, 214, 407, 418, 461-470, 509, 539, 603, 638, 644, 645, 677, 684, 752, 753, 811, 998, 1007, 1017, 1023, 1119
permeameter 647-652, 656-671, 737-738
permittivity 243, 283-296, 304, 308-329, 484-487, 1156
petroleum hydrocarbon 541, 551, 552, 781, 782, 795, 803, 875
pH 475-481, 488-503, 622, 638-639,642, 751-756, 759, 766, 770, 772, 778, 788, 798, 826, 846-848, 852, 854-855, 856, 859, 862-865, 869, 941, 947-949, 954-957, 960-969, 973-980, 985-1004, 1007, 1010-1037, 1043-1045, 1117, 1128, 1148-1150, 1163-1174, 1182, 1188
phenol index 783, 788, 804, 875, 1176-1189
Philip's transient infiltration equation 607
photo ioization detector PID 803-805, 814-817, 822
photogrammetry 27, 44, 73-74, 81-85
photometric turbidity measurement 436
photomosaic 47, 73-74, 82-91
phreatic aquifer → unconfined aquifer
piezometer 681, 682
piezometric head 966, 969, 1002-1009, 1015-1020
piezometric surface 642, 644, 683
pipeline 16, 217, 263, 264, 285, 291, 302-305
pipette method → grain size analysis
plausibility check 960-961
plus-minus method 367
Poisson's ratio 340, 344, 345, 348, 351, 355
polarization 113, 120, 213-215, 291, 295, 305, 343, 358, 394, 817, 819, 820, 1132, 1150, 1153
pole reduction 172
pole-dipole array → array, dc geoelectrical
pollutant → contaminant
pollution → contamination
polycyclic aromatic hydrocarbon 476, 490, 502-503, 541, 552, 781-782, 788, 795, 802-804, 876-899, 994, 999, 1093, 1163, 1170-1190
polymerase chain reaction -PCR 884-890, 906-910
ponding 604-606
pore fluid 213, 219, 244, 337, 437, 214, 439, 461, 469, 544-550, 789, 796, 843, 846, 855, 1022, 1025, 1161
pore water pressure 544-550
pore-size distribution 539, 607
porosity 214, 244, 245, 337, 403-408, 414-417, 423-425, 433-446, 449-456, 461-465, 509, 537, 539, 604, 607, 608, 644-646, 712, 719, 723-729, 730, 734-735, 752, 758, 762, 788-789, 796, 823-827, 843-844, 846-848, 855, 862, 987, 997-1009, 1021-1029, 1105, 1150, 1156, 1157, 1159, 1161, 1162
positioning 18-20, 21, 115, 1063-1066, 1073, 1099
potential gradient 610
precautionary measure 1
precipitation 3, 40, 53, 124, 125, 134, 481, 567-594, 600-619, 642, 684, 753-754, 773-774, 832, 846, 859, 886, 895, 959, 965-974, 981-999, 1008, 1027-1040, 1130, 1148, 1168
preservation 758, 774-778, 793-796, 831, 1086
pressure head 607-611, 649, 660-661, 678-680, 730, 1021
pressure transducer 641, 651, 663, 673, 678, 769, 770
pressure-retaining bailer 765-773
principal component analysis 954, 955
principle of equivalence 173, 191, 212
principle of suppression 212

probability 507, 943-950, 966
profiling 162, 168-173, 192-197, 200, 206, 209, 215-234, 241-242, 255, 259-269, 1055, 1101-1102, 1109-1117, 1128-1129, 1143-1154
proton-precession magnetometer 168, 169, 404
pseudosection 221-224, 230, 266, 276
pulse test 647-649, 671, 677-680, 1105, 1119
pumping rate 442, 536, 638, 682-687, 692, 696, 700-701, 759-777, 813, 815, 846, 1015
pumping test 492, 638, 647, 677, 681-711, 985, 993, 1000, 1019, 1105, 1119
purging 761, 762, 764, 770, 771, 772, 775, 777
push broom scanner → along-track scanner
pycnometer 729

quality assurance/quality control 6, 11-17, 21-22, 43, 44, 85, 134, 174, 194, 197, 225, 266, 267, 303, 312, 382, 387, 393, 409, 415, 447, 494, 777, 822, 845, 866

radar interferometry → radar methods (in remote sensing)
radar methods 84, 97, 111-113, 120, 133, 151, 1091, 1056, 1060-1064, 1068-1072, 1185
radiation temperature 105, 143
radiomagnetotellurics 242, 248, 253, 255, 261
radiometer 1132
radiometric correction 125, 127, 128
radiometric resolution 82, 101, 107, 121, 128, 134
rain 140, 141, 478-480, 503, 568-578, 588-593, 600-619, 627, 646, 751, 762, 773, 787-809, 822, 846-849, 858-862, 869, 875-878, 906, 967-968, 973-975, 981-999, 1018, 1032-1033, 1039-1040, 1064-1079, 1090-1095, 1131, 1150-1151, 1159-1162, 1178-1185
rain gauge 569, 575-578
range resolution 112, 326, 327
reactive transport → transport modeling
real-aperture radar → radar methods (in remote sensing)
receiving stream/lake/pond 42, 140
recharge 421, 479, 498, 568-571, 584, 600, 605, 610-619, 641, 965-974, 1007, 1015-1019, 1105, 1116, 1130, 1148
recombinant DNA -rDNA 889, 909
reconnaissance survey 3, 23, 32, 54, 302, 366
rectification 34, 45, 73-75, 77, 78, 81, 83, 88, 99, 103, 126, 127
redox potential 436, 442, 446, 469, 475, 480-481, 488-497, 503, 622, 638, 642, 750-755, 769-772, 826, 885, 901, 910, 996, 1027, 1128, 1133, 1148
redox process 480, 878, 884, 895, 899, 1027-1028
redox zonation 886, 895, 896, 901
reduction to the magnetic equator/pole 172
reference station 20, 44, 134, 191-197
reflection coefficient 295, 347, 369
reflection seismics → seismic methods
refraction seismics → seismic methods
refraction tomography → seismic methods
regional groundwater system 4, 507, 508
regional planning 42, 64, 1063, 1079
relief displacement 75, 101, 126, 127
remanent magnetization 164-167, 172-173, 180
remediation 1, 14, 141, 143, 524, 541, 876-891, 949, 963, 1016, 1019
remote sensing 3-6, 23-26, 44, 97-115, 134, 508, 966, 1053-1054, 1131, 1135
reporting 21, 175, 227, 269, 369, 383, 416, 687-690
representative elementary volume 824
resistivity → electrical resistivity
resistivity survey 205-234, 243, 245, 272, 976-982, 1000

resolution 24-28, 34, 73, 86, 98-138, 148-150, 208-209, 217-223, 243, 248-252, 255-265, 274-276, 283-284, 291-305, 325-327, 352-362, 367-369, 375-390, 406, 411, 414, 418-420, 432, 437-443, 448-451, 458-459, 489, 531, 1067, 1072, 1104, 1143, 1151
retardation 826-827, 845, 894, 899, 997, 1012, 1013
retardation coefficient 826, 827, 997
retention → contaminant retention
Ricker wavelet 297
Rietveld refinement 866
risk assessment 3, 4, 10, 24-25, 40-46, 99, 100, 823, 941, 949, 961-973, 997-998
root zone 614, 619, 791
rubble → building rubble
runoff 3, 58, 567-571, 582, 590-605, 612-618, 646, 965, 968, 998, 1105

saline water 213, 286, 243, 276, 392, 469
salinity 245, 420, 537, 582, 826
salinity log 436-446, 469, 1128
saltwater intrusion/saltwater rise 243, 1009
sample preparation 713-714, 750, 797-798, 822, 831, 872
sample preservation 774, 795
sample size 509, 659, 786, 787, 788, 793, 942, 1029
sampling 122, 134, 145, 176, 178, 299, 300, 307, 362, 374-377, 382, 389-395, 509, 524-532, 540-543, 556-560, 619-642, 678, 684, 749-797, 807-815, 831-836, 874-893, 906-910, 960-961, 973, 995-999, 1152, 1165-1166, 1178, 1181,
-, soil gas 541, 558, 985, 807, 810, 811,
-, water 475, 499, 560, 624, 628, 642, 753, 759, 765, 768, 769, 772, 777, 973
satellite image 12, 13, 86, 98, 99, 127, 1064-1066, 1073, 1091, 1132
satellite-based remote-sensing system → space-borne imaging system
saturated flow 613, 1004, 1012, 1013
saturated hydraulic conductivity 605, 644, 645, 653-658, 674
saturated zone 481, 492, 556, 607-612, 620, 644, 750-757, 796, 810, 823, 1013, 1016
saturation index 481, 1157
schistosity 1127
Schlumberger array → array, dc geoelectrical
screening 552, 803, 814, 875, 971, 972
sealed soil sampler 557
sealing 593, 612, 627, 630, 771, 795, 813, 815
second-derivative analysis 172
sedimentation method 716
sedimentation method → grain size analysis
seepage 39, 46, 52, 99, 124-130, 567, 591, 612, 625, 752, 758, 973, 981, 990-1001
SEG standard 363, 383
segment gamma ray log 635-636
seismic energy source 337, 338, 349-373, 383-395
seismic methods 337-396, 403, 1053, 1056, 1092, 1100-1116, 1133, 1143-1144, 1150-1153, 1161-1162,
-, reflection seismics 231, 283-285, 289-296, 300, 338-340, 352-353, 365-374, 383-392, 520, 523, 976, 987-992, 1104, 1110-1114, 1143-1144,
-, refraction seismic 293-294, 338-339, 363-370, 377-393, 1100-1104, 1143-1150, 1153, 1162
seismic migration 353, 380
seismic noise 371
seismic receiver 338-339, 353-354, 359-395, 1153
seismic reflection → seismic methods
seismic refraction → seismic methods
seismic resolution 352, 381, 385
seismic velocity 338, 347, 373, 382, 1143
seismic waves/elastic waves 293, 294, 337-369, 373-396, 440, 455, 461,1053, 1110

self-potential 215, 216, 225, 1122
self-potential method 435, 452, 1122
sensitivity, matrix 208, 217, 1160
sewage 765, 803, 810, 964, 1104, 1138, 1174-1185
sewage sludge 148, 803, 1104-1185
shaded relief map 62, 1060-1061, 1091
sheet flow 591
short core sampler 793
side-looking airborne radar → radar methods (in remote sensing)
sieve analysis → grain size analysis
sieve-pipette method → grain size analysis
signal-to-noise ratio 101, 122-128, 208, 217, 219, 252, 267, 338, 357, 359, 374-378, 381-386, 411-412, 476, 500
single-ring infiltrometer 653
sinkhole 39-40, 56, 201, 203, 1092
site suspected to be hazardous 31, 32, 475-477, 507-510, 622-625
skin effect 248-253, 651-652, 657-660, 677-687, 699
sleeve friction 543-550
slowness 349, 527
slug/bail test 647, 649, 671-677, 1105, 1119
Snell's law 349, 363, 367
soil air 480, 481, 558, 656, 754, 813
soil classification 546, 547, 548, 867
soil contamination → contamination
soil gas 541, 558, 568, 749-753, 807-813
soil gas sampling → sampling
soil horizon 751, 786-789
soil moisture 26-29, 36-40, 285, 587, 588, 593, 594, 606, 612, 615, 1103
soil salinization 138, 1092
soil sampling → sampling
soil texture 602, 604, 607, 608, 615, 722, 723, 792
soil type 519-520, 544-546, 583, 593, 603, 618, 653-659, 723, 733, 751, 786, 787
soil water content 582, 604-616, 731-732, 752-754, 965, 1170
solubility 479, 624, 751, 862, 880, 884
sorption 750-757, 773, 783, 802-887, 998, 1008, 1013, 1026, 1151
sorption coefficient 825, 845
sorption isotherm → isotherm
sorptivity 607, 655-658
sounding 206-208, 212-231, 240-242, 255-269, 275-276, 403, 406-425, 508, 509, 510, 546, 770, 787, 794, 813, 1053, 1100-1104, 1109-1110, 1119-1123, 1131, 1143, 1150-1162
sounding rod 794, 813
source/sink term → transport modeling
space-borne system 4, 12, 73, 78, 86, 87, 97-108, 117-120, 125, 128, 138-139, 149, 1092
species 31, 41, 54, 753-758, 773, 816-826, 853-855, 865-867, 887, 889, 960, 994, 1008, 1013, 1021, 1026-1027
specific retention 646, 725, 727
specific storage 646, 1002
specific yield/drainable porosity 615-617, 645-646, 725-729
spectral resolution 62, 98, 101, 108, 110, 116, 121
spectrometer 97-110, 115, 123-136, 817, 821, 822, 859
spin 168, 403, 404
spreading of contaminants 494, 751, 752, 787, 1064, 1104-1105
stability 40, 56, 225, 965, 1007, 1017, 1034, 1039
stacking 215, 252, 259, 299-302, 307, 357, 374-386, 391-393, 409, 412, 415, 416, 418, 420-425, 1143
staff gauge 597
stagnant water 761, 777
starting model → modeling
static correction 364-365, 374, 376-378, 391-396
statistical methods 126, 569-574, 577-596, 600-601, 610, 613-614, 632, 642, 752, 787, 941-953, 956, 958, 960, 1017-1018, 1133, 1163-1166
steady state condition 630-632, 645, 655, 658, 660, 665, 757, 831, 839, 840-843, 991, 993, 1003, 1012, 1018, 1022

steady-state flow 655, 658, 665, 831, 839
stereometric processing 34, 38, 46, 78-80, 83-85
sterile sample 790
stochastic method 966
Stokes' law 712, 716-717
storativity 465, 567-568, 586-588, 604, 615-617, 626, 646, 647, 671, 681-710, 723, 725, 757, 765, 774-775, 787-796, 815, 891, 1002-1007, 1015-1023, 1062, 1072, 1105, 1174-1181
strain 337, 341-345, 367, 878, 906, 1033, 1039, 1040
stratigraphic column 340, 507, 508, 515, 516, 517
stratigraphy 508-516, 541, 793, 965, 967, 1015, 1131
stream and lacustrine sediment sampling → sampling
stream flow 591-596, 613
stream gauge 614, 617
stream hydrograph 617
stress 41, 337, 341-350, 355, 509, 539, 548, 617-618, 638-641, 970, 1100, 1104, 1132
strike 170, 173, 178, 242, 254-255, 260, 267, 433, 444, 446, 465, 1118, 1121, 1150-1151
structure map 203, 965, 977-979
subrosion 40, 92, 310, 1092, 1118
subsidence 3, 40, 92, 113, 114, 133, 151, 152, 201, 495, 507, 510, 614, 634, 965, 975, 1092, 1093, 1118
sulfide precipitation 754
sun elevation angle 77
surface nuclear magnetic resonance 403-426, 1150-1162
surface sealing 3, 612
surface water 25, 31, 42, 61, 100, 109, 130, 567, 627, 642, 664, 684-687, 749-759, 782, 871, 965-968, 994-995, 1005-1006, 1015-1019, 1060, 1161, 1078-1085, 1138, 1174
surveying → land surveying
susceptibility log 435, 451, 456, 467
swath 101-106, 112-122
synthetic-aperture radar → radar methods (in remote sensing)

tachymeter 18-21
tectonic structure 178, 270, 390, 391, 507-511
telemetry 492-495
temperature 26, 35, 44, 104-116, 124-129, 141-146, 168, 192, 299, 301, 346, 431-436, 441-447, 465, 469, 475-503, 568-569, 581-590, 593, 613, 622, 636-638, 641-651, 681-684, 717-718, 728-729, 736-737, 752, 759, 770-775, 800-833, 854-855, 872-875, 955, 965, 974, 995-997, 1022, 1105, 1116, 1117, 1128, 1129, 1133, 1148, 1181
temperature log 436, 447, 537, 636, 1128, 1129
tensiometer 607, 841
tension infiltrometer 656
terrain correction → corrections (in gravity methods)
terrain feature 38-40, 73-74, 133
test tube 809, 811, 812, 833
texture 73, 84, 508-509, 539, 556, 570, 602-615, 659, 722-723, 785, 792, 871, 1161
Theis equation 691
theodolite → land surveying
thermal analysis 821, 863
thermal behavior 99
thermal survey 124, 125
Thiessen method 577, 579, 581
Thornthwaite equation 588
time of travel 946, 960-961, 973, 989-990, 995-999, 1008, 1033, 1036
tomography 206, 224, 227, 300, 310-311, 367-368, 384-390, 539, 1104, 1115-1116, 1144
topographic (terrain) correction → corrections (in gravity methods)
topographic base map → base map
topographic map 12, 74, 80-90, 99, 108, 130, 964, 965, 1060
topography 13, 20-21, 38-44, 62, 73-75, 81-82, 112, 114, 128, 151, 208, 216-218, 225, 231, 267-268, 276, 338,

369, 388-390, 568, 578, 593, 602, 613, 964, 978, 980, 993, 1061, 1078, 1143, 1150-1151
total carbon 538, 788, 798, 803, 863, 865, 871, 875, 989-990
total cyanide 780, 801, 806
total dissolved solids 557, 961
total inorganic carbon 865, 866, 871, 989-996
total organic carbon 501, 871-875, 783, 871, 872, 875, 901, 911, 964, 966-967, 989-999, 1168, 1170, 1179, 1180, 1184, 1185
total organic halides 875
total petroleum hydrocarbons TPH 875
total porosity → porosity
total station → land surveying
toxicity 823, 994, 998, 1174, 1176
trace element 508, 752-754, 765, 812, 814, 818-821, 944, 948, 1163-1165, 1174, 1180
tracer 594, 618, 620, 681, 684, 734, 735, 826, 839, 843, 844, 855, 1120-1121, 1025-1026, 1128
tracer test 5, 445, 681, 734, 1121
transient electromagnetic sounding 214, 219, 239, 242, 251-269, 275-276
transmission coefficient 302, 347-349
transmissivity 645-647, 677, 692-710, 764, 1119
transport equation 830, 1008, 1021-1036
transport modeling 5, 754, 1001, 1008-1013, 1018, 1020-1035, 1041-1044
transport of reactive substances → transport modeling
trench/trenching 508, 510, 519, 520, 521, 522, 523, 789, 792, 794, 814, 1103, 1104, 1105, 1131
turbidity 436, 442, 445, 446, 476, 483, 499, 500, 501, 560, 642, 686-690, 775, 778
Turc equation 589
type of bonding 754

unconfined aquifer 418, 644, 646, 673, 686, 691, 710, 725
unconformity 976
unexploded ordnance 15, 285, 162, 169, 180
univariate statistical methods 941-942, 1165
unsaturated zone/vadose zone 404, 416-418, 481, 492, 527, 556, 591, 603, 607-613, 617-620, 644-645, 655-656, 663, 750, 753, 756, 757, 796, 810, 1013, 1016, 1043, 1151-1158
uplift 510, 978, 987, 1092, 1118

vadose zone → unsaturated zone
value of aquifer 3
van der Waals attraction 1026
variable head test 660-661
vegetation anomaly 40, 58, 100, 130
velocity analysis 376-378, 382, 393
velocity field 365-368, 382-385
vertical coplanar loop configuration 248, 249, 250, 270
vertical electrical sounding 206-234, 412, 416, 982, 1100-1104, 1109, 1123
very low frequency method 242, 248, 253, 254, 255, 259, 260, 261, 264, 267, 269
viscosity 442, 479, 536, 645, 716, 717, 1009
VLF resistivity method 242, 253-259, 269
volatile halogenated hydrocarbon 541, 551, 753, 769-773, 781-788, 798, 805, 807, 781, 782, 805, 877, 886, 1163, 1178
volatile halogenated hydrocarbon VHC 781, 782, 805
volatile organic compond 551, 552, 560, 561, 794, 795, 796, 798, 809, 812, 822, 876, 884, 994, 999-1001
volumetric water content 646, 729, 830
vulnerability 475, 477

Ward's method 956
waste 213, 230, 285, 309, 477, 495-501, 532, 541, 550-562, 752-757, 787, 807-813, 820, 849, 855, 892, 899-900, 964-974, 980-984, 991-1001, 1007, 1075, 1078-1086, 1092-1093, 1119, 1122, 1131, 1133, 1138, 1143, 1163-1168, 1174, 1179-1189
waste incineration plant 58, 1181
wastewater treatment 15, 143, 477, 500-501, 820, 1081
water balance 568-569, 612-618, 959, 965-974, 998, 1016, 1021
water budget 584-586, 612-615, 687, 1019
water content 263, 288, 307, 404-407, 413-426, 437, 438, 444, 461, 509, 537, 582, 604-611, 617, 645-646, 729-733, 750-753, 819, 830-831, 965, 998, 1105, 1127, 1145, 1150-1162
water level 586, 615-620, 641-642, 653-656, 664, 673-676, 681-697, 769-775
water level measurement 685, 993, 1054, 1087-1093, 1153
water level meter 769
water protection area 3, 4, 14, 477, 1006, 1187
water quality 442, 445, 470, 502, 530, 619-625, 634, 642, 686-688, 750, 776, 777, 948, 949, 965, 1001, 1020, 1148
water retention curve 729-733
water table 40-43, 217, 285, 307-309, 410, 418, 420, 519, 524, 532-541, 556, 582-583, 591, 606-620, 632, 644, 652, 660-681, 710, 759-775, 810, 892-899, 1053, 1079, 1092-1093, 1105-1106, 1116, 1147-1156, 1185
water treatment 765, 998
water unit 436, 438, 454, 1125
water-logged area 39, 1093, 1138
watershed 577, 593, 613, 685
waterworks 475, 477, 499, 1103-1105, 1118, 1020
wavefront method 367
wavelength dispersive x-ray spectroscopy 818
wavenumber filtering 375
weather condition 34-36, 45, 54, 98, 125, 226, 815
weather radar 576-578
weathered zone 207, 340, 346, 351-356, 364, 382, 1113, 1121, 1127
weathering 100, 130, 132, 254, 274, 507, 519, 751, 854, 965, 1099-1103, 1113, 1118-1121, 1127, 1131
weir 595-598, 688
well 575, 612-710, 1156
well casing → casing
well completion 433, 448, 449, 467, 468, 525, 760, 769, 775, 985
well development 626-627, 656, 681, 687
well hydrograph 700
well logging → borehole logging
well recovery 698
well screen → casing
well storage 685
well test 508, 638, 681-682, 684-685, 973
well yield 766
Wenner array → array, dc geoelectrical
Wentworth scale → grain size analysis
wet sieving → grain size analysis
whisk broom scanner → across-track scanner
withdrawal of groundwater 558, 968, 614, 673, 682-691, 970, 1022, 1035
World Geodetic System 1984 1066

x-ray diffraction analysis 463, 464, 849-851, 862-869
x-ray fluorescence spectroscopy 541, 554, 555, 556, 818, 866-867

Zeeman effect 169
zone of weakness 392

3-D analytic signal 174
3-D resistivity imaging 206
200-day line 624

Printing: Mercedes-Druck, Berlin
Binding: Stein + Lehmann, Berlin